Excavation
Handbook

OTHER McGRAW-HILL HANDBOOKS OF INTEREST

Baumeister · Marks' Standard Handbook for Mechanical Engineers
Brady and Clauser · Materials Handbook
Brater · Handbook of Hydraulics
Callender · Time-Saver Standards for Architectural Design Data
Conover · Grounds Maintenance Handbook
Considine · Energy Technology Handbook
Crocker and King · Piping Handbook
Croft, Carr, and Watt · American Electricians' Handbook
Davis and Sorensen · Handbook of Applied Hydraulics
Emerick · Handbook of Mechanical Specifications for Buildings and Plants
Emerick · Heating Handbook
Emerick · Troubleshooters' Handbook for Mechanical Systems
Fink and Beaty · Standard Handbook for Electrical Engineers
Foster · Handbook of Municipal Administration and Engineering
Gieck · Engineering Formulas
Harris · Handbook of Noise Control
Harris and Crede · Shock and Vibration Handbook
Havers and Stubbs · Handbook of Heavy Construction
Heyel · The Foreman's Handbook
Hicks · Standard Handbook of Engineering Calculations
Higgins and Morrow · Maintenance Engineering Handbook
King and Brater · Handbook of Hydraulics
La Londe and Janes · Concrete Engineering Handbook
Leonards · Foundation Engineering
Lund · Industrial Pollution Control Handbook
Manas · National Plumbing Code Handbook
Mantell · Engineering Materials Handbook
Merritt · Building Construction Handbook
Merritt · Standard Handbook for Civil Engineers
Merritt · Structural Steel Designers' Handbook
Myers · Handbook of Ocean and Underwater Engineering
O'Brien · Contractor's Management Handbook
Peckner and Bernstein Handbook of Stainless Steels
Perry · Engineering Manual
Rossnagel · Handbook of Rigging
Smeaton · Switchgear and Control Handbook
Stanair · Plant Engineering Handbook
Streeter · Handbook of Fluid Dynamics
Timber Engineering Co. Timber Design and Construction Handbook
Tuma · Engineering Mathematics Handbook
Tuma · Handbook of Physical Calculations
Tuma · Technology Mathematics Handbook
Urquhart · Civil Engineering Handbook
Waddell · Concrete Construction Handbook
Woods · Highway Engineering Handbook

A visionary example of open-pit mining is the spectacular workings of the United States Borax and Chemical Corporation in the midst of the Mojave Desert of southern California.

This awesome amphitheater is 4700 ft west to east, 3300 ft north to south, and 550 ft deep. It is carved out of the desert alluvium and clays and the borate ore. Overburden and the blasted ore are loaded by outsize electric shovels and hauled by monstrous diesel trucks over the boulevardlike haul roads. Overburden is lifted hundreds of feet to waste pile areas beyond the rim of the pit and ore is hauled to the primary crusher at the foot of the conveyor, whence it is raised to the stockpiles of the processing plant.

In 20 yr some 168 million tons, or about 96 million yd^3, of combined overburden and ore have been wasted and processed. Such colossal quantities are enough to fill a freight train of hopper cars stretching more than around the world, or some 32,000 mi. *(United States Borax and Chemical Corporation.)*

Excavation Handbook

HORACE K. CHURCH
Consulting Engineer

McGRAW-HILL BOOK COMPANY
New York   St. Louis   San Francisco   Auckland   Bogotá
Hamburg   Johannesburg   London   Madrid   Mexico
Montreal   New Delhi   Panama   Paris   São Paulo
Singapore   Sydney   Tokyo   Toronto

Library
Vermont Technical College
Randolph Center, Vermont

Library of Congress Cataloging in Publication Data

Church, Horace K
  Excavation handbook.

   Bibliography: p.
  Includes index.
  1. Excavation. 2. Engineering geology.
I. Title.
TA730.C48   624.1'52   80-19630
ISBN 0-07-010840-4

Copyright © 1981 by McGraw-Hill, Inc. All rights reserved.
Printed in the United States of America. No part of this
publication may be reproduced, stored in a retrieval system,
or transmitted, in any form or by any means, electronic,
mechanical, photocopying, recording, or otherwise, without
the prior written permission of the publisher.

1234567890 HD HD 8987654321

*The editors for this book were Harold B. Crawford and
Joseph Williams, the production supervisor was Teresa F.
Leaden, and the designer was Mark E. Safran. It was set in
Gael by ComCom.*

Printed and bound by Halliday Lithograph Corp.

Contents

Preface xv

1. EARTH AND GEOLOGY OF EXCAVATION 1-1

2. ROCKS, ORES, MINERALS, AND FORMATIONS 2-1

    The Grand Canyon—The Classical Record of Rocks 2-3
    Definition of Rock. 2-5
    Classification of Rocks 2-5
        Igneous Rocks 2-5
        Sedimentary Rocks. 2-6
        Metamorphic Rocks 2-7
    Rocks, Ores, and Minerals 2-9
    Formations 2-13
        The Lineage of a Formation 2-14
        Examples of Formations 2-14
        Attitude of Formations—Dip and Strike 2-20
        Correlation of Rock Formations for Excavation 2-22
    Summary 2-26

3. ROCK WEATHERING 3-1

    Forces of Weathering 3-4
        Decomposition by Chemical Action 3-4
        Disintegration by Mechanical Action 3-5
        Combined Effects 3-11
    Rates of Weathering 3-12
    Summary 3-15

4. LANDFORMS AND GEOMORPHOLOGY. 4-1

    Ice 4-3
    Water 4-7
        Valleys and Gorges 4-8
        Floodplains 4-9
        Fanglomerates 4-9

Terraces 4-10
Mudflows 4-10
Mesas and Buttes 4-11
Wind 4-12
Volcanics 4-13
Gravity 4-14
Crustal Movements 4-17
Faults 4-17
Summary 4-19

5. OPEN-CUT EXCAVATIONS 5-1

Airports 5-3
Building Sites and Large-Foundation Excavations 5-4
Canals 5-5
Dams and Levees 5-6
Highways 5-8
Pits and Open Mines 5-10
Quarries 5-11
Railroads 5-12
Sanitary Fills 5-13
Trenches and Small-Foundation Excavations 5-16
Summary 5-17

6. EXAMINATION OF EXCAVATION 6-1

Office Work 6-3
Field Work 6-10
Equipment 6-10
Purpose of Field Reconnaissance 6-11
General Observations 6-12
Long-Range Surveillance 6-14
Walking Centerline—Detailed Observations 6-15
Conclusions from Field Work and Subsequent Excavation
History 6-16
Summary 6-18

7. EXPLORATION OF EXCAVATION 7-1

Manual Means 7-3
Mechanical Means 7-4
Backhoes 7-4
Bulldozers and Angledozers 7-5
Drills 7-6
Instrumental Means—The Seismic Timer 7-26
Refraction Studies 7-26
Ripping and Blasting Zones According to Seismic Velocities 7-32
Uphole Studies 7-32
Velocity-Depth Relationships for the Three Classes of Rock 7-33
Comparison of Methods for Exploration of Excavations 7-36
Summary 7-37

8. COSTS OF MACHINERY AND FACILITIES 8-1

Cost of Ownership and Operation of Machinery and Facilities 8-3
Elements of Hourly Ownership and Operating Cost Tabulation . . . 8-3
Use of Table of Hourly Ownership and Operating Costs 8-34
Production of Machinery 8-34
Continuous Production 8-34
Intermittent Production 8-35

Contents   ix

Direct Job Unit Cost. 8-35
Summary 8-36

9. CLEARING AND GRUBBING 9-1

Principles 9-3
Forest Regions 9-4
Estimating Costs 9-12
  Necessary Data 9-13
  Timber Cruise 9-13
Methods and Machinery 9-14
  Crane Method for Heavy Clearing and Grubbing 9-14
  Brush Rake Method for Light Clearing and Grubbing 9-22
Production and Cost Estimates 9-29
Summary 9-34

10. DEVELOPMENT OF WATER SUPPLY 10-1

Symbols, Formulas, and Weights and Measures 10-3
General Considerations 10-4
Fundamental Equations 10-6
  Bernoulli's Theorem 10-7
  General Equation 10-7
Friction Losses in Pipes 10-8
  Hazen-Williams Formula 10-8
Friction Losses in Hoses 10-9
Friction Losses in Pipe Appurtenances 10-9
  Equivalent Length of Pipe Method 10-9
  Flow-of-Water Method 10-19
  Velocity Method 10-20
  Tables of Losses 10-20
Example of Use of the General Equation 10-22
  Solution for Maximum Elevation, h_{e2}, of the Pump 10-23
  Solution for Mechanical Head, h_m, of the Pump 10-26
Methods and Machinery 10-27
  Pipe 10-28
  Pipe Fittings 10-28
  Valves 10-28
  Nozzles 10-30
  Meters 10-30
  Pumps 10-30
  Water Tanks 10-39
  Reservoirs 10-41
Examples of Water Supply Development 10-41
  Use of Nearby Fire Hydrant 10-41
  Use of Reactivated Line of Water District 10-43
  Use of River Water Elevated above the River 10-45
Productions and Costs 10-49
Summary 10-52

11. FRAGMENTATION OF ROCK 11-1

Ripping 11-5
  Principles 11-5
  Methods and Machinery 11-11
  Productions and Costs 11-16
Blasting 11-23
  Principles 11-23
  Explosives 11-23
  Methods of Blasting 11-37

Contents

Air Noise and Earth Vibration from Blasting 11-59
Safety in the Use of Explosives 11-60
Blasthole Drills and Drilling 11-61
Examples of Successful and Economical Drilling and Use of Explosives 11-133
Summary 11-139

12. LOADING AND CASTING EXCAVATION 12-1

Tractor-Bulldozers and Tractor-Angledozers 12-3
   Resolution of Forces when Bulldozing on the Level over Loose Rock-Earth 12-4
   Resolution of Forces when Bulldozing Downgrade over Loose Rock-Earth 12-6
   Capacities and Specifications 12-7
   Loading by Tractor-Bulldozers 12-7
   Casting by Tractor-Angledozers 12-9
   Summary 12-12
Bucket Loaders 12-12
   Crawler-Mounted Bucket Loaders 12-14
   Wheels-Tire-Mounted Bucket Loaders 12-17
Shovels 12-21
   Motions, Mechanics, and Specifications 12-21
   Operating Cycles 12-24
   Loading 12-29
   Casting 12-30
   Summary 12-31
Backhoes 12-32
   Operating Cycle and Specifications 12-33
   Efficiency 12-34
   Operating Cycle Times 12-35
   Backhoe with Telescoping Boom and Rotating Dipper 12-38
   Production and Cost Estimation for Difficult Work 12-39
Small Backhoes 12-40
   Auxiliary Machinery 12-41
   Summary 12-41
Cranes 12-42
   Uses in Excavation 12-42
   Operation 12-43
   Specifications 12-45
   Operating Cycle Times 12-46
   Auxiliary Machinery 12-49
   Crane Selection and Production and Cost Estimation 12-49
   Summary 12-51
Draglines 12-52
   Operating Cycle Times 12-53
   Production and Cost Estimation for Various Operations 12-54
   Area Mining by Dragline 12-59
   Summary 12-60
Trenchers 12-61
   Ladder Trencher 12-61
   Efficiencies of Trenchers 12-61
   Wheel Trencher 12-62
   Productions of Trenchers 12-66
   Summary 12-66
Belt Loaders 12-67
   Mobile Belt Loaders 12-67
   Movable Belt Loaders 12-72
   Summary 12-78
Wheel-Bucket Excavators 12-78
   Estimating Productions 12-79

Contents   xi

    Linear-Digging–Linear-Traveling Excavator 12-79
    Circular-Digging–Linear-Traveling Excavator 12-82
    Summary 12-84
Summary 12-85

13. HAULING EXCAVATION 13-1

    Types and Characteristics of Hauling Machines 13-3
        Characteristics and Economics of Internal-Combustion Engines . . . 13-3
        Speed Indicators and Work-Cost Indicators for Hauling Machines . . 13-7
    Crawler-Tractor-Bulldozers 13-9
    Crawler-Tractor-Drawn Scrapers and Rock Wagons 13-11
        The Effective Drawbar Horsepower 13-12
        Coefficient of Tractor and Rolling Resistance 13-13
        Dynamics of Haulage by Crawler-Tractors and Scrapers 13-18
        Production and Cost Estimation for Crawler-Tractor-Scrapers . . . 13-20
        Pull-type Rock Wagons 13-22
        Summary 12-25
    Wheels-Tires-Mounted Scrapers and Bottom-Dump, Rear-Dump, and
        Side-Dump Haulers 13-26
        Principles of Haulage by Wheels-Tires Mounted Scrapers and Haulers . 13-26
        Dynamics of Haulage 13-28
        Hauling-Cycle Calculations 13-31
    Scrapers 13-37
        Push-Loaded Scraper 13-37
        Self-Loading Scraper 13-44
        Summary 13-51
    Bottom-Dump Haulers 13-51
        Types and Functions 13-51
        Comparison with Scrapers as Fundamental Haulers 13-53
        Hauling Cycle Times and Average Loads 13-56
        Productions and Costs 13-56
        Summary 13-63
    Rear-Dump Haulers 13-63
        Types and Applications 13-63
        Hauling Cycle Times 13-70
        Average Loads 13-70
        Productions and Costs 13-72
        Summary 13-74
    Side-Dump Haulers 13-74
        Hauling Cycle Times 13-79
        Average Loads 13-79
        Production and Cost Estimates 13-79
        Summary 13-80
    Considerations for Scraper or Hauler Selection 13-80
    Belt-Conveyor Systems 13-84
        Parts of a Belt-Conveyor System 13-91
        Dynamics of a Belt-Conveyor System 13-96
        Productions and Costs of a Belt Conveyor 13-109
        Rear-Dump Truck Haulage 13-111
        Belt Conveyor Haulage 13-112
        Summary 13-113
    Cableway Scraper-Bucket Systems 13-113
        Dragscraper Machine 13-113
        Track Cable Machine 13-122
        Summary 13-127
    Railroads 13-128
        Physical and Economic Criteria 13-128
        Principles and Dynamics of Railroad Haulage 13-131
        Locomotive Selection Procedure 13-134

Production and Cost Calculations for Locomotives and Cars 13-137
Summary 13-138
Hydraulickers 13-138
   Principles of Hydraulicking 13-140
   Machinery and Methods. 13-141
   Production and Costs for Hydraulicking 13-144
   Comparison of Haulage by Hydraulicking and by Scrapers 13-145
Dredges 13-145
   Grapple Dredge 13-145
   Bucket Dredge 13-149
   Self-Propelled Hopper Dredge 13-150
   Suction-Cutterhead-Pipeline Dredge 13-150
   Production and Cost Estimation 13-168
   Summary 13-169
Summary of Haulers. 13-159

14. DUMPING AND COMPACTING EXCAVATION 14-1

Dumping and Placing 14-3
Building Compacted Embankments 14-4
   Spreading and Mixing 14-5
   Principles, Testing, and Specifications for Compaction of Embankments 14-15
   Wetting. 14-20
   Compacting 14-26
Compaction of Trash and Cover Material in a Sanitary Landfill or Dump 14-43
Summary 14-46

15. HAUL ROAD CONSTRUCTION AND MAINTENANCE 15-1

Construction. 15-3
   Grades 15-3
   Alignment. 15-4
   Methods and Machinery. 15-5
   Productions and Costs 15-12
Maintenance. 15-12
   Motor Graders 15-13
   Tow-Type Graders. 15-13
   Water Wagons 15-16
Summary 15-19

16. ALLIED OPERATIONS, MACHINERY AND COMPONENTS, AND FACILITIES 16-1

Wellpoint Systems. 16-3
   Principles, Methods, and Machinery 16-3
   Estimated 1978 Cost of Unwatering Foundation Excavation 16-5
   Alternative Dewatering Methods for Foundation 16-6
Cofferdams 16-7
   Lead Cofferdams 16-8
   Water Cofferdams 16-9
   Principles of Fundamentals of Cofferdam Design 16-9
Pumps for Unwatering for Excavation 16-12
   Centrifugal Suction Pumps. 16-13
   Centrifugal Submersible Pumps 16-13
   Diaphragm Pumps. 16-13
   Magnitude of Large Unwatering Jobs 16-18
Lighting Plants 16-19
   Principles of Illumination 16-22
   Lighting Equipment 16-27
   Estimated Costs for a Floodlighting System. 16-28

Contents   xiii

Weighing Excavation.	16-28
Tires.	16-30
Tire Construction	16-30
Factors Affecting Life of Tires	16-33
Tire Specifications.	16-36
Selection of Tires	16-36
Life of Tires.	16-46
Repairs to Tires.	16-50
Relative Costs of Nylon Bias-Ply Versus Steel Radial-Ply Tires.	16-50
Engines, Engine-Generator Sets, and Motors	16-50
Engines.	16-50
Engine-Generator Sets	16-53
Motors	16-68
Transporting Machinery	16-72
Methods and Machinery.	16-76
Production and Costs for Transporting Machinery	16-81
Servicing and Maintaining Machinery on the Job	16-82
Servicing in the Field	16-82
Maintenance in the Field	16-85
Yards, Shops, and Offices	16-87
Yards.	16-87
Shops	16-87
Offices	16-89
Summary.	16-90

17. CALCULATIONS OF QUANTITIES FOR EXCAVATION   17-1

Areas	17-3
Volumes	17-4
Subdivision Method	17-4
Prismoidal Formula	17-4
Average-End-Area Method.	17-5
Average-End-Area Method with Prismoidal Adjustment	17-15
Caclulating Excavation Volumes According to Manner of Fragmentation	17-15
Excavation Volumes and Weights According to Position or Condition.	17-16
Center of Gravity or Center of Mass of Volume in Excavation or Embankment.	17-17
Importance of Using Centers of Gravity When Estimating Hauling Costs	17-21
Measurement of Haul and the Mass Diagram.	17-23
Making Up Tables 17-1, Calculations for Mass Diagram.	17-25
Plotting Mass Diagram, Figure 17-9, from Data of Table 17-1.	17-25
Use of Mass Diagram for Calculating More Economical Haulage by Substituting Another Borrow Pit	17-25
Use of Mass Diagram for Calculating Centers of Gravity in Cuts and Fills.	17-27
Expedient Method for Approximate Volumes of Residuals, Rippable Rock, and Blasting Rock in Linear Cut of Igneous or Metamorphic Rock	17-28
Summary.	17-32

18. PREPARATION OF BID AND SCHEDULE OF WORK.   18-1

Preparation of Bid	18-3
Example of Bid Preparation	18-4
The Unbalanced Bid.	18-27
Summary of Bid Preparation.	18-29
Schedule of Work.	18-29
The Critical-Path Method (CPM)	18-30
General Observations and the Value of Preplanning and Scheduling.	18-35
Summary.	18-36

APPENDIXES. **A-1**

    1. Approximate Material Characteristics A-3
    2. Rock Clauses A-9
    3. Depreciation Schedule for Machinery and Facilities A-11
    4. Conversion Factors for Systems of Measurement A-17
    5. Formulas Frequently Used in Calculations for Rock Excavation
       Projects A-19
    6. Swell Versus Voids of Materials and Hauling Machine Load Factors.   A-21
    7. Approximate Angles of Repose of Materials. A-23
    8. Bearing Powers of Materials A-25
    9. Abbreviations. A-27

Glossary **G-1**

Bibliography **B-1**

Index follows bibliography.

Preface

The *Excavation Handbook* discusses the methods, machinery, and costs for excavating the materials of the Earth's mantle, or regolith, from their initial locations in situ to their final deposition in construction or their final or near final deposition in mining. These materials range from the earthy to the rocky.

Soft earths are really finely divided particles of rock and they are, petrologically, little rocks. Frequently in the text the material is called rock, rock-earth, earth-rock, or earth, the two intermediate designations meaning that either rock, rock-earth, or earth, earth-rock, predominates in the material.

The word "excavation" means work in an open cut, exposed to the sky, and it does not include underground excavation, as in mining and tunneling, except when tunneling is supplementary to open-cut excavation.

Purpose of the Book

The *Excavation Handbook* may be used for general and specific information, for selecting methods and machinery for work, for estimating productions and costs of machinery for a given task, and for estimates of the total cost of excavation in construction and mining in open cuts.

The book is for the use of all college students in engineering, construction, and geology, for private engineering companies, for public engineering agencies, for construction and mining companies, and for machinery, explosives, and supplies companies. It may also be entertaining to people who have a general interest in the art and science of moving the Earth.

Coverage of the Book

The *Excavation Handbook* includes these aspects and phases of many excavations:

1. Geology of excavation as it relates to rocks, ores, minerals, rock formations, rock weathering, and landforms.
2. Kinds of open-cut excavations, which are indicative of the diversity and size of these cuttings and fillings.
3. Examination and exploration of excavations, which are requisite to a well-considered estimate of costs and to a later successful prosecution of the work.
4. The costs of ownership and operation of the machinery and facilities, into which the productions of machinery are divided in order to secure unit costs for the work.
5. The methods, machinery, productions, and costs for the regular seven operations in excavation.
6. Important correlative operations, the principal secondary machines, and the complements of the major machines.
7. Calculations for quantities of excavation, into which productions of machines are divided to determine time for the work.
8. Preparation of a typical bid for excavation and a typical means for scheduling excavation as a part of the complete project.
9. Appendixes, a glossary, and a bibliography covering the science and practices of excavation.
10. A complete, detailed index.

Acknowledgment

I would like to express my appreciation to the many generous people of the earthmoving industry who contributed to the particulars and completeness of the *Excavation Handbook*. No writer is, in fact, a complete authority on his chosen subject, especially on excavation, with its variables of Nature.

Horace K. Church

CHAPTER 1

Earth and Geology of Excavation

PLATE 1-1

Aggradation or uplift of the Sierra Nevada of California. The Sierra Nevada are a classical example of Nature's aggradation or uplift of the Earth's surface. Mount Whitney towers at an elevation of 14,495 ft along with the eastern edge of the batholith, which is some 350 mi from north to south and some 80 mi from west to east. This orogeny commenced some 180 to 135 million yr ago in the Jurassic period and, although intermittent, the continuous raising of the westward dipping fault block is still in progress. Along with the aggradation is the relentless degradation of the land surface by wind, rain, snow, and ice. Throughout the aeons the uplifting of the massive granite has triumphed over the opposing weathering of the many kinds of surficial rocks, and Mount Whitney is slowly increasing in elevation. *(National Park Service.)*

Excavation is the act of removing, moving, and depositing the veneerlike surface of the Earth's outermost crust or regolith. The material excavated may be in the solid or semi-solid state, that is, rock; in the weathered state, usually a mixture of rock and earth; or in the loose state, earth.

In many cases excavation, because of the different degrees of weathering, may be called rock, rock-earth, earth-rock, or earth. The distinction depends upon the relative amounts of rock and earth. Lithologically, all excavated material may be called rock, the rock differing only in the size of the individual particles. Extremes are found in the material basalt. In the solid state in a lava flow it is basalt rock. In the loose state in a cultivated field it is red clay or earth. Between these extremes is the weathered state, a mixture of rock-earth or earth-rock, depending upon the relative amounts of rock and earth.

In construction of public works such as dams and highways and in the mining of quarries and open pits, the finished grade is rarely at a depth greater than 400 ft below natural or original ground. There are some notable exceptions such as the iron mines of Minnesota and the copper mines of Utah and Arizona, as well as some other large open-pit works.

Such deep excavations, spectacular as they may seem, are a mere 50 thousandths of the Earth's radius. They are in a weathered and solid zone wherein nature's destructive forces have been toiling for millions of years. These ruinous energies are the most important ally of human excavators, because if all excavation were in the rock or solid state much more time and money would have been expended for the trillions of cubic yards of excavation since the beginning of recorded history some 6000 yr ago.

The veneerlike surface of the Earth's crust is a variable and fascinating earth-rock mixture and it is presumptious and dangerous for the excavator to remove it without preliminary study. The greater the understanding of geology and its kindred sciences, the greater the mastery of the complexities of excavation.

Some 5 billion yr ago the planet Earth was formed from a gaseous substance which, through cooling and related processes, became a slowly consolidating molten mass. Some 4 billion yr ago the igneous rock basement complex of Earth had been formed, to be followed by systematic degradation by nature's forces. This relentless debasement resulted in the deposition of both marine and nonmarine or continental sedimentary rocks.

Simultaneously with the formation of the sedimentary rocks, the igneous rocks did not remain dormant but rather continued the orogeny by their intrusions and extrusions. From these rocks both metamorphic and additional sedimentary rocks were derived, not only in the past but in the geologic present.

Nature's powerful processes of building up and tearing down are continuously in evidence. An example of building up or aggradation is provided by the frequent lava flows of the volcanoes of the Hawaiian Islands. The six islands are the result of volcanic eruptions from the bottom of the sea during a period of some 8 million yr. Judging by the degrees of rock weathering, the oldest island is the westernmost Kauai and the youngest is the easternmost Hawaii. Mauna Loa is the perennial active volcano of the island chain, the accumulative lava of Hawaii towering 13,800 ft above the Pacific Ocean. The sea bottom is about the same distance below the ocean surface. The sea-bottom floor area of Hawaii is some 16,000 mi^2, and so Hawaii Island represents about 28,000 mi^3 of lava flows.

Turning to more understandable present-day volcanic aggradation, Figure 1-1 shows the 1955 flow of the less pretentious Kilauea Crater of Hawaii.

The area of the picture is 3400 ft north to south and 3400 ft west to east. The black basalt flow within the limits of the picture is 1.5 million ft^2 with an average 8-ft depth. The flow of 440,000 yd^3 took out the north-to-south road, isolating the farmers and cattlemen. The road was restored five days after the flow. Bulldozers worked atop the flow when temperatures had cooled from 1000°C (1830°F) to 700°C (1290°F), the incandescent molten lava being visible through the cooling cracks of the solidifying lava.

Every decade or so there is a sizable, perhaps 1-million-yd^3 lava flow on Hawaii, a sizable contribution to the land but infinitesimal when contrasted with the 153 trillion (153,000,000,000,000) yd^3 of lava of the island Hawaii.

Aggradation, even in the form of breathtaking molten lava flows, is not as spectacular as degradation in the form of landslides. One such massive downward movement of rock

1–4   Earth and Geology of Excavation

Figure 1-1   Aggradation by flow of lava. Basalt lava flow of Kilauea Volcano, Hawaii, in 1955. The area of the picture is 0.41 mi², 3400 ft by 3400 ft. The volume of the black igneous extrusive rock within the picture is 440,000 yd³. *(Hawaiian Volcano Observatory.)*

was the Wood's Gulch slide of 1974 on the abandoned old California Highway 1 of San Mateo County.

Figure 1-2 illustrates this slide of an estimated 110,000 yd³, which weakens the lateral support for many new homes located on an artificial terrace above the shore.

There were several mutually contributing causes of this slide. First, the Pacific Ocean is constantly battering the shoreline foundation of the old Coast highway. Second, the rock structure is unstable, as the moisture-laden Merced formation of sedimentary sandstones, siltstones, and claystones is poorly indurated and rather steeply dipped. Third, earthquakes along the Wood's Gulch fault and the nearby parent San Andreas fault have further weakened the rock formation over the millenia. And fourth, the highway was excavated within a cliff in a location fraught with natural and artificial hazards such as encroaching home-building sites destructive of natural drainage of the steep slopes.

The slide-prone shoreline extends for 12 mi from Daly City southward to Montara and, wherever possible, the Coast highway is being relocated inland away from the treacherous cliffs of the sea.

In the Wood's Gulch slide only an unused highway was immediately affected. This was not the case in the Orinda slide of 1950 in Contra Costa County, California.

Figure 1-3 emphasizes the tragic possibilities of this 250,000-yd³ slump slide of 800-ft height and 300-ft width. After several days of heavy rains the hillside with its cover of shrubs and trees began to slip at 10:30 A.M. on December 9. Within 15 min the four-lane arterial California Highway 75 was covered and the earth flow continued for about 2 h.

In spite of normal Saturday morning traffic in the San Francisco Bay area, averaging 35,000 vehicles daily, there were miraculously no fatalities and no injuries.

The view shows the emergency excavation for the four-lane detour around the toe of the slide. By Saturday night 9000 yd³ of near-liquid-state sediments had been removed, the detour was paved, and the highway was ready for the Monday morning rush of commuters.

Figure 1-2 Degradation by slide of coastline highway. Massive coastline slide of 110,000 yd³ at Wood's Gulch, San Mateo County, California. This degradation occurred in 1974 on the old Coast highway, previously abandoned because of ill-considered inroads on the instability of nature and the subsequent weakening of a rock structure by excavations for a highway and a huge home site along the coast line. *(California Division of Mines and Geology.)*

The causes of the slide are portrayed graphically in Figure 1-3. First, the area is slide-prone, as the rather steep natural slope of the young 10-million-year-old Orinda formation is made up of soft sandstones, shales, and conglomerates, sometimes steeply dipped. The wet shale beddings act as a lubricant for the sliding actions of the adjacent sandstones and conglomerates. Second, the area is within 3 mi of the Pinole and Wildcat faults, subparallel to the notoriously active Hayward fault, and, accordingly, is weakened structurally. Third, excavating to the back slope of the existing highway had reduced the stability of the natural ground by removing the buttress. Fourth, percolation of the winter rains provided lubricated slip planes in the layered sedimentary rocks.

The regrading of the slide area provided for two important future slide preventatives. The slope of the cut was reduced to a ratio of 2:1, considerably increasing the stability. Some 20,000 lineal feet of horizontal drains were installed in the area of the slide and a total flow of some 135,000 gal/day of water was withdrawn to dry out the rock formation. The slope setback at a ratio below the angle of repose and the dewatering have resulted in stable conditions.

Although degradation is more noticeable than aggradation, the two different and opposing natural processes are in delicate balance so that the mountains and the valleys are eternally in equilibrium.

An example of these adversary forces occurs in the Black Hills of the Wyoming–South Dakota border. Degradation is obvious in the daily weathering of the limestones and redbeds and in the stream erosion. Not apparent is the slow, inexorable uplifting of the sedimentary rock dome by igneous rock intrusion, which commenced some 80 million yr ago. After initial uplifting, ending some 35 million yr ago when the dome towered 7500 ft above the plains, weathering reduced the high elevation to the present 4000 ft. Thus,

1-6 Earth and Geology of Excavation

Figure 1-3 Degradation by slide of mountain highway. The beginning of emergency slide removal of the Orinda slump slide of 1950 in Contra Costa County, California. During the winter rainy season 250,000 yd^3 of liquidlike soft rock avalanched over the arterial highway in 15 min. Four-lane traffic was restored in 2 days by emergency crews of the California Division of Highways. Phenomenally, there were no injuries or deaths to travelers and there was little inconvenience to San Francisco Bay area commuters, as the slide and the construction detouring took place on a weekend. *(San Francisco Chronicle, Dec. 10, 1950.)*

building up has exceeded tearing down by 3500 ft. The differential rate is only $5/8$ in every 1000 yr, but it has been of a positive or growth nature.

Another striking example of the triumph of aggradation over degradation is the story of the Sierra Nevada, as described in Plate 1-1.

It is well for the excavator to have a good understanding of geologic time because it will be advantageous to correlate the same rock formations in terms of the geologic time scale in diverse locations around the world.

The Principle of Uniformitarianism was established brilliantly by the father of modern geology, James Hutton (1726–1779). Some 200 yr later the earthmover finds it safely practical to conclude that he will have blasting in the Hornbrook sandstone formation near Hilt, northern California, because he blasted the similar Rosario sandstone formation at Point Loma, southern California. Both formations are Upper Cretaceous marine sedimentary rocks of thick beddings. They are 700 mi apart but of the same geologic series of rocks, 63 to 99 million yr of age.

The early studies of James Hutton and the subsequent laborings of many geologists have produced the main divisions of geologic time for North America, as set forth in Table 1-1.

The rocks of the Cenozoic, Mesozoic, and Paleozoic groups are fairly accurately dated by fossil correlations in the sedimentary rocks and by isotypes in the igneous and metamorphic rocks.

The rocks of the Proterozoic, Archeozoic, and Azoic groups are likewise precisely dated by isotypes, as fossils do not exist in these groups except for rare instances of algae fossils.

TABLE 1-1 Main Divisions of Geologic Time for North America

Era or group of rocks		Period or system of rocks		Epoch or series of rocks	Scale of time (millions of years)	
					Interval	Age
Cenozoic (age of mammals, grass, and land forests)	63	Quaternary	2	Holocene Pleistocene or Glacial	2	2
		Tertiary	61	Pliocene	11	13
				Miocene	12	25
				Oligocene	11	36
				Eocene	22	58
				Paleocene	5	63
Mesozoic (age of reptiles)	167	Cretaceous	72	Upper Cretaceous	36	99
				Lower Cretaceous	36	135
		Jurassic	45		45	180
		Triassic	50		50	230
Paleozoic (age of fishes, amphibia, and swamp forests)	340	Carboniferous	110	Permian	50	280
				Pennsylvanian	30	310
				Mississippian	30	340
		Devonian	60	Upper Devonian	20	360
				Middle Devonian	20	380
				Lower Devonian	20	400
		Silurian	30	Cayugan	10	410
				Niagaran	10	420
				Oswegan	10	430
		Ordovician	70	Cincinnatian	23	453
				Mohawkian	24	477
				Canadian	23	500
		Cambrian	70	Saratogan	23	523
				Acadian	24	547
				Georgian	23	570
Proterozoic (Algonkian age of scum, algae, and jellyfish)	630	Keweenawan	250		250	820
		Huronian	250		250	1070
		Temiskamian	130		130	1200
Archeozoic (Archean; age of life)	800	Keewatinian	800		800	2000
Azoic (no life)	2500				2500	4500

NOTE: This is a composite table, representing the thoughts of several geologists of these times. The ages are in millions of years, and the age of the earth is thought to be about 4.5 billion yr.

The oldest rocks have been dated at 3.3 billion years of age in Rhodesia, or about three-quarters of the age of the Earth.

SUMMARY

The continual and balancing processes of aggradation and degradation have produced the infinitely complex surface of the Earth. It is this regolith that the excavator must remove. Accordingly, it is fundamental that a prudent excavator must be acquainted with the uniformity and the irregularity of the regolith.

CHAPTER 2

Rocks, Ores, Minerals, and Formations

PLATE 2-1

Aged rocks of the Grand Canyon of the Colorado River. The majestic Grand Canyon of the Colorado River in Arizona is Earth's greatest display of rocks over the longest time. On the Kaibab Trail from the South Rim down to the Colorado River and up to the North Rim is a 1.2-billion-yr-old geologic record of igneous, sedimentary, and metamorphic rocks which encompasses about one-fourth the age of Earth.

Plate 2-1 shows the panorama of this assemblage of rocks, ranging from the oldest black metamorphic rocks of the Inner Gorge of the Colorado River up to the white sedimentary rocks of the South Rim, some 1 mi above the Colorado River. One may cross-refer this picture of the stratigraphy of the Canyon to Figure 2-1 and Table 2-1 of this chapter. The numbers of Plate 2-1 approximate the zones of the principal rock formations of the Canyon.

    1. Archeozoic group of metamorphic and igneous rocks. Vishnu schist with granite intrusions. Perhaps 1200 to 2000 million yr ago.

    2. Proterozoic group of metamorphic, igneous, and sedimentary rocks. Vishnu schist with granite dikes. Bass limestone and Hakati shale. Perhaps 570 million to 1200 million yr ago.

    3. Cambrian system of metamorphic and sedimentary rocks. Shinume quartzite, Bright Angel shale, and Muav limestone. 500 to 570 million yr ago.

    4. Devonian system, Temple Butte limestone, and Mississippian series, Redwall limestone, of sedimentary rocks—310 to 400 million yr ago. Between the Cambrian and Devonian periods there was a long interval of erosion or unconformity lasting some 100 million yr.

    5. Pennsylvanian series of sedimentary rocks. Supai limestone and shale—280 to 310 million yr ago. Between the Mississippian and the Pennsylvanian series there is another unconformity of unknown time interval in the sequence of rocks. Consequently no time for the unconformity is hazarded.

    6. Lower Permian series of sedimentary rocks. Hermit shale. Perhaps 263 to 280 million yr ago. Between the Pennsylvanian and the Lower Permian series of rocks there is another erosion surface of unknown length and so no time for this unconformity is speculated.

    7. Middle Permian series of sedimentary rocks. Coconino sandstone. Perhaps 247 to 263 million yr ago.

    8. Upper Permian series of sedimentary rocks. Toroweap limestone and sandstone in lower zone and Kaibab limestone and sandstone in upper zone, reaching the surface of the land. Perhaps 230 to 247 million yr ago.

Between 230 million yr ago and the present there are no rocks, except for the surficial residual sands of the desert country. Mesozoic and Cenozoic groups of rocks are missing. The Grand Canyon is in the midst of another great period of unconformity and another immense erosion surface is being formed, encompassing much of northern Arizona.

If geologic history customarily repeats itself, some millions of years in the future the sea will again invade the land and it will deposit another series, system, or group of sedimentary rocks. Then the land will rise again and future geologists will mark the sands of the South Rim as the fourth unconformity of the Grand Canyon of the Colorado River. *(U.S. National Park Service.)*

THE GRAND CANYON—THE CLASSICAL RECORD OF ROCKS

The Grand Canyon of the Colorado River in northern Arizona features the Earth's hugest excavation by weathering and river erosion and probably the most complete display of rock types of the last 1.2 billion yr. In the 1-mi-deep and 15-mi-wide Canyon the walls show a variety of igneous, sedimentary, and metamorphic rocks.

Let us go down a mile into the Earth and go back into the aeon of 1.2 billion yr ago.

The scene is the Kaibab Trail, commencing at Yaki Point on the South Rim of the Canyon at elevation 7260 ft and ending at the suspension bridge across the river at elevation 2420 ft. The distance is about 8 mi and the downward grade is about 10 percent throughout the 4860-ft depth.

If one wishes to examine the multicolored rocks in detail and to appreciate the plants, the animals, and the magnificent scenery, a 3-day hike is in order.

Figure 2-1 and Table 2-1 show the rock and rock-formation relationships. Table 2-1 gives the log of the probe into the earth's crust and it may be cross-referred to Table 1-1.

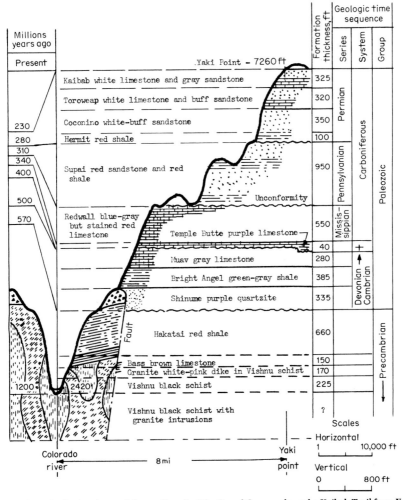

Figure 2-1 Geologic column of the south wall of the Grand Canyon along the Kaibab Trail from Yaki Point to the Colorado River.

2-4 Rocks, Ores, Minerals, and Formations

TABLE 2-1 Geologic Column of the South Wall of the Grand Canyon along the Kaibab Trail from Yaki Point to Colorado River

	Formation thickness	Elevation (ft)
Yaki Point on the south rim, about 230 million yr ago		7260
Paleozoic group of rocks		
Permian series, 230–280 million yr ago		
Kaibab white limestone and gray sandstone	325	
Toroweap white limestone and red-buff sandstone	320	
Coconino white-buff sandstone	350	
Hermit red shale	100	
	1095	
Erosion surface or unconformity of unknown time interval		6165
Pennsylvanian series, 280–310 million yr ago		
Supai red sandstone and red shale	950	
Erosion surface or unconformity of unknown time interval		5215
Mississippian series, 310–340 million yr ago		
Redwall blue-gray but stained red limestone	550	
Devonian system, 340–400 million yr ago		
Temple Butte purple limestone	40	
Erosion surface or unconformity of 400–500 million yr ago, representing missing Silurian and Ordovician systems		4625
Cambrian system, 500–570 million yr ago		
Muav gray limestone	280	
Bright Angel green-gray shale	385	
Shinume purple quartzite	335	
	1000	
Erosion surface or great unconformity of unknown time interval		3625
Proterozoic and Archeozoic or Precambrian group of rocks, 570–2000 million yr ago		
Hakatai red shale	660	
Bass brown limestone	150	
Granite white-pink dike in Vishnu schist	170	
Vishnu black schist	225	
	1205	
Colorado River at the Inner Gorge, about 1200 million yr ago		2420

It is said that the 8-mi hike gives a white-to-black spectrumlike picture of rocks equivalent to a 2400-mi journey from Mexico's Sonora Desert to Canada's Hudson Bay. The range is from white sedimentary Kaibab limestone through pink igneous granite to black metamorphic Vishnu schist.

Down the trail to the Colorado River we pass through at least 970 million yr of many kinds of rocks in about 10 h at a rate of 97 million yr/h. However, if one wished to analyze these same rocks on the surface of the land from the Sonora Desert to Hudson Bay, it would probably require 3 months.

Sixteen hours or so should be devoted to the long uphill hike so as to observe the geology, the landforms, the formations, the rocks, the animals, and the plants. Literally we emerge from the depths of time of the earliest traces of life in the Bass limestone, 470 ft above the river, to the tracks of reptiles in the dune sands of the Toroweap sandstone, 485 ft below Yaki Point. At Yaki Point let us consider Nature's prodigious excavation job during the assumed 10 million yr of erosion of the Canyon.

Let the Canyon be confined to Grand Canyon National Park with a length of 100 mi, an average width of 15 mi from rim to rim, and an average depth of 1 mi. The following statistics may be derived from these dimensions.

Rock volume, eroded and removed ultimately by the river	750 mi^3
or	4,095 billion yd^3
Yearly excavation	410,000 yd^3

Excavating 410,000 yd^3 of rock annually is not impressive to the average earthmover, but so awesome is geologic time that 4.1 trillion yd^3 of rock excavation is mind-boggling. It is equivalent to a trench from Seattle, Washington, to Miami, Florida, with a length of 2600 mi, a width of 1 mi, and a depth of ¼ mi.

After our return to Yaki Point we stand on the 230-million-yr-old Kaibab formation and we have come some 4840 ft vertically upward from the Colorado River. Our upgrade trail has taken us through almost 1 billion yr of rock formations and from a time 1.2 billion yr ago.

Such great time is astounding; let us visualize it by means of a descriptive book commencing 1.2 billion yr ago, each page representing the average threescore years and ten of our lifetimes. The book has some 17,142,857 pages and it is about 1429 ft in thickness. Page 1 describes the Vishnu schist and the last page describes our last view from Yaki Point.

Along about page 3,510,375 we find a description of an algae fossil in the Bass limestone, the beginning of the chapter of life, Genesis. Near page 9,527,000 there is a reference to trilobite fossils in the Bright Angel shale, a definite form of life. An earliest recognizable fish is discussed on page 11,857,143 or thereabouts. This fossil is in the Temple Butte limestone. In the windblown sands of the Coconino sandstone are the fossil tracks of reptiles and amphibians, as described on or about page 13,322,244. The record of the rocks ends about 230 million yr ago at the top of the Kaibab formation and at about page 13,857,143. Thereafter no record exists at Yaki Point. Hence the book has 3,285,714 empty pages except for the advent of humans.

Perhaps humans, immigrants from Asia by way of the land bridge of the Aleutian Islands during the last glacial period, first saw the Canyon about 25,000 yr ago. If such is the case, they would appear on page 17,142,500. There are only 357 pages left and when we leave Yaki Point we are recorded on the last page.

We leave Yaki Point in awe of a mighty chasm, The Grand Canyon of the Colorado River.

DEFINITION OF ROCK

In this text *rock* is defined as material that forms an essential part of the Earth's crust. It includes loose incoherent masses such as beds of sand, gravel, clay, and volcanic ash, as well as firm, hard, solid masses such as granite, sandstone, limestone, quartzite, and schist.

According to this definition and in keeping with the sense of rock excavation, rock also includes the metallic and nonmetallic ores and minerals which are usually contained within the rocks.

One example is the open-pit mining of chromite, in which the disseminated chromium ore is excavated like clay, as in California. Another example is strip mining of subbituminous coal, a sedimentary rock, by huge stripping and loading shovels in Wyoming. A third example is the quarrying of nickel in veins of gabbro at Sudbury, Ontario, by open-pit methods.

The rock excavator should know as much as possible about rocks, ores, and minerals. Success depends on the excavator's familiarity with the materials excavated.

CLASSIFICATION OF ROCKS

By definition, rocks vary from the hard Barre granite of Vermont to the loose granitic sands of California. There are three classes of rocks, containing or not containing ores and minerals, which are forming and have been forming almost from the beginning of the existence of planet Earth.

Igneous Rocks

These are the original rocks, formed initially at varying distances below the Earth's surface. There are two kinds, intrusive and extrusive.

Intrusive, Sometimes Called Plutonic These rocks originated as molten magma and they slowly cooled at great depth, with resultant large crystals. A well-known example is the majestic granite batholith of the Sierra Nevada of California and Oregon, as portrayed in Plate 1-1.

The rock to the left in Figure 2-2 is intrusive granite diorite formed 99 to 180 million

Figure 2-2 Igneous rocks: *(Left)* Intrusive, or plutonic, granite diorite from the Sierra Nevada of California. *(Right)* Extrusive, or volcanic, basalt from the Columbia River plateau of Oregon.

yr ago from the Sierra Nevada. The dark crystals are hornblende and the light crystals are quartz. Depending on the degree of weathering, the rock is soft to hard excavation.

Extrusive, Sometimes Called Volcanic These rocks burst forth at or near the Earth's surface and they cooled rapidly, with consequent minute crystals. An example is the gigantic lava flow of the Columbia River Plateau of Washington, Oregon, and Idaho.

The rock to the right in Figure 2-2 is extrusive basalt of the Columbia River Plateau, 2 to 13 million years old. The holes of the characteristic vesicular structure were caused by air entrapment during cooling, giving the rock a rough appearance. Again, depending on the degree of weathering, basalt may be medium to hard excavation.

Sedimentary Rocks

These rocks are the consolidated and cemented products of the disintegration and decomposition by weathering of the three groups of rocks, igneous, metamorphic, and even sedimentary rocks themselves. They also derive from limy constituents of lake and ocean waters, limestones and coquinas being examples.

The rocks are consolidated by mechanical cementation and by chemical precipitants, sometimes accompanied by pressure. They are found at or near the Earth's surface. If formed on land they are called continental or nonmarine. If formed in the sea they are called marine. Originally they were laid down in horizontal or near horizontal beds with thin to thick bedding. Subsequent crustal deformations may have tilted, folded, and faulted these original strata.

The rock to the left in Figure 2-3 is a soft, punky, well-weathered sandstone of the Del Mar formation of San Diego County, California. This 36- to 58-million-yr-old rock is part of an excavation for dwellings and it is soft excavation for medium-weight tractor-ripper fragmentation.

The rock to the right in Figure 2-3 is hard, unweathered dolomite from the Tule River Indian Reservation of southern California. This bluish-white rock is 230 to 310 million yr old and it requires blasting for fragmentation. It is a form of limestone and these chemically precipitated marine rocks make up hundreds of material pits and quarries of America.

Figure 2-3 Sedimentary rocks. *(Left)* Well-weathered soft sandstone of the Del Mar formation of San Diego County, California. *(Right)* Hard unweathered dolomite from the Tule River Indian Reservation of Tulare County, California.

Metamorphic Rocks

These rocks are formed from existing igneous, sedimentary, and metamorphic rocks. In the last case they have simply reworked themselves. All have been formed by heat, pressure, and attendant gases and liquids below the earth's surface. Two examples are shown in Figure 2-4.

To the left is slate, derived from shale. It is a marine metasedimentary rock of the Mariposa formation with an age of some 146 million yr. It is from a 30-year-old highway cut in Mariposa County, California. Although slightly weathered in the exposed slope, it represents extremely hard ripping for a heavyweight tractor-ripper or else requires blasting. Metamorphic rocks such as this slate are characteristic of the Mother Lode country, which is some 150 mi long and 50 mi wide.

To the right is a granite gneiss from the Marble Mountains of San Bernardino County, California. It is of the Archeozoic group of rocks, dated at about 1450 million yr of age. This rock in its unweathered state requires blasting.

Sometimes the demarcation between igneous and metamorphic rocks is ill-defined, and the process may be reversed. The metamorphic rocks may become molten and cool into igneous rocks. In turn, these may be weathered into sands and become sandstones. And sedimentary shales may likewise be partially metamorphosed to relatively soft shale-slates.

Obviously the igneous, sedimentary, and metamorphic rocks may pass through many cycles, frequently doubling back on themselves. Two examples of metamorphic rocks are shown in Figure 2-5.

To the left is a marble, derived from limestone. This hard rock is from the south portal of the excavation for the San Bernardino Tunnel, California. This Precambrian group of rocks is more than 570 million yr old. The marbles and associated gneisses and schists were blasted in the open-cut excavation.

To the right is finely laminated schist, and the schistose structure is visible above the scale. Schists are formed from igneous and sedimentary rocks or even from reworked metamorphic rocks, and they are generally extremely hard and of great age. This schist is in the foundation excavation for Auburn Dam on the American River, Placer County, California. This schist occurs in an ultrabasic formation made up largely of amphibolites and schists and it is about 160 million yr old. About 20 ft below natural ground or below the weathered rocks it was necessary to blast the rock formation.

The transition from an early stage of metamorphism to a late stage is shown in Figure 2-6. The rock to the left is greenstone, altered from intrusive igneous diabase. The three rocks to the right are serpentine, the one to the extreme right showing accessory asbestos fibers. The rocks are from the Franciscan formation of age about 135 million yr, located on the Redwood Highway, Mendocino County, California.

Figure 2-4 Metamorphic rocks. *(Left)* Extremely hard slate, a metasediment derived from shale. This durable rock is a part of the Mariposa formation of the California Mother Lode country, the formation being made up of slates, phyllites, and metasedimentary sandstones and conglomerates. Along with the Calaveras formation and several others, the Mariposa formation makes up an extremely complex assemblage of rocks covering some 7500 mi^2 of the gold country. *(Right)* Hard granite gneiss from the Marble Mountains of San Bernardino County, California. The rock is a well-weathered cobble, but in its natural state in the unweathered rock formation it would require blasting. The banded rock structure is a property of the gneisses.

2-8   Rocks, Ores, Minerals, and Formations

Figure 2-5 Metamorphic rocks. *(Left)* Extremely hard marble, derived from limestone. Finely seamed structure. From the San Bernardino Mountains, San Bernardino County, California. This not too common rock in excavation is invariably hard and requires blasting in its natural unweathered state. *(Right)* Extremely hard schist, probably derived from metamorphic slates. Finely banded structure as contrasted with the gneisses. The rock is from Placer County in the Mother Lode country of California, and it came from a dam foundation where blasting was in progress.

Figure 2-6 Metamorphic rocks from early state to late stage of change. *(Left)* Greenstone, altered from igneous intrusive diabase. *(Right)* Three rocks of serpentine, altered from the greenstone. The serpentine rock to the right of the group shows accessory asbestos fibers. All rocks are from the Franciscan formation of Mendocino County, California.

The actual three-stage transition is from hard diabase, requiring blasting, to medium-hard greenstone, calling for hard ripping by a heavyweight crawler-tractor-ripper, to soft to medium-hard serpentine, generally fragmented by soft to medium ripping.

A common mineral for boron is ulexite, and Figure 2-7 shows two examples of this sedimentary rock from the open pit of the United States Borax and Chemical Corporation at Boron, California. The ulexite to the left shows the laminations typical of

Figure 2-7 Ulexite, a common mineral of boron. *(Left)* A typical laminated specimen of this sedimentary rock, showing a stratum or layer of fibrous structure. *(Right)* A common example of the white fibrous silken complexion of the ulexite. The minerals are from the open-pit mine of the United States Borax and Chemical Corporation, located in the midst of the Mojave Desert of California.

sedimentary rocks, and the ulexite to the right shows the common loosely intergrown fibrous crystals.

The overburden above the boron ore is several hundred feet in thickness and it consists largely of desert alluvia and clays. It is excavated and wasted several thousand feet beyond the rim of the huge 96-million-yd^3 excavation.

Mining practice is to blast this hard ulexite. It is then loaded out by electric shovels, hauled by diesel-electric trucks to the primary crusher, and elevated by conveyor from the floor of the open pit to the stockpiles and preparation plant near the rim of the pit.

ROCKS, ORES, AND MINERALS

The common rocks, ores, and minerals encountered in construction and in open-pit mining are summarized in Table 2-2.

TABLE 2-2 Rocks, Ores, and Minerals

Igneous Rocks	Formed by crystallization from an originally deep-seated magma.
Intrusive igneous	Formed below the surface by the slow cooling of magma, resulting in medium- to large-size crystals.
Diabase	Dark color. Small crystal matrix with large, elongated white crystals. Medium to hard excavation.
Diorite	Light to dark color. Medium, uniform-size crystals. Soft to hard excavation.
Gabbro	Dark color. Medium-size crystals. Medium to hard excavation.
Granite	Light to dark color. Medium to large crystals. Soft to hard excavation.
Porphyry	Dark color. Fine-grained crystal matrix with coarse light crystals. Medium to hard excavation.
Syenite	Light to medium color. Small crystals. Medium to hard excavation.
Trap rock	Quarry worker's term applied to such intrusive igneous rocks as diabase and gabbro. Dark color. Small- to medium-size crystals. Medium to hard excavation.
Extrusive igneous	Formed near or on surface from fissures or from volcanic eruption. Rapid cooling causes minute crystals and sometimes vesicles or small gas pockets.
Andesite	Medium to dark color. Fine-grained lava flow with thin to thick beddings. Sometimes called felsite. Medium to hard excavation.
Basalt	Dark color. Fine-grained lava flow with thin to thick beddings. Generally characterized by columnar structure of polygonal cooling joints and by vesicles. Medium to hard excavation.
Breccia, volcanic	Light- to dark-colored fine to coarse pyroclastic rocks with different degrees of consolidation due to welding. Originated by violent airborne eruption. Soft to medium excavation.
Obsidian	Dark color. Glassy with no visible grains because of instantaneous cooling. Thin to thick beddings. Medium to hard excavation.
Pumice	Light color. Porous because of release of gases during cooling. Soft to medium excavation.
Rhyolite	Light to medium color. Fine-grained lava flow. Thin to thick beddings. Soft to hard excavation.
Scoria	Light to dark color. Cindery and jagged because of release of gases during cooling. Rhyolite family. Thin to thick beddings. Soft to medium excavation.
Trap rock	Quarryworker's term for extrusive igneous rocks such as basalt and andesite. Medium to dark color. Small crystals. Medium to hard excavation.

TABLE 2-2 Rocks, Ores, and Minerals *(Continued)*

Sedimentary Rocks	Formed by degradation of existing rocks and organic remains and their subsequent deposition by natural forces.
Unconsolidated rocks	
Clays	Light to dark color. Extremely fine size, less than 0.005 mm. More or less plastic. Thin to thick beddings. Soft to medium excavation.
Silts	Light to dark color. Fine size, between 0.005 mm and 0.05 mm. Soft excavation.
Sands	Light to dark color. Fine to medium size, between 0.05 mm and 2.0 mm. Soft excavation.
Gravels	Light to dark color. Medium to large size, between 2.0 mm and 20 cm.
Boulders	Light to dark color. Large to extremely large size, 20 cm to larger than 10 m.
Consolidated rocks	Formed by cementation of deposited rocks by silica, calcium carbonate, or iron oxide, by chemical precipitation from solutions, and by the action of organic agents.
Agglomerate	Cemented angular, as contrasted with rounded, fine to coarse particles. Equivalent to breccia. Light to dark color. Thin to thick beddings. Soft to hard excavation.
Coal, bituminous	Dark color. Compressed organic material associated with limestones, sandstones, and shales. Medium to thick beddings. Medium to hard excavation.
Conglomerate	Cemented, rounded, fine to coarse particles. Light to dark color. Thin to thick beddings. Soft to hard excavation.
Dolomite	Carbonate rock both chemically and organically derived. Light to medium color. Thin to thick beddings. Hard excavation.
Lignite	Brown coal. Medium brown color. Mildly compressed organic matter associated with shales and sandstones. Thin to thick beddings. Soft to medium excavation.
Limestone	Carbonate rock both chemically and organically derived. Differs from dolomite by its lower magnesium carbonate/calcium carbonate ratio. Light to medium color. Thin to thick beddings. Hard excavation.
Limestone, bituminous	Limestone impregnated with bitumen. Medium to dark color. Up to 20% bitumen content. Thin to thick beddings. Medium to hard excavation.
Sandstone	Cemented sand. Light to dark color. Thin to thick beddings. Soft to hard excavation.
Sandstone, bituminous	Sandstone impregnated with bitumen. Medium to dark color. Up to 10% bitumen content. Thin to thick beddings. Soft to hard excavation.
Shale	Cemented clay. Light to dark color. Thin to thick beddings. Soft to hard excavation.
Shale, bituminous	Shale impregnated with bitumen. Medium to dark color. Up to 15% bitumen content. Thin to thick beddings. Soft to hard excavation.
Siltstone	Cemented silt. Light to dark color. Thin to thick beddings. Soft to hard excavation.
Metamorphic Rocks	Formed from existing rocks by heat, pressure, and mineral-bearing hydrothermal solutions, acting together or separately.
Coal, anthracite	Black color. Associated with altered limestones and shales. Hard excavation.

Metamorphic Rocks

Gneiss — Formed from igneous, sedimentary, and older metamorphic rocks. Light to dark color. Coarse banded structure. Medium to hard excavation.

Greenstone — Formed from fine-grained igneous rocks. Light to dark greenish color. Medium to hard excavation.

Marble — Formed from limestone. Light to medium color. Hard excavation.

Quartzite — Formed from sandstone. Light to medium color. Hard excavation.

Schist — Formed from igneous, sedimentary, and older metamorphic rocks. Light to dark color. Fine banded structure, differing from gneiss in its tendency to split into layers because of its foliation. Medium to hard excavation.

Serpentine — Formed from igneous rocks. Light to dark greenish color. Lustrous and waxy. Soft to hard excavation.

Slate, or argillite — Formed from shale with original thin stratifications. Light to dark color. Medium to hard excavation.

Soapstone, or talc — Formed from igneous and metamorphic rocks. Light to dark color. Soft to medium excavation.

Ores and minerals — In this text an ore is a rock containing a mineral. It may be the entire ore body to be excavated or it may be a part of the total excavation. A mineral, metallic or nonmetallic, is any substance of definite chemical composition. It also is considered to be rock, part of the ore.

Aluminum ore — Bauxite, a hydrous aluminum oxide of 55% average aluminum content. Light to dark brown color. Decomposition product of igneous and metamorphic rocks. Occurs in lenses and pockets. Soft to medium excavation.

Asbestos ore — Chrysotile, serpentine, a hydrous magnesium silicate of 100% asbestos content. Light to dark color. Occurs in veins of serpentine. Soft to hard excavation.

Boron ore — Borax, a hydrous sodium borate of 11% average boron content. Light to medium color. Occurs as sediment in extrusive igneous and sedimentary rocks. Thin to thick beddings. Soft to hard excavation.

Chromium ore — Chromite, a ferrous chromic oxide of 47% chromium content. Dark color. Occurs in veins and disseminations in igneous and metamorphic rocks. Medium to hard excavation.

Copper ore — Chalcopyrite, a copper and iron sulfide of 34% copper content. Light to medium color. Occurs in veins and disseminations in igneous and metamorphic rocks. Medium to hard excavation.

Diatomaceous earth ore — Diatomite, a silicon dioxide of 47% silica content. Light color. Occurs in massive sedimentary beds of diatoms, microscopic forms of plant life. Medium to thick beddings. Soft to medium excavation.

Gold ore — Gold, an element, is yellow. Occurs in veins of igneous rocks, chiefly in quartzose rocks. Medium to hard excavation. Occurs also in placer deposits of alluvia as fine gold and nuggets. Soft to medium excavation.

Gypsum ore — Gypsum, a hydrous calcium sulfate of 32% lime content. Light color. Occurs in massive sedimentary beds associated with shales and limestones and in cavities of limestones. Medium to thick beddings. Medium to hard excavation.

Iron ores — Hematite, an iron oxide of 70% iron content. Medium to dark color. Occurs in thick beddings in sedimentary and metamorphic rocks. Soft to hard excavation.

Limonite, a hydrous iron oxide of average 60% iron content. Medium to dark color. Formed as a residual by the decomposition

TABLE 2-2 Rocks, Ores, and Minerals *(Continued)*

Ores and minerals

Iron ores	of iron minerals. Thin to thick beddings. Soft to medium excavation.
	Magnetite, an iron oxide of 72% iron content. Dark color. Magnetic. Occurs in veins and in disseminations in igneous and metamorphic rocks. Medium to hard excavation.
	Taconite, an iron oxide composed of hematite and magnetite, of 71% average iron content. Occurs in ferruginous chert sedimentary rock associated with limestones and sandstones. Thick beddings. Hard excavation.
Kaolin ore	Kaolinite, a hydrous aluminum silicate of 100% kaolin content. Light color. Occurs as thick claylike beds. Soft to medium excavation.
Lead ore	Galena, a lead sulfide of 87% lead content. Dark color. Occurs in veins in igneous, sedimentary, and metamorphic rocks. Medium to hard excavation.
Magnesium ore	Magnesite, a magnesium carbonate of 29% magnesium content. Light color. Occurs in veins in serpentine, as dolomite, and in sedimentary deposits of medium to thick beddings. Soft to hard excavation.
Manganese ore	Pyrolusite, a manganese dioxide of 63% manganese content. Dark color. Associated with sedimentary and metamorphic rocks in thin to thick beddings. Medium to hard excavation.
Mercury ore	Cinnabar, a mercury sulfide of 86% mercury content. Medium red color. Occurs as disseminations in igneous, sedimentary, and metamorphic rocks. Soft to hard excavation.
Molybdenum ore	Molybdenite, a molybdenum disulfide of 60% molybdenum content. Medium to dark color. Occurs as small veins and disseminations in igneous, sedimentary, and metamorphic rocks. Medium to hard excavation.
Nickel ore	Niccolite, a nickel arsenide of 44% nickel content. Light to dark color. Occurs in veins in igneous and metamorphic rocks. Medium to hard excavation.
Nitrate ore	Soda niter, a sodium nitrate of 100% sodium niter content. Light to medium color. Associated with gypsum. Thin to thick beddings. Soft to hard excavation.
Phosphate ore	Apatite, a calcium fluorophosphate or calcium chlorophosphate, or both, of average 18% phosphorus content. Light to medium color. Occurs as veins in igneous rocks with hard excavation and in sedimentary deposits of thin to thick beddings with soft to medium excavation.
Potassium ore	Sylvite, a potassium chloride of 52% potassium content. Light to medium color. Associated with halite. Thin to thick beddings. Soft to medium excavation.
Salt ore	Halite, a sodium chloride of up to 100% purity. White to light color. Occurs in massive sedimentary beds associated with gypsum and potassium ores. Medium to thick beddings. Medium to hard excavation.
Silver ore	Argentite, a silver sulfide of 87% silver content. Dark color. Occurs in veins of igneous and metamorphic rocks, along with gold and nickel. Medium to hard excavation.
Sulfur ore	Occurs as a yellow element in medium to thick beddings in volcanic and sedimentary rocks. Up to 100% purity in native state. Soft to medium excavation.
Tin ore	Cassiterite, a tin oxide of 79% tin content. Light to dark color. Occurs in veins in igneous rocks, in which excavation is medium to

Ores and minerals

	hard. Occurs also as rounded pebbles in stream alluvia, in which excavation is soft.
Uranium ore	Uraninite, a mixture of uranium oxides, averaging about 86% uranium content, and generally associated with other minerals. Light to dark color. Occurs as veins in igneous and metamorphic rocks and as disseminations in sedimentary rocks with medium to heavy beddings. Medium to hard excavation.
Zinc ore	Sphalerite, a zinc sulfide of 67% zinc content. Light to dark color. Associated with galena. Occurs in veins of igneous, sedimentary, and metamorphic rocks. Medium to hard excavation.

NOTE 1: Thicknesses of beddings for some extrusive igneous rocks and for sedimentary rocks, indicative of excavation methods, are given in this table.
    Thin beddings   Up to 1-ft thickness.
    Medium beddings   Between 1- and 3-ft thickness.
    Thick beddings   Greater than 3-ft thickness.

NOTE 2: The excavation characteristics of rocks, given as soft, medium, and hard, are dependent largely on the degree of weathering, as discussed in Chapter 3. In terms of excavation methods, the characteristics are idealized in this table.
    Soft excavation   Excavation without preliminary ripping or with soft ripping by heavyweight crawler-tractor-ripper.
    Medium excavation   Excavation by preliminary medium to hard ripping.
    Hard excavation   Excavation by preliminary extremely hard ripping or blasting.

Rocks are classified as igneous, sedimentary, and metamorphic. These three classifications are somewhat broad, as some igneous and sedimentary rocks may be in an initial stage of metamorphism, which is scarcely discernible.

Ores are classified according to the principal mineral for which they are mined, some ores providing several minerals and native elements. The most common mineral for the element desired is given. Table 2-2 includes metallic and nonmetallic ores most commonly mined in open-cut excavation.

The deposition of ores may be simple or complex. It may be contemporary, the ore being formed at the same time as the host rocks. It may be subsequent, the mineral compounds having been gathered from the host rocks and deposited therein. It may be replacement, the present ore having been deposited after removal of the original ore.

After the ore has been deposited, weathering and further enrichment of the ore body may create several zones within the deposit. An example is a top zone of complete oxidation, a second zone of leaching, a third zone of oxide enrichment, a fourth zone of sulfide enrichment, and a bottom zone of primary ore.

The forms of ore deposits are tabular deposits, veins, pockets, disseminations in the host rock, and residual bodies. All these forms may be excavated by open-cut methods provided the yield of the desired mineral is economical.

Ores usually, but not always, contain gangue. Gangue is any valueless material that is necessarily excavated along with the desired ore or mineral. Generally it is a part of the host rocks. Gangue is removed by initial concentration processes and by final smelting.

The descriptions of ores and minerals have been simplified for the excavator in terms of their occurrences and characteristics for excavation.

FORMATIONS

For the earthmover an understanding of the rock formation is fully as important as knowledge of the rocks that make up the formation. As defined and used by the U.S. Geological Survey, the *formation* is the ordinary unit of geologic mapping, consisting of a large and persistent stratum of some kind of rock.

Actually, in the language of the excavator and the construction and mining geologist, a formation is an individual rock or assemblage of rocks of igneous, sedimentary, or metamorphic origin that is sufficiently homogeneous or distinctive to be given a distinct name. The given name is typically the name of a locality where the rock is exposed and identified.

The Lineage of a Formation

Referring to Table 1-1, the formation is a subdivision of a series of rocks, which in turn is a part of the larger system of rocks. For example, the limestone and dolomite hard-rock excavation of the Osage River basin near Warsaw, Missouri, make up the Theodosia formation. The lineage of the Theodosia formation is as follows:

Theodosia	formation	*from*
Jefferson City	subseries	*from*
Canadian	series	*from*
Ordovician	system	*from*
Paleozoic	group	

The age of the fossil-bearing Theodosia formation is about 489 million yr.

Examples of Formations

The term formation is applied to igneous, sedimentary, and metamorphic rocks and their combinations. For example, the Topanga formation of southern California is an 18-million-yr-old Middle Miocene assemblage of sedimentary sandstones, conglomerates, and shales, with much interbedding of volcanic intrusive and extrusive rocks such as andesites and basalts. The well-known formation is of both continental and marine origin.

In such varied formations the excavator may find troublesome thin dikes and sills of hard igneous rocks separating masses of soft sedimentary rocks. As a result, systematic ripping and blasting of the formation are both sometimes made impossible by the attitude of the formation and the consequent excavation difficulties. Such a transition might be from a 3-ft thickness of shale to a 5-ft thickness of basalt to a 4-ft thickness of sandstone, all strata having extreme dip and the basalt sill requiring costly drilling and blasting.

A unique intrusive igneous rock is the granite formation known as tonalite. By definition it is simply quartz diorite of the granite family. However, when weathered it is characterized by up to room-size, rounded boulders, both on the surface and submerged beneath natural ground in a matrix of "DG," or decomposed granite. In this formation are easily rippable rocks containing "drill-and-shoot" boulders, all at the same plane of excavation. Figure 2-8 shows this approximately 99-million-yr-old troublesome formation in a 1,044,000-yd³ cut for a freeway in Riverside County, California. Prior to excavation

Figure 2-8 Massive boulder formation of granite. Tonalite, a kind of granite distinguished by huge, weathered boulders, in Riverside County, California. In spite of the awesome appearance of this granite formation, this part of the 75-ft-deep cut was excavated by soft to hard ripping by heavyweight tractor-rippers. Grooves left by the ripper teeth are visible along the relatively smooth face of the cut, the flat face being in marked contrast to the rough natural ground. Outsize boulders were blasted before being loaded out to the fill.

the surface of this 75-ft-deep and 3700-ft-long cut resembled the area above the cut slope of the picture. In spite of the formidable appearance of the boulders and outcrops of granite, relatively little yardage of the million-cubic-yard cut required blasting. Below the grade of the cut, the weathered granite with its imbedded boulders is slowly succeeded by jointed semisolid and ultimately solid rock.

Basalt is an extrusive igneous rock, usually occurring in flows with characteristic columnar tabular formation. Figure 2-9 illustrates the outcroppings along the centerline of a 27-ft-deep cut for a proposed highway in Siskiyou County, California. The highway is in the shadow of young Mount Shasta, and one of Shasta's volcanic eruptions some 1 million yr ago produced the lava flow of this formation.

Another distinguishing feature of basalt, although not always present, is its vesicular surface, consisting of small holes left by the frothing cooling rock. Figure 2-10 shows this mottled surface. The area of the picture is about 2 ft^2, or about the size of the boulder in the foreground of Figure 2-9.

This basalt formation will be excavated by medium to hard ripping by a heavyweight tractor-ripper to about 4 or 5 ft above grade, and the remaining small prism of rock will be taken out by either extremely hard ripping or blasting. The choice of ripping or blasting will depend on the quantities involved and on the availability of drilling and blasting machinery and crews.

A not too common extrusive igneous rock of pyroclastic nature is tuff, which sometimes, since it is composed of small particles airborne from an erupting volcano, is mistakenly called a sedimentary rock. This is especially true because it resembles heavy-bedded buff and brown sandstones.

Figure 2-11 shows an existing freeway cut in the Bishop tuff formation of Mono County, California. This excavation involved four major cuts totaling 1.7 million yd^3 with a maximum 53-ft-deep cut in this approximately 2-million-yr-old formation.

The andesitic tuff rings with a pick blow and shows an ugly surface. And yet, all the roadway excavation was ripped out economically. Testimony to this less costly means of excavation is apparent in the smooth, clean-cut slope as contrasted with the jagged natural slope. On the left side of the road there appears to be a rough-cut slope. Actually, this is a natural slope on the other side of the gulch of Crooked Creek.

Figure 2-9 Well-weathered surficial basalt. Weathered basalt lava flow from Mount Shasta, Siskiyou County, California. The young formation, some 1 million yr old, will be ripped to near grade of the 27-ft-deep highway cut. A few feet above grade the blocky basalt will become semisolid jointed rock and it will call for either extremely hard ripping by a heavyweight tractor-ripper or expensive blasting because of the shallowness of the remaining prism of rock.

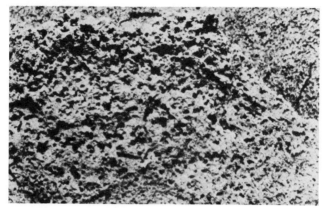

Figure 2-10 Basalt with air cells. Vesicular basalt. The area of the photo is a 2-ft square. The holelike vesicles have been formed by the rapid expansion of gases during the quick cooling of the lava flow. Sometimes after cooling the vesicles are filled by one or more minerals, and in that case the vesicular structure becomes an amygdaloidal rock. Basalt is not always vesicular but the exceptions with smooth surfaces comprise a small percentage of the total basalts.

Figure 2-11 Volcanic tuff. Bishop tuff, Mono County, California. This deceptive igneous pyroclastic formation of 1.7 million yd³ in four cuts up to 53 ft deep was of soft to hard ripping and the yardage was ripped entirely by heavyweight tractor-rippers. The misleading nature of the formation is graphically portrayed in the contrast between the massive rocky natural ground and the ultimate smoothness of the cut slope when finished by the slope board of the crawler-tractor.

A troublesome formation for excavation is scoria, as shown in Figure 2-12. Scoria is a rough, rubblelike extrusive igneous rock, and it is usually formed by the extremely fast cooling of basaltic lava. This approximately 1-million-yr-old scoria formation is in the Owens Valley, Inyo County, California. Like many of the extrusive igneous rock formations of the Owens Valley fault area, it is of the Pleistocene series and among the youngest of California rocks. Further south in the Mojave Desert is the scene of equally spectacular Pleistocene volcanic activity.

A typically soft marine sedimentary formation is the Capistrano of California, consisting of thinly bedded friable sandstones, shales, and conglomerates. Figure 2-13 shows this formation in a highway cut in San Diego County. The first 10-ft depth of residuals was excavated without ripping and the rest of the excavation to grade was ripped easily by a medium-weight tractor-ripper.

Figure 2-12 Cinderlike basaltic lava. Scoria, an extrusive igneous rock formed from basaltic lava. This sharp, angular, abrasive rock is along the centerline of an 18-ft cut in a proposed freeway of Inyo County, California. Seismic studies and subsequent excavation revealed that the rubble merged into solid rock a few feet above finished grade. Accordingly the formation required expensive drilling and blasting commencing some 5 ft above grade. The overlying scoria is of medium weight and hard whereas the country rock, solid basalt, is both heavy and hard.

Figure 2-13 Young sedimentary rock formation with little deformation. The Capistrano formation of southern California is normally made up of sandstones, shales, and conglomerates. In this roadside cut of 20-ft depth are 10 ft of residuals overlying 10 ft of soft shales, sandstones, and conglomerates. Excavation was relatively soft in this San Diego County formation, requiring only mild ripping by a medium-weight tractor-ripper for full fragmentation.

2-18   Rocks, Ores, Minerals, and Formations

In this picture are shown the medium-resistant shales of the upper portion; the middle, more resistant sandstones; and the lower, less resistant conglomerates. The hand pick shows the 8-in thickness of the hardest sandstone. The age of the Capistrano Upper Miocene formation is about 15 million yr.

Sometimes there are variations in the same formation with respect to the kinds of rocks and their respective consolidations, although generally a formation is fairly uniform. This variation in the Capistrano formation is illustrated in Figure 2-14. A 10-ft lens of San Onofre breccia forms an overhanging ledge above the softer sandstone. The breccia is made up of cemented particles of schist and interbedded grit and sandstone. Thus, the excavator would find hard breccia above the soft sandstone. Normally, in a downward sequence of strata the hardness of the formation increases with depth, but this is not necessarily so. The dimensions of this well-weathered river terrace may be visualized with respect to the man shown in the picture. Above the breccia are clay and sand residuals. The excavator would encounter:

1. Fifteen feet of easily excavated residuals.

2. Ten feet of breccia calling for hard ripping by a heavyweight tractor-ripper.

3. Twenty feet of siltstone requiring medium ripping by a medium-weight tractor-ripper.

This bank has been eroded and weathered by the San Juan River and by the winds and the rains.

A thick-bedded sedimentary limestone that forms the overburden of anthracite coal stripping in Pennsylvania is shown in Figure 2-15. This hard limestone of maximum 8-ft thickness requires blasting prior to casting by walking draglines. Only thinly bedded limestones of a relatively soft nature may be successfully ripped by heavyweight tractor-rippers. This limestone formation is of the Pennsylvanian series of the Carboniferous system. The Carboniferous is the anthracite-coal-bearing system. The Pennsylvanian series is about 295 million yr old and it has been contorted by the mountain building of the Appalachian Revolution sometimes into anticlines and rather deep synclines. This orogeny has complicated the art of coal strip mining in many areas of the Appalachian Mountains, resulting in high ratios of overburden

Figure 2-14   Hard stratum interbedded in soft strata of sedimentary rocks. The Capistrano formation with a variant—a 10-ft lens of San Onofre breccia. This breccia is hard and frequently it characterizes the formation in the coastal area. With reference to the man, it is 10 ft thick and it overhangs the less resistant light sandstone. Excavation would involve medium to hard ripping by a heavyweight tractor-ripper.

Figure 2-15 **Massive hard-rock overburden.** Overburden of limestone formation in eastern Pennsylvania anthracite coal field. About 80 ft of overburden is being cast by the walking dragline so as to uncover the coal vein. The hard rock is blasted prior to removal because of the heavy bedding of the limestone strata. Such massive limestone formations of the Carboniferous system of rocks are common in the eastern and central United States, occurring as outcrops in the mountains and as underlying beddings in the plains. *(E. I. du Pont de Nemours & Co., Blasters' Handbook, 1966, p. 309.)*

yardage to recoverable coal tonnage. In this case the limestone overburden is about 80 ft in thickness.

The excavation of sedimentary rocks such as shales, sandstones, and limestones depends on the bedding or lamination thickness and the vertical joints, even though the individual stratum of the rock may be hard and strong. These planes of weakness and their cubic dimensions determine the ease or difficulty of ripping and the necessity for alternate blasting. A generalization for these structural spacings is:

Spacings less than 12 in	Soft to medium excavation
Spacings 12 to 36 in	Medium to hard excavation
Spacings more than 36 in	Hard to extremely hard excavation

The metamorphic rocks are generally hard and durable. A common one is granite gneiss, or altered granite. Banded and wavy structures, due to heat and pressure, distinguish this fine-grained rock. It usually calls for extremely hard ripping or blasting for fragmentation, and its excavation characteristics are akin to those of granite except that it does not weather as rapidly as the coarser-grained granite.

A variable metamorphic rock formation, derived from marine sedimentary rocks, is the Bedford Canyon formation of California. Figure 2-16 shows schist outcroppings of this formation in a rather shallow highway cut adjacent to a proposed 880,000-yd^3 72-ft-deep cut for a freeway. These metamorphosed sandstones, siltstones, and conglomerates are fairly solid up to the grass roots, and the existing highway cut indicated what might be

Figure 2-16 Minimum weathered outcrops of metamorphic rocks. Bedford Canyon formation of metamorphosed sedimentary rocks. This California formation of phyllites, schists, argillites, and graywackes, located in San Diego County, is of the Upper Jurassic system of rocks, about 146 million yr old. It is an inlier of about 2 mi along the centerline of a freeway and it is surrounded by younger granites about 99 million yr old. The contacts between the igneous and the metamorphic formations are customarily in the swales or in the fill areas of the freeway. This picture shows a cut in the existing highway paralleling the proposed freeway cut. Semisolid rock is but a few feet below natural ground.

expected in the rock excavation for the proposed adjacent freeway. The correlation was good as the huge adjacent cut was ripped to an average 35-ft depth and blasted under this level to grade.

Another fairly common metamorphic rock is slate, derived from shale. In America the well-known slates used for commercial purposes are those of Vermont and New York State. Figure 2-17 shows the slate of the Slate River area of Virginia. In the midst of the massive slate there is a small fault, with its brecciated zone between the two hand picks of the picture. In the weakened zone rock has been crumbled and the adjacent faces have been weakened by bending. However, only a few feet from the fault the rock remains massive and hard. From a practical standpoint this 465-million-yr-old Ordovician formation will require blasting.

The beddings of slate vary from less than ⅛ in to a foot or more in thickness, and characteristically there is a minimum thickness of residuals and little weathered slate between the residuals and the solid slate. In the Mother Lode country of California there are many slate and other metamorphic rock formations, which trend generally north to south along the axis of the gold mining activities for a distance of some 160 mi. Such a slate formation is the Mariposa, which, in contrast to the Slate River, Virginia, formation, is generally without minor faults. Although the cleavage planes are apparent, they are not evidence of weakness, as the hard rock requires blasting. In a moderate-size highway cut of 40 ft maximum depth the "rock line" was but a few feet below the grass roots, and some 60,000 yd³ of the total 75,000-yd³ cut was blasted.

Attitude of Formations—Dip and Strike

Sedimentary and extrusive igneous rocks are laid down horizontally or nearly horizontally. Later the formation may be dipped or tipped by deformation of the Earth's crust. While in a horizontal or dipped attitude, the rocks may be converted to metamorphic rocks. It is important for the earthmover to recognize the *dip and strike* of the rock

Figure 2-17 Brecciated fault zone of metamorphic rock. The slate or argillite member of this Slate River, Virginia, rock formation is a hard, tough, metamorphic facies requiring blasting. Nature's weathering process on this hard, durable rock is manifestly slow. As shown between the two hand picks, there is a narrow brecciated zone of the small fault, but otherwise within a short (less than 2 ft) distance the rock is solid. Because of the contortions of metamorphism slates generally occur in steeply dipped attitudes, and with their associated schists, phyllites, and quartzites they present unknown and perplexing problems to the excavator. *(H. Ries and Thomas L. Watson,* Engineering Geology, *Wiley, New York, 1925.)*

formation because it is important in the fragmentation and the loading of the rock. It is also significant from a safety standpoint because of the possibility of dangerous slides even in moderately dipped rock formations.

Figure 2-18 shows the relationship of the dip and strike of a formation to a horizontal plane or datum plane of a body of water.

Three examples of the value of dip-strike information to the excavator are:

1. In the case of fragmentation by ripping, one should not rip at right angles to the strike and away from the direction of dip because the ripper points would slide along the inclined beddings and would not penetrate the stratum efficiently. It would be better to rip in the opposite direction so as to lift the stratum with the ripper points or to rip parallel to the strike. Naturally, the ideal approach for ripping must be modified sometimes to accommodate the physical aspects of the work area.

2. When drilling vertical blastholes, the drill steel is apt to be deflected and wedged in the partings of the strata if the dip is vertical or nearly vertical. It is better to drill off the vertical at an angle to the plane of the dip.

3. When loading soft sedimentary rock by scrapers without preliminary ripping, the problems with the cutting edge of the scraper bowl as it is forced into the stratum are the same as those with the points of the ripper. The solutions are the same, as in the case of the cutting edge the operator is trying to penetrate a stratum, just as the operator is trying to force the ripper points into the stratum.

Information concerning dip and strike is of great significance to the miner investigating the worth of an ore deposit and the cost of the proposed mining work. In open-cut excavation the length of the strike and the width of the deposit on the Earth's surface, together with the dip, determine the amount of ore that can be taken out economically by surface mining.

Correlation of Rock Formations for Excavation

Just as similar machines rip, blast, and load rock in like manner in the same rock formation, so the same types of formations, though many miles apart, behave in pretty much the same way when ripped, blasted, and loaded.

The Principle of Uniformitarianism of geologist James Hutton was enunciated by this Scottish man of vision in 1785 in his revolutionary book *Theory of the Earth*. Hutton stated that the present is the key to the past and that, given time, the processes now at work could have produced all the geologic features of the Earth.

In terms of the excavation of rock formations, this principle may be extended to the theory that rock formations of the same kind and of the same time on the geologic time scale tend to have the same characteristics for excavation. The theory is necessarily a generalization because the spatial aspects of the same rock formations vary according to locality.

This extension of the Principle of Uniformitarianism, supported by many correlations, is an example of the practical, useful application of an accepted theory. In a sense it is a basic assumption to which modifications may be added to complete the understanding of a new and questionable rock formation. The correlation of the unknown formation with a known formation in which the excavator has had experience is made by the use of correlation charts, as shown in Table 2-3. In this table the center section, embracing the Basin Ranges, Mojave Colorado Desert, Peninsular Ranges, and part of the Transverse

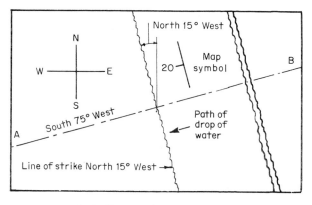

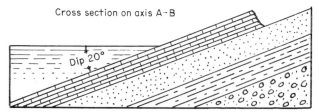

Figure 2-18 Strike and dip of rock formation. The illustration shows a dipped formation of sedimentary rocks along the edge of a body of water representing a horizontal plane.

Strike is the compass direction of the intersection of its bedding plane with a horizontal plane and it is measured in degrees.

Dip of a rock formation is a measure of its tilt or slope in the direction of its dip. The angle of dip is the acute angle which this direction makes with the horizontal plane and it is measured in degrees.

The strike of this formation is North 15° West or South 15° East and it is represented by the edge of the water. The dip is 20° below the horizontal plane of the water toward a compass bearing of South 75° West.

Formations 2-23

TABLE 2-3 Correlation Chart for Sedimentary Rock Formations of Southern California
(Interval of time is 135 million years)

			SOUTHERN COAST RANGES				TRANSVERSE RANGES	
NOTE: The authors' names that appear at the tops of the columns do not indicate responsibility for the contents of the columns (see text).			Santa Maria Basin & Huasna Basin Woodring & Bramlette (50), Taliaferro (43)	Cuyama Valley Dibblee (52), Beck (52), Schwade (54)	Reef Ridge-Kettleman Hills Woodring, et al (40), Stewart (46)	S. J... Hoo... Dibl...	...s Basin-...es Hills ...odford (51), ...oodring, (46)	Santa Monica Mountains Hoots (51), Durrell (54)
QUATERNARY	Pleistocene	Upper	Terrace deposits / Orcutt sand	Fanglomerate	Older alluvium	Fan... ter...	...non-...ce deps. sd.	Alluvial plain and marine terrace deposits
		Lower	Paso Robles formation	Paso Robles formation	Tulare formation	Tular...	sd. silt rl	Marine conglomerate
T E R T I A R Y	Pliocene	Upper	Careaga fm. / Foxen fm.	San Joaquin fm		form...	form...	Pico (San Diego) fm.
		Lower	Sisquoc fm.	Morales fm.	Etchegoin fm. / Jacalitos fm.	Chand... form...		
	Miocene	Upper	Sta. Margarita fm. / Santa Margarita fm. / Quatal fm.		Reef Ridge sh		Sta. Mar. fm. / st. mbr. mbr.	Modelo fm.
		Middle	Monterey fm. / Point Sal fm.	Monterey fm. / Bitter Creek sd / Caliente sh	Monterey fm / McLure sh		Maricopa or Monterey fm.	Topanga fm.
		Lower	Vaqueros fm.	Painted Rock sd / Soda Lake sh	Temblor fm.		Temblor or Vaqueros	Vaqueros fm.
	Oligocene		Lospe fm / Sespe fm.	Simmler fm.			Pleito fm.	Sespe fm.
	Eocene	Upper			Kreyenhagen sh		San Emigd... fm. / Reed sh	
		Middle		Unnamed Eocene beds	Avenal ss		Tejon / Live / Uv	
		Lower						
	Paleocene		?	?				"Martinez fm."
CRETACEOUS	Upper		Asuncion (?) gp. Pacheco (?) gp.	Unnamed Cretaceous beds	Panoche fm			"Chico fm." Trabuco fm.
	Lower		"Shasta" gp.					

Portions of a correlation chart of sedimentary formations in Southern California are shown, with brackets drawn to indicate the Panoche and the Chico formations—See text for discussion.

SOURCE: Charles W. Jennings and Mort D. Turner, *Geology of Southern California*, California Division of Mines and Geology, 1954, chap. 3, plate 1.

Ranges, has been omitted for the sake of simplicity. Such charts are available from a number of private and public agencies, which include:
    1. Geology departments of various colleges and universities.
    2. American Association of Petroleum Geologists, Tulsa, Oklahoma.
    3. State divisions of geology.
    4. U.S. Geological Survey, Washington, D.C.
    5. *History of the Earth*, 2d ed., by Bernard Kummel, W. H. Freeman and Company, San Francisco, Calif., 1970.

Some of these sources give valuable additional information in terms of rock types (igneous, sedimentary, or metamorphic), as well as data on stratification, dip and strike, folding and faulting, and even degree of weathering in some cases.

The use of rock formation correlations is typified in the following three illustrations.

1. An abundant sedimentary rock formation of northern California is the Chico, made up of thickly bedded sandstones and tightly bedded shales. This old formation, Upper Cretaceous dating back some 90 million yr, has the reputation of requiring much blasting in the heavy excavation of the northern California area over a long period of time.

In 1962 seismic studies were made for an 11-million-yd^3 freeway across Pacheco Pass of the California Coast Range. This formation is the Panoche and it also is of the Upper Cretaceous series. An early correlation was made by means of the chart of Table 2-3, linking the Chico and the Panoche formations. The later seismic studies and the excavation experience show that the contractor blasted some 65 percent of the roadway excavation.

In 1965 preliminary studies were made for an 8.5-million-yd^3 rock job for a freeway across Santa Susana Pass, some 260 mi south of Pacheco Pass and some 420 mi south of Chico, the city after which the Chico formation was named. The Santa Susana Pass formation is known as the Chico, and so there was no hesitation in correlating this rock work with that of Pacheco Pass. Seismic studies verified the assumption, as did the excavations for the job, during which 68 percent of total yardage was blasted.

Figure 2-19 shows dramatically the massive nature of the Upper Cretaceous formation at La Jolla, California. Both the miniature sea mount and the main rock structure show few planes of weakness.

2. After a lot of experience igneous rock formations may also be correlated, although there is more deviation in the attributes of two similar formations than in the case of the sedimentaries. In 1959 preliminary seismic studies were made of a proposed freeway excavation a few miles east of Barstow, California. The rock in one big cut was rhyolite,

Figure 2-19 Hard sedimentary rock characteristic of the Upper Cretaceous series. An unnamed Upper Cretaceous sedimentary rock formation at La Jolla, California. This formation, correlative to the Chico some 550 mi northward, is massive and it presents the same excavation problem of drilling and blasting. The Rosario formation of coastal Baja California, another 220 mi southward, correlates with the unnamed La Jolla formation. Straddling the Oregon–California boundary 160 mi north of Chico is the Hornbrook formation of Upper Cretaceous rocks in which blasting experience corresponds to that in other formations of the same geologic time. Between Hornbrook and Rosario there are 900 mi of near perfect correlations among six formations of the Upper Cretaceous series of sedimentary rocks, typical of the practical feasibility of rock and rock-formation correlations in excavation appraisals.

a 19-million-yr-old Miocene rock lava flow. The rock was moderately weathered and the rock line was judged to be about 9 ft below the ground surface. Practically all the 547,000-yd^3 cut was blasted.

In 1960 studies were made for a proposed freeway excavation across Golconda Summit, east of Winnemucca, Nevada. A study of the geology showed the excavation to be in Eocene rhyolite, which, although three times the age of the Miocene rhyolite of Barstow, was probably weathered under the same desert conditions. It was a foregone conclusion that ripping and blasting conditions would be the same. And they were in fact about equal, demonstrating that correlation was feasible over a distance of 400 mi and over a time span of 40 million yr.

Similar correlations prior to excavation have been made in the granite formations of the West. Agreements are closer in this intrusive igneous formation than in the extrusive igneous formations of basalt, andesite, and rhyolite.

3. As with igneous rock formations, geologists have cataloged well the metamorphic rock formations. The time determinations are based largely on associations with sedimentary rocks of like age and, of course, often on radioactive methods of dating. Within reason, correlations are entirely feasible.

In 1967 preliminary studies were made on 18 million yd^3 of freeway excavation in Cajon Pass, San Bernardino County, California. Of this, 6.9 million yd^3 was in Pelona schist, a Precambrian formation more than 500 million yr old. All schist excavation was in the rift zone of the active San Andreas fault and thus was subject to weakening deformations over tens of millions of years. The prediction of the studies and the experience of the contractor indicated that no drilling and blasting should be done, although the cuts ran up to 130 ft in depth.

In 1969 investigations were made on a 9.4-million-yd^3 reach of the California Aqueduct located west of Palmdale, California. About 2.5 million yd^3 of canal excavation was located in the same Pelona schist. And, by coincidence, the aqueduct section was also in the rift zone of the San Andreas fault, paralleling the fault some 8000 ft to the north. Here, too, the contractor's experience corresponded to that of the builder of the Cajon Pass, although there was a modicum of blasting at the bottom of one 60-ft-deep cut. As an observation on the influence of fault action, there have been two major earthquakes of magnitude greater than 6 on the Richter scale in the Cajon Pass during recorded history, whereas there have been none along this particular reach of the California Aqueduct.

Finally, in appraising the worth of correlation of far removed rock formations of identical ages, the equivalence of eastern Pennsylvania limestone at Allentown to western Texas limestone at Sierra Blanca, despite dissimilar physiographic and climatic conditions, should be cited.

In 1968 preliminary seismic studies and observations were made on 55,000 yd^3 of jointed and semisolid limestone for building-foundation excavation at Allentown. The formation was the upper Jacksonburg member of the Pennsylvanian series, some 295 million yr old. Almost all the rock required blasting, in keeping with the quarrying and road-building methods of the area.

Shortly thereafter studies were made for mountaintop microwave stations near Sierra Blanca, Texas. A mountain of white limestone was the site of one of these stations. This rock was also of the Pennsylvanian series. Again, both seismic studies and excavation methods were similar in these two formations, 1800 mi apart.

Thus it is evident that nationwide correlations of rock formations are feasible. Actually, worldwide correlations are entirely workable. Kummel, in his book *History of the Earth*, correlates the relatively soft Miocene marine sedimentary formations of America with the comparable Hofuf, Dam, and Hadrukh formations of Saudi Arabia. There is no reason to believe that the sandstones and shales of the Upper Miocene Dam formation would behave very differently under a two-shank heavy-duty tractor-ripper than the Upper Miocene Capistrano formation of southern California, even though Saudi Arabia and the western United States are some 10,000 mi apart. Of course, one can speculate beyond the Earth in making correlations. The exploration of the planet Mars disclosed the similarity between the basalt of the desert of Mars and the same rock in the Mojave Desert of California, even though an average of 135 million mi of space and untold aeons of time separate these comparable extrusive igneous rock formations.

SUMMARY

Just as the general picture of the geology of the excavation is important to the excavator, so also is the detailed picture of the rocks, ores, minerals, and rock formations equally essential to the understanding of the rock and earth excavation.

The earthmover should fix in mind the nature of the excavation, which is most important in the case of rock work, because in this basic knowledge is the key to the successful use of methods, machinery, workers, and money.

CHAPTER 3

Rock Weathering

PLATE 3-1

Uneven weathering of eroded plutonic rock. Irregular weathering of tonalite, a kind of granite, in an abandoned quarry of San Diego County, California. The blocky rock has many unpredictable highly weathered cooling joints, the largest being above the front of the car. This condition made it impossible for the quarry excavator to develop an economical drilling and blasting program without removing some 50 ft additional depth of rock below the present quarry floor. Such costly pioneering work was not feasible. The exposed face dramatizes the problems of the quarry excavator in this well-weathered granite formation.

Chapter 2 discussed pragmatically the weathered zone of rock in the thin regolith of the Earth's surface. It is this weathered zone that is of special interest to the earthmover because the degree of weathering to the final finished grade will determine the methods and the machinery for the excavation.

The contractor in public works such as highways and dams thinks in terms of the whole mass of excavation. The miners in open-pit mines and the stone excavators in the quarries consider overburden and stone, as well as ores, as materials for the entire excavation. Both these classifications of materials have weathering characteristics.

Figure 3-1 shows the stages of disintegration and decomposition in a steep granite cut beside a highway in Los Angeles County, California.

First is a thin 2-ft layer of residuals, consisting of clays and sands and including small cobbles. Second is a 10-ft layer of weathered granite, or decomposed granite (DG). Last is the semisolid and solid granite with its weakening cooling joints. As depth increases below the grade of the highway, the cooling joints will be fewer and thinner and the granite will become solid.

The geologist speaks of these three layers in terms of the degrees of weathering, as illustrated in Figure 3-2. While the geologist speaks of residuals, weathered rock, and semisolid and solid rock, the earthmover talks of "potato dirt," "hard pan," and "shooting rock."

The cross section of Figure 3-2 is idealized, being simplified into three zones which actually merge into each other imperceptibly. In many excavations the degree of consolidation or apparent specific gravity from loose humus to solid rock bears a nearly straight-line relationship to depth from ground surface to solid rock.

As must always be emphasized when appraising excavation from method and cost criteria, no factor is as important as the degree of weathering of the rock. Two extremes of granite weathering illustrate this truism.

Figure 3-1 Weathered boulders of tonalite formation embedded in decomposed granite. The degrees of weathering in a highway rock cut of granite in Los Angeles County, California. Dimensions may be visualized by the hand pick resting on the boulder in the middle of the picture. At the grass roots is a 2-ft top layer of clay and sand residuals containing some cobbles. Beneath the residuals is a 10-ft layer of weathered granite or DG in which is the embedded large boulder. This cubic-yard-size boulder is an example of spheroidal weathering, which commenced in the approximately cubical pattern of cooling joints. Beneath the weathered zone are the semisolid and the solid granite, which become increasingly more solid with depth and with the narrowing of the cooling joints.

3-4 Rock Weathering

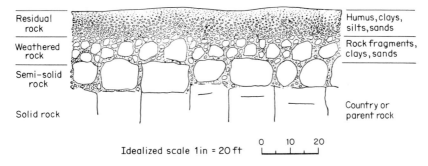

Figure 3-2 Rock excavator's conception of cross section of composite excavation.

Solid "country-rock" granite, relatively free of weakening joints, requires:
  1. Fairly close pattern of drilling or spacing of holes. Rather expensive drilling.
  2. Explosive yield of perhaps 1.5 yd^3 per pound of explosive. Average cost for blasting.
  3. Loading by shovel. Relatively expensive loading out of rock.
  4. Hauling by rear-dump or side-dump trucks. Relatively expensive haulage.
  5. Fill placement and compaction. Relatively expensive.
All five cost elements add up to high unit cost excavation.

In contrast, granitic sands, the ultimate stage in the weathering of granite, call for:
  1. Loading and hauling by push-tractors and scrapers. Relatively inexpensive.
  2. Relatively low cost for fill placement and compaction. These two cost elements add up to low unit cost for excavation.

Overall costs for these two kinds of excavation might be in the ratio of $2.00/yd^3 for the solid granite and $0.75/yd^3 for the derivative sand, the difference being solely due to the difference in degree of weathering.

Weathering is the work of Nature's continual efforts to lower the mountains and raise the valleys and thereby produce the ideal level land or peneplain. That all the land of the Earth is not completely leveled is due to the opposition between the raising of the mountains and the relative or actual lowering of the valleys. It is an ageless battle between creation and destruction.

FORCES OF WEATHERING

The weathering forces are chemical decomposition, the usually invisible destroyer, and mechanical disintegration, the normally obvious devastator.

Decomposition by Chemical Action

Decomposition results from hydration, hydrolysis, oxidation, carbonization, and dissolution, all mutually supporting the inexorable change from trap rock to clay.

Hydration example: Conversion of hard periodotite rock into soft serpentine by addition of water.

Hydrolysis example: Conversion of feldspar and mica from granite to clay by water containing mild acid.

Oxidation example: Breakdown of olivine basalt into soft iron compounds with characteristic brown color and stains.

Carbonization example: The breakdown of arkose or feldspathic sandstone into clays by the action of carbon dioxide and water.

Solution example: The dissolution of limestone by water, resulting in deep fissures and caverns.

When groundwater descends into and dissolves sedimentary rocks, it leaves a red clayey soil mantling the surface and filling the crevices and caverns. Figure 3-3 shows a cross section of this decomposition as exaggerated in a quarry face from which dimension building stone has been taken. Such karstlike topography is common in limestone areas of temperate and tropical regions. The manifest problems confronting the drilling and blasting crews are lost drill holes because of stuck drill rods and bits and

Figure 3-3 Chemical and mechanical weathering of sedimentary rock. Predominantly chemical decomposition of the Bedford limestone formation of southern Indiana. Surface and near-surface dissolution by descending groundwater has left a surface blanket and a subsurface seam- and fissure-filling deposit of red clayey residuals, as exposed in this face of a quarry for dimension building stone. The residual is called *terra rosea*, and it extends to considerable depth below the ground surface. *(William D. Thornbury,* Principles of Geomorphology, *Wiley, New York, 1961, p. 319.)*

lost explosive force because of soft pockets of residuals within the limestone formation.

Reiterating, the five agencies of chemical decomposition processes discussed above can be mutually and simultaneously active, all contributing to slow but insidious weathering.

Disintegration by Mechanical Action

In contrast to the decomposition of the Indiana limestone of Figure 3-3 is the predominantly mechanical disintegration of the Arizona limestone of Figure 3-4.

In this semiarid country with little rain and little ground cover, chemical forces are not significant when compared with mechanical energies. In this monolithic rock there are no topmost residuals and no intermediate weathered zones. Hard rock is at the surface. The limestone is of the Pennsylvanian series and about 295 million yr old, whereas the Indiana limestone is probably of the Mississippian series and 325 million yr old. They may be considered of about the same age, but they have been treated differently by the environments of their geographic locations.

Disintegration results from mechanical forces and these forces are at work at the same time as the energies of chemical decomposition. The mechanical breakup of rock is caused chiefly by the following agents.

Frost Wedging Water is generally present in the joints or interstices of semisolid rock. When water freezes into ice the volume increases by 9 percent. Terrific expansion forces are generated, serving to split the rock further. After the following thaw there is another freeze, and the breakup continues indefinitely. Radiocarbon dating indicates that such wedging of huge angular blocks of granite occurred thousands of years ago. Figure 3-12 shows the results of frost wedging, along with the weathering effects of rains, roots, and other agents.

Temperature Changes Daily and annual temperature changes disintegrate rock. In the United States these changes can vary through 75°F daily in the western deserts and thru 150°F annually in the northern states, causing compression and tension. These

3–6 Rock Weathering

Figure 3-4 Predominantly mechanical weathering in arid climate. Predominantly mechanical disintegration of Arizona limestone near Globe, Arizona. There is no cover of residuals and the hard rock is slightly weakened only by the irregular pattern of thin joints. The exposed face is about 30 ft wide and 40 ft deep. *(U.S. Geological Survey.)*

changes act on rocks in both microscopic and megascopic ways. Minutely, a rock such as granite slowly disintegrates because of the differences in coefficients of expansion of the chief mineral constituents: quartz, feldspar, hornblende, and biotite. As these microscopic particles work against each other over the millennia, the rock slowly disintegrates and forms, ultimately, granitic sand.

The relationship of the different minerals with unlike thermal characteristics is illustrated in Figure 3-5. In this diorite with its little vein of aplite are chiefly light quartz, feldspar, and muscovite and dark biotite and hornblende.

Massively, temperature changes cause the spalling of rocks such as granite in more or less concentric layers. These layers are generally several inches thick. This spalling should not be confused with exfoliation, which resembles spalling but is caused by chemical decomposition.

Cooling In the process of cooling, igneous rocks such as intrusive granite and extrusive basalt contract and the masses pull apart from each other and form cooling joints. These joints are then invaded by chemical and mechanical agents, which further the weathering processes.

Figure 3-6 portrays stages in the weathering of granite, commencing with the formation of the jointing planes and ending with the final production of sand. The cooling joints of the distant rock formation were frost-wedged to produce the rubblelike mass. Accumulative disintegration and decomposition resulted in the boulders of the foreground and the sands of the desert floor.

This transition from solid granite to sand is taking place in the high desert country of Joshua Tree National Monument of southern California. Elevation is about 5000 ft and annual temperature range is about 75°F.

Figure 3-5 **Disintegration due to different thermal characteristics of different minerals.** Diorite from the Sierra Nevada of California. The rock is about 1 ft³ in size. The vein of aplite, largely quartz, and the basic diorite minerals—quartz, feldspar, muscovite, biotite, and hornblende—have different expansion coefficients and these different rates of expansion and contraction slowly loosen the bonds between the different minerals. Eventually the diorite becomes a nearly incoherent mass, which can be reduced to sand by the blow of a sledgehammer.

Figure 3-6 **Aeons of progressive weathering of igneous rock.** Disintegration of granite in Joshua Tree National Monument of southern California. Commencing with the cooling joints in the distant ridge, a combination of the predominantly mechanical disintegration agents ice, rain, and wind has produced desert sand during the 81 million yr of age of this basement complex of the Sierra Nevada. *(U.S. National Park Service.)*

The excavation of the granite formation of Figure 3-6 for highway or railroad building would call for the following methods:

1. The cubic-yard-size weathered boulders of the foreground would be loaded out by rubber-tire tractor-shovel and hauled by rear-dump haulers. They would be incorporated into the fill provided the fill was sufficiently deep.

2. A few feet of the residuals of the desert floor would be loaded by push-tractors and hauled by rubber-tire scrapers.

3. The weathered zone of granite below the residuals and in the distant ridge would be fragmented by tractor-rippers and loaded out and hauled either by shovel and haulers or by push-tractor and scrapers. Choice of machinery would depend upon the degree of weathering.

4. The semisolid and the solid underlying granite would be fragmented by blasting, loaded by tractor-shovel, and hauled by rear-dump haulers.

A careful coordination of these different methods would be of prime importance to both efficiency and low cost.

Figure 3-7 displays vividly the characteristic vertical cooling joints of columnar basalt. This extrusive igneous formation of Le Puy, France, shows the following three zones of in-situ weathering:

1. Residual zone. The upper 10 ft shows the final effects of mechanical disintegration and chemical decomposition. The end result is the topmost foot of red clay.

2. Weathered rock zone. The middle 10 ft indicates the effects of all the factors working in the residual zone plus the breaking down of the hexagonal basaltic columns.

3. Semisolid and solid rock zone. The lower few tens of feet portray vividly this remarkable example of horizontal polygonal and vertical linear jointing of the cooling basalt. The rock is firm and hard.

At the bottom of the wall is a talus slope made up of mechanically weathered fragments of rock. The breakage of these cubic-yard rectilinear blocks has taken place along both vertical and horizontal cooling joints.

Figure 3-7 Unique weathering of columnar basalt. Weathering of columnar basalt near Le Puy, France. Three zones are present: The upper 10 ft are residuals; the middle 10 ft, weathered rock; and the lower few tens of feet, semisolid and solid rock. The individual columns are characteristically hexagonal and about 5 ft in diameter. *(H. Ries and Thomas L. Watson,* Engineering Geology, *Wiley, New York, 1925, p. 88.)*

The fragmentation procedure in such a basalt formation would be: first, no tractor ripping in the residual zone; next, soft to medium ripping in the weathered zone; and finally, extremely hard ripping and blasting in the semisolid and solid rock zone.

Breakage There are several visible and invisible mechanical movements causing rock breakage. One visible one is the fall of rock from a steep vertical face or escarpment or from a ridge. An example is shown in Figure 3-7, in which a talus slope is accumulating at the bottom of the face of columnar basalt.

When a turbulent stream undercuts the soft shale of a gorge, the overlying sandstones or limestones will break up in the gravity fall. This degradation is shown in Figure 3-8. Flexure along a jointing pattern also creates breakage, which accentuates the jointing pattern and sometimes results in definite cubic formations, as shown in Figure 3-8. This sandstone formation along the steep bank of the Green River in Canyonlands National Park in Utah has developed two-dimensional joints on the surface and invisible three-dimensional joints below and parallel to the rock surface. The percolation of river water and rainwater has widened the joints. The resulting cubelike blocks are about 20 ft on each side and weigh some 640 tons.

Near the surface and far below the surface these near cubes will be weathered along the jointing planes of weakness, as illustrated in Figure 3-9. Such spheroidal weathering takes place more commonly in igneous intrusive rocks such as granite. This action is illustrated by the huge embedded granite rock in the center of the picture of Figure 3-1. Spheroidal weathering is nearly nonexistent in igneous extrusive rocks such as basalts and andesites and in the metamorphic rocks.

In Figure 3-9 nearly final weathering has reduced the volume of the original cube

Figure 3-8 Massive cubical pattern of weathering in jointed sandstone. Breakage of falling blocks of sandstone along the gorge of the Green River in Canyon Lands National Park, Utah. Beneath this regional pattern of jointed sandstone, the underlying softer shales have been eroded by the swift water, causing the huge 640-ton blocks of sandstone to fall and break up. Stream degradation will reduce them ultimately to sand. *(U.S. National Park Service.)*

3-10 Rock Weathering

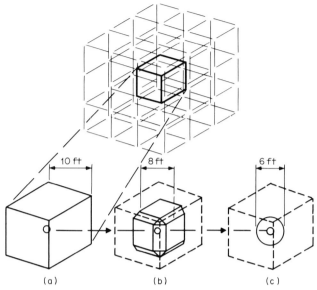

Figure 3-9 The spheroidal weathering of rock. Starting with initial weathering of huge prism of fresh cubically jointed rock made up of average 10-ft cubes; ending with near final weathering, with 6-ft spheres of solid rock surrounded by a well-weathered matrix of clays, sands, and small rock fragments. The matrix is 89 percent and the solid rock is only 11 percent of the original prism. The complete weathering will produce only clays and sands. *(a)* Cube—Initial weathering: beginning of chemical decomposition and mechanical disintegration; 37 yd^3 or 100 percent solid rock; 6 faces, 12 edges, and 8 corners for early weathering. *(b)* Cube-Sphere—Partial weathering: 16 yd^3 or 43 percent solid rock; 26 faces, 48 edges, and 24 corners for continued weathering. *(c)* Sphere—Nearly final weathering: 4 yd^3 or 11 percent solid rock; infinite number of minute area faces, no edges, and no corners; maximum area for weathering.

by 89 percent, although the overall reduction in the linear dimension is only 40 percent. When the 6-ft sphere becomes sand, the weathering will be complete.

An example of breakage by folding, showing dramatically the fracturing of the contorted strata, is apparent in Figure 3-10. An individual stratum of this Chickamauga limestone formation near Ben Hur, Virginia, has a maximum thickness of about 24 in. As the reverse fold becomes more pronounced, the strata will be more highly fractured and a minor brecciated fault will form. Such folds may include the entire volume of a huge cut and so their weakening effect may reduce a sizable yardage of apparently hard rock to rather easily ripped weathered rock.

Faulting fractures rock on both minor and major scales. A minor fault has been shown in Figure 2-17. In contrast to minor faulting there is the major weakening of an entire rock formation or several rock formations due to brecciation by a major fault over tens of millions of years.

Southern California is literally laced by major faults and this large-scale faulting is the explanation for several calamitous earthquakes during the past century. Figure 3-11 delineates this netlike fault system.

Most conspicuous is the famous and notorious San Andreas fault, stretching in its land appearance 660 mi from Point Arena, north of San Francisco, to the Mexican border. Geologists believe that the brecciated or gouge zone of the fault extends ½ to 2 mi on either side of the axis of the fault.

As discussed in Chapter 2, the appraisal of an 18-million-yd^3 freeway across the San Andreas fault in Cajon Pass, California, verified this contention. The excavation was investigated in 1967 by 63 seismic studies. Additionally, researches were made on the intensities and the magnitudes of four major earthquakes in Cajon Pass since 1857. From

Figure 3-10 Weathering by severe minor faulting in rock structure. Contorted strata in the Chickamauga limestone formation near Ben Hur, Virginia. The maximum thickness of an individual stratum is 24 in. Within the slightly reversed fold there is already a small brecciated zone of crumbled rock. As the reversed fold becomes fully developed, a minor fault will occur, producing a sizable brecciated and gouged zone. *(Virginia Geological Survey.)*

these analyses it was deduced that 4,300,000 yd^3 of granites, schists, and gneisses and 600,000 yd^3 of sandstones, all within the fault or rift zone, could be ripped economically. The contractor's excavation experience on the job corroborated the analyses. It was a striking example of the rock-weakening effects of major faults.

Plant Roots and Burrowing Animals Rock structures are weakened by roots and by burrowing. In faces of new cuts of decomposed granite, the tree and shrub roots extend down 30 ft or more in their strivings for moisture. Figure 3-12 shows the combined results of temperature changes, rains, frosts, and tree roots in the breakup or weathering of quartzite.

Combined Effects

In this metamorphic quartzite of Monroe, New York, the total combined effect of all destructive agents has been to reduce hard quartzite, normally requiring blasting, to rubble. Only ripping is required and not much of that if the cut is shallow. Cooperating Nature, by transforming this rock formation from the hardest to just about the softest, has materially reduced the work of the earthmover.

Climate has a profound influence on the weathering of rocks. The effects of different climates on the same basalt rock formation are dramatically illustrated by two pictures of the windward and the leeward sides of the western part of Maui Island, Hawaii. Figure 3-13 shows a vigorous early mature landscape on the windward side of the mountains. Both heavy vegetation and rapid erosion have resulted from the heavy annual rainfall of 200 in. Herein both mechanical disintegration and chemical decomposition are evident, but the chemical weathering has predominated.

Figure 3-14 displays the relatively arid climate of the leeward side of the mountains. Depth of weathering is less, streams are almost nonexistent, and vegetation is either scarce or absent. Mechanical disintegration is more in evidence than chemical decomposition.

Table 3-1 summarizes generally the degrees of weathering according to 11 contribu-

3-12   Rock Weathering

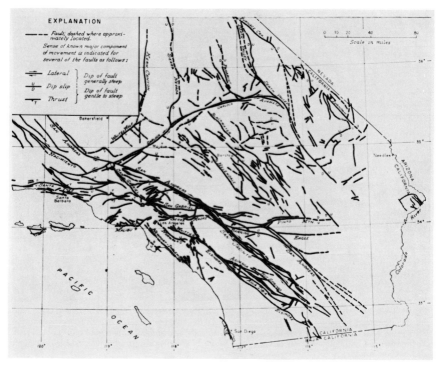

Figure 3-11 Widespread system of faults in 13,200 mi² of California. Major and minor faults of southern California. Notorious faults that have dealt death and destruction during the past century are Imperial, Newport-Inglewood, San Andreas, San Gabriel, San Jacinto, Santa Yñez, Sierra Nevada, and White Wolf. Others have been active before recorded history and presently are dormant. Throughout aeons all have been allies of the excavator by weakening the rock structures within their rift zones, which may extend up to 2 mi on either side of the fault axis. *(California Division of Mines and Geology.)*

tory factors. These influences are observable. Obviously, in using the table one must weigh the factors and use good judgment, as the table is idealized. And of course all factors are interrelated.

RATES OF WEATHERING

There are no real data concerning the linear rate of rock weathering in the natural state. The several thousands of years of recorded history are too brief for conclusions even if there had been observations.

Meager available information centers around a wide range of tombstones and buildings in England and in the New England states during the past 300 yr, around the 8000-yr-old ruins of Yucatan, and around surface rocks that show largely mechanical disintegration during the past 15,000 yr of the last interglacial period. These weatherings in igneous, sedimentary, and metamorphic rocks average about 2 in for the past 10,000 yr. These weatherings are all in exposed buildings and monuments and it is probable that they are the result of mechanical disintegration. Accordingly, the average rate does not reflect the more effective chemical decomposition of rock in the natural state below ground surface.

If chemical forces are twice as effective as mechanical agents, then a hypothetical rate of weathering might be 6 in, or 0.50 ft, in 10,000 yr for the combined weathering process for rocks in situ. If this rather vague rate is applied to the 10-ft cube of unweathered rock of Figure 3-9, the rock should become clay and sand in 100,000 yr.

Rates of Weathering   3–13

Figure 3-12 Cumulative effect of several mechanical agents on weathering. Metamorphic quartzite of Monroe, New York, weathered and broken by temperature changes, frost, rain, and plant roots. Fresh quartzite requires blasting. This well-weathered quartzite requires little or no ripping if the cut is only a few feet deep. *(H. Ries and Thomas L. Watson,* Engineering Geology, *Wiley, New York, 1925, p. 225.)*

Figure 3-13 Strong geologically early weathering in wet climate. Vigorous early mature landscape on the windward side of the mountains of western Maui. In this subhumid climate rock weathering has been advanced by the annual 200 in of rainfall, resulting in chemical decomposition by the rich vegetation and in mechanical disintegration by the abundant streams. *(U.S. Air Force.)*

3-14   Rock Weathering

Figure 3-14   Weak, geologically late weathering in dry climate. Feeble, immature landscape on the leeward side of the mountains of western Maui. In this semiarid climate rock weathering has been reduced by limited rainfall, culminating in little chemical decomposition because of sparse vegetation and moderate mechanical disintegration by the scarce streams. *(U.S. Air Force.)*

Personal observations of surface and subsurface rocks over a long period of time in some 1 billion yd^3 of rock excavation indicate that the rocks most resistant to weathering are metamorphic, that those of average resistance are igneous, and that the least resistant rocks are sedimentary. This is a generalization, as may be inferred from the fact that sedimentary limestone in arid country is more resistant than igneous granite in semiarid, subhumid surroundings, and that granite is more unyielding than metamorphic serpentine in humid lands.

The depth of weathering varies greatly, being dependent upon the factors of Table 3-1. Based on observations of deep cuts in public and private works, the maximum depth appears to be about 400 ft. The rocks and ores of underground mines show weathering at considerably greater depths, but the chemical decomposition and replacement processes are the result of metamorphic actions rather than surface and subsurface weathering.

TABLE 3-1   Observable Factors Affecting Weathering

Factor	Degree of weathering		
	Minimum	Average	Maximum
Rainfall, average annual (in)	Up to 15	15 to 50	50 and up
Temperature, average annual (°F)	Up to 40	40 to 80	80 and up
Humidity, average annual (%)	Up to 40	40 to 80	80 and up
Topography	Level	Hilly	Mountainous
Rock exposure	South	East-west	North
Vegetation	Sparse	Average	Abundant
Residuals, depth	Thin	Average	Thick
Rock outcrops	Heavy	Light	None
Rock grain	Small	Medium	Large
Rock color	Dark	Medium	Light
Rock hardness or durability	Hard	Medium	Soft

SUMMARY

Rock weathering is an important ally of the earthmover in his efforts to excavate rock in the many stages of mechanical and chemical breakup. In all cases, to a greater or lesser degree, Nature has substituted generously for the tractor-ripper and for explosives in the fragmentation of rock.

Weathering studies are all significant to the excavator, as through the millennia Nature may very well have fragmented the rock down to the finished grade without the knowledge or understanding of the earthmover.

CHAPTER 4

Landforms and Geomorphology

PLATE 4-1

Obstacles to rock excavation along a bluff shoreline. Along the shores of submergence into an encroaching sea there are often accentuated promontories and canyons that present serious problems to highway and railroad builders. Plate 4-1 illustrates the highway construction enigma near Hecate Light of the rough Oregon coast. Characteristics are steep sidehill cuts and fills, bridges and tunnels, and ever-present confined working conditions. A compounding perplexity is the unknown stability of the weathered basalt rocks. These 49-million-yr-old lavas are sometimes of columnar jointing, making for dangerous uncertainties in the necessarily steeply sloped sidehill cuts. *(Brubaker Aerial Surveys.)*

Landscapes or geomorphology, physical geology that deals with the form of the Earth, are to the rock formations as the forest is to the trees. Landforms are the volumes of rock to be excavated and their understanding is important to the earthmover.

For example, is the gravelly river terrace of Figure 4-1 simply a comparatively level and deep alluvial fan washed down from the mountains? Or is it a thinly veneered horizontal sedimentary rock deposit? If the former, it is common excavation, moderately consolidated. If the latter, the rock may require blasting after the removal of the alluvium blanket. These questions are paramount for the construction of a highway or a railroad from the narrow valley to the top of the terraces.

The geomorphic processes that shape the earth and as a result control the excavator's methods are degradation, or the breaking down of the earth's crust, and aggradation, or the building up of the crust. Generally the same process is responsible both for degradation at higher elevations and resulting aggradation at lower levels. For the rock specialist, degradation uncovers the existing rock, making it visible, and aggradation obscures the present-age rock. Accordingly, the imponderable is generally the result of building up processes and the resulting landforms.

Degradation or subtraction and aggradation or addition of earth-rock are the result of six agents: ice, water, wind, volcanics, gravity, and crustal movements. All six forces may work individually or collectively and cooperatively.

ICE

Ice action on a large scale involves erosion, transportation, and deposition of earth-rock by glaciers. On a small scale the action involves frost or ice wedging, as discussed in Chapter 3. Although glacial actions are as old as Precambrian groups of rocks, the excavator is interested primarily in the landforms of those of the Pleistocene series, commencing about 2 million yr ago and ending about 18,000 yr ago. The last one, known as the Wisconsin in North America and the Wurm in Europe, influenced greatly the lives of our ancestors. It also released meltwater to form the Great Lakes and to excavate the rapids and basin of Niagara Falls.

Figure 4-2 illustrates the extent of these glaciers or ice sheets. They covered the

Figure 4-1 A possible misleading interpretation of surficial rock formation. Deceptive terraces of Cola Creek, Castle Hill basin, New Zealand. Unless there are exposed outcroppings of subsurface earth-rock structures along the banks of the creek, a cursory examination of the surface gravels of the terraces will not disclose the true nature of deep excavation for a highway or railroad ascending the escarpment. The simple question is whether the gravels extend to the grade of the excavation or are merely a veneer atop a rock formation. *(William D. Thornbury,* Principles of Geomorphology, *Wiley, New York, 1961, p. 159.)*

4–4 Landforms and Geomorphology

Figure 4-2 Coverage of four glacial periods in North America. Areal extent in North America of four glacial periods during the past 1 million yr. This vast and powerful action of moving ice resulted in transportation of incalculable amounts of valuable topsoils from Canada to the north central and northeastern United States, in excavation of the Great Lakes by the meltwaters, and in formation of the present-day glaciers of the Rocky Mountains and the Sierra Nevada. It is presumed that the Northern Hemisphere is presently in the midst of the fourth interglacial period, the Wisconsin in North America and the Wurm in Europe. *(U.S. Geological Survey.)*

northern half of North America, which included many of the northern states. In the 11 western states the glaciers covered much of the high elevations of the Sierra Nevada and the Rocky Mountains, and many mountain glaciers have survived into the present interglacial period.

There are four principal landforms resulting from the deposition of the earth-rock mixtures. All have one characteristic of materials composed of unsorted and unstratified till and drift materials made up of clays, sands, gravels, cobbles, and boulders. These particles vary in proportions and in shape from rounded to angular. The four landforms are *lateral moraines, terminal moraines, drumlins,* and *eskers.* They are pictured and discussed in Figures 4-3, 4-4, and 4-5. Since they are all formed by deposition from melting ice, their landforms sometimes overlap in configuration.

An analysis of these glacial landforms is important to the excavator both prior to construction and during the prosecution of the work. The following example points up the necessity for analysis.

In 1960 freeway building at Conway Summit on the east slope of the Sierra Nevada of California called for an 850,000-yd^3 cut of maximum depth 140 ft. Exposed granite surrounded the cut. Some bidding contractors believed that the gravelly surface material was an overburden atop country granite and they adjusted their bids accordingly. In one case the following analyses were made and conclusions drawn.

Figure 4-3 **The wide gradation of the conglomerate of a glacial moraine.** Glacial ground moraine near Lansing, Michigan, typifying a lateral or terminal moraine. The lateral moraine results from the deposition of materials when the ice melts along the edges of the glacier. The terminal one comes from placement at the end or tongue of the glacier. The eroded bank of till is about 30 ft high and it illustrates the heterogeneous nature of the moraine, ranging from silts to boulders of up to 2-ft^3 volume and laid down unsorted and without stratification. Over the aeons till is sometimes cemented by mineral agents and then it becomes tillite. Thus, a moraine may represent soft to hard excavation for the earthmover, but usually it is soft. *(U.S. Geological Survey.)*

    1. A geology map showed the cut to be part of a terminal moraine dating back about 18,000 yr to the Wisconsin glacial stage. Such youthful age precluded any thought of even mild consolidation.

    2. Limited exploratory borings showed some refusals to drilling, suggesting either bedrock or, more likely, interbedded boulders.

    3. Seismic studies showed no shock-wave velocities greater than 2900 ft/s to finished grade.

    4. The conclusions were:

      a. The gravelly overburden indicated a moraine rather than typical decomposed or weathered granite.

      b. The boring refusals were caused by large boulders and not by country rock or bedrock.

      c. Seismic studies, indicating uniform low velocities to excavation limits, forecast a gravel-boulder deposit with anticipated soft ripping by medium-weight tractor-ripper.

      d. Finally, it was wisely concluded that the huge cut could be taken out by methods for common excavation.

The beginning of a peneplain by the sculpturing of a glacier is shown in Figure 4-6. This aerial photograph shows a section of the glaciated Sierra Nevada of central California in the vicinity of Donner Pass. The eight-lane freeway section of the view is at about elevation 6000 ft and Donner Pass in the eastward distance is at elevation 7008 ft.

At this high elevation the surface of the bare granite is striated by the grooving action of the glaciers and it is strewn with rock particles torn from the mountainside by glacial gouging. The particles range in size from sands to room-size boulders. Along the distant horizon may be seen the remains of *cirques.* Cirques are generally shaped as steep amphitheaters and they are faceted where the tops of the slopes meet the natural slopes of the rugged mountains. Bear Valley floor to the left in the picture is made up of rather

4-6   Landforms and Geomorphology

Figure 4-4 Large rocks carried by glacier a short distance from the mountains. The terminal moraine of a cliff glacier on Mount Lyell of the Sierra Nevada, California. As compared with the two men of the picture, the largest of the massive glacier-borne rocks are room size. They were torn from the cirques and the gorges of the mountains and they were transported a relatively short distance, as attested by their angular form. Excavation of this rubble would call for shovel and trucks, with some blasting of the hugest fragments. *(U.S. National Park Service.)*

Figure 4-5 Wide gradation of well-traveled, well-rounded rocks of a glacial drumlin. Drumlins and eskers are deposits from melting ice within the body of the glacier, rather than at the extremities as in the case of the moraines. Clay, sand, pebbles, and boulders are mixed in this unconsolidated till of Oconomowoc, Wisconsin. It is sometimes called boulder clay and it represents soft excavation. Drumlins are elongated hills or ridges up to 1 mi in length and 200 ft in depth, their axes being parallel to the movement of the glacier. Eskers are rather narrow, sinuous ridges, longer but lower than drumlins. The till or drift material of an esker resembles closely that of a drumlin, as they have a common origin, and they are oriented along the path of the glacier. *(U.S. Geological Survey.)*

Figure 4-6 Glacially sculptured granite of Donner Pass of the Sierra Nevada of California. Ageless leveling of granite and andesite of the great Sierra Nevada of California by aeons of glacial action. Rock particles are torn from the mountainsides by gouging action and are then transported by the ice down the mountainsides, grooving the rock surfaces and removing any loose rock fragments. An almost bald rock surface remains, with little or no surface residuals except in small vales. Such topography always creates abnormally high rock excavation costs. *(California Division of Highways.)*

coarse alluvium with little soil cover. In general the land is topped by a minimum thickness of residuals, which supports only sparse vegetation.

The rock excavation for this magnificent freeway called for much blasting in granite and andesite, as ripping of the rock by heavyweight tractor-rippers extended below natural ground by only a few inches or feet. Cuts ranged in depth from a few feet to several tens of feet, and the freeway alignment is mostly on the ridges. The preponderance of rock excavation, the short construction season, and the remoteness of the work area contributed to high cost to the state and low profit or loss to the excavators.

WATER

Running water has produced and is forming the greatest number of landforms, which make up excavations. It is the most powerful of the forces involved in Nature's eternal efforts to level the land. A trip by airplane across the United States from Los Angeles to New York unveils a myriad of landforms of interest to the earthmover. A few are:

1. The several-hundred-foot-deep waterborne fanglomerates of the San Gabriel Valley call for hundreds of millions of cubic yards of gravel excavation for the aggregates used in the Los Angeles area.

2. The abrupt water-carved topography of the San Bernardino granite mountains presents real rock excavation problems to the roadbuilders.

3. The desiccated floor and the mudflows of the Mojave Desert, the results of flash floods, require continuous cuts and fills in what appears to be a monotonously flat desert.

4–8 Landforms and Geomorphology

4. The nearly vertical rock walls of the Colorado River escarpment made the abutment excavation for Hoover Dam painstaking and costly.

5. The precipitous walls of the Black Canyon of the Gunnison River of Colorado called for tedious sliver-cut excavation for the Denver and Rio Grande Railroad.

6. The easily worked river-gravel terrace deposits of the central states afford building materials for the Mississippi Valley states.

7. The northeast to southwest trending ridges of the Appalachian Mountains call for hard-rock excavations for railroads and highways, unless location engineers can find water gaps and wind gaps.

8. The igneous rock escarpment of the Hudson River Palisades provides an abundance of trap rock for the needs of the trade area.

Running water degrades the landscape in the form of streams, ranging from little brook gullies to big river valleys, as in the combined watersheds of the Conewango Creek, Allegheny River, Ohio River, and Mississippi River, stretching 1100 mi from New York State to Louisiana. On the other hand, water aggrades the same land in this watershed all the way from the little gully outwash of Conewango Creek to the big Mississippi River Delta.

Valleys and Gorges

Figure 4-7 dramatizes the late stage in the development of a mature river gorge. Such a rock exposure in the walls, together with the rock cuts of the highway and railroad, indicates that the excavator will encounter soft, medium, and hard excavation and that there will be some blasting. If excavation is planned in nearby canyons of the tributaries

Figure 4-7 Unmistakable visible harbingers of hard-rock excavation. Mature gorge of the Cheakamus River, British Columbia. A powerful mountain stream is cutting this precipitous waterway, compounding the granite excavation for the sidehill cuts of the highway and the railroad. Exposed faces foretell problems of any future excavation in the Cheakamus gorge and the gorges of nearby tributary streams. *(Geological Survey of Canada.)*

of the Cheakamus River and if the rock formations in these resemble those of Cheakamus Gorge, then the proposed excavation will follow the same pattern for work.

Contrastingly, an earthmover who is considering heavy subsurface excavation for a reservoir in plains country and sees a fairly deep stream with slumping banks in the general area must assume that the reservoir will involve soft excavation. Quite naturally, the earthmover would wisely verify the supposition.

Floodplains

Just as the valley is the result of stream erosion, so the *floodplain* is the product of stream deposition or aggradation. Floodplains are generally fairly level and consist of clays, silts, sands, and gravels. To considerable depths they present no excavation problems.

Fanglomerates

A possibly deceptive landform is the alluvial fanglomerate, which is formed by material transported by water at the end of mountain valleys. The fanglomerate consists of fine to coarse aggregates, ranging from silts at the base to room-size boulders near the apex of the triangular fan.

In a fanglomerate the unknowns to the earthmover are these:

1. Does the fanglomerate exist down to and below the grade of the excavation, or is it a shallow deposit atop the pediment of the mountain rock?

2. If the fanglomerate does, indeed, exist to grade, does the invisible mass resemble the surface material as far as gradation is concerned?

Such questions were raised when the Owens Valley Aqueduct, supplying water to Los Angeles County from Bishop, California, was built. The 300-mi water carrier crosses many fanglomerates of the Owens and Mojave Valleys, the granitic alluvia being derived from the steep east escarpment of the Sierra Nevada. Complete investigations of these and other variables of the open-cut and tunnel excavations were made prior to construction.

Figure 4-8 outlines several fanglomerates from the Funeral Mountains near Furnace

Figure 4-8 Physical features of a typical geological basin-range province. Fanglomerates of Death Valley, southern California. Furnace Creek Inn is built upon a fanglomerate flanked by a fault escarpment. In back of the inn are several small alluvial fans debouching from the canyons of the foothills. Furnace Creek fault is at the base of the distant Funeral Mountains and many major fanglomerates issue from the large canyons. *(California Division of Mines and Geology.)*

4–10   Landforms and Geomorphology

Creek, Death Valley, California. These triangular aprons are being constantly built by the water transportation of well-weathered rock particles from the eroded mountains. Furnace Creek Inn is built upon a recent fault scarp and the major Furnace Creek fault is at the base of the distant mountains.

The unsorted materials of a fanglomerate are strikingly shown in Figure 4-9. The dissected wadi reveals up to cubic-yard-sized boulders. Such boulders would "refuse" an exploratory drill, perhaps erroneously indicating bedrock. Obviously this common excavation would present some problems, calling for shovel loading of trucks, but there would be no blasting in the loose alluvia.

Terraces

Excavation in abrupt, steep-sloped river terraces is generally obvious, as the formation shows up in the exposed bank. This is not the case with flat-sloped terraces, as shown in Figure 4-1. In these the true nature of the innocent-appearing slopes may be obscured by slope wash.

Mudflows

Mudflows are movements of water-saturated materials down streambeds and slopes. They may be mild or they may be of startling and even fatal proportions. A sudden heavy flash flood in the mountains, unknown to the careless camper in a sunny desert arroyo, may give the victim less than a 1-min warning roar before he is swept away.

Clays and silts of desert hillsides and volcanic ash of mountain slopes are common

Figure 4-9   Massive, poorly graded conglomerate of an alluvial fan. Stream channel crossing an alluvial fan or fanglomerate. Bank shows the characteristic texture of fine to coarse materials laid down by the waters. Some of the large boulders are in the middle of the arroyo and they have been moved only by powerful floodwaters. *(U.S. Forest Service.)*

materials of mudflows. Figure 4-10 outlines an ancient mudflow of volcanic breccia in the Mojave Desert near Victorville, California. The area is about 7 mi^2 and the volume of the mudflow is about 50 million yd^3. Such sudden mudflows have reached velocities of 60 mi/h, with devastating results.

Such mudflows in level or gently rolling country generally offer no excavation problems. However, this is not in the case in mountainous terrain, where the flow material may include larger than room-size angular and rounded boulders and where the underlying bedrock or pediment is of unknown and sometimes variable depth.

Mesas and Buttes

Mesas and buttes, typical of the 11 western United States, are formed chiefly by the combined agents of water and wind. Sometimes the ranges, or uplifted horsts of basin-range geologic provinces, appear to be mesas because of their steep escarpments. However, they are the direct result of vertical faulting, modified by water and wind. It is true that horsts, because of their configuration, present the same excavation problems as do the mesas.

Usually mesas and buttes are distinguished by a hard-rock cap that protects the lower, less resistant rocks and preserves the plateaulike surface of the land.

When a highway or railroad line must ascend the escarpment of a mesa from the valley below, the sidehill cuts may cross different rock strata and thereby may present excavation problems because of the varying hardnesses of rock and the tendency for slopes to slide.

Figure 4-11 shows a four-lane highway descending the side of a mesa to the lagoon of the Pacific Ocean near Del Mar, California. The escarpment of this mesa is not steep, having been eroded from the well-weathered sedimentary rock formation. This Del Mar formation, some 47 million yr old, is made up of multicolored mudstones, siltstones, and sandstones. There are oyster interbeds, testifying to its marine origin. It is of medium

Figure 4-10 Outsize but typical mudslide characteristic of desert country. Breccia mudflow in Mojave Desert, near Victorville, California. The magnitude of the mudflow of volcanic rock may be judged by the narrow ribbon of the road in the left foreground. The lobe is about 2½ mi across and about 7 mi^2 in area and totals some 50 million yd^3. Such flows, generally the result of prolonged mountain rains, are usually soft to medium excavation. They are an unsorted mass of many sizes of earth-rock particles, ranging from clays to room-size boulders. *(William D. Thornbury,* Principles of Geomorphology, *Wiley, New York, 1961, p. 93.)*

4–12   Landforms and Geomorphology

Figure 4-11   Easy soft-rock excavation along gentle bluff of eroded mesa. Four-lane highway descending the escarpment of a mesa along the Pacific Coast near Del Mar, California. The moderately sloped bluff, together with the medium-hardness Del Mar marine sedimentary rock formation, presented no excavation problems to the earthmover. The mixture of mudstones, siltstones, and sandstones is sufficiently indurated to allow for a designed backslope of ration about 1:4.

hardness, as indicated by the steep sideslope of the highway, and it calls for medium ripping by a tractor-ripper.

The oceans, with their shores, bays, inlets, coves, and estuaries, shape the land so as to cause problems for the building of excavations and embankments. Plate 4-1 outlines the complexity of highway building near Hecate Light on the Oregon coast. The rock is basalt and andesite and is subject to slides in the cut slopes and in the wave-carved shoreline. Here just about all excavation problems are present. After a railroad or a highway is built in such a shore of submergence, the external cause of failures, undercutting by the waves, will always exist.

WIND

Wind degrades and aggrades. Degradation or erosion of existing surfaces serves the earthmover by uncovering earth-rock formations that would be obscured otherwise by misleading residuals.

Deposition by winds results chiefly in loess and sand dune formation. Loess is a fine silt and dust deposit of the central United States. Deflation by the wind involves pickup from the prairie lands of fine particles, which are then windborne and deposited at considerable distances. Loess has accumulated to depths of 40 ft and it is characterized by near-vertical slopes when eroded by nature or excavated by humans. Being of recent making and having little cementation or pressure, it is not consolidated and it is soft excavation.

Sand dunes are common to shores of lakes and oceans and to desert lands. Figure 4-12 shows barchand dunes in the Imperial Valley of California. These ever-creeping landforms cover and uncover the old plank road, which was built early in this century.

Dunes are obviously soft excavation. They present only a moderate loading problem but generally they offer an aggravating hauling problem to the excavator. Figure 4-12 shows dramatically the haulage enigma because early automobiles could not pass over these sands without the plank or corduroy road, even though the natural road-building material was laid carefully. In this same country the new freeway called for moderate cuts and fills but the single-axle-drive rubber-tire scrapers could not tra-

Figure 4-12 Eternally encroaching sand dunes. Barchand sand dunes obliterating the old narrow plank road across the Imperial Valley of southern California. In the early part of this century, because of the ever-shifting sands this corduroy road was built between El Centro and Yuma, Arizona. Every ¼ mi it was widened to allow passing. It served well until improved road-building methods made it possible to construct and maintain a first-class highway over the treacherous desert. Persistent migratory dunes no longer bedevil the building of freeways across the dunes of the seashores and the deserts. *(U.S. Forest Service.)*

verse the sandy haul roads. The problem of traction and flotation was solved by the use of two-axle- or all-axle-drive scrapers with twin engines, supplemented by liberal wetting of the haul roads.

A similar haulage problem confounded the builder of the Cape Cod Canal of Massachusetts some 40 yr ago. Rubber-tire bottom-dump haulers were used and only single-drive axles were available. Work was possible by liberally wetting the unstable sands of the cuts.

VOLCANICS

Volcanics interest the earthmover almost exclusively in terms of the unknown consolidation of cinder cones and breccia and the questionable hardness of lava flows. The certainty of hard excavation normally requiring blasting is evident in the lava flow of Figure 1-1. This "aa" or rough vesicular basalt flow may be excavated by hard to extremely hard preliminary ripping by a heavyweight tractor-ripper.

Tuffs and breccias differ only in the maximum size of their particles, both being pyroclastic or formed from airborne volcanic tephra. Tuff is of less than 1-in diameter and breccia contains fragments of more than 1-in diameter.

Moderately solid breccia rock is shown in Figure 4-13. The canyon of Rio de las Frijoles in northern New Mexico has been eroded in this thick deposit. Cliff dwellers quarried blocks for their cliff-hanging homes from this rock formation. Such tuff and breccia are hard, but they are usually rippable by heavyweight tractor-rippers. The formidable, massive nature of the formation is accentuated by reference to the man in the picture.

A typical Mojave Desert basalt cinder cone is shown in Figure 4-14. It is a lightweight-

Figure 4-13 **Canyon wall of finely graded medium-hard volcanic breccia.** Eroded pyroclastic breccia of the Canyon del Rio de las Frijoles in northern New Mexico. The massive vertical walls above the man at the left of the picture, suggesting extreme hardness, are to be contrasted with the weathered caverns at the right of the picture, suggesting softness. Excavation properties are in between, as the medium-hard igneous rock formation is usually rippable by heavyweight tractor-rippers. *(U.S. Forest Service.)*

aggregate source for an open-pit operation near Red Cinder Mountain. These cones are usually mildly consolidated and, in spite of a rather forbidding appearance, they are generally soft to medium-hard excavation.

GRAVITY

The land-forming effects of gravity are largely the results of the already discussed actions of ice, water, wind, and volcanics. However, visible examples of the work of gravity are provided by creep, slump, slide, and talus slope, for here are landforms only partially dependent on the aforementioned four agents.

Creep or slump in a hillside or river bank is a gradual process, usually but not always accompanied by a high moisture content and by horizontal slump ridges in the formation of the hillside. Figure 4-15 pictures the gentle movement of a slump that was preceded by creep. The vertical fall of the detached section of the bank is about 4 ft vertically and about 6 ft horizontally. The bank was weakened by the undercutting of the stream in flood stage.

Of significance to the earthmover in all creeps and slumps is an indication of an unstable earth-rock formation. Such indications also mean soft excavation to moderate depth. From a design standpoint they point up the necessity of flat back slopes of cuts and possibly a requirement for retaining walls in the cuts.

Of course slides, big or small, indicate highly unstable natural earth or rock. Sometimes slides occur in rock slopes which have been blasted because of weakening of the natural

Figure 4-14 Volcanic cinder cone of soft to medium-hard aggregate-size scoria. A huge cinder cone of the Mojave Desert in California is the source of lightweight aggregates for this commercial open-pit materials plant. The red volcanic cinders are ripped and bulldozed from the cinder pit *(CP)* to the screening plant *(P)* by a medium-weight tractor and then the sizes are stockpiled in the yard *(SY)*. Excavation is soft to medium hardness. Nearby is a sister cinder cone in which a 75-ft-deep sidehill cut was made for a freeway. The excavation experience of the earthmover was identical to that of the operator of the cinder pit. *(California Division of Mines and Geology.)*

Figure 4-15 Slump in residuals and well-weathered rock formation. Slumping of subsoil and soil in the bank of a stream near Spartanburg, South Carolina. The gentle slump was preceded by a gradual creep of the edge of the bank, which was undercut by the stream in flood stage. While more common in earthy materials, creep and slump are also common in weathered rock formations. *(U.S. Soil Conservation Service.)*

rock by the back-breaking or shattering effects of the blast beyond the designed prism of excavation. Such effects are common in weathered granite and serpentine.

In sedimentary rock formations the slide generally takes place along the moisture-lubricated stratum of clay or shale, the dip of the slide-prone strata sometimes being as low as 15° to the horizontal. Such a slide of dangerous proportions is dramatically shown in Figure 1-3. This huge mass movement was caused by moisture-laden earth-rock, weakened by removal of the natural buttress when the back slope of the highway was excavated. Like creeps and slumps, slides indicate a weak earth-rock formation and resultant soft excavation. In slide-prone country the hazards to construction are manifest. Not only are the hazards indigenous to the immediate work area or right-of-way, but they may endanger also the adjacent property above the cut, as shown in Figure 1-3, and below the fill because of surcharging of the natural ground.

Sometimes there is a monetary gain to the excavator in terms of a profitable unit price for excavation. Many slides in cut faces are caused by too steep back slopes, especially in the grading for building sites. If a slide occurs the earthmover excavates additional earth-rock, which is more easily handled than it would have been in its original state.

Talus slopes are typical of mountainous country. Sometimes they take the form of steep cones, and the instability of the fallen rock fragments is obvious. On the other hand, in the case of the moderate slope of a rock stream the talus slope would be relatively stable. Such a rock stream is shown in Figure 4-16.

The lower slope of the talus rock stream is not especially steep and yet the stream is continuing to flow down into the valley. Were a railroad or a highway to be built across this landform, the back slopes would be unstable and the roadbed would be displaced slowly downstream. Similarly, any fills would be in jeopardy. It is doubtful that any work done by humans could be located economically or safely on such ever-mobile land.

Excavation in this landform is soft and the poorly sorted but generally hard rock particles are a good source for aggregates.

Figure 4-16 Slowly advancing rock stream of graded conglomerate. Rock stream in Silver Basin, San Juan Mountains, Colorado. These streamlike talus slopes extend down into the valleys and they originated as rock falls from glacially eroded steep cliffs of the mountains in the distance. Slowly moving rock streams would make for instability in either excavation or embankment and in any building raised thereon. The inherent instability is synonymous with soft to medium excavation characteristics. Like alluvia and fanglomerates, rock streams are rarely consolidated by any cementing agency. *(U.S. Geological Survey.)*

CRUSTAL MOVEMENTS

These kinetic phenomena in rock formations are the result of internal forces within the Earth's crust or regolith. They are usually the results of ages of work by the pressure within but sometimes they result from swift deformations by earthquakes. They are subsurface activities and the surface manifestations have generally been obliterated by years of weathering. Now and then one sees a recent scarp or ridge resulting from an earthquake fault.

Faults

Figure 4-17 shows transverse cross sections of common vertical faults. As indicated by the legend of Figure 4-17, these sometimes concealed faults are misleading even after careful field investigations.

In the case of the normal fault, a ditching contractor who based the estimate for bidding on borings to the west of the fault would possibly have quadrupled the estimated cost when working to the east of the fault.

On the other hand, by the same prebidding estimate the contractor in the case of the reverse fault would probably find the costs quartered when excavating to the east of the concealed fault.

Of course there are more variations and complexities in the vertical faults, but these normal and reverse faults typify the excavation problems.

Another type of fault is the lateral one, in which the sides of the fault move more or less horizontally with respect to each other. If they are truly lateral and not complicated by vertical movement, they are generally not misleading. Perhaps the greatest number of vertical and lateral faults, some with displacements of 20 ft, are in California. Their extremely extensive distribution is depicted in Figure 3-11. These have created both engineering and excavating problems. Sometimes the age-old movement creates brecciation in rift zones thousands of feet wide. These zones are made up of fractured rocks and clays rather than of the adjacent solid rocks from which they were derived. The salutary effect of of the rift zone on the cost of excavation was discussed in Chapter 3.

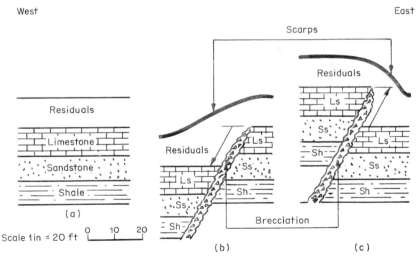

Figure 4-17 Types of vertical faults. *(a)* Unfractured beds. Undisturbed strata prior to faulting. *(b)* Normal fault. Tens of years after the faulting. Rather gentle slope after years of weathering. Fault is concealed by deceptive north-south trending ridge. An excavator digging a 20-ft ditch from west to east would run into about 14 ft of additional rock at the fault line. *(c)* Reverse fault. Tens of years after the faulting. Rather steep slope after years of weathering. Fault is concealed by deceptive north-south trending ridge. An excavator digging a 20-ft ditch from west to east would run into all earth at the fault line. Brecciation is made up of angular rocks torn from the sides of the plane of the fault.

4–18 Landforms and Geomorphology

A deformation equally as important as faulting of the rock formation is *folding*. Figure 4-18 shows the progressive deformation of sedimentary rock strata.

In a general way folds may have nearly vertical, inclined, or nearly horizontal axes. They may be of a minor nature, several inches to several tens of feet, or they may be of major proportions, up to tens of miles in length.

In Figure 4-18 an anticline has been slowly warped into a horizontal or recumbent fold, the upper half being shown in the lower cross section. At this juncture of development the strata could not resist the lateral forces, and the recumbent fold was faulted horizontally and the zone of brecciation created.

Such deformation presents to the earthmover obvious problems in estimating and subsequent excavation. Synclines, anticlines, folds, thrusts, and faults are common in the anthracite coal fields of eastern Pennsylvania. Here the coal, usually interbedded with limestones and shales, has been contorted in such manner as to complicate greatly the open-pit mining of the coal. Figure 2-15 shows such a coal stripping mine in which the rock formation has been only mildly deformed, affording the miner unusually good

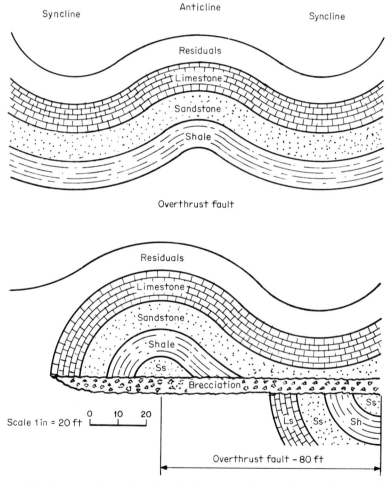

Figure 4-18 The progressive deformation of sedimentary rock strata. The original anticline became a recumbent fold and the recumbent fold is now an overthrust fault as the result of intense deformation. Brecciation is made up of angular rocks torn from sides of the plane of the fault.

Figure 4-19 Extreme deformation and weakening of a sedimentary rock formation. Folding and faulting of a sedimentary rock formation within the rift zone of the San Andreas fault near Palmdale, California. In this 1,146,000-yd^3 cut the sandstones and the shales were so weakened by ages of continuous flexuring that the excavator was able to fragment the rock by only soft ripping by a tractor-ripper. Nature, with infinitely patient work, had substituted for the machines of humans.

working conditions. There is a fairly fixed ratio of near horizontally bedded overburden to coal in terms of cubic yards per ton.

There is one benefit from these deformations. The rocks are necessarily severely fractured in the flexuring. The limestones, which normally might require blasting, may be ripped for scraper operation and may be sufficiently fragmented by Nature for shovel and dragline work.

A classic example of folding and faulting in the rift zone of a major fault is portrayed in Figure 4-19. The rift zone is perhaps 1 mi in width on either side of the San Andreas fault.

This 1,146,000-yd^3 76-ft-deep cut is within the rift zone near Palmdale, Claifornia. The Anaverde formation of sedimentary rocks includes sandstones and shales which were laid down some 8 million yr ago. During this epoch the originally horizontal strata were contorted into these picturesque faults and reverse folds. Because of rather thin bedding and the weakening effects of the continuous flexuring, the rocks were fragmented by soft ripping.

In contrast to the soft ripping in this freeway cut, the same Anaverde formation in a railroad cut a few miles to the east and outside of the rift zone required medium to hard ripping by the same heavyweight tractor-rippers.

This chapter ends the discussion of geology, rocks, ores, minerals, rock formations, rock weathering, and landforms. It would be interesting and intellectually and financially rewarding for the rock excavator to pursue these fascinating subjects far beyond the elementary treatment in this book.

SUMMARY

The body of excavation is always a landform, and the study of landforms, or geomorphology, will assist the earthmover in his appraisal of the excavation. Appearances of landforms are deceptive, especially when they are covered by a mantle of weathered material which may not indicate the nature of the excavation below the mantle. In many cases an understanding of the simple landforms and recognition of them will help in the initial examination of the excavation.

CHAPTER 5

Open-Cut Excavations

PLATE 5-1

Largest open-pit mine of the world. The earth's largest excavation by humans is the Bingham Canyon, Utah, open-pit copper mine of the Kennecott Copper Corporation. This huge bowl averages 2 mi in width and 0.5 mi in depth. The excavation was commenced in 1904, and the historical and logistical statistics are awesomely impressive.

1. A total of 3.6 billion tons of porphyry copper ore and overburden or waste rock has been excavated and hauled to the 20-mi distant smelter and to the nearby waste-pile areas. Of these 3.6 billion tons, ore made up 1.31 billion tons and overburden accounted for 2.29 billion tons as of 1976. The stripping ratio, or ratio of overburden to ore, is 1.75:1.

2. The pit is 2.25 mi west to east, 1.50 mi north to south, and 0.50 mi deep. The average train of locomotive and 16 ore cars carries 1390 tons and is about 565 ft long. To haul 3.6 billion tons the continuous string of trains would stretch 277,000 mi, or once around the Earth, then to the Moon, and finally around the Moon.

3. A close scrutiny of Plate 5-1, showing the "thumbprint" of humans on the earth's surface, reveals 55 benches of average 45-ft height and average 80-ft width. Rail haulage takes place below the 6340-ft level and truck haulage is above this level.

4. Fragmentation of ore and overburden is by blasting, and for every pound of total explosives used 4.30 tons, or about 1.89 yd^3, of material is removed. Annually 15,300 tons of explosives and blasting agents is used, filling a string of box cars about 3 mi long.

5. Machinery inventory for the average daily production of 433,000 tons during 1975 in a normal triple-shift day was:
   a. 100 mi of standard gauge railroad track.
   b. 54 miles of track in 3 mine-tunnel entries.
   c. 18 blast-hole drills of 8- to 12-in drill-hole diameters.
   d. 37 electric power shovels of 6- to 15-yd^3 dipper capacity.
   e. 60 electric and 2 diesel locomotives of 70- to 125-ton classification.
   f. Approximately 1000 ore and overburden hopper cars of average 87-ton capacity.
   g. 97 rear-dump haulage trucks ranging from 65- to 150-ton capacity.

6. On October 13, 1974, a world's record for excavation was set. In 24 h 504,167 tons of ore and overburden was loaded and hauled away. This tonnage is equivalent to 222,000 yd^3 of solid granite. It may be compared with the hard-rock excavation for a good-sized freeway cut or dam spillway.

7. The excavation for the Panama Canal was 268,000,000 yd^3 of rock-earth, or about 502 million tons. Accordingly, it would require approximately the material from seven Panama Canals to fill the colossal open pit of Bingham Canyon.

(Kennecott Copper Corporation and *Utah Power and Light Company.)*

There are nine principal kinds of open-cut excavations, those for: airports, building sites, canals, dams and levees, highways, pits and open mines, quarries, railroads, and trenches.

All these excavations generally combine both common and rock materials, including residuals, weathered rock, and semisolid and solid rock. The rock work is generally of a lesser yardage when compared with the common excavation, but it is of greater unit cost and it presents more problems.

The frontispiece and the accompanying description of the huge open pit of the United States Borax and Chemical Corporation vividly portray open-cut excavation.

AIRPORTS

Ideally, airports are located in flat or gently rolling country for obvious reasons of safety and construction costs. Actually, airports involving sizable rock excavation are a small minority of all airports.

Examples of those having 100 percent common excavation are numerous. Three such examples are cited below.

  1. The Baltimore airport in Maryland was built years ago in the swamps of Chesapeake Bay. The borrow pit for the fill material, some 2 mi away, was common excavation of silts and clays.
  2. The Kansas City airport, Missouri, is built on a flat alluvial plain at the confluence of the Missouri and Kansas Rivers. Grading consisted of a mild leveling process in the sand and gravel river terraces.
  3. The Seattle-Tacoma airport of Washington is built on a plain. Considering its 11,600-ft runways, proportionately little excavation, all common, was involved.

One must turn to hilly and mountainous country for examples of rock excavation, where sometimes the only place for an airport is atop a rocky mountain. Such an airport is the Wheeling airport, West Virginia, built during World War II under emergency winter conditions.

  1. Figure 5-1 shows the necessary bad-weather beginning of this 3-million-yd^3 rock job. Because of snowy and icy conditions and consequent excavation and haulage problems, the residuals and well-weathered shales and limestones were moved by crawler-tractor-drawn scrapers rather than by rubber-tire or wheel-type scrapers. Hauls averaged 2000 ft one way downhill, calling for maximum traction on the uphill empty return.

Figure 5-1 Airport building in icy, snowy wintry weather. World War II winter emergency construction of an airport in the mountains near Wheeling, West Virginia. The 2.5-yd^3 shovel is loading 15-ton rear-dump trucks with frozen weathered limestones and shales. This is the opening cut of the 3-million-yd^3 job. The work is icy and costly and it will not improve until the end of the spring thaws. During the thaws the steepening haul roads will be of mud rather than ice.

5–4 Open-Cut Excavations

The weathered rock was loaded without blasting by the 2½-yd³ shovel into 15-ton rear-dump trucks. Solid rock was blasted before loading. The winter picture shows three of the four trucks shuttling without turning on the initial 300-ft one-way level haul. An average of 4.5 buckets was loaded in 2.0 min for a 7.5-yd³ payload. By elimination of two turnings cycle time was reduced to 4.7 min, and the three-unit fleet production was about 200 yd³h.

After the bad weather the excavator brought in wheel-scrapers for the summer work and the overall production for the job improved greatly.

2. The San Clemente Island airport off the coast of southern California was hewn from rhyolite-dacite volcanic solid rock. The maximum depth of cut was about 65 ft. The weathered rock was ripped to about 45-ft depth and the solid lava flow was blasted to grade.

The runways, stretching completely across the north end of the island, are 11,000 ft long and 500 ft wide and the excavation was 10 million yd³ of moderately weathered and solid rock. Two 2.5-yd³ shovels and eight 22-ton rear-dump rock haulers were used for the loading and hauling of the huge yardage.

BUILDING SITES AND LARGE FOUNDATION EXCAVATIONS

These sites for homes, huge buildings, and plants may run to millions of cubic yards of common and rock excavation. As in most excavation, the common is generally less than the rock work.

One exception to this generalization was the 1967 rock excavation for an additional plant of the Mack Truck Company at Allentown, Pennsylvania. Blasting was used for 140,000 yd³ of the Pennsylvanian series limestone, totaling just about all the excavation. This rock formation typified excavation with minimum residuals and weathered rock.

On the other hand, Figure 5-2 shows a granite rock excavation in which blasting of solid rock was at a minimum when contrasted with the total excavation. The maximum depth of cut in this notoriously well-weathered granite of southern California was 60 ft. Of a total of 1,943,000 yd³, only 198,000 or 10 percent was blasted.

The track drill of Figure 5-2 drills 4½-in-diameter holes at the rate of 610 ft of hole per 8-hour shift. Hole spacing pattern is 7 ft by 7 ft, the average production of the drill rig being 138 yd³/h. Ammonium nitrate explosives were used, the combined powder factor of the explosives and blasting agents being about 0.75 lb/yd³ of granite.

Figure 5-2 Hard-rock excavation for mountain home sites. A 1,943,000-yd³ excavation in granite for a tract of home sites in the San Raphael Hills of Los Angeles County, California. In spite of negligible residuals, the volume of weathered granite was extremely high with respect to the solid rock. Only 198,000 yd³, or 10 percent, of the total yardage was blasted.

After blasting, this granite was loaded out and hauled by wheel scrapers, a common practice in such formidable-appearing solid granite. In five representative housing tracts in the same granite formation, only 269,000 yd^3, or 8 percent, of the total 3,429,000 yd^3 was blasted. This low ratio of "shooting rock" was obtained in spite of the fact that the maximum depth of cut was 125 ft and the average depth was 83 ft.

CANALS

Classic canals excavated out of rock are numerous. Three are described below.

Panama Canal Of some 175 million yd^3 of common and rock excavation, about 130 million yd^3 was in dry work and the remaining 45 million yd^3 was dredged hydraulically. The canal was completed in 1914 and of course heavyweight tractor-rippers and wheel or rubber-tire scrapers would today handle much of the so-called rock which was blasted and loaded by railroad-type shovels of the early twentieth century.

The total bid price for the prism excavation of the canal was $138 million, resulting in a unit price for unclassified excavation of about 79 cents. It is probable that today's gigantic earthmovers could excavate such a canal for a bid price in the range of $0.75 to $1.00 per cubic yard.

St. Lawrence Seaway This mammoth rock excavation with channels, canals, and locks stretching some 100 mi from Ogdensburg, New York, to Montreal, Quebec, involved several million cubic yards of solid beddings of limestone. In addition to fragmentation of the rock by conventional blasting, it was necessary to give the walls of the rock excavation a relatively smooth surface. Preshearing or presplitting was developed to a remarkable degree during the period of the 1950s and the 1960s. In this operation, depending on the kind of rock and the smoothness desired, drill holes of 1½ to 4-inch diameter are spaced on 1-ft to 4-ft centers along the line of the face desired and the explosives are detonated along with the main charges.

Excavation for the canal was in a glaciated area in which residuals and weathered rock were at a minimum. Consequently solid rock excavation predominated, far outweighing the common excavation.

California Aqueduct The main channel of this lengthy canal stretches some 600 mi from Oroville Dam in Butte County to Auld Valley Dam in Riverside County. The open-cut excavation ranged from the residuals and alluvium of the valleys to the solid rocks of the mountains. The rocks varied from soft sedimentary siltstones through medium igneous granites to hard metamorphic marbles. However, of the tens of millions of cubic yards, less than a million cubic yards of igneous and metamorphic rocks was blasted.

Figure 5-3 shows one of these rare exceptions to the general practice of rock ripping as the means of fragmentation. Rock is in the intake channel of the Tehachapi pumping plant in Kern County, California. This 1966 excavation included alluvium overlaying sandstone of the Tejon formation, which in turn led into the country rock, the granite.

The alluvium required no ripping and the sandstone was ripped out without difficulty. However, the granite ranged from medium ripping through hard ripping to blasting. The picture shows the phase of hard ripping, as a heavyweight tractor-ripper and a tractor-dozer are delivering the rock from the upper level of the ridge down to the loading shovel. The 6-yd^3 shovel is loading three 35-ton rear-dump rock haulers. Shortly thereafter, as the ridge was brought down, it became necessary to blast the granite rock. Of the total rock yardage of the intake channel, some 15 percent was blasted. In spite of the small percentage of blasted rock, this case is atypical of the usual case of rock excavation in the aqueduct. This significant distribution between rippable and "drill-and-shoot" rock for all the excavation for the California Aqueduct is typified in the following summary of excavation for the Fairmont to Leona reach, located in the foothills of the San Gabriel Mountains.

1. Length of reach, mi	18
2. Excavation, yd^3	9,400,000
3. Number of cuts through the ridges	114
4. Average yardage of cuts, yd^3	82,000
5. Maximum depth of cut, ft	76
6. Distribution of excavation, west to east, according to rock formations, yd^3.	
a. Granite. All was ripped. 36%	3,400,000

5–6 Open-Cut Excavations

b. Anaverde formation of sandstones and conglomerates. All was ripped.	4%	380,000
c. Alluvium of clays, silts, sands, and gravels. No ripping.	32%	3,100,000
d. Dacite. All was ripped.	1%	30,000
e. Pelona schist. About 50,000 yd³, or 2% of total 2,490,000 yd³ was blasted, 98% being ripped.	27%	2,490,000
Total yd³ of reach		9,400,000

Compared with the 600 mi of other reaches, this one in the pediment formations of the mountain range was considered to be rocky. Yet only 50,000 of the total 6,300,000 yd³ of rock was blasted, and this is but 0.8 percent of the total rock.

DAMS AND LEVEES

The open-cut excavations for dams may be subdivided in a general way into eight classes.

1. *Foundation excavation* or the removal of residuals and weathered rock so as to form a semipervious base for an earth-fill dam and an impervious setting for a concrete dam. Some of this excavation, especially in the case of a concrete dam, may require blasting or "dental" excavation if fault zones of breccia and gouge are present.

2. *Cutoff wall excavation.* This trench work is carried well below foundation excavation and is taken down to bedrock if bedrock is present. If bedrock is not present, as in the case of many earth-fill dams, it is carried down into firm material.

3. *Inlet and outlet portals* for the diversion tunnel. This excavation is characteristic of both types of dams, but more especially in the relatively rough topography of the setting for a concrete dam.

4. *Diversion channel excavations,* in lieu of diversion tunnels, are usually associated with earth-fill dams as a means of dewatering the work area of the dam.

5. *Spillway channel excavation,* as distinguished from the integral spillway of the dam. Since concrete dams are associated with mountainous country, rock excavation is usually present, whereas in the case of earth-fill dams the excavation tends to be common rather than rock.

6. *Power-plant and tailrace excavations,* down in the bed of the stream, may be either rock or alluvia or a combination of both materials.

7. *Borrow pits* located in residuals, alluvia, or weathered rock provide fill materials for earth-fill dams and concrete aggregates for the concrete appurtenances. Usually they

Figure 5-3 Excavation for canal to pumping plant of aqueduct. Rock excavation for the intake channel of the Tehachapi pumping plant, a part of the California Aqueduct. Although the yardage of all the rock blasted in connection with the total tens of millions of cubic yards of rock involved in the building of the canal was insignificant, this rock excavation in the granite pediment of the Tehachapi Mountains accounted for about 15 percent of the rock of the inlet channel. Almost without exception heavyweight tractor-rippers substituted for explosives in the rock excavation for the 600-mi canal.

are common excavation, although the cases of weathered rocks and cemented rocks such as caliche sometimes call for fragmentation by ripping. For concrete dams borrow pits may provide concrete aggregates and sundry materials.

8. *Quarry excavations* provide aggregates and haul-road metal for both earth-fill and concrete dams. They produce rock for semipervious, pervious, and blanket-rock zones for earth-fill dams.

There are many minor examples of rock excavation in the building of dams such as highway and railroad relocations, access roads, haul roads, building sites, and staging area excavations.

The enormous rock excavation quantities for two big dams under construction in 1975 are set forth in the following tabulations.

1. *New Melones Dam, Stanislaus River, California*

Foundation	1,380,000 yd^3
Intake structure	50,000
Slide areas	1,910,000
Spillway	15,000,000
Total rock excavation	18,340,000 yd^3

This dam is the rock-fill type and the total rock excavation will go into rock fill. These quantities do not include a sizable amount of rock removed during the first stage of construction for the diversion tunnel and other appurtenances. The rocks are metamorphic or altered basalts, sandstones, and shales of the Mother Lode country.

2. *Auburn Dam, American River, California*

Foundation, for both main dam and cofferdam	4,620,000 yd^3
Power plant	220,000
Service spillway	1,100,000
Auxiliary spillway	800,000
Tailrace and channel improvement	650,000
Quarry riprap	30,000
Total rock excavation	7,420,000 yd^3

This dam will be a double-arch concrete dam, and the first contract is for the excavation preliminary to the actual building of the dam. The dam is located in a slightly faulted area of the Mother Lode country, and the foundation excavation includes 82,000 yd^3 for fault zones that involve closely controlled lines and grades or "dental" work. The quantities are generally a mixture of residuals and weathered and solid amphibolites and schists. The rock for the power plant and for tailrace and channel improvements and the quarry riprap are solid. All rocks are typically metamorphic in the work area.

The Balsas River Dam on the Río de las Balsas of Mexico is a part of the Infiernillo electric project and it was built during the 1960 to 1970 period. The 900,000-kW generating dam is of the rock-fill type with a crest length of 175 m (574 ft) and a height of 152 m (498 ft). Five million cubic meters (6.5 million yd^3) of quarry rock excavation in a basalt-andesite formation was required.

Four 4-yd^3 electric shovels and two 4-yd^3 diesel shovels loaded out a fleet of 22-ton rear-dump rock haulers with an average production of 15,080 m^3 day of double shifting. The average rate is 157 m^3 (205 yd^3) hourly for each shovel.

In this mixture of weathered and solid igneous rock such production is excellent, indicating good management and good machinery availability. The average hourly productions suggest a shovel performance in keeping with the following deduced tabulation of factors.

Shovel cycle	25 s
Shovel dipper factor, ratio of payload to capacity	0.60
Overall job efficiency	60%

Such quarry excavation on a hillside, as shown in Figure 5-4 is typical of foundation excavations for the abutments of concrete and rock-fill dams, although in foundation excavations generally only weathered rock is removed. During this quarry excavation, the undesirable residuals and well-weathered rock were removed initially and wasted,

5–8 Open-Cut Excavations

Figure 5-4 Dam building in precipitous mountain gorge. Quarry excavation for the dam on the Río de las Balsas of Mexico. Two 4-yd³ shovels are opening up the excavations by means of a "sliver cut" and later an additional four shovels will be used for full-scale quarry production of 15,000 m³ (19,600 yd³) each double-shift working day. In this pioneering work of residual and weathered rock removal, shovel efficiency is necessarily low because of the confined working conditions and the instability of the slide-prone rock formation. *(Bucyrus Erie Company.)*

then the acceptable slightly weathered rock was removed, and finally the desirable solid rock, requiring blasting, was taken out and the operation proceeded efficiently.

HIGHWAYS

Everybody is most familiar with the everyday sight of the work for roads, highways, expressways, and freeways. Excavation ranges from a few thousand cubic yards of common classification per mile for a "turnpike" road across a flat desert playa all the way up to several million cubic yards of rock per mile for an eight-lane freeway through rugged mountains.

In 1 mi of freeway building through the San Raphael Hills of Los Angeles County, 8,140,000 yd³ was removed with a maximum 210-ft depth of cut in this all-granite job. In spite of the great depth in this hard rock only about 1.5 million, or 9 percent, of the total 11.6 million yd³ was blasted.

The magnitude of highway excavation may be inferred from a summary of 506 public and private earth-rock and rock jobs analyzed during the past 20 yr. Of these varied jobs, 301, or 60 percent, were highway building. Since some 1 billion yd³ made up these 506 projects, there was some 600 million yd³ of highway excavation, and this represents only the experience of one engineer.

Figure 5-5 illustrates a typical rock job for a freeway of 3,510,000 yd³, made up of about 1,350,000 yd³ of granite, 1,424,000 yd³ of schist, and about 736,000 yd³ of residuals of granitic and schistose origins. The freeway is located in the mountains of San Diego County, California.

Figure 5-5 Huge saddle cut of freeway construction. Granite rock excavation in San Diego County, California, in 1975. This huge, 1,006,000-yd³ cut is about half excavated. It is 930 ft from top of slope to top of slope and 110 ft in maximum depth. About one-third of the total excavation was ripped to a depth of about 25 ft and hauled by wheel-type scrapers and by rear-dump hauler loaded by wheel-type shovels. The remaining two-thirds of the yardage is being systematically blasted prior to shovel loading. Four track drills are working at the 600-ft-wide elevation. Two 8-yd³ bucket loaders are loading out a fleet of 40-ton rear-dump rock haulers. A rock cut of such size will require about 8 months from the initial pioneering work to the final finishing to rough grade.

Analyses for bid preparations for the contractors disclosed that the rock distribution was in accordance with the following tabulation.

Kind of rock by cuts	Total rock, yd³	Maximum depth of cut, ft	Depth to rock, ft	Cubic yards of rock to be blasted, yd³
Granite:				
Cut 1	344,000	67	25	150,000
Cut 5	1,006,000	110	25	626,000
Schist:				
Cut 2	359,000	56	38	29,000
Cut 3	880,000	72	32	246,000
Cut 4	185,000	39	35	18,000
Residuals (Cuts 1, 2, 3, 4, 5, and 6)	736,000	110	Variable	
Totals	3,510,000			1,069,000

This formidable-appearing excavation with heavy outcrops of granite and schist includes only about 1,069,000 yd³, of "drill-and-shoot" rock, or 30 percent, of the total 3,510,000 yd³.

Among the 301 highway rock excavation jobs studied in the United States during the past 20 yr have been the following two outstanding hard-rock projects involving high percentages of necessary blasting.

    1. Santa Susana Pass Freeway in the Santa Monica Mountains of southern California, totaling 8,533,000 yd³. About 5,800,000 yd³ of thickly bedded sandstone of the Chico formation, or about 68 percent of the total yardage, was blasted.

    2. Pacheco Pass Freeway in the Coast Range of southern California, totaling 11,190,000 yd³. About 5,100,000 yd³ of heavily bedded sandstones and schists, or about 46 percent of the total yardage, was blasted. By coincidence the bulk of the "drill-and-shoot" rock was in the Panoche formation, also an Upper Cretaceous marine formation of sedimentary rocks correlating with the Chico formation of the

5-10 Open-Cut Excavations

Santa Susana Pass Freeway. The correlation of rock formations was discussed in Chapter 2.

It is revealing that the rock excavations with highest percentages of yardage requiring blasting are in the heavily bedded sedimentary rocks rather than in the igneous or metamorphic rocks. Preexcavation analyses and subsequent excavation experiences show this to be true in all parts of the United States. Of some 1 billion yd^3 studied, the three kinds of rock make up these percentages, as detailed in Table 5-1 and summarized below.

Igneous rocks	39%
Sedimentary rocks, thickly bedded	20
Metamorphic rocks	11

The average percentage of "drill-and-shoot" rock in the so-called rock highway job is about 10 percent of the total roadway excavation. This minimum of blasting, as contrasted with the much higher figure of a generation ago, is due to rock fragmentation by modern heavyweight class I tractor-rippers. These ripper- and bulldozer-equipped machines have up to 700-hp, 190,000-lb working weight and 76,000-lb usable drawbar or ripping-shank pull.

PITS AND OPEN MINES

The description of the Bingham Canyon copper mine of Kennecott Copper Corporation, Plate 5-1, dramatizes the biggest rock excavation made by humans. However, on a mundane, less grandiose scale many hundreds of pits and open mines produce a huge variety of building and construction materials and metallic and nonmetallic ores for human needs. These include sands and gravels, bauxite, phosphorite, iron, copper, coal, and a host of other materials and commodities.

Excavation varies from loose sands to ultrahard taconite iron ore. Most of the coals and the ores are beneath overburdens so that there is a dual excavation of waste material and desired product.

The bituminous-coal strip mining of Figure 5-6 illustrates the sequence of overburden removal by casting of the 50-yd^3 stripping shovel, coal loading from the vein by the 8-yd^3 coal shovel, and coal hauling from the mine to the preparation plant by the 25-ton bottom-dump haulers.

In this central Illinois mine the coal vein is 5 ft thick and the highwall to the right of

TABLE 5-1 Percentages of Kinds of Rocks and Depths to "Drill-and-Shoot" Rock as Encountered in About 1 Billion yd^3 of Excavation in 508 Public and Private Works of Excavation

Kinds of rock	Percentage of rock	Depth to rock line, ft
Igneous:	39	
Intrusive, as granite	25	49
Extrusive:		
Thick-bedded, as rhyolite	7	43
Amorphous, as breccia	7	49
Average, weighted		46
Sedimentary:	50	
Nonbedded	9	
Noncemented, as gravel (*)	(5)	(120)
Cemented, as conglomerate	4	52
Bedded:	41	
Up to 1-ft bedding, as shale	6	90
1- to 3-ft bedding, as sandstone	15	71
3 ft and up bedding, as limestone	20	53
Average, except (*), weighted		59
Metamorphic:	11	
Massive, as gneiss	6	41
Laminated, as slate	5	43
Average, weighted		42
Grand total	100	
Grand average, except (*), weighted		55

Figure 5-6 Open-pit areal coal mine. Bituminous-coal stripping mine of central Illinois. The sequence of operations for this typical open-cut mine is as follows.

   1. Sixty feet of limestone and shale overburden, previously blasted, is cast by a 50-yd³ stripping shovel from the highwall to the right to the waste area to the left from which the coal vein has been removed.
   2. An 8-yd³ shovel loads out the 5-ft coal seam, which may or may not have been blasted, into a fleet of six 25-ton bottom-dump coal haulers. The single-shifting coal-loading shovel follows the triple-shifting stripping shovel.
   3. The coal haulers haul on the mucky shale floor of the pit for 0.2 mi and then ramp up to the 0.9-mi main haul road to the preparation plant.
   Every day the stripping shovel casts about 38,000 yd³ of overburden for the 3500-ton coal production of the plant.

the stripping shovel is 60 ft high. The overburden mixture of limestone and shale requires blasting and it bears a volume ratio of 12:1 to the coal or a ratio of 10.5 yd³ of overburden per ton of coal.

The stripping shovel cycles in 55 s and its consistent round-the-clock production is about 1800 yd³/h. It is on triple-shift operation so as to cast about 5400 yd³ for every 500 tons/h of single-shift coal production. The coal haulers are equipped with large single tires so as to have flotation and traction in the muck of the pit floor and on the sometimes soft, spongy haul roads. Haul is 1.1 mi, one way, and net cycle time is 11 min. Travel speed during the wet-season operation averages only 13 mi/h loaded and 22 mi/h empty. Sustained production for the fleet of six coal haulers is 3500 tons of coal per 7-h shift. During the dry season the rolling resistance of the coal haulers will be reduced greatly and cycle time will drop down to 9 min.

QUARRIES

The quarries of the United States produce hundreds of millions of tons of rock products annually. The more important nonmetallics from quarry-type operations are asbestos, bituminous rock, feldspar, fluorspar, gypsum, limestone and dolomite, manganese, perlite, rocks for crushed stone and riprap, shale, sulfur, talc and soapstone, and uranium ores.

Quarries vary greatly in size from the little mica quarry producing 50 tons/day to the big limestone fluxing-stone quarries producing 50,000 tons/day on round-the-clock triple shift operation.

A typical medium-size commercial quarry is shown in Figure 5-7. This granite quarry of San Diego County, California produces some 5000 tons of decomposed granite and

5-12 Open-Cut Excavations

Figure 5-7 Hard-rock quarry. Quarry operations of a medium-size commercial rock producer. In the foreground are two track drills putting down blast holes in the solid granite. In the far background is an upper face of decomposed and well-weathered granite, which does not require blasting but which can be used for plant products. An 8-yd³ tractor-loader-type shovel is loading out the two 35-ton rear-dump rock haulers. A downgrade haul road leads to the primary crusher. Quarry output averages about 5000 tons of decomposed, semisolid, and solid granite from the quarry every 8-h day. *(South Coast Asphalt Products Company.)*

crushed granite every 8-h day. A unique feature of the operation is that the overburden is largely salable. It is made up of decomposed granite, called "DG", and well-weathered granite. These materials, requiring little blasting, are used for road base and simply as excellent random fill for works requiring more embankment.

The major use of the crushed rock is for two concrete plants and for select base materials for road building. One plant is the asphaltic cement concrete plant owned by the company, and the other plant is the Portland cement concrete plant owned by an associated company.

Crushed rock is sold to concrete plants without owned sources for aggregates and to a variety of industries requiring decomposed and crushed rock for road building and for the concrete of home and industrial buildings.

The quarry is located in the same tonalite granite formation as the abandoned quarry of Plate 3-1. Because of more favorable economic conditions and because of perhaps better planning, the present quarry has reached its present stage and it will continue into greater development and rock production.

Figure 5-8 shows the complex of material-handling machinery and plants that process the rock from the primary crusher.

RAILROADS

Because of the generally maximum 1 percent grades of railroads, which call for relatively deep cuts as contrasted with maximum highway grades of about 6 percent, the ratio of rock excavation to common excavation is high in hilly and mountainous country.

Such a new railroad is the Palmdale to Colton cutoff of the Southern Pacific Company across the Mojave Desert and through Cajon Pass of southern California. Its length is 78 mi, of which 24 mi are in the formidable-appearing rocks of Cajon Pass. From north to south through the pass these rocks are sedimentary shales, conglomerates, and sandstones, igneous granites, and metamorphic schists.

The total excavation was 6 million yd³ of which 1.6 million was in the Cajon Pass rock area. In spite of the array of imposing formations, the rocks were well-weathered and the rock requiring blasting did not exceed 50,000 yd³ of the most question-

Figure 5-8 Sequence of quarry work. Sequence of quarry-to-plant rock excavation and distant plant. *Center:* One of two drills putting down holes for blasting. *Foreground:* Blasted granite from the last shot of 32,000 tons. *(Right)* Loading out one of two 35-ton rear-dump rock haulers. *(Left)* Section of the 1600-ft well-maintained haul road with 11 percent downgrade to the primary crusher. *Far background:* Complex of primary crusher, secondary crushers, screening and conveyor equipment, asphaltic cement concrete plant, and Portland cement concrete plant of the efficient multiproduct industry of approximately 5000-ton/day capacity. *(South Coast Asphalt Products Company.)*

able 567,000 yd³ of the Cajon Pass excavation. This 567,000 yd³ was within the rift zone of the San Andreas Fault, and unquestionably the age-old weakening effect of the fault action resulted in minimum "drill-and-shoot" rock. This subject is discussed in Chapters 3, 4, and 5.

Figure 5-9 pictures the rugged nature of the east escarpment of Cajon Pass. The new line leaves Cajon Creek and ascends a steep 5-mi grade to the summit, whence begins the descent to the Mojave Desert.

In 1965 the Santa Fe Railway Company's relocation of its main line around Galisteo Creek Dam, New Mexico, required 900,000 yd³ of rock excavation in 4 mi. Cuts were up to 60 ft in depth. The excavation was in sedimentary rocks made up of heavily bedded Navajo sandstones, medium-bedded Dakota sandstones, and Pierre shales.

About 330,000 yd³ of sandstone, 37 percent of total yardage, was blasted. The rock was loaded by two 6-yd³ wheel-type tractor-shovels and hauled by six 18- to 32-ton rear-dump rock haulers. Residuals and weathered rock were ripped by two medium-weight tractor-rippers, push-loaded by two heavyweight push-tractors, and hauled by six 40-yd³ wheel-type scrapers.

Both percentage of "drill-and-shoot" rock and machinery types were in contrast on these two railroad rock jobs in formidable mountainous areas.

The Cajon Pass job of the Southern Pacific Company required but 9 percent of the questionable yardage to be blasted and all excavation was handled by conventional tractor-rippers and wheel-type scrapers. On the other hand, the excavation of the Galisteo Creek area required considerable blast-hole drilling machinery and both push-tractor–scraper spreads and shovel–rear-dump hauler spreads of machinery.

All the rock of the Santa Fe Railway Company job was questionable and, proportionately, four times as much rock was blasted.

SANITARY FILLS

Hundreds of millions of cubic yards of both common and rock excavation are moved annually in the building of sanitary fills, the euphemistic term for the old trash dumps. These "cut-and-cover" fills are of two types, those in level country and those in hilly or mountainous country. They differ, not in the manner of building the fills, but rather in the methods used for excavation.

5-14 Open-Cut Excavations

Figure 5-9 Railroad building. Parallel relationship of the new Southern Pacific Company line to the existing Santa Fe Railway Company line through Cajon Pass of the San Bernardino Mountains of southern California. The new line is to the left or north and follows through the area which has been pioneered with access roads and cleared and grubbed of chaparral. Cajon Creek is in the middle of the picture at elevation 3680 ft and the summit in the far distance is at elevation 4260 ft. In this approximately 5-mi section of line the ruling grade is 2.2 percent. The rock east of Cajon Creek is the Crowder formation of soft friable fanglomerates, conglomerates, and sandstones, requiring only soft ripping by heavyweight tractor-rippers for all excavation. The rock west of Cajon Creek, in the foreground, is the Punchbowl formation of massive shales, conglomerates, and sandstones, requiring medium to heavy ripping. Southward are several sidehill cuts in granite and schist with a maximum centerline cut of 102 ft, in which only about 50,000 yd³ of hard rock was blasted. *(G. Sheley (ed.), "Building 78.4 Miles of Railroad," Western Construction, April 1967, p. 71.)*

Figure 5-10 shows a dump in level country near Long Beach, California. Fill material, consisting of clays, silts, and sands, is taken from a wide, deep trench by push-tractor-loaded wheel-type scrapers and hauled upgrade to the fill area, where it is bulldozed in a 1.5-ft layer atop the 4-ft compacted layer of trash. Crawler-tractors serve in both bulldozing and compaction operations.

The two 21-yd³ wheel scrapers in Figure 5-10 are hauling about 300 yd³/h of cover material, or enough for about 3600 tons of trash daily. Accordingly, this dump will require some 600,000 yd³ of common excavation annually. Level-country dumps seldom involve rock excavation because they are usually in residuals and alluvia.

After this long trench is excavated, the accumulated trash atop the high wall will be bulldozed into the trench and then another sequence of alternate layers of cover and trash will be added until natural ground level and perhaps a higher elevation are reached. Lines and grades and construction methods for the fill are controlled closely.

As contrasted with the Long Beach sanitary fill in level country, the dump of the Santa Susana Mountains, some 35 mi to the north, calls for the leveling off of siltstone-sandstone ridges for the cover material. The sedimentary rock formation is well weathered and requires only soft to medium ripping by medium-weight tractor-rippers.

Figure 5-11 shows a medium-weight bulldozer cascading cover material to the fill some 80 ft below. This dump, much smaller than the Long Beach fill, requires only three medium-weight combination bulldozers and tractor-rippers for all the work, as the dozing distances on the fill are generally not great. When distances are too long to be economical, a crawler-tractor-drawn scraper is put into service.

About 1000 yd³ of cover material per 8-h day is sufficient for the some 1500 tons of trash handled daily. In this dump the three crawler-tractors handle both the excavation of the cover material and the trash compaction work in the fill.

Figure 5-10 Excavating cover material for a sanitary landfill. Sanitary fill construction in level country. Silts and sands are being excavated from this wide, deep trench for cover for the "cut-and-cover" dump. At the top of the high wall is a pile of accumulated trash. Beyond this pile the alternate layers of 4 ft of compacted trash and 1.5 ft of cover are being placed. After the trench is excavated to grade, it will be filled with the alternate layers of trash and cover and the cover will come from another trench parallel and adjacent to it. In large cities there is about 1 ton of trash annually for every person. A ton of trash compacts to about 2.5 yd^3 and requires about 0.6 yd^3 of excavation. Accordingly, this Los Angeles County sanitary fill can accommodate about 1 million population.

Figure 5-11 Delivery of cover material for a sanitary landfill by gravity. Bulldozing cover material of siltstone-sandstone from a ridge down to a sanitary fill. This dump is located in the Santa Susana Mountains of Los Angeles County, California. Construction methods for the alternate layers of trash and cover are the same as those for an operation in level country, as only the sources of the excavations differ. After a fill is completed and landscaped, the area is generally used for recreational purposes, for example as a park or small golf course. No buildings of more than one story are built because of the low density of the fill. However, a generation of building controlled sanitary fills indicates that the fills are remarkably stable.

Both sanitary fills are privately owned and both are controlled by the governmental body with local jurisdiction over methods. Such strict controls almost always prevent three objections to dumps—odor, rats, and spontaneous combustion. All are eliminated by compacting the alternate layers of trash and earth or rock to as high a density as possible. On compaction 1 yd^3 of loose trash becomes ½ yd,^{3} and 1½ ft of loose cover becomes a 1-ft thickness.

TRENCHES AND SMALL-FOUNDATION EXCAVATIONS

Trench excavations vary greatly in both quality and quantity. A trench for a telephone conduit in loam may be only 18 in deep and 12 in wide, resulting in soft excavation and only 6 yd^3 per station of 100 ft. Such excavation is an easy job for a small sidecasting vertical-boom trencher.

On the other hand a trench for a 6-ft-diameter concrete storm drainpipe in rock country may be 18 ft deep and 12 ft wide, calling for blasting and a minimum of 800 yd^3 per station. And, if the worker handling the powder is not diligent and allows too much backbreak from the charges, the yardage could well run to 1200 yd^3 of rock per 100 ft of length. Such a trench is a hard excavation job for a large, full, revolving 2½-yd^3 backhoe even after good fragmentation.Common trench excavations include those made for the following applications:

1. Government, commercial, and home utilities such as water, gas, electricity, telephone, sewers, and storm drains.

2. Large commercial pipelines such as those used for water, gas, oil, and coal transportation.

3. Farm applications, such as surface-ditch irrigation and drainage and subsurface pipe drainage.

4. Highway construction, such as side and lateral ditches, culverts, and diversion ditches.

Many other kinds of trenches are excavated as a part of the many types of open cuts described in this chapter.

Figure 5-12 illustrates vividly a particularly tough trench excavation. It is part of the San Bernardino Valley Municipal Water District's master distribution system for this California county.

The heavy fanglomerate material is described as "85 percent over 8 inch mechanically locked material." This notation is equivalent to saying that the alluvium is a high-density, tightly cemented conglomerate with up to boulder-size particles. The fanglomerates from the San Bernardino Mountains cover great areas and they extend to considerable depth.

The trench through this troublesome formation was 22 ft deep and 13 to 25 ft wide. The length was 6 mi and so some 500,000 yd^3 was excavated at the rate of about 80,000 yd^3 mi.

The excavation methods were changed occasionally during the work, but they were essentially as follows:

1. The first few feet of depth being characteristically gravels and cobbles, this zone was taken out by a 2½-yd^3 dragline assisted by two medium-weight bulldozers to a depth of 6 ft.

2. A 2-yd^3 backhoe then excavated to an average depth of 20 ft. In this zone oversize boulders were drilled by jackhammers and blasted.

3. To provide for bedding material beneath the 78-in-diameter steel pipe, the backhoe excavated an additional 8 to 12 in below grade.

This trench is an extremely rough example of excavation in cemented conglomerates made up of outsize fanglomerate boulders.

Analyses of many open-cut rock excavations warrant valuable conclusions as to kinds of rock and depths to "drill and shoot" rock in the several classifications. These data are as given in Table 5-1.

The table is based on 506 projects of public and private excavations for which complete seismic studies were made before excavation and in which there were many followups of the experiences of the earthmovers and seismic studies of the work in progress. These correlations during the actual work indicate that the excavator in rock country com-

Figure 5-12 Trench excavation for main pipeline of water system. Tough trench excavation in tightly cemented boulder-size fanglomerate of the San Bernardino Mountains, California. After blasting of the largest of the embedded boulders, the remaining 16 ft of the 22-ft-deep trench is being excavated by the 2-yd³ backhoe. The bulldozer in the trench is leveling bedding material for the 6½-ft-diameter steel pipe. The top 6-ft-deep zone of up to cobble-size alluvium was excavated by a 2½-yd³ dragline assisted by two medium-weight bulldozers. About 0.5 million yd³ of this "mechanically locked" fanglomerate was excavated in 1975. *(San Bernardino Valley Municipal Water District.)*

mences to blast the rock when the shock-wave velocity gets up into the range of 6000 to 7000 ft/s.

Inferences from the table that are most significant to the rock excavator are:

1. Minimum depth to "rock line" or depth to blasting below original ground is in the metamorphic rocks, averaging 42 ft.
2. Intermediate depth is in the igneous rocks, averaging is 46 ft.
3. Maximum depth is in the sedimentary rocks, averaging 59 ft.
4. Depths in the sedimentary rocks vary inversely according to thicknesses of bedding.

Up to 1-ft thickness of stratum	90-ft depth
1- to 3-ft thickness	71-ft depth
3 ft and greater thickness	53-ft depth

5. The average depth to "drill-and-shoot" rock in all kinds of rock is 55 ft.

Of course, the tabular data are averages and as averages they are to be properly interpreted. One knows that the "rock line" is of variable depth even in the same kind of rock and in the same rock formation of the same job, let alone in like rocks and formations of different jobs. Nevertheless, the data are valuable generalizations.

The master curves from which the table derives show that the average 55-ft-deep "rock line" varies from 27- to 111-ft depths, or from -51 percent to $+102$ percent of the average value. This fact lends emphasis to the inconsistency of the weathered zone of the Earth's regolith and the urgency for the earthmover to learn all about the rocks and the rock formations before commencing rock excavation in open cuts.

SUMMARY

In this chapter a few examples of the nine principal kinds of excavation have been given and these may be called medium and large projects. Of course there are many kinds of

5–18 Open-Cut Excavations

small projects involving excavations, such as those for a swimming pool or for a small irrigation ditch on a 160-acre farm. Unfortunately there is no summary of total annual excavation in the United States, but it is probable that it is some 2 billion yd^3 bank measurement. Consequently, excavation is a big business.

CHAPTER 6

Examination of Excavation

PLATE 6-1

Rugged terrain for the examination of rock excavation. Topography of a four-lane freeway through the Antelope Valley of southern California. The rock excavation includes 7.7 million yd^3 in 10.8 mi of mountains and valleys. Nine principal cuts are typified by the cut of Plate 6-1, in which a cross section of the cut and the centerline of the freeway are delineated. The cut is 907,000 yd^3 with a maximum depth of 130 ft, and it is a symmetrical through cut with two 1100-ft benches on each side. Rocks include massive sandstones of the Vasquez formation and medium-bedded sandstones and conglomerates of the Mint Canyon formation. There are minor interbedded basalts in the sedimentary rocks of the Vasquez continental formation. In this cut 565,000 yd^3, or 62 percent of the total roadway excavation, was blasted, the rock line being about 40 ft below natural ground. For the entire job 5.8 million yd^3, or 75 percent, was ripped and 1.9 million, or 25 percent, was blasted.

The shaggy terrain accentuates the problems of walking the centerline of mountain jobs, but it also emphasizes the desirability of field reconnaissance. Such were the experiences and the policy of the efficient excavator of the large rock job of 1960 to 1962. *(California Division of Highways.)*

For both practicability and efficiency the examination of excavation in the office and in the field should precede the exploratory work in the field. Exploration of the excavation is the subject of Chapter 7.

There are two phases of examination, to be conducted in the following order:

1. Office work in terms of securing and analyzing topographical and geological data and historical excavation experiences in the area of excavation.

2. Field work in terms of "walking centerline" with literal down-to-earth examination of the rock job.

The office work should precede the field work because, metaphorically, it is best to view the forest before seeing the trees. A typical office examination of a 17.9 million-yd^3, 12-mi freeway excavation located in the mountains is used as an example.

OFFICE WORK

1. Analyses of plans and specifications provide a general outline of the examination. Cuts and fills are then outlined on detailed plan sheets, together with notations of yardages, depths of cuts, and height of fills.

Cuts and fills are then transferred to a master plan sheet or title sheet showing the entire job. Usually this master plan sheet is a reproduction of a United States Geological Survey (USGS) map of either the 7½-min series with a scale of 2000 feet to the inch or the 15-min series with a scale of 1 mile to the inch. The large-scale, 7½-min-series map is preferable because of greater detail and ease of drafting work on the map. However, since master plan sheets are generally reduced in size, the scale of the sheet may very well be an odd one and it may not be scaled precisely with the regular engineer's scale. In such a case the use of proportional dividers will solve the problem of transfer of distances to the odd-scale map.

2. The master plan sheet is now transferred to the topographical USGS map. If available, the 7½-min-series map should be used, and this series of maps is generally available except for remote areas of the United States.

Now it is possible to visualize the job by means of this topographical map. One sees clearly the relationships of centerline and cuts and fills to the mountains, valleys, and streams. Relationships are established also to the works of humans, such as roads, railroads, towns, dams, canals, mines, quarries, pipelines, electrical transmission towers, and the like.

3. The final step on the working plan sheet is to transfer the local geologic map to the sheet so as to divide up the map according to kinds of rock and rock formations. In keeping with the purpose of examination of rock excavation, step 3 is the most important.

Generally one or more geologic maps are available for the area from the following sources:

1. County: Engineering and environmental departments in the county seat.
2. State: Divisions of geology, public works, and/or environmental engineering in state capitals.
3. Federal: USGS in Washington, D.C., as well as other offices in principal cities.
4. State tourist information bureau in the state capital.
5. National Park Service, Washington, D.C.
6. Geological Society of America, Boulder, Colorado.
7. American Association of Petroleum Geologists, Tulsa, Oklahoma.
8. Libraries of the larger U.S. cities.

Scales of geological maps range from 50 miles to the inch to 2000 feet to the inch. Again, one should use the largest-scale map available so as to ensure precision. When the scale is small, as in the maps of the American Association of Petroleum Geologists, the small area can be enlarged for greater accommodation.

Precision is especially important when there are several rock types and different formations in the excavation area. The maps usually are divided according to the geologic series of rocks but sometimes they are divided into formations, corresponding to the fourth division of geologic time.

Generally the maps are accompanied by a description of the kinds of rocks in the series or formation. This information enables the engineer to anticipate what rocks will be met with in the field reconnaissance and in the rock work.

As has been mentioned, when transferring from the plan sheet to the USGS topograph-

6-4 Examination of Excavation

ical map and from the local geologic map to the same topographical map, it is efficient and precise to use proportional dividers rather than the engineer's scale unless the scales happen to be the same. Of course all completed originals should be copied for use in the field.

Figure 6-1 shows the transfer of the centerline of the freeway to the geologic map of the Cajon Pass, the location of the rock excavation. From this map, which shows the distribution of the rock formations, one can now transfer the contacts between the rock formations to the map for field reconnaissance shown in Figure 6-2.

Figure 6-2 shows the working map or plan for the examination of the freeway. Only the middle section, about 4 mi long, is portrayed. The centerline on either side of the San Andreas Fault includes the rift zone of the fault. It includes 9,850,000 yd^3 of the roughest excavation in the 17,900,000-yd^3 job.

The rough topography of Cajon Pass is accentuated by the 3100-ft high elevation of the 3,471,000-yd^3 cut between stations 186 and 214 and the 2700-ft low elevation of the fill between stations 214 and 224. From top of cut to bottom of fill is a 400-ft vertical drop and 1400 ft laterally, and this will make for difficult pioneering of the excavation. Scrutiny of the master plan sheet, Figure 6-2, discloses other aspects of the field reconnaissance and subsequent field exploratory work.

1. The centerline of the rift zone of the San Andreas fault crosses the freeway centerline at station 235 in the center of a 1,761,000-yd^3 cut of conglomerate-breccia. This longtime faulting accounts for the breccia. The conglomerate is really a fanglomerate from the San Bernardino Mountains to the east.

2. If, as supposed, the rift zone extends 1 mi or so on both sides of the axis of the fault, then one may expect weathered rock formations from at least station 182 to station 288. An investigation of earthquake activity in Cajon Pass during the recorded history of the last century disclosed the following disturbances:

   a. Old Cajon Pass Road, which followed approximately the present alignment of U.S. Highway 66, was covered by slides and debris by a violent earthquake in 1899. Its intensity on the Mercalli scale was 8 to 9, equivalent to 7+ on the presently used Richter scale.

   b. In 1907 there was another powerful quake in the pass. Its magnitude was 6 and it was accompanied by heavy slides in the area of the freeway.

While the recorded seismic activity is less than a century old, one must keep in mind that the San Andreas fault has been active for over 150 million yr. Accordingly, the rift zone has been weakened continually for aeons and it is still being weakened.

These studies made prior to field reconnaissance suggested exactly the findings of the seismic studies for the excavation and the experience of the earthmover when the rock excavation was taken out. There was medium and hard ripping in the rift zone but there was no blasting of roadway excavation.

It is probable that more seismic studies of excavation were made on this freeway than on any other similar rock work in the United States. The California Division of Highways made available to the bidding contractors 39 studies and the writer made 28 studies. Total studies were 67, or one study for every 147,000 of the 9,850,000 yd^3 of questionable rock work.

3. Judging by the paucity of trails and roads through the work area, field reconnaissance must involve jeep travel and considerable hiking over the rough country.

4. The high-walled sidehill cut along the existing Highway 66 at Blue Cut should give indications of the nature of the 3,471,000-yd^3 schist cut to the southeast and on a bearing paralleling the fault. Overbend excavations for the pipeline near the freeway will be indicative of the excavation to the same depths. Likewise, data on the tunnel work to the east of station 237 will be helpful.

5. An important facet of the office work is the correlation of the rock formations of the freeway with the same formations with which the earthmover is familiar. This association of formations has been discussed in Chapter 2. Proceeding upstation, the cuts may be correlated in this manner.

Examination and exploration of the rock excavation for the Cajon Pass Freeway took place prior to and during 1966. The work was done during the 1967–1969 period.

The schist cuts are between stations 224 and 263. This same Pelona schist in sizable cuts

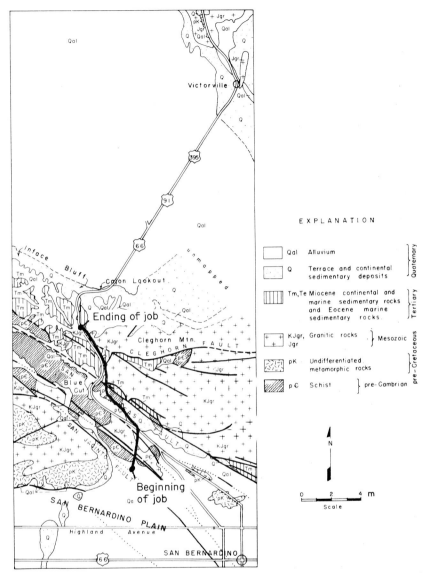

Figure 6-1 Geologic map of the Cajon Pass area, San Bernardino County, California. Centerline of entire Cajon Pass Freeway is superimposed on the geologic map of the area. Areal map shows the distribution of the rock formations. *(Olaf P. Jenkins,* Geology of Southern California, *California Division of Mines and Geology, Geologic Guide no. 1, 1954, p. 48.)*

6–6 Examination of Excavation

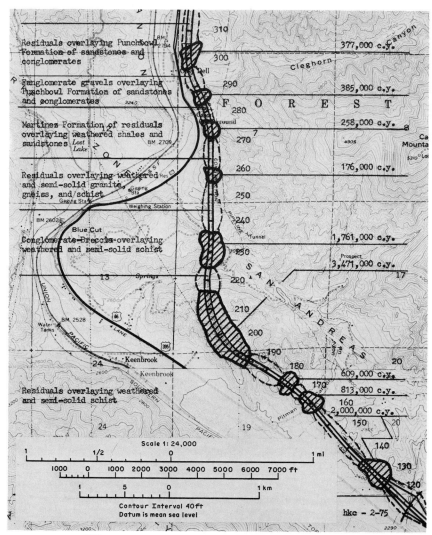

Figure 6-2 Map for field reconnaissance of Cajon Pass Freeway of southern California, middle section. Station 115 to station 318. 9,850,000 yd³ of roadway excavation. *(U.S. Geological Survey, Cajon 7½-min Topographic Quadrangle Map, 1968.)*

was excavated just a few years previously in the Bouquet Canyon Road across the San Andreas fault rift zone just west of Palmdale, California. There was no blasting, as the weathered rock was ripped by tractor-rippers of less weight and horsepower than those of 1966.

The 258,000-yd³ cut is in the Martinez formation of sedimentary rocks. About 1961 up to 145-ft-deep cuts were taken out by hard ripping on Highway 150 through San Marcos Pass, Santa Barbara County, California, in the same Martinez formation.

The two northern cuts, totaling 762,000 yd³, are in the Punchbowl formation of the Cajon beds. Unfortunately there is no correlation for these undivided or unclassified Miocene continental sedimentary rocks in terms of excavation experience. It is known, however, that the soft though massive Miocene rocks are rippable to a depth of 100 ft.

The lower and upper sections of the Cajon Pass Freeway, omitted in this discussion, are similarly correlated with recent work by machinery of the past 5 yr.

Generally an office appraisal of the job would include the history of the excavation for the existing U.S. Highway 66 and for the Southern Pacific Railroad to the east across Cajon Creek. However, the highway and the railroad were built several decades ago when rock ripping was not in vogue and fragmentation was by explosives. Consequently, an investigation of the excavation methods was considered to be not helpful.

In large public works the contracting officer or agency usually supplies the bidders with rather complete materials and soils data as part of the specifications, special provisions, and plans. This valuable information may include logs of test hole borings, as well as cores if obtained, exploration trench or open-pit and quarry data, and surficial geologic rock-type and formation data.

Sometimes in large private works the builder may supply similar information, although usually not as complete as that of the agencies. This collective material on rock excavation results in a better understanding of the job and better balanced bidding. Four examples are explained below.

    1. Figure 6-3 is a portion of a standard soil survey sheet for a mountain highway in the Sierra Nevada of California.

Test holes 4, 5, and 6 were drilled to below grade in the 160,000-yd^3 cut between stations 567 and 585. All three logs record easy drilling for the auger drill in the well-weathered volcanic rock. This information was augmented by private seismic studies of the job. Studies 1 and 2 at stations 572 and 576 showed up to 5300-ft/s shock-wave velocities to 40-ft depth and 9200 ft/s to greater depths. These studies supported the test hole data.

Since the maximum centerline cut was 34 ft, the excavation was fragmented by methods ranging up to hard ripping for a heavyweight tractor-ripper.

    2. Figure 6-4 is part of a geologic plan and profile for a reach of the California Aqueduct through the northern foothills of the San Gabriel Mountains of southern California.

The plan of areal geology shows the 238,000-yd^3 cut between stations 1852 and 1858 to be in Pelona schist, an old metamorphic rock the excavation history of which is well known. Rotary-core-drill hole 33 was put down to 25 ft below the canal invert elevation in the 95-ft-deep cut. Cores from the cut were available to the bidders. A private seismic study, 23, shows velocities and velocity interfaces and it suggests rippability to grade. Bucket-drill hole 34 to a depth of 43 ft was put down through weathered schist to 10 ft below grade, where refusal was met. The California Department of Water Resources' seismic profiles SP-68 and SP-67 are plotted in the smaller cut between stations 1863 and 1869.

There is good conformity between the agency's data and private seismic work. The rock was fragmented by methods ranging up to hard ripping for a heavy-weight tractor-ripper.

    3. Figure 6-5 is part of the geologic map for the spillway of New Melones Dam on the Stanislaus River of Central California.

These some 15 million yd^3 of complex metamorphic rocks were investigated thoroughly by agency and private means. The agency means, as delineated on the complete map for the 6000-ft-long, 400-ft-wide, and maximum 237-ft-deep spillway included core drill holes, exploration drifts, and a sizable, 32-ft-deep test quarry. As is customary, core samples were available to the bidders.

Thirteen private seismic studies of the spillway rock excavation were made for the several bidders. Averages indicated a rock line at 6-ft depth, where a 6000-ft/s shock-wave velocity changed to 15,100 ft/s. The contractor's ripping and blasting experience corroborated the inferences of the agency and private investigators.

    4. Several years ago a huge tract for homes in the San Raphael Hills of Los Angeles County, California was appraised for the granite rock excavation of 1,943,000 yd^3. There were four principal steps in the investigation.

        a. Test holes of 4-in diameter were drilled by track drills to a maximum 70-ft depth of cut. The test holes evaluated the hardness of the rock, as well as determining

6–8 Examination of Excavation

the important drilling speeds according to the depth of drilling in the event that blast-hole drilling should be necessary.

b. Uphole seismic studies were made in the eight drill holes, thus establishing shock-wave velocities in the several zones of depth.

c. Twelve refraction seismic studies were run so as to cover the entire excavation area down to the grade of the finished cuts.

d. The foregoing explorations provided an estimate that 198,000 yd³, or 10.6 percent of total yardage, would require blasting. All data were made available to the bidders and consistent, reasonable bids were obtained.

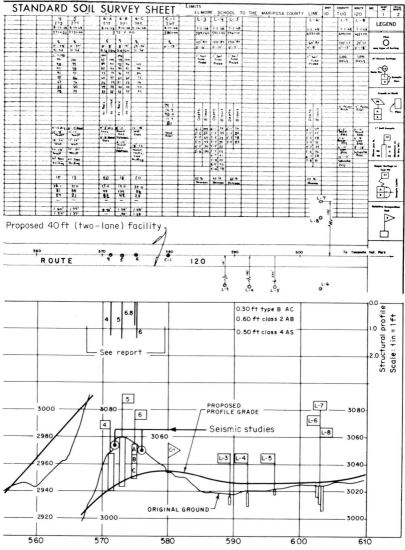

Figure 6-3 Route 120 Highway—FH-39-151 Mariposa and Tuolomne Counties, California. *(California Division of Highways.)*

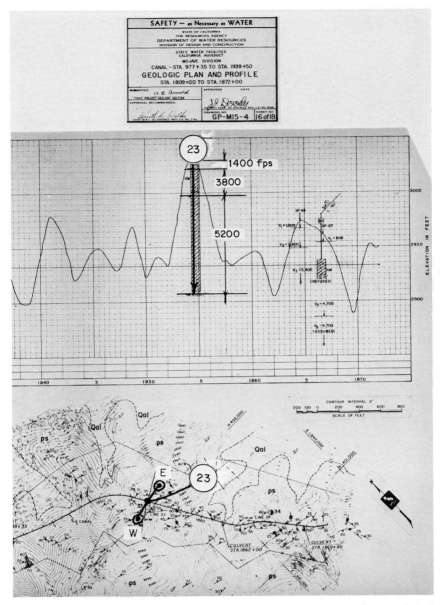

Figure 6-4 Section of Fairmont–Leona Siphon reach, California Aqueduct. Significant symbols of area of illustration: (Qal) alluvium of Quartenary system; (ps) Pelona schist of Precambrian group; (RD) test hole by rotary core drill; (BDP) test hole by bucket drill; (23) seismic study or profile by the writer; (SP-67, SP-68) seismic profiles by Dept. of Water Resources. *(California Dept. of Water Resources.)*

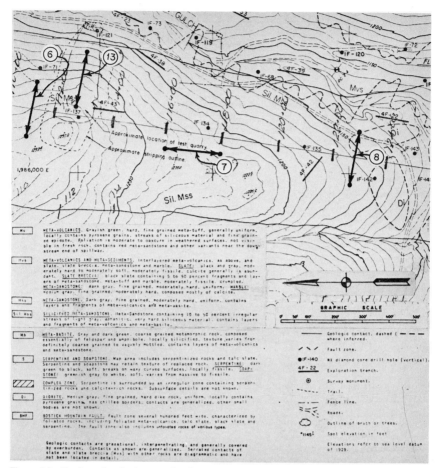

Figure 6-5 Section of geologic map of spillway of New Melones Dam, Stanislaus River, California. Four of the thirteen private seismic studies of the 15-million-yd³ spillway rock excavation are shown. Symbols of the complex rock formations include all within the work for the spillway. Not all are represented in this spillway section. *(U.S. Bureau of Reclamation.)*

This description of office work highlights the necessary prelude to field reconnaissance. A story of field work for another job emphasizes the importance of this correlated office and field effort.

FIELD WORK

Equipment

As in the case of office work, a worker is as good as the available tools, and for "walking centerline" the following items of equipment are suggested.

   1. Brush-turning, hard-texture, loose-fitting clothing suitable for the prevailing weather conditions. In rainy weather "breathing" rain gear is important rather than the sweat-producing and ventilation-lacking clothing made from synthetic fabrics. If brush is thick and if rocks are sharp, let alone if barbed-wire fences are encountered, horsehide gloves should be worn both summer and winter. Boots should be waterproof high cuts, preferably engineer's type boots with over-ankle strap and relatively smooth composition

soles. Lace boots are time-consuming to put on, the hooks get caught in brush and barbed wire, and the boots are relatively heavy.

2. A light pack, of the heavy-duty Boy Scout type, for carrying plans and equipment. Shoulder straps should be broad and there should be a belly band. One's hands should be free when hiking over rough country and climbing sharp and slippery rocks.

3. Lunch pail. The regular plastic pail with 1-pt vacuum bottle is excellent.

4. Two canteens are advisable. In desert hiking with greater than 100°F temperatures, one can consume the water of two canteens in 4 h. The general-issue or "GI" English quart size is good. One of these should be worn on the belt for convenience.

5. Plans, specifications, and master plan sheet should be placed if possible in a three-ring binder so as not to be defaced.

6. Field notebook, slide rule or pocket computer, and engineer's scale.

7. Special equipment. These items are well nigh indispensable.
    a. Camera, instant type.
    b. Binoculars. Wide-angle and seven-power magnification.
    c. Engineer's tape. Nonmetallic of 100-ft length.
    d. Clinometer. Military pocket-type, reading in percentage of grade or inclination.
    e. Pocket altimeter. Precision to ± 10 ft of elevation.
    f. Compass. Boy Scout pocket-type.
    g. Magnifying glass. Pocket-type.
    h. Geologist's hand pick. Belt-type.
    i. Small first-aid packet and snakebite kit.
    j. Revolver, if in order, as protection against "man and beast."

These items will give a pack weight of about 15 lb, and the hands are free for the roughest of hiking.

In addition to the "tools of the trade" the engineer must consider his safety, especially in remote mountain and desert areas. Under extreme conditions it is best to have a fellow hiker in the event of such potentially fatal mishaps as broken limbs and heat prostration. If it is necessary to hike or even travel by jeep alone, always tell associates and the hotel manager what plans are afoot for the day.

Purpose of Field Reconnaissance

The purpose of field reconnaissance is to find out everything possible that relates to the rock excavation. When the entire construction or mining job is considered, quite naturally other matters pertaining to the entire work must be investigated. In this discussion only relevant phases of the excavation are considered. These items include the following major operations:

1. Pioneering and opening up of the job.
2. Clearing shrubs, chaparral, and trees to the lines of plans.
3. Excavation.
4. Haulage of excavation.
5. Fill operations, including placement and compaction.

Related to these five items are such operations as these:

1. Borrow pit and waste area locations.
2. Development of water supply.
3. Building of access roads, sometimes over privately owned lands.
4. Establishment of offices and machinery yard.
5. Relationships with property owners who may or may not have concluded negotiations with the contracting officer or quarry or open-pit area owner.

Thus the field operator must look to many job factors when making a survey. These factors are discussed below in relation to a mountain highway job.

Figure 6-6 is part of a plan–profile sheet for the Cleghorn Road to Crestline Highway on the north slope of the San Bernardino Mountains of southern California. Field investigation of this job, combined with seismic studies of the roadway excavation, was made between February 4 and 11, 1969. It was a rainy, snowy winter period and no field work could be done except during three days of the period, as the average elevation was 3600 ft at the rain-snow line.

Ten seismic lines were run, requiring 10 h, so that the hiking and reconnaissance work

6-12 Examination of Excavation

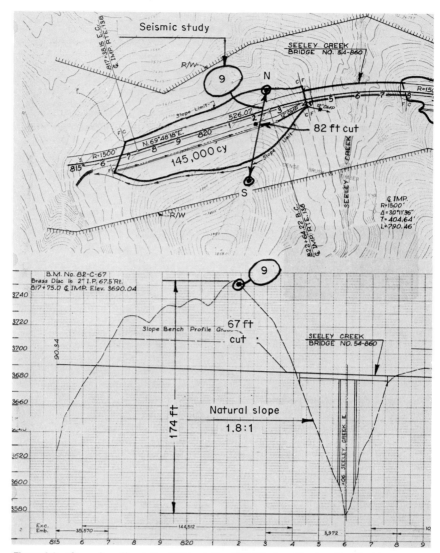

Figure 6-6 Plan and profile, station 815 to station 829, Cleghorn Road to Crestline Freeway, San Bernardino County, California. *(California Division of Highways.)*

required about 20 h for the 5.1-mi job. As shown by Figures 6-6 and 6-7, topography was rough and weather was bad. There was no access by jeep, so that total hiking was about 15 mi, including twice the length of centerline plus lateral distances from the existing highway. This highway was 0.3 mi to the north along the East Fork of the Mojave River and about 300 ft below the elevation of the freeway. Hiking for the work not pertaining to seismic studies was at the rate of 0.8 mi/h, a general average for the unfavorable conditions.

General Observations

The following observations relate specifically to this job, and in principle to all excavation jobs of the same general nature.

Pioneering or Opening Up of the Job Two factors make for expensive work. First, the job is in the San Bernardino National Forest and work is subject to both state and federal government controls. Federal control means minimum environmental damage to existing trails and minimum building of new access roads. Second, the rough topography and the granite rock require ripping out of rock instead of normal bulldozing for the pioneering and access roads.

Clearing Brush and Trees The job is in average mountain growth of flora in granitic soil. However, no burning is permitted, and so a disposal area must be found within reasonable distance or else the chaparral and small trees must be shredded and mulched. Again, the rough country makes clearing quite difficult. There may be merchantable timber so that free-of-charge logging may be a bonus in the form of lowered clearing costs.

Excavation Of paramount importance is the rock excavation. Office work with geology maps showed the earth-rock structure to be thin to thick residuals overlying average weathered granite, so-called "DG," which in turn overlies jointed and semisolid granite.

The granite is of the tonalite variety, characterized by up to room-size weathered boulders on the surface and embedded in a matrix of decomposed granite beneath the surface.

While geologic maps are of primary value in determining the nature of the rock excavation, a good secondary source is an aerial picture, such as that shown in Figure 6-8. This vertical view in San Saba County, Texas, shows definite relationships between vegetation and surface residuals of different rock types. Low mesquite grows on Mississippian Barnet shale (MB). Cedars, live oaks, and other trees grow on Pennsylvanian Marble Falls limestone (PMF), and upon Ordovician Honeycut dolomite and limestone (OH).

Sometimes different shrub and tree types grow on opposite sides of a fault separating two different rock types. For example, north of Stony Gorge Reservoir in Glenn County, California, the Stony Creek fault separates the western unnamed formation of metamorphic serpentines from the eastern Knoxville formation of sedimentary shales, sandstones, and conglomerates. Tall pines and chaparral favor the serpentine soil, while

Figure 6-7 Walking the centerline of proposed mountain freeway. Field reconnaissance hiking in the San Bernardino Mountains of southern California in the wintertime. Elevation is 3600 ft, marking the transition from rain to snow at the snow line. All possible jeep trails have been obliterated by washouts of the mountain ravines and slides of the water-saturated slopes. "Walking the centerline" for about 15 mi was at the rate of ¾ mi/h. By an interesting coincidence, this is just about the same rate which obtains for hiking in the desert in summer heat of 120°F. Physical discomfiture is about equal.

6-14 Examination of Excavation

medium-size manzanita and tall grass prefer the shale, sandstone, and conglomerate residuals. The fault line literally separates the the western dark foliage from the eastern light leafage.

The U.S. Production and Marketing Administration, which furnished the picture of Figure 6-8, has done remarkable work in correlating vegetation with types of soil. It is merely an extension of these principles to correlate vegetation with types of parent or country rock beneath the soil cover or residuals.

Long-Range Surveillance

Job surveillance from the existing road to the north along the East Fork of the Mojave River indicates the following job facts.

1. Canyons are steep, particularly that of Seeley Creek, probably calling for the use of crawler-tractor-drawn scrapers rather than wheel-tractor-drawn units.

2. Canyon streams, tributary to the Mojave River, are running full over boulders of up to cubic-yard size and exposed granite. Initial construction of fills and building of bridge abutments and piers will be difficult and expensive.

3. Brush is from medium to thick and good-size trees are fairly abundant, varying from 6 in to 3 ft in diameter.

4. Cuts on the north slope of the mountains are and will be wet, regardless of season, and the slopes are slide-prone because of saturation of weathered granite. Drill holes for blasting will be wet, probably necessitating use of more costly dynamite instead of ammonium nitrate explosives. Likewise, the wet conditions will cause less job efficiency than that obtainable when working on a sunny slope.

5. About 1.5 mi west of the little hamlet of Cedar Springs is an alluvial flat area along the West Fork that is suitable for offices and yard. However, it is subject to spring flooding and a diversion channel should be excavated.

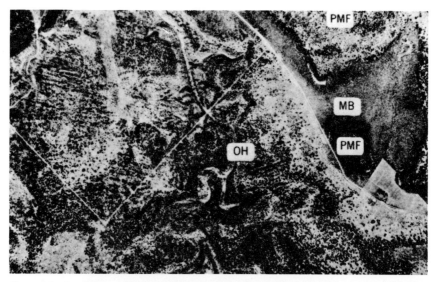

Figure 6-8 Correlations of kinds of rocks with kinds of vegetation. Correlation of vegetation with types of soil in San Saba County, Texas. Mesquite favors the Barnet shale (MB). Cedars, live oaks, and other trees are partial to Marble Falls limestone (PMF) and to Honeycut dolomite and limestone (OH). The residuals supporting the vegetation are the extremely weathered results of the immediately underlying rocks and thus the presence of these rocks may be inferred from the vegetation. However, in a rock excavation there may be only a surface zone of shale supporting the mesquite. The mass of harder rock beneath the shale might be limestone, in which case the deduction that the excavation is softer shale would be erroneous and costly. Such field reconnaissance observations must be a part of the mosaic of examination and exploration of rock work. *(U.S. Production and Marketing Administration.)*

Walking Centerline—Detailed Observations

Walking centerline results in the following notes concerning the cuts and fills according to station limits. All these notes will be analyzed by the office engineers, the estimators, selected job superintendents, and other affected personnel.

Stations 657 to 687

Three minor through cuts, totaling 90,000 yd^3. Maximum centerline cut 24 ft. Scattered to dense brush. Some medium-size timber. Streams running full. Well-weathered rock. Soft to medium ripping for heavy-duty tractor-rippers. Use of wheel scrapers on maximum 3:1 natural slopes.

Stations 687 to 693

One through cut of 32,000 yd^3 with maximum 36-ft cut.

Average brush and small trees. Streams running full. Well-weathered rock. Soft to medium ripping. Possibly crawler-tractor-scrapers on account of 1.6:1 slopes.

Stations 693 to 701

One through cut of 109,000 yd^3 with 80-ft cut.

Medium brush and few medium trees. Full streams. Top of cut and slopes show moderately weathered rock. Blasting is estimated below 40-ft depth, requiring about 15 percent or 16,000 yd^3 of blasting. Soft to hard ripping to 40-ft depth.

Slope is 1.7:1 from top of cut to bottom of fill, 110 ft below. Use tractor-scrapers if rock breaks up satisfactorily or wheel loaders and rear-dump rock haulers if rock fragments are too big for scraper work.

Stations 708 to 714

One through cut of 70,000 yd^3 with 40-ft maximum cut.

Moderate brush and some small trees. Full streams. Cut is moderately weathered. Soft to hard ripping. A 4:1 maximum slope calls for use of wheel-type scrapers.

Stations 720 to 725

One 93,000-yd^3 through cut with maximum 57-ft cut. Sawpit Canyon bridge adjoins station 725 to the east. Rock will move to west down 2.2:1 slope.

Moderate brush and medium-size trees. Sawpit Canyon stream running full over outsize boulders and bedrock. Abutment and two-pier construction will be difficult and costly.

Partially weathered rock outcrops. Soft to medium ripping to 50-ft depth, suggesting minimal blasting. Crawler-tractor-scraper work.

Blasting estimated to be less than 5 percent, or 5000 yd^3. Probably to be hauled by tractor-scrapers.

Stations 730 to 738

One through cut of 184,000 yd^3 and maximum 81-ft cut. Moderate brush and fair amount of medium-size trees. Full streams.

Rock appears to be well weathered with maximum 3:1 slope on centerline. Soft to hard ripping.

Use of wheel scrapers with little likelihood of having to substitute crawler-tractor-scrapers.

Stations 747 to 754

One through cut of 99,000 yd^3 and maximum 67-ft cut.

Dense brush and moderate number of medium-size trees. Full streams of usual 2-ft depth and average 15-ft width.

Average weathered rock and maximum 3.5:1 natural slope. Soft to hard ripping and haulage by wheel-type scrapers.

Stations 760 to 766

One through cut of 134,000 yd^3 and maximum 90-ft cut.

Moderate brush and fair number of medium-size trees. Full streams.

Rock outcrops on top and slope of cut. Steep centerline slope of 1.7:1 down 205 ft vertically. Soft to hard ripping to 60-ft depth. Blasting below 60-ft depth, estimated at 5 percent, or 7000 yd^3. Use crawler-tractor-drawn scrapers.

Stations 768 to 776

Three through cuts, totaling 192,000 yd^3, with maximum 75-ft cut.

Medium brush and average number of medium-size trees. Three feet of water in turbulent boulder-strewn canyons.

Rock is moderately weathered. Soft to hard ripping to 25-ft depth or 75 percent of

6-16   Examination of Excavation

192,000 yd³. Haulage of 145,000 yd³ by crawler-tractor-drawn scrapers with maximum 1.5:1 slope. Remaining 47,000 yd³ to be blasted, loaded by wheel-type shovels, and hauled by rear-dump rock haulers.

Stations 778 to 784
One through cut of 86,000 yd³ with maximum 72-ft cut. Heavy brush and moderate-size trees. Canyons carrying 2 ft of water over boulder-strewn bedrock.

Rock is average weathered, and westerly slope is 1.6:1 down 147 ft to bottom of fill.

Soft to hard ripping. Use crawler-tractor-drawn scrapers.

Stations 789 to 794
One through cut of 93,000 yd³ with maximum 65-ft cut at inside ditch line.

Dense brush and average number of medium-size trees. Easterly Burnt Mill Canyon running 2-ft of rough water. Building of 320-ft bridge over canyon will require costly work for abutments and two piers.

Rock will move to west down maximum 3:1 slope. Soft to hard ripping to 40-ft depth with use of wheel-type scrapers. Blasting estimated at 8 percent or 8000 yd³. Use wheel shovel and rear-dump haulers.

Stations 799 to 809
Two through cuts of total 99,000 yd³ with maximum 44-ft centerline cut.

Moderate brush and medium-growth small trees. Streams running to 2 ft of water. Rock is well weathered. Soft to medium ripping. Haulage by wheel scrapers, as steepest natural slope is 3:1.

Stations 816 to 824
One through cut of 145,000 yd³ with 82-ft maximum cut.

This cut and adjoining westerly Seeley Creek bridge are shown in Figure 6-6. Centerline slope to east is 1.8:1, forecasting costly excavation and bridge building in the steep, rocky gulch.

Heavy brush and average-height trees. Creek running continuous 4-ft depth and 10-ft width. Cut appears rippable to half depth, resulting in 15 percent or 22,000 yd³ of blasting. Soft to hard ripping to 41-ft depth. Maximum centerline slope is 3:1 for westward moving excavation.

Wheel scrapers for ripped rock and wheel loader and rock trucks for blasted rock.

Stations 820 to 854
Five through cuts totaling 83,000 yd³ with maximum 30-ft centerline cut.

Medium brush and small trees. Brooks running full with about 1 ft of water.

Rock is well weathered and slope is maximum 3:1. Soft to medium ripping. Haulage by wheel-type scrapers.

Stations 855 to 865
Two through cuts of total 114,000 yd³ with maximum 40-ft centerline cut.

Heavy brush and timber of average height. Brooks running full with about 1 ft of water.

Centerline slopes average 3:1, and rock is well weathered. Soft to medium ripping. Haulage by wheel-type scrapers.

866 to 873, End of Job
One through cut of 29,000 yd³ with deepest cut of 28 ft.

Dense brush and trees of average height. Small canyons are running about 2 ft of rough water. Well-weathered rock. Easy slopes of maximum 4:1 slope. Soft to medium ripping for heavyweight tractor-rippers. Haulage by wheel-type scrapers.

Conclusions from Field Work and Subsequent Excavation History

The following general conclusions, after hiking centerline through this series of 16 cuts, are of great value to the office engineers and estimators.

1. Clearing should be gone over carefully as there may be some merchantable timber. Tree count is desirable, including tree diameters.

2. Abundant water is available for fill compaction and haul road maintenance either from wells in the river alluvium or from dams across the tributaries.

3. All ripping should be done by heavyweight tractor-rippers so as to take out marginal rock without blasting.

4. On the steepest slopes much rock can be bulldozed to bottoms of fills.

5. Consideration should be given to crawler-tractor-drawn scrapers for short, steep hauls, especially in view of wet, muddy haul roads in early and late summer.

6. Most of the rock will be hauled by wheel-type scrapers. An analysis of the mass diagram for the excavation, showing haul distances, will establish zones of work for the two types of scrapers.

7. Preliminary field estimate for blasting is 105,000 yd^3, or 6.2 percent, of total 1,700,000 yd^3 of roadway excavation. This rough estimate is to be checked by exploratory drilling and by seismic studies, accompanied by final quantity takeoffs from the plans.

8. Building of three bridges, each with two piers, with the longest to be 360 ft and the highest to be 105 ft, will be unusually costly in the rough, wet canyons.

Excavation jobs in the area and in the same granite rock formation were made objects of close study and these examinations resulted in these notes.

1. North portal excavation for the San Bernardino Tunnel, located northerly from station 745. Maximum centerline cut was about 90 ft. Cut was in granite gneisses and was taken out without blasting, although ripping ranged up to hard and extremely hard for heavyweight tractor-rippers. Excavation in 1967 and 1968.

2. First section of the relocated Cleghorn Road, located to the west of job. Work is in progress. It consists of five principal cuts with maximum centerline depth of 110 ft, as compared with 90 ft of this job. Topography is not as rough.

In the cuts where blasting is or has been taking place, the rock line appears to average about 40 ft below natural ground. Cuts show a fair amount of embedded boulders, typical of tonalite granite, which are difficult to handle without blasting and costly to blast by blockholing.

3. Cedar Springs Dam, 2 mi to the north, is being built. Spillway in the same granite rock was excavated to 90-ft depth by soft to extremely hard ripping in the last 30 ft of depth. Presently at the 90-ft depth the use of heavyweight tractor-rippers equipped with single shanks is about ended. The embedded boulders are up to room size, and they are surrounded by a little matrix of decomposed granite. The rock is becoming semisolid. Systematic drilling and blasting has commenced for this 210-ft-deep spillway. The diameter of the blast holes is 6 in and the hole spacing or pattern is 10 ft by 10 ft. Thus far ammonium nitrate is being used instead of dynamite, the holes being fairly dry.

With these field notes covering the reconnaissance of the job, together with rock exploratory work if deemed necessary and with further field investigations of clearing and water supply, the estimators will be ready for bid preparation on those items pertaining to the rock excavation.

Figure 6-9 shows two of the cuts being taken out in the Cleghorn Road to Crestline Freeway. The cut in the foreground is between stations 789 and 794. This 93,000-yd^3 cut with a maximum 65-ft depth has been brought down to about 30-ft depth by tractor-rippers and wheel-type scrapers. Although there was estimated to be 8000 yd^3 of blasting in this cut, all excavation was taken out by the tractor-rippers.

The distant cut is between stations 778 and 784 and is 86,000 yd^3 with the maximum cut of 72-ft depth. It is almost down to grade and it was taken out with wheel-type scrapers instead of crawler-tractor-drawn scrapers as suggested in the notes of the field reconnaissance. Actually, all excavation except blasted rock was hauled by wheel-type scrapers. After the cuts were opened up by bulldozers, it was found possible to lessen the grades of the haul roads by switchbacks, and thus the more economical wheel-type scrapers could be used for the job.

Figure 6-10 illustrates the finished freeway together with the topography of the mountainside job. The cut in the foreground and the three cuts in the distance are between stations 789 and 824. These cuts appear to be sidehill-type rather than through-type. However, all four cuts have small slopes on the downside or outside ditch line and so they are technically through cuts.

The picture is taken toward the west and the Mojave River is to the north. Tributaries run northward and their erosion has caused the ravines and canyons and the many northward trending ridges through which the freeway passes. The growth represents average density of brush and average size of trees.

Reiterating, it is well to investigate every excavation job on foot, whether it be the centerline of a highway or canal or the many lines and grades of a dam or open-pit mine. What may appear formidable from a distance may be easy rock work on closer examina-

6–18 Examination of Excavation

Figure 6-9 **Progress of construction of mountain freeway.** Excavation for the Cleghorn Road to Crestline Freeway in 1969. These two cuts total 179,000 yd^3. All granite rock is being ripped by heavyweight tractor-rippers, in spite of a maximum depth of cut of 72 ft. Wheel-type scrapers are negotiating up to 25 percent grades on switchback haul roads. Location is on the north slope of the San Bernardino Mountains of southern California. Elevation is 3750 ft and weather is bad during the springtime. The Mojave River lies to the north.

tion. For example, in desert areas what appears to be a dark lava flow from a half-mile view may very well be a harmless, gently sloping fanglomerate with a superficial coating of dark desert varnish. Conversely, the apparently easygoing alluvium may be a thin veneer of gravels on a gently sloping granite pediment.

Perhaps the greatest advantage in hiking is the unhurried opportunity to size up a job without interruptions, as well as the therapeutic value of a nice walk. In this hiking it is advisable to walk the job in both directions since hindsight is sometimes better than foresight.

In bidding on a highway job it is said that, because the costs of materials, machinery, and workers are fairly fixed and because methods are fairly uniform, the low bidder must have an "angle." The "angle" generally involves the excavation and especially rock work, be it for roadway, subbase, base, or quarry stone.

Perhaps, as shown in Figure 6-11, midway of the job and 100 yd upstream from centerline an eroded gully barely conceals an ample source of good rip-rap for the 100,000-yd^3 revetment item of the bid schedule. The specified acceptable source may call for a 10-mi-longer haul. Development of the midjob quarry could mean a savings of 1 million yd·mi of haul or possibly $200,000 in haulage costs, giving the perceptive bidder a significant advantage and a greater profit margin.

Such an "angle" could have been discovered only by the hiker on centerline with wide-angle vision.

SUMMARY

Two correlative investigations form the basis of the analysis of excavation. These are examination of excavation in the office and exploration of excavation in the field, the subject of Chapter 7. The one is not truly valuable without the other and they go hand-in-hand, as should always be the case with interrelated efforts.

Figure 6-10 **Completion of construction of mountain freeway.** Completed Cleghorn Road to Crestline Freeway, summer of 1972. Total excavation, 1.7 million yd^3 with 22 major cuts of maximum depth 90 ft in the 5.1-mi mountainside job. The distant high-sloped cuts in ridges trending northward to the Mojave River illustrate the topography of the north slope of the San Bernardino Mountains. The freeway traverses a series of many ridges and ravines, resulting in short, steep hauls for the granite rock formation. The northward exposure and the rather luxuriant growth of brush and trees, coupled with the high elevation, resulted in moisture-laden rock formation and a short construction season in spite of southern California's semiarid climate.

Figure 6-11 **Discovery of source of materials by walking centerline.** A youthful gully being deepened into a deep ravine by weathering of the bedrock. Spalls of metamorphic schist are being removed by hydraulic action. This exposure of solid rock may suggest a source of quarry stone for use in the nearby projected highway. The exposure also indicates the depth of the shallow residuals and weathered schist. Such a discovery only comes from the inquisitive mind of the field reconnaissance man while walking centerline.

6–20 Examination of Excavation

It is preferable for both examination and exploration to be performed by the same individual or, at least on a large project, for all the persons involved to be associated closely. There are unfortunate examples of lack of cooperation between office personnel working on examination and field personnel carrying on exploration. Under such conditions the result is not the best even though the individuals are the finest.

CHAPTER 7

Exploration of Excavation

PLATE 7-1

Part of working areas of foundation explorations for dam. Switchback trails built for the foundation explorations for the south abutment of Auburn Dam, American River, Placer County, California. The outline of the projected dam is 700 ft above the river. The correlated investigations included trenches, tunnels, drill holes, and seismic studies of the complex metamorphic rock formations and required about 10 yr. The unusually complete studies were necessary because of minor faults and shear zones within these typical Mother Lode country rock formations. The concrete arch dam requires solid foundations in this earthquake-prone country.

The excavation quantities for the preliminary foundation excavation for Auburn Dam are given in Chapter 5. About 1.7 million yd^3 is being excavated for the left or south abutment shown in picture.

The rocks are metamorphic schists and amphibolites and metasedimentary siltstones and sandstones. Although the exploration of excavation for the abutment by the contracting officer was largely for the purpose of design of the dam, the investigations fixed the hard-rock line at about 25 ft below original ground. The 25-ft-thick upper zone was made up of residuals and rippable weathered rock.

The outline of the concrete arch dam as shown in Plate 7-1 shows only about one-tenth of the 4000-ft-long crest of the 6.5-million-yd^3 structure. The combination irrigation and hydroelectric dam combines 2.3 million acre·ft of water capacity with 750 MW of ultimate electric power.

Exploration of rock excavation in the field is the sequel to examination in the office and field. These prior activities determine the scope and the methods for the subsequent exploration.

The methods used are manual, mechanical, and instrumental.

1. Manual methods may be as simple as the use of pick and shovel or a sounding rod. More refined methods include hand auger, wash boring, and soil sampler.

2. Mechanical methods include the use of bulldozers and angledozers, backhoes and other excavators, and many kinds of drills.

3. In present-day practice instrumental methods are confined to geophysical testing by seismic timers which includes ground-surface refraction studies and *uphole* studies of shock-wave velocities and interfaces between the different shock-wave velocities.

All these methods have their proper spheres and they may be used individually or collectively. All are forms of sampling and the results may be expressed axiomatically: The greater the sampling extent, the less the error in exploration of rock excavation.

MANUAL MEANS

The use of pick and shovel is almost as old as civilization and it needs no explanation. One simply picks and digs until one finds rock. Early mineral prospectors used this method to most profitable advantage.

The sounding rod is a series of 4-ft-long and ¾-in-diameter pipes with couplings. The first section is fitted with a removable drive head and a drive point. A heavy sledge is used and increased depth is attained by adding lengths of the pipes. No samples of materials can be taken, and so the nature and the degree of consolidation of the residuals and the weathered rock depend on individual judgment. Obviously, the depth of investigation is limited to a few feet, generally less than 10 ft.

Figure 7-1 illustrates the three common kinds of hand tools.

Drilling a Hole with a Simple Open-Spiral Auger without Casing A degree of sampling is possible by examining the materials brought up by the auger and estimating of the depth from which the material was obtained. The degree of consolidation of the earth-rock structure may be judged by the ease or difficulty of augering. Augers range from 2- to 6-in diameter. In sands depths to 100 ft are possible. The maximum depth depends upon the leverage and down pressure applied to the auger. When small rocks are encountered, as in the case of weathered rock, a chopping bit may be substituted for the auger. If the material caves badly, as in the case of alluvium, a casing may be used advantageously. As shown in Figure 7-1, the hand auger can penetrate only partly into the zone of weathered rock.

Wash Boring Drilling a Hole with Casing and with a Chopping Bit Where pressured water is available it is sometimes expedient and efficient to use the simple pipe jet with casing of 2- to 4-in diameter, depending on the size of the flushed particles. Materials are judged by the washings and consolidation is appraised by the driller. With the ordinary two-person crew depths to 75 ft are possible, providing materials and water pressure are favorable. Again, drilling can reach only upper levels of the weathered rock zone.

Wash Boring Drilling a Hole and Sampling with Split-Tube Sampler Dry sampling, in the sense of securing an undisturbed sample, may be used in connection with wash boring. The chopping bit of Figure 7-1 is removed from the bottom of the wash pipe and a split-tube core sampler is substituted. This tube is forced into the earth-rock at the bottom of the hole and the core or sample is brought to the surface.

Manual tools have three desirable features, simplicity, portability, and economy. They have two limitations: (1) the limiting depth is 100 ft under the most favorable circumstances and generally it is considerably less; and (2) penetration is confined to maximum-weathered rock, as indicated in Figure 7-1.

Figure 7-2 displays a variety of manual tools as furnished in a soil sampling kit intended for up to 25-ft depths. It weighs 180 lb and contains 43 items, including 15 soil sampling tools. Samples are 1½-in diameter.

Among the important items are: (1) handle; (2) drive head; (3) nine 2½-ft drill rods with couplings; (4) ship auger; (5) closed-spiral auger; (6) open-spiral auger; (7) Iwan post-hole or sampling auger; (8) probing rod; (9) split-tube sampler; (10) saw-tooth and pocket shoes for split-tube sampler.

7-4 Exploration of Excavation

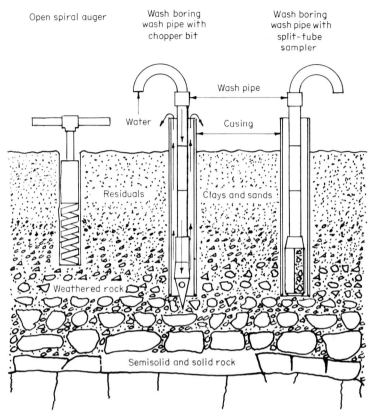

Figure 7-1 Manual tools for exploration of rock formations. No scales are shown, as the illustrative drawing is not in proportion. Spiral augers average 3-in diameter. Casings average 3-in diameter and water pipes average 1-in diameter. Residuals may be taken at 10-ft thickness, weathered rock at 10-ft thickness, and the semisolid and solid rock at indefinite thickness.

Figure 7-3 shows use of the soil sampling kit. The one-person operation is ideal for simple shallow-depth exploration through residuals and well-weathered rock to bedrock. The split-tube sampler is used for qualitative analyses of the earth-rock.

For more than one-person operation, affording up to 6-in-diameter holes and up to 100-ft depths under ideal conditions, there are larger-scale tools than those of the pictured soil sampling kit. These include longer handles for two- or four-person operation, 5-ft drill rods, augers of up to 6-in diameter, heavy chopping bits for weathered rock, and large-diameter split-tube samplers.

MECHANICAL MEANS

Backhoes

Small and medium-size backhoes, either wheel- or crawler-type, are efficient exploratory tools. Such machines can excavate trenches up to 20 ft deep, giving stratigraphic cross sections of residuals, weathered rock, and sometimes even semisolid rock.

Figure 7-4 illustrates exploratory work and initial excavation for a 3000-ft trench down the steep side of Slide Mountain, Nevada. The 5-ft-deep, 2-ft-wide trench is for utilities for a radio communication center atop the nearly 9000-ft-high mountain. The combination exploratory-excavation work was done to ascertain the suitability of the dual-purpose crawler-tractor-mounted-backhoe-bulldozer for the difficult work. The machine was ade-

Figure 7-2 Sampling kit for soil exploration. Manual soil sampling kit, offering a variety of hand augers, soil samplers, and split-tube sampler with accessories. Weight of 180 lb and depth capacity of 25-ft with the furnished drill rods. Simplicity, portability, and economy are the advantages of this assembly for one-person operation. *(Acker Drill Company.)*

Figure 7-3 Spiral auger and split-tube sampler for soil investigation. Use of open-spiral auger and split-tube sampler in the exploration of excavation in connection with highway building. The simple split-tube sampler gives a compact, corelike sample from a given depth of the drilled hole. *(Acker Drill Company.)*

quate for the rough terrain and the highly variable earth-rock, which consisted of weathered granite boulders embedded in sands of decomposed granite.

Hourly production was 15 lineal feet of trench, or 6 yd^3 neat trench measurement. Actual production, with allowance for caving of the sides of the trench, was about 19 yd^3/h.

Exploratory work showed that up to ½-yd^3 boulders persisted to and below the bottom of the 5-ft-deep trench, an example of the difficulty of trench exploration in weathered granite of the Sierra Nevada.

Bulldozers and Angledozers

These machines excavate a trench or sidehill cut, affording ample width and depth to analyze and diagram the exposed rock formation. Such trenches are popular with contracting officers in the case of public works and with designing engineers in the case of private works.

7-6 Exploration of Excavation

Figure 7-4 Rock exploration with tractor-backhoe-bulldozer. Medium-weight crawler-tractor-mounted backhoe-bulldozer trenching for utilities installation on the flank of Slide Mountain, Nevada. The weight of the combination machine is about 26,000 lb, and the diesel engine is 100 hp. The backhoe bucket is 30 in wide with 6.75 ft³ struck capacity. The earth-rock is weathered granite, consisting of large boulders embedded in a matrix of sands and decomposed granite. (*J. I. Case Company.*)

Figure 7-5 illustrates a face of rock exposed by a sidehill bulldozer cut in connection with the exploration of a 1,840,000-yd³ sidehill cut for a freeway in San Diego County, California. The rock is schist of an igneous-metamorphic formation made up of greenstones, tuffs, schists, and granites. The face of the exploratory cut is about 12 ft deep and the rock line of the semisolid formation is at a depth of about 10 ft.

Figure 7-6 is a diagram of the wall section of a 20-ft-deep exploratory sidehill cut dug by a medium-weight bulldozer equipped with a two-shank ripper. This cut called for about 250 yd³ of earth-rock excavation and it was about a 3-h job for the medium-weight tractor.

The face, equivalent to a 60-ft-long 20-ft-high mural, showed the designing engineers what stratigraphy might be expected in the first 20-ft depth of excavation. Several similar sidehill cuts were made in critical locations of the multimillion-cubic-yard excavation for a huge housing project.

The ease or difficulty in ripping and bulldozing the hard rock formation gave indications of probable grading methods and machinery. Correlative to this investigation, one of the interested bidding contractors made additional sidehill cuts and trenches and put down several test holes in the sedimentary rock cuts.

As a result of all these tests, the builder's engineer had excellent data for design work relating to the characteristics of the materials and for estimating the probable costs of unclassified excavation, and the interested bidders could figure closely their grading costs. In short, there resulted ultimately a low cost to the buyer of a lot for his home.

Drills

Prior to discussing various drills for exploratory work, it is well to approximate their efficiency in this special work. Several of the types of drills are used for blasthole drilling, in which, because of repetitive production work, their efficiencies are just about double that for exploratory drilling.

Efficiency may be expressed as the ratio of average drilling speed in feet of hole per hour over a work period of days, weeks, or months to *penetration rate*, or rate while

Figure 7-5 Sidehill cut by tractor-ripper-bulldozer for rock investigation. An exploratory sidehill cut by a heavyweight combination tractor-bulldozer-ripper. The rock is schist of an igneous metamorphic formation of San Diego County, California. The cut face is about 12 ft deep and the rock line, or separation between rippable weathered rock and drill-and-shoot semisolid rock, is about 10 ft deep.

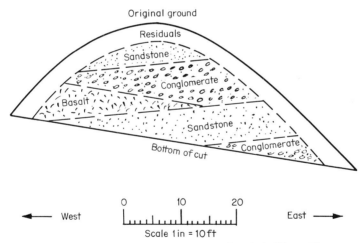

Figure 7-6 A wall section of 20-ft-deep exploratory sidehill cut by bulldozer. This cut was made in the Topanga formation of the Santa Monica Mountains, southern California, in connection with a proposed excavation for a site for homes. The 23-million-yr-old Topanga formation is made up of marine sandstones and conglomerates with volcanic intrusive and extrusive basalt rocks.

The history of the wall cross section, proceeding from bottom to top, is:

1. At least 6 ft of conglomerate was deposited on the bottom of the sea.
2. Next 7 ft of sandstone was laid down on the conglomerate.
3. At least 6 ft of basalt lava flow, tonguing to nothing at the right, was extruded on the sandstone.
4. An 8-ft layer of conglomerate was deposited on the basalt and the sandstone.
5. An unknown depth of sandstone, 4 ft within the wall section, was laid down on the conglomerate and possibly on the tongued basalt.
6. Finally all rocks have been weathered to a 3-ft depth of residuals.
7. Since deposition, the rock formation has been dipped a few degrees to the west.

7-8 Exploration of Excavation

actually drilling in feet of hole per hour. Thus, efficiency is a measure of the working time losses due to moving to the next hole, setting necessary casing, changing drill rods and bits, freeing stuck drill rods and bits, mechanical failures, and the like.

In blasthole drilling, efficiency may be approximated at 67 percent. In exploratory drilling, efficiency, obviously a more variable figure, may be estimated at about 33 percent. The chief reasons for the lower efficiency are the considerably more time required to move to the next hole and drill setup, deeper holes, more time in setting casing, and intermittent drilling while taking samples and making notes.

This example illustrates the relationship of these efficiencies. A combination of medium-weight track drill and a 600-ft^3/min, compressor, putting down a 3½-in-diameter hole in quarry blasting of granite, requires 7.5 min to drill down an 11-ft length of drill rod. The penetration rate is 1.47 ft/min. Whence:

Penetration rate at 100% efficiency	88 ft/h
Blasthole drilling at 67% efficiency	59 ft/h
Exploration drilling at 33% efficiency	29 ft/h

Auger Drill Perhaps midway between the manual auger and the mechanical auger is the portable one-person-operated drill shown in Figure 7-7. This unit, powered by a 10-hp air-cooled gasoline engine, weighs about 300 lb with accessory equipment.

Auxiliary equipment includes augers of up to 3-in diameter and core barrels and bits for ⅞-in-diameter cores, as well as a 1½-hp water pump. A variety of tools is available, including the soil sampling kit of Figure 7-2. Depth capacity is about 50 ft.

When one considers that much exploration for rock takes place within a 50-ft depth, the portability and economy of this simple auger drill are obvious advantages.

Figure 7-7 Portable power-operated auger drill for rock exploration. Portable one-person-operated auger drill, powered by 10-hp air-cooled gasoline engine. Weight with tools about 300 lb. Depth capacity is about 50 ft for both auger and core drilling. The driller is checking the depth of soft residuals down to shales and limestones in fields of Pennsylvania. The variety of tools used includes the soil sampling kit of Figure 7-2. *(Acker Drill Company.)*

The drill of Figure 7-7 is being used in farm country with maximum depth of residuals in order to investigate the depth to shales and limestones. In such work the average drilling speeds during a complete investigation are approximately as tabulated below.

	Speed, ft/h	
Class of rock	Auger drill	Core drill
Soft (residuals)	60–90	
Medium	30–60	3–5
Hard		0–3

A truck-mounted auger drill is efficient and economical for exploratory drilling in soft to medium-hard rock formations with overburden cover of clay-silt-sand residuals. Figure 7-8 shows such a machine exploring for a shallow highway cut in Indiana. Rock formation is soft earthy residuals overlying shales and limestones.

The auger drill has a capacity of up to 200-ft depth and 5- to 36-in hole diameter. Figure 7-9 shows the principal parts of an auger drill assembly and the following five kinds of auger cutterheads and bits for the conveyor-type flight auger:

a. Earth auger for residuals with single flight and up to 24-in diameter.

b. Earth-rock auger for earth and well-weathered rock with double flight and up to 24-in diameter.

Figure 7-8 Carrier-mounted heavyweight auger drill for rock investigation. Truck-mounted auger drill putting down 12-in-diameter hole through soft residuals for exploration of depth of shales and limestones. The flight of spirals is loaded with the sticky clay residuals. When the shales and limestones are encountered in the weathered zone, the bit will be changed to the rock-type. *(Acker Drill Company.)*

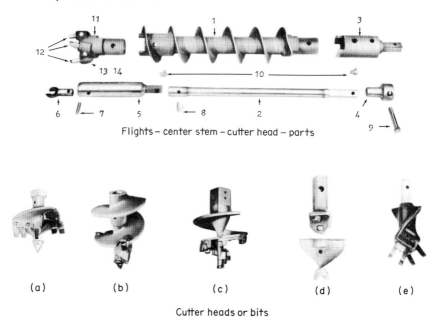

Figure 7-9 Principal parts of auger-drill assembly with five types of bit. *(a)* Earth auger for earth and residuals. *(b)* Earth-rock auger for well-weathered rock. *(c)* Rock auger for weathered rock. *(d)* Fishtail bit for loosely consolidated rocks. *(e)* Finger bit for shales and medium-hard rocks. *(Acker Drill Company.)*

    c. Rock auger for weathered rock with double flight and up to 24-in diameter.
    d. Fishtail bit for clean, straight holes in soft unconsolidated residuals and rocks. Up to about 8-in diameter.
    e. Finger-type head for shales and medium-hard rocks. Up to about 12-in diameter.
    Most heads have alloy-steel replaceable bits, either hardfaced or with carbide inserts. The conveyor flight augers are of slightly smaller diameter than the heads or bits and they are of either solid or hollow-stem construction; the latter type makes it possible to take core samples, as the hollow stem acts as a casing.
    Speed of drilling for this abrasive-type drill varies from tens of feet hourly in soft earth to a few feet hourly in medium hardness of rock. The auger drill is inadequate for work in hard rocks. In the hands of an experienced driller it is a proper tool for determining the depth of the rippable rock. Ranges of hourly drilling speeds for a 5- or 6-in-diameter hole are shown in the table. Drilling efficiency is estimated at 33 percent.

Class of rock	Speed, ft/h
Soft (residuals)	60–120
Medium (well-weathered rock)	0–60

Bucket Drill The bucket drill is an abrasive drill in the auger drill family. The cutting edges of the rotating bucket function as an auger and the bucket, in addition to bringing the cuttings to the surface, serves as a large sampling device. It is a powerful drill for exploratory work, as well as for large-diameter holes for caissons, structure footings, water wells, and the like.
    Figure 7-10 shows a medium-weight bucket drill discharging a bucketful of residual clays. The carrier truck is all-wheel drive, especially designed as a versatile vehicle for

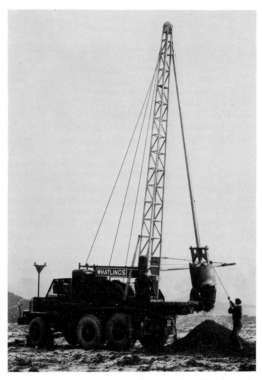

Figure 7-10 Carrier-mounted bucket drill for sampling weathered rock. Bucket drill exploring for fire clay near Dunfermline, Scotland. A 30-in-diameter hole is being put down through residuals of ordinary clay to determine the depth of the desirable fire clay. Qualitative analyses are made throughout the depth of the hole to fix the interface at the top of the fire clay seam. *(Earthdrill, Incorporated.)*

the exploratory drill. The depth of overburden above a fire clay seam is being investigated near Dunfermline, Scotland. As the 30-in-diameter bucket brings up large samples, qualitative analyses may be made to determine the interface separating the ordinary clays from the desired merchantable fire clay.

Bucket drills range from a $\frac{1}{2}$-yd^3, 48-in-diameter bucket with 100-ft depth capacity to a $1\frac{1}{2}$-yd^3, 120-in-diameter bucket with 225-ft depth capacity. Specifications for a suitable bucket drill for exploratory work would include:

Capacity: $\frac{3}{4}$-yd^3 bucket and 150-ft depth.

Power: Diesel engine of 100 hp.

Controls: Hydraulic with variable speeds and crowding or down pressure for the bucket.

Truck chassis: Three-axle all-wheel-drive type. Six large traction-flotation tires. Gross vehicle weight (GVW) rating 45,000 lb.

Like auger drills with somewhat similar cutting bits, the bucket drill is limited to soft to medium-hard rock formations. For example, in hard sandstones the penetration becomes a milling rather than a drilling operation, resulting in drilling speeds as low as 1 ft/h with excessive consumption of bits. Likewise, drilling speeds vary from several feet per hour in soft shales to several inches per hour in hard sedimentary rocks. A popular axiom of the experienced driller is: If the rock formation cannot be drilled by a bucket drill, it cannot be ripped, and blasting is in order.

As an approximate guide for estimating drilling speeds of a medium-weight bucket drill with 24-in-diameter bucket and capacity for drilling to 150-ft depth, the following table is suggested:

7-12 Exploration of Excavation

Class of rock	Speed, ft/h
Soft (residuals)	20–40
Medium (well-weathered rock)	10–20
Hard (weathered rock)	0–10

Figure 7-11 portrays combined bucket-drill and seismic explorations for Santa Susana Pass Freeway in southern California. The massive Chico sandstone formation of this 8.5-million-yd^3 rock job was discussed in Chapter 2.

The heavy-duty bucket drill of the California Division of Highways, called "Big Mo," is shown putting down test holes up against the ledge sandstone in order to determine if a near vertical face separates the residuals from the ledge rock or if there is a pediment or inclined surface of sandstone beneath the residuals. On striking the sandstone pediment drilling speed was down to 1 ft/h of 18-in-diameter hole.

In the foreground is the state seismic crew checking out the Chico formation and securing up to 13,000 ft/s shock-wave velocities a few feet below the original ground. The seismic study is being made in a test area where a heavyweight tractor-ripper with one shank attempted without success to rip economically the heavily bedded sandstone. Only a prohibitively costly production of 80 yd^3/h bank measurement was possible below a 6-ft depth.

The combined drilling, seismic, and ripping tests were made to acquaint both the state and the bidding contractors with the possible pitfalls in this spectacular job. The investigations, made after the first bids were rejected as too high, established the justifications for the initial bids. The bids of the second letting were about equal to the first bids, both initial and final bids being submitted by experienced rock excavators. The low bidder blasted about 68 percent of the total 8.5 million yd^3 of roadway excavation.

Big Mo, like many modern exploratory drills of the abrasive type, is able to work as a bucket drill, an auger drill, a rotary drill, and a core drill. It lacks only the facility for percussive types of drilling.

Jackhammer The jackhammer, a percussive-type drill, is a hard-rock drill and it is not suitable for exploratory work in residuals and well-weathered rock atop the drill-and-

Figure 7-11 Combined drilling and seismic analyses of rock excavation. Combined bucket-drill and seismic explorations are shown for the massive Chico formation of sandstones in the Santa Susana Pass Freeway of southern California. The California Division of Highways made the findings available to the bidders on this 8.5-million-yd^3 rock job. At the particular locations the speed of the bucket drill was reduced to 1 ft/h in the heavily bedded marine formation, and seismic shock-wave velocities up to 13,000 ft/s were secured within a few feet below original ground.

shoot rock at the rock line. This is because the drilled rock or cuttings are forced up and out of the hole by the same compressed air which actuates the drill. Earthy, claylike residuals cannot be forced up the hole because of the tendency to ball up, although this limitation can be partially overcome by placing casing in the hole to the depth of the hard rock. Accordingly, the overburden must be removed down to the semisolid rock before efficient drilling is possible. After this, satisfactory exploratory work may proceed through the semisolid rock into the solid rock.

Equipment for use of a 60-lb jackhammer generally includes: a truck-mounted or pull-type compressor of about 165-ft^3/min capacity; a set of drill steel in 2 ft-multiple lengths up to 20 ft; a supply of 1½-in-diameter tungsten carbide bits; several 20-ft lengths of 1-in-diameter air hose; and miscellaneous air hose fittings.

Such an exploratory drill will put down the following hourly footages of 1½-in-diameter hole with drilling efficiency at 33 percent:

Class of rock	Speed, ft/h
Soft	14–20
Medium	8–14
Hard	0–8

In most blasting work the drilling is done by percussive-type drills mounted on crawlers, particularly in the field of public works construction. In drilling there is a correlation between the drilling speeds of jackhammers and track drills. Track drills put down holes of about 2-in to 6-in diameter in systematic blasting, and in exploratory work the holes are usually of 3½- or 4-in diameters.

Correlation for exploratory drilling by jackhammers and by track drills gives the following table of average drilling speeds at 33 percent drilling efficiency:

	Speed, ft/h	
Class of rock	Jackhammer 1½-in hole	Track drill 4-in hole
Soft	17	62
Medium	11	38
Hard	4	12

The table indicates that the drilling speed of the jackhammer averages about 12 percent that of the track drill.

A distinct advantage of using the jackhammer or track drill for exploratory work is that the operation gives a good estimate or guide for the anticipated blasthole drilling speeds and for the drilling characteristics of the rock formation. It goes without saying that in such appraisals it is necessary to make full and complete notes, and preferably time studies, of the drilling operation.

Figure 7-12 displays the variety of bits used for jackhammers and for small and large track drills. The bits vary in size from 1¼-in diameter for jackhammers to 6-in diameter for large track drills. Different configurations and compressed air outlets are shown in examples (a), (b), (c), (d), and (e). Example (f) illustrates a bit with tungsten-carbide-steel inserts, which increase the life of the bit by a factor of 30 or more as compared with an alloy-steel bit.

Figure 7-13 shows the use of a 57-lb jackhammer for drilling hard granite. Residuals and weathered rock have been removed so as to provide efficient drilling conditions for exploratory work. The semisolid and solid granite is fairly uniform and drilling speed averages about 7 ft/h. The heavy-weight jackhammer has about a 20-ft depth capacity. The driller is checking for the presence of joints or seams and for drilling speeds. These factors will cast light on methods to be used for the drilling and blasting program.

If metamorphic rocks such as schists were being explored, the goals would be the same. However, if sedimentary rocks such as alternate beddings of sandstone and shale were being drilled, then thicknesses of beddings and angles of dips as well as the respective

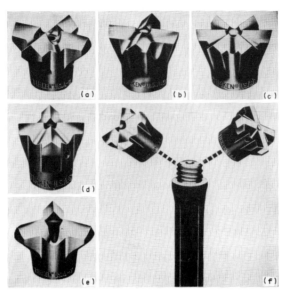

Figure 7-12 Some configurations of bits for percussive drills such as jackhammers and track drills. *(a)* Four-point bit, standard for most drilling, with center air hole for hard rock. *(b)* Four-point bit with modified airways. *(c)* Six-point bit with center air hole. *(d)* Four-point bit with side air hole for soft rock. *(e)* Three-point bit with center air hole. *(f)* Four-point alloy-steel and carbide-steel insert bits with drill steel. Not shown is the single-chisel bit for short holes in soft to medium rock. *(Timken Roller Bearing Company.)*

drilling speeds would be significant for making drilling and blasting plans for the job under investigation.

Track Drill The track drill is perhaps the most popular exploratory drill for the rock excavator because of its availability, portability, depth range, and indications of drilling speeds to be encountered in blast-hole drilling. The last capability is most important because drilling and blasting techniques and costs can be anticipated by exploratory drilling by the same kind of drill that probably will do the subsequent blast-hole drilling.

Figure 7-14 illustrates a track drill putting down 3½-in-diameter hole in weathered andesite of eastern New York State. The reach of the mast from the front of the crawlers is about 9 ft, affording an efficient spread of holes from the one drill stand. Thereby travel time over the rough natural ground and setup time are minimized. Average drilling speed in such a hard-rock formation is about 20 ft/h, the hour including travel and setup times. Figure 7-14 also shows a team of track drill and compressor traveling between exploratory holes for a canal job across Cactus Plain in Arizona. In spite of the soft desert alluvium and steep wadis of the dissected land, the track drill pulling the air compressor averages 2 mi/h. The traction air motors of the drill receive air from the compressor, making the combination a self-contained exploratory unit. In rather soft alluvia the team is able to traverse steep-sided dry streams whose banks have +20 percent grades. In one day the drill is able to put down several exploratory holes in alluvia and schists along the centerline and to travel 8 mi over the rough mountainous desert terrain.

Estimated ranges of hourly drilling speeds for a track drill putting down 3½-in-diameter exploratory holes are set forth in the following tabulation:

Class of rock	Speed, ft/h
Soft	50–75
Medium	25–50
Hard	0–25

Figure 7-13 Percussive jackhammer for rock exploration. Heavyweight jackhammer investigating hard rock excavation. Residuals and weathered rock have been removed and drilling is at rate of about 7 ft/h in semisolid and solid granite. Joints and seams are being checked, as well as drilling speeds. These factors relate to blasthole drilling methods and costs. *(Joy Manufacturing Company.)*

The practical maximum depth of hole for the track drill is about 80 ft unless the drill is of the downhole hammer type, as beyond this depth the inertia of the drill steel lowers the energy delivery to the bit and thereby lowers the drilling speed. At the same time expulsion of the cuttings by the compressed air becomes difficult and finally impossible. Generally this depth limit is adequate because most cuts are shallower than 80 ft and because below this depth the rock formation has become uniform and needs no deeper investigation.

Rotary Drill The rotary drill, equipped with either a drag bit or a tricone bit, is able to penetrate both soft and hard rocks to considerable depths. The drag bit is usually of the fishtail or finger type with removable hardfaced teeth and it is suitable for soft formations. Two- and four-blade bits are used, the two-blade bit giving faster drilling speed but being subject to stalling. At this juncture the four-blade bit of smaller diameter is substituted.

Tricone bits are similar to those used in drilling for oil and gas. Figure 7-15 shows a tricone bit, consisting of three rotating conical cutters mounted in the rotating head. Cuttings from rotary drill bits are removed by compressed air delivered through the hollow drill stem and released through the bit. It is possible, of course, to use *drilling fluid* or *mud*, as in oil-field drilling, but this method for removing cuttings is not normally used in exploratory drilling.

Figure 7-16 shows a rotary drill, equipped with a drag bit, exploring the excavation for a 180-ft-deep, 50-ft-diameter surge chamber. The chamber is part of the Los Angeles Tunnel at Castaic Dam, California. After the 5-in-diameter hole was drilled, uphole seismic studies were made in the Hungry Valley sedimentary rock formation. Shock-wave velocities were correlated with rates of drilling in the several zones of depth. Velocities

(a)

(b)

Figure 7-14 Exploratory drilling by track drill and compressor teams. *(a)* Putting down 3½-in-diameter holes in andesites of eastern New York. Large weathered rocks present traveling difficulties but these are offset partially by the extended boom facility, which makes it possible to drill several holes from the same central drill stand. *(b)* Crossing a wadi of the Arizona desert between exploratory holes for canal excavation. Rock is valley alluvia overlying hard schist. The track drill and towed compressor average 2 mi/h.

ranged from 3100 ft/s near the surface to 6200 ft/s at depth, indicating a mixture of tedious tractor ripping and blasting in the small-diameter confined area of the surge chamber.

Drilling rates for rotary drills, based on 33 percent efficiency, are given in the following tabulation of suggested hourly rate ranges for 5-in-diameter holes.

	Speed, ft/h	
Class of rock	With drag bit	With tricone bit
Soft	130–200	35–50
Medium	60–130	15–35
Hard	—	0–15

Mechanical Means   7-17

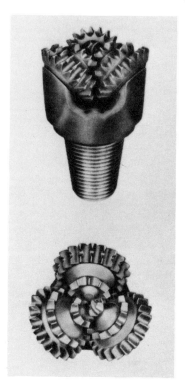

Figure 7-15 Rotary drill bit for investigation of rock excavation. Tricone bit for rotary drilling, showing the three rotating cones. One cone has two cutting elements and the other two have three cutting elements each. Such a bit performs well in either soft or hard rock. *(Joy Manufacturing Company.)*

As in the case of the track drill, the rotary drill using either drag bit or tricone bit furnishes valuable data anticipatory to future drilling and blasting. This is especially true if use of the rotary drill with tricone bit is planned for the blasthole drilling. Derived information is:
    1. Net drilling speeds, or penetration while actually drilling.
    2. Life of bits, generally a costly item of drilling.
    3. Uniformity of rock formation, which usually contributes to stable drilling speeds and fixed powder factor in pounds per cubic yard or per ton of excavation.
    4. Presence of joints, seams, fissures, or pockets of soft residuals, which make for expensive drilling and blasting.
    In short, after exploratory drilling by the kind of drill to be used for blasthole work on the job, a wealth of data is available for a well-calculated cost estimate for production drilling and blasting. If different types of drills are used for exploratory work and anticipated blasthole drilling, either a correlation factor for drilling speeds with the two drills or data from experience in the same rock formation with the particular drill must be used.

    Core Drill   The core drill is an abrasive rotary type for the express purpose of extracting near continuous cores of rock. It is used almost universally by designers of public and private works involving rock excavation. It is also used less extensively by excavators in their appraisals of rock excavation.
    There are two kinds of core drills, according to the nature of the abrasive or cutting material of the bit. More common is the diamond drill, in which the bits are studded with embedded diamonds and which is used to extract cores of up to 6-in diameter. Less common is the shot or calyx drill, in which chilled steel shot is fed down through the drill stem and under the serrated cutting edge of the rotating bit and which yields cores of

Figure 7-16 Carrier-mounted rotary drill for exploration of tunnel excavation. Rotary drill, equipped with 5-in drag bit, exploring excavation for 180-ft-deep surge chamber of the Los Angeles Tunnel of the Castaic Dam. The rock formation is the Hungry Valley, made up of average-thickness beddings of sandstones and shales. The rate of drilling in this medium rock was about 25 ft/h, which was unusually low because of delays with stuck steel and bits.

up to 6-ft diameter. Both drills have similar operating elements, including power unit, drill stem, rotating table, water pump, casing, core barrel, and bit. Such machinery for the larger-diameter shot drill is necessarily much heavier and more powerful then that for the diamond drill.

Diamond Drill. The diamond drill may be truck-mounted or skid-mounted, the truck mounting being more popular. Core diameters range from $7/8$ to 6 in. Commonly used core diameters are: EX, $7/8$ in; AX, $1 1/8$ in; BX, $1 5/8$ in; and NX, $2 1/8$ in.

Figure 7-17 shows several types of bits used in the drilling process. The *casing bit* is used to ream a hole of proper size for setting casing through porous or water-bearing formations which would impede or prevent drilling progress. The *chopping bit* has two functions. When, as sometimes happens, driving standpipe becomes impossible because of materials gathering inside the pipe, the chopping bit serves to break up the material and clear the pipe to the bottom of the drive shoe. The chopping bit also serves to break up any small portion of the core which might have broken off during extraction of the core. The *coring bit,* with diamond surface settings, drills an annular hole which permits the passage of the core barrel down over the core. The *impregnated coring bit* is applicable to most of the rock formations that are too rough for the set coring bit to handle. It is most practical in hard, badly broken, or fractured formations.

Figure 7-18 shows a typical box of core samples, type NX of $2 1/8$-in diameter. The cores were taken by the California Division of Highways in anticipation of a future highway in the Fortuna Mountain area of San Diego County. The highway cut being investigated is 300 ft deep. The rock of the area is granitic and metavolcanic, and these particular cores

Figure 7-17 Bits for diamond core drilling. Four of the principal kinds of bits used in diamond core drilling. *(a)* Casing bit. *(b)* Chopping bit. *(c)* Coring bit. *(d)* Impregnated coring bit. These four bits have variations to the pictured configurations. Other types of bits, usually for some special purpose, are impregnated-casing, pilot-plug, reaming, fishtail, rose, and ferrule. *(Joy Manufacturing Company.)*

are of granite. The cores were taken at depths of 27.5 to 32.5 ft in an exploratory hole of 285-ft depth.

These cores indicate that at an average 32-ft depth the rock is hard and firm and that recovery is practically 100 percent. *Recovery* is that percentage of a given length of core that is unbroken. In this example a break in the core indicates cooling joints in the granite, or perhaps breakage during core extraction. The cause can be determined by close study of any indications of old weathering or young breakage. It is probable that the granite at the horizon of the core samples will require blasting.

When this rock excavation job is up for bidding, these and all other cores will be available for the contractors' inspection. Available also will be logs of all exploratory holes, including auger and bucket drill logs. It is probable that the state will also have available seismic studies of excavation.

Figure 7-19 shows a core drill exploring overburden and rock structure at Lincoln Park in Chicago, Illinois. Test holes were drilled to a 1500-ft depth in order to check the feasibility of constructing underground tunnels for temporary storage of rainwater and sewage overflow. This drill may be tractor-mounted or truck-mounted. It is a multitype drill, affording auger drilling to a depth of 200 ft depth, rotary drilling to 1000 ft, and core drilling to 1500 ft.

Representative performance in four test holes through an average 100 ft of glacial drift made up of hard till, sands, and gravels and through an average 264 ft of limestones and shales is given in Table 7-1. The rig used was a Mobile Drilling Company Model B-50 Explorer, and the maximum hole depth for the project was 591 ft.

7-20 Exploration of Excavation

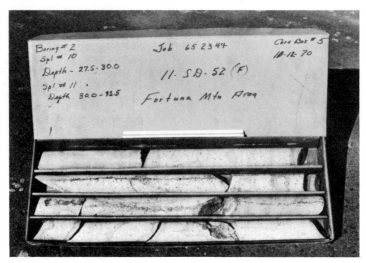

Figure 7-18 Box of typical cores, taken by core drill. Typical box of cores taken by the contracting officer in the investigation of a highway cut in granite. The cores were extracted in October 1970, and the highway had yet not been fully planned as of 1976. Sometimes a decade of rock exploration precedes a large rock excavation project. The cores show good recovery at average 32-ft depth, and they indicate that blasting will be necessary at this depth in the huge 300-ft-deep cut in granite and metavolcanics. *(California Division of Highways.)*

Figure 7-19 Carrier-mounted core drill taking cores from deep rock strata. Core drilling through 100 ft of overburden, consisting of sands and gravels of till, and 362 ft of limestones and shales. Rock exploration is for the feasibility of building deep tunnels for the storage of rain water and sanitary sewage beneath Lincoln Park in Chicago, Illinois. The average drilling speed in overburden by a $5\frac{5}{8}$-in tricone bit was 5.6 ft/h and the speed in limestone and shale with a $2\frac{1}{8}$-in core bit was 12.8 ft/h. *(Mobile Drilling Company.)*

TABLE 7-1

Date:	July 1971
Project:	Metropolitan Sanitary District, Chicago
Contractor:	American Testing & Engineering, Indianapolis, Ind.

Job conditions: Overburden consisted of hardpan, sand and gravel, and free-running sand to average depth of 90 ft. Overburden drilled with 5⅝-in tricone bit and cased with 4-in pipe to rock, then NX casing placed 5 ft into rock in preparation for coring. Rock coring performed using wire line method with NQ rods and core barrel. Clear water was pumped by John Bean Model 435-11 (35 maximum gal/min) at pressures ranging from 200 to 500 lb/in². Coring performed through thin shale and heavy limestone formations to average depth of 360 feet.

		Drill Log Data*		
	Type of Drilling	Depth, ft	Total Time, h	Rate, ft/h
#1 Hole	Overburden	90	9.5	9.5
	Core	263	24.5	10.7
	Total	353	34.0	10.3
#2 Hole	Overburden	104	17.0	6.1
	Core	216	23.0	9.3
	Total	320	40.0	8.1
#3 Hole	Overburden	104	11.0	9.5
	Core	217	17.5	12.4
	Total	321	28.5	11.3
#4 Hole	Overburden	100	33.0	3.0
	Core	362	17.5	20.9
	Total	462	50.5	9.1

*Figures taken from drill log do not include setup time and are based on three 8-h shifts daily.
SOURCE: Mobile Drilling Company.

The averages for the drill log data of Table 7-1 are:

Average rate of drilling overburden with inclusion of setting of casing, tricone bit of 5⅝-in diameter.	5.6 ft/h
Average rate of drilling limestone and shale, core bit of 2⅛-in diameter	12.8 ft/h

Note that drilling time does not include setup times from hole to hole but does include all other necessary times for the drilling operation. It is probable that penetration rate or actual drilling speeds are about 50 percent faster, giving 8.4 ft/h in overburden and 19.2 ft/h for combined limestone and shale rock formation.

Actual penetration speeds, exclusive of all delays to the drilling, for diamond core drills are given approximately in the following table for all sizes of cores from ⅞- to 2⅛-in diameter. Efficiency is 100 percent. The table is adjusted for probable average drilling rates at 33 percent efficiency.

	Speed, ft/h	
Class of rock	100% efficiency	33% efficiency
Soft, as salt and limestone	30–40	10–13
Medium, as limestone, dolomite, graywacke, and greenstone	20–30	7–10
Hard, as granite and gabbro	10–20	3–7

SOURCE: Christensen Diamond Products Company.

Shot or Calyx Drill. The *shot* or *calyx drill* is used for large holes up to about 6 ft in diameter. Occasionally these holes are used for exploratory work, since a worker can be lowered into the hole in order to make visual inspections of the rock structures. However,

7–22   Exploration of Excavation

the drill is used more commonly for large-diameter holes for foundations and penstocks and for communication and ventilating shafts in underground mines.

Figure 7-20 shows a cross section of the workings of a shot drill within a rock formation. Water acts as a lubricant and as a carrier for the cuttings. The cuttings and the intermingled particles of the spent shot settle in the *calyx*. The casing is equipped with a drive shoe for penetrating the residuals and weathered rock and for seating onto the semisolid or solid rock. Below this interface the cores of rock are extracted.

Standard shot drills put holes down to 600-ft depth and extract cores of 3- to 20-in diameter. Specially built drills are sometimes used for unusual rock exploration involving deeper and larger holes of up to 3-ft diameter and 1000-ft depth.

Figure 7-21 illustrates a derrick-type calyx drill exploring a dam foundation in South America. The heavy, large-diameter cores call for massive drills and for powerful drilling

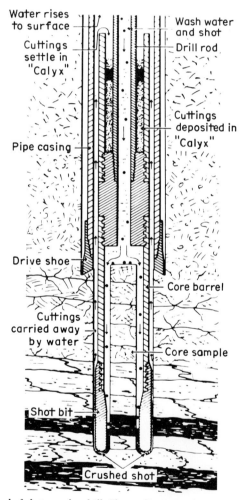

Figure 7-20 Drill head of shot or calyx drill. The workings of a shot or calyx drill exploring rock formation of residuals, weathered rock, and semisolid rock. Water, bearing chilled steel shot, is circulated downward between core and core barrel to the serrated shot-bit cutting edge, whence it is circulated upward with its cuttings and spent shot particles between the core barrel and casing. At the top of the core barrel, the mixture falls into the calyx for deposit. *(Ingersoll-Rand Company.)*

Mechanical Means 7-23

Figure 7-21 Head works of large-capacity shot or calyx drill. Shot or calyx drill exploring dam foundation in South America. The derrick-type drill is extracting 66-in-diameter cores from this combination hole for penstock and examination of rock formation. It is probable that the average drilling speed is about 0.4 ft/h. Such massive and powerful drills are of special construction and they are of a "one-job" nature. *(Ingersoll-Rand Company.)*

machinery. This installation is of a semipermanent, nonportable nature, except for the supporting trackage for lineal movement.

The huge size of some calyx cores is emphasized in Figure 7-22. These 36-in-diameter cores of hard sandstone were extracted for the examination of foundation work for Norris Dam, located on the Clinch River of Tennessee. In this investigation began the now established practice of lowering a geologist into the hole for minute inspection and photographing of the rock structure. It is probable that the sustained rate of drilling in this hard sedimentary rock was about 0.4 ft/h.

At 33 percent efficiency a shot drill will perform at about the drilling speeds tabulated below, according to class of rock and diameter of hole.

	Speed, ft/h, according to hole diameter, in		
Class of rock	3 to 6 in	6 to 15 in	15 to 30 in
Soft	3–6	1.5–3	0.8–1.5
Medium	1.5–3	0.8–1.5	0.4–0.8
Hard	0.8–1.5	0.4–0.8	0.2–0.4

The large-diameter calyx is used by the contracting officer in qualitative investigations of rock pertaining to design rather than by the earthmover in his appraisal of the excavation characteristics. Of course, with the use of modern camera equipment it is possible

7-24   Exploration of Excavation

Figure 7-22 Large-diameter cores from calyx drill. Large-diameter (36-in) calyx cores taken from the rock foundations for Norris Dam, one of the huge dams of the Tennessee Valley Authority. During the examination the chief geologist, Dr. Charles P. Berkey, was lowered down into the hole to make one of the first visual inspections of the walls of a test hole. Anxiety over the seamy nature of strata prompted this personal observation in addition to examination of small-diameter cores. These explorations by the contracting officer gained much valuable data for the selection of methods and machinery for the rock excavation. *(Ingersoll-Rand Company.)*

to scan the rock structure in the walls of a small-diameter hole and thus eliminate the lowering of a geologist except under some unusual investigative problem. However, when exploring variable rock formations, the unusual is sometimes the most valuable investigation.

Cable Drill   The well-known and long-used *cable drill* is a gravity percussive machine, depending on the free fall of a string of drilling tools for the breakage of the rock. Over a period of about a century it has been known as a spudder, a churn drill, and a well drill. Figure 7-23 shows this rig set up for exploratory work. It is set up on cribbing on a sidehill in order to check the depth of overburden and weathered rock overlying merchantable trap rock or basalt in northern New Jersey. The rig is drilling a 5⅝-in-diameter hole and swinging a 1350-lb string of tools, including a set of jars. The hole was cased through about 15 ft of residuals and weathered rock, at which depth solid basalt was encountered. The rate of drilling through some 20 ft of rock was 0.8 ft/h. Inasmuch as the proposed quarry floor was some 100 ft below the top of the suitable rock, the quarry site was regarded as economically feasible.

During drilling the percussive energy comes from the fall of the string of tools through an adjustable 18- to 36-in stroke of the walking beam with a frequency of 40 to 60 blows per minute. In Figure 7-23 the stroke was 36 in and the frequency was 53 blows per minute. Normally, drilling is done without jars and the string of tools is made up of swivel rope socket, drill stem, and bit. When a 5⅝-in-diameter hole is being drilled, such tools weigh about 1000 lb, and with a 30-in stroke and a frequency

Figure 7-23 **Cable drill for rock exploration.** An efficient crawler-mounted cable drill exploring the site for a proposed trap rock or basalt quarry in northern New Jersey. A 5⅝-in-diameter hole is being put down through residuals and weathered rock into the solid basalt. After setting casing through 15 ft of overburden, acceptable rock was encountered. Three more comparable holes were drilled and the site was judged acceptable.

of 50 blows per minute, some 125,000 ft·lb/min of energy is available for breaking the rock.

The commonly used bit is the chisel bit. As with the kindred percussive bits for jackhammer and track drill (Figure 7-12), other types are available to be used according to the nature of the rock structure. Bits may be dressed by forging and sledging on the job or, more efficiently, by work in a shop. Drilling may be with or without casing, depending on the earth-rock formation. Water is fed into the hole during drilling, forming a mixture with the cuttings. Periodically the string of tools is pulled and the mixture is removed with the dart-valve bailer. If samples are desired, a special drill bit with hollow-tube sampler may be substituted for the regular bit. Otherwise, the nature of the rock is judged by the cuttings in the bailer. Casing is driven by the stroking of the tools, the drive head being driven by drive clamps attached to the stem.

The reciprocating action of the spudder beam provides the hammerlike blows for drilling and driving casing. A bull reel carries the drilling cable, a calf reel handles the cable for the casing, and a sand reel contains the bailer cable. Different positions of the crankpin control length of stroke, usually three lengths. Engine throttle controls frequency of blows.

Such a popular-sized rig as shown in Figure 7-23 has a depth capacity of about 700 ft when swinging 900 lb of tools.

Figure 7-24 displays the common drilling tools and some of the many fishing tools used to extract stuck tools or tools which have become loose because of broken cable.

Based on 33 percent efficiency for exploratory drilling, average hourly drilling speeds for holes of 4- to 6-in diameter are as listed in the following tabulation:

Class of rock	Speed, ft/h
Soft	6–12
Medium	3–6
Hard	0–3

Wash-Boring Drill The properties of the cable drill make it an ideal machine for wash boring simply by the addition of a high-pressure water pump. Wash boring is an efficient and economical means of exploring for rock when the overburden is made up of residuals, alluvia, or extremely well-weathered rock. The principles of operation are the same as those for manual work as shown in Figure 7-1.

In the case of the mechanical cable drill a greater quantity of water is introduced into the drill rod or pipe, a cutting bit is affixed to the end of the rod, a driving head is attached to the top of the rod, and almost always the hole is cased.

The spudder beam furnishes reciprocating blows to the drill rod and to the casing pipe if necessary. Bull reel, calf reel, and sand reel suffice for all hoisting and setting work.

An inherent advantage of the use of the cable drill is in exploring through alternate layers of hard and soft rocks, such as lava and alluvium. In such cases the lava may be drilled by cable tools and the underlying alluvium may be cased and wash-bored.

Wash boring is limited to about 100-ft depth because of the abnormal water pressures needed to flush out large particles of material. Hole diameter is limited to about 6 in. In soft soils, such as residuals and alluvia, overall drilling speeds of 20 ft/h are possible. In stiff clays footage drops to about 8 ft/h.

By means of a split-tube sampler attached to the drill pipe samples may be taken at any desired depth of hole.

Although the ubiquitous veteran cable drill may lack certain sophistications of principle, it is an efficient, reliable, simple, and economical drill for much rock exploration.

Many drills are multipurpose for exploratory work, being custom-built for a variety of kinds of drilling. The machine of the California Division of Highways, Figure 7-11, can put down auger, bucket-auger, rotary, and core holes. With its 10-wheel truck chassis, it is suitable for travel over almost all terrains and for drilling in almost all kinds of rock formations.

INSTRUMENTAL MEANS—THE SEISMIC TIMER

Refraction Studies

The principles of refraction seismography have been known and applied for about 75 yr. In the decade 1920 to 1930 they were applied to oil exploration and during the past 25 yr they have been applied successfully to exploration of earth-rock excavation in the thin regolith excavated by human agency.

Principles The theory is simple. Artificially created shock waves travel through an earth-rock structure at velocities proportional to the degrees of consolidation of the residuals, weathered rock, and semisolid and solid rock. In their downward penetration of the crust they are refracted away from the normal or vertical toward the tangential or horizontal according to Snell's law of refraction. It is possible, then, to calculate the shock-wave velocities according to the degrees of consolidation and the depth to the interfaces at which velocities and degrees of consolidation change.

As will be explained later is this chapter, excavation methods and costs bear an empirical relationship to the shock-wave velocities. Expressed arithmetically, cost of excavation is proportional to degree of consolidation of earth-rock, degree of consolidation is proportional to velocity, and therefore cost of excavation is proportional to shock-wave velocity. Much work in correlating excavating machinery and costs with velocity has made this generalization just about axiomatic.

There are two types of seismic timer in common use and acceptance, the single-station and the multistation machine. The single-station unit involves a fixed position for the geophone or pickup element of the timer and variable positions for the point of excitation of the shock wave.

The multistation unit includes a fixed point of excitation and variable positions for the several, usually 12, geophones. Times for the several distances are recorded graphically

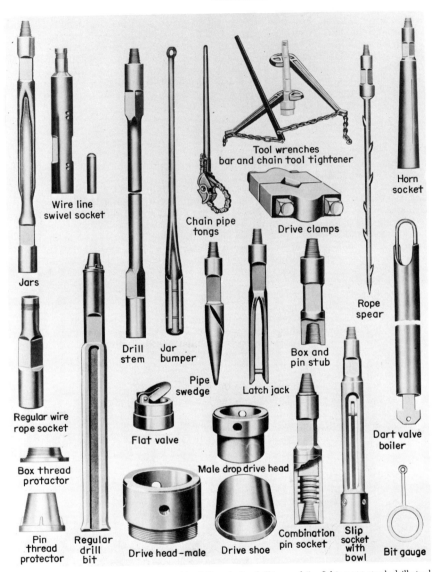

Figure 7-24 Cable drill tools. Some cable drill tools for drilling and for fishing out stuck drill steel or drill steel which is loose by reason of a broken cable. Jars are a unique device for drilling in variable rock formations in which the tools are apt to get stuck. They are set between the cable socket and the stem and the stroking of the spudder beam jars loose the tools. In characteristically difficult rock structures they are used continuously. *(Bucyrus Erie Company.)*

on a photographic negative, from the picture of which times in milliseconds are scaled.

Both types are reliable and they have inherent advantages and disadvantages. As both types produce time elements for the passage of the shock waves from the point of excitation to the geophone, this description of the seismic timer is confined to the use of the single-station unit.

In this discussion of theory, field techniques, and calculations there will be no elaboration on stratigraphic problems sometimes encountered in seismic studies or on their

7-28 Exploration of Excavation

mathematical solution. For such professional treatment the reader is referred to the bibliography and to the instruction manuals of the makers of the several seismic timers.

Snell's law of refraction as applied to seismic work may be expressed verbally as follows: When a shock wave passes through a layer in which its velocity is V_u and strikes an interface separating this upper zone of lower velocity from the lower earth-rock zone of higher velocity V_l at an angle I_u with the normal, it is bent at an angle R_l with the normal in the lower layer, V_l. The relationship among these four elements is expressed by the following derived equation:

$$\frac{\sin I_u}{\sin R_l} = \frac{V_u}{V_l}$$

Figure 7-25 shows ideally the paths of travel of shock waves from points of excitation to the geophone in a three-layer earth-rock structure consisting of residuals, weathered rock, and semisolid and solid rock.

The further development of the above equation results in combination theoretical and practical equations for depths from the ground surface to the interfaces between the layers having the calculated velocities. These equations are set forth in Figure 7-26, a typical study of a 100-ft-deep cut in granite with four layers of different velocities.

Equipment and Methods Figure 7-27 illustrates the equipment needed by the instrument operator and sledge operator when the sledge is used for excitation of the shock wave. The sledge method is suitable for seismic lines of maximum 320 ft for 107-ft depth exploration. Usually explosives are used on longer lines (up to 640 ft and longer) for investigating to 217-ft or greater depths. When explosives are used, a few pounds of stick dynamite, instantaneous electric caps, blaster box, firing line, and an adze for cutting out holes for the explosive charges are added to the two packs.

Figure 7-28 illustrates the setup for a refraction seismic study. The job is a 406,000-yd³ mountain highway in the Mother Lode country of the Sierra Nevada Mountains, California. The highway is 2.04 mi long. The rock formation is the Calaveras, which, along with the Mariposa, is a common metamorphic assemblage of rocks in this fractured and faulted country. It is made up of altered sedimentary rocks consisting of schists, slates, quartzites, phyllites, and conglomerates. These two kindred formations were involved in 11 of the 509 rock excavation jobs used to develop the correlations of Figure 7-33.

Pictured in Figure 7-28 is the beginning of a 320-ft seismic line across the centerline of a 272,000-yd³ cut with maximum 69-ft centerline depth. Two other 320-ft lines and a 200-ft line were run along the 1200-ft length of the centerline. The pickup geophone is spiraled into the residuals under the right foot of the instrument operator. The 100-ft

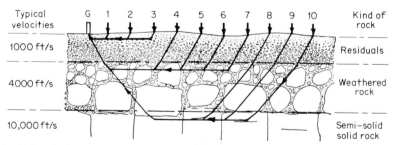

Figure 7-25 Ideal paths of travel of shock waves in an earth-rock structure according to Snell's law of refraction. The seismic timer is actuated by the earliest of the many shock waves arriving from a single point of excitation. The waves from each of the excitation points 1 through 10 are picked up by the geophone G.

Waves from 1, 2, and 3 arrive first through the upper layer of 1000 ft/s. However, of the many waves from 4, that one passing through both the upper layer and the middle layer of 4000 ft/s arrives first. Likewise, waves from 5, 6, and 7 follow the same pattern of travel through the upper and middle layers. Again, however, wave 8 reaches the geophone before any other waves through the upper and middle layers by passing through all three layers, 1000, 4000, and 10,000 ft/s. Of course, waves from 9 and 10 follow the path of wave 8.

As diagrammed, a wave passing from a lower-velocity layer through the interface to a higher-velocity layer is bent farther away from the normal to the interface.

Instrumental Means—the Seismic Timer 7-29

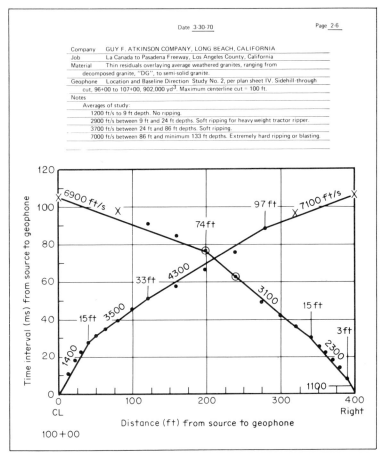

Figure 7-26 Seismic study of excavation. Calculations for depths and velocities below 100+00 centerline.

$$D\text{-}1 = \frac{40}{2}\left(\frac{3500-1400}{3500+1400}\right)^{1/2} = 15 \text{ ft}$$

$$D\text{-}2 = 0.9 \times 15 + \frac{120}{2}\left(\frac{4300-3500}{4300+3500}\right)^{1/2} = 33 \text{ ft}$$

$$D\text{-}3 = 0.1 \times 15 + 0.8 \times 33 + \frac{280}{2}\left(\frac{7100-4300}{7100+4300}\right)^{1/2} = 97 \text{ ft}$$

tape is stretched out from the geophone beyond the sledge operator. The sledge with welded-on steel disk is cable-connected to the seismic timer. The sledge operator's back is toward the geophone so that the shock wave will be driven downward toward the geophone. At the instant of impact a leaf switch, mounted on the sledge handle near the head, will close, opening up the gate of the timer, and time will start recording on the indicator-light panel. When the shock wave reaches the geophone, the gate will close and the time interval in milliseconds will be visible in the lighted bulbs. In this favorable terrain the reversed 320-ft line was run in 35 min. The averages for this study, probing to a depth of 107 ft minimum in this 69-ft deep cut, are listed below:

1300 ft/s to 6-ft depth. No ripping.

3000 ft/s between 6- and 28 ft depth. Soft ripping for heavyweight tractor-ripper.

7-30 Exploration of Excavation

  7500 ft/s between 28- and 48-ft depths. Blasting.
  9600 ft/s between 48- and minimum 107-ft depth. Blasting.

The averages for all four studies in the cut fixed the rock line at an average 29-ft depth, below which blasting would be necessary. A quantity takeoff established that 106,000 yd^3, or 39 percent of the yardage of the cut, would be blasted.

Eight seismic studies were made in the three principal cuts, and the averages of the studies compared favorably with those for an adjacent section of the highway in the same Calaveras formation which was analyzed several years ago. The history of excavation in the adjacent job, along with the studies of both jobs, provided the bidding contractor with a wealth of knowledge for his bid preparation for the unclassified excavation.

The correlation of the studies of these two adjacent jobs, backed up by job experience on one of them, is a good example of the correlative approach to exploration of rock excavation which was discussed in Chapter 2 and charted in Table 2-3. When such methods are used, bidding on questionable rock excavation becomes less hazardous and approaches more closely an exact science.

Figure 7-29 illustrates shock-wave excitation by blasting. A 400-ft reversed line is being run in a 1,484,000-yd^3 through cut, examining the rock excavation to 133-ft minimum depth. Excitation by sledge was carried out to 280 ft of line and thereafter explosives were used at 340- and 400-ft distances. The formation is the Panoche, which was excavated for the Pacheco Pass Freeway in Merced County, California. The formation was discussed in Chapter 2.

The rock line in this cut was established at 30-ft depth. Calculations for estimating unit bid price on the unclassified roadway excavation called for 371,000 yd^3 or 25 percent, to be ripped by heavyweight tractor-rippers and 1,113,000 yd^3, or 75 percent, to be blasted. The rock excavator blasted about 65 percent of a total 11 million yd^3 of the Panoche and Franciscan formations of the job.

Figure 7-27 Equipment for seismic investigation of rock-earth by two-person crew. *(Left)* Pack of sledge operator. Total weight, including sledge, is about 30 lb. Equipment includes: 14-lb sledge with welded-on striking disk; 220- and 340-ft cables on reels; sender for radio link between sledge and seismic timer in lieu of connecting cable; canteen; and two lunch pails. It is sometimes possible, if there is no interference, to use the radio link. However, the cable connection is more satisfactory.

(Right) Pack of instrument operator. Total weight of about 30 lb. Equipment includes seismic timer with radio-link receiver, geophone, 100-ft engineer's tape, geologist's hand pick, case for plans, and canteen. Extra sledge with striking plate for soft ground is shown but it is not a part of the regular gear.

Figure 7-28 Field setup of seismic equipment for refraction study for 320-ft line across the centerline of a 69-ft-deep mountain highway cut of 272,000 yd^3. Geophone under right foot of instrument operator marks the beginning of the right-to-left leg of the line, located 160 ft to the right of centerline. This tranverse line was run in favorable topography and brush and it required 35 min. Two other 320-ft lines and a 200-ft line were made along centerline with steep gradients and much brush, requiring 2 h. Thus the time for the analysis of the rock excavation of the cut required about 2½ h, about normal for mountain work in favorable weather.

Figure 7-29 Seismic studies using explosives for excitation of shock wave. Excitation of shock wave by blasting. The line is 400 ft long, probing to a minimum 133-ft-depth in a 95-ft deep cut of 1,484,000 yd^3. The cut is in the Panoche formation, made up of thin residuals overlying weathered, semisolid, and solid heavily bedded sandstones. The rock line was established at 30-ft depth.

The shot of three sticks of 1¼-in by 8-in cartridges of dynamite is 280 ft from the geophone near the instrument operator. The worker on the blaster box has just pressed the firing button and simultaneously fired the shot and started the seismic timer to record the time interval before the shock wave is picked up by the geophone. By such means depth probes may be extended to 200 ft and more, but generally topography and other limitations prevent examinations of more than 250-ft depth.

Ripping and Blasting Zones According to Seismic Velocities

Commencing some 25 yr ago with the use of the seismic timer, it was necessary to express results in terms of shock-wave velocities through earth-rock structures with respect to two methods of rock fragmentation. Concurrent with the introduction of the seismic timer was the development of medium-weight and heavyweight tractor-rippers as economical tools for rock fragmentation. It was then natural to express rippability of tractor-rippers in terms of shock-wave velocities until a velocity was reached beyond which blasting was necessary. In the ensuing years many correlative studies have been made. Suggested guides for rippability and blasting are listed below.

1. For medium-weight tractor-rippers in the 200- to 300-engine-hp and 60,000- to 90,000-lb working-weight specification ranges:

0 to 1500 ft/s	No ripping
1500 to 3000 ft/s	Soft ripping
3000 to 4000 ft/s	Medium ripping
4000 to 5000 ft/s	Hard ripping
5000 to 6000 ft/s	Extremely hard ripping or blasting
6000 ft/s and higher	Blasting

2. For heavyweight tractor-rippers in the 300- to 525-engine-hp and 100,000- to 160,000-lb working-weight specification ranges:

0 to 1500 ft/s	No ripping
1500 to 4000 ft/s	Soft ripping
4000 to 5000 ft/s	Medium ripping
5000 to 6000 ft/s	Hard ripping
6000 to 7000 ft/s	Extremely hard ripping or blasting
7000 ft/s and higher	Blasting

Uphole Studies

Another kind of seismic study, more precise than the refraction study, is the uphole study. An exploratory drill hole is necessary for this vertical exploration, but, most unfortunately, these are rarely available. And when they are available, they are generally caved, capped permanently at the collar, or backfilled by the one making the exploration. Of some 6000 seismic studies of 1 billion yd^3 of excavation, it has been possible to make only 95 studies of the uphole kind.

The uphole study is based simply on the determination of the time for a shock wave to travel upward along the wall of the hole between two different depths of the hole. The difference between the depths, or the thickness of the layer of the rock formation, is divided by the difference between two seismic travel times for the two elevations and the quotient is the shock-wave velocity. The results are precise.

Experience with up to 340-ft-deep uphole studies indicates definitely better results than with the refraction technique, especially when one realizes that the baseline for a 340-ft-deep refraction study should be at least 1020 ft long, practically impossible in average topography.

Figure 7-30 shows the field layout for an 80-ft deep uphole study in volcanic rock of San Diego County, California. Explosive charges were set at intervals of depth in the hole and detonated by seismic electric caps, and the seismic timer recorded the time from the depth of the charge to the collar of the hole. A stick of dynamite, or a fraction of a stick, was suspended from the two leg wires of the cap, which was embedded conventionally in the stick of dynamite. Deep-hole seismic caps are spooled with leg wires of up to 400-ft length and the wires are marked at 5-ft intervals.

Usually holes are wet, sometimes with clear water and sometimes with light mud. Under such conditions the charges are weighted with 6-in spikes so as to settle to the desired depth. The long legs of the cap may be used for successively shallower shots.

Figure 7-31 is the seismic study sheet for one of 10 uphole studies in granite of the San Raphael Hills, Glendale, California. The study was made in the same tract excavation shown in Figure 5-2. The drilling and blasting contractor put down the exploratory holes by track drill, had uphole studies made, and then bid on the hard-rock drilling and blasting on a lump-sum basis.

Instrumental Means—the Seismic Timer 7-33

Figure 7-30 Field setup of seismic timer equipment for uphole study. Firing cable with reel from blaster box to leg wires of seismic cap of charge. Blaster box connected to seismic cap and to gate which opens the inlet of the seismic timer. Firing button simultaneously fires charge and opens gate of the timer. Engineer's tape, 100-ft length, with weight for measuring depth of hole. If depth of hole, usually unknown, is greater than 100 ft, additional cord is used. Seismic timer with connected geophone located near collar of hole (foreground, left to right).

Hole, with light board from which are suspended leg wires of the cap with explosives charge. In wet holes the charge is weighted with 6-in spikes. Geophone with cable stabilizing geologist's pick (middle ground, left to right).

As a result of these combined drilling and seismic exploratory data, the contractor determined:
    1. Ten depths to and beyond the rock line separating rippable from drill-and-shoot rock.
    2. Rates for drilling in the hard rock.
    3. Seismic data for the hard rock, from which hole diameter and hole spacing pattern were estimated.
From these data, quantities and unit costs for blasting hard rock, necessary for a venturesome lump-sum bid, were determined.

The seismic study sheet shows a slowly bending velocity versus depth curve, typical of weathered granite. The rock line is at 55-ft depth in the 100-ft-deep cut. In the prism of rock represented by the study, a quantity takeoff based on the assumption that the rock line paralleled the original ground revealed the quantity of drill-and-shoot rock.

When making seismic studies it is sometimes expedient to take advantage of an existing exploratory trench, as pictured in Figure 7-32. This trench in weathered granite was excavated by the contracting officer for the New Lake Arrowhead Dam, San Bernardino County, California. Under this fortuitous circumstance there are two advantages.

First, in the walls of this 150-ft-long, 18-ft-deep excavation one may see the weathering of the rock structure. Second, one may run a "below-original-ground" study to corroborate the findings of the nearby parallel study on the original ground. In this case a 120-ft line was run, with depth penetration of 40 ft below the bottom of the trench to a total 58-ft depth. The results of the study agreed with the results of the other four studies of the site 3 excavation of 1,100,000 yd^3 to a maximum 80-ft depth. The rock line separating rippable weathered granite from drill-and-shoot semisolid granite averaged 77-ft depth.

Such exploratory trenches are not uncommon. However, they are characteristically short and of shallow depth, the shortness necessarily controlling the length of the seismic line and therefore the depth penetration.

Velocity—Depth Relationships for the Three Classes of Rock

Figure 7-33 summarizes the velocity-depth relationships of igneous, sedimentary, and metamorphic rock formations. One may cross-refer this figure to Table 5-1, which shows

7-34 Exploration of Excavation

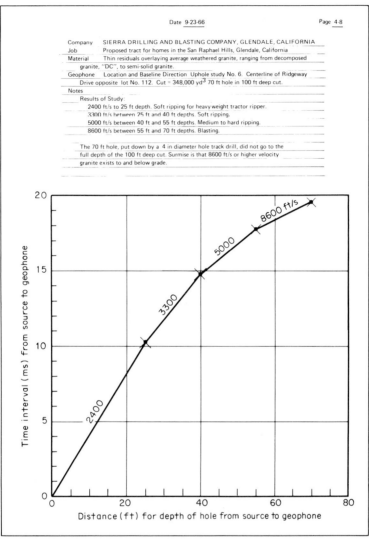

Figure 7-31 Seismic study of excavation. Calculations for average velocities within four zones of depth:

Original ground to 25-ft depth:	25 ft/10¼ ms	=2400 ft/s
25-ft to 45-ft depth:	15 ft/4½ ms	=3300 ft/s
40-ft to 55-ft depth:	15 ft/3 ms	=5000 ft/s
55-ft to 70-ft depth:	15 ft/1¾ ms	=8600 ft/s

the relative percentages of the three basic kinds of rock in some 1 billion yd³ of rock excavation in public and private works.

Chapter 3 contains several figures illustrative of the use of Figure 7-33.

Figure 3-4 pictures minimum-weathered limestone. Entering Figure 7-33 at the ordinate 6000 ft/s, one finds for the curve "Sedimentary—Min W" an abscissa of 31-ft depth where the rock excavator might commence to blast the rock.

Figure 3-6 shows average-weathered igneous intrusive granite. If a highway were

Figure 7-32 Ideal seismic study at bottom of exploratory trench. Refraction seismic study in an exploration trench of borrow pit for dam building. Excavation is in tonalite granite, made up of thin residuals with some surface boulders atop weathered granite with embedded boulders, atop semisolid and solid granite. The findings of the 120-ft seismic line corroborated the findings of a nearby parallel 320-ft line on natural ground. These two studies, together with three other studies in site 3, set the rock line at 77-ft depth in the 1,100,000-yd^3, 80-ft maximum depth rock borrow pit. Trenches present excellent opportunities to inspect rock formations in the weathered zone, as in this case, where the method reveals weakening dikes, cooling joints, and the embedded boulders in the matrix of decomposed granite.

projected through the ridge with a cut of 50-ft depth, one would enter the graph at the abscissa 50-ft depth, intersect the curve "Igneous—Av W," and find the ordinate 6700 ft/s within the questionable zone. The canny rock mover would anticipate some costly drill-and-shoot rock near grade.

Figure 3-12 illustrates well-weathered quartzite. If one were considering a dam-abutment excavation of 80-ft depth in this metamorphic rock formation, one would enter the graph at the abscissa 80 ft, intersect the curve "Metamorphic-Max W," and read the ordinate of 5800 ft/s for the probable shock-wave velocity at grade. The velocity approaches the zone of questionable ripping or blasting, and so ripping near grade would be difficult.

One appreciates that Figure 7-33 is "average" and is merely an indicator to be proved by localized tests and studies of the particular rock formation. In a cut of the same rock formation with the same degree of weathering, two seismic studies may differ materially because of the vagaries of nature.

The advantages of the single-station seismic timer for both refraction and uphole studies are many, as summarized below:

1. *Portability.* As Figure 7-27 emphasizes, the two-person crew carries two packs and can go anywhere they can crawl.

2. *Speed.* It is normal to run an average of eight studies for depth penetrations in the 60- to 100-ft range in eight working hours.

3. *Precision.* Seismic examination being a form of sampling, it is axiomatic that "the more samples, the greater the precision." Much experience based on thousands of studies indicates that errors in velocity and depth determinations are about ± 5 percent for the individual study. In the some 1 billion yd^3 of rock explored, the average was 170,000 yd^3 per study.

4. *Versatility.* Any rock formation may be explored, ranging from alluvium of 1000 ft/s to andesite of 24,000 ft/s and to practical depths over 200 ft.

5. *Economy.* A typical 5-million-yd^3 rock excavation with less than five different rock

7-36 Exploration of Excavation

formations calls for about 25 studies. Four field days and four office days, along with perhaps two travel days, are required, giving a total of 20 labor-days. In 1976 the cost of such an investigation amounted to less than 0.1 cent/yd^3.

COMPARISON OF METHODS FOR EXPLORATION OF EXCAVATIONS

Choice of any or several of the manual, mechanical, and instrumental means for exploration depends on the purpose of the investigation. The contracting officer, the rock mover, and the miner may have different or the same objectives. Table 7-2 is suggested as a guide to the chief capabilities of the several means of rock exploration.

The vast array of exploratory tools offers just about every facility to the inquisitive earthmover. Excavators vary greatly in their completeness of rock investigations.

Some big companies do little exploratory work, relying on vast experience and the data furnished by the contracting officer. Others, more inquisitive, go to great lengths to learn everything possible about the invisible hard rock.

When one considers the great disparity in unit prices for unclassified excavation submitted by experienced bidders, one must conclude that more exploratory work should be done. In April 1975 bids were taken for a budgeted $24 million freeway in granite excavation in San Diego County, California. All seven bidders had worked in the rather common granite formation. Bids ranged from $0.50 to $1.50 per cubic yard of roadway excavation in rather well balanced bids. Admittedly, this is an unusual percentage difference in a unit price bid, but it suggests that more exploratory work would have resulted in more logical conformity of unit prices for the unclassified excavation. There was 2,300,000 yd^3 of roadway excavation, giving a $2.3

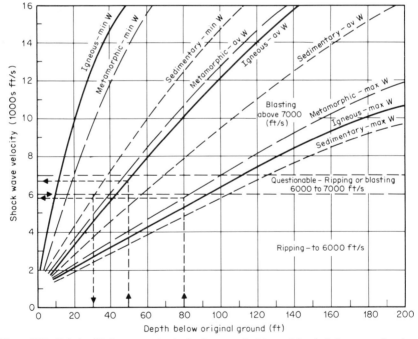

Figure 7-33 Relationship between seismic shock-wave velocities and depths below ground surface for the three basic kinds of rocks with minimum, average, and maximum degrees of weathering. The data for the curves are based on some 6100 seismic studies made in 509 rock excavation jobs totaling about 1 billion yd^3.

TABLE 7-2

Method of exploration	Limitation of rock hardness	Limitation of depth	Purpose of exploration	
			Qualitative	Quantitative
Manual:				
Pick and shovel	Hard	Limited	Yes	Yes
Sounding rod	Medium	Limited	No	Yes
Auger	Medium	Limited	Yes	Yes
Wash boring	Soft	Limited	Yes	Yes
Mechanical:				
Backhoe	Medium	Limited	Yes	Yes
Bulldozer	Medium	Limited	Yes	Yes
Auger drill	Medium	Limited	Yes	Yes
Bucket drill	Medium	Limited	Yes	Yes
Jackhammer	None	Limited	Limited	Yes
Track drill	None	Limited	Limited	Yes
Rotary drill	None	Unlimited	Yes	Yes
Core drill	None	Unlimited	Yes	Yes
Cable drill	None	Unlimited	Yes	Yes
Wash boring	Soft	Limited	Yes	Yes
Instrumental:				
Seismic timer:				
Refraction study	None	Limited	Limited	Yes
Uphole study	None	Unlimited	Limited	Yes

million difference in bid total due to the item. The low bid was $16,729,332 and the high bid was $19,921,313.

SUMMARY

As was explained in the summary of Chapter 6, exploration of excavation is a correlative to examination of excavation. Accordingly, good management regards examination and exploration as of equal importance. In so doing ownership is assured of a good appraisal of the excavation.

CHAPTER 8

Costs of Machinery and Facilities

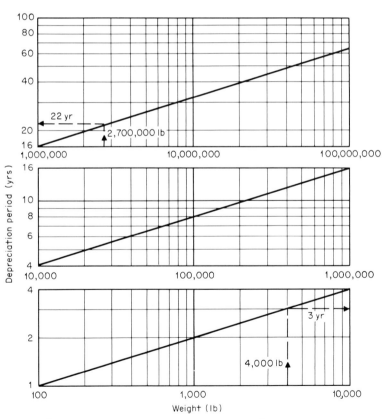

PLATE 8-1

Depreciation period in years according to weight in pounds for excavating machinery powered by engines and motors. The idealized graph is based on the formula:

$$DY = 2^{(\log WP - \log 100)}$$

where DY = depreciation period in years and WP = weight of machine in pounds.

COST OF OWNERSHIP AND OPERATION OF MACHINERY AND FACILITIES

The hourly cost of ownership and operation of rock excavation machinery varies greatly for a given machine or facility because of the range of the following factors:
   1. Total cost of job, depending on discount buying, cost of financing or borrowing, sales and use taxes, and freight to job location.
   2. Salvage or trade-in value.
   3. Period or rate of depreciation in years and hours of use yearly.
   4. Severity of work, as expressed in consumption of parts and replacements of both mechanical equipment and tires.
   5. Fuel and lubricant consumption.
   6. Labor rates, including fringe benefits, for operators, oilers, and maintenance mechanics.

These items are also functions of location, latitude, altitude, weather, and physical characteristics of the job. In the face of these many influencing factors, the estimator uses one or more of the following methods for estimating costs of ownership and operation.
   1. Company records of machinery costs, tempered by adjustment for the particular job conditions.
   2. In the case of a new variation or type of machine, a basic buildup of estimated costs founded on experience with a similar machine.
   3. Local rental schedules for machines, from which costs of ownership and operation may be deduced.
   4. Fundamental buildup in accordance with the material furnished by:
      a. Books and magazines of the construction and mining industries.
      b. Data from the manufacturers of the machines.
      c. Guides and tables for the costs of ownership and operation, as furnished by associations such as the Associated General Contractors of America.

A buildup of costs based on diverse information is illustrated in Tables 8-1 and 8-2, the one for a heavyweight tractor-ripper and the other for an average-size shovel for massive excavation.

In these tables, under "Hourly Cost of Ownership and Operation" is one column headed "Dollars" and one headed "% of $1000s of Cost." The second column translates the dollars of the first column into terms of the cost of the machine. In the case of the tractor ripper depreciation is $17.10 hourly and is 7.1 percent of 240, the number of thousands of dollars of cost.

This method of translating all hourly costs of ownership and operation, except the wages and fringe benefits of operator(s), is used in Table 8-3. The percentage method automatically takes care of inflationary increases in all elements of machinery costs. It is simply based on the principle that all costs are approximately proportional to the total cost of the machine. Errors are in the order of ± 5 percent.

The percentage figures of Table 8-3 are suggested averages for the United States and they are based on:
   1. Average cost of machine, assembled and erected, fob job in 1978.
   2. Average unit prices for diesel fuel, gasoline, electrical energy, lubricants, hydraulic oils, and filters in 1978.
   3. Average unit prices for consumed items such as tires, cutting edges, bits, cables, and the like in 1978.
   4. *Contractors' Equipment Manual, 1974,* issued by the Associated General Contractors of America.
   5. Cost data of construction and mining companies.
   6. Cost estimates of manufacturers of machinery.
   7. Cost data of magazines, books, and other publications.
   8. Cost data gathered by the author over 50 years.

Elements of Hourly Ownership and Operating Cost Tabulation

The elements of Table 8-3 are arranged in working sequence from left to right.
   Machine with Description   Machines, with auxiliary equipment when necessary for work, are arranged alphabetically with notations of capacities and ratings. Capacities and

8-4   Cost of Machinery and Facilities

TABLE 8-1 Estimated Hourly Cost of Ownership and Operation of Heavyweight Tractor-Ripper-Bulldozer of 410-hp Engine and 110,000-lb Weight (1978)

Cost of tractor-ripper-bulldozer, fob job site	$240,000
Less trade-in or salvage value at 20% of cost	48,000
Cost to be depreciated	192,000
Depreciation period of 8 yr of 1400 h/yr or total of 11,200 h of life.	

	Hourly cost of ownership and operation	
	Dollars	% of number of $1000s of cost
Depreciation, $192,000/11,200	17.10	7.1
Interest, 8.0%; taxes, 2.0%; insurance, 2.0%; storage, 1.0%. Total 13% of average investment of $135,000 or $17,600 yearly.		
$17,600/1400	12.60	5.2
Replacement cost escalation 7% of $240,000/1400	12.00	5.0
Total fixed charges	41.70	17.4
Repairs, replacements, and labor:		
Major and minor items	28.80	
Consumed parts, ripper and bulldozer	7.20	
Total	36.00	15.0
Operation, less operator:		
Diesel fuel, 16 gal at $0.40/gal	6.40	
Lubricants, hydraulic oil, filters	1.30	
Oilers and grease truck	2.50	
Total	10.20	4.2
Total hourly cost of ownership and operation, less operator	87.90	36.6

TABLE 8-2 Estimated Hourly Cost of Ownership and Operation of Electric Mining Shovel with 18-yd³ Capacity and 1600-hp Motors (1978)

Cost of shovel, erected, fob job site	$2,400,000
Less trade-in or salvage value at 20% of cost	480,000
Cost to be depreciated	1,920,000
Depreciation period of 20 yr of 4000 h/yr, or total of 80,000 h of life.	

	Hourly cost of ownership and operation	
	Dollars	% of number of $1000s of cost
Depreciation, $1,920,000/80,000	24.00	1.0
Interest, 8.0%; taxes 2.0%; insurance 2.0%. Total 12.0% of average investment of $1,260,000, or $151,200 yearly. $151,200/4000	37.80	1.6
Replacement cost escalation 7% of $2,400,000/4000	42.00	1.8
Total fixed charges	103.80	4.4
Repairs, replacements, and labor	66.00	2.7
Operation, less operator(s):		
Electrical energy, 1069 hp, average consumption 790 kWh @ $0.045	35.60	
Lubricants	3.60	
Total	39.20	1.6
Total hourly cost of ownership and operation, less operator(s)	209.00	8.7

TABLE 8-3 Cost of Ownership and Operation of Excavating Machinery and Facilities

Machinery with description	Weight and power		Depreciation factors			Yearly fixed charges as % of cost				Hourly costs as % of number of $1000s of cost			
	Weight of machine, lb	Delivered horsepower of prime mover(s)	Depreciation period, yr	Hours of use yearly	Depreciation period, total hours	Depreciation	Interest, taxes, insurance, storage	Replacement cost escalation	Total fixed charge	Fixed charges	Repairs, replacements, labor thereto	Operating expenses	Total hourly costs
Air compressors, portable, wheel-tire-mounted:													
Diesel engine:													
125 ft³/min	3,300	40	6	1200	7,200	13.3	7.6	7.0	27.9	23	18	6	47
175 ft³/min	3,800	50	6	1200	7,200	13.3	7.6	7.0	27.9	23	18	7	48
250 ft³/min	4,200	75	6	1600	9,600	13.3	7.6	7.0	27.9	17	18	10	45
365 ft³/min	4,500	110	6	1600	9,600	13.3	7.6	7.0	27.9	17	18	13	48
600 ft³/min	6,800	180	8	1600	12,800	10.0	7.3	7.0	24.3	15	16	14	45
900 ft³/min	10,000	300	8	1600	12,800	10.0	7.3	7.0	24.3	15	16	14	45
1200 ft³/min	14,000	360	8	1600	12,800	10.0	7.3	7.0	24.3	15	16	14	45
1750 ft³/min	23,000	525	8	1600	12,800	10.0	7.3	7.0	24.3	15	16	12	43
2000 ft³/min	28,000	600	8	1600	12,800	10.0	7.3	7.0	24.3	15	16	12	43
Gasoline engine:													
85 ft³/min	2,400	30	6	1100	6,600	13.3	7.6	7.0	27.9	25	20	11	56
125 ft³/min	2,900	40	6	1100	6,600	13.3	7.6	7.0	27.9	25	20	13	58
175 ft³/min	3,400	50	6	1100	6,600	13.3	7.6	7.0	27.9	25	20	16	61
Electric motor:													
900 ft³/min	6,800	300	8	1600	12,800	10.0	7.3	7.0	24.3	15	12	40	67
1200 ft³/min	9,500	360	8	1600	12,800	10.0	7.3	7.0	24.3	15	12	37	64
1750 ft³/min	15,600	525	8	1600	12,800	10.0	7.3	7.0	24.3	15	12	34	61
2000 ft³/min	19,000	600	8	1600	12,800	10.0	7.3	7.0	24.3	15	12	31	58
Air compressors, stationary, skid-mounted:													
Diesel engine:													
900 ft³/min	9,000	300	8	1600	12,800	10.0	7.3	7.0	24.3	15	16	18	49
1200 ft³/min	12,600	360	8	1600	12,800	10.0	7.3	7.0	24.3	15	16	16	47
1750 ft³/min	20,700	525	8	1600	12,800	10.0	7.3	7.0	24.3	15	16	14	45
2000 ft³/min	24,300	600	8	1600	12,800	10.0	7.3	7.0	24.3	15	16	13	44

TABLE 8-3 Cost of Ownership and Operation of Excavating Machinery and Facilities *(Continued)*

Machinery with description	Weight and power		Depreciation factors			Yearly fixed charges as % of cost				Hourly costs as % of number of $1000s of cost			
	Weight of machine, lb	Delivered horsepower of prime mover(s)	Depreciation period, yr	Hours of use yearly	Depreciation period, total hours	Depreciation	Interest, taxes, insurance, storage	Replacement cost escalation	Total fixed charge	Fixed charges	Repairs, replacements, labor thereto	Operating expenses	Total hourly costs
Air compressors, stationary, skid-mounted:													
Electric motor:													
900 ft³/min	6,100	300	8	1600	12,800	10.0	7.3	7.0	24.3	15	12	44	71
1200 ft³/min	8,500	360	8	1600	12,800	10.0	7.3	7.0	24.3	15	12	38	65
1750 ft³/min	14,000	525	8	1600	12,800	10.0	7.3	7.0	24.3	15	12	32	59
2000 ft³/min	21,800	600	8	1600	12,800	10.0	7.3	7.0	24.3	15	12	26	52
Air drills, complete with tools:													
Crawler-mounted:													
3½-in-diameter hole	12,000		6	1600	9,600	13.3	7.6	7.0	27.9	17	35	1	53
4½-in-diameter hole	13,000		6	1600	9,600	13.3	7.6	7.0	27.9	17	35	1	53
5½-in-diameter hole	14,000		6	1600	9,600	13.3	7.6	7.0	27.9	17	35	1	53
Crawler-mounted, with compressor, downhole hammer, and rotary drilling:													
Diesel engine:													
6½-in-diameter hole, 330 ft³/min	43,000	200	8	1600	12,800	10.0	7.3	7.0	24.3	15	22	3	40
7½-in-diameter hole, 330 ft³/min	48,000	300	8	1600	12,800	10.0	7.3	7.0	24.3	15	22	4	41
8½-in-diameter hole, 600 ft³/min	53,000	400	10	1600	16,000	8.0	7.2	7.0	22.2	14	20	5	39
9½-in-diameter hole, 600 ft³/min	58,000	500	10	1600	16,000	8.0	7.2	7.0	22.2	14	20	6	40
Electric motor:													
8½-in-diameter hole, 600 ft³/min	39,700	400	10	1600	16,000	8.0	7.2	7.0	22.2	14	20	12	46
9½-in-diameter hole, 600 ft³/min	43,400	500	10	1600	16,000	8.0	7.2	7.0	22.2	14	20	13	47
Drifters, with feed shell and support column:													
Lightweight	400		3	1200	3,600	26.7	8.6	7.0	42.3	35	50	1	86
Medium-weight	500		3	1200	3,600	26.7	8.6	7.0	42.3	35	50	1	86
Heavyweight	600		3	1200	3,600	26.7	8.6	7.0	42.3	35	50	1	86

Description	C1	C2	C3	C4	C5	C6	C7	C8	C9	C10	C11	C12	C13
Jackhammers:													
Lightweight	30		3	1200	3,600	26.7	8.6	7.0	42.3	35	50	1	86
Medium-weight	45		3	1200	3,600	26.7	8.6	7.0	42.3	35	50	1	86
Heavyweight	60		3	1200	3,600	26.7	8.6	7.0	42.3	35	50	1	86
Air equipment:													
Pipe or hose, with valves and fittings			3	1400	4,200	26.7	8.6	7.0	42.3	30	10	1	41
Bit sharpener and forge			7	1400	9,800	11.4	7.4	7.0	24.8	18	12	1	31
Auger drills, spiral-type, diesel engine, complete with tools:													
Vertical and inclined holes, 3- to 36-in-diameter holes, to 200-ft depth, 140 hp. truck-mounted	35,900	350	10	1200	12,000	8.0	7.2	7.0	22.2	18	15	5	38
Horizontal and inclined holes, 3- to 24-in-diameter holes, to 100-ft length, 150 hp, crawler-mounted	22,700	230	10	1200	12,000	8.0	7.2	7.0	22.2	18	15	5	38
Automobiles and pickup trucks, gasoline engines:													
Lightweight	3,000	100	2	2000	4,000	40.0	9.8	7.0	56.8	28	18	36	82
Medium-weight	4,000	150	3	2000	6,000	26.7	8.6	7.0	42.3	21	15	36	72
Heavyweight	6,000	200	4	2000	8,000	20.0	8.1	7.0	35.1	18	12	36	66
Backhoe excavators, diesel engine:													
Wheel-tire- and crawler-tractor-mounted, with front-end loader:													
½ yd³, 100 hp	26,000	100	8	1400	11,200	10.0	7.3	7.0	24.3	17	14	4	35
1 yd³, 150 hp	40,000	150	8	1400	11,200	10.0	7.3	7.0	24.3	17	14	4	35
Truck-mounted, all diesel engines:													
½ yd³, 250 hp	36,000	250	8	1400	11,200	10.0	7.3	7.0	24.3	17	12	6	35
1 yd³, 350 hp	52,000	350	8	1400	11,200	10.0	7.3	7.0	24.3	17	12	5	34
2 yd³, 450 hp	74,000	450	8	1400	11,200	10.0	7.3	7.0	24.3	17	12	4	33
Crawler-mounted, full revolving type:													
1 yd³, 100 hp	69,000	100	10	1400	14,000	8.0	7.2	7.0	22.2	16	5	3	24
2 yd³, 150 hp	113,000	150	10	1400	14,000	8.0	7.2	7.0	22.2	16	4	2	21
3 yd³, 200 hp	174,000	200	12	1400	16,800	6.7	7.0	7.0	20.7	15	4	2	21
4 yd³, 250 hp	201,000	250	14	1400	19,600	5.7	7.0	7.0	19.7	14	3	1	18
5 yd³, 300 hp	263,000	300	16	1400	22,400	5.0	6.9	7.0	18.9	14	3	1	18

TABLE 8-3 Cost of Ownership and Operation of Excavating Machinery and Facilities *(Continued)*

Machinery with description	Weight and power		Depreciation factors			Yearly fixed charges as % of cost				Hourly costs as % of number of $1000s of cost			
	Weight of machine, lb	Delivered horsepower of prime mover(s)	Depreciation period, yr	Hours of use yearly	Depreciation period, total hours	Depreciation	Interest, taxes, insurance, storage	Replacement cost escalation	Total fixed charge	Fixed charges	Repairs, replacements, labor thereto	Operating expenses	Total hourly costs
Belt conveyors, with rollers, frame, supports, and power:													
Portable, stackers, 100-ft long units. Belt speed 400 ft/min, belt angle 15–26°:													
Diesel engine:													
36-in belt width, 45 hp	12,400	45	8	1400	11,200	10.0	7.3	7.0	24.3	17	14	3	34
48-in belt width, 80 hp	24,300	80	8	1400	11,200	10.0	7.3	7.0	24.3	17	14	3	34
60-in belt width, 125 hp	43,100	125	8	1400	11,200	10.0	7.3	7.0	24.3	17	14	2	33
72-in belt width, 180 hp	61,900	180	8	1400	11,200	10.0	7.3	7.0	24.3	17	14	2	33
Electric motor:													
36-in belt width, 45 hp	12,000	45	10	1400	14,000	8.0	7.2	7.0	22.2	16	12	7	35
48-in belt width, 80 hp	23,000	80	10	1400	14,000	8.0	7.2	7.0	22.2	16	12	6	34
60-in belt width, 125 hp	41,000	125	10	1400	14,000	8.0	7.2	7.0	22.2	16	12	5	33
72-in belt width, 180 hp	59,000	180	10	1400	14,000	8.0	7.2	7.0	22.2	16	12	5	33
Stationary, 100-ft-long units. Belt speed 400 ft/min, belt angle 3°:													
Diesel engine:													
36-in belt width, 15 hp	11,000	15	8	1400	11,200	10.0	7.3	7.0	24.3	17	14	2	33
48-in belt width, 30 hp	20,000	30	8	1400	11,200	10.0	7.3	7.0	24.3	17	14	2	33
60-in belt width, 45 hp	37,000	45	8	1400	11,200	10.0	7.3	7.0	24.3	17	14	1	32
72-in belt width, 60 hp	53,000	60	8	1400	11,200	10.0	7.3	7.0	24.3	17	14	1	32
Electric motor:													
36-in belt width, 15 hp	10,000	15	10	1400	14,000	8.0	7.2	7.0	22.2	16	12	3	31
48-in belt width, 30 hp	18,000	30	10	1400	14,000	8.0	7.2	7.0	22.2	16	12	3	31
60-in belt width, 45 hp	35,000	45	10	1400	14,000	8.0	7.2	7.0	22.2	16	12	2	30
72-in belt width, 60 hp	50,000	60	10	1400	14,000	8.0	7.2	7.0	22.2	16	12	2	30

Belt loaders, diesel engines:										
Movable, stub-type:										
Belt 48 in wide and 60 ft long. 120 hp	52,000	120	10	1200	12,000	8.0	7.2	7.0	22.2	18	16	3	37
Belt 60 in wide and 60 ft long. 180 hp	60,000	180	10	1200	12,000	8.0	7.2	7.0	22.2	18	16	3	37
Belt 72 in wide and 60 ft long. 260 hp	70,000	260	10	1200	12,000	8.0	7.2	7.0	22.2	18	16	4	38
Mobile:										
Pull type, by crawler-tractor:										
Wheel-tire-mounted. Belt 48 in wide and 40 ft long. 120 hp.	27,000	120	10	1200	12,000	8.0	7.2	7.0	22.2	18	16	5	39
Crawler-mounted. Belt 60 in wide and 40 ft long. 300 hp.	90,000	300	10	1200	12,000	8.0	7.2	7.0	22.2	18	16	5	39
Self-propelled type, crawler-tractor-mounted:										
Belt 72 in wide and 40 ft long. Complete with two tractors. Loader hp 550. Tractor hp 820. Total hp 1370.	24,2000	1370	10	1200	12,000	8.0	7.2	7.0	22.2	18	16	5	39
Brush burners, portable, wheel-tire-mounted, gasoline engine:										
Medium-duty, 3 hp, 30-gal/h fuel consumption	500	3	4	1400	5,600	20.0	8.1	7.0	35.1	25	16	285	326
Heavy-duty, 12 hp, 60-gal/h fuel consumption	1,500	12	4	1400	5,600	20.0	8.1	7.0	35.1	25	16	233	274
Bucket drills, truck-mounted, with tools, diesel engines:										
$1/2$-yd^3 bucket, to 48-in-diameter and 100-ft-deep hole, 250 total combined hp	23,000	250	10	1200	12,000	8.0	7.2	7.0	22.2	18	14	5	37
1-yd^3 bucket, to 120-in-diameter and 150-ft-deep hole, 350 total combined hp	36,000	350	10	1200	12,000	8.0	7.2	7.0	22.2	18	14	5	37
$1 1/2$-yd^3 bucket, to 120-in-diameter and 200-ft-deep hole, 450 total combined hp	48,000	450	10	1200	12,000	8.0	7.2	7.0	22.2	18	14	4	36
Buckets:										
Clamshell, heavy-duty, and grapple, heavy-duty:										
2 yd^3	11,500		6	1600	9,600	13.3	7.6	7.0	27.9	17	16	1	34
3 yd^3	16,000		6	1600	9,600	13.3	7.6	7.0	27.9	17	16	1	34
4 yd^3	21,000		6	1600	9,600	13.3	7.6	7.0	27.9	17	16	1	34
5 yd^3	27,000		6	1600	9,600	13.3	7.6	7.0	27.9	17	16	1	34
6 yd^3	34,000		6	1600	9,600	13.3	7.6	7.0	27.9	17	16	1	34

TABLE 8-3 Cost of Ownership and Operation of Excavating Machinery and Facilities *(Continued)*

Machinery with description	Weight and power			Depreciation factors			Yearly fixed charges as % of cost				Hourly costs as % of number of $1000s of cost			
	Weight of machine, lb	Delivered horsepower of prime mover(s)	Depreciation period, yr	Hours of use yearly	Depreciation period, total hours	Depreciation	Interest, taxes, insurance, storage	Replacement cost escalation	Total fixed charge	Fixed charges	Repairs, replacements, labor thereto	Operating expenses	Total hourly costs
Buckets:														
Dragline, heavy-duty:														
2 yd³	6,000		8	1600	12,800	10.0	7.3	7.0	24.3	15	14	1	30	
3 yd³	8,000		8	1600	12,800	10.0	7.3	7.0	24.3	15	14	1	30	
4 yd³	10,000		8	1600	12,800	10.0	7.3	7.0	24.3	15	14	1	30	
5 yd³	12,000		8	1600	12,800	10.0	7.3	7.0	24.3	15	14	1	30	
6 yd³	15,000		8	1600	12,800	10.0	7.3	7.0	24.3	15	14	1	30	
Buildings, portable, wheels-tires-mounted:														
Office, 60 × 12 ft	7,000		5	2000	10,000	16.0	7.8	7.0	30.8	15	2	5	22	
Bunk house, 60 × 24 ft	14,000		5	3800	19,000	16.0	7.8	7.0	30.8	8	2	5	15	
Mess hall, 60 × 24 ft	22,000		5	3800	19,000	16.0	7.8	7.0	30.8	8	2	5	15	
Machine shop, 60 × 24 ft	10,000		5	2900	14,500	16.0	7.8	7.0	30.8	11	2	5	18	
Tool house, 60 × 12 ft	5,000		5	2900	14,500	16.0	7.8	7.0	30.8	11	2	1	14	
Powder magazine, 12 × 12 ft	2,000		5	2000	10,000	16.0	7.8	7.0	30.8	15	1	1	17	
Cable drill, complete with tools:														
Truck-mounted, diesel engines, 3- to 6-in-diameter, up to 700-ft-deep holes, 225 total combined hp	17,000		10	1200	12,000	8.0	7.2	7.0	22.2	18	12	6	36	
Cars, steel, rail, side-dump, quarry-type:														
10 cysm, 14 tons	14,000		15	2000	30,000	5.3	6.9	7.0	19.2	10	12	1	23	
20 cysm, 27 tons	27,000		15	2000	30,000	5.3	6.9	7.0	19.2	10	12	1	23	
40 cysm, 54 tons	54,000		15	2000	30,000	5.3	6.9	7.0	19.2	10	12	1	23	
60 cysm, 81 tons	81,000		15	2000	30,000	5.3	6.9	7.0	19.2	10	12	1	23	

Description											
Chippers, log, portable, wheel-tire-mounted, diesel engine:											
8 tons/h of chips, 130 hp	16,000	130	8	1400	11,200	10.0	7.3	7.0	24.3	18	12	6	36
20 tons/h of chips, 290 hp	45,000	290	8	1400	11,200	10.0	7.3	7.0	24.3	18	12	5	35
Cleaning equipment, steam, portable, wheel-tire-mounted, gasoline engine:											
100 gal/h, water to steam, 3 hp, 2 gal/h fuel oil	500	3	5	1800	9,000	16.0	7.8	7.0	30.8	17	15	75	107
200 gal/h, water to steam, 4 hp, 4 gal/h fuel oil	1,100	4	5	1800	9,000	16.0	7.8	7.0	30.8	17	15	47	79
400 gal/h, water to steam, 5 hp, 8 gal/h fuel oil	2,400	5	6	1800	10,800	13.3	7.6	7.0	27.9	16	15	46	77
600 gal/h, water to steam, 10 hp, 12 gal/h fuel oil	3,600	10	6	1800	10,800	13.3	7.6	7.0	27.9	16	15	43	74
Core drills, complete with tools, diesel engine:											
Diamond-type, 1- to 6-in-diameter and up to 1500-ft-deep hole:											
Crawler-mounted, 50 hp	18,000	50	8	1400	11,200	10.0	7.3	7.0	24.3	18	22	2	42
Truck-mounted, 200 total combined hp	15,000	200	8	1400	11,200	10.0	7.3	7.0	24.3	18	22	5	45
Skid-mounted, 50 hp	4,000	50	8	1400	11,200	10.0	7.3	7.0	24.3	18	35	5	58
Shot or calyx type, 2- to 20-in diameter and up to 600-ft-deep hole:											
Crawler-mounted, 60 hp	21,000	60	8	1400	11,200	10.0	7.3	7.0	24.3	18	22	2	42
Truck-mounted, 210 total combined hp	18,000	210	8	1400	11,200	10.0	7.3	7.0	24.3	18	22	4	44
Skid-mounted, 60 hp	7,000	60	8	1400	11,200	10.0	7.3	7.0	24.3	18	35	6	59
Crane excavators, complete with bucket or grapple, diesel engine:											
Crawler-mounted:											
2 yd³, 40 tons, 150 hp	102,000	150	10	1600	16,000	8.0	7.2	7.0	22.2	14	6	2	22
3 yd³, 50 tons, 200 hp	154,000	200	12	1600	19,200	6.7	7.0	7.0	20.7	13	5	2	20
4 yd³, 60 tons, 250 hp	178,000	250	14	1700	23,800	5.7	7.0	7.0	19.7	12	4	2	18
5 yd³, 70 tons, 300 hp	215,000	300	16	1700	27,200	5.0	6.9	7.0	18.9	11	3	2	16
Truck-mounted:											
2 yd³, 30 tons, 10-ft radius, 400 total hp	62,000	400	10	1400	14,000	8.0	7.2	7.0	22.2	16	6	3	25
3 yd³, 60 tons 12 ft radius, 500 total hp	90,000	500	12	1400	16,800	6.7	7.0	7.0	20.7	15	5	3	23
4 yd³, 90 tons 12 ft radius, 600 total hp	118,000	600	14	1600	22,400	5.7	7.0	7.0	19.7	12	4	3	19
5 yd³, 120 tons 16 ft radius, 700 total hp	147,000	700	16	1600	25,600	5.0	6.9	7.0	18.9	12	3	3	18

TABLE 8-3 Cost of Ownership and Operation of Excavating Machinery and Facilities *(Continued)*

Machinery with description	Weight and power		Depreciation factors			Yearly fixed charges as % of cost				Hourly costs as % of number of $1000s of cost			
	Weight of machine, lb	Delivered horsepower of prime mover(s)	Depreciation period, yr	Hours of use yearly	Depreciation period, total hours	Depreciation	Interest, taxes, insurance, storage	Replacement cost escalation	Total fixed charge	Fixed charges	Repairs, replacements, labor thereto	Operating expenses	Total hourly costs
Crushers, primary, complete, electric motors:													
Jaw-type, with pan-feed:													
48 × 42 in, 220 tons/h, 200 total hp	207,000	200	10	1600	16,000	8.0	7.2	7.0	22.2	14	11	3	28
60 × 48 in, 340 tons/h, 270 total hp	361,000	270	10	1600	16,000	8.0	7.2	7.0	22.2	14	11	3	28
84 × 66 in, 500 tons/h, 330 total hp	469,000	330	10	1600	16,000	8.0	7.2	7.0	22.2	14	11	3	28
Gyratory type:													
36 in, 480 tons/h, 225 hp	180,000	225	15	1600	24,000	5.3	6.9	7.0	19.2	12	15	4	31
54 in, 1250 tons/h, 350 hp	280,000	350	15	1600	24,000	5.3	6.9	7.0	19.2	12	15	4	31
72 in, 2500 tons/h, 500 hp	540,000	500	15	1600	24,000	5.3	6.9	7.0	19.2	12	15	3	30
Cutters, stump, portable, wheels—tires-mounted, gasoline engine:													
10-in depth, 37 hp	3,200	37	8	1400	11,200	10.0	7.3	7.0	24.3	18	12	16	46
24-in depth, 65 hp	5,200	65	8	1400	11,200	10.0	7.3	7.0	24.3	18	12	19	49
Dragline excavators, complete with bucket:													
Crawler-mounted:													
Diesel engine:													
2 yd³, 150 hp	100,000	150	10	1600	16,000	8.0	7.2	7.0	22.2	14	8	2	24
3 yd³, 200 hp	140,000	200	12	1600	19,200	6.7	7.0	7.0	20.7	13	7	2	22
4 yd³, 250 hp	185,000	250	14	1700	23,800	5.7	7.0	7.0	19.7	12	6	2	20
5 yd³, 300 hp	225,000	300	16	1700	27,200	5.0	6.9	7.0	18.9	11	5	2	18
6 yd³, 350 hp	270,000	350	18	1700	30,600	4.4	6.9	7.0	18.3	11	4	2	17
Electric motors:													
2 yd³, 150 hp	98,000	150	10	1600	16,000	8.0	7.2	7.0	22.2	14	7	3	24
3 yd³, 200 hp	137,000	200	12	1600	19,200	6.7	7.0	7.0	20.7	13	6	3	22
4 yd³, 250 hp	181,000	250	14	1700	23,800	5.7	7.0	7.0	19.7	12	5	3	20
5 yd³, 300 hp	221,000	300	16	1700	27,200	5.0	6.9	7.0	18.9	11	4	3	18
6 yd³, 350 hp	265,000	350	18	1700	30,600	4.4	6.9	7.0	18.3	11	3	3	17

Equipment										
Pontoon-mounted, walker-type:										
Diesel engine, electric motors, or diesel-electric drive:										
8 yd³, 700 hp	750,000	700	20	4000	80,000	4.0	6.8	7.0	17.8	4	3	1	8
12 yd³, 1100 hp	1,150,000	1150	20	4000	80,000	4.0	6.8	7.0	17.8	4	3	1	8
18 yd³, 1600 hp	1,700,000	1600	20	4000	80,000	4.0	6.8	7.0	17.8	4	3	1	8
Electric motors-generators-motors, ac to dc:										
18 yd³, 1600 hp	1,700,000	1600	20	7000	140,000	4.0	6.8	7.0	17.8	3	3	2	8
27 yd³, 4200 hp	2,600,000	4200	20	7000	140,000	4.0	6.8	7.0	17.8	3	3	2	8
36 yd³, 5600 hp	3,400,000	5600	20	7000	140,000	4.0	6.8	7.0	17.8	3	3	2	8
54 yd³, 8400 hp	5,100,000	8400	30	7000	210,000	2.7	6.7	7.0	16.4	3	2	2	7
72 yd³, 11,000 hp	6,800,000	11000	30	7000	210,000	2.7	6.7	7.0	16.4	3	2	2	7
108 yd³, 16,800 hp	10,900,000	16800	30	7000	210,000	2.7	6.7	7.0	16.4	3	2	2	7
Dredges, hydraulic, diesel engines, complete:										
12-in pump, 725 hp; auxiliary engine 335 hp	140,000	1060	12	2800	33,600	6.7	7.0	7.0	20.7	7	6	4	17
18-in pump, 1125 hp; auxiliary engine 335 hp	460,000	1460	16	2800	44,800	5.0	6.9	7.0	18.9	7	6	4	17
24-in pump, 2875 hp; auxiliary engine 800 hp	670,000	3680	20	2800	56,000	4.0	6.8	7.0	17.8	6	6	4	16
Engines, clutch, skid-mounted:										
Diesel:										
10 hp	150	10	5	1200	6,000	16.0	7.8	7.0	30.8	26	24	34	84
20 hp	300	20	5	1200	6,000	16.0	7.8	7.0	30.8	26	24	34	84
40 hp	600	40	5	1200	6,000	16.0	7.8	7.0	30.8	26	18	34	78
60 hp	900	60	5	1200	6,000	16.0	7.8	7.0	30.8	26	18	34	78
100 hp	1,400	100	6	1200	7,200	13.3	7.6	7.0	27.9	23	16	34	73
200 hp	2,900	200	6	1200	7,200	13.3	7.6	7.0	27.9	23	16	34	73
300 hp	4,400	300	6	1200	7,200	13.3	7.6	7.0	27.9	23	16	34	73
400 hp	5,800	400	6	1200	7,200	13.3	7.3	7.0	27.9	23	16	34	73
600 hp	8,700	600	8	1200	9,600	10.0	7.3	7.0	24.3	20	14	34	68
800 hp	11,600	800	8	1200	9,600	10.0	7.3	7.0	24.3	20	14	34	68
1000 hp	14,500	1000	8	1200	9,600	10.0	7.3	7.0	24.3	20	14	34	68
1200 hp	17,400	1200	10	1200	12,000	8.0	7.2	7.0	22.2	18	12	34	64
1600 hp	23,200	1600	10	1200	12,000	8.0	7.2	7.0	22.2	18	12	34	64

TABLE 8-3 Cost of Ownership and Operation of Excavating Machinery and Facilities *(Continued)*

Machinery with description	Weight and power		Depreciation factors			Yearly fixed charges as % of cost				Hourly costs as % of number of $1000s of cost				
	Weight of machine, lb	Delivered horsepower of prime mover(s)	Depreciation period, yr	Hours of use yearly	Depreciation period, total hours	Depreciation	Interest, taxes, insurance, storage	Replacement cost escalation	Total fixed charge	Fixed charges	Repairs, replacements, labor thereto	Operating expenses	Total hourly costs
Engines, clutch, skid-mounted:													
Gasoline or gas:													
10 hp	100	10	5	1200	6,000	16.0	7.8	7.0	30.8	26	26	121	173
20 hp	200	20	5	1200	6,000	16.0	7.8	7.0	30.8	26	26	121	173
40 hp	400	40	5	1200	6,000	16.0	7.8	7.0	30.8	26	20	121	167
60 hp	600	60	5	1200	6,000	16.0	7.8	7.0	30.8	26	20	121	167
100 hp	1,000	100	6	1200	7,200	13.3	7.6	7.0	27.9	23	18	121	162
200 hp	2,000	200	6	1200	7,200	13.3	7.6	7.0	27.9	23	18	121	162
300 hp	3,000	300	6	1200	7,200	13.3	7.6	7.0	27.9	23	18	121	162
Explosives loading truck, bulk ammonium nitrate and fuel oil mixture, pneumatic- or auger-type, diesel engine:													
5 tons, 2 axles, 150 hp	14,000	150	8	1600	12,800	10.0	7.3	7.0	24.3	15	12	5	32
12 tons, 3 axles, 250 hp	22,000	250	8	1600	12,800	10.0	7.3	7.0	24.3	15	12	5	32
Floodlights, portable, wheel-tire mounting, 30-ft tower:													
Diesel engine-generator:													
5 kW, 10 hp	2,200	10	5	1200	6,000	16.0	7.8	7.0	30.8	26	20	2	48
10 kW, 20 hp	2,600	20	5	1200	6,000	16.0	7.8	7.0	30.8	26	20	3	49
15 kW, 30 hp	3,100	30	5	1200	6,000	16.0	7.8	7.0	30.8	26	20	4	50
Gasoline engine-generator:													
5 kW, 10 hp	2,200	10	5	1200	6,000	16.0	7.8	7.0	30.8	26	20	5	51
10 kW, 20 hp	2,600	20	5	1200	6,000	16.0	7.8	7.0	30.8	26	20	9	55
15 kW, 30 hp	3,700	30	5	1200	6,000	16.0	7.8	7.0	30.8	26	20	13	59

Fusion piercing drill, diesel engine, complete with delivered oxygen, fuel oil, and water: 12,000 ft³/h of oxygen, 200 hp

	120,000	200	20	4000	80,000	4.0	6.8	7.0	17.8	4	3	5	12
Generators, electric, skid-mounted:													
Diesel engine:													
5 kW, 8 hp	150	8	5	1200	6,000	16.0	7.8	7.0	30.8	26	20	30	76
10 kW, 15 hp	250	15	5	1200	6,000	16.0	7.8	7.0	30.8	26	20	30	76
20 kW, 30 hp	500	30	5	1200	6,000	16.0	7.8	7.0	30.8	26	20	30	76
40 kW, 60 hp	1,000	40	5	1200	6,000	16.0	7.8	7.0	30.8	26	20	30	76
60 kW, 90 hp	1,500	90	5	1200	6,000	16.0	7.8	7.0	30.8	26	20	30	76
100 kW, 150 hp	2,500	150	6	1400	8,400	13.3	7.6	7.0	27.9	20	18	30	68
200 kW, 300 hp	5,000	300	6	1400	8,400	13.3	7.6	7.0	27.9	20	18	30	68
300 kW, 450 hp	7,000	450	6	1400	8,400	13.3	7.6	7.0	27.9	20	18	30	68
400 kW, 600 hp	9,500	600	8	1600	10,800	10.0	7.3	7.0	24.3	15	16	30	61
500 kW, 750 hp	12,000	750	8	1600	10,800	10.0	7.3	7.0	24.3	15	16	30	61
600 kW, 900 hp	14,400	900	8	1600	10,800	10.0	7.3	7.0	24.3	15	16	30	61
800 kW, 1200 hp	19,100	1200	10	1600	16,000	8.0	7.2	7.0	22.2	14	14	30	58
1000 kW, 1500 hp	24,000	1500	10	1600	16,000	8.0	7.2	7.0	22.2	14	14	30	58
1200 kW, 1800 hp	28,900	1800	10	1600	16,000	8.0	7.2	7.0	22.2	14	14	30	58
Gasoline engine:													
5 kW, 8 hp	100	8	5	1200	6,000	16.0	7.8	7.0	30.8	26	22	87	135
10 kW, 15 hp	200	15	5	1200	6,000	16.0	7.8	7.0	30.8	26	22	87	135
20 kW, 30 hp	450	30	5	1200	6,000	16.0	7.8	7.0	30.8	26	22	87	135
40 kW, 60 hp	950	60	5	1200	6,000	16.0	7.8	7.0	30.8	26	22	87	135
60 kW, 90 hp	1,400	90	5	1200	6,000	16.0	7.8	7.0	30.8	26	22	87	135
100 kW, 150 hp	2,200	150	6	1400	8,400	13.3	7.6	7.0	27.9	20	20	87	127
Graders:													
Drag, tow-type, by tractor, wheel-tire-mounted:													
12-ft width, 6 cysm.	12,000		8	1400	11,200	10.0	7.3	7.0	24.3	17	8	2	27
16-ft width, 8 cysm.	13,000		8	1400	11,200	10.0	7.3	7.0	24.3	17	8	2	27
20-ft width, 10 cysm	14,500		8	1400	11,200	10.0	7.3	7.0	24.3	17	8	2	27
Motor grader, 3 axles, with scarifier, diesel engine:													
12-ft blade length, 180 hp	36,000	180	8	1400	11,200	10.0	7.3	7.0	24.3	17	10	7	34
14-ft blade length, 210 hp	45,000	210	8	1400	11,200	10.0	7.3	7.0	24.3	17	10	6	33
16-ft blade length, 240 hp	60,000	240	8	1400	11,200	10.0	7.3	7.0	24.3	17	10	5	32

8–15

TABLE 8-3 Cost of Ownership and Operation of Excavating Machinery and Facilities *(Continued)*

Machinery with description	Weight and power — Weight of machine, lb	Weight and power — Delivered horsepower of prime mover(s)	Depreciation factors — Depreciation period, yr	Depreciation factors — Hours of use yearly	Depreciation factors — Depreciation period, total hours	Yearly fixed charges as % of cost — Depreciation	Yearly fixed charges as % of cost — Interest, taxes, insurance, storage	Yearly fixed charges as % of cost — Replacement cost escalation	Yearly fixed charges as % of cost — Total fixed charge	Hourly costs as % of $1000s of cost — Fixed charges	Hourly costs as % of $1000s of cost — Repairs, replacements, labor thereto	Hourly costs as % of $1000s of cost — Operating expenses	Hourly costs as % of $1000s of cost — Total hourly costs
Loaders, bucket type, diesel engine													
Crawler-mounted, with ripper:													
2 cysm, 140 hp	34,000	140	8	1400	11,200	10.0	7.3	7.0	24.3	17	12	5	34
3 cysm, 190 hp	54,000	190	8	1400	11,200	10.0	7.3	7.0	24.3	17	12	5	34
4 cysm, 280 hp	82,000	280	8	1400	11,200	10.0	7.3	7.0	24.3	17	12	5	34
Wheel-tiremounted:													
2 cysm, 100 hp	21,000	100	8	1400	11,200	10.0	7.3	7.0	24.3	17	12	6	35
3 cysm, 130 hp	27,000	130	8	1400	11,200	10.0	7.3	7.0	24.3	17	12	6	35
4 cysm, 170 hp	36,000	170	8	1400	11,200	10.0	7.3	7.0	24.3	17	12	6	35
5 cysm, 260 hp	50,000	260	8	1400	11,200	10.0	7.3	7.0	24.3	17	12	6	35
6 cysm, 325 hp	74,000	325	8	1400	11,200	10.0	7.3	7.0	24.3	17	12	5	34
8 cysm, 410 hp	106,000	410	8	1400	11,200	10.0	7.3	7.0	24.3	17	12	5	34
10 cysm, 550 hp	139,000	550	8	1400	11,200	10.0	7.3	7.0	24.3	17	12	5	34
Locomotives, rail, diesel engine:													
15 tons, 180 hp	30,000	180	12	2000	24,000	6.7	7.0	7.0	20.7	15	10	6	31
25 tons, 300 hp	50,000	300	12	2000	24,000	6.7	7.0	7.0	20.7	15	10	6	31
45 tons, 540 hp	90,000	540	16	2000	32,000	5.0	6.9	7.0	18.9	14	8	6	28
65 tons, 780 hp	130,000	780	16	2000	32,000	5.0	6.9	7.0	18.9	14	8	6	28
120 tons, 1400 hp	240,000	1400	16	2000	32,000	5.0	6.9	7.0	18.9	14	8	6	28
Monitors, hydraulicking, complete with stands:													
1½-in discharge-nozzle diameter, 3-in intake pipe diameter	4,000		8	1400	11,200	10.0	7.3	7.0	24.3	17	8	1	26
2-in discharge-nozzle diameter, 4-in intake pipe diameter	5,000		8	1400	11,200	10.0	7.3	7.0	24.3	17	7	1	25
3-in discharge-nozzle diameter, 6-in intake pipe diameter	7,000		8	1400	11,200	10.0	7.3	7.0	24.3	17	6	1	24

Motors, electric, with controls, 1800 r/min:													
10 hp	200	10	6	1400	8,400	13.3	7.6	7.0	27.9	20	12	119	151
20 hp	400	20	6	1400	8,400	13.3	7.6	7.0	27.9	20	12	119	151
40 hp	700	40	6	1400	8,400	13.3	7.6	7.0	27.9	20	12	119	151
60 hp	1,100	60	6	1400	8,400	13.3	7.6	7.0	27.9	20	12	119	151
100 hp	1,800	100	8	1400	11,200	10.0	7.3	7.0	24.3	17	10	109	136
200 hp	2,600	200	8	1400	11,200	10.0	7.3	7.0	24.3	17	10	109	136
300 hp	3,400	300	8	1400	11,200	10.0	7.3	7.0	24.3	17	10	109	136
400 hp	3,800	400	8	1400	11,200	10.0	7.3	7.0	24.3	17	10	109	136
600 hp	5,700	600	10	1400	14,000	8.0	7.2	7.0	22.2	16	8	99	123
800 hp	7,600	800	10	1400	14,000	8.0	7.2	7.0	22.2	16	8	99	123
1000 hp	9,600	1000	10	1400	14,000	8.0	7.2	7.0	22.2	16	8	99	123
1200 hp	11,500	1200	10	1400	14,000	8.0	7.2	7.0	22.2	16	8	99	123
Pipe, steel, with fittings, 10 gauge, 100-ft units of length:													
2-in diameter	300		3	1400	4200	26.7	8.6	7.0	42.3	30	5	0	35
4-in diameter	600		3	1400	4200	26.7	8.6	7.0	42.3	30	5	0	35
6-in diameter	900		3	1400	4200	26.7	8.6	7.0	42.3	30	5	0	35
8-in diameter	1200		4	1400	5600	20.0	8.1	7.0	35.1	25	5	0	30
10-in diameter	1500		4	1400	5600	20.0	8.1	7.0	35.1	25	5	0	30
12-in diameter	1800		4	1400	5600	20.0	8.1	7.0	35.1	25	5	0	30
Pumps, complete, with power: Centrifugal, sump, portable, wheels-tires-mounted:													
Diesel engine:													
2 in, 10M, 7 hp	250	7	6	1200	7200	13.3	7.6	7.0	27.9	23	20	6	49
3 in, 20M, 10 hp	400	10	6	1200	7200	13.3	7.6	7.0	27.9	23	20	7	50
4 in, 40M, 20 hp	1000	20	6	1200	7200	13.3	7.6	7.0	27.9	23	20	8	51
6 in, 90M, 40 hp	1800	40	8	1200	9600	10.0	7.3	7.0	24.3	20	16	9	45
8 in, 125M, 80 hp	4200	80	8	1200	9600	10.0	7.3	7.0	24.3	20	16	10	46
10 in, 240M, 120 hp	4900	120	8	1200	9600	10.0	7.3	7.0	24.3	20	16	12	48
Gasoline engine:													
2 in, 10M, 7 hp	200	7	6	1200	7200	13.3	7.6	7.0	27.9	23	20	17	60
3 in, 20M, 10 hp	350	10	6	1200	7200	13.3	7.6	7.0	27.9	23	20	20	63
4 in, 40M, 20 hp	850	20	6	1200	7200	13.3	7.6	7.0	27.9	23	20	22	65
6 in, 90M, 40 hp	1700	40	8	1200	9600	10.0	7.3	7.0	24.3	20	16	25	61

TABLE 8-3 Cost of Ownership and Operation of Excavating Machinery and Facilities (Continued)

Machinery with description	Weight and power		Depreciation factors			Yearly fixed charges as % of cost				Hourly costs as % of number of $1000s of cost			
	Weight of machine, lb	Delivered horsepower of prime mover(s)	Depreciation period, yr	Hours of use yearly	Depreciation period, total hours	Depreciation	Interest, taxes, insurance, storage	Replacement cost escalation	Total fixed charge	Fixed charges	Repairs, replacements, labor thereto	Operating expenses	Total hourly costs
Pumps, complete, with power:													
Centrifugal, sump, portable, wheels-tires-mounted:													
Electric motor:													
2 in, 10M, 7 hp	150	7	6	1200	7200	13.3	7.6	7.0	27.9	23	16	23	62
3 in, 20M, 10 hp	250	10	6	1200	7200	13.3	7.6	7.0	27.9	23	16	26	65
4 in, 40M, 20 hp	500	20	6	1200	7200	13.3	7.6	7.0	27.9	23	16	29	68
6 in, 90M, 40 hp	900	40	8	1200	9600	10.0	7.3	7.0	24.3	20	12	32	64
8 in, 125M, 80 hp	2100	80	8	1200	9600	10.0	7.3	7.0	24.3	20	12	35	67
10 in, 240M, 120 hp	2500	120	8	1200	9600	10.0	7.3	7.0	24.3	20	12	38	70
Centrifugal, submersible:													
Electric motor:													
2 in, 10M, 4 hp	50	4	6	1200	7200	13.3	7.6	7.0	27.9	23	20	17	60
3 in, 20M, 10 hp	100	10	6	1200	7200	13.3	7.6	7.0	17.9	23	20	19	62
4 in, 40M, 20 hp	200	20	6	1200	7200	13.3	7.6	7.0	17.9	23	20	21	64
6 in, 90M, 40 hp	400	40	8	1200	9600	10.0	7.3	7.0	24.3	20	16	23	59
Diaphragm, portable, wheel-tire-mounted:													
Diesel engine:													
2 in, 2M, 3 hp	200	3	6	1200	7200	13.3	7.6	7.0	27.9	23	20	3	46
3 in, 3M, 6 hp	350	6	6	1200	7200	13.3	7.6	7.0	27.9	23	20	5	48
4 in, 6M, 10 hp	500	10	6	1200	7200	13.3	7.6	7.0	27.9	23	20	6	49
Gasoline engine:													
2 in, 2M, 3 hp	150	3	6	1200	7200	13.3	7.6	7.0	27.9	23	20	9	52
3 in, 3M, 6 hp	250	6	6	1200	7200	13.3	7.6	7.0	27.9	23	20	16	59
4 in, 6M, 10 hp	350	10	6	1200	7200	13.3	7.6	7.0	27.9	23	20	18	61
Electric motor:													
2 in, 2M, 3 hp	100	1	6	1200	7200	13.3	7.6	7.0	27.9	23	16	13	52
3 in, 3M, 6 hp	200	2	6	1200	7,200	13.3	7.6	7.0	27.9	23	16	16	55
4 in, 6M, 10 hp	350	3	6	1200	7,200	13.3	7.6	7.0	17.9	23	16	18	57

Description											
Turbine, well-type, stationary, complete, head of 500 ft:											
Diesel engine:											
8 in, 500 gal/min 100 hp	4,000	100	8	1200	9,600	10.0	7.3	7.0	24.3	20	12	9	41
10 in, 1000 gal/min 200 hp	8,000	200	8	1200	9,600	10.0	7.3	7.0	24.3	20	12	9	41
12 in, 1500 gal/min 300 hp	15,000	300	8	1200	9,600	10.0	7.3	7.0	24.3	20	10	7	37
14 in, 2000 gal/min 400 hp	22,000	400	8	1200	9,600	10.0	7.3	7.0	24.3	20	10	7	37
16 in, 2500 gal/min 500 hp	30,000	500	8	1200	9,600	10.0	7.3	7.0	24.3	20	10	6	36
Electric motor:											
8 in, 500 gal/min 100 hp	3,000	100	10	1200	12,000	8.0	7.2	7.0	22.2	18	8	24	50
10 in, 1000 gal/min 200 hp	6,000	200	10	1200	12,000	8.0	7.2	7.0	22.2	18	8	20	46
12 in, 1500 gal/min 300 hp	12,000	300	10	1200	12,000	8.0	7.2	7.0	22.2	18	8	17	43
14 in, 2000 gal/min 400 hp	18,000	400	10	1200	12,000	8.0	7.2	7.0	22.2	18	8	15	41
16 in, 2500 gal/min 500 hp	24,000	500	10	1200	12,000	8.0	7.2	7.0	22.2	18	8	14	40
Turbine, booster-type, stationary, skid-mounted, complete, head of 500 ft:											
Diesel engine:											
8 in, 500 gal/min 100 hp	3,000	100	8	1200	9,600	10.0	7.3	7.0	24.3	20	12	12	44
10 in, 1000 gal/min 200 hp	6,000	200	8	1200	9,600	10.0	7.3	7.0	24.3	20	12	12	44
12 in, 1500 gal/min 300 hp	10,000	300	8	1200	9,600	10.0	7.3	7.0	24.3	20	11	11	42
14 in, 2000 gal/min 400 hp	13,000	400	8	1200	9,600	10.0	7.3	7.0	24.3	20	11	11	42
16 in, 2500 gal/min 500 hp	17,000	500	8	1200	9,600	10.0	7.3	7.0	24.3	20	11	11	42
Electric motor:											
8 in, 500 gal/min 100 hp	2,000	100	10	1200	12,000	8.0	7.2	7.0	22.2	18	8	34	60
10 in, 1000 gal/min 200 hp	4,000	200	10	1200	12,000	8.0	7.2	7.0	22.2	18	8	33	59
12 in, 1500 gal/min 300 hp	6,000	300	10	1200	12,000	8.0	7.2	7.0	22.2	18	8	32	58
14 in, 2000 gal/min 400 hp	9,000	400	10	1200	12,000	8.0	7.2	7.0	22.2	18	8	31	57
16 in, 2500 gal/min 500 hp	11,000	500	10	1200	12,000	8.0	7.2	7.0	22.2	18	8	30	56
Rippers, pull-type, by crawler-tractor, with three shanks, steel, wheel-mounted:											
8,000 lb weight	8,000		8	1400	11,200	10.0	7.3	7.0	24.3	17	6	1	24
14,000 lb weight	14,000		8	1400	11,200	10.0	7.3	7.0	24.3	17	6	1	24
28,000 lb weight	28,000		8	1400	11,200	10.0	7.3	7.0	24.3	17	6	1	24
Rollers or compactors, heavy-duty, unballasted weights:											
Sheepsfoot, tow-type, by tractor:											
Double-drum, 5 ft by 5 ft, 24,000 lb	24,000		8	1400	11,200	10.0	7.3	7.0	24.3	17	8	1	26
Triple drum, 5-ft width by 5-ft diameter, 37,000 lb	37,000		8	1400	11,200	10.0	7.3	7.0	24.3	17	8	1	26

TABLE 8-3 Cost of Ownership and Operation of Excavating Machinery and Facilities *(Continued)*

Machinery with description	Weight and power		Depreciation factors			Yearly fixed charges as % of cost				Hourly costs as % of number of $1000s of cost			
	Weight of machine, lb	Delivered horsepower of prime mover(s)	Depreciation period, yr	Hours of use yearly	Depreciation period, total hours	Depreciation	Interest, taxes, insurance, storage	Replacement cost escalation	Total fixed charge	Fixed charges	Repairs, replacements, labor thereto	Operating expenses	Total hourly costs
Rollers or compactors, heavy-duty, unballasted weights:													
Sheepsfoot, vibratory, tow-type, by tractor, diesel engine:													
Single drum, 6-ft width by 5-ft diameter, 75 hp, 23,000 lb	23,000	75	8	1400	11,200	10.0	7.3	7.0	24.3	17	10	4	31
Sheepsfoot, self-propelled, two-drum or four-wheel, four-drum, with bulldozer, diesel engine:													
Two-drum or four-wheel, 6-ft width by 5-ft diameter, 180 hp, 37,000 lb	37,000	180	8	1400	11,200	10.0	7.3	7.0	24.3	17	12	6	35
Two-drum or four-wheel, 6-ft width by 6-ft diameter, 320 hp, 62,000 lb	62,000	320	8	1400	11,200	10.0	7.3	7.0	24.3	17	12	6	35
Four-drum, 6-ft width by 6-ft diameter, 400 hp, 80,000 lb	80,000	400	8	1400	11,200	10.0	7.3	7.0	24.3	17	12	6	35
Pneumatic, tow-type, by tractor, multiple box, wheels-tires:													
25 tons, ballasted, 22,000 lb	22,000		8	1400	11,200	10.0	7.3	7.0	24.3	17	10	1	28
50 tons, ballasted, 32,000 lb	32,000		8	1400	11,200	10.0	7.3	7.0	24.3	17	10	1	28
75 tons, ballasted, 36,000 lb	36,000		8	1400	11,200	10.0	7.3	7.0	24.3	17	10	1	28
100 tons, ballasted, 45,000 lb	45,000		8	1400	11,200	10.0	7.3	7.0	24.3	17	10	1	28
125 tons, ballasted, 55,000 lb	55,000		8	1400	11,200	10.0	7.3	7.0	24.3	17	10	1	28
Rolling choppers, tow-type, by crawler-tractor, unballasted weights:													
Single-drum:													
8-ft width, 14,000 lb	14,000		8	1400	11,200	10.0	7.3	7.0	24.3	17	8	1	26
12-ft width, 22,000 lb weight	22,000		8	1400	11,200	10.0	7.3	7.0	24.3	17	8	1	26
16-ft width, 46,000 lb weight	46,000		8	1400	11,200	10.0	7.3	7.0	24.3	17	8	1	26

8–20

Tandem drums:												
7-ft width, 19,000 lb weight	19,000		8	1400	11,200	10.0	7.3	7.0	24.3	17	8	1	26
8-ft width, 27,000 lb weight	27,000		8	1400	11,200	10.0	7.3	7.0	24.3	17	8	1	26
10-ft width, 37,000 lb weight	37,000		8	1400	11,200	10.0	7.3	7.0	24.3	17	8	1	26
Rotary drills, complete with tools:												
Crawler-mounted:												
Diesel engine:												
3- to 6-in-diameter hole, to 2500-ft depth	33,000	100	8	1400	11,200	10.0	7.3	7.0	24.3	17	12	2	31
5- to 8-in-diameter hole, to 2500-ft depth	62,000	150	8	1400	11,200	10.0	7.3	7.0	24.3	17	12	2	31
6- to 9-in-diameter hole, to 2500-ft depth	90,000	200	12	4000	48,000	6.7	7.0	7.0	20.7	5	12	2	19
7- to 11-in-diameter hole, to 2500-ft depth	150,000	300	12	4000	48,000	6.7	7.0	7.0	20.7	5	12	2	19
9- to 15-in-diameter hole, to 2500-ft depth	220,000	400	12	4000	48,000	6.7	7.0	7.0	20.7	5	12	2	19
Electric motors:												
3- to 6-in-diameter hole, to 2500-ft depth	32,000	100	8	1400	11,200	10.0	7.3	7.0	24.3	17	10	5	32
5- to 8-in-diameter hole, to 2500-ft depth	59,000	150	8	1400	11,200	10.0	7.3	7.0	24.3	17	10	4	31
6- to 9-in-diameter hole, to 2500-ft depth	86,000	200	12	4000	48,000	6.7	7.0	7.0	20.7	5	10	3	18
7- to 11-in-diameter hole, to 2500-ft depth	152,000	300	12	4000	48,000	6.7	7.0	7.0	20.7	5	10	3	18
9- to 15-in-diameter hole, to 2500-ft depth	209,000	400	12	4000	48,000	6.7	7.0	7.0	20.7	5	10	3	18
Truck-mounted, diesel engines:												
3- to 6-in-diameter hole, to 2500-ft depth, 2 axles, 250 total combined hp	40,000	250	8	1400	11,200	10.0	7.3	7.0	24.3	17	12	3	32
5- to 8-in-diameter hole, to 2500-ft depth, 3 axles, 400 total combined hp	60,000	400	8	1400	11,200	10.0	7.3	7.0	24.3	17	12	3	32
Saws, logging, gasoline engine:												
Chain type, manual:												
Medium weight, 30-in bar length, 2 hp	20	2	3	1400	4200	26.7	8.6	7.0	42.3	30	27	20	77
Heavyweight, 45-in bar length, 4 hp	30	3	3	1400	4200	26.7	8.6	7.0	42.3	30	27	11	58
Circular type, portable, single axle, wheel-tire-mounted 24-in diameter saw, 15 hp	250	10	5	1400	7000	16.0	7.8	7.0	30.8	22	20	42	84

TABLE 8-3 Cost of Ownership and Operation of Excavating Machinery and Facilities *(Continued)*

Machinery with description	Weight and power		Depreciation factors			Yearly fixed charges as % of cost				Hourly costs as % of number of $1000s of cost			
	Weight of machine, lb	Delivered horsepower of prime mover(s)	Depreciation period, yr	Hours of use yearly	Depreciation period, total hours	Depreciation	Interest, taxes, insurance, storage	Replacement cost escalation	Total fixed charge	Fixed charges	Repairs, replacements, labor thereto	Operating expenses	Total hourly costs
Scales, platform, complete with scale house, erected:													
50 tons	33,000		12	1400	16,800	6.7	7.0	7.0	20.7	15	5	1	21
100 tons	45,000		12	1400	16,800	6.7	7.0	7.0	20.7	15	5	1	21
150 tons	67,000		12	1400	16,800	6.7	7.0	7.0	20.7	15	5	1	21
200 tons	76,000		12	1400	16,800	6.7	7.0	7.0	20.7	15	5	1	21
Scrapers, wheels-tires:													
Tow-type, by crawler-tractor, 2 axles:													
15 cysm, 22 tons	25,000		8	1400	11,200	10.0	7.3	7.0	24.3	17	10	1	28
20 cysm, 30 tons	35,000		8	1400	11,200	10.0	7.3	7.0	24.3	17	10	1	28
25 cysm, 38 tons	48,000		8	1400	11,200	10.0	7.3	7.0	24.3	17	10	1	28
Self-propelled type, push-loaded by crawler-tractor(s), two or three axles, diesel engine(s):													
Single engine:													
15 cysm, 22 tons, 350 hp	67,000	350	8	1400	11,200	10.0	7.3	7.0	24.3	17	12	6	35
20 cysm, 30 tons, 400 hp	74,000	400	8	1400	11,200	10.0	7.3	7.0	24.3	17	12	5	34
25 cysm, 38 tons, 450 hp	96,000	450	8	1400	11,200	10.0	7.3	7.0	24.3	17	12	5	34
30 cysm, 45 tons, 500 hp	115,000	500	8	1400	11,200	10.0	7.3	7.0	24.3	17	12	5	34
35 cysm, 52 tons, 550 hp	126,000	550	8	1400	11,200	10.0	7.3	7.0	24.3	17	12	5	34
40 cysm, 60 tons, 550 hp	138,000	550	8	1400	11,200	10.0	7.3	7.0	24.3	17	12	4	33
Twin engines:													
15 cysm, 22 tons, 480 hp	71,000	480	8	1400	11,200	10.0	7.3	7.0	24.3	17	15	7	39
20 cysm, 30 tons, 600 hp	81,000	600	8	1400	11,200	10.0	7.3	7.0	24.3	17	15	7	39
25 cysm, 38 tons, 720 hp	98,000	720	8	1400	11,200	10.0	7.3	7.0	24.3	17	15	7	39
30 cysm, 45 tons, 840 hp	132,000	840	8	1400	11,200	10.0	7.3	7.0	24.3	17	15	7	39
35 cysm, 52 tons, 960 hp	153,000	960	8	1400	11,200	10.0	7.3	7.0	24.3	17	15	7	39
40 cysm, 60 tons, 960 hp	174,000	960	8	1400	11,200	10.0	7.3	7.0	24.3	17	15	6	38

Equipment													
Self-propelled type, self-loaded by integral elevator, 2 or 3 axles, diesel engine:													
Single engine:													
15 cysm, 22 tons, 350 hp	52,000	350	8	1400	11,200	10.0	7.3	7.0	24.3	17	13	7	37
20 cysm, 30 tons, 400 hp	74,000	400	8	1400	11,200	10.0	7.3	7.0	24.3	17	13	6	36
25 cysm, 38 tons, 450 hp	83,000	450	8	1400	11,200	10.0	7.3	7.0	24.3	17	13	6	36
Shovels, full revolving, crawler-mounted:													
Diesel engine:													
2 yd³, 150 hp	117,000	150	10	1600	16,000	8.0	7.2	7.0	22.2	14	8	1	23
3 yd³, 200 hp	168,000	200	12	1600	19,200	6.7	7.0	7.0	20.7	13	7	1	21
4 yd³, 250 hp	210,000	250	14	1700	23,800	5.7	7.0	7.0	19.7	12	6	1	19
5 yd³, 300 hp	275,000	300	16	1700	27,200	5.0	6.9	7.0	18.9	11	5	1	17
6 yd³, 350 hp	330,000	350	18	1700	30,600	4.4	6.9	7.0	18.3	11	4		16
Electric motor:													
2 yd³, 150 hp	115,000	150	10	1600	16,000	8.0	7.2	7.0	22.2	14	7	2	23
3 yd³, 200 hp	165,000	200	12	1600	19,200	6.7	7.0	7.0	20.7	13	6	2	21
4 yd³, 250 hp	206,000	250	14	1700	23,800	5.7	7.0	7.0	19.7	12	5	2	19
5 yd³, 300 hp	271,000	300	16	1700	27,200	5.0	6.9	7.0	18.9	11	4	2	17
6 yd³, 350 hp	325,000	350	18	1700	30,600	4.4	6.9	7.0	18.3	11	3	2	16
Electric motor-generator-motor, ac to dc:													
8 yd³, 650 hp	500,000	650	20	4000	80,000	4.0	6.8	7.0	17.8	4	3	2	9
12 yd³, 950 hp	800,000	950	20	4000	80,000	4.0	6.8	7.0	17.8	4	3	2	9
18 yd³, 1400 hp	1,200,000	1400	20	4000	80,000	4.0	6.8	7.0	17.8	4	3	2	9
27 yd³, 2400 hp	1,900,000	2400	20	7000	140,000	4.0	6.8	7.0	17.8	3	3	2	8
36 yd³, 3500 hp	2,700,000	3500	20	7000	140,000	4.0	6.8	7.0	17.8	3	3	2	8
54 yd³, 5900 hp	4,500,000	5900	30	7000	210,000	2.7	6.7	7.0	16.4	3	3	2	8
72 yd³, 9000 hp	6,900,000	9000	30	7000	210,000	2.7	6.7	7.0	16.4	3	3	2	8
108 yd³, 13,000 hp	12,000,000	13000	30	7000	210,000	2.7	6.7	7.0	16.4	3	3	2	8
Slackline excavators:													
Dragscraper-type, complete with head and tail towers:													
Diesel engine:													
3 yd³, 300-ft span, 225 hp	54,000	225	12	1600	19,200	6.7	7.0	7.0	20.7	13	10	3	26
4 yd³, 400-ft span, 300 hp	62,000	300	12	1600	19,200	6.7	7.0	7.0	20.7	13	10	3	26
5 yd³, 500-ft soan, 375 hp	79,000	375	12	1600	19,200	6.7	7.0	7.0	20.7	13	10	3	26
Electric motor:													
3 yd³, 300-ft span, 225 hp	50,000	225	12	1600	19,200	6.7	7.0	7.0	20.7	13	8	5	26
4 yd³, 400-ft span, 300 hp	60,000	300	12	1600	19,200	6.7	7.0	7.0	20.7	13	8	5	26
5 yd³, 500-ft span, 375 hp	80,000	375	12	1600	19,200	6.7	7.0	7.0	20.7	13	8	5	26

TABLE 8-3 Cost of Ownership and Operation of Excavating Machinery and Facilities *(Continued)*

Machinery with description	Weight and power			Depreciation factors			Yearly fixed charges as % of cost				Hourly costs as % of number of $1000s of cost			
	Weight of machine, lb	Delivered horsepower of prime mover(s)	Depreciation period, yr	Hours of use yearly	Depreciation period, total hours	Depreciation	Interest, taxes, insurance, storage	Replacement cost escalation	Total fixed charge	Fixed charges	Repairs, replacements, labor thereto	Operating expenses	Total hourly costs
Slackline excavators:														
Track-cable scraper-type, complete with crawler-mounted head and tail towers, 1000-ft span:														
Diesel engine:														
6 yd³, 700 total combined hp	418,000	700	12	1600	19,200	6.7	7.0	7.0	20.7	13	8	2	23
8 yd³, 1000 total combined hp	515,000	1000	12	1600	19,200	6.7	7.0	7.0	20.7	13	8	2	23
10 yd³, 1200 total combined hp	770,000	1200	12	1600	19,200	6.7	7.0	7.0	20.7	13	8	2	23
12 yd³, 1400 total combined hp	1,030,000	1400	12	1600	19,200	6.7	7.0	7.0	20.7	13	8	2	23
15 yd³, 1700 total combined hp	1,280,000	1700	12	1600	19,200	6.7	7.0	7.0	20.7	13	8	2	23
Electric motor:														
6 yd³, 700 total combined hp	406,000	700	12	1600	19,200	6.7	7.0	7.0	20.7	13	7	3	23
8 yd³, 1000 total combined hp	500,000	1000	12	1600	19,200	6.7	7.0	7.0	20.7	13	7	3	23
10 yd³, 1200 total combined hp	750,000	1200	12	1600	19,200	6.7	7.0	7.0	20.7	13	7	3	23
12 yd³, 1400 total combined hp	1,000,000	1400	12	1600	19,200	6.7	7.0	7.0	20.7	13	7	3	23
15 yd³, 1700 total combined hp	1,250,000	1700	12	1600	19,200	6.7	7.0	7.0	20.7	13	7	3	23
Tractors:														
Crawler-type, diesel engine, with bulldozer or angledozer, with push block for loading scrapers, or with attached other equipment for land clearing and grubbing:														
50 hp	13,000	50	8	1400	11,200	10.0	7.3	7.0	24.3	17	12	5	34
100 hp	25,000	100	8	1400	11,200	10.0	7.3	7.0	24.3	17	12	5	34
200 hp	44,000	200	8	1400	11,200	10.0	7.3	7.0	24.3	17	12	5	34
300 hp	74,000	300	8	1400	11,200	10.0	7.3	7.0	24.3	17	12	4	33
400 hp	95,000	400	8	1400	11,200	10.0	7.3	7.0	24.3	17	12	4	33
500 hp	135,000	500	8	1400	11,200	10.0	7.3	7.0	24.3	17	12	4	33
600 hp	154,000	600	8	1400	11,200	10.0	7.3	7.0	24.3	17	12	4	33
700 hp	173,100	700	8	1400	11,200	10.0	7.3	7.0	24.3	17	12	4	33

Crawler-type, diesel engine, with bulldozer and three-shank ripper:									
50 hp	14,000	50	8	1400	11,200	10.0	7.3	7.0	24.3	17	15	4	36
100 hp	29,000	100	8	1400	11,200	10.0	7.3	7.0	24.3	17	15	4	36
200 hp	51,000	200	8	1400	11,200	10.0	7.3	7.0	24.3	17	15	4	36
300 hp	86,000	300	8	1400	11,200	10.0	7.3	7.0	24.3	17	15	4	36
400 hp	108,000	400	8	1400	11,200	10.0	7.3	7.0	24.3	17	15	4	36
500 hp	155,000	500	8	1400	11,200	10.0	7.3	7.0	24.3	17	15	4	36
600 hp	172,000	600	8	1400	11,200	10.0	7.3	7.0	24.3	17	15	4	36
700 hp	190,000	700	8	1400	11,200	10.0	7.3	7.0	24.3	17	15	4	36
Wheel-tire type, with bulldozer or push block for loading scrapers, diesel engine:									
50 hp	12,000	50	8	1400	11,200	10.0	7.3	7.0	24.3	17	12	5	34
100 hp	17,000	100	8	1400	11,200	10.0	7.3	7.0	24.3	17	12	5	34
200 hp	37,000	200	8	1400	11,200	10.0	7.3	7.0	24.3	17	12	5	34
300 hp	55,000	300	8	1400	11,200	10.0	7.3	7.0	24.3	17	12	5	34
400 hp	83,000	400	8	1400	11,200	10.0	7.3	7.0	24.3	17	12	5	34
500 hp	109,000	500	8	1400	11,200	10.0	7.3	7.0	24.3	17	12	5	34
Tractor-trailer(s), wheel-tire-type, diesel engine:									
Brush haulers, on-the-road type:									
Rear dumper, 3 axles, 12 tons, 24 cysm, 250 hp	25,000	250	8	1400	11,200	10.0	7.3	7.0	24.3	17	12	12	41
Log haulers, bunks, on-the-road type:									
12 tons, 3 axles, 250 hp	22,000	250	8	1400	11,200	10.0	7.3	7.0	24.3	17	12	12	41
24 tons, 5 axles, 400 hp	35,000	400	8	1400	11,200	10.0	7.3	7.0	24.3	17	12	12	41
Machinery haulers, on-the-road type:									
35 tons, 5 axles, single gooseneck, 250 hp	30,000	250	8	1400	11,200	10.0	7.3	7.0	24.3	17	8	9	34
60 tons, 9 axles, double gooseneck, 400 hp	56,000	400	8	1400	11,200	10.0	7.3	7.0	24.3	17	8	9	34
Rock haulers, off-the-road type:									
Bottom dumpers:									
Single unit, 2 or 3 axles:									
20 cysm, 30 tons, 300 hp	49,000	300	8	1400	11,200	10.0	7.3	7.0	24.3	17	10	7	34
30 cysm, 45 tons, 360 hp	71,000	360	8	1400	11,200	10.0	7.3	7.0	24.3	17	10	7	34
40 cysm, 60 tons, 420 hp	83,000	420	8	1400	11,200	10.0	7.3	7.0	24.3	17	10	7	34
50 cysm, 75 tons, 480 hp	102,000	480	8	1400	11,200	10.0	7.3	7.0	24.3	17	10	6	33
60 cysm, 90 tons, 540 hp	109,000	540	8	1400	11,200	10.0	7.3	7.0	24.3	17	10	6	33
70 cysm, 105 tons, 600 hp	116,000	600	8	1400	11,200	10.0	7.3	7.0	24.3	17	10	6	33

TABLE 8-3 Cost of Ownership and Operation of Excavating Machinery and Facilities *(Continued)*

Machinery with description	Weight and power		Depreciation factors			Yearly fixed charges as % of cost				Hourly costs as % of number of $1000s of cost			
	Weight of machine, lb	Delivered horsepower of prime mover(s)	Depreciation period, yr	Hours of use yearly	Depreciation period, total hours	Depreciation	Interest, taxes, insurance, storage	Replacement cost escalation	Total fixed charge	Fixed charges	Repairs, replacements, labor thereto	Operating expenses	Total hourly costs
Tractor-trailer(s), wheel-tire-type, diesel engine:													
Rock haulers, off-the-road type:													
Bottom dumpers:													
Double units, 4 or 5 axles:													
80 cysm, 120 tons, 420 hp	118,000	420	8	1400	11,200	10.0	7.3	7.0	24.3	17	8	4	29
100 cysm, 150 tons, 480 hp	143,000	480	8	1400	11,200	10.0	7.3	7.0	24.3	17	8	4	29
120 cysm, 180 tons, 540 hp	156,000	540	8	1400	11,200	10.0	7.3	7.0	24.3	17	8	4	29
140 cysm, 210 tons, 600 hp	161,000	600	8	1400	11,200	10.0	7.3	7.0	24.3	17	8	4	29
Rear dumpers:													
Single unit, 2 or 3 axles:													
20 cysm, 30 tons, 300 hp	57,000	300	8	1400	11,200	10.0	7.3	7.0	24.3	17	12	6	35
30 cysm, 45 tons, 360 hp	71,000	360	8	1400	11,200	10.0	7.3	7.0	24.3	17	12	5	34
40 cysm, 60 tons, 420 hp	96,000	420	8	1400	11,200	10.0	7.3	7.0	24.3	17	12	5	34
50 cysm, 75 tons, 480 hp	119,000	480	8	1400	11,200	10.0	7.3	7.0	24.3	17	12	4	33
Side dumpers:													
Single unit, 3 axles:													
14 cysm, 20 tons, 240 hp	50,000	240	8	1400	11,200	10.0	7.3	7.0	24.3	17	12	5	34
20 cysm, 30 tons, 300 hp	78,000	300	8	1400	11,200	10.0	7.3	7.0	24.3	17	12	4	33
30 cysm, 45 tons, 360 hp	106,000	360	8	1400	11,200	10.0	7.3	7.0	24.3	17	12	4	33
Double units, 5 axles:													
28 cysm, 40 tons, 240 hp	75,000	240	8	1400	11,200	10.0	7.3	7.0	24.3	17	12	3	32
40 cysm, 60 tons, 300 hp	117,000	300	8	1400	11,200	10.0	7.3	7.0	24.3	17	12	3	32
60 cysm, 90 tons, 360 hp	159,000	360	8	1400	11,200	10.0	7.3	7.0	24.3	17	12	2	31
Rock haulers, on-the-road type:													
Bottom dumpers:													
Single unit:													
10 cysm, 12 tons, 3 axles, 250 hp	22,000	250	8	1400	11,200	10.0	7.3	7.0	24.3	17	12	13	42
20 cysm, 24 tons, 5 axles, 400 hp	35,000	400	8	1400	11,200	10.0	7.3	7.0	24.3	17	12	13	42

Description													
Double units:													
20 cysm, 24 tons, 5 axles, 250 hp	35,000	250	8	1400	11,200	10.0	7.3	7.0	24.3	17	12	8	37
40 cysm, 48 tons, 9 axles, 400 hp	55,000	400	8	1400	11,200	10.0	7.3	7.0	24.3	17	12	8	37
Rear dumpers:													
Single unit:													
10 cysm, 12 tons, 3 axles, 250 hp	25,000	250	8	1400	11,200	10.0	7.3	7.0	24.3	17	15	12	44
20 cysm, 24 tons, 5 axles, 400 hp	40,000	400	8	1400	11,200	10.0	7.3	7.0	24.3	17	15	12	44
Side dumpers:													
Single unit:													
10 cysm, 12 tons, 3 axles, 250 hp	25,000	250	8	1400	11,200	10.0	7.3	7.0	24.3	17	15	12	44
20 cysm, 24 tons, 5 axles, 400 hp	40,000	400	8	1400	11,200	10.0	7.3	7.0	24.3	17	15	12	44
Double units:													
20 cysm, 24 tons, 5 axles, 250 hp	41,000	250	8	1400	11,200	10.0	7.3	7.0	24.3	17	15	7	39
40 cysm, 48 tons, 9 axles, 400 hp	65,000	400	8	1400	11,200	10.0	7.3	7.0	24.3	17	15	7	39
Water haulers, tankers, sprayers, off-the-road type:													
6,000 gal, 410 total combined hp	63,000	410	8	1400	11,200	10.0	7.3	7.0	24.3	17	10	8	35
8,000 gal, 480 total combined hp	78,000	480	8	1400	11,200	10.0	7.3	7.0	24.3	17	10	8	35
10,000 gal, 550 total combined hp	92,000	550	8	1400	11,200	10.0	7.3	7.0	24.3	17	10	8	35
12,000 gal, 620 total combined hp	107,000	620	8	1400	11,200	10.0	7.3	7.0	24.3	17	10	8	35
14,000 gal, 690 total combined hp	116,000	690	8	1400	11,200	10.0	7.3	7.0	24.3	17	10	8	35
Trenchers, crawler-mounted, diesel engine:													
Ladder type:													
16- to 24-in width, to 11-ft depth, 80 hp	17,000	80	8	1400	11,200	10.0	7.3	7.0	24.3	17	15	4	36
24- to 36-in width, to 15-ft depth, 120 hp	31,000	120	8	1400	11,200	10.0	7.3	7.0	24.3	17	15	4	36
24- to 72-in width, to 25-ft depth, 160 hp	55,000	160	8	1400	11,200	10.0	7.3	7.0	24.3	17	15	3	35
Wheel type:													
16- to 24-inch width, up to 6-ft depth, 60 hp	20,000	60	8	1400	11,200	10.0	7.3	7.0	24.3	17	12	3	32
20- to 36-inch width, up to 7-ft depth, 100 hp	29,000	100	8	1400	11,200	10.0	7.3	7.0	24.3	17	12	3	32
30- to 60-in width, to 8-ft depth, 150 hp	50,000	150	8	1400	11,200	10.0	7.3	7.0	24.3	17	12	3	32
Trucks, diesel engine:													
Brush haulers, on-the-road type:													
Rear dumper, 2 axles, 8 tons, 16 cysm, 250 hp	12,000	250	8	1400	11,200	10.0	7.3	7.0	24.3	17	12	24	53
Rear dumper, 3 axles, 12 tons, 24 cysm, 400 hp	20,000	400	8	1400	11,200	10.0	7.3	7.0	24.3	17	12	24	53

TABLE 8-3 Cost of Ownership and Operation of Excavating Machinery and Facilities *(Continued)*

Machinery with description	Weight and power		Depreciation factors			Yearly fixed charges as % of cost				Hourly costs as % of number of $1000s of cost			
	Weight of machine, lb	Delivered horsepower of prime mover(s)	Depreciation period, yr	Hours of use yearly	Depreciation period, total hours	Depreciation	Interest, taxes, insurance, storage	Replacement cost escalation	Total fixed charge	Fixed charges	Repairs, replacements, labor thereto	Operating expenses	Total hourly costs
Trucks, diesel engine:													
Lubricating trucks and refueling trucks, on-the-road type:													
2 axles, 60 ft³/min compressor, 5 tanks and pumps, 8 reels	18,000	150	8	1400	11,200	10.0	7.3	7.0	24.3	17	12	3	32
3 axles, 90 ft³/min compressor, 11 tanks and pumps, 18 reels	30,000	250	8	1400	11,200	10.0	7.3	7.0	24.3	17	12	3	32
Machinery haulers, flat-bed type, on-the-road type:													
2 axles, 8 tons, 250 hp	11,000	250	8	1400	11,200	10.0	7.3	7.0	24.3	17	10	26	53
3 axles, 12 tons, 400 hp	18,000	400	8	1400	11,200	10.0	7.3	7.0	24.3	17	10	26	53
Rock haulers, off-the-road type:													
Rear dumpers, 2 or 3 axles:													
10 cysm, 15 tons, 200 hp	30,000	200	8	1400	11,200	10.0	7.3	7.0	24.3	17	10	7	34
14 cysm, 20 tons, 250 hp	37,000	250	8	1400	11,200	10.0	7.3	7.0	24.3	17	10	7	34
20 cysm, 30 tons, 350 hp	50,000	350	8	1400	11,200	10.0	7.3	7.0	24.3	17	10	7	34
27 cysm, 40 tons, 450 hp	66,000	450	8	1400	11,200	10.0	7.3	7.0	24.3	17	10	7	34
40 cysm, 60 tons, 650 hp	95,000	650	8	1400	11,200	10.0	7.3	7.0	24.3	17	10	7	34
54 cysm, 80 tons, 850 hp	120,000	850	12	4000	48,000	6.7	7.0	7.0	20.7	5	8	7	20
67 cysm, 100 tons, 1000 hp	143,000	1000	12	4000	48,000	6.7	7.0	7.0	20.7	5	8	7	20
94 cysm, 140 tons, 1400 hp	190,000	1400	12	4000	48,000	6.7	7.0	7.0	20.7	5	8	7	20
120 cysm, 180 tons, 1700 hp	230,000	1700	12	4000	48,000	6.7	7.0	7.0	20.7	5	8	7	20
174 cysm, 260 tons, 2300 hp	310,000	2300	12	4000	48,000	6.7	7.0	7.0	20.7	5	8	7	20
Side dumpers:													
10 cysm, 15 tons, 200 hp	32,000	200	8	1400	11,200	10.0	7.3	7.0	24.3	17	10	6	33
14 cysm, 20 tons, 250 hp	39,000	250	8	1400	11,200	10.0	7.3	7.0	24.3	17	10	6	33
20 cysm, 30 tons, 350 hp	53,000	350	8	1400	11,200	10.0	7.3	7.0	24.3	17	10	6	33

Equipment												
Rock haulers, on-the-road type:												
Rear dumpers:												
6 cysm, 8 tons, 2 axles, 250 hp	13,000	250	8	1400	11,200	10.0	7.3	7.0	24.3	17	10	29	56
10 cysm, 12 tons, 3 axles, 400 hp	24,000	400	8	1400	11,200	10.0	7.3	7.0	24.3	17	10	23	50
Side dumpers:												
6 cysm, 8 tons, 2 axles, 250 hp	15,000	250	8	1400	11,200	10.0	7.3	7.0	24.3	17	10	24	51
10 cysm, 12 tons, 3 axles, 400 hp	26,000	400	8	1400	11,200	10.0	7.3	7.0	24.3	17	10	19	46
Water haulers, tankers, sprayers, 2 axles, off-the-road type:												
4000 gal, 200 hp	30,000	200	8	1400	11,200	10.0	7.3	7.0	24.3	17	10	7	34
6000 gal, 300 hp	45,000	300	8	1400	11,200	10.0	7.3	7.0	24.3	17	10	7	34
8000 gal, 400 hp	68,000	400	8	1400	11,200	10.0	7.3	7.0	24.3	17	10	7	34
Water haulers, tankers, sprayers, on-the-road type:												
2000 gal, 2 axles, 250 hp	12,000	250	8	1400	11,200	10.0	7.3	7.0	24.3	17	10	26	53
4000 gal, 3 axles, 400 hp	20,000	400	8	1400	11,200	10.0	7.3	7.0	24.3	17	10	21	48
Wagons, rock, side-dump, crawler-mounted, crawler-tractor-towed:												
6 cysm	18,000		8	1400	11,200	10.0	7.3	7.0	24.3	17	12	1	30
8 cysm	27,000		8	1400	11,200	10.0	7.3	7.0	24.3	17	12	1	30
Water tanks, gravity flow:												
Portable, power-raised, 2 axles, wheel-tire-mounted, gasoline engine, 8 hp:												
6,000 gal	8,000	8	8	1400	11,200	10.0	7.3	7.0	24.3	17	8	1	26
8,000 gal	10,000	8	8	1400	11,200	10.0	7.3	7.0	24.3	17	8	1	26
10,000 gal	13,000	8	8	1400	11,200	10.0	7.3	7.0	24.3	17	8	1	26
12,000 gal	16,000	8	8	1400	11,200	10.0	7.3	7.0	24.3	17	8	1	26
16,000 gal	20,000	8	8	1400	11,200	10.0	7.3	7.0	24.3	17	8	1	26
20,000 gal	26,000	8	8	1400	11,200	10.0	7.3	7.0	24.3	17	8	1	26
Movable, stand-skid-mounted:												
6,000 gal	6,000	8	8	1400	11,200	10.0	7.3	7.0	24.3	17	5	0	22
8,000 gal	8,000	8	8	1400	11,200	10.0	7.3	7.0	24.3	17	5	0	22
10,000 gal	10,000	8	8	1400	11,200	10.0	7.3	7.0	24.3	17	5	0	22
12,000 gal	13,000	8	8	1400	11,200	10.0	7.3	7.0	24.3	17	5	0	22
16,000 gal	16,000	8	8	1400	11,200	10.0	7.3	7.0	24.3	17	5	0	22
20,000 gal	22,000	8	8	1400	11,200	10.0	7.3	7.0	24.3	17	5	0	22

TABLE 8-3 Cost of Ownership and Operation of Excavating Machinery and Facilities *(Continued)*

Machinery with description	Weight and power		Depreciation factors			Yearly fixed charges as % of cost				Hourly costs as % of number of $1000s of cost			
	Weight of machine, lb	Delivered horsepower of prime mover(s)	Depreciation period, yr	Hours of use yearly	Depreciation period, total hours	Depreciation	Interest, taxes, insurance, storage	Replacement cost escalation	Total fixed charge	Fixed charges	Repairs, replacements, labor thereto	Operating expenses	Total hourly costs
Welding equipment, portable, wheel-tire-mounted:													
Diesel engine:													
100 A, 15 hp	500	15	6	1400	8,400	13.3	7.6	7.0	27.9	20	24	22	66
200 A, 25 hp	1,000	25	6	1400	8,400	13.3	7.6	7.0	27.9	20	24	22	66
300 A, 40 hp	1,500	40	6	1400	8,400	13.3	7.6	7.0	27.9	20	24	22	66
400 A, 50 hp	2,000	50	8	1400	11,200	10.0	7.3	7.0	24.3	17	24	22	63
600 A, 80 hp	3,000	80	8	1400	11,200	10.0	7.3	7.0	24.3	17	24	22	63
800 A, 100 hp	4,000	100	8	1400	11,200	10.0	7.3	7.0	24.3	17	24	22	63
Electric motor:													
100 A, 10 hp	250	10	6	1400	8,400	13.3	7.6	7.0	27.9	20	20	55	95
200 A, 15 hp	500	15	6	1400	8,400	13.3	7.6	7.0	27.9	20	20	56	96
300 A, 20 hp	700	20	6	1400	8,400	13.3	7.6	7.0	27.9	20	20	57	97
400 A, 25 hp	900	25	8	1400	11,200	10.0	7.3	7.0	24.3	17	20	58	95
600 A, 40 hp	1,400	40	8	1400	11,200	10.0	7.3	7.0	24.3	17	20	58	95
800 A, 60 hp	1,800	60	8	1400	11,200	10.0	7.3	7.0	24.3	17	20	58	95
Well-point systems, complete:													
Diesel engine:													
6-in centrifugal pump, 50 well points, 40 hp	11,000	40	5	1200	6,000	16.0	7.8	7.0	30.8	26	24	3	53
10-in centrifugal pump, 150 well points, 120 hp	24,000	120	6	1200	7,200	13.3	7.6	7.0	27.9	23	20	5	48
Electric motor:													
6-in centrifugal pump, 50 well points, 40 hp	9,000	40	5	1200	6,000	16.0	7.8	7.0	30.8	26	20	7	53
10-in centrifugal pump, 150 well points, 120 hp	19,000	120	6	1200	7,200	13.3	7.6	7.0	27.9	23	16	10	49

Wheel bucket excavators, belt-loading, crawler-mounted:										
Linear digging, linear traveling:										
Diesel engine:										
16-ft wheel, twelve 1-yd³ buckets, 500 hp	250,000	500	10	1400	14,000	8.0	7.2	7.0	22.2	16	8	2	26
Circular digging, linear traveling:										
Diesel engine-generator-motor drive:										
18-ft wheel, eight 1 yd³ buckets, 640 hp	450,000	640	20	4000	80,000	4.0	6.8	7.0	17.8	4	4	1	9
25-ft wheel, eight 2 yd³ buckets, 1600 hp	700,000	1600	20	4000	80,000	4.0	6.8	7.0	17.8	4	4	1	9
Combination ac motor-dc generator-dc motor drives and ac-motor drives:										
30-ft wheel, ten 2½ yd³ buckets, 2400 hp	1,620,000	2400	20	4000	80,000	4.0	6.8	7.0	17.8	4	4	1	9

8-32 Cost of Machinery and Facilities

ratings are given specifically rather than within a range, so that the user can approximate closely a given machine.

Weight and Power of Machine These are averages for a given machine as built by one or more manufacturers. In some cases they have been estimated from specifications for a machine of given capacity and rating.

Depreciation Factors These three factors, depreciation period in years, hours of use yearly, and depreciation period in total hours, are estimated for average working conditions of construction and mining in the United States. They may be varied for particular physical working conditions, shifts of work daily, and bookkeeping practices.

Depreciation Period in Years. The years of depreciation vary greatly, even for the same machines working under like conditions. The double logarithmic graph of Plate 8-1 rationalizes a value for years of depreciation according to the weight of the machine. Of all possible determinants weight is the most consistent. The graph approximates averages for excavation machines. The graph, the abscissa of which commences with 100 lb and ends with 100 million lb of weight, shows a doubling of the depreciation period in years for every tenfold increase in weight. The equation is:

$$DY = 2^{(\log WP - \log 100)}$$

where DY = depreciation period in years and WP = machine weight in pounds.

There are two examples shown in the graph, namely:
1. A 4000-lb medium-weight automobile or pickup truck would be depreciated in 3.0 yr.
2. A 2.7-million-lb, 36-yd³ strip-mining shovel would be depreciated in 22 yr.

Hours of Use Yearly. Hours of use yearly is likewise an average figure for the United States. It varies according to length of working season, which is 5 to 12 months, and to number of shifts daily, which varies from one on an average construction job to three for a vast open pit-mine like that shown in Plate 7-1. The hours of use yearly are generally in terms of single shifting, with some double and triple shifting in cases of large, expensive machines used in the massive excavations of construction and mining.

Depreciation Period in Total Hours. This summary figure is the product of depreciation period in years times hours of use yearly.

Yearly Fixed Charges as Percentage of Cost These four charges are expressed as an annual percentage of the cost of the machine. Cost is the total cost of the machine, assembled and erected, at job site and it includes sales taxes and freight.

Depreciation. Depreciation is calculated from the depreciation period in years. All machines are given a trade-in or salvage value of 20 percent of cost. Depreciation is reckoned by the uniform or straight-line method.

Yearly cost in percentage of cost (PC) of machine is given:

$$PC = \frac{100 - \text{percent trade-in or salvage value}}{\text{depreciation-period, yr}}$$

or

$$PC = \frac{80}{\text{depreciation period, yr}}$$

Interest, Taxes, Insurance, and Storage. These four items are functions of the average value of the machine during the depreciation period. Their annual rates, estimated on the basis of average figures for 1978, total 13.0%, broken down as follows: interest, 8.0%; property and use taxes, 2.0%; insurance, 2.0%; storage, 1.0%. Yearly cost in percentage of cost (PC) of machine is given:

$$PC = \frac{0.13 \left(\frac{C + C/Y}{2} \right)}{C}$$

wherein C = cost and Y = depreciation period in years. The equation reduces to:

$$PC = 6.5 \left(1 + \frac{1}{Y} \right)$$

Replacement Cost Escalation. This hedge against increased inflationary cost for the replacement of the machine after the depreciation period is the third element of yearly fixed charges as percentage of cost. It allows for a like-kind replacement. It is estimated at 7 percent yearly of the cost of the machine.

Total Fixed Charges. The total of depreciation, interest, taxes, insurance, storage, and replacement-cost escalation is expressed as percentage yearly of the cost of the machine.

Hourly Costs as Percentage of Thousands of Dollars of Cost These percentage elements of four cost factors are summarized in Table 8-3, which gives factors to be applied to the current cost of a machine in order to arrive at the hourly expense of ownership and operation.

Fixed Charges. For fixed charges, the hourly percentage cost (HPC) is calculated by the equation:

$$\text{HPC} = \frac{\frac{\text{total yearly fixed charges} \times \text{cost}}{\text{hours of use yearly}}}{\frac{\text{cost}}{\$1000}}$$

The equation reduces to:

$$\text{HPC} = \frac{\text{total yearly fixed charges} \times 1000}{\text{hours of use yearly}}$$

Repairs and Replacements, Including Labor. This hourly percentage cost includes both major and minor repairs of a mechanical nature and replacement of consumed items such as tires, cutting edges, bits and drill steel, cables, and the like. These costs may be subdivided approximately as follows:

Parts and replacements, 50%
Labor, mechanics, 25%
Shop and field service, 10%
Repairs by outside agencies, 15%

As tires and tire repairs account for a significant percentage of the cost of ownership and operation, these costs are sometimes set up as a separate item. However, for simplicity's sake and without real loss of precision, they may be included in repairs, replacements, and labor.

Information for the 1978 dollar costs of repairs, replacements, and labor for an hour of operation has been translated into percentages of the thousands of dollars of cost for the 1978 machine.

Operating Expenses. As emphasized, these expenses do not include the hourly wages and fringe benefits of operator(s). For diesel and gasoline engines and for electric motors, including motor-generator drives, they include all costs for fuel and electrical energy, for all lubricants for the prime movers and power trains and assemblies, for hydraulic oils, for filters, and for oilers and grease trucks when they are used.

Approximate formulas for these costs are in terms of the rated or flywheel horsepower of prime movers delivering an average 67 percent of total horsepower. In some cases changes have been made in the 67 percent factor to accommodate special engine and electric-motor configurations of drive. In terms of 1978 fuel and energy costs the general formulas are:

Prime mover	Fuel or energy cost	Hourly cost at 67% load factor
Diesel engine	0.40/gal	$0.024 × rated horsepower
Gasoline engine	0.50/gal	$0.045 × rated horsepower
Gas (LPG) engine	0.52/gal	$0.047 × rated horsepower
Electric motor	0.035/kWh	$0.033 × rated horsepower

These general equations have been modified sometimes in Table 8-3 for special conditions. Examples are:

1. Dragline excavator, pontoon-type, diesel engine, 12-yd³ capacity, is not lubricated by a grease truck with oilers. The dragline oiler, part of a two- or three-person crew, takes care of lubrication. The hourly cost is estimated to be $0.019 times rated engine horsepower.

2. Belt conveyor, stationary, 100-ft segment, 60 in wide, electric motor, is rarely

8–34 Cost of Machinery and Facilities

lubricated and needs no grease truck. Hourly cost is estimated to be $0.029 time rated horsepower.

For Table 8-3 these hourly costs, based on formulas, are translated into percentages of thousands of dollars of 1978 machine cost.

Total Hourly Costs. This fourth element summarizes fixed charges; repairs and replacements, including labor; and operating expenses. All four percentage elements are rounded off to the nearest integer.

Use of Table of Hourly Ownership and Operating Costs

Two examples are:

1. A 1986 estimate for the hourly cost of ownership and operation of a mining shovel, full revolving, crawler-mounted, electric motors, 8 yd^3, 500 hp. Cost (1986): $ 1,700,000. Table 8-3 gives the hourly cost as 9 percent of the thousands of dollars of cost of shovel, exclusive of runner and oiler.

Machine cost hourly, 9% of $1700,		$153.00
Operators' cost hourly, with fringe benefits:		
Runner	$22.00	
Oiler	20.00	42.00
Total hourly costs		$195.00

2. The data of Table 8-3 can be varied at will for particular conditions. Assume that an air drill, crawler-mounted with compressor for both downhole hammer and rotary drilling, powered by electric motors, and drilling 9½-in-diameter holes, is to be used in a dolomite quarry for fluxing stone. Work is round-the-clock triple-shifting throughout the year 1981. These determinants are in order:

a. Depreciation period, yr	5
b. Hours of use yearly, based on about 85% availability	7,400
c. Depreciation period, total hours	37,000

Yearly fixed charges as percentage of cost of $240,000 in 1981 are:

Depreciation, %	16.0
Interest, taxes, insurance (no storage), %	
0.060 (1 + ⅕)	7.2
Replacement cost escalation, %	7.0
Total yearly fixed charges, %	30.2

Hourly costs as percentage of thousands of dollars of cost become:

Fixed charges, $\dfrac{30.2\% \text{ of } 1000}{7400}$	4%
Repairs, replacements, labor	20
Operating expenses	13
Total hourly costs, exclusive of operator(s)	37%

The hourly costs, except for operator(s), are 37 percent of $ 240, or about $89, instead of 47 percent of $240, or about $113, as suggested in Table 8-3.

In conclusion, let it be emphasized that Table 8-3 is a guide for average conditions in the United States, that it does not include the cost of operator(s), and that it can and should be interpreted and modified for particular conditions. These particular conditions are typical of excavations in massive public works such as dams and in huge private works such as open-pit mines.

PRODUCTION OF MACHINERY

Production of machinery is the time rate of performing work. The period of time may be the minute, hour, shift, day, month, year, or duration of the work. Production may be continuous or intermittent.

Continuous Production

Examples of continuous work are the productions of air compressors, engines, generators, motors, and pumps. These machines operate almost without either minor delays (less than a few minutes in duration) or major delays (more than a few minutes in duration) caused usually by mechanical failures.

The efficiency of the machine is expressed as the ratio of the time rate of energy output, horsepower, to the time rate of energy input. An example is the efficiency of an electric motor, expressed as follows:

$$\text{Efficiency} = \frac{\text{output, hp}}{\text{input, hp}}$$

$$\text{Input hp} = \frac{\text{input, kW}}{0.746}$$

Intermittent Production

Examples of intermittent work are the productions of drills, excavators, haulers, and tractors. These machines operate on cycle times, the cycle being the repetitive operation resulting in production. Cycles may be exemplified by that for a shovel, consisting of loading, swinging, dumping, and returning times, and by that for a scraper, made up of loading, hauling, dumping, turning, returning, and turning times.

These cycles are interrupted by minor and major delays. Examples of the minor delays to a shovel are checking grade, handling outsize rocks, minor maintenance, moving the shovel at the face of a cut, operator's delays, and spotting of haulers at the shovel.

Major delays, such as those due to mechanical failures and weather, are not considered to be factors affecting the operating efficiency of the machine.

Efficiencies are reckoned in two ways, each convertible to the other. Both ways express the ratio of available work time for the cycling of the machine to actual work time. Available work time is that time during which the machine can operate on continuous cycles without interruptions due to minor delays.

One well-accepted way employs the available work time in terms of the number of available minutes of the hour. Examples are the 30-min, 40-min, 50-min, and 60-min hours, equivalent respectively to 50, 67, 83, and 100 percent efficiency. As most machines with intermittent productions operate on a cycle, this kind of efficiency is convenient and readily understandable. It is the method used in the production manuals of the manufacturers of machinery, by earthmovers, and in the many examples of production and costs in the text.

An example is provided by the calculations for production of a bottom-dump hauler operating on an estimated 12.5-min hauling cycle and an estimated payload of 25 yd³ bank measurement (25 cybm), and working an estimated 45-min working hour, or at 75 percent efficiency. The estimated hourly production is given by the equation:

$$\text{Hourly production, cybm} = \frac{45.0 \text{ min}}{12.5 \text{ min}} \times 25 \text{ cybm} = 90$$

The other way is expressed by the equation:

$$\text{Hourly production, cybm} = \frac{75\% \text{ of } 60 \text{ min}}{12.5 \text{ min}} \times 25 \text{ cybm} = 90$$

The efficiency of a well-managed job, considering carefully the unavoidable minor delays, may be based on a 30- to 50-min working hour. One extreme is the 30-min working hour for a backhoe loading rock from a small basement excavation into trucks working in a confined area. Another extreme is the 50-min working hour for scrapers removing overburden in an open-pit mine in which working conditions are ideal.

The selection of a reasonable efficiency for a contemplated operation must be based on a combination of theory, practicability, and experience. For the sake of uniformity and ability to compare different operations, most of the efficiencies of machines used in the examples of the text are the 50-min working hour, or 83 percent.

DIRECT JOB UNIT COST

The basic unit of cost used in the text is the *direct job unit cost*. As developed in Chapter 18, this unit cost is the basic one for calculating total job costs and total bid prices.

Direct job unit cost is calculated by dividing total cost of ownership and operation of machine by production of machine for the corresponding period of time. Total cost of ownership and operation of machine is the sum of the cost of ownership and operation of the machine and the cost of operator(s) with fringe benefits. The time period is generally the hour, shift, or day.

Direct job unit cost is used mostly in the text, except where it is desired to include supervision as part of the cost. In this case, the term *total direct job unit cost* is used.

The usual equation for direct job unit cost is:

$$\text{Direct job unit cost} = \frac{\text{total hourly cost of ownership and operation of machine}}{\text{hourly production of machine}}$$

SUMMARY

Calculations for cost of ownership and operation of machinery, production of machinery, and direct job unit cost are the sequential estimates for determining the economics of excavation. Upon them depend the complete cost estimate on the basis of which the feasibility of the work is decided and the bid for the work presented.

Whether the work is small or big, knowledge and thoroughness must guide the calculations, or else one of two unfortunate circumstances will plague the earthmover. Either the estimate will be too low, resulting in a loss for the job, or it will be too high, so that the contractor ends up being an unsuccessful bidder for the work.

CHAPTER 9

Clearing and Grubbing

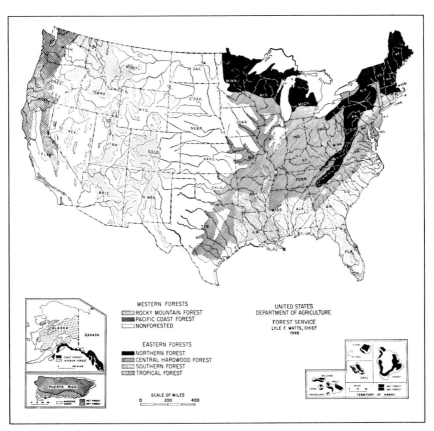

PLATE 9-1

Forest regions of the United States. This map shows the eight natural forest regions of the conterminous United States, Alaska, Hawaii, and Puerto Rico. These huge forests present many diverse problems to the excavator in clearing and grubbing the land prior to excavating for public and private works, such as airports and open-pit coal mines. If excavation is shallow and the forest growth is heavy, the cost for clearing and grubbing is a sizable percentage of the cost for excavation. *(U.S. Forest Service.)*

Chapter 9, as well as the succeeding ensuing chapters through Chapter 15, deal with the principal phases of excavation. These are subdivided into three parts: principles; methods and machinery; and productions and costs.

PRINCIPLES

The century-old definition of clearing and grubbing is presently applicable. Clearing consists of cutting down and removing or burning trees and brush, and it is an aboveground operation. Grubbing includes the removing of stumps and roots and it is a belowground action.

The standard specifications of the California Division of Highways define clearing and grubbing as follows:

Descriptions. This work shall consist of removing all objectionable material from the highway right of way, bridge construction areas, road approaches, material sites within the right of way, areas through which ditches and channels are to be excavated, and such other areas as may be specified in the special provisions. Clearing and grubbing shall be performed in advance of grading operations and in accordance with the requirements herein specified.

Standard specifications for a particular job are usually expanded by special provisions. An example of these provisions for a 1976 freeway in San Diego County, California, is as follows:

Clearing and grubbing. Clearing and grubbing shall conform to the provisions in Section 16, "Clearing and Grubbing," of the standard specifications and to these special provisions. Vegetation shall be cleared and grubbed only within the excavation and embankment slope lines. All existing vegetation, except fruit bearing trees, specified herein, outside the areas to be cleared and grubbed, shall be protected from injury or damage resulting from the Contractor's operations.

All activities controlled by the Contractor, except cleanup and other required work, shall be confined within the graded areas of the roadway.

Nothing shall be construed as relieving the Contractor of his responsibility for final cleanup of the highway as provided in Section 4-1.02, "Final Cleaning Up," of the standard specifications. Fruit bearing trees within the right of way shall be removed and disposed of except between approximate stations RT-LN 2250 and 2270. Attention is directed to "Type BW Fence" of these special provisions regarding grubbing and grading along fence lines.

Vegetable growth, 6 inches or less in diameter, shall be reduced to chips of maximum thickness of ½ inch and distributed uniformly on embankment slopes, except rock areas designated by the Engineer. The layer of chips shall not exceed one inch in depth. Burial or mixing of chips will not be required; however, the chips shall be spread prior to placement of the top soil.

Diseased or insect ridden chips will not be moved from one location to another unless they have been decontaminated. Decontamination shall be performed as directed by the Engineer and will be paid for as extra work as provided in Section 4-1.03D of the standard specifications.

Vegetable growth greater than 6 inches in diameter may be reduced to chips and distributed uniformly on embankment slopes or may be disposed of in embankments as follows:

The material shall be placed outside of the 1:1 inclined plane sloping out and down from the outside edge of the shoulder of the planned roadbed, but not within 10 feet of the finished slope line, measured normal to the slope; except the material shall be placed in areas consisting of coarse rocky material. The disposal area shall not be subject to seasonal fluctuations in moisture content, as determined by the Engineer.

Roots and trunks shall be trimmed to within 2 feet of the stump. Down trees shall be separated into stumps and logs. Single layers of stumps shall be spaced so that compaction equipment can readily pass between them. Tree trunks and trimmed branches shall be placed parallel to each other in a single layer. Each layer of stumps, trunks, and branches shall be covered with a minimum of 3 feet, compacted thickness of embankment material. No material shall be disposed of where it will interfere with any planned work.

Full compensation for chipping, stockpiling, and distributing vegetable growth on embankments shall be considered as included in the contract lump sum price paid for clearing and grubbing, and no additional compensation will be allowed therefor.

These examples of clearing and grubbing definitions by standard specifications and special provisions are not unduly exacting for excavation. They illustrate forcefully the possible complexity and cost for the excavator.

9-4 Clearing and Grubbing

Sometimes the bid-item unit is the acre and sometimes, as in the special provisions example, it is simply a lump sum for the work.

Years ago the term *clearing and grubbing* referred to the removal of trees, brush, and stumps, which are things of nature. As humans encroached on nature, structures built by them, such as homes, pavements, channels, railroads, and the like, had to be removed from the right-of-way. This preparation for excavation, usually involving demolition, is set up as separate bid items such as "removal of existing tank, lump sum bid."

This chapter is concerned only with clearing and grubbing by simple definition, the most important phase of preparation of right-of-way.

FOREST REGIONS

Tree and brush suggest forests. Plate 9-1 shows the forest regions of the United States.

1. The northern forest, located generally in the northeastern United States, consists generally of ashes, birches, cedars, elms, firs, hemlocks, hickories, maples, and pines. These are a mixture of softwoods and hardwoods. Figure 9-1 shows a choice stand of conifers in New Hampshire. Diameter breast high (dbh) is about 45 in and height is about 100 ft.

2. The central hardwood forest is an 800-mi-wide belt extending from Cape Cod, Massachusetts, to southern Texas. Trees are many softwoods and hardwoods, including ashes, basswoods, beeches, birches, catalpas, cedars, cottonwoods, elms, hickories, locusts, maples, oaks, pines, poplars, sycamores, tupelos, and willows. Figure 9-2 illustrates a stand of yellow poplars and beech-maples in Ohio. The tree being studied by the forester is about 27 in dbh and perhaps 70 ft high.

3. The southern forest, Figure 9-3, is a 300-mi-wide arc along the Atlantic and Gulf Coasts from Virginia to eastern Texas. Four important pines characterize this forest—shortleaf, longleaf, slash, and loblolly. Other significant stands are beeches, cypresses, hickories, oaks, and tupelos. Hardwoods and soft pine pulpwoods are also abundant.

4. The Rocky Mountain forest area embraces all 11 western states and shares this vast land with spacious grasslands and deserts. Conifers dominate, along with aspens, cottonwoods, oaks, and soft pulpwoods. Figure 9-4 pictures a mighty stand of Engelmann spruce in Colorado.

5. The Pacific Coast forest is an average 200-mi-wide coastal zone from northern Washington to southern California. Here are found the forest giants, hemlocks, firs, redwoods, spruces, and the giant sequoias, the General Sherman tree in Sequoia National Park being 30 ft dbh. Figure 9-5 shows a venerable stand of Douglas firs and western hemlocks in Oregon ranging up to 17 ft dbh and 300 ft high. Aspens, cottonwoods, eucalypti, laurels, madrones, maples, and oaks are the medium- and small-size trees. Characteristic troublesome brush is chaparral, made up of better-than-head-high mesquites, scrub oaks, and similar thorny shrubs and dwarf trees.

Other regional forests of the United States but outside the 48 contiguous states, are those of Alaska, Hawaii, and Puerto Rico.

6. The Alaska forest is divided into two natural forests, coastal and interior. They occupy about one-half the area of Alaska. Figure 9-6 portrays one of the magnificent coastal stands of hemlocks and spruces along the Seymour Canal. Also along the coast are alders, cedars, cottonwoods, and willows. In the interior are aspens, birches, poplars, spruces, tamaracks, and willows. Thus, Alaska offers many softwoods and hardwoods. The aerial view emphasizes the enormous task of clearing and grubbing for the excavation.

7. The forest of Hawaii covers about 25 percent of the total area of the seven main islands. However, there are fewer types of trees here than in the other seven forest regions. The ohia lehuas of Figure 9-7 are most abundant. The tree marked "2" is about 30 in dbh and is 90 ft high. Koas, or Hawaiian mahogany, mamanis, and kiawes, or mesquites, are native trees. Trees introduced from other lands are ashes, cedars, eucalypti, oaks, and redwoods. The native and imported trees offer a fair variety of softwoods and hardwoods.

8. The tropical forest covers small coastal regions of Florida and Texas, but the true tropical forest is in Puerto Rico. Figure 9-8 displays the luxuriant growth in the Luqillo rain forest. The large tree is a tabonuco of about 36 in dbh and 100 ft height. Common trees are ausubos, cedars, lignum vitaes, mahoganies, marias, and teaks. There is a limited variety of softwoods and hardwoods.

Figure 9-1 **Conifers of the northern forest.** A choice stand of conifers of the northern forest in New Hampshire. These large trees are desirable merchandise and they would defray the cost of clearing and grubbing. The tree to the right of the cruiser is about 45 in diameter breast high (dbh) and about 100 ft high. Brush is negligible and all trees are merchantable—a real boon for the excavator in minimizing the cost of preparation of right of way. *(University of New Hampshire.)*

9–6 Clearing and Grubbing

Figure 9-2 **Poplars and beech maples of the central hardwood forest.** High-quality yellow poplars and beech-maples of a woodland of Holmes County, Ohio, representing the central hardwood forest. The tree being studied by the forester is about 27 in dbh and perhaps 70 ft high. This forest represents the example for estimating the cost of clearing and grubbing discussed under "Production and Costs" in this chapter. It was estimated that clearing and grubbing, including hauling of merchantable logs to a nearby saw mill, would cost $2975 per acre. The excavator decided to give the merchantable logs to the owner of the sawmill in exchange for the clearing and grubbing of the right of way in accordance with standard specifications and special provisions. *(Ohio Agricultural Experiment Station.)*

Figure 9-3 Loblolly and shortleaf pines of the southern forest. An example of a fast-growing stand of loblolly and shortleaf pine in the southeastern United States. Here brush is rather heavy but the trees are merchantable, ranging up to 24 in dbh and perhaps 75 ft height. The country is level and the climate is equable, making for efficient and economical clearing and grubbing of these soft woods. *(International Paper Company.)*

9–8 Clearing and Grubbing

Figure 9-4 A stand of Engelmann spruces of the Rocky Mountain forest in the mountains of Colorado. Such growths of huge trees are typical of the Rocky Mountain forest. These superb specimens average about 30 in dbh and perhaps 125 ft in height. Brush is characteristically light or nonexistent, making for easy felling and removing of trees for the sawmills. *(U.S. Forest Service.)*

Figure 9-5 Firs and hemlocks of the Pacific Coast forest, a luxuriant growth of moss-festooned Douglas firs and western hemlocks in Oregon. Many parts of the Pacific Coast forest are richly favored with abundant moisture, resulting in fast heavy growths. Here brush is heavy and would be troublesome and costly for clearing and grubbing. The salability of these average 36-in-dbh conifers would more than offset the overall expenses. *(U.S. Forest Service.)*

9–10 Clearing and Grubbing

Figure 9-6 **Hemlocks and spruces of the Alaska forest.** The Alaskan coastal forest is a part of the Alaska forest region. This dense, rich stand of large hemlocks and spruces flanks the banks of the natural Seymour Canal northeast of Sitka, and it typifies the great task of clearing and grubbing in the luxuriant growths of the coastal forest. In this example, the great percentage of merchantable timber would be rafted to the nearest sawmill and the work would be simplified. *(U.S. Forest Service.)*

Figure 9-7 A stand of the most abundant native tree, ohia lehua, on the island of Hawaii. The growth is attributed the annual rainfall of about 100 in and to the salubrious climate at 3000-ft elevation. The resultant heavy undergrowth and the wet rough terrain would make clearing and grubbing both difficult and costly. In spite of its relative abundance ohia lehua, the native tree, is little used as sawn lumber. *(Division of Forestry, State of Hawaii.)*

9–12 Clearing and Grubbing

ESTIMATING COSTS

Human inroads upon the forests have lessened the amount of work for clearing and grubbing but, nevertheless, Figures 9-1 through 9-8 depict vividly what the excavator may expect on many jobs and indicate there is still a lot of work for modern Paul Bunyans.

If growth of trees and brush is medium to heavy, a careful estimate of the cost of clearing and grubbing must be made. Otherwise, the excavator may find that this cost has been underestimated and that it may be alarmingly high in relationship to the cost of the excavation.

In Chapter 6 the necessary field work was described. This reconnaissance included a rough appraisal of clearing and grubbing during the hiking along centerline. In the Cleghorn Road to Crestline Freeway in the San Bernardino mountains, growth of trees and chaparral was rather heavy, as shown in Figures 6-7, 6-9, and 6-10. Thus, further study of clearing and grubbing conditions following the original field examination was suggested.

As a matter of interest, the average lump-sum bid of seven bidders was $199,700 for clearing and grubbing. As the right-of-way included some 197 acres, the equivalent average bid per acre was about $1010 in 1969, equivalent to about $1650 in 1978. In contrast to the equivalent 1978 unit bid of $1650 per acre for the heavy clearing and grubbing in the mountains is that for a San Diego County freeway where only eye-high chaparral existed. The 1976 average unit price of four bidders was $290, equivalent to a 1978 price of about $330, only 20 percent of the price in San Bernardino County.

Equally interesting are the ratios of total bids for clearing and grubbing versus total bids for rock excavation. For the heavy work in San Bernardino County the ratio was 17 percent, whereas for the light work in San Diego County the ratio was but 2 percent. Obviously, clearing and grubbing of the San Bernardino County rock job was a more important determinant in bidding the jobs.

In estimating the item of clearing and grubbing one must consider what is below the

Figure 9-8 Tabonucos of the tropical forest of Puerto Rico. Although associated with Puerto Rico, the tropical forest also extends to small areas of the coastal regions of Florida and Texas. Of all of the eight forest regions, the tropical forest symbolizes the greatest luxuriousness. Surrounding these tabonucos is the entwined undergrowth. These wet forests in hilly and mountainous country provide a test for people and machines in the clearing and grubbing of the land. Although tree numbers and sizes appear to be more or less equal to those of Figure 9-2, it is probable that the cost per acre for clearing and grubbing would be double that of the Ohio land, or in the vicinity of $6000/acre, based on 1978 costs. *(U.S. Forest Service.)*

ground as well as what is above the ground. The apparent vegetation may be grouped according to the size classification of the U.S. Forest Service, tabulated below.

	Dbh (diameter breast high), in
Seedlings	0–3
Saplings	3–4
Poles	4–12
Standards	12–24
Veterans	Over 24

Generally, brush and chaparral are equivalent to seedlings and saplings and thus are to be wasted by burning or burying or to be salvaged by chipping. Poles, depending upon size, may be wasted, salvaged, or sold. Almost always, standards and veterans are merchantable and are to be felled, cut to suitable log lengths, and hauled to the sawmill. Depending on size, standards may be chipped if the special provisions call for chip spreading.

Seedlings, saplings, and sometimes nonmerchantable poles are usually removed, along with their roots, by crawler-tractors equipped with bulldozers or brush rakes. In the case of felled standards and veterans, it is well to know definitely what is below the ground or the exact nature of the stump roots, so that grubbing or stumping may be done efficiently. Whether stumps are to be pulled, blasted, or cut or ground in place, the root classification is important. The cost of removal varies from minimum for small lateral roots to maximum for large tap roots. Such roots are illustrated in Figures 9-26 and 9-29.

Tap-root trees include black gums, hickories, white oaks, and yellow pines. Combination tap- and lateral-root trees are numerous and they include ashes, chestnuts, various oaks, white pines, poplars, walnuts, and most fruit trees. Lateral-root trees are most numerous and they include cedars, elders, firs, hemlocks, maples, spruces, and white pines.

Necessary Data

The cost of clearing and grubbing is based fundamentally on the all-important tree count, in which the following data for an average acre are amassed and tabulated: (1) number of trees; (2) average tree diameter, dbh; (3) average height of trees; (4) kind of trees and kind of roots; (5) amount of brush—dense, average, or sparse; (6) green or dead growth; (7) kind of soil; (8) moisture in soil; (9) ground, frozen or not; (10) topography—level, hilly, or mountainous.

These data, together with a forecast of weather and information on accessibility, will guide the estimator as to whether the excavator should perform the work or sublet the clearing and grubbing to a specialist or to a logger. Sometimes the specialist will clear and grub simply for the consideration of the merchantable timber and sometimes there is a compromise, depending on the data of the tree count.

And so it is most important that the owner or prime contractor get complete information and, if the decision is to subcontract the work, enter into a well-conceived contract. Since the initial building of access roads and development of water supply will depend upon the scheduled prosecution of clearing and grubbing, the contract must have a time limit and penalty clause for nonperformance.

Timber Cruise

A typical two-person timber cruise for a tree count is shown in Figure 9-9. The compass person is recording the dbh of the tree, which the cruiser at the tree is determining with a Biltmore stick. Shortly afterward the height of the tree will be measured with a two-scale Abney hand level and a 100-ft measuring tape. The Biltmore stick uses the principle of similar triangles; the Abney level with measured baseline is based on the same principle, with either direct readings or percentage readings for heights.

In such a survey not all trees in all areas are measured. The entire area is sampled in keeping with the desired measure of precision. The results are then tabulated, and average tree volume in cubic feet and in merchantable board feet of lumber is calculated for an average acre.

The need for quick estimation of limited clearing and grubbing of coal-stripping properties in the central states led to development of the graphs of Figure 9-10. By using these, one person is able to cruise small stands of timber and, with reasonable precision, estimate

9–14 Clearing and Grubbing

Figure 9-9 Two foresters running a two-person timber cruise for a tree count. The forester in the foreground is running bearing on the traverse of the survey and is recording the diameter at breast height (dbh) of the tree being measured with a Biltmore stick by the forester of the background. Shortly the height of the tree will be fixed by the use of an Abney hand level and tape for establishing the length of the lateral baseline. These determinations, together with the aforementioned data, will give information for estimating the cost of clearing and grubbing an acre of the forest. *(U.S. Forest Service.)*

the cubic feet and the merchantable board feet by simply measuring the dbh with a Biltmore stick. Having these approximate data, one can negotiate with sawyers and loggers without delay.

These principles give merely a thumbnail sketch of the many intricacies of the art of clearing and grubbing. Starting with the sequoias of the pine family and ending with the catalpas of the bignonia family, there are 27 families and 900 species and varieties of trees in North America north of Mexico. Most of these trees may occur in level, hilly, or mountainous terrain, in dense, average, or sparse growths, in small, medium, or large sizes, and with tap, tap-lateral, or lateral roots. Truly, the complexities of clearing and grubbing are just about limitless.

METHODS AND MACHINERY

The foregoing principles obviously imply that different methods must be used for clearing and grubbing. Basically, the methods are two, each one using different key machinery for the average job.

Crane Method for Heavy Clearing and Grubbing

The crawler-mounted crane equipped with a grapple is used as the loading and casting machine in hilly and mountainous country with medium to heavy growth. Such conditions obtain in Figure 9-4. Here felled trees would be cut to suitable lengths for hauling to the sawmill and, together with stumps, would be cast to the access road.

Auxiliary machinery would include the following units:

1. Several medium- or heavyweight crawler-tractors equipped with bulldozers, brush rakes, tree pushers, tree splitters, and winches, as conditions and experience dictate.
2. A medium-weight crawler-tractor equipped with brush bucket and hooks.
3. Heavy-duty portable chain saws and perhaps circular saws.
4. A brush burner or steel burning box.
5. Brush haulers and log haulers.
6. A miscellany of choker chains and similar tools.

These machines together with several special clearing and grubbing units are shown in Figures 9-11 to 9-25. Grubbing or stumping, the removal of stumps, is done by crawler-tractor, equipped with bulldozers, tree pushers, or tree splitters, or by crawler-tractors equipped with winch and choker chain. Alternate methods are by stump cutter or by blasting. The possible enormity of grubbing is shown dramatically by Figure 9-26, the

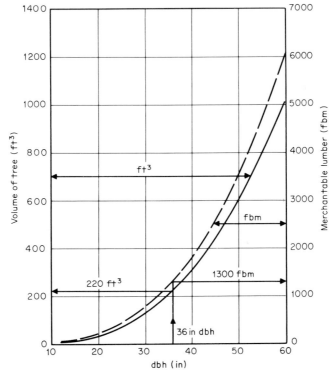

Figure 9-10 Cubic feet and merchantable board feet of trees according to dbh.
NOTES:
  dbh is diameter of tree in inches at breast height of 4½ ft.
  Taper of average tree is 3.2%, or about 3 in per 8 ft of length.
  Tree top of less than 8-in diameter is wasted.
  Merchantable board feet are estimated at 50 percent of the total volume of the tree.
  Approximate equations for curves:
  $$\text{Cubic feet} = 0.0047 \, (\text{dbh})^3$$
  $$\text{Board feet} = 0.0282 \, (\text{dbh})^3$$
  where dbh is in inches.

Figure 9-11 **Heavy-duty chain saw felling 48-in-dbh tree.** The heavy-duty chain saw, measuring up to 48-in bar or cutting edge, has largely replaced the two-person manual cutoff saw and the engine-driven bicycle saw for the felling of trees. The sawyer, who is completing the wedge-shaped undercut of a large tree, will then make the backcut and the tree will fall in the direction of the undercut. A narrow strip of wood left between the undercut and the backcut will prevent the tree from rotating as it falls. The felling of this approximately 48-in-dbh tree will require about 15 min, depending on the hardness of the wood. *(American Forest Product Industries.)*

Figure 9-12 Sawyer cutting slash for disposal with heavy-duty chain saw with 42-in bar. In this heavy brush of seedlings and saplings, the slash will be stacked for burning or burial in the process of clearing and grubbing. Estimating production and costs for this brush cutting depends obviously on the size and density of the growth, and the estimate is largely a matter of judgment of the experienced specialized contractor for clearing and grubbing. It is a facet of costs which defies analysis by conventional time-motion studies and it emphasizes the unique importance of personal know-how. *(McCulloch Corporation.)*

Methods and Machinery 9-17

Figure 9-13 Crawler-tractor tree pusher, front-mounted, and grapple, rear-mounted, on 200-hp crawler-tractor felling a 30-in-dbh tree. The roots will be torn out along with the tree and then the roots will be cleaned out completely by the tractor's bulldozer. The operator can control the direction in which the tree will fall. When a pattern is established for the felling of the trees in the same direction, they may be piled speedily and efficiently for future handling. A tree of this size along with its roots may be cleared and grubbed in about 5 min, depending largely on the nature of the root system. If the tree is merchantable, then the roots may be sawed off closely, thus affording more timber than provided by the conventional method of felling shown in Figure 9-11. The incorporation of tree pusher, bulldozer, and grapple into this crawler-tractor assembly provides versatility in both clearing and grubbing. *(Fleco.)*

Figure 9-14 A V-shaped tree splitter or tree cutter mounted on a 200-hp crawler-tractor. This ambidextrous tool, with its serrated cutting edge, can shear off trees of up to about 15 in-dbh either to the left or to the right by a continuous pass. The larger hardwood tree is being split in the middle, and simultaneously the two halves are being sheared off at ground level. This is a one-pass operation on a tree of about 24 in dbh. As 400-hp and more powerful tractors may be equipped with the tree splitter, it can handle trees of considerable dbh. The tractor works in low gear at about 1 mi/h and the tree is sheared in a fraction of a minute. Tools such as the tree pusher of Figure 9-13 and the tree splitter are especially efficient in large-scale clearing and grubbing projects. *(Fleco.)*

Figure 9-15 A brush rake mounted on a 300-hp crawler-tractor in the midst of clearing brush and poles in a stand of conifers of northern Idaho. This is the key machine for the brush-blade method of clearing and grubbing, perhaps the most popular method because of its reliance on the ubiquitous workhorse of the excavator. All the growth in this area has been cleared by the tractor–brush rake. In this mountainous country its teammate is the crawler-tractor with brush bucket and clamps. As the tractor–brush rake clears and stacks the growth along centerline at about 400-ft centers, the tractor–brush bucket will load and haul the slash to the nearest burn area. If the burn areas are less than 500 ft apart, it is economical for the tractor–brush rake to handle the entire job by dozing less than about 250 ft, one way. In the conditions pictured, the tractor–brush rake can clear and grub about an acre in 8 h. *(Mike Lapinski, "Right-of-Way Clearing—Headache or a Profit," Western Construction, December 1975, p. 29.)*

Figure 9-16 Clearing and grubbing by team of crawler-tractor–brush rakes. In contrast to the working conditions of Figure 9-15 are these for two 300-hp-tractor–brush rakes clearing and grubbing a light stand of brush, seedlings, and saplings in level country of the southeastern United States. Here production is at its highest, as implied in the clearing and grubbing estimates for the two machines. The tractors are working in unison with slightly overlapped rakes, and they are able to clear and grub more than 16 acres daily, or about 8 acres each, instead of the 1 acre handled by the tractor of Figure 9-15. The pictures dramatize the differences in size and amount of growth and in terrain, all of which control production and unit cost. *(Fleco.)*

Methods and Machinery 9-19

Figure 9-17 The crawler-mounted crane with grapple is the key machine for the crane method of clearing and grubbing. This 40-ton crane with 60-ft boom is used in hilly and mountainous country for reaching out and gathering slash, stumps, and trees and depositing them on the access or haul road. The tractor–brush rake and the tractor–brush bucket cannot work efficiently in uphill dozing and haulage. As pictured, the crane is working from the pioneered access road. Of course, the production of the handling crane is a function of the aforementioned variables of size and amount of growth and terrain. Under average conditions, working in conjunction with sawyers and crawler-tractors equipped with brush rakes and brush buckets, the crane can take care of about an acre every 8 h. *(Mike Lapinski, "Right-of-Way Clearing—Headache or a Profit,"* Western Construction, *December 1975, p. 31.)*

Figure 9-18 Action of the crane grapple. The grapple of the crane is shown dropping slash, trees, and stumps for the waiting tractor–brush bucket which will carry the waste to the burn area. This log-handling grapple has two tines. Occasionally, four-tine grapples are used for a preponderance of brush and stump handling. Capacities in tons are matched to the crane capacities. *(Mike Lapinski, "Right-of-Way Clearing—Headache or a Profit,"* Western Construction, *December 1975, p. 31.)*

9-20 Clearing and Grubbing

Figure 9-19 A 200-hp crawler-tractor–brush bucket equipped with hooks. After loading the waste growth, the tractor will turn and head for the burn area some 600 ft away. Production and cost for this simple, straightforward operation are easily calculable according to the following tabulation.

Cycle time	Loading and turning	0.5 min
	Hauling 600 ft at an avg 3 mi/h	2.3
	Dumping and turning	0.3
	Returning at average 4.5 mi/h	1.5
	Net cycle time	4.6 min
Production	Loads hauled per 45-min hour	9.8
	Loads hauled per 8 h	78

It is probable that 78 loads of waste growth would take care of 1 acre of clearing and grubbing. Therefore the costs are:

Loading brush
 Machine, cost of 200-hp crawler-tractor with brush bucket and hooks (1978 cost $115,000).
 Hourly cost of ownership and operation
  Machine (Table 8-3)

Tractor: 34% of $115,000	$39.10
Operator	15.60
Total	$54.70
Direct job cost, 8 × $54.70	$438
Direct job unit cost, per acre. $438/1	$438

(Mike Lapinski, "*Right-of-Way Clearing—Headache or a Profit,*" Western Construction, *December 1975, p. 31.*)

Figure 9-20 A portable circular saw, known as a *bicycle saw*, used to saw felled timber, usually pulpwood, into logs for hauling to the mill. The 30-in-diameter circular saw, driven by a 15-hp gasoline engine, may be adjusted also to the horizontal position for cutting standing timber. The maximum size log or tree handled is about 10-in diameter. The 1978 cost of the saw was about $1600, and the hourly cost of ownership and operation was estimated at 84 percent of $1.6, or about $1.40 (Table 8-3). The consequent total for machine and sawyer was about $16.60 hourly. Of course, sawing costs vary with the diameter of the tree or log and the hardness of the wood. *(Grant W. Sharpe.)*

Figure 9-21 A 200-hp crawler-tractor equipped with brush bucket and hooks is shown loading out an 8-ton brush-hauler truck. When haul distances to the waste areas exceed that of economical haulage by the tractor–brush bucket, it is feasible to use the rear-dump trucks or the rear-dump tractor-trailers. This method is necessary when brush must be wasted in a distant sanitary fill or dump because of environmental or fire hazard prohibition on burning. *(Fleco.)*

9-22　Clearing and Grubbing

Figure 9-22 The crawler-tractor-drawn rolling chopper shown may be used for clearing a right-of-way when the chopped remains of brush, seedlings, and saplings need not be removed completely. Such rights-of-way are for trench excavations for pipelines and other conduits of the utility nature. Here a narrow section of the total width needs to be completely cleared and grubbed for the trench excavation. The rest of the right-of-way need not be grubbed and in the future need be only occasionally cleared of new growth. The rolling choppers, water-ballasted drums with chopping blades, may be used in single, double, or triple units depending on the amount and size of the growth and on the horsepower and weight of the crawler-tractor. Generally, the higher the travel speed, the greater the efficiency of the chopper. *(Fleco.)*

massive root spread of an old Douglas fir. These methods are illustrated by Figures 9-13 and 9-27 to 9-30.

The tree pusher of Figure 9-13 tears out the roots of the tree along with the felling of the tree. The stumping or grubbing of Figures 9-27 to 9-30 is described in the legends accompanying the pictures.

Emphasis must be placed on medium- and heavyweight crawler-tractors as the workhorses for clearing and grubbing. Equipped simply with bulldozers and rear-mounted winches, they can pioneer the access roads, fell trees of considerable girth by bulldozer or choker chain, and skid logs by either drawbar pull or cable winch. With attached brush rake they can gather and stockpile slash and loose rocks, and with attached brush buckets and hooks they can transport and load slash of considerable size. Equipped with tree splitter or tree pusher, they can splinter and fell sizable trees, taking with the trunks the attached roots. When equipped with root plow, they can cut off roots below ground and minimize the handwork of the root pickers. They are the universal tool for stumping. When pulling rolling choppers, the crawler-tractor with bulldozer is especially effective in stands of brush, seedlings, and saplings. There the chopped growth can be raked and burned after a short period of drying out.

As pictured in Figure 9-31, the bulldozer-equipped crawler-tractor may well be the only machinery used in the clearing and grubbing of light to heavy brush. The versatile machines are brushing the chapparal into piles for chipping according to the special provisions of this chapter for this huge freeway job.

Brush Rake Method for Light Clearing and Grubbing

In the common method the crawler-tractor equipped with a brush rake is the key machinery in flat and hilly country with light to medium growth. Such conditions are shown in Figure 9-16. Herein, a team of heavyweight crawler-tractors with brush rakes is just about doing a complete job of clearing and grubbing a light growth of timber. In such work just about the only cramp upon the tractor is its limitation in steep upgrade bulldozing. For

Figure 9-23 Heavy-duty portable log chipper, suitable for making chips for distribution on embankment slopes in accordance with the special provisions of this chapter. This machine, weighing 45,000 lb with a 290-hp engine, can take logs up to 5-ft diameter and 10-ft length. Hourly production is as high as 25 tons, depending on the nature of the wood. Chips of maximum ½-in thickness are required by the special provisions. After suitably sized chips have been produced by the adjustable cutters, they will be hauled to the fill slopes and distributed to a maximum depth of 1 in. Assuming an hourly production of 15 tons of chips and a specified depth of 1 in, the chipper can cover 12,000 ft² of fill slope hourly, ample output for the average excavation job. *(Vermeer of California.)*

9-24 Clearing and Grubbing

Figure 9-24 Heavy-duty brush burner supplying both forced air and fuel oil to a stack of wasted growth. This unit can supply up to 80,000 ft³/min of air and up to 60 gal/h of fuel oil to support combustion. It is especially effective in the burning of green growth. Burning may be accomplished by two other methods, namely, by simply burning the pile without benefit of mechanical assistance, or by use of a steel burn box about 20 ft long, 10 ft wide, and 8 ft high and welded up from ⅜-in plate. The boxes can be equipped for forcing air-entrained fuel oil into the bottom of the box, thus producing very intense heat. When the crane method of clearing and grubbing is used, the box can be moved short distances along the right-of-way for most efficient work by the tractor–brush rakes and tractor–brush buckets. *(Fleco.)*

Figure 9-25 Tractor-trailer log haulers on a private logging road in the state of Washington. Such log haulers of five axles have a nominal capacity of about 24 tons on load-restricted public highways. It is probable that the average load on this private logging road is about 30 tons, or about 7200 board feet of dimension lumber. On this slightly downgrade road the loaded units average about 25 mi/h and the light unit with "piggy-back" trailer averages about 35 mi/h. Gentle curves and good road maintenance are the requisites. *(St. Regis Paper Company.)*

Methods and Machinery 9-25

Figure 9-26 The massive upended root spread of an old-growth Douglas fir, blown down by the November 1958 windstorm of the Pacific Northwest. The lateral root system averages about 25 ft in diameter. While it is an unusual example, nevertheless it emphasizes both the complexity of the roots and the great work of stumping a good-size tree. In this case, after the felling of the tree the stumping would necessarily be a piecemeal undertaking by sawyers, choker setters, and possibly powder operators. *(Washington Department of Natural Resources.)*

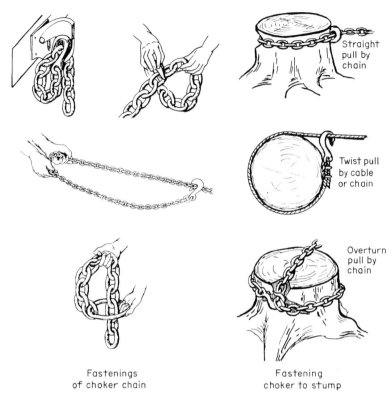

Fastenings of choker chain

Fastening choker to stump

Figure 9-27 Stumping, or grubbing, by chain or cable choker. The pull is either by drawbar or rear-mounted winch of the crawler-tractor. The three common attachments are straight-pull by chain, twist-pull by cable or chain, and overturn-pull by chain. The straight-pull is the easiest attachment. The twist-pull, putting little kink in the line so that either cable or chain may be used, gives an advantageous twist to the stump and roots. The overturn pull by chain is most effective as it provides greatest leverage to the stump and roots. For safety and to eliminate slippage, all stumps should be notched for both chain and cable chokers. As cable tends to kink and break, the chain choker is preferred. The head end of the chain choker may be hooked into the clevis or hook of the winch cable when pull by winch is preferred over use of drawbar.

Figure 9-28 Large-capacity, heavy-duty stump cutter removing a 30-in-diameter stump to a depth of 24 in below ground level. Such stumping requires about 10 min, depending upon the root structure and the nature of the wood. The pictured machine has a cutting wheel diameter of 30 in and is powered by a 65-hp engine. For clearing and grubbing the stump cutter has two advantages over conventional means of stumping. First, it is easily maneuverable into places where heavy machinery cannot work without damage to the area as shown in the picture. Second, it is not necessary to remove the stump as it is reduced to chips. *(Vermeer of California.)*

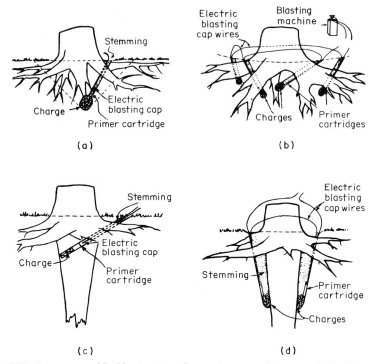

Figure 9-29 Stump removal by blasting. According to the nature of the roots. *(a)* Loading a small lateral-rooted stump beneath the tree trunk. *(b)* Loading a lateral-tap-rooted stump with connected charges. *(c)* Loading a tap-rooted stump in a hole bored into the tap root. *(d)* Loading a large tap-rooted stump by charges placed in holes dug alongside the root. (Blasters' Handbook, E. I. du Pont de Nemours & Co., 1966, p. 379.)

In all cases the amount of the charge depends on the diameter and condition of the stump, the nature of the root system, and the kind of soil—loose or firm. The looser the soil, the more the charge.

The experience of the powder operator is basic in determining the amount of charge. The following table for recommended charges of 1½ in × 8 in cartridges of dynamite is based on firm, dense soil. If the soil is dry and light, the charges will have to be increased materially. The charges are also based on placement either under or alongside the stump. In the case of a tap-rooted stump, only one-fifth as much dynamite will be required if it can be placed in a hole bored in the tap root.

Recommended Charges

Diameter of stump, in, at 1 ft above ground	Number of 1¼ in × 8 in dynamite cartridges		
	Green	Old but solid	Partly rotted
6	2	1½	1
12	4	3	2
18	6	4	3
24	8	6	4
30	10	7	5
36	13	9	6
42	16	12	8
48	20	15	10

SOURCE: *Blasters' Handbook*, E. I. du Pont de Nemours & Co. (1966), p. 381.

The charges are for 1¼-in by 8-in dynamite cartridges, but sometimes only size 1⅛ in by 8 in is available. In such case, multiply the number of cartridges in the table by 1.25.

Figure 9-30 The crawler-tractor root plow is a most effective tool for grubbing in open country with light or medium brush. It is a V-shaped cutting blade which is drawn laterally through the soil by a crawler-tractor, severing the roots below the surface. The brush may then be piled by a bulldozer or brush rake for burning or haulage. This light to medium 150-hp tractor is drawing a 9-ft-wide cutting edge through the roots of mesquite brush in the southwestern United States. This tractor–root plow, traveling at about 1.5 mi/h with slightly overlapping passes, can grub some 8 acres per 8-h day preparatory to cleanup of severed brush. This systematic grubbing of cut-and-fill areas in semiarid country is economical for the excavator, costing about $46/acre for the tractor-root plow work. *(Fleco.)*

ing. For the brush-rake method supplementary machinery, subject to changes according to the job requirements, might include the following units.

1. Heavyweight crawler-tractor with brush bucket and hooks.
2. Heavy-duty chain saws.
3. Steel burning box or brush burner.
4. Brush haulers.
5. Rear-mounted winches with choker chains for the crawler-tractors.
6. Miscellany of small tools such as axes and hand saws.

In this description of methods and machinery there is no discussion of machines used in large-scale logging. These include tractor-mounted and tractor-drawn arches and pans for the skidding of trees and logs and skid-mounted winches for the cable skidding and yarding of trees and logs. Such methods and machines are not typical of average clearing and grubbing for excavation.

PRODUCTION AND COST ESTIMATES

The estimated hourly cost of ownership and operation of machinery is one of the factors determining unit cost, the other being estimated hourly production. In this chapter and the following ones through Chapter 15 the hourly cost is determined as explained in Chapter 8. For the machines used in excavation the values of Table 8-3 are applied, as adjusted to the 1978 costs of the machinery and the 1978 costs for workers with average fringe benefits.

Of the several phases of excavation, clearing and grubbing productions and costs are least documented. It is as though writers of articles, manufacturers' handbooks, and books on earth and rock excavation shrug off this important preliminary work as being insignificant when compared with the all-important, romantic excavation.

That this attitude is ill-conceived is demonstrated by the following not uncommon relationship. Let us assume that a double-track railroad is being built through rolling country with a dense growth of trees averaging 24 in dbh. In a representative 1-mi of right-of-way, there is 53,000 yd^3 of excavation, costing $0.95/yd^3. There is 18.2 acres of

9–30 Clearing and Grubbing

Figure 9-31 Clearing and grubbing a heavy growth of chaparral in a mountain arroyo for freeway building in San Diego County, California. In this picture one 300-hp and one 150-hp crawler-tractor with bulldozer are brushing down average 40 percent grades to a 190-hp crawler-tractor with loading bucket, which is hauling the brush to a nearby brush pile for chipping. The two brushing tractors in the upper left part of the picture are traveling a 600-ft-long triangular loop to doze a total swath of 3500 ft^2 of brush down to the lower access road.

The cycles of the brushing tractors average 8.2 min in this precipitous terrain, and the total average production is about 26,000 ft^2/h. Another 300-hp tractor with bulldozer is working behind the picture taker in more favorable terrain, with an average production of about 32,000 ft^2/h. Thus the combined production for the three brushing tractors, when brushing exclusively, is 58,000 ft^2/h, or 1.33 acres/h.

During a 7-day period of 56 working hours, the four tractors cleared and grubbed, preparatory to chipping, 26.8 acres of cut-and-fill area for rock excavation. The costs for this work are tabulated below. Hourly costs of ownership and operation and of operators are as derived from Table 8-3 for 1978 costs.

Machines
  300-hp crawler-tractor with bulldozer, 33% of $183 = $60.40. Two units
    @ $60.40 $120.80
  150-hp crawler-tractor with bulldozer, 34% of $85   28.90
  200-hp crawler-tractor with brush bucket, 34% of $115   39.10
  Operators, 4 @ $15.60   62.40
  Total cost $251.20
Hourly cost of supervision by foreman with medium-weight pickup truck.
  Salary @ $20.00 and truck @ $3.30   23.30
Total hourly cost $274.50
Total cost, 56 × $274.50 $15,372
Unit cost per acre, $15,372./26.8 acres $574

The lump sum bid for clearing and grubbing, involving some 180 acres, was $220,000, or about $1222/acre. This random sample of the cost under particular conditions is, of course, not representative of all the clearing and grubbing. It is simply an estimate of the cost of one phase. Building of trails, chipping and chip spreading, and removal of trees are some of the items entering into the bid price of $1222/acre.

clearing and grubbing, costing $1050/acre. This seemingly innocuous initial work costs $19,110, or 38 percent of the $50,350 cost of excavation. Obviously, this cost is worthy of much attention.

The reasons for the paucity of reference data for clearing and grubbing are easily explained. First, the methods, machinery, productions, and costs are not studied sufficiently in the field by the writers. Without this accumulation of data the writer, trying to theorize on estimating, is without basic premises. The average estimator, even one who has never time-studied a heavyweight crawler-tractor-bulldozer in the field, can estimate hourly production and unit cost for bulldozing under given conditions with the aid of reliable manufacturers' handbooks. We may call this method safe objective estimating, as dozing cycles, efficiency, payload of blade, and hourly cost of ownership and operation are known fairly precisely. On the other hand, consider the same tractor equipped with a brush rake, clearing and grubbing an acre of mountainous right-of-way as shown in Figure 9-6, where the growth varies for both trees and chaparral, the growth is green, there is a variation in the root structures of the several conifers, topography is irregular, and weather is variable. The functions of the tractor are myriad, including trail blazing, tree toppling, tree skidding, stumping, and brush raking. Little or no book knowledge is available, the printed word dealing with generalities. The cost estimate must be based largely on experience.

Second, this experience is unfortunately never recorded. In this field of clearing and grubbing, the doers, those with the most knowledge, are workers and they have no time to formalize their hard-won experience for the estimator. Additionally, in this special field of endeavor experience is considered to be proprietary and there is a natural reluctance to give it away.

Third, the manufacturers of machinery realize that nature has arrayed a a formidable host of variables to confound the estimator, and they too are reluctant to provide even sketchy handbook data for estimating productions and costs for clearing and grubbing.

Occasionally, when a phase of clearing and grubbing lends itself to conventional analyses of production based on the mechanics of work, one may apply the usual objective methods of estimating. Figure 9-16 exemplifies such a working condition. These two heavyweight tractors, equipped with brush rakes, are clearing and grubbing a light stand of brush, seedlings, and saplings. They are working shoulder to shoulder in this work, although in systematic team work the following tractor would be slightly behind the other one so that the rake would overlap by about 3 ft.

Production and unit cost for 5 acres of work, the area being 1090 ft long and 200 ft wide and wherein brush piles are windrowed along both sides for burning, are estimated in this tabulation.

Two 300-hp crawler-tractors with 12-ft wide and 6-ft high rakes, together with a burn man, make up the work force. The method is to brush laterally to both sides. No supervision is needed for this simple task.

```
Cycle of the two teamed tractors:
  Brushing 220 ft at 2 mi/h                                    1.3 min
  Stacking and turning                                         0.7
  Brushing 200 ft at 2 mi/h                                    1.3
  Stacking and turning                                         0.7
  Net cycle time                                               4.0 min
Hourly production:
  Total width of swaths per cycle, 48–12 ft                    36 ft
  Length of swath                                              200 ft
  Area of swaths per cycle                                     7200 ft²
  Cycles per 50-min hour                                       12.5
  Area cleared and grubbed, 90,000 ft²                         2.07 acres
Unit cost for 1978:
  Hourly cost of ownership and operation of
  300-hp tractor-rake. 1978 cost, $183,000.
    Machine. 33% × $183.                                       $60.40
    (Table 8-3)
    Operator                                                   15.60
    Total                                                      $76.00
  For two machines                                             $152.00
  1 burn man                                                   11.50
  Total                                                        $163.50
  Unit cost per acre, $163.50/2.07                             $79.00
```

When one contrasts the fixed conditions of Figure 9-16 with the variable situations of Figure 9-15, one appreciates the difficulties in estimating by this kind of an analysis based on time-motion considerations.

What, then, are the means for estimating? First, it can be avoided simply by calling in subcontracting specialists and asking for competitive subbids. It is a common and an economical method. Such a specialist has know-how, machinery, and workers, and generally submits a subbid which is as low as the cost for the excavator to do the work.

9-32 Clearing and Grubbing

Furthermore, the building of access roads, clearing and grubbing, and development of water supply are works which must be done swiftly so that excavation can commence. Economy of time and money is a factor favoring subcontracting.

The alternative is to bring all resources into play and to prepare a well-thought-out estimate of productions and costs. The estimate is usually based on the following considerations.

1. *Safe, objective preparation,* based on all available knowledge of the individual productions and resultant unit costs for machines working under the given conditions. This fundamental principle of estimating is the most desirable of all methods.

2. *Relatively safe subjective preparation,* based on previous experiences in similar work. This analogical method is expedient and, when previous experience is weighted with adjustment factors, it is a dependable means for estimating.

3. *Least safe comparative preparation.* This not unpopular means involves simply a study and analysis of a compilation of competitive bids for clearing and grubbing under like or near like conditions. From this examination comes an average unit bid price which may well be not too far astray from the standpoints of competition and profit. It is a fact that almost all bids for excavation are prepared with proper attention to all three of these methods, the one blending into the others almost imperceptibly.

In the discussions of many of the preceding figures in this chapter, there have been some thoughts about the productions of the machines. Here follow additional estimates for productions and unit costs. Because these machines work together as a team, although they sometimes work independently, one must relate their performances to a unit of work, the acre of clearing and grubbing. Also, this acre must be defined in a rather broad sense.

The example of a clearing and grubbing estimate is for a 6-mi freeway through farming country in Ohio. The individual isolated stands of timber, 10 in all, average 860 ft in length, 200 ft in width, and 3.95 acres, and they total 39.5 acres. The stands, through which the right-of-way passes, are typified by Figure 9-2 showing the central hardwood forest of Ohio. Here are average-size trees of up to 30 in dbh and 80 ft in height, ranging from saplings to veterans. Brush is scant or negligible. The country is rolling and hilly. The right-of-way for the four-lane freeway has an average 200-ft width. There is a surveyors' jeep trail along centerline.

The job's special provisions permit brush burning in designated areas adjacent to the stands of timber, except during the summer months of June, July, August, and September. As clearing and grubbing are scheduled for early spring, brush and slash will be burned. A preliminary timber cruise, by which methods and machinery were fixed, gave the following tree count for an average acre.

	Number	*Cubic feet*	*Board feet*
Veterans, averaging 27 in dbh	12	1,080	6,480
Standards, averaging 16 in dbh	27	540	3,240
Poles, averaging 10 in dbh	109	540	3,240
Totals	148	2,160	12,960

Merchantable board feet are estimated at 12,960 per acre, or 511,930 total for the 39.5 acres of clearing and grubbing.

Job conditions call for the crane method and the sequence of work is as follows.

1. Light grading of about 12,900 ft of 24-ft-wide access road as close as possible to centerline and laterally to the existing highway. A few sections of 12-in-diameter culvert pipe in a few small ravines will provide drainage. This light grading, estimated at about 19,400 yd^3 of excavation, will involve clearing and grubbing of the 8600-ft length within the stands of timber. The remaining 4300-ft length is in open country.

2. Systematic clearing and grubbing of the remaining right-of-way and the hauling of merchantable logs to the sawmill.

Estimated productions and unit costs in 1978 for the several operations are based on 2 months of work, and they are:

1. Building of the access roads calls for 12 cuts of average 500-ft length and 2-ft depth, giving a gentle rolling grade, by sometimes following the natural contours of the land. The material is soft residuals and the average haul is 600 ft, one way. All grading is to be done by a 300-hp crawler-tractor, equipped with bulldozer, pulling a 20-cysm wheel-tire-mounted scraper. Spreading on the fill is to be done by the bulldozer and compaction will

be by the tractor-scraper machines. Natural moisture in the cut is adequate for compaction. It is estimated that the hourly production of the tractor-scraper is 276 cybm. The grading of the access roads will require 80 h.

Hourly cost of ownership and operation of the machines, as derived from Table 8-3:
300-hp tractor-bulldozer. 1978 cost = $183,000.
33% of $183 $60.40
20 cysm wheel-tire-mounted scraper. 1978 cost =
$79,000. 28% of $79   22.10
Hourly cost of operator   15.60
Total hourly cost $98.10
Total cost of machines and operators:
80 × $98.10 $7,848.00
Cost of culvert pipe, installed:
Twelve 30-ft sections of 12-in-diameter, 16-gauge corrugated steel pipe, estimated at $16/ft.
360 × $16 $5,760.00
Total cost for building of access road $13,608.00

2. Felling of trees is done by sawyers with heavy-duty chain saws

Sawyers' time is distributed as follows:
474 veterans at 10 min each   79 h
1066 standards at 6 min each   107
4306 poles at 4 min each   288
Total time   474 h
Number of sawyers working 160 h monthly   3
Hourly cost of ownership and operation of heavy-duty chain saw. 1978 cost = $900
Machine: 58% of $0.9 $ 0.60
Sawyer   15.20
Total $15.80
Total machine cost, or total cost: 474 × $15.80 $7,489.20

3. Stumping is done by a 300-hp crawler-tractor-bulldozer with winch and chain choker, axe wielders, and choker setters.

Tractor, and axe wielders; and choker setters' time for:
474 veterans at 6 min each   48 h
1060 standards at 5 min each   89
4306 poles at 4 min each   187
Total time   324 h
Number of axe wielders–choker setters   3
Hourly cost of ownership and operation of 300-hp crawler-tractor-bulldozer-winch. 1978 cost $192,000.
33% of $192 (Table 8-3) $ 63.40
Hourly cost of operator   15.60
Total hourly cost of machine $ 79.00
Hourly cost of 3 axe wielders-choker-setters @ $15.20   45.60
Total hourly cost of machines and workers $124.60
Total cost of stumping work, 324 × $124.60 $40,370.40

4. Cutting merchantable trees into logs and loading logs onto log haulers are done by a 40-ton crane, crawler-mounted, with grapple, and two sawyers with heavy-duty chain saws.

Crane and sawyers' time for sawing and loading:
474 veterans at 5 min each   40 h
1066 standards at 3 min each   54
4306 poles at 2 min each   144
Total time for work   238 h
Hourly cost of ownership and operation of machines:
40-ton crane with grapple (1978 cost $233,000):
Crane: 22% of $233 (Table 8-3) $ 51.30
Operator and oiler   30.80
Total $ 82.10
2 sawyers with heavy-duty chain saws @ $15.20   30.40
Total hourly cost of machines and workers $112.50
Total cost of felling trees, cutting, and loading of logs
238 × $112.50 $26,775.00

9-34 Clearing and Grubbing

5. Brushing and stacking slash by 300-hp crawler-tractor with rake for burning by one worker. Estimated time for acre is 6 h. Total time for 39.5 acres is estimated at 237 h.

Hourly cost of ownership and operation of 300-hp tractor-rake (1978 cost $183,000):	
Machine, 33% of $183 (Table 8-3)	$60.40
Operator	15.60
Total	$76.00
Hourly cost of worker to burn slash	11.50
Total hourly cost of machine and worker	$87.50
Total cost for brushing and stacking slash:	
237 × $87.50	$20,737.50

6. Logs are hauled 10 mi over a highway to the sawmill by rented 24-ton log haulers. Rental at $50.00 hourly, everything furnished. Estimated 98 trips for 2133 tons of logs. Trip estimated at 1.5 h, giving log-hauler rental time of 147 h.

Total cost for log hauling, 147 × $50.00	$7350.00

7. Supervision for clearing and grubbing. Supervisor with pickup truck.

Hourly cost of ownership and operation of medium-weight pickup truck (1978 cost $4500):	
Machine: 72% of $4.5 (Table 8-3)	$3.30
Total cost for pickup truck, 450 h @ $3.30	$1485.00
Supervisor's salary, 2 months @ $3520	7040.00
Total cost for supervision of clearing and grubbing	$8525.00

The cost of the entire operation may be summarized as follows:

Total cost for clearing and grubbing	$117,505.10
Cost per acre, 39.5 acres total	$2,974.80
Cost per 1000 board ft, estimated 12,960 board ft/acre	$229.50

After negotiation, the owner of the sawmill offers either to buy the estimated 511,920 board ft of logs at the mill or to clear and grub the right-of-way according to the specifications in exchange for all timber. The contractor, not being a skilled professional timber man, elects wisely to accept the second offer and signs an equitable contract with the sawmill owner. If all goes well, the rock excavator will be able to profit by the amount of his bid for the item of clearing and grubbing, except for overhead costs.

The foregoing estimate for the cost of clearing and grubbing a hypothetical right-of-way illustrates the importance of this costly item. Such a 6-mi four-lane freeway in hilly country might have 1 million yd³ of roadway excavation, resulting in a prorated cost of $0.118/yd³. If the excavation were bid in at $1.00/yd³, then the cost of clearing and grubbing would be 12 percent of that for excavation.

SUMMARY

Clearing and grubbing, usually a preliminary operation necessary for excavation, may be most costly in relation to the cost of the excavation. For that reason the work warrants a most careful analysis, as under some circumstances the cost of clearing and grubbing may exceed the cost of excavation.

CHAPTER 10

Development of Water Supply

PLATE 10-1

A simple water supply system. The most workaday example of the development of a water supply was this fireman's hose with its nozzle connected to a nearby fire hydrant. The 300,000-yd³ rock excavation for home sites was in the San Gabriel Mountains of southern California. One hose operator applied average 160 gal/min to two small fills. The rental for 320 ft of 4-in-diameter pipe and one 300-ft length of 3-in-diameter fire hose was about $630 for the job, or $0.0021/yd³, in 1961. The 1978 cost was about $0.0041/yd³. Rarely are job conditions and job location as fortuitous for such minimum cost of development of water supply. Sometimes the cost is 1000 times greater for a large dam or freeway, and the development is infinitely more complex.

SYMBOLS, FORMULAS, AND WEIGHTS AND MEASURES

Symbols and Notations

A	= area, ft^2
A_p	= area of pipe, ft^2
A_s	= area of surface, ft^2
C	= any coefficient
C_l	= lift coefficient
D	= diameter of pipe, ft
E	= total energy, ft·lb
E_e	= elevation energy, ft·lb
E_k	= kinetic energy, ft·lb
E_p	= pressure energy, ft·lb
e	= efficiency
e_h	= hydraulic efficiency
e_m	= mechanical efficiency
F	= force, lb
f	= friction factor for flow of water
g	= acceleration of gravity, 32.2 ft/s^2
H	= total head, ft
h_e	= elevation head, ft
h_f	= friction head, ft
h_m	= mechanical head, ft, as from a pump
h_p	= pressure head, ft
h_v	= velocity head, ft
K	= any constant
K_c	= constant for contraction of pipe
K_e	= constant for expansion of pipe
K_{gv}	= constant for flow through gate valve
K_p	= constant for entry into pipe
L	= length, ft
NPSH	= net positive suction head
P	= power, ft·lb/s
p	= pressure
Q	= quantity rate of flow, ft^3/s or gal/min
R	= hydraulic radius, for full flow in pipe $= D/4$
S	= slope of hydraulic grade line or slope of pressure heads
spg	= specific gravity of water, 62.4 lb/ft^3
t	= time, s
V	= volume, ft^3
v	= velocity, ft/s
W	= total weight, lb
w	= unit weight of water, 62.4 lb/ft^3
whp	= water horsepower, 550 ft·lb/s

NOTE: When the general equation, based on Bernoulli's theorem, is used for the solution of a water-flow problem, the several heads are followed by a numerical suffix to designate that point in the system where the head exists. For example, $h_e 2$ means that the elevation head exists at point 2.

Conversion Formulas

There are several handy conversion formulas which eliminate repetitive calculations for some much used values. They are:

Q, $ft^3/s = Q$, gal/min$/449$
h, ft $= 2.31$ lb/in^2
v, ft/s $= 8.03$ $(h_v,\ ft)^{1/2}$
A, $ft^2 = 0.785(D,\ ft)^2$

Q, gal/min $= 449 Q$, ft^3/s
$lb/in^2 = 0.433 h$, ft
h_v, ft $= 0.0155(v,\ ft/s)^2$

Convenient Weights and Measures

Water weighs 62.4 lb/ft^3 and 8.34 lb/gal.
1 ft^3 of water $= 7.48$ gal
1 gal water $= 0.134$ ft^3
1 hp $= 33,000$ ft·lb/min and 550 ft·lb/s

10-4 Development of Water Supply

GENERAL CONSIDERATIONS

According to the usual standard specifications and the special provisions for an excavation job, the development of water supply may be described in relationship to highway building. The standard specifications of the California Division of Highways define it in these terms: "Description. This work shall consist of developing a water supply for all water required for the work, and of furnishing and applying all water for which separate payment is made, including water used in the performance of work otherwise paid for as extra work."

In these standard specifications the bid item for the development of water supply does not include payment for the application of the water for such other bid items as excavation and embankment. The excavator must allow for the costs of and profit on such items in the bid items in which water is used.

Occasionally, however, the bid items include one for the application of water along with the development of water. Such a job was the relocation of Lake Hughes Road high above the east shore of Castaic Reservoir, Los Angeles County, California. Figure 10-1 shows the grading of this mountain highway in 1969.

The length of the rock excavation job was 3.5 mi with 2,003,000 yd³ of unclassified roadway excavation. The uphill grade to the saddle of the highway at elevation 2350 ft was a steady 6 percent.

The source of water was Warm Springs Creek at an elevation of 1750 ft and the maximum vertical lift of water was about 600 ft. Water trucks were loaded by a 6-in centrifugal pump with pipe stand from a sump pit of the stream.

There were five bidders on this rock job and the average of bids pertaining to the development of water supply and application of water to the fills was as follows:

Development of water supply	$229,000
Application of 49,000,000 gallons @ $0.69	34,000
Total	$263,000

The water requirement estimate of the contracting officer was 49 million gal for 2,003,000 yd³ of rock. With an allowance of about 1.5 gal/yd³ for haul-road maintenance and miscellaneous purposes, there was 23 gal or 192 lb of water per cubic yard for compaction. The weathered sandstones and shales weighed about 3400 lb in situ, and so

Figure 10-1 This mountain highway excavation with minimum water requirement for fills and haul-road maintenance used only 21 gal/yd³, or 4.9 percent by weight. The only source of water was a stream at one end of the 3.5-mi, 2,003,000-yd³ job. The high point of centerline was 600 ft above Warm Springs Creek, Los Angeles County, California. Water delivery was by three 3500-gal water trucks, which was a more economical method than building a system of pumps and pipes in the rugged and inaccessible mountainous country. The picture shows fill compaction by a 20-ton sheepsfoot roller and contributory 22-ton rear-dump trucks. Little water was being added to the moisture-laden residuals and well-weathered shales and sandstones.

the applied water, in addition to natural moisture, was 5.3 percent of the total weight of 3592 lb/yd³.

The average bidder felt that the total cost for development of water supply and application of water would be about $0.13/yd³ equivalent to a bid of $0.20 in 1978. Water application was planned to be minimum because of the high lift from stream to fill and the rough topography of the sidehill cuts and fills. Figure 10-1 shows the sidehill fill building, the rock coming from a 500-ft distant sidehill cut. Little water was added to the fill, as the combined compaction of the 20-ton sheepsfoot roller and the 44-ton-GVW trucks met the specification requirements.

The work may be as simple and cheap as the use of a fire hose as shown in Plate 10-1. In keeping with the rough, confining mountainous country, production was low, averaging but 250 yd³/h. The average water application was 160 gal/min, adding about 38 gal/yd³ to the natural moisture of the weathered and semisolid granite.

Figure 10-2 shows another simple reach of a more elaborate water system consisting of four similar reaches. Water for the 12,000-gal tank came 2100 ft downgrade from a main line of the urban water works. Flow through the 8-in line to the tank was about 900 gal/min, whereas the demand of the 10,000- and 4000-gal water haulers was about 700 gal/min. Thus supply and demand were in balance.

Whenever water is available from a nearby existing water system, it is usually economical to develop the water supply accordingly rather than to build a system of pumps, wells, and appurtenances. In this instance the four similar reaches included about 1 mi of 8-in pipe, four meters, about eight gate valves, and four 12,000-gal portable tanks. The system not only served the excavation, 2,740,000 yd³, but also 206,000 yd³ of base courses, 58,000 yd³ of concrete pavement, and 12,000 yd³ of structural concrete, a total of 3,016,000 yd³ of excavation and materials.

A huge dam requires a stupendous amount of water. Perris Dam, an earth-fill structure of Riverside County, California, required about 880 million gal for the 22 million yd³ of excavation. Daily consumption, with double shifting, was about 2 million gal, or about 2080 gal/min. Even a large residential or industrial excavation job with 2500 yd³/h production may require 1200 gal/min. The water consumption at Perris Dam, 2 million gal/day, was equal to that of a city of 14,000 population. Thus are dramatized the size and

Figure 10-2 Portable water tank at end of one of four simple reaches comprising the water system for a rock excavation of 2,740,000 yd³ for an eight-lane freeway in San Diego County, California. Each reach was essentially an 8-in line between the main of the urban water works and a 12,000-gal portable tank. The length of the job was 4.1 mi, and the four tanks were spaced an average of 5500 ft apart. Accordingly, maximum haul was about 2800 ft and average haul was about 1400 ft, economical haul distances for both fill building and haul-road maintenance. This reach served a huge fill, the toe of which can be seen in the background. The water hauler is of 10,000-gal capacity and the loading time is 2.5 min, or 4000 gal/min. The tank served only two haulers, the other one being a 2000-gal unit used largely for haul-road maintenance.

complexity of the development of a water-supply system for a huge rock excavation job. For excavation there are three possible sources of water, as follows:

1. *Surface:*
    a. Relatively static ponds, lakes, and reservoirs. They may be higher or lower than the work.
    b. Flowing creeks, rivers, inlets, bays, oceans, canals, and waterways. They are usually at a lower elevation than the excavations and embankments.
2. *Subsurface.* These sources are in fixed locations and they include:
    a. Existing wells of private and public ownership. They are generally below the elevation of the work.
    b. Natural springs, perched water, and underground reservoirs, as in a karst topography of sinks and caverns. They may be above or below the elevation of the work.
3. *Existing facilities:*
    a. Water systems of towns, cities, water and irrigation districts, and other public agencies.
    b. Water supplies of industries such as railroads, quarries, open pits, and the like. This source is of limited quantity. Existing facilities may be above or below the elevation of the work.

Generally, with the exception of saltwater sources and some huge freshwater bodies, the water is purchased by the excavator. However, if the contracting officer or owner owns or has jurisdiction over the supply, the water may be given or sold to the excavator.

Water requirements for the job depend upon both job and natural factors. Job factors include:

1. Cubic yards of excavation and embankment.
2. Porosity and gradation of the rock placed in the embankment.
3. Specifications controlling the degree of compaction. There may be no compaction, as in the case of a random or waste-area fill.
4. Lengths of access roads, detours, and haul roads to be built and maintained.
5. Miscellaneous water requirements, usually nominal.

Natural conditions are equally important and they include:

1. Natural moisture in the excavation. This may very well provide most of or all the water required by the specifications. Actually, in certain locations and times of the year it is necessary to remove water in the embankment by aerating the fill mechanically with tractor-drawn multiple gangs of disk harrows.
2. Rainfall during the excavation. Rains not only shut down jobs but also frequently delay resumption of the work because of excessive wetness in the cut and fill.
3. High temperature with resultant evaporation increases water consumption over fills and haul roads of large areas.
4. Humidity without rainfall may actually increase moisture in the fill in spite of existing sunshine.

Water requirements are generally specified in terms of gallons per cubic yard of excavation and they vary within large limits. The consumption on a high desert freeway job in Inyo County, California, in which the excavation was porous volcanic Bishop tuff, was 65 gal/yd^3 during four seasons. Water was applied in both cut and fill, the distribution being: in cut, 55 gal, 85 percent; in fill, 8 gal, 12 percent; and for haul-road maintenance, 2 gal, 3 percent. The annual rainfall is 10 to 20 in.

On the other hand, consumption on a similar job in Humboldt County, California, where rainfall is 50 to 60 in annually, was only about 15 gal/yd^3. The excavation was a mixture of residuals and weathered and semisolid sandstone, shale, conglomerate, schist, and intruded basalt of the Franciscan formation. The water use was but 23 percent of that of the Inyo County job.

The design of facilities for the development of water supply is based on the principles of fluid mechanics, just as in the engineering of any water-supply system. Water, for practical purposes, is an incompressible liquid with a nearly fixed weight of 62.4 lb/ft^3, or 8.34 lb/gal.

FUNDAMENTAL EQUATIONS

There are several fundamental theoretical equations for the mechanics and design of water systems with their pipes, pipe appurtenances, pumps, wells, tanks, reservoirs, and

other hydraulic elements. The equations involve the total energy E of a cubic foot of water at a given point in the system. There are three energies in a frictionless flow of water. Together they make up the total energy E. They are:
1. Elevation energy, E_e, a potential for conversion to kinetic energy, E_k.

$$E_e = h_e w$$

where E_e = elevation energy, ft·lb
h_e = height above a given datum elevation, ft
w = weight of cubic foot of water, 62.4 lb/ft³
2. Pressure energy, E_p, a potential for conversion to kinetic energy, E_k.

$$E_p = h_p w$$

where E_p = pressure energy, ft·lb
h_p = height of water above pressure plane, ft
w = weight of cubic foot of water, 62.4 lb/ft³
3. Kinetic energy, E_k

$$E_k = \frac{wv^2}{2g}$$

where E_k = kinetic energy, ft·lb
w = weight of cubic foot of water, 62.4 lb/ft³
v = velocity of water, ft/s
g = acceleration of gravity, 32.2 ft/s²

The continuity principle for flow of water in a closed water system is expressed by the following equation, which holds for any point in the system.

$$Q = A_1 v_1 = A_2 v_2 = A_3 v_3 = \cdots = A_n v_n$$

where Q = time rate of flow, ft³/s
A_1 = area at point 1, ft²
v_1 = velocity of water at point 1, ft/s

Bernoulli's Theorem

Bernoulli's theorem states that in a frictionless or ideal fluid the total energy E at any point equals the total energy E at any other point. When this basic principle is applied to two points, 1 and 2, the equation of equality results:

$$E = E_{e1} + E_{p1} + E_{v1} = E_{e2} + E_{p2} + E_{v2}$$

Substituting the values for the three individual energies in the equation above, there results this equation:

$$E = h_{e1} w + h_{p1} w + \frac{wv_1^2}{2g} = h_{e2} w + h_{p2} w + \frac{wv_2^2}{2g}$$

But $\qquad E = Hw \qquad$ and $\qquad \dfrac{v^2}{2g} = h_v$

Substituting these two values in the above equation and dividing both sides of the equation by w, there results the simple equation:

$$H = h_{e1} + h_{p1} + h_{v1} = h_{e2} + h_{p2} + h_{v2}$$

Thus, in a frictionless or ideal fluid, total head H at any point equals total head H at any other point. It follows that the total head, which is the sum of the elevation, pressure, and velocity heads, is a constant.

In all water systems, there are friction losses h_f taking place throughout the flow of water. In some water systems there is a mechanical head h_m developed by a pump or pumps placed within the system. Such a system is shown schematically in Figure 10-5.

General Equation

One must modify the theoretical equation of equality of the total head H in order to take care of these two additional heads, friction h_f and mechanical h_m. If one considers two points, 1 and 2, between which there is a friction loss h_f and a mechanical gain h_m, a completely practical basic equation may be derived for the solution of water flow problems encountered in the development of water supply.

$$H = h_{e1} + h_{p1} + h_{v1} - h_f + h_m = h_{e2} + h_{p2} + h_{v2}$$

where H = total head, ft
h_{e1}, h_{p1}, and h_{e1} = respective elevation, pressure, and velocity heads at point 1, ft
h_f = total friction head between points 1 and 2, ft
h_m = total mechanical head between points 1 and 2, ft
h_{e2}, h_{p2}, and h_{v2} = respective elevation, pressure, and velocity heads at point 2, ft

FRICTION LOSSES IN PIPES

Friction losses may be divided into three general classifications:
1. Friction head losses within the pipe or pipes. They are major losses.
2. Friction head losses within the hose or hoses. They are major losses.
3. Friction head losses within the pipe appurtenances. These include losses at entrance to and discharge from pipe, at contraction and expansion of pipe, through pipe fittings and bends, and through valves, nozzles, and meters. They are minor losses.

Formulas for the flow of water in pipes as a result of friction losses are empirical and they includes those of Chézy, Darcy, Darcy-Weisbach, Fanning, Manning, and Hazen-Williams. Because of simplicity, flexibility, and precision, the formula most commonly used in the design of water-supply systems is that of Hazen-Williams.

Hazen-Williams Formula

$$v = 1.32 C R^{0.63} S^{0.54}$$

where v = velocity in pipe, ft/s
C = coefficient depending upon the condition and kind of pipe, generally taken as 100 for steel pipe in average condition
R = hydraulic mean radius or area of flow divided by the wetted perimeter, ft. For a pipe it is $0.25D$, ft, or $0.0208D$, in
S = slope of hydraulic grade line as expressed by the sine of the angle of slope

Since diameter of pipe is usually given in inches and since the coefficient C is usually taken at 100, the Hazen-Williams formula may be expressed as:

$$v = 11.5 D^{0.63} S^{0.54}$$

where v = velocity in pipe, ft/s
D = diameter of pipe, in
S = slope of hydraulic grade line or sine of angle of slope

The solution of the Hazen-Williams formula involves raising decimal integers to decimal integer powers. An example of its solution for the head in feet for a 100-ft length of pipe to overcome friction and produce a given flow is given below.

PROBLEM To determine friction head in feet for a flow of 500 gal/min in an 8-in-diameter steel pipe in average condition. 1. The necessary velocity for the given flow of 500 gal/min, based on the equation $Q = Av$, is given by the equation

$$v = Q/A = \frac{500/7.48 \times 1/60}{0.349} = 3.19 \text{ ft/s}$$

Turning to the Hazen-Williams formula and solving for the unknown value of S, we obtain the equation

$$S^{0.54} = \frac{v}{11.5 D^{0.63}} = \frac{3.19}{11.5 \times 8^{0.63}} = 0.0751$$

$S = 0.0084$, which is the sine of angle of slope

2. The necessary head to overcome friction in a 100-ft length of pipe is given by the equation

$$h = 100 \times 0.0084 = 0.84 \text{ ft}$$

The friction head is therefore 0.84 ft per 100 ft of pipe when velocity is 3.19 ft/s.

Obviously, solution for friction head by the Hazen-Williams formula is laborious and error-prone. In order to simplify calculations one uses Table 10-1 or the nomograph of Figure 10-3, both of which are based on the formula. Either may be used, Table 10-1 offering greater precision.

For example, using the nomograph for a flow of 500 gal/min through an 8-in-diameter pipe, one scales a velocity of 3.1 ft/sec and a friction head loss of 8.0 ft/1000 ft of pipe, equivalent to 0.8 ft per 100 ft of pipe. Using the table for the same flow and the same pipe, one finds a velocity of 3.2 ft/s and a head loss of 0.838 ft per 100 ft of pipe. The table also gives the velocity head for the given velocity, a value not given by the nomograph. In this instance a velocity head of 0.16 ft creates the velocity of 3.20 ft/s to maintain the 500-gal/min flow. As the head loss due to friction is 0.84 ft, the total head necessary for flow is 1.00 ft per 100-ft length of pipe. One notes that solution of the formula $Q = Av$ for velocity gives a value of 3.19 ft/s, whereas Table 10-1 gives a value of 3.20 ft/s. This discrepancy, well within the precision limits for estimating, is explained by the use of inside diameters of pipe in the table whereas the formula solution is based on the use of nominal diameters.

Table 10-1 and the nomograph of Figure 10-3 are based on $C = 100$. "Average condition of pipe" assumes a pitted pipe of several years of age. Generally the excavator uses good or new pipe, thus reducing the costly friction losses as well as the possibility of pipe failures. Values of C for pipe in other than average condition are given at the end of Table 10-1. Correction factors K to be applied to the friction head losses of Table 10-1 are given for the respective values of C.

Table 10-2 gives numbers raised to the powers which are used as constants in the flow formulas. Those of the Hazen-Williams formula, raised to the powers 0.63 and 0.54, are included in Table 10-2. The table eliminates calculations by logarithms, by the log-log slide rule, or by the electronic calculator with facilities for power and root calculations.

FRICTION LOSSES IN HOSES

Table 10-3 gives friction losses for 100-ft lengths of hoses of different diameters according to flow Q in gallons per minute. The table is for rubber or rubber-substitute hose in average condition. Generally, the friction head losses are somewhat less than those for the same size pipe in average condition.

FRICTION LOSSES IN PIPE APPURTENANCES

Generally minor losses are not significant when compared with the major losses in pipes and hoses, particularly when the conduits are several thousands of feet long. In such cases they are sometimes neglected because the error in estimating the friction head losses in the conduits may very well be several times the minor losses. However, in conduits of less than 1000-ft length, they may be of the same magnitude as the friction losses in the conduit itself. For example, for a flow of 500 gal/min in a 500-ft length of new 6-in-diameter steel pipe the friction head loss is 8.6 ft. If a gate valve with one-fourth opening were inserted in the line, the additional friction head loss would be 14 ft, according to Table 10-7. For such reasons it is well always to include minor friction head losses. There are three methods for calculating them.

Equivalent Length of Pipe Method

By this method either a nomograph or a table gives a length of pipe of the same nominal diameter as the appurtenance, which length of pipe has the same friction head loss. Figure 10-4 includes a nomograph of friction head losses in pipe fittings, bends, valves, entrances, enlargements, and contractions, expressed in equivalent lengths of pipe. Table 10-4 gives similar friction head losses in terms of equivalent lengths of pipe.

Determinations by the nomograph and the table differ somewhat, as may be expected in the results of many studies by different individuals, both in the laboratory and in the

(Text continues on page 10–19)

TABLE 10-1 Flow of Water in Steel Pipes of Average Condition According to the Hazen-Williams formula
(Friction losses in steel pipe, $C=100$)

	Nominal diameter 2 in				Nominal diameter 3 in		
	Standard wt steel 2.067 inside diameter				Standard wt steel 3.066 inside diameter		
Flow, US gal/min	Velocity, ft/s	Velocity head, ft	Head loss, ft per 100 ft	Flow, US gal/min	Velocity, ft/s	Velocity head, ft	Head loss, ft per 100 ft
5	0.48	0.00	0.120	10	0.43	0.00	0.063
6	0.57	0.01	0.167	15	0.65	0.01	0.134
7	0.67	0.01	0.223	20	0.87	0.01	0.227
8	0.77	0.01	0.285	25	1.09	0.02	0.344
9	0.86	0.01	0.355	30	1.30	0.03	0.481
10	0.96	0.01	0.431	35	1.52	0.04	0.640
12	1.15	0.02	0.604	40	1.74	0.05	0.820
14	1.34	0.03	0.803	45	1.95	0.06	1.02
16	1.53	0.04	1.03	50	2.17	0.07	1.24
18	1.72	0.05	1.28	55	2.39	0.09	1.47
20	1.91	0.06	1.55	60	2.60	0.11	1.74
22	2.10	0.07	1.85	65	2.82	0.12	2.01
24	2.29	0.08	2.18	70	3.04	0.14	2.31
26	2.49	0.10	2.52	75	3.25	0.16	2.62
28	2.68	0.11	2.89	80	3.47	0.19	2.96
30	2.87	0.13	3.29	85	3.69	0.21	3.31
35	3.35	0.17	4.37	90	3.91	0.24	3.67
40	3.82	0.23	5.60	95	4.12	0.26	4.06
45	4.30	0.29	6.96	100	4.34	0.29	4.47
50	4.78	0.36	8.46	110	4.77	0.35	5.33
55	5.26	0.43	10.1	120	5.21	0.42	6.26
60	5.74	0.51	11.9	130	5.64	0.49	7.26
65	6.21	0.60	13.7	140	6.08	0.57	8.32
70	6.69	0.70	15.8	150	6.51	0.66	9.48
75	7.17	0.80	17.9	160	6.94	0.75	10.7
80	7.65	.91	20.2	180	7.81	0.95	13.2
85	8.13	1.03	22.6	200	8.68	1.17	16.1
90	8.61	1.15	25.1	220	9.55	1.42	19.2
95	9.08	1.28	27.7	240	10.4	1.7	22.6
100	9.56	1.42	30.5	260	11.3	2.0	26.2
110	10.5	1.7	36.4	280	12.2	2.3	30.0
120	11.5	2.1	42.7	300	13.0	2.6	34.1
130	12.4	2.4	49.6	320	13.9	3.0	38.4
140	13.4	2.8	56.9	340	14.8	3.4	43.0
150	14.3	3.2	64.7	360	15.6	3.8	47.8
160	15.3	3.6	72.8	380	16.5	4.2	52.8
170	16.3	4.1	81.4	400	17.4	4.7	58.0
180	17.2	4.6	90.5	420	18.2	5.1	63.5
190	18.2	5.1	100.	440	19.1	5.7	69.2
200	19.1	5.7	110.	460	20.0	6.2	75.3

Friction Losses in Pipe Appurtenances

Nominal diameter 4 in				Nominal diameter 5 in			
Standard wt steel 4.026 inside diameter				Standard wt steel 5.047 inside diameter			
Flow, US gal/min	Velocity, ft/s	Velocity head, ft	Head loss, ft per 100 ft	Flow, US gal/min	Velocity, ft/s	Velocity head, ft	Head loss, ft per 100 ft
20	0.50	0.00	0.061	30	0.48	0.00	0.043
30	0.76	0.01	0.128	40	0.64	0.01	0.073
40	1.01	0.02	0.219	50	0.80	0.01	0.110
50	1.26	0.03	0.330	60	0.96	0.01	0.154
60	1.51	0.04	0.463	70	1.12	0.02	0.205
70	1.76	0.05	0.615	80	1.28	0.03	0.262
80	2.02	0.06	0.788	90	1.44	0.03	0.326
90	2.27	0.08	0.980	100	1.60	0.04	0.396
100	2.52	0.10	1.19	120	1.92	0.06	0.555
110	2.77	0.12	1.42	140	2.24	0.08	0.739
120	3.02	0.14	1.67	160	2.56	0.10	0.946
130	3.28	0.17	1.93	180	2.88	0.13	1.18
140	3.53	0.19	2.22	200	3.20	0.16	1.43
150	3.78	0.22	2.53	220	3.52	0.20	1.70
160	4.03	0.25	2.84	240	3.85	0.23	2.00
170	4.29	0.29	3.18	260	4.17	0.27	2.32
180	4.54	0.32	3.53	280	4.49	0.31	2.66
190	4.79	0.36	3.90	300	4.81	0.36	3.03
200	5.05	0.40	4.29	320	5.13	0.41	3.41
220	5.55	0.48	5.12	340	5.45	0.46	3.81
240	6.05	0.57	6.01	360	5.77	0.52	4.24
260	6.55	0.67	6.97	380	6.09	0.58	4.68
280	7.06	0.77	8.00	400	6.41	0.64	5.15
300	7.57	0.89	9.09	420	6.73	0.70	5.64
320	8.07	1.01	10.2	440	7.05	0.77	6.14
340	8.58	1.14	11.5	460	7.38	0.85	6.67
360	9.08	1.28	12.7	480	7.70	0.92	7.22
330	9.59	1.43	14.1	500	8.02	1.00	7.79
400	10.1	1.6	15.5	550	8.82	1.21	9.28
420	10.6	1.7	16.9	600	9.62	1.49	10.9
440	11.1	1.9	18.5	650	10.4	1.7	12.6
460	11.6	2.1	20.0	700	11.2	1.9	14.5
480	12.1	2.3	21.7	750	12.0	2.2	16.5
500	12.6	2.5	23.4	800	12.8	2.5	18.6
550	13.9	3.0	27.9	850	13.6	2.9	20.8
600	15.1	3.5	32.8	900	14.4	3.2	23.1
650	16.4	4.2	38.0	950	15.2	3.6	25.5
700	17.6	4.8	43.6	1000	16.0	4.0	28.1
750	18.9	5.6	49.5	1100	17.6	4.8	33.5
800	20.2	6.3	55.8	1200	19.2	5.7	39.3

Development of Water Supply

TABLE 10-1 Flow of Water in Steel Pipes of Average Condition According to the Hazen-Williams formula *(Continued)*
(Friction losses in steel pipe, $C=100$)

Nominal diameter 6 in				Nominal diameter 8 in			
Standard wt steel 6.065 inside diameter				Standard wt steel 7.061 inside diameter			
Flow, US gal/min	Velocity, ft/s	Velocity head, ft	Head loss, ft per 100 ft	Flow, US gal/min	Velocity, ft/s	Velocity head, ft	Head loss, ft per 100 ft
50	0.56	0.00	0.045	130	0.83	0.01	0.069
60	0.67	0.01	0.063	140	0.90	0.01	0.079
70	0.78	0.01	0.084	150	0.96	0.01	0.091
80	0.89	0.01	0.107	160	1.03	0.02	0.102
90	1.00	0.02	0.133	170	1.09	0.02	0.114
100	1.11	0.02	0.162	180	1.15	0.02	0.126
120	1.33	0.03	0.227	190	1.22	0.02	0.140
140	1.56	0.04	0.302	200	1.28	0.03	0.154
160	1.78	0.05	0.387	220	1.41	0.03	0.183
180	2.00	0.06	0.481	240	1.54	0.04	0.215
200	2.22	0.08	0.584	260	1.67	0.04	0.250
220	2.40	0.09	0.697	280	1.80	0.05	0.286
240	2.67	0.11	0.819	300	1.92	0.06	0.325
260	2.89	0.13	0.950	350	2.24	0.08	0.433
280	3.11	0.15	1.09	400	2.57	0.10	0.554
300	3.33	0.17	1.24	450	2.88	0.13	0.689
320	3.56	0.20	1.39	500	3.20	0.16	0.838
340	3.78	0.22	1.56	550	3.52	0.19	0.999
360	4.00	0.25	1.73	600	3.85	0.23	1.17
380	4.22	0.26	1.92	650	4.17	0.27	1.36
400	4.44	0.31	2.11	700	4.49	0.31	1.56
450	5.00	0.39	2.62	750	4.81	0.36	1.77
500	5.56	0.48	3.19	800	5.13	0.41	1.99
550	6.11	0.58	3.80	850	5.45	0.46	2.23
600	6.66	0.69	4.46	900	5.77	0.52	2.48
650	7.22	0.81	5.17	950	6.09	0.58	2.74
700	7.78	0.94	5.93	1000	6.41	0.64	3.02
750	8.34	1.08	6.74	1100	7.05	0.77	3.60
800	8.90	1.23	7.60	1200	7.69	0.92	4.23
850	9.45	1.39	8.50	1300	8.33	1.08	4.90
900	10.0	1.6	9.44	1400	8.97	1.25	5.62
950	10.5	1.7	10.2	1500	9.61	1.44	6.39
1000	11.1	1.9	11.5	1600	10.3	1.7	7.20
1100	12.2	2.3	13.7	1800	11.5	2.1	8.95
1200	13.3	2.7	16.1	2000	12.8	2.5	10.9
1300	14.4	3.2	18.6	2200	14.1	3.1	13.0
1400	15.6	3.8	21.4	2400	15.4	3.7	15.2
1500	16.7	4.3	24.3	2600	16.7	4.3	17.7
1600	17.8	4.9	27.4	2800	18.0	5.0	20.3
1700	18.9	5.6	30.6	3000	19.2	5.7	23.0

Friction Losses in Pipe Appurtenances   10–13

	Nominal diameter 10 in				Nominal diameter 12 in		
	Standard wt steel 10.027 inside diameter				Standard wt steel 12.000 inside diameter		
Flow, US gal/min	Velocity, ft/s	Velocity head, ft	Head loss, ft per 100 ft	Flow, US gal/min	Velocity, ft/s	Velocity head, ft	Head loss, ft per 100 ft
180	0.73	0.01	0.042	200	0.57	0.01	0.021
200	0.81	0.01	0.051	250	0.71	0.01	0.032
220	0.89	0.01	0.061	300	0.85	0.01	0.045
240	0.98	0.01	0.071	350	0.99	0.02	0.059
260	1.06	0.02	0.083	400	1.14	0.02	0.076
280	1.14	0.02	0.095	450	1.28	0.03	0.095
300	1.22	0.02	0.108	500	1.42	0.03	0.115
350	1.42	0.03	0.143	550	1.56	0.04	0.137
400	1.63	0.04	0.183	600	1.70	0.05	0.161
450	1.83	0.05	0.228	700	1.99	0.06	0.214
500	2.04	0.06	0.277	800	2.27	0.08	0.275
550	2.24	0.08	0.330	900	2.56	0.10	0.341
600	2.44	0.09	0.338	1000	2.84	0.13	0.415
650	2.64	0.11	0.450	1100	3.12	0.15	0.495
700	2.85	0.13	0.516	1200	3.14	0.18	0.581
800	3.25	0.16	0.660	1300	3.69	0.21	0.674
900	3.66	0.21	0.821	1400	3.98	0.25	0.773
1000	4.07	0.26	0.998	1500	4.26	0.28	0.878
1100	4.48	0.31	1.19	1600	4.55	0.32	0.990
1200	4.89	0.37	1.40	1800	5.11	0.41	1.23
1300	5.30	0.44	1.62	2000	5.68	0.50	1.50
1400	5.70	0.50	1.86	2200	6.25	0.61	1.78
1500	6.10	0.58	2.11	2400	6.81	0.72	2.10
1600	6.51	0.66	2.38	2600	7.38	0.85	2.43
1700	6.92	0.74	2.66	2800	7.95	0.98	2.78
1800	7.32	0.83	2.96	3000	8.52	1.13	3.17
1900	7.73	0.93	3.27	3500	9.95	1.54	4.21
2000	8.14	1.03	3.60	4000	11.4	2.0	5.39
2200	8.95	1.24	4.29	4500	12.8	2.5	6.70
2400	9.76	1.48	5.04	5000	14.2	3.1	8.15
2600	10.6	1.7	5.84	5500	15.6	3.8	9.72
2800	11.4	2.0	6.70	6000	17.0	4.5	11.4
3000	12.2	2.3	7.61	6500	18.4	5.3	13.2
3200	13.0	2.7	8.58	7000	19.9	6.2	15.2
3400	13.8	3.0	9.60	7500	21.3	7.1	17.3
3600	14.6	3.3	10.7	8000	22.7	8.0	19.4
3800	15.5	3.7	11.8	8500	24.2	9.1	21.7
4000	16.3	4.1	13.0	9000	25.6	10.2	24.2
4500	18.3	5.2	16.1	9500	27.0	11.3	26.7
5000	20.3	6.4	19.6	10000	28.4	12.5	29.4

SOURCE: Kelly Pipe Company.
NOTE: The value of 100 used for C in the equation

$$V = 1.32 C R^{0.63} S^{0.54}$$

is for steel pipe in average condition after several years of service. It is a conservative value to be used in the design of a water system. However, the value of C may range from a low of 60 for highly tuberculated small pipes to a high of 140 for large pipes in new condition.

To compensate for this widely ranging value of C it is suggested that the following table be used to adjust the head losses in feet per 100 ft of pipe, as given in Table 10-1:

Development of Water Supply

Condition of pipe	C according to pipe condition	K, multiplier for head loss based on $C = 100$
Bad, old, highly tuberculated	60	2.58
	70	1.92
Poor, rough	80	1.51
	90	1.22
Average, pitted	100	1.00
	110	0.84
Good, fairly smooth	120	0.71
	130	0.62
New, smooth	140	0.54

EXAMPLE: Required to find the total friction loss for 4500 ft of new 12-in-diameter pipe with flow Q of 2000 gal/min.
Table 10-1 gives head loss, feet per 100 ft of pipe, of 1.50 ft.
For new pipe $C = 140$. The multiplier for head loss, as given in the table above, is 0.54.
Head loss for 100 ft of new pipe is $0.54 \times 1.50 = 0.81$ ft.
Total head loss in 4500 ft of new pipe is $45.0 \times 0.81 = 36.4$ ft.
The use of new pipe instead of pipe of several years of age results in a head loss reduction, due to friction, of 46 percent.

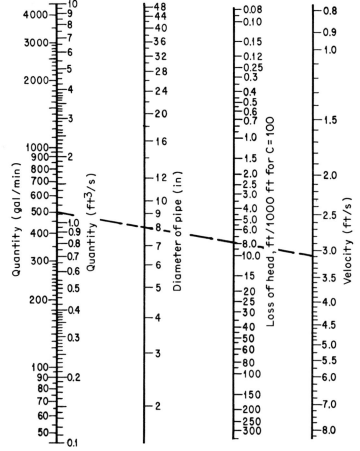

Figure 10-3 Flow of water in steel pipes of average condition according to the Hazen-Williams formula. The line connecting any two of the values determines the other two values. (EXAMPLE: For quantity of 500 gal/min and diameter of pipe of 8 in, loss of head due to friction is 8.0 ft per 1000 feet of pipe and velocity is 3.1 ft/s.)

TABLE 10-2 Powers of Numbers Used as Constants in Flow Formulas

Number	0.5	0.526	0.54	0.555	0.58	0.63	0.65	$\frac{2}{3}$
				Number to given power				
.00010	.0100	.0079	.0069	.00602				
.00011	.0105	.0083	.0073	.00635				
.00012	.0110	.0087	.0076	.00667				
.00013	.0114	.0091	.0080	.00697				
.00014	.0118	.0094	.0083	.00727				
.00015	.0122	.0098	.0086	.00755				
.00016	.0126	.0101	.0089	.00782				
.00017	.0130	.0104	.0092	.00809				
.00018	.0134	.0108	.0095	.00835				
.00019	.0138	.0111	.0098	.00860				
.00020	.0141	.0113	.0101	.00885				
.00022	.0148	.0119	.0106	.00933				
.00024	.0155	.0125	.0111	.00980				
.00026	.0161	.0130	.0116	.0103				
.00028	.0167	.0136	.0121	.0107				
.00030	.0173	.0141	.0125	.0111				
.00032	.0179	.0145	.0130	.0115				
.00034	.0184	.0150	.0134	.0119				
.00036	.0190	.0155	.0138	.0122				
.00038	.0195	.0159	.0143	.0126				
.00040	.0200	.0163	.0147	.0130				
.00042	.0205	.0168	.0150	.0134				
.00044	.0210	.0172	.0154	.0137				
.00046	.0214	.0176	.0158	.0141				
.00048	.0219	.0180	.0162	.0144				
.00050	.0224	.0184	.0165	.0147				
.00055	.0235	.0193	.0174	.0155				
.00060	.0245	.0202	.0183	.0163				
.00065	.0255	.0210	.0191	.0170				
.00070	.0265	.0219	.0198	.0178				
.00075	.0274	.0227	.0206	.0185				
.00080	.0283	.0234	.0213	.0191				
.00085	.0292	.0242	.0220	.0198				
.00090	.0300	.0250	.0227	.0204				
.00095	.0308	.0257	.0233	.0210				
.0010	.0317	.0264	.0240	.0216				
.0011	.0332	.0278	.0253	.0228				
.0012	.0346	.0291	.0265	.0239				
.0013	.0360	.0304	.0277	.0250				
.0014	.0374	.0317	.0289	.0260				
.0015	.0387	.0328	.0299	.0270				
.0016	.0400	.0339	.0310	.0280				
.0017	.0412	.0350	.0320	.0290				
.0018	.0424	.0361	.0330	.0299				
.0019	.0436	.0371	.0340	.0308				
.0020	.0447	.0381	.0350	.0317				
.0022	.0469	.0400	.0368	.0334				
.0024	.0490	.0420	.0385	.0350				
.0026	.0510	.0438	.0402	.0366				
.0028	.0529	.0455	.0419	.0382				
.0030	.0548	.0472	.0435	.0398				
.0032	.0566	.0489	.0450	.0412				
.0034	.0583	.0505	.0464	.0426				
.0036	.0600	.0520	.0479	.0440				
.0038	.0616	.0534	.0493	.0454				
.0040	.0632	.0549	.0507	.0468				
.0042	.0648	.0563	.0520	.0480				
.0044	.0664	.0578	.0534	.0492				
.0046	.0679	.0592	.0548	.0505				
.0048	.0693	.0606	.0560	.0518				
.0050	.0707	.0619	.0572	.0530	.0463	.0354	.0320	.0294
.0055	.0741	.0650	.0601	.0558	.0490	.0377	.0340	.0312
.0060	.0774	.0680	.0630	.0585	.0516	.0399	.0360	.0331

10-16   Development of Water Supply

TABLE 10-2   Powers of Numbers Used as Constants in Flow Formulas *(Continued)*

Number	0.5	0.526	0.54	0.555	0.58	0.63	0.65	⅔
				Number to given power				
.0065	.0806	.0710	.0659	.0611	.0540	.0419	.0380	.0349
.0070	.0836	.0738	.0688	.0638	.0564	.0439	.0399	.0366
.0075	.0866	.0764	.0714	.0662	.0588	.0458	.0417	.0383
.0080	.0894	.0790	.0738	.0685	.0610	.0477	.0435	.0400
.0085	.0922	.0816	.0763	.0708	.0632	.0496	.0451	.0417
.0090	.0948	.0840	.0786	.0730	.0652	.0514	.0468	.0434
.0095	.0974	.0864	.0809	.0752	.0674	.0532	.0485	.0450
.010	.100	.088	.083	.0776	.069	.055	.0501	.047
.011	.105	.093	.087	.0816	.073	.058	.0533	.049
.012	.110	.097	.092	.0856	.077	.061	.0566	.052
.013	.114	.102	.096	.0895	.081	.065	.0596	.055
.014	.118	.106	.099	.0930	.084	.068	.0625	.058
.015	.122	.110	.103	.0965	.088	.071	.0653	.061
.016	.126	.113	.107	.1000	.091	.074	.0681	.063
.017	.130	.117	.111	.104	.094	.076	.0710	.066
.018	.134	.121	.114	.107	.097	.079	.0737	.069
.019	.138	.124	.117	.110	.100	.082	.0762	.071
.020	.141	.128	.121	.113	.104	.085	.0789	.074
.022	.148	.134	.127	.120	.110	.090	.0840	.078
.024	.155	.141	.133	.126	.115	.095	.0888	.083
.026	.161	.147	.139	.132	.121	.100	.0935	.088
.028	.167	.152	.145	.137	.126	.105	.0982	.093
.030	.173	.158	.151	.142	.131	.109	.103	.097
.032	.179	.164	.156	.147	.136	.114	.107	.101
.034	.184	.169	.161	.152	.141	.118	.111	.105
.036	.190	.174	.166	.157	.146	.122	.115	.109
.038	.195	.179	.171	.162	.150	.127	.120	.113
.040	.200	.184	.176	.167	.155	.131	.124	.117
.042	.205	.189	.180	.172	.159	.135	.128	.121
.044	.210	.193	.186	.176	.164	.139	.131	.124
.046	.214	.197	.190	.180	.168	.143	.135	.128
.048	.219	.202	.194	.185	.172	.147	.139	.132
.050	.224	.206	.198	.189	.177	.151	.143	.136
.055	.235	.217	.209	.199	.187	.160	.152	.145
.060	.245	.228	.219	.209	.196	.170	.161	.154
.065	.255	.238	.228	.219	.205	.179	.169	.162
.070	.265	.247	.238	.228	.214	.187	.178	.170
.075	.274	.256	.247	.237	.223	.195	.186	.179
.080	.283	.265	.256	.246	.232	.203	.194	.187
.085	.292	.274	.265	.254	.240	.211	.201	.194
.090	.300	.282	.273	.262	.249	.219	.209	.201
.095	.308	.290	.281	.270	.257	.227	.217	.209
.100	.316	.298	.289	.278	.264	.234	.224	.216
.11	.332	.313	.304	.293	.279	.249	.239	.230
.12	.346	.328	.318	.307	.294	.262	.252	.243
.13	.360	.342	.332	.321	.308	.276	.267	.257
.14	.374	.357	.346	.334	.321	.290	.280	.270
.15	.387	.370	.359	.348	.334	.303	.293	.282
.16	.400	.382	.372	.361	.347	.316	.306	.295
.17	.412	.395	.384	.373	.360	.328	.319	.307
.18	.424	.407	.397	.386	.372	.340	.330	.319
.19	.436	.419	.409	.398	.383	.351	.341	.331
.20	.447	.430	.420	.409	.394	.363	.353	.343
.22	.469	.451	.441	.430	.417	.386	.377	.365
.24	.490	.472	.461	.451	.439	.408	.399	.387
.26	.510	.493	.482	.472	.460	.430	.420	.408
.28	.529	.513	.502	.492	.480	.450	.440	.429
.30	.548	.532	.522	.511	.500	.470	.460	.450
.32	.566	.550	.541	.531	.519	.490	.480	.470
.34	.583	.568	.559	.550	.538	.508	.500	.490
.36	.600	.586	.576	.569	.556	.526	.519	.509
.38	.616	.603	.593	.586	.573	.544	.538	.526
.40	.632	.620	.610	.603	.591	.563	.556	.544

Friction Losses in Pipe Appurtenances

Number	0.5	0.526	0.54	0.555	0.58	0.63	0.65	⅔
				Number to given power				
.42	.648	.636	.626	.620	.607	.580	.572	.562
.44	.664	.651	.642	.636	.624	.596	.590	.580
.46	.679	.666	.658	.650	.640	.616	.609	.599
.48	.693	.681	.674	.665	.655	.633	.624	.616
.50	.707	.697	.690	.680	.671	.649	.640	.632
.55	.741	.734	.725	.718	.709	.688	.682	.675
.60	.774	.768	.760	.754	.746	.727	.720	.715
.65	.806	.800	.794	.787	.782	.765	.759	.754
.70	.836	.832	.827	.821	.818	.803	.797	.792
.75	.866	.863	.860	.856	.850	.838	.834	.829
.80	.894	.892	.889	.884	.880	.871	.870	.865
.85	.922	.920	.918	.914	.910	.904	.902	.900
.90	.948	.947	.945	.943	.940	.936	.935	.935
.95	.974	.974	.973	.972	.970	.968	.968	.968
1.0	1.00				1.00	1.00	1.00	1.00
1.1	1.05				1.06	1.06	1.07	1.07
1.2	1.10				1.12	1.12	1.13	1.13
1.3	1.14				1.17	1.18	1.19	1.20
1.4	1.18				1.22	1.24	1.25	1.26
1.5	1.22				1.27	1.29	1.30	1.31
1.6	1.26				1.32	1.35	1.36	1.37
1.7	1.30				1.36	1.40	1.42	1.43
1.8	1.34				1.41	1.45	1.47	1.49
1.9	1.38				1.45	1.50	1.52	1.54
2.0	1.41				1.50	1.55	1.58	1.60
2.2	1.48				1.58	1.64	1.67	1.70
2.4	1.55				1.66	1.73	1.77	1.80
2.6	1.61				1.74	1.82	1.86	1.90
2.8	1.67				1.81	1.91	1.95	1.99
3.0	1.73				1.89	1.99	2.04	2.09
3.2	1.79				1.96	2.07	2.13	2.18
3.4	1.84				2.03	2.15	2.22	2.27
3.6	1.90				2.10	2.23	2.30	2.36
3.8	1.95				2.17	2.30	2.39	2.44
4.0	2.00				2.23	2.38	2.47	2.52
4.2	2.05				2.30	2.46	2.54	2.60
4.4	2.10				2.36	2.53	2.62	2.69
4.6	2.14				2.42	2.60	2.70	2.77
4.8	2.19				2.48	2.67	2.77	2.85
5.0	2.24				2.54	2.74	2.84	2.93
5.5	2.35				2.69	2.91	3.02	3.12
6.0	2.45				2.82	3.08	3.20	3.30
6.5	2.55				2.95	3.24	3.38	3.49
7.0	2.65				3.08	3.39	3.55	3.67
7.5	2.74				3.21	3.54	3.71	3.84
8.0	2.83				3.33	3.69	3.87	4.00
8.5	2.92				3.45	3.83	4.02	4.17
9.0	3.00				3.57	3.97	4.18	4.33
9.5	3.08				3.69	4.10	4.33	4.49
10.0	3.16				3.80	4.24	4.48	4.65
11.0	3.32				4.00	4.50	4.75	4.93
12.0	3.46				4.21	4.75	5.02	5.22
13.0	3.60				4.42	5.00	5.30	5.51
14.0	3.74				4.62	5.25	5.57	5.79
15.0	3.87				4.81	5.48	5.83	6.06
16.0	4.00				4.99	5.70	6.07	6.33
17.0	4.12				5.17	5.92	6.30	6.59
18.0	4.24				5.34	6.15	6.55	6.85
19.0	4.36				5.50	6.36	6.78	7.11
20.0	4.47				5.67	6.57	7.01	7.37

Maximum error less than 1 percent.
SOURCE: W. Kent, *Kent's Mechanical Engineers' Handbook, Power,* Wiley, New York, 1936, pp. 2–16.

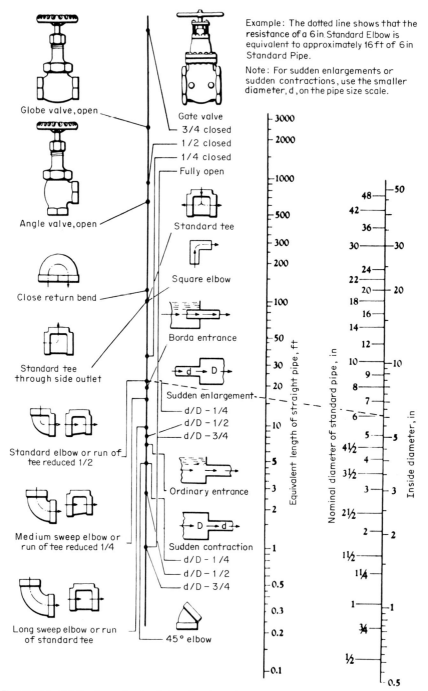

Figure 10-4 Friction losses in pipe fittings, bends, and valves expressed in the equivalent lengths of steel pipe in feet. *(The Crane Company.)*

field. However, the differences are not significant, and together the nomograph and the table include almost all the examples of minor losses. It is suggested that they be used in a supplementary manner.

Flow-of-Water Method

Table 10-5 gives friction head losses for turbine-type meters in terms of flow and size of meter.

TABLE 10-3 Friction Losses for Water Flow in 100-ft Length of Rubber or Rubber-Substitute Hose in Average Condition in Feet of Head

	Inside diameter of hose, in				
Flow, gal/min	2	2½	3	4	5
20	1.6	0.7			
40	5.6	1.8	0.7		
60	11.8	3.2	1.4		
100	28.8	8.1	3.9	1.1	0.2
200	106.1	30.0	13.6	3.2	1.1
300		43.8	21.0	4.8	1.6
400		106.0	48.5	9.2	3.9
500			74.0	17.1	6.0
600			101.0	22.0	8.0
700			130.0	28.1	10.0
800				35.6	13.6
900				46.2	17.5
1000				62.3	22.0

SOURCE: Gorman Rupp Company.

TABLE 10-4 Friction Losses in Pipe Fittings, Bends, and Valves Expressed in the Equivalent Lengths of Steel Pipe of Same Nominal Diameter, Feet

	Diameter of pipe, in							
Item	2	3	4	5	6	8	10	12
90° elbow	6	8	11	14	16	21	26	32
45° elbow	3	4	5	6	8	10	13	15
Tee, side outlet	12	17	22	28	33	44	55	66
Close return bend	13	18	24	31	37	49	62	73
Check valve	21	32	42	53	63	81	105	125
Foot valve	46	64	75	76	76	76	76	76
Gate valve	1	2	2	3	4	4	6	7
Globe valve	55	82	115	135	165	215	280	335
Pressure-altitude valve	63	94	133	155	190	247	322	385

SOURCE: Gorman Rupp Company.

TABLE 10-5 Friction Head Losses in Commonly Used Turbine-Type Water Meters for Meter and Screen

	Meter size, in				
Item	2	3	4	6	10
Flow of water, Q:					
gal/min	160	350	1000	2000	5500
ft^3/s	0.36	0.78	2.23	4.46	12.25
Friction head loss:					
lb/in^2	13.0	10.3	8.4	6.8	6.5
ft	30	24	19	16	15

SOURCE: Rockwell Manufacturing Company.

Velocity Method

This method is based on the fundamental formula for friction head loss as a function of velocity head.

$$h_f = \frac{Kv^2}{2g}$$

where h_f = friction head loss, ft
 K = coefficient for head loss as given in the accompanying table
 v = velocity as defined, ft/s
 g = acceleration of gravity, 32.2 ft/s²

Table 10-6 gives friction head losses for nozzles. Although it is a tabulation according to velocity, it is based on the following empirical formula:

$$h_f = 0.04 \frac{v^2}{2g}$$

The tabulation is given in order to eliminate the calculation by formula.

Tables of Losses

In Table 10-7 friction losses in gate valves are given according to the opening of valve. Values of K_{gv} are given for different valve openings according to the size of the valve. This table provides a refinement over the friction head losses for partially open valve as given in the nomograph of Figure 10-4. Table 10-7 gives two additional openings, ⅜ and ⅛.

Table 10-8 gives friction losses for water flow in entrance to and discharge from pipe, as in the case of a reservoir. Values of K_e are given for four kinds of entrance. Table 10-8 provides a refinement over the head losses as given in Figure 10-4, which gives only ordinary or sharp-cornered entrance and Borda or inward-projecting entrance.

Table 10-9 gives friction losses for water flow through sudden and gradual contractions from large to small pipe. The table provides a refinement over the head losses obtained from Figure 10-4. Figure 10-4 gives only three ratios of small to large pipe diameters, and the friction head losses are assumed independent of velocity in the small pipe.

TABLE 10-6 Friction Losses, Approximate, for Water Flow through Nozzles

Discharge velocity, ft/s	Head loss, ft
10	0.1
20	0.2
30	0.6
40	1.0
50	1.6
60	2.2
80	4.0
100	6.2
120	8.9
140	12.2
160	15.9
180	20.1
200	24.8

NOTE: This table is based on the following empirical equation:

$$h_f = 0.04 \times \frac{v^2}{2g}$$

where h_f = head loss, friction, ft
 v = velocity at discharge, ft/s
 g = acceleration of gravity, 32.2 ft/s²
SOURCE: Robert L. Dougherty, *Fluid Mechanics*, McGraw-Hill, New York, 1965, p. 346.

TABLE 10-7 Friction Losses in Gate Valves According to Opening of Valve

Valve size (nominal diameter of valve), in	Coefficient, k_{gv}					
	Ratio of height of valve opening to diameter of full valve opening					
	1/8	1/4	3/8	1/2	3/4	1
2	140	20	6	3	1	1
4	91	16	6	3	1	1
6	74	14	5	2	1	1
8	66	13	5	2	1	1
12	56	12	5	2	1	1

NOTE: This table gives coefficients for the equation

$$h_f = \frac{K_{gv} v^2}{2g}$$

where: h_f = head loss, friction, ft
K_{gv} = coefficient of table for head loss
v = velocity, ft/s, in a pipe with the same nominal diameter as that of the gate valve
g = acceleration of gravity, 32.2 ft/s²

SOURCE: Leonard C. Urquhart, *Civil Engineering Handbook*, McGraw-Hill, New York, 1962, p. 4-71.

TABLE 10-8 Friction Losses for Water Flow in Entrance to and Discharge from Pipe, as in the Case of a Reservoir

Friction Losses at Entrance

Entrance loss occurs after the water enters the pipe

$$h_f = \frac{K_p v^2}{2g}$$

where h_f = head loss, friction, ft
K_p = coefficient of table for head loss
v = velocity at entrance, ft/s
g = acceleration of gravity, 32.2 ft/s²

Description of entrance	Coefficient, K_p
Inward-projecting	0.78
Sharp-cornered	0.50
Slightly rounded	0.23
Bell-mouth	0.04

Friction Losses at Discharge

Discharge loss occurs after the water leaves the pipe.

When water with a velocity v is discharged into a reservoir which is so large that the velocity within it is negligible, the kinetic energy is dissipated and the discharge loss coefficient becomes 1.0 under this condition. Accordingly,

$$h_f = \frac{v^2}{2g}$$

where h_f = head loss, friction, in ft
v = velocity at discharge, ft/s
g = acceleration of gravity, 32.2 ft/s²

SOURCE: Leonard C. Urquhart, *Civil Engineering Handbook*, McGraw-Hill, New York, 1962, pp. 4-71

TABLE 10-9 Friction Losses for Water Flow through Sudden Contraction from Large Pipe to Small Pipe

Velocity in small pipe, ft/s	Coefficient, K_c								
	Ratio of diameter of small pipe to diameter of large pipe								
	0.0	0.1	0.2	0.3	0.4	0.5	0.6	0.7	0.8	0.9
2	0.49	0.49	0.48	0.45	0.42	0.38	0.28	0.18	0.07	0.03
5	0.48	0.48	0.47	0.44	0.41	0.37	0.28	0.18	0.09	0.04
10	0.47	0.46	0.45	0.43	0.40	0.36	0.28	0.18	0.10	0.04
20	0.44	0.43	0.42	0.40	0.37	0.33	0.27	0.19	0.11	0.05
40	0.38	0.36	0.35	0.33	0.31	0.29	0.25	0.20	0.13	0.06

NOTE: This table gives K_c values for the following equation:

$$h_f = \frac{K_c v^2}{2g}$$

where h_f = head loss, friction, in feet
K_c = coefficient of table for head loss
v = velocity in small pipe, ft/s
g = acceleration of gravity, 32.2 ft/s²

NOTE: For an efficient total contraction angle between 20° and 40°, h_f for water flow through gradual contraction from large pipe to small pipe is given, as an approximation, by the same equation as that for sudden contraction, $K_c = 0.1$.

$$h_f = \frac{0.1 v^2}{2g}$$

where h_f = head loss, friction, ft
v = velocity in small pipe, ft/s
g = acceleration of gravity, 32.2 ft/s²

SOURCE: Leonard C. Urquhart, *Civil Engineering Handbook*, 4th ed., McGraw-Hill, New York, 1962, p. 343.

Table 10-10 gives friction losses for water flow through sudden and gradual expansions from small to large pipe. The table provides a refinement over the head losses as given in Figure 10-4. Figure 10-4 gives only three ratios of small to large pipe diameters, and the friction head losses are assumed independent of the velocity in the small pipe and of the total angle of expansion.

Table 10-11 contains two parts: values for flow of water through nozzles according to sizes and pressures of nozzles; and heights and distances that effective streams can be thrown by nozzles. The data give pertinent information on performances of nozzles of 1- to 3-in diameter so that a nozzle can be selected with suitable pressure for both desired discharge and desired horizontal reach of stream. Table 10-11 may be used in conjunction with Table 10-6 to determine friction head loss in a nozzle. An illustration is the determination of loss in a 1½-in-diameter nozzle discharging 558 gal/min with a horizontal reach of 84 ft. The results obtained are:
1. Discharge, $Q = (558/7.48)(1/60) = 1.24$ ft³/s
2. $v = Q/A = 1.24/0.0123 = 101$ ft/s
3. $h_f = 6.3$ ft, from Table 10-6 by interpolating for 101 ft/s.

An ensuing example of the use of the general equation for the solution of a problem involving the use of the nozzle for wetting a fill illustrates the use of Table 10-6. Plate 10-1 is a picture of the nozzle being used to supply water for fill compaction.

EXAMPLE OF USE OF THE GENERAL EQUATION

In order to illustrate the use of the general equation for the solution of flow problems, the following example is given, as shown schematically in Figure 10-5.

Water is to be taken from a 16-in main line at elevation 1000 ft and pumped 3000 ft horizontally to the top of a temporary reservoir at elevation 1200 ft. The static pressure of the main line varies from 80 to 120 lb/in², the minimum static head being 184.6 ft. Requirement is 1000 gal/min, equivalent to Q of 2.23 ft³/s.

A 3003-ft length of 8-in-diameter steel pipe in good condition is to be used. From Table 10-1: (1) $v = 6.41$ ft/s; (2) $h_v = 0.64$ ft; (3) head loss due to friction is 3.02 ft per 100 ft of

TABLE 10-10A Friction Losses for Water Flow through Sudden Expansion from Small Pipe to Large Pipe

Velocity in small pipe, ft/s	Coefficient K_e for sudden expansion									
	Ratio of diameter of small pipe to diameter of large pipe									
	0.0	0.1	0.2	0.3	0.4	0.5	0.6	0.7	0.8	0.9
2	1.00	1.00	0.96	0.86	0.74	0.60	0.44	0.29	0.15	0.04
5	0.96	0.95	0.89	0.80	0.69	0.56	0.41	0.27	0.14	0.04
10	0.93	0.91	0.86	0.77	0.67	0.54	0.40	0.26	0.13	0.04
20	0.86	0.84	0.80	0.72	0.62	0.50	0.37	0.25	0.12	0.04
40	0.81	0.80	0.75	0.68	0.58	0.47	0.35	0.22	0.11	0.03

NOTE: This table gives coefficients for the equation:

$$h_f = \frac{K_e v^2}{2g}$$

where h_f = head loss, friction, ft
K_e = coefficient of table for head loss
v = velocity in small pipe, ft/s
g = acceleration of gravity, 32.2 ft/s²

TABLE 10-10B Friction Losses for Water Flow through Gradual Expansion from Small Pipe to Large Pipe

Total angle of expansion	Coefficient K_e for gradual expansion								
	Ratio of diameter of small pipe to diameter of large pipe								
	0.1	0.2	0.3	0.4	0.5	0.6	0.7	0.8	0.9
10°	0.04	0.04	0.04	0.04	0.04	0.05	0.03	0.02	0.01
30°	0.16	0.16	0.16	0.16	0.16	0.15	0.13	0.10	0.06

For efficient total-expansion angles of 10° and 30°, h_f is given by the same equation as that for sudden expansion, to which the tabulated K_e values apply.

SOURCE: Leonard C. Urquhart, *Civil Engineering Handbook*, 4th ed., McGraw-Hill, New York, 1962, pp. 4–70.

pipe, based on $C=100$ for average-condition pipe; (4) multiplier for 3.02, based on $C=120$ for good pipe, is 0.71; (5) friction head loss is given by $0.71 \times 3.02 = 2.41$ ft per 100 ft of pipe. A centrifugal pump is to be located suitably within the line so as to overcome elevation head and friction head from entrance to discharge.

Appurtenances are: gate valve at foot of reach; check valve at foot of reach; gate valve at downstream side of pump; gate valve and check valve at upstream side of pump; and three long-sweep elbows and two 45° elbows within the reach of pipe.

It is desired to know: (1) the maximum elevation h_e of the pump location in order to avoid any suction head in the action of the pump; and (2) the mechanical head h_m of the pump needed to raise the water to the top of the reservoir.

Solution for Maximum Elevation, h_{e2}, of the Pump

Applying the general equation between points 1 and 2:

$$h_{e1} + h_{p1} + h_{v1} - h_f = h_{e2} + h_{p2} + h_{v2}$$

At point 1 and between points 1 and 2:

$h_{e1} = 0.0$ ft
$h_{p1} = 184.6 - 0.6 = 184.0$ ft
$h_{v1} = 0.6$ ft
h_f = sum of 6 known minor friction head losses in appurtenances and unknown major friction head loss in pipe (unknown because the length of pipe from entrance to pump is unknown)

TABLE 10-11A Values for Flow of Water Through Nozzles According to Sizes and Pressures of Nozzles

Nozzle pressures by pitot, lb/in²	Size of nozzle, in																			
	1				1⅛				1¼				1⅜				1½			
	Discharge gal/min	Pressure loss 100 ft 2½-in hose, lb/in²	Vertical reach, ft of stream	Horizontal reach, ft of stream	Discharge gal/min	Pressure loss 100 ft 2½-in hose, lb/in	Vertical reach, ft	Horizontal reach, ft	Discharge gal/min	Pressure loss 100 ft 2½-in hose, lb/in²	Vertical reach, ft	Horizontal reach, ft	Discharge gal/min	Pressure loss 100 ft 3-in hose, lb/in²	Vertical reach, ft	Horizontal reach, ft	Discharge gal/min	Pressure loss 100 ft 3-in hose, lb/in²	Vertical reach, ft	Horizontal reach, ft
20	132	4.8	35	37	167	7.3	36	38	206	10.6	36	39	250	5.8	36	40	298	8.1	37	42
25	148	5.8	43	42	187	8.9	44	44	230	13.1	45	46	280	7.2	45	47	333	10.1	46	49
30	162	6.8	51	47	205	10.5	52	50	253	15.5	52	52	307	8.6	53	54	365	11.9	54	56
35	175	7.9	58	51	221	12.1	59	54	273	17.8	59	58	331	9.9	60	59	394	13.7	62	62
40	187	8.9	64	55	237	13.8	65	59	292	20.0	65	62	354	11.2	66	64	422	15.5	69	66
45	198	9.9	69	58	251	15.3	70	63	309	22.2	70	66	376	12.5	72	68	447	17.3	74	71
50	209	10.9	73	61	265	16.8	75	66	326	24.7	75	69	396	13.8	77	72	472	19.1	79	75
55	219	11.9	76	64	277	18.3	79	69	342	27.2	80	72	415	15.1	81	75	494	20.8	83	78
60	229	12.8	79	67	290	19.8	83	72	357	29.6	84	75	434	16.4	85	77	517	22.6	87	80
65	238	13.8	82	70	301	21.3	86	75	372	31.7	87	78	451	17.6	88	79	537	24.3	90	82
70	247	14.8	85	72	313	22.9	88	77	386	33.9	90	80	469	18.8	91	82	558	26.0	92	84
75	256	15.8	87	74	324	24.5	90	79	399	36.1	92	82	485	20.0	93	84	578	27.8	94	86
80	264	16.7	89	76	335	26.1	92	81	413	38.6	94	84	500	21.2	95	86	596	29.5	96	88
85	272	17.7	91	78	345	27.7	94	83	425	40.8	96	87	516	22.5	97	88	614	31.2	98	90
90	280	18.7	92	80	355	29.3	96	85	438	43.1	98	89	531	23.8	99	90	633	32.9	100	91

SOURCE: C.V. Davis and K.E. Sorensen, *Handbook of Applied Hydraulics*, 3d ed., McGraw-Hill, New York, 1968.

Example of Use of the General Equation 10–25

TABLE 10-11B Heights and Distances That Effective Streams Can Be Thrown by Nozzles

Pressure at base		Angle with horizontal, °	1½- to 2-inch diameter		3-in diameter	
lb/in²	Head, ft		Max height, ft	Max distance at max height, ft	Max height, ft	Max distance at max height, ft
50	116	20	15	75	15	75
		50	60	100	62	100
100	231	20	30	140	30	140
		50	110	165	120	180
200	462	20	45	210	50	220
		50	175	180	180	250

SOURCE: Leonard C. Urquhart, *Civil Engineering Handbook*, 4th ed., McGraw-Hill, New York, 1962, pp. 10–45.

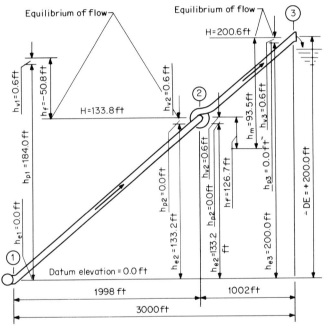

Figure 10-5 General equation applied to solutions of two problems in flow of water through pipe, pipe appurtenances, and pump.

$$h_{e1} + h_{p1} + h_{v1} - h_f + h_m = h_{e2} + h_{p2} + h_{v2}$$

The schematic drawing illustrates the two problems solved in the text.

PROBLEM 1: To find the elevation of the pump, h_{e2}. Point 1 to point 2.

$$h_{e1} + h_{p1} + h_{v1} - h_f = h_{e2} + h_{p2} + h_{v2} = H$$

0.0 ft + 184.0 ft + 0.6 ft − 50.8 ft = 133.2 ft + 0.0 ft + 0.6 ft = 133.8 ft

PROBLEM 2: To find the mechanical head of the pump, h_m. Point 2 to point 3.

$$h_{e2} + h_{p2} + h_{v2} - h_f + h_m = h_{e3} + h_{p3} + h_{v3} = H$$

133.2 ft + 0.0 ft + 0.6 ft − 26.7 ft + 93.5 ft = 200.0 ft + 0.0 ft + 0.6 ft = 200.6 ft

10–26   Development of Water Supply

Known minor losses in equivalent lengths of pipe:
  Figure 10-4:
    Ordinary entrance   12 ft
    Two fully open gate valves @ 4 ft   8
    One long-sweep 90° ell   14
    One 45° ell   10
  Table 10-4:
    One check valve   63
    Total equivalent length pipe   107 ft
    Minor losses, $h_f = 1.07 \times 2.41$ — 2.6 ft

Unknown major loss in pipe:
The unknown loss in the unknown length of pipe may be expressed in terms of h_{e2}, the unknown elevation head of the pump. Thus:

$$h_f = L \times \frac{2.41}{100} \tag{1}$$

$$L = \frac{h_{e2}}{\text{sine angle of slope}}$$

$$= \frac{h_{e2}}{200/3000}$$

$$= \frac{h_{e2}}{0.0666} \tag{2}$$

Substituting (2) in (1):

$$\text{Major loss, } h_f = \frac{h_{e2}}{0.0666} \times \frac{2.41}{100} = -0.362 h_{e2}$$

Total h_f for the reach of pipe from entrance to pump $= -(0.362 h_{e2} + 2.6)$ ft

At point 2:
h_{e2} = unknown = h_{e2} ft
h_{p2} = 0.0 ft
h_{v2} = 0.6 ft

Substituting in the general equation:

$$0.0 \text{ ft} + 184.0 \text{ ft} + 0.6 \text{ ft} - (0.362 h_{e2} + 2.6) \text{ ft} = h_{e2} + 0.0 + 0.6$$

Solving for h_{e2}:

$$h_{e2} = 133.2 \text{ ft}$$

The pump must be located not higher than 133.2 ft above the entrance at point 1. Probably it should be located at the entrance, where electrical energy for the motor logically would be available. The pump motor should be controlled by a water-level-actuated switch at the reservoir.

The length of the line from entrance to pump is given by:

$$L = \frac{133.2}{0.0666} = 2000 \text{ ft}$$

Accordingly, the length of the line from point 2, pump location, to point 3, discharge at reservoir is $3003 - 2000 = 1003$ ft.

Solution for Mechanical Head, h_m, of the Pump

Applying the general equation between points 2 and 3:

$$h_{e2} + h_{p2} + h_{v2} - h_f + h_m = h_{e3} + h_{p3} + h_{v3}$$

At point 2 and between points 2 and 3:

h_{e2} = 133.2 ft
h_{p2} = 0.0
h_{v2} = 0.6
h_f = sum of 5 known minor friction head losses in appurtenances and major friction head loss in 1003-ft length of pipe.

Known minor losses in equivalent lengths of pipe:
Figure 10-4:
 One fully open gate valve   4 ft
 Two long sweep 90° ells @ 14 ft   28
 One 45° ell   10
Table 10-4:
 One check valve   63
 Total equivalent length pipe   105 ft
 Known major loss, length of pipe   1003
 Grand total equivalent length pipe   1108 ft

Total h_f for reach of pipe from pump to discharge into reservoir $= 11.08 \times 2.41 = 26.7$

$h_m =$ unknown $= h_m$ ft

At point 3:

$h_{e3} = 200.0$ ft
$h_{p3} = 0.0$
$h_{v3} = 0.6$

Substituting in the general equation:

$$133.2 \text{ ft} + 0.0 \text{ ft} + 0.6 \text{ ft} - 26.7 \text{ ft} + h_m = 200.0 \text{ ft} + 0.0 \text{ ft} + 0.6 \text{ ft}$$

Solving for h_m:

$$h_m = 93.5 \text{ ft}$$

The centrifugal pump must develop a minimum head of 93.5 ft for the flow of $Q = 1000$ gal/min. The water horsepower, whp, of the pump is given by the equation:

$$\text{whp} = \frac{W h_m}{550} \quad (1)$$

where $W =$ total weight of water pumped per second
     $= 2.23 \text{ ft}^3/\text{s} \times 62.4 \text{ lb} = 139.2 \text{ lb}$
  $h_m =$ height of water lift $= 93.5$ ft

Substituting in equation (1):

$$\text{whp} = \frac{139.2 \times 93.5}{550} = 23.7$$

Assuming a pump efficiency of 55 percent, the brake horsepower of the motor should be a minimum of $23.6/0.55 = 43$ at a continuous-duty rating.

METHODS AND MACHINERY

Choices of methods depend upon the problems presented by the development of water supply. The entire flow or parts of the flow in a large integrated system may move upgrade, essentially level, or downgrade. Despite the complexities, the problems may be solved by application of the general equation involving equality of total head H at different points in the system.

When the flow is upgrade, the system may include pipes and fittings, valves, a pump or pumps, water tanks, and reservoirs. If the flow is essentially level, the system may involve all the machinery of the upgrade flow. For flow downgrade the pump or pumps are usually not necessary. A possible exception would be a reach of pipe in which a saddle or high elevation is between the inlet and outlet and the pressure head is not sufficient to overcome the elevation head.

The following discussion of machinery includes such nonmachinery items as pipe, reservoirs, and the like.

Pipe

Pipe is usually the portable line type, as is illustrated in Figures 10-6, 10-16, and 10-17. It is usually lightweight and it is available in sizes within the range of common use, 4- to 12-in diameter. Pipe section lengths are generally 40 ft. The ends of the pipe sections are grooved and they are held together by a quick gasket coupling, the coupling entering the grooves of the two pipe ends. The sections of pipe are locked together in a positive but flexible grip. Thus the reach of pipe may conform to reasonable changes of alignment, typically essential for a portable installation in rough country.

Table 10-12 gives characteristics of lightweight portable line pipe. A frequently used size is 8-in diameter, with a wall thickness of 10 gauge (0.134 in) and weighing 11.44 lb/ft.

A conveniently handled, efficient pipe diameter for the average flow of water is 8 in. For example, an excavator using a shovel-truck spread and a scraper spread may average 1200 yd^3/h, calling for perhaps 48,000 gal/h, or 800 gal/min. Table 10-1 gives a practical velocity of 5.13 ft/s and a workable friction head loss of 1.99 ft per 100 ft of pipe for average condition pipe, and a low of $0.54 \times 1.99 = 1.07$ ft per 100 ft of new pipe. If the job were building a rock-fill dam of 80-ft crest height, and 1000 ft of new 8-in-diameter pipe were used from diversion tunnel to tanks at crest, the combined elevation and friction heads would equal 91 ft ($h_e + 10.0\ h_f = 80 + 10.0 \times 1.07 = 91$ ft). Table 10-13 shows that a Model 125-M 8-in pump could deliver 800 gal/s against a total head H of 100 ft. The combination of 8-in-diameter pipe and 8-in centrifugal pump is adequate for the job.

On small jobs 4-, 5-, and 6-in-diameter pipes are used and on large jobs 10- and 12-in sizes are frequently used.

Pipe Fittings

Figure 10-4 illustrates an array of the most commonly used pipe fittings. Sometimes these fittings are the threaded type but usually they are fittings adapted for welding to the unthreaded ends of the portable pipe. Exceptions are valve fittings.

Often the excavator makes his own weldments either in the job shop or else at the pipeline. Generally it is more expeditious and cheaper to fabricate on the job rather than to purchase from a supply house and stock fittings in the warehouse.

Valves

There are five kinds of valves used in the water systems of the excavator. The first two types listed below are for manual and the last three for automatic control of water flow.

1. The most commonly used one is the gate valve, in which a gate or disk is raised and lowered to control flow from smallest to full opening. Friction head loss is high for small openings and negligible for full opening.

2. The globe valve and its counterpart, the angle valve, are not used customarily because of the high friction head loss due to their construction. Their use is generally confined to pipes of less than 4-in-diameter pipe in which flow is of the order of 100 to 400 gal/min.

3. The pressure-altitude valve is used to control the shutoff level in tanks and reservoirs by sensing the hydrostatic head. It is a useful and sometimes necessary part of the system, especially when remote tanks and reservoirs are used for storage. Because of its internal complexity its friction head losses are the highest of any of the valves.

4. Check valves, for permitting only one-way flow of water:
    a. Horizontal swing type, the most efficient and simple design, in which the flap is hinged at the top, thus affording minimum friction head loss
    b. Vertical type, in which the valve stem travels in a guide to ensure proper seating and in which friction head losses are average
    c. Horizontal ball type, in which a ball dropping into a seat prevents backward flow of water and in which friction head losses are high

5. The foot valve or flap valve is really a check valve of the swing type. It is a vertical valve at the foot of the line and is used when pumping with a suction head as it prevents loss of water between top of water and pump elevation. Friction head losses are minimal.

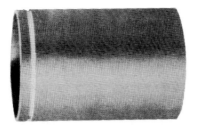 

Grooved ends on standard pipe

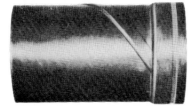 

Grooved nipples welded to light weight pipe

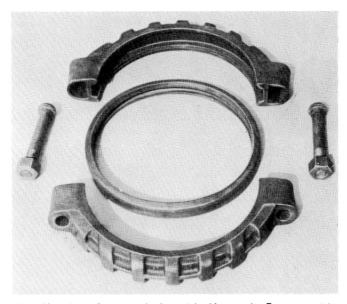

Coupling housing, gasket and bolts ready for assembly

Figure 10-6 Portable line pipe such as is generally used in the development of water supply. The size of the pipe varies from 4-in to 12-in diameter and the average length of a section is 40 ft. The gauge is usually 10 (0.134 in), 12 (0.109 in), or 14 (0.083 in). The sections of pipe with their end grooves are locked together by the coupling and gasket in a positive but flexible manner so that the reach of pipe can be adapted to rolling and moderately rough terrain. The lightness of the pipe contributes to speedy and economical erection of the water-supply system. *(Kelly Pipe Company.)*

Nozzles

Nozzles are used for both wetting, as of cuts and fills, and hydraulicking, as for the movement of earth-rock. Wetting is treated in Chapters 11 and 14 and hydraulicking is described in Chapter 13.

The *monitor* is a type of nozzle, differing from other nozzles with respect to size and operation. Nozzles range up to about 3-in diameter and monitors from about 4- to 10-inch diameter. In operation the nozzle is associated with manual operation, either in hand or mounted on a stand, and the monitor with mechanical or hydraulic operation. Of course these two distinctions overlap in actual use since nozzle and monitor serve the same purpose, creation of a high-velocity stream of water.

The distinction is accentuated in Plate 10-1, showing a 1⅛-in nozzle delivering 160 to 240 gal/min, and in Figure 13-75, showing a 3-in monitor delivering 3800 gal/min.

Meters

Meters in use for water supply systems are of two types: the nutating or wobble type, more common in sizes up to 3 in; and the turbine type, more common in sizes greater than 3 in. Meters record the quantity of water in gallons or in cubic feet, both of these units being used as the basis of cost.

Pumps

Pumps for water supply are almost always of the centrifugal type, reciprocating pumps being confined to older permanent installations. In the centrifugal pump, pressure head is developed by the action of centrifugal force. The casing is designed so that the velocity of the water leaving the periphery of the impeller is gradually converted into pressure at the pump discharge. Centrifugal pumps are of two classifications, volute and turbine.

Volute Pump This pump consists of an impeller rotating inside a volute, or spiral, casing. Taking in water axially through the eye, the pump discharges radially. It may be of the single-suction, one-inlet type or of the double-suction type, the latter having two inlets which neutralize axial thrusts.

The size of a volute is the internal diameter of the flange at discharge, as is the size of the turbine pump. The pump is rated by its capacity in gallons per minute and head in feet at maximum efficiency for a given speed in revolutions per minute.

TABLE 10-12 Dimensions and Weights for Lightweight Electric-Welded Steel Pipe

Inside diameter, in	Wall thickness		Weight, lb/ft	Inside diameter, in	Wall thickness		Weight, lb/ft
	Gauge	in			Gauge	in	
4	14	0.083	3.54	8	14	0.083	7.09
4	13	0.095	4.05	8	13	0.095	8.11
4	12	0.109	4.65	8	12	0.109	9.31
4	11	0.120	5.12	8	11	0.120	10.25
4	10	0.134	5.72	8	10	0.134	11.44
4	9	0.148	6.32	8	9	0.148	12.64
4	8	0.165	7.04	8	8	0.165	14.09
5	14	0.083	4.43	10	14	0.083	8.86
5	13	0.095	5.07	10	13	0.095	10.14
5	12	0.109	5.82	10	12	0.109	11.64
5	11	0.120	6.40	10	11	0.120	12.81
5	10	0.134	7.15	10	10	0.134	14.31
5	9	0.148	7.90	10	9	0.148	15.80
5	8	0.165	8.81	10	8	0.165	17.62
6	14	0.083	5.31	12	14	0.083	10.63
6	13	0.095	6.08	12	13	0.095	12.17
6	12	0.109	6.98	12	12	0.109	13.96
6	11	0.120	7.68	12	11	0.120	15.37
6	10	0.134	8.58	12	10	0.134	17.17
6	9	0.148	9.48	12	9	0.148	18.96
6	8	0.165	10.57	12	8	0.165	21.14

SOURCE: Kelly Pipe Company.

Figure 10-7 shows a 6-in single-stage self-priming volute pump, powered by a 107-hp diesel engine. It is drawing water from a reactivated deep well and discharging into a water tank. The combined static head and friction head total about 70 ft. The pump can deliver about 775 gal/min.

As used in development of water supply, the sizes of single-stage volute pumps vary from 3 in with discharge of 400 gal/min against a 360-ft total head to 12 in with 5000 gal/min discharge against a 200-ft total head. They may be driven by electric motors or by combustion engines, as shown in Figure 10-7.

Figure 10-8 shows a cutaway section and a cross section of a volute pump equipped with a diffuser to make it self-priming. The cutaway section shows the workings of the diffuser, which enables the pump to develop a suction head of about 25 ft. Unless the pump is submerged, as in the case of some electric-motor-driven pumps for special purposes, the pressure of the atmosphere must force the water upward through the intake channel into the vacuum created by the impeller. Accordingly, the elevation of the pump should not be higher than 20 ft above the elevation of the water to provide a practical height of suction head. The cross section shows the action of the diffuser in eliminating the entrained air within the volute.

Figure 10-9 gives performance curves for a single-stage 10-in volute pump with a 50-hp, 1150-r/min electric motor. At constant speed, the highest efficiency is 57 percent at 1700 gal/min against a 55-ft head. The brake horsepower is 42, as may be checked by the equation

$$\text{bhp} = \frac{1700 \times 8.34 \times 55}{0.57 \times 33,000} = 42$$

Figure 10-9 also gives performance curves for a 6-in single-stage centrifugal volute pump according to its speed in revolutions per minute. The advantage of varying speed to secure different capacities at a given total head is desirable. For example, the job requirement of an average 800 gal/min for a total head of 150 ft might vary between largely separated limits. The curves show that 60 percent throttle gives 460 gal/min, 80 percent gives 850 gal/min, and full throttle gives 1100 gal/min. Ability to control rate of discharge makes for a desirable even flow of water.

Table 10-13 gives delivery for self-priming volute pumps in gallons per minute according to size, total head, and suction head. The 3- to 8-inch sizes are adequate for the deliveries and total heads encountered in most water-supply development projects. In cases of greater capacities and higher heads one may use larger pumps or booster pumps of the same type.

Table 10-14 is a more complete summary of characteristics of general-service centrifugal pumps, including both volute and turbine types. Sizes, stages, revolutions per minute, and recommended rated horsepower are given as functions of capacity and head. The range of flow rate in gallons per minute and head in feet embraces more than average requirements for development of water supply. Likewise, pump sizes range from 2- to 12-in, again covering the commonly used volute and turbine pumps. The table is a good guide for the tentative selection of a pump to meet given requirements.

Turbine Pump The turbine pump is associated with deep-well operation and is usually a multistage unit. It is an axial-flow pump, a construction which adapts itself to the confines of the well casing.

Figure 10-10 shows the working parts of a two-stage deep-well turbine pump and the relationship of the pump to all parts of the well and to the stratigraphy of the aquifer. Water passes through the gravel packing, through the screen, and into the suction nozzle. It is then picked up by the first-stage impeller, passes through the diffuser, and enters the intermediate pump case. It is next picked up by the second-stage impeller, passes through the diffuser, and enters the top pump case, from which, now with a great velocity, it enters the discharge column or barrel. In this passage from screen to barrel, static head and velocity head are converted into pressure head.

Figure 10-11 illustrates the setup of an electric-motor-driven deep-well turbine pump. The well is 240 ft deep with 20-in-diameter casing. The aquifer is valley alluvia. The pump is an 8-in, four-stage model and the main-line pipe is of 8-in diameter. The pump serves a farming area of about 160 acres and water is available for rock excavation in the granites of the surrounding hills and mountains. Delivery is about 600 gal/min at 90 lb/in^2 static

10–32   Development of Water Supply

Figure 10-7 A 6-in portable volute pump, powered by a 107-hp diesel engine, supplied water temporarily to a 12,000-gal movable tank. The water table in the valley of the San Luis Rey River, San Diego County, California, was about 20 ft below the valley floor. Pending the installation of electric-motor-driven pumps in additional wells, this interim setup in a reactivated well discharged about 775 gal/min into the tank. The 10,000-gal water hauler made three trips hourly to the nearby fill at a rate of about 500 gal/min delivery, so that the pump capacity was adequate for this initial development of water supply. Later, at a maximum hourly production of about 5700 yd^3 from belt loaders, shovels, and scrapers, the water requirements reached about 3100 gal/min. The rock was a mixture of residuals, weathered, semisolid, and solid granite on this huge, 11,670,000-yd^3 freeway job.

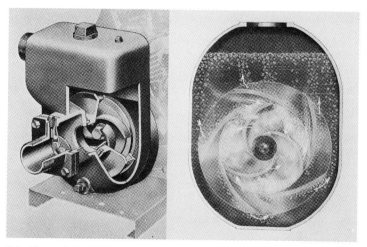

Figure 10-8 The working parts of a centrifugal volute pump of the self-priming type. The cutaway section shows the diffusing veins and the cross section shows the action of these veins in eliminating the entrained water, thus creating a more perfect vacuum to induce maximum suction head. While such a pump may produce a suction head of 25 ft, approaching the ideal 34 ft, it is well to use a practical figure of 20 ft as the limit of suction head. *(Marlow Pumps.)*

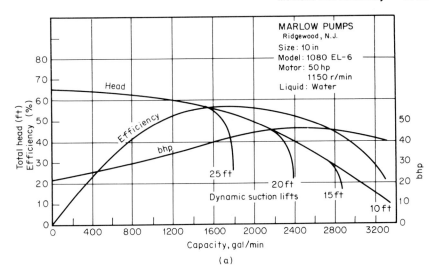

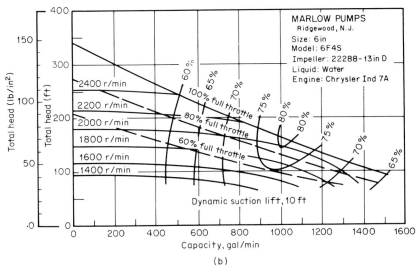

Figure 10-9 (a) Performance curves for a centrifugal pump. (b) Effect on the head of varying the speed of a centrifugal pump. *(Marlow Pumps.)*

TABLE 10-13 Standard Tables for Self-Priming Centrifugal Pumps
(Delivery in gallons per minute according to total head)

Total head, including friction	Height of pump above water 10 ft	Height of pump above water 20 ft	Total head, including friction	Height of pump above water 10 ft	Height of pump above water 20 ft
Model 4-M, 1½ in			Model 40-M, 4 in heavy		
15 ft	67		25 ft	665	
20 ft	66		30 ft	660	475
25 ft	65	47	40 ft	645	465
30 ft	63	47	50 ft	620	455
40 ft	54	44	60 ft	585	435
50 ft	37	34	70 ft	535	410
55 ft	25	25	80 ft	465	365
Model 10-M, 2 in			90 ft	375	300
25 ft	166		100 ft	250	195
30 ft	164	115	110 ft	65	50
40 ft	157	113	Model 90-M, 6 in		
50 ft	145	107	25 ft	1500	
60 ft	122	97	30 ft	1480	1050
70 ft	85	75	40 ft	1430	1020
Model 15-M, 3 in			50 ft	1350	970
30 ft	250	170	60 ft	1225	900
40 ft	230	165	70 ft	1050	775
50 ft	200	155	80 ft	800	600
60 ft	160	138	90 ft	450	365
70 ft	110	100	100 ft	100	100
Model 20-M, 3 in			Model 125-M, 8 in		
30 ft	333	235	25 ft	2100	1570
40 ft	310	230	30 ft	2060	1560
50 ft	275	220	40 ft	1960	1520
60 ft	220	195	50 ft	1800	1450
70 ft	160	155	60 ft	1640	1360
80 ft	90	90	70 ft	1460	1250
Model 30-M, 4 in light			80 ft	1250	1110
30 ft	500	350	90 ft	1020	940
40 ft	495	345	100 ft	800	710
50 ft	475	340	110 ft	570	500
60 ft	450	325			
70 ft	415	300			
80 ft	355	270			
90 ft	250	215			
100 ft	100	100			

NOTE: Shaft or brake horsepower must be taken from manufacturer's performance curves or calculated by the equation:

$$\text{hp} = \frac{\text{gal/min} \times 8.34 \times \text{head}}{\text{efficiency} \times 33{,}000}$$

Model 125-M, 8-inch pump with 800 gal/min, 100-ft head, and assumed 50 percent efficiency:

$$\text{hp} = \frac{800 \times 8.34 \times 100}{0.50 \times 33{,}000} = 40$$

SOURCE: Contractors' Pump Standards.

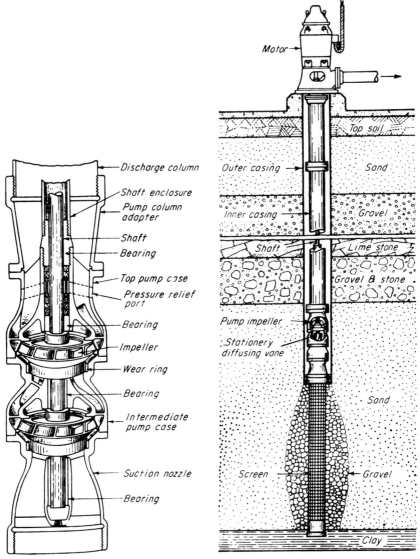

Figure 10-10 The working parts of a two-stage centrifugal-turbine deep-well pump and the parts of the complete working assembly, from the gravel packing at the intake screen to the discharge pipe at the driving electric motor. The stratigraphy of the sediments and sedimentary rock varies from the sand of the aquifer through solid limestone to the surface residuals. Such a two-stage pump of 12-in-diameter classification with 9-in-diameter impeller at 1770 r/min can perform in accordance with these characteristics:

Capacity	1600 gal/min
Total head	102 ft
Efficiency	78%
Shaft or brake horsepower	53 hp

(Johns Manville, Industrial Products Division.)

TABLE 10-14 Ratings of General-Service Centrifugal Pumps

Gal/min		Total head, ft: 40	60	80	100	120	140	160	180	200	240	280	320	360	400	440
100	Size*	2	1½	1½	1½	1½	1½	1½	1½	2	1½	1½	1½	2	2	2
	Stages	1	1	1	1	1	1	1	1	1	2	2	2	2	2	2
	r/min	1750	3500	3500	3500	3500	3500	3500	3500	3500	3500	3500	3500	3500	3500	3500
	hp†	2	3	5	5	5	7½	7½	10	10	15	15	15	30	30	30
200	Size*	2½	2½	2½	2½	2½	2	2	2	2	2½	2	2	2	2	2
	Stages	1	1	1	1	1	1	1	1	1	1	2	2	2	2	2
	r/min	1750	1750	1750	1750	3500	3500	3500	3500	3500	3500	3500	3500	3500	3500	3500
	hp†	5	5	7½	10	10	15	15	15	15	20	25	30	30	40	40
300	Size*	3	2½	2½	2½	2½	2½	2½	2½	2½	2½	2½	2½	2½	2½	2
	Stages	1	1	1	1	1	1	1	1	1	1	1	1	1	1	2
	r/min	1750	1750	1750	3500	3500	3500	3500	3500	3500	3500	3500	3500	3500	3500	3500
	hp†	5	7½	10	15	15	15	20	25	25	30	40	40	40	50	60
400	Size*	3	3	3	3	3	3	3	3	3	3	3	3	3	3	3
	Stages	1	1	1	1	1	1	1	1	1	1	1	1	1	2	2
	r/min	1750	1750	1750	1750	1570	1750	3500	3500	3500	3500	3500	3500	3500	1750	1750
	hp†	7½	10	15	15	20	20	25	30	30	40	50	50	60	75	75
600	Size*	4	4	4	3	3	3	3	3	3	3	4	4	4	4	4
	Stages	1	1	1	1	1	1	1	1	1	1	1	1	1	1	2
	r/min	1750	1750	1750	1750	1750	3500	3500	3500	3500	3500	3500	3500	3500	3500	1750
	hp†	10	15	20	25	30	40	40	40	50	50	75	75	75	100	125
800	Size*	4	4	4	4	4	4	4	4	4	4	4	4	4	4	5
	Stages	1	1	1	1	1	1	1	1	1	1	1	1	1	2	2
	r/min	1750	1750	1750	1750	1750	3500	1750	1750	3500	3500	3500	3500	3500	1750	1750
	hp†	15	20	25	30	40	40	50	60	60	75	75	100	100	125	150
1000	Size*	5	5	4	5	4	4	4	4	5	6	4	4	4	5	6
	Stages	1	1	1	1	1	1	1	1	1	1	1	1	1	2	2
	r/min	1750	1750	1750	1750	3500	3500	3500	3500	1750	1750	3500	3500	3500	1750	1750
	hp†	15	20	30	40	40	50	60	60	75	100	100	125	125	150	200
1500	Size*	6	6	6	6	6	6	6	6	6	6	6	6	6	6	6
	Stages	1	1	1	1	1	1	1	1	1	1	1	1	1	2	2
	r/min	1750	1750	1750	1750	1750	1750	1750	1750	1750	1750	1750	1750	1750	1750	1750
	hp†	20	30	40	50	60	75	75	100	100	125	150	200	200	200	250

Capacity, gal/min														
2000	Size*	8	8	8	8	8	8	8	8	8	8	8	8	
	Stages	1	1	1	1	1	1	1	1	1	1	2	2	
	r/min	1750	1750	1750	1750	1750	1750	1750	1750	1750	1750	1750	1750	
	hp†	25	40	50	75	100	100	125	125	150	200	300	400	
2500	Size*	10	8	8	8	8	8	8	8	8	8	8	8	
	Stages	1	1	1	1	1	1	1	1	1	1	2	2	
	r/min	1150	1750	1750	1750	1750	1750	1750	1750	1750	1750	1750	1750	
	hp†	40	50	60	75	100	125	125	150	200	250	300	350	
3000	Size*	10	10	10	10	10	10	10	10	10	6	8		
	Stages	1	1	1	1	1	1	1	1	1	2	2		
	r/min	1150	1750	1750	1750	1750	1750	1750	1750	1750	1750	1750		
	hp†	40	60	75	100	125	125	150	200	200	250	300		
3500	Size*	10	10	10	10	10	10	10	10	10	10			
	Stages	1	1	1	1	1	1	1	1	1	1			
	r/min	1150	1750	1750	1750	1750	1750	1750	1750	1750	1750			
	hp†	50	75	100	125	125	150	200	200	250	300			
4000	Size*	12	12	10	10	10	10	10	10					
	Stages	1	1	1	1	1	1	1	1					
	r/min	1150	1750	1750	1750	1750	1750	1750	1750					
	hp†	60	100	125	150	200	200	250	300					
4500	Size*	12	12	12	12	12	12	12	12	12				
	Stages	1	1	1	1	1	1	1	1	1				
	r/min	1150	1750	1750	1750	1750	1750	1750	1750	1750				
	hp†	60	125	150	200	200	250	250	300	300				
5000	Size*	12	12	12	12	12	12	12	12	12				
	Stages	1	1	1	1	1	1	1	1	1				
	r/min	1150	1750	1750	1750	1750	1750	1750	1750	1750				
	hp†	75	100	125	150	200	200	250	300	300				

*Diameter, in, of discharge opening, nominal size.

†Recommended motor rating for pumping water.

NOTE: The general-service type of centrifugal pump includes the single-suction volute, the double-suction volute, and the multistage pumps.

This reference table gives size of pump, number of stages, r/min, and recommended motor or engine horsepower according to the capacity in gal/min and the total head in feet. The tabulations should be always cross-referred to the performance curves for the individual pump as given by the manufacturer.

SOURCE: Worthington Pump and Machinery Corporation.

10–37

10-38 Development of Water Supply

Figure 10-11 Permanent well installation available for water supply for excavation. The setup is shown of a 600-gal/min deep-well turbine pump for combined farming irrigation and water supply for nearby rock excavation jobs in the Pauma Valley of southern California. The 8-in-diameter four-stage pump is powered by a rated 100-hp motor and delivery from the 240-ft-deep well is at 90 lb/in² pressure. The total head is 448 ft, and the efficiency for the four-stage pump is about 74 percent. Shaft or brake horsepower of the electric motor is given by the equation:

$$\text{hp} = \frac{600 \text{ gal/min} \times 8.34 \text{ lb} \times 448 \text{ ft head}}{0.74 \text{ efficiency} \times 33,000 \text{ ft·lb}}$$
$$= 92 \text{ hp}$$

pressure. Such delivery is adequate for a production of 7200 yd³ each 8-h working day in the average mixture of residuals, weathered, and semisolid granites.

Figure 10-12 gives performance curves for an 8-in single- or multistage turbine pump. These pumps are available in sizes from 4 to 64 in, and the range of sizes is far more than adequate to satisfy the deliveries and the heads encountered from the smallest to the largest of excavation jobs. The number of stages available for each size of pump varies but generally the stages do not exceed five.

The performance curves are typical of those of the manufacturers and the information includes:
1. On the abscissa, capacity in gal/min
2. On the left ordinate
    a. Feet total bowl head, per stage
    b. Diameters of available impellers
3. On the right ordinate
    a. Feet net positive suction head NPSH required at suction impeller
    b. Submergence over bell, feet
    c. Brake horsepower
4. Graphs
    a. NPSH and submergence over bell
    b. Total head per stage according to size of impeller
    c. Efficiency
    d. Brake horsepower per stage

The performance curves are used as is illustrated in Figure 10-12.

The selection of a pump, and especially a deep-well turbine pump, involves many considerations beyond the use of performance curves. In the example given in Figure 10-12 these include strength and elongation of pump shaft, size of pump barrel, and kinds of bearings and seals. Actually, the engineering and estimating for the entire average or large water system call for a combination of theory and practice. The average excavator does not employ an engineer who is thoroughly familiar with hydraulics or fluid mechanics simply because the excavator does not design a sufficient number of water-supply systems. And so reliance must be placed on the competent pipe, pump, and well supply engineers. Such dependence is axiomatic when one does not know a subject from both theoretical and practical standpoints.

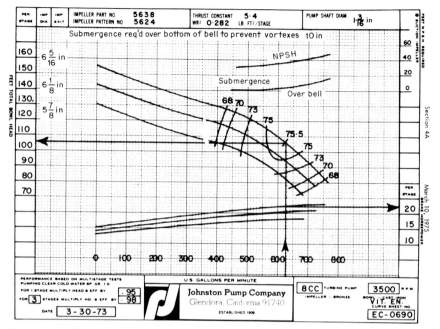

Figure 10-12 The use of performance curves for an 8-in single-stage or multistage centrifugal-turbine deep-well pump is exemplified by this problem of determining the preliminary specifications for a capacity of 625 gal/min with a total head of 300 ft.

1. Tentative examinations of several pump performance curves indicate that pump model 8CC is adequate.
2. At 625 gal/min and with one $6\frac{5}{16}$-in-diameter impeller the pump will handle a total head of 106 ft.
3. Three stages are required, giving:
   a. Total head $= 0.98 \times 3 \times 106 = 312$ ft
   b. Efficiency $= 0.98 \times 75.5\% = 74\%$
4. Shaft or brake horsepower $= 3 \times 22.2 = 67$

(Johnson Pump Company.)

Failure to make use of such services locally resulted in costly litigation between a rock excavator and a pipe supplier on one desert freeway job, involving an 8-mi reach of 8- and 10-in diameter pipe. The rented pipe deteriorated and failed after less than a year of service because of the alkaline salts of the desert water. Both the supplier and the renter accused each other of negligence in not investigating the corrosive nature of the water. Such information was readily available from local experienced miners, ranchers, and engineers of the water district in the area.

Water Tanks

Both portable and semiportable tanks are used frequently on all except the smallest excavation jobs. In addition to these large units, there are the welded, generally smaller tanks which sometimes are simply set into a convenient bank for loading the water haulers.

The tank of Figure 10-13 is a portable unit with its wheels-tires assembly. It is raised and lowered by an integral cable winch so that setup and takedown times are minimal. A float-controlled valve shuts off water when the tank is full. A 10-in-diameter downspout with lever-controlled butterfly valve discharges up to 3000 gal/min.

Portable Tanks A particular design of tank embodies these features: (1) Raising and lowering by two hydraulic jacks, powered by a 4½-hp gasoline engine and pump. (2) Wheels-tires assembly with air brakes. (3) Float valves with 4- to 12-in inlets. (4) Downspouts of 10- to 20-in diameter. Weights and dimensions for several popular-size tanks are:

10–40 Development of Water Supply

Size, gal	Weight, empty, lb	Length, ft	Width, ft	Height, erected, ft
8,000	10,000	25	8	21
12,000	18,000	40	8	21
16,000	21,000	40	9	22

These portable tanks may be moved over highways without permit.

Semiportable Tanks These tanks do not have running gear and they are not self-erecting. In all other respects they are similar to the portable units. Weights and dimensions for several popular tanks are:

Size, gal	Weight, empty, lb	Length, ft	Width, ft	Height, erected, ft
5,000	5,600	16	8	21
6,500	7,000	20	8	21
8,000	8,500	24	8	21
10,000	10,200	32	8	21
12,000	12,500	40	8	21
16,000	15,000	40	9	22
20,000	25,000	40	10	23

The smaller of these tanks may be moved over the highways without permit but in some cases the larger units require permits.

Illustrative of the advantages of using these tanks is this comparison of productions and unit costs for 3500-gal water haulers with and without a 12,000-gal tank when water delivery was 350 gal/min. This condition existed in the setup of Figure 10-13.

Water requirements on the job were neatly met by two water haulers. Without the tank, three units would have been necessary and there would have been a considerable increase in the unit cost for applying water. Table 10-15 is an analysis of the elements of productions and 1978 unit costs. This simply demonstrated saving of $6560 emphasizes the significant economies which may be achieved by analyzing every phase of an excavating job.

Figure 10-13 A portable water tank served by a municipal water system. This 12,000-gal storage tank serves a 180,000-yd³ rock excavation job for a shopping center in San Diego County, California. Water requirements were about 21,000 gal/h or 350 gal/min. This requirement was met by the nearby fire hydrant. Two 3500-gal water haulers wetted the cut, haul road, and fill for the average 703-yd³/h production. If water haulers had loaded directly from the hydrant, three units would have been necessary instead of the two water haulers in service. The hourly saving was $33.50. Since average production was 703 yd³/h, the saving was $0.048/yd³ for the application of water to the cut, haul road, and fill.

TABLE 10-15

	Water hauler loaded by:	
	Tank	Fire hydrant
Water-hauler hauling cycles, min:		
Loading	2.5	10.0
Hauling, wetting fill and haul roads, and returning	17.1	17.1
Total time	19.6	27.1
Hourly productions:		
Loads hauled	3.06	2.22
Gallons hauled	10700	7800
Water haulers required for 21,400 gal/h	2	3
Costs of machinery:		
Water hauler, $56,000		
Water tank, $32,000		
Hourly cost of ownership and operation:		
Water hauler:		
Machine (from Table 8-3), 48% of $56 $26.90		
Operator 10.10		
Total $37.00		
Water tank:		
Machine (from Table 8-3), 26% of $32 $8.30		
Total hourly costs of ownership and operation		
Water haulers	$74.00	$111.00
Water tank	8.30	
Total	$82.30	$111.00
Unit cost for application of water, per yd³ of excavation at average 703 yd³/h	$0.117	$0.158
Saving per yd³	$0.041	
Saving on job of 160,000 yd³	$6560	

Reservoirs

Temporary reservoirs are used extensively in water-supply development projects and they are used for three purposes. First, to ensure an adequate supply of water in the event of a failure in the system. This insurance is most important in a far-flung system of wells, pumps, and pipes. Second, to secure water from a stream or a watershed above the reservoir. Third, to compensate for a marginal rate of supply during the working hours. For example, if a dam job were being worked two shifts and the available water supply of 2000 gal/min were just adequate, the use of a reservoir in which the supply could be stored during 24 h would make 3000 gal/min available during the 16 working hours.

Figure 10-14 shows a typical reservoir during construction. This construction is refined by the inclusion of a waterproof lining and thin asphalt surface. The reservoir is in permeable desert alluvia of sands and gravels. Where they can be located in impervious soils such as clays, a common circumstance, or in semisolid rock, an uncommon situation, far less expensive building is possible.

Usually reservoirs are located near the high points of the water system and the locations are set so as to minimize cross-hauling of the water. It would be uneconomical, for example, to pump water 1000 ft upgrade against 60 ft of elevation head on a highway job only to haul the water 1000 ft downgrade in a water hauler. A healthy combination of design and experience dictates the practical and economical layout of any water system.

EXAMPLES OF WATER SUPPLY DEVELOPMENT

Following are three examples of the development of water supply which show the relationship among principles, methods and machinery, and common sense, with its trial-and-error processes.

Use of Nearby Fire Hydrant

The final plan of the simple system of Plate 10-1 is shown in Figure 10-15. The initial system included these same elements, with the exception that 2½-in hose was used

Figure 10-14 Construction of a reservoir for reserve water-supply system for Mojave Desert freeway. The estimated water requirement was 52 gal/yd^3 yard or 1,040,000 gal for a production of 20,000 yd^3 each 8-h working day. Two deep-well turbine pumps working up to 16 h/day supplied this water and additionally kept the two 1,200,000-gal reservoirs filled for any emergencies. The two reservoirs were located at high points of the two reaches of the system. Double-acting pressure valves allowed for both filling of the reservoirs and back flow into the lines. The reservoir was 120 ft long, 80 ft wide, and 12 ft deep, with about 600,000 gal capacity. It had been excavated by a bulldozer and lined with heavy, waterproof-coated fabric and was being surfaced with a 2-in thickness of asphaltic concrete. About 3000 yd^3 of excavation in desert alluvium, 1500 yd^2 of fabric, and 170 tons of asphaltic concrete were required. The cost was about $3000 in 1960, equivalent to $6000 in 1978. *(Pacific Lining Company.)*

instead of the eventual 3-in hose. The system was based on two requirements: (1) delivery should average 160 gal/min, with the maximum about 240 gal/min; and (2) the stream from the nozzle should have a horizontal reach of about 75 ft. The considerations had to be based on an available static head of 85 lb/in^2, or 196.4 ft, at the hydrant.

The first installation from hydrant to nozzle included these elements: (1) 4-in gate valve; (2) 2½-in meter; (3) a 90° ell; (4) 320 ft of good condition steel pipe, (5) 4-in gate valve; (6) 300 ft of good-condition 2½-in hose; (7) 1⅛-in nozzle with integral ½-turn valve.

The system did not provide enough pressure head at the base of the nozzle for an adequate stream. An application of the general equation will confirm the short supply. As Q must be 240 gal/min, or 0.535 ft^3/s and the reach of the stream must be 75 ft, one commences with these two controls.

Table 10-11 indicates that a 1⅛-in nozzle with 55 lb/in^2 at the base will discharge 277 gal/min 69 ft horizontally. This distance is satisfactory and Q is taken at 277 gal/min, or 0.617 ft^3/s. If the entrance from the hydrant is at point 1 and the discharge from the nozzle is at point 2, one may apply the general equation to determine the necessary pressure head at the hydrant. Velocities and velocity heads for $Q=277$ gal/min, or 0.617 ft^3/s, are as follows.

1. In 4-in pipe (from Table 10-1, interpolated for 277 gal/min): $v=6.98$ ft/s; $h_v=0.75$ or 0.8 ft; $h_f=7.86$ ft per 100 ft pipe.
2. In 2½-in hose: $v=Q/A=0.617/0.0341=18.1$ ft/s.
3. At nozzle discharge: $v=Q/A=0.617/0.00690=89.4$ ft/s; $h_v=124.1$ ft.

At point 1 and between points 1 and 2:

h_{el} = 50.0 ft
h_{pl}, unknown = h_{pl}
h_{vl} = 0.8 ft
h_f = sum of known friction head losses

Entrance to 4-in pipe (Table 10-8):

$h_f = 0.50 \times \dfrac{6.98^2}{64.4}$ = 0.4 ft

Through open 4-in gate valve (Table 10-7):

$h_f = 1.0 \times \dfrac{6.98^2}{64.4}$ = 0.8

Through 2½-in meter (Table 10-5 interpolated for 2½-in meter): = 27.0
Through 320 ft of 4-in pipe, 90° ell, and open gate valve:
  Equivalent length of pipe:
  320 ft of pipe (Table 10-4): 320 ft
  One 90° ell 11
  One open gate valve 2
  Total 333 ft
h_f for good pipe = 0.71 × 7.86 = 5.59 ft per 100 ft of pipe

$h_f = 3.33 \times 5.59$ = 18.7

Sudden contraction from 4-in pipe to 2½-in hose (Table 10-9); $v = 18.1$ ft/s:

$h_f = 0.27 \times \dfrac{18.1^2}{64.4}$ = 1.4

Through 300 ft of 2½-in hose (Table 10-3, interpolating for 277 gal/min):

$h_f = 3.00 \times 40.6$ = 121.8

Through nozzle (Table 10-6, interpolated for $v = 89.4$ ft/s): h_f = 5.0
h_f, total = −175.1 ft

At point 2:
h_{e2} = 0.0 ft
h_{p2} = 0.0 ft
h_{v2} = 124.1 ft

Substituting in the general equation:

$$50.0 \text{ ft} + h_{p1} + 0.8 \text{ ft} - 175.1 \text{ ft} = 0.0 \text{ ft} + 0.0 \text{ ft} + 124.1 \text{ ft}$$

$$h_{p1} = 248.4 \text{ ft}$$

The necessary pressure head of 248.4 ft exceeds the available pressure head of 195.6 ft, and so the friction head losses of the line must be reduced. The obvious reduction is by increasing the size of hose to 3 in. The present friction head loss in the hose is 121.8 ft.

Table 10-3, interpolated for 277 gal/min, gives a loss of 19.2 ft per 100 ft of 3-in hose, giving, for 300 ft, $h_f = 3.00 \times 19.2 = 57.6$ ft. Now total friction head losses are: $h_f = 175.1 - 121.8 + 57.6 = 110.9$ ft.

Substituting in the general equation:

$$50.0 \text{ ft} + h_{p1} + 0.8 \text{ ft} - 110.9 \text{ ft} = 0.0 \text{ ft} + 0.0 \text{ ft} + 124.1 \text{ ft}$$

$$h_{p1} = 184.2 \text{ ft}$$

The necessary pressure head now is less than the available pressure head by 11.4 ft and the requirements are met. The system will adjust itself to the law of continuity, $Q_1 = Q_2$, and both delivery and reach of stream will be increased slightly. Actually, in the trial-and-error installation a 3-in hose was substituted for the original 2½-in hose.

In this reach of 620 ft of pipe and hose, the total friction head losses of 110.9 ft are divided in these proportions.

Major losses in pipe and hose, 68 percent or 75.5 ft
Minor losses in entrance, gate valve, meter, 90° ell, contraction, gate valve, and nozzle, 32 percent or 35.4 ft

Minor losses can be a significant part of the total friction head losses in reaches less than 1000 ft, 32 percent in this case.

Use of Reactivated Line of Water District

During the initial stage of building a deep fill of a freeway rock excavation job in the mountains of San Diego County, California, it was expeditious to use an existing water supply until water was available from abandoned and new wells in the San Luis Rey

10-44   Development of Water Supply

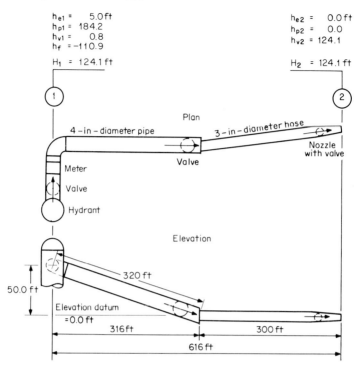

Figure 10-15 General equation. The total head at point 1, less the friction head loss between points 1 and 2, equals the total head at point 2.

$$h_{e1} + h_{p1} + h_{v1} - h_f = h_{e2} + h_{p2} + h_{v2} = H$$
50.0 ft + 184.2 ft + 0.8 ft − 110.9 ft = 0.0 ft + 0.0 ft + 124.1 ft = 124.1 ft

Valley. The source was a reactivated main line in which the static head was estimated to be 90 ft. This system is shown in Figures 10-16 and 10-17. The total reach of 1830 ft of two sizes of steel pipe in good condition is described briefly as follows.

1. From main line at elevation 855 ft to saddle at highest elevation of reach at 912 ft, 57 ft of elevation head lift: total of 1160 ft of 8-in pipe with two 4-in meters, two gate valves, and two 90° ells.

2. From saddle at elevation 912 ft to discharge of pipe stand at elevation 645 ft, 267 ft of elevation head drop: total of 645 ft of 8-in pipe and 25 ft of 6-in pipe with two 45° ells, gate valve, and three 90° ells.

The reach was built largely on the basis of experience, as the rock excavator has developed many water supplies under many job conditions. As has been stated, the estimated available static head of 90 ft at the entrance to the reach was sufficient to raise the elevation of the flowing water higher than the saddle, the critical or controlling point of the reach. When the general equation was applied to point 1, the entrance to the reach, and point 2, the saddle, the following analysis resulted.

At point 1 and between points 1 and 2:

$h_{el} = 0.0$ ft
$h_{pl} = 206.7$ ft
$h_{vl} = 1.2$ ft
$h_f = -128.7$ ft

At point 2:

$h_{e2} = 57.0$ ft
$h_{p2} =$ unknown
$h_{v2} = 1.2$

Substituting in the general equation:

$$0.0 \text{ ft} + 206.7 \text{ ft} + 1.2 \text{ ft} - 128.7 \text{ ft} = 57.0 \text{ ft} + h_{p2} + 1.2 \text{ ft}$$

$$h_{p2} = 21.0 \text{ ft}$$

Thus a pressure head of 21.0 ft is estimated to exist at the saddle, adequate to force the water over the highest elevation of the reach.

Use of River Water Elevated above the River

The two preceding water systems were built on the basis of experience. This system, taking water from a stream and elevating it 402 ft above the river to a freeway rock excavation job, was designed. The equations and design diagrams are given in Figures 10-18 and 10-19.

The job was in high desert country of Inyo County, California, and rock excavation was 1,014,000 yd³. The 1974 bid item for development of water supply was set by the contracting officer at a $20,000 lump sum.

The rock was largely the Bishop tuff formation, a rhyolitic cemented ash with interbedded rhyolite flows. The formation is porous and the country is hot and dry. It was felt that about 60 gal of water per cubic yard would be needed for cut, fill, and haul roads. Excavation was scheduled for a 6-month period, requiring about 850 yd³/h average. The

Figure 10-16 Section of reach from elevation 912 ft at saddle to elevation 630 ft at bottom of pipe stand. Length of 8-in pipe is 645 ft for a drop of 282 ft, giving a descent angle of about 26°, or 49 percent slope. A 10,000-gal water hauler is being loaded under the spout of the pipe stand. Discharge Q from spout loads the 10,000 gal in an average of 6.8 min, giving a Q of 1490 gal/min. The estimated 90-ft static head at the main line was sufficient to overcome friction head losses in 1160 ft of 8-in pipe and the elevation head of 57 ft in order to put 1490 gal/min over the saddle and down to the pipe stand. This temporary reach was abandoned after 4 months of service when the fill elevation was raised and when the planned system of wells in the San Luis Rey Valley was completed. Ultimately the delivery reached 1000 gal/min continuously for the 32,000 yd³ of granite excavated each 18-h working day.

10–46 Development of Water Supply

Figure 10-17 A simple pipe stand serving two large water wagons is shown loading a 10,000-gal water hauler at the end of the temporary reach. Delivery was 1490 gal/min during the actual loading. The two 10,000-gal units furnish water to a rapidly raised fill and to a long haul road. As the hauling cycles of the haulers average 24 min, five loads or 50,000 gal are hauled hourly and average use is 830 gal/min. From the saddle, 40-ft sections of 8-in portable pipe lead to the pipe stand along a steep 49 percent slope. If this reach were permanent, a 16,000-gal tank would take the place of the pipe stand, thereby reducing the loading time from 6.8 to about 2.0 min. The hauling cycle would be reduced to 19.2 min, and on the basis of a 60-min hour efficiency each hauler would deliver 31,200 instead of 25,000 gal/h. Savings in the application of water would be 20 percent. One easily appreciates the economy of using water storage tanks.

water requirement was based on 1060 yd³/h with delivery of a maximum of 1060 gal/min. Two reaches were built.

Reach A was from the Owens River to two 10,000-gal tanks located 182 ft above the river. It included essentially a 15-in pump at the river, 5230 ft of new 8-in pipe, and the two tanks. The tanks served water haulers for the fills and two monitors for the cuts.

Reach B was from the tanks to the two monitors located 220 ft above the tanks. Essentially it included two 15-in pumps, 3000 ft of new 8-in pipe, 400 ft of 4-in hose, and the two 1⅜-in monitors.

Figures 10-18 and 10-19 show, respectively, the layouts of reaches A and B. The general equation will be used to check their designs.

Reach A Delivery at point 2 was established at 1060 gal/min or 2.36 ft³/s. The 180-hp diesel engine of the pump was estimated to deliver 1060 gal/min at three-quarters of full throttle. To check reach A, the theoretical mechanical head h_m of the pump will be determined, and from it the brake horsepower of the engine will be estimated.

At point 1 and between points 1 and 2:

$h_{e1} = 0.0$ ft
$h_{p_1} = 10.0$ ft
h_{v1} (Table 10-1, 8-in pipe, interpolated for 1060 gal/min) = 0.8 ft
h_f = sum of known friction head losses

Entrance (Table 10-1 interpolated for 1060 gal/min):
$v = 6.79$ ft/s (Table 10-8):

$$h_f = 0.78 \times \frac{6.79^2}{64.4} = 0.6 \text{ ft}$$

Equivalent length of 8-in new pipe (Table 10-1 interpolated for 1060 gal/min):
Friction head loss = 3.37 ft per 100 ft of average pipe.

For new pipe, $h_f = 0.54 \times 3.37 = 1.82$ ft per 100 ft of pipe.
5230 ft of pipe   5230 ft
Table 10-4:

One foot valve	76
One open gate valve	4
One pressure valve	247
Four 90° ells @ 21 ft	84
One tee	44
Total equivalent length pipe	5685 ft

$h_f = 56.8 \times 1.82 = 103.4$ ft

$h_f = -104.0$ ft
h_m unknown $= h_m$ ft
At point 2:
$h_{e2} = 222.0$ ft
$h_{p2} = 0.0$ ft
$h_{v2} = 0.8$ ft

Applying the general equation:

$$0.0 \text{ ft} + 10.0 \text{ ft} + 0.8 \text{ ft} - 104.0 \text{ ft} + h_m = 222.0 \text{ ft} + 0.0 \text{ ft} + 0.8 \text{ ft}$$

$$h_m = 316.0 \text{ ft}$$

Water horsepower (whp) $= \dfrac{316.0 \times 2.36 \times 62.4}{550} = 84.6$

Brake horsepower (bhp) of diesel engine of pump $= \dfrac{\text{whp}}{\text{pump efficiency}}$

Assuming pump efficiency of 55 percent,

$$\text{bhp} = 84.6/0.55 = 153.8$$

Percent of rated bhp of engine $= 154/180 = 86$

It is probable that the engine worked continuously at 78 percent of full throttle to deliver 1060 gal/min. Tests showed that delivery at the tanks was 911 gal/min at an estimated three-fourths of full throttle.

Figure 10-18 shows two pressure valves in reach A. Flow to the tanks was regulated by the lower valve with alternate discharge pipe, located at the pump, and by the upper valve, located below the two tanks. When tanks became full, the upper valve closed and the increased pressure head opened the lower valve, discharging the water back into the river. When water was lowered in the tanks by the water haulers and the monitors, the upper valve opened to fill the tanks and the resulting decreased pressure head closed the lower valve, restoring the flow in the line.

Reach B This reach was designed on two premises: that each of two 1⅜-in monitors would deliver 530 gal/min, or 1.18 ft³/s, for a horizontal reach of stream of about 90 ft; and that one or two pumps would deliver the necessary mechanical head. To check reach B, the theoretical mechanical head of the pump(s) and the resultant necessary engine brake horsepower will be determined.

At point 3, and between points 3 and 4:

$h_{e3} = 0.0$ ft
$h_{p3} = 20.0$ ft
h_{v3} (Table 10-1, interpolated for 1060 gal/min in 8-in pipe) $= 0.8$ ft
$h_f =$ sum of known friction head losses
Entrance:
    Table 10-1, interpolated for 1060 gal/min:

$$v = 6.79 \text{ ft/s}$$

    Table 10-8:

10-48   Development of Water Supply

$$h_f = 0.50 \times \frac{6.79^2}{64.4} = 0.4 \text{ ft}$$

Equivalent length of 8-in new pipe (Table 10-1, interpolated for 1060 gal/min)
Friction head loss = 3.37 ft per 100 ft
of average pipe. For new pipe, $h_f =$
$0.54 \times 3.37 = 1.82$ ft per 100 ft of pipe.

3000 ft of pipe	3000 ft
Table 10-4	
Four open gate valves @ 4 ft	16
Three 90° ells @ 21 ft	63
One tee	44
Total equivalent length pipe	3123 ft

$h_f = 31.2 \times 1.82 = 56.7$ ft

h_f = sum of known friction head losses
Two 200-ft lengths of 4-in hose; for each hose, $Q = 530$ gal/min
(Table 10-3, interpolated for 530 gal/min): $h_f = 4.00 \times 18.8 = 75.2$ ft
Two 4-in open gate valves:
$v = Q/A = 1.18/0.0872 = 13.5$ ft/s
$h_f = 1.0 \times \frac{13.5^2}{64.4} = 2.8$ ft
Two valves @ 2.8 ft = 5.6
Two 1⅜-in monitors; for each monitor:
$v = Q/A = 1.18/0.0103 = 115$ ft/s
(Table 10-6, interpolated for 115 ft/s):
$h_f = 8.2$ ft
Two monitors @ 8.2 ft = 16.4

$h_f = -154.3$ ft
h_m, unknown = h_m ft

At point 4:

$h_{e4} = 220.0$ ft
$h_{p4} = 0.0$ ft
$h_{v4} =$ For two monitors = $2 \times \frac{115^2}{64.4} = 410.7$ ft
Applying the general equation:

$$0.0 \text{ ft} + 20.0 \text{ ft} + 0.8 \text{ ft} - 154.3 \text{ ft} + h_m = 220.0 \text{ ft} + 0.0 \text{ ft} + 410.7 \text{ ft}$$

$$h_m = 764.2 \text{ ft}$$

Water horsepower (whp) $= \frac{764.2 \times 2.36 \times 62.4}{550} = 204.6$

Brake horsepower (bhp) of diesel-engine pumps $= \frac{\text{whp}}{\text{pump efficiency}}$

Assuming pump efficiency of 55 percent, bhp = 204.6/0.55 = 372.0

The pumping job called for two 225-hp diesel engines for the two pumps. The first pump was placed at the beginning of the reach and the second was placed near the end of the pipeline. Only 257.1 ft mechanical head was needed to bring water to the end of the pipeline at the tee. The diesel engine of the first pump needed only to develop 123 bhp at 56 percent of full throttle.

In reach B the major and minor friction head losses are proportionately:

Major losses in pipe and hoses, 84 percent	129.8 ft
Minor losses in fittings, valves, and monitors, 16 percent	24.5
Total losses	154.3 ft

In this 3400-ft reach the minor losses are 16 percent, whereas in the 620-ft reach of Plate 10-1 the losses were 32 percent. Generally, minor losses are proportionately less the longer the reach. If the reach is longer than 1 mi, some estimators ignore the minor losses.

Productions and Costs   10-49

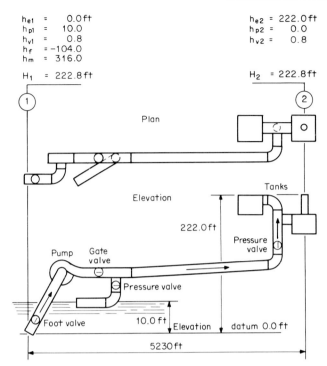

Figure 10-18 General equation for reach A. The total head at point 1, less the friction head loss plus mechanical head gain between points 1 and 2, equals the total head at point 2.

$$h_{e1} + h_{p1} + h_{v1} - h_f + h_m = h_{e2} + h_{p2} + h_{v2} = H$$
$$0.0 \text{ ft} + 10.0 \text{ ft} + 0.8 \text{ ft} - 104.0 \text{ ft} + 316.0 \text{ ft} = 222.0 \text{ ft} + 0.0 \text{ ft} + 0.8 \text{ ft} = 222.8 \text{ ft}$$

Justification is based on the fact that the precision for estimating all losses is not within the percentage value of the minor losses.

The analyses of these three water systems by means of the general equation, based on practical application of Bernoulli's theorem, emphasize the advantages of well-thought-out designs as opposed to building a water system solely by experience and the usual correlative trial-and-error corrections.

PRODUCTIONS AND COSTS

The development of water supply for excavation is necessarily also the development for nearly all other phases of the job. For example, in this chapter a water system was cited which not only served 2,740,000 yd³ of excavation but also 276,000 yd³ of base courses and concrete. Additionally, it provided for haul-road maintenance and for miscellaneous purposes. On the particular job perhaps only 80 percent of the cost of development should have been prorated to excavation.

As is usually the case, this and other prorations of costs depend on bookkeeping methods. For simplicity in this discussion all costs will be assigned to excavation. Also, for illustrative purposes both operating cost and the cost of water will be included in the total cost of development of water supply.

The cost of ownership and operation will be based on Table 8-3. In some cases an interpolation of the table will be made, as is generally done in estimating from tabular data.

Earlier in this chapter three water-supply development projects were described. Below are estimates for these three systems in terms of 1978 costs.

10–50   Development of Water Supply

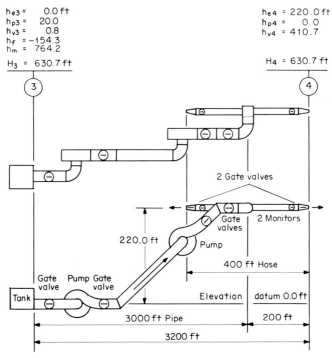

Figure 10-19 General equation for reach B. The total head at point 3, less the friction head loss plus mechanical head gain between points 3 and 4, equals the total head at point 4.

$$h_{e3} + h_{p3} + h_{v3} - h_f + h_m = h_{e4} + h_{p4} + h_{v4} = H$$

0.0 ft + 20.0 ft + 0.8 ft − 154.3 ft + 764.2 ft = 220.0 ft + 0.0 ft + 410.7 ft = 630.7 ft

1. In the simple system of Plate 10-1, the excavation was 300,000 yd³, scheduled at 250 yd³/h hourly for an estimated 1200 working hours, or about 10 months of work. Average water delivery was 160 gal/min, 8000 gal or 1070 ft³ per 50-min working hour.

	Machinery
Hourly cost of ownership and operation by machinery (Table 8-3):	
1. 320 ft of 4-in pipe, cost $813; 35% of $0.81	$0.28
2. Two 4-in gate valves	
One 2½-in meter	
One 90° ell, cost $318; 35% of $0.32	0.11
3. 300 ft of 3-in hose, cost $1491; 41% of $1.49	0.61
4. One 1⅛-in nozzle, cost = $105; 35% of $0.10	0.04
Total	$1.04
Hourly cost of water, 1070 ft³ @ $0.225 per 100 ft³	2.41
Total hourly costs	$3.45
Total job cost, 1200 × $3.45	$4140.
Cost per yd³ of granite excavation	
1. Ownership and operation, $1.04/250	$0.004
2. Water, $2.41/250	0.010
Total	$0.014

2. The temporary system of one reach of Figures 10-16 and 10-17 served about 760,000 yd³ of excavation over a period of 4 months, or about 704 working hours. Average

production was 1080 yd³. Although water delivery actually was 1670 gal/min during the loading of the water haulers for the maximum production of 2500 yd³/h, the average delivery was 720 gal/min, or 43,200 gal/h.

	Machinery
Hourly cost of ownership and operation by machinery (Table 8-3):	
1. 1790 ft of 8-in pipe, cost = $9093; 35% of $9.09	$ 3.18
2. One 8-in gate valve	
Two 3-in meters	
Two 45° ells, weldments	
Two 90° ells, weldments, cost $600; 35% of $0.60	.21
3. 45 ft of 6-in pipe, cost $172; 35% of $0.17	.06
4. One 6-in gate valve	
Three 90° ells, weldments	
One reducer, weldment, cost $325; 35% of $0.32	.11
Total	$ 3.56
Hourly cost of water, 43,200 gal @ $0.45 per 1000	19.44
Total hourly costs	$23.00
Total cost for temporary reach, 704 × $23.00	$16,192.
Cost per cubic yard of granite excavation	
1. Ownership and operation, $3.56/1080	$0.003
2. Water, $19.44/1080	.018
Total	$0.021

3. The complete system of Figures 10-18 and 10-19 provided water for 1,014,000 yd³ of excavation over a period of 6 months or about 1190 working hours, the average hourly production being 852 yd³. The job averaged 65 gal/yd³ of volcanic tuff. Total water consumption was 65,900,000 gal or an average of 923 gal/min. The water was used almost exclusively for excavation in cut, embankment in fill, and on haul roads. Paving was asphaltic concrete and there were no sizable concrete structures, so that little additional water was called for.

	Machinery
Hourly cost of ownership and operation by machinery (Table 8-3)	
1. 8230 ft of 8-in pipe, cost $41,808. 35% of $41.81	$14.63
2. One 8-in foot valve	
Five gate valves	
Two pressure valves	
Seven 90° ells, weldments	
Two tees, weldments, cost $4350, 35% of $4.35	1.52
3. 400 ft of 4-in hose, cost $1990; 41% of $1.99	0.82
4. Two 1¾-in monitors with stands	
Two 4-in gate valves, cost $3500; 35% of $3.50	1.22
5. One 15-in centrifugal pump with 180-hp diesel engine, cost $16,600; 42% of $16.60	6.97
6. Two 15-in centrifugal pumps with 225-hp diesel engines, cost $36,400; 42% of $36.40	15.29
7. Two 10,000-gal movable tanks, cost $28,200; 22% of $28.20	6.20
Total	$46.65
Hourly cost of water, 55,300 gal @ $0.10 per 1000 gal	5.53
Total hourly costs	$52.18
Total cost for complete system, 1190 × $52.18	$62,094.
Cost per cubic yard of tuff excavation	
1. Ownership and operation, $46.65/852	$0.055
2. Water, $5.53/852	0.006
Total	$0.061

The comparison of unit costs per cubic yard among these three different systems, as tabulated below, emphasizes the influences of kinds of rock, physical conditions, and sources of water.

10–52   Development of Water Supply

	Ownership and operation	Water	Total
The simple system	$0.004	$0.010	$0.014
The temporary reach	0.003	0.018	0.021
The complex system	0.055	0.006	0.061

One notes that these examples of costs include no water haulers or operators. They are for delivery of water to the job and do not include the application of water to cuts, fills, or haul roads. These costs are parts of the unit bid prices for excavation and embankment.

SUMMARY

The development of water supply is sometimes considered to be a part of excavation and the costs are therefore a part of excavation costs. However, whether they be separate or integral, the methods and the costs are to be carefully examined. The development may be as simple as hooking into a nearby water main or it may be as complex and costly as drilling wells, installing pumps, and building a long main pipeline with reservoirs.

In many cases the planning of the water supply is equivalent to the designing of water supply for a good-sized city, the volume of water being the same but the distribution being less complex. However, in both cases it is manifest that good engineering is mandatory.

CHAPTER 11

Fragmentation of Rock

PLATE 11-1

Smooth economic transition from ripping to blasting in a huge granite cut of yardage 9,776,000 yd³, length 4700 ft, and maximum centerline cut 235 ft. Three heavyweight tractor-rippers are ripping extremely hard semisolid rock in the background where the cut averages 70 ft below original ground and where scrapers are averaging about 1500 cybm/h. In the foreground two track drills are putting down blastholes in semisolid and solid granite, where the cut averages about 60 ft below original ground. Blasted rock is being bulldozed by two bulldozers to a 10-yd³ bucket loader, which is loading 50-ton rear-dump trucks. Bucket loader production is about 700 cybm/h.

In this area of the cut seismic studies disclosed that weathered rock, in which seismic shock-wave velocity averaged 4100 ft/s, changed to 7800-ft/s semisolid rock at 56 ft average depth. The tractor-rippers are reaching the end of ripping and the blasting is commencing at an average depth of about 65 ft below original ground.

Several factors determine the time for this change, if indeed the change is to be made. Notably they are: (1) the remaining yardage of hard rock, as it may be economical to continue ripping rather than to bring in a drilling and blasting crew; (2) the desired hourly production, as a spread using tractor-rippers and scrapers may produce double the yardage of a spread using drills, shovel, and rear-dump haulers; and (3) a comparison of the total direct job unit costs for the two methods, ripping and blasting. Total costs for the ripping method include ripping, push loading by tractors, hauling by scrapers, and fill building; and those for the blasting method include drilling and explosives, loading by shovel, possibly bulldozing to the shovel by tractors, hauling by rear- or side-dump haulers, and fill building.

Obviously, hardness or degree of consolidation of the rock, although most important, is not the only determinant in the shift from ripping to blasting for fragmentation.

This chapter commences the actual chain of events for excavation, which ends with Chapter 16. It is helpful to examine again Chapters 1 through 7, as they offer provocative ideas for excavation from the cut to the fill or to the plant.

Fragmentation is the breaking up of rock so that it can be loaded or cast efficiently and economically. *Rock* is a mass of material, loose or solid, which makes up an integral part of the earth. In the geologist's terms rock may be *residual, weathered, semisolid,* or *solid.* In terms of the excavator it may be *soft, medium,* or *hard,* and with or without an overburden of earth.

The terms of the excavator and the geologist relate in this manner. Soft rock includes the harder residuals and the softer weathered rock. Medium rock includes the harder weathered rock and the softer semisolid rock. Hard rock includes the harder semisolid rock and the solid rock. The overburden of earth generally comprises the residual rock, although residual may be any form of rock which is not acceptable for its purpose.

There are presently two methods of fragmentation, by medium- and heavyweight tractor-rippers and by blasting. The tractor-rippers are used in the soft and medium rocks and blasting is used in the relatively hard rocks. Figure 11-1 illustrates the generalization of these relationships.

In this idealized cross section of a 102-ft-deep cut, if the largest cross section were located at the high point of a symmetrical cut of 500,000 yd³ and 3000-ft length, the quantities within the three important zones would be approximately as given in this table:

Zone	Cubic Yards	Percentage
1. No ripping	125,000	25
2. Ripping	290,000	58
3. Blasting	85,000	17

1. There is generally a relatively thin blanket of earth or residuals which does not require ripping. Therefore, the major cross section shows a centerline depth of 13 ft, or

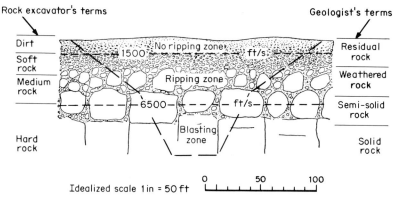

Figure 11-1 Fragmentation by ripping and blasting in a 102-ft-deep railroad cut and their relationship to the nature of the rock formation and to seismic shock-wave velocities. Ripping is generally by medium-weight or heavyweight tractor-rippers. Blasting is generally by drilling holes of appropriate diameter and depth on appropriate areal patterns and by using suitable explosives. The average seismic shock-wave velocity in igneous, sedimentary, and metamorphic rocks which separates the zones for no ripping and ripping is 1500 ft/s. The average velocity in these same three classifications of rocks which separates the zones for ripping and blasting is 6500 ft/s.

These velocities of 1500 and 6500 ft/s are for heavyweight tractor-rippers. For medium-weight tractor-rippers the velocities are, respectively, 1500 and 5500 ft/s. These tractor-rippers are described in Chapter 7 in the section explaining the use of the seismic timer as a means for exploration of rock excavation.

11-4 Fragmentation of Rock

13 percent of the total 102 ft. However, the estimated yardage for this mantle is 25 percent, or 192 percent of the centerline-depth ratio.

2. The rippable rock is 58 percent of the total 500,000 yd^3, although the centerline-depth ratio is 44 percent.

3. The centerline depth of rock requiring blasting is 45 ft, or 44 percent of the total 102-ft depth. However, the estimated yardage for the hard rock is but 17 percent of total yardage.

Such an estimate is based on the uniform weathering characteristics of most rock formations, whose degree of weathering decreases fairly uniformly with increase in depth. Thus the zones or layers of Figure 11-1 may be likened to the concentric layers of an onion.

Exceptions to this generality of uniform weathering are shown in Figures 3-3 and 3-4, both involving limestone formations in which the change from residuals to solid rock is abrupt and the interface is planar and not spheroidal. Another similar example is shown in Figure 7-6. Extrusive igneous and sedimentary rocks have planar interfaces and outcroppings just below the residuals.

Usually there are three bases for the selection of ripping or blasting, along with the ancillary ones discussed in the legend for Plate 11-1.

1. The logs, the cuttings, and the cores of the exploratory drill holes. Such logs are shown in Figure 6-3 and Table 7-1. Cores are illustrated in Figure 7-18. Usually the logs contain notes on drilling speeds in different zones of depth and, properly interpreted, these speeds or penetration rates have a bearing on the selection of methods for fragmentation. In percussion drilling by jackhammer drills, track drills, and cable drills, the cuttings identify both the nature and the hardness of the rock. In core drilling, the recovery, length ratio of solid cores to hole, and the cores are indicative of hardness and uniformity, as shown in Figure 7-18. For best results from exploratory drilling, it is requisite that an engineer work directly with the driller so as to take full notes for considered interpretation.

2. The report of seismic studies. As explained in the seismic timer section of Chapter 7, this method provides a tangible index of the consolidation or hardness of the earth-rock in terms of shock-wave velocity. These finite values are set forth in Chapter 7 in terms of zones for: (1) no ripping; (2) soft ripping; (3) medium ripping; (4) hard ripping; (5) extremely hard ripping or blasting; (6) blasting. Velocity-depth relationships are shown in Figure 6-4 and a typical study is given in Figure 7-26. Chapter 7 covers field exploration of rock excavation.

3. The experience and common sense and imagination of the rock excavator. These are in terms of familiarity with the particular or similar rock formations and the practical and economical methods used in the excavation.

When these three factors are considered harmoniously, the right selection of ripping or blasting emerges.

Fragmentation is usually not a separate bid item in excavation as it is considered to be a part of the excavation item. This is true for both bidding on unclassified excavation, common and rock together, and on classified excavation, common and rock separately. Inclusion in the excavation bid is true universally in public works, such as blasting for the spillway excavation of a dam. Sometimes in private works, such as excavation for a plant site, ripping would be a part of the excavation item but blasting would be a separate item.

The Standard Specifications of the California Division of Highways state: "This work [earthwork] shall consist of all operations necessary to excavate earth and rock. . . ." In public works there is one exception to this definition as it applies to rock, and it is covered by the special provisions for the job. It is *contour blasting*, more commonly known as *line drilling* or *preshearing*. Examples are shown in Figures 11-37 and 11-38. This item is a separate one for drilling only and the unit for bidding is the lineal foot. Contour blasting is becoming popular in both public and private works for aesthetic reasons, for safety, and for engineering considerations.

In foundation excavation for Auburn Dam, California, there were 202,000 lineal ft, or 38 mi, of drilling. Average unit bid price in 1974 was $1.77 per lineal foot.

Hereafter in this chapter ripping and blasting for fragmentation of rock are considered separately, although they are usually simultaneous and well-coordinated operations.

RIPPING

Principles

The ripper has a long recorded history, as it was the first means for fragmenting rock. From the building of the Roman Appian Way in 312 B.C., when a wheel-mounted plow was drawn by oxen, the ripper has evolved into a crawler-tractor-ripper-bulldozer of enormous proportions. As of 1978 the largest weighs 190,000 lb, has a 700-flywheel horsepower diesel engine, and exerts a pull of some 76,000 lb on the tips of the ripper shanks. Since the Romans loosened rock-earth with a pull of some 400 lb, it is obvious that a pull of 76,000 lb, 190 times greater, can fragment rock formations of considerable consolidation.

The tractor-ripper changed the conception of rock-earth classifications. Up to about 1930, when real rock ripping commenced, excavation was classified as *common*, that which could be dug readily, and *rock*, that which required blasting prior to digging. The separation of common, which included loose rocks up to 1 ft dimension, and rock was fairly simple. With the use of rippers an intermediate classification emerged, rippable rock, which in the beginning included only well-weathered rock because of the size and power limitations of the available pull-type rippers and crawler-tractors. As of 1978 tractor-mounted rippers such as a 160,000-lb, 524-hp machine are fragmenting well into the semisolid rocks. Such a tractor-ripper-bulldozer is shown ripping granite in Figure 11-16.

As a result of this development in machines and ripping techniques, the previous classifications of common and rock excavation have been largely eliminated from specifications and the all-embracing term *unclassified excavation* has been adopted. Thus the responsibility for determining the absolute and relative quantities of so-called common and rock excavation belongs to the rock excavator.

For rock excavation the crawler-tractor-mounted ripper is used almost exclusively, although some pull-type rippers drawn by crawler-tractors are used when ripping is done on a nonproduction basis. The ranges of the weights and horsepowers of the medium- and heavyweight machines are tabulated below. The table is based on specifications of the several manufacturers. Specifications for crawler-tractors, without ripper and bulldozer, are given in Table 13-5. Weights are working weights, including fully equipped tractor, ripper with three shanks, and bulldozer. Drawbar or tip-pull at 1.0 mi/h is not based on the theoretical drawbar pull, but rather on a practical 0.40 coefficient of traction in loose rock-earth.

Classification	Working weight, lb	Flywheel horsepower	Drawbar or tip pull, lb, at 1.0 mi/h
Medium-weight	55,000	200	22,000
	83,000	300	33,000
Heavyweight	109,000	410	44,000
	160,000	524	64,000
	190,000	700	76,000

Figure 11-2 shows the size and relationship of the shanks, tips, and shank protectors of a two-shank, adjustable parallelogram-type ripper mounted on a heavyweight tractor. Important parts of the ripper are:

1. *The point*, which enters the rock structure by wedge action, splitting and breaking the stratified rocks and crushing and breaking the amorphous rocks. Figure 11-3 shows a point with its critical angle A and the three common types of points, each point being designed for rocks of different characteristics.

2. *The shank*, which holds the tip and the shank protector, and which, if of the fixed type, mounts into the pocket of the tool bar or, if of the swiveling type, into the pocket of the clevis. There are three types of shanks: straight for blocky and slabby rock; curved for tight, laminated rock; and modified for average rock. The shanks of Figure 11-2 are modified, being used in mixed rock formations.

3. *The tool bar*, the heavy transverse box section, to which are attached the shanks and which is raised, lowered, and inclined by the power assembly.

11–6   Fragmentation of Rock

 4. *The push block,* sometimes supplemented by a lower shelf, for tandem tractor ripping. A tractor-bulldozer pushes against the push block so as to increase the tip pull of the leading tractor-ripper in order to rip out a local prism of extremely hard rock. The projecting push block is shown in Figure 11-5. The ripper with swiveling shanks of Figure 11-2 has no push block, as it is practical to equip only the fixed-shank ripper with push block and shelf.

 5. *The power assembly,* consisting of arms and hydraulic cylinders for raising, lowering, and inclining the tool bar with its attached shanks. Figure 11-4 illustrates the combined actions of this assembly for the most efficient adjustable parallelogram ripper.

The older and currently less used ripper is the hinged type. The tool bar is hinged to the tractor frame or to the track-roller frame so that it swings through an arc of about 30°. This action results in a like change of angle of the tips. Since a principle of ripping is that the tip angle should be constant at any depth of ripping, there are compensating, but limited, adjustments of the shank positions within the shank pockets. However, these adjustments are manual and time-consuming.

Figure 11-5 shows a hinged-type ripper mounted on the track-roller frame. This design applies the pull on the draft arms at a point nearest to ground contact with resultant minimizing of many structural stresses in the tractor-ripper unit. The heavyweight tractor-ripper-bulldozer weighs 93,000 lb with a 280-hp engine.

The ripping speed of a tractor-ripper is usually in the 0.75- to 1.50-mi/h range or in first gear, the number of tips and the depth of ripping being adjusted to the speed. About 1 mi/h makes for "comfortable" ripping with nearly constant speed except in

Figure 11-2 Working parts of the shank assembly of a heavyweight tractor-ripper, as used in granite formation of all degrees of weathering. Sizes of parts may be judged by the 36-in yardstick. The shank is the modified type and the tip is the short type for difficult penetration and heavy shocks. The shanks are mounted in swivel clevises in order to provide free movement through the variable rock formation. Both tip and shank protector are pin-connected to the shank for quick exchange to new parts in the field

Figure 11-3 Three basic kinds of tips for ripping are shown above, along with the critical angle A for tip penetration into the rock formation. The angle may be varied in three ways: by the configuration of the tip itself, by the type of shank, and by the angular adjustment of the shank. Angular adjustment of the shank for best penetration of the tip is possible in the adjustable parallelogram ripper as shown in Figure 11-4. The short tip is used when penetration is most difficult and shock is most severe. The long tip is used for highly abrasive rock when breakage is not a serious factor. The intermediate tip is suggested for applications in which abrasive rock is hard enough to break the tip. These three tips are all reversible and self-sharpening, providing good penetration and long life. *(Caterpillar Tractor Co.)*

Figure 11-4 Ripping characteristics of the adjustable parallelogram-type ripper. *(a)* Backward position of the tip provides an aggressive tip angle for entry into the rock formation. *(b)* The tip having entered and having reached the desired depth for ripping in the formation, its best ripping angle is found by adjusting the shank to the near vertical position. *(c)* For prying out rock particles, the shank is adjusted to a forward position. All these operations are by trial and error when ripping of an unfamiliar formation is begun, but they quickly become adjustments by rote. *(Caterpillar Tractor Co.)*

hard and extremely hard ripping. Such comfortable ripping is shown in Figure 11-6. A heavyweight tractor-ripper-bulldozer is advancing easily through weathered granite with two shanks and about 2-ft ripping depth. This would be described as soft to medium ripping.

"Uncomfortable" ripping is emphasized strikingly in Figure 11-7. The heavyweight machine is on tiptoe, with resultant structural and power transmission assembly stresses and excessive wear and tear in this extremely hard ripping. Higher production and lower unit cost could have been achieved by use of one shank in the hard igneous rock.

In further discussions of principles, the tractor-ripper-bulldozer specifications will be those of the composite heavyweight unit, as this machine is most used in rock excavation. Salient specifications are:

1. Working weight, equipped with three-shank ripper and bulldozer, 125,000 lb
2. Flywheel horsepower, 450 hp
3. Tip pull, based on 40% coefficient of friction between grousers and loose rock, 50,000 lb

Figure 11-5 A massive swing or hinged-type ripper attached to the track-roller frame rather than to the tractor frame. The resulting advantages are a decrease in angular displacement of the shanks, due to the long radius of swing, and a lowering of structural stresses, due to attachment to track-roller frame. The heavyweight tractor-ripper-bulldozer weighs 93,000 lb with a 280-hp engine. The push block and shelf for the pushing bulldozer make it especially adapted for tandem ripping, as shown in Figure 11-10. The weight of the ripper itself with one shank is 21,000 lb, about double that of the swing-type ripper when attached to the tractor frame. *(Shepherd Machinery Co.)*

Figure 11-6 "Comfortable" ripping in weathered granite by a heavyweight tractor-ripper equipped with two swiveling shanks and tips. This is soft-medium ripping in rock of 4000-ft/s seismic shock-wave velocity. The tractor-ripper is traveling in low gear at about 1 mi/h with about 2-ft tip penetration. The production is in the neighborhood of 1200 cybm/h, supplemented by that of another heavyweight unit which both rips and helps to push-load the scrapers. Together they rip about 1600 yd^3/h. Often the operators of tractor-rippers and push tractors cooperate so that the machines provide ample coordinated efforts for ripping and push loading for maximum scraper efficiency. This sensitive teamwork takes place without supervision and causes wonderment in those who are unfamiliar with rock excavation practices.

Figure 11-7 "Uncomfortable" ripping, as contrasted to the "comfortable" ripping of Figure 11-6. The uncomfortable ripping is characterized by low production, extremely high costs for repairs and replacements, and high unit cost for ripping. The comfortable ripping of Figure 11-6 is distinguished by high production and low unit cost. Unless this tractor-ripper-bulldozer, working tiptoe fashion, is ripping out an infrequent knob of hard rock, it appears advisable to change to blasting, as shown in Plate 11-1. *(Caterpillar Tractor Co.)*

For a near constant speed of 1.0 mi/h the depth of ripping depends on the consolidation of the rock formation and the number of points. The depth for a single point may vary from 6 ft in soft ripping for trench excavation to a mere few inches for extremely hard ripping in mass rock excavation. In the case of two or three points the depth may vary from 1 to 3 ft through the range of soft, medium, and hard ripping.

Spacing between furrows, usually resulting from multiple passes over an area, depends upon the excavation method to be used and the gradation requirements for the ripped rock. By excavation method is meant the machines used for loading, casting, and hauling of the ripped rock, because the method will determine the degree of fragmentation desired. The correlative is gradation, depending on the end use of the rock, for example, for compacted fill, waste fill, hydraulicked fill, stockpile, primary crusher, and the like. The closer the spacing, the smaller the ripped particles of rock; the easier the loading, casting, and hauling; and the more uniform the gradation. Conversely, the wider the spacing, the larger the particles; the harder the loading, casting, and hauling; and the less uniform the gradation.

Depth and spacing of furrows go hand in hand: the deeper the furrow, the wider the spacing. This is explained by the cross section of the breakage area of a single point. As the point advances an inverted triangular area is broken, the sides forming about a 45° angle with the horizontal and vertical. It follows that spacing must be such as to accommodate this inverted triangular prism so as to rip all the rock. For a given job both depth and spacing are usually based on experience and trial and error on the job.

Table 11-1 summarizes data for the average use of three, two, and one point on the composite heavy-duty tractor-ripper-bulldozer. The table is idealized but nevertheless it is in keeping with ripping practices. The estimated hourly productions should be compared with those of Figure 11-17, which are based on many observations of actual ripping.

Notable in Table 11-1 is the number of passes of the machine over the overall track width in order to get full area coverage and average depth of one-half the total depth of ripping at the tips. The number of passes varies from 0.7 pass to an average depth of 1.2 ft in soft ripping to 3.6 passes to an average depth of 0.8 ft in extremely hard ripping.

Figure 11-1 presents a generalization on fragmentation by ripping and blasting according to the nature of the rock formation. There are also many generalizations as well as specifics on the subject in Chapters 2 through 7.

The ripping characteristics of igenous, sedimentary, and metamorphic rocks are summarized in the following paragraphs.

TABLE 11-1 Idealized Data for the Performance of a Composite Heavyweight Tractor-Ripper When Ripping in Soft, Medium, Hard, and Extremely Hard Earth-Rock Formations
Pertinent specifications: weight 125,000 lb; engine horsepower, 450; ripper tip or drawbar pull, based on 40% friction coefficient between grousers and ripped rock, 50,000 lb; outside track width, 10.7 ft.

	Soft ripping	Medium ripping	Hard ripping	Extremely hard ripping
Seismic shock-wave velocity, ft/s	1500 to 4000	4000 to 5000	5000 to 6000	6000 to 7000
Number of tips	3	2	1	1
Spacing of tips, ft	5.0	10.0		
Depth of ripping, ft	2.5	2.0	2.0	1.5
Average depth of ripping, ft	1.2	1.0	1.0	0.8
Net width ripped by one pass, ft	15.0	8.0	4.0	3.0
Net cross-sectional area ripped by one pass, ft^2	18.8	8.0	4.0	2.2
Feet of ripping travel hourly by tractor-ripper	3840	3840	3310	3310
Cubic yards ripped hourly	2670	1140	570	270
Number of passes to cover outside track width	0.7	1.3	2.7	3.6

NOTE: Net width ripped by one pass is the total of widths ripped by all tips based on 45° angles of breakage from tips to ground surface. Net cross-sectional area ripped by one pass is the total area of the inverted triangles of breakage by all tips. Feet of ripping travel hourly by tractor ripper is based on: 300-ft length of ripping at 1.00 mi/h for soft and medium ripping and 0.75 mi/h for hard and extremely hard ripping; turning time 0.50 min; 50-min working hour. Number of passes to cover outside track width is 10.7 ft divided by net width ripped by one pass.

Igneous rocks, unless weathered, are difficult to rip because of their lack of laminations into which the ripper point may be forced. Of course, extrusive igneous rocks such as lava flows are stratified but stratification is massive and few cleavage planes exist. Ripping then is a crushing and breaking action, the success of which depends on the degree of weathering and presence of cooling and flexure joints. Igneous rocks make up about 39 percent of rock excavation. The many variations of formations are shown in Figures 2-8, 2-9, 2-11, 2-12, 3-1, 3-6, 3-7, 4-7, 4-10, 4-13, 4-14, 5-2, 5-3, 5-5, 5-7, 6-7, 6-9, 7-4, 7-14, 7-18, and 7-32.

Sedimentary rocks, except for the unconsolidated ones such as alluvia, have bedding planes of weakness. When the individual hard strata do not exceed 2 ft in thickness, these rocks may be ripped. If the strata are soft and punky, greater thicknesses may be ripped. Ripping is a splitting and cleaving action causing both horizontal and vertical separations. Sometimes, even when bedding is less than 2 ft thick, the ripper may rip out huge slabs, which makes more complete breakage by subsequent passes very difficult. Such a ripping condition is shown in Figure 11-8. Sedimentary rocks also have vertical joints of weakness, generally resulting from weathering and flexure, and these contribute to rippability. Sedimentary rocks are usually less hard than igneous rocks.

Sedimentary rocks make up about 50 percent of rock excavation. The many kinds of consolidated and unconsolidated sediments are shown in Figures 2-13, 2-14, 2-15, 2-19, 3-3, 3-4, 3-8, 3-10, 4-3, 4-5, 4-8, 4-9, 4-11, 4-16, 4-19, 5-1, 5-6, 5-9, 5-10, 5-11, 5-12, and 5-16.

Metamorphic rocks are generally less affected by weathering than igneous and sedimentary ones and they are harder to rip. The laminations of such rocks as schists and phyllites are not planes of weakness and they are usually free of joints. The major contribution to ripping is the weathering of these altered igneous and sedimentary rocks. As in the case of igenous rocks, ripping is a crushing and breaking action.

Metamorphic rocks make up about 11 percent of rock excavation. Some kinds are

Figure 11-8 Tractor-ripper-bulldozer with 700-hp engine and weighing 190,000 lb ripping massive altered sandstone. The bedding or stratification of this marginably rippable rock formation is between 8 and 15 ft in thickness, unfavorable for ripping. The dip is a few degrees in the direction of ripping and the strike is at right angles to ripping, both favorable attitudes for ripping. Seismic shock wave velocities are high, ranging from 5400 to 15,000 ft/s, and adverse to ripping. Indicative of the difficult ripping are the use of a single high-riding shank and tip and the visible smoke coming from the tip.

The hourly production of the tractor-ripper ranged from 1375 cybm, excellent in the extremely hard rock, down to a low uneconomical output. When this was reached, blasting was substituted for ripping.

The job is highway construction in Montana. The altered sandstone is a member of the Livingston formation, a 70-million-yr-old division of the Upper Cretaceous, which always presents a formidable obstacle to excavation. *(Caterpillar Tractor Co.)*

shown in Figures 2-16, 2-17, 3-12, 6-11, and 7-5. Table 5-1 gives a complete summary of the percentages of the kinds of rocks encountered in open cut-rock excavation.

Physical characteristics favoring rock rippability are: brittleness, crystallization, faults, fractures, friability, large grain size, joints, low strength, moistness, softness, stratification, weak cementation, and weathered nature. Those attributes not favoring rippability are: amorphousness, dryness, hardness, high strength, massiveness, small grain size, solidity, strong cementation, toughness, and unweathered nature.

Methods and Machinery

Methods are determined by the nature of the earth-rock formation and by the manner of loading or casting and hauling the ripped rock. The disposition is generally by the following means.

1. *Bulldozing.* If rock is dozed to a waste area, it is the end of the work. The tractor-ripper need only reduce the rock to a size for efficient dozing. If, however, the dozing is to a shovel, dragline, stationary belt loader, conveyor, primary crusher, slackline excavator, or hydraulicking monitor, the rock must be reduced to proper size for handling. In many cases multiple passes are necessary and the method is based on experience with, and trial and error in, the rock formation. Soft rocks require minimum ripping and hard rocks call for maximum ripping.

2. *Scraper haulage.* In soft and friable rocks ripping produces good grading from fine to coarse particles. Scraper loading by push-tractor is made easy, and one or two passes usually suffice for adequate fragmentation. In hard and tough rocks ripping does not provide good grading, there not being enough fines within the big particles to ease the loading. Multiple passes are necessary. In medium rocks an average condition exists. The largest ripped rocks should not exceed 2 ft in overall dimension for efficient loading.

3. *Shovel, dragline, backhoe, wheel-bucket excavator, slackline excavator, and other bucket excavators.* The maximum size of ripped rock should be compatible with the size of the bucket, the overall size being perhaps 25 percent of the smallest bucket dimension.

4. *Portable and mobile belt loaders.* These machines for heavy rock excavation range from 36- to 72-in belt width. To minimize wear of belt and side flashing, the ripped rock should be graded and the overall dimension of the largest rock should not exceed 25

percent of belt width. Usually, except in soft rocks, multiple passes of the tractor-ripper are necessary.

Tractor-rippers are most commonly used with scrapers and ripping is done parallel to the loading path of the scrapers. The direction of ripping sometimes depends on the lay of the stratification and joints of the rock formation. For example, if the scrapers were loading northward and the dip of the bedding were northward, westward, or eastward, then one would rip northward to force the tips into the weak bedding planes. Were the dip to the south, the tips would tend to be forced out of the rock along the plane of stratification. In this case the tractor-ripper would rip southward but parallel to the loading direction of the scrapers.

Figure 11-8 shows the heaviest and most powerful of tractor-ripper-bulldozers working in an altered sandstone formation in which dip is in the direction of travel and strike is at right angles to travel. The attitude of stratification is good for ripping, but the massive strata, up to 15 ft thick, and the high shock-wave velocities of the rock, up to 15,000 ft/s, limit severely the performance of the machine. Ultimately, blasting will be necessary.

Figure 11-9 illustrates ripping in the metamorphic Bedford Canyon formation of southern California. Two heavyweight tractor-ripper-bulldozers are ripping and bulldozing to the bucket-type loader. The machine to the right is one of the heaviest of tractor-ripper-bulldozers. Engine-flywheel horsepower is 524 bhp and working weight is 160,000 lb.

In this formation of schists and phyllites, average seismic shock-wave velocity changed from 4800 ft/s, medium ripping, to 12,100 ft/s, blasting, at a 31-ft depth. That the change has taken place is shown at the right side in the picture, where a hard prism of rock has been blasted. These two machines succeeded in ripping to grade and bulldozing to loader with a combined production of 470 cybm/h. There are no bedding or weakened planes in the formation and ripping is equally efficient in all directions.

Figure 11-10 illustrates tandem ripping by two heavyweight machines. In the weathered basalt formation small, 200-ft^2 areas with seismic shock-wave velocities up to 7300 ft/s were encountered and the single unit with simultaneous 25,000 lb horizontal pressure and 31,000 lb vertical pressure was unable to rip out the troublesome areas. Tandem ripping afforded simultaneous 47,000 lb horizontal pressure and 59,000 lb vertical pressure on the single ripper tip. This expeditious tandem ripping avoided blasting these small yardages of extremely hard rock.

Another method of ripping is to continually wet the cuts during the ripping. Figure 11-11 shows two monitors playing streams of water on a cut which is being ripped by two

Figure 11-9 Ripping and bulldozing schists and phyllites to a 7-yd^3 bucket loader. The rock is metamorphic and the formation is variable in consolidation, ranging from medium-weathered to semisolid. To the right is a pile of rock which has been blasted, although it is about the same depth below ground surface as is the rock being ripped successfully. The two heavyweight tractor-ripper-bulldozers have combined production of 470 cybm/h. The heavier of the machines, which has 524 hp and weighs 160,000 lb, is to the right.

Figure 11-10 Tandem ripping to remove a troublesome knob of hardest rock is shown as carried out in a troublesome basalt formation of the Santa Monica Mountains of California. The lead tractor-ripper is equipped with a hinged-type ripper with one tip and the bench-type push block of the tool bar provides for both horizontal push and vertical down pressure from the bulldozer of the push tractor. In this instance the combined simultaneous forces of both tractor-rippers on the single tip provide 47,000 lb horizontal point pull and 59,000 lb downward pressure. Such tandem ripping is recommended as an expediency for the ripping of occasional harder-than-usual knobs or small areas of outcrops but it is not suggested for continuous operation. The reason is operational because the "uncomfortable" ripping makes for difficult teamwork between the two operators and unusual wear and tear on the tractor-rippers.

Figure 11-11 Wetting cut to assist ripping of volcanic tuff is shown as carried out with two monitors in order to increase production of the two heavyweight tractor-rippers and thereby reduce the unit cost for ripping. The rock is rhyolitic tuff of the Bishop tuff formation, Inyo County, California. The formation is soft to hard ripping. The highest average seismic shock-wave velocity is 5900 ft/s for hard rock in which each of the tractor-rippers averages about 600 cybm/h. Water in the amount of 55 gal/yd^3 is added to the porous tuff of little or no natural moisture in this high, semiarid desert country. The mixture of rock and water then contains 7.9 percent moisture on its way to the fill, but much of the water content is lost to evaporation by wind and sun. Aside from increasing production and decreasing unit cost, wetting the cut also lessens wear on critical ripper parts such as tips, shank protectors, shanks, and track grousers. *("Ladd & McConnel Progressing Rapidly on 395 Expressway."* Earth, *David Byrnes (ed.), June–July 1975, p. 10.)*

heavyweight machines and in which scrapers are being push-loaded. The water is supplied at a rate of 55 gal/yd³ to this volcanic tuff of the high desert country. Wetting serves three purposes:

1. For ripping, it acts as a lubricant for the tips and the shanks, easing the passage through the abrasive tuff and lowering wear on these parts. The cost for ripping is lowered because of the resulting higher hourly production and lesser maintenance.

2. For loading the scrapers, it also acts as a lubricant for the particles of ripped rock and for the cutting edge and bowl of the scraper. Loading time and wear are reduced, with resultant savings of money.

3. For water consumption, always a high cost in desert country, the application can be controlled better by the use of the monitors than by conventional spraying water haulers. Additionally, water control is better for maintaining the right optimum moisture on the fill. There is one hazard of the wetting method; namely, the water also lubricates tires and thus increases cutting by sharp-edged rocks. This tendency may become costly and it must be watched closely.

When certain rock formations have no weaknesses to facilitate ripping, it is sometimes expedient to preblast before ripping so as to create seams and fissures in the rock structure. In this method the pattern of drill-hole centers is about 1.5 times as large and the powder factor, expressed in pounds of explosive per cubic yard, is about half as large as for regular blasting. Preblasting was used extensively about 20 years ago because then tractor-rippers were much lighter and less powerful than present machines.

The method is still used, the combination method producing better fragmentation than either ripping or blasting as separate operations. The total unit cost for comparison is the sum of preblasting, ripping, push loading, hauling by scrapers, and fill building. When conventional blasting is used, the total comparative cost is usually the sum of blasting, possible bulldozing to shovel, loading by shovel, hauling by rear-dump haulers, and fill building. Usually preblasting is a transitional method between ripping and blasting and comparative costs should be carefully scrutinized.

Figure 11-12 demonstrates a unique example of all-directional ripping in sedimentary

Figure 11-12 Successful ripping of a hard sedimentary rock formation with a high shock-wave velocity of 7000 ft/s. The Hungry Valley formation of the Tehachapi Mountains of California is made up of thin- to medium-thickness beddings of sandstones, shales, and conglomerates. Here five heavyweight tractor-ripper-bulldozers are working after the day's regular shift, during which eight machines were ripping. Each machine is ripping about 225 cybm/h. The picture is toward the east. The dip of 25° is to the north, the strike being west to east. The excavation moves to the south. Because of bedding and dip, ripping is done in all directions but usually to the south into the dipped strata and in the direction of loading of the scrapers.

rocks. The huge through cut featured 8,750,000 yd³, 320-ft centerline depth, and 1450-ft distance between tops of slopes. Seismic studies disclosed these average shock-wave velocities:

 1400 ft/s up to 21-ft depth: soft ripping
 4000 ft/s between 21- and 55-ft depth: soft to medium ripping
 7000 ft/s between 55- and minimum 184-ft depth: extremely hard ripping or blasting

The entire cut was taken out by ripping in spite of the high velocity of 7000 ft/s below a 55-ft depth. The tractor-ripper-bulldozers are working in all directions. These two seeming anomalies are explained by the rock excavator's conviction that ripping was cheaper than blasting and by the thin to medium bedding of the sedimentary rocks. When pictured, the working grade was about 150 ft above the finish grade. During the regular working day eight machines ripped for a combined hourly production of 1800 yd³.

 Although this chapter emphasizes the use of medium-weight and heavyweight tractor-rippers, all crawler-tractors may be equipped with a ripper. A small tractor with bulldozer-mounted ripper points is shown in Figure 11-13. The machine is being used for lot excavation for a home. It exerts about 7800 lb thrust on the four tips. The same tractor may be equipped with a rear-mounted ripper, in which case the weight is about 14,000 lb.

 Whereas Figure 11-12 features high-production ripping, Figure 11-14 illustrates work under confined limitations. On this rock job, characterized by short and deep cuts and fills, hauling was by two crawler-tractor-drawn scrapers and fill compaction was by sheepsfoot roller with tractor. A motor grader completed the limited machinery inventory, typical of small jobs with little working space.

 The single-tip or -shank tractor-ripper-bulldozer is used extensively in pipeline work. For such trench ripping, the penetration is up to 72 in. Figure 11-15 pictures such work with shank raised clear of the ground and lowered to full ripping position.

 An example of the heavyweight machines is shown in Figure 11-16. The specifications are weight of 160,000 lb, engine horsepower of 524 hp, and tip pull of 64,000 lb. Rock is weathered granite of 5100-ft/s seismic shock-wave velocity. A single shank is being used to the full 4-ft depth for comfortable ripping, resulting in hourly production of 800 yd³.

 Presently there are two developments contributing possibly to still greater capabilities of tractor-rippers. One is the vibration of the shanks to augment penetration by the tips. Vibration is accomplished by eccentrics attached to the shanks, power being furnished

Figure 11-13 A small lightweight tractor-ripper-bulldozer with four points attached to the bulldozer. This machine is typical of the many small machines used in small excavations of loosely consolidated rocks. In this case the rock is a lightly cemented conglomerate. The complete machine weighs 13,000 lb with a 50-hp engine. Normally it is equipped with a rear-mounted ripper, but this mounting is more efficient for the average work of the contractor specializing in excavations for home sites.

11-16   Fragmentation of Rock

Figure 11-14 Swing- or hinged-type ripper in hard granite schist. This heavyweight tractor-ripper with its hinged-type ripper attached to the track-roller frame weighs 94,000 lb and has 280 engine hp. The rock is granite schist and the highway building is on the north escarpment of Lake Isabella, California. The seismic shock-wave velocity in this semisolid igneous rock formation is 6800 ft/s in the zone of extremely hard ripping or blasting. The rocky promontory, jutting into the lake, is the hardest short section of the highway. In the 5000-yd^3 prism the tractor-ripper is averaging only 180 cybm/h. However, the small yardage does not justify bringing in a track drill and compressor to the rather remote area. Often this relatively expensive method of fragmentation by ripping is more economical than blasting.

Figure 11-15 Deep ripping for pipeline trench. Rock ripping for trenches is common. This heavyweight machine is ripping for a pipeline across the Mojave Desert of California. It is a swing-type ripper, as shown in Figure 11-5, equipped with a special, single long shank with 54-in penetration. The rock is a mixture of desert alluvia and well-weathered andesitic lava. The 12-ft-wide, average 6-ft-deep trench was ripped and excavated in two layers. An average 3-ft depth of alluvia was ripped and excavated by backhoe and then the remaining 3-ft depth of weathered lava was handled similarly. When the tractor-ripper encountered unusually hard rock, tandem ripping was used advantageously.

by an engine mounted on the tool bar. The other development is the introduction of expansive gases through the tips of the ripper. The expanding gases tend to split the rock ahead of and surrounding the tips. Presently neither method is being used on a large scale, although both have possibilities for enhancing the mechanical action of the ripper.

Productions and Costs

Production and unit costs for ripping hinge on several variables, and notably they are:

    1. The nature of the rock, that is, its classification as igneous, sedimentary, or metamorphic and as to the kind of rock, as basalt, sandstone, schist and the like.

2. The degree of weathering of the rock.

3. The arbitrary categorization of the rock as soft, medium, hard, or extremely hard ripping, as in Table 11-1 and Figures 11-17 and 11-18.

4. The hourly cost of ownership and operation of the tractor-ripper-bulldozer, as the unit is generally equipped with bulldozer.

One may rightfully suspect that variables 1, 2, and 3, which determine estimated hourly production, may mean different things to different estimators and thus that they are error-prone. Variable 4, into which hourly production is divided to obtain direct job unit cost, can be estimated with reasonable precision.

An arithmetic means of estimating production is to determine the number of tips and the depth of ripping which are applicable to the rock formation. These determinants are based on the nature and the degree of weathering of the rock, and their correct evaluation depends on the experience of the estimator. The calculations are set forth in Table 11-1.

When time and money permit, a practical estimate may be based on actual exploratory ripping of the rock formation, but this ripping must be done to full rippable depth. Otherwise there is an uninvestigated zone of rock.

This method was used in the appraisal of the Chico formation of sandstones, illustrated in Figure 7-11. As described in the text, the heavyweight tractor-ripper succeeded in ripping only 80 yd^3/h. The estimated unit cost was \$0.56/yd^3 in 1966. It was agreed that blasting would be more economical in the massive sandstone of 6600-ft/s seismic shock-wave velocity.

Graphs for hourly production and direct job unit cost of the composite medium-weight and heavyweight tractor-ripper-bulldozers are given in Figures 11-17 and 11-18. The production curves are based on many field correlations between seismic shock-wave velocities and performances of machines over a period of 22 yr. The estimates for unit cost are based on the estimated hourly cost of ownership and operation for the two machines in soft, medium, hard, and extremely hard ripping.

The 1978 cost for the composite heavyweight machine was \$280,000. Table 8-3 gives 36 percent of \$280 for the hourly cost of ownership and operation for a 400- to 500-hp

Figure 11-16 Heavyweight tractor-ripper-bulldozer fragmenting weathered granite. This machine, which weighs 160,000 lb and has 524 engine flywheel hp, exerts a horizontal tip pressure of 64,000 lb, based on a traction coefficient of 0.40 in loose rock-earth and no vertical tip pressure. The vertical tip pressure can be as high as 54,000 lb but with simultaneous horizontal tip pressure of 42,000 lb. The machine is ripping medium-weathered granite of 5100-ft/s shock-wave velocity with hourly production of 800 cybm. One shank is being used in this firm formation of the Laguna Mountains of San Diego County, California. The variable nature of the tonalite formation is displayed in the cut slope of the background. Some 300 ft in back of the machine the rock is being blasted for shovel excavation where the shock wave velocity is 7200 ft/s.

11–18 **Fragmentation of Rock**

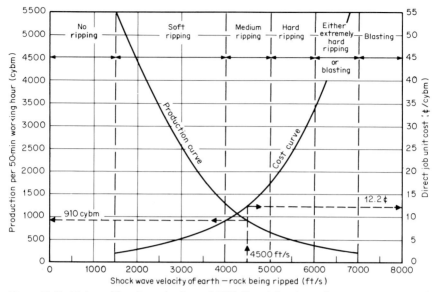

Figure 11-17 Estimated hourly production and 1978 direct job unit cost for ripping by composite heavyweight tractor-ripper with bulldozer weighing 125,000 lb with a 450-hp engine.

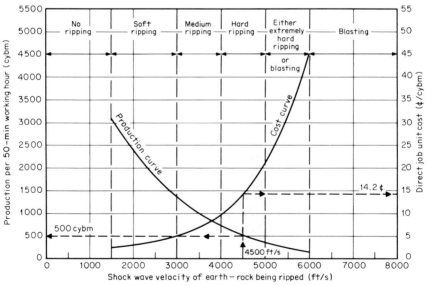

Figure 11-18 Estimated hourly production and 1978 direct job unit cost for ripping by composite medium-weight tractor-ripper with bulldozer weighing 70,000 lb with a 240-hp engine.

crawler-tractor with three-shank ripper and bulldozer. It is based on hard ripping and the total hourly cost is:

Machine, 36% of $280	$100.70
Operator	15.60
Total	$116.30

This cost must be adjusted to soft, medium, and extremely hard ripping in terms of the differences in costs for points and shank protectors, consumed items, and associated cost of repairs, replacements, and labor. The adjustments and the resultant costs are given in the following tabulation, which shows the 1978 estimated costs of ownership and operation of a composite heavyweight tractor-ripper-bulldozer in different kinds of rock ripping.

Item	Soft ripping	Medium ripping	Hard ripping	Extremely hard ripping
Tips @ $40.00 each:				
3 tips	$120.00			
2 tips		$80.00		
1 tip			$40.00	
1 tip				$40.00
Hours of use	72	24	8	3
Hourly cost	$1.67	$3.33	$5.00	$13.33
Shank protectors @ $60.00 each:				
3 protectors	$180.00			
2 protectors		$120.00		
1 protector			$60.00	
1 protector				$60.00
Hours of use	864	288	96	32
Hourly cost	$0.20	$0.42	$0.62	$1.87
Total hourly cost for tips and protectors	1.87	3.75	5.62	15.20
Hourly cost for repairs, replacements, and labor thereto	32.53	33.55	37.43	39.30
Total hourly cost	$34.40	$37.30	$43.05	$54.50
Adjusted hourly cost of ownership and operation	$ 92.10	$ 95.00	$100.70	$112.20
Operator	15.60	15.60	15.60	15.60
Total hourly costs, used for figuring direct job unit cost of Figure 11-17	$107.70	$110.60	$116.30	$127.80

Although most rock ripping is done by heavyweight machines, a great amount is done by medium-weight units in residuals and weathered rocks. The composite medium-weight tractor-ripper-bulldozer has the following controlling specifications:

Working weight, with three-shank ripper and bulldozer	70,000 lb
Flywheel horsepower	240 hp
Drawbar pull at 1.0 mi/h	59,000 lb
Estimated tips pull	28,000 lb

The 1978 cost for the composite machine was $154,000. Table 8-3 gives 36 percent of $154 for the hourly cost of ownership and operation for a 250-hp crawler-tractor with three-shank ripper and bulldozer. The total hourly cost of ownership and operation, based on hard ripping, is:

Machine, 36% of $154	$55.40
Operator	15.60
Total	$71.00

As in the case of the heavyweight unit, when the cost of $71.00 is adjusted for different kinds of ripping, the following estimated total hourly costs of ownership and operation emerge:

11-20   Fragmentation of Rock

Soft ripping	$66.10
Medium ripping	$67.70
Hard ripping	$71.00
Extremely hard ripping	$77.40

Field studies disclose that, for the same seismic shock-wave velocities of earth-rock, the hourly ripping production of the composite medium-weight machine is about 55 percent that of the heavyweight unit. Graphs for the hourly production and direct job unit cost for the medium-weight machine are given in Figure 11-18, again based on many observations. The two illustrations of Figures 11-17 and 11-18 indicate that medium ripping at 4500 ft/s for the heavyweight machine at a cost of $0.122/cybm is hard ripping for the medium-weight unit at a cost of $0.142/cybm.

In the following estimates for direct-job unit cost of ripping, the hourly cost of ownership and operation is for the composite heavyweight tractor-ripper-bulldozer. These hourly costs have been developed previously in this chapter. Hourly productions and unit costs, as functions of seismic shock-wave velocity of the earth-rock being ripped, are set forth in Figure 11-17.

EXAMPLE 1: Estimating productions and unit costs for ripping for a dam. Borrow-pit site 3 for New Lake Arrowhead Dam, San Bernardino County, California, was made up of residuals and weathered and semisolid granite. Maximum cut in the comparatively flat mesalike pit was 80 ft. One of the pit's exploratory trenches is shown in Figure 7-32.

Five seismic studies gave these averages for the maximum 80 ft of cut.

    1200 ft/s up to 12 ft-depth: no ripping
    2700 ft/s between 12- and 40-ft depth: soft ripping
    4700 ft/s between 40- and 77-ft depth: medium ripping
    6600 ft/s between 77- and minimum 107-ft depth: extremely hard ripping or blasting

Quantity calculations gave these yardages and corresponding percentages of the total 1.1 million yd³ in the above four depth zones.

    0 to 12-ft depth: 165,000 yd³, 15%
    12- to 40-ft depth: 385,000 yd³, 35%
    40- to 77-ft depth: 506,000 yd³, 46%
    77- to 80-ft depth: 44,000 yd³, 4%

It is desired to estimate the hourly productions and the unit cost for ripping in the four zones, the machine requirements, and the weighted unit cost for ripping and possibly for blasting in order to fragment the rock for scraper work.

Figure 11-17 and simple calculations yield the following tabulation.

| Zone | Velocity, ft/s | Tractor-ripper-bulldozer data ||| Total cost |
		Hourly cybm Production	Unit cost, $	Machines required	
No ripping, 165,000 yd³; hourly scraper production is 1600 cybm	1200	...	...	0.0	0
Soft ripping, 385,000 yd³; hourly scraper production is 1400 cybm	2700	3000	0.043	0.5	$16,600
Medium ripping, 506,000 yd³; hourly scraper production is 1200 cybm	4700	800	0.140	1.5	$70,800
Extremely hard ripping 44,000 yd³; hourly scraper production is 800 cybm	6600	260	0.490	3.1	$21.600

Total costs amount to $109,000. The unit cost for ripping, weighted for the total 1.1 million yd³ of scraper work, is $0.099.

It is possible that half of the 44,000 yd³ of extremely hard rock might require blasting. After blasting, the rock would require medium ripping to prepare it for scraper work. If blasting is estimated at $0.80/yd³, then the recast unit cost for the work would be according to the following tabulation:

		Total Cost
Soft ripping	385,000 yd³ at $0.043	$ 16,600
Medium ripping	528,000 yd³ at $0.140	73,900
Extremely hard ripping	22,000 yd³ at $0.490	10,800
Blasting	22,000 yd³ at $0.800	17,600
Total cost		$118,900

The unit cost for ripping and possible blasting, weighted for the total 1.1 million yd³ of scraper work, is $0.108.

The machine requirements for ripping in the three zones are 0.5, 1.5, and 3.1, calling for one, two, and four machines, respectively. When, theoretically, these machines would not be ripping, they would assist the push-loading tractors by triple pushing instead of the conventional double pushing.

EXAMPLE 2: Estimating production and unit cost for ripping under "comfortable" conditions in a freeway cut (Figure 11-6)

This machine is actually ripping about 1200 cybm/h, based on the production of the scraper spread. The seismic shock-wave velocity of the weathered granite is 4000 ft/s. Figure 11-17 gives for 4000 ft/s an hourly production of 1300 cybm and a unit cost of $0.092/cybm.

One may use Figure 11-17 to determine the unit cost for an hourly production of 1200 cybm. Enter the left ordinate at 1200 cybm, go right to the production curve, drop downward to the cost curve, go right to the right ordinate, and read $0.098/cybm. The shock-wave velocity on the abscissa is 4100 ft/s.

EXAMPLE 3: Estimating production and unit cost for ripping under "uncomfortable" conditions in a highway cut (Figure 11-7)

Tip penetration is about 1 ft for this tractor-ripper-bulldozer equipped with two shanks. It is probable that the seismic shock-wave velocity in this formidable granite formation is between 6000 and 7000 ft/s. If this assumption is valid, Figure 11-17 gives hourly productions between 380 and 200 cybm and respective unit costs between $0.33 and $0.64 per cybm. These results demonstrate graphically and economically that it is time to go to blasting in this extremely hard granite.

EXAMPLE 4: Estimating production and unit cost for mass ripping in huge 8,750,000-yd³ freeway cut (Figure 11-12)

Five machines are working for 3 h after the completion of the regular 8-h shift in order to keep up the ripping schedule for a shift production of 14,400 cybm. Machine hours total 79, giving a machine-hour production of 182 cybm. Extrapolating the production curve for 182 cybm/h, one travels upward to a unit cost of about $0.68. This high cost for ripping is explained by the high (7000 ft/s) seismic shock-wave velocity in this sedimentary rock formation at the depth being ripped. According to Figure 11-17, for 7000 ft/s, hourly ripping production is 200 cybm and direct job unit cost is $0.64.

By either actual production on the job or by estimated production according to seismic shock-wave velocity, the unit cost suggests that blasting is in order.

EXAMPLE 5: Estimating production and unit cost for marginal cost ripping (Figure 11-14).

As stated in the text, it was not economical to bring in a drilling and powder crew for this 5000-yd³ prism of extremely hard schist in the remote small grading job. Seismic shock-wave velocity was 6800 ft/s. Figure 11-17 for this velocity gives an hourly production of 220 cybm and a unit cost of $0.55. The actual production rate of ripping during the excavation of the 5000 yd³ was 180 yd³/h. The actual unit cost was then about $0.68. Under normal conditions, even though the cost for actual ripping is lower than that for actual blasting, it is more economical to blast. This apparent contradiction is explained by the fact that in extremely hard ripping not only does cost of ripping go up but also the cost of other operations increases dramatically because of accelerated wear and tear. Specifically, maintenance on the undercarriages of the push tractors and on the cutting edges and the bowls of the scrapers increases. In short, the overall unit cost of the entire fragmenting, loading, and hauling of the rock that are the determining factors.

EXAMPLE 6: Estimating production and unit cost of ripping for trench excavation in pipeline work (Figure 11-15).

This ripping is usually a special application of a long single shank. The machine is fragmenting desert alluvia and well-weathered lava to 6-ft depth, the average seismic shock-wave velocity being 3200 ft/s. Dimensional and hourly production data are given in the following tabulation.

Width of trench, ft	12.0
Average depth of trench, ft	
Upper lightly cemented alluvia	3.0
Lower well-weathered lava	3.0
Total	6.0
Average tip penetration, ft	3.5
Average cross-sectional area ripped by one pass, ft³	12.2
Cross-sectional area of trench, ft²	72.0
Number of passes for alluvia and lava zones	6
Average ripping travel speed, mi/h	1.0
Average cycle traveling speed, mi/h	0.9
Length, ft, of trench ripped per 50-min working hour	660
Cubic yards, trench measurement, ripped hourly	1760

NOTE: The sequence of operations was ripping to 3-ft depth by two passes and excavation by bulldozer and backhoe, followed by ripping from 3- to 6-ft depth by four passes and excavation by backhoe.

For a production of 1760 cybm/h, Figure 11-17 gives a unit cost of $0.071/cybm. As there are 2.67 yd³/ft of trench, the ripping cost was $0.19/ft of trench in terms of 1978 costs.

The average seismic shock-wave velocity in alluvia and weathered lava was 3200 ft/s. Obviously, in this special application the relationships between velocities and hourly productions and unit costs, as set forth in Figure 11-17, are less precise than those for mass rock ripping.

EXAMPLE 7: Performance of larger heavyweight tractor-ripper-bulldozer (Figure 11-16)

This machine is in "comfortable" ripping of weathered granite of 5100-ft/s seismic shock-wave velocity. Figure 11-17 gives an hourly production of 620 cybm for the composite heavyweight machine, whereas this big machine's production was 800 cybm. The production is 29 percent higher than 620 cybm. The difference is explained by the combined effect of 160,000 lb weight and 524 engine hp. These are 28 percent and 16 percent higher than the respective specifications of the composite machine.

The bigger machine cost $360,000 in 1978. The estimated hourly expense of ownership and operation in hard ripping is $145.10 (36% of $360 + $15.60). The unit cost for ripping, based on the actual production of 800 cybm, is $0.181/cybm. For 5100 ft/s velocity, Figure 11-17 also gives a unit cost of $0.181 thus demonstrating the reliability of Figure 11-17 for machines in the heavyweight class.

It is an axiom that productions of heavyweight tractor-ripper-bulldozers are proportional to their averaged weight and horsepower ratios when they are comfortably ripping. The advantages of the larger machines are in their greater productions and ability to rip somewhat harder rocks with somewhat higher seismic shock-wave velocities. As an example, studies show that a machine with an average 33 percent advantage in weight and horsepower will comfortably rip rock of 10 percent higher seismic shock-wave velocity. This same rock would be uncomfortable ripping for the smaller machine, and of course it would be considerably more expensive to rip.

BLASTING

Blasting is a dual operation, made up of the interrelated drilling of blastholes and the exploding, or loading into the holes and detonating of the explosives. These are considered to be mutual undertakings, although one usually chooses the right explosives for the blasting and then correlates the methods of drilling to the ideal use of the explosives. For these reasons it is most practical to discuss drilling and exploding together under the term *blasting*.

Two variations of the term *drilling* as used in this chapter need explanations. Drilling includes not only the drilling of holes to receive explosives but also the driving of adits and laterals as chambers to receive explosives when the *coyote* method of blasting is used. In this method preliminary blasting and excavation of rock in the adits and laterals are necessary. Accordingly, in this instance drilling in the sense of providing a hole for explosives includes the dual operations of blasting, that is, both drilling and exploding. Drilling also includes the method of putting down blastholes by jet piercing, in which the rock within the blasthole is not drilled out but rather is disintegrated and removed by intense flame from an orifice.

Principles

The principles of blasting may be developed most clearly and practically by a description of explosives and a discussion of their application or use by means of the drilled blastholes.

Explosives

An inclusive definition of explosives is given by E. I. du Pont de Nemours & Co. as follows:

> The explosives commonly used in commercial blasting operations are, with few exceptions, mixtures of solids or solids and liquids which are capable of rapid and violent decomposition with resultant conversion into large volumes of gas. Decomposition of a *high* explosive, such as dynamite, takes place with extreme rapidity, while in the case of a *low* explosive, such as black blasting powder, it takes place more slowly, the action simulating rapid burning or combustion. High explosives are termed *detonating* explosives whereas low explosives are referred to as *deflagrating* explosives.

Low Explosives Low explosives are black powders composed of very intimate mixtures of sulfur, charcoal, and either potassium nitrate (also known as saltpeter) or sodium nitrate. They burn progressively over a relatively sustained period of time, in contrast to detonating explosives, which decompose practically instantaneously. Black powders are the slowest-acting of all the explosives and they give a shearing and heaving action tending to blast rock into large, firm fragments. Their action derives from a relatively slow development of gas pressure so that they must be loaded carefully and closely confined. Burdens or hole spacings should be well balanced since there is a tendency for the rock to yield at weak points. Black powders are manufactured in granular and pellet forms. The granular powders are of different gradations and usually they are packed in 25-lb heavy containers. Pelletized powders are available in cartridges of 8-in length and diameters ranging from 1½ to 2½ in, packed in 50-lb cases.

Black powders are relatively insensitive to shock and friction, but they ignite at about 572° F (300° C). Ignition may be by flame, spark, hot wire, or hot surface or by such blasting accessories as squibs, electric squibs, igniters, and detonating cord.

Black powders do not have true velocities, as their rate of burning is affected by confinement. In open trains of powder, speed is in terms of less than 1 ft/s, but when the powder is confined in steel pipe, speed may reach 2000 ft/s.

Presently black powder is little used in rock excavation, having given way to high explosives. However, because of its slow heaving action, as contrasted to the fast shattering action of high explosives, it is moderately used in the quarrying of heavy solid stone for riprap and in the blasting of coal in open-pit mines.

High Explosives High explosives are of two types. One type is cap-sensitive and it includes the dynamites—straight dynamite, extra dynamite, and gelatin dynamite. The other type consists of cap-insensitive blasting agents and includes, in a general sense, the nitro-carbo-nitrates in many mixtures and forms for many special purposes. Over a short period of time many formulas for blasting agents have been developed and their chemical compositions have overlapped those of the dynamites. Accordingly, their type classification is generic rather than specific.

Dynamites Dynamites are cap-sensitive mixtures which contain an explosive compound, either as a sensitizer or as the principal means for developing energy, and which, when properly initiated, decompose at detonation velocity. Most but not all dynamites contain nitroglycerin as the explosive compound.

Dynamites can be made in many types and grades, each having different properties and characteristics. Unlike most simple blasting agents, such as ammonium nitrate–fuel oil mixtures, the compositions, properties, and characteristics of dynamites can be modified or tailored to provide the best blasting action for each kind of work. In selecting a dynamite for any specific purpose many factors must be considered. The most important ones involve the rock to be blasted, its density, hardness, toughness, friability, seismic shock-wave velocity, and the like. Other considerations are the degree of fragmentation desired, whether the blastholes are wet or dry, and the uniformity of the rock formation with respect to seams, fissures, gouge and fault zones, and weathering. Each blast presents some combination of these and other conditions and hence a dynamite with the proper properties should be selected. Quite naturally, these requirements and properties hold for all kinds of explosives.

Important properties of explosives, listed in the specifications for the explosives, are strength, density, velocity, and water resistance. *Strength* refers to the energy content of an explosive which, in turn, contributes to the force and power it develops and the work it is capable of doing. The rating designates the percentage by weight of nitroglycerin in the formula. The straight dynamites are the bases with which all other dynamite grades are compared. Dynamites are sometimes graded according to their *bulk strength*. This refers to the strength per cartridge of the explosive, and the bulk strength figure indicates that one cartridge of the dynamite, so marked, has a strength comparable with one cartridge of straight dynamite of the same percentage and size. The total strengths of two explosives are not in proportion to the percentage strengths, which relate to the ingredient nitroglycerin, because there are other ingredients in the mixture. For example, a 60 percent straight dynamite is not 3 times as strong as a 20 percent dynamite but only about 1½ times as strong. Non-nitroglycerin explosives and blasting agents are given strength ratings which designate their energy relative to an equal weight of a grade of straight dynamite.

Density is conveniently expressed by the number of 1¼ in $\times$ 8 in cartridges contained in a 50-lb case. This figure varies from about 85 to 205. In most blasting agents and large-diameter explosives, density is related to that of water and is expressed as grams per cubic centimeter. The purpose of density variations in explosives is to enable the blaster to concentrate or distribute charges at will by varying the strength-volume relationship.

Velocity is a measure of the speed at which the detonation wave travels through a column of an explosive. Unless otherwise stated, the velocity data refer to a 1½-in-diameter column. Velocities range from about 4000 to about 23,000 ft/s. As the velocity increases, the explosive usually produces a greater shattering effect in the harder rocks. Strength and density also influence shattering action so that all three properties must be considered in the selection of an explosive.

Water resistance of high explosives varies widely and this quality must be considered, particularly when the explosive is used under wet conditions. Of the three dynamites, gelatin dynamites are the most resistant. Of the blasting agents, water gels are best for wet conditions. In all cases the intrinsic water resistance of the explosive is most important, as a watertight container or bag can be broken easily in handling and in loading the blasthole.

There are three kinds of dynamites: straight dynamites, extra dynamites, and gelatin dynamites. Straight dynamites consist entirely of nitroglycerin, which by itself is explosive. The most popular strengths are the 50 percent and 60 percent grades. The high-strength straight dynamites are characterized by high velocity, which imparts a quick shattering effect. They are seldom used for general blasting because of high cost and sensitivity to shock.

Extra dynamites, compared with straight dynamites, are, grade for grade, lower in velocity and water resistance and less sensitive to shock and friction. They are considerably more economical. They have less shattering effect, which is advantageous in certain kinds of work.

Gelatin dynamites have an explosive base of nitrocotton-nitroglycerin gel and vary in consistency from thick, viscous liquids to tough, rubberlike substances. They are insoluble in water and tend to waterproof other materials which they coat or enclose. Plasticity makes it possible to load them solidly into blastholes in order to obtain maximum loading density. This combination of excellent features makes them most desirable for hard rocks and ores in rock excavation. The representative characteristics of straight, extra, and gelatin dynamites are given in the following table:

Kind of dynamite	Weight strength or grade, %	Bulk strength %	Density of cartridges per 50 lb 1¼ in × 8 in	Velocity, ft/s	Water resistance
Straight	30	30	102	11,600	Poor
	40 to 50	40 to 50	102 to 104	13,800 to 16,100	Good
	60	60	106	18,200	Excellent
Extra	20 to 35	16 to 29	110	8,000 to 9,500	Very good
	40 to 50	35 to 43	110	10,200 to 11,200	Very good
	60	55	110	12,000	Very good
Gelatin	20 to 60	32 to 66	85 to 96	10,500 to 19,700	Excellent
	75 to 90	70 to 79	101 to 107	20,600 to 22,700	Excellent

SOURCE: E. I. du Pont de Nemours & Co.

The cartridge sizes for packaged dynamites are many in order to facilitate handling and the loading of blastholes for maximum blasting efficiency. In small sizes, ⅞- through 2-in diameters, the 1⅛- and 1¼-in diameters and 8-in lengths are the most popular, although lengths up to 24 in are available. Dynamites of up to 10-in diameter are available, but the common sizes are of 3- to 6-in diameter with lengths of 8 to 36 in.

Figure 11-19 illustrates a few of the many sizes and types of dynamite cartridges. Not only is the dynamite itself tailor-made for a given task, but the containing cartridge is also tailor-made for the particular loading condition.

Blasting Agents A blasting agent is a cap-insensitive chemical composition or mixture which contains no explosive ingredient and which can be made to detonate with a high-strength explosive primer.

As has been stated, the formulas sometimes overlap those of the dynamites. One group of blasting agents, classified as oxidizing materials because they contain no high explosives, is known as nitro-carbo-nitrates. Another group contains non-nitroglycerin high-explosive ingredients and must be classified strictly as high explosives. In these descriptions blasting agents are divided into three classifications in keeping with their blasthole-loading properties and without regard to their specific compositions or formulas. These classifications are: cartridged, or packaged, for placing in the blasthole; free-running for flowing into the blasthole, either manually or mechanically by bulk-loading equipment; and plastic or viscous, bagged for loading into the blasthole manually.

The several advantages of the blasting agents include safety, ease of handling and loading, and economy. Almost any blasting in open-cut rock excavation may be handled by blasting agents, and as of 1978 blasting agents came into use almost exclusively in construction and in quarrying and open-pit mining.

Cartridged Blasting Agents. The performance of these cartridged agents is equivalent to that of the more costly dynamites. There is available a wide range of container sizes of 3- to 9-in diameter and holding 3 to 75 lb. The smaller cartridges are packaged in 50-lb quantities in heavy paper boxes. Figure 11-20 illustrates two kinds of 5-in-diameter 19-lb cartridges. Characteristics of representative cartridged blasting agents are given in the following table:

Brand	Weight strength, %	Cartridge or bulk strength, %	Specific gravity	Equivalent 1¼ × 8 in cartridge count per 50-lb case	Velocity, ft/s	Water resistance
Dynatex B	65	40	1.07	130	12,500	Fair
Dynatex B-WR	65	40	1.07	145	9,850	Fair
Tritex 2	55	40	1.07	120	9,900	Good

SOURCE: Hercules, Inc.

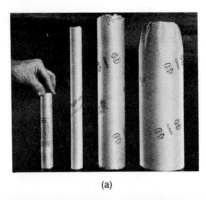

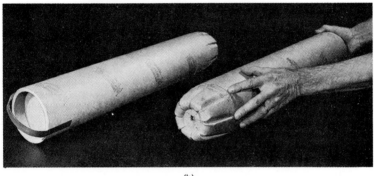

Figure 11-19 Representative dynamite cartridges for straight, extra, and gelatin dynamites. They are packaged in 50-lb heavy paperboard containers. *(a)* Left to right: 1¼ in × 8 in, 1¼ in × 16 in, 3 in × 16 in, and 5 in × 16 in. *(b)* Left to right: 5 in × 20 in with tape bail and tapered end for ease in loading the blasthole. *(E. I. du Pont de Nemours & Co.)*

A comparison between the data of this table and the dynamites shows many similar characteristics, attesting to the logic of substituting the blasting agent for dynamite because of lower unit cost per pound.

Free-Running Blasting Agents. These nitro-carbo-nitrate agents may be poured manually or mechanically loaded into blastholes. They range from the economy grade of ammonium nitrate prills, to which fuel oil must be added, through to the high-grade water-resistant pelletized explosives.

Figure 11-21 shows these two free-running blasting agents. They come usually in 50-lb boxes, although some may come in bulk for mechanical loading.

Prills are available in two forms, ammonium nitrate, AN, and AN/FO, consisting of premixed ammonium nitrate and fuel oil, FO, the latter being the more common kind. To use AN prills, No. 2 fuel oil must be added in the proportion of 6 to 94 percent prills by weight or about 3½ qt of oil per 100 lb of prills.

Mixing is by several methods. If the mixture is manually loaded into the blasthole, one method is either to pour the prills into the dry hole and to add the oil or to add the oil to the prills before the loading, the latter means being preferable. A second method is to use an oil probe injected into the bag for metering the oil. A third most satisfactory method is to premix at a mixing plant and then to either bag the mixture or load it into bulk-loading equipment for pouring into the blasthole.

The average bulk density of prills and oil or the mixture is 0.80 or about 50 lb/ft³. If hole loading is by compressed air injector, the density is 0.95, or about 60 lb/ft³.

Ammonium nitrate prills with added oil are not suitable for wet holes. In dry holes, as

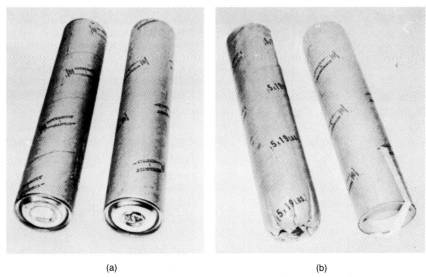

(a) (b)

Figure 11-20 Representative blasting agent cartridges: *(a)* 5-in-diameter 19-lb cartridge with metal ends. *(b)* 5-in-diameter 19-lb cartridge with tapered crimp and fiberboard insert closure. *(E. I. du Pont de Nemours & Co.)*

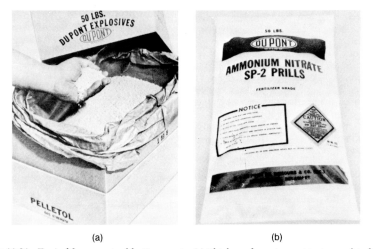

(a) (b)

Figure 11-21 Typical free-running blasting agents: *(a)* A high-grade, water-resistant granulated blasting agent. *(b)* AN prills. When fuel oil is added, AN becomes AN/FO, a low-cost, efficient blasting agent of 0.80 density.

These free-running agents are made with many poured densities, ranging from 0.45 to 1.15. Their average density is about 0.85. They are supplied in 50-lb containers for hand loading and in bulk for mechanical loading. *(E. I. du Pont de Nemours & Co.)*

they are poured they fill the entire volume of the hole, thus compensating for their relatively low specific gravity. Detonation velocities vary from about 10,700 to 13,800 ft/s suggesting their explosive efficiency.

AN/FO is simply the manufactured blasting agent ammonium nitrate premixed with fuel oil. Its proportions and characteristics are those of AN prills with the added oil. Its advantages over AN prills with added oil are uniformity of mixture and convenience in handling and loading.

In addition to the free-running blasting agents AN and AN/FO there are many more of diverse formulas, developed by the several manufacturers for special purposes. These nitro-carbo nitrate agents have different specific gravities, different poured densities, different velocities, and different resistances to water.

Densities range from 0.45 to 1.15, affording column loading of the blasthole for almost any powder factor desired. These additional free-running agents are packaged in 50-lb bags contained in heavy paper boxes.

The Pelletol (Figure 11-21a) is an example of a pelletized, free-running, waterproof agent with the explosive attributes of 60 percent gelatin dynamite. The individual pellets have a specific gravity of 1.5 and their poured density is 1.0. Thus they sink readily and they load with average density into the blasthole.

Plastic or Viscous Blasting Agents. These water-resistant water gels and slurries are forms of nitro-carbo-nitrate agents which have been developed for use in wet blastholes and in coyote tunnels. Although classified as non-cap-sensitive, a few of these agents may be detonated by electric blasting caps under favorable conditions.

Their characteristics are: high density, high bulk strength, high water resistance; good slump, which permits displacement of air and water and thereby high poured density in the blasthole; ease of handling and loading; reliability and safety; and economy.

In partially wet holes, water gels are sometimes used along with AN/FO. The water-resistant water gel is loaded from the bottom of the hole up to a level slightly above the water level, being primed by a detonating fuse. AN/FO is then loaded into the dry length of the hole, being primed by the same detonating fuse. These two blasting agents comprise an efficient, economical team for much blasting work.

Water gels and slurries are packaged in conveniently sized waterproof bags which may be slit for loading. They are available with different specifications and formulas for different jobs, but in general their characteristics are those of the other nitro-carbo-nitrate blasting agents. Figure 11-22 shows the packaging of 3-in-diameter 10-lb bags. Typical ranges of sizes are: 4 in, 15 lb; 5 in, 30 lb; 6 in, 30 lb; 7 in, 50 lb; and 8 in, 50 lb.

Flogels are slurry-type, water-resistant, highly plastic, water-based explosives, which provide high loading densities and excellent explosive-to-rock coupling. Such distinguishing features are common to slurries of the same specifications and formulas. Representative characteristics of three such plastic, viscous blasting agents are tabulated below.

Brand	Energy relative to equivalent value of AN/FO	Density	Detonation velocity of 5-in-diameter cartridge, ft/s
Flogel	1.35	1.40	18,000
Flogel 3	1.25	1.15	16,500
Flogel AL	1.60	1.40	18,000

SOURCE: Hercules, Inc.

There are many manufacturers of explosives in the world. Their tested formulas or mixtures for dynamites and blasting agents vary, so that a great many explosives are available for many conditions of blasting. Because of the complexities of the countless rock formations, the rock excavator must rely on experience, sound advice, and trial and error right on the job to ensure the most efficient and most economical blasting practices.

Initiating Devices Different explosives call for different initiating devices and it is well to choose the device more carefully than the explosive itself, or else a misfire

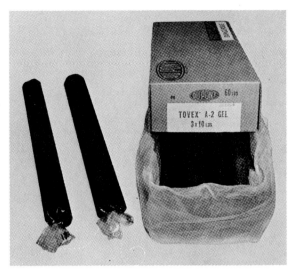

Figure 11-22 A representative waterproof plastic or viscous blasting agent. These water-gel and slurry blasting agents are water-resistant and are adapted to use in wet blastholes. This gel is supplied in 3-in-diameter 10-lb bags, packaged six bags to a 60-lb fiberboard container. Usually, wet holes are wet only in the lower part of the total depth. When such is the case, the water gel is loaded into the lower part and the non-water-resistant AN/FO is loaded into the upper part of the blasthole. The two agents combine for maximum efficiency with maximum economy. *(E. I. du Pont de Nemours & Co.)*

may occur with resultant delays, added costs, and perhaps danger in the blasting. These devices are used to initiate charges of explosives, to supply or transmit flame to start an explosion, or to carry a detonation wave from one charge to another charge of explosives. They are discussed below in relationship to black powders, dynamites, and blasting agents.

Black Powder Initiating Devices. The devices used with black powders may be either electric or nonelectric. There is only one electric device and that is the *electric squib*. It contains an enclosed deflagrating charge which is activated by an electric current. These squibs resemble the electric caps of Figure 11-24. They are considered to be the safest and most effective device for initiating blasting or pellet powder.

There are two nonelectric methods; by safety fuse or by detonating fuse or primers. Safety fuse contains a core or train of potassium nitrate black powder, and standard rates of burning are 30 or 40 s per foot of fuse. By varying the lengths of the fuses, the individual charges of powder may be initiated at different times. The safety fuse may be ignited by a match or by special devices such as the hot-wire fuse lighter or lead-spitter fuse lighter. When lighting several fuses, one of the special lighters is recommended. Safety fuse is usually furnished in 50-ft and 3000-ft coils. It is shown, but with blasting cap, in Figure 11-23.

Special faster-burning safety fuses are available, ranging in speed up to 1.0 ft/s. Such a fuse, illustrated in Figure 11-26, is used in firing a number of charges as in secondary blasting. It is not used for sequential or rotational blasting.

The other nonelectric method, by means of detonating fuse or by primers of high explosives, is shown in Figure 11-25. This means is efficient in large blasts.

Dynamite Initiating Devices. Three methods are used to initiate dynamite explosions. One is by the electric blasting cap of the instantaneous or millisecond (MS)-delay types. Both types contain three chief elements: a bridge which, when heated by an electric current, ignites the ignition mixture; the ignition mixture which detonates the primer charge; and the primer charge. The primer charge initiates the dynamite. Figure 11-24 illustrates electric blasting caps and the manner of their insertion into a cartridge of dynamite.

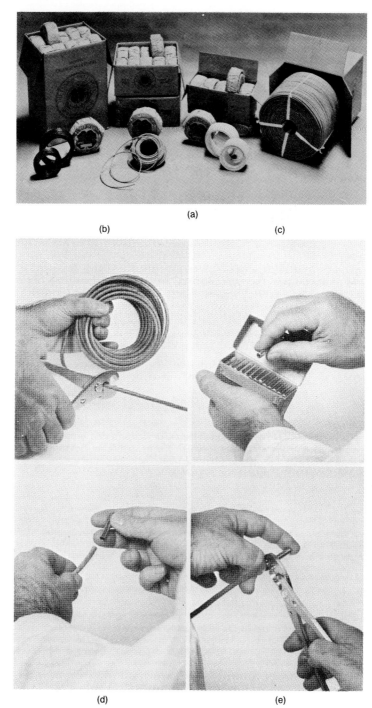

Figure 11-23 Safety fuse and blasting caps. *(a)* Packaging of 50-ft coils of safety fuse in 100-, 500-, 1000-, and 3000-ft total lengths in containers and 3000-ft reel of fuse. *(b–e)* Crimping a blasting cap onto a length of fuse. *(b)* Cutting sufficient length of fuse. *(c)* Removal, one at a time, of cap from box. *(d)* Slipping hollow end of cap on end of fuse. *(e)* Crimping cap to fuse with cap crimper. *(Hercules Inc.; E. I. du Pont de Nemours & Co.)*

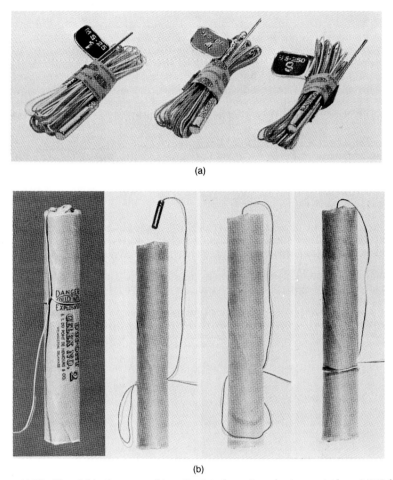

Figure 11-24 Electric blasting caps and insertion into dynamite and primer cartridges. *(a)* MS delay electric blasting caps: (left to right) No. 1 with 25-ms delay; No. 4 with 100-ms delay No. 8 with 250-ms delay. *(b)* Cartridge with inserted cap and the three steps in the insertion and the securing of the cap. The hole for the cap is punched into the end or the side of the cartridge by the pointed handle of a cap crimper or by a wooden awl. No pointed metal object capable of producing a spark or heat should be used. *(E. I. du Pont de Nemours & Co.)*

In Figure 11-24 one must note that the numerals 1, 4, and 8 do not refer to the delays in milliseconds of the caps. The figures are merely designative, their respective millisecond delays being 25, 100, and 250. There are some differences in the relationships between designative numerals and actual millisecond delays of the caps according to the practices of the manufacturers of the caps. For example, No. 8 of Figure 11-24 has a 250-ms delay, whereas in the table on page 11-32 No. 8 has a 240-ms delay. Although these differences are apparently of a small order, nevertheless it is good practice not to mix MS delay caps of different manufacturers.

A second method is by blasting cap, crimped to the end of a safety fuse, as illustrated in Figure 11-23. This cap contains three elements: a charge of ignition powder at the top to pick up the flame of the safety fuse; a middle primer charge to convert burning to detonation; and a bottom base charge of high-velocity explosive. This method is used a great deal in secondary blasting and in multiple-hole blasts, where there may be stray

Relationships Between Delay Period Designations and Nominal Firing Times for Millidet No. 8 Electric Blasting Caps

Delay period designation	Nominal firing time, ms	Delay period designation	Nominal firing time, ms	Delay period designation	Nominal firing time, ms
0	12	6	170	12	400
1	25	7	205	13	450
2	50	8	240	14	500
3	75	9	280	15	550
4	100	10	320	16	600
5	135	11	360	17	700

SOURCE: Hercules, Inc.

electrical currents not favoring the usual electrical initiation. Figure 11-26 illustrates the use of blasting caps with a fast safety fuse.

The third method is by detonating fuse, commonly called Primacord.* This fuse contains pentaerythritol tetranitrate (PETN), a high explosive of 21,000 ft/s detonation velocity. It resembles the safety fuse of Figure 11-23 and it comes in 500- and 1000-ft reels. It is water-resistant and it is not subject to initiation by lightning and stray electrical currents. These properties make the detonation fuse safer than electric caps with their array of surface and in-hole wires. Detonating fuse is initiated by an electric cap or a primer. Delayed action firing may be accomplished by millisecond-delay connectors set within the line of the detonating fuse.

Blasting Agent Initiating Devices. Blasting agents, since they are with few exceptions cap-insensitive, must be detonated by a high-strength explosive primer, sometimes called a *booster*. These primers include the following devices.

1. *Dynamite cartridges,* spaced throughout the depth or even confined to the bottom of the blasthole. These dynamite primers may be initiated by any of the previously described three methods for initiating dynamite.

2. *Detonating fuse,* as previously described for initiating dynamites.

3. *Combined dynamite and detonating fuse.* In this method the fuse connects the dynamite cartridges, which are spaced throughout the depth of the blasthole.

4. *Special primers or boosters,* as shown in Figure 11-25. There are a great number of these devices of different sizes, weights, and formulas. They, in turn, may be initiated by electric blasting caps or a detonating fuse. The special primers for blasting agents should be selected carefully for the particular agent, for the dimensions of the blasthole, and the dryness or the wetness of the blasthole.

Initiating devices are as important as the explosives, since their choice determines the efficiency of the explosive and the resultant unit cost per cubic yard of rock for the blasting work.

Blasting Circuits The word "circuit" implies a closed electrical circuit for the firing of one or more charges of explosives rather than a plan or arrangement for a nonelectric means of firing. However, for the sake of simplicity in this discussion, the term *blasting circuit* is applied to both the fuse and the electric-wire means of firing as used both on the ground surface and in the coyote tunnels and blastholes.

The conductor of a circuit may be a safety fuse with a velocity of 1 to 40 ft/s, a detonating fuse of 21,000 ft/s velocity, or two wires of instantaneous action. All these conductors have been described.

Safety-Fuse Circuit. Figure 11-26 shows a typical trunk-line layout for a high-velocity safety fuse of 1.0 ft/s. The trunk line and leads are being used for secondary blasting in a quarry. The 14 individual charges of dynamite will be detonated by blasting caps at the ends of the lead fuses from the trunk line.

If only one charge were to be fired there would be simply one lead to the charge, the lead being of sufficient length for safety purposes. Cords or lengths of fuse are connected by square knots. Downlines at the holes and similar junctions are connected by half hitches and clove hitches.

*Manufactured by the Ensign-Bickford Co.

Figure 11-25 **These representative primers, or boosters, for the blasting agents** have the following important features: detonating velocity of 25,000 ft/s; sensitivity to electric blasting caps and detonating fuse or Primacord; nearly complete resistance to water; built-in wells for caps and tunnels for fuse; and safe and nonheadache attributes.

They vary in weight from ⅓ to 5 lb and in dimensions from 1⅜- to 5-in diameters and from 4⅜- to 4½-in heights. Thus they encompass the necessary range of weights and dimensions for use in all blastholes.

They are packaged in heavy paper cases varying from 35 to 60 lb. *(Hercules, Inc.)*

Figure 11-26 **This layout for initiation of blasting by safety fuse with blasting caps** is representative for the secondary blasting of oversize rock or boulders resulting from primary blasting in a quarry. The ignition point is at *A* and cross ties are shown at points *B* and *C*. This high-velocity safety fuse burns at a rate of 1.0 ft/s, affording ample time for the safety of the powder operator. Detonation is by blasting caps at the ends of the fuse lines. Explosives are 1¼ in × 8 in 40 percent dynamite cartridges and, assuming the individual rocks to be about 5 ft in diameter, the charge would be about one stick of dynamite for each rock or boulder drilled for the blockhole method of secondary blasting. *(E. I. du Pont de Nemours & Co.)*

Safety fuse is relatively safe and simple. It is made in several grades resistant to both abrasion and water.

Detonating-Fuse Circuit. Circuits for one, two, and three rows of blastholes are shown in Figure 11-27. The Primacord is looped and cross-tied so as to provide a second means of detonation if the main trunk line between holes should be broken accidentally.

Detonating-fuse MS connectors are used for short-interval delays. The initiation is by blasting cap, either electric or with safety fuse, taped to the detonating fuse. Firing by electric blasting cap is preferred because of control of the time of initiation.

Detonating fuse is the safest and most practical method for firing large-blasthole primary blasts and coyote-tunnel blasts. Safety is in terms of immunity to electrical disturbances, including lightning. Practicability is in terms of abrasion and water resistance and simplicity.

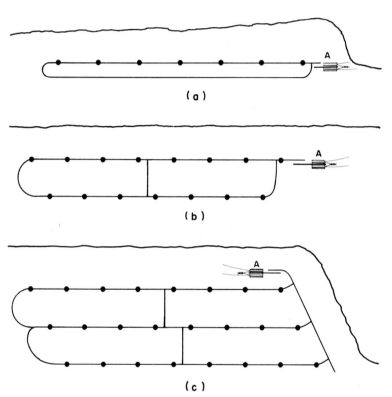

Figure 11-27 Layouts for detonation of blasting by detonation or Primacord fuse. *(a)* Single row of holes connected by Primacord and detonated by electric blasting caps or cap and fuse attached at point A (note that Primacord loop is tied to trunkline with right angle connection). *(b)* Double row of holes connected by Primacord. Note that Primacord from back line is tied to trunkline with right-angle connection (detonated by electric blasting caps or caps and fuse attached at point A). *(c)* Three rows of holes connected by Primacord. Note right-angle connections to trunklines (detonated by electric blasting caps or caps and fuse attached to point A).

These typical layouts are for instantaneous detonations by the fuse, initiated by an electric cap at point A. If delay blasting is desired, MS connectors may be placed in the fuse lines. The most effective delay intervals range from 5 to 35 ms, depending on the kind of rock and the nature of the rock formation. Detonation or Primacord fuse is used usually in lieu of electric circuits and blasting caps when the blastholes or ground are wet and when safety from electrical disturbances and lightning is necessary or mandatory. (*E. I. du Pont de Nemours & Co.*)

Electrical Circuit. Electrical circuits used for firing explosives must be designed to accommodate the characteristics of electric caps within the circuit, so that total resistance of the copper or aluminum wire must be calculated, as well as the energy output of the blasting machine. Thus the powder operator must cope with both conditions and electrical design. In practice, the good powder operator takes care of these two problems without theory and without ado. The cheapness of the electrical circuit as compared with the detonating-fuse circuit must be weighed against the wetness of the work and the dangers of extraneous electrical disturbances.

Four kinds of circuits are shown in Figure 11-28. They are series, parallel, series in parallel, and parallel in series. The series circuit is simply a single path for the current through the individual electric blasting caps.

The parallel circuit provides for the leg wires of the caps to be connected to the two sides of the circuit. It has the disadvantage that the leg wires are sometimes not long enough to span the distance between sides of the circuit.

The series-in-parallel and parallel-in-series circuits are combinations of the series and parallel circuits and they enable one to fire a large number of caps without a large power input to the circuit.

Connections between wires are made by forming a flat loop and then twisting the loop

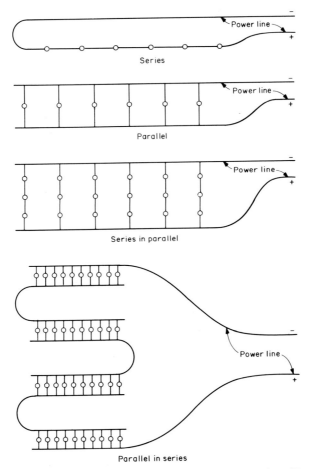

Figure 11-28 Four circuits for connecting electric blasting caps to power line. *(Hercules, Inc.)*

11-36 Fragmentation of Rock

several times to form a low-resistance connection. Cap leg wires are connected to the bus wire by a circumferential twist of the leg wire around the bus wire.

Circuits are tested by galvanometers and voltameters, and of course this testing is a necessary preblasting precaution to avoid costly and perhaps ultimately fatal misfires.

Electrical blasting circuits may be fired by either blasting machines or utility power lines. The use of the power lines calls for a rather sophisticated electrical shooting station and it is of the nature of a special installation. Most commonly either condenser-discharge or generator-type blasting machines are used, as illustrated in Figure 11-29.

The condenser-discharge machine consists of a battery or set of batteries which energizes a condenser. The resultant high-voltage current initiates the blasting caps. They are small and lightweight, and can fire up to 1200 caps in a parallel-series circuit of 450 V. There are two kinds of condenser-discharge machines. One kind discharges one impulse into a single-circuit main line. Delay blasting results from the use of MS-delay electric blasting caps within the circuit.

The other kind, the sequential timer, provides precisely timed separate impulses into a multicircuit main cable. Usually there are 10 circuits with built-in millisecond delays. In a particular machine 10 sequential delays might range from 25 to 250 ms with 25-ms delays between each two successive impulses. Between the machine and the 10 blasthole circuits there is located a protected portable distribution panel with 10 sets of binding posts for the 10 circuits from the blasting machine. To these binding posts are attached the circuits going outward to the blastholes.

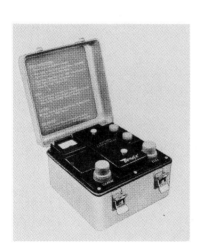

(a)   (b)

Figure 11-29 Two basic types of blasting machines. *(a)* Condenser-discharge type, made up essentially of a battery, condenser, and controls. They are made in several sizes, ranging in capacity up to 1200 electric blasting caps. There are two kinds, for instantaneous and for sequential initiation. Instantaneous initiation involves one pulse of energy to the MS-delay caps. Sequential initiation involves several pulses, separated by designed MS delays within the machine, through a multicircuit main cable to the instantaneous or MS delay caps. *(b)* The long-popular generator type, pushdown operation, made up essentially of a generator actuated by the rack-and-pinion drive of the handle. It is made in several sizes up to a rated capacity of 50 electric caps. It has been largely replaced by the small sizes of the condenser-discharge type. *(E. I. du Pont de Nemours & Co.)*

Either instantaneous or MS-delay caps may be used within the individual blasting circuits, thus affording a wide range of actual delay times. Accordingly, the sequential timer or sequential blasting machine lends itself to more precise control of rock fragmentation and ground vibrations.

The generator machine with its simple push-down handle is simple and reliable. It is time-tested and popular within its capability range. It is made in several sizes with rated series capacities from 10 to 50 caps. The 50-cap machine, when hooked up to a parallel-series circuit, can fire up to 200 caps.

Safe and efficient firing by electrical circuits calls for both theory and experience and these requisites are especially important when firing many heavy charges under adverse physical conditions. The very economy of this means of firing brings with it the necessary care in building the circuit or circuits. Before using the blasting machine all precautions are necessary; these dos and don'ts are set forth in the "When shooting electrically" section of Table 11-9.

Theory of Fragmentation of Rock by Explosives Until a few years ago it was believed that all forces for fragmentation radiated from the charge in the manner shown in Figure 11-30, the breakage being attributed solely to these radiating forces, which attenuate inversely as the square of the distance from the explosive charge. This theory is still held presently, but it is supplemented by the sequential theory of reflection of the shock waves from a free surface, such as the vertical or slanted face of the rock.

This composite action is illustrated in Figure 11-31. It is explained by the reflection theory of rock breakage according to the following sequence of events.

1. Detonation converts the blasthole charge into gas at pressures which far exceed the compressive strength of the rock.

2. The zone, sphere or cylinder, of rock around the gaseous pocket or column of explosive is crushed to shallow depth by pressure that rapidly drops below the level of the rock compressive strength.

3. As crushing stops, the remaining energy continues into the unbroken rock as a compressive wave pulse, or shock front, traveling at shock-wave velocity without breaking the rock.

4. On reaching a free surface, such as a visible or invisible joint or a visible surface, the compressive wave is reflected back as a tension wave.

5. Tension is then exerted on the rock to a measurable depth behind the free surface.

6. Rock that is pulled toward the face by a tensile strain pulse and pushed toward it by a compressive strain pulse at the same time ruptures under sufficient total strain.

7. The process repeats when the remaining portion of the impinging pressure wave reaches a new free face and is reflected as a tension wave. The number of repetitions is predictable from data on both the rock-breaking strain and the pulse-fall strain.

As a result of these theories, the U.S. Bureau of Mines has established a relationship between characteristic impedance of explosives (density times detonation velocity) and characteristic impedance of rock (density times velocity of longitudinal shock wave in the rock). These theories pertain to homogeneous rock. Rocks and rock formations are not uniform, being variously weathered and with joints, bedding planes, cleavages, and other unconformities. Because of these variables it is almost impossible to apply the equation of relationship in a practical manner to a given rock at different locations of a given rock excavation job. The rock excavator must rely on the knowledge and the experience of the powder operators.

Methods of Blasting

There are two general methods of primary blasting, as shown in Figure 11-32: by blastholes and by tunnels or coyotes. These two plans may be combined. For example, in coyote blasting it is usually desirable to maintain a uniform back slope, and blastholes, either vertical or slanted, may then be used in conjunction with the tunnels.

Blasting by Blastholes This is, by far, the more common method because of several important inherent characteristics:

11-38 Fragmentation of Rock

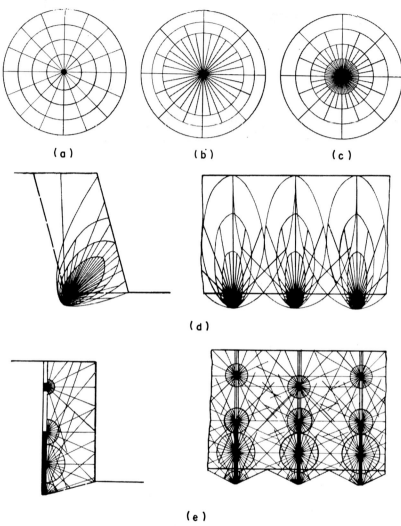

Figure 11-30 Actions of explosives as illustrated graphically. The lines radiating from the centers of the diagrams represent the forces resulting from the explosions. The forces attenuate with distance from the charge. Usually they diminish inversely as the square of the distance from the charge of explosive. *(a)* In the case of steam explosion all lines go to the outer surface of the sphere. *(b)* For black powder, a low explosive, a small portion reaches the outer surface. *(c)* For dynamite, a high explosive, a smaller portion reaches the surface, but there is a concentration of forces immediately around the explosive.

This contrast in forces between black powder and dynamite accounts for the heaving action of the low explosive and the shattering action of the high explosive.

(d) Action of explosives with pocket loading. Pocket loading of a blasthole concentrates the explosive in the bottom of the hole below the finished grade. It is applicable to friable rocks which break up by fall and it is desirable for breaking out completely the toe of the face of rock.

(e) Action of explosives with deck loading. Deck loading and also column loading distribute the forces throughout the depth of the blasthole. These loadings are applicable to hard rock, in which it is important to produce a shattering effect for complete fragmentation.

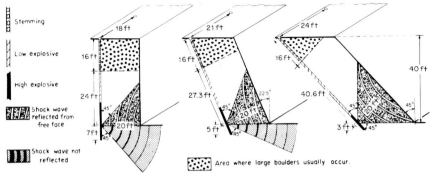

Figure 11-31 Yields of cubic yards of rock per foot of length of blasthole for vertical and slanted or inclined blastholes. This drawing by Professor Boris J. Kochanowsky of Pennsylvania University depicts shock-wave propagation by detonation of explosive at point L in the blasthole. It shows graphically how the useful proportion of the explosive energy available at the toe of the bench is increased by the slanted or inclined drilling and blasting. Where this technique applies, typical data confirm that the benefits include reduced drilling costs, improved powder factor, and greater safety.

In the case of the vertical hole of 47-ft total length, spacing is 18 ft and burden is 20 ft. Total yield to grade is 533 yd^3 or 11.3 yd^3 per foot of total length of hole.

For the slanted hole inclined at 22.5° and of 47.9-ft total length, spacing is 21 ft and burden is 21.6 ft. Total yield to grade is 672 yd^3 or 14.0 yd^3/ft per foot of total length of hole.

In the case of the slanted hole inclined at 45° and of 58.7-ft total length, spacing is 24 ft and burden is 28.3 ft. Total yield to grade is 1006 yd^3, or 17.1 yd^3/ft per foot of total length of hole.

As compared to the vertical hole, the increases in yields per foot of length of hole are 26 percent for the slanted hole inclined at 22.5° and 89 percent for the slanted hole inclined at 45°.

Assuming fragmentation with the same powder factor is satisfactory in all three cases, then drilling costs for the slanted holes would range from 21 to 47 percent below that for the vertical hole. [*Walter E. Trauffer (ed.), Pit and Quarry Handbook, Pit and Quarry Publications, Chicago, 1974, p. A-53.*]

1. The depth of faces or benches may be controlled and correlated to the blasthole pattern so as to harmonize with the nature of the rock, physical features of the work, and nature and size of the loading and hauling machinery. Depths may vary up to 40 ft for construction and up to 180 ft for quarrying and open-pit mining.

2. Good fragmentation is possible by the correlation of pattern and diameter of blastholes and characteristics of explosive for a given rock.

3. The lines and grades of the blasted prism of rock may be controlled, especially by smooth blasting to required contours with presplitting drilling, line drilling, and slant drilling.

4. As contrasted with blasting by coyotes, the charges of explosives are usually smaller, making for better control of possibly harmful vibrations.

5. Blasting by blastholes lends itself to the trial-and-error method of determining expeditiously and economically the best relationship between drilling and blasting practices.

Figure 11-32 shows that blastholes are put down about 8 ft below grade. Figure 11-33 explains the reason for this procedure, which ensures rock breakage down to grade and a smooth floor at grade. The depth drilled below grade usually equals about one-half the spacing distance of the blastholes.

Loading Methods Figure 11-34 illustrates two ways of loading the blastholes with the charges. In *column loading,* the charge is solid from the bottom to within a few feet of the collar of the hole. From the top of charge to the collar, *stemming* is added so as to confine the explosive forces as much as possible. Another important function of stemming is to protect the loaded charge from accidental detonation. By coincidence the length of stemming is generally equal to the subdrilling or distance from grade to the bottom of the hole, and this equality simplifies calculation of the powder factor for the rock excavation above grade, the quantity for payment. Usually stemming is drill cuttings, earth, sand, or fine crushed rock. In column loading with electric-cap detonation, the primer is at the bottom of the hole. With a detonating fuse the fuse extends from the top to the bottom of the hole.

11-40 Fragmentation of Rock

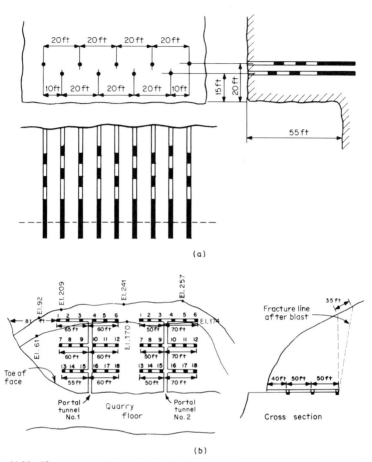

Figure 11-32 The two general methods of blasting. (a) The more common method is by use of blastholes. The holes may be vertical, slanted, or horizontal. They may be column-loaded, deck-loaded, or pocket-loaded, all with stemming. The illustration shows staggered vertical holes with deck loading.

Spacing, parallel to face, is 20 ft. Burden at right angles to face is 5 ft and burden between face and first row of holes is 15 ft. Face is 55-ft height. The holes are 8 ft below grade to allow for complete fragmentation down to grade. Yield is 3.70 yd³/ft of blasthole, but the yield for the row of holes nearer the face is 11.11 yd³/ft because of the greater 15-ft setback.

(b) The less common method is by use of coyote holes or tunnels. This method depends in part on fall of rock for fragmentation and it is applicable to irregular formations. It is usually used for production of rip-rap, ballast, ores, and quarry rock. It is difficult to maintain good lines and grades of back slope and floor. Powder factors are lower than those for the blasthole method. Topographical and environmental conditions must be favorable for these massive charges and resultant great yardages.

In the case shown, involving blasting of rhyolite, the explosives totaled 95,950 lb and the rock totaled 444,000 yd³. The low powder factor was 0.22 lb of explosive per cubic yard of rock, perhaps one-third of that for use of blastholes.

In *deck loading* the total charge is divided into segments separated by stemming. Certain rocks and rock formations of soft and medium nature lend themselves to deck loading. In detonation by electric caps a primer is placed at the bottom of each deck. If detonation is by detonating fuse, again the fuse extends to the bottom of the hole.

Figure 11-35 illustrates conventional column loading with stemming from top of explosives to collar of hole. It also shows stemming in deck loading to distribute the charge and also to prevent the loss of explosive force into a soft mud seam.

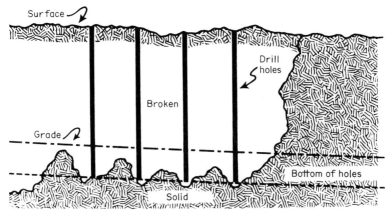

Figure 11-33 Necessary drilling of blastholes below finished grade. If blastholes were carried down only to grade, there would result conical or pyramidal extensions of solid rock above the desired finished grade. These are shown in the space between the plane of hole bottoms and grade. In order to excavate to grade, these extensions would have to be blasted out by the blockhole method, an expensive and time-consuming job. Hence, blastholes are carried down to below grade as illustrated. As rock tends to break out at an angle of about 45° to the plane of the hole bottoms, the holes are carried down to about one-half the distance between the blastholes. (E. I. du Pont de Nemours & Co.)

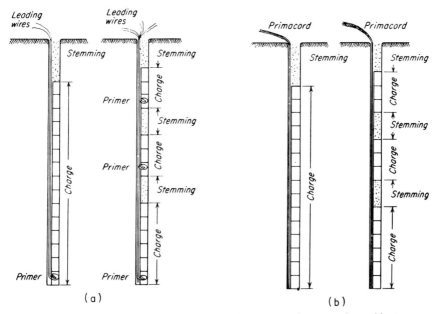

Figure 11-34 Methods of loading blastholes with explosives using alternative electric blasting caps or detonating fuse as initiating devices. (a) Detonation by electric caps using primers: column loading with stemming (left); deck loading with stemmings (right). (b) Detonation by detonating or Primacord fuse: column loading with stemming (left); deck loading with stemmings (right). (R. L. Peurifoy, Construction Planning, Equipment, and Methods, McGraw-Hill, New York, 1965, p. 265.)

11–42 Fragmentation of Rock

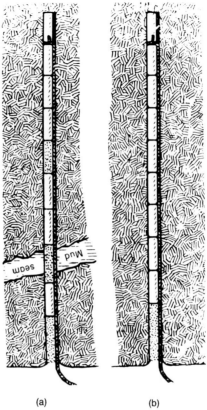

(a) (b)

Figure 11-35 The principal uses of stemming. *(a)* Stemming between the top of the column-loaded explosive and the collar of the hole in order to confine the force of the explosion to fragmentation of the rock formation. *(b)* Stemming between the deck loadings of the explosives so as to distribute the forces between maximum at the bottom and minimum at the top of the hole and to reduce loss of force into the soft mud seam. Stemming is usually the cuttings from drilling, as well as clay, sand, or small rock. All stemming should be damp so as to compact in the blasthole. *(E. I. du Pont de Nemours & Co.)*

When the rock formation is made up of alternating beds of limestone or sandstone and shale, the shale is sometimes of a soft or earthy nature. The same condition exists in cavernous and vuggy limestones. Blocking off of these voids or strata is done by means of stemming.

In the illustration of the mud seam the decked charges diminish from the bottom to the top of the hole so as to correspondingly lessen the explosive forces according to the nature of the rock or the rock formation.

Tamping poles longer than the depth of the hole are used for the freeing and the compaction of explosives during the loading of the blasthole. They are of about 1½-in-diameter straight-grained hardwood with no exposed metal parts except certain approved connectors for joining the sections of pole. Poles are marked or notched at 1-ft intervals for depth measurements.

A third method for loading blastholes is by the pocket or chamber, as illustrated in Figure 11-36. Chambers, prevalent during the common use of black powder, are little used presently. A notable exception is their popularity when the jet-piercing drill is used, as explained in Figure 11-36.

Breaking out the toe of the face is facilitated by the chambering method, as it concentrates explosive forces at the bottom of the hole. Pocket loading may be combined with

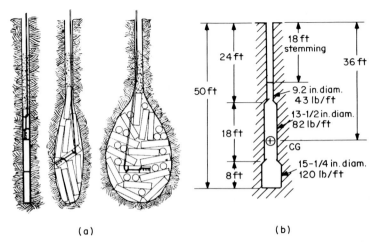

Figure 11-36 Blasting by pocketing or chambering the blasthole. *(a)* By springing the hole by several successive charges of dynamite. The first charge is shown on the left. In the center and on the right are successive loads showing how larger charges are used for each shot until a chamber of the required size is obtained. When springing several 20-ft deep, 5⅝-in-diameter holes in sandstone by three springings, these averages resulted:

For 1¼ in × 8 in cartridges of 40% strength dynamite:

First springing	2 cartridges
Second springing	5 cartridges
Third springing	20 cartridges
Total springings	27 cartridges

The resulting pocket or chamber was charged with 200 lb of explosives. *(Hercules, Inc.)* *(b)* By chambering the hole with jet-piercing drill. The jet-piercing drill is unique among drills in its ability to increase the diameter of or to pocket the drill hole at any point in its depth during the actual drilling operation.

In this instance 27 blastholes were designed to yield 35 yd^3 of hard taconite per foot of blast hole. The powder factor of the AN/FO explosive was about 1.54 lb/yd^3.

The resultant well-fragmented taconite was about 100,000 tons, or about 47,000 yd^3. By pocket loading, a tough 30-ft burden at toe of face was well broken and grade of pit was maintained. *(Union Carbide Corp.)*

either column or deck loading so as to ensure good fragmentation with low powder factor. Such a combination, pocket and column loading, is shown in the example of the jet piercing drill of Figure 11-36.

Yield in Cubic Yards of Rock per Foot of Blasthole. Table 11-2 gives cubic yards per foot of blasthole according to the drilling pattern. This pattern is made up of *spacing,* the distance in feet between rows of holes at right angles to the face, and *burden,* the distance between rows of holes parallel to the face.

One must note that sometimes the burden from the face to the first row of holes is greater than the burden between rows of holes. Figure 11-32 shows this difference, which must be considered when calculating yardages of rock. One must also note that for total yardage to grade the values of Table 11-2 should be multiplied by depth of hole to grade and not by total depth of hole.

In open-pit mining and quarrying, yields are sometimes, and even exclusively, reckoned in tons and the values of Table 11-2 must be multiplied by a factor representing tons per cubic yard bank measurement. These factors may be calculated easily from the "cubic yards in cut" column of Appendix 1, Approximate Material Characteristics.

Densities for Loading Blastholes with Explosives. Table 11-3 gives pounds per lineal foot of blasthole for given densities of explosives in grams per cubic centimeter. Tables 11-2 and 11-3, when used together, provide the means for bringing density of explosives, diameter of blasthole, density of loading, pattern of blastholes, and powder factor into

TABLE 11-2 Cubic Yards of Rock per Lineal Foot of Blasthole According to Pattern of Blastholes

Spacing is distance in feet between blastholes parallel to face.
Burden is distance in feet between blastholes at right angles to face.
For total cubic yards of rock per blasthole, multiply by height of face in feet.

Spacing	Burden														
	4	5	6	7	8	9	10	11	12	13	14	15	16	17	18
4	0.59	0.74	0.89	1.04	1.19	1.33	1.48	1.63	1.78	1.93	2.07	2.22	2.37	2.52	2.67
5	0.74	0.93	1.11	1.30	1.48	1.67	1.85	2.04	2.22	2.41	2.59	2.78	2.96	3.15	3.33
6	0.89	1.11	1.33	1.56	1.78	2.00	2.22	2.44	2.67	2.89	3.11	3.33	3.56	3.78	4.00
7	1.04	1.30	1.56	1.81	2.07	2.33	2.59	2.85	3.11	3.37	3.63	3.89	4.15	4.41	4.67
8	1.19	1.48	1.78	2.07	2.37	2.67	2.96	3.26	3.56	3.85	4.15	4.44	4.74	5.04	5.33
9	1.33	1.67	2.00	2.33	2.67	3.00	3.33	3.67	4.00	4.33	4.67	5.00	5.33	5.67	6.00
10	1.48	1.85	2.22	2.59	2.96	3.33	3.70	4.07	4.44	4.81	5.19	5.56	5.93	6.30	6.67
11	1.63	2.04	2.44	2.85	3.26	3.67	4.07	4.48	4.89	5.30	5.70	6.11	6.52	6.93	7.33
12	1.78	2.22	2.67	3.11	3.56	4.00	4.44	4.89	5.33	5.78	6.22	6.67	7.11	7.56	8.00
13	1.93	2.41	2.89	3.37	3.85	4.33	4.81	5.30	5.78	6.26	6.74	7.22	7.70	8.19	8.67
14	2.07	2.59	3.11	3.63	4.15	4.67	5.19	5.70	6.22	6.74	7.26	7.78	8.30	8.81	9.33
15	2.22	2.78	3.33	3.89	4.44	5.00	5.56	6.11	6.67	7.22	7.78	8.33	8.89	9.44	10.00
16	2.37	2.96	3.56	4.15	4.74	5.33	5.93	6.52	7.11	7.70	8.30	8.89	9.48	10.07	10.67
17	2.52	3.15	3.78	4.41	5.04	5.67	6.30	6.93	7.56	8.19	8.81	9.44	10.07	10.70	11.33
18	2.67	3.33	4.00	4.67	5.33	6.00	6.67	7.33	8.00	8.67	9.33	10.00	10.67	11.33	12.00
19	2.81	3.52	4.22	4.93	5.63	6.33	7.04	7.74	8.44	9.15	9.85	10.56	11.26	11.96	12.67
20	2.96	3.70	4.44	5.19	5.93	6.67	7.41	8.15	8.89	9.63	10.37	11.11	11.85	12.59	13.33
21	3.11	3.89	4.67	5.44	6.22	7.00	7.78	8.56	9.33	10.11	10.89	11.67	12.44	13.22	14.00
22	3.26	4.07	4.89	5.70	6.52	7.33	8.15	8.96	9.78	10.59	11.41	12.22	13.04	13.85	14.67
23	3.41	4.26	5.11	5.96	6.81	7.67	8.52	9.37	10.22	11.07	11.93	12.78	13.63	14.48	15.33
24	3.56	4.44	5.33	6.22	7.11	8.00	8.89	9.78	10.67	11.56	12.44	13.33	14.22	15.11	16.00
25	3.70	4.63	5.56	6.48	7.41	8.33	9.26	10.19	11.11	12.04	12.96	13.89	14.81	15.74	16.67
26	3.85	4.81	5.78	6.74	7.70	8.67	9.63	10.59	11.56	12.52	13.48	14.44	15.41	16.37	17.33
27	4.00	5.00	6.00	7.00	8.00	9.00	10.00	11.00	12.00	13.00	14.00	15.00	16.00	17.00	18.00
28	4.15	5.19	6.22	7.26	8.30	9.33	10.37	11.41	12.44	13.48	14.52	15.56	16.59	17.63	18.67
29	4.30	5.37	6.44	7.52	8.59	9.67	10.74	11.81	12.89	13.96	15.04	16.11	17.19	18.26	19.33
30	4.44	5.56	6.67	7.78	8.89	10.00	11.11	12.22	13.33	14.44	15.56	16.67	17.78	18.89	20.00
31	4.59	5.74	6.89	8.04	9.19	10.33	11.48	12.63	13.78	14.93	16.07	17.22	18.37	19.52	20.67
32	4.74	5.93	7.11	8.30	9.48	10.67	11.85	13.04	14.22	15.41	16.59	17.78	18.96	20.15	21.33
33	4.89	6.11	7.33	8.56	9.78	11.00	12.22	13.44	14.67	15.89	17.11	18.33	19.56	20.78	22.00
34	5.04	6.30	7.56	8.81	10.07	11.33	12.59	13.85	15.11	16.37	17.63	18.89	20.15	21.41	22.67
35	5.19	6.48	7.78	9.07	10.37	11.67	12.96	14.26	15.56	16.85	18.15	19.44	20.74	22.04	23.33
36	5.33	6.67	8.00	9.33	10.67	12.00	13.33	14.67	16.00	17.33	18.67	20.00	21.33	22.67	24.00
37	5.48	6.85	8.22	9.59	10.96	12.33	13.70	15.07	16.44	17.81	19.19	20.56	21.93	23.30	24.67
38	5.63	7.04	8.44	9.85	11.26	12.67	14.07	15.48	16.89	18.30	19.70	21.11	22.52	23.93	25.33
39	5.78	7.22	8.67	10.11	11.56	13.00	14.44	15.89	17.33	18.78	20.22	21.67	23.11	24.56	26.00
40	5.93	7.41	8.89	10.37	11.85	13.33	14.81	16.30	17.78	19.26	20.74	22.22	23.70	25.19	26.67

harmony for efficient economical fragmentation. The following example illustrates the computations.

The height of face in an abandoned limestone quarry is 50 ft. It is planned to reactivate the quarry with its hard and medium bedded limestone. Experiences in nearby similar limestone deposits indicate that a powder factor of 1.0 lb AN/FO per 2.5 tons of limestone may be anticipated. This is equivalent to 4380/5000 or 0.88 lb/yd³ of limestone. Experience also indicates that a 12-ft spacing by 15-ft burden pattern gives good fragmentation for the primary crusher. It is planned to use AN/FO of 0.80 poured density.

Table 11-2, for the tested spacing, gives 6.67 yd³ per foot of blasthole. At 0.88 powder factor, 5.9 lb AN/FO is required per foot of blasthole. Table 11-3, for density of 0.80 and loading of 5.9 lb/ft, gives blasthole diameter of 4¾ in. If the blastholes were drilled 6 ft below grade of quarry floor to ensure a good level floor and explosive were loaded to within 6 ft of collar of hole, good fragmentation at low unit cost for drilling and blasting should result.

Trial-and-Error Spacing for Blastholes. Table 11-4 gives trial-and-error spacing for blastholes at the beginning of blasting operations when experience is lacking in the

						Burden						
19	20	21	22	23	24	25	26	28	30	32	34	36	38	40
2.81	2.96	3.11	3.26	3.41	3.56	3.70	3.85	4.15	4.44	4.74	5.04	5.33	5.63	5.93
3.52	3.70	3.89	4.07	4.26	4.44	4.63	4.81	5.19	5.56	5.93	6.30	6.67	7.04	7.41
4.22	4.44	4.67	4.89	5.11	5.33	5.56	5.78	6.22	6.67	7.11	7.56	8.00	8.44	8.89
4.93	5.19	5.44	5.70	5.96	6.22	6.48	6.74	7.26	7.78	8.30	8.81	9.33	9.85	10.37
5.63	5.93	6.22	6.52	6.81	7.11	7.41	7.70	8.30	8.89	9.48	10.07	10.67	11.26	11.85
6.33	6.67	7.00	7.33	7.67	8.00	8.33	8.67	9.33	10.00	10.67	11.33	12.00	12.67	13.33
7.04	7.41	7.78	8.15	8.52	8.89	9.26	9.63	10.37	11.11	11.85	12.59	13.33	14.07	14.81
7.74	8.15	8.56	8.96	9.37	9.78	10.19	10.59	11.41	12.22	13.04	13.85	14.67	15.48	16.30
8.44	8.89	9.33	9.78	10.22	10.67	11.11	11.56	12.44	13.33	14.22	15.11	16.00	16.89	17.78
9.15	9.63	10.11	10.59	11.07	11.56	12.04	12.52	13.48	14.44	15.41	16.37	17.33	18.30	19.26
9.85	10.37	10.89	11.41	11.93	12.44	12.96	13.48	14.52	15.56	16.59	17.63	18.67	19.70	20.74
10.56	11.11	11.67	12.22	12.78	13.33	13.89	14.44	15.56	16.67	17.78	18.89	20.00	21.11	22.22
11.26	11.85	12.44	13.04	13.63	14.22	14.81	15.41	16.59	17.78	18.96	20.15	21.33	22.52	23.70
11.96	12.59	13.22	13.85	14.48	15.11	15.74	16.37	17.63	18.89	20.15	21.41	22.67	23.93	25.19
12.67	13.33	14.00	14.67	15.33	16.00	16.67	17.33	18.67	20.00	21.33	22.67	24.00	25.33	26.67
13.37	14.07	14.78	15.48	16.19	16.89	17.59	18.30	19.70	21.11	22.52	23.93	25.33	26.74	28.15
14.07	14.81	15.56	16.30	17.04	17.78	18.52	19.26	20.74	22.22	23.70	25.19	26.67	28.15	29.63
14.78	15.56	16.33	17.11	17.89	18.67	19.44	20.22	21.78	23.33	24.89	26.44	28.00	29.56	31.11
15.48	16.30	17.11	17.93	18.74	19.56	20.37	21.19	22.81	24.44	26.07	27.70	29.33	30.96	32.59
16.19	17.04	17.89	18.74	19.59	20.44	21.30	22.15	23.85	25.56	27.26	28.96	30.67	32.37	34.07
16.89	17.78	18.67	19.56	20.44	21.33	22.22	23.11	24.89	26.67	28.44	30.22	32.00	33.78	35.56
17.59	18.52	19.44	20.37	21.30	22.22	23.15	24.07	25.93	27.78	29.63	31.48	33.33	35.19	37.04
18.30	19.26	20.22	21.19	22.15	23.11	24.07	25.04	26.96	28.89	30.81	32.74	34.67	36.59	38.52
19.00	20.00	21.00	22.00	23.00	24.00	25.00	26.00	28.00	30.00	32.00	34.00	36.00	38.00	40.00
19.70	20.74	21.78	22.81	23.85	24.89	25.93	26.96	29.04	31.11	33.19	35.26	37.33	39.41	41.48
20.41	21.48	22.56	23.63	24.70	25.78	26.85	27.93	30.07	32.22	34.27	36.52	38.67	40.81	42.96
21.11	22.22	23.33	24.44	25.56	26.67	27.78	28.89	31.11	33.33	35.56	37.78	40.00	42.22	44.44
21.81	22.96	24.11	25.56	26.41	27.56	28.70	29.85	32.15	34.44	36.74	39.04	41.33	43.63	45.93
22.52	23.70	24.89	26.07	27.26	28.44	29.63	30.81	33.19	35.56	37.93	40.30	42.67	45.04	47.41
23.22	24.44	25.67	26.89	28.11	29.33	30.56	31.78	34.22	36.67	39.11	41.56	44.00	46.44	48.89
23.93	25.19	26.44	27.70	28.96	30.22	31.48	32.74	35.26	37.78	40.30	42.81	45.33	47.85	50.37
24.63	25.93	27.22	28.52	29.81	31.11	32.41	33.70	36.30	38.89	41.48	44.07	46.67	49.26	51.85
25.33	26.67	28.00	29.33	30.67	32.00	33.33	34.67	37.33	40.00	42.67	45.33	48.00	50.67	53.33
26.04	27.41	28.78	30.15	31.52	32.89	34.26	35.63	38.37	41.11	43.85	46.59	49.33	52.07	54.81
26.74	28.15	29.56	30.96	32.37	33.78	35.19	36.59	39.41	42.22	45.04	47.85	50.67	53.48	56.30
27.44	28.89	30.33	31.78	33.22	34.67	36.11	37.56	40.44	43.33	46.22	49.11	52.00	54.89	57.78
28.15	29.63	31.11	32.59	34.07	35.56	37.04	38.52	41.48	44.44	47.41	50.37	53.33	56.30	59.26

particular rock of the particular location. When blasting is carried out continuously in one location, such as in an ore deposit, techniques are fairly well standardized. However, in transient public works, such as a 10-mi reach of canal, one may encounter several different rocks in several different rock formations. Under such diverse conditions trial-and-error blasting methods are used necessarily. Here Table 11-4 is especially valuable as a practical guide for work in unfamiliar surroundings.

Buffer Blasting. Where there is a low face and where shooting out the toe of the face presents a problem, some confinement of the blast is needed for good fragmentation. *Buffer blasting* against a previously blasted pile of rock at the face is an effective means of confinement. The buffer is efficient in blocky and seamy rocks such as limestones and sandstones and it is economical, as usually a lower powder factor is possible. An added advantage is less throw of the blasted rock horizontally, a sometimes unwelcome feature of blasting in a quarry as it requires cleaning up the quarry floor after the blasting work.

Controlled Blasting. *Controlled blasting* is a means of avoiding irregular back break in the face of a rock cut, which sometimes produces loose rock in the face. The practice

TABLE 11-3 Densities for Loading Blastholes with Explosives

Blasthole diameter, inches	Pounds per lineal foot of blasthole for given densities of explosives in grams per cubic centimeter												
	0.40	0.50	0.60	0.70	0.80	0.90	1.00	1.10	1.20	1.30	1.40	1.50	1.60
1	0.1	0.2	0.2	0.2	0.3	0.3	0.3	0.4	0.4	0.4	0.5	0.5	0.5
1¼	0.2	0.3	0.3	0.4	0.4	0.5	0.5	0.6	0.6	0.7	0.7	0.8	0.8
1½	0.3	0.4	0.5	0.5	0.6	0.7	0.8	0.8	0.9	1.0	1.1	1.1	1.2
1¾	0.4	0.5	0.6	0.7	0.8	0.9	1.0	1.1	1.2	1.4	1.5	1.6	1.7
2	0.5	0.7	0.8	1.0	1.1	1.2	1.4	1.5	1.6	1.8	1.9	2.0	2.2
2¼	0.7	0.9	1.0	1.2	1.4	1.5	1.7	1.9	2.1	2.2	2.4	2.6	2.8
2½	0.8	1.1	1.3	1.5	1.7	1.9	2.1	2.3	2.6	2.8	3.0	3.2	3.4
2¾	1.0	1.3	1.5	1.8	2.1	2.3	2.6	2.8	3.1	3.3	3.6	3.9	4.1
3	1.2	1.5	1.8	2.1	2.4	2.8	3.1	3.4	3.7	4.0	4.3	4.6	4.9
3¼	1.4	1.8	2.2	2.5	2.9	3.2	3.6	4.0	4.3	4.7	5.0	5.4	5.7
3½	1.7	2.1	2.5	2.9	3.3	3.7	4.2	4.6	5.0	5.4	5.8	6.2	6.7
3¾	1.9	2.4	2.9	3.3	3.8	4.3	4.8	5.3	5.7	6.2	6.7	7.2	7.6
4	2.2	2.7	3.3	3.8	4.4	4.9	5.4	6.0	6.5	7.1	7.6	8.2	8.7
4¼	2.5	3.1	3.7	4.3	4.9	5.5	6.1	6.8	7.4	8.0	8.6	9.2	9.8
4½	2.8	3.4	4.1	4.8	5.5	6.2	6.9	7.6	8.3	9.0	9.6	10.3	11.0
4¾	3.1	3.8	4.6	5.4	6.1	6.9	7.7	8.4	9.2	10.0	10.7	11.5	12.3
5	3.4	4.2	5.1	6.0	6.8	7.6	8.5	9.4	10.2	11.0	11.9	12.8	13.6
5½	4.1	5.1	6.2	7.2	8.2	9.3	10.3	11.3	12.3	13.4	14.4	15.4	16.5
6	4.9	6.1	7.3	8.6	9.8	11.0	12.2	13.5	14.7	15.9	17.1	18.4	19.6
6½	5.7	7.2	8.6	10.0	11.5	12.9	14.4	15.8	17.2	18.7	20.1	21.5	23.0
7	6.7	8.3	10.0	11.7	13.3	15.0	16.7	18.3	20.0	21.7	23.3	25.0	26.7
7½	7.7	9.6	11.5	13.4	15.3	17.2	19.1	21.0	23.0	24.9	26.8	28.7	30.6
8	8.7	10.9	13.1	15.2	17.4	19.6	21.8	23.9	26.1	28.3	30.5	32.6	34.8
8½	9.8	12.3	14.7	17.2	19.7	22.1	24.6	27.0	29.5	31.9	34.4	36.8	39.3
9	11.0	13.8	16.5	19.3	22.0	24.8	27.5	30.3	33.0	35.8	38.6	41.3	44.1
9½	12.3	15.3	18.4	21.5	24.5	27.6	30.7	33.8	36.8	39.9	43.0	46.0	49.1
10	13.6	17.0	20.4	23.8	27.2	30.6	34.0	37.4	40.8	44.2	47.6	51.0	54.4
11	16.5	20.6	24.7	28.8	32.9	37.0	41.1	45.3	49.4	53.5	57.6	61.7	65.8
12	19.6	24.5	29.4	34.3	39.2	44.1	49.0	53.9	58.8	63.7	68.6	73.5	78.4
13	23.0	28.7	34.5	40.2	46.0	51.7	57.5	63.2	69.0	74.7	80.5	86.2	92.0
14	26.7	33.3	40.0	46.7	53.3	60.0	66.7	73.3	80.0	86.7	93.3	100.0	106.7
15	30.6	38.3	45.9	53.6	61.2	68.9	76.5	84.2	91.8	99.5	107.1	114.8	122.4
16	34.8	43.5	52.2	60.9	69.7	78.4	87.1	95.8	104.5	113.2	121.9	130.6	139.3
17	39.3	49.1	59.0	68.8	78.6	88.5	98.3	108.1	117.9	127.8	137.6	147.4	157.3
18	44.1	55.1	66.1	77.1	88.2	99.2	110.2	121.2	132.2	143.3	154.3	165.3	176.3
19	49.1	61.4	73.7	85.9	98.2	110.5	122.8	135.1	147.3	159.6	171.9	184.2	196.4
20	54.4	68.0	81.6	95.2	108.8	122.4	136.0	149.6	163.2	176.9	190.5	204.1	217.7

TABLE 11-4 Trial-and-Error Spacing for Blastholes of 6-in diameter at the Beginning of Blasting Work

Height of face	Kind of material and distance between drill-hole centers (ft)					
	Granite, gneiss, quartzite	Limestone and dolomite (hard, solid and thickly-bedded)	Limestone (medium thickly-stratified)	Limestone (thinly laminated)	Sandstone	Shale (medium-hard to soft)
25	10 × 12	10 × 12	12 × 14	12 × 14	12 × 14	15 × 18
30	12 × 14	12 × 15	12 × 15	12 × 16	13 × 15	15 × 18
35	12 × 14	12 × 15	14 × 16	12 × 16	14 × 17	16 × 18
40	12 × 14	12 × 16	14 × 18	14 × 18	14 × 18	17 × 21
45	13 × 15	12 × 16	14 × 19	14 × 20	15 × 18	18 × 22
50	14 × 15	14 × 17	14 × 20	14 × 20	15 × 20	19 × 24
55	14 × 16	14 × 18	15 × 21	15 × 22	16 × 20	19 × 24
60	14 × 16	15 × 20	15 × 22	16 × 25	16 × 22	20 × 26
70	15 × 18	16 × 20	16 × 23	16 × 26	20 × 24	22 × 27
80	16 × 20	18 × 22	16 × 25	18 × 28	18 × 25	20 × 28
90	16 × 21	18 × 22	18 × 26	18 × 29	18 × 25	20 × 29
100	16 × 22	18 × 24	20 × 28	20 × 30	20 × 30	20 × 30
120	16 × 25	20 × 25	20 × 29	20 × 30	20 × 30	20 × 32

NOTE 1: For 9-in-diameter holes it has been found that the spacing can usually be increased about 50 percent in each direction and equal or improved fragmentation obtained. This is due to the greater capacity for explosive per 9-in hole.

NOTE 2: These suggested spacings or patterns are for 6-in-diameter blastholes. Spacing and burden may be increased or decreased in proportion to the diameter of the new trial-and-error blasthole in relation to 6 in.

As an example, for sandstone the table suggests 15-ft spacing parallel to the face by 18-ft burden at right angles to the face. If 8-in-diameter holes were to be drilled, the pattern of the blastholes would be increased by 33 percent. Spacing then would be 20 ft and burden would be 24 ft.

Whereas cubic yards of sandstone per foot of 6-in-diameter hole is 10.0, the corresponding figure for the 8-in-diameter hole would be 17.8 yd³; the increase in yield, cubic yards per foot of blasthole, would be 78 percent.

If AN/FO, a blasting-agent explosive of 0.80 density, were to be loaded into the 6-in-diameter hole, the pounds per foot of hole would be 9.8 and the corresponding figure for the 8-in diameter hole would be 17.4. The increase is 78 percent. Hence, for both hole diameters the powder factor is 1.02 yd³ of sandstone per pound of AN/FO.

Assuming fragmentation to be satisfactory in both cases, the possible economy in using 8-in diameter blastholes would be in the lesser footage of drilling for the same yardage of sandstone, 44 percent in this case.

SOURCE: Walter E. Trauffer (ed.), *Pit and Quarry Handbook*, Pit and Quarry Publications, Chicago, 1974.

11-48 Fragmentation of Rock

achieves safety as well as satisfying engineering and aesthetic considerations. There are three principal methods.

1. *In-line drilling* a single row of small-diameter, closely spaced holes is drilled along the neat excavation line. They are of 2- to 3-in diameter and are spaced at 2 to 4 times the hole diameter. They are not loaded with explosives, as their purpose is to create a plane of weakness for a smooth face when the primary blast is detonated. Line drilling is best suited to homogeneous formations in which bedding planes, seams, and joints are minimal.

2. *Smooth or contour blasting* is illustrated by Figure 11-37, which shows a portal excavation. Holes are drilled along the neat excavation line either before or after the primary blasting. The holes are then loaded and fired with MS-delay caps between holes so as to facilitate a shearing action. Table 11-5 is a guide to practices in this work.

3. *Preshearing or presplitting* involves a single row of closely spaced small-diameter holes along the neat excavation line. Figure 11-38 illustrates the half holes remaining in a shale-siltstone-sandstone face after preshearing. Preshearing to grade was specified for two 500-ft deep through cuts for diversion of the Levisa Fork of the Big Sandy River in Kentucky. Holes are 2¾-in diameter with 2½-ft spacing and 4-ft burden. The holes in preshearing are fired together before any adjacent excavation is blasted. Table 11-6 suggests explosive loadings and spacings for preshearing.

Controlled blasting is specified presently on many rock excavation jobs as the aforementioned achievements have been accompanied by overall economy.

Delay Blasting. Delay blasting in the pattern of blastholes is used commonly for several purposes: (1) as it is a form of buffer blasting, it tends to reduce the powder factor;

Figure 11-37 Smooth or contour blasting for a portal excavation. In comparison with the person standing above the ladder, the holes and spacing for the smooth blasting are average, being about 2 in and 2 ft, respectively. These holes were loaded and detonated after primary blasting of the rock of the portal. *(E. I. du Pont de Nemours & Co.)*

Figure 11-38 Presheared or presplit face of shale-siltstone-sandstone cut for river diversion. The half holes of the 2¾-in-diameter preshearing blastholes, spaced on 2½-ft centers, may be seen in the face behind the production drilling rotary rig. These large-area, 500-ft-high faces of two 6-million-yd³ through cuts are being presheared. Such diameter and spacing call for ⅛ to ⅜ lb of explosive per foot of hole. Preshearing holes are being drilled by two heavyweight track drills. *(Dave Williams (ed.), "Slurry Blasting Cuts Costs for New River Canyon," Roads and Streets, September 1975, p. 45.)*

(2) it usually gives better fragmentation; (3) it reduces undesirable ground vibrations and noise of blasting; and (4) it makes possible a greater total charge of explosives and greater total yield of rock.

For MS electric caps, the delay periods are tabulated in the earlier section on dynamite initiating devices. For detonating fuse or Primacord, delay connectors are introduced into the fuse circuit, as described in the same section.

Both delaying devices are marked with the delay in milliseconds. As the shock-wave velocity of initial burden movement is about 1000 ft/s, a good rule of thumb is to limit the delay interval to less than 1 ms per foot spacing. Otherwise the shock wave of the blast of one row of holes would reach the second row of holes before it was detonated and detonation in it would be cut off. However, since cutoffs generally occur in the upper third of the blasthole, the systems can tolerate somewhat longer time intervals.

The common arrangement is to prime all holes in each row parallel to the face with the same MS period, the MS periods increasing backward away from the face. The first row might include MS 25s, the second MS 50s in keeping with the above-mentioned tolerance, the third MS 75s, and the fourth MS 100s. Succeeding rows would have MS 25 intervals. Figure 11-39 shows three additional methods for delay blasting. Part *a* is by a staggered pattern delayed in echelon sequence firing, part *b* is by a square pattern delayed in echelon, and part *c* is by multiple rows with staggered pattern, resembling the aforementioned common arrangement. The ultimate arrange-

TABLE 11-5 Guide for Smooth or Contour Blasting Practices

Hole diameter, in	Spacing, ft	Burden, ft	Explosive charge, lb/ft
1½–1¾	2	3	0.12–0.25
2	2½	3½	0.12–0.25

NOTE: Spacings, burdens, and explosive charges are dependent on the nature of the rock and the rock formation and they represent an average range of values.
SOURCE: E. I. du Pont de Nemours & Co.

TABLE 11-6 Guides for Preshearing or Presplitting Practices

Hole diameter, in	Spacing, ft	Explosive charge, lb/ft
1½–1¾	1–1½	0.08–0.25
2–2½	1½–2	0.08–0.25
3–3½	1½–3	0.13–0.50
4	2–4	0.25–0.75

NOTE: Spacings and explosive charges are dependent on the nature of the rock and the rock formation and they represent an average range of values.
SOURCE: E. I. du Pont de Nemours & Co.

ment is based on the experience of the powder operator, supplemented by a degree of trial and error on the job.

Blasting by Coyotes or Tunnels Sometimes the most economical blasting results are obtained by using a number of large, concentrated charges located in one or more small tunnels driven into the rock. Such a method is illustrated in Figure 11-32. Here two adits or main tunnels were driven into the sidehill at right angles to the face to a length of 240 ft at the grade of the quarry floor. Three laterals or cross-tees were then driven at right angles to the adits for a combined length of 720 ft. Total length of tunnels was 960 ft.

The tunnels were charged as illustrated in Figure 11-32 and detonated by detonating fuse or Primacord as shown in Figure 11-40. Tunnels are kept as small as possible for economy in driving and for minimum expense in backfilling with set rock and smaller rock on top of the explosives. Figure 11-41 shows average dimensions of a typical tunnel, 6 ft high and 4½ ft wide, and two patterns for drilling a round of blastholes. In the case of the draw cut, nine 5-ft holes of about 1½-in diameter are shown, whereas for the burn cut eleven 4½-ft holes of like diameter are shown. If the tunnel of Figure 11-41 were used in the second blasting operation diagramed in Figure 11-32, there would be 960 yd^3 of tunnel rock excavation.

Drilling is usually by light drifter air hammers with their supporting columns. According to the class of rock—soft, medium, or hard—drilling speeds vary greatly from 100 to 0 ft/h. High-strength, high-velocity gelatin dynamites or water-gel blasting agents are usually used, the latter being preferable because of their lack of objectionable fumes in the underground work.

The powder factor depends on the hardness of the rock and varies from 3 to 6 lb/yd^3 in the small tunnels. For example, if the holes of the draw cut of Figure 11-41 were loaded for one-half of their lengths with 1.4-density water gel, 33 lb would be loaded, and 5 yd^3 would be pulled per round of blasting, resulting in a powder factor of 5.0 lb of water gel per cubic yard of rock.

Blasted rock may be removed by several methods or combinations of methods. It may be hand-loaded into wheelbarrows or into a steel skid pan hauled by a mechanical winch. It may be loaded by a small rocker shovel into a steel skid pan or it may be loaded out by a small wheel loader. The loading of explosives and backfilling material may be by roller-wheel conveyors or by the same machines that have been

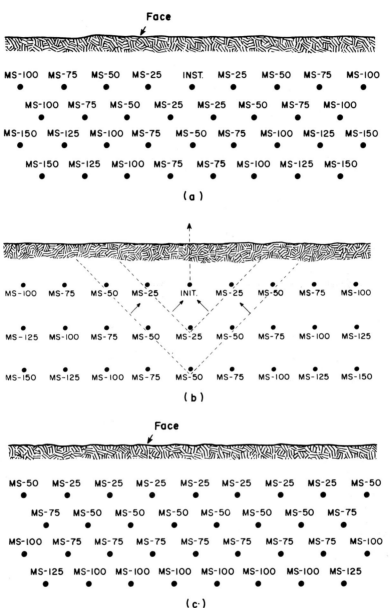

Figure 11-39 Three ways for delay blasting by use of MS-delay electric blasting. *(a)* A staggered pattern delayed in echelon sequence firing. *(b)* A square pattern delayed in echelon. *(c)* An arrangement of MS delays in multiple rows with staggered pattern. These plans are for electric circuits. If detonating of Primacord fuse is used, somewhat similar plans may be arranged by the use of MS-delay connectors placed within the fuse lines. *(E. I. du Pont de Nemours & Co.)*

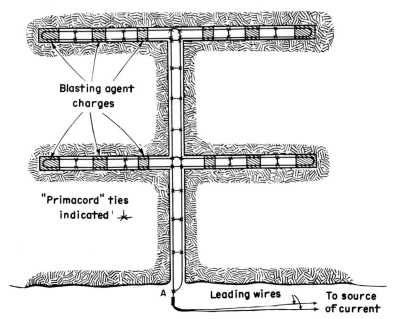

Figure 11-40 Typical loading and detonation methods for coyote or tunnel blasting. Individual charges of blasting agent are placed in the cross-tees in accordance with the planning of the blast. They are separated by stemmings of set rock and smaller rock, and the charges nearest the adit or main tunnel are bulkheaded with stemmings so as to confine the explosive forces. There are no charges or stemmings in the adit. Detonation of blasting agent charges is by detonating or Primacord fuse arranged in parallel lines with cross ties at intervals and between the charges. Initiation is by electric cap at the end of the fuse. (*E. I. du Pont de Nemours & Co.*)

used to remove the blasted rock. The backfill of set rock and smaller rock is hand-placed.

Figure 11-42 illustrates a coyote blast with one adit and one lateral. The rock formation is irregular, made up of both boulders and country rock, and it is an ideal formation for coyote blasting. As a heaving action was desired, 8150 lb ammonium nitrate dynamite of 9800-ft/s velocity was used, primed by 100 lb of straight dynamite. A total of 30,000 tons, or about 14,300 yd^3, of rock was fragmented. The powder factor was 0.58 lb/yd^3, testimony to the relatively low powder factors obtainable by coyote blasting.

Combination Coyote Tunnel and Blasthole Blasting As has been mentioned, in order to shear off a well-defined back wall it is common practice to combine coyote work with the use of blastholes. The method is also used to ensure good fragmentation in the prism of rock well above the level of the cross-tees, as this zone is subject to a heaving action rather than a shattering action by the charges in the cross-tees. In a general sense soft to medium rock formations are adapted to coyote-tunnel blasting, whereas hard rock formations are adapted to the combination of coyote tunnels and blastholes. An example of this efficient and economical method in hard rock is illustrated at the end of the chapter.

The principles of blasting are all dedicated to fragmenting the rock satisfactorily for its use. Figure 11-43 shows good fragmentation in a 100,000-ton or 47,000-yd^3 blast of taconite, a hard iron ore. In this blasting the drilling was by jet-piercing drills. The hole was chambered as shown in Figure 11-36. Twenty-seven holes on 31 ft × 31 ft pattern yielded about 35 yd^3/ft of blasthole. AN/FO explosives were used with a powder factor of 1.54 lb/yd^3, or 0.72 lb/ton. Such excellent fragmentation is routine under day-by-day regular drilling and blasting of the same rock in the same location.

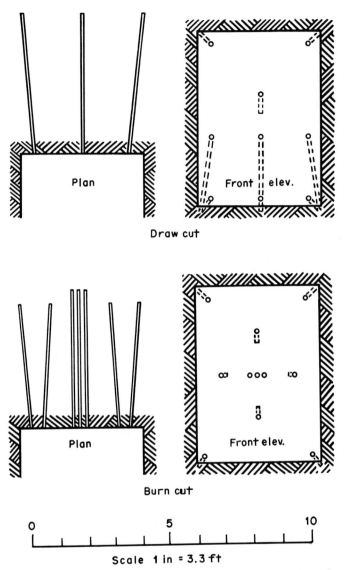

Figure 11-41 Representative patterns of drilling blastholes in tunnels or adits and crosscuts or laterals for the coyote or tunnel method of blasting. Explosives are loaded into the crosscuts or laterals. The crosscuts are backfilled first with set-rock and then with smaller rocks. The charge is usually detonated by detonating or Primacord fuse and primers. *(Ingersoll-Rand Co.)*

11-54   Fragmentation of Rock

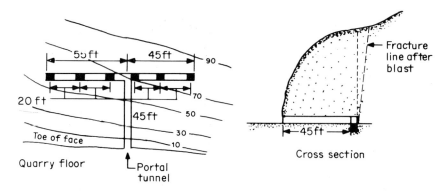

TUNNEL PLAN AT A TEXAS TRAP ROCK QUARRY
Favorable results were obtained in laying down 30,000 tons of rock on the quarry floor with 8,150 pounds of Special No. 1 powder.

The rock lies in partly boulder formation and partly solid. The height of face varies from 50 feet to 85 feet. Eight thousand one hundred and fifty pounds Special No. 1 in $12\frac{1}{2}$ lb. bags and 100 lbs. of 60% Straight Nitroglycerin Dynamite were used. Thirty thousand tons of rock were broken by the blast. The photographs show the formation and result of this shot.

SOLID AND BROKEN
The natural boulder formation of the quarry is shown at the left. On the right is the result of the blast.

Figure 11-42 Coyote tunnel blasting in Texas trap rock quarry. The blasted rock is made up of thin residuals overlying weathered boulders and fragmental rock derived from the parent trap rock. In turn the weathered rock overlies the semisolid and solid country or parent rock. It is an ideal formation for coyote tunnel blasting, as the weathered zone needs only a heaving action rather than an abrupt shattering action for fragmentation. About 14,300 yd^3 of the mixed rocks was blasted by a total of 8250 lb of explosives with a low powder factor of 0.58 lb/yd^3 and good fragmentation. *(Hercules, Inc.)*

Secondary Blasting Secondary blasting, as illustrated in Figure 11-44, is generally insignificant with respect to primary blasting when blastholes are used. However, when the coyote tunnel method is used, good fragmentation by the direct action of the explosive forces is limited because breakage is both by these forces and by gravity. Thus there may be a high percentage, up to 20 percent, of rock requiring secondary blasting.

This excessive secondary blasting is typical of weathered rock containing large weathered boulders and large angular rocks detached from the parent rock. It is also typical of hard rock in the upper prism of the blasted rock where the shattering forces of the coyote blast have attenuated. It is well to reckon with the probability of significant secondary blasting when comparing the overall unit cost for coyote blasting versus that for the use of blastholes.

Loading of Blastholes Loading of explosives in blastholes may be manual or mechanical, depending on the size and uniformity of the work. Manual loading is common in all blasting and mechanical loading is typical of large operations.

Manual Loading. Figure 11-45 illustrates manual work. Holes are wet in this 400,000-yd^3 highway cut of Colorado. The combination of explosives is water gel in the wet lower length of hole, topped by AN/FO in the upper dry portion. As the water level varies in the shale-sandstone formation, the loading of each hole is a little different and manual loading is efficient.

In the foreground the powder operator is dropping a 2 in $\times$ 16 in plastic cartridge of water gel suspended on Primacord. A splash will indicate water level. Thereafter more cartridges are added until the water gel is above water level. AN/FO then will be poured into the hole, as shown in the right background, until the powder factor is attained. From 6 to 8 ft of stemming is added to the 22-ft holes.

Finally the Primacords of all holes are tied into the Primacord trunk line with suitable successive 5-ms delays between rows of holes. The lightning-proof circuit is then ready for detonation by electric blasting cap.

Such painstaking procedures call for manual loading. The four powder operators load

Figure 11-43 Excellent fragmentation for a particular purpose—the goal of all blasting—is illustrated here by the excellent results of a 100,000-ton or 47,000-yd^3 blast of taconite, a hard iron ore, in the Thunderbird mine of Minnesota. A particular fragmentation was desired for both the 6-yd^3 loading shovel in the pit and the 54-in gyratory crusher at the plant. The size of the worker in the upper right corner of the view shows both the size and the uniformity of the blasted taconite. Blasthole spacing was 31 ft by 31 ft for the 50-ft holes and the yield was about 35 yd^3/ft of blasthole. The chambered blastholes were drilled by jet piercing drills of nominal 9.2-in diameter. The powder factor for the AN/FO explosive was 1.54 lb/yd^3 of rock. *(Union Carbide Corp.)*

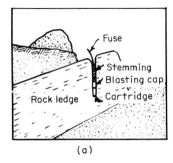
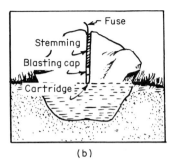

(a) (b)

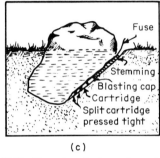
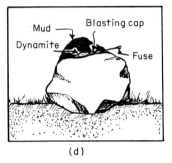

(c) (d)

Figure 11-44 Examples of minor primary blasting and secondary blasting of rock. *(a)* Chip blast. *(b)* Blockhole. *(c)* Snakehole. *(d)* Mudcap.

Often outcroppings of a rock ledge must be blasted as part of the primary blasting in a rock job. After primary blasting there is usually some secondary blasting resulting from incomplete fragmentation by the primary blasting or from the presence of surface boulders or embedded boulders in a matrix of softer rock. Methods for fragmenting these rocks are illustrated in the drawing. Initiation of the explosive, usually dynamite, may be by fuse and cap or by wires and electric cap. Average charges for blasting small ledges and boulders are given in this tabulation.

Thickness of ledge or diameter of boulder, ft	Approximate number of 1¼ in × 8 in cartridge of 40 to 50% dynamite for hard rock			
	Chip blast	Blockhole	Snakehole	Mudcap
1½	¼	¼	1	2
2	¼	¼	1	3
3	½	½	1½	4
4	¾	¾	4	7
5	1	1	6	12

up to 450 holes in 2 days. This is equivalent to loading about 1200 lb of mixed water gel and AN/FO hourly or 300 lb per worker-hour.

Figure 11-46 illustrates manual loading under the same circumstances as those of Figure 11-45, except that electric caps and an electrical circuit are being used instead of Primacord. Production averages about 4800 yd^3 per 8-h shift. Three powder operators are loading 3½-in-diameter, 30-ft-deep holes with 5800 lb of water gel and AN/FO in 8 h. Loading rate is 240 lb per worker-hour.

The average rate for these two examples of manual loading of a mixture of water gel and AN/FO into 3½-in-diameter, average 26-ft-deep holes is 270 lb per labor-hour.

Bulk Loading. Bulk loading of water gel is shown in Figure 11-47. This 20,000-lb-capacity loader places water gel at a preset rate, usually 400 lb/min. Capacities up to 30,000 lb are used.

When reasonable allowances for delays are made, the following schedule should ob-

Figure 11-45 Hand loading blasting agents on highway job. In the wet holes of this sandstone-shale formation of the Colorado Mountains a combination of water gel and AN/FO solves the explosives problem. In the foreground a cartridge of water gel is being dropped into the hole, suspended by a length of Primacord. In the background AN/FO is being poured on top of the water gel in the dry portion of the hole. Primacord is being used both in the hole and in the surface trunk system because of the wet holes and for safety from electrical storms. *(Dave Williams (ed.), "Wet Blasting Made Possible with Small Water Gel Cartridges," Highway and Heavy Construction, April 1976, p. 30.)*

Figure 11-46 Manually loading blasting agents on freeway job. In this example of the use of water gel and AN/FO in wet holes, an electrical circuit with electric caps and primers is the means for detonation. In front of and to the left of the three powder operators are the piles of recently blasted rock. The blasts average about 4800 yd^3, in keeping with a day's production of rock. Here are shown drilling and loading of blastholes, two days' reserve of fragmented rock, and the loading and hauling of the rock. The rock is granite and the 26-ft lift is the last of the 1,327,000-yd^3 cut, of which about 1 million yd^3 required systematic blasting. Location is San Diego County, California.

Figure 11-47 Mechanical loading of blasting-agent water gel. The 20,000-lb-capacity loading truck is pouring water gel into two blastholes at the same time. The combined flow rate is about 400 lb/min and the output of the mechanical loader is about 10,000 lb per working hour while at the drilling site. With allowance for the return trip to the bulk plant for another load after 2 h on the site, the overall production of the loader is in the 6000- to 10,000-lb/h range. *(E. I. du Pont de Nemours & Co.)*

tain for average conditions when loading 6-in-diameter, 30-ft holes on a 12 ft × 15 ft pattern with a powder factor of 1.5 lb/yd^3. Figure 11-47 shows two holes being loaded together.

Pounds of water gel per blasthole	200
Cycle for loading two holes simultaneously, min:	
Loading 200 lb at 400/2 or 200 lb	1.0
Changing holes	0.5
Total	1.5
Loading time per hole with 200 lb	0.8
Holes loaded per 40-min working hour	50
Loading rate hourly, 50 × 200, lb	10,000

The bulk loading truck returns every 2 h to the bulk plant for another load. If this round trip should require 1 h, then the overall production of the truck would be 20,000/3, or 6670 lb per working hour. The 20,000-lb-capacity truck would then handle some 50,000 lb of explosives in 8 h.

There are three types of water-gel bulk loaders. One is a finished product pumper that places water gel in the blastholes. Another is a batch pumper containing a mixing unit that actually batches a tankful of finished product on site. The third is a continuous pumper that carries all ingredients, which are mixed as needed at the blasthole site.

Bulk loading of AN/FO is pictured in Figure 11-48. This 20,000-lb-capacity unit is delivering into the hole by auger. Capacities range from 3 to 12 tons. Discharge rate is from 200 to 600 lb/min, set usually at 400 lb/min. The other delivery system is pneumatic and discharge rates are the same.

Capacities of the AN/FO bulk loader while at the site and for the average working hour are about the same as those for the water-gel loading unit. Prior to bulk loading blastholes with blasting agents such as water gel and AN/FO, the holes must be loaded with initiating and detonating agents.

In the case of water gel these can be Primacord with primers or electric caps with primers. In the case of AN/FO they can be electric caps with primers or, if there is

Figure 11-48 Bulk loading truck augering AN/FO into blast hole. Delivery of mixed fuel ammonium nitrate prills is at a rate of about 400 lb/min during flow. If the blasthole load is 200 lb, the cycle per hole is about 1.0 min. With 40 min/h working efficiency, about 8000 lb of AN/FO can be loaded hourly at the blast hole site. If 1 h is allowed for the round trip back to the bulk plant, 3.5 h are required for handling 20,000 lb of explosives. The overall production rate for the bulk loader is about 6000 lb/h or 100 lb/min. *(Amerind-MacKissic, Inc.)*

danger of electrical disturbances such as lightning, they can be Primacord with primers. These preparations must take place prior to the bulk loading of the holes and the work must be done well in advance of the fast-loading operation. The ends of the Primacord and of the leg wires of the blasting caps must be securely fastened at the collars of the holes so as not to interfere with the work of the powder operators when loading the holes.

To summarize, under the right conditions bulk loading of explosives is considerably cheaper than hand loading. The ideal conditions, provided by large jobs such as freeways, dams, quarries, or open-pit mines, are: one kind of blasting agent is used; the rock structure is uniform, calling for fairly fixed hole diameter, hole depth, and powder factor; and the hole loading program can be well planned. Generally all these conditions do not exist in the average rock excavation job, a circumstance which makes for the popularity of hand loading of blastholes.

Air Noise and Earth Vibration from Blasting

Noise Complaints about blasting are primarily about noise and secondarily about vibration. Nature's weather, flora, and topography and human control of explosive charges determine the level of noise. Humans alone, by control of blasting, are able to minimize vibration.

Weather determines the velocity of sound in different directions and at different altitudes. The variables affecting sound propagation are wind velocity and temperature as a function of altitude; studies of these variables have established that fundamentally favorable conditions are clear to partly cloudy skies, relatively warm daytime temperatures, rapidly changing winds, brief showers, and location to the windward side of blasting. Unfavorable conditions are still air; fog, haze, and smoke; constant temperature; light winds; and location to the lee side of blasting.

These simple factors are taken into consideration when blasting is carried out in areas of high population. Rough topography, such as a hill between blasting and the affected area, reduces noise. Flora in the form of tall trees and shrubs dampens noise.

Of course reduction of the charge in the total quantity or in individual quantities

11-60   Fragmentation of Rock

detonated by delay blasting reduces noise as well as earth vibration. Generally an observance of Nature's favorable conditions and the experience of the powder operator in detonating the charges will lower noise to an acceptable level.

Vibrations Seismic vibrations which are destructive of buildings and other structures are the result of forces originating from blasting and from earthquakes. They have been under intense study for a number of years and seismic instruments are available to measure velocities, amplitudes, and accelerations of these earth vibrations. In a general way these instruments are known as engineering seismographs.

Equations have been developed to correlate potential damage with the kinetics of motion of a particle of earth in the area of possible damage. The intensity of earth vibration is commonly expressed as follows:

1. Energy ratio, ER, is calculated from frequency, F, and amplitude, A, according to the equation:

$$ER = (3.29 \times FA)^2$$

where F = frequency, Hz
A = amplitude, in

2. Particle velocity is measured directly from the velocity reading in inches per second.

Particle velocity and energy ratio are directly proportional. A particle velocity of 1.8 in/s is equal to an energy ratio of 1.0. The direct measurement of particle velocity does not require frequency analysis of complex wave forms, and it is often the desirable method as it can be read directly from calibrated velocity recordings by nontechnical people. An energy ratio of 1.0 or a particle velocity of 1.8 in/s is regarded as allowable or tolerable for structures. Presently most blasting is monitored by engineering seismographs to ensure acceptability of the blasting and to provide records in case of litigation over alleged damages from blasting.

Table 11-7 is suggested as a guide for estimating permissible charges of explosives according to distances from a possibly affected area. It is the guide of the state of New Jersey, and of course it is not applicable to other locations where controls might be more or less stringent. Usually controls are exercised in keeping with the actual on-site problems of vibrations and noises.

Safety in the Use of Explosives

The transportation, storage, and use of explosives are fraught with many dangers. To guard against these dangers rather complete regulations have been promulgated by federal, state, county, and city agencies. In addition, the manufacturers of explosives publish books and pamphlets on safety, which are expressed in practical terms. Finally, the innate common sense of the experienced powder operator is most important in fending off possible disasters.

Transportation of explosives to the work area is usually by common carrier or special transporter of the seller. The explosives are delivered to the magazine for storage on the job, whence they are hauled by the rock excavator's vehicle to the actual site of the blasting work. In all cases the rail car or vehicle is especially designed for the purpose and carries suitable integral signs or placards, such as "Explosives A," "Explosives B," "Oxidizers," and "Dangerous." For the transporters, mechanical excellence and appropriate fire-fighting equipment are mandatory.

The rock excavator is concerned generally only with storage on the job. In the case of a quarry or other permanently located work, the magazines are solidly built structures, usually of concrete or similar construction. In the case of a construction job, such as a canal or dam, the magazines are of a semiportable or portable nature. Such a set of magazines is shown in Figure 11-49. The four magazines can store some 85,000 lb of high explosives, blasting agents, primers, electric blasting caps, and accessories for a huge rock freeway job. Daily consumption on the job is about 8000 lb of explosives.

Magazines contain sometimes not pounds but tons of explosives and they must be located at minimum distances from human-built structures which are appropriate to the amounts of stored explosives. Table 11-8 gives these quantity versus distance relationships in keeping with inhabited buildings, passenger railways, public highways, and other magazines.

TABLE 11-7 Quantities of Explosives Versus Distances for Blasting

Quarry blasting quantity—distance table

Distance, ft, from blast area to nearest structure, neither quarry owned or leased	Maximum quantity of explosives per shot for instantaneous firing or per delay, for delay firing, lb
100	75
200	92
300	116
400	140
500	176
600	209
700	259
800	330
900	403
1000	495
1200	770

Construction blasting quantity—distance table

Maximum quantity of explosives per shot for instantaneous firing or per delay, for delay firing in pounds	Distance from blast area to nearest building or structure in feet (trenching, tunnels, shafts, side hill or through cuts)
See Note 1	Less than 5
¼ lb	6–10
⅕ lb per foot of distance	15–50
¼ lb per foot of distance	60–200
Over 50—See Note 2	

NOTE 1: Total quantity of explosives shall not exceed ¼ pound per shot up to five feet from nearest building or structure.
NOTE 2: Seismic control, to determine the ground constant, shall be required for more than 50 pounds of explosives.
NOTE 3: Quarry blasting includes open-pit blasting such as is done in mining operations. The above two tables are suggested as guides. Quantities may be much greater and quantity versus distance relationships may vary greatly according to topography, weather, and the nearness or farness of people, animals, homes, buildings, structures, and the like. The kind of rock or rock formation is of major importance. Blasting of rock usually calls for strict on-the-site requirements, inspection, and seismic monitoring.
SOURCE: New Jersey Department of Labor and Industry.

TABLE 11-8 American Table of Distances for Storage of Explosives
(as revised and approved by the Institute of Makers of Explosives, September 30, 1955)

Explosives		Distances when storage is barricaded, ft				Explosives		Distances when storage is barricaded, ft			
Pounds over	Pounds not over	Inhabited buildings	Passenger railways	Public highways	Separation of magazines	Pounds over	Pounds not over	Inhabited buildings	Passenger railways	Public highways	Separation of magazines
2	5	70	30	30	6	500	600	340	135	135	31
5	10	90	35	35	8	600	700	355	145	145	32
10	20	110	45	45	10	700	800	375	150	150	33
20	30	125	50	50	11	800	900	390	155	155	35
30	40	140	55	55	12	900	1,000	400	160	160	36
40	50	150	60	60	14	1,000	1,200	425	170	165	39
50	75	170	70	70	15	1,200	1,400	450	180	170	41
75	100	190	75	75	16	1,400	1,600	470	190	175	43
100	125	200	80	80	18	1,600	1,800	490	195	180	44

Fragmentation of Rock

TABLE 11-8 American Table of Distances for Storage of Explosives *(Continued)*
(as revised and approved by the Institute of Makers of Explosives, September 30, 1955)

Explosives		Distances when storage is barricaded, ft				Explosives		Distances when storage is barricaded, ft			
Pounds over	Pounds not over	Inhabited buildings	Passenger railways	Public highways	Separation of magazines	Pounds over	Pounds not over	Inhabited buildings	Passenger railways	Public highways	Separation of magazines
125	150	215	85	85	19	1,800	2,000	505	205	185	45
150	200	235	95	95	21	2,000	2,500	545	220	190	49
200	250	255	105	105	23	2,500	3,000	580	235	195	52
250	300	270	110	110	24	3,000	4,000	635	255	210	58
300	400	295	120	120	27	4,000	5,000	685	275	225	61
400	500	320	130	130	29	5,000	6,000	730	295	235	65
6,000	7,000	770	310	245	68	75,000	80,000	1,695	690	510	165
7,000	8,000	800	320	250	72	80,000	85,000	1,730	705	520	170
8,000	9,000	835	335	255	75	85,000	90,000	1,760	720	530	175
9,000	10,000	865	345	260	78	90,000	95,000	1,790	730	540	180
10,000	12,000	875	370	270	82	95,000	100,000	1,815	745	545	185
12,000	14,000	885	390	275	87	100,000	110,000	1,835	770	550	195
14,000	16,000	900	405	280	90	110,000	120,000	1,855	790	555	205
16,000	18,000	940	420	285	94	120,000	130,000	1,875	810	560	215
18,000	20,000	975	435	290	98	130,000	140,000	1,890	835	565	225
20,000	25,000	1,055	470	315	105	140,000	150,000	1,900	850	570	235
25,000	30,000	1,130	500	340	112	150,000	160,000	1,935	870	580	245
30,000	35,000	1,205	525	360	119	160,000	170,000	1,965	890	590	255
35,000	40,000	1,275	550	380	124	170,000	180,000	1,990	905	600	265
40,000	45,000	1,340	570	400	129	180,000	190,000	2,010	920	605	275
45,000	50,000	1,400	590	420	135	190,000	200,000	2,030	935	610	285
50,000	55,000	1,460	610	440	140	200,000	210,000	2,055	955	620	295
55,000	60,000	1,515	630	455	145	210,000	230,000	2,100	980	635	315
60,000	65,000	1,565	645	470	150	230,000	250,000	2,155	1,010	650	335
65,000	70,000	1,610	660	485	155	250,000	275,000	2,215	1,040	670	360
70,000	75,000	1,655	675	500	160	275,000	300,000	2,275	1,075	690	385

NOTE 1: "Explosives" means any chemical compound, mixture, or device, the primary or common purpose of which is to function by explosion, i.e., with substantially instantaneous release of gas and heat, unless such compound, mixture, or device is otherwise specifically classified by the Interstate Commerce Commission.

NOTE 2: "Magazine" means any building or structure, other than an explosives manufacturing building, used for the permanent storage of explosives.

NOTE 3: "Natural Barricade" means natural features of the ground, such as hills, or timber of sufficient density that the surrounding exposures which require protection cannot be seen from the magazine when the trees are bare of leaves.

NOTE 4: "Artificial Barricade" means an artificial mound or revetted wall of earth of a minimum thickness of three feet.

NOTE 5: "Barricaded" means that a building containing explosives is effectually screened from a magazine, building, railway or highway, either by a natural barricade, or by an artificial barricade of such height that a straight line from the top of any sidewall of the building containing explosives, to the eave line of any magazine, or building, or to a point twelve feet above the center of a railway or highway, will pass through such intervening natural or artificial barricade.

NOTE 6: When a building containing explosives is not barricaded, the distances shown in the table should be doubled.

NOTE 7: "Inhabited Building" means a building regularly occupied in whole or in part as a habitation for human beings or any church, schoolhouse, railroad station, store, or other structures where people are accustomed to assemble, except any building or structure occupied in connection with the manufacture, transportation, storage or use of explosives.

NOTE 8: "Railway" means any steam, electric, or other railroad or railway which carries passengers for hire.

NOTE 9: "Highway" means any public street or public road.

NOTE 10: When two or more storage magazines are located on the same property, each magazine must comply with the minimum distances specified from inhabited buildings, railways, and highways, and in addition they should be separated from each other by not less than the distances shown for "Separation of Magazines," except that the quantity of explosives contained in cap magazines shall govern in regard to the spacing of said cap magazines from magazines containing other explosives. If any two or more magazines are separated from each other by less than the specified "Separation of Magazines" distances, then such two or more magazines, as a group, must be considered as one magazine, and the total quantity of explosives stored in such group must be treated as if stored in a single magazine located on the site of any magazine of the group, and must comply with the minimum distances specified from other magazines, inhabited buildings, railways and highways.

NOTE 11: The Institute of Makers of Explosives does not approve the permanent storage of more than 300,000 pounds of commercial explosives in one magazine or in a group of magazines which is considered as one magazine.

NOTE 12: This table applies only to the manufacture and permanent storage of commercial explosives. It is not applicable to transportation of explosives, or any handling or temporary storage necessary or incident thereto. It is not intended to apply to bombs, projectiles or other heavily encased explosives.

NOTE 13: All types of blasting caps in strengths through No. 8 cap should be rated at 1½ lb of explosives per 1000 caps. For strengths higher than No. 8 cap, consult the manufacturer.

NOTE 14: For quantity and distance purposes detonating fuse up to 60 grains per foot should be calculated as equivalent to nine pounds of high explosives per 1000 ft. Heavier case loads should be rated proportionately.

SOURCE: E. I. duPont de Nemours & Co., *Blasters' Handbook*, Wilmington, Del., 1966.

Figure 11-49 A naturally safe setting for powder magazines. The capacity of the four magazines is about 40 tons of high explosives, blasting agents, primers, detonating fuse, electric blasting caps, and accessories. The location is in an abandoned quarry of San Diego County, California. Headquarters for the round-the-clock guards is 100 yd distant. The following magazines, left to right, contain different explosives, as listed:

1. Semitrailer for water-gel blasting agent, storage of about 18,000 lb.
2. Heavy steel skid-mounted magazines for dynamites, primers or boosters, and detonating fuse or Primacord, storage of about 4000 lb.
3. Semitrailer for AN/FO blasting agent, storage of about 18,000 lb.
4. Heavy steel skid-mounted magazine for electric blasting caps and nonexplosive accessories.

Delivery to the magazines is by the seller's highway vehicle of special construction. About 8000 lb of explosives is used daily in the granite rock excavation for the nearby freeway.

Careless use of explosives accounts for most accidents. Time-tested dos and don'ts offer the soundest advice covering use, as well as transportation, storage, and disposal of all explosives and accessories. Table 11-9 is an inclusive, thought-provoking list of such dos and don'ts for safety.

TABLE 11-9 Dos and Don'ts
Adopted by The Institute of Makers of Explosives, February 1, 1964

Definitions

    1. The term "explosives" as used herein includes any or all of the following: dynamite, black blasting powder, pellet powder, blasting caps, electric blasting caps and detonating fuse.

    2. The term "electric blasting cap" as used herein includes both instantaneous electric blasting caps and all types of delay electric blasting caps.

    3. The term "primer" as used herein means a cartridge of explosives in combination with a blasting cap or an electric blasting cap.

When Transporting Explosives

    1. DO obey all federal, state and local laws and regulations.

    2. DO see that any vehicle used to transport explosives is in proper working condition and equipped with a tight wooden or nonsparking metal floor with sides and ends high enough to prevent the explosives from falling off. The load in an open-bodied truck should be covered with a waterproof and fire-resistant tarpaulin, and the explosives should not be allowed to contact any source of heat such as an exhaust pipe. Wiring should be fully insulated so as to prevent short circuiting, and at least two fire extinguishers should be carried. The truck should be plainly marked so as to give adequate warning to the public of the nature of the cargo.

    3. DON'T permit metal, except approved metal truck bodies, to contact cases of explosives. Metal, flammable, or corrosive substances should not be transported with explosives.

    4. DON'T allow smoking or unauthorized or unnecessary persons in the vehicle.

    5. DO load and unload explosives carefully. Never throw explosives from the truck.

    6. DO see that other explosives, including detonating fuse, are separated from blasting caps and/or electric blasting caps where it is permitted to transport them in the same vehicle.

    7. DON'T drive trucks containing explosives through cities, towns or villages, or park them near such places as restaurants, garages and filling stations, unless it cannot be avoided.

    8. DO request that explosive deliveries be made at the magazine or in some other location well removed from populated areas.

    9. DON'T fight fires after they have come in contact with explosives. Remove all personnel to a safe location and guard the area against intruders.

When Storing Explosives

    10. DO store explosives in accordance with federal, state or local laws and regulations.

    11. DO store explosives only in a magazine which is clean, dry, well ventilated, reasonably cool, properly located, substantially constructed, bullet and fire resistant and securely locked.

    12. DON'T store blasting caps or electric blasting caps in the same box, container or magazine with other explosives.

    13. DON'T store explosives, fuse, or fuse lighters in a wet or damp place, or near oil, gasoline, cleaning solution or solvents, or near radiators, steam pipes, exhaust pipes, stoves, or other sources of heat.

    14. DON'T store any sparking metal, or sparking metal tools in an explosives magazine.

    15. DON'T smoke or have matches, or have any source of fire or flame in or near an explosives magazine.

    16. DON'T allow leaves, grass, brush, or debris to accumulate within 25 ft of an explosives magazine.

    17. DON'T shoot into explosives or allow the discharge of firearms in the vicinity of an explosives magazine.

    18. DO consult the manufacturer if nitroglycerin from deteriorated explosives has leaked onto the floor of a magazine. The floor should be desensitized by washing thoroughly with an agent approved for that purpose.

    19. DO locate explosives magazines in the most isolated places available. They should be separated from each other, and from inhabited buildings, highways, and railroads, by distances not less than those recommended in the American Table of Distances.

When Using Explosives

    20. DON'T use sparking metal tools to open kegs or wooden cases of explosives. Metallic slitters may be used for opening fiberboard cases, provided that the metallic slitter does not come in contact with the metallic fasteners of the case.

    21. DON'T smoke or have matches, or any source of fire or flame, within 100 ft of an area in which explosives are being handled or used.

    22. DON'T place explosives where they may be exposed to flame, excessive heat, sparks, or impact.

    23. DO replace or close the cover of explosives cases or packages after using.

    24. DON'T carry explosives in the pockets of your clothing or elsewhere on your person.

    25. DON'T insert anything but fuse in the open end of a blasting cap.

    26. DON'T strike, tamper with, or attempt to remove or investigate the contents of a blasting cap or an electric blasting cap, or try to pull the wires out of an electric blasting cap.

    27. DON'T allow children or unauthorized or unnecessary persons to be present where explosives are being handled or used.

28. DON'T handle, use, or be near explosives during the approach or progress of any electrical storm. All persons should retire to a place of safety.

29. DON'T use explosives or accessory equipment that are obviously deteriorated or damaged.

30. DON'T attempt to reclaim or to use fuse, blasting caps, electric blasting caps, or any explosives that have been water soaked, even if they have dried out. Consult the manufacturer.

When Preparing the Primer

31. DON'T make up primers in a magazine, or near excessive quantities of explosives, or in excess of immediate needs.

32. DON'T force a blasting cap or an electric blasting cap into dynamite. Insert the cap into a hole made in the dynamite with a punch suitable for the purpose.

33. DO make up primers in accordance with proven and established methods. Make sure that the cap shell is completely encased in the dynamite or booster and so secured that in loading no tension will be placed on the wires or fuse at the point of entry into the cap. When side priming a heavy wall or a heavyweight cartridge, wrap adhesive tape around the hole punched in the cartridge so that the cap cannot come out.

When Drilling and Loading

34. DO comply with applicable federal, state and local regulations relative to drilling and loading.

35. DO carefully examine the surface or face before drilling to determine the possible presence of unfired explosives. Never drill into explosives.

36. DO check the borehole carefully with a wooden tamping pole or a measuring tape to determine its condition before loading.

37. DO recognize the possibility of static electrical hazards from pneumatic loading and take adequate precautionary measures. If any doubt exists, consult your explosives supplier.

38. DON'T stack surplus explosives near working areas during loading.

39. DO cut from the spool the line of detonating fuse extending into a borehole before loading the remainder of the charge.

40. DON'T load a borehole with explosives after springing (enlarging the hole with explosives) or upon completion of drilling without making certain that it is cool and that it does not contain any hot metal, or burning or smoldering material. Temperatures in excess of 150° F are dangerous.

41. DON'T spring a borehole near another hole loaded with explosives.

42. DON'T force explosives into a borehole or through an obstruction in a borehole. Any such practice is particularly hazardous in dry holes and when the charge is primed.

43. DON'T slit, drop, deform or abuse the primer. DON'T drop a large size, heavy cartridge directly on the primer.

44. DO avoid placing any unnecessary part of the body over the borehole during loading.

45. DON'T load any boreholes near electric power lines unless the firing line, including the electric blasting cap wires, is so short that it cannot reach the power wires.

46. DON'T connect blasting caps, or electric blasting caps to detonating fuse except by methods recommended by the manufacturer.

When Tamping

47. DON'T tamp dynamite that has been removed from the cartridge.

48. DON'T tamp with metallic devices of any kind, including the metal end of loading poles. Use wooden tamping tools with no exposed metal parts, except nonsparking metal connectors for jointed poles. Avoid violent tamping. Never tamp the primer.

49. DO confine the explosives in the borehole with sand, earth, clay, or other suitable incombustible stemming material.

50. DON'T kink or injure fuse, or electric blasting cap wires, when tamping.

When Shooting Electrically

51. DON'T uncoil the wires or use electric blasting caps during dust storms or near any other source of large charges of static electricity.

52. DON'T uncoil the wires or use electric blasting caps in the vicinity of radio frequency transmitters, except at safe distances. Consult the manufacturer or The Institute of Makers of Explosives pamphlet on "Radio Frequency Hazards."

53. DO keep the firing circuit completely insulated from the ground or other conductors such as bare wires, rails, pipes, or other paths of stray currents.

54. DON'T have electric wires or cables of any kind near electric blasting caps or other explosives except at the time and for the purpose of firing the blast.

55. DO test all electric blasting caps, either singly or when connected in a series circuit, using only a blasting galvanometer specifically designed for the purpose.

56. DON'T use in the same circuit either electric blasting caps made by more than one manufacturer, or electric blasting caps of different style or function even if made by the same manufacturer, unless such use is approved by the manufacturer.

57. DON'T attempt to fire a single electric blasting cap or a circuit of electric blasting caps with less than the minimum current specified by the manufacturer.

58. DO be sure that all wire ends to be connected are bright and clean.

TABLE 11-9 Dos and Don'ts *(Continued)*
Adopted by The Institute of Makers of Explosives, February 1, 1964

59. DO keep the electric cap wires or leading wires short circuited until ready to fire.

When Shooting with Fuse

60. DO handle fuse carefully to avoid damaging the covering. In cold weather warm slightly before using to avoid cracking the waterproofing.

61. DON'T use a short fuse. Know the burning speed of the fuse and make sure you have time to reach a place of safety after lighting. Never use less than 2 ft.

62. DON'T cut fuse until you are ready to insert it into a blasting cap. Cut off an inch or two to insure a dry end. Cut fuse squarely across with a clean sharp blade. Seat the fuse lightly against the cap charge and avoid twisting after it is in place.

63. DON'T crimp blasting caps by any means except a cap crimper designed for the purpose. Make certain that the cap is securely crimped to the fuse.

64. DO light fuse with a fuse lighter designed for the purpose. If a match is used the fuse should be slit at the end and the match head held in the slit against the powder core. Then scratch the match head with an abrasive surface to light the fuse.

65. DON'T light fuse until sufficient stemming has been placed over the explosive to prevent sparks or flying match heads from coming into contact with the explosive.

66. DON'T hold explosives in the hands when lighting fuse.

In Underground Work

67. DO use permissible explosives only in the manner specified by the United States Bureau of Mines.

68. DON'T take excessive quantities of explosives into a mine at any one time.

69. DON'T use black blasting powder or pellet powder with permissible explosives or other dynamite in the same borehole in a coal mine.

Before and After Firing

70. DON'T fire a blast without a positive signal from the one in charge, who has made certain that all surplus explosives are in a safe place, all persons and vehicles are at a safe distance or under sufficient cover and that adequate warning has been given.

71. DON'T return to the area of any blast until the smoke and fumes from the blast have been dissipated.

72. DON'T attempt to investigate a misfire too soon. Follow recognized rules and regulations, or if no rules or regulations are in effect, wait at least one hour.

73. DON'T drill, bore, or pick out a charge of explosives that has misfired. Misfires should be handled only by or under the direction of a competent and experienced person.

Explosives Disposal

74. DON'T abandon any explosives.

75. DO dispose of or destroy explosives in strict accordance with approved methods. Consult the manufacturer or follow The Institute of Makers of Explosives pamphlet on destroying explosives.

76. DON'T leave explosives, empty cartridges, boxes, liners, or other materials used in the packing of explosives lying around where children or unauthorized persons or livestock can get at them.

77. DON'T allow any wood, paper, or any other materials employed in packing explosives to be burned in a stove, a fireplace, or other confined space, or to be used for any purpose. Such materials should be destroyed by burning at an isolated location out of doors and no person should be nearer than 100 ft after the burning has started.

Blasthole Drills and Drilling

Drilling will be discussed in terms of the methods and machinery used when primary blasting is done by use of blastholes and coyote tunnels and when secondary blasting is done by blastholes. The text deals with kinds of drills and their auxiliary machinery and equipment. Almost all the drills have been discussed in Chapter 7, as they are used in both exploratory and blasthole drilling. It is well to review the exploratory drilling section of Chapter 7, as the critique presented there will serve as a background.

Based on principles of drilling there are three kinds of drills. They are:

1. *Fusion or jet-piercing drills* are unique drills which burn out the blasthole by spalling the rock with an intense flame, produced by combining fuel oil and compressed oxygen in a downhole burner with jets. Its use is in hard, spallable rocks such as basalt, dolomite, granite, quartzite, sandstone, syenite, and taconite. The spalling action may be complemented by rotary drilling to permit drilling rocks with poor spallability; this in effect creates a thermomechanical drill.

2. *Percussive drills* are of two types: the cable drill and the air-hammer drill. In the former, the percussive force comes from repeated blows from a free-falling string of drilling tools which mounts a heavy cutting bit. The action of gravity shatters and pulverizes the rock. In the air-hammer drill, the percussive force comes from the high-frequency blows of a compressed-air hammer, which are delivered either to a drill-rod assembly that mounts the cutting bit or directly to the cutting bit as in the case of the down-the-hole drill. The rock is shattered and pulverized. Air-hammer drills may be subdivided into hand-held jackhammers, manually and mechanically controlled drifters, mechanically controlled large hammers mounted on the mast or guide of a self-contained mobile drill, and mechanically controlled down-the-hole large hammers. Cable drills are adapted to soft, medium, and hard rocks. Their bit is illustrated in Figure 7-24. Air-hammer drills are suited to medium and hard rocks. Their bits are shown in Figure 7-12.

3. *In a rotary drill,* a rotating drill stem mounts two kinds of cutter-head. One is either an auger or a drag bit of the fishtail or finger type, as shown in Figure 7-9. These bits are adapted to soft and medium rocks and they depend on cutting and abrasion for drilling. The other is the roller bit, made up of two or more cones, as shown in Figure 7-15. This bit has both cutting and crushing actions and it is adapted to soft and medium and sometimes to hard rocks.

For drilling in small quantities and to shallow depths in primary blasting, jackhammers are generally used. For drilling performed in the driving of coyote tunnels, jackhammers with jack legs and drifters with automatic feeds and supports are employed.

Compressed Air and Its Uses in Drilling Aside from the cable drill and, to a limited extent, the fusion or jet-piercing drill, all blasthole drills require compressed air for their operation. In many machines it is used for the following many purposes:

1. Actuating the air hammer for both hammering and rotating the bit, and raising and lowering the air hammer

2. Feeding the drifter into the rock face, as in coyote tunnel drilling

3. Continually removing the cuttings from the bottom of the blasthole and intermittently blowing out the hole

4. Actuating motors for traveling the drilling machine

5. Actuating motors for driving various drill assemblies, such as hydraulic pumps

These functions require air compressors over a wide range of capacities at different pressures. Additionally, accessories such as pipe, hoses, and valves and fittings are needed.

Compressors. Compressors used in rock excavation are in the 125- to 5000-ft^3/min range of ratings or capacities. The standard rating in cubic feet per minute is based on the quantity of free air entering the compressor at sea level, 14.7 lb/in^2, and 60°F which it takes in at inlet and compresses to a stated pressure. The stated pressure varies from 100 to 250 lb/in^2, each pressure being for a particular purpose. With outlet pressure of 100 lb/in^2, the working pressure at the hammer in a well-designed distribution system should be in the 90 to 100 lb/in^2 range. The pressure loss is due to friction of air flow, leaks, and cooling of the compressed air.

The modern compressor is of the rotary or reciprocating kind. The reciprocating, or piston, kind is the older and it is being replaced by the rotary kind, which uses vanes, lobes, or screws which revolve within a chamber to compress and transmit the flowing air. The compressor may be single-, two-, or multistage. Single- and two-stage units are used for blasthole drilling. Figures 11-50 and 11-51 show the parts of a typical stationary two-stage rotary screw compressor.

Besides the compressor with its power unit there are the following important auxiliaries, some or all of which may be part of the complete assembly.

1. *Receiver.* This pressure tank receives the compressed air at the desired outlet pressure. It provides storage, equalizes compressed air surges, and removes condensed water and lubricating oil.

2. *Intercooler.* The intercooler is placed between the two stages of a reciprocating compressor to reduce the temperature of the air and to remove moisture from it. Water is the coolant. In the single-stage or two-stage rotating compressor, oil is the coolant. The cooling effect reduces power consumption.

3. *Aftercooler.* This unit is installed between the compressor outlet and the receiver for like reasons. Moisture removal at this point is most important, as moisture freezes during expansion and hinders lubrication of the air hammer.

11–68   Fragmentation of Rock

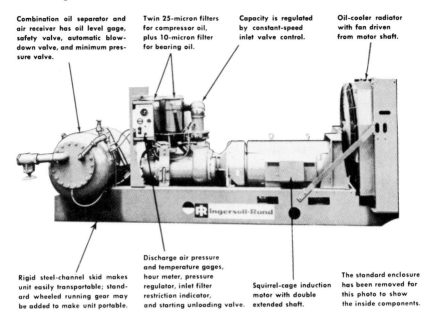

Model	Motor hp	Actual free-air delivery @ 125 psi	Length	Width	Height	Shipping weight
EXL-900	250	900 cfm (25.5 m³ min.)	12'7" (384 cm.)	71" (180 cm.)	63" (160 cm.)	7378 lb. (3347 kg.)
EXL-1200	300	1200 cfm (34 m³ min.)	12'6" (381 cm.)	80" (203 cm.)	71" (180 cm.)	8442 lb. (3829 kg.)

NOTE: Motors normally supplied are 460-3-60; all standard voltage ratings are available on request. Dimensions and weights shown are approximate.

Figure 11-50 Assemblies of a 1200-ft³/min, 125-lb/in², two-stage rotating screw compressor. This stationary compressor with its basic steel frame and standard enclosure, which has been removed for illustration, may be converted readily to a two-axle wheels-tires-equipped portable unit. *(Ingersoll-Rand Co.)*

Compressors may be portable or stationary. Portable ones are characteristic of construction work and both kinds are typical of industrial workings like quarries and open pits. They may be powered by steam engines or turbines, diesel or gasoline engines, or electric motors. Modern usage is confined mostly to diesel engines and electric motors, the selection depending on availability and economy of electrical energy as opposed to diesel fuel. Again one finds diesel engines to be popular in construction and both diesel engines and electric motors to be favored in industrial work.

Portable compressors driven by gasoline and by diesel engines and ranging from 100- to 1600-ft³/min capacity are shown in Figure 11-52. They are of both the single- and the two-axle type. Specifications, ratings, and engine data for the complete line of machines are given in Table 11-10.

A stationary air compressor driven by an electric motor is illustrated in Figure 11-53. Specifications, ratings, and motor data for the manufacturer's complete line of machines are given in Table 11-11.

Air Friction Losses in Pipes, Hoses, and Fittings. The transmission of compressed air from compressor to air hammer is by hose or combination of pipe and hose. Friction losses may be significant, and these should be estimated in all cases except in the shortest of lines. Table 11-12 is for friction losses of air in both pipes and hoses. Table 11-13 is for friction losses of air in pipe and hose fittings in terms of the equivalent length of pipe. Table 11-14 gives recommended pipe sizes for transmission of different quantities of air over different lengths of pipe. Together the three tables are ade-

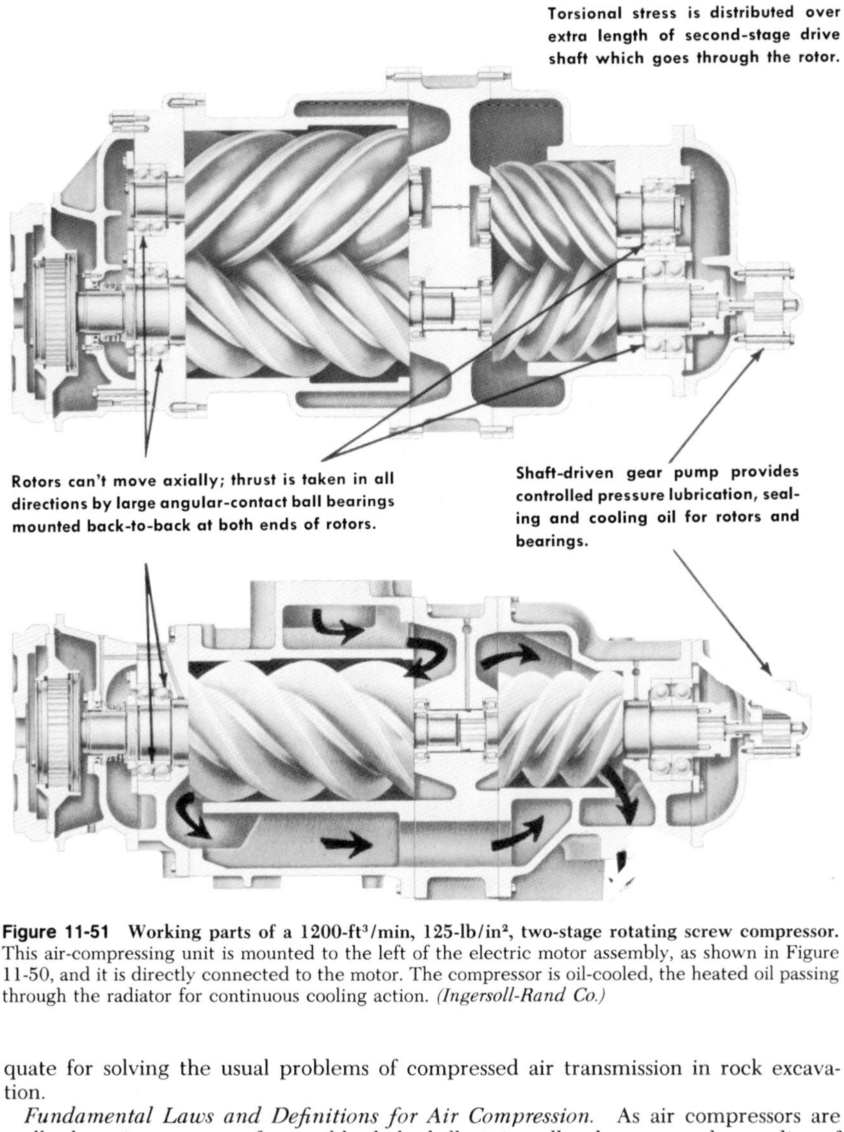

Torsional stress is distributed over extra length of second-stage drive shaft which goes through the rotor.

Rotors can't move axially; thrust is taken in all directions by large angular-contact ball bearings mounted back-to-back at both ends of rotors.

Shaft-driven gear pump provides controlled pressure lubrication, sealing and cooling oil for rotors and bearings.

Figure 11-51 Working parts of a 1200-ft³/min, 125-lb/in², two-stage rotating screw compressor. This air-compressing unit is mounted to the left of the electric motor assembly, as shown in Figure 11-50, and it is directly connected to the motor. The compressor is oil-cooled, the heated oil passing through the radiator for continuous cooling action. *(Ingersoll-Rand Co.)*

quate for solving the usual problems of compressed air transmission in rock excavation.

Fundamental Laws and Definitions for Air Compression. As air compressors are really the prime movers of many blasthole drills, it is well to have an understanding of the fundamental laws and definitions pertaining to air compression. These are enumerated and explained in this glossary.

Absolute pressure Pressure lb/in², measured from absolute zero or sum of gauge and atmospheric pressures. Used in all theoretical and empirical equations involving absolute pressure, volume, and absolute temperature.

Absolute (Fahrenheit) temperature Temperature measured from absolute zero, equal to sum of air temperature, °F, and 460°F. Likewise used in all theoretical and empirical equations involving absolute pressure, volume, and absolute temperature.

Adiabatic compression or expansion Change in volume and pressure of air in which no heat is transferred to or from the air from any outside source.

11-70 Fragmentation of Rock

TABLE 11-10 Specifications for Gasoline- and Diesel-Engine-Driven Portable Air Compressors

Compressor Specifications	Airvane	Airscrew
Type	Oil-flooded sliding vane	Oil-flooded rotary screw
Operating speed range, r/min	1250–2150	1250–2300
Operating temperature range	−20 to +125°F (−29 to +52°C)	
Air service connections	Two ¾ in	
Type of control	0% to 100% demand	
Type oil filter	Full-flow replaceable element	
Type intake air filter	Single stage dry (common with engine) (Two stage dry, RPQ-185)	
Mountings available	Wheel, utility, skid	

Compressor Ratings	Airvane		Airscrew	
	RPV-100	RPV-150	RPS-185	
	RPQ-100	RPQ-150	RPQ-185	
Rated delivery—ft³/min (m³/min)	100 (2.8)	150 (4.2)	185 (5.2)	
Rated operating pressure, lb/in² gage (kg/cm²)	100 (7)	100 (7)	100 (7)	
Operating pressure range, lb/in² gage (kg/cm²)	70–125 (4.9–8.8)	70–125 (4.9–8.8)	70–125 (4.9–8.8)	
Engine Specifications	RPV-100G RPQ-100G	RPV-150G RPV-150D RPQ-150G RPQ-150D	RPS-185G RPQ-185G	RPS-185D RPQ-185D
Type	Gasoline	Gasoline Diesel	Gasoline	Diesel
Make and model	I-H UC-153	I-H UC-200 GM Bedford 220	I-H UC-200	GM Bedford 220
No. of cylinders	4	4 4	4	4
Rated horsepower	49	68 63	68	72

Compressor Specifications	RP-250	RP-365
Type	Oil-flooded sliding vane	
Operating speed range, r/min	900–1800 r/min	
Operating temperature range	−20 to +125°F (−29 to +52°C)	
Air service connections	Two ¾ in	Two 1¼ in
Type of control	0 to 100% demand	
Type oil filter	Full-flow replaceable element	
Type intake air filter	Single stage dry (common with engine)	
Mountings available	Wheel or skid	
Compressor Ratings	RP-250	RP-365
Rated delivery—ft³/min (m³/min.)	250 (7)	365 (10.2)
Rated operating pressure— lb/in² gage	100 (7)	100 (7)
Operating pressure range— lb/in² gage (kg/cm²)	65–125 (4.5–8.8)	65–125 (4.5–8.8)
Engine Specifications	RP-250	RP-365
Type	Diesel	Diesel
Make and model	Hercules D3400	GM4-71
Number of cylinders	6	6
Rated horsepower	90	122

Compressor Specifications	RPS-650 RPQ-650	RPS-800 RPQ-800	RPS-850	RPS-900 RPQ-900	RPS-1050 RPQ-1050	RPS-1200 RPQ-1200	RPS-1600 RPQ-1600
Type			Single-stage, direct drive, oil-flooded, rotary screw				
Operating speed range, r/min	900–1800	1100–2200	900–2100	900–1700	900–1850	1100–1800	900–1900
Operating temperature range:			−20° to +125°F. (−29° to +52°C)				
RPS machines							
RPQ machines			−20° to +120°F. (−20° to +49°C)				
Air service connections	One 2 in or 2½ in	One 2½ in	One 2½ in	One 2½ in	One 2½ in	One 3 in	One 3 in
	One ¾ in	One ¾ in	One ¾ in	One ¾ in	One ¾ in	One ¾ in	One ¾ in
Type of control			Automatic zero to 100% demand				
Type oil filter			Full flow replaceable element with pressure by-pass				
Type intake air filters			Two-stage dry				
Compressor Ratings							
Rated delivery (m³/min) ft³/min	650 (18)	800 (22.4)	850 (23.8)	900 (25.2)	1050 (29.4)	1200 (33.6)	1600 (42.4)
Rated operating pressure, psig (kg/cm²)	100 (7.03)	100 (7.03)	100 (7.03)	100 (7.03)	100 (7.03)	100 (7.03)	100 (7.03)
Operating pressure range, lb/in² gage (kg/cm²)	70–125 (4.9–8.8)	70–125 (4.9–8.8)	70–125 (4.9–8.8)	70–125 (4.9–8.8)	70–125 (4.9–8.8)	70–125 (4.9–8.8)	70–125 (4.9–8.8)
Engine Specifications							
Make and model (diesel)	GM 6V-71N	CAT D33T	GM 6V-71N	GM 8V-71N	GM 8V-71N	CAT D34T3A	GM 12V-71N
Number of cylinders	6	6	6	8	8	6	12
Rated horsepower	205	250	228	277	277	380	425

SOURCE: Joy Manufacturing Co.

Figure 11-52 Representative portable air compressors from 100 to 1600 ft³/min with diesel and gasoline engines. Portable compressors powered by diesel engines range up to 2500 ft³/min at 150 lb/in². When driven by electric motors their range is approximately the same. In both kinds of power two units may be combined to form one integral machine of 5000 ft³/min at 150-lb/in² capacity. *(Joy Manufacturing Co.)*

Figure 11-53 Stationary 260-ft³/min, 125-lb/in² air compressor with 60-hp electric motor. This compact, rotary, screw-type air compressor is one of the 18 stationary units of the stationary rotary screw compressor type built by the manufacturer, as outlined in the specifications of Table 11-11. They range from 180 to 1535 ft³/min and from 100 to 125 lb/in². *(Joy Manufacturing Co.)*

TABLE 11-11 Specifications for Electric-Motor-Driven Stationary Air Compressors

Compressor Specifications

Type:	Oil-flooded rotary screw, nonsymmetrical rotor design
Rated speed:	1750 r/min
Type of control:	Modulating demand, pneumatically operated inlet control valve
Air service connections:	1½ in (RCS-180 through RCS-275)
	2½ in (RCS-335 through RCS-470)
	3 in (RCS-565 through RCS-720)
Cooling system:	Oil to air (standard)
	Water to oil (optional)

Compressor Ratings

Model	Rated Delivery cfm	Rated Delivery m³/min.	Rated Op. Pres. psig	Rated Op. Pres. kg/cm²	Max. Pres. psig	Max. Pres. kg/cm²	Motor HP	Weight lbs.	Weight kg
RCS-180	180	5.1	115	8.0	125	8.7	40	2200	999
RCS-195	195	5.4	100	7.0	110	7.7	40	2200	999
RCS-220	220	6.2	115	8.0	125	8.7	50	2300	1044
RCS-235	235	6.5	100	7.0	110	7.7	50	2300	1044
RCS-260	260	7.3	115	8.0	125	8.7	60	2400	1088
RCS-275	275	7.7	100	7.0	110	7.7	60	2400	1088
RCS-335	335	9.3	115	8.0	125	8.7	75	4200	1905
RCS-360	360	10.2	100	7.0	110	7.7	75	4200	1905
RCS-435	435	12.2	115	8.0	125	8.7	100	4300	1950
RCS-470	470	13.3	100	7.0	110	7.7	100	4300	1950
RCS-565	565	15.8	115	8.0	125	8.7	125	6200	2812
RCS-610	610	17.2	100	7.0	110	7.7	125	6200	2812
RCS-665	665	18.6	115	8.0	125	8.7	150	6600	2993
RCS-720	720	20.4	100	7.0	110	7.7	150	6600	2993
RCS-925	925	25.9	115	8.0	125	8.7	200	7500	3405
RCS-970	970	27.2	100	7.0	110	7.7	200	7500	3405
RCS-1170	1170	32.8	100	7.0	110	7.7	250	8400	3815
RCS-1535	1535	43.0	100	7.0	110	7.7	300	9400	4268

NOTE: Voltages for all models are 460/575 V or 200/230 V.
SOURCE: Joy Manufacturing Co.

TABLE 11-12 Friction Losses of Air in Pipes and Hoses, Expressed in Loss of Pressure

Although these tables are prepared for losses in pipes, they may be used safely for losses in hoses. The losses in pipes and hoses are approximately equal.

(pressure loss, lb/in^2, in 100 ft of pipe for air at 60, 80, 100, and 125 lb/in^2 gauge)

Free air, ft^3/min	Equivalent ft^3/min of compressed air				Nominal pipe diameters								
					½ in				¾ in				
	60 lb	80 lb	100 lb	125 lb	60 lb	80 lb	100 lb	125 lb	60 lb	80 lb	100 lb	125 lb	
10	1.97	1.55	1.28	1.05	.59	.46	.38	.31	.14	.11	.09	.08	
20	3.94	3.10	2.56	2.10	2.23	1.74	1.42	1.17	.53	.41	.34	.28	
30	5.90	4.66	3.84	3.16	4.94	3.84	3.13	2.54	1.14	.90	.74	.60	
40	7.87	6.21	5.13	4.21	8.90	6.93	5.55	4.53	1.99	1.55	1.28	1.05	
50	9.84	7.76	6.41	5.26	14.2	10.7	8.65	7.01	3.08	2.42	2.00	1.62	
60	11.81	9.31	7.69	6.31	…	…	…	…	4.45	3.47	2.84	2.33	
70	13.78	10.87	8.97	7.37	…	…	…	…	6.06	4.73	3.85	3.14	
80	15.74	12.42	10.25	8.42	…	…	…	…	7.96	6.14	5.01	4.08	
90	17.71	13.97	11.53	9.47	…	…	…	…	10.00	7.75	6.40	5.17	
100	19.68	15.5	12.82	10.52	…	…	…	…	12.60	9.62	7.80	6.33	
125	24.60	19.4	16.02	13.15	…	…	…	…	21.0	15.5	12.4	9.8	
150	29.51	23.3	19.22	15.78	…	…	…	…	31.5	23.0	18.1	14.4	

| Free air, ft³/min | Equivalent ft³/min of compressed air ||||| Nominal pipe diameters |||||||||||
| | 60 lb | 80 lb | 100 lb | 125 lb | | 1 in |||| 1¼ in ||||
					60 lb	80 lb	100 lb	125 lb	60 lb	80 lb	100 lb	125 lb
10	1.97	1.55	1.28	1.05	0.05	0.04	0.03	0.02	0.011	0.0086	0.0071	0.0058
20	3.94	3.10	2.56	2.10	0.16	0.13	0.10	0.08	0.040	0.032	0.026	0.021
30	5.90	4.66	3.84	3.16	0.34	0.28	0.23	0.19	0.086	0.068	0.056	0.046
40	7.87	6.21	5.13	4.21	0.59	0.46	0.38	0.31	0.146	0.116	0.096	0.079
50	9.84	7.76	6.41	5.26	0.92	0.73	0.60	0.49	0.22	0.18	0.146	0.120
60	11.81	9.31	7.69	6.31	1.30	1.02	0.84	0.69	0.32	0.25	0.21	0.17
70	13.78	10.87	8.97	7.37	1.75	1.36	1.12	0.92	0.42	0.34	0.28	0.23
80	15.74	12.42	10.25	8.42	2.24	1.76	1.44	1.18	0.55	0.44	0.36	0.30
90	17.71	13.97	11.53	9.47	2.88	2.23	1.85	1.49	0.69	0.55	0.45	0.37
100	19.68	15.5	12.82	10.52	3.45	2.69	2.21	1.81	0.84	0.66	0.55	0.45
125	24.60	19.4	16.02	13.15	5.38	4.18	3.41	2.79	1.31	1.03	0.85	0.69
150	29.51	23.3	19.22	15.78	7.81	5.75	4.91	3.99	1.87	1.47	1.20	0.99
175	34.44	27.2	22.43	18.41	10.8	8.10	6.80	5.45	2.58	2.00	1.64	1.32
200	39.36	31.0	25.63	21.05	14.5	10.9	8.79	7.11	3.31	2.58	2.12	1.73
250	49.20	38.8	32.04	26.31	…	…	…	…	5.30	4.05	3.30	2.67
300	59.0	46.6	38.45	31.57	…	…	…	…	7.51	5.78	4.71	3.83
350	68.9	45.3	44.86	36.83	…	…	…	…	10.3	7.90	6.45	5.15
400	78.7	62.1	51.26	42.09	…	…	…	…	13.7	10.3	8.30	6.74

TABLE 11-12 Friction Losses of Air in Pipes and Hoses, Expressed in Loss of Pressure *(Continued)*
Although these tables are prepared for losses in pipes, they may be used safely for losses in hoses. The losses in pipes and hoses are approximately equal.
(pressure loss, lb/in² in 100 ft of pipe for air at 60, 80, 100, and 125 lb/in² gauge)

Free air, ft³/min	Equivalent ft³/min of compressed air				Nominal pipe diameters							
					1½ in				2 in			
	60 lb	80 lb	100 lb	125 lb	60 lb	80 lb	100 lb	125 lb	60 lb	80 lb	100 lb	125 lb
20	3.94	3.10	2.56	2.10	0.018	0.014	0.012	0.010				
30	5.90	4.66	3.84	3.16	0.039	0.031	0.026	0.021				
40	7.87	6.21	5.13	4.21	0.067	0.053	0.044	0.036				
50	9.84	7.76	6.41	5.26	0.103	0.081	0.067	0.055				
60	11.81	9.31	7.69	6.31	0.15	0.12	0.095	0.078				
70	13.78	10.87	8.97	7.37	0.20	0.16	0.13	0.10				
80	15.74	12.42	10.25	8.42	0.25	0.20	0.16	0.14				
90	17.71	13.97	11.53	9.47	0.31	0.25	0.20	0.17				
100	19.68	15.5	12.82	10.52	0.38	0.30	0.25	0.20				
125	24.60	19.4	16.02	13.15	0.59	0.46	0.38	0.32				
150	29.51	23.3	19.22	15.78	0.83	0.65	0.54	0.44	0.020	0.016	0.013	0.011
175	34.44	27.2	22.43	18.41	1.15	0.90	0.73	0.60	0.030	0.024	0.020	0.016
200	39.36	31.0	25.63	21.05	1.52	1.15	0.95	0.78	0.044	0.033	0.027	0.022
250	49.20	38.8	32.04	26.31	2.35	1.82	1.48	1.20	0.056	0.044	0.036	0.030
300	59.0	46.6	38.45	31.57	3.27	2.55	2.10	1.72	0.072	0.056	0.046	0.038
350	68.9	54.3	44.86	36.83	4.55	3.53	2.86	2.35	0.090	0.070	0.058	0.048
400	78.7	62.1	51.26	42.09	5.86	4.53	3.70	3.03	0.106	0.084	0.069	0.057
450	88.6	69.9	57.67	47.35	7.60	5.80	4.65	3.80	0.165	0.130	0.107	0.088
500	98.4	77.6	64.08	52.61	9.27	7.12	5.79	4.71	0.24	0.19	0.15	0.12
600	118.1	93.1	76.90	63.14	13.7	10.4	8.45	6.85	0.31	0.25	0.20	0.17
700	137.8	108.7	89.71	73.66					0.40	0.31	0.26	0.21
800	157.4	124.2	102.5	84.18					0.62	0.49	0.40	0.33
900	177.1	139.7	115.3	94.70					0.89	0.70	0.57	0.47
1000	196.8	155	128.2	105.2					1.20	0.94	0.77	0.64
1200	236.1	186	153.8	126.3					1.53	1.20	0.99	0.81
1500	295	233	192.2	157.8					1.98	1.55	1.27	1.05
									2.47	1.91	1.56	1.29
									3.53	2.75	2.23	1.83
									4.71	3.67	3.00	2.45
									6.40	4.90	4.00	3.23
									8.10	6.20	5.05	4.10
									9.96	7.62	6.20	5.04
									15.0	11.4	9.05	7.45
									24.1	18.3	14.5	11.7

Free air, ft³/min	Equivalent ft³/min of compressed air						Nominal pipe diameters							
							2½ in				3 in			
	60 lb	80 lb	100 lb	125 lb			60 lb	80 lb	100 lb	125 lb	60 lb	80 lb	100 lb	125 lb
50	9.84	7.76	6.41	5.26			0.013	0.010	0.008	0.007				
60	11.81	9.31	7.69	6.31			0.018	0.014	0.011	0.009				
70	13.78	10.87	8.97	7.37			0.023	0.018	0.015	0.012				
80	15.74	12.42	10.25	8.42			0.030	0.023	0.019	0.015				
90	17.71	13.97	11.53	9.47			0.037	0.028	0.024	0.020				
100	19.68	15.5	12.82	10.52			0.043	0.035	0.029	0.023	0.015	0.012	0.010	0.008
125	24.60	19.4	16.02	13.15			0.068	0.055	0.043	0.036	0.023	0.018	0.015	0.012
150	29.51	23.3	19.22	15.78			0.095	0.074	0.061	0.050	0.032	0.025	0.021	0.017
175	34.44	27.2	22.43	18.41			0.127	0.099	0.081	0.067	0.043	0.034	0.028	0.022
200	39.36	31.0	25.63	21.05			0.163	0.128	0.105	0.086	0.055	0.043	0.036	0.029
250	49.20	38.8	32.04	26.31			0.248	0.195	0.160	0.131	0.082	0.065	0.054	0.044
300	59.0	46.6	38.45	31.57			0.35	0.27	0.23	0.19	0.117	0.092	0.075	0.062
350	68.9	54.3	44.86	36.83			0.47	0.37	0.31	0.25	0.157	0.124	0.101	0.083
400	78.7	62.1	51.26	42.09			0.61	0.48	0.40	0.33	0.20	0.160	0.131	0.108
450	88.6	69.9	57.67	47.35			0.77	0.60	0.50	0.41	0.26	0.20	0.165	0.135
500	98.4	77.6	64.08	52.61			0.95	0.75	0.62	0.50	0.31	0.25	0.20	0.17
600	118.1	93.1	76.90	63.14			1.37	1.08	0.89	0.73	0.45	0.35	0.29	0.24
700	137.8	108.7	89.71	73.66			1.83	1.43	1.18	0.97	0.60	0.48	0.39	0.32
800	157.4	124.2	102.5	84.18			2.37	1.87	1.54	1.25	0.78	0.61	0.60	0.41
900	177.1	139.7	115.3	94.70			3.00	2.35	1.95	1.57	0.98	0.77	0.63	0.54
1000	196.8	155	128.2	105.2			3.70	2.89	2.37	1.94	1.20	0.94	0.78	0.64
1200	236.1	186	153.8	126.3			5.45	4.21	3.45	2.78	1.70	1.37	1.12	0.92
1500	295	233	192.2	157.8			8.54	6.62	5.39	4.38	2.70	2.15	1.73	1.43
2000	394	310	256.3	210.5			15.8	12.0	9.66	7.80	4.85	3.77	3.09	2.52
2500	492	388	320.4	263.1			...	...	...	...	7.80	6.00	4.85	4.00

TABLE 11-12 Friction Losses of Air in Pipes and Hoses, Expressed in Loss of Pressure (Continued)

Although these tables are prepared for losses in pipes, they may be used safely for losses in hoses. The losses in pipes and hoses are approximately equal.
(pressure loss, lb/in^2, in 100 ft of pipe for air at 60, 80, 100, and 125 lb/in^2 gauge)

Free air, ft^3/min	Equivalent ft^3/min of compressed air				Nominal pipe diameters											
					4 in				5 in			
	60 lb	80 lb	100 lb	125 lb	60 lb	80 lb	100 lb	125 lb	60 lb	80 lb	100 lb	125 lb
250	49.20	38.8	32.04	26.31	0.021	0.017	0.014	0.011				
300	59.0	46.6	38.45	31.37	0.030	0.024	0.020	0.016				
350	68.9	54.3	44.86	36.83	0.040	0.032	0.026	0.022				
400	78.7	62.1	51.26	42.09	0.052	0.041	0.034	0.028				
450	88.6	69.9	57.67	47.35	0.065	0.051	0.042	0.035				
500	98.4	77.6	64.08	52.61	0.078	0.062	0.051	0.042				
600	118.1	93.1	76.90	63.14	0.112	0.089	0.073	0.060				
700	137.8	108.7	89.71	73.66	0.150	0.119	0.098	0.081				
800	157.4	124.2	102.5	84.18	0.193	0.154	0.126	0.104				
900	177.1	139.7	115.3	94.70	0.24	0.193	0.159	0.131				
1000	196.8	155	128.2	105.2	0.30	0.23	0.19	0.16	0.095	0.075	0.062	0.051
1200	236.1	186	153.8	126.3	0.43	0.34	0.28	0.23	0.135	0.105	0.088	0.072
1500	295	233	192.2	157.8	0.66	0.52	0.43	0.35	0.21	0.162	0.135	0.110
2000	394	310	256.3	210.5	1.16	0.91	0.75	0.61	0.36	0.28	0.24	0.192
2500	492	388	320.4	263.1	1.82	1.43	1.16	0.95	0.56	0.44	0.37	0.30
3000	590	466	384.5	315.7	2.62	2.05	1.65	1.35	.80	0.63	0.52	0.43
3500	689	543	448.6	368.3	3.55	2.78	2.25	1.84	1.08	0.85	0.70	0.58
4000	787	621	512.6	420.9	4.56	3.55	2.91	2.37	1.41	1.11	0.91	0.75
4500	886	699	576.7	473.5	5.95	4.60	3.70	3.00	1.78	1.40	1.16	0.95
5000	984	776	640.8	526.1	7.35	5.68	4.55	3.70	2.20	1.73	1.43	1.17
6000	1181	931	769.0	631.4	10.6	8.12	6.58	5.34	3.13	2.45	2.01	1.65
7000	1378	1087	897.1	736.6					4.39	3.40	2.79	2.28
8000	1574	1242	1025	841.8					5.75	4.46	3.63	2.89
9000	1771	1397	1153	947.0					7.35	5.65	4.60	3.76
10000	1968	1552	1282	1052					9.04	6.95	5.65	4.59

	Equivalent ft³/min of compressed air				Nominal pipe diameters							
					6 in							
Free air, ft³/min	60 lb	80 lb	100 lb	125 lb	60 lb	80 lb	100 lb	125 lb
---	---	---	---	---	---	---	---	---
1500	295	233	192.2	157.8	0.080	0.063	0.052	0.043
2000	394	310	256.3	210.5	0.140	0.110	0.091	0.075
2500	492	388	320.4	263.1	0.22	0.170	0.140	0.115
3000	590	466	384.5	315.7	0.31	0.24	0.20	0.164
3500	689	543	448.6	368.3	0.42	0.33	0.27	0.22
4000	787	621	512.6	420.9	0.54	0.43	0.35	0.29
4500	886	699	576.7	473.5	0.68	0.54	0.44	0.36
5000	984	776	640.8	526.1	0.84	0.67	0.55	0.45
6000	1181	931	769.0	631.4	1.19	0.94	0.77	0.63
7000	1378	1087	897.1	736.6	1.65	1.30	1.06	0.87
8000	1574	1242	1025	841.8	2.15	1.70	1.39	1.13
9000	1771	1397	1153	947.0	2.75	2.15	1.76	1.43
10000	1968	1552	1282	1052	3.34	2.61	2.14	1.76
11000	2165	1707	1410	1157	4.12	3.23	2.63	2.14
12000	2361	1863	1538	1263	4.93	3.87	3.13	2.55
13000	2558	2018	1666	1368	5.78	4.52	3.69	3.00
14000	2755	2173	1794	1473	6.75	5.25	4.27	3.50
15000	2952	2328	1922	1578	7.67	5.92	4.82	3.93
16000	3149	2484	2051	1684	8.70	6.90	5.56	4.55
18000	3542	2794	2307	1894	11.10	8.70	7.05	5.77
20000	3936	3105	2563	2105	14.2	10.8	8.65	7.04

TABLE 11-12 Friction Losses of Air in Pipes and Hoses, Expressed in Loss of Pressure *(Continued)*

Although these tables are prepared for losses in pipes, they may be used safely for losses in hoses. The losses in pipes and hoses are approximately equal.
(pressure loss, lb/in², in 100 ft of pipe for air at 60, 80, 100, and 125 lb/in² gauge)

Free air, ft³/min	Equivalent ft³/min compressed air				Nominal pipe diameters											
					8 in							10 in			
	60 lb	80 lb	100 lb	125 lb	60 lb	80 lb	100 lb	125 lb			60 lb	80 lb	100 lb	125 lb		
1500	295	233	192	158	0.021	0.016	0.014	0.011								
2000	394	310	256	211	0.035	0.028	0.023	0.019								
2500	492	388	320	263	0.955	0.043	0.036	0.029								
3000	590	466	385	316	0.077	0.061	0.051	0.041			0.017	0.014	0.011	0.009		
3500	689	543	449	368	0.104	0.082	0.068	0.055			0.024	0.019	0.016	0.013		
4000	787	621	513	421	0.134	0.105	0.087	0.071			0.033	0.026	0.021	0.017		
4500	886	699	577	473	0.166	0.133	0.111	0.090			0.043	0.034	0.028	0.023		
5000	984	776	641	526	0.21	0.163	0.136	0.111			0.054	0.042	0.035	0.029		
6000	1181	931	769	631	0.29	0.23	0.190	0.016			0.065	0.052	0.043	0.035		
7000	1378	1087	897	737	0.39	0.31	0.26	0.21			0.094	0.074	0.060	0.050		
8000	1574	1242	1025	842	0.51	0.40	0.33	0.27			0.126	0.099	0.081	0.067		
9000	1771	1397	1153	947	0.66	0.52	0.43	0.35			0.162	0.127	0.105	0.086		
10000	1968	1552	1282	1052	0.79	0.62	0.52	0.42			0.21	0.164	0.135	0.110		
11000	2165	1707	1410	1157	0.98	0.77	0.64	0.52			0.26	0.20	0.165	0.135		
12000	2361	1863	1538	1263	1.16	0.91	0.76	0.62			0.31	0.24	0.20	0.163		
13000	2558	2018	1666	1368	1.37	1.07	0.89	0.72			0.36	0.29	0.24	0.193		
14000	2755	2173	1794	1473	1.60	1.24	1.03	0.84			0.43	0.33	0.28	0.23		
15000	2952	2328	1922	1578	1.77	1.39	1.15	0.94			0.50	0.39	0.32	0.26		
16000	3149	2484	2051	1684	2.06	1.62	1.34	1.09			0.56	0.44	0.36	0.30		
18000	3542	2794	2307	1894	2.64	2.04	1.69	1.37			0.64	0.50	0.42	0.34		
20000	3936	3105	2563	2105	3.18	2.49	2.04	1.67			0.81	0.63	0.52	0.43		
22000	4329	3415	2820	2315	3.97	3.02	2.50	2.03			1.00	0.78	0.64	0.53		
24000	4723	3725	3076	2525	4.65	3.60	3.00	2.42			1.18	0.93	0.76	0.63		
26000	5116	4036	3332	2736	5.50	4.25	3.50	2.84			1.40	1.10	0.90	0.75		
28000	5510	4346	3588	2946	6.40	4.95	4.06	3.28			1.65	1.30	1.06	0.88		
											1.90	1.50	1.23	1.01		

SOURCE: Ingersoll-Rand Co.

TABLE 11-13 Friction Losses of Air in Pipe and Hose Fittings in Terms of the Equivalent Length of Pipe
(These data may be applied to any liquid or gas)

Nominal pipe size Dn std wt	Actual inside diam, in	Gate valve Full open	45° elbow	Long-sweep elbow or run of std. tee	Std elbow or run of tee reduced ½	Std tee through side outlet	Close return bend	Swing check valve Full open	Angle valve Full open	Globe valve Full open	Equivalent resistance of std wt welding elbows Length of straight pipe, ft* 90° Elbows Short radius R/Dn=1	90° Elbows Long radius R/Dn=1½	45° Elbows Short radius R/Dn=1	45° Elbows Long R/Dn=1½
Resistance factor		0.19	0.42	0.6	0.9	1.8	2.2	2.3	5.	10.				
½	0.622	0.35	0.78	1.11	1.7	3.3	4.1	4.3	9.3	18.6	†	0.68	†	0.44
¾	0.824	0.44	0.97	1.4	2.1	4.2	5.1	5.3	11.5	23.1	†	0.91	†	0.58
1	1.049	0.56	1.23	1.8	2.6	5.3	6.5	6.8	14.7	29.4	1.6	1.15	1.01	0.74
1¼	1.380	0.74	1.6	2.3	3.5	7.0	8.5	8.9	19.3	38.6	2.1	1.5	1.33	0.98
1½	1.610	0.86	1.9	2.7	4.1	8.1	9.9	10.4	22.6	45.2	2.4	1.8	1.6	1.14
2	2.067	1.10	2.4	3.5	5.2	10.4	12.8	13.4	29	58	3.1	2.3	2.0	1.5
2½	2.469	1.32	2.9	4.2	6.2	12.4	15.2	15.9	35	69	3.7	2.7	2.4	1.7
3	3.068	1.6	3.6	5.2	7.7	15.5	18.9	19.8	43	86	4.7	3.4	3.0	2.2
4	4.026	2.1	4.7	6.8	10.2	20.3	24.8	26.0	57	113	6.1	4.4	3.9	2.9
5	5.047	2.7	5.9	8.5	12.7	25.4	31	33	71	142	7.7	5.6	4.9	2.9
6	6.065	3.2	7.1	10.2	15.3	31	37	39	85	170	9.2	6.7	5.9	3.6
7	7.024	3.7	8.3	11.8	17.7	35	43	45	98	197	†	†	†	4.3
8	7.981	4.3	9.4	13.4	20.2	40	49	52	112	224	12.1	8.8	7.7	5.7
10	10.020	5.3	11.8	16.9	25.3	51	62	65	141	281	15.2	11.0	9.7	7.1

*For 180° bend multiply values for 90° bend by 1.34.
†Short radius elbows, R/Dn=1, not made in this size and weight.
‡Not made in this size.
Data on fittings based on information published by Crane Co.
Data are based on Fanning coefficient of 0.006, as taken from Chart No. 18 of Catalog 211 of Tube Turns, Inc.
SOURCE: Ingersoll-Rand Co.

TABLE 11-14 Recommended Pipe Sizes for Transmitting Compressed Air at 80 to 125 lb/in² Gage

Volume of air, ft³/min	Length of pipe, ft				
	50–200	200–500	500–1000	1000–2500	2500–5000
	Nominal size pipe, in				
30–60	1	1	1¼	1½	1½
60–100	1	1¼	1¼	2	2
100–200	1¼	1½	2	2½	2½
200–500	2	2½	3	3½	3½
500–1000	2½	3	3½	4	4½
1000–2000	2½	4	4½	5	6
2000–4000	3½	5	6	8	8
4000–8000	6	8	8	10	10

SOURCE: R.L. Peurifoy, *Construction Planning, Equipment, and Methods*, McGraw-Hill, New York, 1956, p. 253.

Aftercooler Heat exchanger placed after the outlet of the compressor to remove heat of compression.

Air compressor A machine to increase air pressure by decreasing the volume of air.

Boyle's law At a constant temperature, the volume occupied by a given weight of air varies inversely as the absolute pressure. The isothermal expansion or contraction is expressed by the equation:

$$P_i V_i = P_f V_f = C$$

where P_i = initial absolute pressure, lb/in²
V_i = initial volume, in³
P_f = final absolute pressure, lb/in²
V_f = final volume, in³
C = a constant

Brake horsepower Actual input horsepower delivered to the compressor by the output shaft of the prime mover.

Capacity Volume of air delivered per minute by an air compressor, measured by the cubic feet of free air taken in by the compressor at 14.7 lb/in² atmospheric pressure and 60°F. The standard rating of an air compressor when qualified by outlet pressure.

Celsius temperature Temperature indicated by the Celsius scale, the freezing point of water being 0°C and the boiling point being 100°C at 14.7 lb/in². Conversion from Fahrenheit temperature is given by the equation:

$$T_c = \frac{T_f - 32}{1.8}$$

T_c = temperature, °C, and T_f = temperature, °F.

Centrifugal compressor One in which compression is effected by a rotating vane or impeller that imparts velocity to the air to give the desired pressure. A form of rotating compressor.

Charles' law At constant pressure the volume of a given weight of air varies in direct proportion to its absolute temperature. This adiabatic expansion or contraction is expressed by the equation:

$$\frac{V_i}{T_i} = \frac{V_f}{T_f} = K$$

where V_i = initial volume, in³
T_i = initial absolute temperature, °F
V_f = final volume, in³
T_f = final absolute temperature, °F
K = a constant

Compression ratio Ratio of absolute outlet pressure of air from a compressor to absolute inlet pressure.

Density of air Weight of unit volume, usually 1 cubic foot of air. Density varies with pressure and temperature. At 60°F and 14.7 lb/in² absolute, weight is 0.0766 lb/ft³. Volume per pound of air is 13.06 ft³.

Discharge pressure Absolute air pressure in pounds per square inch at outlet from compressor.

Double-acting compressor Reciprocating compressor which compresses air at both ends of cylinder.

Diversity factor Ratio of the actual quantity of air required for all uses to the sum of the individual quantities required for each use.

Fahrenheit temperature Temperature indicated by the Fahrenheit scale, the freezing point of water being 32°F and the boiling point being 212°F at 14.7 lb/in² atmospheric pressure. Conversion from the Celsius temperature is given by the equation:

$$T_f = 32 + 1.8 T_c$$

where T_f = temperature, °F, and T_c = temperature, °C.

Free air Air as it exists under atmospheric conditions at any time.

Gauge pressure Air pressure in excess of atmospheric pressure.

Inlet pressure. Absolute air pressure in pounds per square inch at inlet to air compressor.

Intercooler Heat exchanger located between two compression stages to remove heat of compression.

Isothermal compression and expansion Change of pressure and volume of air without any change of temperature.

Load factor Ratio of average delivery of a compressor during a given time to the rated load or capacity, both being in cubic feet per minute at same pressure.

Multistage compressor Compressor having two or more stages to compress air from atmospheric to desired discharge pressure.

Outlet pressure Absolute air pressure in pounds per square inch at discharge from compressor.

Reciprocating compressor Machine which compresses air by means of a piston reciprocating in a cylinder.

Rotary compressor Machine which compresses air by action of rotating vanes, lobes, or screws.

Single-acting compressor Reciprocating compressor which compresses in only one end of the cylinder.

Single-stage compressor Machine which compresses air from atmospheric pressure to desired discharge pressure in a single operation.

Standard conditions Those arbitrarily fixed values which form the basis for calculations involving variations in volume of air by reason of changes in pressure and temperature. They are absolute pressure of 14.7 lb/in² and temperature of 60°F, equivalent to absolute temperature of 520°F.

Theoretical horsepower Horsepower required to compress adiabatically the air delivered by a compressor through the specified pressure range, without provision for friction losses.

Two-stage compressor Machine which compresses air from atmospheric pressure to desired discharge pressure in two operations. The first stage compresses air to an intermediate pressure and the second stage compresses it further to the desired final pressure.

Vacuum Measure of the extent to which pressure is less than atmospheric pressure. A vacuum of 5.0 lb/in² is equivalent to an absolute pressure of 14.7 − 5.0 or 9.7 lb/in².

Volumetric efficiency Ratio of the capacity or rating of an air compressor to its piston or volume displacement.

An air compressor works under conditions somewhere between isothermal and adiabatic, depending on the results of cooling the air between the inlet and outlet of compressor. Combining Boyle's law and Charles' law results in a characteristic equation expressing the relationship between initial and final pressures, volumes, and temperatures when compressing or expanding air under adiabatic conditions. The equation is:

$$\frac{P_i V_i}{T_i} = \frac{P_f V_f}{T_f} = C$$

where P_i = initial absolute pressure, lb/in²
V_i = initial volume, in³ or ft³
T_i = initial absolute temperature, °F
P_f = final absolute pressure, lb/in²
V_f = final volume, in³ or ft³
T_f = final absolute temperature, °F
C = constant

When air is compressed or expanded adiabatically under standard conditions the characteristic equation becomes:

$$\frac{14.7 V_i}{520} = \frac{P_f V_f}{T_f}$$

or

$$0.0283 V_i = \frac{P_f V_f}{T_f}$$

An example of the practical use of the characteristic equation in determining the final temperature of the air in a receiver of a compressor is illustrated in the following calculations, which are based on standard and adiabatic conditions. Premises are:

Initial factors:

P_i = 14.7 lb/in² absolute atmospheric pressure of free air
V_i = 1200 ft³ capacity per minute of compressor
T_i = 520°F, absolute temperature of free air

Final factors:

P_f = 114.7 lb/in², absolute pressure of air in receiver
V_f = 22 ft³, capacity of receiver
T_f = unknown absolute temperature of air in receiver

Substituting in the characteristic equation:

$$0.0283 V_i = \frac{P_f V_f}{T_f}$$

$$0.0283 \times 1200 = \frac{114.7 \times 22}{T_f}$$

$$T_f = 743°F$$

The ambient temperature of the compressed air is 743 − 460 = 283°F. This 283°F temperature of available compressed air is not a practical operating temperature, but the combined effects of the cooling fins on the compressor and the circulating cooling oil within the compressor reduce the value to an acceptable figure.

The theoretical horsepower for compressing air under isothermal conditions is given by the equation:

$$\text{hp} = \frac{4833 \times V_i \times \log_{10}(P_f/P_i)}{33,000}$$

$$= 0.1465 \times V_i \times \log_{10}(P_f/P_i)$$

where V_i = capacity of compressor, ft³/min
P_i = absolute pressure of free air, lb/in²
P_f = absolute pressure of air in receiver, lb/in²

When the values used in the example of the application of the characteristic equation are substituted in the equation for theoretical horsepower, the following horsepower for the 1200-ft³/min compressor results:

$$\text{hp} = 0.1465 \times 1200 \log_{10}\left(\frac{114.7}{14.7}\right)$$

$$= 157$$

The 157 hp is equivalent to 13.1 hp per 100 ft³/min of air compressor capacity. For practical reasons, however, the theoretical horsepower requirement is based on adiabatic conditions because of the power lost to heat in the air compression. The theoretical horsepower for compressing the air under adiabatic conditions is given by the equation:

$$\text{hp} = \frac{n}{n-1} \times 0.0643 V_i \left[\left(\frac{P_f}{P_i} \right)^{n-1/n} - 1 \right]$$

where $n = 1.4$ for air in adiabatic compression
V_i = capacity of compressor, ft³/min
P_i = absolute pressure of free air, lb/in²
P_f = absolute pressure of air in receiver, lb/in²

When the same values used in the characteristic equation are substituted in the equation for horsepower under adiabatic conditions, the horsepower of the 1200-ft³/min compressor is determined as:

$$\text{hp} = \frac{1.4}{1.4-1} \times 0.0643 \times 1200 \left[\left(\frac{114.7}{14.7} \right)^{(1.4-1)/1.4} - 1 \right]$$
$$= 216$$

The 216 hp is equivalent to 18.0 hp per 100 ft³/min of compressor capacity.

Efficiency of Air Compressors. A general range of mechanical efficiencies of air compressors is 90 to 95 percent. Assuming 90 percent efficiency, about 20 hp of prime mover is required for every 100 ft³/min of air compressor rating at 100 lb/in². This 20 hp per 100 ft³/min may be compared with the average horsepowers for 650- to 1600-ft³/min compressors of Tables 11-10 and 11-11. The comparisons:

For compressors driven by diesel engines: 29 hp per 100 ft³/min
For compressors driven by electric motors: 21 hp per 100 ft³/min

Engine ratings are generally higher than those of electric motors for given tasks. The 21 hp for the electric motors is a good check on the 20 hp of the calculations.

Increasing the pressure of delivered air to air hammers adds to their penetration or drilling speed. Over several decades the pressure of compressors has been increased from 100 to 250 lb/in². The increase in penetration as a function of increase in air pressure is expressed by the equation:

$$S_f = (1.00 + 1.5P)S_i$$

where S_f = final penetration speed, in/min
P = percentage increase in pressure
S_i = initial penetration speed, in/min

The equation is for a 0 to 50 percent increase in pressure.

An example is ascertaining the final speed when pressure is increased from 100 to 125 lb/in² and initial penetration speed is 30 in/min.

Solution:
$$S_f = (1.00 + 1.5P)30$$
$$= 41 \text{ in/min}$$

The increase is 37 percent. A significant similar increase is obtained in cubic yards of rock drilled hourly.

Fusion or Jet-Piercing Drill This drill is shown in Figures 11-54 and 11-55 and its unique drilling by the chamber or pocket method is illustrated in Figure 11-36. The "drill" is really not a drill but rather a burning machine which delivers fuel oil, oxygen, and cooling water to a blowpipe mounted in the mast and extending into the blasthole. The drill is equipped with electric motors, hydraulic pumps and jacks, and air compressors for traction, leveling, raising and lowering the blowpipe, and other functions of the drilling operation.

Important specifications of the model JPM-5 jet-piercing drill (Figure 11-54) are in the following table:

Working weight, lb	120,000
Transformer capacity, 150 kVA, hp	200
Hydraulic system, hp	75
Air compressor, ft³/min	13

Hole exhaust system, air, hp	19
Fuel oil system, gal	800
Water system, burner cooling, 22 gal/min at 75 lb/in², gal	150
Standard blowpipe assembly, burner and reamer:	
Hole depth, ft	56
Nominal blowpipe diameter, in	1¼ to 8
(Diameter may be extended selectively to 22 in)	
Average hourly consumption of jet-piercing blowpipe:	
Oxygen, ft³	12,000
Fuel oil, gal	48
Water, gal	850

Figure 11-56 illustrates the working parts of the jet-piercing blowpipe. The blowpipe is made up of a swing joint, kelly, and burning assembly. Oxygen, water, and fuel oil are introduced into the rotating blowpipe through the swing joint. The round, triple-fluted kelly is standard rotary drilling equipment. Oxygen and fuel are carried through separate tubes in the bore of the kelly to the burning assembly. Cooling water is brought to the burner through the remaining space of the kelly. The combustion elements are combined in the combustion chamber and exhausted through three or four diverging nozzles to produce the piercing jet flames at about 4300°F.

The blowpipe with its jets rotates at speeds of 15 to 47 r/min, the speed depending on the characteristics of the rock. The flame jets, when operating under conditions of 12,000 ft³ of oxygen and 48 gal of fuel oil hourly, have an energy equivalent of 600 hp.

Figure 11-54 Fusion or jet-piercing drill. The drill is jet-piercing blastholes in taconite, a hard iron ore of the Mesabi range, Minnesota. Blastholes average 9.2 in diameter and penetration speed averages 17 ft/h between limits of 12 and 25. The drill performs well in rocks with high spall-ability as in the case of the taconite. The drill's ability to pocket and chamber blastholes makes for large drilling patterns and low powder factors in hard rocks. *(Linde Division, Union Carbide Corp.)*

Figure 11-55 Jet-piercing drills and auxiliary machinery. On the lower bench are the auxiliary machines. Oxygen is supplied by one of the 100,000-ft³ semitrailer carriers which are being exchanged. To the left and in back of the carriers is the crawler-mounted 6000-gal water carrier, which is serviced by a water-tank truck of similar capacity. Both oxygen and water carriers suffice for 8 h of work. Fuel oil is carried in the integral 800-gal tank of the drill, more than ample for a shift of work. A number of drills are serviced from a centrally located gaseous oxygen production plant by several of the 100,000-ft³ semitrailer carriers, this integrated system working well for specialized drills putting down blastholes in a particular rock formation, taconite ore. *(Linde Division, Union Carbide Corp.)*

The jets cut and spall the rock, and these cuttings and spallings are expelled continuously from the blasthole by exhaust gases consisting both of combustion products and of steam from the cooling water ejected behind the burner head. Efficiency is a measure of rock spallability, and, when spallability is poor, the action of the flame jets is augmented by addition of a tricone roller bit to form a thermomechanical drill head.

The jet flames produce a hole of somewhat variable diameter because of the changes in spallability at different elevations of the rock. A contourometer averages out the volume of the hole, usually twice that based on the nominal size of the burner, and thus charges of explosives for a desired powder factor may be calculated.

The relationships between nominal blowpipe diameters and average diameters of actual resulting blastholes are given in the following tabulation:

Nominal blowpipe diameter, in	Average blasthole diameter, in
1¼	1¾
4	6⅝
5½	7⅞
6½	9¼
8	10¾

Because of the ability of the jet-piercing drill to make a chamber at the bottom of the blasthole and thereby increase explosive forces at that lowest elevation, spacing and burden may be increased and subdrilling below grade may be decreased with respect to the pattern of the holes. Blasthole patterns up to 37-ft spacing and 37-ft burden have been used successfully in hard rocks. A 30 ft × 32 ft pattern with 50-ft depth of hole is illustrated in Figure 11-36, the drilling of this pattern in Figure 11-55, and the resulting excellent fragmentation in Figure 11-43.

Table 11-15 gives representative hourly penetration speeds when actually drilling. Breakdown of operating functions over a 12-month period is:

11–88 Fragmentation of Rock

Burning or drilling time, piercing and chambering		75.5%
Necessary operations:		
Moving and setup	9.4%	
Change burners and reamers	2.1	
Casing blasthole	1.0	
Total		12.5
Repairs		1.6
Delays, nonfunctional		10.4
Total		100.0%

This sustained high efficiency of 75.5 percent for actual drilling time is typical of huge permanent mining operations in which rock is fairly uniform and management is excellent.

The average penetration speed in the taconite of Table 11-15 is 17 ft/h. The tabulation indicates that 75.5 percent of available time is spent in actual drilling, resulting in an average footage of 13 ft/h. A typical example of drilling for 47,000 yd³ or 100,000 tons of taconite is shown in Figures 11-36 and 11-43. Twenty-seven holes averaged 51 ft in depth for a total of 1377 ft. The hole was pierced at 36.7 ft/h, requiring 1.39 h. Chambering required 0.40 h. Total burning time per hole was 1.79 h. Actual feet of hole hourly while burning $= 51/1.79 = 28.5$ ft/h.

An estimate of total working time in the taconite excavation gives $(1.79 \times 27)/0.755 = 64$ h. Accordingly the jet-piercing drill averaged $47,000/64 = 734$ yd³/h, equivalent to 1562 tons/h.

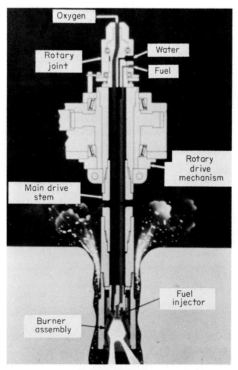

Figure 11-56 Working parts of the jet-piercing blowpipe. The 4300°F jet flame cuts and spalls the rock and the disintegrated rock is expelled from the blasthole by the combined pressure of the fuel-oil combustion gases, oxygen, and steam from the cooling water. Under normal operating consumptions of fuel oil and oxygen, the flame jets have energy equivalent to 600 hp. When spallability is low, a tricone roller bit is added to form a thermomechanical drill head. *(Linde Division, Union Carbide Corp.)*

TABLE 11-15 Average Hourly Penetration Speeds of Jet-Piercing Drills According to Kind of Rock and Kind of Iron Ore Formation (ft of average 9.2-in-diameter hole per hour)

Rock:	ft/h
Conglomerate, Michigan	20
Diorite, Michigan	12
Granite, Massachusetts	20
Quartzite, Michigan	9
Sandstone, New York	20
Schist, Michigan	9
Syenite, Arkansas	18
Iron ore formation:	
Argillaceous, Michigan	14
Specularite	
Hematite, Labrador	32
Magnetite, Michigan	30
Taconite, ferruginous chert	
Lower cherty magnetic, Area A	25
Lower cherty magnetic, Area B	12
Upper cherty magnetic, Area C	23
Lower slaty	12
Oxidized and altered	14

NOTE: These average speeds are penetration rates while actually drilling, and they must be adjusted for moving and setup, changing burners and reamers, setting casing, and like times in order to estimate an overall drilling rate for a working hour. It is suggested that the values of the table be multiplied by 0.67 for estimates.

SOURCE: Linde Division, Union Carbide Corp.

Auxiliary Machinery. For three or more jet-piercing drills on site, gaseous oxygen production plants are installed. The oxygen is charged directly into high-pressure, 2400-lb/in², semitrailer gas transporters of 100,000-ft³ capacity. With a demand of 12,000 ft³ per drill, one transporter serves a drill for an 8-h shift. Figure 11-55 shows two of the transporters during exchange at the drill.

Fuel oil is supplied to the 800-gal tank of the drill by a tank truck, the 800-gal supply being more than enough for an 8-h shift. Water is furnished by a 6000-gal insulated and heated water wagon mounted on crawlers, as shown in Figure 11-54. In turn it is serviced by a water-tank truck.

The crew of the jet-piercing machine consists of a driller and a helper, and the crew for auxiliary machinery amounts to a fraction of a worker according to the nearness of oxygen, fuel oil, and water supplies.

Explosives. A distinction of drilling by the jet-piercing machine, as affecting the use of explosives, is its ability to chamber and pocket the blasthole at will.

Figure 11-36 is a diagram of the unique loading of the variable-diameter hole produced by repeated passes of the blowpipe. With this kind of loading the excellent fragmentation of Figure 11-43 resulted from an AN/FO powder factor of 1.54 lb/yd³, or 0.57 lb/ton of hard taconite.

Explosives and blasting techniques may be selected in accordance with the explosives section of this chapter.

Productions and Costs. An estimate of drill production in taconite has been given in the text. The estimate is based on data of Figure 11-36, illustrating a hole for blasting 47,000 yd³ or 100,000 tons of taconite. The hole pattern was 30 ft × 32 ft, designed to yield about 35 yd³/ft of 50-ft-deep blasthole.

The total drill time was estimated to be 64 h, giving an hourly production of 734 yd³ or 1562 tons. The quantity of AN/FO was estimated to be 72,400 lb, giving a powder factor of 1.54 lb/yd³.

The following tabulation gives data controlling the estimate for the 1978 costs of blasting.

Cost of drilling:
    Machine cost, fob job $1,090,000
    Hourly cost of ownership and operation:

Machine (Table 8-3), 12% of $1090		$130.80
Driller and helper		24.10
Total		$154.90
Direct job cost, 64 × $154.90		$9,914
Unit cost for drilling:		
Per cubic yard, $9914/47000		$0.211
Per ton, $9914/100,000		$0.099
Cost of explosives and accessories:		
AN/FO. 72,400 lb @ $0.125		$9,050
Primers, 81 at 1 lb each; 81 @ $1.50		122
Primacord, 27 holes requiring 27 (50+5)−1485 ft, 1485 ft @ $0.05		74
Electric caps, #6 delays. 27 @ an average of $0.65		18
Trunk-line wire, about 1500 ft @ $0.02		30
Total		$9,309
Unit cost for explosives and accessories:		
Per cubic yard, $9309/47,000		$0.198
Per ton, $9309/100,000		$0.093
Cost of loading explosives by hole-loading truck:		
Machine cost, fob job		$66,000
Hourly cost of ownership and operation:		
Machine (Table 8-3). 32% of $66		$21.10
Driver		11.50
Two powder operators @ $12.10		24.20
Total		$56.80
Average rate of loading AN/FO, per minute of total working time, lb		100
Time for loading 72,400 lb at 6000 lb/h, h		12.1
Total direct job cost, 12.1 × $56.80		$687
Unit cost for loading explosives and accessories:		
Per cubic yard, $687/47,000		$0.015
Per ton. $687/100,000		$0.007

	Total	Unit cost	
Item	amount	yd³	Ton
Drilling	$ 9,914	$0.211	$0.099
Explosives and accessories	9,309	0.198	0.093
Loading explosives and accessories	687	0.015	0.007
Total	$19,910	$0.424	$0.199

To summarize, the fusion or jet-piercing machine is adapted to drilling hard spallable rocks; by means of a supplemental rotating cone bit it can function also in nonspallable hard rocks. This combination thermomechanical bit makes the machine a versatile hard-rock drill.

It is a mass production drill, ideally suited to large-scale, permanently located works such as quarries and open-pit mines, in which the rock is suitable for fusion or jet piercing.

The ability to chamber or pocket the blasthole by multiple passes of the blowpipe enables the powder operator to place explosives in the right place for most economical blasting. This flexibility also makes for larger patterns of holes and thereby reduces unit drilling costs. The pocketing of the hole below desired finished grade results in economical removal of the toe of slope due to concentration of explosives where most needed.

The ability to drill the hardest of rocks economically and to place the explosives most effectively are outstanding advantages of the fusion or jet-piercing drill.Before reading the following discussion of percussion and rotary drills, the reader may wish to examine again those sections of Chapter 7 pertaining to the auger drill, jackhammer, track drill, rotary drill, and cable drill. The descriptions of these machines are applicable to their use in blasthole drilling.

Cable Drill The percussion cable drill is used for small- to medium-scale blasthole drilling where a more expensive drill of either the percussive air-hammer or rotary type is not justified. Also, it is an all-purpose drill for work in the softest to hardest of rocks. The operation of the drill is described in Chapter 8.

Figure 11-57 pictures a cable drill of average size putting down 6-in-diameter blast-

Figure 11-57 Medium-weight cable drill putting down blastholes. The machine is drilling 6-in-diameter blastholes in a small dolomite quarry. The pattern is irregular because of the effort to straighten out the quarry face. Hourly drilling production is 6 ft of hole, hourly, probably resulting in the drilling of 27 yd^3 of rock. It is a one-person operation with an inexpensive drill and with low operating cost. The blasthole work may well be intermittent, supplemented by water-well drilling, as is often the case with cable drills. *(Bucyrus Erie Co.)*

holes in a small dolomite quarry at Washington, Missouri. The principal specifications for this complete machine are given below.

Weight, including string of drilling tools, hand tools, and truck chassis, lb	17,000
Weight of complete string of drilling tools, lb	1500
Horsepower of drill engine	75
Horsepower of truck engine	150
Capacity of truck chassis, tons	2½
Adjustable strokes of drilling tools, in	18, 24, 30
Range of hole diameters, in	4 to 6

For drilling of 6-in-diameter blastholes the string of drilling tools is made up of:

	Weight, lb
Chisel bit, 6-in diameter and 5-ft length	260
Stem, 4½-in diameter and 14-ft length	740
Jars, 4⅜-in diameter	330
Rope socket	170
Total	1500

Strokes per minute vary from 40 to 60. With weight of tools at 1500 lb and stroke frequency at 50/min, the gross energy for drilling is 188,000 ft·lb/min, or 5.7 hp. When drilling the medium-hard cherty dolomite of Figure 11-57, the overall drilling speed averages about 6 ft/h. Such speed in the variable pattern of holes for straightening out the irregular quarry face probably results in a dolomite production of about 27 yd^3/h.

There are available larger sizes of cable drills, but these are used primarily for deep-hole water, oil, and gas drilling, although, of course, they may be used for large-diameter blastholes. However, they are used infrequently because of the widespread popularity of other kinds of drills for big-hole work.

These larger cable drills range up to 36,000 lb when mounted on a semitrailer chassis and they swing up to 6000 lb of tools with up to 46-in strokes. Hole diameters range up to 12 in for blasthole drilling.

Figure 11-57 illustrates the beginning of drilling a blasthole. Because of the weathered rock at the collar of the hole it is being cased. The rock strata are horizontal and jars are not being used between rope socket and stem. If the strata were steeply dipped or cavernous, the jars would be used to loosen a stuck string of tools. The dart bailer is leaning against the frame of the machine.

A representative study for a month of drilling in an average quarry gives this breakdown of time distributions for the operations and the delays of a cable drill.

	Hours	Percent of time
Actual drilling	128.1	72.3
Necessary operations:		
Bailing out hole	12.7	7.2
Changing bits	4.8	2.7
Casing hole and adjusting casing	4.3	2.5
Moving to next hole	17.5	9.9
Total	39.3	22.3
Downtime for repairs	9.5	5.4
Total	176.9	100.0

Although this time study shows a 72 percent drilling efficiency, it is good practice to use a 67 percent efficiency factor for most blasthole drilling. With such an assumption, the average hourly drilling speeds for a cable drill putting down holes of 4- to 6-in diameter are as listed in Table 11-16.

The drilling speeds are slow with respect to those of other kinds of drills. However, both investment and hourly cost of ownership and operation are also low. These factors fix a definite place for the cable drill in small, infrequent, and isolated blasthole drilling.

Bits. There are two kinds of bits. The older kind is the chisel-edge type as shown in Figures 7-24 and 11-57. The modern one is the carbide-insert, or button, kind as illustrated in Figure 11-58. The chisel-edge bit is available in several configurations and steel alloys for different rocks and rock formations. The chisel-edge bit requires sharpening or dressing, sometimes frequently, whereas the button bit cannot be sharpened. However, because of extremely long life ranging up to 1500 ft of hole in hard rocks and because of elimination of time-consuming bit changing, the button bit is more economical than the chisel-edge kind.

The crew sometimes consists of only a driller and sometimes of a driller and a helper or bit dresser, depending on the kind of bit in use. However, even if the chisel-edge bit is used, the helper is sometimes unnecessary if the bit is dressed in a nearby blacksmith shop.

Auxiliary Machinery. Even when the chisel-edge bit, requiring dressing, is used, the backup equipment is minimal. It consists of an average water supply for floating the bit cuttings in a sludge at the bottom of the hole and a convenient forge and other blacksmithing tools for bit sharpening.

Water supply averages about one barrel hourly and it may be brought in by truck or else piped or pumped in from a convenient source. If a blacksmith shop is nearby, which is a rarity, bits may be taken in for dressing. Usually, bit dressing is done right on the job.

TABLE 11-16 Average Hourly Drilling Speeds of Cable Drills Putting Down Blastholes of 4- to 6-in-Diameter, Based on 67% Drilling Efficiency

Class of rock	Speed, ft/h
Soft	12 to 24
Medium	6 to 12
Hard	0 to 6

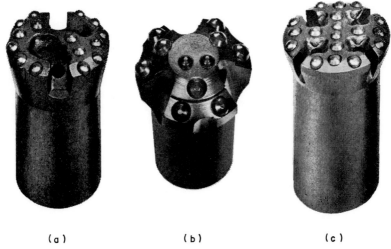

(a) (b) (c)

Figure 11-58 Three types of carbide-insert button bits for cable drills. *(a)* The A-type VPC self-sharpening percussion bit is designed for relatively soft formations and high penetration rates. Available in 2-in through 6-in size range with popular thread design. *(b)* The "TF" VPC self-sharpening percussion bit is engineered for medium to hard formation. Large-diameter buttons result in long life and test penetration. Available in 2½-in through 4½-in size range with popular thread design. *(c)* The "4T" VPC self-sharpening percussion bit is generally used in hard abrasive formations where gage wear is a problem. Available in 2¾-in through 3½-in size range in popular thread designs.

These three configurations are designed for soft, medium, and hard rocks. For use with cable drills they are available in a size range of 4- to 8-in diameter. The different configurations and range of sizes cover the different kinds of rocks and rock formations, as well as average-size blast holes. Like all button bits for other percussive drills actuated by compressed air, these bits have greater life and usually faster drilling speeds than the chisel-edge kind of bit. *(Varel Manufacturing Co.)*

It consists of heating the bit to a glowing temperature for malleability, sharpening the cutting edges by sledge, gauging with circular gauge, and quenching the bit for suitable hardness.

The forge is a simple blacksmith type with or without a blower and it is usually built of steel plate. Bridge anvil, hand forging tools, and slack tub for quenching are modified blacksmith tools. The forge may be fired by coal, coke, gas, or oil. Quenching is by water or oil. The blacksmithing is in the hands of the tool dresser or helper and this special, expert work is done during the drilling.

The use of explosives in the constant-diameter or pocketed blasthole of the cable drill is explained in the explosives section of this chapter.

Production and Costs. An estimate for the production and costs for the primary blasting of Figure 11-57 is tabulated below.

Data controlling the estimate for blasting in the dolomite quarry are:

Drilling:
Approximate desirable tonnage per blast, tons	15,000
Equivalent yardage, yd³, $\dfrac{15{,}000 \times 2000}{4870}$	6160
Height of face, ft	40
Initial trial pattern for 5-in-diameter blasthole, based on experience in local dolomite	12×16
Cubic yards per foot of hole, Table 11-2	7.11
Cubic yards per hole, 7.11×40	284
Number of holes to be drilled, 6160/284	22
Total depth of hole, ft, with 6 ft of subdrilling	46
Estimated average rate of drilling in medium-hard rock, Table 11-16, ft/h	6
Hours of drilling per hole, 46/6	7.7
Cubic yards drilled hourly, 284/7.7	37

Explosives and accessories:

11-94 Fragmentation of Rock

Kind of explosives because of wet holes	water gel
Detonation because of wet holes and electrical storms	Primacord
Initial trial powder factor, lb/yd³	1.4
Pounds water gel per hole, 1.4×284	398
Loading density of water gel	1.2
Pounds of water gel per foot of hole, Table 11-3	10.2
Height of column of explosives in hole, ft, 398/10.2	39
Number of 1-lb primers, 3 per hole	66
Approximate feet of Primacord for layout as in Figure 11-27	1620
Actual quantities of blasted rock for 22 holes:	
Yardage, 22×284, yd³	6250
Tonnage, $\dfrac{6250 \times 4870}{2000}$, tons	15,220

Costs for drilling, explosives and accessories, and hand loading of explosives and accessories, as of 1978, are:

Drilling:	
Machine with tools and equipment, cost	$48,000
Hourly cost of ownership and operation	
Machine (Table 8-3), 36% of $48	$17.30
Driller and helper	24.10
Total	$41.40
Yardage drilled by 22 holes, yd³	6250
Drilling hours, 6250/37	169
Direct job cost, 169×$41.40	$6997
Direct job unit cost, per ton, $6,997/15,220	$0.46
Explosives and accessories:	
Water gel, 1.4×6250=8750 lb; 8750×$0.45	$3938
Primers, 66 @ $1.50	99
Primacord, 1620 ft at $0.05	81
Total	$4118
Direct job unit cost, per ton, $4,118/15,220	$0.27
Hand loading of explosives and accessories: It is estimated that a crew of 3 powder operators can load 22 holes with 8750 lb of water gel in 8 h. Explosives are delivered to the crew by the powder company.	
3 powder operators @ $12.10/h	$36.30
Direct job cost, 8×$36.30	$290
Direct job unit cost per ton, $290/15,220	$0.02
Total direct job unit cost per ton, $11,405/15,220	$0.75

To summarize, the cable drill, operating on the principles of the modern machine, has been in use for at least 100 years. Relative to the compressed-air-actuated percussion drill and the rotary drill for blasthole drilling, it is remarkably inexpensive to own and operate. It can drill all kinds of soft, medium, and hard rocks. It is an economical drill for small to medium-size workings where blasthole drilling is intermittent. The driller sometimes works part-time in the small quarry, supplementing the rock work by drilling water, gas, and oil wells and exploratory holes. Although the design and operation of the cable drill are not sophisticated, it has filled for generations and will continue to fill an important niche in blasthole drilling. Its very simplicity guarantees its demonstrated longevity.

Jackhammer Drill The hand-held jackhammer is a small percussion drill, actuated by compressed air at about 90 lb/in² and imparting about 2000 blows per minute to the bit through the drill rod. It is used in shallow-depth, low-yardage primary blasting and in secondary blasting.

The jackhammer is illustrated in Figure 7-13, where it is working in exploratory drilling, in Figure 11-59, where four jackhammers are being used in primary blasting, and in Figure 11-60, where it is being used in secondary blasting. Figure 11-61 shows the fragmentation of a large rock which had been incompletely fragmented by primary blasting.

Jackhammers used in rock work range in weight from 30 to 70 lb. Typical specifications in the three weight classifications are detailed in Table 11-17.

Jackhammers may be used dry or wet, air or air and water being used for blowing out cuttings and cleaning out the hole. Holes may be drilled to 20-ft depth, but depths are usually confined to less than 10 ft. Drill rod lengths are in 2- or 2½-ft increments. Hole diameters are 1¼ to 2½ in, the average being about 1¾ in.

The approximate ranges of hourly drilling speeds are as listed in Table 11-18. Efficiency

Figure 11-59 Four medium-weight jackhammers drilling for primary blasting. This work is an example of small-yardage, shallow-cut drilling, typical of jackhammers in primary blasting. The cut is about 10 ft below the present grade and the holes will be subdrilled about 2 ft below the finished grade for 12-ft-deep holes. The pattern is 5 ft × 5 ft and the diameter of the holes is 1½ in. Rock excavation is for an access road to a dam. The truck-mounted 365-ft³/min air compressor is adequate for the four jackhammers. *(Ingersoll-Rand Co.)*

Figure 11-60 Heavyweight jackhammer drilling for secondary blasting. Three 5-ft-dimension rocks resulted from the primary blasting of the granite cut in the background because of the character of the jointed igneous rock. One rock has been blasted, one is being drilled, and the third one farther along at the toe of the slope is waiting disposition. A 3-ft, 1½-in-diameter hole receives one or two cartridges of dynamite in accordance with Figure 11-44, the powder factor of 0.3 lb/yd³ being sufficient to shatter the 3½-yd³ rocks. Fragmentation is small enough for acceptance into the rock fill of the freeway. The excellent superintendence of the blasting of this granite cut resulted in negligible secondary blasting, which can amount to as much as 15 percent of total yardage and thereby become an expensive part of blasting cost.

Figure 11-61 Unsatisfactory secondary blasting of 5-yd³ rock. This schist rock was blasted in a manner similar to that of Figure 11-60, the rock excavation jobs being adjacent in San Diego County, California. The worker and the yardstick imply the 6-ft dimension. One-half of the rock has been reduced to fragments acceptable in the rock fill. It is necessary to blockhole again the remaining half of the 2½ yd³. Such experiences underline the high cost of secondary blasting, which in this case has doubled because of the second blasting. This rounded, weathered, metamorphic rock was embedded in a matrix of softer rock. Secondary blasting is typical of seamed, fissured, and jointed rocks and of rock formations with embedded rocks. These conditions are shown in Figures 3-1, 3-3, 3-8, and 3-9.

TABLE 11-17 Representative Specifications for Jackhammers

Specification	Weight classification		
	Light	Medium	Heavy
Weight, lb	36	55	62
Length, in	20	22	24
Cylinder, bore, in	2⅜	2½	2⅝
Cylinder stroke, in	2⅛	2⅝	2¹¹⁄₁₆
Chuck size, hexagonal, in	⅞	⅞	1
Diameter of air hose, in	¾	¾	¾
Diameter of water hose, in	½	½	½
Estimated air consumption, ft³/min, at 90 lb/in²	85	90	105
Average diameter of drill hole, in	1¼	1¾	2¼

TABLE 11-18 Average Hourly Drilling Speeds for Jackhammers Based on 67% Drilling Efficiency

Class of rock	Speed, ft/h
Soft	28–40
Medium	16–28
Hard	0–16

is 67 percent for actual drilling time, 33 percent of available time being consumed by necessary operational delays and other delays lost to repairs and the like.

Auxiliary Machinery. Medium- and heavyweight jackhammers weighing up to 70 lb require 90 to 125 ft³/min in order to work at 90 lb/in². Important specifications of typical small, portable air compressors for use with jackhammers are given in Table 11-10 and they range in capacity from 150 to 365 ft³/min. When two of the heavyweight jackhammers are used, it is well to have an adequate air compressor of 250 ft³/min or larger.

Usually the jackhammer works close to the compressor and is connected directly to the air receiver by a single hose of ¾- or 1-in diameter. The friction losses in hose and fittings are nominal, as shown in Tables 11-12 and 11-13.

Bit and Drill-Rod Life. Different shapes of bits of the chisel type are shown in Figure 7-12. Choice of bit shape depends on characteristics of the rock and is based largely on the experience of the driller. Drill rods of ⅞- and 1-in diameter are used with length increments of 24 or 30 in. The consumption of bits and drill rods is variable and some generalizations are offered as guides in Tables 11-19 and 11-20.

The four medium-weight jackhammers of Figure 11-59 are drilling in medium-hard limestone for primary blasting of an access road to a dam site. Table 11-17 indicates that four medium-weight jackhammers would consume 360 ft³/min when drilling at the same time. However, they do not drill simultaneously, and Table 11-21 gives multipliers to adjust rated air consumptions to actual requirements when several rock drills are working off the same air compressor or compressors, as in the illustration.

For four jackhammers the adjustment averages 12 percent, the multiplier equals 0.88, and the estimated consumption is $0.88 \times 360 = 317$ ft³/min. The truck-mounted 365-ft³/min compressor, closely coupled to the four jackhammers with negligible air friction losses, is adequate for the work.

Explosives. Explosive practices in jackhammer work may be determined from the section on principles of blasting of this chapter. Because of the small hole diameters, 1¼ to 2¼ in, stick dynamite is used normally, as in the following estimate for productions and costs.

Production and Costs. An estimate for the production and costs for the primary blasting of Figure 11-59 is tabulated below.

Data controlling the estimate for blasting in the through cut of medium-hard limestone are as follows for a total yardage of 1780 yd³, of which 740 yd³ is to be blasted.

Explosives and accessories:	
Range of hole depths, ft including 2 ft subdrilling	4–12
Range of patterns of holes, ft	3×3 to 5×5
Average pattern of holes, ft	4×4
Cubic yard per foot of hole, average pattern (Table 11-2)	0.59
Cubic yards per hole, average 6 ft to grade, 6×0.59	3.6
Trial powder factor, lb/yd³	1.5
Pounds explosives per hole, 1.5×3.6	5.4
Average height of column of explosives, ft	6
Pounds explosives per foot of column, 5.4/6	0.9
Explosives, 50% extra dynamite in 1¼ in×8 in cartridges, loading density	1.25
Accessories, detonation by MS-delay electric caps	
Pounds of explosives required, 1.5×740	1110
Drilling:	
Average depth of holes, ft	8
Area to be drilled, ft²	4020
Holes to be drilled, 4020/(4×4)	252
Total footage of holes, ft, 252×8	2016
Estimated average speed of drilling, ft/h (Table 11-18)	16
Drill hours required, 2016/16	126
Hours of drilling for four jackhammers, 126/4	32
Hand loading of explosives and accessories:	

Estimated that two powder operators can load and detonate 242 holes with 1110 lb dynamite in 16 h. The loading rate is 35 lb of cartridge dynamite and accessories per labor-hour.

Costs for drilling, explosives and accessories, and hand loading of explosives and accessories, as of 1978, are as follows:

Drilling:	
Machines, cost	
365 ft³/min skid-mounted compressor, diesel engine	$18,800
Four-wheel drive off-road truck carrier, diesel engine	30,000
Four medium-weight jackhammers, rods, and bits, @ $1100	4,400
Four 50-ft lengths of ¾-in hose and fittings, @ $110	440
Total cost	$53,640
Hourly cost of ownership and operation:	
Machines (Table 8-3)	
365 ft³/min compressor, skid-mounted. 48% of $18.8	$ 9.00

11-98 Fragmentation of Rock

Truck carrier. 53% of $30	15.90
Four medium-weight jackhammers, 86% of $4.4	3.80
Four 50-ft lengths of ¾-in hose. 41% of $0.4	.20
Four drillers @ $12.20	48.80
Total	$77.70
Hours for drilling	32
Total direct job cost, 32 × $77.70	$2486
Direct job unit cost, per yd³, $2486/740	$3.36
Explosives and accessories:	
50% extra dynamite, 1110 lb @ $0.53	$588
Electric caps, MS-delay, 242 @ $0.72	181
Trunk wire, 1200 ft @ $0.02	24
Total	$793
Direct job unit cost, per yd³ $781/740	$1.07
Hand loading of explosives and accessories:	
Hourly cost, 2 powder operators at $12.10	$24.20
Total direct job cost, 16 × $24.20	$387
Direct job unit cost per yd³, $387/740	$0.52
Total direct job unit cost per yd³	$4.95

The high unit cost of $4.95/yd³ accentuates the cost of blasting small-yardage shallow cuts by use of jackhammers and dynamite as contrasted with work with large-hole drills and blasting agents. In general, the unit cost of drilling and loading the holes decreases inversely as the area of the pattern of holes.

The estimated direct job cost of drilling, explosives, and loading explosives in the 740-yd³ rock excavation is $3663.

The heavyweight jackhammer of Figure 11-60 is blockholing 3½-yd³ rocks for secondary blasting in a granite cut. One 1½-in-diameter, 3-ft-deep hole is drilled into each rock and the hole is loaded with two sticks of 1¼ in × 8 in dynamite. The hole is stemmed tightly to the collar. The entire operation requires about 0.5 h. The auxiliary machinery consists of a 250-ft³/min air compressor and a service truck. The driller does the powder work. Detonation is by electric blasting cap.

Figure 11-61 illustrates a not uncommon but expensive unsuccessful secondary blasting which must be repeated for satisfactory fragmentation for the rock fill of the freeway. Any

TABLE 11-19 Estimated Life of Tungsten Carbide Chisel Bits of Jackhammers with 5 to 15 Regrinds

Class of rock	Life, ft of drilled hole
Soft	800–1200
Medium	400– 800
Hard	0– 400

TABLE 11-20 Estimated Life of Alloy Steel Drill Rods for Jackhammers

Class of rock	Life, ft of drilled hole
Soft	6000–9000
Medium	3000–6000
Hard	0–3000

TABLE 11-21 Multipliers for Air Consumption of a Number of Rock Drills Working off the Same Air Compressor or Compressors

Number of drills	1	2	3	4	5	6	7	8	9	10
Multiplier	1.0	2.0	3.0	4.0	5.0	6.0	6.3	7.2	8.1	9.0

form of secondary blasting, be it blockholing, snakeholing, or mudcapping as shown in Figure 11-44, usually results in a splitting instead of a breaking action. As a result fragments are large. The skill and knowledge of both driller and powder operator are important for this costly phase of blasting.

An estimate for the production and costs for the secondary blasting of Figure 11-60 is tabulated below.

Data controlling the estimate for the secondary blasting of Figure 11-60 are as follows for the 3½-yd³ rock to be blasted:

Drilling, loading of explosives, and detonating time, 0.5 h
Explosives and accessories, 50% dynamite, two 1¼ in×8 in cartridges, 0.9 lb. One instantaneous electric blasting cap.
Combination driller–powder operator handles the work.

Costs for drilling, explosives and accessories, and hand loading explosives and accessories (1978) are:

Drilling and hand loading of explosives and accessories:
Machines, cost
  250-ft³/min portable compressor, diesel engine $17,200
  Utility truck   10,000
  Heavyweight jackhammer with drill rods and bits   1,320
  50-ft length ¾-in hose and fittings   110
Hourly cost of ownership and operation:
Machines (Table 8-3):
  Compressor, 45% of $17.2 $ 7.70
  Utility truck, 53% of $10   5.30
  Jackhammer, 86% of $1.3   1.10
  Hose and fillings, 41% of $0.1   . . .
  Driller–powder operator   12.20
  Total $26.30
Hours for drilling and hand loading explosives   0.5
Total direct job cost, 0.5×$26.30 $13.20
Direct job unit cost per yd³, $13.20/3.5 $3.77
Explosives and accessories:
  50% extra dynamite, 0.9 lb @ $0.53 $0.50
  Electric blasting cap   .50
  Total direct job cost $1.00
  Direct job unit cost per yd³, $1.00/3.5 $0.28
Total direct job unit cost per yd³ $4.05

An estimate for the cost of secondary blasting in the case of Figure 11-61 is based on doubling the direct job cost for two blastings. Direct job unit cost equals (2×$14.20)/5 =$ 5.68.

Secondary blasting, as demonstrated in the cases of Figures 11-60 and 11-61, is expensive and can be a sizable percentage of the cost of primary blasting. A hypothetical but realistic comparison is given in the following tabulations.

Premises:
  1 million yd³ of primary blasting @ $1.25 $1,250,000
  50,000 yd³ (5%) of secondary blasting @ $4.05   202,500
  Total cost $1,452,500
Consequences:
  Relationship of cost of secondary blasting to that of primary blasting,
    $202,500/$1,250,000   16%
  Actual unit cost of both primary and secondary blasting, per yd³,
    $1,452,500/1,000,000 $1.45

If the bidder on a 10-million-yd³ rock job underestimated the amount of secondary blasting by 25 percent, the error might well add $250,000 to the cost even if high production methods for secondary blasting halved the $4.05 unit cost of the analysis.

To summarize, the hand-held jackhammer is used for shallow-depth, small-quantity primary blasting and for secondary blasting. The depth of hole is 20 ft maximum and preferably less than 10 ft. The diameter of holes ranges from 1¼ to 2½ in, averaging about 1¾ inch.

In shallow-depth primary blasting, as contrasted with average work, the hole pattern is close and cubic yards per foot of hole is less than one. The unit drilling costs are high. The cost of explosives is also high because of relatively higher unit cost per cubic yard for explosives and for devices such as electric caps and electric circuits.

In secondary blasting, costs per cubic yard are high because such work cannot usually be set up on a production basis and because of the nature of the work. The amount of auxiliary machinery is great when compared with the key piece of machinery, the jackhammer. In Figure 11-60 the jackhammer, costing $1320, is supported by $27,600 worth of other machinery.

When, at first blush, a job is considered to be jackhammer work, the estimator should exhaust the possibilities of substituting mass-production drills such as the track drill with supporting portable compressor. Under favorable circumstances such a substitution might have been made in the jackhammer drilling of Figure 11-59.

Drifters In this text the *drifter* is defined as a manual-mechanical air drill intermediate in size between the small, hand-held jackhammer and the large, completely mechanically operated hammer of big-hole drilling machines equipped with masts. The hammer delivers about 1800 blows per minute through the drill-rod assembly to the bit. Originally the drifter was designed for underground work in drifting and crosscutting in tunnels and mines, which accounts for its name. Presently large air hammers mounted on the masts of huge drills are sometimes called drifters, making the term drifter somewhat ambiguous. Representative specifications for drifters are given in Table 11-22.

The drifter assembly is mounted variously as illustrated in Figures 11-62, 11-63, and 11-64. Figure 11-62 shows the drifter in the typical work of driving a small tunnel. The drifter is mounted on the feed shell, which is supported by the column.

Figure 11-63 illustrates a unique application of a drifter with feed shell. The tractor-compressor-drill is putting down blastholes for the widening of a sidehill highway cut in schist. The 13-ft hinged beams carry a driller's basket, air hose, and complete drifter. The machine is steadied by hydraulic outriggers.

Figure 11-64 pictures a "quad" mounting of drifters on a 200-hp crawler-tractor which pulls two air compressors. This custom-built machine is typical of many ingenious applications of drifters. Three workers operate four heavyweight drifters in the drilling for a pipeline trench. Compact design, efficient operation, and mobility are features of this high-production, low-unit-cost drilling machine.

Drifters may be used dry or wet. Holes may be drilled to about 50-ft depth when the drifter is mounted as in Figure 11-63 for open-cut work. Lengths and increments of length of drill rods depend on the nature of the work. In tunnel work the lengths are adjusted to the length of the round being pulled by one blasting. In open cut work the drill rod length is about 10 ft. Hole diameters range from $1\frac{1}{2}$ to $3\frac{1}{2}$ in.

An approximate range of hourly average drilling speeds, according to hardness of rock, is suggested in Table 11-23. Speeds are based on 40 min of actual hourly drilling time or 67 percent efficiency.

Auxiliary Machinery. When a single drifter is used, the size of the matching air compressor may be determined from the estimated air consumption as listed in Table 11-22. Conservative compressor capacities are in the 165 to 365 ft³/min range. When

TABLE 11-22 Representative Specifications for Drifters

Specification	Weight classification		
	Light	Medium	Heavy
Weight less mountings, lb	110	140	180
Weight including feed shell and column support, lb	400	500	600
Length, in	32	33	35
Cylinder bore, in	3	$3\frac{1}{2}$	4
Cylinder stroke, in	$2\frac{5}{8}$	$3\frac{1}{8}$	$3\frac{3}{8}$
Chuck size, round, in	$\frac{7}{8}$–$1\frac{1}{4}$	$\frac{7}{8}$–$1\frac{1}{4}$	$1\frac{1}{4}$–$1\frac{1}{2}$
Diameter of air hose, in	1	$1\frac{1}{4}$	$1\frac{1}{4}$
Diameter of water hose, in	$\frac{1}{2}$	$\frac{1}{2}$	$\frac{1}{2}$
Estimated air consumption, ft³/min, at 90 lb/in²	155	245	350
Average diameter of drilled hole, in	$1\frac{1}{2}$	$2\frac{1}{2}$	$3\frac{1}{2}$

Figure 11-62 **Lightweight drifter with air-feed shell and pneumatic column.** The original use of the drifter for drilling in the drifts and crosscuts of mines is shown. Drilling is a manual-mechanical operation. Compressed air hoses run to the drifter, to the feed of the horizontal shell or drifter guide, and to the vertical supporting column. A water hose leads to the drifter for wet drilling. *(Ingersoll-Rand Co.)*

TABLE 11-23 Average Hourly Drilling Speeds for Drifters Based on 67% Drilling Efficiency

Class of rock	Speed, ft/h
Soft	70–100
Medium	30– 70
Hard	0– 30

drifters are used in teams, as in tunneling, total air consumption may be based on the total consumption according to the specifications of the drifters as modified by the appropriate multiplier from Table 11-21.

In tunnel drilling, air transmission is usually by trunk-line pipe from the compressor through the length of the adit and then by hoses through the laterals or cross-tees to the drifters. Such an arrangement is shown in Figure 11-40. Air friction losses may be calculated by Tables 11-12 and 11-13.

A step-by-step analysis of the selection of air compressor and pipe-hose air-transmission lines in tunnel work for coyote blasting is outlined below. The working drifter is shown in Figure 11-62, the general arrangement of adit and two cross-tees is shown in Figure 11-40, and the drilling pattern in a round of blasting is illustrated in Figure 11-41. Data and calculations may be referred to these figures.

Figure 11-63 Drifter as integral part of tractor-compressor-drill assembly. This manufacturer's adaptation of the drifter and feed shell to a wheel tractor–compressor assembly provides a portable, far-reaching air drill for scaling and blasthole drilling in sidehill cuts. The machine is steadied by four fore-and-aft hydraulic jacks. Blastholes may be drilled up to a diameter of 2½ in at considerable vertical and horizontal distances from the tractor-compressor unit. *(Schramm, Inc.)*

DATA:
    1. Two medium-weight drifters, each requiring 245 ft^3/min, are to be used (Table 11-22). Required air delivery: $2 \times 245 = 490$ ft^3/min.
    2. Length of pipe from compressor to end of adit, 175 ft. One gate valve, two 90° elbows, and two tees in adit pipe.
    3. One length of 50-ft hose to each drifter in laterals; two gate valves in each lateral hose.

CALCULATIONS:
    1. Adit pipe
        a. Table 11-14: 2-in-diameter pipe for 150 ft of pipe delivering 490 ft^3/min.
        b. Table 11-13: Friction loss in terms of equivalent length of pipe for open gate valve, two elbows, and two tees: $(1.1 + 2 \times 3.1 + 2 \times 5.2) = 18$ ft. Equivalent total length: $175 + 18 = 193$ ft
        c. Table 11-12: Total pressure loss for 490 ft^3/min at 100 lb/in^2: $1.93 \times 1.8 = 3$ lb/in^2.
    2. Lateral hoses
        a. Table 11-22: Drifter hoses are 1¼-in diameter.
        b. Table 11-13: Equivalent length of hose for two open gate valves: $(2 \times 0.74) = 2$ ft. Equivalent total length: $50 + 2 = 52$ ft.
        c. Table 11-12: Total pressure loss for 245 ft^3/min at 100 lb/in$^2 = 0.52 \times 3.2 = 2$
    3. Total friction losses, 5 lb/in^2

4. A 600-ft³/min compressor delivering 600 ft³/min at 100 lb/in² is adequate for the job.

If wet drilling is to be used in the tunnel work, a simple water system of ¾-in-diameter pipe in the adit and ½-in-diameter hoses in the laterals may be installed.

Drifter Bits. Drifter bits may be of either the chisel-edge or button kind, as illustrated in Figure 11-65. Both kinds have tungsten carbide inserts for long life and both have

Figure 11-64 "Quad" heavyweight drifters mounted on medium-weight crawler-tractor. This custom-built high-production drill for putting down blastholes for pipeline trench produces a staggered hole pattern, four holes being drilled together on each of the two rows. The tractor pulls two air compressors so that the assembly is entirely integrated. In essence a three-person crew operates four traveling drifters. The machine is especially adapted to drilling in nearly level rock formations in flat country, a favorable condition for the novel self-contained machines. *(E. I. du Pont de Nemours & Co.)*

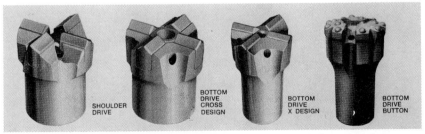

Figure 11-65 Configurations of chisel-edge and button bits as used by drifter and big-hole percussion air hammers. There are other shapes of the button bit, several of which are shown in Figure 11-58 and, like the several shapes of the chisel-edge bit, they are designed for different rocks and for different weatherings of the rock. The chisel-edge bit is a cutting tool and as such is adapted to the harder rocks. The button bit is a crushing tool and is suitable for the softer rocks. However, in drilling a 60-ft hole one may run into soft weathered rock, medium semisolid rock, and hard solid rock, so that the selection of either chisel-edge or button bit for the entire depth of hole must be left to experience and judgment. *(Joy Manufacturing Co.)*

several configurations for different rocks and diverse weatherings of the rocks. The chisel-edge bit may be reground, as shown in Figure 11-66. The bit is of 3½-in-diameter and is being used in medium granite. With an average of 16 grindings, its life averages 1200 ft of hole. The button bit cannot be ground and it averages 1800 ft in the same granite.

A comparison between estimated costs for the two kinds of bits follows:

	Chisel Edge	*Button*
Cost, 1978	$70.00	$121.00
Cost for 16 regrindings @ $1.50	$24.00	—
Total cost of life	$94.00	$121.00
Life, ft of hole	1200	1800
Cost per ft of hole	$0.078	$0.067
Cost per yd³ of granite, 2.07 yd³/ft	$0.038	$0.032
Savings by use of button bit, per yd³		$0.006
Percentage savings by button bit		16

As in the case of jackhammers, bit life and drill-rod life vary between wide extremes. Life of bits depends on several rock characteristics. Chiefly, these are specific gravity, hardness, grain structure, spallability, abrasiveness, and uniformity. Most kinds of rock vary in these features. For example, calcareous sandstone is soft and nonabrasive, whereas siliceous sandstone is hard and abrasive.

Table 11-24 is a suggested generalization for bit life in drifter drilling.

Life of drill rods is approximately a function of bit life. Again, Table 11-25 is a suggested generalization.

Explosives. Explosive practices in drifter work may be determined from the Principles of Blasting, an earlier section of this chapter. With respect to the foregoing example of the use of drifters in tunneling for coyote blasting, an estimate for explosive requirements is in the text accompanying Figure 11-41.

Production and Costs. An estimate for the productions and costs for blasting an adit and two cross-tees similar to those of Figure 11-40 is tabulated below. Data for controlling the estimate for the hard rock are taken from the examples of machinery selection in the preceding section on auxiliary machinery. The layout and the hole pattern for the round

Figure 11-66 Regrinding a 3½-inch-diameter carbide-insert chisel bit. In medium granodiorite of San Diego County these bits averaged 16 regrindings. The bit grinder is driven by an air motor and hourly production averages 10 grindings at a 1978 cost of $1.50 per bit. Five track drills average 3000 ft of hole per 8-h day. The average 90 regrindings for the footage result in 33 ft of hole for each regrinding. Bit life is about 560 ft of hole in this hard phase of drilling, but overall average bit life is about 1200 ft of hole.

TABLE 11-24 Estimated Life of Tungsten Carbide Insert Bits for Drifters

	Life, ft of hole drilled	
Class of rock	Chisel edge bit	Button bit
Soft	1200–1800	1600–2400
Medium	1600–1200	800–1600
Hard	0– 600	0– 800

TABLE 11-25 Estimated Life of Alloy Steel Drill Rods for Drifters

Class of rock	Life, ft of hole drilled
Soft	6000–9000
Medium	3000–6000
Hard	0–3000

of blasting are shown in Figures 11-40 and 11-44. Because of the confines of the tunnels and the limited yardage of 350 yd³ and tonnage of 794 tons of rock, it is assumed that the blasted rock or muck will be hand-loaded into barrows, wheeled to the adit, and hauled through the adit to dumping area by tugger hoist and skid pan. In the coyote blasting some of the rock will then be hauled back to the cross-tees for set rock over the explosive charges.

Drilling:
  Medium-weight drifters working in separate tunnels   2
  Drilling speed in hard rock, Table 11-23, ft/h   15
  Time for drilling round of holes with 1-ft subdrilling, total 54 ft: 54/15   3.6 h
  Cubic yards drilled by average 5 ft round, $6 \times 4.5 \times 5/27$ 5.0
  Total rounds drilled, 350/5.0   70
  Cubic-yards drilled hourly by 1 drifter, 5.0/3.6   1.4
  Cycle for blasting and excavating one round is estimated to be:
    Drilling   1.6 h
    Loading and detonating explosives   1.0
    Ventilating after blast   .5
    Mucking out blasted rock by hand work   3.5
    Total time, h   8.6
  As the combined time for loading and detonating explosives, ventilating, and mucking totals 5.0 h, exceeding the drilling time of 3.6 h, progress in tunnels must be based on a time of 5.0 h per round, as during the 5.0-h period the drifter will be working in another cross-tee. Time for job $=(5.0/2)70=175$ h.

Explosives and accessories:
  Diameter of hole, in   1¾
  Explosives: 50% gelatin dynamite in 1¼ in $\times$ 8 in cartridges detonated by electric blasting caps. Trial powder factor, lb/yd³   5.0
  Dynamite requirements, lb. 5.0×350 1750
  Instantaneous and MS-delay electric blasting caps, 9×70 630
  Trunk-line wire (estimated), ft   2800

The 1978 costs for drilling, explosives and accessories, and hand loading of explosives and accessories are as tabulated below.

Drilling:
  Machines, cost
    600 ft³/min portable compressor, diesel engine $27,000
    Two medium-weight drifters, complete, @ $4900   9,600
    175 ft of 2-in-diameter pipe with valves and fittings   315
    Two 50-ft lengths of 1¼-in diameter air hose with valves and fittings, at $250   500
    175 ft of ¾-in-diameter pipe with valves and fittings   120
    Two 50-ft lengths ½-in-diameter hose with valves and fittings, at $65   130
    Total cost $37,665
  Hourly cost of ownership and operation:
    Machines (Table 8-3)

600-ft³/min portable compressor, diesel engine, 48% of $27	$13.00
Two medium-weight drifters, complete, 86% of $9.6	8.30
175 ft of 2-in-diameter pipe with valve and fittings, 41% of $0.3	0.10
Two 50-ft lengths of 1¼-in-diameter air hose with valves and fittings, 41% of $0.5	0.20
175 ft of ¾-in-diameter pipe with valves and fittings, 41% of $0.1	...
Two 50-ft lengths ½-in-diameter hose with valves and fittings, 41% of $0.1	...
Two drillers at $12.20	24.40
Total	$46.00
Hours for drilling, actual plus delay time	175
Direct job cost, 1750×$46.00	$8050
Direct job unit cost per yd³, $8050/350	$23.00
Explosives and accessories:	
50% gelatin dynamite, 170 lb @ $0.57	$998
Instantaneous and MS-delay electric blasting caps, 630 @ $0.72	454
Trunk wire, 2800 ft at $0.02	56
Total	$1,508
Direct job unit cost per yd³, $1508/350	$4.31
Hand loading of explosives and accessories:	
Hourly cost for one powder operator	$12.20
Hours of loading, actual plus delay time	175
Direct job cost, 175×$12.20	$2135
Direct job unit cost, per yd³, $2135/350	$6.10
Total direct job cost	$11,693
Total direct job unit cost, per yd³, $11,693/350	$33.41

The unit cost of $33.41 does not include the cost of ventilation and mucking out the blasted rock. The volume of air for the layout of Figure 11-40 would vary greatly according to the particular conditions, the practices of the tunnel workers, and the bindings of the local regulations. Cost would vary accordingly. The mucking methods and resultant costs would likewise depend on the special working conditions and the preferences of the experienced tunnel workers.

To summarize, in this text the drifter is considered to be a manual-mechanical air hammer intermediate in size between the smaller, hand-held jackhammer and the larger mechanical hammer mounted on the mast or guide of a heavy integrated drill. The comparison is in terms of weight, bore, and stroke of the air cylinder and cubic feet per minute of air consumption.

Within this context the drifter has two principal uses in open-cut rock excavation: drilling for tunnels in connection with the coyote method of blasting, and, in special applications, as an integral part of a special-purpose drill. However, because of ambiguity in definition and understanding, the drifter often is considered to be any air hammer larger than the jackhammer, with the exception of the down-the-hole air hammer.

Track Drill In this text the track drill is defined as a crawler-mounted machine with a guide or mast carrying a heavyweight, large-capacity air hammer for drilling blastholes of 2- to 6-in diameter. The frequency of blows is about 1600/min. The drill receives air from an attending portable compressor, the compressed air being used by the hammer and air motors actuating the movements of the mast and the crawlers. Of all kinds of blasthole drills, the track drill is the ubiquitous workhorse. It is medium-weight, mobile, and efficient for average rock drilling in construction and mining.

Figure 11-67 accentuates the mobility of the track drill. It is drilling 3-in-diameter blastholes in a sidehill granite cut near Skagway, Alaska. Figure 11-68 illustrates the many possible drilling attitudes of this flexible machine. Figure 11-69 shows the working parts of a track drill.

Track drills may be classified as light-, medium-, and heavyweight, the criteria being in terms of weight of machine, weight of air hammer, air consumption of the hammer, and average diameter of holes drilled. Table 11-26 gives representative specifications for the three classes of track drills. Table 11-27 gives average hourly drilling speeds according to the class of rock.

Performance. A time study of a medium-weight track drill putting down 3½-in-diameter by 30-ft-deep holes in soft, medium, and hard zones of granite of San Diego County, California, gives these results. The drilling is shown in Figures 11-70 and 11-71.

Figure 11-67 The precipitous perch of a heavyweight track drill is seen in this sidehill cut in granite for a highway through West White Pass, Alaska. Blastholes of 3-in diameter are being drilled with a 7-ft × 8-ft production pattern. The wide-reach boom facilitates drilling several holes from the one setup. Much more time is spent in positioning the drill for a day of work than in actual drilling. The procedure for setting up the drill is to first drill a 2½-in-diameter hole by jackhammer for the anchor pin and then to hoist the drill into position with a several-part cable line by the winch of the crawler-tractor. For preshearing and production drilling, 10 heavyweight track drills and 10 supporting compressors of average 1125-ft³/min capacity are being used on the 850,000-yd³ highway job in the hard rock of the rugged mountains. *(Barbara Kalen, "Building a Road Through Historic Area and Solid Granite,"* Western Construction, *November 1975, p. 40.)*

Figure 11-68 Some of the many drilling attitudes of the track drill. Mobility and many possible stances are among the most desirable advantages of the track drill for the many working conditions encountered in blasthole drilling. From one setup several holes may be put down over a large-hole pattern, and holes may be drilled vertically or at any angle upward or downward. The track drill is not only a workhorse but also a mountain goat for the rock excavator. *(Joy Manufacturing Co.)*

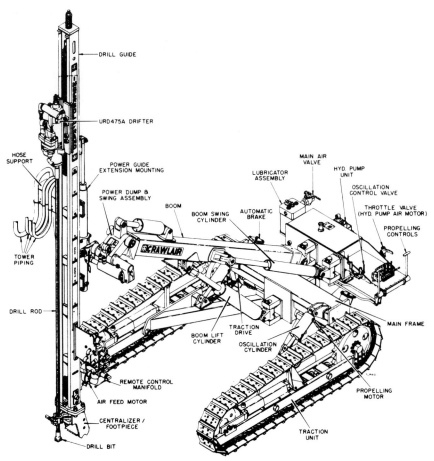

Figure 11-69 Working parts of the track drill, a compressed air–actuated machine. The hammer, mounted on the drill guide, is air-driven and fed by an air motor. The boom is positioned by hydraulic cylinders, the pump being powered by an air motor. The crawlers are driven by separate air motors, which are powerful enough not only to traction the drill on steep, soft terrain but also to pull the supporting compressor at the same time. These characteristics not only make for blasthole drilling efficiency but also for competent exploratory drilling, where the track drill not only determines the consolidation and nature of the rock but also provides data concerning drilling speeds and the abrasive nature of the rock for production blasthole work. *(Ingersoll-Rand Co.)*

TABLE 11-26 Representative Specifications for Track Drills

Specification	Weight classification		
	Light	Medium	Heavy
Weight of track drill, lb	8000	12,000	16,000
Weight of air hammer, lb	190	375	550
Air hammer cylinder, bore, in	3½	4¾	5½
Air hammer cylinder stroke, in	3⅜	3⅝	3⅝
Chuck size, round, in	1–1½	1¼–1¾	1½–2
Diameter of air hose, in	1½	1¾	2
Estimated air consumption @ 100 lb/in^2, ft^3/min	365	750	1050
Average diameter of hole, in	2	3½	5

TABLE 11-27 Average Hourly Drilling Speeds for Track Drills Based on 67% Drilling Efficiency

Class of rock	Speed, ft/h
Soft	100–150
Medium	50–100
Hard	0–50

NOTE: These ranges for drilling speeds are for holes up to 40 ft deep.

Figure 11-70 The convenient controls of a track drill. All controls, with the exception of those for moving the drill, are concentrated on the mast at the driller's position. These levers control positioning of the boom, raising and lowering the air hammer on the mast, feeding the air hammer, and supplying compressed air and water to the air hammer. From one setup of the drill, the driller can put down holes over a wide area, a most important ability in rough terrain.

11-110　Fragmentation of Rock

Figure 11-71 Heavyweight track drills in last 26-ft lift of huge granite cut. These three track drills, one in the background, are the subjects for the preceding time study and estimate for drilling production. They are putting down 3½-in-diameter blastholes in the hard granite with a combined production of about 357 yd³/h. The auxiliary air compressors consist of two of 1050-ft³/min and one of 1200-ft³/min capacity. They are drilling a few feet from the ditch line of the cut, so that at this station blasting work is nearing completion. They are part of a team of five heavyweight track drills, which furnish about 600 yd³/h to the 10-yd³ bucket-type loader in the background. Rock fragmentation is excellent.

	Time, min	*Percent*
Drilling cycle for average hole:		
Drilling:		
Soft rock, 10 ft at 5.0 ft/min	2.0	9
Hard rock, 20 ft at 1.2 ft/min	16.5	74
Average medium rock, 30 ft at 1.6 ft/min	18.5	83
Delays:		
Changing drill rods, twice	0.9	4
Cleaning out hole	0.5	2
Pulling drill rods	1.5	7
Moving drill	1.0	4
Total delays	3.9	17
Total cycle	22.4	100
Production:		
Feet of hole drilled per 50-min working hour, $50/22.4 \times 30$		67
Effective depth of hole due to 4 ft of subdrilling		26
Hole pattern, burden x spacing, ft		7×8
Cubic yards drilled per hole, $\dfrac{7 \times 8 \times 26}{27}$		54
Holes drilled per 50-min working hour, $50/22.4$		2.2
Estimated cubic yards drilled hourly, 2.2×54		119

Seismic studies of this excavation had disclosed these correlative data:
    0- to 6-ft depths, 1700 ft/s. Forecast of soft ripping.
    6- to 30-ft depths, 10,000 ft/s. Forecast of blasting.
    30- to minimum 53-ft depths, 18,000 ft/s. Forecast of blasting.

An excellent method for estimating the production of a track drill is furnished by the Ingersoll-Rand Co. and the practical procedure is set forth in Table 11-28. Although developed for use with Ingersoll-Rand Co. machinery, the table may be used for machinery of other manufacturers by methods of analogy. An example of this

TABLE 11-28 Method for Estimating the Production of a Track Drill
Based on average hourly rate of drilling, ft of hole hourly; drilling pattern of holes, burden by spacing, ft; cubic yards of rock drilled hourly and per shift; and tons of rock drilled hourly and per shift

I-R Crawlair Drill Production Calculator	
Method	Example
To use this calculator it is only necessary to know four things: the drilling equipment used (rig, drill and compressor), the type of rock, the size of the hole, and the number of hours worked per day	Calculate production in cubic yards per day for: CM350 Crawlair with VL140 drifter and DXL750 compressor. Rock: Limestone—Davenport, Iowa Hole Size: 3½ in Working 8 h/day
Step 1. Determine average drilling rate in Barre granite for specified drilling equipment, from **Table 11-28A**.	53 ft/h
Step 2. Convert to drilling rate in rock at job site. Multiply by Drillability Factor from **Table 11-28B**.	Drillability factor for Davenport limestone = 1.79 53 ft/h × 1.79 = 95 ft/h
Step 3. Determine drilling pattern, as follows: Select **Burden** from **Table 11-28C**, for hole size and type of rock. Select **Spacing** from **Table 11-28C**, for same hole size and type of rock.	Burden = 9 ft Spacing = 7 ft
Step 4. Determine cubic yards of rock displaced per hour by multiplying drilling rate (Step 2) by cubic yards displaced per foot of blasthole for pattern determined in Step 3. See **Table 11-28D**.	95 ft/h × 2.33 = 221 yd³/h
Step 5. Determine cubic yards per shift by multiplying cubic yards per hour by hours per shift	221 yd³/h × 8 h = 1768 yd³/shift
Step 6. If you wish to convert to tons per shift, multiply by tonnage factor as follows:	1768 × 2¼ = 3978 tons/shift
Rock	*Tonnage Factor*
Basalt and trap rock	2½
Granite and limestone	2¼
Sandstones	2⅛

NOTE: Although the example in this tabulation is based on the use of Ingersoll-Rand Co. drills and compressors, similar examples may be worked for comparable machinery of other manufacturers. This method by analogy is described in the notes of Table 11-28A.

extended use is given below, where the table is applied to the drilling in Figure 11-71.

DATA:
    1. Heavyweight track drill mounting a 479-lb 5¼-in-cylinder-bore air hammer.
    2. The serving air compressor is rated at 1050 ft³/min.
    3. Blasthole diameter is 3½ in and holes are 30 ft deep with 4 ft of subdrilling below finished grade. Thus, a 26-ft depth of lift is being taken out as far as basis of payment is concerned.
    4. The rock is California granite of San Diego County.

ESTIMATE:
    1. Table 11-28A shows the Ingersoll-Rand Co. drill and compressor which are closest to the specified machines to be the URD 550 hammer and DXL 1050 compressor. For a 3½-in-diameter hole in Barre granite the drilling rate, including setup and change of drill rods, is 83 ft/h.

11-112   Fragmentation of Rock

2. Table 11-28B shows the drillability factor for California granite to be 1.10. Drilling rate, $1.10 \times 83 = 91$ ft/h.

3. The 91-ft/h drilling rate should be adjusted to a 50-min working hour, becoming 76 ft/h. Note that this average drilling speed is within the range of 50 to 100 ft/h for the medium rock of Table 11-27.

4. Table 11-28C, for 3½-in-diameter hole in hard rock, gives a 9-ft burden and a 6-ft spacing. Thus the hole pattern is 6 ft × 9 ft.

5. Table 11-28D gives 2.00 yd³ of rock per foot of hole for the 6 ft × 9 ft pattern.

6. Because of subdrilling, 30 ft of hole is being drilled for the 26-ft lift of rock for which payment is made. Accordingly, for estimating purposes the hourly speed of drilling is $(26/30) \times 76 = 66$ ft/h.

7. The estimated hourly production of the track drill is $66 \times 2.00 = 132$ yd³ of granite.

The schedule beneath Table 11-28B indicates that one may expect a 3-in button-bit life of an average 650 ft in the granite. Assuming the same life for a 3½-in bit, the life in terms of the "pay" yardage of rock is $600 \times (26/30) \times 2.00 = 1040$ yd³. With a 1978 bit cost of \$113, the unit cost for button bits is \$0.109/yd³.

A comparison between the interpretation of the time-study data and calculations based on Table 11-28 for the drilling of Figure 11-71 gives good agreement. It demonstrates the feasibility of using Table 11-28 for estimating productions of track drills.

It is good practice in rock excavation to maintain as nearly a level grade as possible so that drills, shovels, rock haulers, and other accessory machinery may move easily and efficiently. After the removal of overburden or weathered rock this is sometimes not possible, but a level floor is then created after the first lift of blasted rock is removed.

Figure 11-72 shows the unavoidable slope of the rock surface after the removal of ripped rock. In this granite quarry of southern California the hole pattern is 5 ft × 5 ft, visible in the foreground. The depths of the individual 3-in-diameter holes are calculated carefully so that they bottom out in a horizontal plane to ensure a level quarry floor. These two of three medium-weight track drills are served by 900- and 1200- ft³/min compressors to ensure ample air delivery. In soft granite they drill 150 ft/h and in medium granite 80 ft/h. The average powder factor is 0.79 lb/ton. AN/FO with dynamite boosters is used and detonation is by instantaneous and MS-delay electric blasting caps. Together the three drills provide about 5800 tons of granite each 8-h shift, ample for the aggregate plant.

Figure 11-73 emphasizes an aggravatingly uneven drill stand, made at the beginning of drilling by the rough grade left by the tractor-rippers. The rock is complexly weathered schist so that it was impossible to keep a plane floor between rippable and nonrippable rock. The irregularity was typical of the 1,846,000-yd³ freeway cut in California. The subsequent grades after blasting were kept level. Holding the drill firm is somewhat difficult on a spherical surface of bald rock, but when the stand is established the driller will put down several blastholes from the same stand.

Auxiliary Machinery. The lightweight track drill requires about 365 ft³/min of air and the heavyweight unit takes about 1050 ft³/min at 100 or 125 lb/in for fastest drilling. If a down-the-hole air hammer is used, it is sometimes desirable to work at 250 lb/in². The drill is directly connected to the compressor by 50 or 100 ft of hose when greatest mobility is desired.

Figure 11-74 shows five heavy drills coupled to 1050- and 1200-ft³/min air compressors by 100-ft lengths of 2-in-diameter hose. The hoses allow the drills ample latitude for an 8-h shift of drilling.

Figure 11-75 illustrates a three-drill team of heavy track drills with 950-ft³/min compressors advancing in echelon for high-production drilling. The versatile drills are pulling their compressors as they advance in this most efficient work. Two rows of completed holes are visible to the left of the leading machine.

Where mobility of compressor is unimportant, as in the case of quarrying or dam building, air transmission is through a system of trunk-line pipe and individual hoses to the drills. In some cases the compressors are stationary.

Example of Equipment Selection for a Quarry. The following selection of drills, compressors, and air-transmission system for a large quarry is calculated for a stationary compressor layout. Requirement is the drilling and blasting of 16,000 tons of medium-

TABLE 11-28A Drilling Equipment
Average drilling rate, ft/h, in Barre granite
Overall drilling rate 40-ft hole, including setup and steel change

Crawlair Model No.	Drill model no. Drifter	Downhole drill	Compressor Model No.	Press lb/in²	Hole size, in 2¼	2½	3	3½	4	4½	5	5½
LM 100	YD 90M	…	DR 365	100	50	44	…	…	…	…	…	…
LM 300	VL 120	…	DR 600	100	…	58	44	35	23	19	…	…
	…	DHD 24	DR 365	100	…	…	…	…	…	…	…	…
CM 350	VL 120	…	DR 600	100	…	64	49	39	37	…	…	…
	URD 475	…	DXL 850	100	…	75	57	45	46	…	…	…
	URD 475	…	DXL 900	100	…	92	70	55	44	…	…	…
	VL 140	…	DXL 750	100	…	88	67	53	23	19	…	…
	…	DHD 24	DR 365	100	…	…	…	…	33	28	…	…
	…	DHD 24	DXL 600	150	…	…	…	…	62	52	…	…
	…	DHD 24	DXL 750H	250	…	…	…	…	52	44	38	…
	URD 550	…	DXL 1050	100	…	…	80	63	…	…	…	…
ECM 350	VL 120	…	DR 600	100	…	64	48	39	41	21	…	…
	URD 475	…	DXL 850	100	…	82	63	49	51	31	…	…
	URD 475	…	DXL 900	125	…	101	77	60	48	57	…	…
	VL 140	…	DXL 750	100	…	97	74	58	25	48	42	…
	…	DHD 24	DR 365	100	…	…	…	…	36	…	…	…
	…	DHD 24	DXL 600	150	…	…	…	…	68	…	…	…
	…	DHD 24	DXL 750H	250	…	…	…	…	57	48	42	…
	URD 550	…	DXL 1050	100	…	…	88	69	68	57	48	…
CM 1000	URD 550	…	DXL 1050	100	…	…	104	83	110	95	83	…
	VL 170	…	DXL 1200	100	…	…	…	…	62	52	…	…
	…	DHD 24	DXL 750H	250	…	…	…	…	…	…	…	…

TABLE 11-28A Drilling Equipment (Continued)
Average drilling rate, ft/h, in Barre granite
Overall drilling rate 40-ft hole, including setup and steel change

Crawlair Model No.	Drill model no.		Compressor		Hole size, in							
	Drifter	Downhole drill	Model No.	Press lb/in²	2¼	2½	3	3½	4	4½	5	5½
CM 1000	...	DHD 15	DXL 600	150	...	...	...	...	...	...	26	22
	...	DHD 15	DXL 750H	250	...	...	...	...	...	...	49	42

NOTE: Table 11-28A is prepared for Ingersoll-Rand Co. drill and compressor models. In order to reconcile the drill data with those of drills manufactured by other companies, the weights and the bores × strokes for the several drills are tabulated:

Model of Drill	Weight, lb	Bore × Stroke, in
YD 90M	187	3½ × 3⅜
VL 120	187	4¾ × 3⅝
URD 475	377	4¾ × 4⅝
VL 140	430	5½ × 3⅝
URD 550	460	5½ × 3⅝
VL 170	840	6¹¹⁄₁₆ × 3⅝
DHD 24 (downhole drill)	72	2¹¹⁄₁₆ × 6
DHD 15 (downhole drill)	131	3⁷⁄₃₂ × 5

The compressor ratings are standard, the numeral indicating the rating in cubic feet per minute at 100 lb/in². Accordingly, compressors of other manufacturers may be compared readily with the models of Table 11-28A. By means of these reconciliations the drilling speeds for a given combination of drill and compressor may be estimated by means of Table 11-28A.

TABLE 11-28B Drillability of Various Rocks

No.	Rock type	Location	Drill-ability factor	Abrasive index
1	Barre Granite	Barre, Vermont	1.00	1.00
2	Granite	Bulgaria	0.45	2.29
3	Granite	Grand Coulee, Wash.	0.50	2.40
4	Granite Gneiss	Hamburg, N.J.	0.67	1.46
5	Granite	Westchester, N.J.	0.67	1.00
6	Granite	Snettisham Dam, Alaska	0.78	0.74
7	Granite Gneiss	Vancouver, B.C.	0.89	1.03
8	Protogine Granite	Mont Blanc Tunnel, France	0.92	0.86
9	Granite	Newark, N. J.	1.05	1.27
10	Granite	California	1.10	0.54
11	Granite	Dworshak, Idaho	1.11	1.14
12	Granite Gneiss	Denver, Colorado	1.52	1.00
13	Rhyolite	Kirkland Lake, Ontario	.60	0.60
14	Porphyry	Denver, Colorado	.82	2.05
15	Porphyry	Murdockville, Quebec	.89	0.72
16	Diorite	Oregon	0.34	8.20
17	Basalt	New York State	0.56	3.76
18	Pegmatite	Vancouver, B.C.	0.67	2.14
19	Felsite	Denver, Colorado	0.75	0.25
20	Chalcopyrite	New Guinea	0.78	0.36
21	Banded Gneiss	Solna, Sweden	0.89	0.77
22	Magnesite	Vienna, Austria	0.94	0.04
23	Andesite	Mossy Rock, Wash.	1.27	0.28
24	Quartzite	Canada	0.33	1.45
25	Quartzite	Minnesota	0.56	8.60
26	Quartzite	Canada	0.72	3.17
27	Quartzite	New Zealand	0.78	1.70
28	Quartzite	Carters Dam, Georgia	1.00	1.40
29	Quartzite	Capetown, South Africa	1.22	2.70
30	Magnetite	Canada	0.55	3.50
31	Magnetite	Kiruna, Sweden	0.56	1.54
32	Magnetite	Kirkland Lake, Ontario	0.59	1.41
33	Magnetite	Kiruna, Sweden	.67	1.38
34	Taconite	Kirkland Lake, Ontario	.84	4.13
35	Taconite	Kirkland Lake, Ontario	.86	2.91
36	Siderite	Suffern, N.Y.	.89	.55
37	Siderite	Sarajevo, Yugoslavia	.90	.80
38	Magnetite	Kiruna, Sweden	.95	1.21
39	Magnetite	Kiruna, Sweden	1.00	1.23
40	Siderite	Sarajevo, Yugoslavia	1.00	.55
41	Magnetite	Kiruna, Sweden	1.22	.33
42	Hematite (Red)	Sarajevo, Yugoslavia	1.50	.40
43	Hematite (Dark)	Sarajevo, Yugoslavia	2.20	.70
44	Sandstone	Michel, B.C.	.75	2.80
45	Sandstone	New Zealand	2.30	1.20
46	Sandstone	Nova Scotia	2.70	0.14
47	Shale	Michel, B.C.	.75	2.80
48	Shale	Scranton, Pa.	2.00	.00
49	Sandy Dolomite	Hanover, Pa.	.60	0.14
50	Limestone	Washington	.61	0.54
51	Limestone	Vancouver, B.C.	.67	0.44
52	Limestone	Denver, Colorado	.78	0.36
53	Limestone	Washington	.78	0.67
54	Limestone	Washington	.89	0.11
55	Calcite	Hanover, Pa.	.89	0.20
56	Limestone	Millerville, Va.	.89	0.12
57	Limestone	Buffalo, N. Y.	.89	0.09
58	Limestone	Bellefonte, Pa.	.94	0.09
59	Limestone	Tulsa, Okla.	1.12	0.03
60	Limestone	Tulsa, Okla.	1.19	0.10
61	Limestone	Saratoga, N. Y.	1.22	0.01

TABLE 11-28B Drillability of Various Rocks *(Continued)*

No.	Rock type	Location	Drillability factor	Abrasive index
62	Dolomite	Hanover, Pa.	1.70	0.01
63	Limestone	Portsmouth, N. H.	1.77	0.65
64	Limestone	Davenport, Iowa	1.79	0.28

NOTE 1: Rock drillability: multiply factor times known rate in Barre granite.
NOTE 2: Anticipated bit life: divide known bit life Barre granite by abrasive index.
NOTE 3: Based on abrasive index, the average estimated 3-in button bit life should be within the following ranges:

Granites	400–900 ft
Quartzites	200–500 ft
Limestones	1300–4000 ft
Taconite	200–300 ft
Sandstones	200–300 ft
Basalt	100–300 ft
Diorite	100–200 ft
Magnetite	250–700 ft

TABLE 11-28C Determination of Hole Pattern (Burden × Spacing)

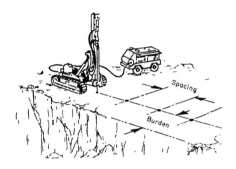

	Burden, ft					
	Hole diameter, in					
General type of rock	2½	3	3½	4	4½	5	5½
Heavy, strong rocks (granite & basalt)	6	7.5	9	10.5	11	12	13.5
Medium rocks (limestone)	6.5	8	9	10.5	12	13	14
Light, weak rocks (sandstone & limestones)	7	9	11	12	13	15	17

	Spacing, ft					
	Hole diameter, in					
General type of rock	2½	3	3½	4	4½	5	5½
Heavy, strong rocks (granite & basalt)	4	5	6	7	7.5	8	9
Medium rocks (limestone)	5	6	7	8	9	10	11
Light, weak rocks (sandstone & limestones)	6	7.5	9	10	11	12.5	14

TABLE 11-28D Cubic Yards of Rock Displaced per Foot of Borehole

Spacing burden on boreholes, ft	Average burden of boreholes, ft														
	4	5	6	7	8	9	10	11	12	13	14	15	16	17	18
4	.59	.74	.89	1.04	1.19										
5		.93	1.11	1.30	1.48	1.66	1.85	2.03	2.22						
6			1.33	1.56	1.77	2.00	2.22	2.44	2.65	2.88					
7				1.81	2.00	2.33	2.70	2.85	3.11	3.37	3.63				
8					2.37	2.65	2.96	3.26	3.55	3.85	4.15				
9						3.00	3.33	3.66	4.00	4.33	4.67	5.00			
10							3.70	4.07	4.44	4.81	5.18	5.55	5.92		
11								4.48	4.88	5.30	5.70	6.11	6.52	6.90	
12									5.33	5.77	6.22	6.66	7.11	7.56	8.00
13										6.26	6.74	7.22	7.70	8.19	8.67
14											7.26	7.77	8.30	8.80	9.33
15												8.33	8.80	9.44	10.00
16													9.48	10.07	10.66
17															11.33
18															12.00

SOURCE: Ingersoll-Rand Co.

Figure 11-72 Drilling on a 5:1 slope in a quarry. Overburden in this granite quarry has been removed, exposing the shield of acceptable rock with an 11° dip from the far drill to the near drill. The depths of the individual holes in the 5 ft × 5 ft hole pattern have been calculated so as to produce a level quarry floor after the blasting of this first lift. Thereafter, in the second lift and in the succeeding several lifts, all holes will be of the same depth in each lift. On this mild slope there is no problem in maintaining a stable drill stand.

Figure 11-73 Unstable drill stand in an unevenly weathered rock formation. The 1,846,000-yd³ freeway cut is in a complexity of metamorphic and volcanic rocks in which schist predominates. The weathering and consolidation of the rocks are variable so that this bald knob remained after the end of ripping and before blasting. With few reservations, systematic blasting began at this horizon and continued down to grade. There were several of these troublesome humps in the first lifts of the blasting. Once the driller positioned the track drill firmly atop the smooth, rounded surface, several holes were put down from the same drill stand. Of all drills for production work, the track drill is best suited to the many operational problems encountered here.

Blasting 11-119

Figure 11-74 Five heavyweight track drills working in the final lift of a 1,327,000-yd³ granite cut. About 860,000 yd³, or 65 percent of this huge cut, was blasted, a 25-ft topmost zone in the 150-ft-deep cut having been ripped. Presently 3½-in-diameter holes are being put down on 7 ft × 8 ft pattern in the 26-ft deep lift to grade. Holes are put down to 30-ft depth, allowing for 4 ft of subdrilling. The combined hourly production of five drills is 675 yd³ of "pay" rock excavation down to finished grade, sufficient for the 10-yd³ bucket-type loader. Ahead of the drills is about 10,000 yd³ of shot rock from the previous buffer blasting.

Figure 11-75 A team of three heavyweight track drills pulling air compressors and advancing in echelon for systematic production drilling. In this large level area of nearly horizontal sedimentary rock, conditions are ideal for most efficient work. Each drill pulls its closely coupled air compressor and advances along the row of blastholes. The holes are about 12 ft deep and no change of drill rod is necessary, making for highly efficient drilling. The amount of actual drilling time relative to available time, under these ideal conditions can be in the 70 to 80 percent range. A startling contrast is the work of Figure 11-67, where this amount is probably in the 20 to 30 percent range. Between these extremes is the average efficiency of about 67 percent. *(Ingersoll-Rand Co.)*

hard thickly bedded limestone per 8-h shift for a 72-in gyratory crusher of the aggregate plant. Considerations and calculations are in the following list:
    1. The physical conditions of the proposed quarry call for the heaviest track drills, serviced by stationary air compressors driven by electric motors. Air transmission is to be by main-line pipe and individual hoses from pipe to drills. The face of the quarry is to be 30 ft.
    2. Machinery
        a. The track-drill hammer has a 5½-in cylinder bore, weighs 690 lb, and requires 1200 ft^3/min at 100 lb/in^2.
        b. The main-line pipe averages 1200 ft with a maximum 1600-ft length.
        c. Individual hoses are of 2-in diameter and 250-ft length.
    3. Established drilling and blasting procedures for the familiar limestone
        a. Efficient hole diameter for the heavyweight track drills is 5 in.
        b. Powder factor 1.1 lb/yd^3.
        c. AN/FO of 0.80 loading density works well in the dry blastholes.
    4. Track drill calculations
        a. Table 11-3, for 0.80 loading density and 5-in-diameter hole, gives loading of 6.8 lb per foot of hole. The yield per foot of hole with a powder factor of 1.1 is 6.8/1.1 =6.2 yd^3. Table 11-2 gives a hole pattern of 12 ft×14 ft for this yield. The trial-and-error initial pattern will be 12 ft×14 ft. The holes are to be subdrilled 6 ft, so that about 6 ft of stemming will be used from the top of AN/FO column to the collar of hole.
        b. Table 11-27 suggests an average drilling speed of 50 ft/h in the medium-hard limestone.
        c. Table 11-28 gives the URD 550 and VL 170 air hammers as closest to the specified one of 690 lb weight and 5½-in cylinder bore. The average air requirement is 1125 ft^3/min. The drilling speed for a 5-in hole in Barre granite averages 66 ft/h for the URD 550 hammer with a 1050-ft^3/min compressor or the VL 170 hammer with a 1200-ft^3/min compressor. The speed of 66 ft/h is equivalent to 55 ft/h for 50-min hour efficiency. The drillability factor for average limestone is 1.04, giving an estimated 57 ft/h for the average drilling speed. This speed correlates with the 50 ft/h of Table 11-27.
        d. The yield for each 30-ft-deep hole to grade is 30×6.2=186 yd^3 for the 12 ft× 14 ft hole pattern. For the 30-ft lift, 36 ft of hole is drilled. The hourly drilling production, then, is (57/36)186=294 yd^3, equivalent to 647 tons of limestone weighing 4400 lb/yd^3.
        e. For a production of 2000 tons/h, four track drills are required with air consumption of about 4800 ft^3/min.
    5. Compressor calculations
        a. Four 1200-ft^3/min compressors as shown in Figure 11-50 and 11-51 are logical units for the central compressor station. Combined delivery is 4800 ft^3/min at 125 lb/in^2. The combined rated horsepower of the electric motors is 1200.
        b. The theoretical horsepower for delivery of 1200 ft^3/min at 125 lb/in^2 is given by the characteristic equation under adiabatic conditions, as set forth in the text section on compressed air. Substituting in the equation:

$$\text{hp} = \frac{1.4}{1.4-1} \times 0.0643 \times 1200 \left[\left(\frac{139.7}{14.7} \right)^{(1.4-1)/1.4} - 1 \right]$$
$$= 243$$

The actual required horsepower, based on 90 percent compressor efficiency, is 270 for each compressor, or 1080 for the four air compressors.
    6. Air transmission line calculations
        a. Air transmission losses through main-line pipe and track-drill hoses may be kept to a minimum in semipermanent and permanent installations. Suggested pipe sizes for tolerable losses are given in Table 11-14. For 4800 ft^3/min and maximum 1600-ft length an 8-in-diameter pipe is recommended.
        b. The loss in this pipe is given in Table 11-12. For 4800 ft^3/min at 125 lb/in^2 it is interpolated at 0.103 lb/in^2 for 100 ft of pipe. Equivalent length of the few valves and fittings of the main line is insignificant. The total friction loss is 16.0× 0.103=1.6 lb/in^2, estimated at 2 lb/in^2. The diameter of the four 150-ft lengths

of hose is established by the diameter of the hammer inlet, which is 2 in. Table 11-13, for tee and gate valve, gives equivalent length of pipe of 11.5 ft. Total equivalent length of hose for each drill is 162 ft. Table 11-12, for 1200 ft³/min at 125 lb/in², gives loss of 7.4 lb/in² per 100-ft length of hose. Loss per length of hose is $1.62 \times 7.4 = 12$ lb/in². The total pressure loss to each track drill is $2+12=14$ lb/in². Accordingly, the estimated pressure at the four air hammers is $125-14=111$ lb/in². This pressure is adequate for desired fast work.

Bits and Drill Rods. The discussion of bits and rods in the previous section covering drifters applies generally to the track drill. Bits vary from 2- to 5½-in diameter and, as in the case of the drifter, may be either chisel-edge, capable of regrinding, or button-type. Both types have different configurations for different kinds of rock and rock formations. Rod lengths vary from 10 ft for lightweight drills to 20 ft for the heavy machines, and diameters vary from 1¼ to 2 in. A heavyweight drill can work in a 30-ft lift with but one change of rods. Estimated lives of bits and drill rods are in keeping with Tables 11-24 and 11-25 for the drifter air hammers.

Explosives. Explosive practices in track drill work may be determined from the section Principles of Blasting in this chapter.

Production and Costs. Often an estimator observes and takes notes on work in progress so as to be guided in estimating a similar project. With binoculars, stop watch, and notebook, along with experience and tact, the estimator can amass much data in a few hours. Such a study could be made of the work shown in Figure 11-74. The following data were developed.

Drilling:	
Heavyweight track drills	5
Air compressors, two 1050-ft³/min, two 1200-ft³/min	5
Diameter of blasthole, in	3½
Hole pattern, ft	7×8
Depth of lift, ft	26
Depth of hole, ft, with 4 ft of subdrilling	30
Average time, min, to put down one hole and move to the next	30
Calculations from above data:	
Number of holes drilled in 8 h, $5 \times \dfrac{60}{30}$	100
Cubic yards per hole, 26-ft lift, $\dfrac{7 \times 8 \times 26}{27}$	54
Cubic yards drilled hourly by 5 drills, 100×54	5400
Explosives and accessories:	
Approximate pounds of explosives in each hole:	
Water gel, 62%	40
AN/FO, 38%	25
Total, 100%	65
Holes loaded for one daily blast	100
Primers, lb, at 1 lb per hole	100

Calculations for one day's blast, based on the above data, are as follows:

Explosives:	
Water gel, lb, 100×40	4000
AN/FO, lb, 100×25	2500
Total, lb	6500
Cubic yards blasted, 100×54	5400
Powder factor, lb/cybm	1.2
Primers, lb, 100×1	100
Electric blasting caps, instantaneous and MS-delay, 100×1	100
Trunk-line wire, ft (estimated)	840
Loading explosives and accessories:	
Powder operators to load 6500 lb explosives in 8 h	3
Powder magazines	4
Explosives truck	1

Costs, 1978, for drilling, explosives and accessories, storage of explosives and accessories, hand-loading of explosives and accessories, and supervision are as tabulated below:

Drilling:	
Machines, cost:	
Five heavyweight track drills	$175,000

11-122 Fragmentation of Rock

Five 100-ft lengths of 2-in-diameter air hose	4,500
Five portable air compressors, diesel engine	
Three 1050-ft^3/min @ $50,000	150,000
Two 1200-ft^3/min @ $59,000	118,000
Total cost	$447,500
Hourly cost of ownership and operation:	
Machines (Table 8-3):	
Five track drills, 53% of $175	$ 92.80
Five hoses, 41% of $4.5	1.80
Three 1050-ft^3/min compressors, 45% of $150	67.50
Two 1200-ft^3/min compressors, 45% of $118	53.10
Five drillers @ $12.20	61.00
Five helpers @ $11.90	59.50
Total	$335.70
Total cost for 8 h, 8 × $335.70	$2686
Direct job unit cost per yd^3, $2,686/5400	$0.497
Explosives and accessories:	
Water gel, 4000 lb @ $0.48	$1920
AN/FO, 2500 lb @ $0.125	312
Primers, 100 @ $1.50	150
Electric blasting caps, instantaneous and MS-delay, 100 @ $1.38	138
Trunk-line wire, 840 ft @ 2.0¢	17
Total	$2537
Direct job-unit cost per yd^3, $2537/5400	$0.470
Storage of explosives and accessories:	
4 powder magazines (2 buildings and 2 heavy-duty trailers @ $5000)	$20,000
Hourly cost of ownership and operation	
Magazines (Table 8-3), 17% of $20	$3.40
Total cost for 8 h, 8 × $3.40	$27
Direct job unit cost, per yd^3, $27/5400	$0.005
Hand loading explosives and accessories:	
Machine, cost	
1 explosives truck, flat-bed, diesel	$20,000
Hourly cost of ownership and operation	
Machine (Table 8-3):	
1 explosives truck, 53% of $20	$10.60
1 truck driver	11.50
3 powder operators @ $12.10	36.30
Total	$58.40
Total cost for 8 h, 8 × $58.40	$467
Direct job unit cost, per yd^3, $467/5400	$0.086
Supervision:	
Machine cost	
1 heavyweight pickup truck	$9600
Hourly cost of ownership and operation	
Machine (Table 8-3):	
1 pickup truck, 66% of $9.6	$ 6.30
Supervisor	20.00
Total	$26.30
Total cost for 8 h, 8 × $26.30	$210
Direct job unit cost per yd^3 $210/5400	$0.039
Summary of direct job unit costs per yd^3:	
Drilling, 45%	$0.497
Explosives and accessories, 43%	0.470
Storage of explosives and accessories, %	0.005
Hand loading of explosives and accessories, 8%	0.086
Supervision, 4%	0.039
Total unit cost, per yd^3	$1.097

In summary, the track drill with its separate air compressor is the most widely used drill for hard-rock drilling in construction and open-pit quarrying and mining. It may be used with an air hammer mounted on the mast or guide or with a down-the-hole hammer for deeper holes.

The drill's agility and stability make for most efficient work in controlled blasting. Line drilling, smooth blasting, and preshearing take place along the neat excavation line, and there topography is generally rough. Figure 11-67 shows an extreme condition near the top of slope of a sidehill highway cut. The track drill had previously put down a row of slanted holes along the slope line to the left of the drill for preshearing of the face of cut.

The track drill is a favorite one for exploratory work as it can pull the air compressor over fairly rough terrain and it can drill to considerable depth with the down-the-hole hammer. Examination of cuttings provides qualitative data on the rock excavation, and the speed of drilling, which is fairly constant for any depth, gives valuable information for estimating drilling and blasting costs for the job being investigated. Rock excavation implies medium to hard rock in generally rough country, in which the advantages of the versatile track drill are evident.

Down-the-Hole Drill The down-the-hole percussion drill uses an air hammer which works in the bottom of the hole. The hammer imparts blows directly to the bit instead of to the top of the drill-rod assembly as in the case of the regular air hammer. By this means energy is not wasted in overcoming the increasing inertia of the assembly as the hole becomes deeper, and a constant drilling speed is maintained in homogeneous rock throughout the depth of the hole.

The down-the-hole hammers work at both 100 and 250 lb/in^2 in order to give high productions with necessarily limited piston displacement and to expel cuttings from deep holes. Blows per minute average about 800 at 100 lb/in^2 and 1300 at 250 lb/in^2.

The hammers may be used with medium- and heavyweight track drills in lieu of the mast-mounted hammers, as indicated in Table 11-28, part A. This adaptation is especially important in deep exploratory drilling, where the regular hammer is limited to about 80-ft depth. The down-the-hole hammer may be used dry or wet.

The modern down-the-hole drill is usually a combination machine for optional rotary drilling. The rotary drill is used for soft to medium rocks and the percussion drill is used for medium to hard rocks. This versatility is important not only for the same degree of hardness throughout the depth of the hole but also for varying hardnesses throughout the depth. For example, in the latter case the rotary drill could be used for soft shale and a shift to the percussion drill could be made when hard limestone is encountered. Typical specifications for such combination machines are given in Table 11-29.

Figure 11-76 shows a medium-weight combination down-the-hole percussive air hammer and rotary drill, truck-mounted. Figure 11-77 illustrates a heavyweight combination machine, crawler-mounted.

Figure 11-78 illustrates four representative sizes of down-the-hole air hammers for 4- to 8-in-diameter holes. These are equipped with button bits but may be equipped also with chisel-edge bits. When they are used with a reamer the blastholes may be enlarged to 10-in diameter. As the hammer must go down into a confining hole, the bore of the cylinder is necessarily limited. When working at 100 lb/in^2, the four drills average 259 ft·lb of energy per blow, but when they are working at 250 lb/in^2, their energy averages 645 ft·lb. Thus, drilling capability is increased by 150 percent by the corresponding increase in working pressure.

Table 11-30 gives manufacturer's specifications for the four down-the-hole air hammers of Figure 11-78.

A study of drilling speeds in Barre granite, as given for regular mast-mounted air hammers and down-the-hole hammers, gives these results:

Regular hammers, working at 100 lb/in^2, average   55 ft/min
Down-the-hole hammers, working at 250 lb/in^2, average   57 ft/min

For all practical purposes the average drilling speeds, as given in Table 11-31, are the same.

TABLE 11-29 Representative Specifications for Combination Down-the-Hole Percussive Air-Hammer and Rotary Drills as Relating to the Down-the-Hole Drilling

Specification	Medium-weight truck-mounted	Heavyweight crawler-mounted
Total weight with tools, lb	40,000	70,000
Engine horsepower, drill	400	400
Compressor rating, ft^3/min, at 250 lb/in^2	600	900
Diameters of blastholes, in	5–6½	5–9

Figure 11-76 Medium-weight combination down-the-hole percussive air hammer and rotary drill. This dual-purpose drill is putting down blastholes in medium limestone of an Illinois quarry. If limestone should turn hard in a lower stratum, drilling can be shifted from rotary to percussive by means of the down-the-hole air hammer. Rotary holes may be drilled up to maximum size for blasthole work. Percussive holes by down-the-hole air hammer may be drilled to about a 8-in diameter. Such a drill is excellent for exploratory drilling where soft-medium-hard rock formations are encountered to considerable depth. *(Bucyrus Erie Co.)*

Table 11-32 summarizes rates of drilling for down-the-hole hammers when drilling in Barre granite. This table is extended from Table 11-28A so as to cover all the manufacturer's down-the-hole hammers. Use of Tables 11-28 and 11-32 for estimating the performance of the medium-weight combination drill of Figure 11-76 is explained below.

DATA:
1. Rock is hard limestone.
2. Diameter of blasthole is 5½ in.
3. Down-the-hole hammer weighs 127 lb.
4. Air compressor is rated at 750 ft^3/min and 250 lb/in^2 air pressure.

CALCULATIONS:
1. Table 11-32 gives DHD 15 hammer and 750 ft^3/min at 250 lb/in^2 compressor as comparable to the above data.
2. Table 11-32, for 5½-in-diameter hole, gives 42 ft/min drilling speed in Barre granite for DHD 15 hammer and 750 ft^3/min compressor at 250 lb/in^2.
3. Table 11-28B, gives 1.04 average drillability factor for 13 limestones. Suggested drilling speed in limestone is $1.04 \times 42 = 44$ ft/min.
4. Estimated speed for 50-min working hour is $0.83 \times 44 = 37$ ft/h.
5. Table 11-28C, suggests a hole pattern of 11×14 ft for the 5½-in-diameter blasthole, and Table 11-28D, gives 5.7 yd^3/ft.
6. Hourly drill production is $37 \times 5.70 = 211$ yd^3, equivalent to $2.25 \times 211 = 475$ tons.

When track drills use down-the-hole hammers, the auxiliary machinery is the same as with mast-mounted hammers, except that drill stems are used in lieu of drill rods. When

Figure 11-77 **Heavyweight combination down-the-hole air percussive and rotary drill.** This largest of the heavyweight combination machines is drilling blastholes in an open-pit mine. For drilling with the down-the-hole air hammer, pictured in the insert, the tower handles a 40-ft drill stem, making it possible to drill out an average depth lift without adding another length of drill stem. Blastholes may be drilled up to 9-in diameter with a generous-size hole pattern. By such efficient means drilling cost per cubic yard is minimized. *(Ingersoll-Rand Co.)*

combination down-the-hole percussive and rotary drills are used, there is no auxiliary machinery except bits. These drills are entirely self-contained, with their own integral compressors. Estimated life of bits is in keeping with those of Table 11-24. Explosive practices for down-the-hole drills may be determined from the Principles of Blasting, discussed in this chapter.

Productions and costs may be estimated exactly as they are for track drills, as explained in the preceding section.

In summary, the down-the-hole air hammer has the following inherent advantages over the more common mast- or guide-mounted hammer.

1. In homogeneous rock there is a nearly constant drilling speed at different depths, whereas the speed of the mast-mounted hammer falls off rapidly after about 40-ft depth because of the inertia of the drill rod.

2. In deep holes the drilling speed of down-the-hole hammers is in excess of that of the conventional hammer.

3. Exploratory drilling may be carried down to depths in excess of 80 ft, the practicable limit for the mast-mounted hammer. This depth capacity is important when cuts in construction and open-pit mining must be explored to depths which often exceed 300 ft.

Finally, for a wide range in diameter and depth of blastholes and exploratory holes, the medium- or heavyweight track drill equipped with either a mast-mounted or a down-the-hole percussive hammer is an ideal drill.

Rotary Drills Rotary drills are associated with soft to medium rock drilling although they are used also in hard rock work. They have ample capacities for the greatest hole

Figure 11-78 Down-the-hole air hammers with button bits. These four hammers may be equipped with chisel-edge bits for unusually hard drilling and they encompass the range of hole diameters for average blasthole work, 4 to 7½ in. Larger hammers are available. At an efficient working pressure of 250 lb/in², they consume 500 to 1270 ft³/min. The hammers may be used for a track drill or for the combination air-percussive and rotary drills. Because of their nearly constant drilling speeds in homogeneous rock at different depths they are good tools for deep-hole exploratory work. *(Ingersoll-Rand Co.)*

diameters and depths and are well suited to many medium and large rock excavation jobs. The combination rotary and down-the-hole percussion and rotary drill is discussed in the preceding section of this chapter and the machines are illustrated in Figures 11-76 and 11-77.

There are two kinds of bits for rotary drilling, the drag and cone types. They are described in the auger drill and rotary drill sections of Chapter 7 and shown in Figures 7-9 and 7-15. The drag bit is associated with soft- to medium-rock drilling and the cone bit is used in medium to hard rock. The drag bit is common to light- and medium-weight drills and both bits are usual with medium- to heavyweight machines.

Drag bits are of the fishtail type for loosely consolidated rocks and of the finger type for rocks of medium weathering. Cuttings from the bottom of the hole may be removed by the flight of spirals or by compressed air.

Cone bits, generally known as Hughes bits, are of two- and three-cone types. They are made in different tooth and overall configurations for different rocks and rock formations. Broadly, the two-cone type is used in soft to medium rocks with inclined bedding and the three-cone type is used in medium to hard rocks of nearly horizontal attitudes. Cuttings are ejected by compressed air. The intricate parts of a three-cone bit are illustrated in Figure 11-79.

Figure 11-80 shows a custom-built rotary drill using a finger bit with spirals for removal of cuttings. It is drilling soft borax ore in the open-pit mine of the United States Borax and Chemical Co. at Boron, California. The finger bit and spiral are shown in Figure 7-9. The fast-drilling machine was designed by and built in the shops of the company for this particular kind of drilling. Blastholes are 4-in diameter and 34 ft deep and are being put down on a 9 ft $\times$ 9 ft pattern. About six holes are drilled hourly, resulting in 8-h-shift production of about 4300 yd³, or 6300 tons, of borax ore.

Figure 11-81 pictures a heavyweight rotary drill being used in an open-pit copper mine of Arizona. Tricone bits of up to 10½-in diameter are used and holes may be drilled up to 55-ft depth with one pass of drill stem. In this medium to hard ore, the drilling speed is about 30 ft/h and the pattern for the 10-in-diameter holes is about 24 $\times$ 30 ft. Hourly production is about 800 yd³ or 1400 tons.

The heaviest drill of this type, built by the manufacturer, puts down up to 17½-in-

TABLE 11-30 Specifications for Down-the-Hole Air Hammers

Drill model	DHD-24	DHD-15	DHD-260	DHD-17
Hole size:				
(in)	4–4½	5–5½	6, 6½, 8	7–7½
(mm)	102–114	127–140	152, 165, 203	178–191
Cylinder bore:				
(in)	2¹¹⁄₁₆	3⁷⁄₃₂	4¼	5
(mm)	68.3	81.8	108	127
Stroke:				
(in)	6	5	6	5⅞
(mm)	152	127	152	149
Blows/min:				
(100-lb/in² air)	945	840	775	755
(250-lb/in² air)	1500	1335	1235	1200
Energy per blow:				
(ft/lb @ 100 lb/in²)	125	150	320	440
(m/kg @ 7.03 kg/cm²)	17.3	20.7	44.22	60.8
(ft/lb @ 250 lb/in²)	315	375	795	1095
(m/kg @ 17.6 kg/cm²)	43.5	51.8	109.4	151.3
Overall length:				
(in)	45⁵⁄₁₆	55¹⁵⁄₃₂	59⁹⁄₁₆	62¹⁷⁄₃₂
(mm)	1151	1409	1513	1588
Outside diameter:				
(in)	3²¹⁄₃₂	4⅜	5⅜	6½
(mm)	93	111	137	165
Net weight:				
(lb)	72	131	192	329
(kg)	32.7	59	87	149
Air consumption:				
(cfm @ 100 lb/in²)	195	240	230	550
(m³/min @ 7.03 kg/cm²)	5.5	6.8	6.5	15.6
(cfm @ 250 lb/in²)	500	580	580	1270
(m³/min @ 17.6 kg/cm²)	14.2	16	16	36

NOTE: These specifications may be cross-referred to Table 11-32 for average drilling rates in Barre granite. In turn, the drilling rates in other kinds of rocks may be estimated, as explained for Table 11-28, both for these drills and compressors and for those of other manufacturers when sizes and weights of drills and capacities of compressors are known.

SOURCE: Ingersoll-Rand.

TABLE 11-31 Average Hourly Drilling Speeds for Down-the-Hole Air Hammers of Track Drills or Combination Percussive and Rotary Drills Working at 250 lb/in² Air Pressure and 67% Drilling Efficiency

Class of rock	Speed, ft/h
Soft	100–150
Medium	50–100
Hard	0–50

diameter holes. This outsize blasthole affords drilling economies as explained in the following analysis. Each foot of hole can be loaded with 83 lb of AN/FO with 0.80 loading density. Assuming a powder factor of 1.0 lb/yd³, 83 yd³ or 2241 ft³ of rock can be blasted per foot of hole. An area of 2241 ft² makes for a square hole pattern of 47 ft × 47 ft. If the machine drills at an average speed of 30 ft/h, reduced to 27 ft/h to allow for subdrilling, 2240 yd³, or about 4480 tons, of copper ore can be drilled hourly. This is equivalent to almost 36,000 tons for each 8-h shift. Such sustained high production is possible in permanently located open-pit mines where rock and operating conditions are stable.

Important specifications for the rotary drill of Figure 11-81 are listed on p. 11-130.

TABLE 11-32   Drilling Rates, in ft/h (m/h), for Down-the-Hole Air Hammers in Barre Granite*

Mounting	DHD Model No.	Compressor Capacity, ft³/min (m³/min)	Compressor Pressure, lb/in² (kg/cm²)	4 (102)	4½ (114)	5 (127)	5½ (140)	6 (152)	6½ (165)	7 (178)	7½ (191)	8 (203)	8½ (216)
CM/ECM 350 Crawlair drill	DHD 24	365 (10.3)	100 (7.03)	23 (7.0)	19 (5.8)								
		600 (17.0)	150 (10.54)	43 (13.1)	36 (11.0)								
		750 (21.2)	250 (17.57)	60 (18.3)	50 (15.2)								
CM 1000 Crawlair drill	DHD 24	600 (17.0)	150 (10.54)	47 (14.3)	39 (11.9)								
		750 (21.2)	250 (17.57)	66 (20.1)	55 (16.7)								
CM 1000 Crawlair drill	DHD 15	750 (21.2)	125 (8.79)	...	...	24 (7.3)	20 (6.1)						
		600 (17.0)	150 (10.54)	...	...	31 (9.4)	26 (7.9)						
		750 (21.2)	250 (17.57)	...	...	50 (15.2)	42 (12.8)						
T3 Drillmaster drill	DHD 15	600 (17.0)	100 (7.03)	...	...	20 (6.1)	17 (5.2)						
		750 (21.2)	250 (17.57)	...	...	50 (15.2)	42 (12.8)						

T3 Drillmaster drill	DHD 260	600 (17.0) 750 (21.2)	100 (7.03) 250 (17.57)	...	...	22 (6.7) 70 (21.3)	19 (5.8) 62 (18.9)	...	15 (4.6) 50 (15.2)
	DHD 17	900 (25.5)	100 (7.03)	...	...	...	...	28 (8.5)	25 (7.6)	22 (6.7)
T4 Drillmaster drill	DHD 15	640 (18.1) 750 (21.2)	150 (10.54) 250 (17.57)	29 (8.8) 50 (15.2)	25 (7.6) 42 (12.8)	...	...	...	...
	DHD 260	640 (18.1) 750 (21.2)	150 (10.54) 250 (17.57)	...	...	39 (11.9) 70 (21.3)	34 (10.4) 62 (18.9)	...	27 (8.2) 50 (15.2)
	DHD 17	900 (25.5)	150 (10.54)	...	...	...	...	44 (13.4)	39 (11.9)	34 (10.4)

*For 40- to 60-ft holes, including setup and steel changing.

NOTE: These drilling rates may be cross-referred to Table 11-28 for estimates of rates for other kinds of rocks. Likewise, the rates for other combinations of drills and compressors of other manufacturers may be estimated as explained in Table 11-28 when sizes and weights of drills and capacities of compressors are known. For conversion of the rates to the 50-min working hour, the values of the table should be multiplied by the factor 0.83.

SOURCE: Ingersoll-Rand Co.

11-130 Fragmentation of Rock

Approximate working weight, lb (electric-motor drive)	157,000
Maximum pulldown or bit loading weight, lb	70,000
Horsepower:	
Rotary drive	50
Air compressor	150
Hoist and propel	50
Miscellaneous	40
Total	340
Compressor capacity, ft^3/min, at 40 lb/in^2	982
Maximum length of drill stem, ft	55
Hole diameter range, in	7–11

Table 11-33 presents representative specifications for light, medium, and heavy rotary drills, and Tables 11-34 and 11-35 give average drilling rates with tricone and drag bits, respectively.

The rotary drill is a self-contained machine and needs no auxiliary equipment except

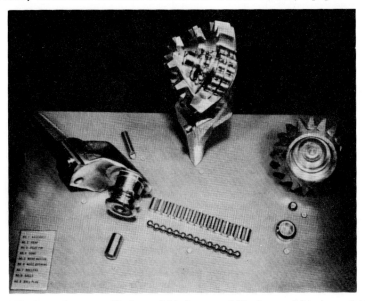

Figure 11-79 Parts of a cone or Hughes rock bit for rotary drills. Cone rock bits are available in a range of diameters for all blasthole drilling. They are of two- and three-cone types for soft, medium, and hard rock drilling. There are different configurations of teeth and cones for different rock formations. Common drilling weights vary from 2000 to 8000 lb/in of bit diameter and rotational speeds vary from 30 to 150 r/min. The bit is rotated by a square or fluted drill stem, known as a kelly, and the drill stem slides freely in the rotary table of the drilling assembly. Air passages within the bit provide for cleaning of the teeth and expelling of cuttings by compressed air. *(Hughes Tool Co.)*

TABLE 11-33 Representative Specifications for Self-Contained Rotary Drills

	Class		
Specification	Light	Medium	Heavy
Approximate working weight, lb	40,000	90,000	205,000
Pulldown or bit loading weight, lb	15,000	60,000	95,000
Horsepower, drilling	200	360	790
Compressor ft^3/min at 40 lb/in^2	600	900	1200
Kind of mounting	Truck	Truck	Crawlers
Maximum length of drill stem, ft	30	30	50
Hole diameter range, in	5–8	6–10	9–15

Figure 11-80 Rotary drill with finger bit and spiral flight putting down blastholes in soft borax ore. The company-designed and company-built special-purpose drill is in the open-pit mine of the United States Borax and Chemical Co. at Boron, California. The drill is one of two which together drill out some 12,600 tons of borax ore each 8-h shift. Soft borax ore, weighing about 2900 lb/yd^3 per cubic yard, is ideal for rotary drilling with finger bit. Hole diameter is 4 in, hole pattern is 9 ft × 9 ft, lift is 30 ft, and depth of holes is 34 ft. About six holes are drilled hourly, giving an average drilling speed of 204 ft/h for each drill.

TABLE 11-34 Average Hourly Drilling Speeds for Rotary Drills Using Tricone Bits Based on 67% Drilling Efficiency

Class of rock	Speed, ft/h
Soft	70–100
Medium	40– 70
Hard	0– 40

NOTE: The ranges of speeds are for holes up to 60 ft in depth.

TABLE 11-35 Average Hourly Drilling Speeds for Rotary Drills Using Drag Bits of Fishtail or Finger Type Based on 67% Drilling Efficiency

Class of rock	Speed, ft/h
Soft	120–240
Medium	0–120

NOTE: The ranges of speeds are for holes up to 60 ft in depth.

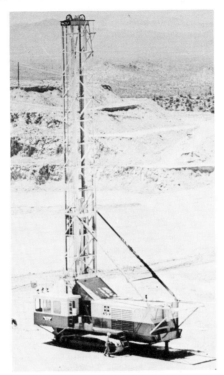

Figure 11-81 Heavyweight rotary drill with tricone or Hughes bit putting down blastholes in medium to hard copper ore. This intermediate-size drill, one of three sizes of heavyweight drills built by the manufacturer, is working in an open-pit copper mine in Arizona. The drills, powered by electric motors or diesel engines, have the following ranges of important specifications which control performance.

Specifications	Small drill	Large drill
Total horsepower, electric motors	240	400
Total horsepower, diesel engine	215	480
Compressor at 40 lb/in^2, ft^3/min	765	1310
Maximum length of drill stem, ft	40	65
Hole diameter range, in	7–9	9–15
Weight, lb	93,000	233,000

(Bucyrus Erie Co.)

the extra drill stems and bits. Usually masts and drill stems are long enough to put down blastholes with one pass for the lift of rock. Bit life estimates for roller bits are presented in Table 11-36. The life of drag bits is high, as they are used in soft and medium rocks. For this reason and because they come in many forms and of different alloy steels no table of suggested bit life is given.

Explosives practices in rotary drill work may be determined from the section Principles of Blasting in this chapter.

An estimate for production and costs of blasthole drilling by a rotary machine identical to that of Figure 11-81 is based on the following data.

    1. Rock is hard limestone overburden of an open-pit coal mine of Ohio.
    2. Diameter of blasthole is 9 in, based on local experience in the limestone.
    3. Average drilling speed is estimated at 30 ft/h.
    4. Depth of lift or height of face is 40 ft.

TABLE 11-36 Estimated Life of Two- and Three-Cone Roller Bits for Rotary Drills

Class of rock	Feet of hole drilled
Soft	2500–4000
Medium	1000–2500
Hard	0–1000

For 6-in-diameter holes and 40-ft face, the pattern indicated by Table 11-4 for hard limestone is 12 ft×16 ft. Adjustment for 9-in-diameter holes gives a 1.5 ft×12 ft by 1.5 ft×16 ft or 18 ft×24 ft pattern. This pattern is also in accord with local practices in quarries and mines.

```
Production:
  Cubic yards per lineal foot of hole (Table 11-2)            16.0
  Cubic yards per hole per 40-ft lift, 40×16.0                 640
  Total depth of hole ft, with 5 ft of subdrilling              45
  Holes drilled hourly, 30/45                                 0.67
  Cubic yards drilled hourly, 0.67×640                         429
Costs, 1978:
  Machine cost                                            $460,000
  Hourly cost of ownership and operation:
    Machine (Table 8-3), 18% of $460                       $ 82.80
    Driller and helper                                       24.10
    Total                                                  $106.90
  Unit cost per yd³, $106.90/429                            $0.249
```

To summarize, the rotary drill is a high-production large-diameter blasthole drill for soft to medium-hard rocks. It is self-contained and is sometimes combined with the down-the-hole percussion air-hammer drill for hard rock work, thereby becoming a versatile drill. It is used largely in massive public works such as dams and in huge private works such as open-pit ore and coal mines and quarries. In these rock excavations big-diameter holes and large hole patterns are feasible and economical.

Examples of Successful and Economical Drilling and Use of Explosives

High Face Although most blasting takes place in lifts or benches up to about 60-ft height of face, there are examples of considerably higher faces, as shown dramatically in Figure 11-82. This blasting of a 150-ft face gave excellent fragmentation of the limestone. Additionally, bringing all rock down to the quarry floor made for safety and economy by keeping shovel loading and truck haulage at the same elevation.

The blastholes were 6-in diameter on a pattern of 25-ft spacing and 35-ft burden and with 5 ft of subdrilling below the grade of the quarry floor. High faces make for generous hole patterns, resulting in this case in a yield of 32 yd³ or about 80 tons of limestone per foot of blasthole.

Total explosives were 62,300 lb and 350,000 tons of rock were brought down, giving an unusually low powder factor of 0.18 lb/ton. The combination of large hole pattern and high powder factor resulted in unusually low unit cost for drilling and blasting.

Delay Detonation Figure 11-83 illustrates the positive effect of delay firing in contrast to instantaneous detonation. At the left of the picture the well-fragmented rock is the result of short-interval MS-delay initiation, whereas at the right the blocky rock is the result of no-delay instantaneous firing. The results of simultaneous firing will be costly shovel loading and considerable expensive secondary blasting.

Buffer Blasting The results of confining explosive forces by buffer blasting is well illustrated in the sequential pictures of Figure 11-84. The upper picture shows the blast-hole pattern for the blasting of the ramp. On both sides of the pattern are the confining prisms of solid rock. The lower picture shows the excellent fragmentation resulting from the confined interacting explosive forces. Probably the good gradation of rock results in about 33 percent voids, and so the pile of blasted rock, along with the invisible volume below grade, is about 1.5 times the original volume.

Sequential Firing and Buffer Blasting Another excellent example of sequential or delayed firing is illustrated in the step-by-step pictures of Figure 11-85. The blast was set

Figure 11-82 Economical drilling and blasting in a 150-ft-high face. *(Top)* A 150-ft face in a Texas limestone quarry. *(Center)* 62,000 lb of explosives in action. *(Bottom)* Result: good fragmentation and plenty of rock. An unusually high face of 150 ft, a modest hole diameter of 6 in, a generous hole pattern of 25 ft × 35 ft, and breakage by fall of rock contributed to low unit cost for drilling and blasting and to excellent fragmentation of the hard limestone rock. A remarkably low powder factor of 0.18 lb/ton was made possible by the combination of circumstances and astute planning of the operation. *(Hercules, Inc.)*

off after the loading of the holes as shown in Figure 11-46. The good fragmentation, resulting in negligible secondary blasting, is the result of both sequential firing and buffer blasting.

Picture *a* shows the area to be blasted, in back of which is the pile of previously shot rock which will act as a buffer. The powder operator is setting the distribution box for sequential firing in order to detonate the five rows of holes parallel to the face at about 17-ms intervals. The first detonation is in the row closest to the face. Data covering the blast are: (1) depth of lift, 26 ft; (2) depth of hole, 30 ft, with 4 ft of subdrilling; (3) cubic yards of "pay" rock to grade, 4805; (4) a total of 77 holes of 3½-in diameter; (5) hole pattern 8 ft × 8 ft; (6) five rows of holes parallel to the face; (7) explosives consisting of 3460 lb of water gels and 2310 lb of AN/FO, totaling 5770 lb; (8) powder factor in the hard granite, 1.2 lb/yd³.

Picture *b* shows the blast. The heights of the plumes from the blastholes show the results of the sequential firing, as the plumes are successively shorter in the rows of holes farther from the face. As the picture was taken, there was no throw of rock of any size, as is demonstrated by the foregrounds of the three pictures.

Picture *c* emphasizes the results of excellent blasting. Fragmentation is good. Secondary blasting will be nil. There are no new rock fragments in the area of future drilling. These results are testimony to good work. The blasted pile of granite will act as the buffer for the next blast.

Coyote or Tunnel Blasting A successful large-scale coyote blast was set off in 1976 in the quarry of the Oregon Portland Cement Co. near Durkee, Oregon. Figure 11-86 is the artist's conception of the coyote shot. The plan is a combination of regular tunnels to shatter and heave the burden up and out and downholes to shear off the back face.

Figure 11-87 is a plan view of the two adits or main tunnels and the six laterals or

Figure 11-83 Delayed detonation as contrasted to instantaneous detonation. These two adjacent quarry blasts show, to the left, the good fragmentation attained by short-interval delay initiation as contrasted, at the right, to the bad fragmentation resulting from instantaneous firing. In such delayed detonation the row of holes next to the face is fired instantaneously, the next row back is fired by perhaps 25-ms delays, the third row by 50-ms delays, and the remaining rows by progressive delays increasing by 25 ms. The exact millisecond intervals between the rows of holes depend upon the nature of the rock and largely upon the experience of the powder operator. *(E. I. du Pont de Nemours & Co.)*

11-136 Fragmentation of Rock

Figure 11-84 Blasting by double buffers. Sequential pictures. The explosive forces will be confined by the solid prisms on each side of the blasthole pattern. Excellent rock fragmentation has been secured by the unusual double-buffer blasting. *(Hercules, Inc.)*

cross-tees, showing loading plan for 281,000 lb of AN/FO. A total of 256 downholes of 5-in diameter were drilled, the loading being about 99,000 lb of explosives. In all, the charge was 380,000 lb of explosives for about 500,000 tons of limestone. The powder factor was 0.76 lb/ton or 1.66 lb/yd^3.

The rich limestone, averaging 98 percent calcium carbonate, is highly faulted and dipped at 40 to 55°. The uplifted faulted blocks are of the Triassic and Carboniferous systems and average about 230 million yr age. The topography and lithology of the limestone influenced the choice of coyote blasting as against the regular lift or bench method using conventional blastholes. Of considerable importance was the immediate bringing of the working level down to a safe and economical elevation.

Coyote tunneling is confined and necessarily costly. In this instance a combination of manual and machine work was used in the tunneling work and the loading of explosives within the tunnels.

The blast and the resulting 0.5 million tons of limestone are shown in Figure 11-88.

(a)

(b)

(c)

Figure 11-85 Sequential firing and buffer blasting. *(a)* Prior to sequential firing of 5770 lb of explosives for a yield of 4805 yd³ of granite. Buffer is ahead of drilled area. *(b)* During the blast. Relative heights of plumes from the blastholes show timings of the sequential firings of the rows of holes. The earliest firing is next to the face. The latest firing is in the backmost row. *(c)* Results of the blast. An inspection by drilling and blasting supervisor, powder operator, and superintendent confirms the fine fragmentation resulting from a well-planned shot in the granite of a freeway excavation in San Diego County, California. The blasted rock lodges against the buffer pile in the background and the new pile will act as buffer for the next blast. Some 15,000 yd³ of blasted rock is in reserve for about three days of excavation work.

11-138   Fragmentation of Rock

Figure 11-86 Plan for large-scale blasting by coyote or tunnel method. The plan is a combination of the coyote method and the downhole method of blasting. The horizontal tunnels are for breaking and heaving the rock and the vertical and inclined downholes are for shearing off the face of the rock. About 500,000 tons of well-fragmented limestone was blasted with 380,000 lb of explosives. The huge prism of rock is about 170 ft deep from portal entrances, 270 ft wide, and 130 ft high at the back face. The faulted and sheared nature of the 230-million-yr-old rock formation is shown in the back face of the artist's drawing. *(Hercules, Inc., The Explosives Engineer, no. 1, 1976, pp. 2, 5, 9.)*

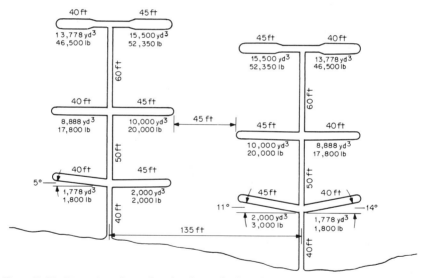

Figure 11-87 Dimensions of tunnels and explosives loading plan for cross-tees. The layout of the two adits and the six cross-tees with their loadings of AN/FO, along with the plan for the downholes and their explosives loadings, was carefully calculated for the nature of the limestone and the desired objectives of the coyote-hole blasting. The foremost objective was excellent fragmentation at minimum cost for tunneling, drilling, and explosives. *(Hercules, Inc., The Explosives Engineer, no. 1, 1976, pp. 2, 5, 9.)*

Figure 11-88 **Result of blasting by combination coyote and downhole methods.** A total of 500,000 tons of well-fragmented hard limestone was brought down by the blasting without objectionable vibrations or rock fly. All objectives were attained in this remarkable example of systematic planning and prosecution of a low-cost operation. *(Hercules, Inc.*, The Explosives Engineer, *no. 1, 1976, pp. 2, 5, 9.)*

Fragmentation was excellent, vibrations were minimal, and there was no fly rock—one of the age-old criteria for a good shot.

SUMMARY

The five foregoing examples of successful blasting methods are a few of many. Other methods are implied or suggested in the text of this chapter. Still others will occur to the readers whether they are experts or novices in the arts of drilling and blasting. And still others may result from discussions with rock experts.

In the fragmentation of rock by ripping and blasting it is well to use all sources of information in planning the work. These wellsprings include books and people. Of all people the practical and experienced driller and powder operator are most important because their knowledge comes firsthand from daily work in one of Nature's most baffling creations, rock and rock formations.

Alternate methods and combinations of alternate methods must be analyzed for the most economical method for rock fragmentation.

CHAPTER 12

Loading and Casting Excavation

PLATE 12-1

Big Muskie is the world's largest full-revolving excavator and the largest mobile land machine. It is a walker dragline, the four huge shoes serving as pads for the raising and backward thrusting of the base supporting the revolving machinery frame. Big Muskie is shown working in a coal stripping mine in southeastern Ohio. The overburden, which the walker dragline casts off the coal seam, is made up of clays, shales, sandstones, and limestones and it is about 170 ft in maximum depth. It is blasted prior to excavation. There are a number of underlying coal seams, the most important one being 36 in thick. Some 100 ft below Big Muskie is a large shovel loading coal into the coal haulers.

Specifications for Big Muskie include the following amazing outsize statistics:

Working weight	27 million lb, or 13,500 tons
Length, extended 310-ft long boom, ft	488
Width, ft	151
Height, ft	222
Total digging and casting reach, ft	575
Digging depth, ft	185
Dragline bucket capacity, yd^3	220
Average bucket load, tons	320
Combined pull of four 5-in-diameter drag cables, lb	2 million
Combined rated load for four 5-in-diameter hoist cables, lb	1.1 million
Electric power through Ward Leonard system: ac motors to dc generators to dc motors. The horsepowers of the motors are, approximately:	
Prime-mover ac motors	30,000
Bucket drag	8000
Bucket hoist	8000
Swing	6200
Walk	7200

The total horsepower for the four dragline movements is 29,400, but all these movements are not simultaneous so that the 30,000-hp prime-mover motors provide ample power to generators.

Estimated production data for Big Muskie:	
Cycle time per bucket of overburden, 180° swing, s	68
Buckets handled in 55-min working hour, 92% efficiency	49
Bucket payload, bank measurement, yd^3	146
Hourly production, cybm	7,150
Daily production, 22.5 h, cybm	161,000
Yearly production, 365 days, 95% availability, cybm	55.8 million
General information on Big Muskie:	
Cost, fob job and erected, estimated for 1978	$55 million
Number of crew of operator, oiler, and ground operator	3
Railroad cars for shipment of components of shovel	407
Maximum number of erection crew	100
Erection time, months	20
Hourly power consumption, equivalent to that of a city of 25,000 people on a monthly basis, kWh	11,000

Big Muskie may be compared with the 59 boom swing shovels used in excavating the Panama Canal. The total dry land rock excavation for the Panama Canal was 130 million yd^3 dug over a 5-yr period of single shifting work. The production of 59 shovels, ranging in dipper size from 1¾ to 5 yd^3, was 26 million yd^3 annually. The corresponding figure for Big Muskie is 19 million yd^3. Accordingly, Big Muskie excavates at a rate 73 percent of that for the 59 shovels used for rock excavation of the Panama Canal, both being on a single-shift comparable basis. *(Bucyrus-Erie Co.)*

Loading and casting are not bid items in a bid for an excavation project because both are considered to be a part of the excavation bid item. The practice is true for both public and private works, although in rare instances in private works casting is considered to be a separate operation calling for a separate bid item.

Loading and casting are done by several machines ranging in size from the 220-yd^3 dragline of Plate 12-1 down to the ¼-yd^3 backhoe of Figure 7-4. These basic types of rock excavators include: (1) tractor-bulldozers; (2) bucket-type loaders; (3) shovels; (4) backhoes; (5) cranes; (6) draglines; (7) trenchers; (8) belt loaders; and (9) wheel-bucket excavators.

All machines, except movable belt loaders and wheel-bucket excavators, both load and cast rock-earth. Bulldozers combine loading with hauling in the sense of pushing the load or bulldozing. When equipped with an angle blade or even with a tilt blade, they cast by moving rock from one side of the travel path to the other side.

Some machines have different configurations, such as crawler- or wheels-tires-mounted backhoes and bucket-type loaders with single or articulated frames.

When loading, all machines have cycles of operation. When casting, with the single exception of trenchers, the machines also have cycles. Thus the productions and the costs may be estimated.

Each machine has inherent characteristics and advantages which fit it for general or particular rock-earth excavation. In the following text each machine will be described with respect to characteristics, operations, productions, and costs.

Just as the tractor-bulldozer of this chapter is considered to be both a loading and a hauling machine, so the slackline excavator of Chapter 13 is a loading and hauling machine. Chapters 12 and 13 have considerable interrelationship.

TRACTOR-BULLDOZERS AND TRACTOR-ANGLEDOZERS

A bulldozer has two definitions: (1) a complete unit made up of a tractor, either crawler- or wheels-tires-mounted, equipped with a front-mounted blade or moldboard with controls for pushing rock-earth; and (2) the front-mounted assembly without the tractor. In this text the complete machine is a tractor-bulldozer and the assembly is a bulldozer.

Wheels-tires-mounted tractor-bulldozers are not used normally in rock-earth because of lesser traction and greater maintenance than the crawler machine. In the following text reference is to the crawler-tractor-bulldozer simply as a tractor-bulldozer.

The bulldozer is raised and lowered by one or two hydraulic cylinders or by cable hoist. The hydraulic unit affords down pressure on the bulldozer, which is an important function not obtainable with a cable hoist except by sheer weight of the blade. The blade may be tilted by a hydraulic cylinder on the left side and it may be angled up to about 25° to a tranverse axis by an angling cylinder. Tilting and angling are optional features.

Tractor transmissions may be mechanical or hydraulic-mechanical with varying ranges of forward and reverse speeds up to 7 mi/h. Ratings are for engine flywheel horsepower and, by graph, for drawbar pull versus travel speed. Drawbar pull or, for bulldozing, bulldozer push is based on tractor performance when traveling on natural ground with rolling resistance of 110 lb per ton of tractor-bulldozer weight, or 5.5 percent of weight. Drawbar horsepower may be calculated by the specification graph for drawbar pull versus speed. The average ratio of drawbar to flywheel horsepower is 0.69. Among tractors the ratio varies little.

In this text the push-pull horsepower is taken as 69 percent of the engine flywheel horsepower, the force corresponding to a given speed of tractor-bulldozer being applied at track-ground contact when the machine is in motion. If this tractive force exceeds the weight of the tractor-bulldozer times the coefficient of friction between tracks and ground, the tracks will slip. Accordingly, allowance for the coefficient of friction is always mandatory.

There are three shapes of blades or mold boards for the bulldozer, and they may be classified according to the plan view when one looks down on the bulldozer.

1. The straight blade has a straight cutting edge from side to side except for relatively small edges or end bits on each side. Physically it is smaller than the other two blades and it is most used of the three shapes because of its versatility for all kinds of work. The angle blade is a straight blade, but with lower height and greater width for side casting. A straight blade with angling and tilting features is illustrated in Figure 12-6.

12–4 Loading and Casting Excavation

2. The straight U blade, sometimes called the modified U blade, is shown in Figure 12-1. Physically it is midway between the straight and the U blade. It has intermediate concavity and it is used both for utility bulldozing and for the mass movement or haulage of rock-earth. In some designs the cutting edge is straight with large edges, and in other configurations the cutting edge is slightly concave with side edges.

3. The U blade is bowl-shaped with side or plate edges. Physically it is the largest of the blades and it is designed for mass movement of rock-earth over relatively long bulldozing distances. The U blade is illustrated in Figures 13-1 and 13-2.

In addition to these three basic shapes there are several blades which are specially built for particular operations, such as coal and iron ore handling, land clearing, and land reclamation.

In this text all bulldozer capacities, productions, and costs, unless otherwise emphasized, are in terms of the intermediate straight U blade. Corresponding data for the straight blade and the U blade would be proportionately lesser or greater and they should be analyzed individually.

The tractor-bulldozer, equipped with a ripper, is shown in Figure 12-1. It is one of the heaviest units with a working weight of 160,000 lb, engine flywheel horsepower of 524, and drawbar horsepower of 362.

The bulldozer load ahead of the huge machine may be calculated by the formula accompanying Figure 12-1, which gives a load of 18.9 cylm or 51,000 lb of average weight rock-earth when bulldozing on the level. Taking this machine as an example, the dynamics and resolution of forces in bulldozing work are set forth in the following examples of level and downgrade work.

Resolution of Forces When Bulldozing on the Level over Loose Rock-Earth

The formula of Figure 12-1 for calculating bulldozer loads in terms of dimensions of bulldozer blade is one of several used for this purpose. It is an empirical, practical formula based on actual loads, loose measurement, as measured in the field.

In theory it is known that the horizontal force of the bulldozer blade, weight of tractor

Figure 12-1 The capacity of a bulldozer, loose measurement, when bulldozing on the level, may be calculated by the approximate formula:

$$\text{Capacity} = \text{area of cross section} \times \text{two-thirds width}$$

$$\frac{1.1H \times 1.6H}{2} \times \text{two-thirds } W = 0.59 H^2 W$$

When H and W are in feet,

$$\text{Capacity in cubic yards} = 0.022 H^2 W$$
$$\text{Bulldozer capacity} = 0.022 \times 7.1^2 \times 17.0 = 18.9 \text{ cylm}$$

bulldozer times 0.40 (coefficient of traction for the machine when working on loose rock-earth), must slightly exceed the horizontal resisting force of the load, weight of load times friction coefficient of load, if the load is to move.

Unfortunately, the friction coefficient varies with the nature of the load and no empirical coefficients have been established. The coefficient represents sliding friction of the load over the ground and along the blade as well as internal friction within the mass of rock-earth when in motion. The coefficient is low in the bulldozing of clay-silt-sand residuals and high in the movement of poorly blasted rock.

The coefficient can be calculated for the load of Figure 12-1 by equating resisting force to moving force, as given by this equation:

$$(18.9 \times 2700)C = 160,000 \times 0.40$$

where 18.9 = bulldozer load, cylm
2700 = weight of average rock-earth, lb/cylm
C = coefficient of friction of load
160,000 = weight of tractor-bulldozer, lb
0.40 = coefficient of traction of tractor-bulldozer

When the equation is solved for the unknown C, C is found to be 1.25. The similarly calculated average coefficient of friction for the bulldozer loads of Table 12-2 is 1.30.

The theoretical bulldozing speed of the machine is 187 ft/min, or 2.1 mi/h. The practical speeds are about 140 ft/min, or 1.6 mi/h.

Resolution of Forces When Bulldozing Downgrade over Loose Rock-Earth

If it is assumed that the friction coefficient of the load is the same 1.25, then the dynamics of downgrade bulldozing, as well as upgrade bulldozing, can be analyzed.

In downgrade work the weights of the tractor bulldozer and the load, which act vertically, must be resolved into components at right angles and parallel to the inclined grade in order to calculate tractive and friction forces and gravitational forces along the grade. Traction and friction forces are lessened as compared with those of level bulldozing but the new gravitational forces assist the tractor bulldozer moving force and lessen the resisting force of the load.

When the machine of Figure 12-1 bulldozes down a 20 percent grade, an algebraic-trigonometric equation for the unknown load gives 88,000 lb. This is equivalent to a loose load of 32.6 cysm. The theoretical speed of the machine is 190 ft/min or 2.2. The practical speeds are about 142 ft/min or 1.6 mi/h.

In these theoretical analyses the movable loads are 18.9 cylm or 51,000 lb for level work and 32.6 cylm or 88,000 lb for 20 percent downgrade work. The load of 32.6 cylm for downgrade bulldozing is far in excess of normal load. The discrepancy is explained largely by the fact that the size of a bulldozer blade for a given tractor is in keeping with all kinds of rock-earth and working conditions. The machine of Figure 12-1 is built for bulldozing everything from wet earth to dry rock on the level, on upgrades, and on downgrades.

The conversion graph for bulldozer loads of Figure 12-2 gives realistic factors for converting loads on level to downgrade and upgrade loads. On a 20 percent downgrade the estimated load for the machine of Figure 12-1 is 1.3 × 18.9 or 24.6 cylm, as contrasted to the theoretical 32.6 cylm.

These discussions provide a study of theoretical and practical calculations of tractor-bulldozer performance. Rock-earth is one of Nature's highly variable substances and the estimator must combine theory and necessary practicability in solving earthmoving problems. One cannot gainsay the worth of empiricism, pragmatism, and experience.

Figure 12-2 contains practical data for estimating the production of a tractor-bulldozer. The conversion graph changes loose loads in level bulldozing to loose loads when working upgrade or downgrade. The conversion factors change loose measurement loads, cylm, to bank measurement loads, cybm, according to the nature of the load.

The cycle-time graph gives times for level or downgrade work according to the length of bulldozing, one way, and combines both loading and hauling operations. This combination work is illustrated in Figures 12-3, 12-4, and 12-5. The cycle-time graph is based on the following criteria:

12-6 Loading and Casting Excavation

Two changes of gears	0.20 min
Bulldozing speed forward, equivalent to 1.2 mi/h	110 ft/min
Backing speeds in reverse, ft/min:	
Level, equivalent to 6.6 mi/h	580
10% upgrade, equivalent to 5.6 mi/h	490
20% upgrade, equivalent to 4.9 mi/h	430
30% upgrade, equivalent to 3.3 mi/h	290
40% upgrade, equivalent to 2.6 mi/h	230

The load formula of Figure 12-1 and the data of Figure 12-2 enable one to estimate the hourly production of the tractor-bulldozer of Figure 12-1 when dozing ripped weathered rock 400 ft down a 20 percent grade.

Cylm bulldozer load by formula, $0.022 \times 7.1^2 \times 17.0$	18.9 cylm
Cylm load on 20% downgrade, 1.30×18.9	24.6 cylm
Cybm load on 20% downgrade, 0.78×24.6	19.2 cybm
Cycle time on 20% downgrade	5.2 min
Cycles or loads per 50-min working hour, 50/5.2	9.6 loads
Cybm bulldozed hourly, 9.6×19.2	184 cybm

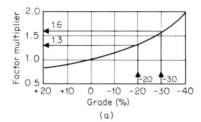

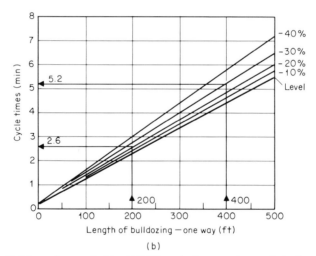

Figure 12-2 Bulldozer charts and table. (a) Bulldozer loads conversion graph, load on level, yd³, to load on given grade, yd³. Load on given grade is equal to factor times load on level. Bulldozer loads conversion factors, loose measurement, cylm, to bank measurement, cybm, are as tabulated here:

Sand-gravel	0.88
Rock-earth, weathered	0.78
Solid rock, well blasted	0.67
Solid rock, poorly blasted	0.60

(b) Crawler-tractor-bulldozer cycles for loading and hauling according to lengths of bulldozing and grades. Cycle times for upgrade bulldozing for less than 200-ft length and for less than +20 percent grades may be reasonably approximated by using the same lengths and corresponding downgrades.

Capacities and Specifications

Often when estimating the capacity of a bulldozer for a tractor one does not have available the dimensions of the blade so as to calculate capacity by the formula: capacity, cylm = $0.022H^2W$, where H = height in feet and W = width in feet. Specifications for tractor-bulldozers disclose that there is a close relationship between blade dimensions for a straight U-shaped blade and the horsepower of the tractor engine. These data suggest the following equation for calculating the capacity of the bulldozer blade in terms of the horsepower of the tractor engine.

$$\text{Capacity, loose measurement, cylm} = 0.027 \text{ hp}$$

From this equation results Table 12-1, the use of which is suggested for approximate estimates.

Bulldozer capacity for level work is calculated by the formula:

$$\text{Capacity, cylm} = 0.022H^2W$$

Typical specifications for machines with straight U blades are listed in Table 12-2, and more complete specifications for crawler-tractors without ripper and bulldozer are given in Table 13-5.

Loading by Tractor-Bulldozers

Tractor-bulldozers are used for several loading operations, of which four common ones are:

    1. Loading rock haulers by bulldozing from cut into a loading chute. This work is practicable when it is possible to bulldoze downgrade or level with minimum upgrade work. This practicability is common to all kinds of bulldozer loading operations.

    2. Hauling and lending loading assistance to the bucket-type loader as illustrated in Figure 12-3. Here the machine is both bulldozing to and helping the loader to get a maximum load.

    3. Hauling and loading rock into a movable belt loader, as shown in Figure 12-4.

    4. Hauling and loading blasted conglomerate rock into a primary crusher, as illustrated in Figure 12-5.

These four operations are combination hauling and loading work, which is called simply *bulldozing*. In this bulldozing the word "hauling" is used in the sense of transportation of the rock-earth, and thus the tractor-bulldozer is a kind of hauler.

Figures 12-1 and 12-2 and Tables 12-1 and 12-2 give data for estimating production of tractor-bulldozers in loading operations. Along with Table 8-3, these figures and tables all combine to help give an estimate for unit cost. Two examples are presented below.

EXAMPLE 1 Two 320-hp machines (Figure 12-4) are bulldozing ripped weathered basalt 200 ft down a 20 percent grade to a movable-belt loader. The bulldozer blade is 5.4 ft high and 14.2 ft wide.

Hourly production for each machine:
Load, cylm, level work (Figure 12-1), cylm = $0.022 \times 5.4^2 \times 14.2$ =	9.1
Load, cylm, down 20% grade. Figure 12-2. 1.3×9.1 =	11.8
Load, cybm, down 20% grade. Figure 12-2. 0.78×11.8 =	9.2
Cycle time, min (Figure 12-2)	2.6
Cycles or loads per 50-min working hour, $50/2.6$ =	19.2
Hourly production, each tractor-bulldozer, cybm, 19.2×9.2 =	177

Costs, 1978:
Machine, 320-hp crawler-tractor-bulldozer	$190,000
Hourly cost of ownership and operation	
Crawler-tractor-bulldozer (Table 8-3). 33% of $190	$62.70
Operator	15.60
Total	$78.30
Direct job unit cost per cybm. $78.30/177	$0.44

EXAMPLE 2 Three 385-hp machines (Figure 12-5) are bulldozing blasted conglomerate 400 ft down a 30 percent grade to a primary crusher. The bulldozer blade is 6.2 ft high and 15.8 ft wide.

12-8 Loading and Casting Excavation

TABLE 12-1 Estimated Bulldozer Capacities, Loose Measurement, for Level Bulldozing According to Tractor Engine Horsepower

Engine horsepower	Capacity, loose measurement, cylm
50	1.4
100	2.7
150	4.0
200	5.4
250	6.8
300	8.1
350	9.5
400	10.8
450	12.2
500	13.5
550	14.8
600	16.2
650	17.6
700	18.9

TABLE 12-2 Representative Specifications for Crawler-Tractor-Bulldozers with Straight U Blades

Tractor engine horsepower	Weight, lb	Blade data		
		Height, ft	Width, ft	Capacity, cylm
105	24,600	3.2	8.7	2.0
195	49,800	4.8	11.8	4.8
300	75,000	5.0	13.9	7.6
410	97,200	6.0	14.4	11.4
524	146,000	7.1	17.0	18.9
700	173,100	7.0	19.8	21.3

Figure 12-3 Bulldozing to and helping to load a bucket loader. The 410-hp tractor-bulldozer-ripper is one of two units which are ripping and bulldozing weathered volcanic rock to the bucket loader and at the same time assisting it in getting the greatest bucket loads. The money-making finesse results from the teamwork of two highly skilled operators. The bulldozer loads an average of about 7.6 cybm and the bucket loads an average of about 4.1 cybm. The bulldozing distance is about 100 ft down a 15 percent grade and the two tractor-bulldozer-rippers are adequate for the loader production of 500 cybm/h. Such efficient triple action of ripping, bulldozing, and assisting loading is typical of this well-managed, high-production rock excavation job in California freeway building.

Figure 12-4 **Bulldozing ripped weathered basalt to a movable belt loader.** Two 320-hp tractor-bulldozer-rippers are both ripping the rock and bulldozing to the belt loader. Distance is 200 ft down a 20 percent grade. Together these machines are able to bulldoze about 354 cybm/h of the ripped volcanic rock to the loader. About one-third of the available time is spent in ripping, so that the actual output to the loader is about 236 cybm/h, adequate for the 42-in belt-width loader under the particular operating conditions of inadequate rock haulers. As additional rock haulers were added for the long haul, a third machine was added and hourly production was boosted to about the capacity of the belt loader, that is, to about 340 cybm/h.

Hourly production for each machine:
Load, cylm, level work (Figure 12-1) cylm $= 0.022 \times 6.2^2 \times 15.8 =$	13.4
Load, cylm, down 30% grade (Figure 12-2), $1.6 \times 13.4 =$	21.4
Load, cybm, down 30% grade (Figure 12-2), $0.67 \times 21.4 =$	14.3
Cycle time, min (Figure 12-2).	5.2
Cycles or loads per 50-min working hour, 50/5.2	9.6
Hourly production, each tractor bulldozer, cybm. 9.6×14.3	137

Costs, 1978:
Machine, 385-hp crawler tractor bulldozer	$239,000
Hourly cost of ownership and operation	
Crawler-tractor-bulldozer (Table 8-3). 33% of $239	$78.90
Operator	15.60
Total	$94.50
Direct job unit cost	
Per cybm, $94.50/137	$0.69
Per ton, 2.03 tons per cybm. $0.69/2.03 =$	$0.34

The comparison between the unit costs of Examples 1 and 2 affords a means of appraising the costs of two different-size machines in somewhat similar downgrade bulldozing. The 320-hp machine bulldozes for $0.0110 per cybm foot of vertical distance. The corresponding figure for the 385-hp unit is $0.0058. The advantage of the latter is in greater power and proportionately larger downgrade load.

Casting by Tractor-Angledozers

Figure 12-6 illustrates an angle bulldozer which is effective in casting rock-earth to one side rather than bulldozing longitudinally as in the case of hauling. Some bulldozers are equipped with both angling and tilting hydraulic cylinders as are shown.

The hourly production of an angle bulldozer is a function of the projected effective area of the blade and the traveling speed of the tractor. The projected area of the blade is given approximately by the equation:

12-10 Loading and Casting Excavation

$$\text{Area} = 0.33 \; HL \cos A$$

where A is the angle of bulldozing. The angle bulldozer of Figure 12-6 is 4.4 ft high and 16.3 ft wide, with a 25° angle of bulldozing. Hence:

$$\text{Projected effective area} = 0.33 \times 4.4 \times 16.3 \times 0.91 = 22 \text{ ft}^2$$

Assuming an average travel speed of 0.5 mi/h or 44 ft/min, the gross yardage angle bulldozed per minute of continuous travel is given by the equation:

$$\text{Cubic yards bank measurement} = \frac{22 \times 44}{27} = 36$$

The angle bulldozer is used mostly in pioneering work on a rock job, such as building access and haul roads and cutting out the tops of slopes along the slope lines. It is not a production machine, being rather a supplementary tool, and an analysis of production and unit cost is not truly meaningful. However, the following example demonstrates principles.

Figure 12-7 shows a cut about to be opened up, and the 320-hp angle bulldozer is cutting out about 800 ft of the slope line of a sidehill cut in weathered granite. Some 980 cybm will be cast down the slope in this pioneering work, to be rehandled in the production excavation of the entire cut.

The initial cut is a triangular cross section of about 11 ft^2 and the second return cut is about 22 ft^2. At an average travel speed of 0.5 mi/h (44 ft/min), the cycle time is:

Cycle times:
First and second cut, 1600 ft at 44 ft/min, min	36.4
Two turnings at 0.5 min each, min	1.0
Total cycle time, min	37.4

Figure 12-5 Long and steep bulldozing of blasted conglomerate to primary crusher. This spectacular bulldozing involves three tractor-bulldozer-rippers which furnish aggregate to the primary crusher of a commercial stone plant of southern California. The bulk of the cemented conglomerate is blasted and the remainder is ripped by the three machines. The bulldozing distance is 400 ft down a 30 percent grade, the maximum vertical distance being 120 ft. The tractor-bulldozer-rippers are rated at 385 hp and weigh about 105,000 lb. When bulldozing only, the three machines provide about 410 cybm, or about 830 tons, hourly for the aggregate plant. The fanlike pattern of the bulldozing may be seen, while in the background center is one of the track drills putting down blastholes in the solid rock formation. *("Hawley Rock Tries New D8K on Work Previously Assigned Only D9G's," D. J. Byrnes (ed.),* Earth, *February–March 1975, p. 5.)*

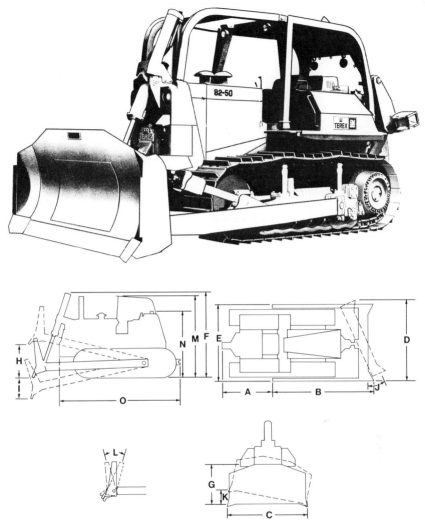

Figure 12-6 A heavyweight tractor with angle bulldozer and ripper. The salient bulldozing specifications of the machine are:

Engine horsepower	372
Weight, lb	96,000
Travel speeds:	
Forward, 0 to 7.0 mi/h, ft/min	0 to 616
Reverse, 0 to 8.6 mi/h, ft/min	0 to 757
Bulldozer blade:	
Height: 5.9 ft. Width: 14.8 ft	
Calculated capacity, cylm	11.3

(General Motors Corp.)

12-12 Loading and Casting Excavation

Hourly production:	
Cycles per 50-min working hour, 50/37.4	1.34
Cybm hourly, $1.34 \times 800 \times (11+22)/27 =$	1310
Costs, 1978:	
Machine, 320-hp crawler-tractor-bulldozer	$190,000
Hourly cost of ownership and operation:	
Crawler-tractor-bulldozer (Table 8-3). 33% of $190	$62.70
Operator	15.60
Total	$78.30
Direct job unit cost per cybm, $78.30/1310	$0.06
Direct job unit cost per foot of slope line, $(980/800) \times \$0.06 =$	$0.07

Summary

The tractor-bulldozer with straight U-type or angle-type blade is an indispensable loading, casting, and hauling machine, hauling being akin to bulldozing. The preceding discussion testifies to the versatile nature of the ever-present tractor-bulldozer.

The loading of hauling units, movable belt loaders, crushers, and excavators by bulldozers with straight U blades is common practice and is production work. Casting with an angle bulldozer, however, is pioneering work and in a sense is a nonproduction job. It includes building of access roads and haul roads, cutting out top of slopes, and the like.

Unit costs for hauling and loading may be estimated precisely. However, casting costs are elusive, the minimum being attained by the excellent supervision of a good foreman. Many times casting costs are submerged in the general rock excavation costs, as would be the case in the example of pioneering the huge sidehill cut.

BUCKET LOADERS

As distinguished from dipper shovels, bucket loaders are not revolving. Another technical difference is that the load is carried in a bucket and not in a dipper.

Figure 12-8 gives operating dimensions and illustrates the bucket action of a medium-size wheels-tires-mounted bucket loader. The machine's nominal capacity is 6 yd^3 and the

Figure 12-7 Angle bulldozer pioneering a cut by cutting out the slope line. The 320-hp tractor–angle bulldozer–ripper is taking out the first of two cuts in establishing the slope line of a huge sidehill cut in weathered granite. The 800-ft-long slope line will take about 40 min of work for the casting of some 980 cybm. The rock job calls for the widening of the existing cut, as may be seen, and the precarious nature of the bulldozing calls for a skilled operator. The machine will travel 1600 ft in making the two passes, and a cross-sectional area of about 33 ft^2 will be removed, the excavation being at the rate of 1.2 cybm per foot of slope line. The cast excavation will be handled a second time in the production excavation of the weathered and solid granite.

Bucket Loaders 12-13

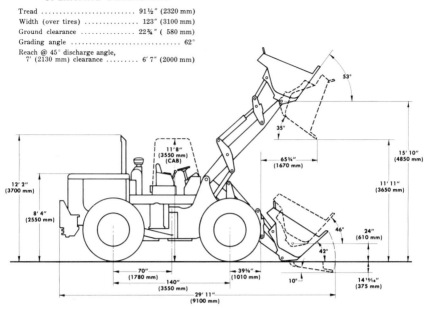

OPERATING DIMENSIONS

```
Tread ........................ 91½"  (2320 mm)
Width (over tires) ............ 123"  (3100 mm)
Ground clearance .............. 22¾" ( 580 mm)
Grading angle ................. 62°
Reach @ 45° discharge angle,
    7' (2130 mm) clearance ........ 6' 7" (2000 mm)
```

Figure 12-8 Operating dimensions and bucket action of bucket loader. The abbreviated specifications for this medium-size loader are:

Nominal capacity, yd³	6
Struck measurement capacity, cysm	5.2
Diesel-engine flywheel horsepower	325
Torque converter	single-stage, single-phase
Transmission travel speeds, forward and reverse, mi/h	
1st gear	0–4.0
2d gear	0–9.0
3d gear	0–22.1
Tire size, in	29.0–29.5 (22 PR)
Breakout force of bucket, lb	50,500
Working weight, lb	70,000

(Caterpillar Tractor Co.)

struck capacity is 5.2 cysm. Engine horsepower is 325 and working weight is 70,000 lb.

The bucket assembly is carried on the frame of the tractor. It consists of two hinged integral arms connected to the frame and bottom of the bucket by four pin connections. Four hydraulic cylinders and two rocker-arm linkages on the arms provide integrated lifting and tilting of the bucket. Tilting of the bucket, in addition to leveling and dumping, provides breakout or prying force on the lip of the bucket so as to load the bucket quickly and efficiently. Bucket capacity ranges in size from ½ to 20 cysm. The machines may be crawler- or wheels-tires-mounted.

Figure 12-9 illustrates one of the largest crawler-mounted bucket loaders. The rock bucket capacity is 4.1 cysm, the engine horsepower is 275, and the working weight with ripper is 83,000 lb.

Figure 12-10 shows one of the largest wheels-tires-mounted loaders. Its rock bucket has a 9.3-cysm capacity, its engine horsepower is 580, and its working weight is 128,000 lb.

The bucket capacities of Table 12-3 are based on the struck measurement capacity of bucket in cysm. Cubic yards loose measurement, cylm, within the bucket vary according to the fill factor for the bucket. It is high for a mixture of rock-earth but low for poorly blasted rock. After the estimated fill factor is applied to the bucket capacity, another factor, a conversion factor, is applied to the loose measurement load to determine the bank measurement load. This second factor compensates for the swell of the rock-earth

12–14   Loading and Casting Excavation

Figure 12-9 Heavyweight bucket loader, crawler-mounted. This large bucket loader has a nominal rock bucket capacity of 5 yd³ and struck capacity of 4.1 cysm. The engine has 275 hp and the working weight with customary ripper is 83,000 lb. The breakout force on the lip of the bucket is 34,300 lb, a most important specification in rock excavation. A slabby limestone formation is being loaded from a ripped cut and cast into the waste pile. The action is just at the beginning of dumping and one can appraise the bucket factor, ratio of load, cubic yards bank measurement, to bucket capacity, struck measurement. In this example of minimum weathered rock bucket factor is about 0.62, halfway between 0.70 for average weathered rock and 0.54 for well-blasted rock. Bucket load is estimated at 0.62 × 4.1, or 2.5 cybm. *(Caterpillar Tractor Co.)*

from bank to loose condition. The product of conversion factor times fill factor is the final bucket factor, representing the ratio of load, cybm, to struck capacity of the bucket, cysm.

Table 12-3 gives average values for the three factors according to the nature of the rock-earth excavation.

Crawler-Mounted Bucket Loaders

Representative specifications for crawler-mounted bucket loader are given in Table 12-4.

Loading   The average loading cycle for crawler-mounted bucket loaders working in rock-earth when loading through an angle of 120° is made up of the following four elements of the approximate times listed, totaling 0.5 min:

   Loading, 7 s (0.12 min)
   Swinging-maneuvering, 12 s (0.20 min)
   Dumping, 3 s (0.05 min)
   Returning-maneuvering, 8 s (0.13 min)
   Total, 30 s (0.50 min)

Hourly production in cubic yards bank measurement per 50-min working hour is expressed by the equation:

$$P = \frac{50}{CT} \times BF \times BC$$

where P = hourly production, cybm
   CT = cycle time, min
   BF = bucket factor (Table 12-3)
   BC = bucket capacity, cysm

By substituting the average cycle time of 0.50 min, the equation becomes:

Figure 12-10 Heavyweight bucket loader, wheels-tires-mounted. This machine has a nominal bucket capacity of 11 yd^3 and struck capacity of 9.2 cysm. The engine horsepower is 580 and the weight is 128,000 lb. Breakout force on the lip of the bucket is 86,000 lb. The bucket loader is handling ripped weathered granite, which is being bulldozed to the loader by the tractor-bulldozer in the foreground. Average bucket load in the variable rock formation is 6.4 cybm, and three buckets are loaded into the 35-ton rock haulers in about 1.8 min. While the formation in the foreground is rippable, the same formation in the background at about the same elevation is being drilled for blasting. Such irregularity in a given formation is not uncommon in intrusive igneous and metamorphic rocks.

TABLE 12-3 Bucket Factors for Average Bucket Loaders According to Nature of Rock-Earth

Nature of rock-earth	Fill factor $\left(\dfrac{\text{loose measurement}}{\text{bucket capacity}}\right)$	Conversion factor $\left(\dfrac{\text{bank measurement}}{\text{loose measurement}}\right)$	Bucket factor $\left(\dfrac{\text{bank measurement}}{\text{bucket capacity}}\right)$
Sand-gravel	1.05	0.88	0.92
Rock-earth, weathered	0.90	0.78	0.70
Rock, well blasted	0.80	0.67	0.54
Rock, poorly blasted	0.60	0.60	0.36

TABLE 12-4 Representative Specifications for Crawler-Mounted Bucket Loaders

Bucket capacities		Engine horsepower	Working weight, ft	Dumping height, ft	Dumping reach, ft	Breakout force, lb
Nominal yd^3	Struck measurement, cysm					
1.0	0.9	62	16,500	8.0	2.7	10,700
1.5	1.2	80	24,500	8.5	3.3	14,200
1.8	1.5	95	27,000	9.1	4.7	15,300
2.0	1.8	130	32,900	9.3	5.3	20,800
3.0	2.5	190	50,700	9.9	6.2	25,800
5.0	4.1	275	83,000	11.3	7.1	34,300

NOTE: Breakout force is the upward force exerted upon the lip of the bucket to loosen or pry out rock-earth when loading the bucket. The average range of speeds forward is 0 to 6.0 mi/h, 0 to 528 ft/min; in reverse it is 0 to 7.0 mi/h, 0 to 616 ft/min.

12-16 Loading and Casting Excavation

$$P = 100 \times BF \times BC$$

Estimation of production and costs for loading rock excavation by crawler-mounted bucket loaders may be illustrated by the example of Figure 12-11, which shows two 4.0-cysm bucket loaders of this type working weathered sandstones and shales of a freeway cut. The hard formation is being ripped by the heavyweight tractor-ripper-bulldozer of the picture. The two machines are loading out double bottom-dump rock haulers of total capacity 24 cysm. The rock haulers are of the heavy off-the-road type and they take a total load of some 68,000 lb. Three buckets are loaded into each of the units for a total load of 17 cybm.

The bucket factor for weathered rock-earth is 0.70 (Table 12-3). By the equation above, production for each bucket loader per 50-min-working hour is calculated to be:

$$\begin{aligned} P &= 100 \times BF \times BC \\ &= 100 \times 0.70 \times 4.0 \\ &= 280 \text{ cybm/h} \end{aligned}$$

Combined production for the two bucket loaders is 560 cybm/h. The loading time for the complete rock haulers is $6 \times 0.50 = 3.0$ min.

Costs, 1978, are estimated to be:

Cost of 4-cysm bucket loader, nominal 5-yd³ capacity	$192,000
Hourly cost of ownership and operation:	
4-cysm loader, bucket type (Table 8-3). 34% of $192	$65.30
Operator	15.60
Total	$80.90
Direct job unit cost per cybm, $80.90/280	$0.29

Casting and Hauling The bucket loader is also a casting and hauling machine, casting and hauling being considered the same kind of operation. The loading cycle becomes the constant portion of the casting or hauling cycle, the variable portion being the part of the total cycle consumed in hauling and returning. The distance traveled is generally less than 300 ft one way. The hauling speed of the crawler-mounted loader averages about 4 mi/h, or 352 ft/min, and the return speed averages about 5 mi/h, or 440 ft/min, on the average level haul road. The casting or hauling cycle time equation is:

Figure 12-11 Two large crawler-mounted bucket loaders working in ripped rock. These machines are loading out ripped sandstone-shale formation in a huge freeway cut of the Santa Monica Mountains, California. The buckets are of 4.0-cysm capacity. The double-bottom dump haulers take a load of six buckets for 17 cybm. The loading time for the complete hauler is estimated at 3.0 min. The hourly production of each loader is estimated to be 280 cybm.

$$CT = 0.50 + \frac{L}{352} + \frac{L}{440}$$
$$= 0.50 + 0.0051L$$

where CT = cycle time, min
L = length of one-way haul, ft

The hourly production in cubic yards bank measurement per 50-min working hour is expressed by the equation:

$$P = \frac{50}{0.50 + 0.0051L} \times BF \times BC$$

where P = hourly production, cybm
L = length of haul, one way, ft
BF = bucket factor (Table 12-3)
BC = bucket capacity, cysm

When casting or hauling upgrade, the average hauling speed of 4 mi/h can be maintained on grades up to +18 percent provided the haul road is in good condition. This maximum grade is fairly steep for casting or hauling, and so one may assume that the equation for hourly production when hauling on the level is applicable to practical upgrade hauls. By reducing the speed to 2 mi/h, or 176 ft/min, the loader can ascend grades up to +43 percent, an impractical and dangerous steepness for working with a heavily loaded suspended bucket. Thus, the crawler-mounted bucket loader has upgrade hauling ability beyond the requirements of normal job conditions.

As an example of production and cost estimation for casting or hauling rock by a crawler-mounted loader let us assume that the bucket loader of Figure 12-11 is to work on a 300-ft one-way haul. The haul is level or up to 18 percent upgrade on a road in average condition. The hourly production in cubic yards bank measurement per 50-min working hour is given by the equation:

$$P = \frac{50}{0.50 + (0.0051L)} \times BF \times BC$$
$$= \frac{50}{0.50 + 0.0051 \times 300} \times 0.70 \times 4.0$$
$$= 69 \text{ cybm/h}$$

Costs, 1978, are estimated to be:

Hourly cost of ownership and operation, as in preceding example of costs for loading $80.90
Direct job unit cost per cybm, $80.90/69 $1.17

The unit cost is relatively high as the loader, because of its characteristics, is not a hauler and it is generally used only in nonproduction hauling.

Wheels-Tires-Mounted Bucket Loaders

These machines are used presently in great numbers in rock excavation because of three factors: easy portability from job to job, high mobility when moving from cut to cut within the job, and high bucket capacity. The economics of their use depends on the exact nature of the rock to be excavated. In practice large bucket loaders, including machines of 20-yd^3 bucket capacity, are used in soft to hard rock formations, the full revolving shovel being used in the hardest of rocks and ores.

These bucket loaders are illustrated in Figures 12-3, 12-8, and 12-10. All sizes of machines work on the same principles, as shown in Figure 12-8, except that in some smaller sizes the frame is a single unit whereas the larger machines always have articulated frames for maximum maneuverability. Table 12-5 contains important specifications for several sizes of articulated bucket loaders used in rock excavation.

Loading By reason of its articulated or hinged frame the wheels-tires-mounted machine has the following fast average loading cycle totaling 0.42 min, when loading through a swing-return angle of 120°.

    Loading, 7 s (0.12 min)
    Swinging-maneuvering, 9 s (0.15 min)
    Dumping, 3 s (0.05 min)

12-18 Loading and Casting Excavation

Returning-manuevering, 6 s (0.10 min)
Total, 25 s (0.42 min)
Hourly production in cubic yards bank measurement per 50-min working hour is expressed by the equation:

$$P = \frac{50}{CT} \times BF \times BC$$

where P = hourly production, cybm
CT = cycle time, min
BF = bucket factor (Table 12-3)
BC = bucket capacity, cysm

By substituting the average cycle time of 0.42 min, the equation becomes:

$$P = 119 \times BF \times BC$$

Production and cost estimation for loading rock excavation by wheels-tires-mounted bucket loaders may be illustrated by the example of Figure 12-10, which shows a 9.2-cysm loader working in ripped weathered granite. Loading is tedious because of embedded boulders in the granite, the loading cycle averaging 0.60 min. Three buckets are loaded into the 35-ton-, 23-cysm-capacity rock haulers.

Production in cubic yards bank measurement per 50-min working hour is given by the equation:

$$P = \frac{50}{0.60} \times 0.70 \times 9.2$$
$$= 537 \text{ cybm}$$

Loading time for each rock hauler is $3 \times 0.60 = 1.80$ min.
Costs, 1978, are estimated to be:

Cost of 9.2-cysm bucket loader, nominal 11-yd³ capacity	$338,000
Hourly cost of ownership and operation:	
9.2-cysm loader, bucket type (Table 8-3). 34% of $338	$114.90
Operator	15.60
Total	$130.50
Direct job unit cost per cybm, $130.50/537	$0.24

Casting and Hauling In these operations the loading cycle becomes the constant of the casting or loading cycle. Distances traveled one way are generally less than 500 ft. The hauling speed averages about 8 mi/h 704 ft/min) and the return speed averages about 12 mi/h (1056 ft/min) over the average good level haul road. The casting or hauling cycle-time equation is:

$$CT = 0.42 + \frac{L}{704} + \frac{L}{1056}$$
$$= 0.42 + 0.0024L$$

where CT = cycle time, min
L = length of haul, one way, ft

The hourly production in cubic yards bank measurement per 50-min working hour is expressed by the equation:

$$P = \frac{50}{0.42 + 0.0024L} \times BF \times BC$$

where P = hourly production, cybm
L = length of one-way haul, ft
BF = bucket factor (Table 12-3)
BC = bucket capacity, cysm

When casting or hauling upgrade, the average hauling speed of 8 mi/h on good haul roads can be maintained on grades up to +9 percent, for which the cycle-time and hourly production equations for level haul roads may be used. On steeper grades hauling speeds may be reduced proportionately, down to about 3 mi/h on a +24 percent grade. The

TABLE 12-5 Representative Specifications for Wheels-Tires-Mounted Bucket Loaders

Specification	Nominal rated capacity, cy1m							
	1.2	2.0	3.0	4.0	6.5	11.0	15.0	20.0
Capacity, cysm	1.0	1.8	2.5	3.2	5.5	9.2	13.4	17.8
Engine flywheel hp	69	100	184	239	380	580	822	1075
Tires, standard	15.50×25 8 PR	17.50×25 12 PR	20.50×25 12 PR	23.50×25 16 PR	29.50×29 22 PR	33.25×35 32 PR	36.00×41 36 PR	36.00×51 42 PR
Dumping height, ft	7.8	8.5	9.6	9.6	11.8	13.3	15.1	17.0
Dumping reach, ft	2.9	4.9	5.9	5.9	7.5	8.8	10.0	11.2
Breakout force, lb	8400	21,000	21,700	33,000	56,100	186,000	125,000	168,000
Working weight, lb	12,800	21,500	32,800	41,300	81,700	128,000	188,000	250,000

NOTE: Breakout force is the upward force exerted on the lip of the bucket to loosen or pry out rock-earth when loading the bucket. Average range of speeds forward and reverse is 0 to 25 mi/h, or 0 to 2200 ft/min. The wide range enables the mobile bucket loader to move rapidly around the job for efficient and economical loading, casting, and hauling.

12-20 Loading and Casting Excavation

average 12-mi/h return speed may be maintained. Under the extreme condition of a +24 percent grade, the cycle-time and production equations become:

Cycle time:
$$CT = 0.42 + \frac{L}{264} + \frac{L}{1056}$$
$$= 0.42 + 0.0047L$$

Production:
$$P = \frac{50}{0.42 + 0.0047L} \times BF \times BC$$

The estimation of production and costs for casting or hauling rock with wheels-tires-mounted loaders may be illustrated by assuming that the loader of Figure 12-10 is to load a belt conveyor of 400-ton/h capacity from a gravel stockpile 500 ft away. The belt conveyor is equipped with an adequate size receiving hopper and feeder. The well-maintained haul road is level. The bucket factor for sand-gravel is 0.92 (Table 12-3).

Hourly production per 50-min working hour is given by the equation:

$$P = \frac{50}{0.42 + 0.0024 \times 500} \times 0.92 \times 9.2$$
$$= 261 \text{ cybm}$$
$$= 1.30 \times 261 = 339 \text{ tons}$$

Costs, 1978, are estimated to be:

Hourly cost of ownership and operation, as in preceding example of loading costs $130.50
Direct job unit cost per ton, $130.50/339 = $0.38

Figure 12-12 illustrates a common use for the wheels-tires-mounted bucket loader. Most of the 359,000-yd³ cut in the Bedford Canyon formation of metasedimentary rocks has been ripped out. The outsize rocks, too large for scraper loading, have been bulldozed to one side for loading into the rock haulers. Loading is by a 7.1-cysm, 325-hp loader. In this cleanup work of large rocks and boulders, five to seven buckets are required to fill the 23-cysm rock haulers. To facilitate loading, the boulders are being wetted down by the water truck in order to lessen the friction between bucket lip and edges and rock.

Figure 12-12 Using a mobile bucket loader for cleanup work in a rock cut. The wheels-tires-mounted bucket loader, by reason of its movability, is an efficient machine for the unavoidable cleanup work in a rock cut which has been ripped and blasted. Here bulldozers have windrowed about 900 yd³ of outsize rocks which are too large to be loaded out by scrapers. The 7.1-cysm bucket loader is working with the rock haulers in removing the troublesome boulders. To facilitate loading the abrasive hard rocks, a water truck is wetting down the pile, thus reducing friction between bucket and rock and reducing wear and the cost of maintenance. The 359,000-yd³ cut in the mountains of San Diego County, California, was fragmented by both ripping and blasting. Only about 0.25 percent of the rock cut resisted complete fragmentation and required the costly cleanup work.

The combined unit cost for wetting, loading, and hauling can be two to four times that for production work. Fortunately, in relation to the total yardage of the cut the approximate 900 cybm of outsize rocks is small, only 0.25 percent. Nevertheless, in a relative sense the cleanup cost can be 1 percent of the $470,000 bid price for the entire cut, amounting to $4700.

Summary Bucket loaders may be crawler- or wheels-tires-mounted, the former ranging from 1- to 5-yd³ bucket capacity and the latter from 1- to 20-yd³ nominal capacity for rock-earth excavation. Both machines are used primarily for loading and secondarily for casting or hauling. The crawler loader is used in rough work, where maneuvering and traction are difficult. The wheels-tires loader is a high-production machine for use where conditions are or can be made favorable to fast operation.

The largest of the wheels-tires-mounted bucket loaders, with a nominal bucket capacity of 20 yd³, can comfortably load 1200 cybm of ripped weathered rock-earth per working hour. The mobility of the wheels-tires machine is important in the frequent movements within a cut due to ripping and blasting work and between cuts in the scheduling of excavation.

Both crawler- and wheels-tires-mounted loaders can cast or haul up practical grades at about the same cycle times as on level hauls because of the conservative traveling speeds.

SHOVELS

Motions, Mechanics, and Specifications

Full-revolving shovels for rock excavation range from 1 to about 200 yd³ in nominal dipper capacity. They are almost always crawler-mounted, as stability is more important than mobility. Among the sizes and types suitable for the rock work, the bucket-type loader, wheels-tires-mounted, is the mobile counterpart of the full revolving shovel.

Power may be supplied by diesel engines or electric motors up to about 12-yd³ capacity, beyond which electric drive is by means of the Ward Leonard system. In this system for large shovels and draglines, the input is high-voltage alternating current to motors which drive dc generators, which in turn drive dc motors of suitable characteristics for the several motions of the shovel or dragline. The efficiency of the Ward Leonard system, that is, the power output/power input ratio, is about 82 percent. Representative specifications for shovels with diesel and Ward Leonard drive systems are listed in Tables 12-6 and 12-7, respectively.

The several motions in the operating cycle of a shovel and their actuating forces are:

1. Crowding and retracting of the dipper stick and dipper when loading and dumping. This reversing action is effected by two cables attached to the ends of the dipper handle and actuated by two cable drums of the machinery deck. The motions may also be powered by hydraulic or electric motors through a pinion on the shovel boom and a rack on the dipper handle.

2. Hoisting of the dipper handle and dipper. This motion is by a two-part or multipart cable between the dipper sheave and boom-point sheave, actuated by the cable drum of the machinery deck.

3. Swinging and returning are mechanically actuated through an internal swing pinion engaging the inside mounted gear of the turntable.

4. Tripping of the dipper door is by cable from the dipper handle and is actuated by a cable drum of the machinery deck.

5. Propelling or traveling in double-crawler machines is by power train through a vertical propeller shaft and subsequent gearbox and chain-sprocket assembly to crawlers. In multicrawler machines the sets of crawlers are propelled by electric motors.

The dipper is loaded by combined crowding and hoisting. After loading of the dipper, hoisting and swinging are a combined operation. After dumping, the returning and lowering of the dipper are also a combined motion.

Figure 12-13 illustrates a 4½-yd³ shovel loading weathered granite into 18-cysm rock haulers. Rock excavation is for a mountain freeway and this size of shovel is used often in this work.

Figure 12-14 shows an 8-yd³ electric shovel being used in a molybdenite mine. The ore

TABLE 12-6 Representative Specifications for Crawler-Mounted and Diesel Engine-Driven Shovels

Nominal dipper capacity, cysm	Working weight, lb	Horsepower for shovel service	Working dimensions with 45° boom angle, ft		
			Cutting radius, ft	Digging depth, ft	Dumping height, ft
1	47,000	110	28	8	15
2	119,000	146	36	9	18
3¼	168,000	205	39	10	21
4¼	210,000	256	41	11	22
5	275,000	300	43	12	23
6	367,000	360	49	13	24

TABLE 12-7 Representative Specifications for Crawler-Mounted Shovels Driven by Electric-Motors with the Ward Leonard System

Nominal dipper capacity, cysm	Working weight, lb	Horsepower for shovel service	Working dimensions with 45° boom angle, ft		
			Cutting radius, ft	Digging depth, ft	Dumping height, ft
7	459,000	530	49	7	24
9	568,000	550	50	7	25
13	721,000	900	56	8	26
15	970,000	1126	60	9	28
20	1,203,000	1565	64	9	32

Figure 12-13 Shovel with 4½-yd³ dipper loading weathered granite. This picture typifies the following aspects of a well-managed shovel excavation job. First, the loading arrangement gives a low-average swing-return angle of 75°. Second, the rock hauler to the left is waiting for a load and so there is no delay to the shovel. Third, capacities of dipper and rock haulers are balanced. Hauler capacity is 18 cysm and dipper capacity is 4.5 cysm. The ratio is 4.0, midway in the range of 3.0 to 5.0 for a balanced relationship. Fourth, a cleanup tractor-bulldozer is present.

In undisturbed weathered rock, dipper hoist pull and speed are important for speedy loading. Hoist pull is 74,000 lb; speed is 99 ft/min; 222 hp is exerted; and upward force per dipper tooth is 14,800 lb, sufficient to fragment the rock. The shovel was used later in excavation of blasted granite.

In terms of size of dipper and transportability to and from the job, the 4½-yd³ machine weighing 210,000 lb is a good selection for the rough rock job of the Tehachapi Mountains of California.

Figure 12-14 Mining shovel with 8-yd³ dipper loading blasted molybdenite. In this molybdenum ore mine of British Columbia, the high efficiency of the shovel operation may be inferred from the well-blasted ore, the minimum angle of swing return, the ample supply of rock haulers, and the clean floor of the pit maintained by the tractor-bulldozer. The 35-ton-capacity rock haulers average 32 tons per load with an average four-dipper load from the shovel. The depth of face or lift averages 33 ft. The hourly production of the shovel is estimated to be 1280 tons of ore. The hourly production of the rotary drill in the background is 60 ft of 9-in-diameter blastholes on 20 ft × 20 ft pattern, giving ample hourly tonnage of 1910. *(Endako Mines.)*

has been blasted. Rock haulers are of 35-ton capacity and they receive four dippers, or 32 tons, from the mining shovel.

Figure 12-15 typifies the tandem or follow-through method of loading rock haulers with minimum swing-return angle and maximum efficiency and safety for the loading work. The shovel has an 11-yd³ capacity and the rock haulers have a 50-ton or 35-yd³ capacity.

Shovels having up to about 20-yd³ dipper capacity with diesel-engine or electric-motor drive are built with little modifications from accepted standard designs of the manufacturers. However, above this capacity shovels are used mostly in particular quarrying and mining work, for which design considerations must be built into the shovel. For this reason these huge, custom-built shovels, as well as draglines, are not considered in detail in this text.

Shovels of up to about 5-yd³ capacity may be converted to backhoes, cranes, or draglines. This conversion is effected by changing the front of the shovel, which consists of boom, dipper handle, and dipper, to that of a backhoe, crane, or dragline. Sometimes there are additional minor modifications.

Figure 12-16 shows a 4¼-yd³ convertible shovel which may be changed to a 4½-yd³ backhoe, 4-yd³ clamshell and 70-ton crane, or 5-yd³ dragline. Figures 12-17 and 12-18 give specifications and working range dimensions for a convertible shovel of 3¼-yd³ dipper capacity.

12-24 Loading and Casting Excavation

Operating Cycles

Shovels up to about 20-yd^3 capacity are used for both loading and casting. Larger machines are used for casting. These huge shovels of up to 200-yd^3 capacity are not included in this discussion of operating cycles because of their different operating characteristics. They are custom-built, special-purpose machines.

Dipper loading usually takes place by a combination of crowding and hoisting and the travel of the dipper up the face of the cut or highwall generally ends at about the elevation of the shipper shaft or dipper-handle pivot shaft. In shovels of up to 20-yd^3 capacity this height averages 22 ft, a height which by coincidence is suitable for loading most rock haulers.

Casting takes place at dipper heights up to average maximum dumping height, which by coincidence is also about 22 ft. Accordingly, loading and casting cycles for the same swing-return angle are about the same except for a negligible difference in dumping times.

For shovels of up to 20-yd^3 capacity, hoisting speeds average 130 ft/min and swing-return speeds, with allowance for acceleration and deceleration, average 3.0 r/min. Average cycle times for various swing-return angles are listed in Table 12-8, and dipper and bucket factors are given in Table 12-9.

Figure 12-15 Effective shovel loading of tandem or follow-through rock haulers. When working space permits, as in this wide freeway cut, the follow-through passage of the rock haulers makes for both efficiency and safety. The average swing angle is 75°, and in this shovel the operator's position to the left of the machine affords good visibility and control for the cycle. Rock haulers always move forward except for backing up at the fill. The shovel of 11-yd^3 capacity loads the 50-ton, 33-cysm haulers with a four-dipper load. Rock is blasted sandstones, shales, and conglomerates of the Hungry Valley formation of the Tehachapi Mountains of California.

Of much interest is the coincidence that this same formation, with a 7000-ft/s seismic shock-wave velocity, is being ripped by another company excavating an adjacent freeway job. Two major companies elected to use two different methods for fragmenting the same rock formation in the extremely hard ripping or blasting classification, as a result of an understandable difference of opinion.

Figure 12-16 Shovel convertible to backhoe, crane, or dragline. This shovel with 4¼-yd³ dipper, like many machines of up to 5-yd³ capacity, may be converted to backhoe, crane, or dragline. The 5-yd³ dragline of Figure 12-37 is the counterpart of the shovel, and both machines are working in the same dam construction on the Balsas River of Mexico. Conversions are made by exchanging the fronts of the machines, exchanging the crawlers if necessary, and making minor changes in the deck machinery. The complete changeover is usually not a major project.

The shovel is loading weathered basalt of excavation for the dam abutment. The rock hauler has a capacity of 30 tons or 20 cysm and takes a five-dipper load from the well-matched shovel. The capacity ratio is 20.0/4.2 = 4.8. *(Bucyrus-Erie Co.)*

TABLE 12-8 Estimated Average Cycle Times for Shovels Loading or Casting Rock Excavation

Angle of swing return, °	30	60	90	120	150	180
Time, min	0.26	0.31	0.37	0.42	0.48	0.53

TABLE 12-9 Dipper and Bucket Factors for Average Loads of Shovels, Backhoes, and Draglines Based on Dipper or Bucket Capacity, Struck Measurement, cysm

Nature of rock-earth	Fill factor $\left(\dfrac{\text{loose measurement}}{\text{dipper or bucket capacity}}\right)$	Conversion factor $\left(\dfrac{\text{bank measurement}}{\text{loose measurement}}\right)$	Dipper or bucket factor $\left(\dfrac{\text{bank measurement}}{\text{dipper or bucket capacity}}\right)$
Rock-earth, average weathered	0.85	0.78	0.66
Rock, well blasted	0.68	0.67	0.46
Rock, poorly blasted	0.45	0.60	0.27

12-26 Loading and Casting Excavation

A shovel cycle when loading blasted rock into rock haulers includes the following elements, evaluated for a typical 0.34-min cycle.

Loading, including crowding and hoisting dipper, 50%	0.17 min
Swinging to dumping position including hoisting, 24%	0.08
Dumping, 10%	0.03
Returning to digging position, 16%	0.06
Total, 100%	0.34 min

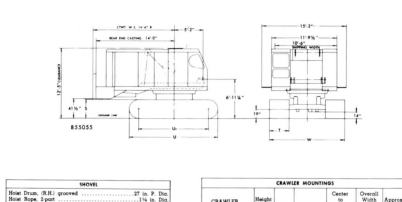

SHOVEL
- Hoist Drum, (R.H.) grooved .... 27 in. P. Dia.
- Hoist Rope, 2-part .... 1¼ in. Dia.
- Boom Point Sheaves (2) .... 40 in. P. Dia.
- Padlock Sheave (1) .... 26 in. P. Dia.

HOE
- Hoist Drum, grooved (L.H.) .... 26 in. P. Dia.
- Hoist rope, 3-part .... 1 in. Dia.
- Drag drum, grooved (R.H.) .... 26 in. P. Dia.
- Drag rope, 2-part .... 1⅛ in. Dia.

CLAMSHELL
- Closing drum, (R.H.) grooved .... 26 in. P. Dia.
- Closing rope, 1-part .... 1 in. Dia.
- Holding drum, (L.H.) grooved .... 26 in. P. Dia.
- Holding rope, 1-part .... 1 in. Dia.
- Boom point sheaves .... (2)27 in. P. Dia.

DRAGLINE
- Hoist drum, (L.H.) grooved .... 26 or 32 in. P. Dia.
- Hoist rope, 1-part .... 1 in. Dia.
- Drag drum, (R.H.) grooved .... 26 in. P. Dia.
- Drag rope, 1-part .... 1⅛ in. Dia.
- Boom point sheaves .... (1)42 in. P. Dia.

CRANE
- Hoist drum, (R.H.) grooved .... 24 in. P. Dia.
- Hoist rope .... 1 in. Dia.
- Aux. hoist drum, (L.H.) grooved .... 26 in. P. Dia.
- Auxiliary hoist rope .... ¾ in. *Dia.
- Boom point sheaves .... (3)27 in. P. Dia.

*1 in. diameter rope optional.

CRAWLER MOUNTINGS

CRAWLER MOUNTINGS	Height Tread Belts S	Width Treads T	Overall Length U	Center to Center Tumblers U₁	Overall Width Tread Belts W	Approx. Bearing Area (Sq. Ft.)
Standard Frames:						
33 in. Treads (Std.)	40 in.	33 in.	15 ft. 11 in.	12 ft. 7 in.	12 ft. 9 in.	73
42 in. Treads	40 in.	42 in.	15 ft. 11 in.	12 ft. 7 in.	13 ft. 6 in.	93

POWER SPECIFICATIONS

Make	GM	Cummins
Model	8V-71N	NT-855-C
Type	Diesel	Diesel
Type of Drive	Tor. Conv.	Tor. Conv.
Cylinders	8	6
Bore x Stroke, inches	4¼ x 5	5½ x 6
Displacement, cu. in.	568	855
Rated for excavator service at full load speed of output shaft:		
Output shaft speed, RPM	1250	1250
Output shaft net H.P.	205	205
Fuel tank capacity, gals.	167	167
Crankcase capacity, qts.	29	36
Cooling system capacity, gals.	18¾	18.5
Starting	12V-Elec.	24V-Elec.
Altitude range, feet	0-4,000	0-12,500

WEIGHTS (LBS.)

	Crane 60 Ft. Boom	Dragline 60 Ft. Boom	Clamshell 60 Ft. Boom	Hoe	Shovel
Net domestic, approx.	145,850	145,750	144,850	172,700	166,650
Working, approx.	148,450	152,100	154,050	174,300	168,250
Export shipping, approx.	147,600	151,200	153,150	174,200	167,350
Ships option tons	127.0	135.0	133.0	129.0	124.0

Hook block and buckets included in working weight and export shipping weight, but not in domestic net weight.

LINE PULLS AND SPEEDS

Drums	Pull in lbs.	Speed F.P.M.	Drums	Pull in lbs.	Speed F.P.M.
24 in. Pitch Dia.	37,700	170	27 in. Pitch Dia.	33,500	191
26 in. Pitch Dia.	34,800	184	32 in. Pitch Dia.	28,350	226

Swing Speed 3.61 RPM   Propel Speed 0.98 MPH

Speeds and line pulls based on standard diesel engine, torque converter drive operating at full load speed of the output shaft. When torque converter is operating at full stall, line pulls are approximately 220% of those shown in table.

Figure 12-17 61-B Series Two Standard Crawler specifications. *(Bucyrus-Erie Co.)*

The calculated cycles for average loading and casting are tabulated below:

	Loading 90° angle	Casting 180° angle
Constants:		
Loading	0.17 min	0.17 min
Dumping	0.03	0.03
Total	0.20 min	0.20 min

UPPER WORKS

Revolving Frame:
One-piece heat treated alloy steel casting. Deep box section design for strength and rigidity with integral boom foot lugs. Rear end casting is bolted to revolving frame. Cast alloy steel machinery side frames bolted to revolving frame. Shear plugs relieve studs of shear loads and maintain alignment.

Main Machinery:
Single main shaft with side-by-side mounted drums plus brakes, clutches and hoist gear. Swing clutches mounted on separate transmission shaft. Boom hoist is located to the rear of the transmission shaft.

Chain Drive:
Multiple strand chain drive between engine and transmission shaft, enclosed and running in oil.

Transmission Shaft:
Heat treated alloy steel shaft, mounted on anti-friction bearings, made in two short sections to facilitate field servicing. Each section mounted on anti-friction bearings and are connected together by a flexible coupling. The bevel gears are mounted on anti-friction bearings and run in oil. Cast steel gear case is recessed into and bolted to the revolving frame. Gear case supports the transmission shaft and encloses the bevel gears.

Swing-Propel Clutches:
Internal expanding, two shoe type clutch. Air controlled with each shoe operated by a separate air cylinder. Clutches are mounted on the transmission shaft.

Swing-Propel Machinery:
A pinion at the bottom of the vertical transmission shaft engages the horizontal swing gear which meshes with the horizontal propel gear. All gears have machine cut hardened teeth.

Air controlled, quick-shift spline type clutches engage either the horizontal swing or propel gear to select either swing or propel motion. Clutches are interlocked to prevent simultaneous engagement of the gears.

Truck Frame:
Heat treated alloy steel casting to accommodate ball bearing swing circle. Box design with heavy "X" bracing. Oil tight housing for propel bevel gears is an integral part of the truck frame. Integral horizontal propel shaft bearing mountings that are line bored. Towing eyes provided at the front of the truck frame.

Swing Circle:
The swing circle consists of two (2) independent rows of precision balls and spacers. It is permanently adjusted at the factory and requires only occasional lubrication from easy accessible fittings.

Axles:
Heat treated cast steel "I" beam section axles with turned ends, that fit into the crawler side frames. Axles are bolted to the truck frame. Shear plugs are installed to relieve the studs of shear loads and to maintain alignment.

Crawler Side Frames:
Heat treated, single unit, box section alloy steel casting. Standard and long length crawler side frames are available.

Hoist Shaft:
Heat treated alloy steel hoist shaft, mounted on anti-friction bearings. Hoist gear is a steel casting with machine cut hardened teeth and is driven by a pinion on the transmission shaft. Cast iron clutch housings turn on anti-friction bearings on the shaft. Split type cast steel drum laggings bolt to housings.

Drum Clutches and Brakes:
Brake and clutch housings are concentric separated by angular cooling ribs. Clutches are internal expanding band type, air set. Brakes are external contracting band type, air operated by foot pedals. Mechanical and mechanical with air assist brakes optional.

Boom Hoist:
Independent power controlled lowering boom hoist that provides positive control of boom both up and down by air controlled clutches. Boom hoist brake spring set-air released. Single lever control. Air operated safety pawl engages a rachet on the boom hoist drum.

Swing Brake:
Friction type, spring set-air released with two V-block shoes. Air assist set for positive holding, actuated when machine is in propel, is standard. Brake drum is secured to the top of the vertical swing shaft.

Controls:
All main functions are air controlled, supplied by a 30 CFM compressor. Air control console is standard. Drum, swing and boom hoist clutches are actuated by graduated type control valves. Drum brake controlled by treadle type air valve. Auxiliary functions controlled by poppet type valves. Manual control of engine and torque converter output shaft governors. Single stick clamshell control is standard on clamshell.

Counterweight:
Cast iron counterweight to match front end attachment requirement.

LOWER WORKS

The standard crawler side frame assembly, complete with 33" treads, may be retracted to provide 10' 6" shipping clearance. It is necessary to remove the propel drive chains before the crawlers can be retracted.

Driving and Take-up Tumblers:
Driving tumblers are heat treated alloy steel castings, straddle mounted on movable shafts for adjusting the tension of the propel chain. Removable screw jack and large adjusting bolts are used to tension chain drive and spacer bars are used to maintain adjustments. Rim and driving lugs are hardened.

Take-up tumblers are heat treated alloy steel castings, straddle mounted on movable shafts for adjusting the tension of the crawler belts. Adjustment of take-up tumblers is similar to that for the driving tumblers. Piston ring type seals provided for driving and takeup tumbler bearings.

Idler Rollers:
Lower idler rollers are single roller style, alloy steel castings with bushing type bearings. Bearing cap and mounting lugs are integral and are bolted to the bottom of the crawler side frame. Bearings have piston ring type seals.

Lubrication:
All gears are shielded. A lubricating guide attached to the right machinery side frame is used to lubricate the deck gearing. Other lubricating fittings are easily accessible or grouped in centralized locations.

Cab:
All steel with offset operator's compartment that is separated from main cab with a door. Adequate doors for access to machinery and ventilation. For shipping, the operator's compartment is removed to obtain 10'6" shipping width.

Power Unit:
Diesel engine, torque converter drive with twin lever control is standard.

Independent Propel:
Air controlled, single speed independent propel is available as optional equipment. Cannot be used with shovel front end.

Third Drum:
Air controlled third drum is available as optional equipment. May be used with or without independent propel and power controlled load lowering. When used with power controlled load lowering, an air operated spline clutch is provided to select either third drum or power controlled load lowering function. Single line pull and speed, based on engine operating at full load speed, 10,000 # @ 171 F.P.M.

Power Controlled Load Lowering:
Air controlled, power controlled lowering of right rope drum is available as optional equipment. Can be installed when machine is equipped with third drum and independent propel. Air operated spline clutches are provided to select either power controlled load lowering or third drum function.

Power controlled lowering of left drum is available for hoe or crane with standard propel.

Upper idler rollers are double roller style steel castings.

Crawler Treads:
Heat treated alloy steel castings with hardened roller path and driving lugs. Single roller path for lateral flexibility. Two full floating pins, widely spaced, connect treads together to form endless belts. 33" treads standard, 42" treads optional.

Propel and Steering:
Swing clutches provide power for propelling after engaging the propel gear on the vertical propel shaft. Horizontal propel shaft carries steering jaw clutches, brakes, and sprockets for propel chain drive. Steering jaw clutches are spring set—air released. An interlock prevents simultaneous disengagement of the steering jaw clutches.

Propel brakes have V-belt linings. **Brakes** are spring set, with air assist for extra holding power, and air released. Propel brakes also serve as digging brakes.

Lubrication:
Propel bevel gears are enclosed and run in oil. Other lubrication fittings easily accessible or grouped in centralized locations.

Figure 12-17 (cont.)

12-28 Loading and Casting Excavation

Hoisting speed of dipper, ft/min	130	130
Average swing-return speed, r/min	3.0	3.0
Cycle:		
Loading and dumping constant	0.20 min	0.20 min
Swinging and returning:		
90° each	0.17	
180° each		0.33
Total	0.37 min	0.53 min

These estimated average cycles are based on the following assumptions:
1. Loading time, made up of crowding and hoisting, is 0.17 min.
2. Average swing-return speed, allowing for acceleration and deceleration, is 3.0 r/min.
3. Dumping time is 0.03 min.

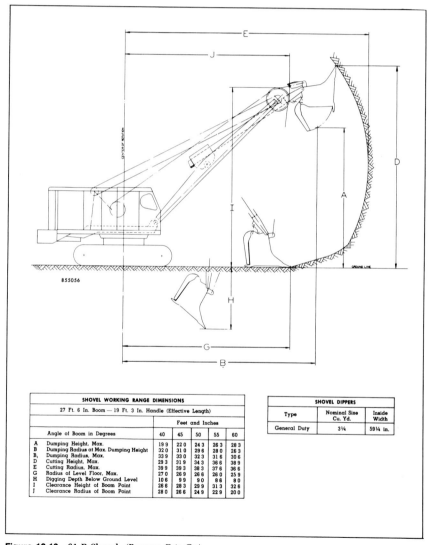

SHOVEL WORKING RANGE DIMENSIONS					
27 Ft. 6 In. Boom — 19 Ft. 3 In. Handle (Effective Length)					
	Feet and Inches				
Angle of Boom in Degrees	40	45	50	55	60
A Dumping Height, Max.	19 9	22 0	24 3	26 3	28 3
B Dumping Radius at Max. Dumping Height	32 0	31 0	29 6	28 0	26 3
B₁ Dumping Radius, Max.	33 9	33 0	32 3	31 6	30 6
D Cutting Height, Max.	29 3	31 9	34 3	36 6	38 9
E Cutting Radius, Max.	39 9	39 3	38 3	37 6	36 6
G Radius of Level Floor, Max.	27 0	26 9	26 6	26 0	25 9
H Digging Depth Below Ground Level	10 6	9 9	9 0	8 6	8 0
I Clearance Height of Boom Point	26 6	28 3	29 9	31 3	32 6
J Clearance Radius of Boom Point	28 0	26 6	24 9	22 9	20 0

SHOVEL DIPPERS		
Type	Nominal Size Cu. Yd.	Inside Width
General Duty	3¼	59¼ in.

Figure 12-18 61-B Shovel. *(Bucyrus-Erie Co.)*

4. In hoisting and swinging simultaneously through the swing angle, the larger of the times is used for the swing time.

Loading

The basic determinant in figuring shovel production in loading average rock-earth is the operating cycle. The actual available time for the cycling depends on the minor and major delays, which in turn depend largely on how good the management of the job is. Average hourly delays to shovel work, based on good management, are summarized in the following tabulation:

Delays to shovel operation	Time distribution	
	min	%
Minor delays of less than 15 min duration:		
Insufficient rock haulers	4.2	7.0
Spotting of rock haulers at shovel	1.0	1.7
Shovel moves at face of cut	1.4	2.3
Shovel repairs and maintenance	1.4	2.3
Shovel handling of outsize rocks and stumps	0.7	1.2
Checking grade	0.3	0.5
Shovel operator's and other delays	0.8	1.3
Subtotals	9.8	16.3
Major delays of more than 15 min duration:		
Moving to next cut and opening up cut	2.4	4.0
Shovel repairs and maintenance	3.0	5.0
Subtotals	5.4	9.0
All minor and major delays, totals	15.2	25.3

The tabulation does not include delays for weather and wet grade, which can be a sizable percentage of total available time. However, these delays do not affect the efficiency on which production should be based. As 25.3 percent of time is spent in unavoidable delays, the efficiency of the average shovel should be based on the 45-min working hour, the actual efficiency for estimating purposes being adjusted to the particular job.

The estimation of loading productions and costs is illustrated by the following two examples.

1. The $4\frac{1}{2}$-yd^3 shovel shown in Figure 12-13 is loading average-weathered granite from a 24-ft-high face and the average swing-return angle is 75°.

Production:
Cycle from Table 12-8, min 0.34
Dipper factor from Table 12-9, weathered rock-earth 0.66
Dipper load, cybm, 0.66×4.5 .... 2.97
Production for 45-min working hour:
$P = (45/0.34)2.97$ cybm 393
Costs, 1978:
$4\frac{1}{2}$-yd^3 shovel $643,000
Hourly cost of ownership and operation:
$4\frac{1}{2}$-yd^3 shovel (Table 8-3). 18% of $643 .... $115.70
Operator and oiler 31.00
Total $146.70
Direct job unit cost per cybm, $146.70/393 .... $0.37

2. The 8-yd^3 electric shovel of Figure 12-14 is loading well-blasted molybdenite from a 33-ft-high face and the average swing-return angle is 60°.

Production:
Cycle from Table 12-8, min 0.31
Dipper factor from Table 12-9, well-blasted rock 0.46
Dipper load, cybm, 0.46×8 .... 3.68
Production per 50-min working hour:

$$P = \frac{50}{0.31} \times 3.68 \text{ cybm} \qquad\qquad 594$$

Hourly production of ore weighing 4300 lb/cybm:

12-30   Loading and Casting Excavation

$$P = \frac{4300}{2000} \times 594 \text{ tons} \qquad 1280$$

In practice a dipper load of ore weighs 8.0 tons, 1 percent more than the estimated load.

$$\text{Estimated load} = 3.68 \times \frac{4300}{2000} = 7.91 \text{ tons}$$

Costs, 1978:
8-yd³ electric shovel	$1,380,000
Hourly cost of ownership and operation	
8-yd³ electric shovel (Table 8-3). 9% of $1380	$124.20
Operator and oiler	31.00
Total	$155.20
Direct job unit cost per ton, $155.20/1280	$0.12

Casting

In casting rock-earth excavation, the shovel cycles for machines of up to about 20-yd³ capacity are approximately equal to those for loading. Accordingly, Tables 12-8 and 12-9 may be conservatively applied to casting work.

The casting shovel works at higher efficiency than the loading shovel. Two minor delays included in the preceding summary of average hourly delays to shovel work are not characteristic of casting. These delays, which total 8.7 percent, are those due to insufficient rock haulers and to spotting of rock haulers at shovel. The adjusted total delays amount to 16.6 percent, giving a comparative efficiency of 83.4 percent, equivalent to a 50-min working hour instead of the suggested 45-min hour for the loading shovel.

Figure 12-19 illustrates casting in the pioneering of a sidehill cut in road building to a coal mine in Kentucky. The shovel has a 3-yd³ capacity. Rock is a weathered limestone

Figure 12-19 Casting in sidehill cut for access road to coal mine. The shovel, of 3-yd³ dipper capacity, is casting rock-earth of weathered limestone through a swing-return angle of 150°. The output per 50-min working hour is estimated to be 206 cybm. Casting is common practice in the building of access roads in mountainous country. Later, as access roads give way to haul roads and eventually to highways, cut-and-fill work calls for the loading of rock haulers. A shovel of 3-yd³ capacity is well suited for casting work of this nature by reason of its casting radius of about 34 ft and working weight of about 150,000 lb. *(Northwest Engineering Co.)*

formation with a working face about the height of the shipper shaft of the shovel. The swing-return angle averages 150°.

Production and cost estimates for the casting work shown in Figure 12-19 are as tabulated below.

Production:
Cycle from Table 12-8, min   0.48
Dipper factor from Table 12-9, weathered rock-earth   0.66
Dipper load, cybm, 0.66×3   1.98
Production for 50-min working hour:

$$P = \frac{50}{0.48} \times 1.98 \text{ cybm} \qquad 206$$

Costs, 1978:
3-yd³ shovel $454,000
Hourly cost of ownership and operation:
  3-yd³ shovel (Table 8-3). 21% of $454 $ 95.30
  Operator and oiler   31.00
  Total $126.30
Direct job unit cost per cybm, $126.30/206 $0.61

Shovels of larger than 20-yd³ capacity with long booms and long dipper handles are used in casting overburden in open-pit rock-earth excavation. Figure 12-20 illustrates the work of a stripping shovel in the contour mining of a coal seam. Such casting to the spoil bank calls for work in a high face or highwall, an ultimately high spoil bank, and a swing-return angle of 180°.

Figure 12-21 shows the action of a huge coal-stripping shovel in a mine of Illinois. The electric shovel has a dipper capacity of 140 cysm and the average dipper load is 250 tons, or about 114 cybm, of the blasted rock-earth formation. The casting cycle averages 55 (0.92 min). The casting distance from highwall to spoil bank is about 350 ft.

The following estimate of hourly production and costs for this 9250-ton custom-built machine is offered as a matter of interest and not as an example of precise estimating. The actual hourly cost of ownership and operation and actual hourly production data are proprietary information, without which an estimate of production and costs can be only suggestive.

The production estimate is based on a 250-ton average load of the shovel dipper, equivalent to 114 cybm of limestone shale weighing 4380 lb/cybm. The average cycle time is 0.92 min for a swing-return angle of 180°. It is assumed that the shovel operates on a 55-min working hour. Hourly production is given by the equation:

$$P = \frac{55}{0.92} \times 114 = 6820 \text{ cybm}$$

The 1978 costs for this shovel are:

140-yd³ electric stripping shovel, erected $36,000,000
Hourly cost of ownership and operation:
  140-yd³ shovel (Table 8-3). 8% of $36,000. $2880.00
  Operator, oiler, and groundman   44.20
  Total $2924.20
Direct job unit cost per cybm, $2,924.20/6820 $0.43

To illustrate the possible difference between the estimated unit cost, based on Table 8-3, and the actual unit cost, the mere elimination of replacement cost escalation would reduce the 8 percent factor to about 6.5 percent. The unit cost would then be $0.34 instead of $0.43, a 23 percent reduction due to a difference in policy for replacement of the machine.

Summary

Full-revolving crawler-mounted shovels are a logical choice for many hard-rock and ore excavations where long life and sustained high production are necessary under exacting but fixed working conditions. These circumstances exist in many public works such as canals, dams, and freeways and in private works such as quarries and open-pit metal and nonmetal mines. For these workings, shovels of up to about

12-32 Loading and Casting Excavation

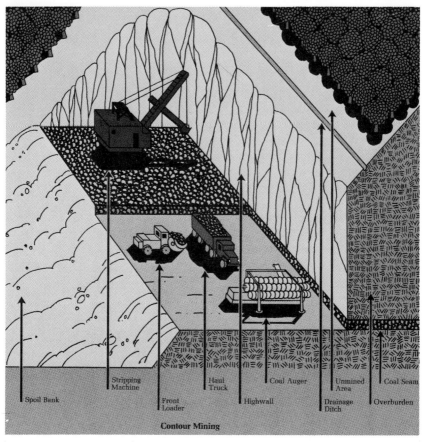

Figure 12-20 Sidehill contour mining of coal and the arrangement of the key machinery. The shovel lends itself to contour mining and it is equally efficient in area mining, as shown in Figure 12-21 and in the area mining diagram of Figure 12-41. In this illustration a front-end loader or bucket loader is loading out the coal seam. Special coal loading shovels of up to 30-yd³ dipper capacity are also used for loading. When the overburden becomes too deep for the capacity of the stripping shovel, the coal auger will be used to auger out as much of the seam as possible. Often the coal seam has been mined out underground almost to the grass roots, and then the amount of coal for augering is reduced considerably. Occasionally, for the same reason, the coal for conventional contour mining is much less than had been anticipated. *(Pittsburg & Midway Coal Mining Co.)*

20-yd³ dipper capacity are of generally fixed specifications, but the larger machines are custom-built for particular applications. In both small and large shovels great life and low maintenance result in low fixed charges and low operating expense, resulting in low direct job unit cost.

BACKHOES

Backhoes are used for trenching and for other subsurface excavation where it is expedient to keep the excavator at original ground level. Backhoes for rock excavation range from 1 to 5 cysm in dipper capacity. They are mechanically and cable- or hydraulically operated and they are usually crawler-mounted. Figures 12-17 and 12-22 give specifications for a 3½-yd³ full-revolving crawler-mounted backhoe which is mechanically and cable-operated on drag and hoist. The backhoe can be converted from the shovel of Figure 12-18.

Operating Cycle and Specifications

There are five actions in the operation of a backhoe, resembling those of the shovel.

    1. Loading the dipper toward the machine by the drag cable of the dipper and the hoist cable of the boom. In hydraulic backhoes this combined action is by one or two hydraulic cylinders on the boom or on the boom and dipper handle.

    2. Hoisting by the hoist cable on the boom or by two hydraulic cylinders between the boom and the machinery deck.

    3. Swinging and returning are mechanically actuated through an internal swing pinion engaging the inside mounted gear of the turntable.

    4. Dumping is by extension of the dipper handle so as to allow fallout of the rock-earth.

    5. Propelling or traveling is by power train through a vertical propeller shaft and subsequent gearbox and chain-sprocket assembly to the crawlers.

The loading of the dipper is by combined dragging and hoisting. After loading the dipper, hoisting and swinging are usually a combined operation. After dumping, the returning and lowering of the dipper are usually combined motions.

Figures 12-17 and 12-22 give detailed specifications and working-range dimensions for a $3\frac{1}{2}$-yd^3 full-revolving crawler-mounted backhoe, and representative specifications for a range of sizes are summarized in Table 12-10.

The backhoe is both a loading and a casting machine. In trenching work casting predominates and in area work, such as foundation excavation, loading prevails over casting.

Figure 12-21 Coal stripping shovel of 140-yd^3 dipper capacity. The shovel is stripping blasted limestone shale overburden of about 75-ft depth and casting the rock some 350 ft laterally to the spoil bank. The coal mine is located in central Illinois. The estimated load of the 140-yd^3 dipper is 114 cybm, or 250 tons. The cycle time for the machine on an 180° swing-return angle averages 55 s, or 0.92 min. When working on a 55-min hour basis, or 92 percent efficiency, the shovel is able to cast about 6800 cybm/h. These large shovels operate three shifts daily with 22.5 h working time, and they cast 365 days annually. With estimated 95 percent availability, this 18.5-million-lb machine casts about 53 million cybm of rock annually. This is approximately equivalent to an 1130-ft cube of solid rock. *(Bucyrus-Erie Co.)*

12–34 Loading and Casting Excavation

Efficiency

The efficiency of a backhoe is controlled by far more variables than that of any other loading and casting machine. In addition to the causes for delay enumerated for the shovel are these, peculiar to backhoe work: (1) confined and exacting, or "dental," excavations for trenches and foundations; (2) tedious work around existing utilities such as water and gas mains; (3) careful work around workers within the area of excavation; (4) limited choice of spoil bank areas when casting and of locations for rock haulers when loading.

An arbitrary selection of an efficiency factor is not businesslike, as the choice of factor must be based on experience in the particular work. The limits may vary from a low of the 20-min working hour, 33 percent, to a high of the 50-min working hour, 83 percent. Illustrative of such a variation are the two backhoes of Figures 12-23 and 12-24. The

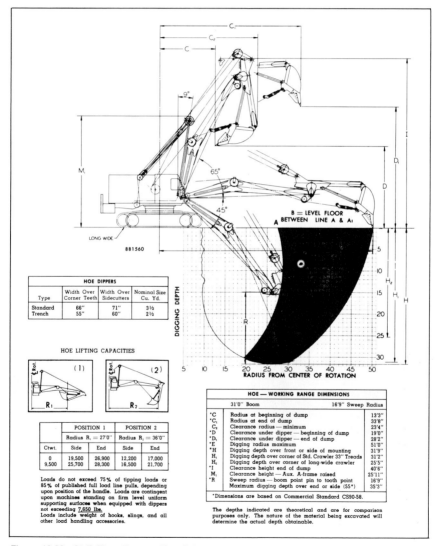

Figure 12-22 61-B Hoe. *(Bucyrus-Erie Co.)*

TABLE 12-10 Representative Specifications for Crawler-Mounted and Diesel-Engine-Driven Backhoes

Specification	Nominal dipper capacity, cysm					
	1	2	3	4	5	6
Working weight, lb	59,000	113,000	174,000	201,000	264,000	352,000
Horsepower for backhoe work	110	146	205	250	300	360
Range of dipper sizes for rock excavation, cysm	⅞–1	1¾–2	2½–3	3¼–4	4¼–5	5¼–6
Digging radius, ft	38	47	51	56	60	64
Digging depth, ft	23	29	32	36	40	44
Dumping height, ft	12	17	19	21	22	24

¾-yd³ backhoe of Figure 12-23 is casting well-blasted volcanic rock on a 0.35-min cycle with an hourly production of 49 cybm. The 2½-cysm backhoe of Figure 12-24 is casting unblasted average- to minimum-weathered granite on a 1.60-min cycle with an hourly production of 52 cybm. Productions are about equal and yet the larger backhoe has 3.3 times the capacity of the smaller machine. A combination of better operating conditions and better management favors the smaller backhoe.

The gravity of working inefficiently is demonstrated by the following comparison of the hourly productions and costs for these two contrasting trench excavations by the efficient ¾-cysm machine of Figure 12-23 and the inefficient 2½-cysm machine of Figure 12-24.

	¾-cysm Backhoe	2½-cysm Backhoe
Production:		
Cycle time, min	0.35	1.60
Dipper factor, Table 12-9	0.46	0.66
Production per 50-min working hour:		
$\dfrac{50}{0.35} \times 0.46 \times 0.75$, cybm	49	
$\dfrac{50}{1.60} \times 0.66 \times 2.5$, cybm		52
Costs, 1978:		
Backhoes	$146,000	$370,000
Hourly cost of ownership and operation		
Backhoes (Table 8-3). 24% of $146	$35.00	
21.5% of $370		$79.60
Operator	15.60	15.60
Total	$50.60	$95.20
Direct job unit cost per cybm	$1.03	$1.83

The unit cost of $1.03 for the ¾-yd³ backhoe does not include the cost of blasting, nor does the unit cost of $1.83 for the 2½-yd³ machine include greatly increased maintenance costs due to the wear and tear induced by the unblasted granite. When these two critical additions are made it is probable that the total unit cost of excavation by the smaller backhoe would be $0.25 to $0.50 less than that of the 2½-yd³ machine.

Operating Cycle Times

The 0.35-min cycle of the backhoe of Figure 12-23, applicable to casting in shallow-depth trenching, is:
  Loading, 7 s (0.11 min)
  Hoisting and swinging, 7 s (0.11 min)
  Dumping, 2 s (0.03 min)
  Returning and lowering to loading position, 6 s (0.10)
  Total, 22 s (0.35 min)

Backhoes of up to 6-yd³ capacity have the following average speeds, which control their loading and casting cycles:

12-36 Loading and Casting Excavation

Figure 12-23 Efficient casting of trench excavation by ¾-yd³ backhoe. The good management of this trench excavation is apparent in the following respects: (1) The minimum-weathered volcanic rock has been blasted to suitable size for full dipper loading. (2) The swing-return angle averages 60°. (3) Grade stakes, to the right of the trench, are set conveniently. (4) The grade checker is always present to prevent both undercutting and overcutting of the desired finished grade.

The 125-hp, 48,000-lb backhoe is casting an estimated 49 cybm/h while working a 50-min working hour for an efficiency of 83 percent.

Figure 12-24 Inefficient casting of trench excavation by 2½-yd³ backhoe. The poor management of this trench excavation is apparent in two respects: (1) The minimum-weathered granite, although of a semisolid nature, has not been blasted. As the depth of the trench increases, the dipper literally is milling out the rock rather than excavating. The length of the operating cycle is abnormal. (2) The walls of the trench are not kept, and cannot be kept, to neat lines, nor can an even grade be maintained.

The 200-hp, 140,000-lb backhoe is casting only about 52 cybm/h because of the long, 1.6-min cycle. The cycle for a well-managed job should be about 0.8 min, or 48 s.

Drag speed for loading, 91 ft/min
Hoist speed, also a component of loading, 60 ft/min
Swing-return speed, average 3.0 r/min

Usually loading time is about 0.10 min and dumping time is 0.3 min, giving a cycle constant of 0.13 min. In trench excavation, hoist takes place to the top of the trench and combined hoist and swing occur to the dumping point; after dumping the sequence follows in reverse. In open-cut work such as foundation-excavation, hoist and swing from loading to dumping are combined, and swing and lowering to loading position are also combined. For comparison, the different cycles for a 15-ft-deep excavation with 90° swing-return angle and with 15-ft dumping height are tabulated below.

Cycle time, min	Trench Cut	Foundation Cut
Loading and dumping, constant	0.13*	0.13*
Hoisting 15 ft	0.25*	0.25
Swinging 90°	0.08	0.08
Hoisting 15 ft	0.25*	
Hoisting 30 ft		0.50*
Returning 90° and lowering to ground level	0.08*	
Returning 90° and lowering to grade of cut		0.08*
Lowering 15 ft within trench, estimated	0.08*	
Total	0.79	0.71

NOTE: The time elements marked by * control the two cycles. For example, in the trench cut after the dipper reaches ground level at the top of the trench, it takes more time to hoist the remaining 15 ft than it does to swing the 90°. Thus, hoist time controls. In the foundation cut, it requires more time to hoist the full 30 ft than to swing 90°, and hoist time controls, as in the trench cut.

However, the difference in these two cycles is only 11 percent, 0.08 min, or 6 s. For this reason and in the interest of simplicity it is assumed that loading and casting cycles for both trench and foundation cuts are the same for the same hoisting distance and the same swing-return angle. Table 12-11 for cycle times is calculated with allowances for equalizing adjustments.

The cycle times of Table 12-11 are based on the following factors and time estimates.

1. Within reasonable estimating precision, cycle times for trenching and foundation excavation are equal.

2. Cycle-time constant includes: loading, 0.10 min; dumping, 0.03 min; lowering dipper into trench after return swing, average 0.04 min, equalizing trenching and foundation work cycles. Total constant, 0.17 min.

3. Hoisting speed, 60 ft/min.

4. Swing-return speed, average 3.0 r/min.

5. When hoisting and swinging simultaneously, the larger of the times is used for calculating cycle times.

Dipper factors for backhoes are given in Table 12-9. They are the same as for shovel dippers.

Figure 12-25 illustrates foundation excavation by a 1¼-yd³ hydraulic backhoe. It is working in well-blasted limestone and it is loading out trucks. The depth is about 15 ft and the total hoist is 30 ft, with swing-return angle of 150°. The efficiency is the 40-min working hour, 67 percent. An estimate of production and costs is tabulated below.

Production:
Cycle time, (Table 12-11)   0.82 min
Dipper factor (Table 12-9)   0.46

Hourly production, $\dfrac{40}{0.82} \times 0.46 \times 1.35$ 28 cybm

Costs, 1978:
1¼-yd³ backhoe $290,000
Hourly cost of ownership and operation:
  Backhoe (Table 8-3). 25% of $290 $72.50
  Operator   15.60
  Total $88.10
Direct job unit cost per cybm, $88.10/28 $0.31

Backhoe with Telescoping Boom and Rotating Dipper

Figure 12-26 illustrates a backhoe with telescoping boom or dipper handle and rotating dipper, two unique features of this particular machine. The backhoe is the largest of three sizes, ranging in weight from 36,000 to 74,000 lb. Dipper capacities range from 21 to 48 in width for rock excavation, roughly equivalent to ½ to 1¼ cysm capacity. Representative specifications for the three sizes are listed in Table 12-12. The pictured machine is excavating for a building foundation in Texas. The rock is hard blue shale and some blasting is necessary. The depth of excavation is 13 ft. The rock dipper is 48 in wide and about 1¼ yd^3 in capacity. The excavation is 72 ft long, 64 ft wide, and 13 ft deep and has a volume of 2220 yd^3. The excavation was completed by loading out trucks in 40 h, and hourly production averaged 56 cybm. The backhoe is wheel-tires-mounted, but all three machines are manufactured with both crawler and wheels-tires mountings.

TABLE 12-11 Estimated Average Cycle Times for Backhoes Loading or Casting Rock Excavation

Total hoist distance, ft	Angle of swing return, °					
	30	60	90	120	150	180
10	0.37	0.40	0.42	0.45	0.49	0.52
20	0.53	0.56	0.58	0.61	0.65	0.68
30	0.70	0.73	0.75	0.78	0.82	0.85
40	0.87	0.90	0.92	0.95	0.99	1.02
50	1.03	1.06	1.08	1.11	1.15	1.18
60	1.20	1.23	1.25	1.28	1.32	1.35

Figure 12-25 A foundation excavation by a 1¼-yd^3 hydraulic backhoe. Although this machine is called a hydraulic backhoe, it is only partially actuated by oil pumps and cylinders. As pictured, the boom is raised and lowered by two cylinders, the dipper handle is moved by one cylinder, and the dipper is controlled by one cylinder. Swing and return and propulsion are by mechanical power trains. The combination hydraulic-mechanical backhoe is an efficient, compact, closely coupled machine.

The backhoe is loading well-blasted limestone into trucks. Estimated cycle time in the 15-ft deep excavation with a 150° swing-return angle is 0.82 min. Estimated hourly production at 67 percent efficiency is 28 cybm at a unit cost of $0.31/cybm. *(Bucyrus-Erie Co.)*

Production and Cost Estimation for Difficult Work

A difficult trench excavation is shown in Figure 5-12. The rock is a cemented and mechanically locked fanglomerate from the San Bernardino Mountains of California. The trench excavation for the 6½-ft-diameter steel pipe cannot be kept to neat excavation lines because of the caving walls. The trench is 13 to 25 ft wide and 22 ft deep. The length

TABLE 12-12 Representative Specifications for Three Sizes of Backhoes with Telescoping Boom, Mounted on Three-Axle Wheels-Tires Carriers

Specification	Nominal capacity, cysm		
	½	1	2
Working weight, lb	36,000	52,000	74,000
Engine-flywheel or brake horsepower:			
Backhoe	120	150	200
Carrier, 3-axle	160	200	250
Range of dipper sizes for rock-earth work	⅓–½	½–1	1–2
Digging radius, ft	28	32	35
Maximum digging depth, ft	20	22	24
Dumping height, ft	17	19	17

Figure 12-26 A foundation excavation by a 1¼-yd³ hydraulic backhoe with telescoping boom and wrist-action dipper. This wheels-tires- or carrier-mounted backhoe is actuated by hydraulic pumps and motors for all motions except traveling. Its unique features are the telescoping boom and wrist action of the dipper, which add to the versatility of the machine. It is carrying a 1¼-yd³ dipper although the nominal rating is 2 yd³. It is shown loading out 2220 yd³ of blasted rock at the rate of 56 cybm/h in a 13-ft-deep excavation which required 40 h. The backhoe excavated both unblasted and blasted blue shale. *(Warner Swasey Co.)*

12–40   Loading and Casting Excavation

of the trench in the fanglomerate is 30,000 ft and yardage is 465,000 cybm. A 2½-yd³ backhoe has excavated to a 16-ft depth and the 2-yd³ backhoe of Figure 5-12 is casting the remaining 97,000 yd³ of harder fanglomerate. Explosives are being used to break up the outsize boulders.

An estimate for the production and costs of rock excavation by the pictured backhoe is tabulated below. The average hoisting distance is 25 ft and the swing-return angle averages 90°. The fanglomerate is equivalent to well-blasted rock. The efficiency is about 67 percent, equivalent to the 40-min working hour.

Production:
    Estimated cycle time (Table 12-11)   0.66 min
    Dipper factor for well-blasted rock (Table 12-9)   0.46
    Hourly production, $\dfrac{40}{0.66} \times 2.0 \times 0.46$ 56 cybm

Costs, 1978:
    2-yd³ crawler-mounted backhoe $290,000
    Hourly cost of ownership and operation:
        Backhoe (Table 8-3). 22% of $290 $63.80
        Operator and oiler   29.80
        Total $93.60
    Direct job unit cost per cybm, $93.60/56 $1.67

Small Backhoes

There is a class of small backhoes, both crawler- and wheels-tires-mounted, which is used for miscellaneous small rock excavations and for exploratory work but which is not used in large-quantity work. Figure 12-27 pictures a wheels-tires-mounted combination backhoebucket loader, as these machines are generally built. It is placing drainage pipe along the top of the slope after having excavated the shallow trench in weathered volcanic rock. The machine weighs 12,000 lb, the diesel engine has 60 hp, the dipper has a rock-

Figure 12-27 A complete pipe-laying job by a small tractor–backhoe–bucket loader. The wheels-tires-mounted tractor–backhoe–bucket loader is a useful machine on any rock excavation job. This machine has excavated a shallow trench along the slope line of 4:1 natural grade. It is being used as a crane to place the drainage pipe and it will later drift in the excavation for backfill. The dipper capacity is ¼-yd³ and the bucket capacity is ⅝ yd³. The lifting capacity, as shown in the picture at about a 10-ft radius, is some 1500 lb. The 14° slope is rather steep for such work by a wheels-tires-mounted machine and its expeditious completion is testimony to the worth of this ever-present jack-of-all-trades machine.

excavation capacity up to $3/8$ yd^3, the digging reach is 18 ft, and the digging depth is 14 ft. The tractor-backhoe-bucket loader is a useful tool for rock excavation and miscellaneous cranelike work.

Figure 7-4 illustrates a crawler-mounted combination tractor-backhoe-bulldozer digging a utility trench in weathered granite. Because of the severity of the work this 13-ton 100-hp backhoe is equipped with only a $1/4$-yd^3 dipper and it is crawler-mounted because of the rough country.

The smaller backhoes of less than $1/2$-yd^3 capacity for rock work have faster cycles than do the large-production backhoes. Their average cycles are:
  Medium digging in maximum weathered rock, 15 s (0.25 min)
  Medium-hard digging in average weathered rock, 20 s (0.33 min)
  Hard digging in minimum weathered rock, 25 s (0.42 min)

Auxiliary Machinery

Because of the differences in consolidation of rock-earth and in width of trench, dippers for the same backhoe are of different lip widths and configurations and of different capacities. Several dippers, buckets, and other accessories are pictured in Figure 12-28.

Figure 12-28 Several dippers, buckets, and grapples for hydraulic backhoes. Hydraulic operation of the dipper of this type of backhoe has made possible such auxiliary equipment as these and other special-purpose useful tools. These attachments, left to right, are: arm lifter, backhoe dipper, clamshell bucket, loader bucket, shaped backhoe dipper for trenching, air percussion hammer, clamshell bucket, and rock grapple. Dippers, buckets, and grapples incorporate wrist-action motion so that these tools are dexterous. *(Adjusta-Buckets, Inc.)*

Summary

The backhoe is an excavator for subsurface work and it may load rock haulers or cast excavation to one side. It is used in trench excavation and in area excavation, as in the case of building foundation work. It may be mounted on crawlers or on wheels-tires, machines of up to 2-yd^3 capacity being mounted in either manner and machines of more than 2-yd^3 capacity being crawler-mounted for desirable stability in larger backhoes.

The motions may be all mechanical, mechanical and hydraulic, or all hydraulic with the exception of traveling motion. Many backhoes with cable-actuated booms and dipper handle-dipper assembly may be converted to shovels, cranes, and draglines. Hydraulic backhoes are usually not convertible into other kinds of excavators.

When loading or casting, several variables control production of backhoes:
  1. Nature of rock-earth, whether in natural weathered state, ripped, or blasted
  2. Hoisting height from grade of excavation to dumping elevation and, when trenching, hoisting height from grade of excavation to ground elevation, during which hoisting the swinging motion cannot take place
  3. Angle of swing return
  4. Dipper factor
  5. Interferences with work, including: confines of excavation; existing utilities within the prism of excavation; presence of workers within excavation area; and inconvenience of location for loading or casting

These above factors also largely control the efficiency of the usual backhoe work.

12-42 Loading and Casting Excavation

CRANES

Uses in Excavation

The crane, equipped with a clamshell or orange-peel bucket, is rarely a loading or casting machine for excavation in the sense of high production. Its uses are in subaqueous excavation and in the rehandling of materials, as in the case of stockpiled aggregates. In open-cut excavations it is a machine for special jobs, among which are the following:

1. Unloading, erecting, disassembling, and loading machinery for excavation.

2. Clearing and grubbing, as shown in Figures 9-17 and 9-18, where grapples are being used for handling trees, stumps, and slash.

3. Loading and casting excavation which is either too large for or inaccessible to loading and casting machinery with high production ability. Figure 12-29 illustrates the use of a grapple for the unloading and the placing of a 22-ton rock in a jetty.

4. Secondary fragmentation of rock in the cut. If there is enough secondary breakage after primary blasting, it is sometimes more expedient and cheaper to break up the oversize rock with a drop or swing ball, as shown in Figure 12-30, than to use secondary blasting.

5. Placing rock in the fill. Riprap in the upstream face of an earthfill dam or in a jetty or breakwater may be placed by a crane. Figure 12-31 illustrates the placing of mat stone of a jetty by means of a rock sling. Figure 12-32 shows the placing of cap stone or riprap on the same jetty by means of a rock-handling, heavy-duty clamshell bucket. All rock is being placed by a crawler-mounted crane of about 8-ton capacity at a 30-ft radius.

Figure 12-29 Crane with grapple for unloading and placing riprap in jetty. The 75-ton crawler-mounted crane is building a jetty, as shown in Figure 12-35. The grapple is rated at 12 to 17 tons per load. The riprap rock being lifted weighs about 22 tons. The maximum opening of the tines of the grapple is 9.0 ft, ample for this amount of riprap rock. The gross load of the crane as pictured is about 57,000 lb, which the crane can handle comfortably at a 40-ft radius. Such a working radius is adequate for the casting and careful placement of the riprap. *("Grapples Handle 20-Ton Rocks on Breakwater Project," Gene Sheley (ed.), Western Construction, January 1960, p. 5.)*

Figure 12-30 Medium-weight drop ball for use in secondary fragmentation of rock. When the volume of rocks to be broken up after primary blasting is significant, it is sometimes economical to use a crane with a drop ball for the secondary fragmentation rather than to use explosives by blockholing. This medium-size, pear-shaped ball weighs about 8000 lb and is capable of shattering a good-sized rock or boulder when dropped from a height of 50 ft. Drop balls are made in pear, spherical, and cylindrical shapes and weigh up to about 12,000 lb. *(Frederick Iron and Steel Co.)*

Cranes may be crawler- or wheels-tires-mounted. When wheels-tires-mounted they are described as truck- or carrier-mounted. Cranes are rated in terms of a percentage of tipping load for a given radius from the center pin. For crawler-mounted cranes the percentage of tipping load is 75 percent and for wheels-tires mounted machines it is 85 percent. The nominal high ratings of both machines are for minimum workable radii.

Operation

The motions of a cable-operated crane are six:

1. Raising or lowering the boom to obtain a proper boom angle for the task is accomplished by a multipart boom cable from the cable drum on the deck frame over the top of the gantry frame to the top of the boom.

2. Digging, or closing the jaws of the clamshell bucket or grapple and hoisting, is carried out by a cable passing from the cable drum of the machinery deck over the boom-point sheave to the bucket.

3. Hoisting and dumping are effected by a cable passing from the cable drum of the machinery deck over the boom-point sheave to the bucket.

4. Swinging and returning of the crane are effected by means of a pinion gear within the geared circle of the truck frame.

5. Dumping of the clamshell bucket or grapple is effected by applying the brake to the hoist cable and releasing the brake of the digging cable.

6. Propelling or traveling is accomplished in crawler-mounted cranes by a power train through the vertical propeller shaft to the crawlers, and in carrier-mounted cranes by a conventional automotive drive from the carrier engine. In carrier-mounted cranes without engines, an uncommon type of crane in rock excavation, the power train is through the vertical propeller shaft to the axles, as in the crawler-mounted machine.

In clamshell-bucket and grapple work, the load is kept from turning by means of a tagline cable which extends from the back of the bucket jacket or from the grapple frame

12-44   Loading and Casting Excavation

Figure 12-31 Rock sling of crane placing mat stone in jetty building. In the building of the Cape May Canal through southern New Jersey, 5500 tons of mat stone is being placed for jetties by means of the illustrated rock sling. A 50-ton crawler-mounted crane working atop the completed jetty handles the rock sling. The sling is dumped by holding the hoist cable and releasing the closing or digging cable. The sling is built of meshed steel to resist abrasion by the sharp-edged granite. In the foreground is the cap stone or full-size riprap which has been placed on the mat stone. *(U.S. Corps of Engineers.)*

Figure 12-32 Heavy-duty clamshell rock bucket placing cap stone or riprap in jetty building. This picture is sequential to that of Figure 12-31, as cap stone or riprap is being placed atop the mat stone. The quantity of riprap is 9400 tons. The 2½-yd^3 bucket is equipped with teeth to grasp firmly the granite rocks. The riprap averages about 10,000 lb and the bucket weighs about 14,000 lb, giving a crane load of 24,000 lb. The 50-ton crane can handle the load comfortably at a 40-ft radius. *(U.S. Corps of Engineers.)*

to a spring-loaded drum or counterweighted sheave assembly within the lattice section of the boom.

Specifications

Figure 12-33 gives capacities of a nominal 51-ton crawler-mounted crane. It is a convertible machine and its standard specifications are given in Figure 12-17. The 51-ton rating is at a 14-ft radius with a 60-ft boom and counterweight W-2. The crane is Class 14–234+. Class 14 means that the maximum load is at a 14-ft radius. The designation 234+ refers to the maximum allowable load of 23,400 lb at a 40-ft radius and with a 60-ft boom and counterweight W-2.

Tables 12-13 and 12-14, respectively, give representative specifications for crawler- and carrier-mounted cranes of various ratings, including those of the capacities used generally in rock excavation. Both crawler-mounted and truck- or carrier-mounted cranes are made in 4- to 300-ton capacities, these machines being cable-actuated. Hydraulic cranes range up to 80-ton capacity, crawler- and carrier-mounted, and these are also used in rock excavation. A typical carrier-mounted crane is shown in Figure 12-34.

TABLE 12-13 Representative Specifications for Crawler-Mounted Cranes

Specification	Nominal rating, tons					
	20	41	51	63	75	83
Radius for nominal tons rating, ft	10	12	14	16	20	20
Nominal clamshell bucket capacity, cysm	1	2	3	4	5	6
Weight of bucket, lb, with nominal capacity load of rock-earth at 2900 lb/cylm	8600	17,300	24,800	32,600	41,500	51,400
Estimated maximum radius for clamshell or grapple work, ft	30	30	35	35	40	40
Engine hp for crane work	110	146	205	250	300	360
Working weight with bucket, lb	59,000	106,000	164,000	207,000	242,000	290,000

TABLE 12-14 Representative Specifications for Truck- or Carrier-Mounted Cranes

Specification	Nominal rating, tons				
	25	40	50	70	80
Data for nominal-ton rating					
Radius, ft	10	10	12	12	12
Boom length, ft	45	45	60	60	70
Nominal capacity of clamshell bucket, cysm	1	2	3	4	5
Weight of bucket, lb with nominal capacity load of rock-earth at 2900 lb/cylm	8,600	17,300	24,800	32,600	41,500
Estimated maximum radius for clamshell and grapple work, ft	50	45	45	40	40
Engine horsepowers					
Crane	90	100	115	150	167
Truck or carrier	150	160	246	290	290
Number of axles	3	3	4	4	4
Working weight with bucket, lb	65,000	91,000	119,000	132,000	145,000

12-46   Loading and Casting Excavation

Operating Cycle Times

Cycle times of cranes are a function of hoisting height, swing-return angle, and lowering to dumping or releasing position, the loading and dumping or releasing being considered constant.

In rock excavation the handling of riprap and miscellaneous large rocks or boulders usually does not involve a high hoist. The closing of a bucket or a grapple takes only a few seconds, perhaps averaging 0.10 min. The unloading may be by careful placing or simply casting, and these operations ought to be conservatively estimated at 0.33 and 0 min, respectively.

Hoisting and lowering speeds for cranes in the 20- to 80-ton capacity range average about 150 ft/min. Swing-return speeds, with allowances for acceleration and decelera-

51 TON CRAWLER CRANE (Class 14-234+)
+ WITH MINIMUM 60 FT. BOOM

MAXIMUM ALLOWABLE LOADS IN POUNDS — CRANE SERVICE

Boom Lgth. in Feet	Rad. in Ft.	Boom Ang. in Degs.	Boom Point Pin Height (Ft. In.)	Standard Crawlers 33-42 In. Treads			
				CTWT. W-1		CTWT. W-2	
				Col. 1	Col. 2	Col. 1	Col. 2
60	14	*82	66' 6"	86,900	98,500	102,000	115,600
	15	*81	66' 3"	77,000	88,100	91,400	103,600
	20	76	65' 6"	50,500	57,200	59,500	67,500
	25	71	63' 6"	36,900	41,900	43,700	49,500
	30	66	61' 6"	28,800	32,700	34,200	38,800
	35	60	59' 0"	23,400	26,600	27,900	31,700
	40	55	55' 9"	19,600	22,200	23,400	26,600
	50	42	46' 9"	14,500	16,400	17,500	19,800
70	15	*82	76' 3"	77,300	87,700	91,000	103,100
	20	78	75' 3"	50,100	56,800	59,100	67,100
	25	74	74' 0"	36,500	41,500	43,200	49,100
	30	69	72' 6"	28,400	32,300	33,800	38,400
	35	65	70' 3"	23,000	26,200	27,500	31,300
	40	60	67' 9"	19,200	21,800	23,000	26,200
	50	50	60' 9"	14,100	16,000	17,100	19,300
	60	38	50' 6"	10,800	12,300	13,300	15,000
80	25	76	84' 6"	36,200	41,100	43,000	48,700
	30	72	83' 0"	28,100	31,900	33,500	38,000
	35	68	81' 3"	22,700	25,800	27,200	30,900
	40	64	79' 0"	18,900	21,400	22,700	25,800
	50	56	73' 3"	13,800	15,600	16,800	18,900
	60	47	65' 3"	10,500	11,900	13,000	14,600
	70	36	53' 9"	8,300	9,400	10,300	11,700
100	30	76	103' 9"	27,400	31,100	32,800	37,200
	35	73	102' 6"	22,000	25,000	26,500	30,100
	40	70	100' 9"	18,200	20,600	22,000	25,000
	50	63	96' 3"	13,100	14,800	16,100	18,100
	60	57	90' 6"	9,850	11,100	12,300	13,800
	70	50	83' 0"	7,600	8,600	9,600	10,900
	80	42	73' 3"	5,950	6,700	7,700	8,700
	90	32	60' 0"	4,650	5,200	6,250	7,000
110	30	77	114' 0"	27,000	30,700	32,400	36,800
	35	74	112' 9"	21,600	24,600	26,100	29,700
	40	72	111' 3"	17,800	20,200	21,600	24,600
	50	65	107' 6"	12,700	14,400	15,700	17,700
	60	60	102' 3"	9,450	10,700	11,900	13,400
	70	54	95' 9"	7,200	8,200	9,200	10,500
	80	47	87' 6"	5,550	6,300	7,300	8,300
	90	40	77' 0"	4,250	4,900	5,850	6,600
	100	30	62' 3"	3,250	3,650	4,650	5,250
120	30	78	124' 3"	26,700	30,300	32,100	36,400
	35	76	123' 3"	21,300	24,200	25,800	29,300
	40	73	121' 3"	17,400	19,800	21,200	24,700
	50	68	118' 3"	12,300	14,000	15,300	17,450
	60	63	113' 9"	9,100	10,300	11,500	13,100
	70	57	108' 0"	6,850	7,750	8,900	10,100
	80	52	100' 9"	5,200	5,900	7,000	7,900
	90	45	92' 9"	3,950	4,450	5,500	6,200
	100	38	80' 6"	2,950	3,350	4,350	4,950
	110	29	65' 3"	2,150	2,400	3,400	3,850
130	35	77	133' 6"	20,900	23,800	25,500	28,900
	40	74	132' 3"	17,100	19,400	20,900	23,800
	50	70	129' 0"	12,000	13,600	15,000	17,000
	60	65	124' 3"	8,750	9,900	11,200	12,700
	70	60	119' 6"	6,500	7,350	8,550	9,700
	80	55	113' 3"	4,850	5,500	6,650	7,500
	90	49	105' 6"	3,550	4,050	5,150	5,850
	100	43	96' 6"	2,550	2,900	4,000	4,550
	110	36	83' 9"	1,800	2,000	3,050	3,450
	120	29	68' 6"	1,100	1,250	2,300	2,600
140	35	78	143' 9"	20,600	23,400	25,100	28,500
	40	76	142' 6"	16,800	19,000	20,600	23,400
	50	71	139' 6"	11,700	13,200	14,700	16,600
	60	67	136' 0"	8,450	9,550	10,900	12,300
	70	62	131' 6"	6,200	7,000	8,250	9,350
	80	58	125' 3"	4,550	5,150	6,350	7,200
	90	53	118' 3"	3,250	3,700	4,850	5,500
	100	47	110' 0"	2,250	2,600	3,700	4,200
	110	42	99' 9"	1,450	1,650	2,750	3,100
	120	35	87' 0"	800	900	2,000	2,250

820944K1   820945K1

The above ratings apply only to machines that are level and standing on hard level uniform supporting surfaces. Loads must be freely suspended. The radii specified are loaded radii. Weights of hooks, hook blocks, slings, and all other load handling equipment except hoist rope shall be considered part of the load. Proper care must be exercised by the operator at all times to prevent shock or side loadings on the boom. Ratings apply only to machines having booms in first class condition built and recommended by Bucyrus-Erie Company.

1. Loads in Column One do not exceed 75 percent of tipping load with the boom in the least stable position in accordance with U.S. Department of Commerce Commercial Standard CS-90-58.
2. Loads in Column Two do not exceed 85 percent of tipping loads with the boom in the least stable position.

* Special deflecting roller for hoist line is required when boom angle exceeds 78 degrees.

BOOM

Construction:

Standard boom is 60 ft. long, two sections, fabricated from alloy steel angles, all welded construction. Sections connected with butt type bolted joints. Insert sections 10 and 20 ft. long are available. Upper section of boom has an open throat design. Two boom point sheaves are standard on clamshell. Three boom point sheaves are standard on crane. One boom point sheave on dragline. Boom point sheaves are mounted on anti-friction bearings.

Suspension:

Eight-part tackle with mast and pendant suspension is standard for boom lengths thru 130 ft. Ten-part tackle with two-part intermediate suspension is standard for 140 ft. boom.

Maximum Length:

Maximum length of boom for crane service is 140 ft. Maximum boom and jib combination are, for standard crawlers with W-1 counterweight, 120 ft. + 30 ft. and standard crawlers with W-2 counterweight, 120 ft. + 50 ft. Maximum boom or boom and jib combination that can be lifted off the ground unassisted are:

Standard Crawler (33 or 42 in. Treads)	Boom		Boom & Jib	
	Side (Ft.)	End (Ft.)	Side (Ft.)	End (Ft.)
Ctwt. W-1	120	140	100 + 30	120 + 30
Ctwt. W-2	140	140	110 + 50	120 + 50

Maximum length of boom that can be carried over the end of crawlers with the mast lowered to a minimum height of 19 ft. 11 in. is 110 ft. Maximum length of boom to which a jib may be attached is 120 ft.

Maximum Angle:

Maximum boom angle is 82 degrees. Special deflecting roller for hoist line is required when boom angle exceeds 78 degrees. Hinged type boom stop is standard on crane.

JIB

Construction:

Jibs are fabricated from alloy steel angles, all-welded construction. Jib lengths of 15, 20, 30, 40 and 50 ft. long are available.

Loads:

Use jibs for crane service only. Allowable loads on main boom sheaves, when jib is attached to boom, must be reduced as follows:

15 ft. jib . . . . 2100 pounds
20 ft. jib . . . . 2250 pounds
30 ft. jib . . . . 2500 pounds
40 ft. jib . . . . 2800 pounds
50 ft. jib . . . . 3000 pounds

Jibs may be used straight in line with boom or at angles with centerline of boom but not to exceed angles shown on jib sketch.

HOIST TACKLE

Parts:

Suggested parts of hoist tackle are as follows (loads in lbs.):

Loads over	27,000	54,000	81,000	108,000	135,000
Parts of line	2	3	4	5	6

The maximum allowable load for a single part line on the auxiliary hoist is 16,500 lbs.

Figure 12-33   61-B Crane. *(Bucyrus-Erie Co.)*

Cranes   12-47

tion, average about 3.0 r/min. A typical cycle when placing riprap in a seawall is tabulated below. Hoist from rock hauler is 10 ft, the average swing-return angle is 90°, and the total distance of hoist and lowering to seawall position for rock averages 30 ft. There is no interference, so that hoist, swing, and lowering may be a combined operation and the return to loading position is the same.

Operation:	Time, min
Loading and placing, estimated	0.43*
Hoisting and lowering to seawall, 30 ft at 150 ft/min	0.20*
Swinging 90°, at 3.0 r/min	0.08
Returning 90°, at 3.0 r/min	0.08*
Total	0.71

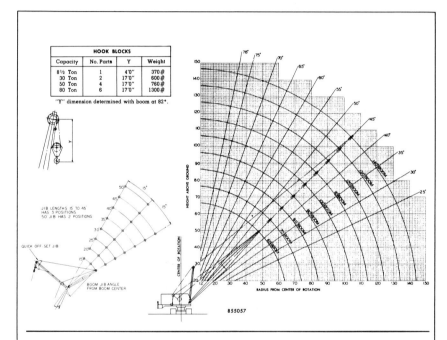

HOOK BLOCKS

Capacity	No. Parts	Y	Weight
8½ Ton	1	4'0"	370#
30 Ton	2	17'0"	600#
50 Ton	4	17'0"	760#
80 Ton	6	17'0"	1300#

"Y" dimension determined with boom at 82°.

JIB LENGTHS 15 TO 45 HAS 3 POSITIONS. 50' JIB HAS 2 POSITIONS.

QUICK OFF-SET JIB

BOOM JIB ANGLE FROM BOOM CENTER

855057

MAXIMUM ALLOWABLE LOADS — JIB (LBS.)

Boom Length In Feet	Boom Angle In Degrees	15' Jib			20' Jib			30' Jib			40' Jib			50' Jib	
		Offset Angle †			Offset Angle †			Offset Angle †			Offset Angle †			Offset Angle †	
		0°	15°	30°	0°	15°	30°	0°	15°	30°	0°	15°	30°	0°	15°
60 to 100	78	*20,000	*20,000	*15,000	*20,000	*20,000	*12,500	*17,200	*14,800	*10,900	*12,600	*10,900	*6,500	*9,600	*8,000
	75	*20,000	*20,000	*14,000	*20,000	*20,000	*11,500	*16,700	*13,800	*9,800	*11,900	*9,800	*6,200	*9,000	*7,000
	70	*20,000	18,500	*13,000	19,200	17,200	*10,500	*15,900	*11,500	*8,600	*11,000	*8,600	*5,300	*8,000	*6,000
	65	15,100	14,100	*12,000	14,400	13,200	*9,800	13,100	*10,000	*7,400	*10,000	*7,400	*5,000	*7,000	*5,000
	60	11,900	11,200	10,800	11,200	10,400	*9,000	10,200	*9,000	*6,400	*9,000	*6,400	*5,000	*6,000	*4,500
	55	9,600	9,100	8,890	9,050	8,500	8,200	8,100	7,500	*5,500	7,200	*5,500	*4,500	*5,000	*4,000
	50	7,900	7,600	7,400	7,400	7,050	6,800	6,600	6,200	*4,500	5,900	*4,500	*4,000	*4,500	*3,500
	45	6,600	6,400	6,200	5,950	5,800	5,500	5,200	*4,000	4,800	*4,000	*4,000	*4,300	*3,000
	40	5,600	5,500	5,400	5,250	5,000	5,000	4,600	4,400	*4,000	4,000	3,800	*3,500	3,550	*3,000
	35	4,800	4,700	4,700	5,000	4,412	4,300	3,950	3,800	3,700	3,400	3,200	3,150	3,000	2,800
110 to 120	78	*20,000	*20,000	*15,000	*20,000	*20,000	*12,500	*17,200	*14,800	*8,000	*12,600	*10,900	*6,500	*9,600	*8,000
	75	*16,500	*16,500	*14,000	*16,000	*15,500	*11,500	*14,500	*13,800	*8,000	*11,900	*9,800	*6,200	*9,000	*7,000
	70	12,000	11,200	10,600	11,500	10,500	9,800	10,600	*9,400	*7,400	9,750	8,350	*5,300	*6,000	*6,000
	65	8,250	7,800	7,500	7,850	7,350	6,950	7,200	6,500	6,050	6,500	5,750	*5,000	6,000	*5,000
	60	5,800	5,550	5,350	5,500	5,150	4,950	4,950	4,550	4,300	4,400	3,950	3,650	4,000	3,500
	55	4,050	3,950	3,800	3,850	3,650	3,500	3,400	3,150	3,000	-2,950	2,650	2,500	2,600	2,300
	50	2,800	2,750	2,650	2,600	2,500	2,400	2,250	2,100	2,000	1,850	1,700	1,600	1,550	1,400
	45	1,850	1,800	1,800	1,700	1,600	1,600	1,350	1,300	1,250	1,050	—	—	—	—
	40	1,150	1,100	1,100	1,000	—	—	—	—	—	—	—	—	—	—

The above loads do not exceed 75% of tipping.   731432K1

†Maximum offset (Angular) from centerline of boom to centerline of jib.

*Indicates that the maximum load is limited by factors other than tipping.

Machine equipped with standard crawlers, 33" or 42" links, and counterweight W-2.

Figure 12-33   (cont.)

12-48 Loading and Casting Excavation

NOTE: Elements marked * control the cycle, as hoisting and lowering time is greater than swinging time. The estimated average cycle times presented in Table 12-15 for cranes placing average riprap or large rocks are based on the following assumptions:
1. Loading time, 6 s (0.10 min).
2. Unloading or placing time, 20 s (0.33 min).

Figure 12-34 Crane of 70 tons nominal capacity, carrier-mounted. This crane can handle a 4-yd^3 clamshell bucket or grapple with an 8-ton rock at a 42-ft radius from the center pin. The pattern of the outriggers is 18 ft longitudinally and 22 ft laterally about the pin, affording good stability. The specially designed carrier provides ruggedness and mobility both on the job and between the rock excavations.

The carrier-mounted crane offers excellent mobility and good stability, and the crawler-mounted crane offers good mobility and excellent stability, so that the rock excavator has a choice for either a particular or general service. *(Koehring Co.)*

TABLE 12-15 Estimated Cycle Times for Cranes Placing Average Riprap or Large Rocks

Total hoisting and lowering distance, ft	Time, min for swing-return angle of:					
	30°	60°	90°	120°	150°	180°
10	0.53	0.56	0.59	0.65	0.71	0.77
20	0.59	0.62	0.64	0.67	0.71	0.77
30	0.66	0.69	0.71	0.74	0.77	0.80
40	0.73	0.76	0.78	0.81	0.84	0.87
50	0.79	0.82	0.84	0.87	0.90	0.93
60	0.86	0.89	0.91	0.94	0.97	1.00
80	0.99	1.02	1.04	1.07	1.10	1.13
100	1.13	1.16	1.18	1.21	1.24	1.27

3. Constant time, loading+unloading, 26 s (0.43 min).
4. Hoisting and lowering speeds, 150 ft/min.
5. Swing-return speeds, with allowance for accelerations and decelerations, 3.0 r/min.
6. When hoisting or lowering and swinging or returning at the same time, the larger of the times is used because it controls the cycle time.

Table 12-16 presents estimated cycle times for cranes loading or casting excavation or casting riprap. These estimated average cycle times are based on the following assumptions:

1. Loading time, 6 s (0.10 min).
2. Unloading or dumping time, 2 s (0.03 min).
3. Constant total time for loading and unloading, 8 s (0.13 min).
4. Hoisting and lowering speeds, 150 ft/min.
5. Swing-return speeds, with allowance for accelerations and decelerations, 3.0 r/min.
6. When hoisting or lowering and swinging and returning at the same time, the larger of the times is used because it controls the cycle time.

Auxiliary Machinery

Cranes use the following principal devices at their cable ends for rock-earth loading.

Grapples Grapples are illustrated in Figures 12-29 and 12-35. Representative specifications for heavy-duty grapples are listed in Table 12-17.

Drop Balls and Rock Slings Drop or swing balls are made in three shapes: pear as illustrated in Figure 12-30, ball, and cylindrical. They are made of wear-resistant nickel-steel alloys and they range in weight from 1500 to about 12,000 lb.

Rock slings are illustrated in Figure 12-31. These are made from hemp matting, steel mesh, and chain mail and they are generally fabricated to the desired size.

Heavy-Duty Clamshell Buckets Heavy-duty clamshell buckets with teeth are shown in Figure 12-32. The nominal rating for clamshell buckets is given approximately in cubic yards struck measurement, measured along the side plates of the bucket. Representative specifications are given in Table 12-18. A similarly acting bucket is the orange-peel type, the several triangular spades of which close into a hemisphere.

Crane Selection and Production and Cost Estimation

Conditions for selection of crane and for estimating productions and costs for placing and casting riprap for a jetty, illustrated in Figure 12-35, are as follows:

1. Specifications call for up to 22-ton riprap, the smallest rock being ½ ton.

TABLE 12-16 Estimated Cycle Times for Cranes Loading or Casting Excavation or Casting Riprap

Total hoist, ft	Time, min, for swing-return angle of:					
	30°	60°	90°	120°	150°	180°
10	0.23	0.26	0.29	0.35	0.40	0.47
20	0.29	0.32	0.34	0.37	0.41	0.47
30	0.36	0.39	0.41	0.44	0.47	0.50
40	0.43	0.46	0.48	0.51	0.54	0.57
50	0.49	0.52	0.54	0.57	0.60	0.63

TABLE 12-17 Representative Specifications for Heavy-Duty Rock Grapples

Specification	Nominal capacity, tons				
	2	3	4–7	7–12	12–17
Maximum tine opening, ft	5.9	6.9	7.2	7.7	9.0
Capacity, lb	4000	6000	14,000	24,000	34,000
Weight, complete, lb	3100	4300	5100	8600	13,100
Gross load for crane, lb	7100	10,300	19,100	32,600	47,100

12-50   Loading and Casting Excavation

2. The crane will work atop the jetty so that a crawler crane will be used for maximum stability.

3. Riprap will be unloaded from rock haulers backed onto the completed section of the jetty by crawler-tractor.

4. The total hoist and lowering distances for placing are 40 ft.

5. The hoisting distance for casting is 10 ft.

6. Tonnages of riprap for placing and casting are equal.

Figure 12-35   Casting maximum size 22-ton rock for riprap of a harbor jetty. A 75-ton crawler-mounted crane is casting this outsize rock for the riprap of a jetty or breakwater at Half Moon Bay, California. The riprap, ranging in size from one-half to 22 tons, is trundled out to the crane by a crawler-tractor-drawn 33-ton-capacity rock wagon. The rocks average about 11 tons. When casting on a 180° swing-return angle, the crane cycle is about 0.47 min, or 28 s. When placing, the crane cycle is about 0.87 min or 52 s. The crane requires 5.4 min to unload 33 tons of riprap, the actual unloading time being estimated at 2.0 min and the time waiting for the exchange of rock haulers being estimated at 3.4 min. The average rate, then, for placing and casting is estimated to be 244 tons per working hour. ("Grapples Handle 20-Ton Rocks on Breakwater Project," Gene Sheley (ed.), *Western Construction, January 1960, p. 6.*)

TABLE 12-18   Representative Specifications for Heavy-Duty Clamshell Buckets Equipped with Standard Jacket and Bottom Teeth

Specification	Nominal capacity, yd^3					
	1	2	3	4	5	6
Length of opening, ft	7.1	9.6	11.0	11.9	13.1	14.1
Width of opening, ft	3.2	4.2	5.4	6.2	6.2	6.6
Capacity, cysm	1.09	2.30	3.60	4.00	5.07	6.02
Weight, complete, lb	5700	11,500	16,100	21,000	27,000	34,000
Load, lb, for nominal rating, based on rock-earth at 2900 lb/cylm	2900	5800	8700	11,600	14,500	17,400
Gross load for crane, lb	8600	17,300	24,800	32,600	41,500	51,400

NOTE: The similarly acting orange-peel bucket, nominal capacity for nominal capacity, has approximately the same capacity in cysm and the same weights—empty, load, and gross load—as the clamshell bucket. Thus, these values may be used for estimating purposes for the orange-peel as well as the clamshell bucket.

The selection is determined as follows:

1. A 22-ton sandstone rock in cubic form is 6.6 ft on each edge. The grapple should have an opening length greater than 6.6 ft. The specifications for grapples indicates that the 12- to 17-ton-capacity unit with 29 percent overload will handle the riprap, as the average size is about 11 tons, less than the 17-ton capacity of the grapple.

2. The gross load for the crane is 13,100 plus 44,000, or 57,100 lb.

3. Table 12-13 indicates that the 75-ton crane can handle a gross load of 42,000 lb at a 40-ft radius. Such a gross load allows for handling a 14.4-ton rock, heavier than the average riprap. By raising the boom slightly, the 75-ton crane can handle the maximum 22-ton rock at about 30-ft radius. Accordingly, the 75-ton crane is the logical selection.

Production and cost estimates for placing and casting riprap by the 75-ton crane are tabulated below.

Production:
Average size of riprap, tons	11
Placing cycle (Table 12-15): hoisting and lowering of 40 ft, 180° swing-return angle, min	0.87
Casting cycle (Table 12-16): hoisting of 10 ft, 180° swing-return angle, min	0.47
Average cycle, equal tonnages for placing and casting	0.67
Time for unloading 33-ton rock hauler with average 3 lifts, min	2.00
If the jetty averages 300 ft in length and there is a good supply of rock haulers, the exchange time for haulers averaging 2 mi/h would be about 3.4 min, giving a total crane cycle time for 33-ton load, min $=2.00+3.40=$	5.40
Production per 40-min working hour, tons $= \dfrac{40}{5.4} \times 33 =$	244

Costs, 1978:
75-ton crawler-mounted crane	$653,000
Hourly cost of ownership and operation:	
75-ton, 5-yd^3 crane (Table 8-3). 16% of $653	$104.50
Operator and oiler	30.80
Total	$135.30
Direct job unit cost per ton, $135.30/244	$0.55

In order to estimate the production of a crane when loading or casting excavation, the cycle time may be estimated by means of Table 12-16. If efficiency is assumed to be the 50-min working hour, the number of buckets handled hourly may be calculated by the equation:

$$\text{Buckets per hour} = 50/\text{cycle time, min}$$

The payload per bucket, in pounds, is the allowable gross load for the crane under the given working conditions, less the weight of the clamshell or orange-peel bucket selected for the work. Figure 12-33 includes a table of allowable loads for a particular crane, any given crane having a similar table. Table 12-18 gives representative weights for heavy-duty clamshell buckets. The resulting payload in pounds divided by the unit weight of the excavation per cubic yard bank measurement gives the payload in cubic yards bank measurement. The unit weight of the material, bank measurement, may be obtained from Appendix 1. Buckets handled hourly times the payload in cubic yards bank measurement gives the estimated hourly production of the crane.

It is evident that the selection of the bucket with respect to both weight and capacity determines the payload because the total weight of the bucket and payload is fixed closely by the capacity of the crane. All relationships must be worked out harmoniously whether the excavation is earth or rock.

Summary

Cranes, crawler-mounted or truck- or carrier-mounted, are used for a few production tasks and many nonproduction chores in excavation. Among these are: clearing and grubbing; unloading, erecting, disassembling, and loading heavy machinery; loading and casting troublesome rocks and boulders; and placing and casting riprap in the fill. For these purposes cranes range in size from 1 to 6 yd^3 in bucket capacity and from 20 to 90 tons in lifting capacity. The chief tools used by cranes are rock grapples, rock clamshell buckets, rock slings, and drop balls for fragmenting rock.

12-52   Loading and Casting Excavation

The choice of mounting depends on whether the emphasis is on stability, an inherent characteristic of the crawler-mounted crane, or on mobility, a characteristic of the truck- or carrier-mounted machine. For placing riprap on a jetty one would choose crawler mounting. On the other hand, for placing riprap on the upstream face of a dam, where the crane would work from the grade of the existing fill, one would choose a carrier-mounted crane.

When working in clearing and grubbing or in placing or casting riprap, the crane is working directly on a bid item. When unloading parts of a huge coal-stripping dragline from a flatcar, it is not working on a bid item, and this chore is chargeable to miscellaneous work. In many respects the crane is indispensable on medium and large rock jobs.

DRAGLINES

Draglines are used for both loading and casting, with casting predominating. Generally the excavation is below the level of the dragline. When digging from lower-level excavation, the machine stays up where it is high and dry, an advantage in casting and especially in loading because the rock haulers do not work well under wet or other unfavorable conditions.

Draglines in rock work range in bucket capacity from 1 to 220 yd^3. The nominal bucket capacity is struck measurement over the top and ends of the sideplates. The machines may be either crawler-mounted, as in Figure 12-37, or base-mounted, with walker shoes or pontoons, as in Figure 12-38. The base-mounted dragline with feet is called a walker.

Draglines of up to about 6-yd^3 bucket capacity are crawler-mounted and diesel-engine-driven, although electric-motor drive is available. These machines have five motions, as listed below.

1. Loading is by pulling the bucket toward the machine with a drag cable attached to the bucket chain. The cable passes through two fairlead sheaves at the base of the boom and thence to the drag-cable drum of the machinery deck.

2. Hoisting by cable from the bucket-chain hoist over the boom point sheave and thence to the hoist-cable drum of the machinery deck.

3. Swinging and returning is by a gear train consisting of the pinion gear and internal gear of the circle of the truck frame.

4. Propelling or traveling is accomplished by power train of gearboxes from the vertical propeller shaft to the crawlers.

5. Loading and dumping of the bucket, especially when loading, are controlled by both drag and hoist cables by means of an equalizing dump cable. This fixed cable runs from the end of the drag chain up to and over the dump sheave at the top of the hoist chain and thence down to the top of the bucket arch.

Figures 12-17 and 12-36 give specifications and allowable load data for a nominal 5-yd^3 dragline, crawler-mounted and diesel-engine-driven, as shown in Figure 12-37. The dragline is convertible from the basic shovel of Figure 12-16. Tables 12-19 and 12-20 give specifications for draglines with nominal bucket capacities of 1 to 20 yd^3.

Above about 6-yd^3 capacity, draglines are base-mounted with walker feet. Up to about 18-yd^3 capacity they may be diesel-engine- or electric-motor-driven. Above 18-yd^3 capacity they are driven by the Ward Leonard system, consisting of a sequence from prime-mover ac motors to dc generators to dc motors, which drives the drag, hoist, swing-return, and propel functions of the dragline.

The loading, hoisting, swinging-returning, and dumping actions of the large walker draglines are identical to those of the smaller crawler-mounted machines. However, the propelling mechanism is uniquely different. When cycling, the machine rests on the base or tub, which is of ample diameter for stability. Movement is by use of two or four shoes or pontoons, which are a part of the revolving frame or machinery deck.

Most walkers have two shoes, one on each side, as shown in Figure 12-38. The frames of the shoes work about a camshaft, which is part of the revolving frame. A revolution of the cam raises the base and steps the machine backward about 7 ft. Walking speed is about $\frac{1}{8}$ mi/h. The walking of Big Muskie, Plate 12-1, is different. This machine has four shoes instead of two and is raised by four hydraulic lifting cylinders between the revolving frame and the shoes and then moved backward by four push cylinders. Each step moves the dragline 14 ft, and the travel speed is $\frac{1}{6}$ mi/h.

Operating Cycle Times

Crawler-mounted machines of up to 6-yd^3 bucket capacity are used for both loading and casting. Cycle times are usually less than 0.50 min for total hoists less than 40 ft and 180° swing-return angle. Walker draglines are used almost exclusively for casting and their cycles for a hoist of 200 ft and 180° swing-return angle can be as high as 1.30 min.

Estimated cycle times for dragline loading and casting with swing-return angles of 30 to 180° are summarized in Table 12-21, and the calculated elements of two cycles for typical loading and casting operations are tabulated below.

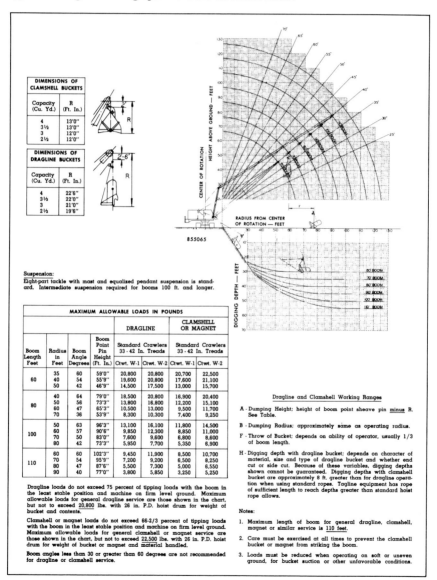

Figure 12-36   61-B Dragline and Clamshell. *(Bucyrus-Erie Co.)*

12-54 Loading and Casting Excavation

Figure 12-37 Dragline of 5-yd³ bucket capacity loading weathered basalt. This dragline weighs 169,000 lb and is powered by a 246-hp diesel engine. The boom is about 75 ft long. The sustained loading production is 1460 m³ each 8-h shift, or 239 cybm each working hour. The dragline is one of two identical machines engaged in low-elevation excavation on the dam construction. The 900,000-kW hydroelectric dam is being built on the Balsas River of Mexico. It is a rock fill structure of 6.5 million yd³, 574 ft crest length, and 498 ft crest height. Four similar convertible shovels are being used in rock excavation, one of which is illustrated in Figure 12-16. *(Bucyrus-Erie Co.)*

Loading: Depth of cut, 30 ft; loading height, 25 ft; total hoist, 55 ft; swing-return angle 120°
Casting: Depth of cut, 50 ft; casting height, 50 ft; total hoist, 100 ft; swing-return angle, 180°

Constants:	Loading	Casting
Loading, min	0.10	0.10
Dumping, min	0.03	0.03
Total, min	0.13	0.13
Hoisting, ft/min	200	200
Swinging-returning, with allowance for acceleration and deceleration, r/min	3.0	3.0

Cycle:	Time, min	Time, min
Loading and dumping	0.13*	0.13*
Hoisting:		
55 ft	0.28*	
100 ft		0.50*
Swinging:		
120°	0.11	
180°		0.17
Returning:		
120°	0.11*	
180°		0.17*
Total time	0.52	0.80

*Elements which control the cycle times, as times for hoisting are greater than times for swinging.

Production and Cost Estimation for Various Operations

Figure 12-37 shows a 5-yd³ dragline loading foundation excavation for a dam into 30-ton rock haulers. The dragline is one of two similar machines, each of which loads 1460 m³

TABLE 12-19 Representative Specifications for Draglines Mounted on Long Crawlers

Specification	Nominal bucket capacity, yd³					
	1	2	3	4	5	6
Working weight with bucket, lb	58,000	103,000	160,000	178,000	236,000	316,000
Engine or electric motor horsepower, approximate	110	146	205	250	300	360
Suggested loading specifications for average work with boom at 45° angle:						
Length of boom, ft	40	50	60	60	70	70
Loading-dumping radius, ft	32	40	47	48	55	56
Maximum load, lb	8700	14,200	18,500	24,200	28,500	35,200
Maximum load data:						
Bucket capacity, yd³	1½	2½	3½	5	6	7
Bucket weight, lb	5,000	7,000	9,000	12,000	15,000	18,000
Load weight, lb	3,700	7,200	9,500	12,200	13,500	17,200
Total weight, lb	8,700	14,200	18,500	24,200	28,500	35,200
Bucket load, cybm	1.0	1.9	2.5	3.2	3.6	4.6
Suggested casting specifications for average work with boom at 35° angle						
Length of boom, ft	50	60	70	80	90	100
Loading-dumping radius, ft	45	54	63	70	80	89
Maximum load, lb	5200	10,000	12,200	16,700	19,800	22,100
Maximum load data:						
Bucket capacity, yd³	¾	1¾	2¼	3¼	4	4½
Bucket weight, lb	3,000	5,500	6,500	8,500	10,000	11,000
Load weight, lb	2,200	4,500	5,700	8,200	9,800	11,100
Total weight, lb	5,200	10,000	12,200	16,700	19,800	22,100
Bucket load, cybm	0.6	1.2	1.5	2.2	2.6	2.9

NOTE: The rock-earth load of weathered rock for a bucket of nominal capacity is based on Table 12-9, which gives a bucket factor of 0.66; the corresponding weight of weathered rock-earth is 3750 lb/cybm.

TABLE 12-20 Representative Specifications for Walker Draglines, Diesel-Engine- or Electric-Motor-Driven

Specification	Nominal bucket capacity, yd³			
	8	12	16	20
Working weight with bucket, lb	850,000	1,270,000	1,700,000	2,110,000
Engine or electric motors horsepower, approximately	750	1,130	1,500	1,870
Boom length range for casting, ft	100–130	130–170	160–210	185–240
Suggested casting specifications for average work with boom at 35° angle:				
Suspended load of nominal capacity bucket, lb				
Bucket	19,200	27,600	36,000	44,400
Rock-earth load at 3750 lb/cybm	19,800	29,700	39,600	49,500
Total	39,000	57,300	75,600	93,900
Bucket load, cybm	5.3	7.9	10.6	13.2
Boom length, conservative, ft	160	170	175	185
Loading-dumping radius, ft	145	153	158	166
Digging depth, ft	80	85	88	92

NOTE: The rock-earth load of weathered rock for a bucket of nominal capacity is based on Table 12-9, which gives a bucket factor of 0.66; the corresponding weight of weathered rock-earth is 3750 lb/cybm.

12-56 **Loading and Casting Excavation**

Figure 12-38 Walker dragline of 4-yd³ bucket capacity casting overburden from atop phosphorite. This dragline is both a casting and a loading machine. After casting the earth-clay overburden from the lenses of phosphorite, as shown in the picture, the machine will load out bottom-dump haulers with the phosphate ore for transportation to the preparation plant. This kind of selective mining is tedious work, as each bucket of overburden or ore is chosen by the inspector, who may be seen immediately under the right-hand corner of the large section of pillared limestone. When casting, the dragline works on a 1.38-min cycle and casts about 2.6 cybm on each pass. When loading, the cycle is about the same and the bucket load is 3.0 tons, the lightweight phosphorite weighing about 2300 lb/cybm. Only two haulers are used in this well-managed Tennessee mine, so that casting and loading operations may be balanced for high efficiency of both the walker dragline and the bottom-dump haulers.

TABLE 12-21 Estimated Cycle Times for Dragline Loading and Casting

Total hoist, ft	Time, min, for swing-return angle of:					
	30°	60°	90°	120°	150°	180°
10	0.21	0.25	0.29	0.35	0.41	0.47
20	0.26	0.29	0.31	0.35	0.47	0.47
30	0.31	0.34	0.36	0.39	0.42	0.47
40	0.36	0.39	0.41	0.44	0.47	0.50
50	0.41	0.44	0.46	0.49	0.52	0.55
60	0.46	0.49	0.51	0.54	0.57	0.60
80	0.56	0.59	0.61	0.64	0.67	0.70
100	0.66	0.69	0.71	0.74	0.77	0.80
120	0.76	0.79	0.81	0.84	0.87	0.90
140	0.86	0.89	0.91	0.94	0.97	1.00
160	0.96	0.99	1.01	1.04	1.07	1.10
180	1.06	1.09	1.11	1.14	1.17	1.20
200	1.16	1.19	1.21	1.24	1.27	1.30

The estimated average cycle times are based on these assumptions:
1. Loading time is 0.10 min, or 6 s.
2. Hoisting speed is 200 ft/min.
3. Swing-return speeds are 3.0 r/min allowing for accelerations and decelerations.
4. Dumping time is 0.03 min, or 2 s. When hoisting and swinging occur at the same time, the larger of the times is used, as it controls the total cycle time.

of weathered igneous rock each 8-h shift. Hourly production for each dragline is 239 cybm.

The total hoist is 35 ft and the swing-return angle is 180°. Table 12-21 gives a 0.48-min cycle time. An estimate for hourly production, based on a 40-min working hour, or 67 percent efficiency, is given by the equation:

$$P = \frac{40}{0.48} \times 0.66 \times 5 = 275 \text{ cybm}$$

This estimate is 17 percent higher than the actual production, which means that the efficiency is less than 67 percent. There is probably waiting for rock haulers as there is no extra hauler at the loading site. One is shown returning in the far distance, and a loaded one may also be seen in the distance. As the hauler behind the dragline has not received the first bucket of rock, it is evident that the dragline was waiting for this hauler.

Figure 12-38 provides a good example of selective open-pit mining. The phosphate rock is a mixture or earth, clay, and fine-grained phosphorite, the desired ore. The phosphate rock occurs in the weathered zone between the limestone pillars. The phosphorite occurs in inclined veins or lenses, usually at the bottom of the phosphate rock mixture. The double duty of the 4-yd³ walker dragline is first to cast aside the overburden with the aid of the inspector down in the pit, who directs the operation, and second, to load out the 12-yd³ bottom-dump haulers.

The total hoist is about 60 ft and the average swing-return angle is 120°. Table 12-21 gives a cycle time of 0.54 min, 39 percent of the actual 1.38-min cycle. This great disparity is explained by the physical nature of the selective mining work. It is definitely not a production job and costs are excessive, as estimated below.

Production:
Cycle, actual, min   1.38
Estimated efficiency, 50-min working hour, %   83
Bucket load, actual, tons   3.0
Bucket load, cybm, calculated at 2300 lb/cybm   2.6

Hourly production, tons = $\frac{50}{1.38} \times 3.0$ 109

Hourly production, cybm = $\frac{50}{1.38} \times 2.6$ 94

Costs, 1978:
Walker dragline, 4-yd³ capacity, electric motors $566,000
Hourly cost of ownership and operation:
  Dragline, equivalent to dragline, crawler-mounted,
    electric motors (Table 8-3). 20% of $566 $113.20
  Operator and oiler   31.00
  Total $144.20
Direct job unit cost per ton, $144.20/109 $1.32
Direct job unit cost per cybm, $144.20/94 $1.53

If the average production-cycle time of 0.54 min were applied to the work, the following figures would emerge:

$$\text{Hourly production, tons} = \frac{50}{0.54} \times 3.0 = 278$$

Direct job unit cost per ton = $144.20/278 = $0.52

The penalty for this selective mining, as opposed to regular open-pit production mining, is $0.80 per ton of phosphate ore.

Figure 12-39 illustrates a 3½-yd³ dragline recasting rock excavation or overburden for strip coal mining in Kentucky. The rock is a weathered limestone formation. The 70-ft boom is at a 40° angle and the bucket is at a 25° angle at the start of loading. The total hoist is 20 ft, the swing-return angle is 180°, and the actual cycle time is 25 s, or 0.42 min. Efficiency is probably about 83 percent. Estimates for hourly production and costs are as tabulated below.

12-58 Loading and Casting Excavation

Figure 12-39 Dragline of 3½-yd³ bucket capacity recasting limestone formation. Recasting is common practice for the removal of overburden in open-pit mining for both small and large draglines. This machine is handling a limestone formation in a strip coal mine of Kentucky. The operator is commencing to load the bucket with an entry angle of 25°, an efficient angle in the loose overburden. The known cycle is 25 or 0.42 min for the 180° swing-return angle and 20-ft lift. The hourly production is estimated to be 275 cybm. Such a 3½-yd³-capacity dragline, equipped with long, wide crawlers and weighing about 150,000 lb, is a versatile, mobile machine for general construction and open-pit mining, as it may be converted to shovel, backhoe, or crane. *(Northwest Engineering Co.)*

Production:
Cycle time, min (Table 12-21)   0.47
    Actual cycle time, min, to be used in estimate   0.42
    Bucket factor (Table 12-9)   0.66
    Hourly production, cybm. $\dfrac{50}{0.42} \times 0.66 \times 3.5$ 275

Costs, 1978:
    3½-yd³ dragline, crawler-mounted $384,000
    Hourly cost of ownership and operation:
        3½-yd³ dragline (Table 8-3). 21% of $384 $ 80.60
        Operator and oiler   31.00
        Total $111.60
    Direct job unit cost per cybm, $111.60/275 $0.41

As this rock has been cast previously, the unit cost of $0.41 is only a part of the total cost for casting.

Figure 12-40 shows a 40-yd³ walker dragline with a 285-ft boom working in a bituminous coal stripping mine of Alabama. The average depth of blasted sandstone-shale overburden is 170 ft. Three seams of coal are being uncovered with thicknesses of 11, 23, and 21 in, totaling 4.6 ft.

The average cycle time of the dragline is 70 s, or 1.17 min, for an average swing-return angle of 180°. The multiple-seam stripping requires careful scheduling of stripping so as to avoid rehandling the overburden. The overburden depth in the picture is about 90 ft. For 150-ft average hoist and 180° swing, Table 12-21 gives an estimated cycle time of 1.05 min.

The 40-yd³ dragline bucket carries an estimated load of 116,000 lb or about 27 cybm. The bucket weighs about 86,000 lb and the total suspended load is about 202,000 lb. The sustained production of the dragline averages 11,450,000 cybm annually for round-the-clock, 365-day-per-year service. Machine availability is about 85 percent. The annual hours of service are about 7450, giving an average hourly production of 1540 cybm.

An approximate estimate of costs for the walker dragline as of 1978 is:

Figure 12-40 Walker dragline of 40-yd³ bucket capacity stripping overburden from three coal seams. This walker dragline is stripping overburden from three coal seams, the total depth in the picture being about 90 ft. The depth for all work averages 170 ft. The average swing-return angle is 135° and the average cycle in blasted sandstone-shale is 70 s. The working weight of the machine is about 3.7 million lb. On a round-the-clock operation and about 85 percent availability, the machine works about 7450 h annually and casts some 11,450,000 cybm/yr at a rate of 1540 cybm/h. The 1978 direct job unit cost for casting is estimated at $0.41 per cybm of blasted overburden. The open-cut bituminous coal mine is located in Jefferson County, Alabama. *(The Drummond Co.)*

Cost of 40-yd³-capacity walker dragline	$7,370,000
Hourly cost of ownership and operation:	
40-yd³ walker dragline (Table 8-3). 8% of $7370	$589.60
Operator, oiler, and groundman	44.20
Total	$633.80
Direct job unit cost per cybm, $633.80/1,540	$0.41

Area Mining by Dragline

Plate 12-1 and Figure 12-40 show the relative positions of blasthole drills, draglines, coal-loading shovels, and coal haulers in typical open-pit coal mines where the area mining method is used. Figure 12-41 shows by an area-mining diagram the several operations for extracting the coal and then reclaiming the spoil banks for farming, grazing, foresting, and other commercial purposes.

Sometimes area mining involves the use of shovels for stripping the overburden, as shown in Figure 12-21. And sometimes contour mining, Figure 12-20, calls for use of draglines as in Figure 12-40. In many contour- and area-mining operations both shovels and draglines are used, depending on topography, nature and depth of overburden, and presence of multiseam beddings of coal.

Essentially the same overburden stripping methods are used to mine metals and nonmetals other than coal in open-pit mines and quarries where stratification of the desired material is fairly level and depth and topography of the overburden are favorable. Examples of such workings are open-pit lead and zinc mines of the tristate area of Missouri, Kansas, and Oklahoma, the kaolin pits of Georgia, and the iron mines of Minnesota. In many cases the stripping of overburden by draglines and

12-60 Loading and Casting Excavation

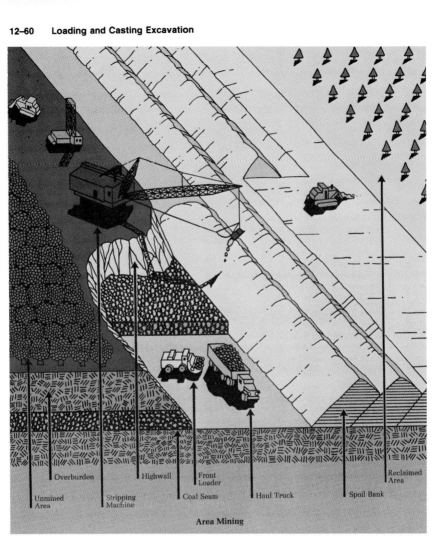

Figure 12-41 Area mining of coal and the arrangement of the key machinery. The walker dragline is used commonly in area mining, as is the stripping shovel as shown in Figure 12-21. The sequential operations in this cross-sectional view of strip coal mining are shown by the pictured machines: (1) clearing of the unmined area by the tractor bulldozer; (2) blasthole drilling of overburden by the crawler-mounted rotary drill; (3) casting of the blasted overburden by the walker dragline; (4) loading of the coal seam into coal haulers by the bucket loader; (5) leveling of the spoil bank by tractor-bulldozer; (6) foresting of the reclaimed area. *(Pittsburg & Midway Coal Mining Co.)*

shovels precedes the opening up of drifts or adits for underground mining operations in hilly or mountainous country.

Summary

Draglines are used both for loading rock haulers and for casting. In loading, both draglines and haulers usually work at elevations higher than the grade of digging so as to avoid wet and rough working conditions. In casting, the same unfavorable conditions are avoided and, more importantly, excavation may be cast up to about 600 ft laterally and 300 ft vertically by the largest of walker draglines.

For rock work, draglines range up to about 220 yd^3 in bucket capacity. They may be mounted on crawlers, wheels-tires, or base shoes as in the case of walker machines. Power may be supplied by a diesel engine, electric motors, a diesel engine–

generator–electric motor combination, or the Ward Leonard system of ac motors–dc generators–dc motors.

Boom lengths and bucket sizes may be varied for adjustment to desired radius of operation and desired weight of bucket and load. In bucket sizes smaller than 6 yd^3, the same basic machine, including revolving frame, may usually be converted to a shovel, backhoe, or crane. The change is neither complicated nor time-consuming, as it involves only a change in the attachments or fronts and changes in the laggings of the cable drums of the deck machinery.

TRENCHERS

Trenchers, also known as ditchers, are used for digging trenches for utilities such as water, gas, sewer, and telephone lines, for oil and gas pipelines, for drainage pipes, and for miscellaneous subsurface linear excavations.

They may be used in some weathered rock formations provided these can be blasted fine enough for efficient digging. This fine blasting is often not possible except in well- and average-weathered formations. The digging action of the machine is such as to excavate to neat lines and grades, thus eliminating expensive undercutting or overcutting. Digging in suitable rock-earth is fast as compared with the speed of a backhoe. The machines can dig trenches up to 25 ft deep and to 6 ft wide, but in most rock-earth excavation the maximum economical width is about 4 ft. Trenchers can both load and cast by means of a side conveyor. There are two kinds of trenchers, both using a series of removable buckets equipped with sidecutter teeth.

Ladder Trencher

The ladder trencher has a series of removable buckets mounted on the chain-sprocket assembly of a hinged or sliding boom, which is lowered into the trench. Excavation is by the feed or travel of the crawler-mounted trencher as it advances along the ditch line. The excavated material is transferred from the bucket line to the side conveyor at the discharge of the bucket assembly. The two types of ladder trenchers, hinged- and sliding-boom, are illustrated in Figures 12-42 and 12-43. The hinged-boom type is more commonly used for trenching, the sliding-boom or vertical-ladder type being used for shallow, narrow trenches for small utilities. Table 12-22 gives representative specifications for the hinged-boom type.

Figure 12-44 illustrates a large ladder trencher digging a hard soapstone formation for a wide sewer trench. The depth varies from 16 to 42 ft, necessitating benching the cut because of the 27-ft depth capacity of the machine. The trencher is equipped with a crumbling shoe at the foot of the boom, which serves also to carry the grade of the curved invert. The trencher is loading out 12- and 14-yd^3-capacity trucks by means of the side conveyor and a supplementary longitudinal conveyor. The specifications for this trencher are those of the 15-ft nominal-depth-capacity machine of Table 12-22.

In the trench of Figure 12-44, of 15-ft depth and 6.3-ft width, the trencher is excavating 150 ft of trench every 8 h. Hourly production is 66 cybm. The hourly production at 100 percent efficiency may be estimated from average loading time of 1.5 min for the 12-ton nominal-capacity trucks. Soapstone weighs 4550 lb/cybm. Hourly production at 100 percent efficiency is given by the equation:

$$P = \frac{60}{1.5} \times \frac{24{,}000}{4{,}550} = 211 \text{ cybm}$$

Accordingly, the estimated efficiency for the trencher is 66/211=31 percent.

The 1978 cost estimate for this trenching work is tabulated below:

Cost of ladder-type trencher	$148,000
Hourly cost of ownership and operation:	
Ladder trencher, 25 ft depth capacity (Table 8-3). 35% of $148	$51.80
Operator and groundman	30.60
Total	$82.40
Direct job unit cost for excavating and loading per cybm, $82.40/66	$1.25

Efficiencies of Trenchers

Trenchers are subject to the same time-consuming delays as are backhoes when they are engaged in trenching work. Among the causes of these delays are: topography; extent of

12-62   Loading and Casting Excavation

Figure 12-42   Medium-weight ladder trencher of the hinged-boom type. This medium-size ladder trencher is cutting a ditch for drainage pipe and is casting excavation which will be used for backfill after the laying of the pipe. Important specifications for the trencher include the following: (1) digging depth, 15 ft; (2) bucket widths, 24, 30, and 36 in, or with sidecutters, to 42 in; (3) bucket-line speeds, 97, 166, and 234 ft/min; (4) digging feed speeds (variable), 0 to 30 ft/min; (5) travel speeds, 1.4, 2.4, and 3.4 mi/h forward, 0.8 mi/h reverse; (6) conveyor belt width, 24 in; (7) conveyor belt speeds, 506, 583, and 806 ft/min; (8) diesel engine horsepower, 100; (9) working weight with 30-in buckets, 37,000 lb.

The capacity of the belt at 506-ft/min speed is about 541 tons/h. If bucket width, bucket line speed, and digging feed were adjusted to this belt speed for average-weathered rock-earth, capacity at 100 percent efficiency would be about 290 cybm/h. *(Koehring Co.)*

shoring for sides of trench; vegetation, such as concealed stumps and roots; obstructions due to buried utilities, sidewalks, paved streets, and building foundations; presence of workers in the trench or the working area; delays to positioning of haulers if loading; difficulty in casting to available locations; and congestion of working area due to other construction activities. These delays in total can reduce the efficiency to surprisingly low figures. The efficiency must be estimated for a given job and it may vary from 20 to 90 percent. Accordingly, estimates must be made carefully, with generous contingency allowances for unanticipated delays.

The work illustrated in Figure 12-42 is under favorable conditions in a farming area. Probably the ladder trencher is able to work on a 50-min working hour, or 83 percent efficiency. The trenching of Figure 12-43 is also being done under excellent working conditions and efficiency should be in the neighborhood of 83 percent. On the other hand, the ladder trencher of Figure 12-44 is known to work at an efficiency of only 31 percent, equivalent to a 19-min working hour. The trenching by the wheel machine of Figure 12-46 is a casting job and conditions are good for a high efficiency rate. This rate has been estimated at 67 percent, equivalent to a 40-min working hour. These four examples of trenching thus represent an efficiency range of 31 to 83 percent in terms of job conditions and not in terms of machine capabilities.

Wheel Trencher

The wheel trencher has a series of removable buckets mounted on the circumference of a rotating wheel. Excavation from the discharging buckets is moved laterally by the side conveyor for loading or casting. The hinged frame of the wheel is lowered into the trench and the rock-earth is excavated by the feed or travel of the trencher as it moves along the line of the ditch. The digging action of the wheel machine is

Figure 12-43 Medium-weight ladder trencher of the sliding-boom type. This trencher of the sliding- or vertical-boom type is casting rock-earth from a light fill for a utility line. Trenches may be excavated up to 8-ft depth and 24-in width. Important specifications for the machine are: (1) bucket width, 12 to 24 in; (2) bucket line speeds, 96 to 225 ft/min; (3) digging feed speeds, 0.5 to 13.8 ft/min; (4) travel speeds, 1.4 and 3.2 mi/h; (5) conveyor belt width, 18 in; (6) conveyor belt speed, 400 ft/min; (7) engine horsepower, 55; (8) working weight, 23,000 lb.

The capacity of the belt at 400 ft/min is about 232 tons/h. If bucket width, bucket line speed, and digging feed were adjusted to this belt speed for average-weathered rock-earth, capacity at 100 percent efficiency would be about 124 cybm/h. *(Barber Greene Co.)*

TABLE 12-22 Representative Specifications for Ladder Trenchers of the Hinged-Boom Type

Specification	Nominal depth capacity, ft		
	11	15	25
Bucket widths, without sidecutters, in	16, 20, 24	24, 30, 36	24, 30, 36, 42, 48
Bucket line speeds, ft/min	97, 166, 233	97, 166, 234	71, 134, 185
Digging feed speeds, variable, ft/min	0 to 30	0 to 30	0 to 50
Travel speeds, mi/h:			
Forward	1.4, 2.4, 3.4	1.4, 2.4, 3.4	1.0, 1.9, 2.7
Reverse	0.8	0.8	0.6
Conveyor belt:			
Width, in	20	24	28
Speed, ft/m	369, 528, 800	506, 583, 806	350, 467, 700
Diesel engine horsepower	80	100	164
Working weight, lb	17,000	37,000	55,000

the same as that of the ladder trencher, being a combination of cutting, milling, and crumbling. Figure 12-45 illustrates the working parts of a large wheel trencher of 8.5 ft nominal depth capacity, and Figure 12-46 shows the same machine excavating hard sandstone. Table 12-23 lists specifications for this machine and two smaller wheel trenchers.

In Figure 12-46 excavation in the average-weathered sandstone is proceeding at 5 to 12 ft/min, averaging 8.5 ft/min during actual digging in the 6-ft-deep and 4.3-ft-wide trench. The hourly production at 100 percent efficiency is 487 cybm. The wheel trencher is casting under favorable conditions and efficiency is probably about 67 percent. At this efficiency, the hourly production is 326 cybm.

12–64 Loading and Casting Excavation

Figure 12-44 Large ladder trencher of hinged-boom type digging hard soapstone formation. The trencher of nominal 25-ft depth capacity is digging a 15-ft-deep, 6.3-ft-wide trench by a combination of cutting, milling, and crumbling work. Every 8 h, 150 ft of trench is excavated, resulting in an average production rate of 66 cybm/h. Based on actual loading time for trucks, the 100 percent efficiency rate for excavation is 211 cybm/h. The efficiency per working hour is thus 31 percent. Trenching efficiency is typically low because of the many unavoidable delays caused by obstacles of human construction and confined working conditions. This work in an urban setting in Texas is representative. *(Lorraine Smith, "Modifications Toughen Trencher's Bite,"* Construction Methods and Equipment, *October 1972.)*

TABLE 12-23 Representative Specifications for Wheel Trenchers

	Nominal depth capacity, ft		
Specification	6.0	7.0	8.5
Bucket widths, in	16–24	20–36	36–66
Wheel speeds, ft/min	152, 259, 366	146, 260, 369	0–243
Digging feed speeds, variable, ft/min	0–30	0–26	0–78
Travel speeds, mi/h:			
Forward	1.4, 2.4, 3.4	to 4.0	to 4.0
Reverse	0.6	to 4.0	to 4.0
Conveyor belt:			
Width, in	24	28	36
Speed, ft/min	369, 528, 800	Up to 748	250–1100
Diesel engine horsepower	80	100	165
Working weight, lb	18,000	30,000	50,000

Figure 12-45 Working parts of a large wheel trencher. The specifications for this machine are those for the 8.5-ft nominal depth capacity machine of Table 12-23, and it is the trencher shown in Figure 12-46. Important assemblies and parts for the trencher are: (1) main frame and boom of the digging wheel, hinged to tractor frame; (2) digging wheel with removable buckets of different sizes or widths; (3) crumbling shoe frame and shoe; (4) side conveyor and hoist behind digging wheel; (5) tractor frame or main carriage, on which the main frame of the digging wheel and the side conveyor are mounted; (6) diesel-engine prime mover, which drives several hydraulic pumps and motors for controls and for actuating several of the assemblies. *(American Hoist and Derrick Co.)*

Figure 12-46 Large wheel trencher casting average-weathered sandstone. This trencher is excavating for a gas loop line around Oil City, Pennsylvania. The trench through average-weathered sandstone is 6 ft deep and 4.3 ft wide. The casting work is being done under rural and favorable conditions, making for high efficiency. When casting at 100 percent efficiency, the lineal progress ranges from 5 to 12 ft/min, averaging 8.5 ft/min, which is equivalent to casting 487 cybm/h. At an estimated efficiency of 67 percent the production per working hour is 326 cybm. The trencher specifications are those for the 8.5-ft nominal-depth-capacity machine of Table 12-23. *(American Hoist and Derrick Co.)*

12–66   Loading and Casting Excavation

The 1978 estimate for costs of excavation by the wheel trencher is:

Cost of wheel-type trencher	$136,000
Hourly cost of ownership and operation:	
Wheel trencher, 30- to 60-in width (Table 8-3). 32% of $136	$43.50
Operator and groundman	30.60
Total	$74.10
Direct job unit cost for casting per cybm, $74.10/326	$0.23

The estimated productions and costs for the large trenchers of Figures 12-44 and 12-46 when working in somewhat similar rock formations present an interesting contrast. The contrast is largely in terms of efficiency of operation. In Figure 12-44, the efficiency is 31 percent. Production is 66 cybm/h at a unit cost of $1.25. In Figure 12-46, the efficiency is 67 percent. The production rate is 326 cybm/h and the unit cost is $0.23. The low-production–high-cost job is one of loading in urban surroundings, whereas the high-production–low-cost trenching is one of casting in a favorable rural setting.

Production of Trenchers

All other factors being equal, the rate of digging and elevating a rock-earth formation depends on the horsepower for the task. The digging consumes more horsepower than the elevating. To elevate each working hour 500 cybm of average-weathered rock-earth weighing 3750 lb/cybm to a 25-ft total height for loading consumes only 24 theoretical hp of the 164 engine hp of the largest trencher of Table 12-22. It is logical to estimate the production of a trencher on the basis of rated engine horsepower and nature of the rock-earth excavation, and Table 12-24 has been prepared on this basis.

Applying Table 12-24 to the ladder trencher of Figure 12-44, the controlling data are: engine horsepower, 164; minimum-weathered rock-earth; and 31 percent efficiency. Interpolated production at 100 percent efficiency is 197 cybm/h; at 31 percent efficiency it is 61 cybm/h.

TABLE 12-24   Approximate Hourly Productions, cybm, for Ladder and Wheel Trenchers at 100 Percent Efficiency

Rock-earth formation	Engine horsepower of trencher			
	50	100	150	200
Alluvium, sand-gravels, lightly cemented	210	420	630	840
Weathered rock-earth:				
Maximum weathering	180	360	540	720
Average weathering	120	240	360	480
Minimum weathering	60	120	180	240

Summary

Ladder and wheel-type trenchers are used in alluvia and weathered rock-earth for subsurface excavations up to 25 ft deep and 6 ft wide. Under favorable conditions the ladder machine may be equipped with a somewhat longer boom so as to increase its depth capacity by a few feet. Its chief advantage over backhoes and draglines for similar work is that it does not cycle and is a continuously operating machine. Accordingly, engine power is applied to continuous digging, providing maximum production and minimum unit cost. Another advantage is that it excavates to neat lines and grade, eliminating both undercutting and overcutting. However, unless excellent small fragmentation results from blasting, it is not generally adapted to trenching in hard rock formations.

Hourly production is sometimes greatly reduced by unavoidable delays in tedious trenching caused by both natural and man-made obstructions. Efficiency may vary from 20 percent in unfavorable urban areas when loading rock haulers to 90 percent in rural surroundings when casting. Of paramount importance when estimating productions and costs is the well-considered selection of an overall efficiency factor to apply to the theoretical output of the trencher. The difference between 20 percent and 90 percent efficiency is a 78 percent reduction in direct job-unit cost for loading or casting.

BELT LOADERS

There are two kinds of belt loaders for rock-earth excavation, mobile and movable. The mobile machine, illustrated in Figure 12-47, moves while it loads or casts. The movable loader, shown in Figure 12-53, is stationary when it loads, but is movable or portable for relocation in another cut.

The mobile loader is adapted to work in flat or gently rolling country and in formations of horizontal stratification, such as alluvia and sedimentary rocks. It is used in ripped rock formations but generally it is not used in blasted rocks. The mobile loader digs by means of the cutting disk or cutting edges and elevates the rock-earth by means of the belt conveyor for loading or casting. The digging force is furnished by the pulling crawler-tractor or tractors. The elevating force is provided by the separate engine of the loader.

The movable loader is simply a belt conveyor, usually equipped with a feeder at the foot of the conveyor. It is adapted for work in hilly and mountainous country. As it is not a digging machine but rather is loaded by crawler-tractor-bulldozers, it may work in any rock-earth formation—natural, ripped, or blasted—which can be loaded onto the belt.

Both loaders have the following general common specifications. The inclination of belt does not exceed 22° or 1.0:0.4 slope, and the belt is of rugged multi-ply construction with widths of 36 to 72 in and lengths of 30 to 70 ft. Engine horsepower for the conveyor ranges from 100 to 600.

Mobile Belt Loaders

There are two types of mobile belt loaders, of which one is pulled by a crawler-tractor or tractors and the other is self-contained, its tractors being an integral part of the complete machine. Mobile loaders may be used either for loading haulers or for casting, although the amount of casting work is negligible.

Figure 12-47 illustrates a wheels-tires-mounted mobile loader pulled by a 235-hp crawler-tractor. A 36-in-diameter disk plow cuts into the 4-ft-high bank of firm valley alluvium and deflects the rock-earth onto the belt. The belt is 48 in wide and 30 ft long. The engine horsepower is 120 and the working weight of the machine is 27,000 lb. The machine is loading bottom-dump semitrailer haulers of 50,000-lb capacity carrying 19 cylm or about 17 cybm of excavation. In this free-flowing alluvium, the belt speed is 400 ft/min and the loading rate is about 22 cylm/min, equivalent to 19 cybm/min. The cross-sectional area of the 4-ft-high cut is 5.0 ft^2, and the travel distance for the 17-cybm load is 92 ft.

Figure 12-47 Mobile belt loader excavating desert alluvium for freeway building. This loader with a 48-in-wide belt and 120-hp engine is pulled by a 235-hp crawler-tractor. The loading rate is 19 cybm/min, or 1140 cybm/h. The hourly production is 650 cybm, giving 57 percent overall efficiency. The lost 43 percent of available time is unavoidable in the exchange of hauling units and the turns at the ends of the 800-ft-long cut. A hauler is loaded out in 0.86 min with 17 cybm of the silt-sand-gravel alluvium in a distance of 92 ft. The alluvium is the toe of a vast fanglomerate eroded down from the San Jacinto Mountains, which may be seen in the far distance. Such a loosely consolidated formation in flat country lends itself to low-cost loading by mobile belt loader and low-cost hauling by on-the-road–off-the-road bottom-dump haulers. Cost on the fill is also low because of the ease of leveling and compacting the long windrows dumped by the haulers.

12–68   Loading and Casting Excavation

The work of Figure 12-47 is freeway building in the San Gabriel Valley of southern California. The typical long shallow cuts and fills make for long hauls in both off-the-road and on-the-road haulage. In this mixed hauling the rugged bottom-dump haulers are efficient, as they are heavy enough for construction haul roads and light enough for use on highways.

Estimated production and 1978 costs for the mobile loader shown in Figure 12-47 are tabulated below.

Production:
  Loading cycle:
    Loading 19 cylm at rate of 22 cylm/min   0.86 min
    Exchanging hauling units   0.33
    Turning at end of 800-ft cut, 1.0 min prorated to each of 8 loads, 1.0/8   0.12
    Total   1.31 min
  Loads per 50-min working hour, 50/1.31   38
  Cybm loaded hourly, 38 × 17   650
Costs, 1978:
  Belt loader, mobile, pull-type, wheels-tires-mounted $ 73,000
  Crawler-tractor-bulldozer, 235 hp   134,000
  Total $207,000
  Hourly cost of ownership and operation (Table 8-3):
    Loader, 39% of $73 $28.50
    Tractor, 34% of $134   45.60
    Tractor-loader operator   15.80
    Total $89.90
  Direct job unit cost for loading per cybm, $89.90/650 $0.14

Figure 12-48 illustrates another of the mobile loaders pulled by one or two crawler-tractors. This heavy machine weighs 90,000 lb and has a 300-hp engine, 60-in belt width and 40-ft length. The loader may be pulled by one or two heavyweight crawler-tractors with a total of 400 to 800 hp, the choice of total horsepower being largely in terms of the consolidation of the rock-earth excavation.

Production of Mobile Belt Loaders   The mobile belt loader embodies two functions, carried on by two power trains. The digging function, by which the rock-earth is delivered to the elevating belt conveyor, is powered by the pulling crawler-tractor(s) or by the integrated tractors in the case of the self-contained loader. The elevating function, by which the rock-earth is loaded into the haulers or cast, is powered by the loader engine in the manner of a belt conveyor. The production of the machine is in terms of the power

Figure 12-48   Large mobile belt loader to be pulled by one or two crawler-tractors. This machine is especially suited to the mass excavations of large public works and open-pit mining operations such as dams, canals, and nonmetal mines in loosely consolidated rock formations. When pulled by two 300-hp crawler-tractors under favorable conditions, the rate for loading can reach 60 cylm/min and the average hourly production can be as high as 1800 cybm. The rather special nature of huge mobile belt loaders calls for expertise in estimating and in application. *(General Motors Corp.)*

Belt Loaders   12-69

of the crawler-tractors for the work of digging. It is both logical and practical to estimate the production in terms of the total rated engine horsepower of the crawler-tractors which either pull or carry the mobile belt loader, as this horsepower controls the time rate of digging and loading the accessory conveyor. Table 12-25 has been prepared on this basis.

The productions of Table 12-25 are based on the digging ability of the tractor(s) pulling or carrying the belt loader. These productions must be reconciled with the carrying capacity of the associated belt conveyor of the loader. These conveyors are designed with sufficient capacities to accommodate the maximum digging capacities of the machines. Engine horsepower, belt speed, and belt width are selected to harmonize with the capability of the loader as a digging machine. Capacities of belt conveyors are given customarily in terms of tons of rock-earth carried hourly according to a belt speed of 100 ft/min, the width of belt in inches, and the weight of rock-earth in pounds per cubic foot. Additionally, tons of rock carried hourly is a function of the cross-sectional area of the load on the belt conveyor.

Tables 12-26, 12-27, and 12-28 supplement Table 12-25. They are prepared in a manner similar to conventional tables for belt conveyors but are calculated conservatively and specifically for use with both mobile and movable belt loaders. Table 12-29 gives factors for converting loose to bank measurement for various types of rock-earth.

Example of Use of Tables for Estimating the Production of a Mobile Belt Loader

The large machine of Figure 12-48 is applied to the job of the smaller loader of Figure 12-47. When the machines are working under like conditions, the determinants for figuring hourly production are:

1. Tractor engine horsepower is 410. From Table 12-25 the rate of loading haulers with alluvium is 41 cybm/min.

2. The bottom-dump haulers have a capacity of 47 cysm and they haul 52 cylm. Table 12-29 indicates that the load is $0.88 \times 52 = 46$ cybm. The time for loading each hauling unit is $46/41 = 1.12$ min.

3. Table 12-27 indicates that a 60-in-wide belt carrying alluvium may have a speed of 700 ft/min. Table 12-28 indicates that the belt can carry a yardage of 61 cylm/min. The belt carries $41/0.88 = 47$ cylm/min, so the capacity is adequate.

4. Exchange time for haulers is 0.33 min and prorated turning time at the end of 800-ft cut is 0.12 min.

5. The 50-min working hour is used, equivalent to 83% efficiency.

On the basis of the preceding data the hourly production is calculated as shown in the following tabulation.

Hourly production:	
Hauling cycle times, min:	
Loading 46 cybm into hauler	1.12
Exchanging hauling units	0.33
Prorated turning at end of 800-ft cut, 1.0 min for 8 loads, 1.0/8	0.12
Total	1.57
Loads per 50-min working hour, 50/1.57	32
Cybm loaded hourly, 32×46	1470
Estimated costs, 1978:	
Belt loader, mobile, pull-type, crawler-mounted	$240,000
Crawler-tractor-bulldozer, 410 hp	251,000
Total	$491,000
Hourly cost of ownership and operation (Table 8-3):	
Loader, 39% of $240	$ 93.60
Crawler-tractor, 33% of $251	82.80
Tractor-loader operator	15.80
Total	$192.20
Direct job unit cost for loading per cybm, $192.20/1470	$0.13

12-70 Loading and Casting Excavation

TABLE 12-25 Approximate Productions, cybm/min, for Mobile Belt Loaders According to Total Engine Horsepower of the Crawler-Tractor(s) at 100% Efficiency

Total engine hp	Total engine horsepower of the crawler-tractor(s)									
	100	200	300	400	500	600	700	800	900	1000
Alluvium, cemented sands-gravels	10	20	30	40	50	60	70	80	90	100
Weathered rock-earth:										
Maximum weathering	9	18	27	36	45	54	63	72	81	90
Average weathering	7	14	21	28	35	42	49	56	63	70
Minimum weathering	4	8	12	16	20	24	28	32	36	40

TABLE 12-26 Areas of Cross Sections of Belt Loads with 20° Angle of Repose for Both Mobile and Movable Belt Loaders

Belt width, in	Area, ft^2
36	0.80
42	1.12
48	1.48
54	1.89
60	2.36
66	2.85
72	3.40

TABLE 12-27 Recommended Maximum Belt Speeds for Mobile and Movable Belt Loaders According to Nature of Rock-Earth

Nature of rock-earth	Belt speed, ft/min, for various belt widths (in)						
	36	42	48	54	60	66	72
Alluvium—silts, sand, gravels	700	700	700	700	700	700	700
Weathered rock:							
Not ripped	500	550	600	600	600	650	650
Ripped	400	400	400	450	450	500	500
Rock, well-blasted	400	400	400	400	400	400	400

TABLE 12-28 Loading or Casting Production Rates for Conveyors of Mobile and Movable Belt Loaders

Belt width, in	Maximum rock size, in	Production rate, cylm/min, for various belt speeds, ft/min				
		300	400	500	600	700
36	18	9	12	15	18	21
42	21	12	17	21	25	29
48	24	16	22	27	33	38
54	27	21	28	35	42	49
60	30	26	35	44	52	61
66	33	32	42	53	64	74
72	36	38	50	63	76	88

NOTE: The production rates in cylm/min are based on the data of Table 12-26 and the given belt speeds of this table.

TABLE 12-29 Factors for Converting Cubic Yards Loose Measurement to Bank Measurement

Nature of rock-earth	Factor or multiplier
Alluvium—silts, sands, gravels	0.88
Rock-earth, weathered, either not ripped or ripped	0.78
Rock, well-blasted	0.67

The unit cost is about 7 percent less than that of the smaller loader. Such a difference may be expected when one substitutes a larger machine of equal quality. However, in this instance the larger haulers, balanced to the larger loader, could not travel on the highways for the on-the-road work. Thus, the experienced excavator chose the right machinery for the job in spite of the advantage of the larger machines for the off-the-road phase of the job.

Figure 12-49 illustrates a heavy mobile belt loader with two integrated heavy crawler-tractors, and Figure 12-50 shows an almost identical loader excavating cemented alluvium for the semipervious fill of Perris Dam, southern California. The alluvium usually requires ripping, but this is not necessary for the multiengine loader. The estimate for

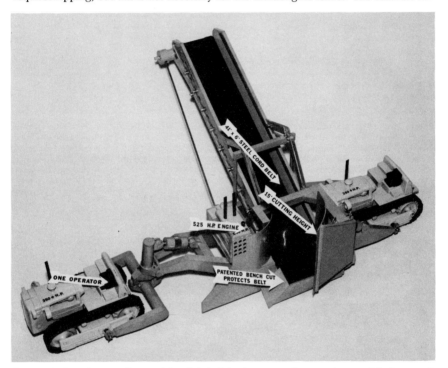

Figure 12-49 A heavy and powerful mobile belt loader mounted on two heavyweight integrated crawler-tractors. The important 1978 specifications for this complete machine include the following:

Loader exclusive of tractors:
    Conveyor belt:
        Width, in   72
        Length, ft   40
        Maximum belt speed, ft/min   700
        Engine horsepower   525
    Cutting edges, lengths:
        Vertical, ft   15
        Horizontal, ft   3
    Controls for hydraulic lift and tilt cylinders, from tractor
    Working weight, lb   92,000
Integrated tractors:
    Total working weight, lb   165,000
    Total engine horsepower   820
Complete machine:
    Working weight, lb   257,000
    Total engine horsepower   1345

(Holland Co.)

12-72 Loading and Casting Excavation

Figure 12-50 Heavy mobile belt loader with two integrated heavy tractors loading alluvium for semipervious fill of dam. The specifications for this machine are those given for the loader of Figure 12-49 except that the total horsepower of the tractors is 770 instead of 820. The location of this work is Perris Dam of southern California, and the estimated performance of the machine when working under the conditions of this picture is as follows: (1) belt speed, 600 ft/min; (2) loading rate, loose measurement, 76 cylm/min; (3) loading rate, bank measurement, 67 cybm/min; (4) loading time for 66-cybm load of bottom-dump hauler, 0.99 min; (5) hourly production in 2000-ft-long cut of cemented alluvium, 2380 cybm; (6) 1978 direct job-unit cost for loading, $0.13; (7) monthly production for two 8-h shifts for 22 days, based on 90 percent availability, 754,000 cybm.

production and costs of loading under these conditions is tabulated below, based on the following known criteria:
    1. The bottom-dump haulers have a 68-cysm capacity and haul an average load of 75 cylm or 66 cybm.
    2. The loading time for 66 cybm is 1.22 min and the exchange time for the haulers is 0.26 min.
    3. The belt speed, adjusted to the loading rate, is about 600 ft/min.

Production:
  Loading cycle times, min:
Loading 66 cybm	1.22
Exchanging hauling units	0.26
Turning at end of 2000-ft cut, 0.9 min prorated to each of 20 loads	<u>0.04</u>
Total	1.54
Loads per 50-min working hour, 50/1.54	32
Cybm loaded hourly, 32×66	2110
Costs, 1978:	
Belt loader, mobile, self-propelled type, 1320 hp	$736,000
Hourly cost of ownership and operation:	
Loader (Table 8-3). 39% of $736	$287.00
Tractor-loader operator	<u>15.80</u>
Total	$302.80
Direct job unit cost for loading per cybm, $302.80/2110	$0.14

Figure 12-51 shows a cut in a shale formation which was excavated by the same heavy mobile belt loader at Fort Peck Dam, Montana.

Movable Belt Loaders

These belt loaders are known also as *stub loaders*. They are set into the hillside cut and the unripped, ripped, or well-blasted rock is bulldozed into the receiving hopper of the belt conveyor. Figure 12-4 shows such an operation of ripping, bulldozing, and loading weathered rock. Figure 12-52 illustrates a movable belt loader with a 60-in-wide belt conveyor, and Table 12-30 gives representative specifications for three belt widths.

Figure 12-51 Average weathered bear paw shale of Montana excavated by mobile belt loader. The 15-ft-deep cut of soft-medium sedimentary rock was taken out with 11 lifts by the loader of Figures 12-49 and 12-50. The shale at Fort Peck Dam averages about 4100-ft/s seismic shock-wave velocity in the cut, which indicates medium ripping for a heavyweight crawler-tractor. The loader dug weathered rock without prior ripping. Loading time for the hauler payload, estimated at 58 cybm, was 1.17 min. From the available data an estimate of production and unit cost for 1978 by customary procedures is tabulated below:

Loading cycle for average 58-cybm load of hauler, min	1.54
Loads per 50-min working hour, 50/1.54	32
Hourly production, cybm, 32×58	1860
Hourly cost of ownership and operation	$302.80
Direct job-unit cost for loading per cybm, $302/1860	$0.16

(Holland Co.)

The cut is usually in a hillside or in an area at a higher elevation than the top of the receiving hopper. Bulldozing starts on the natural slope and ends at the elevation of the hopper with level bulldozing or perhaps a slight upgrade, although bulldozing upgrade is objectionable from both production and cost standpoints. Bulldozing rarely exceeds a 400 ft one-way distance. The cut area is fan-shaped, the loader being at the downgrade apex of the triangle, which has an apex angle of about 135°. Two typical side-by-side cuts are shown in Figure 12-53.

After a cut is finished, the belt loader is moved by a crawler-tractor and set up in the next prepared location, the move requiring about 8 h. The cost of moving the belt must be prorated to the yardage of the cut, so that it is advantageous to take out as sizable a cut as possible.

As the combined action of ripping, bulldozing, and conveyor loading generally produces a well-graded, free-flowing rock-earth mixture, the haulers are mostly of the bottom-dump type. The team of belt loader and matched haulers gives a high-production–low-unit-cost method for loading, hauling, and building of fill.

Productions of Movable Belt Loaders The movable belt loader, like its mobile counterpart, works on a cycle. The average cycle for the two loaders of Figure 12-53 when loading the set of bottom-dump haulers is:

Loading first unit	0.42 min
Exchanging units	0.16
Loading second unit	0.42
Exchanging haulers	0.22
Total	1.22 min

Each loader's actual hourly production is 1230 cybm in the maximum-weathered granite, equivalent to 1580 cylm. The loading times for the two bottom-dump units, carrying 28 cybm, is 0.84 min. Accordingly, the efficiency which emerges based only on actual loading times is as follows:

Actual hourly loader output, cybm	1230
Theoretical hourly output at 100% efficiency, $(60/1.22) \times 28 =$	2000
Efficiency of belt conveyor, %, $1230/2000 =$	62

This efficiency appears to be low until one considers that the complete cycle contains two unavoidable delays for exchange of hauling units. The loading time, on which efficiency must be based, is the total cycle time of 1.22 min. Actual belt-loader efficiency is given by the equation:

$$E = \frac{1230}{(60/1.22) \times 28} = 89\%$$

Both efficiencies are high for the average work of movable belt loaders. The production estimate for a loader, like that for any cycling machine, should be based on the calculated loading cycle and a well-considered efficiency factor. In a well-managed job the attainable efficiency should be the 50-min working hour, or 83%.

Figure 12-52 Movable belt loader with 60-in-wide belt conveyor. The principal specifications for this medium-size movable belt loader are:

Conveyor:	
Width, in	60
Length, ft	60
Loading height, ft	16
Overall dimensions, ft:	
Length	66
Width	11
Height	18
Diesel engine horsepower	185
Working weight, lb	62,000
Estimated nominal production rate at 100% efficiency, cylm/min:	
400-ft/min belt speed	35
500 ft/min belt speed	44
600 ft/min belt speed	52

The production rates in cubic yards loose measurement are based on the data of Table 12-28 and the belt speeds given in this table. *(Athey Products Corp.)*

Figure 12-53 Two movable belt loaders, with 60-in belt width, working in weathered granite. These two machines are loading 2460 cybm/h of ripped weathered granite in this huge 10-million-yd^3 freeway cut of southern California. The production is sustained by a high efficiency, 89 percent. Each double-bottom dump-hauler is loaded out with 28 cybm in 1.00 min. The exchange time between complete haulers is 0.22 min, giving a loader cycle of total 1.22 min. Each loader makes 43.9 loads hourly out of a possible 49.2 loads at 100 percent efficiency. The daily production of the two loaders working 18 h is about 44,000 cybm. The soft, well-weathered granodiorite, an ideal rock for belt-loader work, and the excellent management combined to bring about this unusual output.

TABLE 12-30 Representative Specifications for Movable Belt Loaders

Specification	Nominal belt width, in		
	48	60	72
Conveyor:			
Length, ft	50–60	50–60	60–70
Loading height, ft	13.4–15.6	13.4–15.6	15.6–18.2
Overall dimensions, ft:			
Length	56–66	56–66	66–76
Width	10	11	12
Height	15.1–17.6	15.1–17.6	17.2–20.2
Diesel engine horsepower:			
50-ft length	100	165	
60-ft length	130	185	265
70-ft length			290
Working weight, lb	54,000	62,000	74,000
Estimated nominal production at 100% efficiency, cylm/min:			
400-ft/min belt speed	22	35	50
500-ft/min belt speed	27	44	63
600-ft/min belt speed	33	52	76

NOTE: The production rates in cylm/min are based on the data of Table 12-28 and the given belt speeds of this table.

Tables 12-27, 12-28, and 12-29 are convenient for estimating hourly production of movable belt loaders, as outlined in the following tabulation.

Example of Use of Tables for Estimating Production of Movable Belt Loader Tables 12-27 to 12-29 are convenient for estimating hourly production of movable belt loaders, as outlined in the following example involving the loader of Figure 12-53. The determinants for estimating hourly production in cubic yards bank measurement are:

    1. The rock is maximum-weathered granite. Table 12-29 gives a factor of 0.78 for conversion from loose to bank measurement yardage.

12-76 Loading and Casting Excavation

2. Table 12-27 gives a suggested belt speed of 450 ft/min for a 60-in-wide belt working in ripped weathered rock.

3. Table 12-28 gives a production 40 cylm/min for a 60-in-wide belt at a speed of 450 ft/min.

4. The rock hauler is a set of two bottom-dump units, each of 18 cysm capacity and carrying an estimated load of 14 cybm.

From these data the hourly production in bank measurement is estimated as follows:

Loose measurement load of each hauler unit, cylm. 14/0.78	18
Loading time for each hauler unit, min. 18/40	0.45
Estimated loading cycle times, min:	
Loading first unit	0.45
Exchanging units, 10 s	0.17
Loading second unit	0.45
Exchanging haulers, 15 s	0.25
Total	1.32
Loads hauled per 50-min working hour, 50/1.32	38
Hourly production, cybm 38×28	1060

The theoretical hourly output, based on the capacity of the belt conveyor, is 2400 cylm, equivalent to 1870 cybm. The efficiency of the conveyor is 1060/1870, or 57 percent, whereas the efficiency of the belt loader is 83 percent.

The estimated 1978 costs for loading by the machines of Figure 12-53 are:

Costs, 1978:	
Movable belt loader, 60-in belt	$162,000
Hourly cost of ownership and operation:	
Loader (Table 8-3). 37% of $162	$59.90
Operator and groundman	29.00
Total	$88.90
Direct job unit cost per cybm, $88.90/1060	$0.083

To this unit cost must be added the unit cost for moving the loader to the next location after the cut has been taken out. The work requires a crawler-tractor-bulldozer and a wheels-tires-mounted bucket loader for about 6 h. Generally the move is made after the day's work.

In Figure 12-53 the yardage to be removed is estimated at 47,000 cybm. The unit cost for moving the loader is estimated as shown in the following summary:

Costs, 1978:	
300-hp crawler-tractor bulldozer	$183,000
170-hp wheels-tires-mounted bucket loader	94,000
Hourly cost of ownership and operation (Table 8-3):	
Tractor, 33% of $183	$ 60.40
Operator	15.60
Loader, 35% of $94	32.90
Operator	15.60
Two laborers @ $11.50	23.00
Total	$147.50
Total cost for 6 h work, 6×$147.50	$885.00
Direct job unit cost for moving loader per cybm, $885.00/47,000	$0.019

The combined direct job unit cost for loading and then moving the belt loader is $0.102/cybm, the moving cost being 19 percent of the total cost.

Figure 12-54 illustrates a unique use of the movable belt loader to lower the cost for extreme downgrade haulage. In this mountainous country the grading of a tract site for homes called for bringing a ridge down vertically some 310 ft to the ravine. Obviously, hauling cost by wheels-tires-mounted scrapers would be excessive as a suitable switchback haul-and-return road would total about 2400 ft in length.

The excavator decided to set in a movable loader with a 48-in belt-width at an elevation 200 ft below the beginning of the cut and to use gravity instead of the scrapers for the downgrade movement of rock-earth. The weathered sandstone-shale was ripped and bulldozed to the edge of the cut, whence it cascaded down the natural steep slope to the loader. The well-graded sedimentary rock was then hauled to the fill by the scrapers.

The excavator then wished to anticipate the elevation at which to change to conventional scraper operation. The cost of the belt loader method increased as the ridge was

Figure 12-54 Unique use of movable belt loader to eliminate costly downgrade haulage. Three bulldozers are loading a 48-in belt-width loader located some 200 ft below the cut, from which point scrapers are used for the downgrade haul to the fill. The sandstone-shale ridge of the Santa Monica Mountains of California is being taken down as part of a huge grading job for homes. If wheels-tires scrapers were to be used, they would haul down the steep trail to the left of the picture and then wind back to the cut by a 1600-ft long return haul road. Instead, gravity will bring the rock-earth down to the loader until the cut is brought down to an elevation where the conventional scraper operation is cheaper than combination bulldozing–gravity fall–loader work. The change will take place at about 60 ft below the present cut, or about one-third of the height of the present fall of the rock-earth.

In many instances of rock excavation, simple gravity may be substituted for costly machines, with considerable savings in money and total time for work.

brought down because the length of bulldozing through the transverse ridge to the brow of the cut increased with increasing depth of cut. At the same time the cost of alternative scraper haulage would decrease as the elevation of cut lessened. Both these costs were expressed algebraically in terms of D, the vertical distance below the top of the cut. Equating the unit costs and solving for D gave 60 ft as the point below which scraper work would be more economical. The excavator changed from belt loader to conventional scraper operation at that elevation.

Many problems in rock excavation, both physical and fiscal, may be solved by rather simple imaginative mathematics. Such methods enable the rock mover to anticipate future actions and to make provisions for the necessary machinery for most efficient and least costly work.

Figure 12-55 illustrates a 60-in belt-width loader working in weathered limestone-shale in Ohio. The excavation is rugged work for the belt conveyor, as may be appreciated by scrutiny of the large angular rocks. The bottom-dump haulers are of 13-cysm capacity and they haul about 14 cylm. On the average, 20 haulers are serving the loader on a 2-mi one-way haul.

An estimate for production and costs for the loading is given below.

```
Production:
   Loading rate at 450 ft/min belt speed (Table 12-28), cylm/min        40
   Loading cycle times, min:
      Loading 14 cylm, 14/40                                          0.35
      Exchanging haulers, estimated at 15 s                           0.25
      Total                                                           0.60
   Loads per 50-min working hour, 50/0.60                               83
   Cubic yards bank measurement per load. (Table 12-29) 0.78×14         11
   Hourly production, cybm. 83×11                                      910
```

12-78 Loading and Casting Excavation

Costs, 1978:	
Movable belt loader, 60-in belt	$162,000
Hourly cost of ownership and operation:	
Loader (Table 8-3). 37% of $162.	$59.90
Operator and groundman	29.00
Total	$88.90
Direct job unit cost per cybm, $88.90/910	$0.10

Peak daily production for this well-managed job is 1006 loads per 10-h day; 101 loads/h give 1110 cybm. The direct job unit cost for such a high-production day is $0.08/cybm, or 20 percent less than the estimated unit cost. Such variation is not uncommon, especially in long-haul work.

Summary

Mobile and movable belt loaders are high-production excavators for many different rock-earth formations, ranging from soft residuals and alluvia through undisturbed and ripped weathered rocks to well-blasted rocks. Mobile machines work in consolidations up through ripped rocks and movable machines handle all degrees of consolidation up to well-blasted formations.

Mobile loaders are associated with excavation in flat or gently rolling country, where cuts are long and rock haulers function well. Movable loaders are suitable for hilly and mountainous terrain, where cuts are short and haulers are stationary at the loading location. Additionally, in movable loader work the associated bulldozing to the loader takes place under favorable downgrade conditions.

When working conditions indicate that there is really a choice of the type of loader to be used, a most thorough analysis of productions and costs ought to be made, as offsetting factors may suggest the use of one or the other loader when productions and costs are pretty much the same.

WHEEL-BUCKET EXCAVATORS

Wheel-bucket excavators are used largely in the mass movement of rock-earth in huge public works such as dams and freeways and in open-pit workings such as coal and ore mines. The principles of operation are similar to those of the wheel trencher when

Figure 12-55 Peak hourly production of 1110 cybm for movable belt loader in hard rock. Hard limestone-shale of Ohio is being loaded into bottom-dump haulers by the loader with a 60-in belt width. The minimum-weathered limestone has been ripped into large abrasive sizes and the loading service is markedly severe. Nevertheless, peak production reaches 1006 loads in a 10-h working day. The 20 haulers for the long (2-mi) one-way haul are of 13 cysm capacity and they haul an estimated 11 cybm. Top production is then 1110 cybm/h; the preceding conservative estimate is 910 cybm/h, based on putting out 83 loads hourly during a 50-min working hour. It is probable that the higher production comes from a lesser actual loading cycle and a higher efficiency. *(Athey Products Corp.)*

loading, as discussed previously in this chapter. A rotating wheel with circumferentially mounted rock buckets digs and then discharges onto a belt conveyor, which loads the rock haulers. Rarely is the machine used for casting. There are two kinds of excavators and they differ in the manner of digging by the wheel buckets.

One kind is the linear digging, linear traveling machine. It is illustrated in Figures 12-56 and 12-57. As it advances into the cut, the wheel excavates to a depth of about half the diameter of the wheel-bucket assembly.

The other kind is the circular digging, linear traveling kind. It is illustrated in Figures 12-58 and 12-59. It also proceeds in a straight line into the cut, but the wheel-bucket assembly moves in an arc. The assembly is part of the revolving deck of the machine. The loading belt conveyor is also a part of the revolving deck but it is in a fixed position during the loading, only the wheel-bucket assembly with ladder conveyor moving through the arc. The machine digs to a depth of about two-thirds of the wheel diameter.

Wheel-bucket excavators are used in alluvia, residuals, and weathered rocks. They may be also used in well-blasted rock, provided the fragmentation is small enough to allow the buckets to secure a reasonable loose load, so that a good bucket factor is attainable.

Estimating Productions

As the machine excavates by means of rotating buckets, the loading capacity in cubic yards loose measurement per minute for use in estimating the total loading cycle may be calculated by the formula:

$$P = RPM \times B \times FF \times BC$$

where P = production, cylm/min
 RPM = speed of bucket wheel, r/min
 B = number of buckets
 FF = fill factor (Table 12-3)
 BC = bucket capacity, cysm

The estimated nominal loading capacities in cubic yards loose measurement per minute, as given in the specifications for the machines, are based on the above formula. The fill factor FF is taken as 0.90 for average-weathered rock-earth. Using the machine of Figure 12-56 as an example, the specifications give: $RPM = 5.6$ average; $B = 12$; and $BC = 1.0$. Substituting in the formula:

$$P = 5.6 \times 12 \times 0.90 \times 1.0 = 60 \text{ cylm/min}$$

Linear-Digging–Linear-Traveling Excavator

The salient specifications for the linear-digging–linear-traveling wheel-bucket excavator shown in Figure 12-56 are as follows.

Overall dimensions, ft:	
Length	52.0
Width	60.7
Height	18.9
Digging wheel:	
Diameter, ft	16.0
Width, ft	9.5
Capacity of each of 12 buckets, cysm	1.0
Speed of wheel, r/min	5.1–6.0
Conveyors, width and length, ft:	
Receiving from wheel buckets	4.5×16.7
Elevating	5.0×31.0
Discharging to either of side-by-side haulers	5.0×10.0
Maximum belt speed, ft/min	1100
Crawler mounting, length and width, ft	10.8×14.2
Maximum travel or feed speed ft/min	185
Diesel engine horsepower	500
Working weight, lb	235,000
Estimated nominal loading capacity, cylm/min	60

SOURCE: Barber Greene Co.

The estimate of production and costs for loading by this wheel-bucket excavator in the work shown in Figure 12-56 is outlined below.

Figure 12-56 Linear digging bucket-wheel excavator loading desert alluvium. The 500-hp, 235,000-lb machine is loading an estimated 2090 cybm/h of lightly cemented alluvium. Loading time for the 60-cysm haulers is 1.28 min and exchange time is 0.07 min, the total cycle time being 1.35 min. With allowance for turning at the ends of the 0.5-mi cut, the adjusted cycle is 1.37 min. After it has finished loading the adjacent hauler, the hinged discharging conveyor will be switched to the other hauler with minimum delay. In this efficient work 36 haulers are loaded out hourly, each hauler carrying 58 cybm.

The southern California desert is flat and the underpasses of the freeway provide fill for the corresponding overpasses of the lateral highways, which are spaced every mile. The grading is made up of long cuts and fills, a conditions which is ideal for work by the bucket-wheel excavator.

Figure 12-57 Linear digging bucket-wheel excavator loading bituminous coal. The machine is loading out a 9-ft thick seam of coal at the Navajo Mine of New Mexico. The coal has not been blasted, so loading is severe. Because of the usual narrow width of the working area, the haulers are being loaded in tandem. Operating and estimated data for the work are:

Actual hourly production of the machine, tons	1800
Equivalent hourly production of the machine, cybm	1580
Capacity of coal hauler, tons	120
Hourly cost of ownership and operation, as for the similar machine of Figure 12-56	$216.10
Direct job unit cost per ton	$0.12
Direct job unit cost per cybm	$0.14

The relatively high unit cost per cybm, as compared with the $0.10 cost for the work of Figure 12-56 is explained readily by the different nature of the two excavations, unblasted coal as opposed to lightly cemented alluvium. *(Barber Greene Co.)*

Figure 12-58 Circular digging wheel-bucket excavator loading weathered basalt. The excavator is loading unblasted hard basalt for land reclamation at Tokyo Bay, Japan. The manufacturer builds several sizes, rated nominally for average hourly production in average-weathered rock. Ratings are 650 to 4500 cybm. The rating of this machine is 1500 cybm/h. Actual hourly production in this extremely hard digging is 950 to 1050 cybm. The severity of the work is apparent in the size of the large abrasive rocks. The digging wheel with its ladder belt moves in an arc while loading. The ladder belt discharges onto the discharge belt, visible in the background, which in turn elevates the rock-earth to the waiting 35-ton rock hauler. The discharge belt is in a fixed position while loading. Both wheel bucket–ladder belt assembly and discharge-belt assembly are mounted on the revolving frame of the machine. The 640-hp 430,000-lb machine is mounted on crawlers. *(Mechanical Excavators, Inc.)*

Figure 12-59 Circular digging wheel-bucket excavator loading valley alluvium. The largest of wheel-bucket excavators is loading silts, sands, and gravels for 40 million yd³ of semipervious fill of the San Luis Dam in California. The 2400-hp, 1,625,000-lb machine is loading 4500 cybm/h into the 70-cysm bottom-dump haulers. The cycle-time-saving double-discharge chute is completing the loading of the machine to the right, and the one to the left is in position for a load. Additional haulers are pulling in from the fill so as to provide ample supply and prevent delays to production. The bucket wheel is 30 ft in diameter with 10 buckets of 2½-cysm capacity. The belt conveyors are 84 in wide. The estimated nominal loading capacity is 112 cylm/min, based on 5 r/min of the wheel. At maximum 8 r/min the capacity would increase to 179 cylm/min. *(Bucyrus-Erie Co.)*

12-82   Loading and Casting Excavation

The job is a desert freeway and the rock-earth is lightly cemented alluvium, made up of clay, silt, sand, and gravels. The alluvium is being loaded into 60-cysm bottom-dump haulers, which are loaded out with minimum exchange time by means of the hinged discharging conveyor. Actual loading time for a 58-cybm load is 1.28 min and exchange time is 0.07 min, giving a total cycle time of 1.35 min. The cut of the wheel-bucket assembly is 6 ft high and 9.5 ft wide, giving 2.1 cybm per foot of feed or travel. The machine moves 28 ft at 22 ft/min to load the hauler.

Production:
Loading cycle times, min:
Loading 58 cybm 1.28
Exchanging haulers, 4 s 0.07
Turning at end of ½-mi cut, prorated to 94 loads, 1.7/94 ... 0.02
Total 1.37
Loads per 50-min working hour, 50/1.37 36
Hourly production, cybm. 36×58 2090
Costs, 1978:
Wheel-bucket excavator, linear-digging, 500 hp $712,000
Hourly cost of ownership and operation:
Excavator (Table 8-3). 26% of $712 .... $185.10
Operator and oiler 31.00
Total $216.10
Direct job unit cost per cybm, $216.10/2090 .... $0.10

This unit cost of $0.10 may be compared interestingly with the $0.13 unit cost for the mobile belt loader of Figure 12-48, which was estimated for work in similar desert alluvium.

Circular-Digging–Linear-Traveling Excavator

The important specifications of circular-digging–linear-traveling wheel-bucket excavators as shown in Figure 12-58 are tabulated below.

Specification	Nominal capacity, cybm/h	
	1500	3500
Dimensions, ft:		
Length	90	100
Width at crawlers	24	27
Height	27	31
Working weight, lb	430,000	715,000
Digging wheel:		
Diameter, ft	18	25
Buckets:		
Number	8	8
Capacity, cysm	1	2
Discharges per minute	70	60
Digging speed, r/min	8.7	7.5
Swing speed at digging radius, ft/min	0–145	0–145
Digging radius, ft	44	45
Ladder belt:		
Width, in	54	84
Length, ft	32	34
Speed, ft/min	650	650
Discharge belt:		
Width, in	54	84
Length, ft	42	60
Speed, ft/min	700	700
Discharge chutes	2	2
Travel speeds, forward and reverse, ft/min	40	30
Drives: diesel engine–hydraulic pumps–hydraulic motors; diesel engine–generator; generator–motors		
Total horsepower of engines	640	1600
Estimated nominal loading capacity, cylm/min	63	108

SOURCE: Mechanical Excavators, Inc.

The estimate of production and costs for loading this wheel-bucket excavator in the work shown in Figure 12-58 is outlined below.

The actual production is 950 to 1050, averaging 1000, cybm/h. The unblasted hard extrusive igneous rock is being loaded into 35-ton rock haulers.

Costs, 1978:
Wheel-bucket excavator, circular digging, 640 hp	$1,225,000
Hourly cost of ownership and operation:	
Excavator (Table 8-3). 9% of $1225	$110.20
Operator and oiler	31.00
Total	$141.20
Direct job unit cost per cybm, $141.10/1000	$0.14

Figure 12-59 shows a circular-digging wheel-bucket excavator of large capacity being used to excavate and load valley alluvium for the fill of San Luis Dam, California. Some 40 million cybm is involved in the mass movement of clays, sands, and gravels. The output of the machine is approximately 3000 cybm/h. Details of this custom-built wheel-bucket excavator are shown in Figure 12-60, and its more important specifications are given in Figure 12-61. Supplementary data are:
1. Speed of digging wheel, 2 to 8 r/min
2. Length of discharge belt, 63 ft
3. Estimated nominal loading capacity, based on average 5 r/min of wheel, 112 cylm/min

The estimate of production and costs for loading by wheel-bucket excavator in the work shown in Figure 12-59 is summarized below.

Actual hourly production averages 4500 cybm. The valley alluvium is being loaded into 70-cysm bottom-dump haulers carrying loads of about 65 cybm.

Figure 12-60 **Buckets, digging ladder conveyor, and stacker conveyor of wheel excavator.** The flow of the valley alluvium is from the wheel-bucket assembly to the digging-ladder conveyor to the stacker conveyor to the double discharge chute to the waiting hauler. The wheel-bucket assembly and the digging ladder swing through an arc, while the excavator advances parallel to the cut bank. The stacker conveyor is in a fixed position. The hauler advances as it is being loaded in such a manner as to secure a balanced load.

The average hourly production of the machine is 4500 cybm or 5100 cylm, and its nominal capacity at 100 percent efficiency is calculated to be 6700 cylm. The overall efficiency of the wheel-bucket excavator is about 76 percent. *(Bucyrus-Erie Co.)*

12–84   Loading and Casting Excavation

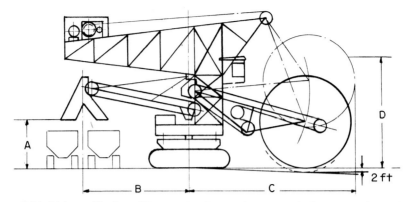

Figure 12-61 Main specifications of Bucyrus-Erie Company's 684-WX wheel excavator, electrically driven.

General:
    Number of buckets   10
    Rated bucket capacity   2½ yd^3
    Wheel diameter over lips   30 ft
    Working weight, approximate   1,625,000 lb
    Ballast required (furnished by purchaser)   125,000 lb
    Width of crawler treads   72 in
    Overall length of mounting   32 ft, 3⅜ in
    Total effective bearing area   345 ft^2
    Maximum stacker swing   180°
    Width of belts   84 in
    Length of digging ladder   38 ft
Working ranges:
    A—Dumping height, approximate   14′-4′
    B—Dumping radius, approximate   63°-0′
    C—Cutting radius, approximate   60°-0′
    D—Cutting height, approximate   40°-0′
Electrical equipment:
    Wheel-drive motor (blown)   (1) {375 hp @ 230 V / 750 hp @ 460 V} dc 75°C Continuous
    Propel motor   (1) {300 hp @ 230 V / 600 hp @ 460 V} dc 75°C Intermittent
    Swing motors   (2) {25 hp @ 230 V / 50 hp @ 460 V} dc 75°C Continuous
    Digging ladder conveyor   (1) 300 hp 440-3-60 ac 55°C Continuous
    Digging ladder
      hoist motor (blown)   (1) {187½ hp @ 230 V / 375 hp @ 460 V} dc 75°C Continuous
    Stacker conveyor motor   (1) 300 hp 440-3-60 ac 55°C Continuous

(*Bucyrus-Erie Co.*)

Costs, 1978:
    Wheel-bucket excavator, circular-digging, 2400 hp $5,400,000
    Hourly cost of ownership and operation:
      Excavator, 9% of $5,400 $486.00
      Operator and oiler   31.00
      Total $517.00
    Direct job unit cost per cybm, $517.00/4500 $0.11

Summary

Wheel-bucket excavators are loading machines of considerable capacity for high-production work in a variety of rock-earth formations. They range in size from 650 to 4500 cybm/h in nominal capacity.

    The huge machines, weighing up to 1,625,000 lb, require disassembly for shipment and erection on the job. For this reason and for other economic considerations they are used only on large projects involving quantities greater than 5 million yd³ or tons. Painstaking analysis of their applicability and unit cost for the work must precede their selection and use, as they are costly and somewhat inflexible in their application to many kinds of rock excavation.

SUMMARY

The diversity and the variations of the nine kinds of loading and casting machines of this chapter suggest the great latitude available in selecting the right machine for a particular project. When carefully analyzed, one machine must necessarily be the best one for the job and it is the work, sometimes of a painstaking nature, of the estimator-engineer to choose the machine best suited for loading, casting, or both.

CHAPTER 13

Hauling Excavation

PLATE 13-1

Haulage by twin-engine scrapers in freeway excavation. *(Top)* Beginning of work, 1965. Fragmentation is by tractor-rippers. Haulage is by twin-engine scrapers. The maximum cross section of the cut is outlined in the picture. Rock is a weathered sedimentary formation of shales and sandstones of the Tehachapi Mountains of California. *(Bottom)* Finished freeway, 1968. Top of slope to top of slope is 1530 ft. Width of eight-lane roadway is 220 ft. Depth of cut at the centerline is 320 ft. Length of the through cut is 2300 ft. The massive volume of cut is 8,750,000 yd^3.

The rough grading for this spectacular cut required 18 months. At the beginning of excavation, as shown in Plate 13-1, the haul was 1700 ft, one way, with a difference of elevation from top of cut to bottom of fill of −530 ft. The downgrade haul was on a −31 percent grade. Twin-engine, all-wheel-drive scrapers were used for these initial excessive grades.

Cycle times for the twin-engine scrapers averaged 10.9 min. The scraper load was about 33 cybm, and each scraper hauled about 150 cybm/h. A scraper spread, consisting of two 385-hp push tractors and nine scrapers, hauled about 1350 cybm/h, or 12,150 cybm each 9-h shift. During the 18 months there were usually two spreads at work with a combined production of about 25,000 cybm/day.

As the cut and fill approached finished grade, the hauling cycle was reduced to 6.0 min and only six scrapers were required for a production of 1350 cybm/h for each spread. The average number of scrapers for a sustained production of 25,000 cybm/day was 15.

TYPES AND CHARACTERISTICS OF HAULING MACHINES

Hauling, like loading and casting, is generally not a separate bid item in an excavation project because it is considered to be a part of the total bid item for excavation. This total bid item includes fragmentation, loading or casting, hauling, haul-road construction, and fill building. The practice holds for both public and private works. Sometimes in private works hauling is bid as a separate item.

Haulage of rock-earth is done by a variety of machines. Haulage is defined as the transportation of materials, and the hauler may vary all the way from a tractor-bulldozer, pushing or bulldozing the load ahead of the bulldozer, to a hydraulic monitor, which provides the real carrier in the form of water. Sometimes, as in the case of a self-loading wheels-tires-mounted scraper or a slackline excavator, the machine both loads and hauls the rock-earth.

The machines of this chapter, to be considered haulers in a literal or a figurative sense, are:

1. Crawler-tractor-bulldozers
2. Crawler-tractor-drawn wheels-tires-mounted scrapers and crawler-tractor-drawn crawler-mounted rock wagons
3. Wheels-tires-mounted scrapers, push-loaded by tractors
4. Wheels-tires-mounted scrapers, self-loaded
5. Bottom-dump haulers
6. Rear-dump haulers
7. Side-dump haulers
8. Belt conveyors
9. Cableway systems
10. Railroads
11. Hydraulickers
12. Dredges

The characteristics, operations, productions, and costs for each of these machines will be discussed in detail.

Haulage is the most important phase of excavation and it is usually the most costly, with the possible exception of rock fragmentation by drilling and blasting.

Although there is customarily no unit bid price for hauling, there is often a qualifying unit bid item for overhaul on a job, and this unit bid price is generally in public works such as road building, where hauls are sometimes long. The unit bid price for excavation may include a so-called free haul distance beyond which the excavator is paid on the basis of a unit price per cubic-yard station or per cubic-yard mile. The cubic-yard station is the haul of a cubic yard through a 100-ft distance, and the cubic-yard mile is the haul through a 1-mi distance. The free haul distance is normally 1000 ft, although it may be as low as 300 ft. The 300-ft distance was used years ago when haulage was by horse- or mule-drawn slip scrapers and wagons. One purpose of the overhaul bid item is to allow flexibility in length of haul to accommodate conditions arising during the prosecution of the work.

Of all the phases of excavation, haulage calls for the greatest expenditure of horsepower. As internal-combustion engines furnish most of this energy, it is appropriate to discuss the attributes of these prime movers here.

Characteristics and Economics of Internal-Combustion Engines

Machinery used for haulage is generally powered by diesel, gasoline, or gas engines, and most machines use diesel engines. Steam engines are rarely used. Gasoline engines are used much more than gas engines. These same three engines are used largely in other machinery for excavation.

The engines are either four-cycle, each piston delivering one power stroke for each two revolutions of the crankshaft, or two-cycle, each piston delivering one power stroke per crankshaft revolution. Four-cycle engines may be naturally aspirated, air entering the cylinder under atmospheric pressure, or pressure-aspirated, air entering the cylinder under greater than atmospheric pressure by means of blower, supercharger, or turbocharger, as the pressurizing device may be called.

Naturally Aspirated Four-Cycle Engine As is the case with all engines when working under altitude and temperature conditions other than standard, the rated flywheel or brake horsepower changes. The standard conditions are a barometric pressure at sea level

13–4 Hauling Excavation

of 29.4 in of mercury and 60°F. Some engine specifications fix the standard temperature at 85°F, but the difference in standards does not appreciably influence the change from rated flywheel horsepower.

The equation for actual flywheel horsepower at the location of the job is:

$$HP = HP_s \frac{P}{P_s}\left(\frac{T_s}{T}\right)^{1/2}$$

where HP = horsepower at location of job
HP_s = rated horsepower under standard conditions
P_s = atmospheric pressure under standard conditions, 29.4 in mercury
T_s = absolute temperature under standard conditions (460°F + 60°F) = 520°F
P = atmospheric pressure at location of job, in of mercury
T = absolute temperature at location of job (460 + observed temperature at location of job, °F)

Calculation of the actual engine flywheel horsepower at the location of the job may be illustrated by the following example. The criteria are:

1. Job elevation is 10,000 ft, where barometric pressure is 20.6 in of mercury, P.
2. Temperature at job location is 0°F or 460°F absolute, T.
3. Engine rating is 400 flywheel hp under standard conditions, HP_s.

Applying the equation:

$$HP = 400 \times \frac{20.6}{29.4}\left(\frac{520}{460}\right)^{1/2}$$
$$= 400 \times 0.701 \times 1.06 = 297 \text{ hp at job location}$$

The solution shows that loss of power due to increase in altitude is 30 percent, or 3 percent per 1000 ft above sea level. It shows also that gain of power due to decrease in temperature is 6 percent for the 60°F drop, or 1 percent per 10°F change of temperature.

The theoretical equation and empirical data provide the following two simple accepted rules for calculating flywheel engine horsepower losses and changes due to differences in elevation and temperature.

1. *Elevation:* For every 1000 ft increased elevation above 1000 ft, decrease the rated horsepower of naturally aspirated engines by 3 percent.

2. *Temperature:* For every 10°F rise in temperature above 60°F, decrease the rated horsepower by 1 percent. For every 10°F fall in temperature below 60°F, increase the rated horsepower by 1 percent. The rule applies to both naturally aspirated and pressure-aspirated engines.

Applying these rules to the rated 400-flywheel-hp engine, the following figures are obtained.

1. Loss in horsepower due to increase in elevation, HP_e:

$$HP_e = \frac{10{,}000 - 1000}{1000}\ (3\% \times 400) = 108 \text{ hp}$$

2. Gain in horsepower due to decrease in temperature, HP_t:

$$HP_t = \frac{60}{10}\ [1\% \times (400 - 108)] = 18 \text{ hp}$$

3. Calculated engine flywheel horsepower at location of job:

$$HP = 400 - 108 + 18 = 310 \text{ hp}$$

Pressure-Aspirated Four-Cycle Engine When an engine is pressure-aspirated, the horsepower depends on the weight of air forced into the cylinder relative to the weight of naturally aspirated air under standard conditions, the loss in engine horsepower in driving the pressurizing mechanism, and the effect of temperature change. Accepted empirical rules are used to calculate engine flywheel horsepower according to rated engine horsepower and elevation and temperature changes.

For elevation change the rules are:

1. No horsepower loss up to 5000 ft above sea level and sometimes up to elevations above 5000 ft.

2. From 5000 ft or from the higher elevation to 10,000 ft, 1 percent loss for every 1000 ft of increased elevation.

3. Above 10,000 ft, 3 percent loss for every 1000 ft of increased elevation.

Because of these relatively smaller losses as compared with naturally aspirated engines, one can readily appreciate the usefulness of pressure-aspirated engines. In all cases it is suggested that the engine specifications of the manufacturer be consulted, as different pressure-aspirated engines have different characteristics.

The change in horsepower due to change of temperature is given by the same empirical rule as that for the naturally aspirated engine.

When the above rules are applied to a 400-hp engine under the same criteria as those for the example of a 400-hp naturally aspirated engine, these comparative figures emerge:

1. Loss in horsepower due to increase in elevation, HP_e:

$$HP_e = \frac{10{,}000 - 5000}{1{,}000} (1\% \times 400) = 20 \text{ hp}$$

2. Gain in horsepower due to decrease in temperature, HP_t:

$$HP_t = \frac{60}{10} [1\% \times (400 - 20)] = 22 \text{ hp}$$

3. Calculated engine flywheel horsepower at location of job:

$$HP = 400 - 20 + 22 = 402 \text{ hp}$$

One sees immediately the advantage of the pressure-aspirated engine under like working conditions. It provides a 30 percent increase in horsepower over its naturally aspirated counterpart.

Pressure-Aspirated Two-Cycle Engine As has been implied, there is no naturally aspirated two-cycle engine, as a blower must be used to force scavenging and combustion air into the cylinder after the power stroke. Likewise, accepted empirical rules are used to calculate engine flywheel horsepowers according to rated engine horsepower and elevation and temperature changes. For elevation changes the rules are:

1. No horsepower loss up to 1000 ft above sea level.

2. Above 1000 ft, 1 percent loss for every 1000 ft increase elevation.

The change in horsepower due to change in temperature is given by the same empirical rule as that for the naturally aspirated engine.

When these rules are applied to a 400-hp engine under the same criteria as those for the example of the 400-hp naturally aspirated engine, these comparative figures result:

1. Loss in horsepower due to increase in elevation, HP_e:

$$HP_e = \frac{10{,}000 - 1000}{1000} (1\% \times 400) = 32 \text{ hp}$$

2. Gain in horsepower due to decrease in temperature, HP_t:

$$HP_t = \frac{60}{10} [1\% \times (400 - 32)] = 22 \text{ hp}$$

3. Calculated engine flywheel horsepower at location of job:

$$HP = 400 - 32 + 22 = 390 \text{ hp}$$

The two-cycle engine, always pressure-aspirated, provides 26 percent more horsepower than the naturally aspirated engine of the same rated horsepower. Many detailed specifications for both naturally and pressure-aspirated engines give horsepower deration tables for elevations above sea level. These tables are approximately in keeping with the empirical rules which have been given in the text.

Engine Flywheel or Brake Horsepower Rating under Standard Conditions Generally the horsepower rating of the engine specifications is determined by a list of the accessories which are mounted on the engine and are necessary to the performance of the engine with the particular machine. The accessories determine the usable horse-

13-6 Hauling Excavation

power at the flywheel. They include all or part of the following: fan, air cleaner, water pump, lubricating oil pumps, fuel pump, generator or alternator, and muffler. Sometimes one or more accessories are not included in the description. The fan is an example and, when this is the case, the large horsepower necessary for the fan must be deducted from the specified flywheel horsepower.

It is imperative that one check closely the list of engine accessories to be sure that none is omitted, for otherwise actual flywheel horsepower may be 10 percent less than the specified valve.

The engines operate approximately in the speed range of 1200 to 2400 r/min. The lower-speed engines are generally of large bore and stroke or high displacement, while the higher-speed ones are usually of small bore and stroke or low displacement.

The fuel consumption of diesel, gasoline, and gas engines varies with respect to fuel characteristics, chiefly specific gravity and number of British thermal units or heat units per pound or gallon of fuel, and the load factor of the engine.

The load factor is the average percentage of engine full horsepower consumed during a period of time, usually the life of the machine. A low load factor would be 30 percent for a crane-clamshell machine casting outsize rocks, whereas a high load factor would 80 percent for a tractor-bulldozer continuously mass bulldozing to a stationary belt loader. For estimating purposes in this text, a conservatively high 67 percent load factor is used except where stipulated. For example, if a 300-hp diesel engine consumes 0.42 lb of fuel oil each flywheel horsepower-hour when working at full load, then the engine consumes $0.67 \times 300 \times 0.42$, or 84 lb of fuel oil each working hour. As diesel fuel weighs 7.09 lb/gal, the engine consumes 84/7.09, or 11.8 gal/h.

Table 13-1 is suggested as a guide for fuel consumption of the three engine types. From these data, consumptions and costs may be figured for an engine of known rated flywheel or brake horsepower in a machine with known or estimated load factor.

The economics of engines and engine selection are illustrated in the two tables of comparative data on page 13-7, the first of which gives the hourly costs of ownership and operation of engines of 300 rated flywheel hp driving electric generators at a 90 percent load factor.

TABLE 13-1 Fuel Data for Internal-Combustion Engines

	Kind of fuel		
Item	Diesel oil No. 2	Gasoline average	Gas liquid butane
Physical properties:			
Specific gravity	0.85	0.71	0.58
Pounds per gallon	7.09	5.92	4.84
British thermal units, Btu:			
Per pound	18,500	18,000	21,500
Per gallon	131,000	113,000	104,000
Fuel consumption:			
At full-rated flywheel horsepower:			
Pound per brake hp·h	0.42	0.52	0.44
Gallon per brake hp·h	0.059	0.088	0.091
At different load factors or percentages of full power			
Gallon per full-rated flywheel or brake hp per hour:			
10 (% load factor)	0.006	0.009	0.009
20	0.012	0.018	0.018
30	0.018	0.026	0.027
40	0.024	0.035	0.036
50	0.030	0.044	0.046
60	0.035	0.053	0.055
70	0.041	0.062	0.064
80	0.047	0.070	0.073
90	0.053	0.079	0.082
100	0.059	0.088	0.091

Item	Diesel fuel No. 2	Gasoline average	Gas liquid butane
Costs, 1978:			
Engine, clutch, skid-mounted	$18,800	$10,400	$10,900
Fuel, per gallon	$0.40	$0.50	$0.52
Fuel consumption hourly at 90% load factor, gal (Table 13-1):			
300×0.053	15.9		
300×0.079	...	23.7	
300×0.082	...	...	24.6
Hourly cost of ownership and operation, 1978			
Fixed charges (Table 8-3):			
23% × $18.8	$4.32		
23% × $10.4	...	$2.39	
23% × $10.9	...	...	$2.51
Repairs and replacements, including labor (Table 8-3):			
16% × $18.8	$3.01		
18% × $10.4	...	$1.66	
18% × $10.9	...	...	$1.74
Fuel:			
15.9 gal @ $0.40	6.36		
23.7 gal @ $0.50	...	11.85	
24.6 gal @ $0.52	...	...	12.79
Total hourly cost	$13.69	$15.90	$17.04

The hours of work necessary to absorb the difference in cost of diesel and gasoline engines by difference in hourly costs of ownership and operation are computed as tabulated below for the engines of the preceding example.

	Kind of engine	
Item	Diesel engine	Gasoline engine
Cost of engine, clutch, skid-mounted, 1978	$18,800	$10,400
Increase in cost of engines	$8400	0
Hourly costs of ownership and operation	$13.69	$15.90
Increase in hourly costs of ownership and operation		$2.21
Hours of work to absorb difference in costs of engines:		
$8400/$2.21		3800

One senses that a lower load factor raises the number of working hours needed to absorb the added engine cost by virtue of the fuel cost savings of the diesel engine. For example, with a 67% load factor, 9440 working hours are required.

In excavation the use of diesel engines as opposed to gasoline and gas engines is almost universal. Some exceptions are automobiles, pickup trucks, lightweight trucks, and tractor-trailers for on-the-road haulage, as well as small power tools.

The use of liquefied gases such as butane and propane is feasible under certain job conditions. Generally they find application in permanently located open-pit mines and quarries and in huge public works such as dams, where working conditions are stable and supply and distribution of liquefied gases are dependable. Assuming these favorable circumstances, economics would determine the choice of fuels.

Speed Indicators and Work-Cost Indicators for Hauling Machines

These indicators are helpful for initial and suggestive appraisals of many machines. They may be used to compare prospective machines for the same work, for it is obvious that

13-8 Hauling Excavation

several machines may do the same work albeit at different production rates and unequal direct job unit costs.

Speed Indicators Relevant speed indicators may be calculated for all wheels-tires- and crawler-mounted machines. They are measures of productivity, as the average travel speeds, both laden and unladen, are important factors in determining the hauling cycle of the machine.

When a machine is in motion at a constant speed on a level haul road, its prime mover must overcome the force of *rolling resistance*, which is calculated by multiplying the weight of the machine by a factor r of rolling resistance. These factors are given in Table 13-3. When hauling upgrade, the resisting forces are both rolling resistance and *grade resistance*, which is figured by multiplying the weight of the machine by a factor g, which is the percent of grade. These factors are given in Table 13-4. When hauling downgrade the same rolling resistance must be overcome, but the force exerted by the grade is an assisting force rather than a resisting one. If the grade assistance force exceeds the rolling resistance force, brakes must be applied to maintain constant speed.

The power applied to a machine bears a relationship to the brake horsepower of the prime mover and it is expressed as a percentage p of the brake horsepower. The equation for speed S, using the factors r, g, and p, is:

$$S = \frac{p \times \text{brake hp} \times 33{,}000}{(r+g) \times \text{machine weight}} \quad \text{ft/min}$$

The machine weight may be laden, gross vehicle weight (GVW), or unladen, tare weight. For wheels-tires- and crawler-mounted machines, the constants or factors may be considered to be equal when the machines are operating under the same working conditions. Thus it may be assumed that the equation, as an indicator of relative speeds, may be reduced to:

$$\text{Laden speed indicator} = \frac{\text{brake hp}}{\text{GVW, 1000s of lb}}$$

The use of speed indicators is demonstrated in the discussion of Figure 13-10. The wheels-tires-mounted hauler travels on the average 98 percent faster than the crawler-mounted hauler when working under like conditions.

Work-Cost Indicators These indicators may be calculated for all machines when the weight of the machine is considered to be the weight of the moving parts.

The total work done by a machine divided by the pay work done is the *work indicator*. It is also the *cost indicator* because cost bears a fairly fixed relationship to the amount of work done. Work is expressed in foot-pounds of energy. The forces described in deriving the speed indicators, when multiplied by the distance of the haul road d for both hauling and returning, give the number of foot-pounds of work. The *work-cost indicator* is expressed by the equations:

$$\text{Work-cost indicator} = \frac{\text{total work hauling} + \text{total work returning}}{\text{pay work hauling}}$$

$$= \frac{[(r+g) \times \text{GVW} + (r+g) \times \text{tare weight}] \times d}{[(r+g) \times \text{load weight}] \times d}$$

The terms "total work hauling," "total work returning," and "pay work returning" include work done when making two turnings and a dumping, as all these elements of the hauling cycle take place when the machine is in motion. No work is performed during the loading element of the hauling cycle, except in negligible cases such as movement of the hauler when being loaded by a moving belt loader. Thus the work-cost indicator is a measure of work or energy expended during the functioning of the hauler in motion.

Again, as the constants or factors may be considered to be approximately equal, the equation becomes:

$$\text{Work-cost indicator} = \frac{\text{GVW} + \text{tare weight}}{\text{load weight}}$$

The use of work-cost indicators is demonstrated in the discussion of Figure 13-10. The wheels-tires-mounted rear-dump hauler shown in this figure can do work at 50 percent of the direct job unit cost of the crawler-tractor- and crawler-mounted side-dump hauler.

The indicators, when applicable, accompany the specifications of the hauling machines of this chapter.

CRAWLER-TRACTOR-BULLDOZERS

As discussed in the section of Chapter 12 covering crawler-tractor-bulldozers, the bulldozer is also a hauling machine. The principles of haulage are discussed and performance data are offered in that section. It is suggested that the reader review the section in order to appraise the worth of the tractor-bulldozer as a hauling machine in the sense of moving rock-earth over surprising distances economically. Figure 13-1 illustrates the pioneering of a haul road through a cut in weathered schist by a 514-hp, 160,000-lb tractor-bulldozer-ripper. The machine is working in confined conditions and both ripping and bulldozing productions are limited. The machine rips and then bulldozes the fragmented rock to the end of the cut. A water wagon is lubricating the hard, sharp rock to assist in both ripping and bulldozing. The excavation for the 200-ft-long, 50-ft-wide haul road through the shallow cut is about 1500 cybm. The ripped rock is being bulldozed in one direction on an average -2 percent grade. The actual time for the combination work is about 8 h.

The estimated 1978 costs for the work of Figure 13-1 are:

Cost of 514-hp tractor-ripper-bulldozer, 1978	$405,000
Hourly cost of ownership and operation:	
Tractor-ripper-bulldozer (Table 8-3), 36% × $405	$145.80
Operator	15.60
Total cost	$161.40
Direct job unit cost for ripping and bulldozing per cybm, $\dfrac{8 \times \$161.40}{1500}$	$0.86

Figure 13-2 illustrates a large U-shaped bulldozer especially designed for mass dozing, as is shown in Figure 13-3. In the work of Figure 13-3 four machines are bulldozing 150 ft down a -10 percent grade to a hydraulicking monitor. The actual cycle time is 1.80 min, and the known hourly production is 1450 cybm of ripped weathered formation of sandstones-shales. For purposes of comparison with the known production, an hourly production estimate for the job is tabulated below.

Figure 13-1 Tractor-ripper-bulldozer opening up a haul road through a shallow rock cut. As preparation for long-haul work by rear-dump trucks, a 50-ft-wide haul road is being ripped and bulldozed out of this 200-ft-long shallow cut. The rock is weathered schist of San Diego County, California and is being excavated at the rate of 1500 cybm in 8 h by the 514-hp tractor-ripper-bulldozer. The rocks are large, sharp, and abrasive and it is economical to keep the cut wet by the water wagon. The water lubricant speeds up both ripping and bulldozing and reduces wear on ripper points and bulldozer cutting edges. The same lubrication method is shown in Figure 11-11, where the rhyolitic tuff is being wetted for both tractor-rippers and scrapers.

13-10 Hauling Excavation

(a)

(b)

Figure 13-2 Specially designed U-shaped bulldozer for mass hauling by tractor-bulldozer. *(a)* The tractor engine is 335 hp, the tractor-bulldozer weighs 80,000 lb, and the bulldozer blade is 16.9 ft wide and 6.7 ft high. The average load, when bulldozing on level ground, is about 17 cybm. *(b)* Bulldozer load is on a -20 percent grade. The average load is about 22 cylm, equivalent to 17 cybm of average weathered rock-earth. On this grade the machine bulldozes a load at about 1.5 mi/h and returns in reverse at about 5.0 mi/h. The hauling cycle is fast when working on an ideal -20 percent grade, varying from 1.5 min for a 100-ft distance to 6.0 min for a 500-ft distance, one way. On the long, 500-ft haul, about 142 cybm may be bulldozed hourly. *(Shepherd Machinery Co.)*

Estimated bulldozer loads:
  Loose measurement (Figure 12-1):
   $\text{Cylm} = 0.022 H^2 W = 0.022 \times 6.7^2 \times 16.9 = 16.7$
  Bank measurement (Figure 12-2): cybm $= 0.78 \times 16.7 = 13.0$

Estimated cycle time, min (Figure 12-2) 150-ft one-way distance, -10% grade	1.9
Loads hauled per 50-min working hour 50/1.9	26
Hourly production, cybm:	
Per machine, 26×13.0	340
For four tractor-bulldozers, 4×340	1360

Figure 13-3 Haulage by mass bulldozing to a hydraulicking monitor. Four tractor-bulldozers of the type illustrated in Figure 13-2 are serving the monitor to the right of the picture. Bulldozing or hauling distance is 150 ft down a −10 percent grade. Cycle time is 1.8 min and hourly production is 1450 cybm of ripped sandstone-shale. This Los Angeles County, California, memorial park project involves 25 million yd³ over a 10-yr period, with 20 million yd³ being bulldozed and carried hydraulically to the fill and the remaining 5 million yd³ being hauled by scrapers. Average daily 8-h production over the 10-yr period for both bulldozer-monitor and scraper haulage is 13,000 cybm.

The production estimate of 1360 cybm/h is 6 percent below the actual average of 1440 cybm/h.

The cost for bulldozing based on an achieved hourly production of 360 cybm per machine and 1978 costs is estimated as follows:

Cost of 335-hp tractor-bulldozer	$202,000
Hourly cost of ownership and operation:	
Tractor-bulldozer (Table 8-3), 33% × $202	$66.70
Operator	15.60
Total cost	$82.30
Direct job unit cost per cybm	$0.23

The summarizing comments of the tractor-bulldozer section of Chapter 12 apply equally to this section on hauling by the same tractor-bulldozer.

The speed and work-cost indicators for the tractor-bulldozers of Table 12-2 are given below, based on the following average values:

1. Brake horsepower of tractor, 372 bhp
2. Weight, tare, 94,300 lb
3. Bulldozer blade capacity, 11.0 cylm
4. Load weight of bulldozer at 2900 lb/cylm, 31,900 lb
5. GVW, 126,200 lb

Speed indicators:

$$\text{Laden:} \quad \frac{\text{Brake hp}}{\text{GVW, 1000s of lb}} = 2.95$$

$$\text{Unladen:} \quad \frac{\text{Brake hp}}{\text{Tare weight, 1000s of lb}} = 3.94$$

$$\text{Work-cost indicator:} \quad \frac{\text{GVW} + \text{tare weight}}{\text{Load weight}} = 6.91$$

The high work-cost indicator implies that the tractor-bulldozer is not an efficient rock hauler, but one must temper this judgment by the fact that the machine is both a loader and a hauler in the sense of moving rock-earth over a distance.

CRAWLER-TRACTOR-DRAWN SCRAPERS AND ROCK WAGONS

Figure 13-4 illustrates a tractor-scraper being loaded by the total forces of a pulling tractor and a push-loading tractor. These machines work admirably under the following operating conditions:

    1. Rock-earth mixture in which large, hard abrasive rocks predominate, as is the case with the work of Figure 13-8

13-12 Hauling Excavation

2. Steep haul roads, as in the case of opening up cuts in rough country
3. Short haul distances, less than 1000 ft one way
4. Poor haul roads, deeply rutted because of weather conditions, which are hard to maintain for self-powered wheels-tires-mounted scrapers
5. Poor weather conditions such as sustained rain, snow, and ice

Figure 13-5 illustrates a tractor–side-dump wagon being loaded by a shovel. These machines work well under the same conditions as those cited for tractor-scrapers, except that the operator must be wary of hauling on downgrades in excess of about -7 percent because the wagon has no brakes. It may very well overrun the tractor if the downgrade force of the wagon exceeds its rolling resistance. When such a condition exists in the operation of the tractor-scraper combination, the operator may drop the scraper bowl to the ground so that the cutting edge brakes the entire tractor-scraper combination.

The Effective Drawbar Horsepower

The horsepower may be calculated by two methods, one of which utilizes a graph or table expressing drawbar pull as a function of speed of travel. The graph or table is a part of the tractor specifications. Figure 13-6 shows such a graph for a 410-flywheel-hp tractor with a working weight of 83,300 lb for drawbar work. The drawbar pull is based on the tractor working on level ground with a rolling resistance of 110 lb per ton of tractor weight. This resistance is 5.5 percent of the weight and it is equivalent to working on a poorly maintained, rutted haul road.

As the graph gives drawbar pull in pounds according to speed in miles per hour, the drawbar horsepower may be calculated by the formula:

$$DHP = \frac{P \times (S \times 5280/60)}{33,000}$$

$$= 0.00267 PS$$

where DHP = drawbar horsepower
P = drawbar pull, lb
S = speed, mi/h

Figure 13-4 Crawler-tractor and specially designed large-capacity scraper. This outsize scraper is a lengthened, enlarged version of the standard machine of the manufacturer designed for adaptation to particular working conditions. These are in terms of material, easily loaded friable weathered rock, uphill return grades up to $+30$ percent, and one-way hauls up to 1500 ft. The scraper is carrying a load of 10.0 cylm of sandstone-shale in the La Puente Hills of southern California. The engines of the pulling and the push-loading tractors are rated at 335 hp. The scraper weighs 42,000 lb and the average load weighs 108,000 lb. The payload of the scraper is about 31 cybm. On a 1000-ft one-way haul, tractor-scraper production is about 190 cybm each working hour. *(Shepherd Machinery Co.)*

Figure 13-5 Crawler-tractor and side-dump rock wagon. The tractor-wagon is on a suitable short (600 ft) one-way haul with a −8 percent grade. It is receiving the last of five dippers from a 3-yd³ shovel. The tractor engine has 131 flywheel hp. Each of three machines hauls 63 cybm/h, giving a total production of 189 cybm/h. The blasted sharp-cornered limestone rocks with few fines create a rough haul road both in cuts and in fills. Such rock characteristics, along with rains and snows in early spring and late fall, mandated the use of crawler machines for this highway construction in mountainous Pennsylvania. *(Athey Products Corp.)*

When the formula is applied to the values of the graph of Figure 13-6, the following relationships emerge:

Gear	Speed, mi/h	Drawbar Pull, lb	Drawbar Horsepower
1	1.0	100,000	267
2	3.0	36,000	288
3	5.0	20,000	267
Average			274

The average tractor drawbar hp/engine flywheel hp ratio is $274/410 = 0.67$.

In the second method, the average drawbar horsepower may be calculated simply by the equation:

$$\text{Drawbar horsepower} = 0.69 \times \text{rated engine horsepower at full load}$$

The factor 0.69 is the average of a number of factors which are calculated from the performance graphs of drawbar pull versus ground speed included in the crawler-tractor specifications. The factor 0.69 is used in the text. The loss of 31 percent of the engine flywheel horsepower is chargeable to mechanical-hydraulic losses in the power train from flywheel to crawlers and to the rolling resistance of the tractor when tested at a standard rolling resistance of 110 lb/ton.

Coefficient of Traction and Rolling Resistance

When a crawler-tractor is pulling a scraper, a wagon, or any load, it may be limited in pounds of pull because of slippage of the tracks before full engine power is developed. The limitation is significant when working under conditions of rain, snow, or ice.

The *coefficient of traction* is that factor by which the component of the tractor weight at right angles to the surface of the ground is multiplied in order to calculate the maximum pull of the tractor. Then one must keep in mind that this force, when hauling uphill or downhill, will be lessened or increased by the effect of the sheer weight of the tractor.

transmission

Planetary-type power shift with 21" (530 mm) diameter, high-torque-capacity oil clutches. Special modulation system permits unrestricted speed and direction changes under full load.

Single-stage torque converter with output torque divider. Connected to transmission by double universal joint for unit construction to provide servicing ease.

Gear	Forward Speed mi/h	(km/h)	Reverse Speed mi/h	(km/h)
1	0-2.5	(4.0)	0-3.1	(5.0)
2	0-4.3	(6.9)	0-5.4	(8.7)
3	0-6.7	(10.8)	0-8.2	(13.2)

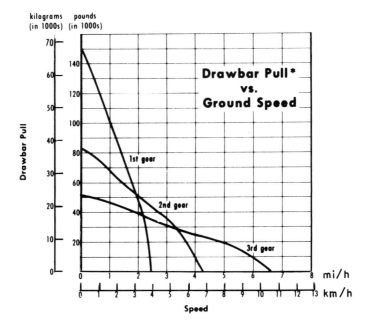

*Usable pull will depend upon traction and weight of equipped tractor.

Figure 13-6 Drawbar pull versus ground speed for a crawler-tractor. The graph and the table for the forward and reverse speeds are from the specifications for the model D9H Caterpillar tractor. Engine flywheel horsepower is 410 and working weight for drawbar work is about 83,300 lb. The calculated average drawbar horsepower for the three forward-speed gears is 274, and the ratio of calculated drawbar horsepower to engine flywheel horsepower is 0.67. Similar calculations for a number of crawler-tractors having a range of flywheel horsepowers yield an average ratio of 0.69. *(Caterpillar Tractor Co.)*

The component of the tractor weight parallel to the ground subtracts from or adds to the maximum tractive effort according to whether the work is uphill or downhill.

Table 13-2 gives acceptable values for coefficients of traction or, as they are sometimes called, *coefficients of friction*, of crawler-tractors.

The following example illustrates the use of Table 13-2.

The tractor of Figure 13-7 is hauling upgrade on a firm earth haul road and the

TABLE 13-2 Coefficients of Traction or Coefficients of Friction for Crawler-Tractors

Surface of travel road	Coefficient
Asphalt pavement	0.45
Clay-loam, dry	0.90
Clay-loam, wet	0.70
Clay-loam, rutted, dry	0.70
Concrete pavement	0.45
Earth, firm	0.60
Earth, loose	0.60
Gravel, compacted	0.50
Gravel, loose	0.25
Ice	0.25
Quarry pit, dry	0.55
Rock, loose, ripped or blasted, dry	0.40
Rock-earth, loose	0.50
Sand, dry	0.35
Sand, wet	0.50
Snow, dry, packed	0.20

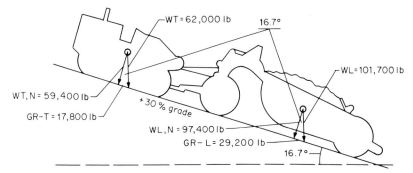

Figure 13-7 Dynamics of upgrade haulage by crawler-tractor and wheels-tires-mounted scraper.

Rolling resistance for firm-earth haul road	2%
Drawbar horsepower of tractor	207 hp
WT, weight of tractor	62,000 lb
WT, N (resultant of weight normal to grade) = 62,000 cos 16.7°	59,400 lb
Maximum drawbar pull = coefficient of traction × 59,400 = 0.60 × 59,400	35,600 lb
There is no rolling resistance calculation for the tractor as it is considered in the calculations for drawbar pull and drawbar horsepower	
GR-T (grade resistance of tractor) = 62,000 sin 16.7°	17,800 lb
WL, weight of load, scraper and rock-earth	101,700 lb
WL, N (resultant of weight normal to grade) = 101,700 cos 16.7°	97,400 lb
Rolling resistance of load = 2% × 97,400	1,900 lb
GR-L (grade resistance of load) = 101,700 sin 16.7°,	29,200 lb
Total rolling and grade resistances of load = 1900 + 29,200	31,100 lb
Maximum 35,600 lb drawbar pull of tractor is adequate for total 31,100 lb resistance of load	
Total rolling and grade resistances of tractor and laden scraper which tractor must overcome on +30% grade = 17,800 + 31,100	48,900 lb

$$\text{Travel speed of tractor} = \frac{207 \times 33{,}000}{48{,}900} = 140 \text{ ft/min or } 1.59 \text{ mi/h}$$

coefficient of traction is 0.60. The tractor weighs 62,000 lb and the drawbar hp is 207. The grade is +30%, equivalent to a 16.7° angle. The component force of the tractor weight at right angles to the road surface is equal to 62,000 cos 16.7°, or 62,000 × 0.958 = 59,400 lb. The maximum possible drawbar pull of the tractor before slippage of tracks is 60% ×59,400, or 35,600 lb.

13-16 Hauling Excavation

When a tractor hauls a load, the drawbar pull must overcome the rolling resistance of the scraper or wagon and its load, that is, gross vehicle weight (GVW). Rolling resistance may be expressed in pounds per ton of GVW, or more simply as a percentage of GVW. Here again, the rolling resistance factor is applied to that component of GVW at right angles to the surface of the travel road. Table 13-3 gives rolling resistances for both wheels-tires-mounted and crawler-mounted rock-earth haulers.

The use of Table 13-3 may be illustrated for the example shown in Figure 13-7.

The weights of the scraper and load are:

Tare or unladen weight	35,000 lb
Load weight, 23 cylm at 2900 lb/cylm	66,700
GVW, laden weight	101,700 lb

When the scraper is working on a level firm earth haul road the rolling resistance is 2.0% × 101,700, or 2000 lb. However, when the machine is working on the same haul road with +30 percent grade, the 2.0 percent factor is applied only to that component of the 101,700-lb vertical force which is normal to the surface of the travel road. The rolling resistance thus equals 2.0% × 101,700 cos 16.7°, that is, 2.0% × 101,700 × 0.958, or 1900 lb. One perceives that even on a +30 percent grade the maximum drawbar pull of the tractor and the rolling resistance of the laden scraper are 96 percent of the corresponding values for a level haul road. For this reason, it is customary to ignore the effect of all practical grades when figuring drawbar pulls of tractors and rolling resistances of tractors and their pulled rock-earth haulers.

When a tractor hauls a load on an upgrade haul, it must overcome the grade resistances of its own weight and that of the laden rock-earth hauler. Table 13-4 gives angles of grades and natural trigonometric functions of the angles according to percent grade. As will be demonstrated later, the table may be used for precise calculations of maximum drawbar pull, rolling resistance, and grade resistance for a particular tractor haulage problem.

Practicable and economical grades for upgrade and downgrade haulage range from 0 to 10 percent. Questionable grades range from 10 to 25 percent. Impracticable and uneconomical grades range from 25 to 40 percent. Rarely do grades exceed 25 percent and then they are usually used only in unavoidable cases. Within the range of estimating precision for rock excavation, which is about ±5%, one may safely assume that for grades less than 30 percent the following two assumptions are reasonable:
1. Sine = tangent = percent of grade
2. Cosine = 1.000

TABLE 13-3   Rolling Resistances of Wheels-Tires- and Crawler-Mounted Rock-Earth Haulers

	Percentage of GVW	
Surface of travel road	Wheels-tires mounted	Crawler-mounted
Asphalt pavement or concrete pavement	1.5	3.0
Dirt, hard, well-maintained, dry	2.0	3.0
Dirt, not firmly packed, some loose material, well-maintained, dry	3.0	4.0
Dirt, poorly maintained	4.0	4.0
Dirt, compacted fills	4.0	4.0
Dirt, ploughed	8.0	6.5
Dirt, deeply rutted	16.0	11.0
Gravel, well compacted, dry	2.0	3.0
Gravel, not firmly compacted, some loose material, dry	3.0	4.0
Gravel, loose	10.0	7.5
Mud, with firm base	4.0	4.0
Mud, with soft spongelike base	16.0	11.0
Rock-earth in cuts	8.0	6.5
Rock-earth in compacted fills	4.0	4.0
Sand, loose	10.0	7.5
Sand, wet	5.0	2.5
Snow, packed	2.5	3.0
Snow, fresh, loose to moderate depth	4.5	4.5

TABLE 13-4 Elements for Calculating Traction Forces or Drawbar Pulls, Rolling Resistances, and Grade Resistances

Percent of grade	Equivalent angle,°	Natural trigonometric functions of angle		
		Sine	Cosine	Tangent
1	0.6	0.010	1.000	0.010
2	1.1	0.020	1.000	0.020
3	1.7	0.030	1.000	0.030
4	2.3	0.040	0.999	0.040
5	2.8	0.050	0.999	0.050
6	3.4	0.060	0.999	0.060
7	4.0	0.070	0.998	0.070
8	4.6	0.080	0.997	0.080
9	5.2	0.090	0.996	0.090
10	5.7	0.100	0.995	0.100
12	6.8	0.119	0.993	0.120
14	8.0	0.139	0.990	0.140
16	9.1	0.158	0.988	0.160
18	10.2	0.177	0.984	0.180
20	11.3	0.196	0.981	0.200
25	14.0	0.242	0.970	0.250
30	16.7	0.288	0.958	0.300
35	19.3	0.330	0.944	0.350
40	21.8	0.371	0.928	0.400
45	24.2	0.410	0.912	0.450
50	26.6	0.447	0.894	0.500

The following example illustrates the use of Table 13-4, based on the dynamics of Figure 13-7.

	Theoretical Solution	Practical Solution
Maximum drawbar pull or traction force, lb:		
$0.60 \times 62{,}000 \cos 16.7°$		
$0.60 \times 62{,}000 \times 0.958$	35,600	
$0.60 \times 62{,}000$		37,200
Grade resistance of tractor, lb:		
$62{,}000 \sin 16.7°$		
$62{,}000 \times 0.288$	17,800	
$30\% \times 62{,}000$		18,600
Resistance forces of laden scraper, lb:		
Rolling resistance:		
$2\% \times 101{,}700 \cos 16.7°$		
$2\% \times 101{,}700 \times 0.958$	1,900	
$2\% \times 101{,}700$		2,000
Grade resistance:		
$101{,}700 \sin 16.7°$		
$101{,}700 \times 0.288$	29,200	
$30\% \times 101{,}700$	...	30,500
Total resistance forces	31,100	32,500
Total resistances of tractor and laden scraper, lb	48,900	51,100
Excess of maximum traction force of tractor over total resistance forces of laden scraper, lb	4,500	4,700
Travel speed of tractor with 207 drawbar hp, ft/min		
$\dfrac{207 \times 33{,}000}{48{,}900}$	140	
$\dfrac{207 \times 33{,}000}{51{,}100}$	...	134

The example emphasizes the feasibility of using the time-saving, practical solution for the dynamics of upgrade or downgrade haulage on grades up to 30 percent. Errors are conservatively estimated at 0 to 4.5 percent. The practical solution may be used for all hauling machines, both crawler- and wheels-tires-mounted.

When grades exceed 30 percent, errors by the practical method become large, and the

13-18 Hauling Excavation

use of the cosines and sines of the angles of slope is recommended, as illustrated in Figure 13-7. For example, on a 50 percent slope, or 26.6° angle of slope, the following errors are incurred:

1. For calculations involving the approximation that sine equals tangent equals percent grade, the error is +11 percent, as the tangent is 0.50 and the grade is 50 percent, whereas sin 26.6°=0.447. Of course, the error is conservative.

2. For calculations involving the approximation that cosine equals 1.000, the error is +12 percent, as cos 26.6°=0.894. Again, the error is conservative for estimating.

Table 13-5 gives typical specifications for crawler-tractors with winches and hydraulic systems for drawbar work pulling scrapers and wagons. When using this table, one must be careful not to use drawbar pulls in excess of the weight of the tractor times the coefficient of traction or friction, because slippage of the tracks will take place without development of full drawbar horsepower. For example, the 300-hp tractor weighs 62,000 lb for drawbar work. If it were pulling a scraper over dry sand, the coefficient of traction would be 0.35 (Table 13-2) and the maximum drawbar pull would be 0.35 × 62,000, or 21,700 lb. This possible pull is less than the 25,900-lb pull at 3 mi/h and more than the 19,500-lb pull at 4 mi/h (Table 13-5). The maximum drawbar pull of 21,700 lb would be developed at the interpolated value of 3.6 mi/h.

Dynamics of Haulage by Crawler-Tractors and Scrapers

Figure 13-7 illustrates the dynamics of upgrade haulage by resolution of forces for a crawler-tractor pulling a laden wheels-tires-mounted scraper. The vertical force, or weight of the machine, is represented by the hypotenuse of a right triangle of forces. It is resolved into two components: the component normal to the surface of the grade is a factor in the calculation of traction force and rolling resistance force, and the component parallel to the surface of the grade is the grade resistance. The angle between the vertical and normal forces equals the angle of slope, as the sides of the two angles are at right angles to each other. Thus, the components may be calculated easily by the trigonometric functions of the angle:

Force normal to surface=weight (vertical force)×cosine of angle
Force parallel to surface=weight (vertical force)×sine of angle

TABLE 13-5 Representative Specifications for Crawler Tractors with Winch and Hydraulic System for Drawbar Work Pulling Scrapers and Wagons

Specification	Nominal flywheel horsepower					
	105	195	300	410	524	700
Drawbar hp., 69%×flywheel hp	72	135	207	283	362	483
Working weight, lb	22,000	46,000	62,000	84,000	124,000	145,000
Travel speeds, Power shift, torque converter transmission, mi/h:						
Forward gears:						
1st	0–2.2	0–3.0	0–2.5	0–2.5	0–2.6	0–2.4
2d	0–3.8	0–6.2	0–4.3	0–4.5	0–4.2	
3d	0–6.3	...	0–6.6	0–6.7	0–6.5	0–7.2
Reverse gears, total ranges	0–7.6	0–7.0	0–8.2	0–8.2	0–7.4	0–8.6
Calculated drawbar pull at given forward speeds:						
1 mi/h	27,000	50,600	77,600	106,100	135,800	181,100
2 mi/h	13,500	25,300	38,900	53,200	68,100	90,600
3 mi/h	9,000	16,900	25,900	35,400	45,200	60,400
4 mi/h	6,800	12,700	19,500	26,600	34,000	45,300
5 mi/h	5,400	10,100	15,500	21,200	27,200	36,200
6 mi/h	4,500	8,500	13,000	17,800	22,600	30,200
Dimensions, ft:						
Length	13.4	16.6	18.3	20.0	21.0	22.0
Width	8.6	8.4	9.2	10.9	11.0	12.0
Height	8.6	10.5	11.2	12.3	12.8	14.9

NOTE: When complete tractor specifications are available which give the data of Figure 13-6 for travel speeds and drawbar pulls, one should use these precise figures for estimating the work capabilities of crawler tractors.

The calculations of Figure 13-7 are precise but they are also unnecessarily cumbersome. Laden upgrade hauls rarely exceed +10 percent and unladen hauls rarely exceed +30 percent in the interest of economy and safety. The tedious trigonometry can be eliminated for all grades up to +30 percent, as explained in connection with Table 13-4.

The mechanical operations of the scrapers of Table 13-6 are actuated by two cables from two drums mounted at the rear of the pulling crawler-tractor. The drums are driven by a power takeoff from the tractor engine.

1. One cable lowers the bowl with its cutting edge to secure a load and then raises the bowl to the load-carrying position.

2. The second cable has a dual function. It raises the apron to secure a load and then lowers the apron after the load is secured so as to maintain a full load. It also powers the load ejector at the rear of the bowl, which acts as a bulldozer to push out the load. The single cable first raises the apron and then, at the end of travel of the apron, moves the ejector forward, dumping the rock-earth over the cutting edge and under the raised apron. The ejector is returned to the loading position by coiled springs in the push-rod case behind the ejector.

Several scrapers are actuated by hydraulic cylinders or by combination hydraulic cylinders and cables. In the scraper of one manufacturer, five double-acting cylinders are used, two for raising and lowering the bowl with its cutting edge, two for raising and lowering the apron, and one for actuating the ejector. These machines provide down pressure on the cutting edge when loading and this action facilitates the loading of rock-earth. This ability is lacking inherently in the cable-actuated scrapers of Table 13-6, as they must

TABLE 13-6 Representative Specifications for Pull-Type Scrapers, Wheels-Tires-Mounted

Specification	Capacity, cysm		
	14.0	21.0	27.0
Capacity, loose measurement with crowned load at 4:1 slope, cylm	15.5	23.3	30.0
Dimensions, ft:			
Length	33.0	38.2	41.5
Width	10.8	11.8	12.9
Height	9.9	10.8	13.0
Tires:			
Front	20.5×25—12PR	23.5×25—20PR	29.5×29—22PR
Rear	26.5×25—16PR	29.5×29—22PR	33.3×33—26PR
Weights, lb:			
Tare, working	25,000	35,000	48,000
Load, rated	42,000	63,000	84,000
GVW	67,000	98,000	132,000
Suggested crawler tractor:			
Flywheel hp	200	300	400
Drawbar hp	138	207	276
Working weight, lb	39,000	65,000	84,000
Total weights, tractor and scraper, lb:			
Tare, working	64,000	100,000	132,000
Load, rated	42,000	63,000	84,000
GVW	106,000	163,000	216,000
Speed indicators:			
Laden: $\dfrac{\text{Brake hp}}{\text{GVW, 1000s of lb}}$	1.89	1.84	1.85
Unladen: $\dfrac{\text{Brake hp}}{\text{Tare weight, 1000s of lb}}$	3.12	3.00	3.03
Work-cost indicators: $\dfrac{\text{GVW+tare weight}}{\text{Load weight}}$	4.05	4.17	4.14

depend on sheer weight for the penetration of the cutting edge. However, this apparent shortcoming is not serious because of the heavy weight of the bowl and because efficient operation calls for the prior ripping of rock-earth or the prior well-blasting of rock before loading.

Bowl factors for average loads of scrapers according to nature of earth-rock and manner of loading are listed in Table 13-7. The use of this table may be illustrated for the scraper of Figure 13-4, which has a capacity of 36.0 cysm. The rock-earth is weathered sandstones-shales and the scraper is being push-loaded for full capacity, as may be seen in the overflowing load.

$$\text{Cubic yards loose measurement} = 1.11 \times 36.0 = 40.0 \text{ cylm}$$
$$\text{Cubic yards bank measurement} = 0.78 \times 40.0 = 31.2 \text{ cybm}$$

Or, alternatively,

$$\text{Cubic yards bank measurement} = 0.87 \times 36.0 = 31.3 \text{ cybm}$$

Production and Cost Estimation for Crawler-Tractor-Scrapers

The hauling cycle for wheels-tires-mounted crawler-tractor-scrapers is divided into three components: loading average rock-earth; hauling and returning, and turning and dumping. Loading by pulling tractor alone takes 1.2 min, and loading by pulling tractor assisted by push tractor takes 0.8 min. Hauling and returning times may be calculated from the data of Tables 13-2, 13-3, and 13-4. One must keep in mind that average speeds for the typical short hauls of crawler-tractors are generally about 90 percent of estimated tractor travel speed. Also, when hauling or returning on excessive downgrades, the travel speed must be adjusted to the grade. This is accomplished by dropping the cutting edge of the bowl to the ground in order to provide braking. When the grade assistance exceeds the rolling resistance, care must be exercised to avoid jackknifing. The sum of turning and dumping times may be approximated as 1.2 min.

The loading time for the push tractor is made up in rock-earth, of approximately 0.8 min loading time and 0.4 min exchange time, giving a total cycle time of 1.2 min. With a 50-min hour working efficiency, the push tractor loads out some 42 scraper loads hourly.

The payloads for both methods of loading vary according to the nature of the rock-earth as, in most cases, this characteristic controls the ratio of cubic yards bank measurement to cubic yards loose measurement in the bowl.

When sands and gravels or average-weathered rock-earth are loaded with the help of a push tractor, the crown of the load is on an average 4:1 slope, giving a loose load of about 1.11 times the struck measurement capacity of the scraper bowl. On the other hand, when these materials are self-loaded by only the pulling tractor, the loose load is about 0.83 times the struck measurement of the bowl. Poorly blasted rock, loaded by either method, has the lowest ratio of loose load to capacity of bowl. The preceding data may be used to estimate productions and calculate direct job unit cost by the use of Table 8-3. This is illustrated by the following example of the work shown in Figure 13-8.

TABLE 13-7 Scraper Bowl Factors for Average Loads of Crawler-Tractor-Drawn and Wheels-Tires-Mounted Scrapers Based on Capacity, cysm

Nature of rock-earth	Fill factor $\left(\dfrac{\text{loose measurement}}{\text{bowl capacity}}\right)$		Conversion factor $\left(\dfrac{\text{bank measurement}}{\text{loose measurement}}\right)$	Bowl factor $\left(\dfrac{\text{bank measurement}}{\text{bowl capacity}}\right)$	
	Push-loaded	Self-loaded		Push-loaded	Self-loaded
Sand-gravel alluvia	1.11	0.83	0.88	0.98	0.73
Rock-earth weathered	1.11	0.83	0.78	0.87	0.65
Rock well blasted	0.80	0.60	0.67	0.54	0.40
Rock poorly blasted	0.60	0.45	0.60	0.36	0.27

Figure 13-8 Haulage by four wheels-tires-mounted crawler-tractor-scrapers. A typical use of four machines is illustrated here for a short, steep haul in rough ripped sandstone-shale of the San Gabriel Mountains of California. Tractor-rippers unearth outsize rocks, as illustrated. The tractor engines have 235 flywheel hp. The scrapers have 18 cysm capacity and haul 15.7 cybm.

The haul is 600 ft one way, with a maximum -24 percent grade from the end of the level cut to the beginning of the level fill. Scrapers are push-loaded by a crawler-tractor with an engine of 235 flywheel horsepower. The hourly production of the four tractor-scrapers is estimated to be 628 cybm.

Criteria:

1. Tractors: weight, 55,000 lb; engine flywheel hp, 235; tractor drawbar hp, 162; maximum speed, 6.5 mi/h.
2. Scrapers: weight, 39,000 lb; capacity, 18.0 cysm. Scrapers are being push-loaded by a crawler-tractor of 235 engine flywheel hp.
3. Rock-earth is ripped weathered sandstone-shale weighing 4260 lb/cybm.
4. Haul is 600 ft, one way, made up of: 200 ft, level, in cut; 200 ft, -24% grade, on haul road; and 200 ft, level, in fill.
5. Haul throughout length is well maintained and compacted.
6. Efficiency for estimating production is the 50-min working hour, or 83%.

Production:

1. From Table 13-7, the bowl factor is 0.87. The estimated payload is $0.87 \times 18.0 = 15.7$ cybm.
2. The estimated gross load (GVW) for the crawler-tractor is $39,000 + (15.7 \times 4260) = 105,900$ lb.
3. From Table 13-3, the rolling resistance of the scraper and load is $4\% \times 105,900 = 4200$ lb.
4. By use of Table 13-4, the maximum grade assistance of the laden scraper is determined as 24 percent, giving $(24\% - 4\%) \times 105,900 = 21,200$ lb as the net thrust of the laden scraper. This sizable thrust through the drawbar to the tractor is excessive, and the cutting edge of the scraper bowl must be lowered to act as a brake when hauling down the 200-ft section of -24 percent grade.
5. The maximum resistance when returning empty up the $+24\%$ grade is:

Tractor grade resistance, $24\% \times 55,000 =$ 13,200 lb
Unladen scraper grade and rolling resistances,
$(24\% + 4\%) \times 39,000 =$ <u>10,900</u>
Total resistance   24,100 lb

6. Table 13-2. The coefficient of traction for the crawler-tractor on loose earth is 0.60. The maximum tractive force is $0.60 \times 55,000 = 33,000$ lb. The force is adequate for pulling the unladen scraper up the maximum $+24$ percent grade.
7. The theoretical hauling speed in 400 ft of level cut and fill is given by the equation:

$$S = \frac{162 \times 33,000}{4200} = 1370 \text{ ft/min} = 14.4 \text{ mi/h}$$

The estimated average hauling speed $= 90\% \times 6.5 = 5.8$ mi/h.

8. The average speed for the 600-ft haul is estimated to be 5.8 mi/h.

13-22 Hauling Excavation

9. The return speed in 400 ft of level fill and cut is estimated to be 5.8 mi/h.
10. The theoretical return speed on a +24 percent grade is given by the equation:

$$S = \frac{162 \times 33,000}{24\% \times 55,000 + (24\% + 4\%)\ 39,000} = 222 \text{ ft/min} = 2.5 \text{ mi/h}$$

11. The estimated return speed on the +24% grade is $90\% \times 2.5 = 2.2$ mi/h.

Hauling cycle, min:	
Loading by push tractor	0.8
Hauling 600 ft at 5.8 mi/h	1.2
Returning:	
400 ft in level fill and cut at 5.8 mi/h	0.8
200 ft up +24% grade at 2.2 mi/h	1.0
Two turnings and dumping	1.2
Total net cycle time	5.0
Hourly production per 50-min working hour, cybm:	
Each tractor-scraper, $\frac{50}{5.0} \times 15.7$	157
Four machines, 4×157	628
Costs for push loading and hauling, 1978:	
Machines:	
Push-loading tractor	$134,000
Tractor pulling scraper	122,000
Scraper	89,000
Push-loading cost:	
Hourly cost of ownership and operation:	
Push-loading tractor (Table 8-3), $34\% \times \$134$	$45.60
Operator	15.60
Total hourly cost	$61.20
Direct job unit cost per cybm, $61.20/628	$0.10
Hauling cost:	
Hourly cost of ownership and operation:	
4 tractors (Table 8-3), $4 \times (34\% \times \$122)$	$165.90
4 scrapers (Table 8-3), $4 \times (28\% \times \$89)$	99.70
4 operators, $4 \times \$15.60$	62.40
Total hourly cost	$328.00
Direct job unit cost per cybm $328.00/628	$0.52
Total direct job unit cost, push loading and hauling	$0.62

Pull-Type Rock Wagons

Characteristics The pull-type wagon, representative specifications for which are given in Table 13-8, is dumped to either side by a centrally located telescoping hydraulic hoist. Hydraulic pumps and controls are mounted on the pulling tractor. The angle of discharge for the bathtub-type body is 45°. Figure 13-9 illustrates the dumping action of the wagon.

The crawler-mounted wagon is not equipped with brakes, so that care must be taken on downgrade haul roads lest the wagon jackknife. The worst danger is risked when a tractor is pulling two loaded wagons on steep downgrades. Grade assistance force and rolling resistance force are balanced under average working conditions on about a -5 percent grade. The greater the rolling resistance, the greater the negotiable downgrade for safe work.

Table 13-9 gives body factors for average loads of wagons according to the nature of rock-earth. The configuration of the relatively shallow-depth, bathtub-type body gives a high ratio of loose load to body capacity, struck measurement. The bathtub-type body with its generous top dimensions offers a good target for the loading shovel.

The following example illustrates the use of Table 13-9 for the wagon of Figure 13-5, which has a capacity of 8.0 cysm. The rock is well-blasted limestone.

        Cubic yards loose measurement = $1.41 \times 8.0 = 11.3$ cylm
        Cubic yards bank measurement = $0.67 \times 11.3 = 7.6$ cybm

TABLE 13-8 Representative Specifications for Pull-Type Wagons, Side Dumping, Crawler-Mounted

Specification	Capacity, cysm		
	6.0	7.0	8.0
Capacity, loose measurement, with crowned load at 4:1 slope, cylm	8.4	9.8	11.5
Dimensions, ft:			
Length	19.6	20.6	22.8
Width	11.2	11.7	11.7
Height, overall and loading	6.3	6.8	7.0
Tracks, crawler, two:			
Width, each, in	18	20	24
Length, ground contact, each, in	41	47	47
Total supporting area, sq ft	10.3	13.1	15.8
Rated capacity, tons	20	20	30
Weights, lb:			
Tare, working	18,300	22,600	26,900
Load, average estimated for rock-earth weighing 2900 lb per cylm	24,400	28,400	33,400
GVW, estimated	42,700	51,000	60,300
Suggested crawler-tractor:			
Pulling one wagon:			
Flywheel hp	150	175	200
Drawbar hp	104	121	138
Weight, lb	30,000	35,000	39,000
Pulling two wagons:			
Flywheel hp	200	250	300
Drawbar hp	138	172	207
Weight, lb	39,000	52,000	65,000
Total weights, tractor and two wagons, lb:			
Tare, working	75,600	97,200	118,800
Load, estimated	48,800	56,800	66,800
GVW, estimated	124,400	154,000	185,600
Speed indicators, tractor and two wagons:			
Laden: $\dfrac{\text{Brake hp}}{\text{GVW, 1000s of lb}}$	1.61	1.62	1.62
Unladen: $\dfrac{\text{Brake hp}}{\text{Tare weight, 1000s of lb}}$	2.65	2.57	2.53
Work-cost indicator, tractor and two wagons:			
$\dfrac{\text{GVW}+\text{tare weight}}{\text{Load weight}}$	4.10	4.42	4.56

TABLE 13-9 Wagon Body Factors for Average Loads of Crawler-Tractor-Wagons, Based on Capacity, cysm

Nature of rock-earth	Fill factor $\left(\dfrac{\text{loose measurement}}{\text{body capacity}}\right)$	Conversion factor $\left(\dfrac{\text{bank measurement}}{\text{loose measurement}}\right)$	Body factor $\left(\dfrac{\text{bank measurement}}{\text{body capacity}}\right)$
Rock-earth weathered	1.41	0.78	1.10
Rock well blasted	1.41	0.67	0.94
Rock poorly blasted	1.41	0.60	0.85

13-24 Hauling Excavation

Figure 13-9 Rock haulage by three crawler-tractor–side-dump crawler-wagons sets. Each set of machines is made up of a tractor of 131 engine flywheel hp and two wagons of 7 cysm capacity. The haul is 1300 ft one way down an average −4 percent grade of the soft haul road to the fill. The wet early springtime work is highway building in the Allegheny Mountains of Pennsylvania. Rock is blasted hard, abrasive limestone. On the soft haul road and fill, beset by frequent rains, the supporting 13.1 ft² of the wagon's crawlers gives a desirable low pressure of 27 lb/in² for the loaded wagons. *(Athey Products Corp.)*

Or, alternatively:

$$\text{Cubic yards bank measurement} = 0.94 \times 8.0 = 7.5 \text{ cybm}$$

Production and Cost Estimation The methods of estimating for crawler-tractor-drawn, crawler-mounted wagons are similar to those for the crawler-tractor-scrapers with two exceptions. First, the loading times are variable and they depend on the capacity of the loader, its operating characteristics, and the controlling job factors. These are discussed in Chapter 12. Second, rolling resistances are different from those of the wheels-tires-mounted scrapers, as shown in Table 13-3. For hard to medium surfaces they are larger and for medium to soft surfaces they are smaller, which emphasizes the advantage of the crawlers for poor haul-road conditions.

As the wagons are top-loaded, one may assume a conservative loose-measurement load based on a crown of 4:1 slope. Table 13-9 indicates that the loose measurement capacity of the wagon averages 1.41 times the struck measurement capacity. As has been mentioned, the crawler-mounted wagons have no brakes and great care must be exercised by the operator when hauling or returning on significant downgrades. The near constant time for two turnings and dumping may be estimated at 1.5 min, there being little difference whether the tractor is hauling one or two wagons in a well-managed job.

In Figure 13-9 the crawler-tractor-wagons are working under conditions calling for their particular operating characteristics.

The estimation of production and calculation of direct job unit cost for crawler-tractor-wagons is outlined below for the wagons of Figure 13-4.

Criteria:

1. The tractor has 131 engine flywheel hp, 90 drawbar hp, and weighs 39,000 lb. Its maximum travel speed is 4.9 mi/h.

2. Each wagon has a capacity of 7.0 cysm and weighs 22,600 lb.

3. The rock is well-blasted limestone and weighs 4380 lb/cybm.
4. The wagons are loaded by a 2-yd³ shovel with a total of 10 dippers in 4.8 min.
5. The haul is 1300 ft long, one-way, down a -4 percent grade. The haul road is deeply rutted.
6. Job efficiency is the 50-min working hour, or 83 percent.

Production:
 1. From Table 13-9, the body factor is 0.94. The estimated payload is $2\times 0.94 \times 7.0 = 13.2$ cybm.
 2. The estimated GVW for the tractor is $2\times 22{,}600 + 13.2 \times 4380 = 103{,}000$ lb.
 3. From Table 13-3, the rolling resistance of the laden wagons is $11\% \times 103{,}000$ lb, or 11,300 lb.
 4. By use of Table 13-4, the grade assistance or push of laden wagons is obtained as $4\% \times 103{,}000$, or 4100 lb.
 5. The net drawbar pull for the tractor is $11{,}300 - 4100 = 7200$ lb.
 6. The grade assistance of the tractor is $4\% \times 39{,}000$, or 1600 lb.
 7. The theoretical hauling speed is given by the equation: $\dfrac{90 \times 33{,}000}{7200 - 1600}$ $=530$ ft/min$=6.0$ mi/h. The estimated hauling speed is $90\% \times 4.9 = 4.4$ mi/h.
 8. The theoretical returning speed of the tractor-wagons up the $+4$ percent grade is given by the equation:

$$\dfrac{90 \times 33{,}000}{4\% \times 39{,}000 + 2(11 + 4\%) \times 22{,}600} = 354 \text{ ft/min} = 4.0 \text{ mi/h}$$

The estimated returning speed is $90\% \times 4.0, = 3.6$ mi/h.

Hauling cycle, min:
Loading	4.8
Hauling 1100 ft at 4.4 mi/h	2.9
Returning 1100 ft at 3.6 mi/h	3.5
Two turnings and dumping	1.5
Net cycle time	12.7

Hourly production per 50-min working hour, cybm

Each set of tractor-wagons, $\dfrac{50}{12.7} \times 13.2 =$	52
Three sets of machines, 3×52	156

Costs for hauling, 1978:
Machines:
Tractor	$100,000
Two wagons @ $58,000	116,000

Hourly cost of ownership and operation:
Tractor (Table 8-3), $34\% \times \$100$	$34.00
Two wagons (Table 8-3), $30\% \times \$116$	34.80
Operator	15.60
Total hourly cost	$84.40
Direct job unit cost per cybm, $84.40/52	$1.62

The unit cost is high but the operating conditions are severe in the early spring when alternate thawing and freezing, as well as rain, bring poor working conditions.

Summary

Haulage by crawler-tractors pulling wheels-tires-mounted scrapers and crawler-mounted rock wagons is relatively costly under good operating conditions but may be economical under unavoidable bad circumstances.

The machines are almost unlimited in their ability to cope with all kinds of rock-earth and all kinds of weather conditions. Their unusual versatility is especially valuable in mountainous country, where both grade and rolling resistances may bog down any other kind of rock-hauling machinery. These considerations carry great weight when opening up a rock excavation job in the late winter or early spring and finishing the work in late fall or early winter.

An index of the relatively high direct job-unit costs when working under good conditions may be inferred from the high work-cost indicators for both tractors-scrapers and tractors-wagons.

WHEELS-TIRES-MOUNTED SCRAPERS AND BOTTOM-DUMP, REAR-DUMP, AND SIDE-DUMP HAULERS

These rock-earth haulers are used in both off-the-road and on-the-road haulage. They may have two axles or multiaxles, and they may have one, two, or three engines. They are illustrated by the figures listed below.

Off-the-road machines:
  Scrapers, Figures 13-15 and 13-21
  Bottom-dump haulers, Figures 13-27 and 13-29
  Rear-dump haulers, Figures 13-30 and 13-36
  Side-dump haulers, Figures 13-41 and 13-42
On-the-road machines:
  Scrapers, Figure 13-23
  Bottom-dump haulers, Figure 13-26
  Rear-dump haulers, Figure 13-31
  Side-dump haulers, Figure 13-40

These modern machines are the successors of a long line of automotive trucks and tractor-trailers and tractor-drawn scrapers and rock wagons. Figure 13-10 illustrates strikingly the development of rock haulers by two manufacturers over a period of two generations. The crawler-tractor-mounted side-dump wagon and the rear-dump rocker hauler are parallel examples, as the crawler-tractor and the wheels-tires tractor are built by one manufacturer and the crawler-wagon and the wheels-tires wagon are built by another company.

The accompanying tabulated comparison of the crawler-mounted and wheels-tires-mounted machines of Figure 13-10 portrays almost a half century of progress in the development of rock haulers.

Item	Crawler-mounted machine	Wheels-tires-mounted machine
Engine flywheel or brake hp	140	330
Capacity:		
Body, cysm	7	22
Load of rock, cybm	6.6	17.0
Weights, lb:		
Tare	61,600	60,700
Load, at 4400 lb/cybm	29,000	74,800
GVW	90,600	135,500
Ratio of load weight/tare weight	0.47	1.23
Estimated cycle time, min, for loading, hauling, two turnings and dumping, and returning on level, 1000-ft one-way haul	9.3	6.1
Production per 50-min working hour, cybm	35	139
Speed indicators:		
Laden: $\dfrac{\text{Brake hp}}{\text{GVW, 1000s of lb}}$	1.55	2.44
Unladen: $\dfrac{\text{Brake hp}}{\text{Tare weight, 1000s of lb}}$	2.27	5.44
Work-cost indicator: $\dfrac{\text{GVW} + \text{tare weight}}{\text{load weight}}$	5.25	2.62

The modern machine has much faster traveling speeds (57 percent more when laden and 140 percent more when unladen) than the special-purpose crawler machine. Likewise, its work-cost indicator is 50 percent of that of the crawler machine, which suggests that it can haul rock at 50 percent of the direct job unit cost of the crawler machine. On the basis of another commonly used index of relative efficiencies of haulers, the ratio of load weight to tare weight, the modern machine has a 162 percent advantage.

Principles of Haulage by Wheels-Tires-Mounted Scrapers and Haulers

These machines, powered by integral engines, are essentially automotive rock-earth haulers. They are designed similarly with respect to engine, torque converter, trans-

(a)

(b)

Figure 13-10 A half century of development of rock haulers. *(a)* A crawler-tractor with 140-flywheel-hp engine pulling a 7-cysm crawler-mounted rock wagon at a maximum speed of 5.3 mi/h with a load of 6.6 cybm. *(b)* A wheels-tires-mounted tractor with 330-flywheel-hp engine pulling a 22-cysm wheels-tires-mounted rocker wagon at a maximum speed of 32 mi/h with a load of 17.0 cybm. *(Athey Products Corp. and Caterpillar Tractor Co.)*

mission, drive line, differential, axles, and wheels. In some machines, such as the heavy-duty off-the-road haulers, three reductions in speed or increases in torque by means of transmission, differential, and final drive are incorporated in the axle-wheel assembly.

The driving wheels, at the contact between the tires and the surface of the road, must overcome rolling resistance and, on upgrades, grade resistance by means of traction force. Traction force is the product of the coefficient of traction or friction times the weight on the driving wheels. These factors are summarized in Tables 13-10 and 13-11 and the preceding Table 13-4.

Dynamics of Haulage

Wheel Horsepower The *wheel horsepower* of a hauler is that horsepower for doing work at the area of contact between driving wheels and the road surface. It is engine brake or flywheel horsepower less all the power transmission losses in the power train from flywheel to final gearing of axle or wheel. It is also power train efficiency times flywheel horsepower. The average efficiency for a rock-earth scraper or hauler is 76 percent.

Tractive Force The tractive force of a hauler must be considered versus the algebraic sum of the forces of rolling resistance and grade resistance or grade assistance. When a scraper or hauler is working on level haul road, the tractive force, in terms of traction coefficient times weight on drive wheels, must exceed rolling resistance, in terms of rolling resistance factor times weight on all wheels, or there is no movement.

TABLE 13-10 Coefficients of Traction for Wheels-Tires-Mounted Scrapers and Haulers

Surface of travel road	Coefficient
Asphalt	0.90
Clay loam, dry	0.55
Clay loam, wet	0.45
Clay loam, rutted	0.40
Concrete	0.90
Earth, firm	0.60
Earth, loose	0.45
Gravel, compacted	0.65
Gravel, loose	0.35
Ice	0.10
Quarry pit	0.65
Rock, loose, ripped	0.45
Sand, dry	0.20
Sand, wet	0.40
Snow, dry, packed	0.20

TABLE 13-11 Rolling Resistances for Wheels-Tires-Mounted Scrapers and Haulers (Percentage of Weight of Laden or Unladen Machine)

Surface of travel road	Percent of weight
Asphalt	1.5
Concrete	1.5
Dirt, hard, dry, well-maintained	2.0
Dirt, not firmly packed, dry, some loose material, well-maintained	3.0
Dirt, poorly maintained	4.0
Dirt, compacted fill	4.0
Dirt, ploughed	8.0
Dirt, deeply rutted	16.0
Gravel, dry, well-compacted	2.0
Gravel, dry, not firmly packed, some loose material, well-maintained	3.0
Gravel, loose	10.0
Mud, with firm base	4.0
Mud, with soft, spongelike base	16.0
Rock-earth in cuts	8.0
Rock-earth in compacted fills	4.0
Sand, loose	10.0
Sand, wet	5.0
Snow, dry, packed	2.5
Snow, fresh, loose to moderate depth	4.5

NOTE: If actual inches of tire penetration of surface are known or determinable, a practical value for the rolling resistance RR of a machine is given by the equation: RR, % of GVW of machine, $= (2.00\% + 1.75\% \times$ inches of tire penetration).

When the machine is working upgrade, the tractive force, in terms of traction coefficient times component of weight on driving wheels normal to the surface, must exceed the sum of rolling resistance and grade resistance. Rolling resistance is now the product of the rolling resistance factor times the component of the weight normal to the surface, and grade resistance is the component of the weight of the machine parallel to the surface. When working downgrade, the resolution of forces is the same as that for upgrade haulage, except that the force due to the grade assists tractive force and the total forces which the traction force must overcome is the algebraic sum of rolling resistance and grade assistance.

In all three cases—level, upgrade, or downgrade hauls—the tractive and the resisting or assisting forces are parallel to the surface of the road.

Upgrade Haulage. Figure 13-11 illustrates the dynamics of upgrade haulage by an all-wheel-drive, twin-engine loaded scraper. Similar resolutions of forces may be applied to all machines, whether two-axle or multiaxle or single-engine, twin-engine, or multiengine.

As in the case of crawler-tractor- drawn scrapers and wagons, the calculations may be conservatively simplified by assuming that on all grades up to +30 percent the following approximations are justified:

1. The force normal to the road surface equals the vertical force or weight.
2. Force parallel to road surface equals the vertical force or weight times percentage of grade.

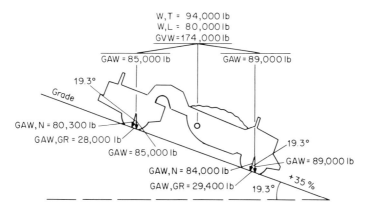

Figure 13-11 Dynamics of upgrade haulage by all-wheel-drive twin-engine loaded scraper.

W, T = tare weight of scraper
GVW = gross vehicle weight
W, L = weight of load
GAW = gross axle weight
GAW, N = component force, normal to road surface, of GAW, for calculating maximum traction force and rolling resistance force
GAW, GR = component force, parallel to surface, of GAW, grade resistance
Rolling resistance of poorly maintained, rutted dirt haul road 4%
Coefficient of traction for the haul road 0.40
Front axle: $GAW, N = 85{,}000 \times \cos 19.3° =$ 80,300 lb
     $GAW, GR = 85{,}000 \times \sin 19.3° =$ 28,000 lb
     Maximum traction force $= 0.40 \times 80{,}300 =$ 32,100 lb
     Rolling resistance force $= 0.04 \times 80{,}300 =$ 3,200 lb
     Grade resistance force =   <u>28,000 lb</u>
     Total resistances force =   31,200 lb
Rear axle: $GAW, N = 89{,}000 \times \cos 19.3° =$ 84,000 lb
     $GAW, GR = 89{,}000 \times \sin 19.3° =$ 29,400 lb
     Maximum traction force $= 0.40 \times 84{,}000 =$ 33,600 lb
     Rolling resistance force $= 0.04 \times 84{,}000 =$ 3,400 lb
     Grade resistance force =   <u>29,400 lb</u>
     Total resistances force =   32,800 lb

As total maximum traction force, 65,700 lb, exceeds total resistance force, 64,000 lb, the loaded scraper can ascend the +35 percent grade.

$$\text{Travel speed of 463-wheel-hp scraper} = \frac{463 \times 33{,}000}{64{,}000} = 239 \text{ ft/min} = 2.7 \text{ mi/h}$$

13-30   Hauling Excavation

As the grade of Figure 13-11 is +35 percent, it is recommended that the trigonometric functions be used.

The calculations of Figure 13-11 are concerned with two questions:
1. Can the laden scraper ascend the grade?
2. If it can ascend the grade, what is its travel speed?

A simple way of answering the first question by figuring maximum gradability is to equate the maximum tractive force to the total resistances and to solve the resulting equation for the unknown angle of grade A. The following equation results:

$$\text{Traction force} = \text{rolling resistance force} + \text{grade resistance force}$$

$$0.40(174{,}000 \cos A) = 0.04(174{,}000 \cos A) + (174{,}000 \sin A)$$

whence
$$\frac{\sin A}{\cos A} = 0.360$$
$$\tan A = 0.360$$
$$A = 19.8°$$
$$\text{Percentage of grade} = \tan A = 36\%$$

As 36 percent exceeds the grade of 35 percent, the laden scraper can ascend the 35 percent grade.

To answer the second question, one notes that the 35 percent grade corresponds to a grade angle of 19.3°. The maximum speed of the laden machine for this grade may be calculated from the equation:

$$S = \frac{\text{wheel hp} \times 33{,}000}{\text{total resistance forces}}$$

$$= \frac{463 \times 33{,}000}{64{,}000} = 239 \text{ ft/min} = 2.7 \text{ mi/h}$$

Downgrade Haulage. The dynamics of downgrade haulage are the same as those for upgrade work except that the grade creates an assistance to motion rather than a resistance. On many downgrade hauls over good haul roads, braking action is necessary unless rolling resistance offsets grade assistance. For example, if rolling resistance is as low as an attainable 2.0 percent of weight, then, except for the compression effect of the engine, brakes must be applied on any grade greater than -2 percent.

Machines may be braked by mechanical brakes, usually hydraulic or air, by retarders, or by brakes assisted by retarders. Retarders in modern haulers are a part of the transmission assembly. They are simply dynamic braking devices consisting of shaft-mounted paddles revolving within a chamber of oil. Filling the chamber with oil slows the paddle wheel and brakes the paddle-wheel shaft. The shaft through the power train brakes the wheels. Heat generated in the oil is dissipated through the retarder's cooling system.

Manufacturers' specifications for the machines give retarder charts along with the engine charts. The charts give rim-pull retardations in pounds as a function of travel speed. From these one may calculate the horsepower absorptions by retarders at the wheel-road contact. The wheel horsepower absorption approximates the flywheel horsepower of the engine.

The following table, based on an analysis of retarder and engine performance charts for a large-capacity (50-ton) rear-dump truck or rock hauler with a 576-flywheel-hp engine, gives the horsepower absorbed by the retarder at the wheels and the horsepower expended by the engine at the wheels according to travel speed.

	Gear and travel speeds		Horsepower at wheels	
Gear	mi/h	ft/min	Absorbed by retarder	Delivered by engine
1st	5	440	586	467
2d	7	616	560	429
3d	10	880	587	480
4th	15	1320	600	480
5th	22	1936	645	440
6th	30	2640	600	480
Averages			596	463
Percent of 576 engine flywheel hp			103	80

On an upgrade +10 percent haul road with 3 percent rolling resistance, the total resistance of the 177,000-GVW truck is 13 percent of 177,000, or 23,000 lb.

$$\text{Travel speed} = \frac{463 \times 33,000}{23,000} = 664 \text{ ft/min} = 7.5 \text{ min}$$

Travel speed would be in second gear.

On a downgrade −10 percent haul road with 3 percent rolling resistance, the total resistance to be overcome to prevent increase in speed would equal 7 percent of 177,000, or 12,400 lb. The maximum speed for full horsepower absorption by the retarder is given by the equation:

$$\text{Travel speed} = \frac{596 \times 33,000}{12,400} = 1586 \text{ ft/min, or } 18.0 \text{ mi/h}$$

Travel speed would be in fourth gear.

Retarders have the capacity for maximum safe downgrade haulage. If a travel speed of 18 mi/h were deemed too high for safety, the speed could be lowered to 12 mi/h, or 1056 ft/min. Travel would be in third gear, and horsepower absorption by the retarder would equal $\frac{1056 \times 12,400}{33,000}$, or 419 hp.

Hauling Cycle Calculations

The hauling cycle includes the following elements.

Loading time. Scrapers are loaded by push tractors or they are self-loaded. This element will be discussed later in the ensuing section on scrapers. Bottom-dump, rear-dump, and side-dump haulers are top-loaded by the several loading machines described in Chapter 12. Loading times for each type of machine are taken up in the respective following sections.

Turning and dumping times. These also will be described for each type of machine separately.

Hauling and returning times. Since these times are based on the common characteristics of scrapers and haulers, they are discussed below for all four machines.

Hauling and Returning Times—General Method There are several methods for calculating hauling and returning times. One method is to rely on experience and, if sufficient accumulated data are available, this method is reliable. Some methods are complex, involving tedious calculations for accelerations and decelerations, complicated nomographs, and generally unnecessary mathematics. Others are too simple, as they are based on average travel speeds which do not provide for transitions from one segment of the haul road to another totally different segment.

Precise calculations are based on several well-known fundamental formulas of dynamics. These formulas are reviewed below because they are basic and their understanding is helpful to a complete appreciation of the many factors involved in the play of forces, speeds, and powers of hauling machines.

The fundamental formulas are:

$$d = vt \quad \text{or} \quad t = \frac{d}{v}$$

$$v = at \quad \text{or} \quad t = \frac{v}{a}$$

$$d = \frac{at^2}{2}$$

$$f = ma$$

$$w = \frac{m}{a}$$

$$f = \frac{wa}{g} \quad \text{or} \quad a = \frac{fg}{w}$$

$$W = fd$$

$$KE = \frac{mv^2}{2} = \frac{wv^2}{2g}$$

where d = distance, ft
v = speed or velocity, ft/min

13–32   Hauling Excavation

$t =$ time, min
$a =$ acceleration or deceleration of machine, ft/min²
$f =$ force, lb
$m =$ mass, lb
$w =$ weight of machine, GVW or tare weight, lb
$g =$ acceleration of gravity, ft/min² (g=116,000 ft/min², or 32.2 ft/s²)
$W =$ work, ft·lb
$KE =$ kinetic energy, ft·lb

Below is an example of the use of several of these formulas in calculating the time for a rock hauler to ascend a ramp to the crest of a dam, the ramp being a segment of the haul road from borrow pit to dam. The calculations give time and distance for a laden bottom-dump hauler when ascending a ramp after entering the ramp at a given speed. The haul up the ramp to the crest of the dam is part of the haul road from the borrow pit to the dam.

Criteria:
Ramp:
  Length, one way, ft   1200
  Grade, %   +8.0
  Grade resistance factor, %   8.0
  Rolling resistance factor for haul road; gravel not firmly
    packed, some loose material, well maintained (Table 13-11)   3.0
  Sum of resistance factors, %   11.0
Speed of entry to ramp (16.7 mi/h), ft/min   1472
Engine horsepower:
  Flywheel   609
  At wheels   463
GVW of laden hauler, lb   346,000

The problem is solved readily by an equation for the kinetic energy and work between the entry to the ramp and the point on the ramp where the maximum attainable speed is reached. The equation is:

KE at entry + work of engine − work of grade and rolling resistances =

                                                                   KE at maximum attainable speed

Speed of entry   1472 ft/min
Sum of grade and rolling resistance forces,
  $f = 11\% \times 346{,}000$ 38,100 lb
Maximum attainable speed on ramp,
  $v = \dfrac{463 \times 33{,}000}{38{,}100}$ 401 ft/min

The equation for kinetic energy and work becomes:

$$\frac{346{,}000 \times 1472^2}{2 \times 116{,}000} + (t \times 463 \times 33{,}000) - 38{,}100 d = \frac{346{,}000 \times 401^2}{2 \times 116{,}000}$$

$$15{,}279{,}000 t - 38{,}100 d = -2{,}991{,}000$$

where $t =$ deceleration time, min
          $d =$ deceleration distance, ft
but $d =$ average velocity × time

$$= \left(\frac{1472 + 401}{2}\right) t = 936 t$$

Substituting in the above principal equation:

$$15{,}279{,}000 t - 35{,}661{,}000 t = -2{,}991{,}000$$

$$t = 0.147 \text{ min}$$

$$d = 936 \times 0.147 = 138 \text{ ft}$$

Time for traveling remaining distance at maximum attainable speed:

$$t = \frac{1200-138}{401} = 2.898 \text{ min}$$

Total time to ascend 1200-ft ramp $= 0.147 + 2.898 = 3.045$ min

These calculations are time-consuming, they assume a knowledge of dynamics and algebra, and they are error-prone. Accordingly, they are impractical.

Hauling and Returning Times—Simple Precise Method The following simple precise method for hauling and returning times is part of the manual *Production and Cost Estimating of Material Movement with Earthmoving Equipment, 1970* of General Motors Corp.; the data are reproduced almost verbatim except for a few additional explanatory notes and for the method of calculating maximum attainable speeds. These speeds are calculated by using wheel horsepower instead of the rim-pull of the performance chart for a particular machine. This method conforms to the system of this text. Both methods result in the same precision measure of estimating the hauling cycles.

The General Motors Corp. method for determining hauling and returning times takes into account that the travel time of an earthmover over a particular section of a haul road can be estimated by dividing the length of the road section by the estimated speed. The equation is:

$$\text{Travel time, min} = \frac{\text{length of section, ft}}{88 \times \text{average speed, mi/h}}$$

In the equation, 88 is a factor to convert miles per hour to feet per minute.

The maximum speed of the machine over a particular haul road section under specific conditions may be calculated with the aid of Tables 13-4, 13-10, and 13-11. However, since the machine may not operate at its maximum speed over the entire length of the road section, the maximum speed must be reduced to a practical average to compensate for the acceleration or deceleration of the earthmover.

The average speed is determined by multiplying the maximum speed on a given road section by a speed factor, as listed in Table 13-12. A speed factor of 1.00 indicates that the machine will maintain maximum speed along the entire road section, while a speed factor less than or greater than 1.00 indicates the machine will travel more slowly or faster, respectively, than the maximum speed.

Several variables determine the speed factor. Among the most important are weight-to-power ratio, initial or final speed, length of road section, and delay factors.

Weight/Engine Horsepower Ratio. The weight-to-power ratio is calculated by means of this formula:

$$\text{Weight-to-power ratio} = \frac{\text{earthmover weight, GVW or tare weight}}{\text{engine flywheel horsepower}}$$

The smaller the weight-to-power ratio, the faster the machine will travel and the greater its acceleration. The speed factor will be higher. And, conversely, the greater the ratio, the lower the speed factor.

Speed and Grade—Momentum. Initial speeds and final speeds affect the earthmover momentum over the road section. A machine which enters or leaves a road section at or near maximum speed will have a speed factor closer to 1.00 than one which starts from a stop or one which must slow down at the end of the road section. An earthmover which enters a haul road section at a higher speed than the maximum speed on that section will have a speed ratio greater than 1.00. The speed factors shown in the following tables should be tempered by these considerations.

Acceleration and momentum affect only the beginning and the end of each haul road section but the length of the section is also important. The longer the section, the more time will be spent at maximum speed and the speed factor will be nearer to 1.00. Machines in motion reach their maximum speed very quickly when entering an uphill haul road section. If the entrance speed is faster than the table speed, the speed factor will be greater than 1.00.

TABLE 13-12 Speed Factors
(for conversion of maximum speed to average speed)

Haul road length, ft	Level haul unit starting from 0 mi/h	Unit in motion when entering haul road section		Uphill grade factor
		Level	Downhill grade	
		Under 300 lb/hp		
0–200	0–0.40	0–0.65	0–0.67	1.00
201–400	0.40–0.51	0.65–0.70	0.67–0.72	(Entrance speed
401–600	0.51–0.56	0.70–0.75	0.72–0.77	greater than
601–1000	0.56–0.67	0.75–0.81	0.77–0.83	maximum at-
1001–1500	0.67–0.75	0.81–0.88	0.83–0.90	tainable speed
1501–2000	0.75–0.80	0.88–0.91	0.90–0.93	on section)
2001–2500	0.80–0.84	0.91–0.93	0.93–0.95	
2501–3500	0.84–0.87	0.93–0.95	0.95–0.97	
3501 and up	0.87–0.94	0.95–	0.97–	
		300–380 lb/hp		
0–200	0–0.39	0–0.62	0–0.64	1.00
201–400	0.39–0.48	0.62–0.67	0.64–0.68	(Entrance speed
401–600	0.48–0.54	0.67–0.70	0.68–0.74	greater than
601–1000	0.54–0.61	0.70–0.75	0.74–0.83	maximum at-
1001–1500	0.61–0.68	0.75–0.79	0.83–0.88	tainable speed
1501–2000	0.68–0.74	0.79–0.84	0.88–0.91	on section)
2001–2500	0.74–0.78	0.84–0.87	0.91–0.93	
2501–3500	0.78–0.84	0.87–0.90	0.93–0.95	
3501 and up	0.84–0.92	0.90–0.93	0.95–0.97	

		380 and up, lb/hp		
0–200	0–0.33	0–0.55	0–0.56	1.00
200–400	0.33–0.41	0.55–0.58	0.56–0.64	
401–600	0.41–0.46	0.58–0.65	0.64–0.70	(Entrance speed
601–1000	0.46–0.53	0.65–0.75	0.70–0.78	greater than
1001–1500	0.53–0.59	0.75–0.77	0.78–0.84	maximum at
1501–2000	0.59–0.62	0.77–0.83	0.84–0.88	tainable speed
2001–2500	0.62–0.65	0.83–0.86	0.88–0.90	on section)
2501–3500	0.65–0.70	0.86–0.90	0.90–0.92	
3501 and up	0.70–0.75	0.90–0.93	0.92–0.95	

NOTE: Average speed in a section of haul road:

Average speed = speed factor × maximum attainable speed

Travel time on section of haul road:

$$\text{Travel time, min} = \frac{\text{length of section, ft}}{88 \times \text{average speed, mi/h}}$$

Total travel time on the haul road:

Travel time, min = sum of travel times on sections, min

SOURCE: General Motors Corp., *Production and Cost Estimating of Material Movement with Earthmoving Equipment, 1970*.

13-36   Hauling Excavation

Reading the Speed Factor Charts.   The procedure is:
1. Determine the weight-to-power ratio, as explained, and choose the proper table.
2. Choose the proper column (starting from stop, downhill, level, or uphill grade).
3. Pick out the proper speed factor opposite the length of the haul road section being examined.
4. Use a speed factor of 1.00 for a machine entering a haul road section at near the maximum speed for the section with no speed limitation at end of section.
5. Speed factor on final section of both hauling and returning should be taken from "Unit starting from stop" column because unit is coming to a stop.
6. Speed factors are for machines equipped with power-shift and torque converter transmissions. For manual-shift transmissions the lower speed factors should be used.
7. Delay factors. There are often hazards or obstructions in the haul road which slow down machine speed. Time allowances must be made when these conditions exist. This adjustment is by the use of smaller speed factors and it is a matter of experience.
   Intermittent factors (consider delay time or slowdown on each section of haul road):
      One-way haul roads
      Delay at passing points
      Sharp curves
      Multiple curves or switchbacks
      Blind corners
      Bridges
      Underpasses
      Railroad crossings
      Cross traffic
   Continuous factors (consider delay time or slowdown over entire haul road):
      Extremely variable and high rolling resistances
      Wet or slippery haul roads
      Unskilled operators
      Long downgrade hauls
8. The use of these charts should be tempered by the experience of the estimator to obtain precise results. Often the results may be compared with the known hauling and returning times for the same or similar machines which have worked on similar haul roads with similar characteristics.

Example of Use of Speed Factors.   The example involves hauling semipervious fill material, alluvium, from the borrow pit to the fill on top of a dam.

Criteria:
  Bottom-dump hauler, as shown in Figure 13-27; engine flywheel hp 609
  Estimated wheel hp = 76% × 609 = 463; estimated capacity of retarder, 609

Working or tare weight	126,000 lb
Load weight, 76 cylm of alluvium rock at 2900 lb/cylm	220,000
Gross vehicle weight (GVW)	346,000 lb
Haul road:	
Borrow pit, level, rolling resistance 4%	1,000 ft
Main haul road, level, rolling resistance 3%	9,000
Ramp to grade of dam fill, +8% grade, rolling resistance 3%	1,200
Fill at top of dam, level, rolling resistance 4%	1,500
Total length, one way	12,700 ft

The total haul road is in excellent condition and well maintained. There are no job conditions to prevent maximum efficiency of the bottom-dump hauler.

Solution:

Hauling: Weight-to-power ratio = $\dfrac{346{,}000}{609}$ = 568

Borrow pit: Maximum speed = $\dfrac{463 \times 33{,}000}{0.04 \times 346{,}000}$ =   1104 ft/min = 12.5 mi/h
             Average speed = 0.53 × 1104 =   585 ft/min = 6.6 mi/h
             Time = 1000/585 =   1.71 min

Main haul road: Maximum speed = $\dfrac{463 \times 33{,}000}{0.03 \times 346{,}000}$ =   1472 ft/min = 16.7 mi/h
             Average speed = 0.93 × 1472 =   1369 ft/min = 15.6 mi/h
             Time = 9000/1369 =   6.57 min

Up ramp:	Maximum speed = $\dfrac{463 \times 33{,}000}{(0.08+0.03)346{,}000}$ =		401 ft/min = 4.6 mi/h
	Average speed = 1.00×401		401 ft/min = 4.6 mi/h
	Time = $1200/401$ =		2.99 min
Fill atop dam:	Maximum speed = $\dfrac{463 \times 33{,}000}{0.04 \times 346{,}000}$ =		1104 ft/min = 12.5 mi/h
	Average speed = 0.75×1104 =		828 ft/min = 9.4 mi/h
	Time = $1500/828$ =		1.81 min
	Total hauling time at avg speed of 11.0 mi/h		13.08 min
Returning:	Weight-to-power ratio = $\dfrac{126{,}000}{609}$ =		207
Fill atop dam:	Maximum speed = $\dfrac{463 \times 33{,}000}{0.04 \times 126{,}000}$ =		3032 ft/min = 34.4 mi/h
	Average speed = 0.88×3032 =		2668 ft/min = 30.3 mi/h
	Time = $1500/2668$ =		0.56 min
Down ramp:	Maximum speed, if retarder at full capacity is used:		
	$\dfrac{609 \times 33{,}000}{(0.08-0.03)126{,}000}$ =		3190 ft/min = 36.2 mi/h
	The speed is too high; assume:		1760 ft/min = 20.0 mi/h
	Average speed = 1.00×1760 =		1760 ft/m = 20.0 mi/h
	Time = $1200/1760$ = 0.68 min		
Main haul road:	Maximum speed = $\dfrac{463 \times 33{,}000}{0.03 \times 126{,}000}$ =		4042 ft/min = 45.9 mi/h
	Average speed = 0.95×4042 =		3840 ft/min = 43.6 mi/h
	Time = $9000/3840 = 2.34$ mi/h		
Borrow pit:	Maximum speed = $\dfrac{463 \times 33{,}000}{0.04 \times 126{,}000}$ =		3032 ft/min = 34.4 mi/h
	Average speed = 0.67×3032 =		2031 ft/min = 23.1 mi/h
	Time = $1000/2031$ =		0.49 min
	Total returning time, average speed = 35.6 mi/h		4.07 min
	Total hauling and returning time, average speed = 16.8 mi/h		17.15 min

The precision measure of the speed factor method is typified by a comparison of calculated times for ascending the ramp in the two examples.

1. Calculation by speed factor method, 2.99 min
2. Calculation in example of application of dynamics, part of preceding discussion, 3.04 min

The difference is 1.7 percent, well within the customary ± 5 percent precision measure. The comparison emphasizes the practicability of the speed factor method.

SCRAPERS

There are two kinds of wheels-tires-mounted scrapers: those push-loaded by crawler-tractors and those self-loaded by an integral ladder-type elevator. They may have two axles or three axles, the more common ones having two axles. The push-loaded scraper, under certain conditions, is able to load itself. One condition occurs in loading down steep grades, particularly in the opening up of a new cut. Under these circumstances the scraper's loading action is one of both bulldozing and loading. The other condition occurs in the loading of an all-wheel-drive scraper, and in suitable materials this scraper is able to get about two-thirds of a normal load. Under both conditions the normally push-loaded scraper gets less than a full load.

Push-Loaded Scraper

Figure 13-12 illustrates the push loading of a scraper by two crawler-tractors. The rock-earth is ripped, well-weathered granite. The large scraper has 32.5-cysm and 34.5-cylm capacity based on crowned load at 4:1 slope. The payload is estimated to be 26.9 cybm. The loading time by two heavyweight push tractors is 0.60 min. The scraper has front-axle drive, the axle carrying 52 percent of GVW of about 210,000 lb, or 109,000 lb.

Scrapers' functions may be cable-controlled, as described in the preceding section covering crawler-tractor-drawn scrapers, or controlled by hydraulic cylinders. The scrapers of Tables 13-13 and 13-14 are hydraulically operated. Bowl, apron, and ejector are

13–38   Hauling Excavation

Figure 13-12 Large scraper being loaded conventionally by two heavyweight tractors. The rock is well-weathered granite of the Laguna Mountains, California, and it is being ripped by the heavyweight tractor equipped with two-shank ripper. The hourly production of this scraper spread, made up of one tractor-ripper, two push tractors, and five scrapers, is estimated to be 1030 cybm.

Scraper, equipped with 1-ft-high sideboards:	
Engine flywheel hp	550
Working weight, lb	115,000
Capacity:	
Cubic yards loose measurement	34
Cubic yards bank measurement	27
Loading time, min	0.60
Push tractors:	
Engine flywheel hp	385
Working weight, lb	90,000
Loading cycle, min:	
Loading	0.60
Exchanging scrapers	0.70
Total cycle	1.30
Estimated hourly production:	
Loads per 50-min working hour, 50/1.30	38
Cubic yards bank measurement hourly, 38×27	1030
Tractor-ripper bulldozer:	
Engine flywheel hp	385
Working weight, lb	110,000

individually controlled. Two double-acting cylinders raise and lower the bowl. One double-acting cylinder raises and lowers the apron. One double-acting cylinder within the push-rod case behind the ejector pushes and retracts the load-ejector plate.

Hauling Cycles Push-loaded scrapers are loaded by one, two, or, rarely, three crawler-tractors. Figure 13-12 illustrates the common practice of using two push tractors. Figure 13-13 shows the uncommon use of three tractors. The reason for this infrequent method is that the third tractor is actually a tractor-ripper equipped with a bulldozer. Ripping is completed for the time being and the operator is simply and efficiently keeping the machine busy. A minimum loading time of 0.42 min results instead of the customary 0.60 min.

Figure 13-14 shows the end of loading a twin-engine scraper. The bowl is full to overflowing. The operator is lowering the apron and raising the bowl and is about to pull away from the first push tractor, which will give the loaded scraper an accelerating push after loading is completed.

The twin-engine scraper with its all-wheel-drive contributes more than twice the traction force of its equal-capacity single-engine single-axle counterpart. The resolution of the total forces for loading is set forth in the following tabulation.

Item	Single-engine, single-axle-drive scraper	Twin-engine, two-axle-drive scraper
Scraper:		
Capacity, cysm	32.0	32.0
Engine(s) wheel horsepower	418	722
Weights, lb:		
Tare:		
Total	121,000	143,000
On driving wheels	74,000	143,000
GVW:		
Total	225,000	248,000
On driving wheels	115,000	248,000
Average on driving wheels during loading	94,000	196,000
Average traction force based on 0.45 coefficient of traction, lb	42,000	88,000
Two push tractors:		
Total engine drawbar horsepower	566	566
Total weight, lb	194,000	194,000
Total traction force for loading scraper, based on 0.60 coefficient of traction, lb	116,000	116,000
Scraper and two push tractors:		
Total traction force, lb	158,000	204,000
Traction force advantage, %		29
Theoretical percent reduction in loading time for equivalent load of 27.7 cybm		23

The comparison shows that during the loading operation the twin-engine two-axle-drive scraper is 2.1 times as powerful as its counterpart. However, the skilled operator

TABLE 13-13 Representative Specifications for Wheels-Tires Scrapers, Push-Loaded and with Single-Engine, Single-Axle Drive

Specification	Capacity, cysm			
	14.0	21.0	28.0	32.0
Capacity, crowned load with 4:1 slope, cylm	15.5	23.3	31.1	35.5
Engine flywheel or brake hp	330	415	550	550
Transmission:				
Gears forward	8	8	8	8
Maximum speed:				
mi/h	32	32	32	34
ft/min	2820	2820	2820	3000
Tires, standard	29.5×29 28 PR	29.5×35 34 PR	33.3×39 38 PR	37.5×39 36 PR
Dimensions, ft:				
Length	41.6	44.4	48.8	50.3
Width	11.1	12.5	13.2	14.2
Height	11.9	12.8	13.6	14.1
Weights, lb:				
Tare, working	66,000	81,000	113,000	121,000
Load, rated	48,000	72,000	94,000	104,000
GVW	114,000	153,000	207,000	225,000
Speed indicators:				
Laden: $\dfrac{\text{Brake hp}}{\text{GVW, 1000s of lb}}$	2.89	2.71	2.66	2.44
Unladen: $\dfrac{\text{Brake hp}}{\text{Tare weight, 1000s of lb}}$	5.00	5.12	4.87	4.55
Work-cost indicator:				
$\dfrac{\text{GVW}+\text{tare weight}}{\text{Load weight}}$	3.75	3.25	3.40	3.33

TABLE 13-14 Representative Specifications for Wheels-Tires Scrapers Push-Loaded and with Twin-Engine Two-Axle Drive

Specification	Capacity, cysm			
	14.0	18.0	24.0	32.0
Capacity, crowned load with 4:1 slope, cylm	15.5	20.0	24.0	35.5
Engine flywheel or brake hp	288	520	610	808
Transmission:				
Gears forward	6	6	4	6
Maximum speed:				
mi/h	23	31	30	35
ft/min	2020	2730	2640	3080
Tires, standard	29.5×29 28 PR	29.5×29 34 PR	33.3×33 38 PR	37.5×39 44 PR
Dimensions, ft:				
Length	39.6	42.0	48.6	52.2
Width	11.1	11.9	11.9	13.2
Height	10.3	12.6	12.5	15.2
Weights, lb:				
Tare, working	55,000	76,000	98,000	138,000
Load, rated	47,000	60,000	80,000	104,000
GVW	102,000	136,000	178,000	242,000
Speed indicators:				
Laden: $\dfrac{\text{Brake hp}}{\text{GVW, 1000s of lb}}$	2.82	3.82	3.43	3.34
Unladen: $\dfrac{\text{Brake hp}}{\text{Tare weight, 1000s of lb}}$	5.24	6.84	6.22	5.86
Work-cost indicator $\dfrac{\text{GVW} + \text{tare weight}}{\text{Load weight}}$	3.34	3.53	3.45	3.65

Figure 13-13 Rare use of three push tractors for loading scraper Three 410-engine-flywheel-hp tractors are loading about 30 cybm in 0.42 min. Rock is ripped average-weathered granite. The tractor trio musters about 180,000 lb thrust for forcing the maximum load of rock into the bowl. The load slightly exceeds that of two push tractors, as, although the loads are equal in loose volume, the 50 percent increase in thrust compresses the loose rock to a greater degree.

The cut is a mammoth, 9,770,000-yd³, 4300-ft-long prism of residual, weathered, and solid granites. The saddle cut in the distance will be 235 ft in centerline depth and 1100 ft wide between the tops of the slopes. The work is the grading of a freeway in the Laguna Mountains of southern California.

Figure 13-14 Ideal full-to-overflowing load of twin-engine two-axle-drive scraper. The scraper is being loaded by two heavyweight push tractors. The rock is ripped well-weathered granite. Loading data are:

Estimated load, cybm	27.7
Cycle of push tractors, min:	
Loading	0.60
Exchanging scrapers	0.55
Total time	1.15
Capacity of push tractors in 50-min working hour:	
$\dfrac{50}{1.15} \times 27.7$, cybm	1200
Engine flywheel hp:	
Scraper, two engines	950
Tractors, two	820
Total hp	1770
Estimated total tractive forces available for loading, based on weights of machines when scraper is half loaded and on coefficients of traction, lb	204,000

does not slip the tires when loading but applies as much traction force as possible. On a well-managed job the loading operation is watched closely so as to eliminate this nonproductive waste of energy and costly wear and tear on tires.

The comparison demonstrates also that two matched push tractors supply most of the loading force, 2.8 times the average maximum for the single-engine, single-axle-drive scraper and 1.3 times that for the twin-engine, two-axle-drive machine. This is the dynamic reason for depending on the tractors for their full force and on the scraper for about 50 percent of its possible full force when loading.

Loading time depends on several additional variables, among which are:
1. Size and characteristics of scraper
2. Condition of the loading area
3. Grade (a downgrade reducing the loading time for push-loaded scrapers)
4. Skill of the operator
5. Most importantly, the horsepower of the push tractor(s)

The loading times are based on the following ranges of flywheel horsepower for the push tractor(s). The times may be adjusted for horsepowers outside of the given ranges.

Range of Scraper Capacities, cysm	Flywheel Horsepower Range of Tractor(s)
10–15	200–400
15–25	400–600
25–35	600–900
35–45	900–1200
45–55	1200–1500

13-42   Hauling Excavation

The hauling cycle times for all wheels-tires scrapers may be calculated by use of Table 13-15 for loading, Table 13-16 for two turnings and dumping, and Table 13-12 for hauling and returning.

Bowl factors for average loads of push-loaded wheels-tires scrapers may be obtained from Table 13-7, which gives fill factors, conversion factors, and bowl factors for push-loaded scrapers according to the nature of the rock-earth.

Productions and Estimated Costs Here follow interesting production data and estimated direct job-unit costs for single-engine, single-axle-drive and twin-engine, two-axle-drive scrapers on the same long downhill haul for freeway construction. The rock is ripped weathered granite. Both kinds of scrapers are push-loaded by two tractors, each with 385 engine flywheel hp. The haul is 4200 ft, one way, down an average -12 percent grade. The haul road is hard and well maintained, with a rolling resistance of about 2 percent.

Item	Single-engine, single-axle-drive scraper	Twin-engine, two-axle-drive scraper
Scraper:		
Capacity:		
cysm	37.6	37.6
cylm	41.7	41.7
cybm	32.5	32.5
Total engine flywheel hp	550	950
Weights, lb:		
Tare, working	123,000	145,000
Load at 2900 lb/cylm	121,000	121,000
GVW	244,000	266,000
Hauling cycle, min:		
Loading	0.8	0.6
Hauling, 4200 ft down a -12% grade:		
At average 25 mi/h	1.9	
At average 27 mi/h	...	1.8
Two turnings and dumping	0.8	0.7
Returning, 4200 ft up a $+12\%$ grade:		
At average 9 mi/h	5.4	
At average 11 mi/h	...	4.3
Total cycle time	8.9	7.4
Hourly production, cybm:		
Loads per 50-min working hour:		
50/8.9	5.6	
50/7.4	...	6.8
Cybm hauled hourly:		
5.6×32.5	182	
6.8×32.5	...	221
Costs, 1978:		
Scraper	$337,000	$412,000
Hourly cost of ownership and operation Scrapers (Table 8-3):		
34%×$337	$114.60	
39%×$412	...	$160.70
Operator	15.60	15.80
Total cost	$130.20	$176.50
Direct job unit cost per cybm:		
$130.20/182	$0.72	
$176.50/221	...	$0.80

The hauling cost for the twin-engine, two-axle-drive scraper is 11 percent higher than that for the single-engine machine. Tables 13-13 and 13-14 indicate that the ratio of total work to paywork, that is, the work-cost indicator, is 3.33 for the single-engine scraper and 3.65 for the twin-engine machine. The cost of doing work is generally proportional to the amount of work done. The tables imply that work by the twin-engine scraper is 11 percent more costly and that the estimated direct job unit cost is also 11 percent more than with the single-engine scraper.

TABLE 13-15 Loading Times for Wheels-Tires Scrapers According to Nature of Rock-Earth

Nature of rock-earth	Loading time, min		
	Push-loaded scraper		Self-loaded, single-axle drive scraper
	Single-axle drive	All-axle drive	
Gravel-sand alluvia	0.8	0.7	1.2
Residuals, silts and clays	0.6	0.5	1.0
Average-weathered rock, ripped	0.8	0.7	1.2
Rock, well blasted	1.2	1.1	Not recommended

TABLE 13-16 Two-Turning and Dumping Times for Wheels-Tires Scrapers, min

Job conditions	Drive of scraper	
	Single-axle	All-axle
Good: large cut and fill; hard	0.7	0.6
Average: medium-size cut and fill, hard to soft	1.0	0.9
Poor: small cut and fill; soft	1.3	1.2

When both scrapers can work with equal efficiency, the single-engine machine is generally less costly. However, Figure 13-15 pictures a hauling condition in which only the twin-engine, two-axle-drive scraper can ascend the steep grade to get back to the cut. Physically it is the only scraper for the job.

The logical applications for multiengine, all-wheel-drive scrapers, either laden or unladen, are on steep haul roads and in soft conditions. These circumstances may present both traction and flotation difficulties, generally surmountable by the machines.

Figure 13-15 illustrates a steep grade being successfully climbed on the return haul. Figure 13-16 shows a short, steep wet haul road from cut to fill wherein both traction and

Figure 13-15 Twin-engine, two-axle-drive scrapers opening up a sheer granite cut. The laden scraper to the left is descending the "slot" of a −48 percent grade at 8 mi/h. The bowl is lowered so that the cutting edge can act as a brake, supplementing the action of the retarder of the machine. The water wagon is wetting the return ramp as far as it can ascend, which is about 75 ft ahead of its position. At that point the grade is +22 percent, the limit of gradability on the ramp for a single front-axle-drive machine whether it be a laden water wagon or an empty scraper. The unladen scraper to the right is ascending the +35 percent grade at 3 mi/h. The granular decomposed granite of the ramp has a low coefficient of traction, about 0.40. The huge cut has been partially taken out, as shown to the right of the picture. The original centerline depth of the cut was 235 ft. The work is freeway construction in San Diego County, California.

13-44 Hauling Excavation

Figure 13-16 Grading work for all-wheel-drive scrapers. This is a typical short, steep, wet haul road during springtime in the rough mountains near Clarkia, Idaho. The rock is well-weathered schists, gneisses, and granites, well saturated by snow and rain runoff waters. The twin-engine, two-axle-drive scrapers are suited ideally to the abrupt steep grades in cuts and fills. Traction and flotation hazards will extend well into the summer.

The haul averages about 500 ft in length, one way, beginning with a -15 percent grade. The traction coefficient is about 0.30 and the rolling resistance is about 10 percent. Total cycle time on the illustrated haul is about 5 mins and traveling speeds average about 5 mi/h. The work exemplifies forest highway building in the northwestern area of the United States. (Western Construction, Gene Sheley (ed.), November 1976, front cover.)

flotation of the scraper must be maximum. Figure 13-17 illustrates an unusual circumstance in which rock must be returned up a steep grade from fill to cut because of an error in balancing cut-and-fill quantities.

In these three examples only the large-tire, all-wheel-drive scrapers among the wheels-tires machines can operate efficiently and economically. The crawler-tractor-drawn scraper, if used as an alternative machine, can work efficiently in all three cases but it cannot work economically except in work such as that of Figure 13-16.

Self-Loading Scraper

The self-loading scraper, typical specifications for which are listed in Table 13-17, is suitable for a variety of rock-earth, ranging from soft loams to ripped well-weathered and well-fragmented soft rock. It is not suited to medium-weathered rock with large fragments, as shown in Figure 13-21, because the elevator by its very nature cannot load outsize rocks or boulders.

Figure 13-18 illustrates the self-loading of the scraper and Figure 13-19 shows the construction of the loading elevator. The machines of Figure 13-18 are two of a spread of four scrapers loading and hauling a ripped, well-weathered, and soft sandstone-shale formation. Figure 13-20 illustrates the speedy dumping and spreading of the rock-earth.

The advantages of self-loaded scrapers over push-loaded machines are:

1. The scraper is self-contained, as it both loads and hauls, in contrast to the other machine, which requires a push tractor for loading.

2. Accordingly, in work such as excavating for home or industrial building sites the machines may be worked individually at any location. This selfness is important in scattered excavations of small quantity.

3. It may be used both in rough grading and in fine or finish grading, as is illustrated in Figure 13-22.

The relative disadvantages of self-loaded scrapers are:

1. Greater hourly maintenance because of the wear and tear on the elevator assembly, especially if there are loading difficulties in rock.

2. Inability to load all kinds of rock-earth formations. This ominous limitation can be irritating and costly when variable rock weathering is encountered on the same job. A common example is in a cut made up of both uniformly well-weathered rock, which is workable, and irregularly weathered rock containing embedded boulders of more than head size, which is not workable.

The scraper's functions are operated hydraulically. A hydraulic motor drives the elevator at loading speeds between 183 and 287 ft/min and at reverse speeds between 105 and 236 ft/min. Occasionally the elevator is reversed during the unloading operation. The bowl with its cutting edge is raised and lowered by two double-acting hydraulic cylinders. Unloading is by the sliding floor of the bowl and by the gate at the ejector end. These are activated together by two double-acting hydraulic cylinders.

Occasionally the self-loading scraper is used in a marginal job in a rock-earth formation where the work can be done more properly by a push-loaded scraper. Such a job is illustrated in Figure 13-21. In this widening of a desert freeway, the rock-earth is a mantle of desert alluvium overlying weathered gneisses. The alluvium grades from sand to 2-ft-diameter boulders. The weathered gneisses require ripping. Regularly, push-loaded scrapers would be used in this rather difficult rock, as loading by the elevator is slow and frequently costly because of breakage of the elevator assembly. Loading time averages 1.8 min. Figure 13-21 shows the sidehill cut of alluvium and ripped gneiss and the confined fill area. The fill area is within the surrounding natural slopes with their large surface boulders, typical of the alluvium. The severity of the work for self-loading scrapers is umistakable.

The self-loading scraper is used extensively in fine or finished grading, as the precise cutting facility enables the operator to hold the grade elevation to within 0.1 ft. Figure 13-22 illustrates fine grading in anticipation of the placing of subbase material for freeway building.

Use of Small Machines on Highway Some self-loading scrapers are small enough to work on streets and on highways, especially if unpaved, when payload is lessened to

Figure 13-17 Unique upgrade haulage by twin-engine, two-axle-drive scraper. Upgrade hauls are far less common than downgrade work, except in open-pit mines and quarries, in dam building, and in other special grading projects. In this example, involving the grading of a tract for homes in the San Gabriel Mountains of southern California, a mistake was made in the balancing of cut-and-fill quantities. It became necessary, because of space limitations, to back-haul weathered granite from fill to cut through about a 75-ft vertical lift.

The grade is +28 percent and, because of extreme summer heat and resultant overheating of the engine and torque converter, it is necessary to reduce the payload to about 33 percent of capacity. The upgrade travel speed is 3.9 mi/h. Under these operating conditions a single-engine, single-axle-drive scraper would be inadequate, as the limiting grade would be about +20 percent because of the low traction coefficient of the decomposed granite haul road.

13–46 Hauling Excavation

TABLE 13-17 Representative Specifications for Self-Loading Wheels-Tires Scrapers

Specification	Nominal capacity, cylm, based on SAE rating with crowned load at 1:1 slope		
	11.0	23.0	35.0
Capacity:			
Cubic yards struck measurement	9.5	19.4	31.4
Crowned load with 4:1 slope, cylm	9.9	20.3	32.3
Engine flywheel or brake hp	144	310	475
Transmission:			
Gears forward	5	6	6
Maximum speed:			
mi/h	25	24	32
ft/min	2200	2110	2820
Dimensions, ft:			
Length	31.2	40.8	47.0
Width	9.1	10.7	12.0
Height	9.7	12.5	12.8
Tires, standard	23.5×25 20 PR	29.5×29 28 PR	37.5×33 36 PR
Weights, lb:			
Tare, working	36,000	69,000	104,000
Load, rated	26,000	53,000	84,000
GVW	62,000	122,000	188,000
Speed indicators:			
Laden: $\dfrac{\text{Brake hp}}{\text{GVW, 1000s of lb}}$	2.32	2.54	2.53
Unladen: $\dfrac{\text{Brake hp}}{\text{Tare weight, 1000s of lb}}$	4.00	4.49	4.57
Work-cost indicator:			
$\dfrac{\text{GVW} + \text{tare weight}}{\text{load weight}}$	3.76	3.60	3.48

Figure 13-18 Suitable rock-earth excavation for self-loading scrapers. In this ripped sandstone-shale formation of excavation for home sites, four scrapers are moving 12,600 cybm on a 300-ft one-way haul every 8 h. Loading time is 1.1 min for a 22-cybm load. The machines are rated at 72,000 lb load and they are carrying about 80,000 lb, a common 11 percent overload for earthmoving machinery under good working conditions. The engine flywheel hp is 415 and the scraper's working weight is about 90,000 lb. The actual work-cost indicator is 3.25, as compared with the average 3.61 of Table 13-17 because of overloading. When such overloading on short hauls at low speeds does not result in an appreciable increase in maintenance, the practice is justified, as it affords a lower direct job unit cost.

Scrapers 13-47

Figure 13-19 Integral elevator assembly for loading the self-loading scraper. When loading the bowl, the cutting edge is dropped and the cut rock-earth is raised and forced into the bowl by the ladder conveyor or elevator. In loading, the operator coordinates depth of cut, speed of elevator, and travel speed with the ease or difficulty in digging. This scraper has an average loose measurement capacity of 28 cylm. In ideal soft rock-earth the loading time is as low as 1 min. In difficult operations the time may be as high as 2 min. An average loading time is 1.2 min.

A typical loading action for this machine involves the following pertinent values:
1. Loading speed, 1.0 mi/h (88 ft/min)
2. Cross-sectional area cut by cutting edge of bowl, 6.2 ft^2; width, 10.5 ft; depth, 0.6 ft
3. Travel distance for 23-cybm load, 99 ft
4. Loading time, 1.1 min

Figure 13-20 High-speed dumping and spreading of load by self-loading scraper. The scraper is traveling at 10 mi/h (880 ft/min). It is laying down an even carpet of pulverized sandstone 11 ft wide, 0.7 ft thick, and 100 ft long in 7 s (0.11 min). The dumping and turning on the fill total 18 s (0.30 min). The rapidly advancing ejector is forcing some of the load over the sides of the bowl. The next scraper will continue the carpet across the 450-ft width of the fill, and then water wagons and compactors will follow through with efficient consolidation of the fill. The 1.4-million-yd^3 excavation is in a young sedimentary rock formation of the southern California shoreline.

(a)

(b)

Figure 13-21 Formidable excavation for self-loading scrapers. *(a)* The sidehill cut of the Mojave Desert freeway job. The confined cut is in alluvium and ripped weathered gneiss, both characterized by outsize boulders and fragmented rocks which impose an average loading time of 1.8 min. All four scrapers are loading and hauling tediously and expensively under these unfavorable conditions. *(b)* The enclosed small fill where dumping is impeded by the large boulders of the load and by the lack of maneuverability. The big boulders are visible in the natural slope to the right.

Figure 13-22 Fine or finishing grading by self-loading scraper. This medium-size machine is fine-grading a cut of decomposed or well-weathered granite prior to the placing of subbase for a California freeway. The close control of cutting edge and ladder elevator of the self-contained loading-hauling scraper lends itself to this exacting work. In this 312,000-yd^3 cut some 500 yd^3 of granular and pulverized granite is being drifted to the low areas of the 1600-ft-long cut and to the adjacent fill. The job requires about 8 h of work.

conform to permissible axle loadings. This versatility classifies the machine as a temporary on-the-road scraper. As it is a self-contained machine and serves for both loading and hauling, it may be used in many kinds of small, scattered excavations involving separated cuts and fills.

Figure 13-23 illustrates such a machine of 9.5-cylm nominal SAE rated heaped capacity and 23,750-lb rated load capacity. Its engine flywheel hp is 152, and it is 32.9 ft long, 8.0 ft wide, and 8.2 ft high. Its front tires are 14.00×17.5 and its driving and scraper tires are 23.5×25. The dimensions are suitable for on-the-road hauling. However, the maximum axle loading of 27,890 lb on the scraper axle precludes use for on-the-road work when carrying the rated load. Data for loads when the payload is reduced by 24 percent to get a temporary permit for on-the-road haulage with a maximum axle loading of 24,000 lb are tabulated below. The 24,000-lb load is 20 percent in excess of the average permissible 20,000-lb axle load, but often it is possible to secure a temporary permit for such loading. Although the possibly permissible payload is 24 percent less than the rated payload, it is often feasible to work under these circumstances, as the total direct job unit cost if generally less than that for a loader and strictly legal hauling machines.

Comparison of axle loadings for a wheels-tires self-loading scaper according to rated off-the-road payload, legal on-the-road payload, and the possibly permissible on-the-road payload.

Item	Rated payload, lb, off-the-road	Legal payload, lb, on-the-road	Possibly permissible payload, lb, on-the-road
Weights of scraper, lb:			
Unladen:			
Front axle	6,940	6,940	6,940
Driving axle	15,430	15,430	15,430
Scraper axle	11,980	11,980	11,980
Tare weight	34,350	34,350	34,350
Laden:			
Front axle	7,640	7,300	7,480
Driving axle	22,570	19,020	20,810
Scraper axle	27,890	20,000	24,000
GVW	58,100	46,320	52,290
Payload:			
lb	23,750	11,970	17,940
cybm (at 3400 lb/cybm)	7.0	3.5	5.3
Percentage of rated payload	100	50	76

Hauling Cycle Times The total time consists of the times for loading, for two turnings and dumping, and for hauling and returning, as discussed previously. These operations are shown, and data for them are given, in the following figures and tables.

Figures 13-18, 13-21, 13-22, and 13-23 illustrate loading, and Table 13-15 gives average loading times according to the nature of the rock-earth. Figure 13-20 illustrates the dumping and spreading of a load. Table 13-16 gives average times for the two turnings and dumpings according to job conditions.

Information on hauling and returning times is provided by the following:

Table 13-12 and its accompanying explanations provide data for calculating hauling and returning times for the wheels-tires scraper.

Table 13-4 and Figure 13-11 with its explanations provide information for calculating gradability,

Table 13-10 gives coefficients of traction.

Table 13-11 provides rolling resistance factors.

The method of calculations of times is identical to that of the example of the bottom-dump hauler, used in explaining the method which includes Table 13-12.

Average Loads Table 13-7 gives fill factors, conversion factors, and bowl factors by which loads may be calculated according to the nature of the rock-earth.

13-50 Hauling Excavation

Figure 13-23 Small, self-loading scraper sometimes adaptable to on-the-road haulage. This off-the-road machine with a rated load capacity of 23,750 lb has a GVW of 58,100 lb. The rated load is equivalent to 7.0 cybm of weathered rock-earth. The maximum load of 27,890 lb is on the scraper axle, in excess of the average legal load of 20,000 lb. Such a legal axle load allows only 3.5 cybm, or only 50 percent of the rated load. Sometimes temporary permission is given to exceed the legal load by some 20 percent if conditions warrant the exemption. If such permission is granted, the scraper can carry a payload of 5.3 cybm, or 76 percent of rated load. Often this lessening of payload is economically feasible for the advantageous on-the-road work involving both loading and hauling by the individual scraper. *(John Deere Co.)*

Productions and Costs for Different Hauls by Identical Machines Scrapers identical for this comparison of short and long hauls are illustrated in Figure 13-18, showing a short (300-ft) one-way haul, and Figure 13-20, showing a long (2100-ft) one-way haul. The rock-earth is ripped well-weathered siltstones, sandstones, and shales of the Linda Vista marine formation of San Diego County, California. The work is by the same contractor, well known for maximum job efficiency.

Important scraper specifications, bearing on the productions and costs, are:

Nominal capacity, cylm	32.0
Estimated capacity, cylm	28.5
Actual payload, cybm	22.3
Engine flywheel or brake hp	415
Weights, lb:	
Tare, working	90,000
Payload, 22.3 cybm at 3600 lb/cybm	80,000
GVW	170,000
Work-cost indicator, $\dfrac{\text{GVW} + \text{tare weight}}{\text{load weight}}$	3.25

Hauling conditions in the cut, on the haul road, and in the fill are excellent, the average rolling resistance being about 3 percent. Tabular data for productions and costs for short and long hauls are given below.

Item	Short haul	Long haul
Haul:		
Length one way, ft	300	2100
Grade, %	−6	−6
Difference in elevation between cut and fill, ft	−18	−126
Hauling cycle, min:		
Loading	1.1	1.2
Hauling:		
300 ft at 9 mi/h	0.4	
2100 ft at 22 mi/h	...	1.1
Dumping and turning	0.6	0.3
Returning:		
300 ft at 5 mi/h	0.7	
2100 ft at 17 mi/h	...	1.4

Turning	0.5	0.3
Total time	3.3	4.3
Production per 50-min working hour:		
Loads hauled hourly:		
50/3.3	15.2	
50/4.3	...	11.6
Cybm hauled hourly:		
15.2×22.3	339	
11.6×22.3	...	259
Costs, 1978:		
Scraper	$258,000	$258,000
Hourly cost of ownership and operation:		
Scraper (Table 8-3), 36% × $258	$ 92.90	$ 92.90
Operator	15.60	15.60
Total	$108.50	$108.50
Direct job unit cost per cybm:		
$108.50/339	$0.32	
$108.50/259	...	$0.42

There is an interesting contrast in loading and hauling unit cost per 100-ft station. For a short haul it is $0.107 and for a long haul it is only $0.020. Years ago overhaul was reckoned beyond the freehaul distance of 300 ft. In this instance, the overhaul costs ($0.42−$0.32)/(21−3), or $0.0055 per station cybm.

Summary

The choice between push-loaded and self-loaded scrapers depends chiefly on the nature of the rock-earth and the kind of job. Soft to medium formations favor the self-loading scraper and medium to hard rock-earth favors the push-loaded machine.

When the job requires excavations in several locations at the same time, as in cuts for confined home sites, the self-sufficient self-loading scraper is more economical. When mass excavation is confined to one cut, the push-loaded scraper is preferred. Obviously, the economical working conditions may overlap, and it is apparent that careful analysis of the job and close estimation of costs are requisite to a well-thought-out selection of the kind of scraper.

When conditions are favorable for both machines, the direct job unit cost for loading and hauling by push tractors and push-loaded scrapers or by self-loaded scrapers is about the same. In such circumstances the earthmover chooses the machine in keeping with his average work. If he excavates soft friable rocks for home and plant sites, he should choose self-loading scrapers. If his work varies from building-site excavations to highways in hard rock formations, he should choose push tractors and push-loaded scrapers. Many earthmovers have both push-loaded scraper spreads and self-loading scraper complements so as to fit the machines precisely to the job at hand. This dexterity allows great flexibility in arriving at the lowest possible estimated unit cost for the excavation.

BOTTOM-DUMP HAULERS

Types and Functions

There are two kinds of bottom-dump haulers, a heavyweight machine suitable only for off-the-road haulage and a medium-weight hauler suitable for both off-the-road and on-the-road work.

An off-the-road hauler is illustrated in Figure 13-24. Its GVW is 111,000 lb and its maximum axle load is 49,000 lb. Specifications for these machines are given in Table 13-18.

A medium-weight combination off-the-road and on-the-road nine-axle hauler is shown in Figure 13-25. The GVW is 190,000 lb and the axle loading for off-the-road work is adjusted to the sturdiness of the machine and to the severity of the work. In this instance it is 22,500 lb. When hauling on the road, the maximum axle loading is reduced to an average of 20,000 lb, with a resulting reduction in payload from 34 to 30 cybm. The specifications for this machine are given in Table 13-19.

Figure 13-24 Bottom-dump haulers in borrow pit for dam construction. This off-the-road hauler weighs 43,000 lb and its engine flywheel hp is 250. Its struck measurement capacity is 21.0 cysm and the average load of conglomerate is 18 cybm. The loading shovel has a 4½-yd³ dipper capacity and fills the hauler with five dippers in 2.0 min. This ratio of 1:5 for shovel dipper and hauler capacities gives a good balance of machines, as acceptable ratios vary from 1:3 to 1:6. The desirable ratios are 1:4 and 1:5. Another earmark of efficiency is the waiting hauler, which ensures high shovel performances. The job is the building of Casitas Dam in Ventura County, southern California.

Figure 13-25 Double bottom-dump haulers being loaded for long off-the-road haulage. These machines are part of a fleet of four units hauling 600 cybm of well-weathered granite hourly. The haul for a San Diego County, California, freeway is 7000 ft one way over a well-maintained, hard, gently undulating haul road. The hauling cycle is 11.1 min, and the average travel speeds are 20 mi/h when laden and 29 mi/h when unladen. Engine flywheel hp is 375. Tare weight is 60,000 lb, load weight is 130,000 lb, and GVW is 190,000 lb. The work-cost indicator is 1.92, a remarkably low figure for a rock-earth hauler.

The 5.8-cysm loader fills the double bottoms with eight buckets in 3.0 min, the unusually fast loading time being made possible by the assisting bulldozer. The payload is 34 cybm. Under these favorable loading and hauling conditions, the selection of these well-built, nine-axle rock haulers is a demonstrably good choice of available machines.

TABLE 13-18 Representative Specifications for Off-the-Road Bottom-Dump Haulers with Two or Three Axles

Specification	Nominal capacity, cysm			
	22.0	27.0	42.0	60.0
Capacity:				
Crowned load with 4:1 slope, cylm	24.6	30.2	47.0	67.2
Tons, rated	33	41	63	100
Engine flywheel or brake hp	330	415	415	600
Transmission:				
Gears forward	8	8	9	9
Maximum speed:				
mi/h	32	34	42	42
ft/min	2820	2990	3700	3700
Number of axles	2	2	3	3
Dimensions, ft:				
Length	42.4	44.8	53.2	61.1
Width	11.7	12.1	13.5	16.7
Height	10.0	11.3	13.0	13.7
Height, loading	9.4	9.7	12.2	13.3
Tires, drive and trailer, standard	29.5×29 28 PR	29.5×35 34 PR	18.0×33 24 PR	21.0×35 32 PR
Weights, lb:				
Tare, working	58,000	65,000	90,000	132,000
Load, rated	66,000	82,000	126,000	200,000
GVW	124,000	147,000	216,000	332,000
Speed indicators:				
Laden: $\dfrac{\text{Brake hp}}{\text{GVW, 1000s of lb}}$	2.66	2.82	1.92	1.81
Unladen: $\dfrac{\text{Brake hp}}{\text{Tare, weight, 1000s of lb}}$	5.69	6.38	4.61	4.55
Work-cost indicator:				
$\dfrac{\text{GVW+tare weight}}{\text{Load weight}}$	2.76	2.58	2.42	2.32

Another medium-weight combination five-axle hauler is shown in Figure 13-26. Its GVW is 90,000 lb and its maximum axle loading is 20,000 lb for on-the-road haulage. When hauling off-the-road, the axle loading is increased to about 22,500 lb or possibly to a greater loading so as to increase payload. Specifications for this machine are given in Table 13-19.

In off-the-road use of the combination off-the-road and on-the-road haulers, the payloads are adjusted to the ruggedness of the machines and to the severity of the hauling conditions.

The functions of a bottom-dump hauler are simple, as the load is dropped by gravity. The doors of the bottom of the body may be hinged along the sides of the body, as illustrated in Figures 13-24 and 13-27, or they may be of the clamshell type, as shown in Figures 13-25 and 13-26. They may be actuated hydraulically or pneumatically. The doors of Figure 13-25 are air-actuated and those of Figure 13-26 are controlled by hydraulic cylinders. In both cases the rate of dumping may be managed closely so as to spread and compact the discharged material quickly and economically.

Comparison with Scrapers as Fundamental Haulers

Bottom-dump haulers are the most efficient and most economical hauling machines when used in rock-earth suitable to their characteristics and when compared with other machines on the basis of pure hauling. By "pure hauling" is meant the work of the machines when in motion, hauling and returning, making two turnings, and dumping. Their superior dynamics and economics in pure hauling, as compared with scrapers which are able to haul the same kinds of rock-earth, are revealed in the following tabulation.

13-54 Hauling Excavation

Figure 13-26 Double bottom-dump haulers traveling at high speeds on expressway. The haul of 4.9 mi one way, is for freeway excavation. The freeway is parallel and close to the expressway, and the well-weathered granite is from a large cut at one end of the freeway under construction. Only 2000 ft of the total 25,900 ft of haul is off-the-road haulage. In spite of six delaying traffic lights, the hauling cycle time is only 11.1 min. Laden travel speeds are 19 mi/h average and 35 mi/h maximum and unladen speeds are 23 mi/h average and 40 mi/h maximum. The payload is 16 cybm and the loose load is about 20 cylm, visibly reduced in the sideboarded bodies. The engine flywheel hp is 275, the tare weight is 33,000 lb, the load weight is 57,000 lb, and GVW is 90,000 lb. The work-cost indicator is 2.16, low in keeping with the small ratio of payload to tare weight for these combination on-the-road–off-the-road bottom-dump haulers.

The expressway leads into Escondido, California, and the haulers pass through a moderately congested area, where the six traffic lights are located.

Figure 13-27 Dumping load of bottom-dump hauler in a long narrow windrow for a dam fill. The 67.5-cysm hauler is laying down a windrow of semipervious material for earth-fill dam building. The windrow is 170 ft long, 8 ft wide, and 3 ft high, making for ideal leveling and compacting. Dumping time is 6 s at a dumping speed of about 10 mi/h. The machine's load is about 62 cybm of desert residuals. The engine flywheel hp is 609 and the GVW is 340,000 lb. The important work-cost indicator of this businesslike hauler is 2.13, which is expressed in a low direct job unit cost for the uphill haulage characteristic of an earth-fill dam.

The total rock-earth excavation for Perris Dam, Riverside County, California, is approximately 22 million yd^3. Crest length is 11,400 ft and maximum crest height above the gently sloping valley is 110 ft. *(General Motors Corp.)*

TABLE 13-19 Representative Specifications for Combination Off-the-Road and On-the-Road Bottom-Dump Haulers Comprising Multiaxle Tractor–Semitrailer–Full Trailer

Specification	Five-axle hauler		Nine-axle hauler	
	On-the-road use	Off-the-road use	On-the-road use	Off-the-road use
Capacity:				
Nominal, cysm	20.0	20.0	40.0	40.0
Crowned load with 4:1 slope, cylm	22.8	22.8	44.8	44.8
Weight, lb:				
Legal on-the-road, with maximum 20,000-lb axle loading:				
Tare, working	31,000		60,000	
Load	59,000		112,000	
GVW	90,000		172,000	
Off-the-road, with maximum 22,000-lb axle loading:				
Tare, working		31,000		60,000
Load		67,000		128,000
GVW		98,000		188,000
Capacity, cybm, based on 3800 lb/cybm	15.5	17.6	29.5	33.4
Engine, flywheel or brake hp	250	250	375	375
Transmission:				
Gears forward	15	15	15	15
Maximum speed:				
mi/h	60	60	50	50
ft/min	4400	4400	5280	5280
Dimensions, ft:				
Length	61.6	61.6	81.9	81.9
Width	8.0	8.0	8.9	8.9
Height	8.5	8.5	9.7	9.7
Height, loading	8.1	8.1	9.7	9.7
Tires:				
Tractor, singles front and duals rear	11.00×22 14PR all tires	11.00×22 14PR all tires	11.00×22 14PR all tires	11.00×22 14PR all tires
Trailers, all duals				
Speed indicators:				
Laden: $\dfrac{\text{Brake hp}}{\text{GVW, 1000s of lb}}$	2.78	2.55	2.18	1.99
Unladen: $\dfrac{\text{Brake hp}}{\text{Tare weight, 1000s of lb}}$	8.06	8.06	6.25	6.25
Work-cost indicator:				
$\dfrac{\text{GVW}+\text{tare weight}}{\text{Load weight}}$	2.05	1.92	2.07	1.94

Item	Push-loaded scraper (Table 13-13)	Bottom-dump hauler (Table 13-18)
Nominal capacity, cysm	21.0	22.0
Engine flywheel or brake hp	415	330
Weights, lb:		
Tare, working	81,000	58,000
Load, rated	72,000	66,000
GVW	153,000	124,000
Speed indicators		
Laden: $\dfrac{\text{Brake hp}}{\text{GVW, 1000s of lb}}$	2.71 (100%)	2.66 (98%)
Unladen: $\dfrac{\text{Brake hp}}{\text{Tare weight, 1000s of lb}}$	5.12 (100%)	5.69 (111%)
Work-cost indicator: $\dfrac{\text{GVW}+\text{tare weight}}{\text{Load weight}}$	3.25 (100%)	2.76 (85%)

Hauling Excavation

One observes that the bottom-dump hauler has an average 4 percent faster speed indicator and a 15 percent lower work-cost indicator when compared with the scraper of about the same capacity. Another index of relative efficiency and economy is the ratio of rated load to tare weight, 89 percent for the scraper and 114 percent for the bottom-dump hauler.

Bottom-dump haulers are used extensively in many open-pit mines, both for the removal of overburden and for the hauling of metal-bearing and non-metal-bearing materials such as copper ore and coal, respectively. Frequently the bodies and bottom-dump doors are designed especially for shock loads, abrasive materials, and outsize rocks and ores. The machines haul everything from light diatomite, weighing 900 lb/cylm, to heavy hematite, weighing 4300 lb/cylm.

Figures 13-28 and 13-29 illustrate machines working in the open-pit coal mines. The bituminous coal weighs 1370 lb/cylm. The comparatively light coal of uniform gradation up to the 8-in maximum size makes for ideal and economical haulage by bottom-dump machines.

Table 13-20 gives constant times for the two turnings and dumping of bottom-dump haulers for use in calculating hauling cycles, and Table 13-21 gives body factors for calculating payloads in bottom-dump haulers in order to figure productions of the machines.

The use of Table 13-21 may be illustrated by the example of the bottom-dump hauler shown in Figure 13-24, which has a capacity of 21.0 cysm and is hauling a load of conglomerate residuals or weathered rock-earth.

1. The loose load is 1.12×21.0, or 23.5 cylm.
2. The bank-measurement load is 0.78×23.5, or 18.3 cybm.
3. Alternatively and simply, the bank measurement load is 0.87 (body factor)×21.0, or 18.3 cybm.

Hauling Cycle Times and Average Loads

As with scrapers the hauling cycle includes loading, two turnings and dumping, and hauling and returning. The machines are top-loaded by any of the following loaders: tractor-bulldozers, by use of loading chute; bucket-type loaders; shovels; draglines; trenchers; belt loaders; and wheel-bucket excavators. Chapter 12 gives data on these loading machines, from which loading times may be calculated. Figure 12-55 illustrates a popular loader for bottom-dump haulers, the movable belt loader.

The estimated times for two turnings and dumping, which are constant or fixed time elements in the hauling cycle, are given in Table 13-20. Tables 13-3, 13-10, 13-11, and 13-12 provide data for calculations for hauling and returning times of wheels-tires-mounted bottom-dump haulers. The estimated loads of bottom-dump haulers in cubic yards loose measurement and cubic yards bank measurement may be calculated by use of Table 13-21.

Productions and Costs

EXAMPLE 1: Actual hauling cycle and estimated production and unit cost for the double bottom-dump hauler of Figure 13-25 are presented below.

The criteria are:
 1. Material: rock-earth, ripped well-weathered granite
 2. Loader: bucket type, nominal bucket rating, 6.5 cylm
 3. Haul road: off-the-road; level with gentle undulating grades; well maintained; hard surface; 7000 ft, one way
 4. Haulers:
  a. Tractor; tandem-drive rear axle; 375 engine flywheel hp
  b. Semitrailer and full trailer, two and four axles, respectively
  c. Total capacity, 40.0 cysm; estimated payload, 34 cybm; tare working weight, of complete machine, 60,000 lb

Figure 13-28 Medium-capacity bottom-dump coal haulers in pit of Illinois strip coal mine.

Salient specifications:
  Capacity:
    Cubic yards struck measurement   40
    Cubic yards loose measurement, crowned load with 4:1 slope   45
    Tons, rated   30
  Engine flywheel or brake horsepower   250
  Maximum travel speed, 35 mi/h   3080
  Tires, singles, tractor-driving and trailer 18.0×25—24PR
  Weights, lb:
    Tare, working   41,000
    Load, rated   60,000
    GVW   101,000
  Speed indicators:

    Laden: $\dfrac{\text{Brake hp}}{\text{GVW, 1000s of lb}}$ 2.48

    Unladen: $\dfrac{\text{Brake hp}}{\text{Tare weight, 1000s of lb}}$ 6.10

  Work-cost indicator: $\dfrac{\text{GVW} + \text{tare weight}}{\text{load weight}}$ 2.53

The average payload is 33 tons, 10 percent in excess of the rated load, which corresponds to typical practice when haul-road conditions are good. The work-cost indicator is low, 2.53, which is usual for these well-designed haulers.

TABLE 13-20 Two-Turning and Dumping Times for Bottom-Dump Haulers

	Times, min, for:	
Job conditions	Single trailer, 2 or 3 axles	Two trailers, multiple axles
Good: Hard surface; large cut and fill.	0.4	0.6
Average: Hard to soft surface; medium size cut and fill.	0.7	0.9
Poor: Soft surface; small cut and fill.	1.0	1.2

13–58 Hauling Excavation

Figure 13-29 Large-capacity bottom-dump coal hauler in pit of Illinois strip coal mine.

Salient specifications:
 Capacity:
  Cubic yards struck measurement   123
  Cubic yards loose measurement, crowned load with 4:1 slope   138
  Tons, rated   100
 Engine flywheel or brake horsepower   550
 Maximum travel speed (43 mi/h), ft/min   3780
 Tires, duals, tractor-driving and trailer 21.0×49—36PR
 Weights, lb
  Tare, working   130,000
  Load, rated   200,000
  GVW   330,000
 Speed indicators:

 Laden: $\dfrac{\text{Brake hp}}{\text{GVW, 1000s of lb}}$ 1.67

 Unladen: $\dfrac{\text{Brake hp}}{\text{Tare weight 1000s of lb}}$ 4.23

 Work-cost indicator: $\dfrac{\text{GVW} + \text{tare weight}}{\text{load weight}}$ 2.30

The average payload is 100 tons, equaling the rated load. The speed indicators show that the big hauler is less speedy than the medium hauler of Figure 13-28. On the other hand, the work-cost factor of the big machine is 9 percent less than that of the medium hauler, suggesting a correspondingly lower direct job-unit cost for the coal haulage. *(Athey Products Corp.)*

TABLE 13-21 Body Factors for Average Loads of Bottom-Dump Haulers, Based on Body Capacity, cysm

Nature of rock-earth	Fill factor $\left(\dfrac{\text{loose measurement}}{\text{body capacity}}\right)$	Conversion factor $\left(\dfrac{\text{bank measurement}}{\text{loose measurement}}\right)$	Body factor $\left(\dfrac{\text{bank measurement}}{\text{body capacity}}\right)$
Sand-gravel alluvia	1.12	0.88	0.98
Rock-earth, weathered	1.12	0.78	0.87
Rock, well blasted	1.12	0.67	0.75

NOTE: Hardness and maximum size of well-blasted rock must be compatible with ruggedness, area of bottom discharge opening, and rear-axle clearance of the bottom-dump hauler. The bottom-dump hauler is not recommended for poorly blasted rock or for outsize rocks and boulders of weathered formations.

Actual hauling cycle, min:
Loading 8 buckets, 36 cybm	3.0
Hauling 7000 ft at average 20 mi/h	4.0
Dumping and turning	0.7
Returning 7000 ft at average 29 mi/h	2.8
Turning	0.5
Total net cycle time	11.1

Estimated hourly production per 50-min working hour, cybm:
Loads hauled hourly, 50/11.1	4.50
Cybm hauled hourly, 4.50×36	162

Direct job unit cost, 1978:
Cost of hauler	$144,000
Hourly cost of ownership and operation:	
Bottom-dump hauler (Table 8-3), 37% of $144	$53.30
Operator	11.80
Total hourly cost	$65.10
Direct job unit cost per cybm, $65.10/162	$0.40

The unit cost is at the rate of $0.057 per cybm-station or $0.30 per cybm-mile.

EXAMPLE 2: The actual hauling cycle and estimated production and unit cost calculations are given below for the double bottom-hauler of Figure 13-26.

The criteria are:
1. Material, rock-earth: ripped, well-weathered granite
2. Loader, bucket type: nominal bucket rating, 6.5 cylm
3. Haul road: off-the-road and on-the-road; level, with well-maintained, hard off-the-road surface

Off-the-road, one way	2,000 ft
On-the-road, six traffic lights on streets, one way	23,900 ft
Total length, one way, 4.9 mi	25,900 ft

4. Haulers

Tractor: single-axle drive, 275 engine flywheel hp. Semitrailer and full trailer: one and two axles, respectively; total capacity 24.0 cysm; estimated payload, 16 cybm or 57,000 lb. Working tare weight of complete machine, 33,000 lb. GVW, 90,000 lb, conforming to maximum 20,000-lb axle load on the highway.

Actual hauling cycle, min:
Loading 4 buckets, 16 cybm	1.8
Hauling 25,900 ft at average 19 mi/h and maximum 35 mi/h	15.5
Dumping and turning	1.3
Returning 25,900 ft at average 23 mi/h and maximum 40 mi/h	13.0
Turning	0.3
Total net cycle time	31.9

NOTE: In this hauling cycle the average speeds are 65 percent of the maximum speeds. On all hauls, off-the-road and on-the-road, one can estimate closely the average hauling speeds, laden and unladen, by multiplying maximum speeds by ⅔.

Estimated hourly production per 50-min working hour, cybm:
Loads hauled hourly, 50/31.9	1.57
Cybm hauled hourly, 1.57×16	25

Direct job unit cost, 1978:
Cost of hauler	$84,000
Hourly cost of ownership and operation:	
Bottom-dump hauler (Table 8-3), 37% × $84	$31.10
Operator	11.80
Total hourly cost	$42.90
Direct job unit cost per cybm, $42.90/25	$1.72

The unit cost is at a rate of $0.35 per cybm-mi. This cost is 12 percent more than that for the off-the-road haulage by a similar machine carrying more than twice the payload, namely the machine of Figure 13-25. There are two possible explanations: (1) The larger of two similar haulers generally offers lower unit cost. (2) In the case of the smaller machine, its hauling cycle time is lengthened by stopping at the 12 traffic lights encountered when hauling and returning.

The following example is of particular interest, as it compares the performances and and costs for two bottom-dump haulers, typified by those of Figures 13-25 and 13-27, for

13–60 Hauling Excavation

a long haul job. The analysis is based on many of the machinery concepts of this text and it demonstrates the tedious but necessary examination involved in selecting the most economical and practical machine for a particular task.

The analysis contained in the letter reproduced below is complete except for the machinery specifications accompanying the complete presentation. The two bottom-dump haulers are described adequately in the text. The 992B is a bucket-type loader of nominal 10-cylm capacity. The D9H is a crawler-tractor equipped with a U-shaped bulldozer and having 410 engine flywheel hp. Its work is to bulldoze the moderately cemented alluvium to the loader so as to get maximum production.

EXAMPLE 3:

HORACE K. CHURCH, C.E.
EARTHMOVING CONSULTANT
3434 DON PORFIRIO DRIVE
RANCHO CARLSBAD
CARLSBAD, CALIFORNIA 92008

November 27, 1978

Mr. Carl G. Malcomb
Malcomb Constructors Incorporated
Suite 202, 18445 Burbank Boulevard
Tarzana, California

Dear Carl:

Herewith is the production and cost analysis for the Mission Oaks Tract excavation job. Yardage of the job is about 225,000 yd.³ Material is the San Pedro formation, made up of sands, gravels, and clays, moderately cemented. The borrow pit is a terrace deposit located at elevation 335 ft and the fill is in an adjoining flood plain at elevation 115 ft. The downgrade haul is 7800 ft, one way, on an average -2.8% grade. There are three segments of the haul road:

1. From center of pit to Mission Oaks Boulevard, -4%	1200 ft
2. From M.O.B. to Santa Rosa Road, -5.5%	3100
3. From S.R.R. to end of fill, level	3500
Total length, one way, average -2.8%	7800 ft

The salient specifications, estimated productions, and estimated direct job unit costs for the two considered bottom-dump haulers are tabulated hereinafter. The estimated hourly costs of ownership and operation are conservative, as they contain an annual 7 percent replacement cost escalation.

	Peterbuilt 375 hp tractor, 2 Fruehauf DES-M2-M20/28 bottom-dump haulers	Caterpillar 660B tractor, Athey PW 660B bottom-dump hauler
Specifications		
Capacity		
Struck measurement, cysm	40.0	55.0
With 13-in sideboards		66.6
Crowned load measurement		
4:1 slope, cylm	44.8	74.6
Engine or brake flywheel hp	375	550
Transmission		
Gears forward	15	8
Maximum speed		
mi/h	60	43
ft/min	5280	3780
Number of axles	9	3

Bottom-Dump Haulers

	Peterbuilt Fruehauf	Caterpillar Athey
Loading height, ft	9.7	13.7
Weights, lb		
Tare, working	60,000	126,000
Load		
44.8 cylm at 2700 lb	121,000	
74.6 cylm at 2700 lb		201,000
GVW	181,000	327,000
Travel speed indicators:		
Laden, hp/GVW in 1000s lb	2.07	1.68
Unladen, hp/tare weight in 1000s lb	6.25	4.37
Work-cost indicator:		
$\dfrac{\text{GVW}+\text{tare weight}}{\text{Load weight}}$	1.99	2.25

Notes: 1. Loader 992B dumping clearance is 15.1 ft, allowing but 1.4 ft for crowning out load in sideboarded PW 660B. Marginal!
2. Heaviest loaded axle of 660B-PW660B is trailer axle with load of about 49% of GVW or 160,000 lb. Excessive!
3. Travel speed indicators infer Peterbuilt-Fruehauf travels much faster than Caterpillar-Athey, laden and unladen on average haul road.
4. Work-cost indicator implies that Peterbuilt-Fruehauf does 12% less total work to accomplish a given amount of pay work in hauling the load on an average haul road.

	Peterbuilt Fruehauf	Caterpillar Athey
Estimated Productions		
Horsepower at driving wheels,		
76% × flywheel hp	285	418
Resistances, % of weight		
Rolling	4	4
Grade, average 2.8%	3	3
Total upgrade	7	7
Necessary tractive force upgrade empty, lb		
7% × 60,000	4200	
7% × 126,000		8800
Average travel speed returning upgrade		
285 × 33,000/4200, ft/min	2240	
418 × 33,000/8800, ft/min		1580
mi/h	25	14
Hauling cycle, min		
Loading 6 buckets from 992B	2.3	
8 buckets from 992B		4.0
Hauling 7800 ft at 30 mi/h	3.0	3.0
Dumping and turning	.7	.5
Returning 7800 ft at 25 mi/h	3.5	
14 mi/h		4.9
Turning	.5	.4
Accelerations, decelerations, and stopping		
average 2 stops at streets	1.0	1.0
Total cycle time	11.0	13.8
Hourly production per 50-min working hour		
Loads hauled hourly, 50/11.0	4.55	
50/13.8		3.62
Payload, 33% swell factor, cybm		
44.8/1.33	33.7	
74.6/1.33		56.1
Cybm hauled hourly per hauling unit		
4.55 × 33.7	153	
3.62 × 56.1		203
Hourly production of 992B loader, cybm		
50/2.3 × 33.7	733	
50/4.0 × 56.1		701
Hauler requirements for one loader		
733/153 = 4.79	5	
701/203 = 3.45		4
Suggested requirements haulers for two loaders	11	9
9-h shift production for 2 992B		

Hauling Excavation

loaders, cybm
2×9×733   13,200
2×9×701   12,600

Estimated direct job unit costs
Machine costs
  Peterbuilt Fruehauf $135,000
  Caterpillar Athey $326,000
  Cat 992B loader, general-purpose
    bucket   439,000   439,000
  Cat D9H, U blade   242,000   242,000

Hourly cost of ownership operation
  Peterbuilt Fruehauf
    Machine, 37% × $135   50.00
    Operator   11.00
    Total   61.00
  Caterpillar Athey
    Machine, 33% × $326   107.60
    Operator   14.30
    Total   121.90
  Cat 992B loader
    Machine, 34% × $439   149.30   149.30
    Operator   14.40   14.40
    Total   163.70   163.70
  Cat D9H, U-blade
    Machine, 34% × $242   82.30   82.30
    Operator   14.30   14.30
    Total   96.60   96.60

Direct job unit costs
  Hauling
    11 Peterbuilt Fruehauf @ $61.00/h =
    $671 $671/1466 $0.46
    9 Caterpillar Athey @ $121.90/h,
    = $1097; $1097/1402 $0.78
  Loading
    2 Cat 992B @ $163.70/h = $327
    $327/1466   0.22
    $327/1402   0.23
  Bulldozing to loader, if necessary
    2 Cat D9H @ $96.60/h = $193
    $193/1466   0.13
    $193/1402   0.14

Total for bulldozing to loader, loading,
and hauling $0.81 $1.15

REMARKS:

1. There is a substantial saving by the use of the Peterbuilt Fruehauf type of off-the-road tractor hauling two-bottom dump haulers. This saving is in keeping with the use of these machines on long hauls with well-maintained haul roads.

2. There are two conditions which, physically, discourage the use of Caterpillar Athey machines: the high loading height of the Athey wagon and the extreme axle loading of the trailer axle.

3. The direct job unit costs are based on ownership of the machines and they allow for about a 10 percent reserve or extra hauling unit. The cost of the operator is an estimate of current wages with fringe benefits.

If there are any questions about this report, please get in touch with me.

Very truly yours,
Horace K. Church

The hauling cost for the Peterbuilt Fruehauf machine, tractor with two trailers, is $0.46/cybm according to the above analysis, and the figure for the Caterpillar Athey hauler, one tractor with one trailer, is $0.78/cybm. The saving in favor of the Peterbuilt Fruehauf machine is 41 percent. The higher travel-speed indicators and lower work-cost indicator for the more economical hauler forecast such a saving. In appraising this unusual saving by one of two similar machines, one must bear in mind that: (1) the hauling conditions are ideal for an off-the-road–on-the-road hauler such as the Peterbuilt Fruehauf machine, and (2) the Caterpillar Athey hauler is one of the largest and heaviest of off-the-road machines and that it is built ruggedly for rough service.

Summary

Because of the inherent low ratio of tare weight to load weight and low work-cost indicator, bottom-dump haulers are economical haulers when rock-earth is favorable and when length of haul is suitable. Simplicity of design makes for low maintenance. Maneuverability is good. As they are top-loaded, they receive a generous crowned load.

As compared with scrapers for the same rock-earth excavation, they show savings on hauls above a range of 2500 to 5000 ft, one way. Where combined loading and hauling unit costs are equal, the distance depends on whether the haul is level, upgrade, or downgrade. The bottom-dump hauler is essentially a machine for long haul of rock-earth suitable for dumping through the bottom of the body.

REAR-DUMP HAULERS

Types and Applications

These machines, sometimes known as end dumps, are the workhorses of rock haulage. They handle all kinds of excavation under just about all varieties of hauling conditions. Their enviable all-purpose reputation in the United States started with the 3½-ton, two-axle Bulldog Mack, and continues at present, about 50 yr later, with the 2000-hp, 240-ton, two-axle rear-dump hauler of Wabco.

Rear-dump haulers are built in different configurations for both off-the-road and on-the-road haulage. Basic designs are:

1. The two-axle truck, sometimes called load-on-back or, in small sizes, bobtail truck, may be rear-axle or two-axle drive.

2. The three-axle truck, sometimes called load-on-back, tandem-drive, or ten-wheeler, may be driven by the tandem rear axles or by all three axles.

3. The two- or three-axle tractor, similar to the chassis of the two- and three-axle trucks, is used to pull a semitrailer of one or two rear axles.

Some machines are built expressly for off-the-road haulage. Another group is intended for a combination of off-the-road work with maximum load and on-the-road work in which the load conforms to the legal axle load, numbers and arrangements of axles, and length.

Figure 13-30 illustrates a three-axle, tandem-drive, 75-ton capacity machine, which was selected for traction and flotation characteristics under the wet and icy conditions in an open-pit anthracite coal mine. Figure 13-31 illustrates a five-axle on-the-road rock hauler made up of a three-axle tractor and a frameless rear-dump trailer with two axles.

There is a variant of the two-axle hauler which has an articulated frame instead of the conventional one-piece frame. It is known as a rocker and it is illustrated in Figures 13-10 and 13-32. In this text it is classed as a two-axle, off-the-road rear-dump hauler. Excellent maneuverability in close quarters and balanced axle loadings are among its desirable features.

The hauler of Figure 13-30 is being loaded with seven dippers of well-blasted limestone-shale from the 8-yd^3 shovel. The haul is 8000 feet, one way, averaging $+10.6$ percent grade. While of extreme length and grade, these long, uphill hauls are characteristic of anthracite coal stripping operations in eastern Pennsylvania. On this particular haul the maximum grade is $+15$ percent. Under extreme weather conditions of rain and snow, the coefficient of traction can reach a low of 0.20. When loaded, 80 percent of GVW is on the two rear driving axles. If rolling resistance is 4 percent and extreme grade resistance is 15 percent, then total resistance is 19 percent. The resistance of the laden hauler is 19 percent of GVW, that is, 19 percent of 260,000 lb, or 49,400 lb. The tractive effort can be as low as 80 percent $\times 260,000 \times 0.20$, or 41,600 lb, and the laden hauler would be immobilized. One sees easily the necessity for the tandem drive-truck under

13–64 Hauling Excavation

Figure 13-30 Three-axle rear-dump hauler of 75-ton capacity used in stripping overburden from coal in open-pit mining.

Important specifications:
  Capacity:
    Cubic yards struck measurement   44
    Cubic yards loose measurement, crowned load, 4:1 slope   51
    Tons, rated   75
  Engine flywheel or brake horsepower   700
  Weights, lb:
    Tare, working   110,000
    Load, rated   150,000
    GVW   260,000
  Speed indicators:

  $\text{Laden: } \dfrac{\text{Brake hp}}{\text{GVW, 1000s of lb}}$ 2.69

  $\text{Unladen: } \dfrac{\text{Brake hp}}{\text{Tare weight 1000s of lb}}$ 6.36

  $\text{Work-cost indicator: } \dfrac{\text{GVW + tare weight}}{\text{load weight}}$ 2.47

(Mack Truck Co.)

the working conditions. Indeed, many all-axle-drive rock haulers are used in this service. In this case of 19 percent total resistance and 20 percent tractive effort, even an all-axle-drive hauler would have only a 5 percent margin to overcome the wet and icy conditions of winter in the vicinity of Tamaqua, Pennsylvania.

The hauler of Figure 13-31 is dumping blasted granite on a freeway fill. The load is 11.0 cybm. The haul, 18,900 ft one way, is off-the-road–on-the-road, and the on-the-road haulage is about 17,400 ft. The net cycle time is 23.2 min. Average hauling speed is 20 mi/h and returning speed averages 25 mi/h. The hauler is a good selection for the job. The semitrailer is frameless and this unique construction allows the machine to haul about a ton more payload than the five-axle hauler of Table 13-24. The important result is a 5 percent reduction in the work-cost indicator.

The bottom-dump hauler of Figure 13-26 and the rear-dump hauler of Figure 13-31 are hauling over the same expressway with its traffic lights. Average hauling speeds are about the same and GVWs and tare weights are about the same. The rocker of Figure

Figure 13-31 Rear-dump hauler of 25-ton capacity used in on-the-road haulage in freeway construction.

Salient specifications:
Capacity:
Cubic yards struck measurement 20
Cubic yards loose measurement, crowned load, 4:1 slope 23
Tons, rated 25
Engine flywheel or brake horsepower 325
Weights, lb
Tare, working 36,000
Load, rated 50,000
GVW 86,000
Speed indicators:

Laden: $\dfrac{\text{Brake hp}}{\text{GVW, 1000s of lb}}$ 3.78

Unladen: $\dfrac{\text{Brake hp}}{\text{Tare weight, 1000s of lb}}$ 9.03

Work-cost indicator: $\dfrac{\text{GVW}+\text{tare weight}}{\text{Load weight}}$ 2.44

13-32 is being loaded with six dippers of well-blasted basalt from the $3\frac{1}{2}$-yd^3 shovel. Haul is 700 ft, one way, on an average -8 percent grade. Three rockers are serving the shovel, for a production of 250 cybm/h.

The huge sidehill cut for the freeway in the Santa Monica Mountains of California is being taken out in several 25-ft-deep lifts, the last lift being illustrated. When the top lift was being excavated and the lowest layer in the fill was being dumped, working space was scarce. It was in this initial phase of rock work that the maneuverability of the articulated rockers offered savings in both cycle time and unit cost for haulage. These savings were apparent throughout the prosecution of the work, as confined quarters are typical of sidehill rock cuts in the mountains.

Figure 13-33 illustrates one of several 50-ton rear-dump rock haulers wisely chosen for an unusual all-year, all-weather mountainside project. Oil shale of a pilot mining project is being hauled from an underground mine at an 8200-ft elevation to a retort facility at the 6000-ft level near Anvil Points, Colorado. The haul road, with many tortuous curves, is 6 mi long and 22 ft wide, with an average -6.9 percent grade and a maximum -9 percent grade. The cycle time for the haulers is about 65 min. The hauler has two equally loaded driving axles, both axles having dual tires. The particular haulers were chosen for the exacting work because of the following desirable characteristics: maximum tractive effort and retarder braking action by reason of all-wheel drive; maximum steering control because of drive on steering wheels; maximum flotation because of all-around dual tires; and minimum stress and strain because of equal axle loadings. Important specifications for this unique 50-ton machine are in the following tabulation. The dimensions of the machine are given in the lower picture of Figure 13-33.

13-66 Hauling Excavation

Capacity:
Cubic yards struck measurement 33.3
Cubic yards loose measurement, crowned load, 4:1 slope 38.3
Tons, rated 50
Engine flywheel or brake horsepower 537
Transmission:
Gears forward 10
Maximum speed:
mi/h 36
ft/min 3170
Tires, eight, all duals ..... 18.00×25—32PR
Weights, equal on the two axles when laden, lb:
Tare, working 75,000
Load, rated 100,000
GVW 175,000
Speed indicators:

Laden: $\dfrac{\text{Brake hp}}{\text{GVW, 1000s of lb}}$ 3.07

Unladen: $\dfrac{\text{Brake hp}}{\text{Tare weight 1000s of lb}}$ 7.16

Work-cost indicator: $\dfrac{\text{GVW} + \text{tare weight}}{\text{Load weight}}$ 2.50

Figure 13-32 Rockers, two-axle rear-dump haulers with articulated frames, hauling blasted basalt in freeway building.

Important specifications:
Capacity:
Cubic yards struck measurement 22
Cubic yards loose measurement, crowned load, 4:1 slope 25
Tons, rated 33
Engine flywheel or brake horsepower 300
Weights, lb
Tare, working 60,000
Load, rated 66,000
GVW 126,000
Speed indicators:

Laden: $\dfrac{\text{Brake hp}}{\text{GVW, 1000s of lb}}$ 2.38

Unladen: $\dfrac{\text{Brake hp}}{\text{Tare weight, 1000s of lb}}$ 5.00

Work-cost indicator: $\dfrac{\text{GVW} + \text{tare weight}}{\text{Load weight}}$ 2.82

Figure 13-33 Rear-dump hauler of 50-ton capacity with two equally loaded axles and all-wheel drive. *(a)* Leaving mine portal at 8200-ft elevation for 6-mi downgrade haul to retort facility at 6000-ft elevation. The sharp exit turn is typical of the 22-ft-wide steep sidehill haul road, especially hazardous in all-weather, all-year work. Rain, snow, and ice demand a rock hauler with the features of this balanced, all-wheel-drive machine. *(b)* Configuration and dimensions of the compact and versatile hauler. *(International Harvester Co.)*

Representative specifications are given in Table 13-22 for off-the-road two-axle rear-dump haulers, in Table 13-23 for on-the-road–off-the-road rear-dump haulers with two or three axles, and in Table 13-24 for on-the-road–off-the road three-axle and five-axle rear-dump haulers consisting of tractor and semitrailer.

The principles of operation of rear-dump haulers are simple. The bodies have either end chutes or tailgates for dumping the load. The chute or inclined lip construction, sometimes called the Boulder Dam body, is common to both off-the-road and on-the-road machines, and the tailgate style is generally used for on-the-road work so as to prevent spillage on the road. The bodies are raised by one or two telescoping hydraulic hoists to

13-68 Hauling Excavation

TABLE 13-22 Representative Specifications for Off-the-Road Rear-Dump Haulers with Two Axles

Specification	\multicolumn{4}{c}{Nominal capacity, tons}			
	13	35	75	170
Capacity:				
Struck measurement, cysm	8.7	23.3	45.6	91.0
Loose measurement, crowned load, 4:1 slope, cylm	10.0	26.8	52.4	104.6
Engine flywheel or brake hp	154	400	750	1492
Transmission:				
Gears forward	5	6	6	MG set electric drive
Maximum speed:				
mi/h	34	32	34	34
ft/min	2990	2820	2990	2990
Dimensions, ft:				
Length	21.6	27.8	30.4	38.0
Width	9.2	12.7	16.0	20.9
Height	10.6	13.3	15.5	15.7
Height, loading	7.9	10.7	12.7	17.4
Tires:				
Front, two, and rear, four	12.00×24 14PR	18.00×25 32PR	24.00×35 48PR	36.00×51 50PR
Weights, lb:				
Tare, working	25,000	61,000	105,000	221,000
Load, rated	26,000	70,000	150,000	340,000
GVW	51,000	131,000	255,000	561,000
Speed indicators:				
Laden: $\dfrac{\text{Brake hp}}{\text{GVW, 1000s of lb}}$	3.02	3.05	2.94	2.66
Unladen: $\dfrac{\text{Brake hp}}{\text{Tare weight 1000s of lb}}$	6.16	6.55	7.14	6.75
Work-cost indicator:				
$\dfrac{\text{GVW} + \text{Tare weight}}{\text{Load weight}}$	2.92	2.74	2.40	2.30

TABLE 13-23 Representative Specifications for On-the-Road or Off-the-Road Rear-Dump Haulers

Specification	Two axles	Three axles
Capacity:		
Tons, nominal	7.5	13.0
Struck measurement, cysm	6.0	10.0
Loose measurement, crowned load, 4:1 slope, cylm	6.9	11.5
Engine flywheel or brake hp	190	325
Transmission:		
Gears forward	10	12
Maximum speed:		
mi/h	60	60
ft/min	5280	5280
Dimensions, ft:		
Length	19.3	25.0
Width	8.0	8.0
Height	8.5	9.1
Height, loading	7.5	8.0

TABLE 13-23 Representative Specifications for On-the-Road or Off-the-Road Rear-Dump Haulers (Continued)

Specification	Two axles	Three axles
Tires:		
Front, two	11.00×22	11.00×22
Rear, four for two-axle hauler and	14PR	14PR
eight for three-axle hauler	all tires	all tires
Weights, lb:		
Tare, working	13,000	24,000
Load, rated, on-the-road	15,000	26,000
GVW	28,000	50,000
Speed indicators:		
Laden: $\dfrac{\text{Brake hp}}{\text{GVW, 1000s of lb}}$	6.79	6.50
Unladen: $\dfrac{\text{Brake hp}}{\text{Tare weight, 1000s of lb}}$	14.62	13.54
Work-cost indicator: $\dfrac{\text{GVW} + \text{tare weight}}{\text{Load weight}}$	2.73	2.85

NOTE: When working in off-the-road haulage, payload is increased by about 15 percent.

TABLE 13-24 Representative Specifications for On-the-Road or Off-the-Road Rear-Dump Haulers, Tractor and Semitrailer

Specification	Three axles	Five axles
Capacity:		
Tons, nominal	14.0	24.0
Struck measurement, cysm	11.0	18.0
Loose measurement, crowned load,		
4:1 slope, cylm	12.6	20.7
Engine flywheel or brake hp	190	325
Transmission:		
Gears forward	10	12
Maximum speed		
mi/h	60	60
ft/min	5280	5280
Dimensions, ft:		
Length	24.1	35.7
Width	8.0	8.0
Height	8.5	9.1
Height, loading	7.5	8.6
Tires:		
Tractor, two front and four rear for three-axle	11.00×22	11.00×22
hauler, two front and eight rear for five-axle	14PR	14PR
hauler	all tires	all tires
Semitrailer, four for three-axle hauler		
and eight for five-axle hauler		
Weights, lb:		
Tare, working	22,000	38,000
Load, rated, on-the-road	28,000	48,000
GVW	50,000	86,000
Speed indicators:		
Laden: $\dfrac{\text{Brake hp}}{\text{GVW, 1000s of lb}}$	3.80	3.78
Unladen: $\dfrac{\text{Brake hp}}{\text{Tare weight, 1000s of lb}}$	6.36	8.55
Work-cost indicator: $\dfrac{\text{GVW} + \text{tare weight}}{\text{Load weight}}$	2.57	2.58

NOTE: When working in off-the-road haulage, payload is increased by about 15 percent.

the dumping angle, which averages about 65° with the horizontal. The dumping time ranges from 0.25 to 0.50 min.

Bodies of off-the-road machines are equipped with a canopy to protect the cab, and sometimes on-the-road haulers are also equipped with this safety feature. In cold climates the bodies are equipped with exhaust heating ducts to prevent the load from freezing to the interior of the body.

Trucks and tractors of rear-dump haulers are designed for use of conventional gasoline, gas, and diesel engines, except for the largest off-the-road haulers, which may have the MG power-transmission system. This system is made up of a diesel engine prime mover, which drives a dc generator, which in turn drives dc electric motors within the wheels. Such a hauler is the 170-ton rear-dump hauler of Table 13-22.

Hauling Cycle Times

Rear-dump haulers are top-loaded by any of the following machines: tractor-bulldozers, by use of loading chutes; bucket-type loaders; shovels; draglines; trenchers; belt loaders, movable and mobile types; and wheel-bucket excavators. Figure 13-32 illustrates the most popular one for hard-rock excavation, the shovel, and Figures 13-34 and 13-35 show loading by bucket loader. Loading times may be figured from the data for loading given in Chapter 12.

The times for two turnings and dumping, estimated for fixed times for the hauling cycle, are given in Table 13-25. Hauling and returning times may be calculated from the data provided in Tables 13-3, 13-10, 13-11, and 13-12 for calculations for wheels-tires-mounted rear-dump haulers. Figure 13-36 shows the hauler of Figure 13-35 traveling downgrade on a well-maintained haul road.

Average Loads

The estimated loads in loose and bank measurement may be calculated by use of Table 13-26. The table is based on the struck-measurement capacity of the hauler. An example of the use of Table 13-26 is provided by the 44-cysm, 75-ton rear-dump hauler of Figure 13-30, which is shown hauling well-blasted rock. The body factor is 0.77. The load, bank measurement, is 0.77×44, or 34 cybm, equivalent to 75 tons of limestone-shale rock.

Figure 13-34 A common loading arrangement for rear-dump haulers in rock cut. This backup method requires a few more seconds than the follow-through mode of Figure 12-15 with its 180° turn. Additionally, the average angle of swing is about 15° greater, important in bucket-loader operation but not significant in shovel work, where hoist time rather than swing time controls the cycle time.

The 8.5-cysm bucket loader is filling the 50-ton, 33-cysm hauler with four buckets of well-blasted granite in 1.9 min. The haul is to a crushing and screening plant for Perris Dam, Riverside County, California. The ultimate placement of some of the processed rock is in the pervious fill for the dam. Much of the crushed rock is handled by the bottom-dump haulers, as illustrated in Figure 13-27. A rear-dump hauler is chosen for the run-of-quarry rock and a bottom-dump hauler is selected for the long upgrade haul of crushed rock, each machine having its own economical application.

Figure 13-35 The efficient loading and exchange of 50-ton-capacity rear-dump haulers. The trucks are loaded with 28 cybm of well-blasted granite by the 8.5-cysm bucket loader. The truck is receiving the final fourth bucket and it will leave to the left of the waiting hauler. The waiting truck will be in exact loading position without delay to the cycling loader. The loading time is 2.0 min and the average swing angle is 90°.

Two heavyweight tractor-bulldozers are drifting well-blasted rock to the loader. In the background is the pattern of blastholes for the next round of blasting. The result of this total synchronization of blasting, bulldozing, and loading is a sustained production of about 507 cybm/h for the loader.

The complete hauling cycle of the 50-ton truck of this picture and Figure 13-36 is 13.5 min on the 8100 ft, one way, downgrade haul. The low cycle time is largely the result of the excellently built and maintained haul road of Figure 13-36.

TABLE 13-25 Times for Two Turnings and Dumping for Rear-Dump Haulers

Job conditions	Times, min, for:		Tractor with semitrailer
	Two-axle truck	Three-axle truck	
Good: hard surface, large cut and fill	1.0	1.2	1.5
Average: hard to soft surface, medium-size cut and fill	1.3	1.6	2.0
Poor: soft surface, small cut and fill	1.6	2.0	2.5

TABLE 13-26 Body Factors, for Average Loads of Rear-Dump Haulers Based on Body Capacity, cysm

Nature of rock-earth	Fill factor $\left(\dfrac{\text{loose measurement}}{\text{body capacity}}\right)$	Conversion factor $\left(\dfrac{\text{bank measurement}}{\text{loose measurement}}\right)$	Body factor $\left(\dfrac{\text{bank measurement}}{\text{body capacity}}\right)$
Sand-gravel alluvia	1.15	0.88	1.01
Rock-earth weathered	1.15	0.78	0.90
Rock, well blasted	1.15	0.67	0.77
Rock, poorly blasted	1.15	0.60	0.69

13-72   Hauling Excavation

Figure 13-36 Rear-dump hauler of 50-ton capacity traveling down 5 percent grade at maximum 27 mi/h. This hard, smooth, well-maintained haul road presents about 2 percent rolling resistance. In downgrade hauling the retarder compensates for about 6200 lb of accelerating force. When the hauler is returning upgrade unladen, the resisting force, 2 percent rolling plus 5 percent grade, is about 5700 lb. The maximum upgrade speed is also 27 mi/h, and the 576-flywheel-hp engine is delivering about 416 hp at the wheel-road contact.

The top view of the laden truck, as well as the picture of Figure 13-37, emphasizes the importance of using a conservative 4:1 slope for the crowned load of a hauler. In both illustrations the actual slopes are about 2:1, but the corners of the bodies are not filled. Accordingly, the 4:1 slope is a practical compensatory factor for figuring loose-measurement loads.

The same 50-ton-capacity rock trucks shown in Figure 13-35 travel on this haul road for freeway building in the Laguna Mountains of southern California.

Turning of the rear-dump hauler at loading point may be by two methods. The better one in terms of time and safety is the follow-through method of Figure 12-15. The less efficient backup method of Figure 13-34 is used more commonly, since there is generally not enough space to make the 180° turn for the follow-through method. This confinement is especially true in a bench cut or shelf at the beginning of a sidehill cut.

Productions and Costs

EXAMPLE 1: The actual hauling cycle and estimated production and costs for on-the-road–off-the-road tractor and semitrailer of Figure 13-31 may be determined as follows:

```
Criteria:
  Rock, well-blasted granite
  Loader, bucket-type, nominal rated capacity, cysm                       6.5
  Haul road, ft:
    Off-the-road, −2% grade, well maintained                            4,500
    On-the-road, level, expressway with five traffic lights            13,900
    Off-the-road, level, fill, well maintained                            500
    Total length, one way, 3.58 mi                                     18,900
  Haulers
    Tractor: three axles; tandem-axle drive; engine flywheel hp 300
    Semitrailer: two axles; nominal capacity, 25 tons; actual load,
      23.6 tons, or 10.4 cybm; tare working weight, 36,000 lb; GVW,
      83,200 lb
Actual hauling cycle, min:
  Loading by three buckets                                                1.5
  Hauling 18,900 ft at average 20 mi/h                                   10.8
  Turning and dumping                                                     2.0
  Returning 18,900 ft at average 25 mi/h                                  8.6
  Turning                                                                 0.4
  Net cycle time                                                         23.3
```

Estimated hourly production per 50-min working hour:
Loads hauled hourly 50/23.3   2.1
Cubic yards hauled hourly, cybm 2.1×10.4   22
Direct job unit cost, 1978:
Hauler cost $92,800
Hourly cost of ownership and operation:
Tractor-trailer rear-dump hauler (Table 8-3), 44% × $93 $40.90
Operator   11.80
Total hourly cost $52.70
Direct job unit cost per cybm, $52.70/22 $2.40

EXAMPLE 2: The actual hauling cycle and estimated production and costs of the quarry trucks of Figures 13-37 and 13-38 are calculated as follows:

Criteria:
Rock, well-blasted granite
Loader, bucket type, nominal rated capacity, 6.5 cysm
Haul road, 1600 ft one way, down average −11% grade, average maintenance
Dumping into receiving hopper of 30-in gyratory crusher
Two haulers: tare working weight, 64,000 lb; payload of well-blasted granite, 80,000 lb; GVW, 144,000 lb; engine flywheel hp, 415
Actual hauling cycle times, min:
Loading by four buckets   1.9
Hauling 1600 ft downgrade at average 12 mi/h   1.5
Turning and dumping   0.7
Returning 1600 ft upgrade at average 10 mi/h   1.8
Turning   0.4
Net cycle time   6.3
Estimated hourly production per 50-min working hour:
Loads hauled hourly, 50/6.3   7.9
Tons hauled hourly, 7.9×40   316
Estimated direct job unit cost, 1978:
Hauler cost $181,000
Hourly cost of ownership and operation:
Rear-dump hauler (Table 8-3), 34% × $181 $61.50
Operator   15.60
Total hourly cost $77.10
Direct job unit cost per ton $0.24
Direct job unit cost per cybm $0.54

Summary

Rear-dump haulers are the most versatile machines for handling all kinds of rock under the greatest variety of hauling conditions. For that distinction they must be built ruggedly, they must dump by raising the body, and they must have excellent flotation and traction characteristics. These desirable features are always present in the off-the-road haulers and they are incorporated, as much as possible, into the on-the-road machines.

When selecting haulers for rock ranging from well-weathered to solid, it is well first to consider the rear-dump hauler and then to appraise the economic suitability of scrapers, bottom-dump haulers, and side-dump haulers for the different consolidations of the rock formation and for the different anticipated haul-road conditions. Comparative analyses of productions and costs will determine the right machine or machines for the job.

SIDE-DUMP HAULERS

These special-purpose rock haulers are generally custom-built and they are typified by the model views of Figures 13-39 and 13-40. They may be load-on-back trucks with two or three axles or they may be made up of a tractor with semitrailer or with semitrailer and full trailer. They are used almost exclusively in quarries and open pits, in which the rock and the haul-road conditions are fixed, as illustrated respectively in Figures 13-41 and 13-42. In these applications the configuration, capacity, and engine horsepower are tailored to fit the

Figure 13-37 A short (1600-ft) one-way haul with average −11% grade in quarry work. A short haul over an undulating downgrade haul road of maximum −15 percent grade imposes a low, 12-mi/h average hauling speed and a correspondingly low 10 mi/h average returning speed. The truck, rated at 35 tons capacity, is hauling 40 tons and GVW is 144,000 lb. Tare working weight is 64,000 lb and engine flywheel hp is 415.

The haul road receives average maintenance with about 3 percent rolling resistance. The tractive effort on a maximum +15 percent grade is 18% of 64,000 lb, or about 11,500 lb. At 10 mi/h on the short (250-ft) segment of the haul road, the engine delivers 307 hp at the wheels or 74 percent of rated engine flywheel hp. The hauling cycle averages 6.3 min and the estimated production of each of the two trucks is 316 tons of granite hourly, giving a total of 632 tons/h for the aggregate plant of San Diego County, California. *(South Coast Asphalt Co.)*

Figure 13-38 Beginning of body hoist to 55° dumping angle in 24 s. The time for combined turning and dumping into a fixed crusher–receiving hopper is understandably less than that for turning and dumping on a rock fill in a grading operation. Turning and backing up to a fixed block or chock are routine, resulting in a 42-s, or 0.70-min, total time. The only incidental delay is an occasional wait for a delay in belt-conveyor operation from the crusher.

The truck load is 80,000 lb, 14 percent over the nominal 35-ton capacity rating. Quarry measurement is generally in tons, the unit corresponding to that of plant production, and the 40-ton load is equivalent to 18 cybm of granite. Two of these trucks on the 1600-ft one-way haul supply ample rock for the average plant production of 5000 tons per 8-h day. The site of the plant is northwestern San Diego County, California, where the pediment of the granite formation commences. *(South Coast Asphalt Co.)*

Figure 13-39 Side-dump quarry trucks of 16-ton capacity with twin bodies. This machine is built for limestone rock haulage in a southern Illinois quarry. In these three haulers, each body has a capacity of 8 tons, matching the size of the primary crusher, and the body is dumped by an overhead electric hoist at the crusher. Important specifications and data for the hauler are:

Capacity:	
Total tons (8 tons in each body)	16
Cubic yards struck measurement, two bodies	12
Cubic yards loose measurement, two bodies, crowned load, 4:1 slope	15
Engine flywheel or brake hp	200
Bodies, hinges on right side; dumping of bodies by overhead electric hoist	
Transmission:	
Gears forward	5
Maximum speed:	
mi/h	35
ft/min	3080
Dimensions, ft:	
Length	25.3
Width	10.0
Height	10.3
Height, loading	8.3
Tires:	
Front	12.00×25-16PR
Rear, duals	14.00×25-20PR
Weights, lb:	
Tare, working	33,000
Load, rated	32,000
GVW	65,000
Speed indicators:	
Laden: $\dfrac{\text{Brake hp}}{\text{GVW, 1000s of lb}}$	3.08
Unladen: $\dfrac{\text{Brake hp}}{\text{Tare weight 1000s of lb}}$	6.06
Work-cost indicator: $\dfrac{\text{GVW} + \text{tare weight}}{\text{Load weight}}$	3.06

This custom-built rock hauler has a high work-cost indicator. The reason is the high ratio of tare weight to payload, 1.03, caused by the unusual configuration of the twin bodies on the same load-on-back chassis. The special construction is made to accommodate the hauler to the primary crusher, and it may or may not be an economical adaptation. *(Euclid Road Machinery Company.)*

13-76   Hauling Excavation

Figure 13-40 Combination side-dump rock hauler of 24-ton on-the-road or 30-ton off-the-road capacity. This multipurpose twin-body machine is built for limestone haulage in quarries and on highways of Illinois. Important specifications and data for the hauler are as tabulated below.

Specification	On-the-Road Haulage	Off-the-Road Haulage
Capacities		
Tons:		
12 tons in each body	24	
15 tons in each body		30
Cubic yards struck measurement	22	22
Cubic yards loose measurement, crowned load,		
4:1 slope	28	28
Diesel engine, flywheel brake hp	275	275
Bodies: twin underbody hoists and hinges on both sides permitting dumping to either side; downfolding side gates.		
Transmission:		
Gears forward	12	12
Maximum speeds:		
mi/h	40	40
fpm ft/h	3520	3520
Dimensions, ft:		
Length	68.0	68.0
Width	8.5	8.5
Height	9.4	9.4
Height, loading	7.6	7.6
Tires:		
Tractor:		
Front	9.00×20—12 PR	9.00×20—12 PR
Rear, duals	11.00×22—14 PR	11.00×22—14 PR
Trailers, all duals	11.00×22—14 PR	11.00×22—14 PR
Weights, lb:		
Tare, working	41,000	41,000
Load, rated	48,000	60,000
GVW	89,000	101,000
Speed indicators:		
Laden: $\dfrac{\text{Brake hp}}{\text{GVW, 1000s of lb}}$	3.09	2.72
Unladen: $\dfrac{\text{Brake hp}}{\text{Tare weight, 1000s of lb}}$	6.71	6.71
Work-cost indicator $\dfrac{\text{GVW} + \text{tare weight}}{\text{Load weight}}$	2.71	2.37

The design and construction of this flexible rock hauler are compromises between on-the-road and off-the-road limitations of the ratio of tare weight to payload. The ratio for the on-the-road machine is 0.85, and the ratio for the off-the-road hauler is 0.68. *(Gar Wood Industries.)*

Figure 13-41 Side-dump quarry rock hauler of 40-ton capacity, with tractor and semitrailer. This single-body machine is built especially for hauling limestone from a 6-yd³ shovel to the primary crusher in this quarry at Presque Isle, Michigan. The body is dumped to one side by the overhead electric hoist. Salient specifications and data are as follows:

Capacity:	
Tons, rated	40
Cubic yards struck measurement	22
Cubic yards loose measurement, crowned load, 4:1 slope	28
Engine flywheel or brake hp	336
Body, dumping to one side by overhead electric hoist	
Transmission:	
Gears forward	5
Maximum speeds:	
mi/h	28
ft/s	2460
Dimensions, ft:	
Length	47.9
Width	12.8
Height	11.5
Height, loading	9.2
Tires:	
Tractor	
Front	13.00 × 25—16PR
Rear, duals	16.00 × 25—20PR
Semitrailer, duals	16.00 × 25—20PR
Weights, lb:	
Tare, working	65,000
Load, rated	80,000
GVW	145,000
Speed indicators:	
Laden: $\dfrac{\text{Brake hp}}{\text{GVW, 1000s of lb}}$	2.32
Unladen: $\dfrac{\text{Brake hp}}{\text{Tare weight, 1000s of lb}}$	5.17
Work-cost indicator: $\dfrac{\text{GVW} + \text{tare weight}}{\text{Load weight}}$	2.62

This rock hauler and the one of Figure 13-40 have large rated capacities. Likewise, their weights and work-cost indicators are approximately the same. Both machines are of high quality and they are well engineered for the same purpose. If the hauls are the same length and if they are similar with respect to grades and maintenance standards, and if the loading shovels are of equal capacity, it is quite likely that the direct job unit costs for haulage are about the same. *(Easton Car and Construction Co.)*

13-78   Hauling Excavation

Figure 13-42 Side-dump hauler of 20-ton capacity dumping copper ore into mine chute. This three-axle, tandem-axle-drive, single-body machine is one of four serving a 2-yd^3 shovel in the small open-pit mine of the United Verde Copper Company at Jerome, Arizona. The rich copper ore weighs 6000 lb/cybm. Important specifications and data are as follows.

Capacity:
   Tons, rated   20
   Cubic yards struck measurement   7
   Cubic yards loose measurement, crowned load, 4:1 slope   8.9
Engine flywheel or brake hp   150
Body: Downfolding side gates; dumping to either side;
   underbody hoists
Transmission:
   Gears forward   15
   Maximum speeds:
      mi/h   35
      ft/min   3080
Dimensions, ft:
   Length   24.0
   Width   9.0
   Height   9.5
   Height, loading   7.8
Tires, 10: front, singles, rear, duals 11.25×24—14PR
Weights, lb:
   Tare, working   30,000
   Load, rated   40,000
   GVW   70,000
Speed indicators:

   Laden: $\dfrac{\text{Brake hp}}{\text{GVW, 1000s of lb}}$ 2.14

   Unladen: $\dfrac{\text{Brake hp}}{\text{Tare weight, 1000s of lb}}$ 5.00

   Work-cost indicator: $\dfrac{\text{GVW} + \text{tare weight}}{\text{Load weight}}$ 2.50

These four ore haulers are a good selection for hauling the ore to a chute which feeds rail cars at a lower level of the mine. Haul varies from 200 to 1800 ft one way, with some heavy grades from the shovel to the chute. The laden speed indicator is average, and the work-cost indicator of 2.50 is good in view of the severe mining service, which demands a suitable ratio of tare weight to payload. *(White Motor Company.)*

particular task. Accordingly, specifications are not fixed as they are for scrapers and bottom-dump and rear-dump haulers.

Rarely are the machines used in public works construction except in the case of mammoth projects involving quarry and open-pit excavations. Once in a while they are used economically in overburden excavation incidental to quarry and open-pit work.

The principles of operation, except for the dumping phase of the hauling cycle, are similar to those of the rear-dump haulers.

Hauling Cycle Times

The machines are top-loaded normally by either bucket loader or shovel, typical in quarries and open pits. Of course, other loading machines may be used, as described in Chapter 12. Loading times may be calculated from the data for loading as given in Chapter 12.

Estimated times for two turnings and dumping are about equal to those for rear-dump haulers, given in Table 13-25. Hauling and returning times may be figured by the use of Tables 13-3, 13-10, 13-11, and 13-12 for the wheels-tires-mounted machines.

Average Loads

The bodies of side-dump haulers are wide and long with respect to depth, whether they be rock wagon–bathtub type, as in Figure 13-39, or downfolding side-gate type, as in Figure 13-40. Such a configuration gives a high ratio of loose load volume to struck measurement volume, 1.27. Table 13-27 gives body factors for the side-dump rock haulers.

An example of the use of Table 13-27 is provided by the 20-ton side-dump hauler of Figure 13-42, which has a 7-cysm capacity and is hauling well-blasted copper ore. The body factor is 0.85. The load, bank measurement, is 7×0.85, or 6.0 cybm. The copper ore weighs 6000 lb/cybm. The load then is 6.0×6000 lb, or 36,000 lb. The payload is 18.0 tons.

Important specifications and relevant data for four typical side-dump rock haulers accompany Figures 13-39 through 13-42.

Production and Cost Estimates

The data accompanying Figure 13-42, which illustrates copper ore haulage in an open-pit mine of Arizona, may be used as an example of production and cost calculations for a side-dump hauler.

Criteria:
Well-blasted rich copper ore, weighing 6000 lb/cybm and 4500 lb/cylm
Shovel, nominal rated dipper capacity   2.0 cysm
Hauls: hard pit floor with average maintenance; hauls upgrade
  One way; average +10% grade; shuttling operation   300 ft
  One way; average +2% grade; turning operation   1400 ft
Weight, lb, of four haulers:
  Tare weight, working   30,000 lb
  Actual load   36,000 lb
  GVW   66,000 lb

Item	Short haul	Long haul
Hauls:		
Length, one way, ft	300	1400
Average grade, %	+10	+2
Difference in elevation, ft	+30	+28
Actual hauling cycle times, min:		
Loading 7 to 9 dippers	3.5	3.5
Hauling:		
Average speed 2 mi/h	1.8	
Average speed 8 mi/h	...	2.0
Dumping	0.4	0.4
Turning	...	0.6
Returning:		

13-80 Hauling Excavation

Shuttling in reverse, average speed, 4 mi/h	1.0	
Average speed, 9 mi/h	...	1.9
Turning	...	0.5
Net cycle time	6.7	8.9
Estimated hourly production per 50-min working hour:		
Loads hauled hourly:		
50/6.7	7.5	
50/8.9	...	5.6
Tons hauled hourly:		
7.5×18	135	
5.6×18	...	101
Estimated direct job unit cost, 1978		
Hauler cost	$89,000	$89,000
Hourly cost of ownership and operation:		
Side-dump truck (Table 8-3), 33% of $89	$29.40	$29.40
Operator	11.80	11.80
Total hourly cost	$41.20	$41.20
Direct job unit cost per ton:		
$41.20/135	$0.31	
$41.20/101	...	$0.41
Direct job unit cost per ton-station (100 ft)	$0.10	$0.03

The analysis accentuates the worth of shuttling the haulers on the short hauls. Thereby the cycle has been reduced from an estimated 7.8 min by the conventional turning method to 6.7 min. The unit cost has been reduced from an estimated $0.36 to $0.31, a saving of 14 percent.

Summary

Side-dump rock and ore haulers are generally custom-built with special designs for work in quarries and open pits, where the rock and the hauling conditions are fairly constant. Moreover, the machines are built extremely rugged so as to have long life expectancy and high availability in the demanding work.

Because they are designed to withstand the heavy shock loads from large shovel dippers and loader buckets, their ratio of tare weight to load weight is high. For example, these ratios for the rear-dump rock haulers of Table 13-22 average 0.80, whereas the average ratio for the four side-dump machines considered in this section is 0.90. Similarly, the work-cost ratios are relatively high, the average being 2.59 for the rear-dump haulers and 2.70 for the side-dump haulers.

CONSIDERATIONS FOR SCRAPER OR HAULER SELECTION

The machines commonly used for haulage of rock-earth in construction and in open-pit quarrying or mining are tractor-bulldozers, scrapers, bottom-dump haulers, rear-dump haulers, and side-dump haulers. As has been explained, these machines are available in different capacities, different engine horsepowers, and different configurations. To state categorically that a given machine will give the lowest direct job unit cost for a given job without an analysis based on both theory and practice is foolhardy.

TABLE 13-27 Body Factors for Average Loads of Side-Dump Haulers Based on Body Capacity, cysm

Nature of rock-earth	Fill factor $\left(\dfrac{\text{loose measurement}}{\text{body capacity}}\right)$	Conversion factor $\left(\dfrac{\text{bank measurement}}{\text{loose measurememt}}\right)$	Body factor $\left(\dfrac{\text{bank measurement}}{\text{body capacity}}\right)$
Sand-gravel alluvia	1.27	0.88	1.12
Rock-earth, weathered	1.27	0.78	0.99
Rock, well blasted	1.27	0.67	0.85
Rock, poorly blasted	1.27	0.60	0.76

The machine is really part of a team of machines which mutually produce the overall direct job-unit cost. For example, when comparing scraper haulage with bottom-dump haulage, one can select the right machine with respect to the cost of pure haulage, but the ancillary costs for loading, haul-road construction and maintenance, and building of fill must also be considered. To illustrate, the 1978 loading cost for a scraper production of 1200 cybm/h by two 410-hp push tractors is $0.16/cybm. The loading cost for two 410-hp tractor bulldozers and a 60-in mobile belt loader for production of 1200 cybm/h when loading bottom-dump haulers is $0.23/cybm. However, the bottom-dump hauler has a lesser unit cost for pure haulage and so at some length of haul the cost curves cross. Below this critical length of haul the scraper method is the more economical and above this length of haul the bottom-dump hauler method is thriftier. The length of haul where loading and hauling costs are equal must be determined, as well as the other related costs, before the right machinery selection can be made.

Such a painstaking analysis was made for the machinery selected for the excavation of 11,670,000 yd³ for a California freeway job, as illustrated in Plate 12-1 and in Figures 12-53, 13-13, 13-15, 13-35, and 13-36. The average haul from the controlling largest 9,776,000-yd³ cut to the fill was 6800 ft, one way, with an average −6.8% grade. The average difference in elevation between cut mass and fill mass was 460 ft. The yardage of the cut was subdivided into six categories according to the degree of consolidation. This division was based on seismic analyses and on exploratory drilling. For each category the most economical machinery was selected. These categories and selections are as tabulated below.

Category	Machinery	Yardage, yd³	Percent of total yardage
Residuals, clays and sands; no ripping	Movable belt loaders, tractor-bulldozers, bottom-dump haulers Push-loading tractors, scrapers	716,000	7
Maximum-weathered granite; soft ripping	Movable belt loaders, tractor-ripper-bulldozers, bottom-dump haulers Tractor-ripper-bulldozers push-loading tractors, scrapers	2,877,000	29
Average-weathered granite; medium ripping	Machinery as for maximum-weathered granite	1,119,000	12
Minimum-weathered granite; hard ripping	Tractor-ripper-bulldozers, push-loading tractors, scrapers	1,097,000	11
Semisolid granite; extremely hard ripping	Machinery as for minimum-weathered granite	492,000	5
Solid granite; blasting	Drills and compressors, tractor-ripper-bulldozers, bucket loaders, rear-dump haulers	3,475,000	36
Totals		9,776,000	100

For a desired daily production of about 80,000 cybm, single and double shifting, the earthmoving machinery and ancillary equipment on the job as of a given day in December 1976 were

In cuts:
  Two-scraper spreads:
    Single-engine scraper spread:

13-82 Hauling Excavation

   Two heavyweight tractor-ripper-bulldozers
   Two heavyweight push-loading tractors
   Eight 40-cysm-capacity scrapers
  Twin-engine scraper spread:
   Two heavyweight tractor-ripper-bulldozers
   Two heavyweight push-loading tractors
   Seven 40-cysm-capacity scrapers
 Two bottom-dump hauler spreads:
   Six heavyweight tractor-ripper-bulldozers
   Two 8-cylm-capacity bucket loaders
   Two 60-in-belt-width movable belt loaders
   Sixteen 40-cysm-capacity double-unit bottom-dump haulers
  Two rear-dump hauler spreads:
   Nine heavyweight track drills
   Nine 1200-ft^3/min portable air compressors
   Four heavyweight tractor-ripper-bulldozers
   Two 10-cylm-capacity bucket loaders
   Three 35-ton-capacity rear-dump haulers
   Four 50-ton-capacity rear-dump haulers
 On haul road:
  Two 14-ft-blade motor graders
  One 16-ft drag scraper with heavyweight wheels-tires tractor
  Two 3500-gal-capacity six-wheel sprinkler water wagons
 On fill:
  Three 35-ton rock-tamping wheel compactors
  One 14-ft-blade motor grader
  One heavyweight tractor-bulldozer
  One medium-weight tractor-bulldozer
  Two 10,000-gal four-wheel sprinkler water haulers
  Four 12,000-gal portable water tanks
 Miscellaneous equipment:
  Three electric-motor-driven deep-well pumps of about 500 gal/min capacity with reaches of 8-in-diameter pipe
  Eight 20-kW diesel-engine light plants
  Two fuel trucks
  One grease truck
  Two maintenance and repair trucks
  One truck crane

Neglecting miscellaneous equipment, there were 100 machines for the excavation. In addition, there were 10 machines down for minor and major repairs, giving a machine availability of 91 percent on the particular day. This high availability for double shifting can be attributed to an important facet of the management, which provided completely overhauled and new machinery for this big demanding job.

A typical day's production was approximately in keeping with the following tabulation:

By two scraper spreads, 15 scrapers, 18 h	33,300 cybm
By two bottom-dump hauler spreads, 16 haulers, 9 h	19,200
By two rear-dump hauler spreads, 9 haulers, 18 h	20,700
Total, 40 scrapers and haulers in action	73,200 cybm

Generalities for proper use of off-the-road hauling machines, based on characteristics and work-cost indicators, are summarized below.

Crawler-tractor-bulldozers: Work-cost indicators average 6.90. In spite of this highest of hauler indicators, they are efficient, economical, self-contained machines for short-haul work in all kinds of rock-earth with the exception of poorly blasted rock, which does not provide good footing for the crawlers. In downgrade work under favorable circumstances, they work well on one-way hauls up to 500 ft in length. Relatively low productions and the physical impossibility of working as a team on most jobs preclude their use for high productions on long hauls. It must be emphasized that the tractor-bulldozer both loads and hauls in the bulldozing operation and this unique ability must be recognized in appraising the unit cost for work.

Crawler-tractors pulling wheels-tires-mounted scrapers: Work-cost indicators average 4.19, a high figure which indicates that hauls must be short and must be justified by the physical condition of the haul road and the rock-earth. The machines handle all kinds of rock-earth except poorly blasted rock, although well-blasted rock is loaded with difficulty. They are useful on hauls with steep downgrades and poor traction and flotation conditions. The economic haul is up to about 1000 ft, one way, under circumstances unfavorable to wheels-tires-mounted scrapers.

Crawler-tractors pulling crawler–rock wagons: Work-cost indicators average 4.36 for a tractor pulling two rock wagons. This is a high figure and indicates that hauls must be short and must be justified by the physical condition of the haul road and rock-earth. Being side dumping, the machines handle all kinds of rock-earth. They may be used on hauls up to 1000 ft in length, one way, and on reasonable downgrades where there is no likelihood of jackknifing. Jackknifing comes about because there are no brakes on the rock wagons and there is a tendency to jackknife on downgrades of more than 5 percent, particularly if two wagons instead of one are being pulled. The machines are at their best when working under bad traction and flotation conditions. The economic haul is up to about 1000 ft, one way, under working conditions unfavorable to wheels-tires-mounted rear-dump haulers. In a sense, they complement the crawler-tractor pulling a wheels-tires-mounted scraper, as they handle the blasted semisolid and solid rock which must be top-loaded and which is beyond the loading and hauling ability of the tractor-scraper.

Scrapers, push-loaded: Single-engine and twin-engine scrapers have an average work-cost indicator of 3.46. This indicator is lower than the 4.19 for the crawler-tractor-scraper and suggests greater productions and lower costs. The machines haul all rock-earth which can be loaded comfortably by push-loading tractors. All residuals and weathered rocks, either unripped or ripped according to degree of consolidation, may be handled. Well-blasted and poorly blasted rock are generally not loaded because of difficulties which result in high maintenance and low availability. The economic haul, as compared with bottom-dump and rear-dump haulers, is lower than a 2500- to 5000-ft range, one way, depending on the manner of loading the top-loaded machines and whether hauls are level, upgrade, or downgrade.

Scrapers, self-loaded: Single-engine scrapers have an average work-cost indicator of 3.61. This is higher than for the push-loaded scraper, indicating that the machine is slightly less efficient and more costly as a pure hauler. However, it loads itself and functions as an entity and therein lies its advantage for certain work. It handles comfortably residuals and also maximum-weathered rock if it has been ripped. Because of the action of its self-loading elevator it cannot load materials harder than soft friable rocks. Its range of economic haul corresponds to that for the push-loaded scraper, being up to 2500 to 5000 ft one way, and is slightly lower when compared with bottom-dump and rear-dump haulers.

Bottom-dump haulers: The work-cost indicators average 2.37, lowest for any scrapers or haulers. Clearly, it is an economical machine when used under the right working conditions. It can handle all rock-earth except poorly blasted rock. Its economic range of haul is from 2500 to 5000 ft, one way, and up, depending on manner of loading and whether the haul is level, upgrade, or downgrade.

Rear-dump haulers: The work-cost indicators average 2.59, higher than for bottom-dump haulers and lower than for scrapers. This indicator implies correctly that the rear-dump machine is built to handle all kinds of rock-earth on all lengths of haul and under most kinds of hauling conditions. That it alone, with the exception of the side-dump hauler, can handle poorly blasted rock makes it an all-purpose machine. When scrapers are able to handle the same rock-earth, its economic range of haul as compared with scrapers is about the same as that for the bottom-dump hauler. When compared with the bottom-dump hauler in hauling rock-earth acceptable to the latter, the haulage costs are higher.

Side-dump haulers: Work-cost indicators for the off-the-road machines average 2.72, 5 percent higher than the 2.59 average for the rear-dump haulers of Table 13-22. The average indicator for the side-dump hauler suggests a slightly less efficient and a correspondingly more costly machine. However, this supposition is not altogether true. Practices, both physical and financial, are not the same in a quarry as they are in dam construction, for example.

13-84 Hauling Excavation

1. The same machine is depreciated over a longer period of time both in years and in hours in quarry service. In many cases the work is carried on by double or triple shifting, and it is the practice to give a longer life to all machines, many of which are ruggedly built and specially designed for the particular service. Accordingly, the hourly fixed charges are lessened.

2. The rock excavation and the haul roads do not change and the operators are generally longtime employees. Thus, the machines are less subject to unusual strains and stresses and there are less repairs and replacements.

3. Regular maintenance and preventive maintenance are handled in a more orderly manner in the central shops than in the average on-site temporary shops of dam construction. In addition, the facilities are better and the mechanics are apt to be more experienced.

These three factors tend to lower both fixed charges and operating expenses. Of equal importance is the rhythm of the haulers in the continuous and unchanging work of a quarry or an open-pit mine, which tends to increase production over that of a public works job such as dam or freeway building. Interesting, also, is the fact that one finds far more custom-built machines of all kinds in private works than in public works, and this truism vouchsafes that the special machines offer economies.

BELT CONVEYOR SYSTEMS

Belt conveyor systems, simple or complex, are used widely in rock haulage and there are two kinds. *Stationary belt conveyors* are installed permanently and semipermanently in works like quarries and open-pit mines and semipermanently in construction projects such as dam building. A semipermanent system for hauling rock used in dam building is illustrated in Figure 13-43. A permanent system for haulage, storage, and reclaiming of borax ore from an open-pit mine is illustrated in Figures 13-44 through 13-48. *Movable belt conveyors* are of two types: conveyors on temporary supports to be moved as excavation proceeds, illustrated in Figures 13-49 and 13-50, and *stackers*, which are literally conveyors that stack materials, illustrated in Figures 13-51 and 13-52.

Figure 13-43 Beginning of a flight of a 7250-ft-long conveyor system for dam building. The 36-in-wide belt supplies 1050 tons of crushed rock hourly for the building of Shasta Dam, a concrete structure of northern California. The system elevates the aggregate 280 ft by 14 flights or lifts. The belt speed is about 450 ft/min, and the total horsepower of driving electric motors is about 680.

Prior to building this system for the job of several years duration, alternative methods and machinery for rock haulage were considered. Among these was haulage by off-the-road bottom-dump haulers working over an expensive mountain haul road. The conveyor system was selected on the basis of ultimate direct job unit cost. *("Shasta, Queen of the Dams," Gene Sheley (ed.),* Western Construction, *November, 1975, p. 33.)*

Figure 13-44 Beginning of a 2000-ft-long conveyor system for haulage of borax ore from open-pit mine to plant. This segment of the system includes the following elements:

1. Dumping hopper for average 80-ton load of rear-dump trucks
2. Pan-type feeder, 60 in wide with 60-hp electric motor drive, for primary crusher
3. Hammermill-type 54-in×70-in primary crusher, 600-hp electric motor drive, rating of 1500 tons/h
4. Belt feeder for inclined conveyor, 60 in wide, 40-hp electric-motor drive
5. Portion of 42-in-wide inclined conveyor from bottom to top of pit. Length, 1000 ft; vertical lift, 300 ft; slope, 32%; angle, 17.5°; 600-hp electric-motor drive at head pulleys; belt speed, 600 ft/min; capacity, 1487 tons/h of ore

Figures 13-45 through 13-48 show successive segments of the conveyor system. *(United States Borax and Chemical Corp.)*

13–86 Hauling Excavation

Figure 13-45 Dumping 80 tons of borax ore into hopper at foot of conveyor system. The lightweight ore weighs 74 lb per cubic foot loose measurement, or 2000 lb/cylm. It is already well fragmented by blasting and by the crushing action of the large bucket loader, and the 80-ton load of the truck is reduced to a maximum 10-in size in about 4 min by the huge hammermill crusher. The upgrade haul from the ore body to the crusher averages about 2000 ft one way. On the average, some 16,000 tons of ore is crushed and elevated by the conveyor system to the plant during the double-shift day. *(United States Borax and Chemical Corp.)*

Figure 13-46 The 300-ft vertical lift of the borax ore by the conveyor system. A portion of the 1000-ft-long flight of the conveyor passes through a tunnel under one of the overburden haul roads before reaching the next horizontal 1000-ft flight. There are no gravity or spring-tensioning belt tighteners on this long inclined flight. After many costly trials and experiments it was found that a simple system of mechanical wedges at the tail pulley kept the proper tension and proper alignment for trouble-free and economical operation. *(United States Borax and Chemical Corp.)*

Figure 13-47 Stacker stockpiling borax ore along 1000-ft level flight of conveyor system. This segment of the conveyor system includes the following elements:

    1. Part of the 1000-ft long level conveyor from top of inclined flight to plant: width, 42 in; belt speed, 600 ft/min; electric-motor drive, 100 hp.

    2. Stacker: length of conveyor, 75 ft; stacker height, 40 ft; belt width, 42 in; electric-motor drive for belt and for travel along main conveyor, 75 hp. The stacker has an integral tripper by means of which the main conveyor dumps the ore onto the lateral conveyor of the stacker.

The maximum size of ore on the main conveyor is about 10 in, with occasional larger chunks. The ore has a fairly uniform gradation as it comes from the crusher. *(United States Borax and Chemical Corp.)*

Belt Conveyor Systems 13–87

Figure 13-46

Figure 13-47

13-88 Hauling Excavation

Figure 13-48 Loading stockpiled borax ore into conveyor system for haulage to plant. The traveling receiving hopper is being loaded near the end of the system at the plant. The driving electric motor has 60 hp. The combination of two conveyor flights, stacker, and receiving hopper gives complete flexibility for providing the plant with adequate borax ore for continuous operation. The total power of the electric motors in the system from the truck dumping hopper to the end of the level conveyor is 1535 hp. The installed cost in 1961 was about $1.2 million. The estimated cost in 1978 is about $3.8 million.

The frontispiece of this book shows an aerial view of the pit and the plant. The general layout of the conveyor system in visible, extending from the truck dumping end of the pit westward to the plant. The complete haulage of the ore follows this logical plan: (1) from variable location of loading shovel to fixed location of crusher by flexible truck haulage; (2) from crusher upgrade to fixed locations of stockpiles and plant by the conveyor system, one of the most economical methods for vertical and horizontal haulage. *(United States Borax and Chemical Corp.)*

Figure 13-49 Movable belt-conveyor system for excavating a 2-million-yd^3 borrow pit. The 100-ft deep pit of sand-gravel alluvium is being excavated for freeway fill building in the Los Angeles basin, California.

Illustrated are:

1. Receiving hopper for the four tractor-bulldozers at foot of lateral conveyor.

2. Lateral conveyor, 100 ft long and 42 in wide, to foot of inclined conveyor. Another similar lateral conveyor will be installed from another area when this borrow-pit area is exhausted. Drive is by a 50-hp electric motor.

3. Inclined main conveyor, 500 ft long and 42 in wide, to surge pile at rim of pit. Drive is by a 340-hp diesel engine.

4. Movable 60-in-wide loading conveyor for loading stockpiled material into on-the-road bottom-dump haulers. Drive is by a 180-hp diesel engine. The loading conveyor is shown in Figure 13-50.

The hourly production of this system is 1500 tons, the basis of borrow-pit payment, or 950 cybm. The total electric-motor and diesel-engine power is 570 hp. The total cost for the system as of 1978 is estimated to be $390,000.

Figure 13-50 Movable belt loader for loading stockpiled alluvium from borrow pit. The 60-in-wide belt loader is conveying sand-gravel from the surge pile or stockpile to the waiting on-the-road double-bottom-dump haulers. The surge pile enables the conveyors of Figure 13-49 and the belt loader to work independently and thus to ensure continuous production of 1500 tons/h.

Prior to installation of the conveyor system, the costs of loading by bucket loaders and hauling by these same bottom-dump haulers from the bottom of the pit had been estimated as a possible alternative method of excavation. The analysis showed an 18 percent saving for the conveyor system. When rock-earth excavation must be moved upgrade, when volumes of excavation are sufficiently large, and when operating conditions are propitious, conveyor systems are usually the most economical means for haulage of mass excavation. The work-cost indicator for the conveyor system is considerably less than for haulage by other machines.

Figure 13-49

Figure 13-50

Figure 13-51 Small stacker with grizzly and receiving hopper in gravel pit. The glacial till in this Central states borrow pit is moderately cemented and contains outsize weathered cobbles and boulders which are too large for the embankment. The 2½-yd³ shovel and the 30-in-wide portable stacker with grizzly and loading hopper perform the dual task of fragmentation and elimination of oversize rock. The approximate 400-cybm/h capacity of the stacker is ample for the output of the shovel.

The conveyor is equipped with a swing chute at the head pulley and the operator is now shifting loading from the rear-dump truck to the waiting bottom-dump hauler on the other side. The shovel is able to maintain a high hourly production of about 175 cybm. *(Barber Greene Co.)*

The Parts of a Belt Conveyor System

Belt The belt is the carrier of the rock-earth. It runs in troughing rollers when carrying the load and on return idlers when returning empty, as illustrated in Figure 13-53. As the belt must handle a variety of rock-earth, it comes in several materials and constructions for specific uses, and from these a logical selection may be made. Basically, there are two choices for strength of belt. These are cotton-ply, or duck, carcass and synthetic fiber carcass.

Cotton-ply construction is shown in Figure 13-54. The belt comes in different numbers of plies and in different weights of plies, expressed in ounces per ply. Tensile strength is a function of the number and weight of plies and the width of the belt.

The construction of a typical synthetic fiber belt is illustrated in Figure 13-55. A synthetic fiber is stronger than a cotton fiber of the same cross-sectional area, and so the synthetic fiber belt provides greater strength, less weight, and more flexibility with fewer fibers. This fact may be illustrated by the following tabulated comparison between cotton

Figure 13-52 Large self-propelled stacker casting overburden in phosphate open-pit mine. The overburden in this phosphate mine of Florida is being cast to the spoil bank by the huge crawler-mounted stacker. The overburden is being excavated and loaded into the receiving hopper by a wheel-bucket excavator of nominal 1500 cybm/h capacity, which is also the nominal capacity of the stacker. The specifications of the excavator are given in the wheel-bucket excavator section of Chapter 12. The lateral distance of the haul from face of cut to spoil bank is about 205 ft. In this earthy overburden the hourly production of the two matched machines, built by the same manufacturer, is about 1900 cybm. Important specifications for the stacker are tabulated below.

Capacity, cybm/h:	
Theoretical	3000
Nominal	1500
Prime mover, diesel-engine, flywheel or brake hp	640
Conveyor belt:	
Length, ft	160
Width, in	42
Speed, ft/min	750
Driving mechanisms:	
Belt, mechanical from diesel engine, estimated required hp	200
Propel, by two hydraulic motors, total hp	350
Swing, by two hydraulic motors, total hp	50
Hoist, by two hydraulic motors, total hp	36
Crawlers, two, mounting:	
Length, each, ft	29
Width, each, in	56
Dimensions, ft:	
Length	185
Width	30
Height	40
Length of boom	150
Weight, lb	500,000

(Mechanical Excavators, Inc.)

13–92 Hauling Excavation

Figure 13-53 Construction and completion of stationary belt conveyor system. In the left picture, the flight of the conveyor system is completed except for installation of the medium-width belt. In the view to the right the system is hauling crushed rock to the distant stockpile. This last flight, like the preceding flights, is a complete unit with its own receiving hopper and prime mover. Each flight receives its load into the receiving hopper above the tail pulley and hauls the material to its head pulley, whence the load drops into the hopper of the next flight. As illustrated, the last flight is dumping on the top of the stockpile. A system can work over any terrain provided its slope does not exceed a practical limitation, as shown in Table 13-31.

Conveyor systems haul rock-earth over vast distances, as exemplified by this 7-mi-long, 30-in-wide belt installation for the haulage of crushed rock at Bull Shoals Dam, Arkansas. The system contained 21 flights, averaging 1800 ft length, over rough topography. *(Hewitt-Robins Corp.)*

fabric and synthetic fiber 42-in-wide belts, both with top cover of ⅜-in thickness and bottom cover of ⅛-in thickness. The specified tensile strength for both belts is 12,000 lb.

Item	*Cotton Ply*	*Synthetic Fiber*
Tensile strength required for belt, lb	12,000	12,000
Thicknesses of covers, in:		
Top cover	⅜	⅜
Bottom cover	⅛	⅛
Number of plies	5	3
Ounces per ply per square yard of belt	48	
Weights, per linear foot of belt, lb:		
Carcass	7.1	3.9
Covers	<u>10.6</u>	<u>10.6</u>
Total weight	17.7	14.5
Total belt thickness, in	0.860	0.688
Rated tensile strength of belt, lb	12,180	13,230

The synthetic fiber belt offers: (1) 9 percent more tensile strength; (2) 18 percent less weight; and (3) 20 percent less thickness, a measure of flexibility when traveling over troughing rollers and pulleys.

Pulleys The head, or driving, pulleys and tail pulleys are at the ends of the conveyor belt. They are of large diameter. The diameter of the head or drive pulley equals approximately the belt width and the tail pulley diameter equals approximately 83 percent of

STANDARD

Multiple layers of suitable duck of the same thickness and ply across the entire Belt. Standard is suitable for nearly all types of Conveyor service.

SHOCK PAD

Substantially the same as Standard but with a reinforced top *Cover* consisting of an abrasion-resistant tire-tread stock on the top surface, backed by a thick pad of resilient rubber. The pad yields to sudden, extreme impacts and pressures—protecting the cover from puncture or breakage and preventing rupture of the carcass.

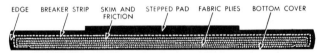

STEPPED PAD ... Stacker-type Cover

Substantially the same as Standard construction except that it is moulded with the *Cover* having an additional thickness standing out in relief in the center Belt area. It is recommended when the loading of abrasive material is concentrated in the center of the Belt with only slight abrasion at the cover edges. Stepped Pad Belting is not recommended for two-pulley or internal drives.

STEPPED PLY

A smooth-top construction having a heavier cover in the center of the Belt than at the edges. This is accomplished by moving the middle portion of one or two of the top fabric plies to the sides and filling in the extra space with cover stock. A Stepped Ply Belt has more crosswise flexibility and troughs more easily. The extra thickness of cover stock gives longer life to the Belt under loading conditions where abrasion is concentrated in the center Belt area.

Figure 13-54 Construction of conveyor belts, cotton or duck fabric carcass. *(Hewitt-Robins Inc.)*

the belt width. The surface of the drive pulley may be smooth or bare, or if greater friction and drive force are desired, the surface may be lagged with a grooved rubber surface vulcanized to the pulley surface.

The drive may be by a single pulley or by tandem pulleys so as to increase the area in contact with the belt for greater friction and driving force. A single pulley may have up to a 240° arc of contact and tandem pulleys may have up to a 500° arc.

In some conveyors the drive pulley is at the tail end of the conveyor belt, in which case the roles of the pulleys are reversed. In other conveyors the drive pulley is internal, the force being exerted on the returning section of the belt. The head pulley drive is the common construction. Table 13-40B gives recommended diameters of pulleys according to belt widths and belt tensions.

Troughing Rollers These rollers or idlers carry the loaded belt. They are shown in Figure 13-53. Table 13-41 gives recommended maximum spacings for the rollers according to weight of materials and widths of belt. The roller assembly may be made up of three rollers, the usual construction, or of five rollers. They are equipped with antifriction bearings. Rollers of 4-, 5-, and 6-in diameters are used, in accordance with the data of Table 13-36.

Return Idlers The flat return idlers carry the returning section of the belt conveyor. These idlers are one piece, of the same width as the belt, and are generally of the same diameter as the troughing idlers. They too are equipped with antifriction bearings. The recommended spacing for return idlers is 10 ft, as noted in Table 13-41.

Supporting Structure This foundation for pulleys, rollers and idlers, accessory machinery, and prime mover with reduction transmission may be as simple as the steel truss

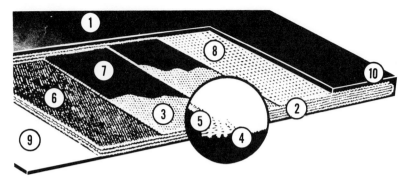

Figure 13-55 Construction of conveyor belts with synthetic fiber carcass.
  1. *Top Cover:* Covers of natural or synthetic rubber and blends are designed to protect the carcass from abrasion and impact. They carry none of the tension load, thus not affecting the strength or tension rating of the belt.
  2. *Carcass:* The strength portion of the belt, consisting of multiple plies of solid woven, rubber-impregnated fabric, bonded together with friction and skim coats.
  3. *Ply:* A single fabric layer of the carcass which provides the load support and tension strength in ply-belt construction.
  4. *Warp:* Lengthwise yarns in a woven fabric which provide the tension-carrying strength of the fabric. Warp yarns are woven over and under the fill yarns.
  5. *Fill:* The transverse or crosswise threads running perpendicular to the warp yarns. The fill provides the lateral strength necessary in a belt and provides for the fastener holding ability of the belt.
  6. *Friction Compound:* A tough, resilient rubber-adhesive compound impregnated in each ply of fabric, which serves as the means of bonding the plies together.
  7. *Skim Coat:* An extra layer of rubber compound between plies which increases flexing life and protects against ply separation.
  8. *Breaker Fabric:* A specially woven fabric placed between the cover and carcass to improve adhesion and provide better resistance to cover gouging or tearing.
  9. *Bottom Cover:* Usually a thinner gauge than the top cover, providing wearing surface against pulleys and supporting idlers.
  10. *Edge:* The rubber-covered edge protects the belt-fabric carcass from moisture and wear.
  (Goodyear Tire and Rubber Co.)

of Figure 13-51 or as complex as the railroadlike base for conveyor, stacker, and receiving hopper of Figures 13-47 and 13-48.

Auxiliary Machinery The accessories include the the following principal equipment items:

Grizzlies. The *grizzly* is used to remove oversize rocks before the load is dropped into the receiving hopper at the foot of the conveyor belt. It is illustrated in Figure 13-51, where outsize cobbles and boulders are shown being chuted off the grizzly. As shown, the grizzly is a simple steel or wood structure of inclined bars spaced in keeping with the maximum size of material to be used. In this case the oversize rocks above 8-in dimension are being stockpiled to the side of the hopper for later removal.

Feeders. In order to deliver material uniformly to the conveyor belt a *feeder* is used. In Figure 13-45 the primary crusher discharges onto a 60-in-wide belt feeder, which both feeds and protects the 42-in-wide inclined main conveyor belt. The generally used kinds of feeders are illustrated in Figure 13-56.

Trippers. As shown in Figure 13-47, deflection of the load along the length of the conveyor belt is sometimes necessary and a tripper, incorporated into the stacker, passes the load into a hopper which feeds the stacker conveyor belt. A simple tripper is illustrated in Figure 13-57.

Belt Takeups. As the belt elongates with use, a belt takeup is necessary. A simple device is a screw adjustment for the tail pulley. A second mechanism is spring loading for the tail pulley, the tail pulley shaft sliding on a lubricated runway. Another method, used for wide belts, is the gravity takeup. The returning belt passes downward and upward around a weighted pulley, which travels vertically in a sliding mechanism, and

(a)

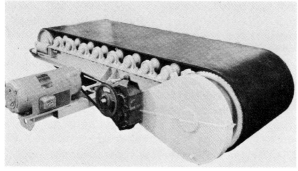

(b)

Figure 13-56 Two kinds of feeders for conveyor belts. *(a)* The *apron feeder* is made up of uniform overlapped steel pans attached onto chains or joined by integral links to form an endless conveying medium. It is used for hard, sharp, abrasive rocks. *(b)* The *belt feeder* is simply a short conveyor with a durable belt. It is used for soft to medium rock-earth. Other kinds of feeders, less used, are the reciprocating unit and the rotary-vane machine, both of which feed the rock-earth at a uniform rate. For any given material the most satisfactory feeder may be selected, and the following material characteristics should be considered: maximum size; hardness; sharpness; wetness; and gradation. *(Iowa Manufacturing Co.)*

this action maintains constant belt tension. All these devices must be frequently adjusted for good belt alignment.

Holdbacks. When a conveyor works upgrade, a braking device on the driving pulley is necessary to arrest downward or reverse motion in the event of failure of the driving mechanism. The holdback is an automatic mechanical device which prevents backward rotation of the head pulley. It must resist a force equal to the load force downward less all friction forces which resist downgrade movement.

Brakes. When a conveyor works downgrade, the load force may exceed the resisting friction force and braking is necessary. A common brake is the combination motor-generator electric prime mover, which functions as a motor when friction force exceeds load force and as a generator or brake when load force exceeds friction force. When an internal combustion engine drives the conveyor belt, either a mechanical brake or a hydraulic retarder may be used.

Prime Movers. Either engines or electric motors may serve as prime movers. For stationary conveyor systems electric motors are most common. For movable systems diesel and gasoline engines are commonly used, depending on the horsepower of the engine, and when electrical energy is available, motors are used. The three movable belt

Figure 13-57 **Tripper for unloading the belt conveyor anywhere along its length.** The automatic belt-propelled tripper is carried on rails along the length of the conveyor belt. The loaded belt passes upward along the inclined rollers, over the top of the pulley, downward and under the bottom pulley, and thence to the rollers. The load, dumped from the top pulley, may enter a hopper to be discharged onto a stacker or it may be deflected simply by a discharge chute. A tripper may be moved manually by a cranking device or it may be moved mechanically by an engine or motor or by the belt itself. *(Hewitt-Robins Inc.)*

conveyors of Figures 13-49 and 13-50 and the other lateral belt conveyor are driven by both engines and motors. The chief considerations for selection are availability of current and hourly cost of ownership and operation. If electrical energy is at hand, the selection is usually motors.

The Dynamics of a Belt Conveyor System

The dynamics are explained in terms of the design for the first flight of the system of Figures 13-44 through 13-48. Figure 13-46 shows this first flight.

Capacity Capacity is the first consideration of design, and a capacity of 1500 tons/h is desired at 100 percent efficiency of the belt conveyor system. Maximum speeds for belts according to the nature and the condition of the material are given in Table 13-28. Sometimes these speeds may be exceeded on the basis of experience gained before or after the system has been installed. However, the approximate speeds of the table should be used in the original design, with provisions for later changes. With these considerations in mind, a speed of 600 ft/min is selected for the borax ore.

Table 13-29 may be used in lieu of calculations for capacities in tons per hour for different widths of belts and different weights of materials. It is based on Table 13-30 with a 20° angle of repose. The borax ore weighs 74 lb per cubic foot loose measurement. The required tonnage per 100 ft/min of belt speed is 1500(100/600)=250 tons/h. Table 13-29 indicates that the belt width should be 42 in. Interpolating for 42-in-wide belt carrying ore weighing 74 lb/ft^3 gives 247 tons/h for the speed of 100 ft/min. At 600 ft/min, the capacity is (600/100)247=1482 tons/h, and this capacity is satisfactory.

Loading. Table 13-30 illustrates the loading of a belt and the cross-sectional areas of the load for three angles of repose or slope. An average angle of repose for a cross section of rock-earth is 20°. The areas are based on a load of width less than that of the belt, as expressed by this equation:

$$\text{Effective width} = W - 2(0.05W + 1)$$

where $W=$ width of belt, in

TABLE 13-28 Recommended Maximum Speeds for Conveyor Belts, ft/min

Kind and condition of material handled	Width of belt, in.										
	14	16	18	20	24	30	36	42	48	54	60
Unsized coal, gravel, stone, ashes, ore, or similar material	300	300	350	350	400	450	500	550	600	600	600
Sized coal, coke, or other breakable material	250	250	250	300	300	350	350	400	400	400	400
Wet or dry sand	400	400	500	600	600	700	800	800	800	800	800
Crushed coke, crushed slag, or other fine abrasive material	250	250	300	400	400	500	500	500	500	500	500
Large lump ore, rock, slag, or other large abrasive material	...	...	...	...	350	350	400	400	400	400	400

NOTE: This table may be cross-referred to Table 12-24. Recommended maximum belt speeds for mobile and movable belt loaders according to nature of earth-rock, ft/min.
SOURCE: Hewitt-Robins Inc.

TABLE 13-29 Carrying Capacities of Troughed Conveyor Belts, tons/h, for a Speed of 100 ft/min

Width of belt, in	Max lumps Sized, in	Max lumps Unsized, in	Weight of material, lb/ft³ 30	50	90	100	125	150	160	180	200
14	2	2½	9	15	28	31	39	46	49	56	62
16	2½	3	13	21	38	42	52	63	67	75	83
18	3	4	16	27	48	54	67	81	86	97	107
20	3½	5	20	33	60	67	83	100	107	120	133
24	4½	8	30	50	90	100	125	150	160	180	200
30	7	14	47	79	142	158	197	236	252	284	315
36	9	18	70	117	210	234	292	351	374	421	467
42	11	20	100	167	300	333	417	500	534	600	667
48	14	24	138	230	414	460	575	690	736	828	920
54	15	28	178	297	534	593	741	890	948	1,070	1,190
60	16	30	222	369	664	738	922	1,110	1,180	1,330	1,480

NOTE: The capacity of a 54-in-wide belt conveyor in tons per hour when carrying copper ore weighing 170 lb/ft³ at a belt speed of 400 ft/min is given by the equation:

$$\text{Capacity} = \frac{400}{100}\left[948 + \frac{10}{20}(1070 - 948)\right]$$

$$= 4036 \text{ tons/h}$$

SOURCE: Hewitt-Robins Inc.

TABLE 13-30

Cross-Sectional Areas of Loads on a Conveyor Belt

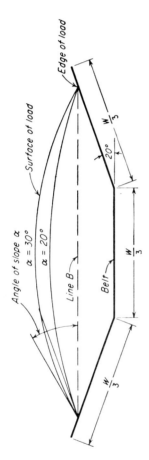

Areas of Cross Sections of Materials for Loaded Belt Conveyors

Width of belt, in.	$0.05W+1$, in.	Area of level load, ft²	Area of surcharge, ft², for angle of repose,°			Total area, ft², for angle of repose,°		
			10	20	30	10	20	30
16	1.8	0.072	0.029	0.059	0.090	0.101	0.131	0.162
18	1.9	0.096	0.038	0.078	0.118	0.134	0.174	0.214
20	2.0	0.122	0.048	0.098	0.150	0.170	0.220	0.272
24	2.2	0.185	0.072	0.146	0.225	0.257	0.331	0.410
30	2.5	0.303	0.118	0.238	0.365	0.421	0.541	0.668
36	2.8	0.450	0.174	0.351	0.540	0.624	0.801	0.990
42	3.1	0.627	0.241	0.488	0.749	0.868	1.115	1.376
48	3.4	0.833	0.321	0.649	0.992	1.154	1.482	1.825
54	3.7	1.068	0.408	0.826	1.264	1.476	1.894	2.332
60	4.0	1.333	0.510	1.027	1.575	1.843	2.360	2.908

NOTE: The carrying capacity of a 42-in belt, moving 100 ft/min, loaded with sand weighing 100 lb/ft³, with a 20° angle of repose, will be 100 ft/min × 100 lb × 1.115 ft² × 60 min ÷ 2,000 lb/ton = 334.5 tons/h. The carrying capacity of this belt for other speeds may be obtained by multiplying 334.5 by the ratio of the speed to 100 ft/min.

SOURCE: Hewitt-Robins Incorporated.

13-100 Hauling Excavation

Thus, a 42-in-wide belt has an effective width of:

$$42 - 2(0.05 \times 42 + 1) = 36 \text{ in}$$

When this 42-in-wide belt is loaded with rock-earth at a 20° angle of repose, Table 13-30 gives a total load cross section of $0.627 + 0.488$, or 1.115 ft^2.

If the conveyor belt carries borax ore weighing 2000 lb/cylm at a speed of 600 ft/min, the hourly capacity in tons is given by the equation:

$$C = \frac{1.115 \times 600 \times 60 \times (2000/27)}{2000} = 1487 \text{ tons/h}$$

Inclination. Table 13-31 gives the relationships of inclined length, horizontal length, and vertical lift in feet for degrees of angle of inclination. The table gives also maximum

TABLE 13-31 Inclined Lengths of Belt Conveyors According to Horizontal and Vertical Distances or Degrees of Angles of Inclination

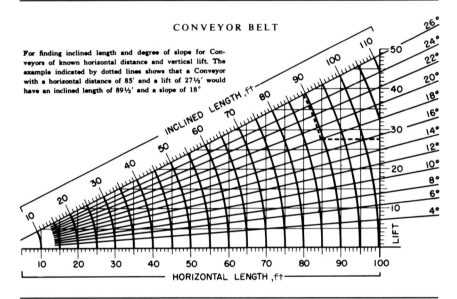

NOTE: The chart is limited to horizontal lengths up to 100 ft. For the greater horizontal lengths use trigonometric functions of angles.

Example: For a horizontal length of 900 ft and a lift of 150 ft, it is required to find the angle of inclination and the inclined length.

Solution: Tangent of angle = $\frac{150}{900}$ = 0.167; angle = 9.5°

Inclined length = $\frac{900}{\cos 9.5°}$ = $\frac{900}{0.986}$ = 913 ft

The maximum safe inclinations of troughed belt conveyors for handling the various materials of rock excavation, in degrees of angle of inclination, are as tabulated below.

Alluvia (sands and gravels):	
Bank run	18
Screened	15
Residuals or earth (clays, silts, and sands, intermixed)	20
Weathered rock-earth, unripped or ripped, intermixed	20
Rock, crushed	22
Rock, well blasted	20

SOURCE: Hewitt-Robins Inc.

safe angles of inclination for various materials of rock-earth excavation. The angles average 19°, which is a safe angle for all rock-earth.

Power A conveyor system is divided into flights and these flights may be different, each requiring a different horsepower. If this is the case, the designer will have several calculations for horsepower. In this example only one upgrade flight is being analyzed.

The total power to drive the flight is the algebraic sum of the horsepowers required for: (1) moving the empty belt; (2) moving the load horizontally; (3) moving the load vertically; (4) turning all pulleys; and (5) driving the head pulleys.

Moving the Empty Belt. Moving the empty belt over the load-carrying troughing rollers and the return idlers involves a friction loss which varies with the moving parts, that is, the troughing and return idlers and belt, and with the friction factors for the different diameters of the idlers. It is a function of:

L = length of conveyor, ft
S = belt speed, ft/min
C = idler friction factor, Table 13-32
Q = weight of moving parts per foot of length of conveyor, Table 13-32

The values of Q in Table 13-32 are average figures for belts and idlers normally used in transporting rock-earth. They are sufficiently precise for preliminary design. Later, as will be explained, the right belt for the tensile strength requirement will be chosen and its weight will be checked against the value of Table 13-32.

The weight is the sum of carcass, top-cover, and bottom-cover weights. *Carcass weight* is the weight per ply per inch width per foot times the number of plies times the belt width in inches, which gives the number of pounds per foot of belt length. *Cover weight* is the sum of the weights of the top, or load-carrying cover, and the bottom or pulley cover, which taken per inch width per foot times the belt width, gives pounds per foot of belt length.

For canvas fabric belts Table 13-33 gives the approximate weights, thicknesses, and strengths of an individual ply of carcass according to weight of cotton fabric. The weight of cotton fabric is given in ounces per ply per square yard of ply and it includes the bonding material of the plies. The use of Table 13-33 may be illustrated by the following example of a cotton carcass belt of 42-in width and 5-ply, 48-oz construction:

Weight per foot of length = $5 \times 42 \times 0.0395$ = 8.30 lb
Thickness = 5×0.086 = 0.43 in
Strength, tensile = $5 \times 42 \times 58$ = 12,180 lb

For synthetic-fiber belts, carcass weights and thicknesses vary with plies and weights of plies, but they also vary according to the particular synthetic fiber used. The Plylon conveyor belt of the Goodyear Tire and Rubber Co. is representative of a good synthetic fiber belt for strength and life of belt in rock-earth excavation. Table 13-34 gives weights, thicknesses, and strengths of carcass according to number of plies of the Plylon belt. The use of Table 13-34 may be illustrated by the following example of a synthetic fiber carcass belt of 42-in width, with 5-ply, 900-lb-strength construction:

Weight per foot length = 42×0.233 = 9.79 lb
Thickness = 0.50 in
Strength, tensile = 42×900 = 37,800 lb

Cover weights may be taken as the same for both cotton and synthetic fabric belts because top and bottom covers are selected for wear and tear resistance and not for tensile strength. Recommended gauges or thicknesses for top and bottom covers in rock-earth excavation are given in Table 13-35.

The use of Table 13-35 may be illustrated for a 42-in-wide belt carrying sharp and abrasive borax ore of larger than 3-in size. The belt selected is with a top cover three-eighths in thick and bottom cover one-eighth in thick, a total thickness of ½ in. The weight per foot is $42 \times 0.0157 \times 16$ = 10.6 lb.

The total weight per foot of the five-ply, 42-in-wide synthetic fiber belt used to illustrate Table 13-34 is 20.4 lb, with the carcass at 9.8 lb and covers at 10.6 lb.

For a conveyor belt installation calling for a specified tensile strength of belt, the synthetic fabric belt, as compared with the cotton fabric belt, is lighter, thinner, and less costly to own and operate. Because it is lighter, less horsepower is consumed. Because it is thinner, there is better flexing over pulleys and load-carrying troughing rollers.

TABLE 13-32A Representative Values of Weight of Moving Parts, Q, Per Foot of Conveyor

| Width of belt, in | Idlers, 5-in-diameter steel pulleys ||||| Weight of belt, lb/ft | Weight of conveyor, lb/ft ||| Q, lb/ft |
| | Troughing || Return ||| | Idlers || Belt | |
	Wt of revolving parts, lb	Spacing	Wt of revolving parts, lb	Spacing			Troughing	Return		
14	18	5'0"	9	10'0"	2.8	3.6	0.9	5.6	10.1
16	20	5'0"	11	10'0"	3.3	4.0	1.1	6.6	11.7
18	22	5'0"	12	10'0"	4.1	4.4	1.2	8.2	13.8
20	24	5'0"	14	10'0"	4.6	4.8	1.4	9.2	15.4
24	26	5'0"	17	10'0"	7.0	5.2	1.7	14.0	20.9
30	31	4'6"	21	10'0"	8.5	6.9	2.1	17.0	26.0
36	36	4'6"	25	10'0"	11.3	8.0	2.5	22.6	33.1
42	40	4'0"	29	10'0"	17.0	10.0	2.9	34.0	46.0
48	45	3'3"	34	10'0"	23.8	13.8	3.4	47.6	64.8
54	74	2'9"	54	10'0"	29.2	26.9	5.4	73.2	105.5
60	80	2'3"	60	10'0"	32.5	35.6	6.0	74.0	115.6

TABLE 13-32B Friction Factors, C, for Conveyor-Belt Idlers Equipped with Antifriction Bearings

Diameter of idler pulley, in	Friction factor
4	0.0375
5	0.036
6	0.030
7	0.025

SOURCE: Hewitt-Robins Corp.

TABLE 13-33 Approximate Weights, Thicknesses, and Strengths of Single-Ply of Cotton Fabric Belts According to Weight of Cotton Fabric

Weight of cotton fabric, oz	Values per inch width per foot length of belt		
	Weight of ply, lb	Thickness of ply, in	Strength of ply, lb
28	0.0231	0.052	25
32	0.0260	0.057	30
36	0.0295	0.065	35
42	0.0344	0.075	45
48	0.0395	0.086	58

TABLE 13-34 Weights, Thicknesses, and Strengths of Carcasses of a Representative Synthetic Fiber Belt

Number of plies	Values per inch width per foot length of belt		
	Weight, lb	Thickness, in	Strength, lb
2	0.067	0.125	150
2	0.075	0.141	210
3	0.092	0.188	315
4	0.142	0.266	420
3	0.150	0.281	540
4	0.192	0.406	720
5	0.233	0.500	900
6	0.283	0.594	1080

TABLE 13-35 Recommended Thicknesses and Weights of Covers for Conveyor Belts in Rock-Earth Excavation

Nature of rock-earth	Thickness of top or load cover, in	Thickness of bottom or pulley cover, in	Total weight, lb, per in width per ft length
Moderately abrasive residuals	1/16–1/8	1/32	0.047–0.078
Abrasive:			
Sand-gravel			
Weathered rock			
Crushed rock:			
To 3-in size	1/8–3/16	1/16	0.094–0.126
Over 3-in size	3/16–1/4	1/16	0.126–0.157
Abrasive and sharp:			
Ripped rock			
Well-blasted rock:			
To 3-in size	3/16–1/4	3/32	0.141–0.173
Over 3-in size	1/4–1/2	1/8	0.188–0.314

NOTE: Cover weight equals 0.0157 lb per 1/32-in thickness per in width per ft length of belt.

The horsepower to move an empty belt is given by the equation:

$$P = \frac{LSCQ}{33,000}$$

where P = horsepower
  L = length of conveyor, ft
  S = belt speed, ft/min
  C = idler friction factor (Table 13-32B)
  Q = weight of moving parts per foot of conveyor length (Table 13-32A)

Hauling Excavation

The power for a 42-in-wide belt with a length of 1000 ft, speed of 600 ft/min, and with 6-in-diameter troughing rollers may be calculated as follows:
 C, from Table 13-32B, is 0.030 for 6-in-diameter rollers or idlers
 Q, from Table 13-32A, is estimated at 46.0 lb per foot of conveyor
Substituting in the equation for horsepower:

$$P = \frac{1000 \times 600 \times 0.030 \times 46.0}{33,000} = 25.1 \text{ hp}$$

Table 13-36 may be used instead of the equation, giving (600/100)5.07, or 30.4 hp, for 5-in-diameter idlers. Applying the -17 percent correction factor for 6-in-diameter idlers, the hp becomes 25.2, and this figure checks closely with the calculated 25.1 hp.

It has been calculated that the 42-in-wide synthetic fiber belt weighs 20.4 lb/ft, whereas the value of Table 13-32 is 17.0 lb. The corrected value of Q then becomes $(46.0 - 17.0 + 20.4) = 49.4$ lb. The horsepower equation then gives 26.9 hp, a negligible difference of 1.8 hp. For practical purposes, Table 13-32 may be used in horsepower calculations for the moving of the empty belt.

Moving the Load Horizontally. Horizontal movement of the belt incurs a friction loss, which varies according to: L, the length of conveyor in ft; S, the belt speed in ft/min; C, the idler friction factor (Table 13-32B); and W, the weight of the load in lb/ft of belt. The equation for horsepower to move the load horizontally is:

$$P = \frac{LSCW}{33,000}$$

The horsepower for the 42-in-wide belt carrying borax ore and weighing 74 lb/ft^3 may be calculated as follows: From Table 13-30, the total cross-sectional area for 20° angle of slope is 1.115 ft.^{2} Then $W = 1.115 \times 74 = 83$ lb. Substituting in the equation for horsepower:

TABLE 13-36 Horsepower Required to Move Empty Conveyor Belts at a Speed of 100 ft/min

Length of conveyor, ft	Width of belt, in									
	14	16	18	20	24	30	36	42	48	54	60
50	0.05	0.06	0.07	0.08	0.11	0.14	0.18	0.25	0.35	0.54	0.63
100	0.11	0.13	0.15	0.17	0.23	0.28	0.36	0.51	0.70	1.14	1.25
150	0.16	0.19	0.22	0.25	0.34	0.42	0.53	0.76	1.05	1.71	1.88
200	0.22	0.25	0.30	0.33	0.45	0.56	0.71	1.01	1.40	2.28	2.50
250	0.27	0.32	0.37	0.42	0.56	0.70	0.89	1.27	1.75	2.85	3.13
300	0.33	0.38	0.45	0.50	0.68	0.84	1.07	1.52	2.10	3.42	3.76
400	...	...	0.60	0.66	0.90	1.12	1.43	2.03	2.80	4.56	5.01
500	...	...	...	0.83	1.13	1.40	1.79	2.53	3.50	5.70	6.26
600	...	...	...	1.00	1.35	1.68	2.14	3.04	4.20	6.84	7.51
800	...	...	...	...	1.80	2.25	2.86	4.05	5.60	9.12	10.00
1000	...	...	...	...	2.26	2.81	3.57	5.07	7.00	11.40	12.50
1200	...	...	...	...	...	3.37	4.29	6.08	8.40	13.70	15.00
1400	...	...	...	...	...	3.93	5.00	7.09	9.80	16.00	17.50
1600	...	...	...	...	...	4.49	5.72	8.10	11.20	18.30	20.10
1800	...	...	...	...	...	5.05	6.43	9.12	12.60	20.50	22.60
2000	...	...	...	...	...	5.62	7.15	10.10	14.00	22.80	24.90
2200	...	...	...	...	...	...	7.86	11.10	15.40	25.10	27.60
2400	...	...	...	...	...	...	8.58	12.20	16.80	27.40	30.10
2600	...	...	...	...	...	...	9.29	13.20	18.20	29.60	32.60
2800	...	...	...	...	...	...	10.00	14.20	19.60	31.90	35.00
3000	...	...	...	...	...	...	10.70	15.20	21.00	34.20	37.60

NOTE: The power values given in the table are based on the use of 5-in-diameter idlers. For 4-in-diameter idlers, increase the values by 4 percent. For 6-in-diameter idlers, decrease the values by 17 percent. The horsepower required to move the empty conveyor belt at a speed different from 100 ft/min is calculated by multiplying the values of the table by the ratio of the desired speed, in ft/min, to 100.

SOURCE: Hewitt-Robins Corp.

$$P = \frac{1000 \times 600 \times 0.030 \times 83}{33,000} = 45.3 \text{ hp}$$

Table 13-37 expresses the horsepower in terms of capacity in tons hourly and length of belt. The capacity of the 42-in belt under consideration has been calculated at 1487 tons/h. For a 1000-ft length, extrapolation of the table to 1487 tons/h gives 49.1 hp for 5-in-diameter idlers. For 6-in-diameter idlers the horsepower is 40.7, a reasonable check on the calculated 45.3 hp.

Moving the Load Vertically. Lifting or lowering the load vertically is a nonfriction component of the motion. If material is lifted, foot-pounds of work will be required; if material is lowered, foot-pounds of work are contributed. The corresponding power varies according to H, the algebraic net change in elevation, ft, and T, the material moved hourly, tons/h.

The equation for horsepower is:

$$P = \frac{H \times (2000T/60)}{33,000}$$

The horsepower required for the 42-in-wide belt to lift 1487 tons/h of borax ore through a 300-ft change of elevation may be calculated by substituting in the equation the values $H = 300$ ft and $T = 1487$ tons/h:

$$P = \frac{300 \times 2000(1487/60)}{33,000} = 450.6 \text{ hp}$$

Table 13-38 gives horsepower in terms of net lift in feet and capacity in tons per hour. The value for the 300-ft lift extrapolated to 1487 tons/h is 450.6 hp, as is to be expected, in confirmation of the calculated value.

Turning All Pulleys. Turning of all pulleys involves a friction loss. Table 13-39 expresses this loss as a percentage of the power requirement to: (1) move the empty belt; (2) move the load on the belt horizontally; and (3) move the load vertically.

In the example of the 42-in-wide belt, the total horsepower requirement for these three elements of the operation is 521.0. The conveyor flight is 1000 ft long and it is inclined at 17.5°. Table 13-39 gives a loss of 3 percent for babbitted bearings or 1.5 percent for antifriction bearings. In the latter case, the power to turn all pulleys is estimated to be 1.5 percent of 521.0, or 7.8 hp. The total horsepower to drive the loaded belt and turn the pulleys equals 521.0+7.8, or 528.8 hp.

Driving the Head Pulleys. Driving the head pulleys from the prime mover is accompanied by a friction loss. The speed-reducing mechanism may be gears, chain drives, or belt drives. The loss may vary from 5 to 15 percent of the horsepower to drive the head pulleys depending on the kind of speed reducer. A conservative average loss is 10 percent. The loss in power for the 42-in-wide belt of the example is then about 10 percent of 528.8, or 52.9 hp.

Total Necessary Shaft Horsepower of Prime Mover. The total necessary shaft horsepower for the conveyor flight of the borax ore example may be summarized as follows:

The criteria are:
1. Material, borax ore weighing 2000 lb/cylm or 74 lb/ft³ loose measurement
2. Length of conveyor, 1000 ft
3. Vertical lift of conveyor, +300 ft
4. Width of conveyor, 42 in
5. Speed of conveyor, 600 ft/min
6. Capacity of conveyor at 100 percent efficiency, 1487 tons/h

The horsepower distribution is:
1. Moving the empty belt, 25.1
2. Moving the load horizontally, 45.3
3. Lifting the load vertically, 450.6
4. Turning all pulleys, 7.8
5. Transmission or reduction mechanism from prime mover output shaft to head pulleys, 52.9

The sum of these components gives a total brake horsepower of prime mover of 581.7.

Tension Factors and Necessary Number of Drive Pulleys The number of drive pulleys, usually head pulleys, needed for driving the conveyor belt depends on the

TABLE 13-37 Horsepower Required to Move Loads Horizontally on Conveyor Belts

Length of conveyor, ft	Load, tons/h												
	50	100	150	200	250	300	350	400	500	600	700	800	900	1,000
50	0.09	0.18	0.27	0.36	0.46	0.55	0.64	0.73	0.91	1.1	1.3	1.5	1.6	1.8
100	0.18	0.36	0.55	0.74	0.91	1.1	1.3	1.5	1.8	2.2	2.6	2.9	3.3	3.6
150	0.27	0.55	0.82	1.1	1.4	1.6	1.9	2.2	2.7	3.3	3.8	4.4	4.9	5.5
200	0.36	0.73	1.1	1.5	1.8	2.2	2.6	2.9	3.6	4.4	5.1	5.8	6.6	7.3
250	0.46	0.91	1.4	1.8	2.3	2.7	3.2	3.6	4.6	5.5	6.4	7.3	8.2	9.1
300	0.55	1.1	1.6	2.2	2.7	3.3	3.8	4.4	5.5	6.6	7.7	8.8	9.9	10.9
400	0.73	1.5	2.2	2.9	3.6	4.4	5.1	5.8	7.3	8.7	10.2	11.6	13.1	14.6
500	0.91	1.8	2.7	3.6	4.6	5.5	6.4	7.3	9.1	10.9	12.7	14.5	16.4	18.2
600	1.10	2.1	3.2	4.2	5.3	6.4	7.4	8.5	10.6	12.7	14.8	17.0	19.1	21.0
800	1.40	2.7	4.1	5.5	7.5	8.2	9.5	10.8	13.7	16.4	19.1	22.0	25.0	27.0
1000	1.70	3.3	5.0	6.7	9.2	10.0	11.7	13.3	16.7	20.0	23.0	27.0	30.0	33.0
1200	2.0	3.9	5.9	7.9	10.8	11.8	13.8	15.7	19.8	24.0	28.0	32.0	36.0	39.0
1400	2.3	4.5	6.8	9.1	12.4	13.7	15.9	18.1	23.0	27.0	32.0	36.0	41.0	45.0
1600	2.6	5.2	7.7	10.3	12.9	15.5	18	21	26	31	36	41	46	52
1800	2.9	5.8	8.7	11.5	14.4	17.3	20	23	28	35	40	46	52	58
2000	3.2	6.4	9.6	12.7	15.9	19.1	22	25	32	38	45	51	57	64
2200	3.5	7.0	10.5	13.9	17.4	21.0	24	28	35	42	49	56	63	70
2400	3.9	7.6	11.4	15.2	18.9	23.0	27	30	38	46	53	61	68	76
2600	4.1	8.2	12.3	16.4	20.0	25.0	29	33	41	49	57	65	74	82
2800	4.4	8.8	13.2	17.6	22.0	26.0	31	35	44	53	62	70	79	88
3000	4.7	9.4	14.1	18.8	23.0	28.0	33	37	47	56	66	75	85	94

NOTE: The power factors given in the table are based on the use of 5-in-diameter idlers. For 4-in-diameter idlers, increase the values by 4 percent. For 6-in-diameter idlers, decrease the values by 17 percent.

SOURCE: Hewitt-Robins Corp.

TABLE 13-38 Horsepower Required to Lift a Load through a Vertical Distance

Net lift, ft	Load, tons/h												
	50	100	150	200	250	300	350	400	500	600	800	1000
5	0.3	0.5	0.8	1.0	1.3	1.5	1.8	2.0	2.5	3.0	4.0	5.1
10	0.5	1.0	1.5	2.0	2.5	3.0	3.5	4.0	5.1	6.1	8.1	10.0
15	0.8	1.5	2.3	3.0	3.8	4.5	5.3	6.1	7.6	9.1	12.0	15.0
20	1.0	2.0	3.0	4.0	5.1	6.1	7.1	8.1	10.0	12.0	16.0	20.0
25	1.3	2.5	3.8	5.1	6.3	7.6	8.8	10.0	13.0	15.0	20.0	25.0
30	1.5	3.0	4.5	6.1	7.6	9.1	11.0	12.0	15.0	18.0	24.0	30.0
40	2.0	4.0	6.1	8.1	10.0	12.0	14.0	16.0	20.0	24.0	32.0	40.0
50	2.5	5.1	7.6	10.0	13.0	15.0	18.0	20.0	25.0	30.0	40.0	51.0
75	3.8	7.6	11.0	15.0	19.0	23.0	27.0	30.0	38.0	45.0	61.0	76.0
100	5.1	10.0	15.0	20.0	25.0	30.0	35.0	40.0	51.0	61.0	81.0	101
125	6.3	13.0	19.0	25.0	32.0	38.0	44.0	51.0	63.0	76.0	101	126
150	7.6	15.0	23.0	30.0	38.0	45.0	53.0	61.0	76.0	91.0	121	152
200	10.0	20.0	30.0	40.0	51.0	61.0	71.0	81.0	101	121	162	202
300	15.0	30.0	45.0	61.0	76.0	91.0	106	121	152	185	242	303
400	20.0	40.0	61.0	81.0	101	121	141	162	202	242	323	404
500	25.0	51.0	76.0	101	126	151	177	202	252	303	404	505

SOURCE: Hewitt-Robins Corp.

TABLE 13-39 Percentage of Shaft Horsepower Required to Overcome Pulley Frictions for Belt Conveyors with Head Drive and Babbitted Bearings

Length of conveyor, ft	Slope of conveyor, °				
	0	1–6	6–11	11–16	16–20
20	112	93	53	35	28
30	76	63	36	25	19
50	45	38	22	15	13
75	30	25	15	12	9
100	22	19	11	8	7
150	15	14	9	7	6
200	14	11	8	6	5
250	12	10	7	5	5
300	11	8	6	5	4
400	9	6	5	4	4
500	7	6	5	4	3
600	6	5	4	3	3
700	5	4	4	3	3
800	4	4	3	3	3
1000	4	4	3	3	3
2000	4	4	3		
3000	4	3	3		

NOTE: For antifriction bearings use one-half the above-listed percentages for babbitted bearings.
SOURCE: Hewitt-Robins Corp.

effective tension or driving force of the driving pulley, which equals the tight-side tension minus the slack-side tension. The equation is:

$$T_e = T_1 - T_2$$

where T_e = effective tension force, lb
T_1 = tight-side tension force, lb
T_2 = slack-side tension force, lb

The coefficient of friction between the belt and a bare metal-surfaced pulley is about 0.25. When the surface is lagged with rubberlike fabric, the coefficient is about 0.35. Tension on the slack side of the belt should not exceed the amount required to prevent slippage between pulley and belt.

For a driving pulley the effective tension, T_e, to transmit a given horsepower is derived from the power equation:

$$P = \frac{(\pi DN)T_e}{33,000}$$

whence

$$T_e = \frac{33,000 P}{\pi DN}$$

where T_e = effective belt tension force, lb
P = horsepower input to pulley, hp
D = pulley diameter, ft
N = r/min of pulley
πDN = S, belt speed, ft/min

The required effective tension or driving force for the 42-in-wide belt for the borax ore of the foregoing example is given by the equation:

$$T_e = \frac{521.0 \times 33,000}{600} = 28,660 \text{ lb}$$

The minimum tension on the slack side of the belt is the sum of tail-pulley tension and the component of the belt weight along the axis or length of the belt. The tail-pulley tension is fixed empirically at 20 lb per inch of belt width, equivalent in our example to 42×20, or 840 lb. The belt has been estimated to weigh 17.0 lb/ft, totaling 17,100 lb for the 1000-ft length. The component W of the 17,100-lb weight along the axis of the belt is given by the equation:

$$W = 17,100 \sin 17.5° = 5140 \text{ lb}$$

The minimum tension on the slack side of the belt, then, equals 840+5140, or 5980 lb.

The maximum tension on the tight side of the belt equals 28,660+5980 or 34,640 lb. The tension factor, F, of the equation of Table 13-40 equals 34,640/28,660, or 1.21. Table 13-40A indicates that tandem-drive bare pulleys with a 400° arc of contact are adequate for the drive.

Table 13-40B indicates also that no 42-in-wide cotton fabric belt can provide 34,640 lb maximum tension. However, the data of Table 13-34 indicate that a 5-ply Plylon belt with tensile strength of 900 lb per inch of width provides 37,800 lb tensile strength. A belt with these necessary characteristics would be selected. The weight per foot of this belt with a three-eighths-in-thick top cover and one-eighth-in-thick bottom cover has been calculated to be 20.4 lb, as against the assumed weight of 17.0 lb/ft in the calculations for horsepower to move the empty belt. However, it has been demonstrated that the horsepower increase due to increase in belt weight is only a negligible 1.8 hp.

Productions and Costs of a Belt Conveyor

The design and building of a large, custom-built conveyor system should finally be placed in the hands of a specialist. There are two reasons for this.

First, both practicable design and good construction are based on theory and experience. A competent engineer can design a system in keeping with theory but might lack experience in this particular field of design.

Second, reference data in handbooks are usually outdated by at least a decade because of technological advances in design and construction of machinery. Prior to 1950 belting incorporated laminated plies of cotton duck fabric. Typical constructions comprised 4 to 10 plies, as set forth in Figure 13-54 and Table 13-40. In the ensuing years synthetic fibers of greater weight–strength characteristics have replaced cotton fibers, and this has resulted in lighter weight, greater tensile strength, lesser thickness, and more flexibility. Likewise, more abrasion-resistant belt covers have been developed. Generally only a specialist can keep abreast of these money-saving improvements in belt conveyor components.

However, the engineer for the owner sometimes prepares a feasibility study for a conveyor system prior to the final design. Such a study is prepared below for the borax ore conveyor system analyzed in the preceding section on system dynamics.

In this open-pit excavation, off-the-road rear-dump haulers had been used to haul ore from the loading bucket shovel at the bottom of the pit to the primary crusher at the rim

TABLE 13-40A Tension Factors for Driving Pulleys of Belt Conveyors

Arc of contact, deg	Bare pulley	Lagged pulley
	Single-pulley drive	
200	1.72	1.42
210	1.70	1.40
215	1.65	1.38
220	1.62	1.35
240	1.54	1.30
	Tandem drive	
360	1.26	1.13
380	1.23	1.11
400	1.21	1.10
450	1.18	1.09
500	1.14	1.06

Formulas: The ratio T_1/T_e is defined as the pulley tension factor F, as given by the table. Therefore,
$T_1 = FT_e$
$T_e = T_1 - T_2$
T_e = effective tension or driving force between pulley and belt, lb
T_1 = tension on tight side of belt, lb
T_2 = tension on slack side of belt, lb

TABLE 13-40B Allowable Working Tensions in Pounds and Pulley Diameters for Conveyor Belts with Cotton Fabric Carcasses

No. of plies	Weight per ply, oz	Width of belt, in								Diameter of pulley, in		
		16	18	20	24	30	36	42	48	Head, drive, tripper	Tail, takeup, snub	Bend
3	32	1440	1,620	...	...	...	...	...	...	16	12	12
3	36	...	...	1800	2160	...	...	...	...	20	16	12
3	42	...	...	2200	2640	3300	...	...	...	20	16	12
3	48	...	...	...	3840	...	...	...	...	24	20	16
4	28	1600	1800	2000	2400	3000	4320	...	...	20	16	12
4	32	1920	2160	2400	2880	3600	4680	...	...	20	16	12
4	36	...	...	2600	3120	3900	4800	...	...	24	20	16
4	42	...	...	...	...	4800	5760	6720	...	24	20	16
4	48	...	...	...	...	6450	7750	9020	...	30	24	20
5	28	2000	2250	2500	3000	3750	4500	...	...	24	20	16
5	32	...	2700	3000	3480	4500	5400	...	...	24	20	16
5	36	...	...	3400	4080	5100	6120	7140	...	30	24	20
5	42	...	...	...	...	6600	7920	9240	10,560	30	24	20
5	48	...	...	...	...	8700	10,400	12,180	13,920	36	30	24
6	28	...	...	3000	3600	4500	5400	...	...	30	24	20
6	32	...	...	...	4320	5400	6480	7560	10,080	30	24	20
6	36	...	...	...	...	6300	7560	8820	12,900	36	30	24
6	42	...	...	...	...	...	9720	11,340	17,300	36	30	24
6	48	...	...	...	...	...	13,000	15,120	...	42	36	30
7	28	...	...	...	...	5250	6300	8820	10,080	36	30	24
7	32	...	...	...	...	6300	7560	10,300	11,780	36	30	24
7	36	...	...	...	...	...	8820	13,200	15,140	42	36	30
7	42	...	...	...	...	...	...	17,640	20,180	42	36	30
7	48	...	...	...	...	...	...	...	...	48	42	36
8	32	...	...	...	...	...	8640	10,080	11,520	42	36	30
8	36	...	...	...	...	...	...	11,760	13,450	48	42	36
8	42	...	...	...	...	...	...	...	17,300	48	42	36
8	48	...	...	...	...	...	...	...	23,050	54	48	42
9	32	...	...	...	...	...	...	11,340	12,900	48	36	30
9	36	...	...	...	...	...	...	13,200	15,140	54	48	36

SOURCE: Hewitt-Robins Corp.

TABLE 13-41 Recommended Maximum Spacings for Troughing Idlers

Width of belt, in	Weight of material, lb/ft³		
	30–70	70–120	120–150
14	5 ft 6 in	5 ft 0 in	4 ft 9 in
16	5 ft 6 in	5 ft 0 in	4 ft 9 in
18	5 ft 6 in	5 ft 0 in	4 ft 9 in
20	5 ft 6 in	5 ft 0 in	4 ft 9 in
24	5 ft 6 in	5 ft 0 in	4 ft 9 in
30	5 ft 0 in	4 ft 6 in	4 ft 3 in
36	5 ft 0 in	4 ft 6 in	4 ft 3 in
42	4 ft 6 in	4 ft 0 in	3 ft 9 in
48	4 ft 0 in	3 ft 3 in	3 ft 0 in
54	4 ft 0 in	2 ft 9 in	2 ft 6 in
60	4 ft 0 in	2 ft 3 in	2 ft 0 in.

NOTE: Recommended spacing for return idlers is 10 ft 0 in.
SOURCE: Hewitt-Robins Inc.

of the pit, whence a belt conveyor carried the crushed ore to stockpiles or direct to the plant. It was felt that an inclined belt conveyor from a primary crusher on the floor of the pit to the rim of the pit would provide lower costs than upgrade haulage by the rock haulers. At the rim of the pit the ore would be transferred to the existing horizontal conveyor. The relative direct job-unit costs developed here are estimated for 1978 rather than for the year of the analysis.

The cost of vertical lift by a conveyor belt is to be compared with the corresponding cost for only the upgrade haulage and downgrade return of the hauler. The unit cost for the rest of the hauling cycle, that is, loading, two turnings, and dumping, is common to both methods of haulage since the ore must be hauled to the receiving hopper of the crusher of the belt conveyor.

Rear-Dump Truck Haulage

Criteria:	
Capacity, tons	80
Engine flywheel or brake hp	1000
Weights of truck, lb:	
Tare or working	120,000
Load	160,000
GVW	280,000
Haul road:	
Difference in elevation	+300
Length, ft	3750
Grade, %	+8.0
Rolling resistance, % of weight	+3.0
Haul-road construction and maintenance, estimated at 20% of hauling cost	
Cost, 1978, of rear-dump truck	$360,000
Hauling cycle:	
Travel speed upgrade	
Total resistance, grade and rolling, % of weight	11
Maximum travel speed: $\dfrac{(76\% \text{ of } 1000)33,000}{11\% \text{ of } 280,000}$, ft/min	814
Speed factor (Table 13-12):	
280,000 lb/1000 hp = 280 lb/hp; factor = 1.0	
Average travel speed = 1.0 × 814, ft/min	814
Travel speed downgrade	
Safe speed, with retarder action (estimated to be 20 mi/h), ft/min	1760
Hauling cycle times (hauling and returning elements only), min:	
Hauling upgrade, 3750/814	4.61
Returning downgrade, 3750/1760	2.13
Total net cycle time	6.74
Production per 50-min working hour	
Loads hauled hourly, 50/6.74	7.42
Tons hauled hourly, 7.42 × 80	594

13–112 Hauling Excavation

Direct job-unit cost, 1978:
Hourly cost of ownership and operation
Rear dump hauler (Table 8-3), 20% × $360	$72.00
Operator	11.80
Total hourly cost	$83.80
Direct job unit cost per ton, $83.80/594	$0.141
Allowance for haul-road construction and maintenance, 20% × $0.141	0.028
Total direct job unit cost per ton	$0.169

Belt Conveyor Haulage

Criteria:
Capacity, based on 50-min working hour, tons	1239
Two electric motors, total horsepower	600
Belt conveyor:	
Width, in	42
Length, ft	1000
Vertical lift, ft	300

Direct job unit cost, 1978:
Estimated cost of belt conveyor, installed	$399,000
Hourly cost of ownership and operation:	
Belt conveyor (Table 8-3), 31% × $399	$123.70
Operator	15.40
Total hourly cost	$139.10
Direct job unit cost per ton, $139.10/1239	$0.112

The belt conveyor haulage saves $0.057/ton, or 34 percent as compared with truck haulage. The saving corroborates the axiom that, when loading and dumping points are fixed, the belt conveyor is a most economical hauling machine.

The work-cost indicator of a belt conveyor is low. The equation is:

$$\text{Work-cost indicator} = \frac{\text{GVW} + \text{tare weight}}{\text{load weight}}$$

In this case the "gross vehicle weight" is the total weight of the load and the belt along the 1000-ft length of the belt, that is, 83,000 plus 20,400, or 103,400 lb. The tare weight is that of the returning empty belt for the 1000-ft length, 20,400 lb. The load weight is that carried on the 1000-ft length of belt, 83,000 lb. Substituting in the equation:

$$\text{Work-cost indicator} = \frac{103{,}400 + 20{,}400}{83{,}000} = 1.49$$

The corresponding work-cost indicator for the truck is given by the same equation, where: GVW is 280,000 lb; tare weight is 120,000 lb; and load weight is 160,000 lb. Substituting in the equation:

$$\text{Work-cost indicator} = \frac{280{,}000 + 120{,}000}{160{,}000} = 2.50$$

The belt conveyor does 40 percent less total work for a given amount of paywork. As is to be expected, the unit cost for haulage is much less than that for truck haulage, being coincidentally 34 percent less in the cost analysis.

Remarks In the analysis of relative costs by truck and belt conveyor haulages, Table 8-3 was used for the estimated hourly costs of ownership and operation. Table 8-3 is a conservative tabulation of costs to be used in estimating in the United States, and the figures are high with respect to those of a company like the United States Borax and Chemical Corp. However, the costs are proportionate, so that the relative, although not the absolute, costs may be used safely for comparative purposes.

During the 18 yr of operation of the complete conveyor system, three new belts have been installed on the inclined flight considered in these analyses. The belts are similar to the 5-ply belt of Table 13-34. The belts have averaged 6 yr or 24,000 h of service, double shifting, and 24 million tons of borax ore haulage. The 1978 cost for the 2000-ft-long, 42-in-wide belt is about $83,000 and the maintenance on the belt for 6 yr approximately equals the cost. The hourly costs, then, are:

Depreciation, $83,000/24,000	$3.46
Repairs, including labor, $83,000/24,000	3.46
Total hourly cost	$6.92

The unit cost for depreciation and for repairs and related labor per ton of borax ore raised vertically 300 ft is $6.92/1000 = $0.0069.

Summary

Belt conveyor systems are used widely in open-cut excavations for public and private works. Both stationary and movable machines offer high productions and low unit costs for haulage when conditions are favorable. They handle all kinds of rock-earth from clay-silt-sand residuals through well-blasted rock. With receiving hopper and feeder they provide a continuous flow of material, which augurs high efficiency and low cost. Fixed loading and dumping points of the stationary belt conveyor are requisite to most competent operation. Movable belt conveyors may or may not have both fixed positions, but their mobility more than offsets this handicap.

The economy of belt conveyors relative to other kinds of haulage is explained theoretically by the low weight of moving parts as contrasted to the weight of the load. The resultant work-cost indicator is the lowest for any machine for haulage.

CABLEWAY SCRAPER-BUCKET SYSTEMS

There are two kinds of systems, depending on the way of handling the scraper or bucket. One is the dragscraper machine, and the other is the track cable machine, which is sometimes called a slackline excavator.

Dragscraper Machine

As the name implies, the bucket of the dragscraper is dragged, both laden and unladen, on the surface of the ground. The only bucket used is the bottomless type, which is known as a *crescent bucket* because of its shape.

Figure 13-58 illustrates a dragscraper machine. The inhaul and pullback cables are shown, as well as the tail block and bridle trolley for lateral movement of the line of travel of the scraper bucket.

The three-drum hoist arrangement of Figure 13-59 shows the setup of cables for the machine of Figure 13-58. The two-drum hoist arrangement does not provide for the

Figure 13-58 Dragscraper machine excavating materials pit. The 4-yd³ bottomless crescent scraper is hauling desert alluvium and well-weathered sedimentary rock a one-way distance of up to 500 ft. The tail block is attached to the bridle trolley, the block being shifted to change the scraper's line of travel. The head works are a mast and three-drum hoist. The prime mover may be a 200-hp electric motor or a 280-hp diesel engine. The schematic layout for the machinery is shown by *b* in Figure 13-59.

The hourly capacity of the machine on the 500-ft haul when working at 100 percent efficiency is about 100 cylm, equivalent to about 77 cybm in this well-weathered formation. *(Sauerman Brothers.)*

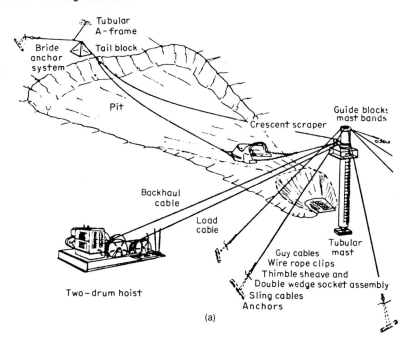
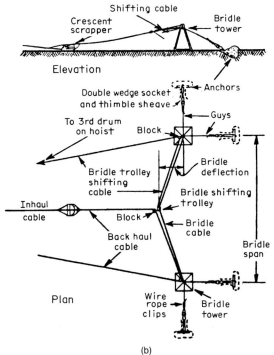

Figure 13-59 Schematic layouts of two- and three-drum hoist operation of Sauerman tower-type dragscrapers. (a) Typical dragscraper machine with two-drum hoist. (b) Rapid-shifting bridle arrangement used with three-drum hoist. (Sauerman Brothers.)

lateral movement of the tail block, this necessary movement being supplied by manual or mechanical movement of the bridle anchor system. The mast-type dragscraper machine is used when there are infrequent changes in position of the head tower. Such an operation would be appropriate in a deep pit and where a permanent dumping point is used, such as the receiving hopper of the illustration.

Figure 13-60 illustrates the configuration of the crescent bucket. In this operation the bucket is being used as part of a track cable machine, as it will be elevated above the ground for the return to the loading point. The crescent bucket may be used with both dragscraper and track cable machines, whereas the dragline-type bucket can be worked only with the track cable or slackline excavator.

Figure 13-61 illustrates a skid-mounted dragscraper machine with a two-drum hoist, one for inhaul and one for pullback. These machines have 1- to 4-yd³-capacity crescent buckets. The portable type can be assembled and dismantled in the field and is suitable for permanent or temporary work. Mobility is important when several pits must be used for one job. Figure 13-62 shows three-drum operation of a skid-mounted dragscraper, which facilitates work in shallow pits, as the tail block can be shifted over the width of the pit. Figure 13-63 illustrates two-drum hoist assemblies with electric-motor and diesel-engine prime movers. Such power assemblies with both two- and three-drum hoists range from 20 to 350 hp for machines of 1- to 5-yd³ bucket capacity.

Specifications Tables 13-42, 13-43, and 13-44 give the principal specifications for dragscraper machines of 1- to 5-yd³ crescent-bucket capacity.

Figure 13-60 Configuration of bottomless crescent scraper for dragscraper and track cable machines. The 12-yd³ scraper has dumped a load into the hopper and is being supported on the track cable trolley for return to the digging position. The lower cable is the pullback line. The three-drum cable setup is shown in the schematics of Figure 13-65. The bottomless scraper is pulled on the surface to the hopper. Such a huge scraper weighs about 12,000 lb and hauls some 36,000 lb of alluvium in each pass of the bucket. On a 1000-ft one-way haul, the hourly production at 100 percent efficiency is about 156 cylm, or some 137 cybm. *(Sauerman Brothers.)*

13-116 **Hauling Excavation**

Figure 13-61 **Skid-mounted dragscraper with 3-yd³ crescent bucket.** This machine is driven by an electric motor of 125 hp and it can operate on a span up to 500 ft in length. Most cable excavators haul materials from wet or submerged pits under adverse conditions. They are uniquely, and almost exclusively, suited to these adverse conditions. The tail anchorage for this machine is a crawler-tractor-bulldozer, which may be seen in part at the extreme left side of the picture. The bulldozer has a load in front of the blade to help resist the pull of the tail pulley. The material is sand-gravel alluvium. On this 200-ft haul the dragscraper handles about 126 cybm, or some 189 tons, in a 50-min working hour. *(Sauerman Brothers.)*

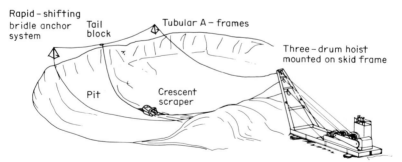

Figure 13-62 Schematic layout of three-drum hoist operation of dragscraper with skid-mounted machinery deck. Typical installation of 1- to 4-yd³ skid-mounted dragscraper machine equipped with three-drum diesel hoist. *(Sauerman Brothers.)*

(a)

(b)

Figure 13-63 Electric-motor- and diesel-engine-driven double-drum hoists for dragscrapers. *(a)* Small electric hoist equipped with two-speed front drum and manual lever controls. Front-drum gearing handles digging action at low speed, conveys at high speed. *(b)* Torque converter diesel-operated 1-yd³ hoist with manual pneumatic controls mounted on operator's platform.

Electric-motor and diesel-engine prime movers of dragscrapers may drive two- or three-drum hoists. Power ranges from 20 hp for ½-yd³ capacity to 350 hp for 5-yd³ capacity. The three-drum hoist is used for track cable or slackline excavators. The corresponding prime movers range in power from 20 hp for ½ yd³ to 410 hp for 3½ yd³ capacity. Large-capacity machines, illustrated in Figures 13-60 and 13-64, are specially built and require appropriate power for the two- and three-drum hoists. *(Sauerman Brothers.)*

TABLE 13-42 Specifications of Sauerman ½- to 5-yd³ Dragscraper Machines

Machine no.	Code word	Kind of power	hp	Machine size, yd³	Length of span, ft	Approx. shipping weight, lb
		Equipped with 2-drum hoists				
ASG-1302	AXCGS	Gas	30	½	200	7,925
ASE-1302	AXCEC	Electric	20	½	200	7,170
ASD-1302	AXCDA	Diesel	30	½	200	8,750
ASG-2002	AXDGU	Gas	45	¾	300	11,285
ASE-2002	AXDEE	Electric	30	¾	300	10,140
ASD-2002	AXDDB	Diesel	45	¾	300	11,615
ASG-2602	AXFGW	Gas	60	1	300	11,985
ASE-2602	AXFEH	Electric	40	1	300	10,850
ASD-2602	AXFDC	Diesel	60	1	300	12,500
ASG-4002	AXGGY	Gas	90	1½	400	16,290
ASE-4002	AXGEJ	Electric	60	1½	400	14,485
ASD-4002	AXGDE	Diesel	90	1½	400	16,170
ASG-5402	AXHGE	Gas	120	2	400	21,825
ASE-5402	AXHEL	Electric	75	2	400	20,070
ASD-5402	AXHDF	Diesel	120	2	400	21,265
ASE-3015	AXKEP	Electric	150	3	400	32,400
ASD-3015	AXKDH	Diesel	225	3	400	35,360
ASE-4020	AXLEM	Electric	200	4	500	35,855
ASD-4020	AXLDO	Diesel	280	4	500	37,750
ASE-5020	AXMEP	Electric	250	5	500	48,810
ASD-5020	AXMDQ	Diesel	350	5	500	48,015

		Equipped with 3-drum hoists			
ASG-313	AXCCT	Gas	30	200	12,775
ASE-313	AXCED	Electric	20	200	12,105
ASD-313	AXCDI	Diesel	30	200	13,685
ASG-320	AXDGV	Gas	45	300	16,165
ASE-320	AXDEG	Electric	30	300	14,970
ASD-320	AXDDJ	Diesel	45	300	16,500
ASG-326	AXFGX	Gas	60	300	17,075
ASE-326	AXFEI	Electric	40	300	15,900
ASD-326	AXFDK	Diesel	60	300	17,140
ASG-340	AXGGZ	Gas	90	400	21,900
ASE-340	AXGEK	Electric	60	400	20,185
ASD-340	AXGDL	Diesel	90	400	21,400
ASG-354	AXHGS	Gas	120	400	28,440
ASE-354	AXHEM	Electric	75	400	26,685
ASD-354	AXHDO	Diesel	120	400	27,400
ASE-3003	AXKEQ	Electric	150	400	41,800
ASD-3003	AXKDN	Diesel	225	400	44,700
ASE-4003	AXLER	Electric	200	500	49,775
ASD-4003	AXLDP	Diesel	280	500	51,665
ASE-5003	AXMES	Electric	250	500	66,335
ASD-5003	AXMDR	Diesel	350	500	65,515

SOURCE: Sauerman Brothers.

13–120 Hauling Excavation

TABLE 13-43 Specifications of Sauerman 1- to 4-yd³ Skid-Mounted Dragscraper Machines

Machine no.	Code word	Kind of power	hp	Scraper size, yd³	Length of span, ft
		Equipped with 2-drum hoists			
ASG-2602S	ACFGA	Gasoline	60	1	300
ASE-2602S	ACFEY	Electric	40	1	300
ASD-2602S	ACFDX	Diesel	60	1	300
ASG-5402S	ACHGE	Gasoline	120	2	400
ASE-5402S	ACHEZ	Electric	75	2	400
ASD-5402S	ACHDU	Diesel	120	2	400
ASE-3015S	ACKEV	Electric	150	3	400
ASD-3015S	ACKDW	Diesel	225	3	400
ASE-4020S	ACLER	Electric	200	4	500
ASD-4020S	ACLDS	Diesel	280	4	500
		Equipped with 3-drum hoists			
ASG-326S	ADFGU	Gasoline	60	1	300
ASE-326S	ADFET	Electric	40	1	300
ASD-326S	ADFDO	Diesel	60	1	300
ASG-354S	ADHGQ	Gasoline	120	2	400
ASE-354S	ADHEP	Electric	75	2	400
ASD-354S	ADHDN	Diesel	120	2	400
ASE-3003S	ADKEK	Electric	150	3	400
ASD-3003S	ADKDL	Diesel	225	3	400
ASE-4003S	ADLEM	Electric	200	4	500
ASD-4003S	ADLDI	Diesel	280	4	500

SOURCE: Sauerman Brothers.

TABLE 13-44 Specifications for Crescent Scrapers or Dragscrapers with Teeth and Roller-Bearing Carriers

Scraper and carrier no.	Code word	Size, yd³	Type of carrier	Recommended size of cables, in — Load	Track	Approx. weight scraper carrier, lb
SC-1050	JAYUS	1	Two-wheel	⅝	¾	1,290
SC-1060	JAYVA	1½	Two-wheel	¾	⅞	1,590
SC-1070	JAYWE	2	Two-wheel	⅞	1	2,870
SC-1090	JAYYB	3	Three-wheel	1	1⅛	3,140
SC-1100	JAYZU	4	Three-wheel	1⅛	1⅛	4,450
SC-1110	JAYFG	5	Three-wheel	1⅛	1¼	4,880
SC-1120	JAYGJ	6	Four-wheel	1¼	1⅜	6,900
SC-1130	JAYKL	8	Four-wheel	1¼	1⅜	8,940
SC-1140	JAYLY	10	Four-wheel	1⅜	1½	9,700
SC-1230	JAYGT	12	Four-wheel	1½	1½	11,840
SC-1240	JAYHM	15	Four-wheel	1½	1⅝	16,000
SC-1250	JAYIP	20	Four-wheel	1⅞	1⅞	22,500

SOURCE: Sauerman Brothers.

Productions Representative hourly productions, cubic yards loose measurement, at 100 percent efficiency are given in Tables 13-45 and 13-46. As is noted in the tables, capacities are given in cubic yards loose measurement, weighing 3000 lb. In order to convert these productions to cubic yards bank measurement according to the nature of the material, the conversion factors of Table 13-47 should be used. The nominal capacity of the crescent scraper corresponds to the average load, loose measurement, because the crescent bucket is continually digging and transporting over the ground surface and thus maintains a heaped load at all times.

An example of the use of the tables is provided by the 12-yd³ crescent scraper of Figure

TABLE 13-45 Rated Handling Capacities of Sauerman Dragscraper Machines in Cubic Yards Loose Measurement

	Machines in sizes from ½ to 5 yd³				
	Capacity in cylm/h at average haul of:				
Machine size, yd³	100 ft	200 ft	300 ft	400 ft	500 ft
½	40	24	18	14	
¾	62	34	25	20	16
1	82	48	35	27	21
1½	130	75	53	42	33
2	172	98	72	55	44
3	293	172	120	93	76
4	391	230	161	123	100
5	488	288	200	155	126

	Machines in sizes from 6 to 15 yd³				
Length of haul, ft	Dragscraper machine size, yd³				
	6	8	10	12	15
100	420	560	700	840	1050
200	282	375	470	564	705
300	215	280	350	420	525
400	168	224	280	336	420
500	138	184	230	276	345
600	120	160	200	240	300
700	108	144	180	216	270
800	96	128	160	192	240
900	...	...	140	168	210
1000	...	...	130	156	195

NOTE: Handling rates are based on normal digging, free-caving material with an average weight of 3000 lb yd³. Dragscraper machines, 1½ through 5-yd³, show capacities when equipped with torque-converter diesel power. Handling capacities are at 100 percent efficiency.
SOURCE: Sauerman Brothers.

TABLE 13-46 Rated Capacities of Sauerman Skid-Mounted Dragscraper Machines

Mach. size, yd³	Capacity, cylm/h at average haul of				
	100 ft	200 ft	300 ft	400 ft	500 ft
1	82	48	35	27	21
2	172	98	72	55	44
3	293	172	120	93	76
4	391	230	161	123	100

NOTE: Handling rates are based upon normal digging, free-caving material with an average weight of 3000 lb yd³ and torque-converter diesel-powered machines. Handling capacities are at 100 percent efficiency.
SOURCE: Sauerman Brothers.

TABLE 13-47 Factors for Converting Rated Hourly Capacities of Dragscrapers as Given in Tables 13-45 and 13-46 from Cubic Yards Loose Measurement to Cubic Yards Bank Measurement

Nature of rock-earth	Conversion factor $\left(\dfrac{\text{bank measurement}}{\text{loose measurement}}\right)$
Alluvium: silts, sands, and gravels	0.88
Residuals: clays, silts, and sands	0.78
Well-weathered rock, either unripped or ripped	0.67

13-60, which is shown hauling alluvium a one-way distance of 1000 ft. Table 13-45 gives an hourly production of 156 cylm at 100 percent efficiency. The conversion factor is 0.88. The hourly production is then 0.88×156, or 137 cybm. At the efficiency of the 50-min working hour, 83 percent, the hourly production is 114 cybm.

This example emphasizes the relationships among rated capacity loose measurement, capacity bank measurement, capacity at 100 percent efficiency, and capacity at average working efficiency. One must allow for all these variables in arriving at the practicable output of excavation machinery. In this instance the well-defined hourly output specified by the manufacturer has had to be reduced by 27 percent in order to arrive at an hourly production figure for estimating direct job unit cost.

Estimated Productions and Costs The calculations are illustrated by the following examples.

EXAMPLE 1. In Figure 13-58 the machine is excavating a mixture of alluvium and well-weathered sedimentary rock in the materials pit. The 4-yd^3 capacity scraper is hauling 500 ft, one way.

Hourly production:	
At 100% efficiency (Table 13-45)	100 cylm
At 50-min working-hour efficiency, $83\% \times 100$	83 cylm
Average conversion factor for alluvium	
and well-weathered rock (Table 13-47), 0.78; $0.78 \times 83 =$	65 cybm
Direct job unit cost, 1978:	
Cost of complete machine, erected	$188,000
Hourly cost of ownership and operation	
Slackline excavator, dragscraper type	$48.90
(Table 8-3), $26\% \times \$188$	
Operator	15.60
Total hourly cost	$64.50
Direct job unit cost per cybm, $64.50/65	$0.99

EXAMPLE 2. In Figure 13-61, the skid-mounted electric-motor-driven dragscraper with 3-yd^3 crescent scaper is hauling sand-gravel alluvium a distance of 200 ft.

Hourly production:	
At 100% efficiency (Table 13-46),	172 cylm
At 50-min working-hour efficiency. $83\% \times 172$	143 cylm
Conversion factor (Table 13-47), 0.88; $0.88 \times 143 =$	126 cybm
Direct job unit cost, 1978:	
Cost of complete machine, erected	$132,000
Hourly cost of ownership and operation	
Slackline excavator, dragscraper type, equivalent,	
machine (Table 8-3), $26\% \times \$132$	$34.30
Operator	15.60
Total hourly cost	$49.90
Direct job unit cost per cybm, $49.90/126	$0.40

As is to be expected, when the similar machines of Examples 1 and 2 are hauling similar materials, the unit cost per station-yard is about the same. For Example 1 it is $0.99/5.00, or $0.20, and for Example 2 it is $0.40/2.00, also $0.20.

Track Cable Machine

This complete machine has headworks, tailworks, hoists, and the track-cable assembly for carrying the laden and unladen dragline-type bucket or the unladen crescent scraper. The machine is also known as a *slackline excavator*.

Figures 13-64 through 13-68 illustrate complete machines, tower and cable layouts, and dragline bucket operation. Operating characteristics and uses are similar to those of the dragscraper machine. These machines are associated with large-scale operations and not with work similar to that of Figure 13-61, which shows a skid-mounted dragscraper excavating and hauling in a small, shallow pit.

Specifications Tables 13-48 and 13-49 give principal specifications for track cable machines up to 3½-yd^3 capacity. As has been mentioned, the machine may use a bottomless crescent scraper (Table 13-44) or a dragline-type bucket (Table 13-49).

Productions Production depends largely on whether the machine uses the conventional dragline bucket or the crescent scraper.

With Dragline Bucket. Table 13-48 includes estimated hourly productions at 100 percent efficiency in cubic yards loose measurement. Table 13-47, which gives conversion factors for dragscrapers, may also be used to convert the hourly productions of Table 13-48 from loose measurement to bank measurement.

With Crescent Scraper. Tables 13-45 and 13-46, along with the supplemental Table 13-47, may be used to give approximate production data for the track cable machine

Figure 13-64 Two-tower track cable machines with 8-yd^3 crescent scrapers. These large cable excavators with rail-mounted head and tail towers are removing overburden from an ore body. The overburden is being dumped in a spoil pile in front of the head towers. The haul is about 600 ft long. The hourly production of each machine at 100 percent efficiency is about 160 cylm, equivalent to about 125 cybm in the overburden of residuals.

Machines like these are custom-built without standard specifications. The drive is by three-drum hoist and the cable arrangement is shown in Figure 13-65. The total horsepower of the electric motors for each machine is 400, and for an alternative diesel engine the horsepower would be about 600.

The excavation is dry and the machines would be ideal for haulage of the same overburden from beneath a water table. This is severe work in which the cable excavator excels over other excavating and hauling machines. *(Sauerman Brothers.)*

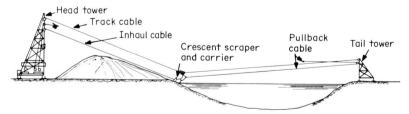

Figure 13-65 Schematic layout of tower machines and three-drum hoist operation of dragscraper. The sectional elevation shows working range of the crescent. The excavated material is deposited on the pile in front of the head tower. *(Sauerman Brothers.)*

13-124 Hauling Excavation

Figure 13-66 Open-pit mining of lead-zinc ore by track cable machine. The machine, with a dragline-type bucket of 2½-yd³ capacity, is digging and hauling rich soft ore in the tristate mining district of Missouri. Ore is hauled an average 500-ft distance to the storage bin. The maximum depth of the pit is 165 ft and the height of the head mast is 83 ft, giving a total lift of about 225 ft to the top of the bin. Figure 13-67 illustrates the dumping action of the bucket for a slackline excavator. Figure 13-68 illustrates the cable layout for the work.

The ore weighs 2970 lb/cylm. The machine averages 75 tons/h, being delayed considerably by the capacity of the washing plant. At 100 percent efficiency its capacity is about 140 tons/h, and at 50-min working-hour efficiency its capacity is about 117 tons/h.

The prime mover is a 200-hp electric motor driving a two-drum hoist. Three split-end eye bolts are embedded firmly in the limestone highwall for the tail anchors. As the bucket returns by gravity to the loading position, a two-drum hoist is used instead of the customary three-drum unit.

The oversize rock from the grizzly atop the hopper is stockpiled by a dragscraper machine. This machine, equipped with a 4-yd³ crescent scraper, was used initially to remove 60 ft, or 530,000 tons, of overburden prior to commencing excavation and haulage of the ore body.

In summary, the combination of dragscraper and track cable machines is working efficiently and economically in this selective mining of lead-zinc ore. *(Sauerman Brothers.)*

when it drags the scraper on the haul and elevates it on the track cable for the return to the loading point.

Estimated Productions and Costs   The example is the open-pit mining of lead-zinc ore, as illustrated and described in Figures 13-66 and 13-68. In Figure 13-66 the attending crawler-tractor-bulldozer is used to maintain a pile of ore ahead of the bucket during times of peak performance, 140 tons/h. The average production is 75 tons/h. It is assumed that the bulldozer works half of the time in the pit and that half of its hourly expense of ownership and operation is charged to the complete loading and hauling operation. The direct job unit cost is based on the average production for the combined operation, 75 tons/h.

Estimated costs, 1978:
  Cost of complete track cable machine, erected $318,000
  Cost of 200-hp crawler-tractor-bulldozer $112,000
  Hourly cost of ownership and operation:

	Track Cable Machine	Crawler-Tractor-Bulldozer
Slackline excavator (Table 8-3), 26% × $318	$82.70	
Operator	15.60	
Crawler tractor bulldozer (Table 8-3), 34% × $112	...	$38.10
Operator	...	15.60
Total hourly cost:	$98.30	$53.70
Direct job unit cost per ton:		
$98.30/75	$1.31	
(50% × $53.70)/75	...	$0.36

The total estimated direct job unit cost is $1.67/ton of ore. If there were no plant delays, the production would be about 117 tons/h, requiring full-time work of the bulldozer. The direct job unit cost would then be ($98.30+$53.70)/117, or $1.30/ton of ore.

The work-cost indicator for cableway scraper-bucket systems is low. For example, the indicator for the dragscraper machine of Figure 13-58 may be calculated in the following manner. The tare or working weight, that is, the total weight of the 4-yd^3 crescent scraper and 500-ft-long cable system, is 6700 lb. The load weight of 4 cylm of alluvia and well-weathered sedimentary rock at 3300 lb/cylm is 13,200 lb. The GVW is 13,200+6700, or 19,900 lb.

Figure 13-67 Dumping action of dragline bucket of track-cable machine. The dumping point is predetermined by a dump button affixed to the track cable. When the traveling carrier reaches this stopping button, the dump block of the chain assembly moves forward and away from the traveler block. This action pulls the dump chain around its sheave and raises the back of the bucket for dumping. The slackline excavator is shown hauling outsize rock from a river bottom deposit to a surge pile for later land haulage to its destination. *(Sauerman Brothers.)*

TABLE 13-48 Specifications and Rated Hourly Capacity for Sauerman Slackline Excavators

Machine no.	Code word	Kind of power	Approximate hp	Size of bucket, yd³	Length of span, ft	Handling capacity in cylm/h when digging 30 ft below mast on an average haul of						Approximate shipping weight, lb
						50 ft	200 ft	250 ft	300 ft	400 ft	500 ft	
UCE-2131	CYCED	Electric	20	½	400	25	22	20	17	...	...	16,040
UCD-2131	CYCDY	Diesel	30	½	400	25	22	20	17	...	...	17,390
UCE-2201	CYDEF	Electric	40	¾	500	39	37	36	33	...	...	22,795
UCD-2201-TC	CYDDA	Diesel	92	¾	500	57	54	51	46	...	...	23,020
UCE-2261	CYFEG	Electric	60	1	600	51	50	46	43	37	...	38,005
UCD-2261-TC	CYFDI	Diesel	140	1	600	83	80	70	65	55	...	39,255
UCE-2401	CYGEH	Electric	100	1½	700	...	79	76	73	66	57	54,740
UCD-2401-TC	CYGDA	Diesel	140	1½	700	...	97	91	84	72	61	53,970
UCE-2541	CYHEJ	Electric	150	2	800	...	...	94	90	80	70	64,930
UCD-2541-TC	CYHDI	Diesel	207	2	800	...	...	114	110	98	84	67,330
UCE-2651	CYJEA	Electric	200	2½	900	...	...	113	112	105	95	92,200
UCD-2651-TC	CYJDM	Diesel	320	2½	900	...	...	150	147	125	107	93,600
UCE-2911	CYLET	Electric	250	3½	1000	...	...	...	156	140	129	154,940
UCD-2911-TC	CYLDS	Diesel	410	3½	1000	...	...	...	182	161	143	157,840

NOTE: Handling capacities are at 100 percent efficiency and in cubic yards loose measurement, weighing 3000 lb/yd³.
SOURCE: Sauerman Brothers.

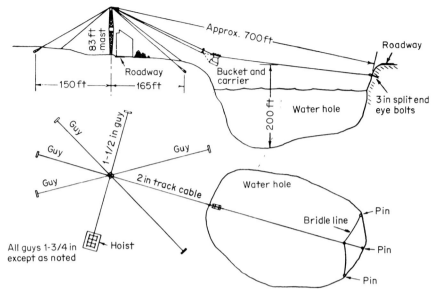

Figure 13-68 Schematic layout of open-pit mining of lead-zinc ore by slackline excavator. The slackline dumps the ore into an elevated hopper, a grizzly keeping out oversize. *(Sauerman Brothers.)*

TABLE 13-49 Representative Specifications of Dragline-Type Buckets for Slackline Cableway Operation

Nominal size, yd³	Capacity, cysm	Weight of bucket, lb		
		Light-duty	Medium-duty	Heavy-duty
1	1.9	2300	3000	3800
1½	1.7	3100	3800	4600
2	2.2	4000	4900	5500
2½	2.7	4400	5700	6600
3	3.3	5600	6700	8000
3½	3.8	6100	7600	9000
4	4.4	6800	8500	10000

$$\text{Work-cost indicator} = \frac{\text{GVW} + \text{tare weight}}{\text{load weight}} = \frac{19{,}900 + 6700}{13{,}200} = 2.02$$

One must interpret the 2.02 indicator with the knowledge that the machine is both a loader and a hauler, and accordingly the indicator is remarkably low.

Summary

Dragscraper and track cable machines are used for open-pit or borrow-pit excavations either in the dry or in the wet, and sometimes the wet loading and hauling are beneath the surface of the water. This subaqueous ability is shared by crane-clamshells, draglines, and backhoes for loading and casting and by dredges for loading and hauling. The cableway machines both load and haul under the most adverse conditions, and this unique property gives them versatility exceeding that of most other machines. The cableway scraper-bucket machines handle a variety of rock-earth, the dragline bucket of the track-cable machine being able to handle well-blasted rock under favorable conditions.

Operating expenses, especially maintenance, are low. Only the scraper or bucket and the inhaul and pullback cables make contact with the excavation. Wear and tear are confined almost exclusively to a few exposed working parts. Partially offsetting these economies are the high fixed charges with respect to the hourly capacity of the machine.

13-128 Hauling Excavation

There are many rock-earth excavations which are obviously proper work for cableway scraper-bucket machines. There are others which should be considered physically and financially because they merit alert consideration.

RAILROADS

Physical and Economic Criteria

When railroads are used for haulage they may be worked in primary or secondary applications. In this text primary haulage is defined as movement of material directly from the cut, with the exception of delivery of material from minor operations such as grizzlies and primary crushers. Secondary haulage is described as movement of material after processing, as by an aggregate plant. Haulage in this text means primary work, and when railroads are so used they are engineered and built almost always in keeping with the following criteria:

1. A fixed or nearly fixed loading point. An example of a fixed point is a loading hopper into which rock-earth has been dumped by a shovel, this operation calling for little or no track shifting. A nearly fixed point is a loading shovel in a quarry rock cut, where lateral shifting of track is necessary, as illustrated in Figure 13-69.

2. Fairly long hauls, preferably in excess of 2 mi in one direction.

3. Nearly level and close to straight hauls, with grades not exceeding +2.0 percent and maximum curvature of 10°. *Curvature* is the central angle subtended by a chord of 100-ft length. For 10° the radius of curvature is 573 ft.

4. A fixed dumping point such as a surge bin or the primary crusher of a cement mill.

Figure 13-69 Loading a train consisting of locomotive and 10 quarry cars. The large-capacity electric shovel is loading alternate cars with 12 tons of limestone to give half of a train load, or a total of 60 tons. The train will go to another loading shovel at another quarry location for the remaining 60 tons, giving a total train load of 120 tons. This method of loading blends the variable limestones for industrial uses. The limestone stratum averages 20 ft in thickness and it is covered by some 60 ft of overburden, which is cast aside by a walker dragline of large capacity. *(Plymouth Locomotive Works.)*

These four desirable conditions are physical, but the overriding requisite is economic. The total cost of the railroad, including right-of-way and all construction, must be amortized over many years or over a vast amount of yardage or tonnage of material. Likewise, the anticipated life of the work must be in keeping with the long depreciation periods for locomotives and cars, 12 to 20 yr or more. Figures 13-69 to 13-71 illustrate phases in the haulage of limestone between quarry and plant in eastern Ohio. The country is gently rolling and the trackage at the top of the limestone ledge is almost level. The total trackage is in a long loop of total length about 5.0 mi, giving a one-way haul of about 2.5 mi.

Figure 13-72 illustrates the loading operation in haulage of dolomite or fluxing stone in a large quarry in Michigan. Here the 5.2-mi one-way haul is level and practically straight except for the wide curves at the shovel and primary crusher.

Both these examples of railroad haulage are in areas of the Central states where the dip of beddings of sedimentary rocks is almost negligible, making for ideal topographic conditions for the use of railroads.

Typical specifications for locomotives and railroad cars of the types used in quarrying are listed in Tables 13-50 and 13-51, respectively.

Principles and Dynamics of Railroad Haulage

In rock-earth excavation by railroads, the following interrelated factors must be considered:

1. The hourly production from the loading machine, as in the case of simple excavation, or from an intermediate production machine, as in the case of a primary crusher located in the quarry rather than at the plant

2. The size of the bucket or dipper of the shovel, which predetermines the capacity of the cars so as to achieve balance between loader and hauler

3. The topography of the land, which determines grades and curves and thereby sizes of trains and accelerations and speeds of the locomotive

4. The degree of refinement in the roadbed, ballast, and track, which fixes train resistances and speeds

These and other important considerations determine the selection of rolling stock and their performance and costs of haulage.

Roadbeds Good drainage with resultant firmness of roadbed is necessary for economical operation. Proper, well-laid ballast and superelevated spiral curves complete the picture for fast speeds. Rail transportation should not be over snaky, uneven tracks, cheaply and hastily laid, with little or no attempt to tamp properly around the ties so as to get evenness and rigidity of rails.

Rail Size The size of rail is expressed in terms of pounds per yard of length, a 60-lb rail weighing 60 lb/yd. The standard rail length is 33 or 39 ft. Rails of 16 to 45 lb are used on short lines where small cars are used; 35- to 80-lb rails are common sizes on narrow-gauge railroads; and 80-lb and heavier rails are used on standard gauge for heavy rolling stock and high speeds.

Rail Capacity The load which may be carried over the steel rails is dependent generally on the condition of the roadbed and the proper spacing of the crossties, which are sometimes known as sleepers. To get the full benefit of the carrying capacity of the rails, crossties must be placed properly and close enough together to allow the rail to carry its capacity and to provide a wide safety margin without bending or springing.

Ties or sleepers are spaced usually from 16 to 24 in apart. The determining factor is not the weight of the locomotive or cars but the permissible weight carried per wheel. Assuming a fair roadbed and crossties properly spaced, rails of 10 to 30 lb will carry about 200 lb per pound weight of rail, 30- to 60-lb rails will carry 250 lb per pound weight of rail, and rails weighing above 60 lb will carry 275 to 350 lb per pound weight of rail.

For example, on a fair track with good crossties and 40-lb rail, the load limit per wheel is 40×250 or 10,000 lb, that is, 5 tons maximum load per wheel. Therefore, a 0-4-0 classification 20-ton locomotive is the largest four-wheel size permissible. On the same track a six-wheel, 0-6-0 30-ton locomotive or an eight-wheel, 0-4-4-0 double-truck 40-ton locomotive may be used.

Rail Gauge The gauge of a railroad or track is the distance between rails measured from inside to inside of rail head. Those most commonly used for rock-earth haulage are 30, 36, 42, and 56½ in. The standard gauge is 56½ in, that is, 4 ft 8½ in, the others being known as narrow gauges. In the metric system the gauges commonly used for construc-

13-130 Hauling Excavation

tion and open-pit work are 0.75 and 1.00 m. Gauges are widened on curves by about one-thirty-second or one-sixteenth in for each degree or fraction of a degree of curvature over 8°. The spread seldom exceeds 1 in.

Train Resistance Before a car or train can be set in motion, a certain resistance must be overcome. This is known as *train resistance* and it is measured in pounds per ton of gross load. It is due to the following components:

1. Friction of car-wheel journals. This will vary considerably from the low value for a smooth, well-lubricated axle with antifriction bearings to a high value for a brass shell bearing and roughly turned axle lubricated by waste soaked in grease.

2. Flange friction due to wheel flanges on edge of rail. This will vary and it increases greatly on rough tracks, causing cars to sway and flanges to bear against first one rail and then the other.

3. Condition of track. Various tables show that the recommended figures for train resistance range from 6 to 60 lb/ton, but for the majority of work in rock-earth haulage one should keep in mind the higher figures.

It is recommended that Table 13-52 be used for train resistance.

Grade Resistance The percent of grade represents the unit rise or fall per 100 linear units of travel. A 4 percent grade means that in traveling 100 ft horizontally the train has climbed a vertical rise of 4 ft. Grade resistance may be taken as uniform and it may be assumed to be 20 lb/ton of gross weight for each percent of grade.

Curve Resistance In the United States, railroad curves are expressed in degrees of the central angle subtended by a chord of 100-ft length. One degree of curvature is equal to a radius of 5730 ft. To obtain the approximate radius of a curve in feet, one divides

Figure 13-70 Main line of railroad in quarry. The level section of track is part of a 5.0-mi loop of line. The gauge is 38 in. Seven locomotives weighing up to 30 tons and 70 side-dump quarry cars of average 12-ton capacity serve the limestone quarry. One of the loading shovels is visible in the distance. The passing trains are switching and the cars will be made up into a complete train of 10 cars and locomotive. *(Plymouth Locomotive Works.)*

Figure 13-71 Dumping a train of cars into the primary crusher. The last of 10 cars is being handled by an overhead hoist mechanism. The locomotive weighs 30 tons, the cars weigh a total of 60 tons, and the limestone rock weighs 120 tons. The total weight of the train is thus 210 tons. The locomotive is driven by a diesel engine. The power train to the two axles is through a hydraulic torque-converter, transmission, drive shafts, and sprockets and roller chains to the axles. The engine horsepower is about 600. *(Plymouth Locomotive Works.)*

TABLE 13-50 Representative Specifications for Locomotives Used in Rock-Earth Excavation

Specification	Nominal weight rating, tons			
	15–25	25–45	45–70	70–120
Range of diesel-engine flywheel or brake hp	210–350	320–580	400–1000	650–1400
Drive train	Hydraulic torque converter, transmission, sprockets and chains	Hydraulic torque converter, transmission, sprockets and chains	Hydraulic torque converter, transmission, universal drive shafts, spur gears	Hydraulic torque converter, transmission, universal drive shafts, spur gears
Wheels, all driving:				
Number	4–6	4–6	8	8
Diameter, in	24–33	33	36–40	33
Track gauge, range, in	23–66	30–66	36–66	36–66
Wheelbase, range, ft	5.9–9.1	7.0–9.0	7.2	7.2
Trucks, center-to-center, ft			19.5	19.5
Dimensions, ft:				
Length, range	15.5–16.0	19.7–21.5	35.3	34.9
Width, range	8.9	9.0	8.0–10.0	7.2–10.0
Height	8.9	10.0	11.0	11.0

13-132 Hauling Excavation

Figure 13-72 Large locomotive hauling large cars and large shovel in dolomite quarry. The output of the primary gyratory crusher of this huge quarry in Michigan averages about 4000 tons/h. Several standard-gauge trains serve the crusher. The largest of the trains and shovels are illustrated here.

Shovel information:
- Dipper capacity, nominal — 16 cysm
- Dippers loaded per car — 2
- Average dipper load — 18 tons
- Shovel cycle, 41 s — 0.68 min
- Time for loading train with 468 tons of rock — 17.7 min
- Shovel output per 50-min working hour — 1320 tons

Locomotive and 13-car train information:
- Diesel engine hp — 1400 hp
- Length of level haul, one way — 5.2 mi
- Weights,
  - Tare weight
    - Locomotive — 72 tons
    - 13 cars, 18 tons each — 234
    - Total — 306 tons
  - Load weight, 13-car load, 18 tons each — 468 tons
  - GVW — 774 tons
- Round trip or cycle time, net — 57.7 min
- Hourly production, per 50-min working hour — 407 tons

Work-cost indicator: $\dfrac{\text{GVW} + \text{tare weight}}{\text{Load weight}} = \dfrac{774 + 306}{468}$ — 2.31

TABLE 13-51 Representative Specifications for Steel Side-Dump Railroad Quarry Cars

Specification	Nominal capacity, cysm			
	10	20	30	40
Capacity:				
Crowned load with 4:1 slope, cylm	11.5	24.2	35.7	47.1
Tons, based on crowned load at 2900 lb/cylm	16.7	35.1	51.8	68.3
Track gauge, in	36.0	56.5	56.5	56.5
Number of axles	2	4	2	4
Dimensions, ft:				
Length	16.0	30.3	38.7	47.1
Width	9.2	10.0	10.0	10.0
Height, overall and loading	7.2	8.0	8.7	9.3
Weights, lb:				
Tare or working	14,000	27,000	45,000	54,000
Load	33,400	70,200	103,600	136,600
GVW	47,400	97,200	148,600	190,600

TABLE 13-52 Train Resistance According to Types of Cars, Track Condition, and Types of Curves

Train resistance, lb/ton	Type of cars	Condition of track	Type of curve
6–10	Large, modern, with first-class bearings	Excellent	None
10–15	Large, standard, with good bearings	Good	None
20–25	Medium-size, first class	Good	Easy
30–35	Small	Fair	Easy
40–45	Small, hard running	Poor	Easy
50–60	Small, poor condition	Poor	Sharp

SOURCE: Plymouth Locomotive Works.
NOTE: One must keep in mind that the smaller the cars in a train, the greater the resistance per ton of load. Contrary to common belief, the largest modern quarry car offers the least train resistance per ton of gross weight.

5730 by the number of degrees of curvature. In the metric system, the radius is less per degree since the chord used is 20 m (65.62 ft) long. To convert from the U.S. Customary to the metric system, multiply the radius by 0.656.

The resistance of curves is expressed usually in pounds per ton of gross load per degree of curvature and it is estimated at 0.50 to 1.00 lb. A conservative figure for average rock-earth railroad haulage is 1.00 lb.

Acceleration Resistance The tractive force required to give to 1 ton (2000 lb) a linear acceleration of 1.0 mi/(h)(s) is 91.2 lb. To accelerate a train of weight W tons at an acceleration rate of a in miles per hour per second requires a tractive force of $(91.2aW)$ lb. On account of the accompanying angular acceleration of the rotating parts, an additional tractive force is required, which raises the acceleration constant, 91.2 lb, by an average of 8 percent for average weights of locomotives and cars. For estimating purposes the constant is increased by 10 percent to become 100. A given linear acceleration of a in miles per hour per second then requires an accelerating force of $100a$ lb/ton.

Air Resistance While air resistance, sometimes called speed resistance, is a factor to be considered in high-speed express service, it can be neglected for speeds less than 30 mi/h, a maximum speed for rock-earth railroad haulage. This fact is demonstrated below.

An accepted formula for the force of air resistance is:

$$f = KAv^2$$

where f = force, lb
K = a constant, 0.0033
A = frontal exposed area of train, ft²
v = speed, mi/h

For a locomotive and train with a 100-ft² frontal area, speed of 30 mi/h, and weight of 500 tons, the air resistance R_a in pounds per ton is given by the equation:

$$R_a = \frac{0.0033 \times 100 \times 30^2}{500} = 0.59 \text{ lb/ton}$$

Within the precision limits in estimating an average 40-lb/ton total resistance for a train, the 0.59 lb/ton for air resistance is negligible.

In calculations for the performance of trains hauling rock-earth at practicable speeds, air or speed resistance is not considered.

Total Resistance The total resistance to motion of the complete train is given by equation:

$$R = W(r + 20G + 1.00g + 100a)$$

where R = total resistance of train = needed tractive force, lb
W = weight of locomotive and cars, tons
r = train resistance, lb/ton (Table 13-52)
G = grade, %

g = curvature, °
a = acceleration of train, mi/(h)(s)

The following example of the use of the equation is illustrated by Figure 13-69.

W, weight of locomotive and loaded cars, 210 tons

r, train resistance (estimated on basis of Table 13-52), 30 lb/ton

G, ruling grade, +1.0% for a short stretch of a few hundred feet leading into the curve approaching the the crusher at the plant)

g, greatest degree of curvature (that of the curve leading into the crusher) 14,409 ft radius.

a, desired acceleration, is 0.10 mi/(h)(s), an acceptable value for the service.

The necessary tractive force F may be maximum at three stages of the train's progress along the railroad line:

1. Gathering speed on level, straight track after leaving shovel:

$$F = W(r + 100a) = 210(30 + 100 \times 0.10) = 8400 \text{ lb}$$

2. Ascending grade leading to crusher; no curve and no acceleration:

$$F = W(r + 20G) = 210(30 + 20 \times 1.0) = 10{,}500 \text{ lb}$$

3. Negotiating curve approaching crusher; no grade and no acceleration:

$$F = W(r + 1.00g) = 210(30 + 1.00 \times 14) = 9240 \text{ lb}$$

The maximum F, or the necessary locomotive tractive force, thus equals 10,500 lb.

Locomotive Selection Procedure

The importance of selecting the right size and type of power haulage unit for performing within a specified range of service is self-evident. On the timing, capacity, and general efficiency of haulage machinery may depend the success or failure of the operation, or at least much of its profitability.

In such a specialized field as locomotive engineering it must be said that the recommendations of the manufacturers' engineers should carry great weight. It is also evident, however, that the earthmover should be able to follow generally sound measures for gaining an approximate idea as to the locomotive which would serve the purpose. The following data are intended to help in the first analysis for locomotive selection.

The principal considerations have been enumerated. They are in terms of the track and the train of loaded cars to be hauled. The ability of a locomotive with sufficient horsepower to start a given load without slipping the wheels is dependent on its weight, and the factor to be considered is known as the *starting drawbar pull*, which is expressed in pounds. It is often referred to as the drawbar pull available at the coupler. The ability of a given locomotive to haul the same given load at a certain specified speed is determined by the horsepower of the locomotive.

Coefficients of Adhesion Tractive force is limited by the coefficient of adhesion or friction between the locomotive wheels and the rails. Coefficient of adhesion is expressed as the percent of the weight of the locomotive supported by the drivers. Table 13-53 gives coefficients of adhesion between steel driving wheels and steel rails according to the condition of the rails.

TABLE 13-53 Coefficients of Adhesion of Locomotives According to Condition of Rails

	Percentage of locomotive weight supported by drivers	
Condition of rails	Unsanded rails	Sanded rails
Most favorable conditions	35	40
Clean, dry rail	25	30
Dry rail	18	24
Slippery, moist rail	15	20
Dry snow–covered rail	11	15

NOTE: Reduce coefficients by 20 percent if cast-iron chilled wheels are used.
SOURCE: Plymouth Locomotive Works.

For average purposes it is well to use 25 percent of the weight of the locomotive on its drivers as its drawbar pull. A diesel-engine-driven locomotive has a greater capacity than a steam-driven locomotive. This advantage is due to the torque characteristics of the engine and to the uniform power application to the driving wheels. When a torque converter is in the power train, the power impulses are still further smoothed out, resulting in more relative capacity.

Weight of Locomotive The general equation for the weight of a locomotive to move a train of a given weight by overcoming all resistances associated with rock-earth haulage is:

$$WL = \left(\frac{5W}{pq}\right)(r + 20G + 1.00g + 100a)$$

where WL = weight of locomotive, tons
W = weight of cars, laden and unladen, tons
p = coefficient of adhesion, % (Table 13-53)
q = weight on drivers, % of weight of locomotive
r = train resistance, lb/ton (Table 13-52)
G = grade, %
g = curvature, degrees
a = acceleration, mi/h/s

The following example of the use of the equation is based on the haulage illustrated by Figures 13-69, 13-70, and 13-71.

It has already been determined that the greatest total resistance occurs when the laden train approaches the crusher at the plant on a 1.0 percent upgrade without curvature and with no acceleration. The weight W of laden cars is 180 tons. The coefficient of adhesion P is for the worst condition of the rails, dry snow-covered, and is 15 percent for sanded rails. The weight on drivers q for the 0-4-0 type of locomotive is 100 percent. Train resistance r was estimated to be 30 lb/ton. There is no grade and no acceleration. Substituting in the general equation:

$$WL = \left(\frac{5 \times 180}{15 \times 100}\right)(30 + 20 \times 1.0 + 0 + 0) = 30.0 \text{ tons}$$

Horsepower of Locomotive The general equation for the horsepower of a locomotive, delivered at driving wheel-rail contacts, to move a train of a given weight by overcoming all resistances associated with rock-earth haulage is:

$$hp = \left(\frac{2.67vW}{1000}\right)(r + 20G + 1.00g + 100a)$$

where hp = horsepower of locomotive delivered at driving wheel-rail contact
v = attained or desired speed, mi/h
W = weight of locomotive and cars, laden or unladen, tons
r = train resistance, lb/ton (Table 13-52)
G = grade, %
g = curvature, °
a = acceleration, mi/(h)(s)

The following example of the use of the equation is based on the haulage illustrated by Figures 13-69, 13-70, and 13-71. Maximum horsepower is required when both desirable speed and total resistance are simultaneously greatest for a certain segment of trackage. In this example the requirement is maximum when the loaded train is being accelerated to a desirable, practicable speed of 20 mi/h on a straight, level main line of trackage. The values to be used in the equation are: $v = 20$ mi/h; $W = 210$ tons; $r = 30$ lb/ton; and $a = 0.10$ mi/(h)(s), an acceptable figure for quarry haulage. There is no grade and no curvature. Substituting in the general equation:

$$hp = \left(\frac{2.67 \times 20 \times 210}{1000}\right)(30 + 0 + 0 + 100 \times 0.10) = 449$$

Assuming a power train efficiency of 0.76 for the locomotive, the flywheel or brake horsepower of the diesel engine is $449/0.76 = 590$ hp.

13-136   Hauling Excavation

Hauling Cycle Calculations for Locomotive and Cars of Figure 13-72 The criteria are:

1. Length of haul one way equals 5.2 mi, or 27,460 ft.
2. Haul is over an oval-shaped loop from the shovel to the primary crusher at the plant to the shovel.
3. The track is level, with two 10° curves at the shovel and crusher and it is in excellent condition, with an estimated train resistance of 20 lb/ton.
4. The average coefficient of adhesion for all-weather operation is estimated at 17 percent.
5. Weights of train of cars
    a. Tare or working weight, 13 cars at 18 tons each, 234 tons
    b. Load of dolomite rock at 36 tons per car, 468 tons
    c. GVW or total laden weight, 702 tons
6. Maximum desirable speeds
    a. Laden, 20 mi/h, or 1760 ft/min
    b. Unladen, 30 mi/h, or 2640 ft/min
7. Assumed accelerations
    a. Laden, to 20 mi/h, 0.05 mi/(h)(s)
    b. Unladen, to 30 mi/h, 0.12 mi/(h)(s)
8. Assumed decelerations, laden and unladen, are based on stopping by train resistance, with little application of brakes.

Weight of Locomotive Substituting the above values in the general equation for weight of a locomotive:

$$WL = \left(\frac{5 \times 702}{17 \times 100} \right)(20+0+1.00 \times 10 + 100 \times 0.05) = 72 \text{ tons}$$

Horsepower of Locomotive In the complete trackage there are two segments which call for high horsepowers of the locomotive. These horsepowers are calculated below.

1. For the laden train of Figure 13-72 when leaving the shovel and accelerating to 20 mi/h on the straight level track, the total weight of the locomotive and cars is 774 tons. Substituting this weight and the values of the variables in the general equation for horsepower of locomotive:

$$hp = \left(\frac{2.67 \times 20 \times 774}{1000} \right)(20+0+0+100 \times 0.05) = 1033$$

2. For the same train when leaving the crusher unladen and accelerating on the curved level track, the total weight of the locomotive and cars is 306 tons. Substituting this weight and the values of the variables in the general equation:

$$hp = \left(\frac{2.67 \times 30 \times 306}{1000} \right)(20+0+1.00 \times 10 + 100 \times 0.12) = 1029$$

The higher horsepower is required for the laden train leaving the shovel. Assuming a power train efficiency of 0.76 for the locomotive, the required flywheel or brake horsepower of the diesel engine is $1033/0.76 = 1360$ hp.

A 72-ton locomotive with a diesel engine of 1400 flywheel or brake horsepower is suitable for the work. In locomotive service, which is hard and usually continuous, one must select the engine on the basis of horsepower rating at continuous sustained load.

	Min
Hauling cycle, calculated	
Loading time. Shovel cycle is actually 0.68 min. 26 dippers.	17.7
Hauling time, min	
Acceleration	
$t = \dfrac{v}{a} = \dfrac{20}{0.05} = 400 =$	6.7
$d = \dfrac{s}{2} \times t = \dfrac{1760}{2} \times 6.7 = 5900$ ft	
Deceleration	
Decelerating force = train resistance × tons	
$20 \times 774 = 15,500$ lb	

$$a = \frac{f}{m} = \frac{\frac{15{,}500}{774 \times 2000}}{32.2} = 0.32 \text{ ft/s}^2$$

$$t = \frac{v}{a} = \frac{\frac{20 \times 5280}{3600}}{0.32} = 92 = \qquad 1.5$$

$$d = \frac{s}{2} \times t = \frac{1760}{2} \times 1.5 = 1320 \text{ ft}$$

Travel at 20 mi/h
$d = 27{,}460 - (5900 + 1320) = 20{,}240$ ft
$t = 20{,}240/1760 = \qquad\qquad\qquad\qquad\qquad 11.5$
Total hauling time at average 15.8 mi/h = $\qquad\qquad\qquad\qquad$ 19.7

Dumping time. Crusher output = 4000 tons/h. $\frac{468}{4000} \times 60 = \qquad$ 7.0

Returning time
  Acceleration
$$t = \frac{v}{a} = \frac{30}{0.12} = 250 \text{ sec} = \qquad 4.2$$

$$d = \frac{s}{2} \times t = \frac{2640}{2} \times 4.2 = 5540 \text{ ft}$$

  Deceleration
    Decelerating force = train resistance × tons
         $30 \times 306 = 9180$ lb

$$a = \frac{f}{m} = \frac{\frac{9{,}180}{306 \times 2000}}{32.2} = 0.48 \text{ ft/s}^2$$

$$t = \frac{v}{a} = \frac{\frac{30 \times 5280}{3600}}{0.48} = 92 \text{ s} = \qquad 1.5$$

$$d = \frac{s}{2} \times t = \frac{2640}{2} \times 1.5 = 1980 \text{ ft}$$

Travel at 30 mi/h
$d = 27{,}460 - (5540 + 1980) = 19{,}940$ ft
$t = 19{,}940/2640 = \qquad\qquad\qquad\qquad\qquad 7.6$
Total returning time at average 23.5 mi/h $\qquad\qquad\qquad\qquad\quad$ 13.3
Total net cycle time $\qquad\qquad\qquad\qquad\qquad\qquad\qquad\qquad\qquad$ 57.7

Production and Cost Calculations for Locomotive and Cars

The production calculations, like the cycle time calculations, are based on Figure 13-72.

Net cycle time, min	57.7
Loads hauled per 50-min working hour, 50/57.7	0.87
Tons hauled hourly by one train, 0.87 × 468	407
Tons hauled hourly by three trains serving shovel with estimated capacity of 1320 tons/h	1221

NOTE: The primary crusher with a capacity of about 4000 tons/h is served by several additional shovels and trains.

Estimated direct job unit cost, 1978:
  Cost of machines:

Locomotive, 72 tons	$ 328,000
13 cars of 40 tons capacity @ $62,000	806,000
Total for complete train	$1,134,000

  Hourly cost of ownership and operation (Table 8-3):

Locomotive, 28% of $328	$ 91.80
13 cars, 23% of $806	185.40
Operators: engineer, fireman, and brakeman	43.40
Total hourly cost	$320.60

  Direct job unit cost:

Per ton, $320.60/407	$0.79
Per ton-mile, $0.79/5.2	$0.15

The low unit cost for rock haulage, $0.15/ton·mi, is in keeping with the low work-cost indicator of the locomotive-car-hauling machine:

13-138   Hauling Excavation

$$\text{Work-cost indicator} = \frac{\text{GVW} + \text{tare weight}}{\text{load weight}} = \frac{774 + 306}{468} = 2.31$$

The cost for track shifting and track maintenance is not included in the direct job unit cost. In such a large-scale quarrying operation this cost is nominal with respect to the unit cost of $0.79/ton. It is probable that a track crew of a foreman, six trackmen, and a medium-weight crawler-tractor-bulldozer with operator can maintain and shift all trackage for the several trains with a combined output of 4000 tons/h.

At an estimated hourly cost of $165.00 (1978), the direct job unit cost is $0.04/ton, or about 5 percent of the cost of haulage.

Summary

Railroad haulage offers low unit cost for rock-earth excavation under the following conditions, typical of huge jobs involving multimillions of cubic yards or tons.

1. Easy upgrades, that is, less than 2.0 percent, as shown in the Bingham Canyon open-pit copper mine, Plate 5-1, where 62 electric-motor- and diesel-engine-driven locomotives and about 1000 ore and overburden cars are used.

2. Fairly fixed loading and dumping points, as illustrated in Figures 13-69, 13-71, and 13-72.

3. Long-term excavations such as a 3- to 5-yr period for dam building or an uncompleted life of 74 additional yr for the copper mine of Bingham Canyon.

4. Double and triple shifting operation so as to afford long depreciation periods in hours for the huge investment in custom-built machinery.

5. Excellent management to deal with the logistics of a railroad and the maintenance of rolling stock and trackage.

HYDRAULICKERS

Hydraulicking is the use of water for the excavation of rock-earth. Water of high velocity from monitors with nozzles is used to break up the formation, and the same water provides transportation or haulage either in open flow or in pipes. The first of these actions is illustrated in Figure 13-73 and the sequential haulage in Figure 13-74.

Nature has used this method for hundreds of millions of years for both the degradation and the aggradation of the earth's mantle of rock-earth. These two processes correspond to taking out a cut and building a fill. In most cases this natural hydraulicking laid down the same sedimentary deposits which humans are excavating hydraulically at the present time.

In the United States hydraulicking was commenced, and machinery and methods were developed, during the last century in the placer gold mining fields of the Western states. The method was used both for removal of overburden and for open-pit mining of the gold-bearing gravels. In later years this economical method for excavation was used in such work as overburden removal in the iron mining of the Cuyuna Range of Minnesota and both overburden removal and ore excavation in the phosphate deposits of Florida.

Currently the monitor is used for highway building in Mexico, dam building in Italy, overburden removal from limestone in Missouri, mining of silica sand in Illinois, phosphate mining in Netherlands Guiana, and for many other unique and economical earthmoving projects.

Principles of Hydraulicking

The method of excavating is to break down the material, usually alluvia or weathered rock or ore, in the cut by means of a powerful stream of water which is discharged from the nozzle of a *monitor*, sometimes called a *giant*. The high-velocity stream is directed against the base of the bank and it erodes and washes out the rock-earth, forming a slurry or liquidlike mud, as illustrated in Figures 13-73 and 13-75. Because this fragmentation is an abrading action, the velocity of the water stream is more important than the quantity of water delivered. In other words, velocity is necessary for excavation and quantity is effectual for the later transportation of the slurry.

After the rock-earth has been removed from the bank, the resulting slurry may flow out of the cut in a natural sluiceway or it may simply form a pond or sump alongside the cut. The sluiceway requires a minimum slope of about -4 percent grade for suitable flow of the slurry; the coarser and denser the slurry, the greater the angle of slope needed.

The solid material in the slurry may make up 10 to 40 percent of the weight of the slurry in terms of dry weight. The exact percentage depends on the nature of the rock-earth and the desired method of transporting the slurry after excavating.

The natural sluiceway may carry the slurry directly to the fill, as in the case of the ripped sandstone-shale formation of Figures 13-73 and 13-74. The sluiceway may also carry the slurry to an adjacent pond or sump, whence it may be pumped to a distant plant, as shown in Figure 13-76. Another method of haulage is to haul or transport the slurry to a fill or spoil bank by means of a flume.

Machinery and Methods

The monitor is a simple machine, consisting of a nozzle which increases the velocity of the delivered water and produces a jet stream in the desired direction. The range of direction is about 360° horizontally and 110° vertically. The direction may be controlled manually by an operating lever in line with the nozzle or it may be controlled hydraulically by cylinders, which in turn are actuated by remote lever valves of a console. The monitor is mounted in a turret or swivel head and the whole assembly with base is securely fixed to the stand. Figure 13-73 illustrates a skid-mounted stand, to be moved at intervals by a crawler-tractor. Figure 13-76 shows three stands, integral parts of the complete slurrifier.

The nozzle diameters in current use vary from 2 to 4 in, although in past massive excavations of placer gold deposits, diameters up to 10 in have been used effectively. The jet or muzzle velocity of the stream ranges up to 150 ft/s. Working pressures at the base of the nozzle range up to 200 lb/in². Discharges range up to 6800 gal/min for a 5-in-diameter nozzle.

The pumps, pipes, and appurtenances which deliver water to the monitor comprise the

Figure 13-73 Hydraulicking weathered sandstone-shale formation from cut to fill. During a 10-yr period at Rose Hills Memorial Park, Whittier, California, 20 million cybm has been flowed from a hillside cut through a natural sluiceway to a broad fill. One of the four tractor-bulldozers serving the monitor may be seen in the background.

The horizontal and vertical movements of this monitor are actuated by custom-built hydraulic rams, which are controlled remotely by valves located on the console of the operator. A total of 1450 cybm is sluiced hourly. The flow is made up of 82 percent solids and 18 percent water by weight. The total water comes from both natural moisture of the rock-earth and the delivery of the monitor. The nozzle is of 2-in diameter, the pressure is 125 lb/in², and delivery is 1350 gal/min. The attainable horizontal distance of the stream is 150 ft, although the downward-directed stream is kept below 75 ft to yield best results in breakup and saturation of the bulldozed rock-earth.

Figure 13-74 Braided natural sluiceway from cut to fill. This figure is sequential to Figure 13-73, completing the haulage by hydraulicking. The 2500-ft-long and average 600-ft-wide fill is about 75 ft in maximum depth. The final hydraulic gradient is about −4 percent, commencing at about −20 percent the beginning of the operation. The hydraulic fill will be topped with a 6-ft blanket of select soil for growing trees and shrubs for the memorial park.

The 20 million yd³ of fill will be placed in about 10 yr at a rate of some 11,000 yd³ per working day. This mass haulage is being done at the remarkably low cost of about $0.06/cybm, based on 1978 estimated costs. This unit cost does not include costs for bulldozing the sandstone-shale formation to the monitor and for topping the hydraulic fill with select soil.

supportive machinery. These may be separate from the monitor, as in the simple setup of Figure 13-73, or they may be integral, with the exception of the main line supply pipe, as illustrated in the self-contained slurrifier of Figure 13-76.

When hydraulicking in mountainous country where reservoirs may be built to afford static heads of 600 ft, the development of water supply is simple. Generally, however, this favorable condition does not exist and it is necessary to develop a water supply. This subject has been covered in Chapter 10 and it is well to review this chapter as a part of the study of hydraulicking.

Table 13-54 gives theoretical discharges of nozzles in gallons per minute according to diameter and the pressure at the base of the nozzle. This table, supplied by a manufacturer, supplements Table 10-11. It gives higher pressures and greater diameters of nozzle tips and it is more applicable to the hydraulicking of rock-earth.

As has been mentioned, the slurry from the monitor may be carried in natural sluiceways, by pipes, or by flumes. In natural sluiceways and in flumes, sands and gravels and well-weathered rocks are handled most easily. In order to maintain a moderate gradient the maximum particle size should not exceed 8 in, maximum cobble size. In pipes the slurry passes through centrifugal pumps and the greatest size must be consistent with pump and pipe diameters. Along with this limitation in size, the hardness and abrasion characteristics of the solids in the slurry must also be considered.

The natural sluiceway, illustrated in Figure 13-74, requires an average gradient of −4 percent, which should be steeper for coarse material and may be gentler for fine sediments. Nature's gradients for a fanglomerate or wash plain from the mountains illustrate the range of gradients. Near the mountains, where large boulders and cobbles are deposited, the gradient may be −20 percent. Two miles or so from the mountains the gradient may be −1 percent, where the fine silts are deposited.

Figure 13-75 Hydraulicking phosphate ore after overburden removal by dragline. The monitor is forming a slurry of phosphate ore for pumping to the plant. The boom of a walker dragline used in the overburden removal is visible in the background. The 3-in-diameter nozzle can discharge 3800 gal/min at 200 lb/in^2. At 3800 gal/min with a slurry containing 38 percent of solids by dry weight, the hourly production of the monitor is about 580 tons of phosphates at 100 percent efficiency.

The monitor is one of two units, mounted on a substantial working platform, with a combined excavation of about 9000 tons of ore every 8-h shift. *(Canaris Corp.)*

TABLE 13-54 Theoretical Discharge of Nozzles in U.S. Gallons per Minute

		Diameter of tips, in				
	lb/in^2	1	1⅛	1¼	1½	1¾
	100	290	370	455	660	904
	125	326	415	511	740	1011
Model A	150	357	455	561	815	1109
	175	363	461	566	820	1114
	200	390	486	591	846	1139
	lb/in^2	1⅛	1¼	1½	1¾	2
	100	375	460	665	909	1190
	125	420	516	745	1016	1325
Model H	150	460	566	820	1114	1450
	175	465	571	825	1119	1455
	200	491	596	851	1144	1480
	lb/in^2	2	2¼	2½	2¾	3
	100	1196	1512	1870	2255	2690
	125	1338	1690	2090	2520	3005
Model M	150	1466	1853	2290	2760	3295
	175	1582	2000	2473	2985	3560
	200	1691	2140	2645	3190	3800

Model	Intake diameter, in	Discharge diameter, in
A	3	1¾
H	4	2
M	6	3

NOTE: Maximum discharge is recommended.
SOURCE: Canaris Corp.

Figure 13-76 Large highwall slurrifier hydraulicking phosphorite and sand. This huge machine mounts three monitors and a battery of centrifugal pumps. The excavation rate is 1020 yd³/h, and water consumption is 11,200 gal/min. After the slurry is ponded, the pumps of the slurrifier, together with centrifugal booster pumps stationed about 1 mi apart, pump the slurry about 6 mi to the plant. The average density of the slurry is 37 percent solids by dry weight. A 1-week record of the slurrifier is tabulated below:

Working time, 86%		144.4 h
Delays:		
Unavoidable, 9%		
Servicing, 5%	9.0 h	
Operational, 1%	1.8	14.8
Instrumentation, 3%	4.0	
Avoidable, 5%		
Mechanical, 5% (This delay is not chargeable to the slurrifier, being caused by failure of a booster pump)		8.8
Total available time, 100%		168.0 h

During the around-the-clock operation 146,662 yd³ was delivered to the plant of Surinam Aluminum Co., located in Netherlands Guiana. *(Canaris Corp.)*

The water consumption by sluiceway is the lowest for any of the three carriers. It is probable that solids make up from 50 to 85 percent by weight of the slurry, depending on the gradation of the rock-earth and the desired hydraulic gradient.

Some factual data are available on gradients of flumes and water consumption in the hydraulic mining for gold in California. These data are interesting in that water was measured by the miner's inch, which unit is equivalent to 1.53 ft³/min, or 11.4 gal/min. The gradients of flumes and water consumption in cubic feet per cubic yard of alluvia are given in Table 13-55. The table suggests that an 89 percent increase in gradient results in a 63 percent decrease in water requirement. A massive compilation of data on water duty indicates that the placer gold miners used from 270 to 2700 ft³ of water per cubic yard of material when using flumes.

A 1912 table of the Union Iron Works Co. of San Francisco gives water requirements for effective hydraulicking as listed in Table 13-56. The original table was in terms of feet of head and cubic feet per minute; these units have been changed to the currently used pounds per square inch unit of pressure and gallons per minute unit of quantity per unit time. This table provides a valuable extension of Table 13-54, as it includes nozzle diameters of 4 and 5 in.

Production and Costs for Hydraulicking

The work shown in Figures 13-73 and 13-74 may be used as an example to illustrate the production and cost calculations. Water is taken from a city water main where pressure is 90 lb/in². A two-stage centrifugal pump, powered by a 140-hp diesel engine, delivers water through 2000 ft of 8-in-diameter pipe to the monitor. The nozzle of the monitor is of 2-in diameter and it is located 100 ft higher than the city water main.

TABLE 13-55 Grades of Flumes and Water Consumption for Hydraulicking Gravel

	Grade of flume, %	Water, ft³ / Gravel, yd³
North Bloomfield Claim 8: Flume, 72 in wide and 32 in deep, 16.0 ft² Gravel, 11,022,000 yd³	−3.4	500
La Grange Co., 5 claims Flume, 48 in wide and 30 in deep, 10.0 ft² Gravel, 2 million yd³	−1.8	1350
Averages	−2.6	925

TABLE 13-56 Water, gal/min, Required for Effective Hydraulicking According to Pressure and Nozzle Diameter

Pressure, lb/in²	Nozzle diameter, in				
	2	2½	3	4	5
43	998	1406	2079	3650	5610
65	1122	1743	2528	4488	7016
87	1182	2020	2917	5161	8026
108	1459	2244	3254	5782	8976
130	1571	2468	3590	6343	9874
152	1683	2693	3800	6844	10659
173	1795	2865	4099	7293	11220

SOURCE: Union Iron Works Co.

The production, on which the direct job unit cost is based, is 1450 cybm/h of the ripped sandstone-shale formation. The slurry is hydraulicked an average distance of 1800 ft one way, with a final hydraulic gradient of −4 percent.

Estimated direct job unit cost, 1978:
Hourly cost of ownership and operation of machinery (Table 8-3):

Monitor, 2-in-diameter nozzle, complete with stand: cost $11,700; 25% × $11.7	$ 2.90
Pipe, 2000 ft of 8-in diameter: cost, $10,160; 35% × $10.2	3.60
One 8-in-diameter foot valve	0.50
Two 8-in-diameter gate valves Three 90°, 8-in-diameter ells, weldments: total cost $1540; 35% × $1.5	
One 8-in two-stage centrifugal pump driven by 140-hp diesel engine, skid-mounted: cost $18,000; 42% × $18.0	7.60
Total for machinery	14.60
Operator and two laborers	38.40
Total hourly cost of ownership and operation	$53.00
Hourly cost of water, 81,000 gal @ $0.45 per 1000 gal	36.40
Total hourly costs	$89.40
Direct job unit cost per cybm, $89.40/1450	$0.0617
Direct job unit cost per station-cybm, $0.0617/18	$0.0034

These unit costs are remarkably low when compared with other kinds of haulage, particularly when the monitor functions also as an excavator.

Comparison of Haulage by Hydraulicking and by Scrapers

Hydraulicking is a special kind of haulage, calling for operating conditions which do not exist generally. The requisites are:

    1. An abundant supply of water at a reasonable cost and an environment in which used water may be absorbed into the soil or wasted in an acceptable manner

    2. For natural sluiceways or flumes, a topography which permits a suitable hydraulic gradient for the slurry

13-144   Hauling Excavation

3. A nonrigid requirement for fill construction, which does not call for normal specifications of compaction
4. A working area which permits use of sluiceways, flumes, or pipelines
5. In rock excavation, a material which lends itself to hydraulicking by size and gradation in sluiceways and flumes, and by size, gradation, hardness, and abrasion in pipes and pumps

When these and other conceivable requirements are met, the earthmover may use this unique, inexpensive haulage. Water may be had from wells on the property, thus making this commodity proprietary, and gravity is free of charge.

Without imagination, the 25-million-yd³ cut-and-fill job of Figures 13-73 and 13-74 would be done by scrapers at a 1978 cost of about $0.46/yd³. Instead, 20 million yd³ is being hydraulicked at about 13 percent of the cost of a scraper operation.

The work-cost indicator for hydraulicking cannot be calculated by the general equation

$$\text{W-C indicator} = \frac{(\text{GVW} + \text{tare weight})}{\text{load weight}}$$

because hydraulicking does not involve the work of a hauling machine. However, an indicator relative to that of the push-loaded scrapers, average 3.46, may be calculated by comparing the energy required to haul 1 cybm of rock-earth by hydraulicker and by scrapers. The work illustrated in Figures 13-73 and 13-74 is used for the comparison.

By Hydraulicker   The energy used in hauling 1450 cybm/h, or 24.2 cybm/min, through the 1800-ft distance is that used to raise 1350 gal/min through the total dynamic head from the original source of water to the monitor. This is estimated to be 208 ft from the source to the water main and 226 ft from the main to the monitor, totaling 434 ft. The water horsepower is $\frac{1350 \times 8.34 \times 434}{33,000} = 148$. The brake horsepower of the prime movers for centrifugal pumps is $148/0.50 = 296$ bhp. The energy for the 1800-ft haul is $\frac{296 \times 33,000}{24.2} = 404,000$ ft·lb/cybm.

By Scraper   The energy required for the push-loaded scraper operation is as follows:

Engine brake hp, rated power of scraper engine	415
Engine average hp delivered, estimated at 67% × 415	278
Capacity:	
Struck measurement, cysm	21.0
Estimated payload, cybm (Table 13-7), 0.87 × 21.0	18.3
Estimated hauling cycle times, min, on hard, well-maintained haul road of 1800-ft length, one way, and with −4% grade	
Loading	0.8
Two turnings and dumping	0.7
Hauling down 4% grade at average 15 mi/h	1.4
Returning up 4% grade at average 11 mi/h	1.8
Net cycle time	4.7
Estimated production per 50-min working hour:	
Loads hauled, 50/4.7	10.6
Cybm hauled, 10.6 × 18.3	194
Cybm/min hauled, 194/60	3.23
Energy per cybm for the 1800-ft haul:	
$\frac{278 \times 33,000}{3.23} =$	2,840,000 ft·lb

Hydraulicking expends only 14.3 percent as much energy as does hauling by scraper for the same cubic yard bank measurement, on the same 1800-ft one-way length of haul. If this factor of 14.3 percent is applied to the average work-cost indicator for the scrapers, 3.46, then the implied work-cost indicator for this particular application of the hydraulicker is 14.3 percent of 3.46, or 0.49.

The reflection of this extremely low work-cost indicator appears in the low cost for hydraulicking in the example, $0.062/cybm. The ratio of this unit cost to the estimated cost of $0.46 for the scraper is 13.5 percent. In this case the work-cost indicators are in approximate proportion to the unit costs.

However, this relationship is entirely accidental because the energy consumption for hydraulicking in a natural sluiceway depends entirely on the work needed to get water of suitable pressure to the monitor. If water came from a higher reservoir with sufficient elevation head, there would be no use of costly energy, as gravity would supply all the energy needed.

DREDGES

A dredge is a floating machine for loading and hauling materials from beneath the surface of the water or from beneath the existing water table in water-bearing materials. They are classified in this text as hauling rather than loading or excavating machines because the hauling element is usually more important and more costly than the loading element for the most popular dredge, the hydraulic suction cutter-head machine.

The basis of payment for this work is the cubic yard bank measurement or, as it is called by dredge operators, the cubic yard apparent volume. Sometimes in industrial operations, such as that of a gravel plant, where weight measurement is used, the basis of payment or output is expressed in tons.

There are two classes of dredges, mechanical and hydraulic, and these two classes are subdivided into three types each. Figure 13-77 illustrates these six dredges with accompanying descriptions of their manners of operation. These descriptions are supplemented by the following information.

Grapple Dredge

This machine is the counterpart of the land-based crane with grapple, clamshell, or orange-peel bucket, as discussed in Chapter 12. Figure 13-78 illustrates a large grapple dredge equipped with a clamshell bucket. Figure 12-29 shows the land use of a grapple which may also be used for subaqueous work. In Figure 13-78 the dredge is loading rock into a barge for disposal in the open sea. The grapple dredge is both a loading and a casting machine. It may be equipped with buckets of up to 12-yd^3 capacity and it may excavate to great depths.

Production These dredges are custom-designed and custom-built, or else they are designed and built as modifications of existing machines. Their hourly productions depend on their individual specifications and on the rock-earth being dredged. Their principles of production, from which their outputs may be calculated, are contained in the crane section of Chapter 12. For example, the production of the machine of Figure 13-78 may be figured as follows:

Criteria:
1. Material: rock and heavy gravels up to 18-in size, classified as rock-gravel alluvia
2. Nominal capacity of clamshell bucket, 6.0 cysm
3. Total hoist, 60 ft
4. Average swing-return angle, 120°

Estimated hourly production:
1. Cycle time (Table 12-16) extrapolated for 60-ft hoist, 0.63 min
2. Buckets cast or loaded per 50-min working hour, 50/0.63=79
3. Bucket load for nominal 6-yd^3 capacity bucket, 17,400 lb
4. Bucket load based on 2900 lb/cybm for rock-gravel mixture, 17,400/2900=6.0 cybm
5. Hourly production, 79×6.0=474 cybm
NOTE: The actual timed cycle for the loading is 37 s, or 0.62 min.

Costs, 1978 Chapter 8 gives no estimated costs of ownership and operation of the large custom-built dredges. For an approximate estimate it is suggested that one use the value for the crane excavator with bucket, crawler-mounted and with diesel engine, of 5-yd^3 bucket capacity. The hourly cost is estimated to be 16 percent of the thousands of dollars of cost of the dredge.

The erected dredge is estimated to cost $2.8 million as of 1978. Hence, the hourly cost of ownership and operation is estimated to be:

Dredge, 16% × $2800	$448.00
Leverman, engineer, and deckhand	50.50
Total hourly cost	$498.50
Estimated direct job unit cost per cybm, $498.50/474	$1.05

Dipper Dredge

This machine is the counterpart of the land-based shovel. Figure 13-79 illustrates a large dredge of 10-yd^3 dipper capacity and with a 50-ft-long dipper handle. It is excavating hard

MECHANICAL

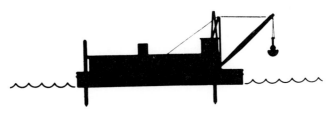

THE GRAPPLE DREDGE

In this type, the work is done by a clamshell bucket suspended from a derrick mounted on a barge. It is most suitable for excavating medium-soft materials in confined areas near docks and breakwaters.

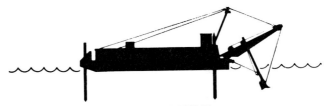

THE DIPPER DREDGE

A powerful dipper bucket mounted at the forward end gives the dipper dredge its main advantage: strong "crowding action," produced as the bucket is forced into material being moved. This permits efficient removal of rock and other hard materials. For its size, a dipper dredge can handle larger pieces, thus reducing blasting needs.

THE BUCKET DREDGE

Buckets mounted on an endless chain do the work here. Each bucket digs, conveys and dumps its own load. A continuous work cycle makes the bucket dredge an efficient mechanical dredge, when used in operations such as sand and gravel production.

Figure 13-77 The two classes of dredges. *(Ellicott Machine Corp.)*

weathered rock from a depth of 25 ft below the water level and casting or loading it into the barge.

The dipper sizes for these dredges range up to 15 yd³ and the dipper handles up to 75 ft in length. They are necessarily boom-swing-type machines like the original land-based railroad-type shovels. They are suitable for the hardest of rock formations, unblasted or blasted. As opposed to other kinds of dredges, the dipper dredge is hoisted onto the spuds when excavating so as to ensure maximum stability.

Production Hourly productions depend on the specifications for these custom-built dredges, as well as on the depth of digging and the kind of excavation. However, performances follow the principles for the shovels described in Chapter 12.

Table 12-8 gives cycle times for shovels, but these values should be increased by

HYDRAULIC

THE PLAIN SUCTION DREDGE

In this type of dredge, a suction pipe is lowered to the surface to be worked. A powerful dredge pump draws up the material, mixed with water, and discharges through a pipeline. Units of this type are used for digging soft, free-flowing materials.

THE SELF-PROPELLED HOPPER DREDGE

Resembling an ocean-going ship, this vessel functions in a way similar to a plain suction dredge. Material is gathered from the bottom by dragged suction heads, then pumped into storage hoppers. When filled, the dredge proceeds to a deep-water dumping area, where the hopper doors in its hull bottom are opened for rapid discharge.

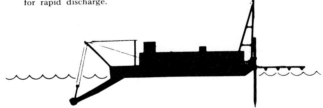

THE CUTTERHEAD PIPELINE DREDGE

The most versatile and widely used excavating unit for transporting waterbound solids. A rotating cutter loosens the material, which is then sucked through the dredging pump, discharged via a pipeline at the stern. These dredges can dig and pump all types of alluvial materials, also clay, hardpan and other compacted deposits.

Figure 13-77 (Cont.)

about 75 percent for dipper dredges because of the greater hoisting heights. If the dredge of Figure 13-79 has an average swing-return angle of 90°, the cycle time is about 1.75×0.37, or 0.65 min. Estimated production per 50-min working hour is calculated as listed below:
1. Cycle time, 0.65 min
2. Dippers loaded hourly, $50/0.65 = 77$
3. Nominal bucket rating, 10 cysm
4. Bucket load of weathered rock (Table 12-9), $0.66 \times 10 = 6.6$ cybm
5. Hourly production, $77 \times 6.6 = 508$ cybm

Costs, 1978 Again, Chapter 8 gives no estimated hourly cost of ownership and operation for custom-built dredges. For an approximate estimate it is suggested that one use

13-148 Hauling Excavation

Figure 13-78 Grapple dredge equipped with clamshell bucket of 6-yd³ capacity. The dredge is deepening a basin of the harbor at San Diego, California to an adequate depth for shipping. The dredge displaces 700 tons and draws 5 ft of water. It is loading some 480 cybm/h of rock and heavy gravels into the barge. The hull, with full revolving machinery deck and winches, together with the emerging bucket, is shown in the lower picture.

the value for shovels with electric motors, 12-yd³ dipper capacity. The hourly figure is 9 percent of the thousands of dollars of cost of the machine.

The erected dredge is estimated to cost $4.1 million as of 1978. Hence, the hourly cost of ownership and operation is estimated to be:

Dredge, 9% × $4100	$369.00
Leverman, engineer, deckmate, and deckhand	67.00
Total hourly cost	$436.00
Estimated direct job unit cost per cybm, $436.00/508	$0.86

Bucket Dredge

Bucket dredges are used in both construction and mining. In mining they are known as placer dredges for the extraction of gold, platinum, and tin from alluvia. They are known also as elevator, ladder, and chain-bucket dredges. The maximum depth of digging is usually 65 or 70 ft below low water, but some placer mining dredges have been built for greater than 100-ft depths.

The ladder or boom mounts an endless chain on which the digging buckets are mounted. Chain speeds are 40 to 60 ft/min, bucket spacings on the chain are 3 to 6 ft, and bucket capacities range from ½ to 20 cubic foot struck measurement. These three factors, which control production, are balanced to the desired output and nature of the material.

Production The production of a bucket dredge in cubic yards bank measurement per minute is given by the equation:

$$P = \left(\frac{S}{D}\right)(BF)\left(\frac{BC}{27}\right)$$

where P = production, cybm/min
S = speed of chain, ft/min
D = spacing of buckets, ft
BF = bucket factor (Table 12-3)
BC = bucket capacity, cysm

The production of a typical large dredge when digging, elevating, conveying, and dumping gravel is calculated from the following criteria: $S = 50$ ft/min; $D = 5$ ft; BF (from Table 12-3) = 0.92; $BC = 15$ ft³.

Substituting in the above equation:

$$P = \frac{50}{5} \times 0.92 \times \frac{15}{27} = 5.11 \text{ cybm/min}$$

Production during a 16-h day at 83 percent efficiency is 4072 cybm.

Costs Hourly costs of ownership and operation are high in relation to those of other dredges of equal purchase price because of the expense of maintaining the chain-bucket assembly. However, because of its uniform flow of material and resultant high hourly delivery, the bucket dredge offers low unit cost when working in suitable rock-earth. It is able to excavate alluvia and well-weathered rock but, like all chain-bucket excavators, it has unusually high maintenance costs when applied to consolidated materials.

Figure 13-79 **Dipper dredge equipped with dipper of 10-yd³ capacity.** This huge dredge is excavating weathered rock and loading the barges, which are towed to sea for disposal of the waste rock. The 50-ft-long dipper handle affords a digging depth of 25 ft at low water level. The attending bumboat is visible astern of the dredge. The dredge has accommodations for crews for double or triple shifting, which is customary in dredging work. The cycle time is about 0.65 min. About 6.6 cybm is the average dipper load, and at 83 percent efficiency the machine loads some 500 cybm/h.

Dipper dredges are suitable for the hardest of unblasted and blasted rock formations and are able to dig consolidated materials beyond the abilities of all other types of dredges. *(Bucyrus-Erie Co.)*

Self-Propelled Hopper Dredge

This huge ship is the largest, most costly, most complicated, and most imposing of the dredges. It is a combination of the suction dredge and a floating bottom-dump hopper for transporting the dredged materials. The operation and the production of the integral suction dredge are described in the ensuing section for the suction cutterhead hydraulic dredge.

The seagoing, self-propelled hopper dredge is a special vessel used mostly in the deepening of waterways wherein the dredged materials must be transported to disposal areas at considerable distance from the waterway. These areas for wasting are generally at sea. In these special operations estimates for productions and costs for the specially designed and built ships are made largely on the basis of the proprietary knowledge and experience of the personnel of the dredging industry.

The overall cost of ownership and operation for this vessel of large tonnage is calculated in the same manner as that for a coasting freighter. As the dredge is both a loading and a hauling machine, the total direct job unit cost depends on the cost of ownership and operation of the ship for the complete passage and on the quantity of dredgings carried. These determinants are fixed by an array of variables, among which are: dredging time, depending on materials and depth of dredging; time under way to and from the disposal area, which depends on traffic and weather; and major delays peculiar to the operations of a vessel working in crowded inland waterways.

Suction-Cutterhead-Pipeline Dredge

These hydraulic dredges (typical specifications for which are given in Table 13-57) are used extensively in rock-earth excavation. With their pipelines for transportation, they dig to depths of 60 ft below low-water elevation, elevate to heights greater than 100 ft

TABLE 13-57 Representative Specifications for Suction-Cutterhead Hydraulic Dredges

Specification	Nominal rating: pump discharge diameter, in		
	12	18	24
Centrifugal pump diameters, in:			
Suction	14	20	26
Discharge	12	18	24
Impeller	36	46	72
Prime movers:			
Driving centrifugal pump:			
Diesel engine, flywheel-brake hp	725	1125	2875
Electric motor, rated generally for selective duty, approximate hp	500	1125	2000
Driving auxiliaries, winches, and cutterhead:			
Diesel engine–generator–electric motors or diesel engine–hydraulic pump–hydraulic motors, flywheel-brake hp	140	380	670
Two electric motors, approximate total hp	140	380	670
Cutterhead, average diameter, in	36	54	72
Winches, driven by hydraulic or electric motors:			
Ladder	1	1	1
Swing	2	2	2
Spuds	2	2	2
Dimensions, approximate, ft:			
Hull:			
Length	60	90	113
Width	18	28	28
Depth	5	6	7
Overall:			
Length	92	127	155
Width	18	28	28
Height	18	27	36
Draft	2.5	3.5	4.0
Ladder, length, average	40	50	60
Dry weight, approximate, lb	140,000	460,000	670,000

above the centrifugal pump, and transport to distances of 2 mi. They handle materials ranging from soft clays to medium-hard rock formations.

Figure 13-80 illustrates a suction-cutterhead-pipeline dredge equipped with an 18-in-diameter centrifugal pump, a popular size of dredge for construction and mining work, and Figure 13-81 shows the principal working parts of a similar dredge with a 12-in centrifugal pump. Figure 13-82 illustrates the convenient, efficient console and cabin for the leverman or operator of a modern dredge.

The essential feature of a suction dredge is the centrifugal pump, which draws water and suspended rock-earth or solids through the suction pipe and then pumps the slurry through the discharge pipe. The mixture may be discharged through a pipeline floating on pontoons or submerged on the bottom and thence to land, where the pipeline may lie on natural ground or fill or upon a supporting structure.

In soft materials such as silts and sands, the loose material is drawn up readily by a plain suction head. In medium materials such as gravels and well-weathered rock, the rock-earth can be agitated sufficiently by water jets attached to the suction head to break up the material. In hard materials such as medium-weathered rock, a rotary cutterhead surrounding the suction head is used to fragment the relatively hard formation.

Output The output of a suction dredge may be expressed in two ways, depending on the significance or the purpose of the quantity.

1. As a basis for production or payment, it may be expressed in terms of an apparent cubic yard, which generally represents 1 cybm of average dredged materials. The apparent cubic yard contains voids averaging about 28 percent, and it equals approximately 1 cybm, which also has about 28 percent voids. However, the actual cubic-yard bank measurement may not have 28 percent voids in the bank. For example, 1 cybm of medium-weathered rock may have only 5 percent voids in the bank before the cutterhead fragments the rock. In this case the apparent cubic yard still has about 28 percent voids and 72 percent solids, but it represents 0.72/0.95 or 0.76 cybm. Accordingly, one must analyze the volume relationships when dealing with some minimum-weathered and semisolid rock formations so as to estimate closely the actual production in relation to the production in terms of apparent volume.

2. As a basis for figuring the dynamics of pumping, or alternatively as a means of estimating production, output may be expressed in terms of the delivery of the pump in gallons per minute and the percentage by apparent volume of the solids contained in the slurry or mixture. Instead of using gallons per minute as a determinant, output may be expressed in terms of pipe diameter and velocity of flow.

Principles of Operation The prime mover(s) of the dredge may be steam engines, diesel engines, or electric motors. Direct drive may be used, or the prime movers may drive generators or hydraulic pumps which in turn drive the auxiliary machinery. Except in the smallest dredges, the centrifugal pump has its own prime mover, as the pump duty is constant and demanding.

The individual operations of the dredge are:

1. The centrifugal pump is driven by the prime mover through a gear reduction box or belts.

2. Jetting around the suction head is by a high-pressure pump with electric or hydraulic drive.

3. The cutterhead surrounding the suction head may have electric or hydraulic drive.

4. Raising and lowering the suction-pipe assembly or ladder is accomplished by winch cable, electrically or hydraulically driven.

5. Swinging the dredge: This lateral movement of the ladder when digging is by swinging about the starboard working spud at the stern, and it is accomplished with two winch cables. The cables either carry fluke-type anchors embedded in the bottom or are attached to deadmen on the banks. The winches are electrically or hydraulically driven.

6. Advancing the dredge is accomplished by means of the starboard working spud, the port walking spud, and the swing cables. The two spuds are dropped alternately so that the dredge rotates correspondingly to starboard and to port and at the same time advances. The spuds are raised and lowered by the two winch cables.

7. For traveling short distances, the dredge may be winched. For long distances it may be towed or, in the case of a portable dredge, it may be dismantled, transported, and reerected.

Figure 13-80 Modern suction-cutterhead hydraulic dredge. This medium-size dredge is recovering sands-gravels for an aggregate plant located near Wichita, Kansas. The dredge works 9 to 12 h daily and 5 days weekly, totaling about 2500 h annually. The 18-in pump and machinery are driven by electric motors of the following horsepowers: centrifugal pump, 1250 hp; cutterhead, 125 hp; and five-drum winch, 60 hp. The slurry is pumped to a stockpile through 2500 ft of 18-in-diameter pipe against a total dynamic head of water of 142 ft. The approximate production statistics for the dredging work are:

Delivery of pump, gal/min	12,000
Solids by weight of slurry	20%
Hourly production, 500 cybm, in tons	660
Annual production, 1,250,000 cybm, in tons	1,650,000

(Dredgemasters International.)

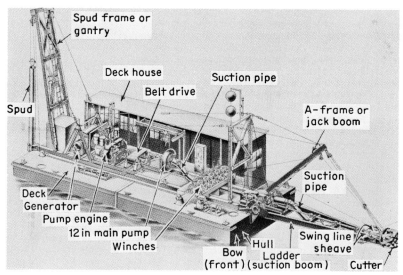

Figure 13-81 Working parts of a suction-cutterhead hydraulic dredge. The 12-in centrifugal pump is driven by its individual diesel engine. The cutterhead and five winches are driven by electric motors, which receive current from the auxiliary diesel engine–generator set. The five winches handle the ladder, swing cables, and spuds. The pump is set above water level in this dredge, although it is frequently located at or below water level in order to avoid priming of the centrifugal pump. The most important part of the dredge is the centrifugal pump, as on it, along with its prime mover, depend production and availability for the usual two-shift or around-the-clock operation of the dredge. *(Ellicott Machine Corp.)*

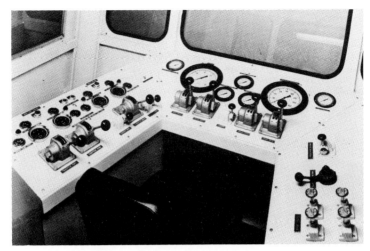

Figure 13-82 Operating console of the lever operator in a modern suction-cutterhead hydraulic dredge. The remote control levers and dial indicators cover the entire range of dredge operations for the machine of Figure 13-80, conveniently located in this manner for the many related operations of dredge work. Those to the left are for the engine of the centrifugal pump and for the engine of the hydraulic pumps used for auxiliary operations. Those to the front are for the five winches for the ladder of the suction pipe, the swing cables, and the spuds. Those to the right are for forward speed and reverse of the five winches and for forward speed and reverse of the cutterhead. Along with the handy controls, the 360° visibility from the wide windows makes for complete mastery by the lever operator.

The complete crew of the dredge consists of the leverman or operator, mechanic, and deckhand. The number of pipeline handlers varies according to the length and complexity of the pipeline. *(Dredgemasters International.)*

A dredge is equipped normally with five winches although small dredges sometimes use only three; the use of only three winches necessarily lowers efficiency.

When a suction dredge is used for rock excavation there are two key parts on which successful operation depends, the cutterhead and the centrifugal pump.

Cutterhead The cutterhead fragments the rock by a rotary slicing and abrading action. Rotation is counterclockwise and cutting takes place on the starboard side, where the blades move upward against the bank. Figure 13-83 illustrates a plain head for soft materials, a high-pressure jet attachment for soft to medium formations, and cutterheads for hard rock formations. Figure 13-84 illustrates a hard-rock cutterhead equipped with removable alloy steel points for exceptionally hard and abrasive rocks.

Cutterheads are made for all sizes of suction dredges used in medium and hard materials. Speeds can be varied from about 5 to 20 r/min and the rotation can be reversed in order to free any clogging or stalling foreign material. The ratio of cutterhead diameter to suction head diameter varies according to the material, but the cutterhead size is generally about three times that of the suction head. No matter what the size and type of cutterhead, its speed is adjusted to the nature of the material so as to give maximum production. The horsepower for the cutterhead averages about 10 percent of that for the pump. In extreme cases when digging minimum-weathered rock, the horsepower may range up to 4500 and just about equal those of the centrifugal pumps.

Centrifugal Pump The more important of the two key parts of the suction dredge is the centrifugal pump. The principles of operation, characteristics, construction, and performance of single-stage centrifugal pumps for water are set forth in Chapter 10. The principles of operation and the characteristics are the same for the pumps used in dredging. However, the construction and performances are different because the service is far more severe when pumping a mixture of 20 percent solids of fragmented rock made up of large, hard, and abrasive particles. The severe service requires a pump with the following attributes:

13-154 Hauling Excavation

1. Wide clearances, which permit passage of high proportion of solids in occasional sizes up to 33 percent of the diameter of the pump
2. Replaceable volute liners and vane tips
3. Extra-heavy shafts and bearings to take the shock loads induced by jamming rocks or foreign materials such as stumps or logs
4. Generally heavier construction and better metallurgy than those of its counterpart for pumping water

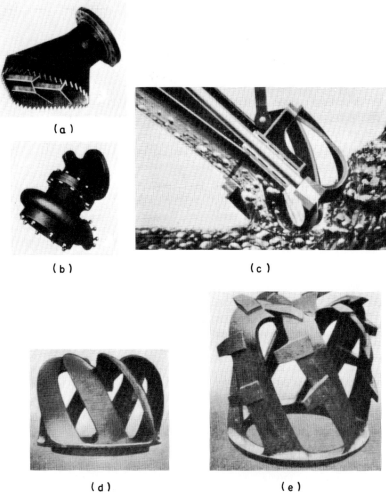

Figure 13-83 Suction head, jetting attachment for suction head, and cutterheads for suction dredges. The suction head (a) is used for soft materials such as clays, silts, and sands, which need no agitation other than by the serrated edges of the head. The high-pressure jet (b) is an attachment for the suction head. It is used in soft to medium materials which need agitation for loosening. The high-pressure supply line is attached to the suction pipeline to the pump. The cutterheads (c–e) are used in medium to hard formations that need to be fragmented by slicing and abrading to an acceptable size for lifting to the pump and for passing through the pump while suspended in the slurry or mixture. (c) Artist's view of the cutter of a hydraulic pipeline dredge. (d) A spiral cutter with seven blades. (e) For special work cutters may be provided with teeth of the so-called shovel type welded to the blades. *(Ellicott Machine Corp.)*

These requirements result in a heavier and more expensive pump and they also result in a less efficient pump, averaging about 50 percent. Discharge or pipe velocities range from 10 to 16 ft/s, 12 ft/s being a good average all-purpose velocity.

Figure 13-85 illustrates a specially designed, rugged centrifugal pump for dredging rock-earth and rock, and Figure 13-86 gives performance curves for this pump.

The rating of the centrifugal pump, that is, the diameter of the discharge opening in inches, is also the nominal rating of the dredge. Usually the diameter of the pipeline is also the diameter of the pump discharge. Table 13-60 gives the hourly output of the dredge or pump in terms of cubic yards bank measurement or apparent volume according to amount of solids, pipe diameter, and velocity of flow. In order to arrive at gallons per minute of flow, one may cross-refer Table 13-60 to Table 13-59." Delivery rate in gallons per minute may be calculated readily by the formula:

$$\text{gpm} = 2.45 D^2 v$$

where gpm = quantity, gal/min
D = diameter of pipe, in
v = velocity of flow, ft/s

Dredges in regular use for rock-earth work range up to a 27-in size. A dredge of this size handles 1252 cybm/h when: (1) pumping through 27-in pipe; (2) pumping at 12 ft/s, the average velocity for dredge work; and (3) pumping 20 percent of solids.

Pipeline The pipeline may be as simple as a short length of pipe from pump to barge or as complex as a 2-mi line with elbows, Y's, and valves from the dredge to the hydraulic fill for a dam. The pipelines are classified as floating, submerged, or shore lines.

Floating pipe is the heavier type, with thicknesses from 0.25 to 0.50 in. Lengths of sections vary from 20 to 60 ft. The metal is high-carbon or alloy steel. The individual sections are supported on floating pontoons, usually two, three, or four, and they are connected by ball joints. The ball joints serve to take care of wave action as well as the long-radius curvature of the line. Turns are through long-radius elbows. The transition

Figure 13-84 Large-diameter cutterhead with removable alloy steel teeth. As with all rock excavating machinery, the ability and capacity of the dredge to handle harder materials have increased with improvements in metallurgy. An example is this 43-in-diameter cutterhead with removable teeth. However, the removable teeth or points of such cutterheads may have to be changed every 30 to 45 min when excavating igneous rocks of minimum weathering. This short and expensive life parallels that of the teeth of crawler-tractor-rippers under extreme conditions. Nevertheless, in both cases it is sometimes economical to excavate without blasting under the prevailing job conditions. This is especially true when cutterheads can eliminate highly expensive subaqueous blasting. *(Dredgemasters International.)*

13–156 Hauling Excavation

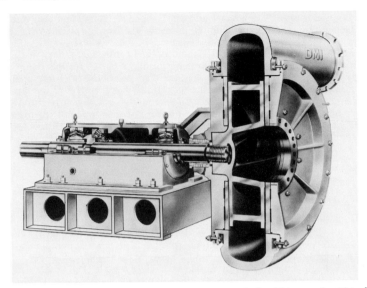

Figure 13-85 Heavy-duty 20-in centrifugal pump for suction dredge. This pump has 20-in-diameter suction, 18-in-diameter discharge, and 46-in-diameter impeller. It is used in the dredge of Figure 13-80.

Among the important specifications are: (1) fully adjustable impeller; (2) heavy-duty antifriction radial and thrust bearings; (3) single sealed bearing cartridge; (4) oversize $7^{3}/_{16}$-in-diameter forged-steel shaft; (5) interchangeable side liners; (6) choice of metals, high-Brinell Duralite or manufacturer's Permalloy, and (7) variable discharge angle through 360°.

Figure 13-86 gives the performance curves for this centrifugal pump. *(Dredgemasters International.)*

from floating pipe to shore pipe is by means of two ball joints. The pontoons are spaced along the section of pipe to secure adequate load distribution and minimum bending of the loaded pipe. They may be custom-fabricated from steel or they may be simply built of steel drums and wooden decking. The floating pipe serves as a walkway from shore to dredge.

One typical section of pipe and pontoons, consisting of 14-in pipe of 0.375-in thickness supported by four pontoons, has the following weights:

Pipe, 50-ft length at 55 lb/ft	2750 lb
Ball joint, one per length	150
Four pontoons, each 10 ft long, 36-in diameter, made of 0.25-in-thick steel, with decking	4320
Unladen weight	7220 lb
Weight of dredgings in pipe, at 77 lb/ft³, 50×82	4100
Total or laden weight	11,320 lb

The displacement of water by the total weight equals $11{,}320/62.4 = 181$ ft³. The volume of the pontoons is 282 ft³, so that the pontoons have ample volume to support the total weight, their capacity being 27,597 lb.

Submerged pipe is similar to floating pipe in general construction. It is placed on the floor of the body of water so as not to interfere with navigation.

Shore pipe is a lighter type because it must be handled frequently unless the dredgings are used at a central point, such as the aggregate plant of Figure 13-80. The thickness varies from 10 gauge, or 0.13 in, to 0.375 in. An occasional practice is to convert worn floating pipe to shore use. To facilitate handling, short lengths of 20 to 40 ft are used. The sections are connected by slip joints to simplify the frequent changes, and the joints provide some flexibility. In lieu of slip joints the pipes may have collars, over which a ring with wedge tightener serves as a connector. Shore pipe may lie on the ground and fill or it may be supported by sleepers or cribbing, depending on the lay of the land. The kind of fill or end use of the dredgings deter-

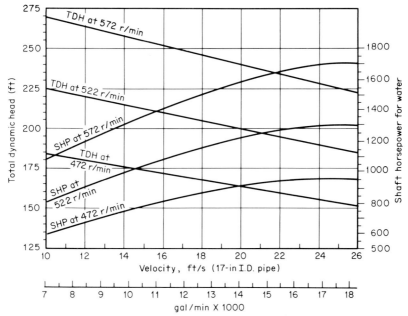

Figure 13-86 Performance curves for centrifugal pump of suction dredge: 20-in-diameter suction, 18-in-diameter discharge, 46-in-diameter impeller, 17-in-I.D. pipeline. This pump, shown in Figure 13-85, is the pump of the dredge of Figure 13-80. When pumping water at 18 ft/s, or 12,700 gal/min, at 522 r/min, the total dynamic head is 205 ft and the required shaft horsepower is 1150, according to the curves. However, the theoretical water horsepower required is given by the equation:

$$\text{hp} = \frac{12{,}700 \times 7.48 \times 205}{33{,}000} = 590$$

Hence, pump efficiency equals 590/1150=51%. *(Dredgemasters International.)*

mines the complexity of the pattern of shore pipeline and the necessary appurtenances for the pipe.

All pipe is placed and shifted by pipeline handlers. Floating pipe and submerged pipe are handled by a hoist mounted on the bumboat or scow. Shore pipe is handled entirely manually or with the help of a crawler-tractor with bulldozer or bucket.

Pipes wear mostly on the bottom and so they are rotated through 120° occasionally to distribute wear and increase life. An empirical equation for average life is:

$$Y = \frac{2D(133t - D)}{133P}$$

where Y = useful life, yr
D = diameter of pipe, in
t = thickness of pipe, in
P = average yearly production, million cybm

If a 24-in-diameter pipe of 0.50-in thickness carries 4 million cybm annually:

$$Y = \frac{2 \times 24[(133 \times 0.50) - 24]}{133 \times 4.0} = 3.83 \text{ yr or 3 yr, 10 months}$$

Dynamics of Dredging by Centrifugal Pump The dynamics are the same as those for the centrifugal pumps of Chapter 10, except for one most important difference. Instead of water, a mixture, or slurry, of rock-earth and water is being pumped. This increases the unit weight of water, 62.4 lb/ft³, in proportion to the amount of solids being pumped. Table 13-58 gives weight per cubic foot of slurry according to percentage by apparent volume.

TABLE 13-58 Weight and Specific Gravity of Mixture of Solids and Water According to the Apparent Volume of Solids Being Dredged (Percent of Solids)

The apparent volume of solids is the bank measurement of the rock-earth being dredged. The table is based on: average percentage of voids in the apparent volume or bank measurement volume, 28% for average rock-earth dredged; average specific gravity of rock-earth in the solid state, 2.66 or 166 lb/ft.^{3}

	Solids			Water		Mixture	
Apparent volume (percent of solids), %	Actual solid volume, %	Weight per ft^3 of mixture, lb		Actual volume, %	Weight per ft^3 of mixture, lb	Weight per ft^3 of mixture, lb	Specific gravity
4	2.9	4.8		97.1	60.6	65.4	1.05
6	4.3	7.2		95.7	59.7	66.9	1.07
8	5.8	9.6		94.2	58.8	68.4	1.10
10	7.2	12.0		92.8	57.9	69.9	1.12
12	8.6	14.3		91.4	57.0	71.3	1.14
14	10.1	16.7		89.9	56.1	72.8	1.17
16	11.5	19.1		88.5	55.2	74.3	1.19
18	13.0	21.5		87.0	54.3	75.9	1.22
20	14.4	23.9		85.6	53.4	77.3	1.24
22	15.8	26.3		84.2	52.5	78.8	1.26
24	17.3	28.7		82.7	51.6	80.3	1.29

Here a distinction must be drawn. In figuring the output of a dredge, the apparent volume is actually the absolute volume of the particles plus the volume of voids among the particles. These voids amount to an average of 28 percent for the materials or rock-earth which are normally dredged. It is also true that these same materials contain about 28 percent voids in the bank before they are dredged. Hence an apparent volume of 1 yd^3 of slurry approximately equals 1 cybm. Accordingly, an apparent volume of 1 yd^3 or 1 cybm contains approximately 72 percent rock-earth in the solid state.

If a slurry contains 20 percent solids by apparent volume, the actual percentage distributions and weights of 1 ft^3 of slurry, based on an average density of 166 lb/ft^3 for solid rock, are as follows:

Weight of 20% apparent volume of rock, $0.72 \times 0.20 \times 166$	23.9 lb
Weight of water, $[(0.28 \times 0.20) + 0.80] 62.4$	53.4
Weight of slurry	77.3 lb
Weight ratio of 1 ft^3 of slurry to 1 ft^3 of water (specific gravity of slurry), 77.3/62.4	1.24

Total Dynamic Head. The total dynamic head against which the centrifugal pump of a dredge works is the sum of suction, velocity, static, and friction heads. It is first calculated for water and this value is then multiplied by the specific gravity of the slurry in order to obtain the actual head against which the pump works.

Suction Head. Suction head is shown in inches of mercury on a vacuum gauge located near the pump in the suction line. The gauge reading may be converted to feet of head of water by multiplying by 1.13. Several factors affect suction head: height of pump above water level or, conversely, depth of pump below water level; resistance of water due to sudden change of velocity from 0 to an average of 12 ft/s; entrance head loss; and friction in the suction pipe. Suction head, however, may be estimated at 16 ft.

Velocity Head. For flow velocities varying from 10 to 16 ft/s velocity head ranges from 1.6 to 4.0 ft. The suggested average velocity head for estimating purposes, based on an average 12-ft/s flow velocity, is 2.4 ft.

Static Head. Static head is the elevation in feet of the highest point of the discharge line above the pump. If the discharge is full clear to the end, the elevation at the end of the pipe is taken for figuring the static head.

Friction Head. Friction head is the sum of losses in feet of head through the pipe and pipe appurtenances such as valves and fittings. It varies with the condition of the pipe, with velocity, and with pipe diameter. Table 13-59 gives values for these losses in pipes of average condition with velocities up to 20 ft/s and diameters up to 48 in. Friction equivalents in feet of straight pipe are given for losses of head in valves and fittings. The table is based on the Hazen-Williams formula for friction losses of water in pipes.

The formula is discussed in Chapter 10. It is shown there that, when the coefficient C has a value of 100 for pipe in average condition, the slope of the hydraulic grade line or the sine of angle of slope is given by the equation:

$$S^{0.54} = \frac{V}{11.5 D^{0.63}}$$

where $v =$ velocity of flow, ft/s
  $D =$ diameter of pipe, in

It is shown also that the friction head per 100 ft length of pipe equals $100S$.

The value of 100 for C, on which Table 13-59 is based, is for steel pipe in average condition after several years of service. It is a conservative estimate. Accompanying Table 13-59 is a tabulation giving values of C for pipe in different conditions, together with multipliers K by which the friction head losses of the table should be multiplied. Because of the excellent smooth, scoured inside surface of dredge pipe, it is suggested that the multiplier K for the head losses of Table 13-59 be taken as 0.54.

Table 13-59 includes friction losses in pipes, valves, and fittings of up to 48-in diameter, ample for dredge work. Friction losses for any inside diameter of pipe may be calculated by the formula:

$$S^{0.54} = \frac{v}{11.5 D^{0.63}}$$

For example, the friction head loss per 100 ft of 28-in-diameter pipe at a flow velocity of 12 ft/s based on a K value of 0.54 for dredge pipe, is figured in this manner:

1. $S^{0.54} = \dfrac{12}{11.5 \times 28^{0.63}}$ $S = 0.0222$

2. The friction head loss for 100 ft of average pipe is $100 \times 0.0222 = 2.22$ ft. This same friction loss may be calculated by interpolating between 24- and 30-in diameters in Table 13-59.

3. The friction head loss for 100 ft of dredge pipe is $0.54 \times 2.22 = 1.20$ ft.

The great savings in horsepower requirements for the diesel-engine prime mover of the centrifugal pump and in diesel fuel cost (1978) which can be achieved by selection of appropriate C and K valves are shown in the following comparisons (page 13–163), based on friction head losses in 2 mi of 30-in-diameter pipe, a flow of 3535 ft³/min, a weight of slurry of 77.3 lb/ft³, and a production of 1580 cybm/h.

TABLE 13-59A Friction Losses in Feet of Head for 100-ft Length of Pipe Based on the Hazen-Williams Formula with C Value of 100

The formula and its modification for figuring friction losses in pipe are discussed in Chapter 10. This table gives the velocity head in feet, velocity in feet per second, capacity in gallons per minute, and friction loss in feet of head per 100 ft of steel pipe in average condition. For average condition of pipe the coefficient C of the formula has a value of 100. For pipe in a different condition the value for friction loss in the table is multiplied by a multiplier K. The values of K for pipe in different conditions are given in this tabulation.

Condition of pipe	C according to pipe condition	K (multiplier for head loss based on $C=100$)
Bad, old, highly tuberculated	60	2.58
	70	1.92
Poor, rough	80	1.52
	90	1.22
Average, pitted	100	1.00
	110	0.84
Good, fairly smooth	120	0.71
	130	0.62
New, smooth	140	0.54

NOTE: For smooth, well-scoured dredge pipe it is suggested that the value of K be 0.54.

TABLE 13-59 (Continued)

Velocity head, ft	Velocity, ft/s	\multicolumn{2}{c}{1}		1¼		1½		2		2½		3		4		5	
		Capacity, gal/min	Friction head, ft	Capacity, gal/min	Friction head, ft	Capacity, gal/min	Friction head, ft	Capacity, gal/min	Friction head, ft	Capacity, gal/min	Friction head, ft	Capacity, gal/min	Friction head, ft	Capacity, gal/min	Friction head, ft	Capacity, gal/min	Friction head, ft
0.015	1	2.45	1.10	4.67	0.78	5.5	0.62	9.8	0.48	5.3	0.37	22	0.30	39	0.21	61	0.17
0.062	2	4.90	3.95	9.34	2.65	11.1	2.23	19.6	1.75	30.6	1.33	44	1.08	78	0.77	122	0.60
0.140	3	7.35	8.45	14.0	5.70	16.5	4.75	29.4	3.70	45.9	2.82	66	2.3	117	1.64	184	1.26
0.250	4	9.79	14.5	18.7	9.75	22.0	8.08	39.2	6.30	61.2	4.81	88	3.9	156	2.79	245	2.14
0.39	5	12.2	21.9	23.4	14.3	27.6	12.3	48.9	9.55	76.5	7.30	110	6.0	196	4.21	306	3.24
0.56	6	14.7	30.7	28.0	20.4	33.2	17.1	58.7	13.3	91.8	10.2	132	8.4	236	5.42	367	4.55
0.76	7	17.2	41.0	32.7	27.8	38.6	22.8	68.5	17.7	107	13.6	154	11.1	275	7.87	428	6.06
0.99	8	19.6	52.4	37.3	35.4	44.0	29.1	78.3	22.7	122	17.4	176	14.3	314	10.1	490	7.77
1.12	8½	20.8	58.5	39.6	39.6	46.9	32.7	83.2	25.4	130	19.4	187	16.0	334	11.3	521	8.70
1.26	9	21.0	65.0	42.0	44.0	49.7	36.2	88.1	28.3	138	21.6	198	17.7	352	12.6	550	9.65
1.40	9½	23.3	72.0	44.3	48.2	52.4	40.0	93.0	31.3	145	24.0	209	19.6	372	13.9	581	10.7
1.55	10	24.5	79.2	46.7	53.0	55.3	44.0	97.9	34.3	153	26.3	220	21.5	391	15.2	612	11.7
1.72	10½	25.7	86.6	49.0	58.0	58.1	48.0	103	37.6	161	29.1	231	23.5	411	16.7	642	12.9
1.88	11	27.0	94.6	51.3	63.3	60.5	52.3	108	41.0	168	31.8	242	25.5	431	18.2	673	14.0
2.06	11½	28.2	102.5	53.6	68.5	63.6	57.0	113	44.4	176	34.4	253	27.6	451	19.8	704	15.2
2.24	12	29.4	110	56.0	74.5	66.6	61.5	118	48.1	184	37.1	264	30.0	471	21.3	735	16.5
2.42	12½	30.6	119	58.3	80.3	69.0	66.3	122	52.0	191	40.0	275	32.3	490	23.0	766	17.8
2.62	13	31.8	128	60.7	87.0	71.5	71.4	128	56.0	199	43.0	286	34.4	510	24.8	796	19.1
3.05	14	34.3	147	65.3	99.4	77.4	82.0	137	64.0	214	49.3	308	40.0	549	28.4	857	22.0
3.50	15	36.7	167	70.0	113	82.5	93.2	147	71.1	230	55.9	330	45.4	587	32.2	918	24.9
3.98	16	39.2	188	74.7	128	88.4	105	157	82.3	244	62.9	352	51.2	628	36.3	980	28.0
4.50	17	41.7	210	79.3	142	94.0	117	167	92.4	260	70.2	374	57.4	666	40.7	1041	31.3
5.05	18	42.1	235	84.0	158	99.4	130	176	103	276	78.0	396	63.8	704	45.2	1102	35.0
5.60	19	46.5	260	88.7	175	104.7	144	186	114	291	86.0	418	70.5	743	50.1	1163	38.8
6.20	20	49.0	285	93.4	192	110.6	158	196	125	305	94.5	440	77.6	782	55.0	1224	42.6

Inside diameter of pipe, in

13–160 Hauling Excavation

Velocity head, ft	Velocity, ft/s	Inside diameter of pipe, in											Velocity, ft/s
		6		8		10		12		14			
		Capacity, gal/min	Friction head, ft	Capacity, gal/min	Friction head, ft	Capacity, gal/min	Friction head, ft	Capacity, gal/min	Friction head, ft	Capacity, gal/min	Friction head, ft		
												15	16
												Capacity, gal/min	Friction head, ft	Capacity, gal/min	Friction head, ft	

Velocity head, ft	Velocity, ft/s	6 Capacity, gal/min	6 Friction head, ft	8 Capacity, gal/min	8 Friction head, ft	10 Capacity, gal/min	10 Friction head, ft	12 Capacity, gal/min	12 Friction head, ft	14 Capacity, gal/min	14 Friction head, ft	15 Capacity, gal/min	15 Friction head, ft	16 Capacity, gal/min	16 Friction head, ft	Velocity, ft/s
0.015	1	88	0.14	156	0.10	244	0.07	351	0.06	480	0.05	551	0.05	627	0.04	1
0.062	2	176	0.49	312	0.35	488	0.27	704	0.22	960	0.18	1102	0.17	1254	0.15	2
140	3	264	1.02	470	0.73	734	0.57	1058	0.46	1440	0.38	1652	0.35	1881	0.33	3
0.25	4	352	1.73	627	1.24	979	0.97	1410	0.78	1920	0.65	2203	0.58	2505	0.56	4
0.39	5	440	2.60	783	1.89	1224	1.45	1763	1.16	2400	0.99	2755	0.85	3140	0.85	5
0.56	6	528	3.65	940	2.65	1468	2.04	2115	1.64	2880	1.38	3304	1.27	3762	1.18	6
0.76	7	617	4.88	1097	3.51	1713	2.70	2468	2.17	3360	1.84	3855	1.70	4390	1.57	7
0.99	8	706	6.23	1253	4.50	1956	3.46	2820	2.80	3840	2.36	4407	2.15	5021	2.01	8
1.12	8½	748	7.00	1331	5.03	2081	3.88	2996	3.12	4070	2.63	4688	2.42	5334	2.12	8½
1.26	9	793	7.75	1410	5.60	2203	4.30	3173	3.47	4320	2.94	4958	2.72	5645	2.50	9
1.40	9½	837	8.58	1488	6.20	2326	4.75	3350	3.83	4560	3.25	5232	3.03	5960	2.77	9½
1.55	10	880	9.42	1567	6.81	2448	5.22	3525	4.21	4800	3.54	5508	3.30	6270	3.04	10
1.72	10½	925	10.6	1646	7.42	2571	5.73	3701	4.59	5040	3.93	5783	3.69	6585	3.30	10½
1.88	11	970	11.5	1723	8.10	2693	6.26	3878	5.00	5280	4.30	6058	3.97	6900	3.61	11
2.06	11½	1013	12.5	1800	8.80	2815	6.80	4054	5.41	5510	4.67	6335	4.30	7215	3.93	11½
2.24	12	1058	13.6	1880	9.53	2938	7.34	4230	5.88	5760	5.05	6610	4.70	7540	4.27	12
2.42	12½	1101	14.6	1959	10.3	3060	7.91	4406	6.33	5990	5.41	6885	5.00	7850	4.60	12½
2.62	13	1145	15.7	2037	11.1	3182	8.52	4583	6.80	6240	5.87	7160	5.40	8151	4.95	13
3.05	14	1234	18.0	2193	12.8	3426	9.80	4935	7.80	6720	6.71	7711	6.20	8775	5.70	14
3.50	15	1321	20.4	2350	14.5	3671	11.1	5288	8.86	7190	7.64	8262	7.06	9420	6.48	15
3.98	16	1410	23.0	2506	16.4	3916	12.5	5640	10.0	7690	8.60	8814	7.94	10010	7.30	16
4.50	17	1500	25.8	2662	18.3	4162	14.0	5992	11.3	8160	9.65	9365	9.00	10670	8.18	17
5.05	18	1586	28.5	2820	20.3	4406	15.6	6346	12.5	8640	10.7	9916	10.0	11290	9.10	18
5.60	19	1675	31.4	2976	22.2	4650	17.3	6698	13.9	9120	11.8	10466	11.0	11910	10.1	19
6.20	20	1764	34.5	3134	24.3	4896	19.0	7050	15.2	9590	13.0	11016	12.0	12540	11.1	20

13–162 Hauling Excavation

TABLE 13-59 (Continued)

Velocity head, ft	Velocity ft/s	\multicolumn{3}{c}{18}			\multicolumn{3}{c}{20}			\multicolumn{3}{c}{24}			\multicolumn{3}{c}{30}			\multicolumn{3}{c}{36}			\multicolumn{3}{c}{42}			\multicolumn{3}{c}{48}			Velocity ft/s
		Capacity, gal/min	Friction head, ft	Capacity, gal/min	Friction head, ft	Capacity, gal/min	Friction head, ft	Capacity, gal/min	Friction head, ft	Capacity, gal/min	Friction head, ft	Capacity, gal/min	Friction head, ft	Capacity, gal/min	Friction head, ft	
0.015	1	794	0.04	973	0.03	1412	0.03	2203	0.02	3170	0.02	4319	0.01	5640	0.01	1
0.062	2	1587	0.14	1956	0.12	2825	0.10	4406	0.07	6340	0.06	8637	0.05	11280	0.04	2
0.14	3	2379	0.29	2940	0.25	4230	0.20	6610	0.15	9520	0.13	12950	0.11	16920	0.09	3
0.25	4	3173	0.49	3920	0.43	5640	0.34	8810	0.27	12690	0.21	17270	0.18	22560	0.15	4
0.39	5	3965	0.73	4900	0.65	7050	0.52	11010	0.40	15860	0.32	21590	0.28	28200	0.23	5
0.56	6	4758	1.03	5800	0.91	8460	0.73	13220	0.57	19040	0.46	25910	0.38	33850	0.32	6
0.76	7	5552	1.37	6850	1.21	9870	0.97	15420	0.75	22210	0.60	30230	0.41	39840	0.44	7
0.99	8	6345	1.75	7830	1.55	11280	1.25	17620	0.96	25380	0.77	33540	0.65	45120	0.55	8
1.12	8½	6740	1.96	8320	1.74	11480	1.40	18720	1.08	26970	0.87	36700	0.73	47940	0.62	8½
1.26	9	7140	2.19	8810	1.93	12690	1.55	19830	1.20	28550	0.96	38860	0.81	50760	0.68	9
1.40	9½	7535	2.41	9300	2.13	13400	1.73	20930	1.33	30140	1.07	41020	0.89	53580	0.76	9½
1.55	10	7930	2.66	9790	2.34	14100	1.87	22030	1.46	31730	1.17	43180	0.90	56400	0.84	10
1.72	10½	8330	2.89	10280	2.57	14800	2.03	23130	1.59	33310	1.29	45340	1.07	59229	0.92	10½
1.88	11	8725	3.14	10770	2.80	15510	2.22	24230	1.63	34900	1.40	47500	1.17	62500	1.00	11
2.06	11½	9120	3.40	11260	3.03	16220	2.42	25340	1.88	36490	1.52	49660	1.27	64860	1.08	11½
2.24	12	9520	3.70	11750	3.29	16920	2.61	26440	2.02	38070	1.65	51820	1.37	67680	1.17	12
2.42	12½	9915	3.98	12240	3.54	17630	2.82	27540	2.10	39640	1.78	53980	1.48	70500	1.27	12½
2.62	13	10310	4.30	12730	3.80	18330	3.03	28640	2.28	41210	1.91	56140	1.60	73320	1.36	13
3.05	14	11105	4.93	13710	4.37	19740	3.50	30840	2.70	44420	2.20	60460	1.82	78960	1.56	14
3.50	15	11897	5.60	14690	4.98	21150	3.97	33050	3.08	46700	2.49	64770	2.08	84600	1.78	15
3.98	16	12690	6.30	15670	5.60	22560	4.50	35250	3.47	50760	2.88	69090	2.32	90240	2.00	16
4.50	17	13480	7.05	16650	6.30	23970	5.00	37450	3.89	53930	3.13	73410	2.60	95880	2.25	17
5.05	18	14280	7.85	17620	7.00	25380	5.60	39650	4.32	57110	3.50	77730	2.90	101520	2.50	18
5.60	19	15070	8.70	18600	7.71	26790	6.17	41860	4.79	60280	3.88	82050	3.20	107160	2.77	19
6.20	20	15860	9.53	19580	8.48	28200	6.78	44060	5.27	63450	4.26	86370	3.50	112800	3.05	20

Inside diameter of pipe, in

TABLE 13-59B Friction Losses in Valves and Fittings Expressed as Equivalent Length in Feet of Straight Pipe of Same Diameter

Valve or fitting	Diameter of pipe, in												
	6	8	10	12	14	15	16	18	20	24	30	36	42	48
Elbow, 90°, long radius	12	16	20	24	28	30	32	36	40	48	60	72	84	96
Elbow, 90°, short radius	18	24	30	36	42	45	48	54	60	72	90	108	126	144
Elbow, 45°	9	12	15	18	21	23	24	27	30	36	45	54	63	72
Elbow, 22½°	6	8	10	12	14	15	16	18	20	24	30	36	42	48
Tee	30	40	50	60	70	75	80	90	100	120	150	180	210	240
Y-branch	22	30	38	45	52	56	60	68	75	90	112	135	158	180
Gate valve	12	16	20	24	28	30	32	36	40	48	60	72	84	96
Globe valve	90	120	150	180	210	225	240	270	300	360	450	540	630	720
Angle valve	30	40	50	60	70	75	80	90	100	120	150	180	210	240
Check valve	24	32	40	48	56	60	64	72	80	96	120	144	168	192
Foot valve	24	32	40	48	56	60	64	72	80	96	120	144	168	192
Flap valve	24	32	40	48	56	60	64	72	80	96	120	144	168	192
Rubber sleeve connection	15	20	25	30	35	38	40	45	50	60	75	90	105	120

SOURCE: Morris Machine Works.

	Factors	
Item	$C=100$	$C=140$ $K=0.54$
Friction head loss per 100 ft of pipe, ft	2.02	1.09
Total friction head loss for 2 mi of pipe, ft	213	115
Slurry horsepower	1764	952
Brake horsepower of prime mover, estimated	3920	2116
Hourly diesel fuel consumption, estimated, gal	231	125
Hourly diesel fuel cost at $0.40/gal	$92.40	$50.00

A savings of $42.80 results in a unit cost saving of $0.027/cybm by reason of a difference in the condition of the pipe.

Table 13-60 gives productions of suction dredges in cubic yards bank measurement, equivalent approximately to cubic yards apparent volume, according to pipe diameter and velocity of flow. The production may be expressed in terms of flow in gallons per minute, either arithmetically or by cross-reference to the values of Table 13-59.

Sometimes, as in the ensuing example for calculating power requirements for a centrifugal pump, the output is reckoned in terms of percentage of solids by weight of slurry. In such cases Table 13-60 is not applicable and calculations must be made for the weight of a cubic foot of slurry. In the example of slurry containing solids at 20 percent of weight of slurry, the weight of a cubic foot of slurry W is given by the equation:

$$\frac{0.20W}{166.0} + \frac{0.80W}{62.4} = 1.0$$

where 166.0 = weight of 1 cubic foot of solids
62.4 = weight of 1 cubic foot of water
W = 71.4 lb per cubic foot of slurry

Table 13-58 indicates that the specific gravity of the slurry is 1.14 and that 20 percent by weight is equivalent approximately to 12 percent by apparent volume.

Power Requirements for Centrifugal Pump The required slurry horsepower for a pump is given by the equation:

$$\text{Shp} = \frac{AvWH}{550}$$

TABLE 13-60 Material in Cubic Yards Bank Measurement Pumped Hourly by Dredges, According to Pipe Diameter and Velocity and Percentage of Solids

Pipe diameter, in	Velocity of flow, ft/s	Material pumped per hour, cybm Amount of solids by apparent volume (% of total volume pumped)										
		4	6	8	10	12	14	16	18	20	22	24
6	10	10	16	21	26	31	37	42	47	52	58	63
	12	13	19	25	31	38	44	50	57	63	69	75
	14	15	22	29	37	44	51	59	66	73	81	88
	16	17	25	34	42	50	59	67	75	84	92	100
8	10	19	28	37	47	56	65	74	84	93	102	112
	12	22	33	45	56	67	78	89	100	112	123	134
	14	26	39	52	65	78	91	104	117	130	143	156
	16	30	45	60	74	89	104	119	134	149	164	179
10	10	29	44	58	73	87	102	116	131	145	160	174
	12	35	52	70	87	105	122	140	157	175	192	209
	14	41	61	81	102	122	142	163	183	204	224	244
	16	47	70	93	116	140	163	186	209	223	256	279
12	10	42	63	84	105	126	147	168	188	209	230	251
	12	50	75	100	126	151	176	201	226	251	276	302
	14	59	88	117	147	176	205	235	264	293	322	352
	16	67	100	134	168	201	235	268	302	335	368	402
16	10	74	112	149	186	223	261	298	335	372	409	447
	12	89	134	179	223	268	313	357	402	447	491	536
	14	104	156	208	261	313	365	417	469	521	573	625
	16	119	179	238	298	357	417	476	536	596	655	715
20	10	116	174	233	291	349	407	465	523	582	640	698
	12	140	209	279	349	419	489	558	628	698	768	838
	14	163	244	326	407	489	570	651	733	814	896	977
	16	186	279	372	465	558	651	744	838	931	1024	1117
24	10	168	251	335	419	503	586	670	754	838	921	1005
	12	201	302	402	503	603	704	804	905	1006	1106	1206
	14	235	352	469	586	704	821	938	1055	1173	1290	1407
	16	268	402	536	670	804	938	1072	1206	1340	1474	1608
28	10	228	342	456	570	684	798	912	1026	1140	1254	1368
	12	274	410	547	684	821	958	1094	1231	1368	1505	1642
	14	319	479	638	798	958	1117	1277	1436	1596	1756	1915
	16	365	547	730	912	1094	1277	1459	1642	1824	2006	2189
32	10	298	447	596	744	893	1042	1191	1340	1489	1638	1787
	12	357	536	715	893	1072	1251	1429	1608	1787	1965	2144
	14	417	625	834	1042	1251	1459	1668	1876	2084	2293	2501
	16	476	715	953	1191	1429	1668	1906	2144	2382	2620	2859

where $Shp =$ slurry horsepower
$A =$ area of discharge pipe, ft^2
$v =$ velocity of flow, ft/s
$W =$ weight of 1 cubic foot of slurry, lb
$H =$ total dynamic head, ft

The application of the above formula may be demonstrated by reconciling the approximate data of Figure 13-80 with the 1250 hp of the electric motor driving the centrifugal pump. The criteria for the calculations are:

Discharge diameter of pump, in	18
Output of pump, gal/min	12,000
Percentage of solids by weight of slurry	20
Amount pumped hourly, approximate, equivalent to 500 cybm, tons	660
Pipe and pipe appurtenances:	
Diameter, in	18
Length, ft	2500
Appurtenances:	
Two 90° elbows, short radius	
Three 22½° elbows	
One 45° elbow	
One flap valve	
Two gate valves	
Height of discharge point at stockpile above pump, ft	40

The values to be substituted in the slurry horsepower formula are obtained as follows:

A for 18-in-diameter pipe, ft^2	1.77
v for velocity of flow (from Table 13-59 for 18-in pipe and 12,000-gal/min) flow, ft/s	15
W for weight of 1 ft^3 of slurry containing 20% solids by weight with specific gravity of 2.66, lb	71.4
H for total dynamic head, ft:	
Suction head, estimated	16.0
Velocity head, estimated	2.4
Static head	40.0
Friction head:	83.5*
Equivalent total length of pipe, ft:	
Length of pipe	2500
Appurtenances (Table 13-59):	
One 90° elbow, short radius	54
Two 22½° elbows, at 18 ft each	36
One 45° elbow	27
One flap valve	72
Two gate valves, at 36 ft each	72
Total equivalent length of pipe	2761
H, total dynamic head, ft	141.9

*The friction head is calculated as follows. The value for a 100-ft length of the equivalent length of pipe (from Table 13-59) is 5.60 ft. This figure is for pipe in average condition, and in the Hazen-Williams formula it is based on $C=100$. For dredge pipe the value of C is 140, and the correction factor for 5.60 ft is 0.54. Then the friction head equals $0.54 \times 5.60 \times 27.6 = 83.5$ ft.

Substituting in the formula:

$$Shp = \frac{1.77 \times 15 \times 71.4 \times 141.9}{550} = 489 \text{ hp}$$

The required brake horsepower for the electric motor driving the centrifugal pump is a function of slurry horsepower, efficiency of pump, and efficiency of the gear reduction box between the motor and the pump. Dredge pump efficiency varies between 50 and 70 percent, a conservative figure being 50 percent, and a conservative efficiency for the gear box is 90 percent. Hence, approximate brake horsepower is given by the equation:

$$bhp = \frac{Shp}{0.50 \times 0.90}$$
$$= \frac{489}{0.50 \times 0.90} = 1087 \text{ hp}$$

The calculations reveal that the 1250 hp of the electric motor is adequate.

Table 13-61 gives the approximate brake horsepower of the prime mover per foot of total dynamic head according to flow in gallons per minute, percentage of solids by

13-166　Hauling Excavation

TABLE 13-61　Brake Horsepower of Prime Mover for Centrifugal Pump of Dredge Required per Foot of Total Dynamic Head

Pumping rate, gal/min	Percentage of solids and specific gravity of mixture of solids and water										
	4 1.05	6 1.07	8 1.10	10 1.12	12 1.14	14 1.17	16 1.19	18 1.22	20 1.24	22 1.26	24 1.29
1,000	0.6	0.6	0.6	0.6	0.7	0.7	0.7	0.7	0.7	0.7	0.7
2,000	1.2	1.2	1.2	1.3	1.3	1.3	1.3	1.4	1.4	1.4	1.4
3,000	1.8	1.8	1.8	1.9	1.9	2.0	2.0	2.0	2.1	2.1	2.2
4,000	2.4	2.4	2.5	2.5	2.6	2.6	2.7	2.7	2.8	2.8	2.9
5,000	2.9	3.0	3.1	3.1	3.2	3.3	3.3	3.4	3.5	3.5	3.6
6,000	3.5	3.6	3.7	3.8	3.9	3.9	4.0	4.1	4.2	4.3	4.3
7,000	4.1	4.2	4.3	4.4	4.5	4.6	4.7	4.8	4.9	5.0	5.1
8,000	4.7	4.8	4.9	5.0	5.1	5.2	5.4	5.5	5.6	5.7	5.8
9,000	5.3	5.4	5.5	5.7	5.8	5.9	6.0	6.1	6.3	6.4	6.5
10,000	5.9	6.0	6.2	6.3	6.4	6.6	6.7	6.8	7.0	7.1	7.2
11,000	6.5	6.6	6.8	6.9	7.1	7.2	7.4	7.5	7.7	7.8	8.0
12,000	7.1	7.2	7.4	7.5	7.7	7.9	8.0	8.2	8.4	8.5	8.7
13,000	7.7	7.8	8.0	8.2	8.3	8.5	8.7	8.9	9.0	9.2	9.4
14,000	8.2	8.4	8.6	8.8	9.0	9.2	9.4	9.6	9.7	9.9	10.1
15,000	8.8	9.0	9.2	9.4	9.6	9.8	10.0	10.2	10.4	10.6	10.8
16,000	9.4	9.6	9.9	10.1	10.3	10.5	10.7	10.9	11.1	11.3	11.6
17,000	10.0	10.2	10.5	10.7	10.9	11.2	11.4	11.6	11.8	12.1	12.3
18,000	10.6	10.8	11.1	11.3	11.6	11.8	12.0	12.3	12.5	12.8	13.0
19,000	11.2	11.4	11.7	12.0	12.2	12.5	12.7	13.0	13.2	13.5	13.7
20,000	11.8	12.0	12.3	12.6	12.8	13.1	13.4	13.7	13.9	14.2	14.5

gpm											
21,000	12.4	12.6	13.0	13.2	13.5	13.8	14.0	14.3	14.6	14.9	15.2
22,000	13.0	13.3	13.6	13.8	14.1	14.4	14.7	15.0	15.3	15.6	15.9
23,000	13.5	13.8	14.2	14.5	14.8	15.1	15.4	15.7	16.0	16.3	16.6
24,000	14.2	14.4	14.8	15.1	15.4	15.7	16.1	16.4	16.7	17.0	17.4
25,000	14.7	15.0	15.4	15.7	16.1	16.4	16.7	17.1	17.4	17.7	18.1
26,000	15.3	15.6	16.0	16.4	16.7	17.1	17.4	17.8	18.1	18.4	18.8
27,000	15.9	16.2	16.6	17.0	17.3	17.7	18.1	18.4	18.8	19.1	19.5
28,000	16.5	16.9	17.2	17.6	18.0	18.4	18.7	19.1	19.5	19.9	20.2
29,000	17.1	17.4	17.9	18.3	18.6	19.0	19.4	19.8	20.2	20.6	21.0
30,000	17.7	18.1	18.5	18.9	19.3	19.7	20.1	20.5	20.9	21.3	21.7
31,000	18.3	18.7	19.1	19.5	19.9	20.3	20.7	21.2	21.6	22.0	22.4
32,000	18.8	19.3	19.7	20.1	20.5	21.0	21.4	21.9	22.3	22.7	23.1
33,000	19.4	19.9	20.3	20.8	21.2	21.6	22.1	22.6	23.0	23.4	23.9
34,000	20.0	20.5	20.9	21.4	21.8	22.3	22.7	23.2	23.7	24.1	24.6
35,000	20.6	21.1	21.6	22.0	22.5	23.0	23.4	23.9	24.4	24.8	25.3
36,000	21.2	21.7	22.2	22.7	23.1	23.6	24.1	24.6	25.1	25.5	26.0
37,000	21.8	22.3	22.8	23.3	23.8	24.3	24.8	25.3	25.8	26.2	26.8
38,000	22.4	22.9	23.4	23.9	24.4	24.9	25.4	26.0	26.4	26.9	27.5
39,000	23.0	23.5	24.0	24.5	25.0	25.6	26.1	26.6	27.1	27.7	28.2
40,000	23.6	24.1	24.6	25.2	25.7	26.2	26.8	27.3	27.8	28.4	28.9

The table is based on: specific gravities as given at the top of the table; pump efficiency, 50%; efficiency of gear reduction box between prime mover and pump, 90%. The equation for brake horsepower per foot of total dynamic head is:

$$\text{hp} = \frac{\text{gpm} \times 8.34 \times \text{sp gr} \times 1}{33,000 \times 0.90 \times 0.50} = 0.000562 \times \text{gpm} \times \text{sp gr}$$

where gpm is flow rate in gal/min and sp gr is specific gravity.

volume, and specific gravity of slurry. The specific gravity of the slurry, weighing 71.4 lb/ft³, is 1.14. Table 13-61 gives 7.7 hp per foot of total dynamic head for a flow of 12,000 gal/min and a specific gravity of 1.14. Thus, the brake horsepower is 141.9×7.7=1093 hp. The figure is a good check on the 1087 hp calculated by the formula.

When calculating brake horsepower one may use either the formula for slurry horsepower, corrected for efficiencies, or Table 13-61.

Production and Cost Estimation

The hourly production of the dredge shown in Figure 13-80 is 660 tons or 500 cybm (apparent yd³). The dredge for this work in sands and gravels has the general specifications of the nominally rated 18-in dredge of Table 13-57.

Hourly cost of ownership and operation, 1978:
    Dredge:
        Approximate cost of dredge, erected at job, 1978 $1,200,000
        Hourly cost of ownership and operation:
            Dredge (Table 8-3), 17%×$1200 $204.00
            Leverman, engineer, and deckhand   50.50
            Total hourly cost $254.50
    Pipe, appurtenances, and pontoons:
        Cost, 1978:
            Pipe, 2500 ft of 18-in diameter and ³⁄₁₆-in thickness, @ $21.50/ft $53,750.00
            Appurtenances:
                14 ball joints for floating pipe @ $270   3,780.00
                Valves and fittings, estimated at 5% of cost   2,690.00
                  of pipe.
            Pontoons, 24 for twelve 40-ft lengths of pipe @ $650   15,600.00
        Total cost $75,820.00

The estimated life for pipe, appurtenances, and pontoons is tabulated below:

$$\text{Pipe } T = \frac{2 \times 18 \times [(133 \times 0.375) - 18]}{133 \times 1.25} \qquad 7 \text{ yr}$$

Ball joints, 1.5×7 10 yr
Valves and fittings, 2.0×7 14 yr
Pontoons, 2.0×7 14 yr

NOTE: The estimated life T of pipe is given by the formula of the previous section covering pipeline. The values used in the formula are: D, diameter of pipe, 18 in; t, thickness of pipe, 0.375 in; P, average yearly production in millions of cybm, 1.25.

The hourly cost of ownership and operation of the pipeline is estimated in the following table:

Item	Pipe	Ball Joints	Valves and Fittings	Pontoons
Depreciation factors:				
Depreciation period, yr	7	10	14	14
Hours of use yearly	2500	2500	2500	2500
Depreciation period, h	17500	25000	35000	35000
Yearly fixed charges as a percentage of cost:				
Depreciation	11.4	8.0	5.7	5.7
Interest, taxes, insurance, and storage	7.4	7.2	7.0	7.0
Replacement cost escalation	7.0	7.0	7.0	7.0
Total fixed charges	25.8	22.2	19.7	19.7
Hourly cost as percentage of $1000s of cost:				
Fixed charges	10	9	8	8
Repairs and replacements, including labor, estimated	2	3	1	1
Total hourly cost	12	12	9	9
Cost, 1978	$53,750	$3780	$2690	$15,600
Hourly cost of ownership and operation:				

12% × $53.8	$6.46			
12% × $3.8		$0.50		
9% × $2.7			$0.20	
9% × $15.6				$1.40
Total hourly cost of ownership and operation of pipeline				$ 8.56
Hourly cost for average one pipeline handler				11.80
Total hourly cost for pipeline				$20.36

The hourly expense of ownership and operation and direct job-unit costs are summarized below.

Item	Machinery	Labor	Total
Hourly cost of ownership and operation:			
Dredge	$204.00	$50.50	$254.50
Pipeline	8.56	11.80	20.36
Total hourly cost	$212.56	$62.30	$274.86
Direct job unit cost:			
Per ton, 660 tons/h	$0.32	$0.10	$0.42
Per cybm, 500 cybm/h	$0.43	$0.12	$0.55

The unit cost is for excavating below the surface of the water to some 40 ft depth, hauling or transporting 2500 ft up a +1.6 percent grade, and stockpiling the sand gravels.

The work-cost indicator for a suction dredge, which is the ratio of total work to pay-work, varies over a wide range. It depends on the percentage of solids by weight carried in the slurry. In the foregoing example of production and costs for the 18-in dredge, the solids amount to 20 percent by weight and the work-cost indicator is simply 1.00/0.20, or 5.00. If, however, clays and silts instead of sands and gravels were being pumped, the percentage of solids could be as high as 24 percent by apparent volume, in which case the indicator, as represented by 1 ft^3 of slurry, would equal 80.3 lb/28.7 lb, or 2.80. On the other hand, if the heaviest gravels were being dredged, the percentage of solids by apparent volume could be as low as 10 percent. Then the indicator would equal 69.9 lb/12.0 lb, or 5.83.

One must keep in mind that, no matter what the work-cost indicator may be, a dredge excavates, hauls, and builds a fill under conditions for which it alone is generally fitted.

Summary

Except for draglines and backhoes for loading and casting only and cableway scraper-bucket systems for loading and hauling from beneath the surface of water, the dredge is the only machine for combined loading and hauling of subaqueous excavation.

Of the several types of dredges, the suction-cutterhead-pipeline dredge is the most versatile. It can excavate well into the range of soft and medium-weathered rocks, eliminating in many cases troublesome and costly blasting. A dredge with a 30-in centrifugal pump, when pumping 20 percent solids by apparent volume, can deliver 2100 cybm/h. The output of slurry is at the rate of 33,100 gal/min. When pumping 1 mi with a lift of 53 ft, a prime mover of 4300 bhp is required.

Such a huge dredge, working around the clock 21.5 h/day, will excavate and haul 45,000 cybm/day and about 13 million cybm/yr. This great output is handled by a dredge crew of four workers and a shore gang of two or three workers.

SUMMARY OF HAULERS

Twelve types of haulers or combined loaders and haulers have been discussed in this chapter. Some are versatile and some are specialized, so that the selection of the right hauler for a given job often calls for meticulous analyses of methods, machinery, productions, and costs. Basically, however, all haulers do work for direct job-unit costs in proportion to their work-cost indicators, provided they are suited to the work. These indicators have been approximated for the 12 machines in accordance with the general equation:

$$\text{Work-cost indicator} = \frac{(\text{GVW} + \text{tare weight})}{\text{load weight}}.$$

13–170 Hauling Excavation

The indicator represents the ratio of total work done to amount of paywork accomplished. The work-cost indicators for the various types of machines included in the 12 categories are:

Haulers:
    Crawler-tractors pulling scrapers   4.12
    Crawler-tractors pulling wagons   4.36
    Scrapers, push-loaded   3.46
    Bottom-dump haulers:
        Off-the-road type   2.52
        On-the-road type   2.00
    Rear-dump haulers:
        Off-the-road type   2.59
        On-the-road type   2.68
    Side-dump haulers:
        Off-the-road type   2.64
        On-the-road type   2.71
    Conveyor systems   1.49
    Railroads   2.31
Haulers and loaders, combined:
    Crawler-tractor-bulldozer   7.12
    Scrapers, self-loaded   3.61
    Cableway scraper-bucket systems   2.02
    Hydraulickers   0.49
    Dredges   4.35

Commencing with these cursory assessments, the estimator may proceed with the cost estimates for the one or more machines which are suitable for the work. The haulers must also be evaluated for the relationship of their costs to those for fragmentation, loading, and fill building or other end use of the rock earth. The cost for hauling must fit neatly into the mosaic of total cost for the rock-earth excavation.

CHAPTER 14

Dumping and Compacting Excavation

PLATE 14-1

Dumping, spreading-mixing, and wetting-compacting rock-earth. These four customarily used machines are getting satisfactory compaction in the embankment from an average-weathered sandstone-shale formation of the Santa Monica Mountains, California.

    1. The twin-engine scraper, one of four being used, is placing an 8-in-thick lift of material. It also contributes mightily to compaction by its weight and its kneading action when traveling over the fill.

    2. The crawler-tractor-bulldozer is spreading and mixing the rock-earth and is aiding compaction by the compression of its tracks.

    3. Water for compaction is being added to the material by the water wagon, and the heavy machine likewise assists in compaction by its loaded tires.

    4. The key compactor is the triple-drum sheepsfoot roller, drawn by the crawler-tractor-bulldozer which helps in spreading and mixing.

    The fill building, controlled by specifications for minimum relative density of fill, is part of an excavation for a huge site for homes. One appreciates that all machines, although typed for a given task, assist in building a satisfactory fill by means of their combined compactive efforts.

DUMPING AND PLACING

After rock-earth is hauled or cast, it may be dumped or placed in several ways according to its use.

1. It may be dumped for random fill, as in building an embankment without specifications for compaction. Such construction was acceptable prior to about 1925, when fills were brought to line and grade and were then compacted by natural settlement. After settlement they were brought up to finish grade and lines. Presently this simple method is used for spoil areas for overburden and for waste dumps in mining and in industry. It is also used in some fills where topography makes it impossible, at least initially, to use the layer or lift method of compaction.

2. It may be dumped onto a stockpile for present or future use as shown in Figure 13-80 or into a primary crusher as illustrated in Figure 13-71. In these and similar cases in mining and industry no machines are needed for the direct dumping.

3. It may be placed as riprap in dam, jetty, breakwater, or seawall building. Jetty construction is shown in Figure 12-35. Figure 14-1 illustrates a seawall, built for the protection of homes along the California coast. The rock was placed in a manner similar to that of Figure 12-35, except that a carrier-mounted crane of 50 tons capacity, equipped with a rock grapple, placed the riprap. A 12-ton-capacity three-axle on-the-road rock hauler delivered the rock.

4. It may be dumped, spread-mixed, wetted, and compacted to satisfy the usually rigid specifications for building an embankment. This method, used during the past 50 years of the development of machinery for embankment construction, is illustrated in Plate 14-1. Of all methods for dumping and placing, it requires the most machinery.

Figure 14-1 Seawall of riprap for beach front protection. Excavation for the seawall and placing of riprap were by a carrier-mounted 50-ton crane equipped with a rock bucket or grapple. The granite rock was brought in by 12-ton rock haulers from a quarry located 50 mi away. Placement was carried out during a few hours on either side of low tide when the sand was firm for the work of the crane and haulers.

The seawall is 400 ft long, 10 ft wide at the top, 40 ft wide at the base, and 20 ft high from the base; 6 ft of this height is above the beach level and 14 ft is below in the previously excavated trench. The beach level fluctuates seasonally by about 4 ft on either side of the present level.

Some 9700 tons of riprap was placed to protect the seaside homes of Oceanside, California. The picture illustrates some damage by the extremely high tides and battering waves of the heavy storm of early 1978.

BUILDING COMPACTED EMBANKMENTS

One perceives that dumping and placing of rock-earth is complex only when embankment construction is controlled by specifications, which define the manner of placing and compacting the materials. Generally the specifications and the basis of payment for private works are similar to those for public works. In dam building, levee work, or some special-purpose work, there is a unit bid price for embankment construction. Generally, however, embankment building is a part of excavation as far as unit bid price is concerned.

Typical of specifications for compacting embankment and for the usual basis of payment for constructing embankments are the following of the California Division of Highways:

19-6.02 **Compacting**—Embankments shall be constructed in compacted layers of uniform thickness and each layer shall be compacted in accordance with the requirements herein specified with the following 2 exceptions:

(1) Sidehill embankments, where the width including bench cuts for bonding existing and new embankments is too narrow to accommodate compacting equipment, may be constructed by end dumping if permitted by the Engineer, until the embankment, including benching, is wide enough to permit the use of compacting equipment, after which the remainder of the embankment shall be placed in layers and compacted as specified.

(2) Where embankments are to be constructed across low, swampy ground which will not support the weight of hauling equipment, the lower part of the embankment may be constructed by dumping successive loads in a uniformly distributed layer of a thickness not greater than that necessary to support the equipment while placing subsequent layers, after which the remainder of the embankment shall be constructed in layers and compacted as specified.

Unless specified herein, or in the special provisions, or directed by the Engineer, the construction of dikes, the placing and compacting of approved material within the right of way where unsuitable material has been removed, and the filling of holes, pits and other depressions within the right of way, shall conform to all of the requirements herein specified for compacting embankments. Trenches, holes, depressions and pits outside of areas where embankments are to be constructed shall be graded to provide a presentable and well-drained area.

Embankments shall be constructed so that each layer shall have a cross fall not to exceed one foot in 20 feet.

The loose thickness of each layer of embankment material before compaction shall not exceed 0.67-foot, except as provided in the following paragraph for rocky material. Each layer shall be compacted in accordance with the following requirements:

The relative compaction of each layer within 2.5 feet of finished grade shall not be less than 95 percent. The relative compaction of the layers of embankment below a plane 2.5 feet below finished grade shall not be less than 90 percent.

When embankment material contains, by volume, over 50 percent of rock larger than 0.5-foot in greatest dimension, the embankment below a plane 3 feet below finished grade, may be constructed in layers of a loose thickness, before compaction, not exceeding the maximum size of rock in the material. When embankment material contains, by volume, between 25 percent and 50 percent of rock larger than 0.5-foot in greatest dimension, the loose layers of embankment shall not exceed 3 feet in thickness. When embankment material contains, by volume, up to 25 percent of rock larger than 0.5-foot in greatest dimension the loose layers of embankment shall not exceed 0.67-foot in thickness in the area between the rocks larger than 0.67-foot. The interstices around the rock in each layer shall be filled with earth or other fine material, after which the layer shall be compacted until there is no visible evidence of consolidation of the material being compacted.

At locations where it would be impractical to use mobile power compacting equipment, embankment layers shall be compacted to the specified requirements by any method that will obtain the specified compaction.

At the time of compaction, the moisture content of embankment material shall be such that the specified relative compaction will be obtained. Embankment material which contains excessive moisture shall not be compacted until the material is dry enough to obtain the required compaction. Full compensation for any additional work involved in drying embankment material to the required moisture content shall be considered as included in the contract price paid for excavating or furnishing the material and no additional compensation will be allowed therefor.

Embankments shall be maintained to the grade and cross section shown on the plans until the acceptance of the contract.

19-6.03 **Payment**—Full compensation for constructing embankments; doing necessary plowing or benching; constructing all dikes; placing and compacting approved material where unsuitable and unstable embankment foundation material has been removed; filling and compacting holes, pits and other depressions; backfilling excavations resulting from the removal of structures and other facilities; placing selected material where required; placing topsoil excavated from within the project limits on slopes; placing selected material and topsoil in stockpiles; all as shown on the plans, and as specified in these specifications and the special provisions, and as directed by the Engineer, shall be considered as included in the contract price paid per cubic yard for excavating the material or the contract price paid for furnishing and placing the material, as the case may be, and no additional compensation will be allowed for such work, except for applying water.

Applying water will be measured and paid for as provided in Section 17.

In this text *compaction* is both a general and a specific term. Generally, it embraces all the operations for constructing an embankment governed by specifications. These are combined spreading-mixing, wetting or drying, and compacting. Specifically, it is simply the compacting or densifying operation which is done by *compactors*, machines built expressly for compacting.

Spreading and Mixing

The amount of work to spread and mix material on the fill depends largely on the kind of haulers used. In order of ease of spreading and mixing, haulers may be considered in this sequence:

1. Crawler-tractor-bulldozers, hydraulickers, and dredges both haul and spread so that no auxiliary machinery is required except for final dressing of the embankment.

2. Scrapers, either crawler-tractor-drawn or self-contained, also need little or no spreading machinery, as they can lay down a lift to the desired thickness.

3. Bottom-dump haulers require some assistance by tractor-bulldozers or motor graders, as they usually leave a windrow which needs leveling. An exception is the machine equipped with a V-shaped spreader or plough mounted in the rear, which serves to level the dumped windrow.

4. Rear-dump and side-dump haulers usually dump over the edge of the fill and a tractor-bulldozer is required to maintain grade to the edge of the embankment. Sometimes dumping takes place on the fill and then a tractor-bulldozer or a motor grader is needed for spreading.

Occasionally the material is cast to the fill and in these rare instances a tractor-bulldozer or a motor grader is used for the spreading operation.

Crawler-Tractor-Bulldozers For spreading-mixing fill materials, the most powerful machines are crawler-tractor-bulldozers of medium or heavy weight. For this work they range in engine flywheel-brake horsepower from 200 to 300 and in weight from 44,000 to 74,000 lb. Larger machines are used occasionally but regular work does not warrant their use. Representative specifications are given in Table 12-2.

When spreading-mixing, tractor-bulldozers pass through the dumped material with a squarely set blade, that is, a blade at right angles to the travel path, drifting the material to both sides of the blade. Travel speeds average about 2 mi/h and one machine can accommodate about 600 to 900 cycm/h, depending on the size of the tractor. On large jobs this work is generally shared by a motor grader or a tow-type grader assigned to both spreading-mixing and haul-road maintenance.

The machines used in spreading-mixing also contribute to compaction, although in a limited way because of the low pressure, about 13 lb/in^2, of the tracks. This dual function is illustrated in Figure 14-2 where a lightweight crawler-tractor-bulldozer is sloping a bank, or spreading, and is helping compaction at the same time.

When crawler-tractor-bulldozers are used to maintain lines and grades of a random fill, the work is not demanding, as is the spreading-mixing operation. The hourly capacity of the machines is about double that for spreading-mixing, or 1200 to 1800 yd^3 embankment measurement for the sizes of tractors so far mentioned.

Wheels-Tires-Mounted Tractor-Bulldozers Mobility contributes to the efficiency of these automotive-type machines, particularly when the haulers are dumping at widespread locations on a large embankment. For spreading-mixing work, their engine horsepowers and weights are comparable with those of their crawler-tractor counterparts. Their representative specifications are given in Table 14-1.

The hourly capacities of wheels-tires-mounted tractor-bulldozers in spreading-mixing on a fill and in simply maintaining lines and grades in a random fill are about the same as those of crawler-tractor-bulldozers of the same horsepower and weight. A machine of 170 engine flywheel hp can handle about 500 cycm/h when spreading-mixing and about 1000 cycm/h when working on a random fill. The corresponding hourly productions of a 300-hp machine are 900 and 1800 cycm, respectively.

14–6 Dumping and Compacting Excavation

Figure 14-2 Lightweight crawler-tractor-bulldozer shaping and compacting slope of fill. The machine is bulldozing upgrade, as the slope has raveled because of rains and has become slightly flatter than the specified 2:1 slope. In the resloping, the tractor is also compacting the eroded slope. This costly second dressing of the slope is necessary for final acceptance by the state of California of the completed freeway in the granite foothills of San Diego County.

One appreciates that the crawler-tractor is a contributing but not an efficient compactor. The pressure contributed by this machine weighing 30,000 lb is 9 lb/in^2, a small fraction of the great pressures exerted by the regular compactors. However, the compactive effort is sufficient for this finishing work.

TABLE 14-1 Representative Specifications for Wheels-Tires Tractor-Bulldozers

Specification	Nominal rating, hp	
	170	300
Diesel engine flywheel-brake hp	170	300
Bulldozer blade, tilt-tip type:		
Dimensions, ft:		
Height	3.4	4.7
Width	12.0	12.2
Capacity:		
Volume, cylm	3.1	5.9
Weight, lb, at 2900 lb/cylm	9000	17,100
Transmission speeds, mi/h:		
Forward:		
1st	4.1	0–3.4
2d	7.2	0–7.6
3d	12.0	0–18.5
4th	19.8	
Reverse:		
1st	4.9	0–3.4
2d	8.6	0–7.6
3d	14.2	0–18.5
4th	23.3	
Tires, four	23.50×25	29.50×29
	12 PR	22 PR
Dimensions, ft:		
Length	21.2	26.2
Width	9.1	10.1
Height	11.2	11.9
Weights, lb:		
Tare, working weight	36,000	64,000
Load of bulldozer blade	9,000	17,100
GVW	45,000	81,100
Work-cost indicator: $\dfrac{\text{GVW} + \text{tare weight}}{\text{load weight}}$	9.00	8.49

Figure 14-3 Medium-weight wheels-tires tractor-bulldozer spreading-mixing decomposed granite. In this easy rock work the task of the tractor-bulldozer is minimal. After the well-weathered granite has been ripped and push-loaded into the scraper, it is the equivalent of free-flowing sand. The scraper lays down a 9-in-thick and 12-ft-wide lift, so that the machine makes merely one or two passes or coverages for mixing and compacting over the already well-spread and well-mixed layer of material. The tractor-bulldozer of 170 engine flywheel hp and 36,000 lb weight is handling about 1200 cybm/h.

Both the tractor-bulldozer and the scrapers are providing all compactive effort on this fill, as they are the equivalent of pneumatic-tires compactors. They provide, respectively, about 28 and 72 percent of the work for compaction. There is no compactor, as such, on the embankment and yet the relative density of the fill material is above the required 95 percent.

The work is part of a freeway building project in the Laguna Mountains of southern California.

Figure 14-3 illustrates the spreading-mixing of free-flowing decomposed granite hauled by scrapers. The scrapers are laying down a lift of about 9-in thickness, so that the tractor-bulldozer is mixing rather than spreading.

Figure 14-4 shows the spreading of large blasted granite rocks which have been rear-dumped by 35-ton-capacity rock haulers and the fast movement of the tractor-bulldozer to another part of the wide embankment. The engine flywheel-brake horsepower is 300 and the weight is 64,000 lb. About 800 cybm/h is being handled from two 6-cysm bucket loaders. In this blasted rock this hourly production is equivalent to some 1000 cycm/h.

The work-cost indicators average 8.74, whereas those for crawler-tractor-bulldozers average 7.12. Although the indicators imply that the latter machines are more economical, it is felt sometimes that the greater mobility of the automotive-type machine, as shown in Figure 14-4, gives the wheels-tires tractor-bulldozer an advantage. Offsetting this advantage are the greater coefficient of traction and the greater resistance to wear and tear of sharp rocks of the crawler-tractor machine. Selection depends largely on weather conditions and nature of the rock-earth encountered in the average work of the earthmover.

Productions and Costs for Spreading-Mixing by Tractor-Bulldozers Although some estimates of hourly productions have been given for tractor-bulldozers when spreading-mixing, definite formulas for hourly productions are hazardous because of different procedures in building the embankment and because of the overlapping and sharing of the work with other machines.

However, the following generalized equations may be applied to both crawler- and wheels-tires tractor-bulldozers.

1. In loose materials such as residuals and maximum and average-weathered rock-earth:

    a. Spreading-mixing materials for compacted embankments

$$\text{cybm} = 3.0 \times \text{hp}$$

    b. Spreading materials at edge of random fills

$$\text{cybm} = 6.0 \times \text{hp}$$

14-8 Dumping and Compacting Excavation

2. In tight materials such as minimum weathered rock-earth and well-blasted rock.
   a. Spreading-mixing materials for compacted embankments

$$\text{cybm} = 1.5 \times \text{hp}$$

   b. Spreading materials at edge of random fills

$$\text{cybm} = 3.0 \times \text{hp}$$

3. In tight materials such as poorly blasted rock, there is no compacted embankment unless fine materials are mixed with the rock mechanically or hydraulically, thereby densifying it:
   a. Spreading materials at edge of random fills

$$\text{cybm} = 3.0 \times \text{hp}$$

In the above equations, cybm is the number of cubic yards bank measurement handled per 50-min working hour, and hp is the engine flywheel-brake horsepower of the tractor. The number of cubic yards compacted measurement (cycm) may be calculated by use of Appendix 1.

An example of cost estimation for spreading rock over the edge of a fill by the machine of Figure 14-4, based on 1978 cost, is tabulated below. The heavyweight wheels-tires tractor-bulldozer is spreading 800 cybm/h of blasted granite. There is no mixing or wetting-compacting in this embankment construction.

Cost of machine	$219,000
Hourly cost of ownership and operation:	
Machine, 300-hp wheels-tires tractor-bulldozer (Table 8-3), 34% × $219	$74.50
Operator	15.60
Total hourly cost	$90.10
Direct job unit cost per cybm, $90.10/800	$0.113

The cost of embankment building is part of the unit bid price for unclassified excavation per cubic yard bank measurement.

Motor Graders On an embankment the motor grader is used for two purposes: (1) it spreads and mixes the dumped loads; and (2) it dresses or fine-grades the fill to the required lines and grades. Fine grading is the last operation on the embankment and, along with the dressing of the cut or borrow pit, it is sometimes considered as work distinct from excavation. Of course the motor grader, in proportion to its weight, also acts as a pneumatic-tires compactor. Figure 14-5 illustrates the working parts of a motor grader.

Figure 14-6 illustrates a heavyweight motor grader spreading-mixing an unusually heavy load. The excavation in the cut has been undercut and it is necessary to backfill to about a 1-ft thickness of lift. Figure 14-7 shows a medium-weight motor grader routinely working in well-weathered granite which has been placed by twin-engine all-wheel-drive scrapers. Table 14-2 gives representative specifications for medium- and heavyweight motor graders of the type generally used in rock-earth excavation.

As in the case of tractor-bulldozers, it is difficult to estimate the hourly production of the motor grader when mixing-spreading because of the overlapping of work of machines on the fill and, in the case of the motor grader, the sharing of work on haul-road maintenance. Motor graders, because of their construction and operating characteristics, are confined to the spreading-mixing of free-flowing materials if they are to work efficiently. These materials are residuals and maximum- and average-weathered rock-earth.

A generalized equation for production when spreading-mixing materials for compacted embankment is:

$$\text{cybm} = 3.0 \times \text{hp}$$

where cybm is the number of cubic yards bank measurement handled per 50-min working hour and hp is the engine flywheel-brake horsepower of the grader. The number of

Building Compacted Embankments   14–9

Figure 14-4 Heavyweight wheels-tires tractor-bulldozer spreading blasted granite. The mobile tractor-bulldozer is spreading 800 cybm/h of large rocks on the acre of rough embankment. Rear-dump haulers of 35-ton capacity unload systematically across the 250-ft-wide fill. In the upper view the machine is finishing-spreading at the west side. In the lower view work is commencing at the east side. The highly mobile, heavy-duty machine, which has an engine flywheel horsepower of 300 and weighs 64,000 lb, is ideally suited for the difficult work.

14-10 Dumping and Compacting Excavation

Figure 14-5 Working parts of a motor grader. The motor grader is powered by a 160-flywheel-brake-hp diesel engine and its working weight is 31,000 lb. The moldboard, or blade, is 13.0 ft long and 2.0 ft high and it can be rotated on the circle through 360°. It may be shifted by an average of 3.0 ft to each side. The circle carrying the moldboard is raised, lowered, and tilted by two moldboard arms. The maximum angle for bank cutting is 90°. The moldboard and circle movements enable the machine to sidecast off the moldboard for the particular kind of work.

The front wheels are of the leaning type, 16.5° left or right, designed to withstand the lateral thrust of the steering wheel. Steering is mechanical with hydraulic power assist. The two rear axles are tandem-driven. The eight forward and reverse speeds provided through a power-shift transmission range from 2.7 to 27.3 mi/h. All controls are actuated hydraulically by levers on the operator's console. *(Fiat-Allis.)*

TABLE 14-2 Representative Specifications for Medium- and Heavyweight Motor Graders

Specification	Nominal rating, moldboard width, ft		
	12	14	16
Diesel engine, flywheel-brake hp	135	180	250
Moldboard dimensions, ft:			
Length	12.0	14.0	16.0
Height	2.0	2.2	2.6
Transmission, speeds, mi/h			
Forward and reverse			
1st	2.5	2.5	2.4
2d	4.0	3.6	3.3
3d	6.4	4.8	4.5
4th	10.1	6.9	6.5
5th	16.2	10.4	9.8
6th	25.5	14.7	13.8
7th	...	20.1	18.6
8th	...	28.8	26.9
Ripper-scarifier, shanks	3	3	4
Tires, six	15.50×25 12 PR	20.50×25 12 PR	23.50×25 12 PR
Dimensions, ft:			
Length	31.2	35.2	38.7
Width, with square blade	12.0	14.0	16.0
Height	10.9	11.6	12.2
Weight, working, lb	32,400	46,000	60,800

(a)

(b)

Figure 14-6 Heavyweight motor grader spreading excessively large load of weathered granite. The 35-ton load of rock-earth, dumped from the departing rear-dump truck, is being leveled by three passes from the motor grader, which has an engine of 250 flywheel-brake hp and weighs 61,000 lb. This infrequent heavy duty is to backfill an undercut grade of the existing cut. *(a)* The second pass through the 4-ft-high pile. *(b)* The completion of the spreading by the third pass. The rock hauler has spread 25 cylm to the 4-ft height, too great for efficient work by the machine. Completion of the task emphasizes the power and weight of this largest of the motor graders.

14–12 Dumping and Compacting Excavation

Figure 14-7 Medium-weight motor grader systematically spreading-mixing weathered granite. The machine is comfortably spreading-mixing 540 cybm/h, equivalent to about the same yardage compacted measurement. The material is being laid down in 8-in lifts by scrapers, one of which can be seen entering the cut to the right of the motor grader. A study reveals that the machine can handle about 20 percent more material, or 650 cybm/h. This capacity is sufficient for the scraper production in this confined, short-haul job of grading a small industrial site.

The motor grader has an engine of 135 flywheel-brake hp and weighs 32,000 lb. The moldboard is 12.0 ft wide and 2.0 ft high.

The location is just east of the contact between westward sandstones and eastward granites along the foothills of San Diego County, California.

cubic yards compacted measurement (cycm) may be calculated by use of Appendix 1.

An example of cost estimation for spreading-mixing for compacted embankment, based on 1978 costs, is tabulated below. The medium-weight motor grader shown in Figure 14-7 is spreading-mixing 540 cybm/h of average-weathered granite.

Cost of machine	$90,300
Hourly cost of ownership and operation:	
Machine, 12-ft motor grader (Table 8-3), 34% × $90.3	$30.70
Operator	15.80
Total hourly cost	$46.50
Direct job unit cost per cybm, $46.50/540	$0.086

The cost of embankment building is part of the unit bid price for unclassified excavation.

Tow-Type Graders The tow-type grader, sometimes called the dragscraper, is a bottomless scraper pulled usually by a wheels-tires tractor and sometimes by a crawler-tractor. Its work is dual, as it is used both for spreading-mixing and for haul-road maintenance.

Figure 14-8 illustrates an 18-ft-wide machine being pulled by a wheels-tires tractor-bulldozer. The two machines are complementary, as each both spreads and mixes the material for the compacted fill. The combination also contributes greatly to compaction because the well-weathered sandstone formation breaks down into sand in the fill, an ideal material for compaction by pneumatic-tires compactors.

On this fill self-loaded scrapers lay down the specified 8-in thickness of lift for the compacted 6-in thickness. In this fast-moving job production averages 1970 cybm/h. A 95 percent relative density of fill is secured by the following machines, carefully routed over the fill for distribution of compactive effort:

Twelve 23-cysm scrapers, two being on the fill at any given time
One 300-hp crawler-tractor pulling a heavyweight double-drum sheepsfoot compactor
One 300-hp self-propelled tamping-foot compactor
One 300-hp wheels-tires tractor-bulldozer pulling a tow-type grader
One 14-ft motor grader
Two 330-hp, 8000-gal water wagons

Building Compacted Embankments   14–13

Figure 14-8 Tow-type grader pulled by wheels-tires tractor-bulldozer. The effective combination of a 300-hp wheels-tires tractor-bulldozer and 18-ft-wide tow-type bottomless grader is spreading and mixing well-weathered sandstone which has been laid down by scrapers. At the same time the machines are contributing to compaction, as their combined weight is about 78,000 lb. The pulverized soft sandstone is ideal for compaction by the pneumatic tires of the two machines. Occasionally the machines make a complete pass over the total 4200-ft circular haul road for maintenance, so that 30-mi/h speeds can be safely supported. Like motor graders, tow-type graders are often cast in this dual role. A 14-ft motor grader and the tow-type grader spread-mix 1970 cybm/h for the compacted fill.

As all these machines compact to varying degrees, one appreciates the impossibility of determining precisely the absolute or relative compactive effort of each.

Figure 14-9 illustrates the working parts of a tow-type bottomless grader, and Table 14-3 gives representative specifications for tow-type graders.

The tow-type grader is generally pulled by a wheels-tires tractor, as shown in Figure 14-8. Sometimes, depending on the nature of the material, the tractor operator lowers the bulldozer so that both tractor and grader are spreading-mixing.

Like the motor grader, the tow-type grader, by construction and operating characteristics, is best suited for work in free-flowing materials. Such materials are residuals and maximum- and average-weathered rock-earth. Although the grader is used for spreading-

Figure 14-9 Working parts of tow-type grader. The 16-ft-wide grader is being towed by a two-axle wheels-tires tractor. It is bottomless and the struck capacity is 7.9 cysm. The operation is hydraulic, being controlled by the operator of the tractor. The bowl is raised and lowered by a hydraulic ram connecting the top of the back of the bowl with the hinged axle. The adjustable cutting depth is up to 6 in below the grade of the wheels-tires. The grader may be equipped with a hydraulically controlled scarifier, or ripper, extended across the full width of the grader behind the wheels-tires assembly. *(Southwest Welding and Manufacturing Co.)*

14-14   Dumping and Compacting Excavation

TABLE 14-3   Representative Specifications for Tow-Type Bottomless Graders

Specification	Nominal rating, width, ft		
	16	18	20
Cutting edge, width, ft	16.0	18.0	20.0
Capacity, cysm	7.8	8.8	9.8
Bowl dimensions, ft:			
Width	16.0	18.0	20.0
Length	4.2	4.2	4.2
Height	3.7	3.7	3.7
Cutting depth range, ft	0–0.5	0–0.5	0–0.5
Tires, two	18.00×25 20 PR	18.00×25 20 PR	18.00×25 20 PR
Dimensions, ft:			
Length	17.5	17.5	17.5
Width	16.0	18.0	20.0
Height	5.4	5.4	5.4
Weight, lb	13,200	13,700	14,200

NOTE: The grader may be equipped with a rear-mounted scarifier controlled by two hydraulic cylinders. The flywheel-brake horsepower of the engine of the towing tractors varies from 250 to 550. Generally the tractors are of the wheels-tires type and they are usually equipped with bulldozers.

mixing, one considers it primarily for the maintenance of haul roads of the cut, the main line, and the fill. When smoothing the haul road in the fill, the grader also mixes and spreads.

As in the case of the motor grader, the hourly production for spreading-mixing materials for compacted embankment is expressed by the generalized equation:

$$\text{cybm} = 3.0 \times \text{hp}$$

where cybm is the number of cubic yards bank measurement handled per 50-min working hour and hp is the engine flywheel-brake horsepower of the towing tractor. The number of cubic yards compacted measurement (cycm) may be calculated by use of Appendix 1.

An example of production and cost estimation for spreading-mixing for compacted embankment, based on 1978 costs, is provided by the work of spreading-mixing well-weathered sandstone as shown in Figure 14-8, which is shared by a 14-ft motor grader of 180 engine flywheel-brake hp. The corresponding horsepower of the tractor-bulldozer pulling the tow-type grader is 300. Because it is impossible to separate the actual work of these two machines, one must allocate the known total hourly output, 1970 cybm, in proportion to the horsepowers of the machines to arrive at an approximation of productions. The tabulation of outputs and costs follows:

Item	180-hp, 14-ft motor grader	300-hp tractor and tow-type grader
Hourly production, cybm (distribution of total of 1970 cybm)	740	1230
Costs, 1978:		
Machines:		
Motor grader	$128,000	
Tractor		$219,000
Tow-type grader		22,000
Hourly cost of ownership and operation:		
Machines (Table 8-3):		
14-ft motor grader, 33% × $128	$42.20	
300-hp wheels-tires tractor-bulldozer, 34% × $219		$74.50
18-ft tow-type grader, 27% × $22		5.90

Operator	15.80	15.60
Total hourly cost	$58.00	$96.00
Direct job unit cost per cybm	$0.078	$0.078

NOTE 1: The cost of embankment building is part of the unit bid price for unclassified excavation.
NOTE 2: If the assumption that productions for the two machines are about in proportion to horsepowers is valid, then the direct job-unit costs are equal.

Final Dressing Spreading an embankment sometimes involves final dressing for acceptance of the job by the contracting officer, as illustrated in Figure 14-2. This sloping work may be done by crawler-tractor-bulldozer, small dragline-type excavators, or backhoes, as illustrated in Figure 14-10. This is the same type of backhoe illustrated in Figure 12-26.

Principles, Testing, and Specifications for Compaction of Embankments

Compaction is the means of increasing the density or unit weight of the mixed-spread rock-earth in order to:
 1. Give it greater strength to support heavier loads
 2. Reduce future settlement by reduction of voids of air and water
 3. Reduce permeability by limiting the flow of water through the material

The increase in density is accomplished by mixing, adding or subtracting water, and compacting.

 1. Mixing takes place in the normal previous work of fragmenting, loading, dumping, and spreading the material on the embankment. The mixing serves to give a uniform gradation to the mixture of rock-earth, resulting in a minimum percentage of voids.

 2. Water is added by water wagons or by other means if the percentage of water for best compaction exceeds the natural moisture of the excavated material. On the other hand, if natural moisture exceeds the desired amount, water is removed by natural aeration, or by manipulating the rock-earth so as to expose it to the open air. Tractor-drawn disk harrows, disk plows, and moldboard plows are used. Figure 14-11 shows the use of a heavy moldboard plow drawn by a crawler-tractor for manipulating material in a borrow pit. In rock-earth excavation water is generally added. The water acts as a lubricant, enabling the particles to slide into the voids and thereby increase the apparent density of the mixture.

 3. Actual compaction takes place on the embankment or fill by the use of several kinds of compactors. The choice of compactor ideally depends on the nature of the material. A generalized range of applications for the various types of compactors is given in Table 14-6. Collectively, their compactive efforts come from pressure, kneading, impact, and vibration. Each type of compactor has two or more of these actions. As will be discussed, every machine passing over the embankment acts as a minor or major compactor and contributes to increasing the density of the fill.

Laboratory Tests of Material before Construction of Embankment Before a job begins and after the cuts or borrow pits have been selected, certain laboratory tests are performed on representative samples of the material to determine optimum moisture-density relationships. Figure 14-12 illustrates the relationships, by weights and volumes, among air, water, and soil in 1 cylm of natural rock-earth as taken from the cut or borrow pit for sampling. The results of the laboratory tests guide the engineer in establishing desired densities for compaction. Testing for moisture-density relationships is usually done by one of two common methods adopted by the American Association of State Highway and Transportation Officials (AASHTO) and known as the Standard AASHTO and the Modified AASHTO. These tests are also known as Proctor tests.

There is an optimum moisture content for maximum density of a given material and given compactive effort. It is determined by compacting a series of five or more different samples, all from the same large sample. The same process is used for each of the five samples but the moisture content is varied. If it is deemed desirable to make more complete tests, additional large samples of the material may be taken.

The series is begun with the soil in a damp condition somewhat below the probable optimum moisture content. After the first sample is compacted, its wet unit weight is determined and a small portion is removed to determine moisture content. This portion

14–16   Dumping and Compacting Excavation

Figure 14-10 Backhoe of 1-yd³ capacity spreading and sloping bank of highway fill. After this fill was completed, there was some raveling, as shown by the loose rock at the toe of the slope. This backhoe with telescoping boom and rotating dipper is an ideal machine for this combined resloping and dressing. The dipper reaches to the bottom of the slope to pull back loose rock and the grade is held easily to 0.1-ft elevation. The dipper is 5 ft wide and the operating cycle is about 0.5 min. In 1 h about 360 ft of length of fill are being sloped, for an area of about 900 yd².

is placed in a drying oven to be dried thoroughly. The weight of the moisture lost is determined and it is expressed as a percentage of the weight of the dry sample. A second sample with an increased moisture content is then compacted and the process is repeated. Additional samples are processed in the same manner until the wet unit weight decreases or until the soil becomes too wet by as judged visually.

After the tests on the several samples are completed, a curve is plotted of dry density as a function of percent moisture. Maximum density, 100 percent, is found by drawing a horizontal line from the peak of the curve to the left ordinate of the graph. Optimum moisture for maximum density is found by drawing a vertical line through the peak of the curve to the abscissa.

Figure 14-13 illustrates the compaction procedures of the two tests and shows a typical moisture-density curve for the more common Standard AASHTO test. The two tests show a considerable difference in total compaction effort, 12,400 ft·lb for the standard test, generally used in embankment construction, and 56,200 ft·lb for the more stringent modified test, used for embankments supporting the heaviest of loads, such as multistory buildings.

Field Tests during Construction of Embankment   Periodic field testing is done to ensure that desired compaction densities are being maintained throughout a particular embankment construction job. These tests also indicate the effectiveness of the compaction machinery and construction methods being used.

A common way of testing is the sand-cone method. The test consists of five basic steps. The field setup for part of these steps is illustrated in Figure 14-14.

  1. A test site is selected on the compacted embankment at least 30 ft away from the operating machinery, which might cause vibrations and thus disturb the calibrated sand.

  2. A steel base plate with a hole of 4-in diameter is placed on the site, and a hole is dug through this opening to the depth of the completed layer or lift being tested. The excavated material is preserved.

  3. The material from the excavated hole is weighed.

  4. After weighing a volume of calibrated sand greater than the volume of the hole,

the sand is poured into the inverted cone atop the hole in the plate until the hole in the embankment, the hole in plate, and the cone are full. The weight of the sand equals the original weight less the weight left in the original container. The total volume of sand equals this weight divided by the specific gravity or unit weight of the sand. The volume of sand in the hole or volume of material removed from the hole equals the total volume of sand less the known volumes of the hole in the plate and the cone. The density of the embankment equals the weight of the material removed divided by the volume of the hole.

5. The moisture content is determined by laboratory analysis of the sample and is first expressed as a percentage of the wet weight. The dry unit weight is calculated by dividing the wet unit weight by 1.000 plus the percent moisture content, expressed as a decimal.

Figure 14-11 Aerating a wet borrow pit for reduction of natural moisture in material. The heavyweight crawler-tractor is towing a heavy-duty moldboard plow through a wet borrow pit in order to aerate the material by manipulation. The weathered rock is for the semipervious zone of an earthfill dam. With a natural moisture content considerably higher than optimum for the required relative density of the fill, the embankment cannot be compacted satisfactorily unless the material is partially dried.

Such processing of fill material is costly but it is sometimes necessary. Cherry Valley Dam is on the west slope of the Sierra Nevada of California. The elevation is about 3800 ft and the dam site is wet from the spring runoff of rain and snow from the eastward range of mountains. *(Southwest Welding and Manufacturing Co.)*

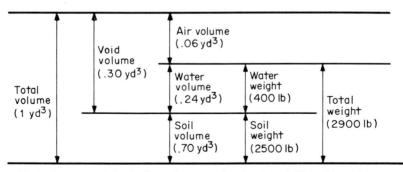

Figure 14-12 Air–water–solid soil relationships in 1 cylm of rock-earth. The cubic yard represents a typical sample taken from a cut or borrow pit to be used for moisture-density tests prior to construction of an embankment. The natural moisture of the sample is high, 16 percent of the weight of the dry soil. Nevertheless, if the moisture-density curve of Figure 14-13 should result from the compaction tests, 98 percent of maximum density is attainable. It is probable that more than 95 percent of maximum density, an ordinary requirement, is attainable by compacting even with the high natural moisture content. *(Hyster Co.)*

14–18 Dumping and Compacting Excavation

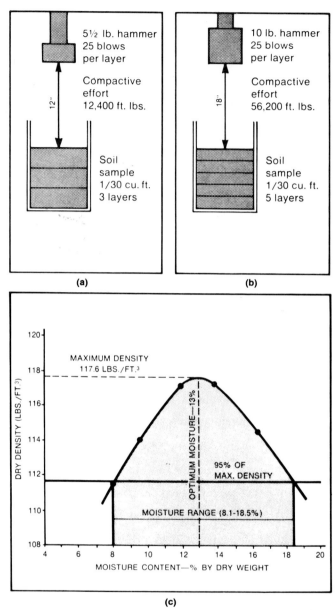

Figure 14-13 Laboratory tests for determining optimum moisture for maximum density of rock-earth in embankment. The Standard AASHTO test *(a)* is used for average embankments. The more severe Modified AASHTO test *(b)*, requiring greater compactive effort from the compactors, is used for embankments supporting extremely heavy loads, where no subsidence is to be tolerated. The moisture-density curve of the Standard AASHTO test *(c)* illustrates the range of permissible moisture content, from 8.1 to 18.5 percent, to obtain 95 percent relative density. Such a range embraces dampness to wetness. *(Hyster Co.)*

Figure 14-14 Field-laboratory setup for making a moisture-density test of compacted embankment. The test for moisture versus density of the embankment is being made in a sidehill highway fill in Idaho. The left picture shows the field laboratory and the field test. Within the wagon are scales and oven. The right picture illustrates the steel base plate with a hole through which material is being excavated, the cone and bottle of preweighed sand, and miscellaneous tools. Large jobs sometimes warrant a laboratory at the engineer's office, where work other than taking the sample may be performed with more convenient facilities.

As a result of the field test the engineer can determine the relative dry density for the particular moisture content. If the relative density is not satisfactory, the necessary changes in compaction methods and machinery can be made without delay so as to ensure a workmanlike job. *(Southwest Welding and Manufacturing Co.)*

Specifications and Machinery Selection for Wetting-Compacting The excavator, before bidding on or selecting wetting-compacting machinery for a job, must determine from the plans and specifications the limits concerning the placement, preparation, and compaction of the material for the embankment. Typical specifications, those of the California Division of Highways, have been given earlier in this chapter. There are four general types of specifications used by the contracting officers to establish minimum standards of compaction for embankments.

　　1. *Method-only specification.* The contracting officer describes in detail how compaction is to be effected but states nothing about the attained results. Typical of this method is a specification calling for P passes or coverages at a given speed in feet per minute by a specific kind of compactor on a specified thickness of lift with a specified range of moisture content. This is a rigid requirement and neither the excavator nor the contracting officer can use other methods or machinery which might very well give a better embankment at a lower cost.

　　2. *Method-and-end-result specification.* Typical of this specification is the requirement that the excavator must obtain 95 percent relative density by the AASHO test with a minimum number of passes P by a specific kind of compactor on a specified thickness of lift with a specific range of moisture content. This is a most rigid and restrictive requirement. It may be most costly since no change in specifications can be made until it is proven in the field that the original requirements are in error.

　　3. *Suggested-method-and-end-result specification.* This arrangement allows the excavator to take advantage of his experience and that of others, and at the same time it offers guidelines to be followed at the beginning of the job. Additionally, it permits changes in methods and machinery during the course of the job. The end result is in terms of the standard tests for compaction in the embankment.

　　4. *End-result specification.* Here only the desired end result is required in terms of the standard tests. The excavator has full choice in selecting the compactors for the job. The contracting officer must employ strict testing procedures and thereby ensure that compaction requirements are fulfilled. The specifications of the California Division of Highways are of the nature of end-result requirements.

14-20 Dumping and Compacting Excavation

In keeping with the continuous development of compactors, there is a growing tendency of the contracting officers to permit the use of new compacting machines, even though the specifications call for standard machines, provided desired compaction is achieved. Presently nearly 80 percent of the contracting officers in the United States call for a given type of compactor to be used in getting the desired density. The specifications for the compactors are set forth in many cases. The remaining 20 percent of the agencies specify end result, allowing the excavator considerable leeway in choice of methods and machinery.

In this text the machines for wetting-compacting are the several kinds of water wagons, hydraulic monitors, and compactors. When appraising the work of compactors, one must keep in mind that they contribute only part of the total compactive effort or effect of all the machines working on the embankment. Thus, when one states that a sheepsfoot compactor gets the specified 95 percent relative compaction in six coverages or passes, one is really saying that, while the compactor is making the six passes, it and several other machines are getting the desired results. This truism is stated in the legend of Plate 14-1.

Referring to this plate, the following table is an estimate of the relative contributions of the four machines on the embankment to the total compactive effort. It is based on the premise, not altogether true but substantially correct, that effort is proportional to the weight of machine. The purpose of the table is to put the compactor in true perspective.

Machine	Effective weight, lb	Percentage of total weight
Compactor:		
Crawler-tractor-bulldozer	74,000	
Triple-drum roller	37,000	
Total	111,000	32
Scraper (on the job there is always one scraper on the embankment)	121,000	35
Crawler-tractor-bulldozer	74,000	22
Water wagon (on the job the water wagon is on the embankment two-thirds of the time; prorated weight)	36,000	11
Total values	342,000	100

In this case the compactor contributes about one-third of the total compactive effort. There have been instances in which the compactor has been eliminated by directing the haulers methodically over the entire area of the embankment and satisfactory relative density has been attained.

Wetting

The first step in estimating wetting requirements is determination of the amount of water to be added per cubic yard bank measurement of average-weathered rock when the initial laboratory tests give the following data:

1. Weight with contained natural moisture, 3750 lb
2. Weight of contained natural moisture, 150 lb
3. Dry weight, 3600 lb
4. Moisture as percentage of dry weight, 4.2%
5. Moisture-density curve of Standard AASHTO test, as shown in Figure 14-13.

The moisture-density curve indicates that 95 percent relative density may be attained with minimum 8.0 percent moisture by dry weight. Economy dictates that the excavator should attempt to get satisfactory compaction by commencing with minimum total water, perhaps 9 percent of the dry weight. The following figures result:

1. Total weight of water per cybm (0.090×3600), 324 lb
2. Weight of contained natural moisture (0.042×3600), 150 lb
3. Weight of water to be added per cybm, 174 lb
4. Volume of water to be added per cybm, 21 gal

Translated into water requirements for a scraper spread excavating 1500 cybm/h, the water consumption is 31,500 gal/h. This is equivalent to four loads from an 8000-gal water wagon. As water wagons generally keep haul roads wetted down for maintenance purposes in addition to their work on the embankment, it is probable that two 8000-gal water wagons would be needed for the job.

When rainfall is abundant and humidity is high, the natural moisture in the material for the embankment must sometimes be lessened because it may exceed the upper limit of the permissible moisture range. In Figure 14-13 this upper limit for 95 percent relative density is 18.5 percent. Generally, the more rocky the material, the lower the natural moisture. Likewise, the more earthy the material, the higher the natural moisture.

There are two ways for removing moisture before the excavation is brought to the embankment. First, it may be aerated with manipulation by disks or plows drawn by tractors so as to dehydrate the rock-earth. This common method is illustrated in Figure 14-11. Second, drier material may be imported and mixed with wet rock-earth on the embankment. There are three objections to the mixing method: under strict material specifications, as in earthen dam building, the added material must blend or be similar; mixing is costly; and mixing is delaying to job progress. Of course, aeration is also costly and delaying.

Water Wagons These machines are off-the-road and off-the-road–on-the-road types and they range in capacities from 1500 to 14,000 gal. Representative specifications for the two types are given in Tables 14-4 and 14-5, respectively. Both types are equipped with pumps driven by separate engines, by engine power takeoffs, and by hydraulic motors and with front-, side-, and rear-mounted spray heads.

Figure 14-15 illustrates an 8000-gal off-the-road water wagon wetting an embankment for home sites. It receives water from a 10,000-gal portable tank located 1000 ft

TABLE 14-4   Representative Specifications for Off-the-Road Water Wagons

Specification	Nominal capacity, gal		
	6000	10,000	14,000
Diesel engine, flywheel-brake hp	330	415	550
Transmission:			
Gears forward	8	8	8
Speed range, mi/h	0–32	0–32	0–32
Tires, four	26.50×29	29.50×35	33.50×39
	24 PR	34 PR	38 PR
Pump assembly:			
Diesel engine, flywheel-brake hp	150	150	150
Pump, centrifugal:			
Suction diameter, in	6	6	6
Capacity, gal/min	1800	1800	1800
Piping, diameter, in	4	4	4
Spray heads, air-actuated (two front, two side, two rear), diameter, in	3	3	3
Dimensions, ft:			
Length	42.2	47.9	51.7
Width	10.7	11.8	12.2
Height, loading	11.2	13.0	13.7
Weights, lb:			
Tare, working	65,000	94,000	111,000
Load	50,000	83,400	116,800
GVW	115,000	177,400	227,800
Speed indicators:			
Laden: $\dfrac{\text{Brake hp}}{\text{GVW, 1000s of lb}}$	2.87	2.34	2.41
Unladen: $\dfrac{\text{Brake hp}}{\text{Tare weight, 1000s of lb}}$	5.08	4.41	4.95
Work-cost indicator:			
$\dfrac{\text{GVW}+\text{tare weight}}{\text{Load weight}}$	3.60	3.25	2.90

TABLE 14-5 Representative Specifications for Off-the-Road–On-the-Road Water Wagons

Specification	Nominal capacity, gal	
	2000	4000
Diesel engine, flywheel-brake hp	190	325
Transmission:		
Gears forward	10	12
Speed range, mi/h	0–60	0–60
Axles	2	3
Tires:		
Front, two	11.00×22	11.00×22
Rear, four for 2000-gal and eight for 4000-	14 PR	14 PR
gal machines	all tires	all tires
Pump assembly, engine-power-takeoff drive, either mechanically or hydraulically driven:		
Pump, centrifugal:		
Suction diameter, in	6	6
Capacity, gal/min	1200	1200
Piping, diameter, in	3	3
Spray heads, air-actuated, two front, two side, two rear, diameter, in	3	3
Dimensions, ft:		
Length	20.3	26.0
Width	8.0	8.0
Height, loading	8.5	9.1
Weights, lb:		
Tare, working	13,000	20,000
Load	16,700	33,300
GVW	29,700	53,300
Speed indicators:		
Laden: $\dfrac{\text{Brake hp}}{\text{GVW, 1000s of lb}}$	6.40	6.09
Unladen: $\dfrac{\text{Brake hp}}{\text{Tare weight, 1000s of lb}}$	14.62	16.25
Work-cost indicator: $\dfrac{\text{GVW} + \text{tare weight}}{\text{Load weight}}$	2.56	2.20

NOTE: When these water wagons are used in on-the-road haulage, the loads are adjusted to conform to the limitations for axle loadings.

away. Figure 14-16 shows two similar 10,000-gal machines wetting down a fill for a freeway. Water is taken from the pipe stand at the end of a reach of the water supply system. Figure 14-17 illustrates a 3500-gal three-axle off-the-road–on-the-road water wagon using all four front and rear spray heads to wet down an embankment for several buildings. A machine of the same type is shown being loaded by a portable tank in Figure 10-13.

Production of Water Wagons. As for all haulers, the production of water wagons may be calculated easily. It is a function of the hauling cycle time and the capacity of the wagon in gallons and is expressed in terms of gallons per hour. The net hauling cycle time in minutes is the sum of times for loading, turning at the loading and discharging points, discharging on the embankment or haul road, and returning. The efficiency on a well-managed job is the 50-min working hour, or 83 percent. Accordingly, the production per working hour is expressed by the equation:

$$\text{Production, gal/h} = \frac{50}{\text{cycle time, min}} \times \text{capacity, gal}$$

Loading may be by water tank or by pipe stand connected to a water line.

Water tanks empty by gravity through 10- to 20-in-diameter downspouts. For a given size of downspout the rate of discharge depends on the average head of water in the tank during discharge and the friction losses at the entry to the downspout and through the controlling butterfly valve. The loading rate from tank to water wagon

Figure 14-15 Water wagon of 8000 gal capacity wetting a confined embankment. This single machine takes care of a production rate of 1580 cybm/h of damp well-weathered sandstone-siltstone. A portable tank for water supply is located 1000 ft away. The logistics are:

Hauling cycle, min:	
Loading 8000 gal of water	2.5
Hauling and returning at average 8 mi/h	3.0
Wetting at average 870 gal/min	9.2
Total cycle time	14.7
Loads hauled per 50-min working hour, 50/14.7	3.4
Water consumption for both haul roads and embankment, gal, 3.4 × 8000	27,200
Estimated gallons per cybm (95% × 27,200)/1580	16
Percentage of water added to natural weight of rock, 16 × 8.3/4070	3.3

The work is grading for home sites at Loma Santa Fe in southern California.

averages about 4300 gal/min per square foot of area of the downspout. Thus a 10,000-gal water wagon requires about 1.3 min for loading from a tank with an 18-in-diameter downspout.

The loading rate by a water line depends on the line pressure, the pipe diameter, and the attendant friction losses in the valves and fittings of the pipe stand. These may be determined and calculations for estimated flow in gallons per minute may be made. Usually, this flow rate is determined simply by a water meter and timing device at the discharge point. When an intermediate water tank is used, its filling rate must be determined so as to ensure that this rate will not be less than the demands of the water wagons.

The time for several turnings at the loading and discharging points may be estimated conservatively at 3.0 min, being as low as 1.5 min when working on a long fill for a freeway or a dam. As the water wagon is a hauler, the hauling and returning times may be calculated by using Table 13-12.

The time for wetting the embankment and the haul roads varies a great deal. On a small fill the water wagon may not discharge water continuously, but on a large fill the wetting may be without any delays. In the latter case the unloading time may simply be the capacity of the wagon divided by the rated capacity of the pump in gallons per minute. An average pump capacity of the large off-the-road water wagons is 1800 gal/min. The average pump capacity of the smaller off-the-road–on-the-road machines is 1200 gal/min. In many cases the production of the water wagon is geared to furnishing the water required for satisfactory compaction and maintenance of haul roads rather than to the capability of the water hauler.

The following example for estimating the production and costs for wetting by the 3500-gal water wagon of Figure 14-17 illustrates the balance between water requirements and the actual ability of the machine insofar as production is concerned. The production data and calculations are as follows:

14-24 Dumping and Compacting Excavation

Figure 14-16 Two water wagons of 10,000-gal capacity wetting freeway embankment. Weathered granite is being placed at a rate of 2410 cybm/h. The rock-earth contains about 6 percent natural moisture. The two water wagons perform according to the data in this table:

Hauling cycle, min:	
Loading 10,000 gal of water	6.7
Hauling and returning, average 3000 ft at 5 mi/h	3.8
Wetting at average 1050 gal/min	9.5
Total cycle time	20.0
Loads hauled per 50-min working hour, 50/20.0	2.5
Water consumption for both haul roads and embankment, gal, $2 \times 2.5 \times 10{,}000$	50,000
Estimated gallons per cybm $(95\% \times 50{,}000)/2410$	19
Percentage of water added to natural weight of rock-earth, $19 \times 8.3/4540$	3.5

The work location is in the Laguna Mountains, San Diego County, California.

Figure 14-17 Off-the-road–on-the-road water wagon of 3500-gal capacity wetting fill. Water is taken from a portable water tank located 2000 ft from the fill. The water wagon delivers 9000 gal/h to the embankment for a production of 540 cybm/h by the two scrapers. The low hourly production is the result of the confining nature of both cut and fill. By adding water in the amount of 17 gal/cybm to the damp weathered granite, a relative density of 96 percent is attained regularly. The high moisture content of the natural material is about double that of the added water from the water wagon.

The specifications of the water wagon are similar to those of the 4000-gal machine of Table 14-5 except for the important feature that all three axles are driven, which results in a facility noticeable in wet conditions.

The job location is the granite pediment of coastal southern California.

Criteria:
Haul road, level, average condition:
  Length, one way, ft — 2000
  Estimated rolling resistance, % of weight of water wagon — 6
  Diesel engine, flywheel-brake hp — 275
Weights, lb
  Laden — 51,000
  Unladen — 20,000
Hauling cycle, min:
  Loading from tank with 12-in-diameter spout, $\dfrac{3500}{(0.78 \times 4300)}$ — 1.0
  Turnings at water tank and on embankment — 3.0
Hauling (Table 13-12):
  Weight/power ratio, 185
  Maximum speed, $\dfrac{0.76 \times 275 \times 33{,}000}{0.06 \times 51{,}000} = 2250$ ft/min
  Average speed, $0.75 \times 2250 = 1690$ ft/min
  Time, $2000/1690 =$ — 1.2
Returning (Table 13-12):
  Weight/power ratio, 73
  Maximum speed, $\dfrac{0.76 \times 275 \times 33{,}000}{0.06 \times 20{,}000} = 5750$ ft/min
  Average speed, $0.75 \times 5750 = 4310$ ft/min
  Time, $2000/4310 =$ — 0.5
Discharging at average 50% of 1200-gal/min pump capacity, $\dfrac{3500}{0.50 \times 1200}$ — 5.8
Total net cycle time — 11.5
Hourly production:
  Loads hauled per 50-min working hour, 50/11.4 — 4.3
  Gallons hauled, 4.3×3500 — 15,000

The estimated hourly production of 15,000 gal is 6000 gal or 67 percent in excess of the actual delivery of water to the embankment, but the estimate is good because it represents provision for the water requirement which would have existed if abnormal rains had not increased the natural moisture in the rock-earth. With the pump working at 75 percent of rated capacity, 900 gal/min, the water wagon can operate on a net hauling cycle of about 10 min and it would provide 17,500 gal per average working hour. Under existing conditions the machine could take care of 1000 cybm/h, almost twice the actual production of the job. Such a comparison points up the flexibility of water wagons.

Cost calculations for the same example, based on 1978 costs, are as follows:

Cost of machine	$51,000
Hourly cost of ownership and operation:	
3500-gal water wagon (Table 8-3)	$24.50
48% × $51	
Operator	11.20
Total cost	$35.70
Water cost:	
9000 gal/h @ $0.50 per 1000 gal	$4.50
Total cost of machinery, operator, and water	$40.20
Direct job unit cost per cybm, $40.20/540	$0.074

Monitors or Nozzles The use of these devices for wetting fills, their specifications, and their dynamics are discussed in Chapter 10. They are illustrated in Plate 10-1 and in Figure 11-11, where water is being applied to the ripped rock in a cut before haulage to the fill. Figure 14-18 illustrates the use of a monitor for the initial wetting of a confined fill. As the fill for the freeway becomes higher and wider, the monitor will not have enough range and then wetting will be done by water wagons.

An example of production and costs for hydraulic monitors is given in the productions and costs section of Chapter 10. The water sources used and machines for bringing water to water wagons and monitors include wells, reservoirs, pumps, pipes, valves and fittings, and water tanks. All these are used in the development of water supply, and their characteristics, dynamics, and productions and costs are discussed in Chapter 10. It is

14-26 Dumping and Compacting Excavation

Figure 14-18 Wetting an embankment by monitor during initial stage of construction. The beginning of fill building for this freeway in rugged mountains is in a gulch. The confined fill is physically and economically suited to use of the monitor. As the fill widens with increase in height, the monitor's range will be inadequate and water wagons will be used. This change is illustrated in Figure 13-8 where on the same job the water wagons are used on a wide fill which also commenced in a gulch.

In this small fill the restricted production by three crawler-tractor-drawn scrapers is only some 450 cybm/h, requiring about 270 gal/min from the monitor. The 1½-in-diameter nozzle is adequate. Water is supplied by a 6-in-diameter pipeline from a temporary reservoir located about ½ mi away and about 300 ft higher than the fill.

suggested that this section on water wagons and monitors be considered complementary to Chapter 10.

Compacting

Every machine which travels on an embankment actually is a compactor, from a 3000-lb pickup truck to a 330,000-lb-GVW bottom-dump hauler. However, in this text a *compactor* is defined as a machine especially designed to increase the density of an embankment.

The compactive effort of the machine may be applied in one or more of the following ways: pressure, kneading, impact, and vibration. Different compactors apply different forces. Compactors used commonly in rock-earth embankments achieve results by the characteristic compressing devices from which they get their names. Those generally used are the *sheepsfoot, tamping-foot, vibratory,* and *pneumatic-tires* types.

Table 14-6 summarizes the generalized applications of compactors for rock-earth according to the nature of the material, and Table 14-7 summarizes the generalized operating characteristics of compactors. Table 14-6 shows that different compactors are able to work in the same materials, but they may not work with the same efficiency or economy. Many factors and conditions influence the selection of the right compactor for a given job, and not the least of these is the experience of the excavator.

Dynamics of Compaction Machinery The general equation for the hourly production of a compactor working at 100 percent efficiency is:

$$\text{cycm} = \frac{[W \times (S \times 5280) \times (T/12)]/27}{P}$$

$$= \frac{W \times S \times T \times 16.3}{P}$$

where cycm = cubic yards compacted measurement per hour
W = compacted width per coverage, ft
S = travel speed, mi/h
T = thickness of compacted lift, in
P = number of passes to achieve desired compaction

If the hourly production in cubic yards bank measurement is desired, it is obtained by multiplying the cycm value by the ratio cybm/cycm. If the material shrinks from cut to

Building Compacted Embankments

TABLE 14-6 Generalized Range of Applications for Compactors According to Materials

Kind of compactor	Residuals			Weathered rock-earth, ripped			Semisolid and solid rock, blasted	
	Clays	Silts	Sands	Maximum weathered	Average weathered	Minimum weathered	Well blasted	Poorly blasted
Sheepsfoot	x	Pressure-kneading			x			
Tamping foot	x	Pressure-kneading-impact-vibration					x	
Vibratory:								
Footed drum			x	Pressure-kneading-vibration		x		
Smooth drum					x	Vibration		x
Pneumatic tires				x	Pressure-kneading		x	

Compactor	Sands	Alluvia Gravels	Cobbles
Vibratory, smooth drum	x	Vibration	x
Pneumatic tires	x	Pressure-kneading	x

NOTE: The ranges are approximate, based on manufacturers' data and writer's observations. In compaction work the overlapping of ranges of different compactors is common, proving that these idealized ranges are suggestions rather than fixed recommendations.

TABLE 14-7 Generalized Operating Characteristics of Self-Propelled and Tow-Type Compactors

Kind of compactor	Thickness of lift, in		Average working speed, mi/h	Average number of coverages
	Loose measurement	Compacted measurement		
Sheepsfoot:				
Self-propelled	7–9	6	4–6	8–16
Tow-type	7–9	6	1½–3½	8–16
Tamping foot:				
Self-propelled	7–9	6	5–8	6–12
Tow-type	7–9	6	2–4	6–12
Vibratory, tow-type	12–28	10–24	1–3	2–4
Pneumatic tires, tow-type	7–9	6	2–4	3–6
	14–18	12	2–4	4–8

NOTE 1: Self-propelled compactors have drums in tandem and one pass gives two coverages. Tow-type compactors may have a single row of drums, in which case one pass gives one coverage, or two rows of drums, in which case one pass gives two coverages. In all cases the configuration of the compactor must be analyzed for the pass-coverage relationship.

NOTE 2: The average number of coverages is based on 95 percent relative density of embankment when tested under the Standard AASHTO test. One must always recognize that the number of coverages depends on the nature of the material, the stringency of the specifications, and many other factors affecting the work of good compaction.

fill, as is the case for most rock-earth, the ratio is more than unity. Conversely, if the material swells, as is the case with semisolid and solid rock, the ratio is less than unity. This factor may be calculated by using values for weight per cubic yard in cut and in fill, as given in Appendix 1.

Self-propelled compactors are built with engines of adequate horsepower to give them a range of suitable speeds when compacting all materials. Tow-type compactors require crawler and wheels-tires tractors of sufficient weight for tractive effort and sufficient horsepower for suitable speeds in materials with different rolling resistances. The necessary weight of the tractor is given by the equation:

14-28   Dumping and Compacting Excavation

$$WT = \frac{WC \times RR}{CT}$$

where WT = weight of tractor, lb
  WC = weight of compactor, lb
  RR = rolling resistance of compactor, % of weight
  CT = coefficient of traction of tractor

The necessary engine flywheel-brake horsepower of a crawler-tractor is given by the equation:

$$hp \times 0.69 = \frac{WC \times RR \times (S \times 88)}{33,000}$$

$$hp = 0.0039 \times WC \times RR \times S$$

where hp = engine flywheel-brake hp of tractor
  0.69 = drawbar hp/engine hp ratio
  WC = weight of compactor, lb
  RR = rolling resistance of compactor, % of weight
  S = travel speed, mi/h

The necessary engine flywheel-brake horsepower of a wheels-tires tractor is given by the equation:

$$hp \times 0.76 = \frac{WC \times RR \times (S \times 88)}{33,000}$$

$$hp = 0.0035 \times WC \times RR \times S$$

where hp = engine flywheel-brake hp of tractor
  0.76 = drawbar hp/engine hp ratio
  WC = weight of compactor, lb
  RR = rolling resistance of compactor, % of weight
  S = travel speed, mi/h

Conservative values for variables of the foregoing equations are given in Table 14-8. The values are based on Tables 13-2, 13-3, 13-10, and 14-7 and on manufacturers' data.

TABLE 14-8 Suggested Values for Members of Equations for Dynamics of Compaction Machinery

Kind of compactor	S, travel speed, mi/h	RR, rolling resistance of compactor, % of weight	CT, coefficient of traction of tractor
Sheepsfoot:			
Self-propelled	4.0	30	
Tow-type:			
By crawler-tractor	2.5	30	0.50
By wheels-tires tractor	4.0	30	0.40
Tamping foot:			
Self-propelled	6.0	25	
Tow-type:			
By crawler-tractor	3.0	25	0.50
By wheels-tires tractor	6.0	25	0.40
Vibratory:			
Tow-type:			
By crawler-tractor	2.0	20	0.50
By wheels-tires tractor	2.0	20	0.40
Pneumatic tires:			
Tow-type:			
By crawler-tractor	2.5	10	0.50
By wheels-tires tractor	5.0	10	0.40

NOTE: T, the thickness of the compacted lift in inches, is generally set forth in the specifications for compacted embankment. P, the number of passes to achieve the desired compaction, is determined by the number of coverages defined by the specifications or it is estimated.

The use of the preceding equations and of Table 14-8 is illustrated below by an analysis of the crawler-tractor-drawn sheepsfoot compactor of Figure 14-21.

Hourly capacity of compactor:
W, compacted width per coverage 20.3 ft
S, travel speed, 2.5 mi/h
T, thickness of compacted lift 6 in
P, number of passes for specified 100% relative density 6

Substituting in the general equation:

$$\text{cybm} = W \times S \times T \times 16.3$$
$$= \frac{20.3 \times 2.5 \times 6 \times 16.3}{6}$$
$$= 830 \text{ cycm/h}$$

At a 50-min working-hour efficiency, the capacity is 690 cycm/h. Actually, three compactors average 1880 cycm/h, or 630 cycm per machine.

Minimum weight of crawler-tractor:
WC, weight of compactor 155,000 lb
RR, rolling resistance of compactor, estimated 30%
CT, coefficient of traction of tractor 0.50

Substituting in the equation for WT, weight of tractor:

$$WT = \frac{WC \times RR}{CT}$$
$$= \frac{155,000 \times 30\%}{0.50}$$
$$= 93,000 \text{ lb minimum}$$

The crawler-tractor with special C-frame for push-pull operation weighs 145,000 lb, affording a good 36 percent safety factor.

Necessary tractor-engine flywheel-brake horsepower for compactor:
WC, weight of compactor 155,000 lb
RR, rolling resistance of compactor, estimated 30%
S, travel speed 2.5 mi/h

Substituting in the equation for the flywheel-brake horsepower of the crawler-tractor:

$$\text{hp} = 0.0039 \times WC \times RR \times S$$
$$= 0.0039 \times 155,000 \times 30\% \times 2.5$$
$$= 453 \text{ hp}$$

Actually, the engine horsepower is 510, giving an ample 23 percent reserve power. Furthermore, the rolling resistance of 30 percent, or 600 lb per ton of compactor weight, may be too conservative an estimate. If RR should equal 25 percent or 500 lb/ton, the necessary horsepower would be reduced to 378 and there would be 35 percent reserve power.

Sheepsfoot Compactors The tow-type sheepsfoot roller was the first compactor, built about 1925 when studies of compaction of embankments commenced. Originally the feet were shaped like those of sheep because it was observed that a flock of sheep in their systematic grazing compacted the soil well. From this prototype developed today's self-propelled modified sheepsfoot compactor, as well as its counterpart equipped with the tamping foot. The original sheepsfoot was patterned after that of a sheep, with the cylindrical foot capped by a circular bearing surface of greater diameter. This construction tended to break up the compacted material as the foot was withdrawn, and so the foot was modified to give it a wedgelike configuration, as shown in Figure 14-19. Both sheepsfoot and tamping-foot compactors are built in tow-type and self-propelled types and their compaction characteristics are alike. The modified sheepsfoot and the tamping foot are illustrated in Figure 14-19.

14–30   Dumping and Compacting Excavation

Figure 14-19 Sheepsfoot and tamping-foot self-propelled compactors equipped with bulldozer. The left picture illustrates a modified sheepsfoot compactor and the right picture shows a tamping-foot machine. The manufacturer builds both machines to the same specifications except for weights and drum and foot construction. Common important specifications are: (1) Each drum must be driven by a separate diesel engine of 165 engine flywheel hp having an integral transmission with three speeds forward and a maximum speed of 12 mi/h. (2) The frame must be articulated. Important dissimilar specifications are:

Specification	Sheepsfoot Compactor	Tamping-Foot Compactor
Working weight, with water ballast, lb	64,000	66,000
Drums, two, dimensions, in:		
Diameter, drum	60.0	65.5
Diameter, overall	79.8	79.5
Width	78.0	80.0
Feet:		
Number, total	144	240
Number in contact with ground, total	16	16
Length, in	9.8 or 9.5	6.7
Cap face area, in^2	7.5 or 12.0	30.2
Pressure of 16 feet, lb/in^2	533 or 333	137

(Hyster Co.)

The sheepsfoot compactor works in a wide range of rock-earth from clays to average-weathered rock containing rock and earth in about equal amounts. It compacts by weight and kneading action from the bottom to the top of the lift. When used in alluvia and minimum-weathered rocks, it displaces rather than compacts and is not efficient.

The feet have an average length of 9.4 in and 8.8 in^2 average cap areas, exerting an average pressure of 388 lb/in^2. For comparative purposes this pressure is calculated on the basis that one row of feet of the drum is in contact with the material at one time. The high unit pressure of the sheepsfoot compactor appeals to many contracting officers, especially for materials in the range of operation of the machine.

By reason of the foot construction the sheepsfoot compactor will not "walk out" of the material, and for that reason it actually compacts the lift below the one on which the drum is riding. Accordingly, travel speed is limited to a maximum of 6 mi/h and lift thickness usually does not exceed 6 in compacted measurement. Generally 6 to 10 coverages are required for satisfactory compaction unless the method-only specification for compaction is used by the contracting officer, in which case as many as 16 coverages or passes may be mandatory. Because of their greater speed and two-directional movement, self-propelled machines are more efficient than tow-type compactors.

Figure 14-20 emphasizes the flexibility obtained when two different kinds and types of compactors work efficiently in the same well-weathered sandstone-shale formation.

Figure 14-20 Flexibility in compaction abilities of tow-type sheepsfoot compactor and self-propelled tamping-foot compactor. The two different compactors with different feet are working as a team to compact an embankment of ripped well-weathered siltstone-sandstone rock. Hourly production is about 1600 cybm. The fills for the home sites are small and confined, so that the self-propelled machine, with twice the effective speed of the tow-type compactor, is not able to maintain this operating advantage. At the same time the tow-type compactor is diverted from high production on the surface of the fill to low production on the slopes. However, the machines complement each other and the result is good compaction at low unit cost. An estimate is that the self-propelled machine is compacting about 75 percent of the total yardage and the tow-type compactor is taking care of the remaining 25 percent. Table 14-6 indicates that both machines are suitable for compacting the maximum-weathered formation of the coastal area of San Diego County, California.

They are a self-propelled tamping-foot machine and a tow-type sheepsfoot roller pulled by a crawler-tractor. Figure 14-21 illustrates a large-capacity "push-pull" sheepsfoot compactor working in the silt-loam zone of an earthfill dam.

Representative specifications for heavy-duty tow-type double-drum and triple-drum sheepsfoot and tamping-foot compactors are given in Table 14-9. Representative specifications for a sheepsfoot self-propelled compactor and its tamping-foot counterpart are given in the legend of Figure 14-19. The same basic self-propelled machine may be equipped with either type of foot.

The self-propelled, articulated, sheepsfoot compactor of Figure 14-19 is provided with modified sheepsfeet, which, as illustrated in the figure, bear little similarity to the sheepsfeet of the tow-type compactor of Figure 14-21. The higher travel speeds of the self-powered compactor call for the modified sheepsfeet, so that the feet may leave or be withdrawn from the lift without disturbing the material, as would be the case with the sheepsfeet of the tow-type compactor. Most self-powered compactors presently in use have either modified sheepsfeet or tamping feet, tamping feet being more common. Production and cost calculations for compaction by tow-type sheepsfoot compactors are illustrated below for the machine of Figure 14-21.

Criteria:

1. The known hourly production for each of the three compactors on the job is 627 cycm.

2. The material is silt-sand, weighing about 3560 lb/cycm and 3060 lb/cybm. The calculated average hourly production of each compactor is 729 cybm.

14–32   Dumping and Compacting Excavation

Direct job unit cost, 1978:
Cost of machines:
  Compactor, including tractor C-frame $150,000
  Crawler-tractor, 510 hp $330,000
Hourly cost of ownership and operation:
  Machines (Table 8-3):
    Compactor, 26% × $150 $ 39.00
    Crawler-tractor, 33% × $330   108.90
    Operator   15.60
    Total hourly cost $163.50
Direct job unit cost:
  Per cycm, $163.50/627 $0.261
  Per cybm, $163.50/729 $0.224

The relatively high compaction cost is due to the specifications for compaction, which call for 100 percent relative density, necessitating 12 coverages by each drum of the compactor.

Tamping-Foot Compactors   These machines are built in both tow and self-propelled types. In addition to the foot of Figure 14-19 there are several other tamping-foot configurations. They are known as *rock feet* and *wedge feet* and they have the common purpose of compacting by pressure, kneading, impact, and vibration. The shape of the foot permits the compactor to "walk out" of the material, leaving a uniformly dense and partially sealed lift. Such a shape also allows the compactor to travel at higher speeds than those of its sheepsfoot counterpart.

The average tamping foot, at 7.6 in long, is shorter than the sheepsfoot and its cap area, which averages 23.0 in^2, is greater than the sheepfoot's. Its pressure averages 170 lb/in^2, as contrasted to 390 for the sheepsfoot.

Figure 14-22 illustrates a medium-heavyweight tamping-foot compactor rebuilding a

Figure 14-21 Heavy high-capacity "push-pull" tow-type sheepsfoot compactor. One of three dual triple-drum sheepsfoot compactors which together are compacting 1880 yd^3 embankment measurement. Material is silt-sand of the semipervious zone of an earthfill dam, part of the Pueblo Dam project, Colorado. The embankment is brought to 100 percent relative density by three round trips of the machine. Important specifications are:

Weight of complete tow-type compactor, ballasted, lb	160,000
Six drums, dimensions, ft:	
Diameter, overall	6.7
Width of coverage by three drums	20.3
Feet:	
Length, in	10
Area of cap, in^2	9
Total number	864
Theoretical number on embankment at one time	36
Foot pressure, lb/in^2	494
Compacted depth of lift, in	6
Coverages by six passes to secure 100% relative density	12
Crawler-tractor:	
Engine flywheel-brake hp	510
Average travel speed, mi/h	2.5
Drawbar pull at 2.5 mi/h, lb	52,000

(Southwest Welding and Manufacturing Co.)

TABLE 14-9 Representative Specifications for Heavy-Duty Sheepsfoot and Tamping-Foot Tow-Type Compactors

Specification	Sheepsfoot type		Tamping-foot type	
	Two-drum	Three-drum	Two-drum	Three-drum
Weight, lb:				
Empty	23,400	36,800	24,800	38,600
With water ballast	32,700	50,000	34,100	51,800
With wet sand ballast	43,100	62,500	44,600	64,300
Drums, dimensions, in:				
Diameter	60	60	60	60
Diameter, overall	79	79	79	79
Width	60	60	60	60
Feet:				
Total number	240	360	240	360
Theoretical number on embankment at one time	12	18	12	18
Length, in	9.2	9.2	9.2	9.2
Area of cap, in^2	7.0	7.0	7.0	7.0
Foot pressure, water ballast, lb/in^2	389	397	406	411
Construction	Removable tips	Removable tips	Solid wedge tips	Solid wedge tips
Dimensions, ft:				
Length	15.2	18.0	15.2	18.0
Width	11.9	18.2	11.9	18.2
Height	7.9	8.1	7.9	8.1
Width of coverage by drums, ft	11.4	17.3	11.4	17.3

NOTE: Feet with 8.0 and 9.0 in^2 of cap area are optional equipment.

Figure 14-22 Tamping-foot compactor with bulldozer rebuilding eroded embankment. The medium-heavyweight compactor is working to bring a sidehill fill up to grade after it has suffered minor surface erosion from a prolonged rainstorm. The average-weathered granite is being laid down by scrapers. The loose lift is 8 in thick and the compacted lift is 6 in thick. Four machine passes, or eight coverages, are adequate for satisfactory compaction.

The work is not on a production basis because of the physical limitations of the eroded area, 200 ft long and 40 ft wide. About 300 cycm/h is being handled, as against about 1200 cycm/h when working systematically on a large embankment. The bulldozer allows the operator to bring the embankment up to exact lines and grade.

The location is a mountain freeway in the Laguna Mountains of southern California.

14-34 Dumping and Compacting Excavation

sidehill embankment of a freeway. Figure 14-23 shows a self-powered medium-heavyweight machine compacting a fill for an industrial building. Figure 14-24 illustrates a tow-type tamping-foot compactor, hauled by the winch of a pipelayer, working on the slope of a massive embankment. Table 14-10 gives representative specifications for self-propelled compactor-bulldozers with tamping feet.

Production and cost calculations for compaction by a self-powered tamping-foot compactor are illustrated below for the example of Figure 14-23.

Criteria:
    1. The known hourly production on the job is 540 cybm/h.
    2. The material is average-weathered rock, weighing about 3940 lb/cycm and 3750 lb/cybm. The calculated hourly production is 510 cycm/h.

Direct job unit cost, 1978:
  Cost of machine:
    Compactor $167,000
  Hourly cost of ownership and operation:
    Machine (Table 8-3), compactor, 35% × $167 $58.40
    Operator   15.60
    Total hourly cost $74.00
  Direct job unit cost:
    Per cybm, $74.00/540 $0.137
    Per cycm, $74.00/510 $0.145

These unit costs, as contrasted with the almost double unit costs of the work of Figure 14-21, reflect first, a relative density requirement of 95 instead of 100 percent, and second, the resultant need for only 6 instead of 12 coverages for the required relative density.

Vibratory Compactors There are two kinds of vibratory compactors: tow-type and self-propelled. In rock-earth excavation the tow-type, illustrated in Figure 14-25, is more common. There are two types of drums, footed and smooth, and both are illustrated in Figure 14-25. The ranges of operation of these compactors in different materials and their means of compaction are given in Table 14-6, and their generalized operating characteristics are given in Table 14-7.

The dynamics of compaction involve the rearrangement of particles to decrease the voids among the particles and thus increase the density of the embankment. The rear-

Figure 14-23 Tamping-foot compactor with bulldozer working on embankment for industrial buildings. The medium-heavyweight machine is shuttling on a 300-ft-long fill of average-weathered granite. The material is being laid down in 9-in-thick lifts by scrapers. The compactor with its bulldozer both mixes and compacts the material, which has rocks of up to 12-in dimension intermixed.

Scraper production is 540 cybm/h, equivalent to about 510 cycm/h. The machine travels at an average of 5 mi/h and secures 96 percent relative density by 6 passes, or 12 coverages. The motor grader of Figure 14-7 is the key machine for spreading-mixing and the water wagon of Figure 14-17 provides wetting.

Figure 14-24 Tow-type tamping-foot compactor hauled by winch of pipelayer. The light-duty double-drum compactor weighs 14,000 lb, requiring a single-line pull of about 12,000 lb. The pipelayer diesel engine has 105 flywheel-brake hp and the winch hauls the compactor up the slope at about 150 ft/min and returns it downslope at about the same speed.

The three 36-ft-long slopes, separated by 8-ft-wide benches, are being compacted separately. The 8-ft-wide compactor requires 0.24 min for one pass or coverage up or down the 36-ft section of the slope. As four coverages are required for the compacting and sealing of the surface, 400 lineal ft of length of section of slope are worked each 50-min working hour. Thus, about 1600 yd² of surface of slope is handled hourly.

The work is the grading of sites for homes in the average-weathered granite of the Verdugo Mountains, Los Angeles County, California.

rangement results from dynamic forces, applied in the case of the footed drum by pressure, kneading, and vibration and in the case of the smooth drum by vibration.

The drum of the compactor is vibrated by means of an eccentrically loaded, engine-powered shaft within the drum. The drum shaft is cushioned in the frame by vibration insulators. The frequency of vibration may be varied to suit different materials and it ranges from 1000 to 4500 oscillations per minute. The *amplitude* is the total vertical distance of the up-and-down drum movement due to the forces imparted to the drum by the vibrating mechanism. The *dynamic force* is the centrifugal force produced by the rotating offset weight on the eccentric shaft of the vibrating mechanism. The *dynamic impact* is the centrifugal force plus the static weight of the compactor. It is a measure of the compacting ability of the machine.

Productions of Vibratory Compactors. The production in a given material is a function of the travel speed, width of coverage of the drum or drums, number of coverages, thickness of lift, and total dynamic impact of the compactor. The practicable speed of the compactor varies from 1 to 3 mi/h, and the number of coverages for satisfactory compaction ranges from two to four. The thickness of lift may be up to 4 ft for rocks up to 1 yd³ in size provided enough smaller material is present to fill, or nearly fill, the voids. Whenever large rocks are used in the embankment, the lift thickness should be about 1 ft more than the greatest dimension of the rocks.

The hourly production of a vibratory compactor may be figured by the general equation given previously under Dynamics of Compaction Machinery. When using the equation for a given compactor working in a given material, one must approximate the travel speed, thickness of lift, and number of passes largely in keeping with experience. There are few empirical data available for estimating the hourly production of vibratory compactors.

Table 14-11 gives representative specifications for tow-type machines as generally used

14–36 Dumping and Compacting Excavation

(a)

(b)

Figure 14-25 Heavyweight tow-type vibratory compactors. *(a)* A smooth-drum machine for granular materials without cohesion. Such rock-earth includes average- and minimum-weathered rock and blasted rock. *(b)* A footed-drum machine for granular materials containing up to 10 percent of cohesive or clayey material. Such rock-earth includes silty and sandy residuals and maximum- and average-weathered rock.

These two heavyweight compactors have an average 51,300-lb total dynamic impact and together they efficiently compact a wide range of embankment materials. *(Southwest Welding and Manufacturing Co.)*

TABLE 14-10 Representative Specifications for Self-Powered Tamping-Foot Compactors with Bulldozers

Specification	Medium-weight	Medium-heavy-weight	Heavy-weight
Diesel engine flywheel-brake hp	170	300	400
Transmission:			
Gears forward and reverse	4	3	3
Maximum speeds, mi/h:			
Forward	19	17	20
Reverse	22	17	20
Frame construction	Articulated	Articulated	Articulated
Drums, four, dimensions, in:			
Diameter, drum	40.5	51.0	61.5
Diameter, overall	53.5	68.5	78.5
Width	38.0	44.5	48.0
Feet:			
Number, total	240	260	360
Theoretical number on embankment at one time	20	20	24
Length, in	6.6	7.5	8.0
Cap area, in^2	18.0	29.8	29.8
Pressure, compactor ballasted, lb/in^2	119	125	129
Dimensions, ft:			
Length	20.5	23.3	26.1
Width	11.9	13.7	15.2
Height	11.2	11.9	12.5
Width of coverage by one pass, ft	12.7	14.8	16.0
Working weights, lb:			
Empty	39,900	66,300	79,100
With water ballast	43,000	74,400	92,100

TABLE 14-11 Representative Specifications for Tow-Type Vibratory Compactors

	Nominal average total dynamic impact, lb					
	27,000		51,300		69,000	
Specification	Smooth drum	Footed drum	Smooth drum	Footed drum	Smooth drum	Footed drum
Drum, dimensions, in:						
Width	60	60	72	72	78	78
Diameter	48	...	66	...	66	
Diameter, overall	...	58	...	72	...	82
Feet:						
Length, in	...	5	...	8	...	8
Cap area, in^2	...	18	...	7	...	7
Number	...	120	...	112	...	136
Centrifugal vibrating force, lb:						
At 1800 r/min	18,000	18,000				
At 2000 r/min	...	...	36,100	36,100		
At 1500 r/min	...	...	...	...	47,000	47,000
Static weight, lb	8,000	10,000	15,400	15,000	21,500	22,500
Total dynamic impact, lb	26,000	28,000	51,500	51,100	68,500	69,500
Diesel engine flywheel-brake hp	39	39	50	50	75	75
Dimensions, ft:						
Length	13.5	13.5	16.0	16.0	16.0	16.0
Width	6.1	6.1	7.3	7.3	7.3	7.3
Height	4.3	4.8	6.0	6.0	6.0	6.0
Weight, lb	9300	11,300	16,900	16,500	23,000	24,000

14–38 Dumping and Compacting Excavation

in rock-earth excavation. Figure 14-26 illustrates the compaction of a canal lining and Figure 14-27 shows the tedious compaction of a slope of embankment. Both jobs are typical of the adaptability of vibratory compactors for different materials and diverse working conditions.

Production and cost calculations for compaction by a vibratory compactor are given below for the example of Figure 14-26.

Criteria:

1. The known average hourly production of the compactor drawn by the crawler-tractor is 250 cycm.

2. The estimated hourly production of the compactor when working without unavoidable delays is about 780 cycm.

Direct job unit cost, 1978:
Costs of machines:
Compactor $49,000
70-hp crawler-tractor 41,000
Hourly cost of ownership and operation:
Machines (Table 8-3):
Compactor, $31\% \times \$49$ $15.20
70-hp crawler tractor, $34\% \times \$41$ 13.90
Operator 15.60
Total hourly cost $44.70
Direct job unit cost:
Cost per cubic yard embankment measurement
With unavoidable delays: $44.70/250 $0.179
With 83% efficiency: $44.70/780 $0.057

NOTE: The unavoidable delays to the compaction, due to slowness in delivering the select blended material for the canal lining, increase the normal unit cost threefold. Such disparities are not unusual in special work of this nature.

Pneumatic-Tires Compactors For rock-earth work these compactors are built as illustrated in Figure 14-28. Four or five large, weighted, oscillating tires provide com-

Figure 14-26 Vibratory compactor with footed drum in canal building. The compactor weighs 16,500 lb and its total dynamic impact is 51,100 lb. The towing crawler-tractor has a 70 flywheel-brake hp engine and weighs 15,000 lb. The material is blended silty clay and sand. A relative density of 100 percent plus is obtained with an average of two coverages over an 8-in thickness of compacted lift.

Hourly production varies with ability to deliver the blended materials for the canal lining. Maximum daily production is 2000 cycm, or an average hourly rate of 250 cycm. At a speed of 2.0 mi/h and at 50-min working-hour efficiency, the hourly production of the compactor would be 780 cycm, as calculated by the general equation, indicative of the unusual delay in waiting for material.

The canal job is located near Pasco, Washington. *(Southwest Welding and Manufacturing Co.)*

Figure 14-27 Vibratory compactor with footed drum working on slope of embankment. The picture demonstrates the necessary exactitude in compacting a slope. The work is slow and costly, involving an additional compaction of the embankment in order to seal the surface of the slope. The material is a well-weathered siltstone-sandstone formation, which has been reduced to silts and sands by ripping, loading, dumping, and spreading-mixing.

As the fill is brought up in horizontal lifts of 12-in thickness, satisfactory 95 percent plus relative density is secured by two coverages. As the fill depth increases in 5-ft increments, the compactor seals off and dresses the slopes in about 8-ft widths of slope. The hourly production of the compactor when working on the regular lifts is about 1200 yd^3 embankment measurement. A small percentage of total time is spent in the slope work.

The work is grading of sites for homes in the Puente Hills of the Los Angeles Basin. *(Southwest Welding and Manufacturing Co.)*

paction by pressure and kneading. Each tire is mounted in a box assembly which is weighted by ballast, usually granular material. For average work the machines with ballast weigh between 50 and 125 tons. They are towed by wheels-tires tractor or by crawler-tractor. The former is shown in Figure 14-29 and the latter is illustrated in Figure 14-31.

The machine compacts from the top to the bottom of the lift. The pressure on the embankment in pounds per square inch equals approximately the air pressure of the tire in pounds per square inch except for the stiffening effect of the tire walls. This rigidity raises the ground pressure somewhat above the air pressure, as some flattening of the tire is prevented. The relationship between tire contact area and ground contact pressure determines the kneading action which densifies the material.

One senses that the forces of pressure and kneading are present also in the large tires of the wheels-tires scrapers and haulers, and this is the reason why the alert foreman directs these machines over a definite pattern on the embankment so as to compact as well as haul the material. Historically the idea for the pneumatic-tires compactor came from the awareness that scrapers and haulers are excellent compactors.

Tables 14-6 and 14-7 set forth the range of operation and the operating characteristics of the pneumatic-tires compactor. This type of machine is suited to materials ranging from maximum-weathered rock-earth to well-blasted rock and to alluvia. It is not efficient in residuals, with their high plasticity, because of its tendency to force aside the material and mire down as is the case in the operation of Figure 14-30.

The thickness of the compacted lift is generally governed by specifications, which indirectly control the thickness of the loose lift. Travel speeds range from 2 to 4 mi/h. The number of coverages varies from three to eight for the desired relative density, but these too may be covered by the specifications.

14-40 Dumping and Compacting Excavation

Figure 14-28 Pneumatic-tires compactor of 50-ton capacity. This compactor, nominally rated at 50 tons with ballast, is the lightest of four compactors built by the manufacturer. The others have 75-, 100-, and 125-ton capacities. Together they encompass just about all the requirements for different jobs with different materials of embankments. The upper machine is towed by a wheels-tires tractor, integrated by the manufacturer. The lower machine, basically identical, is modified by drawbar construction for use with a crawler-tractor. *(Southwest Welding and Manufacturing Co.)*

Figure 14-29 Compacting pattern and oscillation of tires of pneumatic-tires compactor. The nominally rated 75-ton compactor is drawn by a tractor having a 335 engine flywheel-brake hp and weighing 30,000 lb. The ballasted boxes, each mounted on a single large tire, are free to move 12 in vertically so as to conform to surface irregularities and to seek out soft spots in the embankment, while providing uniform pressure. The ground pressure thereby obtained approximately equals the recommended tire air pressure of 85 lb/in^2.

The material is semipervious fill for dam building, ideal for the compactor as it is granular. The compacting pattern indicates a maximum rolling resistance of about 5 percent, or 7500 lb. The tractor drawbar pull is about 12,000 lb, and the speed for efficient compaction is about 5 mi/h.

The dam construction is located in the Sierra Nevada of California. *(Southwest Welding and Manufacturing Co.)*

Figure 14-30 Marginal efficiency of pneumatic-tires compactor working in embankment of residuals. The compactor was chosen for freeway building in which materials varied from residuals to maximum-weathered rock-earth. The nominally rated 75-ton compactor is drawn by a crawler-tractor having a 235 engine flywheel-brake hp and weighing 55,000 lb.

The machine is compacting a fill approaching a bridge. The material is predominantly damp clay residual from the topmost zone of the cut. The compactor becomes mired occasionally because of the poor bearing capacity and high plasticity of the clay. The rolling resistance reaches 15 percent, or 22,500 lb. At the same time the traction coefficient for the crawler-tractor drops to about 0.30, or 16,500 lb drawbar pull. It is necessary to reduce ballast in the compactor so that rolling resistance is less than drawbar pull. Thereby compaction is enabled to proceed without expensive delays due to bogging down.

As the cut is brought down, the material changes to weathered rock and the difficulties are ended. The location is the Arroyo Grande River, California.

14–42 Dumping and Compacting Excavation

Important specifications for the 50-ton-capacity pneumatic-tires compactor of Figure 14-28 are tabulated below, and typical specifications for multiple-box machines rated at 50 to 125 tons are given in Table 14-12.

Specification	With wheels-tires tractor	With crawler-tractor
Compactor:		
Rated capacity, ballasted, tons	25–60	25–60
Tires, four:	18.00×25	18.00×25
	24PR	24PR
Maximum load per tire, lb	30,000	30,000
Rolling width of coverage, ft	9.8	9.8
Working weight, lb:		
Empty	32,450	32,450
With ballast, rated capacity	100,000	100,000
Tractor, suggested:		
Diesel engine flywheel-brake hp	250	200
Tires	26.50×25	
	24PR	
Working weight, lb	24,000	46,000
Compactor and tractor:		
Dimensions, ft:		
Length	33.2	38.5
Width	11.2	11.2
Height	12.0	11.0
Working weight, rated capacity, lb	124,000	146,000

NOTE: The wheels-tires tractor is suggested by the manufacturer. The selection of the crawler-tractor was based on a compactor rolling resistance of 10 percent, or 10,000 lb, and a drawbar pull of the tractor equal to 12,700 lb at 4 mi/h.

TABLE 14-12 Representative Specifications for Pneumatic-Tires Compactors of the Multiple-Box Type

Specification	Nominal rating, ballasted, tons			
	50	75	100	125
Capacity, ballasted, tons:				
Working range	25–50	35–75	50–100	75–125
Maximum, rated	50	75	100	125
Number of boxes and tires	4	4	4	5
Tires:				
Size	18.00×25	21.00×25	21.00×25	21.00×25
	24PR	24PR	44PR	44PR
Maximum load, lb	30,000	39,000	56,400	56,400
Air pressure, maximum/minimum, lb/in^2	95/80	85/70	150/120	150/120
Width of coverage, ft	9.8	10.5	10.5	14.0
Dimensions, ft:				
Length	23.5	23.7	25.7	24.6
Width	11.2	11.9	11.9	16.3
Height	7.5	7.8	9.8	8.2
Ground clearance, ballasted	1.5	1.5	1.5	1.3
Empty weight, with gooseneck, lb	32,450	36,450	45,000	55,400
Suggested tractors:				
Wheels-tires tractor:				
Engine flywheel-brake hp	330	330	415	415
Tires	26.50×29	26.50×29	29.50×35	29.50×35
	24PR	24PR	34PR	34PR
Working weight, lb	33,300	33,300	58,000	58,000
Travel speed, maximum, mi/h	5	5	5	5
Crawler-tractor:				
Engine flywheel-brake hp	200	200	300	400
Working weight, lb	44,000	44,000	74,000	95,000
Travel speed, maximum, mi/h	4	4	4	4

Estimated production and cost calculations for compaction by a sheepsfoot compactor and a pneumatic-tires compactor working in the same material of embankment are presented below for the work of Figure 14-31. The material is maximum- and average-weathered rock-earth for the semipervious zone of an earthfill dam.

Criteria	Pneumatic-tires compactor	Sheepsfoot compactor
Weight of compactor, ballasted, lb	150,000	43,000
Estimated rolling resistance, %:		
Table 13-11	10	
Average value	...	30
Necessary drawbar pull of tractor, lb:		
10% × 150,000	15,000	
30% × 43,000	...	12,900
Weight of tractor, lb	45,000	45,000
Engine flywheel-brake hp	191	191
Estimated drawbar hp, 69% × 191	132	132
Estimated maximum travel speed, mi/h:		
$\dfrac{132 \times 33,000}{15,000} = 290$ ft/min	3.3	
$\dfrac{132 \times 33,000}{12,900} = 338$ ft/min	...	3.8
Estimated average travel speed at 80% of maximum speed, mi/h	2.6	3.0
Estimated hourly production based on general equation: cycm = $\dfrac{W \times S \times T \times 16.3}{P}$		
W = compacted width per coverage, ft:		
Table 14-12	10.5	
Table 14-9	...	11.4
S = travel speed, estimated, mi/h	2.6	3.0
T = thickness of compacted lift, specified, in	6	6
P = number of passes to achieve desired relative density, estimated (Table 14-7)	6	12
Hourly production at 100% efficiency, cycm	445	279
Hourly production at 50-min working-hour efficiency, cycm	369	231
Estimated direct job unit cost, 1978:		
Costs of machines:		
Pneumatic tires compactor	$ 62,000	
Sheepsfoot compactor	...	$ 56,000
Crawler-tractor	124,000	124,000
Hourly cost of ownership and operation:		
Machines (Table 8-3):		
Pneumatic tires compactor, 28% × $62	$17.40	
Sheepsfoot compactor, 26% × $56	...	$14.60
Crawler-tractor with bulldozer, 34% × $124	42.20	42.20
Operator	15.60	15.60
Total hourly cost	$75.20	$72.40
Direct job unit cost per cycm:		
$75.20/369	$0.204	
$72.40/231	...	$0.313

COMPACTION OF TRASH AND COVER MATERIAL IN A SANITARY LANDFILL OR DUMP

The construction of sanitary landfills for the disposal of trash is generally covered rigidly by specifications so as to prevent nuisances and hazards to public health and safety. Odors, vermin, and spontaneous combustion, as well as objectionable appearance, are the most prevalent results of haphazard development of dumps.

Figure 14-31 Flexibility of different kinds of compactors working in the same material. Table 14-6 sets forth ideally the application ranges of compactors in different materials. The practicability of the table and the flexibility of sheepsfoot and pneumatic-tires compactors are demonstrated in this picture of dam building. The materials in an earthfill dam vary through all the kinds listed in Table 14-6 with the addition of crushed rock and riprap. These two compactors, singly and together, handle all these materials except the riprap, which is placed but which is not compacted. In this picture both machines are working in the semipervious zone, which consists of maximum- and average-weathered rock-earth.

The tow-type two-drum sheepsfoot compactor weighs 43,000 lb ballasted, and the crawler-tractor has an engine flywheel-brake hp of 191 and weighs 45,000 lb. The tow-type pneumatic-tires compactor weighs 150,000 lb ballasted and is pulled by a similar tractor.

The work is construction of the Cherry Valley Dam in the Sierra Nevada Mountains of California. *(Southwest Welding and Manufacturing Co.)*

The specifications call for a depth of compacted trash averaging 4 ft, to be covered by a depth of compacted cover averaging 1 ft. Both trash and cover may be compacted by crawler-tractor-bulldozers, as shown in Figure 14-32. When specifications are more severe or when other considerations warrant higher density, special landfill compactors are used to consolidate the trash and the cover material. The machines used for the trash resemble self-propelled tamping-foot compactors, except that their drums are equipped with angular chopping feet instead of the conventional tamping feet. The compactor used for the cover is the one best suited for the cover material, as set forth in Table 14-6.

Wetting of both trash and cover is advisable and is sometimes mandatory under the specifications. Trash requires about 50 percent moisture by weight for most effective compaction. As it averages about 800 lb/cycm in the dry state, water in the amount of about 60 gal/cycm must be added. Cover requires water in keeping with the desired relative density of the lift. The water wagons may be any of those already discussed.

Sanitary fills are of three kinds. In the *separate trench method* deep trenches are filled with trash and covered. The cover material comes from successive trench excavations, as shown by Figure 5-10.

The *progressive slope method* resembles the work of Figure 14-32, where cover material is being bulldozed from the neighboring sharp ridges. In a regular progressive slope method the bulldozer excavates the cover material from an adjacent area, which is then backfilled with trash and cover from the next adjacent area.

The *area method*, typical of hilly and mountainous country, involves simply the filling of depressions with trash and imported cover. Figure 14-32 illustrates a small-capacity sanitary landfill of this kind. Here the source of the imported cover is as illustrated in Figure 5-11, being the conveniently located weathered rock of the surrounding sharp ridges. Figure 14-33 shows a large area-method landfill.

Figure 14-32 Building a sanitary landfill with three medium-weight crawler-tractor-bulldozers. In this area fill, the cover of well-weathered siltstone-sandstone formation is being bulldozed from the surrounding ridges, as illustrated in Figure 5-11. The two similar machines of this picture spread and consolidate trash and cover material. The trash is compacted to a 3-ft thickness and the cover is put down in a layer of 1-ft compacted thickness. No water is being added.

An average of 525 tons of trash is handled every 7 h in this low-cost work. The 1978 direct job unit cost is estimated $2.31 per ton of trash. The fill is located in the Santa Susana Mountains, Los Angeles County, California.

Figure 14-33 Large-area-method sanitary landfill in hilly country. Specifications for this landfill call for a high relative density of both trash and cover material compaction. Trash is being dumped at two levels. The cover material is a clay-silt residual being imported by two wheels-tires scrapers from a borrow pit 1800 ft away. The landfill accommodates about 900 tons of trash hourly, or some 7000 tons per 8-h day.

The major machinery is:
  Three heavyweight crawler-tractor-bulldozers of 365 engine flywheel-brake hp
  One 8000-gal water wagon of 345 engine flywheel-brake hp
  Two 23-cysm-capacity wheels-tires scrapers of 345 engine flywheel-brake hp
  One 12-ft moldboard motor grader of 115 engine flywheel-brake hp

The approximate 1978 direct job unit cost, including the cost of water, is $6.10 per ton of trash. The landfill is located in the Puente Hills of Los Angeles County, California.

SUMMARY

The end use of excavated rock-earth determines the amount of work and cost for its dumping and placing. The simplest operation is dumping, with little or no use of auxiliary machinery. Examples are the dumping of overburden over the edge of a spoil bank, for which only a tractor-bulldozer may be needed, and the dumping of quarry rock into the receiving hopper of a primary crusher. In both cases the end use of the rock may require no machinery and entail no costs.

A complex deposition procedure is the dumping, spreading-mixing, and wetting-compacting of material for the embankment of an earthen dam. When hourly production is 2000 cycm and when specifications are customarily stringent, major machines on the fill may include two tractor-bulldozers, two motor graders, two water wagons, and four compactors. Such an array of machinery can result in a direct job unit cost of $0.35/cycm (1978 costs).

In all cases dumping a load from the hauling machines calls for well-considered analysis, planning, and estimating. It is fully as important as the preceding operations of fragmenting, loading and casting, and hauling the excavation. In embankment construction for an airport, dam, or freeway the cost may be one-fourth of the total cost for all work.

CHAPTER 15

Haul Road Construction and Maintenance

PLATE 15-1

Two-way four-lane haul road for massive rock-earth excavation. On this 7100-ft, average 100-ft-wide, well-maintained haul road with an average -7% grade, 187 laden and 187 unladen scrapers, bottom-dump haulers, and rear-dump haulers pass hourly under this observation point. The traffic density is one machine every 10 s, in addition to maintenance machines.

During an average 18-h working day 73,200 cybm is hauled, or 4070 cybm/h. During peak production in the famous Culebra cut of the Panama Canal, where rock excavation was handled by locomotives and cars, 71,800 cybm was hauled in 8 h, 9000 cybm/h. Thus, in this 9.8 million yd^3 freeway cut, rock is being handled at 45 percent of the rate in the 147.2 million yd^3 Culebra cut. Both cuts contained residuals, weathered rock, and solid rock.

The well-built and well-maintained haul road for this huge cut affords maximum production with a 100 percent safety record. The excellent maintenance is secured by a motor grader, a tow-type grader pulled by wheels-tires tractor, a 4000-gal water wagon, and a 10,000-gal water wagon.

The location is in the Laguna Mountains of San Diego County, California.

Modern haul roads in construction and mining are built to the standards of highways when materials and topography make such building possible and when economic conditions warrant such caliber. Over the past 50 yr travel speeds of automotive-type off-the-road haulers have increased about fivefold from 10 to 50 mi/h. Although much of this increase is due to improvements in design, it is true that attention to construction and maintenance of haul roads accounts for an equal share of the increase in travel speeds.

The reasons for good haul roads are economic. They make for greater travel speeds with lesser shock loads, which are translated into higher production and lower maintenance for the wheels-tires scrapers and haulers. Travel speeds are affected by grades, alignment, smoothness of surface, and hardness of surface.

CONSTRUCTION

Haul roads may be temporary or permanent and their refinements may depend on their exact status. However, the haul roads for a 50-million-yd^3 dam, although temporary, would be superior to the permanent ones for a long-term gravel pit with an annual production of 500,000 yd^3. One rightly associates temporary haul roads with public works construction and permanent ones with private works such as open-pit mines and quarries.

Some haul roads for construction become part of the finished work, as shown in Plate 15-1. This is true of lineal construction such as canals, highways, and railroads. In such cases the cuts and fills for the haul road become integral parts of the finished cuts and fills. Under such circumstances the cost of haul road construction is a part of the cost of excavation, including the embankment building.

In areal grading, such as for an airport or dam, the cuts and fills of the haul road may not be parts of the cuts and fills for the finished earthwork. An example is the haul road from a remote borrow pit to a dam site, in which case the cuts and fills are not a part of the dam and appurtenances and very probably will have to be obliterated as required by the specifications. In this case the cost of the haul road construction could be charged to general expense or to the bid item referring to the material being taken from the pit.

The materials for surfacing a haul road are generally those of the excavation. Sometimes, as in the case of a hard rock job such as a quarry, crushed stone is used for the surfacing because the quarry-run blasted rock may be too big for surfacing. Oiled roads and asphalt-cement concrete paving do not stand up under the concentrated wheel and axle loadings of huge off-the-road machines, which are sometimes 20 times those of on-the-road haulers. Some permanent haul roads in open pits and quarries are paved with Portland cement concrete. In these instances specifications for base course and pavement must be in keeping with design practices for the given wheel and axle loadings. In this text surfacing is considered to be carried out with natural on-the-job or imported materials, compacted by using water only as a binder.

Generally haul road construction and maintenance in public works is not a bid item. Accordingly, the cost is either a part of the bid item for excavation or a part of the general expense of the job, to be prorated according to the bookkeeping practices.

Grades

The more moderate the grades, both unfavorable and favorable, the faster the average travel speed. Usually grades in public works are partially or entirely fixed by the topography and the nature and dimensions of the works, but by close planning they can be kept to a minimum. In private works the source of the rock-earth and the location for disposition are fairly fixed and there is more freedom for the design of economical grades. Such circumstances exist in the open-pit mine of Figure 15-9. Under these conditions one keeps in mind a maximum grade of ± 8 percent as being conducive to economical all-weather operation.

As in railroad location, the excavator avoids adverse grades whenever possible. An adverse grade on a haul road is one which must be ascended to the summit only to be followed by a descending grade or which must be descended to the bottom only to be followed by an ascending grade. In both cases more work is expended by the hauler than by traveling over a straight grade from the beginning to the end of the two contiguous grades.

Modern machines are equipped with retarders, a kind of brake which has been discussed in Chapter 13, and these devices are effective on economical downgrades for both laden and unladen scrapers and haulers. These hydraulic assemblies have removed both the danger and the high maintenance which formerly accompanied the use of friction shoe–drum brakes on long, steep grades.

In a discussion of grades it is well to comment on the energy and time required to lift a laden hauler through a given difference in elevation over two haul roads of different grades but equal rolling resistance.

Criteria:
Engine flywheel-brake hp   400
Horsepower at wheel-road contact, $76\% \times 400$ 304
GVW, lb   200,000
Rolling resistance, %   3

Item	+4% Grade	+8% Grade
Difference in elevation, ft	100	100
Length of grade, ft	2,500	1,250
Resisting forces at constant speed, lb:		
Due to grade:		
$4\% \times 200,000$	8,000	
$8\% \times 200,000$		16,000
Due to rolling resistance, $3\% \times 200,000$	6,000	6,000
Total force	14,000	22,000
Work expended going up grades, ft·lb:		
$2500 \times 14,000$	35,000,000	
$1250 \times 22,000$	...	27,500,000
Time for ascending grades, min:		
$\dfrac{35,000,000}{304 \times 33,000}$	3.50 At 8.1 mi/h	
$\dfrac{27,500,000}{304 \times 33,000}$	...	2.75 At 5.2 mi/h
Percent time saving on +8% grade	...	21

The tabulation shows that doubling the grade and halving the length of the haul road saves 21 percent in travel time, which gives an approximate saving of 21 percent in direct job unit cost for haulage up the steeper ramp. Further analysis shows that, if grade resistance is neglected, travel times for the same increase in elevation are equal. This observation is obvious, as energies expended in overcoming gravity for the same weight through the same vertical distance by the same force must result in equal times for ascent.

Alignment

Although there may be options for alignment in public works, it is generally set by the nature of the work, as are the grades. In private works there is more latitude and the choice may be more favorable to long-range economy.

In all cases one should strive for long-radius curves, compounded or spiraled, with superelevations. Such construction combines speed and safety. If attainable, curves should be kept to a 500-ft minimum radius, equivalent to an 11° curve. Often that radius is not possible, as shown in Figure 15-8, where the radius for the laden rear-dump haulers is 50 ft. Such small radii are typical of work at the loading and dumping points and they are really parts of the turnings rather than elements of the main-line haulage road. However, they often occur as switchbacks on main-line haulage when descending a slope or escarpment.

Travel speeds of 30 mi/h for wheels-tires haulers are not uncommon and superelevation is necessary. It is given ideally by the equation:

$$E = 0.067 \frac{S^2}{R}$$

where E = maximum elevation, ft, per ft of roadbed or surfacing
S = travel speed, mi/h
R = radius of curve, ft

Where haul roads are reasonably dry all year round, E may have a maximum value of 0.12. However, for all-weather roads subject to mud, snow, and ice, the value of E should not exceed 0.06.

The use of the ideal equation and the modification of the value of E for practical purposes are shown by the following example, in which it is desired to travel at 30 mi/h through a curve of 500-ft radius, 11°, in all-weather work. Theoretically, balancing centrifugal force against opposing force due to superelevation, the equation gives:

$$E = 0.067 \times \frac{30^2}{500} = 0.12 \text{ ft per ft of roadbed width}$$

A superelevation of 12 percent is too great, as on a firm, wet clayey surface a hauler will slide dangerously on a -8 percent grade. Using the suggested value of 0.06 for E in the equation and solving for S:

$$S = \left(\frac{0.06 \times 500}{0.067} \right)^{1/2} = 22 \text{ mi/h}$$

Safety and economy are secured by a dynamic balance of forces and a speed reduction of 27 percent on the curve. As a matter of interest, if the curve went through a right-angle turn, that is, if it subtended a 90° central angle, the loss in travel time would be but 7 s.

Sight distance must be adequate for safety. Depending on density of traffic and topography, this distance varies between 130 and 475 ft for different travel speeds.

Table 15-1 gives suggested design controls for haul roads. The table is for two-way traffic on two lanes. In large-scale haulage there are sometimes four lanes, two lanes for each direction to accommodate slow and fast traffic. Such a road is that shown in Plate 15-1 which has a traffic density of 374 machines hourly. In this instance the criterion for use of Table 15-1 is an average hourly traffic density of 187 machines. Even though Table 15-1 embraces a maximum density of only 100 machines hourly, it is usable because the maximum travel speed is 22 mi/h on the long haul road with an average 7 percent downgrade.

Methods and Machinery

Haul roads for public works such as highways and dams are generally within the right-of-way and mostly within the prisms of the cuts and fills of the works. Accordingly they are

TABLE 15-1 Suggested Minimum and Desirable Design Standards for Haul Roads with Two-Way, Two-Lane Traffic of Wheels-Tires Haulers

Design control	Average hourly traffic, expressed as number of machines					
	Under 10		10 to 40		40 to 100	
	Minimum	Desirable	Minimum	Desirable	Minimum	Desirable
Speed, mi/h:						
Flat topography	30	40	35	45	40	50
Rolling topography	20	30	25	35	30	40
Mountain topography	10	20	15	25	20	30
Sharpest curve, °:						
Flat topography	14	10	11	7	9	6
Rolling topography	25	16	18	11	14	9
Mountain topography	56	38	36	18	25	14
Maximum grade, %:						
Flat topography	8	5	8	5	7	5
Rolling topography	12	7	10	7	8	6
Mountain topography	15	10	12	9	10	7
No passing sight distance, ft:						
Flat topography	280	355	315	415	350	475
Rolling topography	205	280	240	315	275	350
Mountain topography	130	215	165	240	200	275
Width of roadbed, ft	50	60	50	60	50	60

NOTE: The table is based on the standards of the American Association of State Highway Officials, 1949, for highway traffic. It has been modified for application to off-the-road–on-the-road haulers used in rock-earth excavation.

15-6 Haul Road Construction and Maintenance

being continually obliterated during the course of the work. At the end of the rough grading the surviving ones are usually obliterated as set forth in the specifications. When haul roads are outside the immediate work area, as is the case with those from remote borrow pits, they may have sizable cuts and fills calling for fairly heavy grading.

Roads for open pits and quarries in private works may be entirely within the work area, as shown in Figure 13-37. On the other hand the main haul road may be entirely outside of the workings, as explained in Figure 13-33, in which case the road is made up of short sidehill cuts and fills.

The construction of these roads may call for much of the machinery described in Chapters 9 to 14. These chapters contain specifications, productions, and costs for machines used in haul road construction and maintenance. Chapter 8 gives data for calculation of costs.

Figure 15-1 presents four sequential pictures of haul-road construction for a Rocky Mountain highway up and over a saddle.

The machine most used for building haul roads, especially in the initial work, is the crawler-tractor-bulldozer. After the machine cuts out the top of the slope of the cut, as shown in Figure 12-7, it then widens the cut sufficiently to accommodate the scraper or the hauler. This widening is shown in Figures 15-2 and 15-3.

In maximum- to average-weathered rock-earth, downgrade haul roads from cut to fill may be commenced by all-wheel-drive scrapers without help from push-loading crawler-tractors. Such a cut is shown in Figure 15-4, where a large four-wheel-drive machine, along with two others, is creating its own haul road and hauling to the fill at the same time.

Figure 15-1A Pioneering the haul road up the sidehill slope to the saddle of the pass. The medium-weight crawler-tractor-bulldozer is working up the slope in order to put in an access road to the saddle of the 329-ft deep 1.3 million yd³ cut. Track drills will climb the road to open up the hard metamorphic rock structure. Afterwards the narrow access road will be widened to serve as the haul road from the saddle down to Ten Mile Creek.

The location of the job shown in these sequential pictures is near Vail, Colorado, in the Rocky Mountains, where the elevation is 9200 ft and the construction season is usually short, commencing in July and ending in September. On this job, however, excavation was continued through the winter in anticipation of paving the following summer. (D. M. Williams (ed.), "The Hardest Part Is Getting Started," Highway and Heavy Construction, September 1977, p. 38.)

Figure 15-1B End of pioneer road built almost to the saddle by crawler-tractor-bulldozer. Beyond the end of the access road, just short of the saddle, the hard metamorphic rock is nonrippable and will be blasted. Encountering such a hard saddle is not uncommon in rock excavation, or else the ridge would have disappeared by age-long weathering. In the foreground is a track drill ready to be set up for drilling. The supply line for compressed air is visible along the ledge of rock, the compressors being located at the bottom of the cut some 1000 ft away. *(D. M. Williams (ed.), "The Hardest Part Is Getting Started,"* Highway and Heavy Construction, *September 1977, p. 37.)*

Figure 15-1C Track drills commencing work in saddle cut. The heavyweight track drills are putting down 3-in-diameter, 30-ft-deep holes on an 8 ft × 8 ft pattern. It is probable that hourly production is 80 ft of hole, or 2.7 holes, in this tedious initial drilling. When the drills are putting down holes in later systematic drilling on benches, the rate of drilling will be 50 ft of hole, or 119 cybm, hourly on the 8 ft × 8 ft tested pattern. *(D. M. Williams (ed.), "The Hardest Part Is Getting Started,"* Highway and Heavy Construction, *September 1977, p. 39.)*

Figure 15-1D Completing excavation of a bench of the single 1.3-million-yd³ cut. The cut is about 67 percent finished and the rock is being hauled 2 mi to an embankment along Ten Mile Creek. As the 27-ft deep lifts or benches were taken out, the steep haul roads within the prism of cut were eliminated and the clean face of the cut resulted. The present gentle downgrade haul road is visible in the lower left section of the picture and the loading of the rock is in the far distance.

The massive hard cut was taken out between July 1976 and July 1977. The all-year-round average rate of excavation was about 108,000 yd³/month. Peak summer production during the 10-h shift was about 480 yd³/h of blasted rock. The excellent haul road was maintained by a large water wagon and a medium-weight motor grader so as to afford a high average travel speed of about 30 mi/h for the 35-ton-capacity rear-dump rock haulers on the 2-mi, one-way haul road. *(D. M. Williams (ed.), "The Hardest Part Is Getting Started," Highway and Heavy Construction, September 1977, p. 38.)*

Figure 15-2 Heavyweight tractor-ripper-bulldozer building a ramp from cut to fill. The ridge of granite is being excavated to eliminate a sharp-radius curve in the dangerous mountain highway shown. The rock will be hauled down the ramp or haul road to the nearby fill. At present the rock, blasted at a level above the highway, is being hauled around the promontory. The top of the new sidehill cut is visible to the left of the machine.

The new haul road will eliminate a sharp switchback and reduce the one-way haul distance by 300 ft. As the cut lowers from natural ground to the finished grade, the rock-earth will become nonrippable and blasting will be commenced.

The highway relocation project is in Bear Ridge, San Diego County, California.

Figure 15-3 Start of haul-road construction at top of high ridge. A four-lane expressway is being replaced by an eight-lane freeway across the Tehachapi Mountains of California. Excavation of a huge sidehill cut is commencing about 200 ft above the existing expressway. A small sidehill cut of the present thoroughfare is visible between the tractor-ripper-bulldozer and the chaparral to its right side. After the machine pioneers a 20-ft-wide sidehill cut down the slope, scrapers will commence the long haulage to the fill in the distant valley.

There have been four stages in the 50 mi of highway construction across the mountains from the San Fernando Valley to the San Joaquin Valley:

1916–1936: Continuous work for a final 18-ft-wide road following the main ridge, from which originated the name Ridge Route.

1936–1950: Three-lane highway following closely the valleys and canyons paralleling the main ridge, called the Alternate Ridge Route.

1950–1970: Four-lane expressway, the rebuilding of the Alternate Ridge Route.

1970–1978: Eight-lane freeway, closely following the alignment of the original Ridge Route and also called the Ridge Route.

During the 62 yr of road building some 110 million yd^3 of rock and earth has been excavated to produce a marvel of highway engineering.

Productions and Costs

Haul road construction is simply a rock-earth excavation job and the productions and costs for clearing and grubbing, development of water supply, fragmentation, loading and casting, hauling, and fill building may be calculated as explained in the preceding chapters.

MAINTENANCE

After the haul road is completed, the all-important maintenance begins. Good maintenance affords high travel speeds and safety and is the key to lowest haulage costs. High travel speeds result from smoothness and hardness. Smoothness reduces vibrations and impacts to the haulers and thereby reduces expensive repairs and replacements for the machine. If a haul road is rough, shock stresses in the machine, including the tires, increase as the square of the speed. Accordingly, if the tolerable speed is 15 mi/h and if an attempt is made to increase the speed to 30 mi/h, shock stresses are increased fourfold. Presumably, the cost of repairs and replacements would also be quadrupled. If the haul road were perfectly smooth, shock loads and maintenance cost for the haulers would be the same for 15-mi/h and 30-mi/h speeds.

Hardness reduces indentation of tires and thereby lowers rolling resistance and makes higher speeds possible. The equation of Table 13-11 indicates that a hauler with 2.0 in indentation will haul at 1.6 times the speed of one with 4.0 in indentation. Smoothness and hardness result from continuous light or fine grading, wetting, and the compacting effected by the haulers.

Safety and, as a result, higher speed, may be achieved by application of a dust palliative. Although palliatives such as calcium chloride, oil emulsions, light oils, and light tars have been applied to surfaces, fresh water is the oldest and most common agent and its use

Figure 15-4 All-wheel-drive scraper creating haul road and simultaneously loading-hauling. The 24-cysm scraper is one of three which are commencing excavation without benefit of haul road and with no push-loading tractors. A maximum-weathered siltstone-sandstone formation and a steep downgrade enable the scraper to maintain about 67 percent of the payload secured normally with a push tractor, equivalent to about 14 cybm. In spite of the reduced payload, this operation is economical in the upper third of the cut because the machines are eliminating push-loading tractors and haul-road building. As the elevation is lowered into the middle third of the cut, push tractors will be set to work and a haul road will be established.

The work is grading for home sites in the Granada Hills of Los Angeles County, California.

is pretty much standard practice among rock-earth excavators. The simultaneous use of motor graders, tow-type graders, and water wagons is the usual method for securing smoothness, hardness, and safety.

Motor Graders

The 12-ft and the 14-ft motor graders of Table 14-2 are generally used, but predominantly the 14-ft-rated machine. Figure 15-5 illustrates a 14-ft machine maintaining the beginning of the haul road in a cut.

The dynamics of maintenance of a haul road by motor graders, as observed on a number of rock-earth jobs, are given in Table 15-2. In the table hourly production is given in area of haul road and in length of two-lane and four-lane haul roads. These values are for one coverage hourly.

Frequency of complete coverage varies from hourly to every 4 h, depending on the material, hardness, and wetness of the haul road. The average is about every 2.5 h, and the hourly productions of Table 15-2 must be multiplied by a suitable factor for the frequency of coverage. For example, the 14-ft motor grader on an average job would maintain 2.5×2400, or 6000 ft, of a two-lane haul road of 50-ft width.

Generally motor graders are used on the entire length of the haul road, in the cut, on the main line, and in the fill. However, practices vary and sometimes the machine is not used in the fill. Table 15-2 is essentially a guide rather than an arbitrary adviser because of the many variables affecting the performance of the motor grader when maintaining haul roads. This limitation is also true for Table 15-3, which gives average performance of tow-type graders.

Estimate of Production and Costs It is desired to know the cost of maintaining a haul road in terms of the hourly production of the rear-dump haulers using the haul road. The pertinent criteria and production and cost data are tabulated below for a typical situation.

Criteria:
    Motor grader to be used, 14 ft with 180 flywheel-brake hp engine

Haul-road length, ft	3800
Roadbed or haul road width, ft	50
Estimated frequency of coverage, based on 3 coverages each 8-h shift, h	2.7
Estimated hourly production of the rock haulers, cybm	460
Hourly production per 50-min working hour	
Roadbed length covered by one pass hourly, ft (Table 15-2)	2400
Roadbed length covered by one pass every 2.7 h, ft, 2.7×2400	6480
Time required to maintain 3800 ft of haul road by one pass, h, 3800/6480	0.59
Cost of motor grader, 1978	$126,000
Hourly cost of ownership and operation, 1978:	
Machine, 14-ft motor grader (Table 8-3) $33\% \times \$126$	$41.60
Operator	15.80
Total hourly cost	$57.40
Cost of coverage, $0.59 \times \$57.40$	$33.90
Production of the rock haulers during 2.7 h of one coverage, cybm. 2.7×460	1240
Direct job unit cost per cybm of rock-earth hauled, $33.90/1240	$0.027

To this cost of motor-grader maintenance must be added that for water-wagon maintenance, which is discussed in the section on water wagons.

Tow-Type Graders

All three sizes of the machines of Table 14-3 are used, but chiefly the 18-ft grader. Figure 15-6 illustrates an 18-ft grader towed by a tractor with an engine of 340 flywheel-brake hp. The machine is maintaining a 2300-ft-long, 50-ft-wide haul road by three passes at hourly frequency.

Table 15-3 gives the dynamics of haul-road maintenance by three sizes of graders towed by wheels-tires tractors of suggested engine horsepowers. The production values are for complete coverage hourly for 50-ft-wide two-lane roads and 100-ft-wide four-lane roads. The production of any tow-type machine depends on the power and weight of the pulling tractor, the machine having no inherent output. The productions of the table are based on the use of the suggested wheels-tires tractors. As production is a direct function of horsepower, the production can be adjusted proportionately for a tractor of different horsepower.

As in the case of motor graders, frequency of complete coverage varies from hourly to

15–12 Haul Road Construction and Maintenance

Figure 15-5 Medium-weight motor grader maintaining haul road in huge cut. In addition to sharing the maintenance of the main line of the haul road with a tow-type grader, this 14-ft motor grader takes care of 2000 ft of 100-ft-wide haul road within the cut. All told, the machine takes care of an equivalent total of 4000 ft of 100-ft-wide haul road.

Most rock spillage causing tire damage occurs in the cut during actual loading and on the first 200 ft of travel before the crowned load settles in the bowl of the scraper or in the body of the hauler. One observes that the pit-floor area shown is remarkably clean, as it has been swept bare of sharp damaging granite rock.

The location is a freeway construction project in the Laguna Mountains of San Diego County, California.

TABLE 15-2 Average Performances of Motor Graders in Maintenance of Haul Roads

Item	Nominal rating of moldboard width, ft		
	12	14	16
Diesel engine flywheel-brake hp	135	180	250
Moldboard dimensions, ft:			
Length	12.0	14.0	16.0
Width of pass, moldboard set at 45° angle:			
Maximum	8.5	9.9	11.3
Average, 2.0 ft overlap	6.5	7.9	9.3
Engine hp per foot of average width of pass	21	23	27
Average depth of cut-fill, in	1.0	1.0	1.0
Average travel speed, including one productive turn:			
ft/min	270	300	350
mi/h	3.1	3.4	4.0
Roadbed maintained with one complete coverage per 50-min working hour:			
Area, yd^2:			
Length of pass, ft	13,500	15,000	17,500
Average width of pass, ft	6.5	7.9	9.3
Square yards of roadbed	9800	13,200	18,100
Length, ft:			
50-ft-wide roadbed for two-lane, two-way traffic	1800	2400	3200
100-ft-wide roadbed for four-lane, two-way traffic	900	1200	1600

NOTE: This table is suggested as a guide and not as a dictum, as there are many variables affecting the performance of a motor grader in haul road maintenance.

Figure 15-6 Medium-size tow-type grader maintaining haul road from cut to fill. Like the motor grader, the tow-type machine skims the surface to fine-grade cuts and fills of 1-in average depth. However, the cutting edge is at right angles to the direction of travel, and it does not drift the material from one side of the roadbed to the other as does the motor grader.

The 18-ft grader is being towed by a tractor of 340 engine flywheel-brake hp. The average cut-fill is about 1.0 in. The haul road is 2300 ft long and 50 ft wide, and the net time for three passes of a complete coverage is 22.5 min at an average speed of 310 ft/min, or 3.5 mi/h. The coverage is hourly, and during the remaining 37.5 min the machine spreads and mixes material on the fill. The rate of maintenance per 50-min working hour is 5100 ft of 50-ft-wide haul road.

The material is well-weathered granite of a freeway in the Merriam Mountains of southern California.

TABLE 15-3 Average Performances of Tow-Type Graders in Maintenance of Haul Roads

Item	Nominal rating, width, ft		
	16	18	20
Suggested engine flywheel-brake hp of wheels-tires tractor	300	350	400
Width dimensions, ft:			
Maximum	16.0	18.0	20.0
Width of pass, 2.0 ft overlap	14.0	16.0	18.0
Engine hp per foot of pass width	21	22	22
Average depth of cut-fill, in	1.0	1.0	1.0
Average travel speed, including one productive turn:			
ft/min	310	320	330
mi/h	3.5	3.6	3.8
Roadbed, maintained with complete coverage per 50-min working hour			
Area:			
Length of pass, ft	15,500	16,000	16,500
Average width of pass, ft	14.0	16.0	18.0
Square yards of roadbed	24,100	28,400	33,000
Length, ft:			
50-ft-wide roadbed for two-lane, two-way traffic	4300	5100	5900
100-ft-wide roadbed for four-lane, two-way traffic	2150	2550	2950

NOTE: This table is suggested as a guide and not as a dictum, as there are many variables affecting the performance of a tow-type grader in haul road maintenance.

15–14 Haul Road Construction and Maintenance

every 4 h, depending chiefly on material, hardness, and wetness of the surface. The average is about every 2.5 h, and the values for hourly productions of Table 15-3 are to be multiplied by the appropriate factor for frequency of coverage.

Estimate of Production and Costs The estimate is based on Figure 15-6.

Criteria:
18-ft grader towed by wheels-tires tractor with diesel engine of
340 flywheel-brake hp:

Haul-road length, ft	2300
Roadbed width, ft	50
Frequency of coverage	Hourly
Net time for complete coverage, min	22.5
Hourly production of scrapers using haul road, cybm	1100
Hourly production per 50-min working hour	
Coverages per hour, 50/22.5	2.22
Hours for one coverage, 1.00/2.22	0.45
Cost of machines, 1978:	
Tow-type grader, 18-ft	$22,000
Wheels-tires tractor, 340-hp	$114,000
Hourly cost of ownership and operation, 1978:	
Machines (Table 8-3):	
Wheels-tires tractor, 340-hp, 34% × $114	$38.40
Tow-type grader, 18-ft, 27% × $22	5.90
Operator	15.60
Total hourly cost	$59.90
Direct job unit cost:	
Cost of complete coverage, 0.45 × $59.90	$27.00
Unit cost per cybm of rock-earth hauled hourly, $27.00/1100	$0.025

The unit cost of $0.025/cybm compares favorably with $0.027 for the motor grader although there are differences in length of haul road, frequency of coverage, and cubic yardage hauled hourly over the haul roads.

Water Wagons

The application of water to a haul road accomplishes two purposes: first, it acts as a dust palliative and thereby reduces the operator's discomfort and increases safety; second, it helps to compact the surface of the roadbed and thereby decreases rolling resistance and increases production.

The amount of water applied per square yard of surface hourly varies greatly according to material of surface, natural moisture of material, humidity, rainfall, evaporation, and density of traffic. Because of these variables or imponderables, it is impossible to come by any empirical equation giving rate of application per square yard of surface. On most jobs the length of the haul road varies and thus another variable is present when an attempt is made to calculate costs of wetting as a function of cubic yards of material passing over the haul road.

In tropical and temperate regions there may be little or no water requirement. In desert and steppe regions the haul roads may require as much as 25 gal per cubic yard of excavation passing over the haul road. It is sufficient to say that water application changes daily according to the judgment of the competent foreman and superintendent.

Water wagons may be used for combination haul-road maintenance and embankment construction or they may be used exclusively for haul-road maintenance. On small jobs the machine is used in combination service, as are the wagons of Figures 14-15 and 14-17. On large jobs the machine may be worked exclusively on the haul road or in combination work, the kind and size being selected for the characteristics of the haul road. Some selections are illustrated in Figures 15-7 to 15-9.

As for the selection of on-the-road or off-the-road water wagons, one should always consider that haul-road maintenance is on-the-road work. It follows logically that, if the machine is to be used exclusively on the haul road, it is probable that an on-the-road–off-the-road water wagon would provide lower costs.

Chapter 14 contains specifications for water wagons and Chapter 8 provides data for the costs of ownership and operation.

Production The water wagon is simply a water hauler and the hauling cycle may be estimated in the following manner, similar to that used for rock-earth haulers.

Figure 15-7 Off-the-road 10,000-gal and on-the-road 3500-gal water wagon maintaining a haul road for freeway construction. These two machines take care of a four-lane two-way haul road of 6000-ft length, made up of the cut and the main line of the road. Combined hourly capacity is 33,800 gal for an area of 66,600 yd² of roadbed and excavation production of about 5100 cybm/h.

The water application rate is 0.51 gal/yd²/h and 6.6 gal/cybm hauled. When interpreting these rates one must keep in mind these important factors: average temperature, 90° F, reaching 105° F; average relative humidity, 40 percent; rainfall, none; haul road material, porous, well-weathered granite; traffic density, 374 haulers hourly.

The location is in the Laguna Mountains of San Diego County, California.

Loading time: From the pipe stand loading time in minutes equals capacity in gallons divided by flow rate in gallons per minute. From the movable storage tank, the approximate time in minutes equals the capacity in gallons divided by 4300 gal/min per foot of area of tank downspout.

The hauling and returning time, when not discharging, may be calculated by methods discussed in connection with Table 13-12.

The time for two turnings may be taken as averaging 0.9 min.

The discharge time is a function of the capacity of the centrifugal pump of the water wagon or the desired flow rate per square yard of roadbed in gallons per minute. Observations indicate that the average rate of discharge is approximately 0.67 of the pump capacity, the travel speed of the wagon when discharging being adjusted to the width of

15-16   Haul Road Construction and Maintenance

Figure 15-8 On-the-road–off-the-road 2000-gal water wagon maintaining haul road for freeway building. The machine is wetting down a two-way haul road 7000 ft long and 50 ft wide. The production of the rear-dump haulers is 664 cybm/h. About 6000 gal of water is applied hourly by three trips of the wagon, corresponding to 0.15 gal/yd^2/h and 9.0 gal/cybm hauled. The climate and weather conditions are identical with those of Figure 15-7. However, the material of the roadbed is weathered schists and phyllites of less porosity than the well-weathered granite. The averages for the work of Figures 15-7 and 15-8 are 0.33 gal/yd^2/h and 7.8 gal/cybm hauled hourly.

The maneuverability and traction of the two-axle load-on-back wagon fit it admirably into the confined work and the steep grades of the ramps. This freeway location is near Poway, San Diego County, California.

the roadbed and the desired rate of flow per square yard of roadbed. Pump capacities are given in the specifications for water wagons, Tables 14-4 and 14-5.

One must bear in mind that when a water wagon is assigned to maintaining a haul road it may not work to capacity simply because the haul road does not need such capacity for best results from wetting. The machine should be selected for greatest anticipated requirements and then the excess capacity may be used in wetting the embankment for compaction.

Estimate of Production and Costs   The example which will be used is the work of Figure 15-8. The haul road is 7000 ft long and 50 ft wide, and it is felt that two complete coverages hourly will be adequate for maintenance. For the contemplated 2000-gal water wagon, four trips hourly would supply 8000 gal and the rate of application would be 0.21 gal per square yard of roadbed hourly. In this work of entering a busy main haulage road and blending with heavy traffic which cannot be delayed, a 40-min working hour at 67 percent efficiency will be used to compensate for unavoidable traffic delays.

The criteria used are as follows:

   1. The portable tank is shown in Figure 10-2, the water wagon in the picture being a 10,000-gal machine used in the building of fills. Because of possible low pressure in the supply line the loading time for the water wagon is set conservatively at 2.5 min. The tank location is 500 ft west of the midpoint of the south to north main haulage road of 7000-ft length.

   2. Average rate of discharge of centrifugal pump of water wagon is taken at 67 percent of rated capacity of 1200 gal/min or at 800 gal/min.

   3. Haul road is approximately level except for short ramps in the cut. The rolling resistance is estimated to be 4 percent. It is planned to wet one-half of the haul road at a time by one load of 2000 gal. Lengths are as follows:

a. Tank to main line haulage road, 500 ft
   b. Average length of haul road wetted by one coverage of 2000 gal, 3500 ft
   c. Return distance to tank, 4000 ft
4. Water wagon data for use of Table 13-12:
   a. Diesel engine, 192 flywheel-brake hp
   b. Weights: GVW, 29,600 lb; tare weight, 13,000 lb
   c. Weight-to-power ratio: laden, 154; unladen, 67
   d. Diesel engine wheel hp, 76% × 192 = 146 hp
   e. Hauling speed, laden:

$$\text{Maximum speed} = \frac{146 \times 33{,}000}{0.04 \times 29{,}600} = 4070 \text{ ft/min} = 46 \text{ mi/h}$$

Average speed (Table 13-12)
53% × 4070 = 2160 ft/min = 25 mi/h

The average speed is too high because of the probable stop at the entrance to the main haul road. Use 12 mi/h.
   f. Returning speed, unladen:

$$\text{Maximum speed} = \frac{146 \times 33{,}000}{0.04 \times 13{,}000} = 9260 \text{ ft/min} = 105 \text{ mi/h}$$

Average speed (Table 13-12):

$$91\% \times 9260 = 8430 \text{ ft/min} = 96 \text{ mi/h}$$

The average speed is unreasonable for returning over a haul road. Use 20 mi/h. These theoretical-empirical unrealistic speeds illustrate the need for common sense and experience in adjusting the impracticable values determined by certain kinds of data.

Production is estimated in terms of the time for two complete coverages of the haul road, which is determined as follows:

Hauling cycle times, min, for one coverage of one-half of haul road:	
Loading under portable tank	2.5
Two turnings	0.9
Discharging 200 gal at 800 gal/min (travel speed when discharging averages 16 mi/h)	2.5
Hauling 500 ft at 12 mi/h	0.5
Returning unladen 4000 ft at 20 mi/h	2.3
Net cycle time	8.7
Net time for two coverages of complete haul road, min, 4 × 8.7	34.8
Gross time for two coverages of complete haul road, based on 40-min working hour, min, (60/40) × 34.8	52.2

The water wagon can comfortably maintain two coverages of the complete haul road hourly.

Costs, 1978:

Cost of water wagon	$32,000
Hourly cost of ownership and operation:	
Machine; water wagon, 2000-gal capacity (Table 8-3), 53% × $32	$17.00
Operator	11.20
Total hourly cost	$28.20
Cost of water, hourly, 8000 gal @ $0.50 per 1000 gal	4.00
Total hourly cost for water wagon and water	$32.20
Direct job unit cost, based on estimated production of 660 cybm hauled hourly, $32.20/660	$0.049

On this work it was observed that a 14-ft motor grader spent half of its working time maintaining the 7000-ft section of haul road, the other half being spent on an adjacent haul road for scrapers. The hourly cost of ownership and operation of the motor grader is $57.40. Accordingly, the combined total cost for haul road maintenance is

$$\frac{(\$32.20 + \$57.40/2)}{660}$$

which equals $0.092/cybm of rock-earth hauled over the road.

Figure 15-9 Maintaining haul roads in open-pit mine of the Mojave Desert with a 7200-gal water wagon. Wetting miles of wide haul roads in the desert requires lots of water. This huge amphitheater is the open-pit mine shown in the frontispiece. Borax ore and overburden are being taken out at the rates of 17 million tons annually and 2910 tons, or 1860 cybm, hourly over haul roads 15,900 ft long and 75 ft wide. The roadbeds of clayey silts and sands are smooth and hard with rolling resistance of about 3 percent.

The water wagon, two-axle load-on-back type with a 415-hp engine, has a 7200-gal capacity and it is one of three wagons with a combined capacity of 23,200 gallons. They average two loads hourly, applying 46,400 gal/h. The application is at the rate of 0.35 gal/yd^2/h, corresponding to 16 gal/ton, or 25 gal/cybm, of ore and overburden hauled over the haul roads. The location is Boron, California. *(D. Byrnes (ed.), "High Class Water," The Earth, April-May 1976, p. 4.)*

Figure 15-10 Excellent mountain haul road built to standards of a state highway. This haul road, built for earthen dam construction, connects the borrow pit for impervious material to the dam. It is 11,100 ft long and 45 ft wide with a uniform −6 percent grade. Because of the fine maintenance by motor graders and water wagons the 12 scrapers maintain the following production:

Hauling cycle times, min:	
Loading, two turnings, and dumping	2.0
Hauling at average 34 mi/h	3.7
Returning at average 16 mi/h	7.9
Net cycle time	13.6
Average delay due to bunching on long haul	2.3
Gross cycle time	15.9
Hourly production:	
Loads hauled per machine, 60/15.9	3.77
Cybm hauled per machine, 3.77×21.2	80
Cybm hauled by fleet of scrapers, 12×80	960

The location is on the west slope of the Sierra Nevada at North Fork, California, where Mammoth Pool Dam is being built.

IMPORTANCE OF GOOD HAUL ROADS

Mountain haul roads are expensive but sometimes the quantities of rock-earth to be hauled justify high design standards which equal those for a first-class highway. Such a highway for dam construction is illustrated in Figure 15-10.

Occasionally, especially for emergency construction, the haul road already exists, and such a one is shown in Figure 15-11. In this instance the haulers are subject to the regular legal restrictions for wheel and axle loadings. For compliance and for economy these haulers are three-axle load-on-back rear dump trucks with allowable GVW of 50,000 lb, resulting in payload of about 26,000 lb or about 7.0 cybm measurement of average-weathered granite.

The importance of a good haul road is strikingly illustrated by Table 15-4, which presents a cost comparison of a good with a hypothetical bad haul road.

SUMMARY

Construction and maintenance of a good haul road should be as carefully planned and executed as those of a good highway. The road should and can be the solid foundation

15–20   Haul Road Construction and Maintenance

(a)

(b)

Figure 15-11 Existing mountain highway used as haul road for emergency repairs of slide. The haul road is 2200 ft long and 40 ft wide with a uniform −8 percent grade. Three three-axle rear-dump trucks are being used in this tedious work with characteristic low production. The net hauling cycle times average 9.8 min but gross time, including unavoidable delays, is about 15 min. Hourly production for the three trucks averages only 80 cybm of average-weathered granite. *(a)* A descending laden truck and the slide area in the distance just above the truck. Laden speed is 9 mi/h and returning unladen speed is 10 mi/h. *(b)* The congestion and the unavoidable delays at the fill for dumping, bulldozing, wetting, and compacting the layers of the embankment are illustrated. The location is the Santa Margarita Mountains of San Diego County, California.

for low hauling costs. While close attention to haul roads is characteristic of large works, it is neglected sometimes in planning for small rock-earth excavations. This omission may be due to ignoring the fact that a small amount of money spent on the haul road will be more than offset by a large amount saved in hauling costs.

The virtually perfect haul road construction and maintenance of Plate 15-1 may be used as an example for a contrast between the good haul road illustrated and a hypothetical bad counterpart. The tabulation presented in Table 15-4 sets forth the considerations

TABLE 15-4 Savings of a Good Haul Road as Contrasted to a Hypothetical Bad Haul Road

Item	Actual good haul road	Hypothetical bad haul road
Criteria:		
Haul road:		
Length, ft	7100	7100
Width, ft	100	100
Grade, average, %	−7.0	−7.0
Average travel speed of scrapers, bottom-dump haulers, and rear-dump haulers, mi/h:		
Actual	18	
Estimated		12
Average hourly production hauled over road, cybm	4070	4070
Hauling cycle times, min:		
Loading, two turnings, and dumping	3.2	3.2
Hauling and returning 14,200 ft:		
At avg 18 mi/h	9.0	
At avg 12 mi/h	. . .	13.5
Net cycle time	12.2	16.7
Direct job unit costs, estimated per cybm:		
Relative, in terms of hauling cycles:		
12.1-min cycle	$0.093	
16.7-min cycle	. . .	$0.127
$\left(\dfrac{16.7 \times \$0.093}{12.2}\right)$		
Estimated decrease because of lesser costs for maintenance of bad haul road, $98.00/4070	. . .	0.024
Estimated increase because of greater costs for maintenance of haulers	. . .	0.014
Comparative totals	$0.093	$0.117
Savings by good haul road (20%) per cybm	$0.024	
Savings for 9.8 million cybm passing over haul road	$235,000	

NOTE: The model of a good haul road is the long, wide boulevard of Plate 15-1.

and the estimated savings made possible by the good haul road.

Whether the job be little or big, a good haul road is the most important consideration for low haulage cost. In the example of Table 15-4 the unit bid price for excavation is $1.22/yd^3. The calculated saving by the good haul road is $0.024/yd^3. The total saving is $235,000, or 2.0 percent of the total bid price for the 9.8 million yd^3 of rock-earth passing over the haul road.

CHAPTER 16

Allied Operations, Machinery and Components, and Facilities

PLATE 16-1

Complete lubrication truck for machines in dam construction. Rock excavation machinery runs mostly by fuel oil and the parts float literally on lubricating oils and greases. The components of the large machines are generally oiled and greased every shift. This heavy-duty lubricating truck threads its way through a maze of ridges and creeks in order to service 91 machines, most of which are serviced every shift. The two-operator crew services a heavyweight tractor-ripper-bulldozer in 10 min. The machine carries an air compressor, 8 pumps, 11 tanks, 8 hose reels, and many storage boxes for special lubricants. It does not refuel the machines, although it does carry a small supply of fuel oil for emergency use.

The location is Pyramid Dam on Piru Creek, Los Angeles County, California. *(Lincoln–St. Louis–McNeil Corp.)*

WELLPOINT SYSTEMS

Often the overburden atop rock excavation is water-bearing. The overburden may be alluvia, such as permeable silts, sands, and gravels, or it may be semipermeable residuals, such as mixed clays and weathered rock-earth. In either case, the wet materials must be removed to get down to the elevation of the semisolid and solid rock. In open-cut excavation there are three principal means for handling the flow of water so that systematic excavation may proceed.

1. As the excavation proceeds, small sumps may be created and the collecting water may be pumped out of the excavation area. This method creates clutter and messiness in the work area and it is generally not satisfactory except for minimum amounts of percolating water.

2. The saturated material or simply water may be held back by *cofferdams*, as described in the following section. This method is applicable to both permeable and semipermeable materials.

3. Water may be removed from permeable materials around the periphery of the excavation prism by a *wellpoint system*, which prevents water from filtering into the mass of excavation. This method is used commonly for foundation excavations, such as those for buildings and dams.

Principles, Methods, and Machinery

The principles are those of a well and pumping system. A series of in-line small-diameter wells is driven around the periphery of the prism of excavation, from which percolating water is to be shielded. The wells remove the water before it can filter into the prism. The principles are illustrated in Figures 16-1 through 16-4.

The parts of the wellpoint through which the water is drawn are shown in Figure 16-1. The particular configuration is for permeable materials such as sand and gravel alluvia.

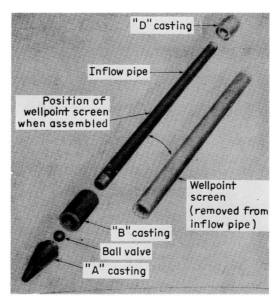

Figure 16-1 Parts of a wellpoint. The point A has an opening through which water is pumped to jet the wellpoint assembly into position below the water table. The ball valve prevents water and material from entering the inflow pipe during the unwatering of the material. The inflow pipe is perforated to receive water through the surrounding screen. The casting D connects to a riser pipe leading up to the valve and header or manifold pipe. The wellpoint screen excludes material during the unwatering. The internal diameter of the assembly varies from 1½ to 2 in. The flow of water varies from about 3 to 30 gal/min. *(Griffen Wellpoint Corp.)*

16-4 Allied Operations, Machinery and Components, and Facilities

The wellpoint is jetted into position by pumping water through the opening at the bottom of the point.

When unwatering semipervious materials such as silts and well-weathered rocks, it is necessary to surround the wellpoint by a cylinder of sand or fine gravel so as to prevent clogging of the screen. This is done by jetting down a large pipe of about 10-in diameter, removing the material within the pipe by flushing, inserting the wellpoint with its riser, backfilling the pipe with sand-gravel, and then pulling the large pipe or casing, leaving the wellpoint surrounded by filter material. The diameter of the inflow pipe and the riser is 1½ or 2 in and the point is of somewhat greater diameter.

The wellpoints around the excavation are spaced on from 2- to 6-ft centers, depending on the anticipated rate of flow of water through the material. Figure 16-2 illustrates the arrangement of risers, crossovers, valves, header, pump, and discharge line.

The wellpoints are driven down below the low level of saturation so as to ensure satisfactory unwatering of the material during excavation. The length of the wellpoint assembly, including riser pipe, should not exceed 24 ft because of the limitation of the suction depth of the centrifugal pump, if the intake of the pump is assumed to be at the same elevation as the top of the riser. When the depth of material to be unwatered does exceed 24 ft, one or more additional independent stages are added, as shown in the three stages of Figure 16-4. The wellpoints of a series are equipped with valves for individual control and are connected to a header pipe, which leads to the intake of the pump.

The schematic operation of a wellpoint system is shown in Figure 16-3. A system may include a few tens of wellpoints or a few hundreds. The one of Figure 16-4 includes 230 wellpoints for an average depth of about 60 ft from water table to bedrock.

The rate of flow from a wellpoint varies within wide limits, ranging from 3 to 30 gal/min. The maximum flow rate for the 230 wellpoints of Figure 16-4 is 3000 gal/min, the average flow through fine-medium-coarse sand being 13 gal/min per wellpoint. The pump for a stage may be engine- or motor-driven and it runs on an around-the-clock basis. As availability must be 100 percent, because percolation is continuous, a standby pump is necessary.

Like several other special operations attending rock excavation, the design, installation, and working of a wellpoint system are for the experienced people of this specific

Figure 16-2 Arrangement of elements of wellpoint system. Risers, coming up from the wellpoints, are connected to the crossovers which join the header pipe. Each crossover is equipped with a valve to regulate the flow of water from the individual wellpoint. The header carries a series of close-centered short risers for use with the crossovers, thus allowing for any desired spacing of the wellpoints.

The centrifugal pump, as well as its counterpart in the distance, has its large-diameter inlet crossover pipe from the header, control valves, and discharge pipe. The pumps are mounted on wheels so that a standby pump may be moved readily into position for mandatory continuous pumping. *(John W Stang Corp.)*

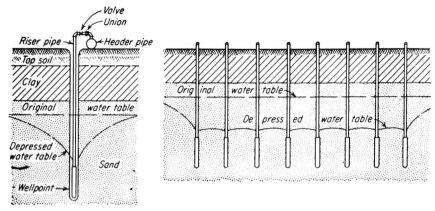

Figure 16-3 Schematic elevation views of work of wellpoint system. The elevation views of the depressed water table show its undulating profile. If the wellpoints were spaced farther apart, the average level of the lowered water table would be higher. The spacings are a function of the permeability of the material, and the practicable and economical distances are determined by experience in the particular or a related material. Short spacings, 2 to 4 ft, are for semipermeable materials such as residuals, silts, and well-weathered rock-earth. Long spacings, 4 to 6 ft, are for permeable sands and gravels. *(Griffen Wellpoint Corp.)*

field of endeavor. As has been stated, the dynamics are those of a water system. Customarily, many premises of design must be tempered by experience. If the excavator's workers are not completely practiced, the proposed wellpoint system should be turned over in its entirety to the professionals in the field of design and installation of such systems.

Estimated 1978 Cost of Unwatering Foundation Excavation

The cost calculations are illustrated below for the work of the three-stage wellpoint system of Figure 16-4.

Criteria:
  Time for unwatering:
    Days of 24-h continuous operation   133
    Total hours   3192
  Estimated excavation of fine to coarse sand, unwatered and removed, cybm   187,000
Cost of machinery, 1978:
  Pipe with fittings:
    Wellpoints, 2-in diameter, 230 @ $66.00 $ 15,180
    Risers and crossovers, 2-in diameter, 6440 ft @ $1.80   11,592
    Headers and crossovers
      8-in diameter with 2-in-diameter short risers, 2 ft c to c, 670 ft @ $9.10   6,097
      10-in diameter with 2-in-diameter short risers, 2 ft c to c, 330 ft @ $14.40   4,752
    Discharge:
      8-in, 1200 ft @ $5.10   6,120
      10-in, 900 ft @ $7.90   7,110
    Valves:
      Gate or key, 2-in, 230 @ $10.00   2,300
    Gate:
      8-in, 4 @ $510.00   2,040
      10-in, 2 @ $800.00   1,600
    Foot:
      8-in diameter, 4 @ $260.00   1,040
      10-in diameter, 2 @ $410.00   820
    Subtotal for pipe with fittings $ 58,651
  Centrifugal pumps, diesel-engine-driven, skid-mounted:
    Four 8-in, 80-hp @ $18,000 $ 72,000
    Two 10-in, 120-hp @ $21,400   42,800
    Subtotal for pumps   114,800
  Total cost of machinery $173,451

16–6 Allied Operations, Machinery and Components, and Facilities

Hourly cost of ownership and operation:
 Machines
  Pipe with fittings (Table 8-3):
   Wellpoints, 2-in diameter, 35% × $15.2 . . . . $ 5.30
   Riser and crossover pipe with valves, 2-in diameter, 35% of $13.9 . . . . . 4.90
   Header, crossover, and discharge pipe with valves:
    8-in diameter, 30% × $15.3 . . . . 4.60
    10-in diameter, 30% × $14.3 . . . . 4.30
   Subtotal for pipe with fittings 19.10
  Centrifugal pumps, diesel-engine-driven, skid-mounted (Table 8-3):
   Working:
    Two 8-in, 2 (46% × $18.0) 16.60
    One 10-in, 48% × $21.4 . . . . 10.30
   Standby:
    Two 8-in, 2 (20% × $18.0) 7.20
    One 10-in, 20% × $21.4 . . . . 4.30
   Subtotal for pumps 38.40
  Total for machines . . . . $ 57.50
 Operators:
  3 pumpmen @ $13.80 and 4 pipelayers @ $11.80 . . . . $88.60
 Assembly and disassembly of wellpoint system:
  Estimated at 40% of cost of pipe with fittings, $58,651, prorated over
   3192 h of operation . . . . $ 7.40

$$\frac{(40\% \times \$58.651)}{3192}$$

 Total hourly cost $153.50
 Total cost for unwatering, 3192 × $153.50 . . . . $489,972
 Direct job unit cost per cybm of estimated foundation excavation,
  $489,972/187,000 . . . . $2.62

Alternative Dewatering Methods for Foundations

In addition to wellpoint systems there are two other means for stabilizing wet material surrounding a prism of excavation. One is mechanical, by freezing, and the other is electrical, by electroosmosis.

Freezing When the surrounding material is alluvium such as silts, sands, and gravels containing freely percolating water, a vertical wall may be frozen. This barrier is equivalent to a cofferdam. The elements of the system are:

1. A series of freezing units around the periphery of the prism. Each consists of a vertical pipe for delivery of the brine to the desired depth and an enclosing larger pipe for the return of the brine to the surface. The small-diameter injection pipe is equipped with a valve to regulate the flow of brine in order to maintain a uniform rate of freezing.

2. A header pipe around the excavation area to deliver brine from the refrigerating plant to the injection pipes.

3. A header pipe around the prism to return the brine from the return pipes to the refrigerating plant.

4. Pumps for circulation of the brine.

5. Refrigerating plant.

The system has been applied to deep excavation within a small area, such as mine shafts through water-bearing sands and gravels, but it has not been applied widely to rock-earth excavation. It is a highly specialized method requiring expert design and installation.

Electroosmosis The stabilization or hardening of the barrier involves a complex phenomenon in the field of electrophysics and electrochemistry. Metal electrodes are placed in the soil and direct current is passed between the electrodes until the soil is hardened by the drawing of free water from anode to cathode. The anodes are usually steel piling, rods, old rails, and the like. The cathodes are generally wellpoints for pumping out the discharging water. The anodes and cathodes are placed on opposite sides of the barrier to be dewatered. The method is applicable to fine-grained soils, such as clay-silts, which are too fine for successful operation of a wellpoint system.

The theory, design, installation, and operation of an electroosmosis system must be undertaken by companies with experience in the particular field. Like the freezing

Figure 16-4 Three-stage wellpoint system for unwatering foundations of a dam. The large system is unwatering approximately 187,000 yd³ of fine to coarse sand excavation. The work requires 133 days, or 3192 h, of round-the-clock operation. The crew averages seven pumpmen and pipemen.
The wellpoint system is made up of the following important components:

Number of 2-in-diameter wellpoints, on 4.3-ft centers	230
Average length of 2-in-diameter risers with crossovers, ft	28
Total length of header and crossover pipe to pumps, ft:	
8-in	670
10-in	330
Total length of discharge pipe, ft:	
8-in	1200
10-in	900
Number of valves:	
Gate or key, 2-in diameter	230
Gate:	
8-in diameter	4
10-in diameter	2
Foot:	
8-in diameter	4
10-in diameter	2
Centrifugal pumps, diesel-engine-driven:	
8-in:	
Working	2
Standby	2
10-in:	
Working	1
Standby	1

The maximum inflow for three pumps is 3000 gal/min, or 13 gal/min per wellpoint. The location is Grayson Dam, Grayson County, Kentucky. *(Moretrench American Corp.)*

method for creating a cofferdamlike barrier around the excavation, it is rarely used in rock-earth excavation.

COFFERDAMS

A cofferdam is a watertight enclosure, from which wet material or water is removed to expose the impermeable bottom for excavation or for another kind of work. The word cofferdam derives from the French word *coffre,* meaning "box" (originally "basket"), and the Dutch word *dam,* meaning "weir." There are two kinds of cofferdams, those for

16-8 Allied Operations, Machinery and Components, and Facilities

holding back land and those for holding back water, and both are used in rock-earth excavation.

Land Cofferdams

Figure 16-5 illustrates a small cofferdam for restraining wet material. The sheet piles may be of wood or steel. Generally the structure is removed after the excavation is completed.

There are several types of both wooden and steel piles. Wooden piles may be single planks with butt joints, single with tongue and grooved joints, single with splined joints, or double or triple with lapped joints. The sheet piles are generally 2 in $\times$ 12 in or 4 in $\times$ 12 in. The type and size depend on the desired strength and the degree of water tightness desired for the cofferdam. Steel piles may be straight web, arch web, deep arch web, or Z-type, all being interlocking.

The design for a land cofferdam is basically the same as that for a dam, the lateral pressure ranging from a minimum for stable material with little moisture to a maximum for unstable saturated material. For saturated material, such as liquid mud with immersed rocks, the angle of repose can approach 0° and the weight can reach 120 lb/ft^3. In such a free-flowing mixture the pressure against the cofferdam would be 1.92 times that for water only. The dynamics of design are similar to those for a water cofferdam, as illus-

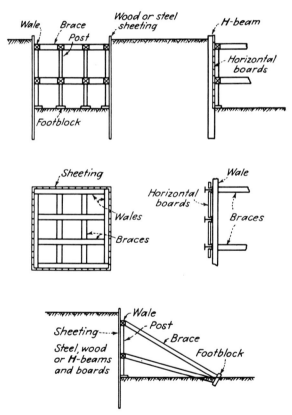

Figure 16-5 Small land cofferdams. These simple cofferdams are used in water-bearing soils for building foundations, bridge abutments, and the like. Wooden or steel sheet piles or horizontal boards with steel H-beams may be used. The upper drawing illustrates the horizontal bracing used for small cofferdams with a greatest horizontal dimension of about 75 ft. The lower drawing shows the angle or knee bracing used for structures of greater lateral dimensions. Commencing with first lift of excavation and upper bracing, successive lifts and bracings are carried on alternately until the desired depth for the work is reached.

trated in Figure 16-7, except that the horizontal force on the face of the cofferdam is a function of the cosine of the angle of repose and the weight of the material.

When the cofferdam is small, up to about 75 ft in diameter or greatest lateral dimension, the bracing is carried horizontally across the enclosure to the opposite side, as illustrated in the upper part of Figure 16-5. However, if a larger area is being excavated, the bracings are angled down to the footblocks, as shown in the lower part of the drawing.

The construction of a cofferdam starts with driving or jetting the piling all around the excavation prism to the desired depth. If steel piling is used, the entire ring of interlocking piles is set up before driving and then a few piles are driven at a time until full depth is reached. If the piles were driven without the preliminary setup, it would be impossible to make the closure with the final pile. After all piles have been driven, excavation is commenced and carried down to a practicable and safe depth before the first bracing is set. Thereafter alternate excavation and setting of bracing proceed until full depth is reached.

Water Cofferdams

Figure 16-6 illustrates a small water cofferdam. Except for shallow depths of water, less than 10 ft, where wooden piles may be used, steel piles are employed. The construction procedure is as follows: First, all piles are driven to the desired depth; next the top frame of horizontal bracing is placed down to about the water level; and then the cofferdam is unwatered. As the water level drops, the steel piles bend inward, as shown by the curved lines of Figure 16-6. This curvature tightens the interlocks of the piles and reduces the leakage of water. As the water level drops, additional frames of bracing are added if necessary. Finally, the material is excavated and bracing is installed alternately until the desired grade is reached, as in the case of the land cofferdam.

When large cofferdams are to be built several kinds may be used, depending on the nature of the material on which the cofferdam is to rest, the depth of water, and the purpose of the cofferdam. Five kinds are shown in Figures 16-7 and 16-9.

Unless a cofferdam is small, as for inconsiderable building foundations, bridge piers and foundations, and the like, the design and construction should be meticulous. Such careful construction is typified by Figure 16-9, showing construction of the upstream cofferdam for a huge dam. Such a cofferdam in itself is sometimes a sizable rock-fill dam.

Fundamentals of Cofferdam Design

Cofferdams are essentially either retaining walls, as shown in Figures 16-5, 16-6, and 16-7 c, or dams, as shown in Figure 16-7b, d, e and Figure 16-10. As such, the stresses in their members are statically determinate and their designs are not unduly complicated. For example, in Figure 16-5 the sheeting and wales are continuous beams, the braces are horizontal or inclined struts or columns, and the posts are lightly loaded columns. The

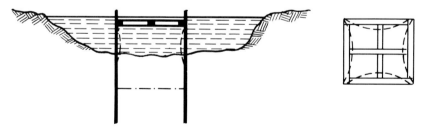

Figure 16-6 Small water cofferdam. This cofferdam is the kind used for works such as small shallow bridge piers and abutments, where excavation is in an open cut and where it is not necessary to use caissons.

Down to a depth of about 10 ft wooden piles or sheeting may be used and at greater depths steel piles are used. The dashed lines of the drawing shows the bending of the steel piles and if the bending is excessive additional horizontal bracing may be installed. After unwatering, alternate excavation and setting of bracing proceed until the excavation is brought down to the desired grade.

Like its small land-cofferdam counterpart, the size is limited by the practicability of the length of span of horizontal bracing. Beyond this span of about 75 ft, cofferdams similar to those of Figures 16-7 and 16-9 are used, as well as rock-fill dams.

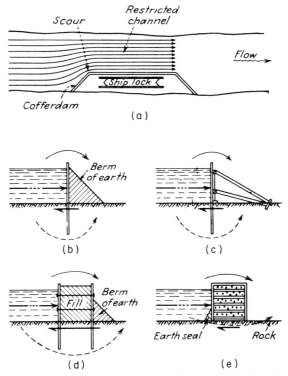

Figure 16-7 Four types of water cofferdams for ship lock construction. The selection of the right cofferdam for this work of diverting the stream and unwatering the area for working depends on several factors. Principal among these are the depth and rate of flow of stream to be anticipated during the period of construction, the nature of the streambed, and the width of the cofferdam if it has an effect on the width of the restricted channel or the working area. If the water is shallow and the streambed is earthen, types *b* and *c*, with adequate depth of piling to prevent seepage, are applicable. If the water is deep and the streambed is earthen, type *d* with sufficient depth of piles is adequate. If the water is shallow or deep and the streambed is rock, type *e* is the one to be used.

These four types, together with that of Figure 16-9, suggest the possibilities of economies in cofferdam selection, which should be investigated completely before commencing the costly construction.

stresses in all these members, based on the forces exerted against the face of the cofferdam, may be readily calculated. An example of calculations for a dam-type cofferdam is given in Figure 16-8. The design calculations for this dam of rectangular profile are similar to those for the circular-type cellular cofferdam of Figure 16-9.

The forces acting on a cofferdam are the lateral pressure of the earth and water being restrained, the weight of the dam, and the upward reaction of the underlying earth. The cofferdam with its structural parts should be designed to withstand these forces, as well as the effects of scouring water, flowing water, and seepage of water under the cofferdam.

The selected height for a cofferdam is a most important consideration, as the cost varies roughly as the square of the height. For land cofferdams the height is fixed. For water cofferdams there is a most economical height, depending on the following considerations for different heights.

1. Probable frequency of floods overtopping the cofferdams during the time of construction (from data available in hydrological records of public agencies)
2. Cost of cofferdam
3. Cost of repairs to construction work as a result of overtopping of the cofferdam
4. Combined costs of cofferdam and repairs to work as a result of flooding

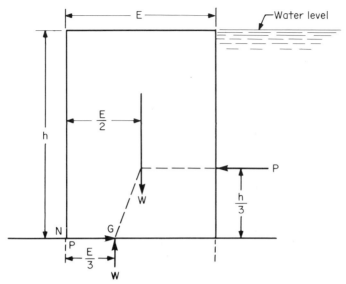

Figure 16-8 Resolution of forces acting on cofferdam of dam type.
W = weight of 1 linear foot of cofferdam filled with earth or with rock-earth. W = approximately $(E \times h \times 100)$ lb.
P = static pressure of water against 1 linear foot of cofferdam, acting at the center of pressure located at point ⅓ h, ft

$$P = \frac{h^2 \times 62.4}{2} = (h^2 \times 31.2) \quad \text{lb}$$

G = economy-safety location where the resultant of forces W and P passes through the base of the cofferdam and where the center of the resisting upward force, also W, passes through the base. From the point of rotation, N, the distance to G equals ⅓ E.
Taking the algebraic sum of all three moments about N, the point of rotation, and equating the sum to 0 at equilibrium:

$$\left[(E \times h \times 100)\frac{E}{2}\right] - \left[(h^2 \times 31.2)\frac{h}{3}\right] - \left[(E \times h \times 100)\frac{E}{3}\right] = 0$$

$$E = 0.79h$$

The value of E is conservative as it neglects the stabilizing effects of the weights of cofferdam parts and the resistance of the skin friction of the piling. It is not conservative in that it also neglects the possible additional pressure of moving water against the face of the cofferdam and the possible uplift of seeping water beneath the cofferdam.

A suggested practical equation is simply: $E = h$.

After these data are amassed, one selects that height of cofferdam which gives the lowest combined cost for the cofferdam and repairs to work as a result of flood damage during the period of construction.

Figure 16-9 illustrates the stages of constructing a circular-type cellular cofferdam. Figure 16-10 shows the completed cofferdam, made up of some 100 cells, interconnecting dikes, and floodgates.

Another kind of cellular cofferdam is the diaphragm type. It is made up of two parallel walls consisting of equal length arc segments placed opposite each other, with the outside of the arcs facing upstream and downstream. The opposite intersections of the arcs are connected by transverse straight diaphragms. The filing is sheet steel, Y-piles being driven at the intersections of the arcs so that the piles of the diaphragms may be connected. The cells are filled with suitable material but they must be filled uniformly to avoid bending the piles of the diaphragms because of the unequal lateral pressure of the backfilled material.

16-12   Allied Operations, Machinery and Components, and Facilities

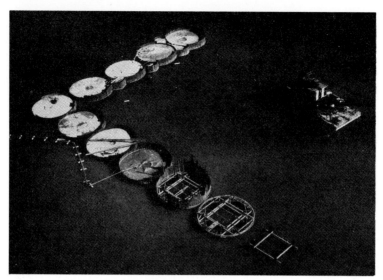

Figure 16-9 Stages of construction of circular-type cellular cofferdam. The cells are 50 ft in diameter. The steel-sheet piling averages 60 ft in length, driven to refusal into the basalt-rock riverbed. Construction proceeds clockwise in the picture in this sequence:

1: The H-piles for the prefabricated circular template are driven in squares and braced with struts and supports.

2: The template for the sheet piles is placed over the square of H-piles.

3: The sheet piles are driven around the template. The template is lifted out of the cell and placed around the square of step 1 to continue another cycle of construction.

4: The tops of the sheet piles are cut off, with the higher side upstream. The cell is backfilled with dredged sands and gravels.

5-8: The cells are completed except for connecting arcs.

9-11: The cells are completed with the addition of connecting arcs and completion of backfills. The cells are higher on the upstream side to afford greater protection from spring floods.

The location is McNary Dam across the Columbia River east of Umatilla, Oregon. *(Guy F. Atkinson Co.)*

Just as dam design and construction are costly and painstaking, so also are the planning and building of cofferdams. Whether they are small or large, for land or for water, their cost is a sizable percentage of that for the work which they protect.

Sometimes the engineers, estimators, and builders of excavation companies are not familiar with this special field of endeavor. In such cases, and especially if sophisticated design and construction are involved, one should use outside expertise. This experienced help can come from foundation engineering companies and from the steel companies furnishing the steel piling. The alternative is simply to sublet the work to a pile-driving company.

The important considerations for the best cofferdam are safety and low cost and these requisites are not necessarily incompatible, as a well-thought-out design combines both objectives.

PUMPS FOR UNWATERING FOR EXCAVATION

These small to medium-size pumps are of two kinds: centrifugal, including suction and submersible types; and diaphragm, for handling water containing large quantities of mud, small rocks, sludge, and trash. In some cases large centrifugal pumps are used to unwater flooded open pits and mines in which the quantities of water are abnormal.

Performances of and specifications for single-stage centrifugal pumps are given in

Figure 16-10 Large circular-type cellular cofferdam for dam building. The huge closure for unwatering the work area of some 60 acres includes 100 cells and a section of dike around the periphery of the work area. The picture is taken from a point facing in a northerly direction. The east side of the cofferdam is upstream. The completed spillway of the dam is in the upper right-hand corner of the picture to the northeast. The partly completed powerhouse is in the lower left-hand corner of the picture to the southwest. The dam remains to be completed.

The individual cells and their connecting arcs are built of steel-sheet piles. The cell diameter is 50 ft, and the cells are backfilled with alluvium. They are designed on the principles of gravity dams, as illustrated in Figure 16-8. The phases of cell construction are illustrated in Figure 16-9. The cellular cofferdam is a solid integral structure, which will withstand overtopping by floodwaters without damage. Floodgates are installed to permit flooding of the work area before overtopping of the cofferdam. Thereby excessive damage to the work area is prevented.

The location is McNary Dam across the Columbia River east of Umatilla, Oregon. *(R.L. Peurifoy, Construction, Planning, Equipment, and Methods, McGraw-Hill, New York, 1956, p. 406.)*

Chapter 10. Performances and specifications for submersible centrifugal pumps and for diaphragm pumps are given in Tables 16-1 and 16-2, respectively.

Centrifugal Suction Pumps

Figure 16-11 illustrates the unwatering of a quarry. These pumping operations are common in the springtime, when many quarries start up after shutdowns during the cold winter months.

Centrifugal Submersible Pumps

These pumps are submerged to the bottom, where they operate under a static or flooding head, and thus they do not work by suction. They are powered by electric motors. Figure 16-12 illustrates two large-capacity pumps unwatering for dam construction and for quarry work. Figure 16-13 shows the sump and discharge stream of a submersible pump working in an open-pit mine.

Diaphragm Pumps

Diaphragm pumps, which lift water by suction, are used to handle either clean water or water containing mud, sands, small rocks, and miscellaneous trash. The greatest dimension of the contained material may be up to two-thirds of the size of the intake of the pump. The diaphragm pump is self-priming and is sometimes equipped

TABLE 16-1 Representative Specifications and Performances of Centrifugal Submersible Pumps

Specification	Nominal discharge diameter, in			
	2	4	6	8
Pump:				
Discharge, diameter, in	2	4	6	8
Impeller, diameter, in	11 11/16	23 1/4	28	28
Speed, r/min	3450	1750	1750	1750
Motor, electric, ac, 60 cycle:				
Horsepower	2	25	60	95
Speed, r/min	3450	1750	1750	1750
Phase	Single	Three	Three	Three
Voltage	115–230	230–460	230–460	230–460
Complete machine:				
Cable length, ft	50	50	50	50
Strainer:				
Open area, in^2	38	48	48	48
Openings, size, in	1/4	3/8	1	1
Dimensions, in:				
Height	23	46	57	57
Diameter	16	32	39	39
Weight, approximate, lb	50	150	450	800
Performance in gal/min for total head of:				
20 ft	130	930	2160	2800
40 ft	100	870	1950	2720
60 ft	70	780	1720	2570
80 ft	. . .	680	1480	2370
100 ft	. . .	550	1100	2120
120 ft	. . .	360	400	1800
140 ft	. . .	. . .	. . .	1250
160 ft	. . .	. . .	. . .	600

Figure 16-11 Two 4-inch centrifugal pumps unwatering a quarry. These pumps are driven by 20-hp electric motors and are working under a suction head of 10 ft and a total head of 30 ft. The final suction head will be 25 ft, giving an average suction head of 18 ft and an average total head of 38 ft. At 1750 r/min and 55 percent efficiency, each pump delivers 550 gal/min, so that the delivery is 1100 gal/min for both pumps. Working around the clock the pumps can pump out an acre-ft of water, 3,260,000 gal, in 50 h. If the area to be unwatered in this quarry is 200 ft × 300 ft, with a water depth of 15 ft, the unwatering time will be 204 h, or 8.5 days.

The quarry is located near Springfield, Ohio. *(Gorman-Rupp Co.)*

TABLE 16-2 Representative Specifications and Performances of Diaphragm Pumps

Specification	Nominal discharge diameter, in		
	2	3	4
Pump, single-acting:			
Strokes per minute	60	60	52
Length of stroke, in	2½	2¹³⁄₁₆	3¾
Usable diameter of diaphragm, in	7	11½	12¾
Displacement, in^3	71	221	381
Prime mover:			
Diesel or gasoline engine, hp	3	6	10
Electric motor, ac, 60-cycle:			
Horsepower	½	1½	3
Speed, r/min	1750	1750	1750
Phase and voltage	Single 115–230 Three 230–460	Single 115–230 Three 230–460	Single 115–230 Three 230–460
Complete machine:			
Weight, wheels-tires-mounted, lb			
Diesel engine	230	350	500
Gasoline engine	150	200	400
Electric motor	150	200	350
Performances, gal/min versus total head:			
5-ft suction head with discharge head of:			
5 ft	26	78	125
15 ft	22	66	113
25 ft	18	64	104
15-ft suction head with discharge head of:			
5 ft	21	68	97
15 ft	19	64	74
25 ft	17	60	70
25-ft suction head with discharge head of:			
5 ft	18	56	82
15 ft	16	60	66
25 ft	15	56	52

(a)

(b)

Figure 16-12 Large centrifugal submersible pumps in typical construction and quarry work. *(a)* The 8-in pump, powered by a 95-hp electric motor at 1750 r/min, is dewatering an interception trench ahead of a construction area for Dillon Dam, Colorado. With a low total head the output is 2750 gal/min, sufficient to unwater 0.051 acre·ft of water hourly. *(b)* The 6-in pump, powered by a 95-hp electric motor at 1750 r/min, is dewatering a quarry. The total head is 113 ft. Discharge is through a manifold with four 4-in-diameter hoses. The pump handles 1600 gal/min, sufficient to remove 0.029 acre·ft of water hourly. The simplicity, compactness, and low weight of these electric-motor-driven pumps fit them for both emergency and sustained work under a variety of adverse working conditions. *(Gorman-Rupp Co.)*

Pumps for Unwatering for Excavation 16-17

Figure 16-13 Intake sump and discharge channel of centrifugal submersible pump. The 8-in pump is unwatering an open-pit mine. While almost all open-cut mines are susceptible to flooding by surface or groundwater, those which are located in flat or rolling country are most liable. Typical of these are quarries and coal stripping mines in the central states, where springs and perched water tables bring groundwater and rains and melting snows bring surface water. Natural or artificial sumps provide resting places for the pumps which discharge into downgrade watercourses, as shown in the picture, or to a streambed located at a higher elevation. The pump is working under a low total head of about 20 ft. With its 95-hp motor, its capacity is approximately 2800 gal/min. The location is a coal stripping mine in southeastern Ohio. *(Gorman-Rupp Co.)*

with a suction accumulator, which provides more efficient operation and smoother pumping flow. Because of their ability to handle many materials along with the water, they are used extensively under most difficult conditions for dewatering operations.

Figure 16-14 illustrates the operating principles and working parts. Parts are few, as the flexible diaphragm acts as a complete piston, the other working parts being the plunger rod, a gear reduction box with eccentric for the plunger rod, and suction and discharge check valves. Figure 16-15 illustrates a 4-in pump with electric-motor drive working in a mudhole, a most difficult task.

Estimate of Production and Costs for Dewatering by Diaphragm Pump The example is the pump of Figure 16-15. The criteria for the work are:
1. 4-in pump, powered by a 3-hp electric motor
2. Suction head, 5 ft; discharge head, 5 ft

Production (from Table 16-2) is 125 gal/min.
Costs, 1978:
    Cost of machinery:
        Pump $1990
        Suction hose, 4-in, 50 ft   66
        Discharge hose, 4-in, 100 ft   88
        Total cost $2144
    Hourly cost of ownership and operation (Table 8-3):
        Diaphragm pump, 4-in, electric-motor-driven, $57\% \times \$2.0$ $1.10
        Hose, 4-in diameter, $41\% \times \$0.2$ 0.10
        Total hourly cost $1.20

    Direct job unit cost per 1000 gal, $\dfrac{\$1.20}{(125 \times 60)/1000}$ $0.16

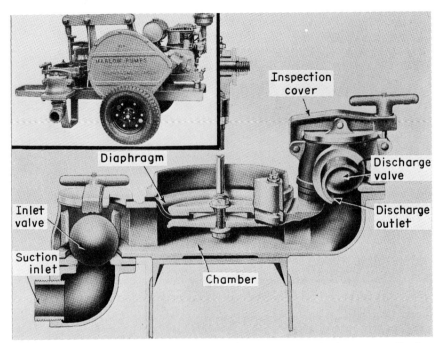

Figure 16-14 Operation, working parts, and approximate production of diaphragm pump. The flow of water is from left to right. When the diaphragm is on the upstroke or on suction, the valve on the suction side opens and the valve on the discharge side closes. On the downstroke or discharge the suction valve closes and the discharge valve opens. The stroking action is by the plunger rod, actuated by a crank arm or eccentric. The pump works at about 60 strokes per minute. As the displacement of the diaphragm is constant, the output is fixed except for loss by slippage through the valves. As the pump handles all kinds of materials, which cause valves to seat improperly, the slippage varies considerably. An estimate of average efficiency is 85 percent.

The approximate output of a diaphragm pump is given by the equation:

$$\text{gpm} = 0.0023 \times \text{spm} \times S \times D^2$$

where gpm = output, gal/min
spm = strokes per minute
S = stroke, in
D = usable diameter of diaphragm, in

(Marlow Pumps.)

Assuming that the area and depth of the dewatering work are (50 ft × 50 ft) and 5 ft, there are 93,500 gal. The time for unwatering is 12.5 h and the cost is $15.00.

Magnitude of Large Unwatering Jobs

Unwatering by centrifugal suction, centrifugal submersible, and diaphragm suction pumps is troublesome, time-consuming, and costly and it interferes with the normal excavation schedule. Accordingly, the work must be done in an efficient manner with well-selected machinery. Of equal importance is the provision of a contingency cost, as part of the estimate for rock excavation, to take care of the anticipated unwatering during the period of work.

Large unwatering jobs for open-pit quarrying and mining may run into hundreds of millions and hundreds of billions of gallons. Such work requires batteries of large centrifugal pumps and indeed may be the controlling factor in determining the feasibility of the entire project. Such a dewatering job is illustrated in the water hole of Figure 13-68, which was dewatered for the open-pit mining of Figure 13-66. If it

Figure 16-15 Diaphragm pump handling water-mud-trash from area for foundation. The 4-in pump with its 3-hp electric motor is dewatering a flooded area before foundation work commences. The suction head is about 5 ft and the discharge head to the nearby water course is about 5 ft. The delivery is about 125 gal/min.

The all-purpose diaphragm pump is the little workhorse for pumping in construction and mining, as it is assigned to the tough jobs for all-material and all-weather tasks. The pumps are self-priming, simply built, and with but one part subject to hard wear. This part is the easily replaceable rubber diaphragm.

The location is a building site in Ohio. *(Gorman-Rupp Co.)*

were desired to unwater this pit in one month, the following table summarizes the size of the pumping operation and the capacity of the necessary machinery would be as tabulated below.

Approximate volume to be pumped	64,000,000 gal
Time for pumping, based on 55-min working hour	39,600 min
Necessary output of pumps	1616 gal/min
Minimum total head of pumps	60 ft
Maximum total head of pumps	200 ft
A pump selection based on Table 10-14 for 200 ft total head:	
two single-stage 5-in centrifugal pumps, powered by 75-hp electric motors mounted in hull	
Combined capacity at 200 ft total head	2000 gal/min
Combined horsepower at 200 ft total head	150 hp

The total cost for the work as of 1978, including a pump operator, would be about $14,000, or $0.22 per 1000 gal of dewatering.

LIGHTING PLANTS

Night work calls for adequate lighting for all phases of excavation in which machinery is performing exacting work or operating at high speed. Safety should be the prime concern in illumination.

There are several kinds of plants. Illumination for permanent and semipermanent works may be as simple as adequate lamps on poles or as complex as the battery of clustered lamps which furnishes the lighting for the dam excavation of Figure 16-16. For the more common temporary, as well as semipermanent, illumination, portable light towers are used. These are illustrated in Figures 16-17 and 16-18. The excellent lighting of Figure 16-19 is furnished by a nearby portable light tower.

Floodlights are rarely designed or even assembled for particular work in excavation because they are moved frequently among different locations with diverse lighting requirements and working conditions. One or more standard plants are used, and these are set up expeditiously in the most favorable locations to give the best illumination over the greatest area.

Figure 16-16 **Floodlighting foundation excavation for a dam.** For this long-term fixed-position work a stationary battery of sixteen 1000-W lamps gives excellent illumination for the rock excavation. It is probable that the level of illumination in the open cut at the center of the picture is 3 fc, adequate for the exacting work. An alternative method is the use of four or five portable light towers, strategically located to give the same good illumination. Either method is good and the choice is a matter of convenience and economics. The location is Mammoth Pool Dam, North Fork, California. [*Walter E. Trauffer (ed.),* Pit and Quarry Handbook, *Pit and Quarry Publications, Chicago, 1974, p. D-11.*]

Figure 16-17 **One of eight portable tower-light plants for 12,000 ft of freeway building.** The plant is made up of a 40-ft tower carrying four 1500-W metal halide lamps and a 20-kVA generator driven by a 38-hp diesel engine. The engine-generator has ample capacity for six lamps. The eight plants are located appropriately in the cut, on the haul road as shown, and in the fill. The plant of the picture covers about 600 ft of the haul-road length.

The total illumination is adequate, as during the 18 months of work there were no accidents chargeable to insufficient illumination. The entire 12,000-ft length of the cut, haul roads, and fill are not lighted at all times, as all areas of the cuts and fills are not active at the same time. The easy, swift portability of the plants is important when locations of work change frequently.

The location is a freeway building project in the Laguna Mountains of San Diego County, California.

246,000 SQ. FT. OF LIGHT

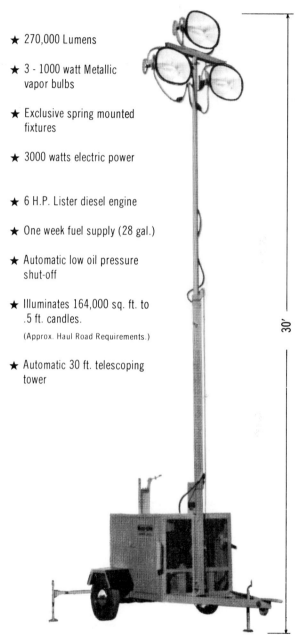

★ 270,000 Lumens

★ 3 - 1000 watt Metallic vapor bulbs

★ Exclusive spring mounted fixtures

★ 3000 watts electric power

★ 6 H.P. Lister diesel engine

★ One week fuel supply (28 gal.)

★ Automatic low oil pressure shut-off

★ Illuminates 164,000 sq. ft. to .5 ft. candles.

(Approx. Haul Road Requirements.)

★ Automatic 30 ft. telescoping tower

Figure 16-18 Three-lamp portable tower-light plant. The manufacturer offers four models of this modular-built machine, which have one, two, three, and four lamps with total lumen ratings of 90,000, 180,000, 270,000, and 360,000, respectively. All plants are powered by diesel engine–ac generator sets of 120 to 220 V. Each lamp of the light plants is rated at 90,000 lm and the lightgrid of Figure 16-20 is applicable to the individual lamps of all four machines. The lamps are of 1000 W power and of the multi-metal vapor kind. *(Allmand Brothers, Inc.)*

16-22 Allied Operations, Machinery and Components, and Facilities

Figure 16-19 Floodlighting for a rock cut in highway construction. The adequate illumination comes from a three-lamp portable tower plant of 270,000 lm capacity located about 200 ft from the bucket loader and about 300 ft from the far tractor. The intensity of illumination on the vertical surfaces normal to the direction of the light beam of the nearer machines is about 6 fc, while that on the corresponding surfaces of the far tractor is about 1 fc. The lighting of the plant is supplemented by the headlights and the working lights of the machines so that safety and speed may prevail over the large working area. The freeway cut is located in the Laguna Mountains of San Diego County, California.

Although mostly standard portable light towers, requiring no design considerations, are used, it is well to discuss the simple principles of illumination in order to help in the selection and application of adequate lighting plants, whether completely preassembled or designed and built for a particular purpose.

Principles of Illumination

  Definitions Visible light is radiant energy which, by its action on the organs of vision, enables them to perform their function of sight.
  Illumination (E) is the surface light density, as luminous flux or power per unit area, on an intercepting surface at a point. It varies directly as the luminous intensity, lumens or candle power, of the source. The unit is the lumen per square foot or the footcandle.
  The *candlepower* (cp) is the amount of light emitted by a sperm candle of 7/8-in diameter and burning at rate of 120 grains, 7.776 grams, per hour, or it is the light emitted by the new candle. The new candle, for calculating purposes, is the equivalent of the sperm candle.
  The *footcandle* (fc) is the illumination produced by the light of one candle at a distance of 1 foot. It equals one lumen per square foot.
  The *lumen* (lm) is the standard unit of luminous flux, one lumen being the light sent out from a unit light source of one absolute candle through a unit solid angle called a *steradian*. A steradian is the solid angle subtended at the center of a sphere by a portion of the surface equal in area to the square of the radius of the sphere. A source having uniform candlepower in all directions would emit 4π lumens.

  Calculations for Illumination The computations are involved, as they deal with many variables such as:
  1. The lumen or candlepower rating of the lamp
  2. The configurations of the lamp and the lamp fixture, which determine the flux density in the direction of the propagation of light
  3. The distance from the lamp to the surface and the inclination of the received beam of light to the central beam of light emitted from the lamp
  4. The inclination of the surface with respect to the received beam from the lamp
  5. The efficiency of the lamp and fixture, which depends largely on age and cleanliness
Because of these complexities, resulting in error-prone calculations, and for practicability, it is suggested that the earthmover use the lightgrids and isofootcandle diagrams of the manufacturers of lamps and lighting plants, as illustrated in Figures 16-20 and 16-21. The

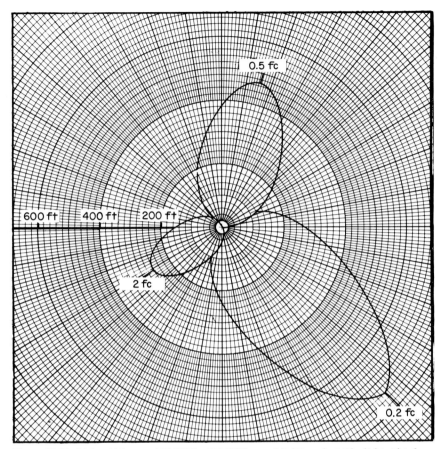

Figure 16-20 Lightgrid for rated 90,000-lm 1000-W lamp of lighting plant. The lightgrid is for a surface normal to the direction of the lamp beam. This surface corresponds approximately to the near vertical surfaces of the work and the excavating machinery for which the illumination is desired.

Lightgrids with different values for footcandles result from different angles of the lamp beam, which change the length of the lightgrid. In the enclosed areas of the lightgrids the values of footcandles are higher than those for the perimeters. The aiming point of the lamp is approximately one-third the length of the lightgrid away from the lamp mast. The illumination at the aiming point is approximately 30 times the footcandles at the perimeter of the lightgrid. *(Allmand Brothers, Inc.)*

footcandles of the lightgrids and isofootcandle diagrams may be compared with the following desirable ranges of values for the work areas.

Areas	Footcandles
Close work by machinery, such as establishing lines and grades, or the servicing and maintenance of machinery	4.0–6.0
General work	2.0–4.0
Work in cuts and fills, such as rough excavation, with the light supplemented by headlights and working lights of machinery	0.5–2.0
Work on haul roads, with light supplemented by headlights and working lights of machinery	0.2–0.5

There are two general equations, the inverse square law and Lambert's cosine law, which are helpful in preliminary estimates for illumination and for the interpretation of lightgrids and isofootcandle diagrams.

16–24   Allied Operations, Machinery and Components, and Facilities

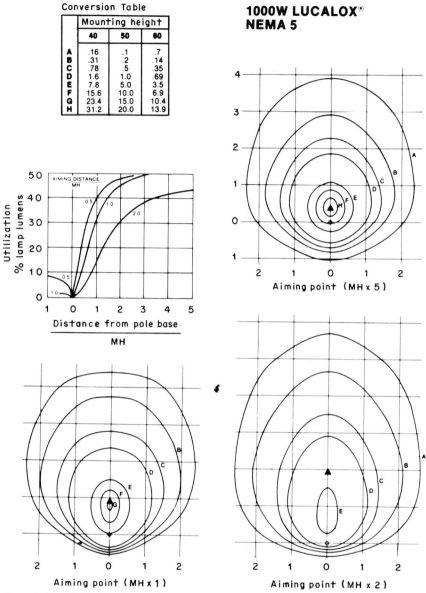

Figure 16-21 Floodlight isofootcandle diagrams and utilization data. The 1000-W metal halide lamp is the metal vapor kind, with a rating of 100,000 lm. The 1000-W Lucalox lamp is the high-pressure sodium kind with a rating of 140,000 lm. The coordinates of the diagrams are in terms of mast heights. The conversion tables are in terms of footcandles on a horizontal surface for the isofootcandle diagrams and for different mast heights.

When illuminating the work and machinery of excavation, the illumination of near vertical surfaces is most important. It is given by the equation:

$$FC_v = \frac{FC_h D}{MH}$$

1000W METAL HALIDE NEMA 5

Conversion Table

	Mounting height		
	40	50	60
A	.16	.1	.7
B	.31	.2	.14
C	.78	.5	.35
D	1.6	1.0	.69
E	7.8	5.0	3.5
F	15.6	10.0	6.9
G	23.4	15.0	10.4
H	31.2	20.0	13.9

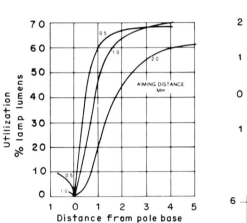

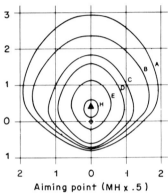

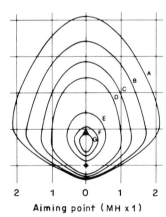

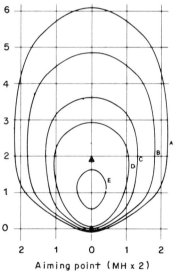

where FC_v = footcandles on vertical surface
FC_h = footcandles on horizontal surface
D = distance from mast to surface, ft
MH = mast height, ft

For example, the metal halide lamp with 40-ft mast gives 0.16 fc on a horizontal surface at 6.1×40, or 244 ft, from the mast. Illumination on a vertical surface is given by:

$$FC_v = \frac{0.16 \times 244}{40} = 0.98 \text{ fc}$$, which represents an acceptable illumination.

(General Electric Co.)

The *inverse square law* states that for a given lumen or candlepower rating of a lamp, the illuminations in footcandles on two surfaces normal to the direction of the same beam of light are inversely proportional to their respective distances from the lamp. For two points E_1 and E_2 at respective distances d_1 and d_2 from the lamp, the equation is:

$$\frac{E_1}{E_2} = \frac{d_2{}^2}{d_1{}^2}$$

Because the two surfaces must be in line with the lamp, this relationship is not true for the lightgrid of Figure 16-20 and the isofootcandle diagram of Figure 16-21, as points on the ground cannot be in line with a beam from the lamp. However, an empirical equation for illuminations at ground level beyond the aiming point may be expressed approximately by an inverse cube rule:

$$\frac{E_1}{E_2} = \frac{d_2{}^3}{d_1{}^3}$$

The inverse cube rule may be developed by analyses of two isofootcandle diagrams of Figure 16-21 for an aiming point at twice the mounting height of 40 ft.

Lambert's cosine law states that the illumination on a surface inclined at an angle θ to the surface normal to the axis of the beam varies according to the cosine of the angle. The equation is:

$$E_\theta = E_n \cos \theta$$

where E_θ = illumination on surface inclined at angle θ, lm or fc
E_n = illumination on surface normal to axis of beam, lm or fc
θ = angle of inclination of surface to that surface normal to axis of beam, °

The use of the inverse cube rule and Lambert's cosine law with reference to Figure 16-20 is set forth in the following examples.

EXAMPLE 1 The lightgrid in the upper right-hand quadrant gives an illumination of 0.5 fc at a distance of 470 ft from the lamp, the inclined distance being equal approximately to the horizontal distance. This illumination is for a surface normal to the beam and it is approximately correct for the illumination on a vertical surface. This near equality holds for surfaces beyond 100 ft from a lamp with mast height up to 40 ft. The aiming point of the lamp is approximately one-third of 470 ft, or 157 ft from the pole. Applying the empirical inverse cube rule, the following illuminations obtain for different distances along the axis of the lightgrid from 470 to 157 ft from the lamp mast.

Distance from Lamp Mast, ft	Calculations	Illumination, fc
470		0.5 (known)
400	$0.5 \times \dfrac{470^3}{400^3}$	0.8
300	$0.5 \times \dfrac{470^3}{300^3}$	1.9
200	$0.5 \times \dfrac{470^3}{200^3}$	6.5
157 (aiming point)	$0.5 \times \dfrac{470^3}{157^3}$	13.4

EXAMPLE 2 If one wishes to read a set of plans spread out on the ground at the aiming point of the lamp, 157 ft from the lamp mast, the illumination may be estimated by use of Lambert's cosine law. At this point the tangent of the angle of the beam with respect to the surface of the ground is 30/157, or 0.19, and the angle is 11°. The angle θ is the complement of this angle, or 79°. Substituting in the equation:

$$E_\theta = E_n \cos \theta$$
$$E_\theta = 13.4 \cos 79°$$
$$= 2.6 \text{ fc}$$

This value, 2.6 fc of illumination, is an acceptable level for limited reading of plans.

Lighting Equipment

Lamps Four kinds of lamps with their different characteristics are used in lighting plants. The lamps are available in convenient 1000- and 1500-W sizes for batteries of one to six lamps in portable-tower lighting plants and much greater numbers of lamps in semistationary and stationary systems. Table 16-3 summarizes the characteristics of these lamps. Lumens per watt is a measure of electrical energy consumption per unit of illumination, varying from 21 to 143. Life in hours is a measure of replacement cost per hour of operation.

It is most important to keep lamps and fixtures clean, as illumination decreases rapidly with accumulation of the dust and dirt of an excavation job. A lamp efficiency of 67 percent calls for a 50 percent higher lamp rating to give the same number of footcandles of illumination obtainable at 100 percent lamp efficiency, and results in a corresponding increase in energy consumption.

Portable Tower Lighting Plants The towers and trailers are generally built to the same specifications and dimensions. The engine-generator sets and lamps of different capacities are added to the tower-trailer assemblies. Representative specifications are given in Table 16-4.

TABLE 16-3 Lamps Used in Lighting Plants

Lamp	Watts	Lumens	Lumens/watt	Life, h
Incandescent tungsten halide (quartz); clear light	1000 1500	21,000 33,000	21 22	3000 3000
High-intensity-discharge mercury vapor; good color	1000 1500	57,500 89,000	58 59	24,000 24,000
High-intensity-discharge metal halide; clear light	1000 1500	100,000 155,000	100 103	10,000 10,000
High-intensity-discharge sodium vapor; yellowish light	1000 1500	130,000 215,000	130 143	20,000 20,000

TABLE 16-4 Representative Specifications for Portable-Tower Lighting Plants

Tower and trailer assemblies complete, without engine-
generators and lamps:
  Tower: Structural steel, three-section, hydraulically operated
    by one hydraulic cylinder
  Trailer: Structural steel, completely housed, lamp storage
    box, fenders, all legally required equipment for
    highway travel, 17-gal fuel tank,
    2 tires, 670×15, 6 PR
  Dimensions:
    Tower height, ft
    Travel height, ft
    Overall width, ft
    Approximate weight, lb
	
	25.7
	6.7
	6.0
	1630

Engine–ac generator sets:
  5.0- and 7.5-kW, 110/220-V generators driven by 15- and 25-
    hp gasoline engines
  6.0- and 12.0-kW, 110/220-V generators driven by 15- and
    30-hp diesel engines
  Approximate weight, lb
	
	1570

Lamps and fixtures:
  Standard: 3 or 4 high-intensity tungsten halogen lamps, each
    1500 W and 90,000 lm capacity
  Optional: 3 or 4 high-intensity mercury vapor lamps,
    Approximate weight, lb
	
	100

Complete machine, approximate weight, lb	3300

Estimated Costs for a Floodlighting System

The cost calculations are illustrated for the example of the floodlighting by eight portable-tower lighting plants, as illustrated by Figures 16-17 and 16-19.

Criteria:
Total length of cut, haul road, and fill, partially but satisfactorily illuminated, ft   12,000
Traffic density of earthmoving machines and miscellaneous vehicles, hourly
    During 9 h of illumination   222
    During 9 h without illumination   374
Rock-earth excavation, cybm/h
    During 9 h of illumination   2730
    During 9 h without illumination   4850
    Average during 18 h of work   3790
Costs, 1978:
    Cost of 8 portable lighting plants, @ average $12,300 $98,400
Hourly cost of ownership and operation (Table 8-3):
    Average 15-kW floodlights, portable, diesel engines, 50% × $98.4 $49.20
    Cost of moving, estimated at 5% × $49.20   2.50
    Operator, estimated at 33% of hourly rate of $18.40 for electrician   6.10
    Total hourly cost $57.80
Direct job unit cost per cybm:
    During 9 h of illumination, $57.80/2730 $0.021
    During 18 h of work, (9×$57.80)/(18×3790) $0.008

For proration to cost of excavation, one uses the figure $0.008/cybm. For 1978 costs, this figure of $0.008/cybm when double shifting is a fair job unit cost. The relatively high cost of $57.80/h for floodlights for the 12,000-ft length is offset by the high 3790-cybm/h production. On the average job with an average production of 1500 cybm, three or four plants are adequate and the unit cost is about the same.

WEIGHING EXCAVATION

Excavated material may be measured in cubic yards or in tons. The more commonly used unit is the cubic yard, either bank measurement, loose measurement, or compacted measurement. These are volume measurements, easily calculated. They may be converted closely to weights by the approximate values of material characteristics given in Appendix 1.

The less common unit is the ton, generally of 2000 lb but sometimes of 2240 lb, the long ton. Its uses are in borrow pits, in open-pit mines, and in quarries, where the basis of payment for excavation is the ton and payloads of scrapers and haulers must be weighed. Scales for this purpose are illustrated in Figure 16-22, which shows residuals from a borrow pit being weighed prior to placement in the embankment of a freeway.

Standard scales as shown in the figure are built with up to 200 tons capacity and adequate length, and larger custom-built scales are available. Scales work on the lever-fulcrum principle, but the means for weight determination may be as simple as the dial-and-beam indicators or as complex as the electronic-dial scale with readout printing of weights or the more sophisticated electronic solid-state indicator with a variety of processed data. Table 16-5 lists typical specifications for standard movable-platform scales.

Basically the procedure used with all devices is simple. The laden machine is weighed for GVW, a predetermined tare weight is subtracted, and the remainder is the payload. For the scraper of Figure 16-22 the arithmetic is:

       GVW   237,000 lb
       Tare weight   121,700
       Payload   115,300 lb

The estimation of production and costs for weighing materials is illustrated below for the work of the scales of Figure 16-21.

Criteria: The output of the scraper spread is 45 loads hourly and the total hourly payload tonnage averages 2720 tons.

Weighing Excavation 16-29

Figure 16-22 Weighing excavation from a borrow pit with standard scales. The 150-ton-capacity scales are weighing residuals from a borrow pit, the material being destined for a distant freeway embankment. Such means of measurement and payment is commonplace in highway construction.

The scale house contains a dial-and-beam indicator, which gives the weigh master duplicate printed readouts of GVW, preestablished tare weight, and resultant payload. In this instance they are 237,000, 121,700, and 115,300 lb, respectively. The tare weight is the average working weight of the scraper, based on periodic weighings. This procedure allows for any changes in weight of accumulated material such as frozen residuals and for such changes as the addition or subtraction of sideboards.

The production of a set of scales is in terms of the excavation weighed hourly. It takes about 45 s to weigh a machine across the scales when the weigh master gives the operator a weight ticket. If the ticket is not needed, as is sometimes the case, the time is about 30 s. Accordingly, in a 50-min working hour, productions in terms of scrapers or haulers are 67 machines with and 100 machines without delivery of weight ticket.

The hourly production in terms of tons weighed is the sum of the payloads for the hour. *(Murphy Scale and Equipment Co.)*

TABLE 16-5 Representative Specifications for Standard Movable-Platform Scales

Specification	Nominal capacity, tons			
	50	100	150	200
Platform dimensions, ft:				
Length	60	60	70	70
Width	10	12	14	14
Weights, complete except for scale house and foundations, lb	39,000	54,000	70,000	76,000
Common specifications: Steel-plate deck, structural-steel-plate girder platform frame, structural-steel H-beam main frame Lever-fulcrum with double-link suspension bearings and hardened steel inserts at bearing points Optional dial-and-beam indicators with weight-ticket multicopy readouts or electronic solid-state indicator with readouts for data processing				

NOTE: These scales are available in portable configurations in which the pedestals rest on individual timber or concrete foundations rather than on the main frame and the decking is timber rather than steel plate.

Costs, 1978:

Cost of 150-ton-capacity scales, for job site	$59,600
Hourly cost of ownership and operation:	
Machine (Table 8-3):	
Scales, platform, complete with scale house, erected, 21% × $59.6	$12.50
Dismantling of scales, estimated at 5% × $59,600 for 2110 h of work	1.40
Weigh master	12.40
Total hourly cost	$26.30
Direct job unit cost per ton of material weighed, $26.30/2720	$0.010

TIRES

Most excavation machinery is mounted on wheels-tires, as opposed to crawlers. Of 114 major machines used on one major rock job for freeway building, 84 units, or 74 percent, were wheels-tires-mounted, the remaining 26 percent being crawler-mounted. Part of the 1977 tire inventory, amounting to a total of about $350,000, is shown in Figure 16-23. The inventory amounted to about $4200 per machine and the figures suggest the importance of tire costs on a rock job.

The value of tires for a machine averages about 8 percent of the cost of the machine, and the cost of repairs and replacements averages about 12 percent of the total cost for ownership and operation. Understandably, tires merit both physical and financial consideration.

Tires in common use range in size from on-the-road $10.00 \times 20 - 12PR$ size, supporting a 5000-lb load, to 36.00×51—50PR size, carrying a 92,000-lb load.

Tire Construction

Figure 16-24 gives nomenclature used in specifying the dimensions of a tire, as set forth by the Tire and Rim Association. Two kinds of tires are in general use: the bias-ply tire has a carcass made up of nylon plies, illustrated in Figure 16-25, and the radial tire has a carcass made up of steel-cable plies, shown in Figure 16-26.

Tires are also classified into two types on the basis of their section height/width ratios. For average work the standard tire, with a section height/section width ratio of 1.00, is employed. The standard tire, as used on a rear-dump hauler, is designated as 18.00×25

Figure 16-23 Part of inventory of huge, on-the-job tires for large rock excavation project. These twenty-eight 37.5×39—44PR wide-base tires are the largest used on the job and they are for fifteen 40-yd³ scrapers. Seventy-four major machines are wheels-tires-mounted.

The maximum capacity of the tire is 60,390 lb, its weight is 2829 lb, and its 1978 cost is $5400. These tires, part of the $350,000. inventory, are worth $151,000. In spite of the great cost, it is necessary to carry a full inventory as most modern earthmovers cannot move without tires.

The location is a California freeway job on which tire wear is average.

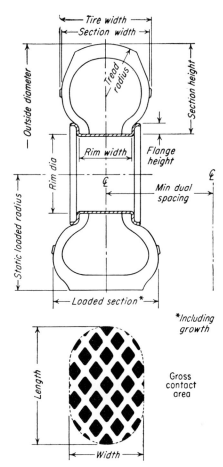

Figure 16-24 Nomenclature for dimensions of tires. In addition to the nomenclature of the cross-sectional view there are other important terminologies for a tire and they are:

Deflection Deflection is the difference between outside radius and static-loaded radius.
Ply rating (PR) The PR is an index of the strength of the tire and most often it is not the number of plies in the tire.
Inflation pressure The capacity of a tire increases with increase in pressure. The pressure should be taken when the tire is at atmospheric temperature and not when it has been heated by service.
Gross contact area The area in square inches equals approximately tire load in pounds divided by tire pressure in pounds per square inch. It is slightly less than this quotient because of the effect of sidewall stiffness.
Standard and wide-base tires A standard tire is shown, its ratio of section height to section width being 1.00. The wide-base tire has a section height less than section width for greater traction and flotation, the ratio averaging about 0.85. The range of pressures is 50 to 90 lb/in^2 for standard and 30 to 70 lb/in^2 for wide-base tires.

Flotation and traction increase and rolling resistance decreases with lowering of tire pressure, but tire costs increase. Under the same operating conditions, flotation and traction decrease and rolling increases with raising of tire pressure, but tire costs decrease. *(Goodyear Tire and Rubber Co.)*

16–32 Allied Operations, Machinery and Components, and Facilities

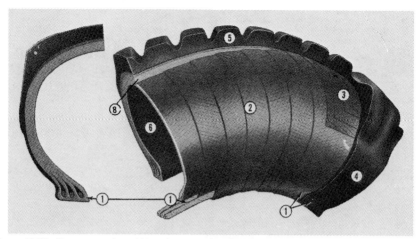

Figure 16-25 Cutaway section of bias-ply tire with nylon-ply carcass.

1. The *tire beads* are bundles of steel wire which are forced laterally by pressure to wedge the tire firmly on the tapered bead seat of the rim. The nylon plies tie into the bead bundles.

2. The *body plies* are the layers of rubber-cushioned nylon cord which make up the tire carcass. Alternating plies of cord cross the tread centerline at an angle or bias. The *ply rating* is an index of strength and is not the actual number of plies in the tire.

3. *Breakers or tread plies*, if used, are confined to the tread area of the tire and are for increased carcass strength and protection to the body plies.

4. *Sidewalls* are protective layers of rubber covering the body plies in sidewall areas.

5. The *tread* is the wearing part of the tire which contacts the ground. It also provides traction and flotation.

6. The *inner liner* is the sealing medium which retains the air in the tire and, combined with the O-ring seal and rim base, eliminates the need for inner tubes and flaps.

7. *Tubes* and *flaps* (not shown) are used occasionally when tire life may be improved thereby.

8. The *undertread* is the protective cushion of rubber between the tread and the body ply. *(Caterpillar Tractor Co.)*

—32PR. For work in soft materials, where flotation and traction are important, the wide-base tire, with a section height/section width ratio of about 0.85, is used. The wide-base tire, as used on a scraper or bottom-dump hauler, is designated as 23.5×25—24PR. In both designations the first figure is the tire width in inches, the second figure is the rim diameter in inches, and the third figure is the ply rating. Note that ply rating does not necessarily mean number of plies, as it is a relative rating indicating the strength of the plies.

The type of tire (standard or wide-base), the tread design, and the ply rating are the most important considerations in the selection of tires for particular work or average work of a machine.

Uses of a standard tire are shown in Figure 13-30, which depicts typical rock haulage over a hard haul road by rear-dump truck, and in Figure 13-26, showing typical on-the-road–off-the-road haulage by bottom-dump machines. Uses of wide-base tires are illustrated in Figure 13-16, showing typical rock-earth haulage by scrapers under soft conditions, and in Figure 13-28, representing haulage over a hard coal vein in the cut and over a soft haul road to the coal preparation plant. The selection of wide-base tires for the coal haulers represents a compromise between the hard and the soft sections of the whole haul road, it being felt that good flotation and traction on the ascending ramp and the long main haulage road are more important than the advantages of a standard tire on the hard coal vein.

Tread designs are basically of three kinds for off-the-road haulage and one kind for on-the-road–off-the-road work. There are many variations of these kinds for fixed conditions of haulage. Figure 16-27 illustrates three off-the-road tire treads and Figure 16-28 shows an on-the-road–off-the-road tire tread.

Factors Affecting Life of Tires

Neglecting accidental failures such as by cuts and bruises, tires deteriorate because of ply separation in the carcass. The separation results from loss of strength due to heat generated by flexing of the plies. The heat reverses the vulcanizing process and causes the rubber to revert to a plastic state.

Flexing results from overloading, overspeeding, and underinflation. Figure 16-29 illustrates the effects of these factors on normal tire life. Another source of heat is external, namely a high ambient temperature. A striking example of the effects of overloading, overspeeding, and high ambient temperature, shown in Figure 16-30, took place during the building of Imperial Dam on the Colorado River near Yuma, Arizona.

In 1936 the largest bottom-dump hauler for off-the-road haulage used $13.50 \times 24 - 14$PR standard tires with an allowable maximum load of 7800 lb but with a designed load of 8250 lb, a 6 percent overload. These tires were the largest manufactured at the time. The machine had been operating well at a maximum speed of 15 mi/h at a maximum 100°F temperature. However, because of optimism on the part of both manufacturer and owner, the capacity of the machine was increased from 10 to 14 tons, resulting in a tire load of 10,500 lb, a 35 percent overload. The speed was increased to 21 mi/h, which in itself was probably not a harmful change. However, Nature raised the temperature in the desert to 120°F, a 20 percent increase over the previous maximum temperature.

While it was difficult to assign values to the factors operating mutually to lower the life of the tires, it was felt, on the basis of a graph similar to that of Figure 16-29, that overloading caused a loss of 50 percent, and that high temperature, possibly combined with high travel speed, resulted in another progressive 50 percent loss of life. When these losses were compounded, a theoretical loss of 75 percent in tire life could be expected. Normally, tire life was about 5000 h, but at the dam it dropped to about 1200 h.

The remedy was to take off the top set of 12-in sideboards, reducing the tire load to

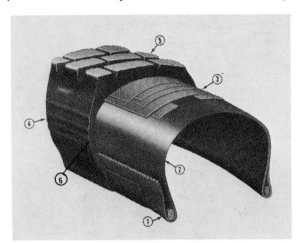

Figure 16-26 Cutaway section of radial-ply tire with steel-ply carcass.
 1. This *bead* consists of a single bead bundle of steel cables or steel strip, spiraled like a clock spring, at each rim interface.
 2. The *radial carcass* consists of a single layer or ply of steel cables laid archwise on the radian of the section from bead to bead.
 3. The *belts* which underlie the tread area around the circumference of the tire consist of several layers or plies of steel cables. The cable in each belt crosses the tread centerline at an angle, the angle being reversed from that of the preceding belt.
 4. *Sidewalls* are protective layers of rubber covering the radial carcass in sidewall areas.
 5. The *tread* is the wearing part of the tire which contacts the ground. It also provides traction and flotation.
 6. The *undertread* is a protective cushion of rubber between the tread and the steel belts.
(Caterpillar Tractor Co.)

(a)   (b)   (c)

Figure 16-27 Three basic tire treads for off-the-road haulage. *(a)* A rock lug tire, standard base, used over hard surfaces and having high pressure and low rolling resistance. Machines with this tire are bucket loaders, rear- and side-dump haulers, bottom-dump haulers, and tractor-bulldozers.

(b) A rock-earth directional-bar tire, wide-base, used over soft surfaces and having low pressure, good flotation and traction, and high rolling resistance. Machines with this tire are bucket loaders, scrapers, bottom-dump haulers, and tractor-bulldozers.

(c) A rock-earth button tire, standard or wide-base, generally used as an all-purpose tire for scrapers and bottom-dump haulers. It is used primarily on all wheels of crawler-tractor-drawn scrapers.

One perceives much overlapping in the uses of these tires. Actually, the earthmover selects the same tire for both driving and trailing wheels and selects the tread best suited to his average work. *(Goodyear Tire and Rubber Co.)*

Figure 16-28 A basic tire tread for on-the-road–off-the-road haulage. This rock-earth tire tread is a compromise for work in both hard and soft conditions at high and low speeds. It is a standard tire. For interchangeability it is used on both driving and trailing wheels. It is also used on the steering wheels, usually in a smaller size. There are many variations of this tread design with different standard configurations for emphasis on either wearing or traction qualities or for a compromise between wearing and traction characteristics. *(Goodyear Tire and Rubber Co.)*

Tires 16-35

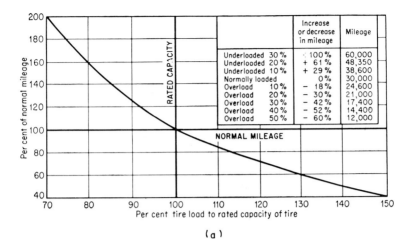

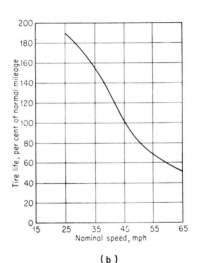

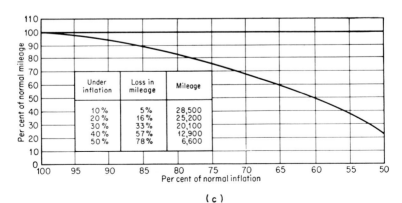

Figure 16-29 The three important factors affecting life of tires. *(a)* Effect of load on tire. *(b)* Effect of speed of vehicle. *(c)* Effect of inflation of tire. *(Goodyear Tire and Rubber Co.)*

Figure 16-30 Example of tire failures on overloaded and overspeeded machine working at high temperatures. In an effort to increase production and to lower hauling costs, 10 of these bottom-dump haulers with 13.50×24—14PR tires were modified at the factory in two respects. Double sideboards were added, increasing the 10-ton nominal capacity to 14 tons, a 35 percent overload. The transmission was changed, increasing the speed from 15 to 21 mi/h, a 27 percent increase. A neglected fact was that normal maximum ambient temperature would increase from 100 to 120°F where the machines would be operating for a long time.

On the 3-mi gentle-downgrade desert haul road, top travel speed was maintained both laden and unladen. Average tire life decreased from 5000 to 1200 h with considerable loss of working time due to tire changes. The remedy was to reduce the load to 11 tons and the speed to 15 mi/h. Tire life increased to an average of 3300 h in spite of the 14 percent overload and persistent high temperatures.

The location was the Imperial Dam on the Colorado River near Yuma, Arizona, in 1936.

8900 lb, a 14 percent overload, and to decrease the travel speed to 15 mi/h. Thereby, tire life increased to about 3300 h, which was considered good under the circumstances.

Tire Specifications

There are two kinds, two types, and many tread designs for applications in rock-earth excavation. The many selections available include both general-purpose and special-purpose tires, and the offerings of the manufacturers require many pages of specifications for tires, rims, and accessories. The specifications also require and deserve the close scrutiny of the user and, almost always, the advice of trustworthy, knowledgeable experts in the field of tire applications.

The specifications listed in Tables 16-6 and 16-7 cover the two basic kinds of tires, bias-ply with nylon-ply carcass and radial-ply with steel-ply carcass, and the two basic types of tires, standard and wide-base, which have been discussed.

There are many treads with different configurations for wear, flotation, and traction. Some tires have treads designed as a compromise among all desirable features, and these are called all-purpose tires, as they are considered suitable for a variety of conditions encountered in rock-earth excavation.

Tables 16-6 and 16-7 are representative of the many useful data furnished by the manufacturers. They contain specifications for standard and wide-base tires for on-the-road–off-the-road and for off-the-road service.

Selection of Tires

The tires of a machine are selected either for a given job, during which they will receive most of their wear or will wear out, or for average work, during which they will be used on several jobs of varying natures over their lives. A given job would be a dam, quarry, or open-pit mine and average work would be a highway job followed by a building site and then excavation for a coal-stripping job.

The given job is represented by the overburden removal work of Figure 13-30. The large truck will be used continually in long-haul upgrade work in all weather conditions. The owner wisely selected a lug-tread tire, combining good wearing traits, traction, and flotation in that order. Representative specifications are given in Table 16-6.

TABLE 16-6 Specifications for Standard-Type Hard-Rock Lug-Tread Tires *(Continued)*

This type and tread of tire is shown in Figure 16-28. It is made both in nylon bias-ply and in steel radial kinds with different treads for different working conditions in rock-earth excavation. It is primarily a tire for wear and it is secondarily a tire for flotation and traction. It is for combination on-the-road and off-the-road service with a maximum speed 50 mi/h.

Part A

Size	Load range	Ply rating	Rim	Min. dual spac. no. chain	Inflated dimensions					Static loaded radius	Loaded dimensions		
					Inf Lb	Load	Ovl width	Ovl diam.	Loaded sect. and gr.		Rev per mi	Gross cont. area	Weight
8.25-20ML	F	12	6.50T	11.6	90	3950	9.4	39.3	10.1	18.8	528	43	99
			7.00T	11.8	90	3950	9.6	39.3	10.3	18.8	528	43	99
9.00-20ML	F	12	7.00T	12.6	85	4520	10.3	40.8	11.3	19.4	512	60	122
			7.50V	12.8	85	4520	10.5	40.8	11.5	19.4	512	60	122
10.00-20ML	F	12	7.50V	13.4	75	4760	11.0	42.3	11.9	20.0	496	69	129
			8.00V	13.6	75	4760	11.2	42.3	12.1	20.0	496	69	129
10.00-20ML	G	14	7.50V	13.4	90	5300	11.0	42.3	11.9	20.0	496	69	139
			8.00V	13.6	90	5300	11.2	42.3	12.1	20.0	496	69	139
10.00-22ML	F	12	7.50V	13.4	75	5070	11.0	44.3	11.9	21.0	473	77	145
			8.00V	13.6	75	5070	11.2	44.3	12.1	21.0	473	77	145
10.00-22ML	F	14	7.50V	13.4	90	5640	11.0	44.3	11.9	21.0	473	77	156
			8.00V	13.6	90	5640	11.2	44.3	12.1	21.0	473	77	156
11.00-20ML	F	12	8.00V	14.1	75	5190	11.7	43.3	12.9	20.6	482	77	154
			8.50V	14.3	75	5190	11.9	43.3	13.1	20.6	482	77	154
11.00-20ML	G	14	8.00V	14.1	90	5780	11.7	43.3	12.9	20.6	482	77	163
			8.50V	14.3	90	5780	11.9	43.3	13.1	20.6	482	77	163
11.00-22ML	F	12	8.00V	14.1	75	5520	11.4	45.3	12.9	21.5	462	86	165
11.00-22ML	F	12	8.50V	14.3	75	5520	11.6	45.3	13.1	21.5	462	86	165
11.00-22ML	G	14	8.00V	14.1	90	6140	11.4	45.3	12.9	21.5	462	86	175
			8.50V	14.3	90	6140	11.6	45.3	13.1	21.5	462	86	175
11.00-24ML	F	12	8.00V	14.1	75	5860	11.8	47.4	13.0	22.5	441	91	180
			8.50V	14.3	75	5860	12.0	47.4	13.2	22.5	441	91	180
11.00-24ML	G	14	8.00V	14.1	90	6520	11.8	47.4	13.0	22.5	441	91	189
			8.50V	14.3	90	6520	12.0	47.4	13.2	22.5	441	91	189
12.00-20ML	H	16	8.50V	15.0	95	6790	13.1	45.2	14.3	21.4	464	94	192
			9.00V	15.2	95	6790	13.3	45.2	14.5	21.4	464	94	192
12.00-24ML	H	16	8.50V	15.0	95	7640	13.0	49.6	13.9	23.4	424	97	235
			9.00V	15.2	95	7640	13.2	49.6	14.1	23.4	424	97	235

TABLE 16-6 Specifications for Standard-Type Hard-Rock Lug-Tread Tires (Continued)

Part A

Size	Load range	Ply rating	Rim	Min. dual spac. no. chain	Inflated dimensions					Loaded dimensions				Weight
					Inf Lb	Load	Ovl width	Ovl diam.	Loaded sect. and gr.	Static loaded radius	Rev per mi	Gross cont. area		
12.00-25ML	H	16T	8.50	15.0	95	7640	12.7	49.5	13.8	23.5	422	98	235
13.00-24ML	J	18	9.00V	16.2	100	9120	13.8	51.4	15.2	24.1	412	105	243
			10.00W	16.4	100	9120	14.2	51.4	15.6	24.1	412	105	243
13.00-25ML	J	18T	10.00	16.7	100	9120	14.2	51.4	15.6	24.1	412	105	258
13.00-25ML	J	18T	8.50	16.0	100	9120	13.8	51.4	15.2	24.1	412	105	258
14.00-24ML	L	20	10.00W	17.7	100	10710	15.3	54.4	16.6	25.7	387	137	340
14.00-24ML	N	24	10.00W	17.7	100	10710	15.3	54.4	16.6	25.7	387	137	386
14.00-25ML	L	20T	10.00	17.0	100	10710	15.0	54.4	16.3	25.7	387	133	384

Part B: This second part of the specifications covers large tires and it is for off-the-road service with maximum speed of 30 mi/h.

Size	Ply rating	Rim	Min. dual spac. no. chain	Inflated dimensions					Loaded dimensions				Gross cont. area
				Inf lb	Load	Ovl width	Ovl diam.	Loaded sect. and gr.	Static loaded radius	Rev per mi		
18.00-49T	20	13.00-2½	24.2	50	20930	20.1	88.2	23.1	40.8	230	340
18.00-49T	24	13.00-2½	24.2	60	23290	20.1	88.2	23.1	40.8	230	340
18.00-49T	28	13.00-2½	24.2	70	25500	20.1	88.2	23.1	40.8	230	340
18.00-49T	32	13.00-2½	24.2	80	27560	20.1	88.2	23.1	40.8	230	340
21.00-25TT	20	15.00-3	26.3	40	16130	23.2	69.4	25.8	30.9	305	315
21.00-25TT	24	15.00-3	26.3	50	18390	23.2	69.4	25.8	30.9	305	315
21.00-25T	28	15.00-3	26.3	60	20460	23.2	69.4	25.8	30.9	305	315
21.00-35T	28	15.00-3	27.6	60	24270	23.6	79.4	26.2	36.7	257	362
21.00-35T	32	15.00-3	27.6	70	26570	23.6	79.4	26.2	36.7	257	362
21.00-35T	36	15.00-3	27.6	80	28710	23.6	79.4	26.2	36.7	257	362
21.00-35T	40	15.00-3	27.6	90	30750	23.6	79.4	26.2	36.7	257	362
21.00-49T	24	15.00-3	27.6	50	26410	23.3	93.9	26.6	43.4	217	428
21.00-49T	28	15.00-3	27.6	60	29390	23.3	93.9	26.6	43.4	217	428
21.00-49T	32	15.00-3	27.6	70	32170	23.3	93.9	26.6	43.4	217	428
21.00-49T	36	15.00-3	27.6	80	34770	23.3	93.9	26.6	43.4	217	428
21.00-49T	40	15.00-3	27.6	90	37250	23.3	93.9	26.6	43.4	217	428

Tire	Ply	Rim											
24.00-25TT	24	17.00-3½	29.8	45			22330	26.0	74.4	29.0	33.6	280	428
24.00-25T	28	17.00-3½	29.8	50			25130	26.0	74.4	29.0	33.6	280	428
24.00-29TT	24	17.00-3½	29.8	45			23920	25.7	77.4	29.0	35.2	268	412
24.00-29T	36	17.00-3½	29.8	65			29670	25.7	77.4	29.0	35.2	268	412
24.00-35T	36	17.00-3½	31.3	65			32550	26.3	84.6	28.9	38.8	243	485
24.00-35T	42	17.00-3½	31.3	75			35380	26.3	84.6	28.9	38.8	243	485
24.00-35T	48	17.00-3½	31.3	85			38040	26.3	84.6	28.9	38.8	243	485
24.00-49T	36	17.00-3½	31.3	65			39050	26.3	98.2	29.5	45.3	208	530
24.00-49T	42	17.00-3½	31.3	75			42450	26.3	98.2	29.5	45.3	208	530
27.00-33TT	30	22.00-4	34.5	50			33480	30.3	88.1	33.7	40.0	233	530
27.00-33TT	36	22.00-4	34.5	60			37250	30.3	88.1	33.7	40.0	233	530
27.00-49T	36	19.50-4	35.1	60			45580	29.25	104.3	33.2	47.1	198	605
27.00-49T	42	19.50-4	35.1	70			49900	29.25	104.3	33.2	47.1	198	605
27.00-49T	48	19.50-4	35.1	80			53930	29.25	104.3	33.2	47.1	198	605
30.00-51T	40	22.00-4½		60			56380	33.1	111.5	37.1	51.0	183	910
33.00-51T	50	24.00-5		70			71550	35.3	117.2	39.5	53.7	174	960
36.00-51T	42	26.00-5	46.5	55			75750	38.9	123.6	43.8	55.7	168	1294

Part C

Tire Size	Tire load limits at various inflation pressures												
	40	45	50	55	60	65	70	75	80	85	90	95	100
8.25-20	2640	2830	3000	3180	3340/10	3500	3660	3800/12					
9.00-20	3850	3050	3240	3440/10	3630	3790/12							
10.00-20	3170	3400	3600	3820	4020/12	4210	4400/14						
10.00-22	3390	3630	3850	4070	4290/12	4510	4700/14						
11.00-20	3560	3830	4070	4300	4510/12	4730	4950/14						
11.00-22	3760	4040	4230	4540	4780/12	5000	5220/14						
11.00-24, 25	3960	4240	4580	4850	5100/12	5350	5590/14						

TABLE 16-6 Specifications for Standard-Type Hard-Rock Lug-Tread Tires (Continued)
Part C

Tire Size	Tire load limits at various inflation pressures												
	40	45	50	55	60	65	70	75	80	85	90	95	100
12.00-20, 21	4010	4300	4570	4850	5090	5330/14	5580	5800/16					
12.00-24, 25	4510	4830	5120	5440	5720	6000/14	6270	6520/16	6780/18	7040	7280/20		
13.00-24, 25	5350	5750	6100	6470	6790	7120/16	7420	7730/18					
14.00-20, 21	5610	6000/12	6390	6770	7110/16	7460	7780	8100/20					
14.00-24, 25	6260	6710	7120	7540	7930/16	8310	8690	9050/20	9380	9730	10060/24		
16.00-21	7420	7960	8490	8950	9420	9860/20							
16.00-24, 25	8210	8800	9370/16	9930	10400	10920/20	11380	11820	12320/24	12780	13240/28		
16.00-57	14710	15800	16760	17730	18650	19550/20							
18.00-24, 25	10690	11450/16	12180	12900/20	13560	14170	14790/24	15380	15950/28	16560	17140/32	17660	18210/36
18.00-33	12570	13460	14390	15260	15950	16710	17550/24	18180	18870/28	19640	20300/32	21050/36	
18.00-49	16490	17660	18820	19950/20	20930	21920	22960/24	24040	24800/28	25700	26650/32	27410	28220/36
21.00-24, 25	13640/16	14570	15510/20	16610	17400/24	18270	19120/28						
21.00-35	16600	17780	18960	20090	21080	22080	23120/28	24020	24970/32	25880	26830/36	27600	28410/40
21.00-49	20760	22220	23690	25080	26360/24	27600	28900/28	30030	31220/32	32350	33540/36	34500	35520/40

Tire Size															
24.00-25	17660/18														
24.00-29		18920													
24.00-33	19100		20460	20940	20130/24 21760/24	21290	22400/28								
24.00-35	19560			22330	23010	23650/24	24210	25370	26500	27590	28650/36				
24.00-49	21250	22770		24220	25610		26940	28240/30 34930/30	29500	30700	31880/36 39440/36	30030	34160/42 42250/42	35260	36330/48
27.00-33	26290	28170		29960	31680		33330	33650	36480	37980		40860			
27.00-49	25300	27100/24		28900	30620/30		32130	35240/36 44410/36		46230	48010/42	49740	51470/48		
30.00-33	31990	34290		36360	38580		40580	42540							
30.00-51	31020	33160		35330	37450		39300	41160							
36.00-51	37630	40320		42880	45340		47700/36	49900							
	55560	59520		63300	66930		70430/42								

NOTE: The underscored figure denotes maximum load and inflation for tire with ply rating shown below figures.

SOURCE: Goodyear Tire and Rubber Co.

TABLE 16-7 Specifications for Wide-Base-Type Sure-Grip Lug-Tread Tires

This type and tread of tire is shown in Figure 16-27B. It is made in nylon bias-ply and in steel radial kinds with different treads for different working conditions. It is primarily a tire for flotation and traction and secondarily a tire for wear. It is for off-the-road service with a maximum speed 30 mi/h.

Part A

Size	Ply rating	Rim	Min. dual spac. no. chain	Inf lbs.	Loaded dimensions			Loaded sect. and gr.	Inflated dimensions		Revs per mi	Gross cont. area
					Load	Ovl width	Ovl diam.		Static loaded radius			
20.5-25	16	17.00-2	24.0	40	11390	21.0	58.5	23.1	26.8		354	237
20.5-25	20	17.00-2	24.0	50	12990	21.0	58.5	23.1	26.8		354	237
23.5-25	12	19.50-2½	27.3	25	11150	24.4	64.1	26.3	28.6		329	320
23.5-25	16	19.50-2½	27.3	35	13580	24.4	64.1	26.3	28.6		329	320
23.5-25	20	19.50-2½	27.3	40	14680	24.4	64.1	26.3	28.6		329	320
23.5-25	24	19.50-2½	27.3	50	16730	24.4	64.1	26.3	28.6		329	320
26.5-25	14	22.00-3	30.6	25	14190	27.4	68.9	29.8	30.9		305	423
26.5-25	20	22.00-3	30.6	35	17280	27.4	68.9	29.8	30.9		305	423
26.5-25	24	22.00-3	30.6	45	20020	27.4	68.9	29.8	30.9		305	423
26.5-25	26	22.00-3	30.6	50	21300	27.4	68.9	29.8	30.9		305	423
26.5-29	18	22.00-3	30.6	30	16900	28.2	72.8	30.2	32.9		286	388
26.5-29	22	22.00-3	30.6	40	19970	28.2	72.8	30.2	32.9		286	388
26.5-29	24	22.00-3	30.6	45	21400	28.2	72.8	30.2	32.9		286	388
26.5-29	26	22.00-3	30.6	50	22770	28.2	72.8	30.2	32.9		286	388
29.5-25	16	25.00-3½		25	17830	30.1	73.0	33.1	31.9		295	556
29.5-25	22	25.00-3½		35	21710	30.1	73.0	33.1	31.9		295	556
29.5-25	28	25.00-3½		45	25160	30.1	73.0	33.1	31.9		295	556
29.5-29	16	25.00-3½		25	18990	29.8	76.6	32.7	34.8		270	580
29.5-29	22	25.00-3½		35	23120	29.8	76.6	32.7	34.8		270	580
29.5-29	28	25.00-3½		45	26790	29.8	76.6	32.7	34.8		270	580
29.5-29	34	25.00-3½		55	30140	29.8	76.6	32.7	34.8		270	580
29.5-35	28	25.00-3½		45	29160	30.0	82.3	32.6	37.5		251	501
29.5-35	34	25.00-3½		55	32810	30.0	82.3	32.6	37.5		251	501
29.5-35	28	27.00-3½		45	29160	30.6	82.1	33.0	37.4		252	501
29.5-35	34	27.00-3½		55	32810	30.6	82.1	33.0	37.4		252	501
33.25-35	32	27.00-3½		45	34710	33.5	88.2	37.3	39.0		239	756
33.25-35	38	27.00-3½		55	39050	33.5	88.2	37.3	39.0		239	756
33.5-33	26	28.00-4		35	30820	33.4	87.6	36.8	38.4		243	700
33.5-33	32	28.00-4		45	35710	33.4	87.6	36.8	38.4		243	700

Tire Size	PR	Rim	PR	Load	Width	OD	SW	SLR	Rev	Wt
33.5-33	38	28.00-4	55	40180	33.4	87.6	36.8	38.4	243	700
33.5-33	44	28.00-4	65	44160	33.4	87.6	36.8	38.4	243	700
37.5-33	30	32.00-4½	40	40470	38.1	93.5	41.8	40.7	229	769
37.5-33	36	32.00-4½	45	43370	38.1	93.5	41.8	40.7	229	769
37.5-33	42	32.00-4½	55	48800	38.1	93.5	41.8	40.7	229	769
37.5-33	54	32.00-4½	70	56250	38.1	93.5	41.8	40.7	229	769
44.5-45	18	37.00-5	20	43820	46.6	117.7	51.5	50.8	184	1376
44.5-45	28	37.00-5	30	55550	46.6	117.7	51.5	50.8	184	1376
44.5-45	38	37.00-5	40	65740	46.6	117.7	51.5	50.8	184	1376
44.5-45	48	37.00-5	55	79190	46.6	117.7	51.5	50.8	184	1376

Part B

Tire load limits at various inflation pressures

Tire Size	25	30	35	40	45	50	55	60	65	70	75
20.5-25	8650	9640 / 12	10540	11390 / 16	12210	12990 / 20	13740	14450 / 24			
23.5-25	11150 / 12	12420	13580 / 16	14680 / 20	15730	16730 / 24					
26.5-25	14190 / 14	15810 / 16	17280 / 20	18680	20020 / 24	21300 / 26	22530 / 28				
26.5-29	15170	16900 / 18	18470	19970 / 22	21400 / 24	22770 / 26	24140	25400	26620 / 34		
29.5-25	17830 / 16	19870	21710 / 22	23480	25160 / 28	26820	28350 / 34	29840	31270 / 40		
29.5-29	18990 / 16	21150	23120 / 22	25000	26790 / 28	28500	30140 / 34	31770	33300 / 40		
29.5-35	20670	23030	25170 / 22	27210	29160 / 28	31020	32810 / 34	34580	36240 / 40		
33.25-35	24600 / 20	27410	29950 / 26	32390	34710 / 32	36920	39050 / 38				
33.5-33	25310 / 20	28200	30820 / 26	33320	35710 / 32	37980	40180 / 38	42140	44160 / 44		
33.5-39	27270	30390	33210 / 26	35910	38480 / 32	40930	43290 / 38	45390	47570 / 44		

TABLE 16-7 Specifications for Wide-Base-Type Sure-Grip Lug-Tread Tires (Continued)
Part B

Tire Size	\multicolumn{11}{c}{Tire load limits at various inflation pressures}										
	25	30	35	40	45	50	55	60	65	70	75
33.5-33	25310/20	28200	30820/26	33320	35710/32	37980	40180/38	42140	44160/44		
33.5-39	27270	30390	33210/26	35910	38480/32	40930	43290/38	45390	47570/44		
37.5-33	30740	34250/24	37430	40470/30	43370/36	46130	48800/42	51400	53860	56250/54	
37.5-39	32990	36750	40170/28	43430	46540/36	49510	52370/44	55160	57800	60360/52	
37.5-51	37310	41570	45430	49120	52640/36	55990	59230/44	62360	65350	68250/52	
44.5-45	49930	55550/28	60790	65740/38	70420	74900	79190/48	83330	87330	91200	

NOTE: Underscored figure denotes maximum load and inflation for tire with ply rating shown beneath figure.
SOURCE: Goodyear Tire and Rubber Co.

Average work is exemplified by the highway building of Figure 12-55. The bottom-dump haulers have been and will be used on many jobs, such as highways, levees, and gravel pits in Ohio. The logical choice was wide-base chevron-tread tires, combining traction, flotation, and good wearing quality. In this case the trailing tire is better for traction than the driving tire, but it is probable that it was removed from the driving position and replaced by the better-wearing tire with more closely spaced chevrons for the rock job. Representative specifications are given in Table 16-7.

Selection for on-the-road–off-the-road tires is limited to the standard kind, as shown in Figure 16-28. A typical use is illustrated in Figure 13-31. Tread selection is generally similar to that of Figure 16-28, which emphasizes good wear rather than flotation and traction. Representative specifications are given in Table 16-6.

In addition to selection of tread configuration, one may also choose extra tread depth, thus providing greater life as well as increase in traction.

After kind, type, and tread design have been selected, size and ply rating are chosen in keeping with load. Size, within narrow limits, is controlled by the tire clearances of the machine. Thus the capacity for the given load is obtained by suitable choice of ply rating; the greater the ply rating, the greater the capacity. However, tire loadings are also a function of air pressure. Tables 16-6 and 16-7 give representative values for tire capacities according to size, ply rating, and pressure in pounds per square inch.

In selecting a tire, one must keep in mind the effects of load, speed, and inflation pressure on the life of the tire, as set forth in Figure 16-29. As is demonstrated in Figure 16-30, ambient temperature also affects tire life.

The Ton–Mile-per-Hour Rating System An important guide for the selection of a tire is the *ton–mile-per-hour* rating system. This is a rating based on the fact that tire failure, neglecting accidents to the tire, is a function of temperature level above the tolerance of the tire at which separation of plies and related failures occur.

The generation of heat depends on three factors: the load on the tire, or flexure per revolution; the speed of travel, or flexures per unit of time; and the temperature of the air surrounding the tire. The equation for the ton–mile-per-hour rating is:

$$\text{Ton-mi/h rating} = \text{mean tire load} \times \text{hourly average speed}$$

Mean tire load is the average load, tons, on the tire for the laden and unladen machine. Some machines, such as motor graders, operate unladen. Rarely, a machine such as a truck may operate laden in both directions. These variations must be taken into account when figuring mean tire load. Hourly average speed is the maximum average speed during any hour of operation. It is the greatest number of miles traveled in 1 h of work.

Table 16-8 gives ton-mph ratings for different sizes and types of tires according to ambient temperatures. The table emphasizes certain facts:
1. Ton-mi/h rating decreases with increase in ambient temperature.
2. Ton-mi/h rating decreases with increase in tread depth.
3. Ton-mi/h rating is greater for steel radial-ply tires than that for nylon bias-ply tires.

An example of the use of Table 16-8 is provided by the self-loading scrapers of Figures 13-18 and 13-20, which are working on both a short haul of 300 ft one way and a long haul of 2100 ft one way. The tires are wide-base 33.25 × 35—32PR with a mean tire load of 18 tons. The ambient temperature reaches 100°F. Tabulations for calculating tons-mi/h are:

Item	300-ft Haul	2100-ft Haul
Tons	18	18
Speed, mi/h	1.73	9.22
Tons-mi/h	31	166
Allowable tons-mi/h at 100°F (from Table 16-8):		
Nylon bias ply	190	190
Steel radial ply	430	430

Nylon bias ply is adequate for both hauls and steel radial ply is more than adequate. Actually the owner switched from nylon bias ply to steel radial ply after the original tires were run out, thereby increasing life from an average 2500 to an average 4000 h. Average tire cost for the four tires of the machine was reduced from $6.27 to $4.51 hourly, a saving of 28 percent.

TABLE 16-8 Ton–Mile-per-Hour System of Rating Large Tires for Rock-Earth Excavation

The mi/h values are a measure of the ability of the tire to absorb heat, which is the result of tire flexure and which causes tire failures exclusive of those caused by accidents to the tire. After the ton-mi/h rating is calculated for a given tire working under given conditions of load and speed, the value is compared with that of the table for a given temperature to appraise the ability of the tire to perform satisfactorily.

$$\text{Ton-mi/h rating} = \text{mean tire load} \times \text{hourly average speed}$$

Mean tire load is the average load on the tire in tons for laden and unladen conditions. Some machines work unladen and, rarely, some machines work always laden. These variations must be taken into account when figuring mean tire load. Hourly average speed is the maximum average speed during any 1 h of operation. It is the number of miles traveled during the hour or the overall travel speed.

After the ton-mi/h value is estimated or calculated for actual work, it may be used to select tires for original equipment of a new machine or to choose better tires for a machine in use.

Ton-mi/h (ton-km/h) Ratings

Tire size	Type	Ambient temperature, °F (°C)			
		60° (16°)	80° (27°)	100° (38°)	120° (49°)
18.00-25	Std. tread (E-3)	150 (219)	140 (204)	125 (182)	110 (161)
	Extra tread (E-4)	120 (175)	110 (161)	100 (146)	90 (131)
	Radial S.C.	210 (306)	195 (285)	175 (255)	160 (233)
18.00-33	Std. tread (E-3)	175 (255)	155 (226)	140 (204)	120 (175)
	Extra tread (E-4)	150 (219)	140 (204)	125 (182)	110 (161)
	Radial S.C.	250 (365)	225 (328)	200 (292)	180 (263)
21.00-35	Std. tread (E-3)	200 (291)	185 (270)	165 (241)	145 (211)
	Extra tread (E-4)	170 (248)	150 (219)	135 (197)	115 (168)
	Radial S.C.	300 (438)	270 (394)	240 (351)	205 (299)
24.00-35	Std. tread (E-3)	210 (306)	195 (285)	175 (255)	160 (233)
	Extra tread (E-4)	185 (270)	165 (241)	150 (219)	130 (190)
	Radial S.C.	335 (489)	305 (445)	275 (401)	240 (351)
26.5-29	Std. tread (E-3)	175 (255)	155 (226)	140 (204)	120 (175)
	Radial S.C.	340 (496)	310 (452)	280 (409)	245 (358)
29.5-29	Std. tread (E-3)	185 (270)	165 (241)	150 (219)	130 (190)
	Radial S.C.	400 (584)	360 (525)	320 (467)	280 (409)
29.5-35	Std. tread (E-3)	215 (314)	200 (292)	180 (263)	160 (234)
	Radial S.C.	530 (773)	450 (657)	370 (540)	280 (409)
33.25-35	Std. tread (E-3)	225 (328)	210 (306)	190 (277)	170 (248)
	Radial S.C.	530 (773)	480 (700)	430 (628)	380 (555)
33.5-39	Std. tread (E-3)	245 (356)	215 (314)	190 (277)	160 (233)
	Radial S.C.	570 (833)	510 (745)	445 (650)	380 (555)
37.5-39	Std. tread (E-3)	275 (401)	250 (365)	230 (336)	205 (299)
	Radial S.C.	580 (846)	520 (759)	455 (664)	390 (569)
37.5-51	Std. tread (E-3)	290 (423)	260 (379)	230 (336)	195 (285)
	Radial S.C.	690 (1008)	620 (905)	550 (803)	480 (700)
36.00-51	Std. tread (E-3)	500 (730)	440 (642)	390 (569)	350 (511)

NOTE 1: Ton-mi/h (t-km/h) rating = mean tire load × workday average speed.
NOTE 2: Consult your tire supplier for their current tire design and ton-mph rating.
NOTE 3: "Std. tread" means nylon bias-ply tire. "Extra tread" means nylon bias-ply tire with extra thick tread. Radial S.C. means radial steel ply tire.
SOURCE: Caterpillar Tractor Co.

The experience of the owner emphasizes the worth of the tons-mph analysis and suggests its use before tires for new machines are selected. All in all, it is manifest that tire selection is not an exact science and that the excavator should use all available help to get the best performance and economy from his expensive tires.

Life of Tires

There are three methods for estimating the life of tires in hours under given working conditions.

1. By the experience of the owner. This time-honored method is based on company records for identical or similar tires in specific work or in average work. It is the most dependable system because it is based on factual evidence.

2. By the combined experience of the owner and of manufacturers of tires and

machines using tires. This method is trustworthy for estimating purposes. Table 16-9 gives values for average tires used on most important machinery.

3. By using several factors affecting tire life as cumulative multipliers to be applied to an ideal life of 6000 h for the average tire. This method is reliable when no experience, as by methods 1 and 2, is available. It is also valuable as a check on the estimates of methods 1 and 2. It may be called a mathematical method. Table 16-10 is for this method.

The use of Table 16-10 may be illustrated for quarry hauling as shown in Figures 5-7, 13-37, and 13-38. From Table 16-10 these eight factors are assigned:

1. Tire maintenance, average	1.0
2. Maximum speed of machine, 10 mi/h	1.2
3. Curves of road bed, moderate	1.0
4. Nature of haul roads and work areas: hard-packed rock-earth, maintained	1.0
5. Tire loads: 10 percent overload	1.0
6. Wheel position on machine: rear-dump hauler	0.7
2 front	0.9
4 driving	0.6
Weighted average	0.7
7. Grades of roadbed, driving wheels only, firm surface:	0.9
10 percent maximum. Weighted average for 6 tires	
8. Miscellaneous working conditions, none	1.0

$$\text{Tire life, } h = (1.0 \times 1.2 \times 1.0 \times 1.0 \times 1.0 \times 0.7 \times 0.9 \times 1.0) \times 6000$$
$$= 0.76 \times 6000 = 4600$$

The actual average life of six 18.00×33—32PR nylon bias-ply tires of the 35-ton capacity truck is 4700 h. The range of life indicated by Table 16-9 for the existing favorable conditions is 3500 to 4500 h, averaging 4000 h. Thus, within reasonable estimating

TABLE 16-9 Estimates for Life of New and Retreaded Tires According to Working Conditions

	Tire life	
Machine and working conditions	New	Retreaded
Motor grader:		
Favorable	4500–6000	3400–4500
Average	3000–4500	2200–3400
Unfavorable	1500–3000	1100–2200
Bulldozer or bucket loader:		
Favorable	2500–3500	1900–2600
Average	1500–2500	1100–1900
Unfavorable	500–1500	400–1100
Scrapers:		
Twin-axle drive:		
Favorable	3000–4000	2200–3000
Average	2000–3000	1500–2200
Unfavorable	1000–2000	800–1500
Single-axle drive:		
Favorable	3500–4500	2600–3400
Average	2500–3500	1900–2600
Unfavorable	1500–2500	1100–1900
Bottom-dump haulers:		
Favorable	4500–6000	3400–4500
Average	3000–4500	2200–3400
Unfavorable	1500–3000	1100–2200
Rear-dump or side-dump haulers:		
Favorable	3500–4500	2600–3400
Average	2500–3500	1900–2600
Unfavorable	1500–2500	800–1900

NOTE 1: The working conditions are those for the traveled roadbeds in the cuts, main-line haulage roads, and in the fills. The average life includes failures due to accidents such as cuts, bruises, and blowouts.

NOTE 2: The table is based on the combined experiences of users and of the manufacturers of tires and machines equipped with tires.

TABLE 16-10 Estimates for Life of New Tire in Hours—Based on Eight Tire-Life Factors and a Basic Life of 6000 h

Tire life, h = 6000 (product of 8 factors)

Determining condition	Factor value
1. Tire maintenance, including installation:	
Excellent	1.1
Average	1.0
Poor	0.7
2. Maximum speed of Machine, mi/h:	
10	1.2
20	1.0
30	0.8
40	0.5
3. Curves of roadbed:	
None	1.1
Moderate	1.0
Severe:	
Single wheels	0.8
Dual wheels	0.7
Tandem wheels	0.6
4. Nature of haul roads and work areas:	
Snow, packed, and no exposed roadbed	3.0
Pavement, asphalt or concrete	1.0
Hard-packed rock-earth, maintained	1.0
Soft earth or sand, maintained	1.0
Soft earth or sand with some rock	0.8
Mud, ordinary	0.8
Mud, abrasive, or with some rock	0.5
Gravel, well maintained	0.9
Gravel, poorly maintained	0.7
Crushed stone, well maintained	0.8
Crushed stone, poorly maintained	0.6
Rock, soft:	
Ripped or well blasted	0.8
Poorly blasted	0.6
Rock, hard:	
Ripped or well blasted	0.6
Poorly blasted	0.4
5. Tire loads:	
50% underloaded	1.2
20% underloaded	1.1
Recommended by Tire and Rim Association	1.0
10% overloaded	1.0
20% overloaded	0.8
40% overloaded	0.5
6. Wheel position on machine:	
Front, steering, not driving	0.9
Driving:	
Motor grader, tandem drive	0.7
Bulldozer	0.6
Bucket loader	0.6
Scraper	0.6
Bottom-dump hauler	0.7
Rear-dump hauler	0.6
Side-dump hauler	0.6
Rear-dump hauler, tandem drive	0.7
Side-dump hauler, tandem drive	0.7
Trailing	1.0
7. Grades of roadbed, driving wheels only:	
Level	1.0
Firm surface:	
6% maximum	0.9
10% maximum	0.8
15% maximum	0.7
25% maximum	0.4

Determining condition	Factor value
Loose or slippery surface:	
6% maximum	0.6
10% maximum	0.6
15% maximum	0.4
8. Miscellaneous working conditions, as tire hazards:	
Favorable	1.5
None	1.0
Unfavorable	0.8
Very unfavorable	0.6

SOURCE: Goodyear Tire and Rubber Co.

precision, Tables 16-9 and 16-10 give results fairly close to the actual tire life, the errors being −15 and −2 percent, respectively, and the average being a conservative −8 percent.

Recapping of tires is common, although some excavators do not feel that recapping is feasible economically. According to industry averages, 75 percent of all tires can be recapped, provided this is done before the tread is worn down to the undertread. The following two examples represent extremes of total life of tires when recapped and illustrate experiences under hard and soft working conditions in the cut, on the haul road, and in the fill or at dumping point.

The first case is that of a quarry owner in California who, using 35-ton-capacity rear-dump trucks with 18.00×33—32PR bias-ply tires, recaps the tires once and then runs out the tires. Conditions are average for tire wear. The life of the original tire is 4700 and the life of the recap is 3800 h, giving a total life of 8500 h.

The second example involves the owner of a kaolin pit in Georgia who uses 24-cysm, all-wheel-drive scrapers for overburden removal in soft clays. The 33.5×33—38PR bias-ply tires are recapped three times and then run out. The conditions are favorable for extended tire wear. The life of the original tire is 4300 and the life of three retreads is 9700 giving a total life of 14,000 h.

Recapping cost averages 50 percent of new tire cost. It follows that, if the life of a recapped tire is greater than 50 percent of the life of the original tire, recapping is economical. The life of a recapped tire ranges between 75 and 85 percent of that of a new tire. The extreme cases of tire recapping in the quarry and in the kaolin pit result in the following comparisons and costs for 1978.

Item	Quarry	Clay Pit
Length of haul, one way, ft	1600	2000
Part of cycle time during which wear occurs, min	4.1	4.0
Load, tons	40	40
Relative hourly productions, based on part of cycle time and proportionate 30-min working hour, station tons	4680	6000
Cost for tires:		
Originals, 6 @ $2120	$12,720	
Recappings, 6 @ $848	5090	
Total for tires	17,810	
Repairs estimated at 15% × $17,810	2,670	
Total tire cost for 8500 h	$20,480	
Originals, 4 @ $3500		$14,000
Recappings, 12 @ $1400		16,800
Total for tires		30,800
Repairs estimated at 12% × $30,800		3,700
Total tire cost for 14,000 h		$34,500
Cost per station ton of haulage:		
$20,480/(8500×4680)	$0.00050	
$34,500/(14000×6000)		$0.00041

The difference of 18 percent in station-ton unit cost for tires reflects the difference between one recapping of tires in the hard quarry service and three recappings of tires in the relatively easy clay-pit service. In rock excavation the average practice is to recap once and then run out and discard the tire. Two recappings are sometimes practicable but rarely are three recappings possible.

Repairs to Tires

The estimated hourly cost of repairs is expressed as a percentage of the cost of the tire and its recappings divided by the total hours of service. It is an average percentage for both a new and a recapped tire. The percentages are:

General Working Condition	Percent of Total Hourly Cost of New and Recapped Tire
Favorable: work in residuals and maximum-weathered ripped rock-earth	12
Average: work in average- and minimum-weathered ripped rock-earth	15
Unfavorable: work in semisolid and solid blasted rock	17

Relative Costs of Nylon Bias-Ply versus Steel Radial-Ply Tires

The comparison of relative hourly costs will utilize actual costs based on the operation of the self-loading scrapers of Figures 13-18 and 13-20. The machines are used in soft sedimentary rocks on short and long hauls. All conditions are favorable to low tire costs.

The company first used nylon bias-ply tires and then changed over to steel radial-ply tires. New tires are mounted on the driving wheels of the tractor and old tires are shifted to the trailing wheels of the scraper. Because of the severe service on the driving and loading wheels of the tractor, which weakens tire carcasses, the company does not recap tires but runs them out.

The following tabulation gives comparative hourly 1978 tire costs for a scraper.

Item	Nylon Bias-Ply Tire	Steel Radial-Ply Tire
Tire size	33.25×35—32PR	33.25×35—32PR
Service load at 30 mi/h, lb	35,710	34,780
Actual maximum load carried, lb	38,600	38,600
Overload, %	8	11
Hours of service	2500	4000
Tire costs:		
4 tires:		
@ $3500	$14,000	
@ $4030		$16,120
Repairs, estimated at 12% of new cost	1,680	1,930
Total cost for tires	$15,680	$18,050
Hourly cost of tires:		
$15,680/2500	$6.27	
$18,090/4000		$4.51
Savings by steel radial-ply tires, 28%		$1.76

ENGINES, ENGINE-GENERATOR SETS, AND MOTORS

Engines

In rock excavation both gasoline and diesel engines are used extensively as prime movers for moving and stationary machinery. Largely because of savings in fuel costs, diesel engines predominate, especially when engine ratings above 150 hp are called for. Natural- and liquefied-gas engines and oil engines are used occasionally, as are steam engines, when fuels are available and when economic considerations warrant their use.

This discussion concerns gasoline and diesel engines as commonly used. Chapter 13 contains a section on the characteristics, fuel consumption, and economics of internal-combustion engines. Although the section is a prelude to the discussion of rock-hauling machines, it is applicable to all machinery and it is suggested that it be reviewed here.

Gasoline Engines These engines are used chiefly in horsepowers up to about 300, intermittent rating, for both automotive and general work. Figure 16-31 illustrates the cross section and performance curves for an engine with a maximum continuous gross

Engines, Engine-Generator Sets, and Motors 16–51

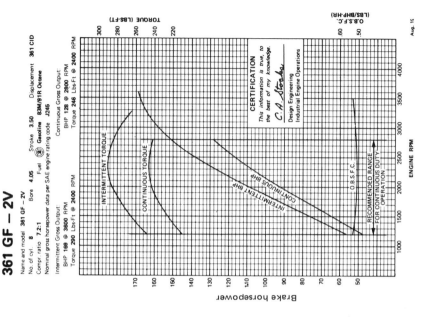

361 CID 8-cyl. Gasoline

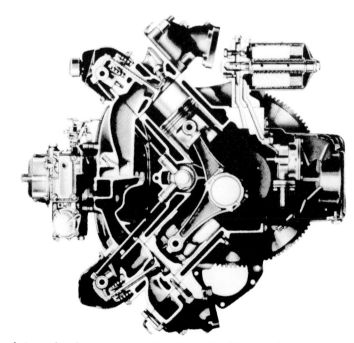

Figure 16-31 Sectional view and performance curves of gasoline engine. The versatile eight-cylinder, V-type engine may be used for automotive service, materials-handling machinery, or general work in rock excavation. For intermittent work, the gross output is 169 bhp at 3600 r/min and for continuous work the gross output is 128 hp at 2800 r/min. The torque and fuel consumption curves are characteristic of engines of these horsepower ratings. The tests are in accordance with those specified in SAE Standard J245, the conditions being 500 ft altitude and 85°F dry air temperature. *(Ford Motor Co.)*

output of 128 hp at 2800 r/min. Table 16-11 gives specifications for a large number of in-line and V-8 engines of a manufacturer.

Gasoline engines are of the four-cycle type, naturally aspirated. Fuel consumption averages 0.52 lb/bhp·h. If the engine of Figure 16-31 operates at a 67 percent load factor in intermittent service, it averages 113 bhp and consumes 113×0.52, or 59 lb, of fuel hourly, equivalent to 10.0 gal/h. Fuel data are given in Table 13-1.

Gas engines, using natural gas or liquid gas such as butane, are built with horsepowers up to 1000 for use in rock excavation. They are generally turbocharged and aftercooled or intercooled. They are discussed in Chapter 13. The terms *aftercooled* and *intercooled* mean the same thing, namely, that the horsepower of an engine has been increased by placing an aftercooler between the turbocharger and the intake port of the cylinder. Thereby the intake air is cooled before entry into the cylinder and greater density is attained as well as the greater volume of air produced by the turbocharger. The result is an increase of power because of the greater weight of oxygen available for combustion. Figure 16-32 illustrates the improved performance curves of a diesel engine resulting from intercooling or aftercooling. The aftercooler, through which the pressurized air passes, is cooled by the cooling system of the engine.

Diesel Engines Rock excavation machinery diesel engines range from 6 to 1600 hp. Both two- and four-cycle engines are built for natural aspiration, turbocharging, or intercooling. In small and medium sizes up to about 400 hp they are built both water-cooled and air-cooled.

Naturally aspirated and pressure-aspirated or turbocharged engines are discussed in Chapter 13. The increase in continuous horsepower provided by a turbocharged as against a naturally aspirated averages about 45 percent. Intercooling further increases the horsepower by about 21 percent. Thus by these two changes a naturally aspirated diesel engine may have its continuous horsepower increased by $[(1.45 \times 1.21) - 1.00]/1.00$ or 75 percent.

Diesel engines for automotive service are characterized by a high ratio of horsepower to weight. They are of the two- or four-cycle and of in-line or V-type design. They may be naturally aspirated or turbocharged with or without intercooling. A typical engine is illustrated in Figure 16-33 and its performance curves are shown in Figure 16-34. The specifications for some of the automotive engines of one manufacturer are given in Table 16-12.

An air-cooled engine of modular design is illustrated in Figure 16-35 and specifications for the complete family of small to medium engines are given in Table 16-13. Modular design is facilitated by the simplicity of air cooling and lends itself to ease of servicing by interchangeability of parts within the family of engines. Modular design is also typical of some medium and large water-cooled diesel engine families.

A large engine of a complete line of two-cycle diesel engines and its application to large off-the-road rear-dump trucks are illustrated in Figure 16-36. The specifications for 13 models of these engines, ranging from 64 to 1600 gross hp, are given in Table 16-14. The extensive range of models provides engines for on-the-road and off-the-road automotive-type machines, for other kinds of moving machines, and for stationary installations.

Production and Costs of Engines. The hourly cost of ownership and operation of an engine is generally a part of that of the complete machine of which the engine is the prime mover, as indicated in Chapter 8. Additionally, however, in Table 8-3 engines are treated individually for average costs.

Examples of hourly cost of ownership and operation are given in the section on internal combustion engines of Chapter 13. When a large engine is to be used in a stationary operation, as in a compressor plant of a large quarry, a complete study of operating conditions is made. The output, on which unit cost is based, is in terms of average cubic feet of air delivered per minute. It is on these bases that the engine is selected and the direct job unit cost estimated. An example of an estimate follows. It is desired to select

Engines, Engine-Generator Sets, and Motors 16–53

TABLE 16-11 Specifications of Gasoline Engines According to Displacement of Engine

LITER/CID	IRRIGATION	CONSTRUCTION	AGRICULTURE	GENERATOR SETS	AUTOMOTIVE	MAT. HANDLING	INDUSTRIAL	MARINE	POWER UNIT AVAIL	TYPE	COMP. RATIO	GROSS INTERMITTENT[1] BHP/RPM[4]	GROSS CONTINUOUS[2] BHP/RPM[4]	GROSS INTERMITTENT TORQUE FT. LBS.[4]	LENGTH	WIDTH	HEIGHT	WEIGHT
FORD GAS ENGINES																		
1.1L/67	●	●	●			●	●		No	I-4	8.0	32 @ 3600	21 @ 2800	48 @ 2800	24.16	20.08	24.50	240
1.6L/98	●	●	●			●	●		No	I-4	8.0	46 @ 3600	33 @ 2800	74 @ 2400	24.16	20.08	25.18	260
104								●	Yes	V-4	7.8	49 @ 3600	36 @ 2800	82 @ 2400	23.48	24.50	25.21	276
172	●	●	●	●		●	●		Yes	I-4	7.5	60 @ 2800	51 @ 2800	138 @ 1200	31.30	20.85	26.67	535
192	●	●	●	●		●	●		Yes	I-4	7.5	63 @ 2800	53 @ 2800	161 @ 1300	31.30	20.85	26.67	535
192	●	●	●	●		●	●		Yes	I-4	7.5	69 @ 2800	58 @ 2800	149 @ 1700	31.30	20.85	26.67	535
200					●				No	I-6	8.3	105 @ 4000	74 @ 2800	166 @ 2200	35.58	22.02	26.00	337
250					●				No	I-6	8.0	108 @ 3500	85 @ 2800	205 @ 2800	35.90	25.10	27.65	422
300	●	●	●	●		●	●		Yes	I-6	7.9	124 @ 3600	101 @ 2800	241 @ 2000	37.88	24.50	28.93	473
300	●	●	●			●			No	I-6	8.0	110 @ 3600	83 @ 2800	200 @ 2800	37.88	24.50	28.93	473
302	●	●	●			●	●		No	V-8	8.0	166 @ 4400	—	248 @ 2800	30.28	24.50	28.22	541
302					●				No	V-8	8.0	160 @ 4200[3]	—	254 @ 2200	30.28	24.50	28.22	480
302					●				No	V-8	7.9	185 @ 4200[3]	—	262 @ 3000	30.28	24.50	28.22	480
302					●				No	V-8	7.9	208 @ 4200[3]	—	277 @ 3400	30.28	24.50	28.22	480
330								●	Yes	V-8	7.4	156 @ 3600	114 @ 2800	262 @ 2000	35.53	32.01	33.62	720
351M					●				No	V-8	8.0	177 @ 4200	—	265 @ 2400	28.98	26.59	26.60	698
351W					●				No	V-8	8.0	197 @ 4400	—	290 @ 2500	30.16	26.55	28.78	601
351W				●					No	V-8	8.0	220 @ 4200[3]	—	334 @ 2800	30.16	26.55	28.78	569
351W				●					No	V-8	8.0	235 @ 4200[3]	—	330 @ 3200	30.16	26.55	28.78	569
361	●	●	●			●	●		Yes	V-8	7.2	169 @ 3600	128 @ 2800	290 @ 2400	35.53	32.01	33.62	730
391	●	●	●			●	●		Yes	V-8	7.2	192 @ 3600	148 @ 2800	333 @ 2400	35.53	32.01	33.62	721
460					●				No	V-8	8.0	157 @ 3200	128 @ 2800	310 @ 1400	31.63	25.85	30.18	713
460				●					No	V-8	8.0	244 @ 4400	—	375 @ 2600	31.63	25.85	30.18	776
460				●					No	V-8	8.0	320 @ 4600[3]	—	424 @ 3200	31.63	25.85	30.18	713
460				●					No	V-8	9.0	330 @ 4600[3]	—	458 @ 3200	31.63	25.85	30.18	713
477	●	●	●			●	●		Yes	V-8	7.3	215 @ 3400	175 @ 2800	359 @ 2100	40.21	34.07	35.81	1032
534	●	●	●			●	●		Yes	V-8	7.3	246 @ 3200	205 @ 3000	461 @ 2400	40.21	34.07	35.81	1032
534	●								No	V-8	7.3	203 @ 3200	161 @ 2800	368 @ 2400	40.21	34.07	35.81	1032

NOTE: Gross intermittent bhp and torque per SAE J245 (500 ft altitude and 85°F dry air); gross continuous bhp per SAE J245 (500-ft altitude and 85°F dry air); marine horsepower at 29.92 Hg and 60°F dry air

SOURCE: Ford Motor Co.

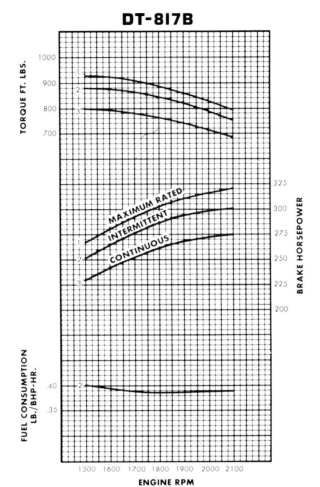

ALTITUDE CAPABILITY: No deration required up to 10,000 ft. altitude. Thereafter, derate 3% per 1000 ft. from curve #1.

Figure 16-32 The effect of intercooling on the performance of a turbocharged diesel engine. Performance curves: *(Above)* turbocharged engine; *(Opposite page)* turbocharged and intercooled engine.

The four-cycle diesel engine has six cylinders; bore and stroke are 5.375 and 6.000 in, respectively; and displacement is 817 in^3.

Item	DT-817B Turbocharged	DTI-817B, Turbocharged and Intercooled	Change by Intercooling, %
Brake horsepower at 2100 r/min:			
Maximum	320	420	+31
Intermittent	300	375	+25
Continuous	275	325	+18
Torque at 1500 r/min, ft·lb:			
Maximum	930	1210	+31
Intermittent	880	1100	+25
Continuous	800	940	+18
Fuel consumption at 1800 r/min, lb/bhp·h	0.39	0.36	−8

Significant are the increases in brake horsepower and torque and the decrease in fuel consumption. *(International Harvester Co.)*

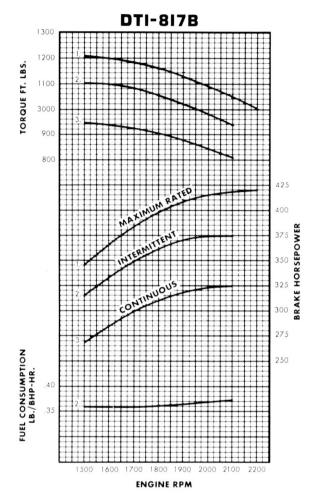

ALTITUDE CAPABILITY: No. deration required up to 7500 ft. altitude. Thereafter, derate 3% per 1000 ft. from curve #1.

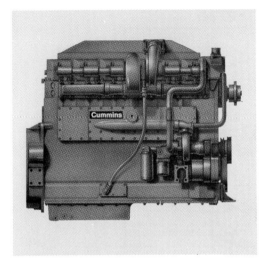 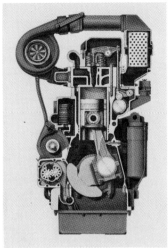

Figure 16-33 An automotive-type turbocharged, aftercooled diesel engine. The engine is Model KTA 600. The specifications for this low weight-to-power ratio, compact, high-horsepower engine are:

Intermittent bhp, without engine fan or alternator, at 2100 r/min	600
Maximum torque at 1600 r/min, lb·ft	1660
Number of cylinders, in-line	6
Bore and stroke, in	$6¼ \times 6¼$
Displacement, in^3	1150
Cycle	4
Dimensions, in:	
Length	62
Width	34
Height	51
Weight, with standard accessories, lb	3650
Weight-to-power ratio, lb/hp	6.1

When equipped with fan and alternator, the intermittent bhp at 2100 r/min is about 540 and the maximum torque at 1600 r/min is about 1490 ft. lb. *(Cummins Engine Co.)*

a diesel engine for a 2000-ft^3/min-rated compressor and to minimize the cost of powering the compressor for each 100 ft^3/min average delivery of air.

Criteria:
    1. The compressor works an average of 1600 h annually.
    2. The average delivery is estimated to be $67\% \times 2000$ or 1340 ft^3/min.
    3. The required rated brake horsepower of the engine for full 2000-ft^3/min delivery is estimated to be 29(2000/100) or 580, from the empirical formula of the section on air compression in Chapter 11. A 600-bhp engine is selected.
    4. The operator cost is estimated at 25 percent of the hourly cost of the compressor operator.

Costs, 1978:	
Cost of 600-hp diesel engine, clutch, skid-mounted	$35,900
Hourly cost of ownership and operation:	
Machine:	
Diesel engine (Table 8-3), $68\% \times \$35.9$	$24.40
Compressor operator, estimated at $25\% \times \$15.40$	3.80
Total hourly cost	$28.20
Direct job unit cost per 100 ft^3/min of average amount of air delivered, ($28.60/60)/(1340/100)	0.0356

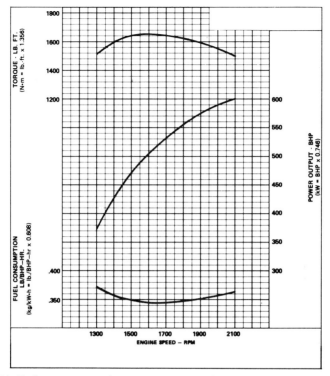

Figure 16-34 Performance curves for automotive-type turbocharged, aftercooled diesel engine. These performance curves for the Model KTA 600 engine of Figure 16-33 are for intermittent horsepower, typical of automotive service. The performances are in keeping with SAE J816b standards of 500-ft altitude and 85°F dry-air temperature. The engine is equipped with a water pump, an oil pump, a fuel system, an unloaded compressor, and an air cleaner. Fan and alternator are not included. Thus, one must adjust the values of the curves to compensate for the necessary fan and alternator. A conservative adjustment is a 10 percent decrease. Hence:

Intermittent bhp at 2100 r/min, 90% × 600	540
Maximum torque at 1600 r/min, lb·ft, 90% × 1660	1490
Fuel consumption at 2100 r/min, lb/bhp, (600/540) × 0.364	0.40

The curves are characteristic of diesel engines, especially the fairly flat torque curve, which indicates the lugging ability of the engine. *(Cummins Engine Co.)*

Engine-Generator Sets

For this practical and nontheoretical discussion of engine-generator sets and motors, the following definitions, units, and simple principles of electricity are useful.

Electricity The phenomena accompanying the movement of electrons and protons which give rise to a field of force possessing potential energy.

Ampere, I The practical unit of current flowing in a current of 1 ohm resistance when volt is impressed on it.

Volt, E Unit of electromotive force which, when applied to a circuit of one ohm resistance, causes a current of one ampere to flow.

Ohm, R The unit of resistance of a circuit in which a current of one ampere flows when subjected to an electromotive force of one volt.

TABLE 16-12 Specifications of Automotive-Type Four-Cycle Diesel Engines

Model	Displacement, in³	Speed range, r/min	Maximum intermittent hp	Maximum torque, ft·lb	Overall dimensions, in			Weight, lb
					Length	Width	Height	
NTC 250	855	1200–2100	250 at 2100 r/min	850 at 1300 r/min	60	34	50	2650
NTC 290	855	1200–2100	290 at 2100 r/min	930 at 1300 r/min	60	34	50	2750
NTC 350	855	1300–2100	350 at 2100 r/min	1065 at 1400 r/min	60	34	51	2795
NTC 400	855	1300–2200	400 at 2100 r/min	1150 at 1500 r/min	60	35	50	2850
KT 450	1150	1200–2100	450 at 2100 r/min	1350 at 1500 r/min	62	34	49	3430
KTA 600	1630	1300–2100	600 at 100 r/min	1660 at 1600 r/min	62	34	51	3650

NOTE 1: The above six engines are part of a line of 15 similar engines for automotive-type machines.
NOTE 2: Model designations: Letters indicate series of engines and numeral is the maximum intermittent horsepower. TC or T, turbocharged engine; TA, turbocharged and aftercooled engine. All engines are four-cycle six-cylinder in-line type.
NOTE 3: Engines are tested under SAE standard J1816b, 500-ft altitude and 85°F dry-air temperature. The tests include all engine accessories except fan, alternator, and optional equipment. A conservative adjustment for horsepower and torque values to accommodate fan and alternator is a reduction of 10 percent in the values.
NOTE 4: For large off-the-road scrapers and haulers the company offers the model VTA-1710-CS engine of 755 flywheel hp at 2100 r/min and maximum torque of 2200 ft·lb at 1550 r/min.
SOURCE: Cummins Engine Co.

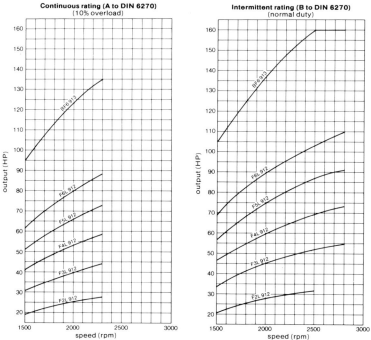

Figure 16-35 Four-cycle four-cylinder air-cooled diesel engine—performance curves for a group of similar modular engines. The engine, Model F4L 912, is rated at 41 continuous bhp, 73 intermittent bhp, and 80 maximum bhp. Its displacement is 230 in^3, and its speed range is 1500 to 2800 r/min. It is one of six engines with two, three, four, five, and six cylinders of the FL 912-913 series of modular engines. The horsepower curves give performances for the six engines in terms of continuous and intermittent ratings. Table 16-13 gives specifications for the complete line of modular engines with largely interchangeable parts. Such modular construction simplifies service and minimizes the inventory of replacement parts. *(Deutz Corp.)*

TABLE 16-13 Specifications of Small to Medium-Size Four-Cycle Air-Cooled Diesel Engines

Model	Optimal range, cont.- max. interm.	Speed range, r/min	Max bhp	Displacement, in³	Approx. main dimensions, in Length	Width	Height	Approx. wt, lb
F1L 208*	3–8	1500–3600	9	25	18	16 15/16	21 1/4	132
F1L 210*	7–14	1500–3000	16	41	18 1/4	21 3/8	24 15/16	176
F1L 411D*	6.5–14	1500–3000	15.7	42.5	16 7/16	20 5/8	26 5/8	242
F2L 411D*	13–28	1500–3000	31.5	85	21 3/8	20 5/8	26 5/8	322
Direct injection for optimum fuel economy								
F2L 912	19–32	1500–2500	35.5	115	23 15/16	22 15/16	30 3/8	518
F3L 912	30.5–55	1500–2800	60	172.5	27 3/4	26 1/8	32	595
F4L 912	41–73	1500–2800	80	230	31 7/8	26 1/8	31 5/8	661
F5L 912	51–92	1500–2800	100	287.5	37 1/4	25 15/16	33	837
F6L 912	61–110	1500–2800	120	345	42 5/16	26 1/8	32	903
BF6L 913	95–160	1500–2800	175	373.5	45 1/8	28 3/16	36 1/4	1069
F6L 413	93–170	1500–2650	185	517.5	39	40 1/4	33 3/4	1311
F8L 413	125–230	1500–2650	250	690	49 13/16	40 1/4	33 3/4	1668
F10L 413	155–285	1500–2650	310	862	56 9/16	40 1/4	35 5/16	2039
F12L 413	187–340	1500–2650	370	1035	62 5/8	40 1/4	35 5/16	2403
Two-stage combustion system for cleanest possible emissions (indoor and out underground)								
F2L 411W	12–24	1500–3000	29.5	85	21 3/8	20 1/8	26 5/8	322
F2L 912W	17–27	1500–2500	33.5	115	23 15/16	22 15/16	30 3/8	518
F3L 912W	27–43	1500–2500	50.5	172.5	27 3/4	26 1/16	32	595
F4L 912W	36–57	1500–2500	67	230	31 7/8	26 1/16	31 5/8	661
F5L 912W	45–72	1500–2500	84	287.5	37 1/4	25 15/16	33	837
F61 912W	54–86	1500–2500	101	345	42 5/16	26 1/8	32	903
F6L 714	58–150	1000–2300	165	580	43 1/4	48 5/8	39 3/4	1742
F8L 714	77.5–200	1000–2300	220	772.5	50 5/8	48 5/8	39 3/4	2061
F10L 714	96–250	1000–2300	275	964	58 3/8	48 5/8	39 3/4	2315
F12L 714	116–300	1000–2300	330	1159.5	64 7/8	48 5/8	39 3/4	2977
BF12L 714	165–370	1000–2300	390	1159.5	75 1/2	48 5/16	39	3153

*Also available for hand-starting. Liquid-cooled engines also available up to 9680 hp.

NOTE: These engines, ranging from 3 to 370 hp, continuous rating, are prime movers for many rock excavation machines. These include on-the-road and off-the-road haulers, bucket loaders, motor graders, air compressors, pumps, and generator sets. Air cooling provides simplicity of design and resultant modular construction, which simplifies the servicing problems of the buyer and the seller of the engines.

SOURCE: Deutz Corp.

Ohm's law A fundamental equation of electrical science for the relation among potential, current, and resistance. It is expressed algebraically in three ways:

$I = \dfrac{E}{R}$ Current varies directly with electromotive force and inversely with resistance.

$E = IR$ Electromotive force varies directly with current and resistance.

$R = \dfrac{E}{I}$ Resistance varies directly as electromotive force and inversely with as current.

Watt, W The unit of power to keep current flowing continuously is the product of voltage and amperage: $W = EI$.

Kilowatt, kW Customary unit of power for machines: $kW = 1000 W$.

The relationship between watts and horsepower, the conventional unit of mechanical power, is given by the equations: 1 hp = 746 W = 0.746 kW; 1 kW = 1.34 hp.

The foregoing equations concerning power apply only to direct current, dc, and they must be modified when applied to alternating current, ac. Whereas the effective power in a dc circuit may be calculated simply by the equation $W = EI$, E being measured by voltmeter and I by ammeter, the effective power in an ac circuit may not be given by the equation unless E and I are in phase. This condition is rarely found in practice.

The power of an ac circuit is given by the equation:

$$W = EI \cos \theta$$

where $\cos \theta$ is the *power factor* by which the apparent power, EI, must be multiplied to obtain the effective power. The angle θ is that angle by which voltage and current are out of phase. The power factor depends on the relative amounts of resistance, inductance, and capacity of the ac circuit, machines such as induction motors and transformers contributing to a low power factor. The average power factor is 0.80, the out-of-phase angle being 37°. The ideal power factor is 1.00, voltage and current being in phase. The ac synchronous motor can be made to have a power factor of 1.00. There are two ways of rating ac generators, or alternators: by kW output, based on a 0.80 power factor, and by kVA output, based on a 1.00 power factor.

A generator is a machine for converting mechanical power into electrical power and a motor is a machine which converts electrical power into mechanical power. Both machines work on about the same fundamental principles.

When a loop or coil of wire is revolved in a magnetic field, an electromotive force is induced of which the value at any instant depends on the speed of rotation, the magnetic field, and the size of the coil. The resulting current is alternating because the direction of the current through the coil is reversed twice during each revolution. This is the fundamental principle of the generator.

If the current is collected from each end of the coil by a ring and brush, the brushes being connected to an outside circuit, there results an alternating current. The assembly is an ac generator or alternator.

If the current is collected from each end of the coil by a commutator, a ring split into halves connected to the ends of the coil with two opposite brushes which are connected to an outside circuit, there results a dc current. This is because each brush connects first with one end of the coil and then with the other end as the current is reversed in the coil. The assembly is a dc generator or dynamo. The magnetic fields of both ac and dc generators are created by dc electromagnets.

When a loop or coil of wire is in a magnetic field and current is passed through it, the coil will move. The force exerted is proportional to the current and the strength of the field. This is the fundamental principle of the motor. The dc motor is built in the same manner as the dc generator. If the dc current passes through a commutator into the ends of the coil, the coil will carry an ac current and move in the magnetic field. The ac synchronous motor is built like an ac generator. If the ac current passes through a ring and brush at each end of the coil, the coil will carry an ac current and it will move in the magnetic field.

The practical application of these principles to generators and motors, together with their construction, is described in a simple manner below.

Figure 16-36 Large two-cycle turbocharged diesel engine powering 120-ton-capacity trucks. The 1000-flywheel-hp engines power twenty-eight 100- and 120-ton-capacity rear-dump trucks. The engine produces average 750 wheel hp at tire-ground contact, or 75 percent of flywheel horsepower. The average weights for the trucks are:

Tare weight	160,000 lb
Pay load	220,000
GVW	380,000 lb

The well-balanced engine and truck combination, when laden, travels at about 30 mi/h on a hard level grade and at about 6 mi/h up the 8 percent grades from the pit. The fleet of 28 trucks hauls 150,000 tons of overburden daily from the copper mine, averaging 5400 tons/day and 225 tons/h per truck. Continuous round-the-clock loading is by large electric shovels with 12-yd³ dippers.

The Caujone pit is operated by the Southern Peru Copper Corp. and it is located at a 12,500-ft elevation in the Andes Mountains. At this elevation the engine produces about 660 wheel hp instead of the calculated 750 wheel hp. *(Detroit Diesel Allison.)*

TABLE 16-14 Specifications of Industrial-Type Two-Cycle Diesel Engines

Model	Displacement, in³	Engine output, hp $\left(\dfrac{hp}{r/min}\right)$ Rated	Continuous	Maximum torque, ft·lb	Overall dimensions, in Length	Width	Height	Weight, lb
2-71	142	65 2000	48 1800	184 at 1400 r/min	32	27	41	960
3-53	159	98 2800	70 2400	205 at 1800 r/min	33	27	35	965
4-53	212	136 2800	93 2400	282 at 1800 r/min	39	27	37	1110
4-71	284	152 2100	117 1800	400 at 1600 r/min	42	29	42	1780
6V-53	318	210 2800	140 2400	445 at 1500 r/min	39	40	37	1485
6-71	426	228 2100	175 1800	600 at 1600 r/min	41	39	48	2010
6V-92	552	270 2100	225 1800	737 at 1400 r/min	41	39	47	1960
8V-92	736	360 2100	300 1800	983 at 1400 r/min	48	39	51	2345
16V-71	1136	608 2100	466 1800	1600 at 1600 r/min	79	45	58	4600
16V-92	1472	720 2100	600 1800	1966 at 1400 r/min	79	44	58	4606
12V-149T	1792	1000 1900	675 1800	2915 at 1500 r/min	91	63	69	9095
16V-149T	2389	1325 1900	900 1800	3880 at 1400 r/min	108	54	68	10,540
16V-149TI	2389	1600 1900	1080 1800	4595 at 1600 r/min	104	64	66	10,865

NOTE 1: The above 13 engines are part of a line of 30 similar engines for excavation machinery.
NOTE 2: Model designations: First numeral is number of cylinders; V denotes V-type engine; second numeral is series of engines; T denotes turbocharged engine; I denotes intercooled engine.
NOTE 3: *Rated power* is intermittent rating. Engine can be operated for about 1 h, followed by 1h of operation at or below the continuous rating.
NOTE 4: *Continuous power* is that at which engine can be operated without interruption or load cycling.
NOTE 5: Engines are tested under SAE standard J816b at 500-ft altitude and 85°F dry-air temperature. Tests include all accessories including fan and alternator, designated as fan to flywheel ratings.
SOURCE: Detroit Diesel Allison.

Engine-generator sets are machines for converting various kinds of energy to electricity or changing the form of electrical energy. Examples are: a diesel engine–generator set which converts the potential energy of fossil fuel to kilowatts; a steam turbine-generator set which converts nuclear energy to kilowatts; or an ac motor–dc generator set which converts alternating to direct current. In rock excavation the generally used machine is the diesel engine–ac generator set. Diesel-engine–ac and –dc generating sets are the subjects of this section.

The power output of the engine is in flywheel-brake horsepower, and this output corresponds to the input to the generator if the two machines are directly connected, which is the usual case. The output of the generator is expressed by the equation:

$$kW_g = \frac{bhp_e}{1.34} \times E$$

where kW_g = output of generator, kW
  bhp_e = output of engine or input of generator, bhp
  1.34 = ratio of 1 kW to 1 hp
  E = efficiency of generator

Practically, at an average 90 percent efficiency for 50- to 1000-kW generators working at three-fourths of full load, the equation reduces to:

$$kW_g = 0.67 \text{ bhp}_e$$

Engine-generator sets are used in rock excavation under the following conditions:
  1. When a large machine, such as a huge shovel or hauler, is equipped with an engine-generator-motor set which is an integral part of the machine
  2. When electrical energy is not available from the utility company
  3. When electrical energy is available but economic considerations warrant the use of the set rather than buying current from the utility company, especially if a transmission line must be built wholly or partly at the expense of the owner

Diesel Engine–DC Generator Sets A number of coils of wire wound on an iron or steel core, together with the commutator, comprise the armature of the generator. In this case the armature is called also a *rotor*. The magnetic field is produced by two, or any multiple of two, electromagnets or poles which surround the armature. These poles and the supporting frame are called the *field*. In this case the field is also called the *stator*. There are three kinds of generators, classified according to the method of exciting the field:
  1. *Series*, in which the field is in series with the armature. The voltage depends on the load.
  2. *Shunt*, in which the field is in parallel with the armature. The voltage may be controlled independently of the load by means of a variable resistance or rheostat placed in series with the field.
  3. *Compound*, in which the field is both in series and in parallel with the armature. This kind is commonly used as it affords approximately constant voltage at different loads.

Engine–dc generator sets are little used today in rock excavation. Most dc motors, being of large size, are powered by motor–generator sets receiving ac power to the prime mover.

Diesel engine–dc generator sets resemble their ac counterparts. General specifications are somewhat the same and the efficiencies of the generators are about the same. Table 16-15 gives efficiencies for dc and ac generators and Table 16-16 gives specifications for diesel engine–ac generator sets.

TABLE 16-15 Approximate Efficiencies of Generators as Determined by Test Performances

DC generators, compound wound with commutating poles

kW	Voltage	Amp.	Efficiency, %		
			½ load	¾ load	¼ load
5	125	40	78.0	81.0	82.0
10	125	80	83.5	87.0	88.0
25	125	200	86.5	88.5	89.5
50	125	400	86.0	89.0	90.0
100	125	800	88.7	89.3	89.0
200	125	1600	89.1	89.6	89.4
400	250	1600	91.7	91.9	91.7
1000	250	4000	92.1	92.6	92.1

AC generators, high-speed horizontal-coupled type

Capacity, kVA	Poles	Speed, r/min	Excitation kW at 125 V	Load		
				½	¾	Full
				Efficiency at 80% power factor		
50	6	1,200	2.25	83.2	85.5	86.1
100	6	1,200	3.00	87.8	89.2	89.6
250	8	900	5.5	89.3	90.8	91.1
400	10	720	7.75	91.6	92.7	92.9
1,100	10	720	13.5	92.8	94.0	94.2
2,000	10	720	20.0	93.5	94.7	95.1
4,000	8	900	22.5	93.9	95.2	95.7

SOURCE: Westinghouse Electric Corp.

Diesel Engine–AC Generator Sets When the armature of a dc generator is equipped with two sets of rings and brushes instead of the commutator, the machine becomes an ac generator, or alternator. The field is excited by direct current, which may be furnished by a rectifying commutator on the shaft of the armature or may be separately furnished by a generator driven by the alternator.

The ac generators may be divided into two kinds:

1. *Revolving armature:* This kind is in keeping with the principle of the revolving coil of wire. The armature is the rotor. Current is delivered from the brushes of the collecting rings.

2. *Revolving field:* The field is the moving member or rotor and the armature is the stator. In generating electric current by causing an inductor to cut magnetic lines, it makes no difference whether the cutting of the magnetic lines is effected by moving the inductor across the magnetic field or by moving the magnetic field across the inductor.

TABLE 16-16 Specifications for Diesel Engine–AC Generator Sets

Specification	Model					
	3304 T	3406 T	3412 T	348 TA	349 TA	399 TA
Engine:						
Aspiration	TC	TC	TC	TC-AC	TC-AC	TC-AC
Cycle, or strokes	4	4	4	4	4	4
Cylinders	4-IL	6-IL	12-V	12-V	16-V	16-V
Displacement, in^3	425	893	1649	1786	2382	3928
Flywheel-brake hp, without fan:						
60 Hz:						
Standby	120	305	532	890	1100	1310
Prime	95	270	486	775	980	1195
50 Hz:						
Standby	93	265	450	735	900	1100
Prime	93	225	412	670	820	990
Generator, three-phase:						
Ratings, kW, 80% power factor:						
60 Hz, 1800 r/min:						
Standby	75	200	350	620	750	
Prime	60	175	320	530	670	
50 Hz, 1500 r/min:						
Standby	60	175	300	510	620	
Prime	60	150	275	465	560	
60 Hz, 1200 r/min:						
Standby	...	...	...	...	...	930
Prime	...	...	...	...	...	850
50 Hz, 1000 r/min:						
Standby	...	...	...	...	...	775
Prime	...	...	...	...	...	700
Efficiency, 60 Hz, prime	85	87	88	92	92	95
Complete machine, skid-mounted:						
Dimensions, in:						
With radiator:						
Length	95	131	150			
Width	32	40	50			
Height	55	68	76			
Without radiator:						
Length	...	...	...	127	147	83
Width	...	...	...	62	61	60
Height	...	...	...	66	69	79
Weight, dry, lb	2645	6380	9167	12,350	15,220	22,700

NOTE 1: Voltages are from 120/208 to 300/600, varying somewhat with different models. When engines are equipped with a fan, flywheel-brake hp and generator outputs are reduced by about 4 percent.

NOTE 2: TC, turbocharged; AC, aftercooled; IL, in-line cylinders; V, cylinders in V; *standby*, continuous service during interruption of normal service; *prime*, continuous service.

NOTE 3: These six sets are part of a complete line of 21 sets of the manufacturer. These sets range in output for 60 Hz and prime service from 50 to 850 kW, 80% power factor.

SOURCE: Caterpillar Tractor Co.

16–66   Allied Operations, Machinery and Components, and Facilities

For generators of medium and large size, the armature should be stationary and the field should revolve, as this simplifies both circuitry and construction. This is the construction of the generators for the sets of Figure 16-38 and Table 16-16.

There are three chief types of ac generators based on the number of circuits in the armature.

1. *Single-phase.* One circuit in the armature, with two terminals. Single-phase alternators are used largely for lighting circuits.

2. *Two-phase.* Two circuits in the armature generating two electromotive forces 90° apart, with three or four terminals. Two-phase alternators are used for lighting circuits and for providing power for machines.

3. *Three-phase.* Three circuits in the armature generating electromotive forces 120° apart, with three or four terminals. Three-phase alternators are used for generating power for machines.

Two- and three-phase generators are known also as *polyphase alternators.*

In the smallest sizes, up to about 5 kW, the generator is single-phase. Three-phase generators of up to about 1000 kW, or 1250 kVA, capacity are used in excavation. Generators are 50 or 60 Hz (cycles per second), 60 Hz being commonly used. Voltages range from 110/220 to 220/440. Speed varies from 3600 r/min for small generators to 1200 r/min for the large machines.

Figure 16-37 illustrates a small diesel engine–ac generator set, such as may be used for a lighting plant. Figure 16-38 shows a medium-size set, skid-mounted. Such a set may be fully enclosed and mounted on wheels-tires for portability. This set may be used to supply current for field offices and shops for dam building in a remote area.

A preliminary cost estimate, based on 1978 costs, of the feasibility of substituting diesel engine–ac generator sets for power purchased from a utility company is presented below for the case of a quarry and aggregate plant for dam construction. The utility company estimates that for the 5-yr period of construction the average cost of delivered energy, including the prorated cost for a transmission line to the dam site, will be about $0.120/kWh.

Criteria for the estimate are:

1. Years of operation   5
2. Months of operation   40
3. Hours of operation, double shifting   14,080
4. Hours of operation annually   2816
5. Peak demand, kW   3000
6. Average demand, kW   1800
7. Average cost of diesel fuel during 5-yr period, estimated, per gallon $0.72

For a peak load of 3000 kW, four sets are selected, model 399TA, 60 Hz, of Table 16-16. The 1978 cost, fob dam site, is four machines at $128,000, or $512,000. The salvage value of the sets at completion of the job is estimated to be 33 percent of $512,000, or $169,000, leaving a $343,000 cost to be depreciated over the 5-yr construction period. The cost of the powerhouse is estimated to be $50,000 and the cost of erection of sets is estimated at $10,000. The total cost to be depreciated, with no salvage value for the powerhouse, is then $343,000+$50,000+$10,000, or $403,000. The total investment is $572,000. The average investment over the 5-yr period is 60 percent of $572,000, or $343,000.

Estimated hourly cost of ownership and operation:
Fixed charges:
  Interest, 8.0%; taxes, 2.0%; insurance, 2.0%; $14.60
  total, 12.0%. Average investment is $343,000:

$$\frac{12\% \times \$343{,}000}{2816} = \$14.60$$

Depreciation:
  Value to be depreciated over 14,080 h is   28.60
  $403,000; $403,000/14,080 = $28.60
  Total fixed charges: $43.20
Operating expenses:
  Repairs and replacements, including labor: hourly $71.70
  cost as a percentage of $1000s of cost, 14%
  (Table 8-3), 14% × $512 = $71.70

Fuel oil: The average demand is 1800 kW. At a conservative 90% generator efficiency, the average flywheel-brake hp of the engines is given by the equation:

$$bhp = \frac{1800 \times 1.34}{0.90} = 2680$$

At 0.42 lb or 0.059 gal/bhp·h, fuel consumption is $0.059 \times 2680 = 158$ gal/h; $158 \times \$0.72 = \$113.80$ 113.80
Oil, lubricants, and filters are estimated at 10% of fuel cost; $10\% \times \$157.50 = \$15.80$ 15.80
Total operating expenses $201.30
Operator   15.40
Total hourly costs $259.90
Direct job unit cost, per kWh of average demand $259.90/1800 = $0.144 $0.144

The preliminary estimate of $0.144/kWh is $0.024 higher than the preliminary quotation of the utility company, $0.120. The estimated power consumption for the 5-yr job is 25.3 million kWh. The difference in the total cost of power, favoring the utility company, amounts to $607,000.

Figure 16-37 **Small diesel engine–ac generator set.** This set is used to provide current to a portable lighting plant, as illustrated in Figure 16-17. Important specifications are:

Engine:	
Flywheel-brake hp	12
Cylinders, air-cooled	2
Displacement, in³	60
Speed, r/min	1800
Fuel tank capacity, gal	10
Generator:	
kVA	5.0
kW, Capacity at 100% power factor	5.0
Frequency, Hz	60
Voltage	110/220
Complete machine:	
Dimensions, in:	
Length	46
Width	34
Height	36
Weight, dry, lb	940

Small portable sets are built with up to about 15 kW capacity. *(Allmand Brothers Inc.)*

Figure 16-38 Medium-size diesel engine–ac generator set. The set is completely assembled with radiator-fan, all accessories, and all controls for immediate operation. Important specifications are:

Engine:
   Four-cycle, six cylinders inline, turbocharged and aftercooled
      Nominal rating, continuous service, 275 hp at 1800 r/min
      Displacement 638 in^3
Generator:
   Revolving field, brushless exciter, three-phase, 60 Hz, 1800 r/min, 120/208, 240/480, 300/600 V

Ratings	Kind of service	
	Standby	Prime
Capacity at 80% power factor, kW	175	150
kVA	219	188
Efficiency of generator, %	90	90
Engine brake hp at 1800 r/min	262	223
Complete machine:		
Dimensions, in:		
Length		115
Width		34
Height		57
Weight, wet, lb		4700

Standby service continuous during interruption of normal service. Prime service continuous. *(Caterpillar Tractor Co.)*

Nevertheless, a painstaking additional estimate of the owner's contemplated diesel engine–generator sets would be made, involving such matters as tax considerations, careful estimates of salvage values, respective dependabilities of the two systems, and availability of diesel fuel over the 5-yr period. Along with this estimate, further negotiations with the utility company would take place. Ultimately a well-considered decision would be made.

Motors

A motor is a machine which converts electrical energy into mechanical energy and it is not to confused with an internal combustion engine, which is sometimes erroneously called a motor. In excavation both dc and ac motors are used, the ac motors predominating. All motors are available in different constructions, such as open, protected, and enclosed types, in order to fit into operating conditions which sometimes are characterized by much dust, dirt, mud, snow, ice, and water.

DC Motors DC motors are built similar to dc generators. The armature, instead of giving current, receives alternating current through the commutator. The magnetic field of the field magnets causes the armature to rotate. Like dc generators, dc motors are classified with respect to the type of armature winding. There are three general kinds of motors, namely, series, shunt, and compound.

A *series* motor has its field and armature in series, and its characteristic is large torque and slow speed, which makes it ideal for machines requiring high torque at starting. A *shunt* motor has its field in parallel with the armature. The motor is in common use for general purposes as its speed decreases only a few percentage points from no load to full load. A *compound* motor has both series and parallel fields, and it is either accumulative or differential according to whether the series field strengthens or weakens the total field when the load increases. The accumulative form is ordinarily used, for it combines the high starting torque feature of the series motor with the constant speed characteristics of the shunt motor.

When starting, dc motors are not connected directly to the line because its resistance is low and excessive current would flow and would injure the motor. A motor is started by reducing the line voltage to a low value by inserting a resistance, a starting rheostat, in series with the armature. The resistance is cut out gradually as the motor comes up to speed. Table 16-17 gives data for dc motors of up to 200 hp, which are used extensively in excavation. Sometimes larger motors are used, an example being the 750-hp hoist motor for an electric shovel of 15-yd^3 dipper capacity.

As electrical energy is almost always distributed as alternating current, a converter must be used to change from the ac to the dc for the motor. The converter is a machine which employs mechanical rotation to change electrical energy from one form to another. It may be compared with a separate ac motor driving a dc generator and it is known as a motor-generator set. When it is completed with a dc motor, it is the Ward Leonard system, as described under Plate 12-1. An alternate source of direct current is the engine–dc generator set, as described in the previous section.

When dc power is produced by a converter, there is a significant loss of power through the two machines. The efficiency of the two machines is the product of their individual efficiencies. If the efficiency of a 500-hp ac motor is 95 percent and the efficiency of the directly connected dc generator is 92%, both at full load, the efficiency of the converter equals 95% × 92%, or 87 percent. The power output of the dc generator would be 325 kW.

In the case of the complete Ward Leonard system, ac motor–dc generator–dc motor, the efficiency from input to ac motor to output of dc motor is about 82 percent. Loss for each of the three machines averages about 6 percent. The efficiencies of the commonly used compound-wound dc motors are given in Table 16-17. The average efficiency at full load in the 25- to 200-hp range is 90 percent.

The contemplated use of dc motors with their desirable characteristics, as opposed to ac motors with their simplicity and mostly continuous alternating current, warrants a careful study of all factors affecting the cost of power for mobile and stationary machines.

AC Motors Because of the availability of alternating current these motors are used extensively as prime movers for machines. There are two kinds, synchronous and induction.

A *synchronous* motor is essentially an ac generator or alternator operated as a motor. It must be brought up to synchronous speed by external means so that it will run at line frequency before it can be connected to the line. It operates only at synchronous speed. These motors are built in large sizes, greater than 200 hp, for service during which it is not necessary to stop the motor frequently.

Synchronous motors are classified as low speed, less than 450 to 500 r/min, and high speed, more than 450 to 500 r/min. The full-load efficiencies for low-speed motors in the 100- to 700-hp range vary from 91 to 95 percent, averaging 92 percent. The full-load efficiencies for high-speed motors in the 100- to 1000-hp range vary between 92 and 96 percent, averaging 94 percent.

The usual type of synchronous motor with rotating field is started by applying voltage to the stator winding with the field circuit closed. The motor then starts up as an induction motor, producing torque to bring the motor and its load up to within a few percentage of synchronous speed. At that speed the dc field is excited and the motor is brought up to synchronous speed.

The *induction* motor is the most used ac motor. They include small, single-phase motors for low horsepower and large, two- or three-phase, or polyphase, motors, used generally in excavation. Their desirable characteristics are similar to those of the dc shunt motor, as they have reasonable starting torque and operate at approximately constant speed. Polyphase motors range from 1 to 1000 hp and there are two types in general use, squirrel cage and slip-ring.

The *squirrel-cage motor* has an armature or rotor made up of heavy copper bars short circuited on each other and with no external connections. Current is led through the field coils only, and the armature is rotated by currents induced by the varying field set up through the field coils. The motor has no wearing parts except the bearings and there is no sparking because there are no brushes. It is simple in construction and reliable in

TABLE 16-17 Data of DC Motors Used in Excavation

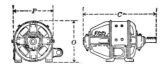

A. Dimensions and weights of open-type general-purpose constant-speed motors

HP	Approx. wt., lb	C	O	P	Approx. wt., lb.	C	O	P	Approx. wt., lb.	C	O	P	Approx. wt., lb.	C	O	P
	3450 r/min				1750 r/min				1150 r/min				850 r/min			
5	...	...	...	...	...	...	...	...	...	...	...	...	450	32	19	19
7½	...	...	...	...	...	...	...	...	450	32	19	19	500	34	19	19
10	...	...	...	...	450	32	19	19	500	34	19	19	650	38	21	21
15	450	32	19	19	500	34	19	19	650	38	21	21	725	40	21	21
20	500	34	19	19	650	38	21	21	725	40	21	21	1000	43	24	24
25	500	34	19	19	725	40	21	21	950	43	24	24	1000	43	24	24
30	650	38	21	21	725	40	21	21	1000	43	24	24	1325	49	28	27
40	725	40	21	21	950	43	24	24	1000	43	24	24	1750	54	31	30
50	950	43	24	24	1000	43	24	24	1325	49	28	27	1950	55	31	30
60	...	...	...	...	1000	43	24	24	1750	54	31	30	2600	63	35	34
75	...	...	...	...	1325	49	28	27	1950	55	31	30	2800	64	35	34
100	...	...	...	...	1950	55	31	30	2600	63	35	34	2800	64	35	34
125	...	...	...	...	2600	63	35	34	2800	64	35	34	3450	66	36	34
150	...	...	...	...	2800	64	35	34	3500	68	36	34	4600	78	41	39
200	...	...	...	...	3500	68	36	34	4600	78	41	39	5300	80	41	39
	690 r/min				575 r/min				500 r/min				450 r/min			
2	...	...	...	...	...	...	...	...	450	32	19	19	450	32	19	19
3	450	32	19	19	450	32	19	19	450	32	19	19	500	34	19	19
5	450	32	19	19	500	34	19	19	650	38	21	21	650	38	21	21
7½	500	34	19	19	650	38	21	21	725	40	21	21	725	40	21	21
10	650	38	21	21	725	40	21	21	950	43	24	24	1000	43	24	24
15	950	43	24	24	1000	43	24	24	1200	48	28	27				
20	1000	43	24	24	1325	49	28	27	1750	54	31	30				
25	1325	49	28	27	1750	54	31	30	1950	55	31	30				
30	1750	54	31	30	1950	55	31	30	2600	63	35	34				
40	1950	55	31	30	2600	63	35	34	2600	63	35	34				
50	2600	63	35	34	2800	64	35	34	3100	64	36	34				
60	2800	64	35	34	3100	64	36	34	3450	66	36	34				
75	3100	64	36	34	3450	66	36	34	3450	66	36	34				
100	3450	66	36	34	4500	73	41	39	5100	75	41	39				
125	4500	73	41	39	5100	75	41	39	6000	76	43	42				
150	5300	80	41	39	6000	76	43	42	6700	80	43	42				
200	6500	78	43	42	6900	86	43	42								
	400 r/min				350 r/min				300 r/min							
1½	450	32	19	19	450	32	19	19	500	34	19	19				
2	450	32	19	19	500	34	19	19	500	34	19	19				
3	500	34	19	19	650	38	21	21	650	38	21	21				
5	725	40	21	21	725	40	21	21	950	43	24	24				
7½	950	43	24	24	1000	43	24	24								

B. Test Performances of Compound-Wound Motors

	115 V		230 V		550 V	
HP	Amps	Efficiency at full load	Amps.	Efficiency at full load	Amps.	Efficiency at full load
1	8.4	77.0	4.25	77.0	1.77	73.0
2	16.2	79.0	8.2	78.0	3.20	82.0
5	41.5	79.5	20.0	80.0	8.2	81.0
10	75.4	85.0	37.5	86.0	15.6	86.5
25	112.0	87.0	72.0	87.5	38.1	88.5
50	...	...	181.0	88.5	76.0	89.5
100	...	...	352.0	90.5	148.0	91.0
200	...	...	700.0	91.5	295.0	92.0

SOURCE: Westinghouse Electric Corp.

operation. The motors have been standardized with different operating characteristics for various applications.

The squirrel-cage motor with its load may be started directly across the ac line or at a reduced voltage by means of a compensator. A compensator is a single winding transformer with several taps, to which the motor is connected successively during the starting operation.

The *slip-ring motor* has a rotor or armature with regular polyphase distributed winding, into the phases of which an external variable resistance is connected through collector rings on the shaft. By such construction the motor is adapted to variable-speed service. Starting and speed variation are accomplished by means of the variable resistance. The starting torque is higher than that of the squirrel-cage motor, thus permitting heavy loads to be started slowly and smoothly.

Table 16-18 gives weights and dimensions and test performances for ac general-purpose squirrel-cage induction motors used in excavation. In the range of 20 to 200 hp the average efficiency for 1200- to 1800-r/min motors at full load is 92 percent. Squirrel-cage and slip-ring polyphase induction motors are illustrated in Figure 16-39. Figure 16-40 shows the installation of a 300-hp three-phase induction motor of the squirrel-cage type.

NEMA Design or Classification of Electric Motors There are many variations in the basic kinds of ac and dc motors, which make it difficult for the earthmover to select the right motor for a given job unless he is an expert in the field of motors and their applications. As assistance, the National Electrical Manufacturers Association (NEMA) provides Table 16-19 of NEMA designs and applications for ac polyphase induction motors. The notes accompanying the table are of further assistance in the selection of both ac and dc motors. In all cases one should consult the manufacturers of motors or the experts in the field of motors because the right choice of motor and controls determines the performance of the motor and the complete machine for which it furnishes power.

Production and cost estimations for an ac squirrel-cage polyphase induction motor are outlined below for the example of the 300-hp motor driving a 50 in × 60 in jaw crusher, as illustrated in Figure 16-40. The crusher setup is shown in Figure 13-38. It is desired to estimate the 1978 unit cost per ton of rock crushed for driving the crusher by the electric motor.

Criteria:
 Estimated average output of motor, hp 225
 Estimated efficiency of motor, % 94
 Cost of electrical energy per kWh $0.055
 Cost of motor and controls $10,900
 Average hourly output of crusher, tons 500
 Hourly cost of ownership and operation of motor:
  For a 300-hp motor Table 8-3 indicates that the hourly cost, excluding the cost of electrical energy, is 27% × $10.9, or $2.90.
  The cost of energy is given by the equation:

  Hourly cost = (kWh input to the motor) × $0.055

16-72 Allied Operations, Machinery and Components, and Facilities

Substituting in the equation,

$$\text{Hourly cost} = \frac{0.746 \times 225}{0.94} \times \$0.055 = \$9.80$$

Hourly cost of ownership and operation = $2.90 + $9.80 = $12.70
Direct job unit cost per ton of rock crushed, $12.70/500 = $0.025

NOTE: It is probable that the owner of a permanent plant would depreciate the motor over a longer period than the 8 yr of Table 8-3. In such a case the 27 percent factor would be lessened and the estimated hourly cost of ownership and operation would be lowered to produce a lower direct job unit cost.

TRANSPORTING MACHINERY

Machines may be self-transported or carried by train, ship, or highway carrier. Rarely, they are carried by airplanes and helicopters. Self-transportation over the

(a)

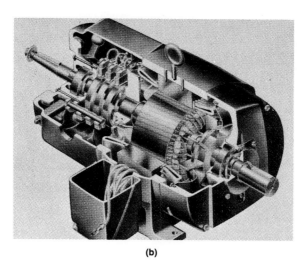

(b)

Figure 16-39 Constructions of alternating-current polyphase induction motors. *(a)* Squirrel-cage explosion-proof motor with tube-type air-to-air heat exchanger, used for adverse operating conditions. The characteristics are simple construction, low maintenance, good starting torque, constant speed, and high efficiency. *(b)* Slip-ring protected motor, used for average working conditions. The characteristics are relatively simply construction, average maintenance, excellent starting torque, variable-speed operation, and average efficiency. *(Allis Chalmers Co.)*

TABLE 16-18 Data of AC General-Purpose Squirrel-Cage Induction Motors Used in Excavation

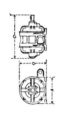

A. Approximate Weights and Dimensions of General-Purpose Squirrel-cage, Induction Motors
(Open type, 60 cycles. Dimensions in inches.)

HP	1800 r/min Wt.,lb	A	C	O	1200 r/min Wt.,lb	A	C	O	900 r/min Wt.,lb	A	C	O	720 r/min Wt.,lb	A	C	O	600 r/min Wt.,lb	A	C	O	514 r/min Wt.,lb	A	C	O	450 r/min Wt.,lb	A	C	O
25																									1550	28	46	29
30																					1550	28	46	29	1800	28	50	29
40																					1800	28	50	29	2000	31	48	32
50													1800	28	50	29	1800	28	50	29	2000	31	48	32	2400	31	54	32
60									1800	28	50	29	2000	31	48	32	2000	31	48	32	2400	31	54	32	3000	31	57	32
75	1800	28	44	30	1550	28	46	29	2000	31	48	32	2400	31	54	32	2400	31	54	32	3000	31	57	32	4000	39	57	40
100	2000	28	47	30	1800	28	50	29	2400	31	54	32	3000	31	57	32	3000	31	57	32	4000	39	57	40	4500	39	62	40
125	2100	28	47	30	2600	31	49	33	3000	31	57	32	4000	39	57	40	4000	39	57	40	4500	39	62	40	5000	39	64	40
150	2600	31	49	33	2600	31	49	33	3000	31	57	32	4000	39	57	40	4500	39	62	40	5000	39	64	40	5500	39	68	40
200	3100	31	52	33	3100	31	52	33	4000	39	57	40	4500	39	62	40	5000	39	64	40	5500	39	64	40	6500	50	67	50
250	3400	31	55	33	3400	31	55	33	4500	39	62	40	5000	39	64	40	5500	39	68	40	6500	50	67	50	7000	50	67	50
300	3400	31	55	33	3800	31	58	33	5000	39	64	40	5500	39	68	40	6500	50	67	50	7000	50	67	50	8000	50	70	50
350	3400	31	55	33	4500	43	53	43	5000	39	64	40	7000	50	67	50	7000	50	67	50	8000	50	70	50				
400	3800	31	58	33	5200	43	55	43	6000	43	65	43	7000	50	67	50	7000	50	67	50	8000	50	70	50				
500	5200	43	55	43	6000	43	65	43	6500	43	65	43	8000	50	70	50	8000	50	70	50	9000	50	74	50				

TABLE 16-18 Data of AC General-Purpose Squirrel-Cage Induction Motors Used in Excavation *(Continued)*
B. Test Performances

		Revolutions per minute					
		514	600	720	900	1200	1800
HP	Voltage	Approximate full-load efficiency					
20	220, 440, 550	85.6	86.5	87.5	88.3	89.0	89.4
	2200	84	85	86	86.5	87.5	87.8
25	220, 440, 550	86.5	87.3	88.2	88.9	89.6	89.8
	2200	85.4	86.1	87	87.7	88.4	88.8
30	220, 440, 550	87.2	87.9	88.7	89.4	90	90.3
	2200	86.5	87	87.8	88.5	89.1	89.4
40	220, 440, 550	88.2	88.8	89.6	90.1	90.6	90.9
	2200	87.8	88.2	89	89.6	90.1	90.4
50	220, 440, 550	88.9	89.6	90.3	90.6	91.2	91.4
	2200	88.8	89.2	89.8	90.3	90.8	91
60	220, 440, 550	89.5	90.1	90.6	91.1	91.5	91.7
	2200	89.4	89.8	90.5	90.8	91.3	91.4
75	220, 440, 550	90.2	90.8	91.1	91.5	92	92.1
	2200	90.2	90.8	91.1	91.4	91.8	91.9
100	220, 440, 550	91	91.4	91.8	92.1	92.5	92.6
	2200	91	91.4	91.8	92.1	92.5	92.5
125		91.6	92	92.2	92.5	92.9	92.9
150		92	92.3	92.6	92.9	93.1	93.2
200		92.7	93	93.1	93.4	93.6	93.6
250		93.2	93.4	93.6	93.7	93.9	93.9
300		93.5	93.7	93.8	94	94.1	94.2
350	220	93.8	94	94.1	94.2	94.3	94.3
400	440	94.1	94.2	94.3	94.4	94.5	94.5
450	550	94.3	94.4	94.5	94.6	94.6	94.6
500	2200	94.5	94.6	94.6	94.7	94.8	94.8
600		94.8	94.8	94.9	95	95	95
700		95	95	95.1	95.2	95.1	95.1
800		95.2	95.2	95.2	95.3	95.3	95.3
900		95.4	95.4	95.4	95.4	95.4	95.4
1000		95.5	95.5	95.5	95.6	95.5	95.5

SOURCE: Westinghouse Electric Corp.

Figure 16-40 Alternating-current polyphase squirrel-cage induction motor. The 300-hp motor is driving a 50 in × 60 in primary jaw crusher handling an average of 500 tons of granite rock hourly. With a 6-in closed setting of the crusher jaws, about 225 hp is required, giving the 300-hp motor an ample capacity margin. The motor is rated at 300 hp, 460 V, 240 A, and 1185 r/min. It is three-phase, six-pole, 60 Hz. The NEMA classification is Class B. The location is a quarry and aggregate plant of San Diego County, California. *(South Coast Asphalt Co.)*

TABLE 16-19 NEMA Design or Classification for AC Polyphase Motors

NEMA DESIGN	STARTING TORQUE	BREAK DOWN TORQUE	STARTING CURRENT	SLIP	APPLICATION
A (curve: % full load torque vs % synchronous speed)	NORMAL	HIGH	NORMAL	LOW	For infrequent starts on moderately easy-to-start loads, requiring slightly more than full load starting torque. The motor's relatively high pull-out torque will sustain occasional short duration peak loads. *Examples:* Conveyors started unloaded, compressors started unloaded, pumps, lathes, grinders, drill presses, machine tools.
					Design A covers a wide variety of motors similar to Design B except that their maximum torque and starting current are higher. Motors of this design are not regularly offered but built to order for special applications.
B (curve)	NORMAL	HIGH	LOW	LOW	For loads were slightly more than full load torque and low slip required to sustain occasional overloads. *Examples:* Conveyors, pumps, machine tools, blowers, grinders, drill presses.
					Design B Motors are the standard general-purpose design. They have low starting current, normal torque, and normal slip. Their field of application is very broad and includes fans, blowers, pumps, and machine tools.
C (curve)	HIGH	NORMAL	LOW	LOW	For infrequent starts on hard-to-start loads requiring high starting (breakaway) torque. Not suitable for long accelerating times such as encountered in bringing high inertia loads up to speed; limited capacity for peak loads. *Examples:* Conveyors started loaded, compressors started loaded, pulverizers, electric stairways.
					Design C Motors have high breakaway torque, low starting current, and normal slip. The higher breakaway torque makes this motor advantageous for "hard-to-start" applications, such as plunger pumps, conveyors, and compressors.
D SLIP 5-8% AND 8-13% (curve)	HIGH	—	LOW	HIGH	
					Design D Motors have a high breakaway torque combined with high slip. 5 - 8% and 8 - 13% Slip Design D Motors are recommended for high inertia machinery, where it is desired to make use of the energy stored in a flywheel under heavy fluctuating load conditions. They are used for multi-motor conveyor drives where motors operate in mechanical parallel.
D SLIP 13% OR GREATER (curve)					For hard to start pulsating loads. Motor has no sharply defined breakdown torque point. Load peaks cause appreciable speed reduction which allows the inertia of a driven machine to carry a part of the peak load, thus reducing current peaks in the line. Also reversing service. *Examples:* Hoists, cranes, elevators, centrifuges and machines with large flywheels, multi-motor conveyor drives where motors operate in mechanical parallel.
					13% or Greater Slip Design D Motors also have high starting torque, but they are limited to short time duty. By developing a high breakaway torque and a high running horsepower in the smallest possible frame size, these motors find use on cranes, hoists, elevators, and for auxiliary movement of machine tools.

NOTE: Types and NEMA classifications of motors for various applications are as tabulated below:

TABLE 16-19 NEMA Design or Classification for AC Polyphase Motors *(Continued)*

Application of motor	Power or current supply	
	Polyphase ac	dc
Air compressors, reciprocating:		
Up to 100 hp	Sq-In Class A	Gp Cs Cw
Above 100 hp	Syn LS	Et LS
Air compressors, centrifugal	Sq-In Class A	Gp Cs Cw
Conveyors	Sq-In Class C	Gp Cs Cw
Cranes	Ch Wr	Ch Sr
Crushers	Sq-In Class B	Gp Cs Cw
Draglines	Sq-In or Wr-In Ch	Ch Sr
Dredges	Syn HS or LS	Gp Cs Cw
Drills	Sq-In Class A	Gp Cs Cw
Excavators, wheel-bucket	Ch Wr	Ch Sr
Generators, dc or ac:		
Up–5 kW	Sq-In Class A	Gp Cs Sh
5–20 kW	Sq-In Class B	Gp Cs Sh
20–40 kW	Sq-In Class A	Gp Cs Sh
Above 40 kW	Syn HS	Gp Cs Sh
Hauling machines using diesel engine–dc generator–dc motor wheel drive		Ch Sr
Hoists	Sq-In or Wr-In Ch	Ch Sr
Pumps, centrifugal:		
Up to 100 hp	Sq-In Class A	Gp Cs Sh
Above 100 hp	Syn HS or LS	Gp Cs Sh
Shovels	Sq-In or Wr-In Ch	Ch Sr
Slackline excavators	Sq-In or Wr-In Ch	Ch Sr

KEY: Ch—crane and hoist type; Cs—constant speed; Cw—compound wound; Et—engine type; Gp—general purpose; HS—high speed; In—induction; LS—low speed; Sh—shunt wound; Sq—squirrel cage; Sr—series wound; Syn—synchronous; Wr—wound rotor.
SOURCE: National Electrical Manufacturers Association.

highways is limited to wheels-tires-mounted machines with suitable dimensions, axle and wheel loadings, and distribution of the loadings. Train, ship, and air transportation is a well-established method, the regulations of which are controlled by the agencies. For most machinery the common method is by highway carriers, and this section is concerned only with this means of transportation. Carrying over the highways may be by the owner's haulers or by those of transportation companies. Sometimes, if unusually heavy loads are involved, the transportation companies may have the only suitable haulers.

Methods and Machinery

These depend on the weight and the size of the machine and the governing rules pertaining to axle and wheel loadings and axle arrangements for the highways to be traveled. Table 16-20 gives typical axle loadings and axle distributions for the California state highway system. If the vehicle conforms to these standards, the movement does not require a permit. However, many movements involve great loads and overwidths, and then a special multiaxle vehicle and a permit are required.

A not unusual case is illustrated in Figure 16-41, where a 91,000-lb scraper is being hauled by a highway tractor with lowboy trailer. By permit, a rear-axle loading of 31,000 lb is allowed and the tractor-lowboy carries the remaining 60,000 lb. Figure 16-42 illustrates a tilt-type trailer which meets legal requirements when it is loaded at rated capacity. Its two axles support 40,000 lb GVW, giving it a rated capacity of 15 tons. Figure 16-43 shows a nine-axle lowboy-type trailer which requires a permit. Its rated capacity is 112,000 lb.

Three-axle and five-axle machinery haulers, consisting of tractor and semitrailer, are either flatbed or gooseneck-lowboy types. The flatbed hauler is used generally for low machines where loading and unloading take place at height of bed. The lowboy machine is used for high heavy machinery where loading and unloading are done near ground level. Specifications for gooseneck-lowboy haulers are given in Table 16-21.

TABLE 16-20   California Legal GVW Loads According to Number of Axles and Axle Spacings, 1978

CHAPTER 5. WEIGHT
Article 1. Axle Limits

Maximum Weight on Single Axle or Wheels

35550. (a) The gross weight imposed upon the highway by the wheels on any one axle of a vehicle shall not exceed ( )[1] *20,000* pounds and the gross weight upon any one wheel, or wheels, supporting one end of an axle, and resting upon the roadway, shall not exceed ( )[2] *10,500* pounds. ( )[3], *except that the gross weight imposed upon the highway by the wheels on any front steering axle of a motor vehicle shall not exceed 12,500 pounds.*

(b) The gross weight limit provided for weight bearing upon any one wheel, or wheels, supporting one end of an axle shall not apply to vehicles the loads of which consist of livestock.

(c) The following vehicles are exempt from the front axle weight limits specified in this section:
 (1) Trucks transporting vehicles.
 (2) Trucks transporting livestock.
 (3) Dump trucks.
 (4) Cranes.
 (5) Buses.
 (6) Transit mix concrete trucks.
 (7) Motor vehicles that are not commercial vehicles.
 (8) Vehicles operated by any public utility furnishing electricity, gas, water, or telephone service.
 (9) Tractors with a front axle at least four feet to the rear of the foremost part of the tractor, not including the front bumper.
 (10) Trucks transporting garbage, rubbish, or refuse.

Amended Ch. 268, Stats. 1959. Effective Sept. 18, 1959.
Amended Ch. 831, Stats. 1970. Effective Nov. 23, 1970.
Amended Ch. 169, Stats. 1971. Operative May 3, 1972.
Amended Ch. 651, Stats. 1975. Effective January 1, 1976. Superseding Ch. 132.

The 1975 amendment added the italicized material and at the point(s) indicated deleted the following:

"(b) The gross weight limit provided for weight bearing upon any one wheel, or wheels, supporting one end of an axle shall not apply to vehicles the loads of which consist of livestock.

(c) The gross weight on the rear axle only of a bus shall not exceed 20,500 pounds.

(d) Vehicles having not more than two axles, designed by the manufacturer for the collection and transportation of garbage, rubbish, or refuse and which are used regularly for such collection and transportation by any person or any governmental entity engaged in the business of, or in providing the service of, collecting, transporting, and disposing of garbage, rubbish, or refuse may exceed the weight limitation imposed by this section when the excess weight is not more than 2,000 pounds and is on the rear axle only. The exemption granted by this subdivision shall only apply to such vehicles when engaged in the collection and transportation of garbage, rubbish, or refuse. If the weight on the rear axle is in excess of the provisions of this subdivision, the allowed load in pounds on the rear axle shall be 18,000 pounds maximum for the purpose of determining the amount of fine for such violation as specified in the table in Section 42030.

Subdivision (c) or (d) shall not be applicable to any highway when it would operate to prevent the State from receiving federal funds for highway purposes."

Computation of Allowable Gross Weight

35551. (a) The total gross weight in pounds imposed on the highway by any group of two or more consecutive axles shall not exceed that given for the respective distance in the following table:

Distance in feet between the extremes of any group of 2 or more consecutive axles	2 axles	3 axles	4 axles	5 axles	6 axles
4	34,000				
5	34,000				
6	34,000				
7	34,000				
8	34,000				
9	39,000	42,500			
10	40,000	43,500			
11	40,000	44,000			
12	40,000	45,000	50,000		
13	40,000	45,500	50,500		
14	40,000	46,500	51,500		

TABLE 16-20 California Legal GVW Loads According to Number of Axles and Axle Spacings, 1978 *(Continued)*

Distance in feet between the extremes of any group of 2 or more consecutive axles	2 axles	3 axles	4 axles	5 axles	6 axles
15	40,000	47,000	52,000		
16	40,000	48,000	52,500	52,500	
17	40,000	48,500	53,500	53,500	
18	40,000	49,500	54,000	54,000	
19	40,000	50,000	54,500	54,500	
20	40,000	51,000	55,500	55,500	55,500
21	40,000	51,500	56,000	56,000	56,000
22	40,000	52,500	56,500	56,500	56,500
23	40,000	43,000	57,500	57,500	57,500
24	40,000	54,000	58,000	58,000	58,000
25	40,000	54,500	58,500	58,500	58,500
26	40,000	55,500	59,500	59,500	59,500
27	40,000	56,000	60,000	60,000	60,000
28	40,000	57,000	60,500	60,500	60,500
29	40,000	57,500	61,500	61,500	61,500
30	40,000	58,500	62,000	62,000	62,000
31	40,000	59,500	62,500	62,500	62,500
32	40,000	60,000	63,500	63,500	63,500
33	40,000	60,000	64,000	64,000	64,000
34	40,000	60,000	64,500	64,500	64,500
35	40,000	60,000	65,500	65,500	65,500
36	40,000	60,000	66,000	66,000	66,000
37	40,000	60,000	66,500	66,500	66,500
38	40,000	60,000	67,500	67,500	67,500
39	40,000	60,000	68,000	68,000	68,000
40	40,000	60,000	68,500	70,000	70,000
41	40,000	60,000	69,500	72,000	72,000
42	40,000	60,000	70,000	73,280	73,280
43	40,000	60,000	70,500	73,280	73,280
44	40,000	60,000	71,500	73,280	73,280
45	40,000	60,000	72,000	76,000	80,000
46	40,000	60,000	72,500	76,500	80,000
47	40,000	60,000	73,500	77,500	80,000
48	40,000	60,000	74,000	78,000	80,000
49	40,000	60,000	74,500	78,500	80,000
50	40,000	60,000	75,500	79,000	80,000
51	40,000	60,000	76,000	80,000	80,000
52	40,000	60,000	76,500	80,000	80,000
53	40,000	60,000	77,500	80,000	80,000
54	40,000	60,000	78,000	80,000	80,000
55	40,000	60,000	78,500	80,000	80,000
56	40,000	60,000	79,500	80,000	80,000
57	40,000	60,000	80,000	80,000	80,000
58	40,000	60,000	80,000	80,000	80,000
59	40,000	60,000	80,000	80,000	80,000
60	40,000	60,000	80,000	80,000	80,000

(b) In addition to the weights specified in subdivision (a), two consecutive sets of tandem axles may carry a gross load of 34,000 pounds each, providing the overall distance between the first and last axles of such consecutive sets of tandem axles is 36 feet or more.

(c) The distance between axles shall be measured to the nearest whole foot. When a fraction is exactly six inches, the next larger whole foot shall be used.

(d) Nothing contained in this section shall affect the right to prohibit the use of any highway or any bridge or other structure thereon in the manner and to the extent specified in Article 4 (commencing with Section 35700) and Article 5 (commencing with Section 35750) of this chapter.

SOURCE: Doris Alexis, director, "Motor Vehicle Code, 1978," Department of Motor Vehicles, Sacramento, Calif., p. 466, Div. 15.

Figure 16-41 Haulage of machinery by carry-tow method. The 91,000-lb scraper is being hauled with a permit over county and state highways. The permit allows an axle loading of 31,000 lb for the rear axle of the scraper, leaving 60,000 lb to be carried by the five-axle tractor–lowboy trailer hauler. The approximate weight distributions in pounds are:

	Tare Weight	Load Weight	GVW
Machinery hauler:			
Tractor:			
Front axle	8,000	2,000	10,000
Rear axles	14,000	26,000	40,000
Lowboy trailer:			
Rear axles	8,000	32,000	40,000
Total	30,000	60,000	90,000
Scraper rear axle		31,000	31,000
Total	30,000	91,000	121,000

Transportation is over county and state highways in San Diego County, California. *(Templeton Co.)*

Figure 16-42 Tilt-type trailer of 15 tons capacity. The tilt-type trailer is a simple machine for easy loading, hauling, and unloading of small and medium-size machinery. It is built in single-, double-, and triple-axle forms with respective capacities of about 8, 15, and 22 tons. The deck is hinged at the back of the main frame so that when the load, under its own power or by winch, moves up the inclined deck beyond the hinge point, the deck drops to the carrying position and is latched to the main frame. When unloading, the procedure is reversed. The tare weight is approximately one-third of the payload, resulting in a most economical carrier. The three-axle machine can carry loads up to that of a 100-hp crawler-tractor-bulldozer. *(Utility Trailer Manufacturing Co.)*

TABLE 16-21 Representative Specifications for Gooseneck-Lowboy-Type Machinery Haulers

	Rated capacity, lb		
Specification	30,000	60,000	112,000
Number of axles, one steering	3	5	9
Tractor:			
Number of axles	2	3	3
Engine flywheel-brake hp	250	325	450
Transmission:			
Gears forward	10	12	12
Maximum speed:			
mi/h	60	60	60
ft/min	5280	5280	5280
Tires:			
Front, 2	10.00×22	10.00×22	10.00×22
Rear, 4 for 2-axle and	12 PR	12 PR	12 PR
8 for 3-axle trailer	All tires	All tires	All tires
Weight, lb	10,000	15,000	17,000
Trailer:			
Number of axles	1	2	6
Tires, all	10.00×22	10.00×22	10.00×22
	12 PR	12 PR	12 PR
Weight, lb	8,000	13,000	39,000
Complete machine:			
Dimensions, ft:			
Length	40.7	52.8	95.5
Width	8.0	8.0	10.0
Height	8.5	9.1	9.4
Length of bed	26.5	39.8	51.5
Weights, lb:			
Tare, working	18,000	28,000	56,000
Payload, rated on-the-road	30,000	60,000	112,000
GVW	48,000	88,000	168,000
Ratio of payload to tare weight	1.67	2.14	2.00
Speed indicators:			
Laden: $\dfrac{\text{Brake hp}}{\text{GVW, 1000s of lb}}$	5.21	3.69	2.68
Unladen: $\dfrac{\text{Brake hp}}{\text{Tare weight, 1000s of lb}}$	13.89	10.48	8.04
Work-cost indicator:			
$\dfrac{\text{GVW} + \text{tare weight}}{\text{Load weight}}$	2.20	2.09	2.00

NOTE: The three- and five-axle haulers may travel without permits with legal loads, but the nine-axle hauler requires a permit because of excessive number of axles, length, and width. Different states and different agencies governing use of the highways have different standards. The data of the above specifications represent average standards.

Figure 16-43 **Large-capacity double-gooseneck-type machinery hauler.** This nine-axle hauler is moving a heavyweight tractor-ripper-bulldozer, weighing about 108,000 lb, over state highways of Washington. The state allows 179,200 lb GVW for this multiaxle machine, resulting in 56,100 lb tare weight and 123,100 lb allowable payload. The prime mover is a 450-hp tandem-drive tractor. The tare weights for the complete machine are:

Tractor	16,600 lb
Trailer, complete	39,400
Total machine	56,000 lb

With an allowable load of 123,100 lb, the ratio of payload to tare weight is 2.20, indicating a highly efficient hauler. The high-powered, 450-hp, engine of the tractor affords good travel speeds over both level and mountainous terrain. *(Royal Industries, Peerless Division.)*

Production and Costs for Transporting Machinery

Estimates, based on 1978 costs, are tabulated below for the example of the work of Figure 16-43. The criteria are:

Engine flywheel-brake hp	450
Wheel hp, $76\% \times 450$	342
Weights, lb:	
Tare, working	56,000
Payload	108,000
GVW	164,000

Haul, over paved highway through valley and mountains:
250 mi; level; 2% rolling resistance:
  Travel speeds
    Laden:

$$\text{Maximum: } \frac{342 \times 33{,}000}{2\% \times 164{,}000} \qquad\qquad 3440 \text{ ft/min}$$
$$39 \text{ mi/h}$$

    Average, estimated: $75\% \times 39$ 29 mi/h
    Unladen:

$$\text{Maximum: } \frac{342 \times 33{,}000}{2\% \times 56{,}000}, \text{ theoretical} \qquad 10{,}080 \text{ ft/min}$$
$$115 \text{ mi/h}$$

    Average, practical, estimated: $75\% \times 60$ 45 mi/h

150 mi; undulating grades, averaging 4%; 2% rolling resistance; 6% total resistance:
  Travel speeds:
    Laden, downgrade and upgrade speeds estimated to be equal:

$$\text{Maximum: } \frac{342 \times 33{,}000}{6\% \times 164{,}000} \qquad\qquad 1150 \text{ ft/min}$$
$$13 \text{ mi/h}$$

    Average, estimated: $75\% \times 13$ 10 mi/h
    Unladen, downgrade and upgrade speeds estimated to be equal:

$$\text{Maximum: } \frac{342 \times 33{,}000}{6\% \times 56{,}000} \qquad\qquad 3360 \text{ ft/min}$$
$$38 \text{ mi/h}$$

Average, estimated: 75% × 38	28 mi/h
Hauling cycle times, h:	
Loading crawler-tractor-ripper-bulldozer	1.0
Hauling:	
250 mi at 29 mi/h	8.6
150 mi at 10 mi/h	15.0
Unloading	1.0
Returning:	
150 mi at 28 mi/h	5.4
250 mi at 45 mi/h	5.6
Net cycle time	36.6
Allowance for stops, meals, and driving breaks	3.7
Total cycle time	40.3
Costs:	
Machines:	
Machinery hauler	$127,000
Pickup truck, pilot car	7,000
Hourly cost of ownership and operation (Table 8-3):	
Machinery hauler, 34% × $127	$43.20
Driver and helper	23.80
Pickup truck, 72% × $0.7	5.00
Pilot	11.20
Total hourly cost	$83.20
Net cost for transportation:	
Total, 40.3 × $83.20	$3353
Per ton·mi, $\dfrac{\$3353}{400 \times 54}$	$0.155
Gross cost for transportation:	
Machines and three workers	$3353
Three nights' lodging and four days' meals for three workers estimated to be: 3 × 3 × $30.00	270
Permit	3
Total	$3626
Unit cost per ton·mi, $\dfrac{\$3626}{400 \times 54}$	$0.168

NOTE: The estimate provides for dead heading back to home base. If a load can be picked up en route to home base, the overall unit cost per ton mile would be lessened. This favorable payload sometimes exists.

SERVICING AND MAINTAINING MACHINERY ON THE JOB

Servicing in the Field

This is the periodic replenishing of fuels, water, antifreeze mixtures, hydraulic oils, lubricating oils and greases, filters, air for tires, and miscellaneous items for the machines. These are called *consumed items,* as distinguished from parts. Servicing may include also cursory inspection of the machine so as to implement the all-important preventive maintenance. For example, a worn fan belt may call for field maintenance, but a crack in the frame can call for either field or shop work.

Servicing machines vary from the extreme of a pickup truck for a small, owner-operated small bucket loader to the other extreme of the large, efficient lubricating vans of Plate 16-1 and Figure 16-44. These large vans generally handle all items except fuel oil, which is dispensed by large refueling tankers. The size and the complexity of the servicing machine depend on the requirements of the job.

It is common for two spreads of rock-earth machines to be serviced by one lubricating van, except for refueling. Servicing of one machine by two equipment greasers requires about 10 min. When single shifting, the machines are serviced at the end of the shift. When double shifting, the machines may be serviced during the midshift lunch periods of the two shifts, at the ends of the shifts, and if necessary during pullouts from the line during the shifts.

Sometimes, when few machines are being used, they are refueled by the lubricating

Figure 16-44 Servicing one complete spread on a rock-earth fill for a dam. After the lubricating van tends this spread, it will move to another one which is handling rock for the fill. Fueling is done by a fuel oil tanker. In all, some 85 machines are being serviced every shift of the double-shift operation.

The van provides full service, with tanks for limited fuel oil, water, antifreeze mixture, crankcase oils, hydraulic oil, and greases. An engine-mounted compressor furnishes air for operating the pumps and for the tires of the machines. There are five heavy-duty pumps for oils and greases. Eight reels and nozzles are available for the various tanks. Two equipment greasers of the van can handle, except for refueling, six machines hourly.

The location is Perris Dam, Riverside County, California. *(Lincoln-St. Louis-McNeil Corp.)*

van, which carries a limited supply of fuel oil, but mostly, because of the great gallonages involved, fuel oil is pumped by regular refueling tankers. The refueling tankers may be operated by the owner or by the fuel oil supplier. Their capacities range from about 1000 to 4000 gal and they are generally equipped with two 50-gal/min pumps and 1½-in hoses and nozzles. They load the average machine, requiring about 130 gal per 8-h shift, in about 3 min net time. The total loading time is about 5 min, so that 12 machines of an average spread may be refueled in about 30 min by two operators.

In the case of permanently located work, such as quarries and open-pit mines, the machine may be brought to a central service station. It is a matter of relative cost whether to bring the machine to the services or to take the services to the machine.

Table 16-22 gives representative specifications for a large-capacity lubricating van.

Estimate of Production and Costs for Servicing (Exclusive of Refueling) Servicing costs for a machinery spread may only be approximated because of the different ways to perform the work. The same is true of refueling, especially since the work may be done either by the owner or by the fuel oil supplier. The following analysis is offered as an explanation of principles and methods.

The following criteria are for the example of a spread of three track drills with three compressors, a bucket loader, two tractor-ripper-bulldozers, six rear-dump rock haulers, one compactor, one motor grader, and one water wagon. A lubricating truck, as illustrated in Figure 16-44, services 12 of the machines after the single shift. The track drills and the compressors are lubricated by the chuck tenders. The lubricating truck requires 20 min for the round trip from shop to machines to shop. An average of 10 min is required to service each machine. The efficiency is the 50-min working hour.

```
Production:
  Operating cycle times, min:
    Travel time to and from machines                                 20.0
    Servicing 12 machines, 12×10                                    120.0
    Total time                                                      140.0
  Actual time for servicing, h, 140/50                                2.8
Costs, 1978:
  Machine, lubricating truck                                      $90,000
  Hourly cost of ownership and operation
    Lubricating truck (Table 8-3), 32% × $90                       $28.80
    Equipment greasers, 2                                           30.40
    Total hourly cost                                              $59.20
  Direct job unit cost
    Total cost of servicing, 2.8 × $59.20                         $165.80
    Actual machine hours of work serviced, 12×8                        96
    Direct job unit cost per machine hour of work, $165.80/96       $1.73
```

NOTE: The cost does not include the materials for servicing, nor does it include the cost of refueling by the separate tanker.

Estimate of Production and Costs for Refueling Service Refueling production and costs are estimated approximately in the following manner. Two fuel operators using two hoses from opposite sides of the refueling tanker, require 38 min to refuel 15 machines of the previous example of servicing, including the 3 compressors. The two operators refuel two machines in 5 min. The operating cycle totals 58 min, 20 min for travel time and 38 min for actual refueling.

The refueling tanker is equivalent in its rate of hourly cost of ownership and operation to the large lubricating truck, 32 percent of the thousands of dollars of cost. The 1978 cost of the 3500-gal refueling tanker, a tandem-drive truck, is $54,000.

Production:	
Operating cycle, servicing 15 machines	58 min
Actual time for servicing, 50-min hour efficiency, 58/50	1.2 h
Hourly cost of ownership and operation:	
Refueling tanker, 3500-gal capacity, 32% × $54	$17.30
Fuel operators, 2	30.40
Total hourly cost	$47.70
Direct job unit cost	
Total cost of servicing, 1.2 × $47.70	$57.20
Actual machine hours of work serviced, 15 × 8	120
Direct job unit cost per machine hour of work, $57.20/120	$0.48

NOTE: The combined direct job unit cost for complete servicing is estimated to be $1.79 + $0.48, or $2.27 per machine hour of work.

TABLE 16-22 Representative Specifications for Large-Capacity Lubricating Truck

Servicing equipment:	
Size of body, ft:	
Length	23
Width	8
Capacity of tanks, gal:	
Diesel fuel	300
Gasoline	150
Water	80
Antifreeze mixture	80
Hydraulic oil	150
Engine oil:	
10 weight, series 1	150
series 3	300
30 weight, series 1	150
series 3	300
Gear oil	300
Waste oil, vacuum system	300
Total	2260
Air compressor, diesel-engine-driven, ft^3/min	95
Delivery pumps	8
Chassis pumps, lb/in^2	400
Reels, hoses, and nozzles	8
Electric hoist	1
Lighting system	1
Truck chassis, tandem-axle drive:	
Diesel engine flywheel-brake hp	327
Transmission speeds	10
Wheelbase, in	254
Tires, 10	10.00 × 22—12 PR
Complete machine:	
Dimensions, ft:	
Length	31.8
Width	8.0
Height	8.5
Weights, approximate, lb:	
Tare	30,000
Payload	16,000
GVW	46,000

Maintenance in the Field

Field maintenance is generally in terms of minor repairs and replacements rather than major work, which takes place in the job shops of the company. However, if the climate is good throughout the year it is not uncommon to make major repairs in the field. This practice is aided by modular repair assemblies, which can replace those requiring major repairs with a minimum of work.

Maintenance involves such items as minor adjustments to machines, minor welding, and replacements of belts, cables, cutting edges, points, teeth, tires, and the like. The work may take place at any time, but preventive maintenance generally occurs during the midshift period, between shifts, or after the day's work.

Maintenance trucks may be simple or complex, ranging from a pickup truck with simple tools to a well-equipped two- or three-axle truck, as illustrated in Figure 16-45. This truck has a crew of two mechanics. Because of excellent year-round climate, both minor and major repairs are made in the field without protective covering. Figure 16-46 illustrates a small two-axle truck with gas and electric welding outfits. Figure 16-47 shows a two-axle tire maintenance truck of a tire company. Tires may be changed by either a truck of the owner or one of the supplier, depending on the size of the job and the resulting inventory of tires and the availability of good service from the tire company.

Figure 16-48 illustrates a supply yard for field servicing of machinery. It is strategically located at the center of the work areas. The supply of tires, part of which is shown in Figure 16-23, is located at one end of the job because of the space limitations in the sidehill supply yard.

By the nature of the operation, field service may take place at any time the machinery is working. Field maintenance is generally confined to good weather. In bad weather, if the ailing machine does not provide cover for the mechanics, it must be brought in for shop repairs.

Some owners prefer to bring machines into the shop for both minor and major repairs on the assumption that they can be made faster, better, and more cheaply. This practice is typical of quarries and open-pit mines whether weather conditions are favorable or unfavorable. The same practice is true generally when public works are being prosecuted all year round in all kinds of weather, as would be the case in several years of dam

Figure 16-45 Changing a heavy transmission in the field during working hours. The well-equipped field maintenance truck performs tasks of both minor and major nature. Among other facilities it carries gas and electric welding outfits, an air compressor, various power tools, and a hoist. The crew consists of two mechanics, one being visible beside the undercarriage of the tractor and the other being invisible within the tractor. The large transmission has been repaired at the site.

The total time for the repairs is 6.2 h and the travel time to and from the shop is 0.5 h, bringing the total work time to 6.7 h. The estimated hourly cost of ownership and operation for the maintenance truck, including the two mechanics, is $48.00. The cost for the repairs to the transmission is 6.7 × $48.00, or about $322.00, exclusive of the cost of parts.

The location is a freeway under construction in San Diego County, California.

Figure 16-46 Preventive maintenance for tractor-ripper-bulldozer during lunch hour. The tractor operator reported an incipient crack in the structural-steel tool bar of the ripper assembly. The welder is repairing the crack during the midshift lunch hour. The small two-axle maintenance truck is equipped with gas and electric welding outfits, an air compressor, and miscellaneous tools, making it ideal for the all-important preventive maintenance during working hours. The location is a freeway under construction in San Diego County, California.

Figure 16-47 Changing a huge but average-size tire during working hours. A 33.5×39—38PR tire weighing 2180 lb is being mounted on the 33-yd^3-capacity scraper. The damaged tire is leaning against the service truck. The complete change requires 50 min. The truck is equipped with a hydraulic hoist and an air compressor. It is the special-purpose machine of the tire supplier. Tire services may be performed by the owner, by the supplier, or by both during the job.

The hourly cost of ownership and operation of the service truck, including the serviceman, is about $22.50. If the work were done by the owner, the total time for the job, including travel time, would be about 1.5 h. The total cost for the change is then about $34.00. The operator of the scraper lends minor help during the tire changing.

The location is a highway building project at Poway Junction, San Diego County, California.

Figure 16-48 Yard for supplies and service and maintenance trucks for freeway construction. The yard is located approximately in the middle of the 3-m-long job. Shown clockwise from the lower left-hand corner of the picture are the following elements for service and maintenance:
  Supply sheds and warehouses for parts
  Fuel oil tanks, being filled by supply tankers
  Company refueling trucks
  Lubricating truck
  Office building
  Water wagon and fuel-oil and gasoline pumps
  Lubricating truck
  Drums of lubricants
  Supplier's delivery truck
  Small service truck
  Truck crane
Presently the maintenance trucks are busy in field repairs along the job. Because of lack of space in this narrow sidehill yard, new tires are stored at one end of the job. The location is in the San Luis Rey River Valley, San Diego County, California.

building. The best practice is a matter of judgment in accordance with the nature of maintenance and repairs, logistics, and weather conditions.

YARDS, SHOPS, AND OFFICES

In public works construction the yards, shops, and offices may be at the job site if the project is large or remote, or at the home office if the job is nearby. In private works, such as quarries, gravel pits, and open-pit mines, the facilities are located at the work site.

Yards

Yards are places to store machinery, construction materials, and supplies. In excavation there is usually little of construction materials and yards may be as simple as that of Figure 16-48 for a vast 12-million-yd^3 job. Figure 16-49 illustrates a yard located at the home office of an excavator specializing in soft-rock jobs, not requiring blasting, within 50 mi of the home office. This kind of operation lends itself to storage and minor and major repair work at one location even though several jobs may be in progress. Naturally enough, the permanent yard tends to become a catch-all repository for old and unused machinery, but the so-called boneyard is a welcome source of materials for the rebuilding of machines and often for the custom building of special-purpose machines.

Shops

Shops and shop facilities range from the medium-size one of Figure 16-50 located, along with the yard of Figure 16-49 at the home office of the company, to the large, well-equipped ones which are part of the plant shown in the frontispiece. In the case of the

Figure 16-49 Typical yard at home office of rock excavator. Presently there are few machines in the yard, as the two spreads of self-loading scrapers are busy within 25 mi of the home office. In the yard from left to right are:
    Unused refueling truck for fuel oil and gasoline
    Unused scraper cannibalized for parts
    Unused tanks
    Unused miscellaneous small machinery
    Bulldozer push arms
    Scraper awaiting repairs
    Flatbed, sideboarded hauler of supplies
    Unused tanks
    Enclosed trailer for storage of parts and supplies
Generally yards are not orderly, but the master mechanic knows the exact location of every machine or assembly which may be used in repairing or rebuilding an existing machine or in the custom building of a machine for a special purpose. Presently a complete lubricating truck is being built to the company's requirements in the adjacent shop. The location is San Marcos, San Diego County, California. *(Templeton Co.)*

Figure 16-50 Centrally located shop at home office of excavator. The medium-size company, working locally within a radius of 50 mi, makes minor and major repairs in the home office shop. Additionally, many special machines such as lubricating and service trucks are built in the shops.
    In the illustration, the complete undercarriage of a 140-hp, 29,000-lb crawler-tractor-bulldozer, including the track frame, sprockets and idlers, rollers, tracks, and shoes, is being rebuilt. In the right background is a medium-size self-loading scraper awaiting major repairs. A crew of five workers is kept busy maintaining two scraper spreads, totaling some 25 major machines, in excellent working shape.
    Good parts service from the local dealers facilitates quick minor and major repairs. An alternate method is to send assemblies to the dealers for repairs, but this owner feels that repairs in his own shop are both expeditious and economical. The shops adjoin the company's yard, as shown in Figure 16-49. Both facilities are located at San Marcos, San Diego County, California. *(Templeton Co.)*

shop of Figure 16-50, all repairs to machinery of the medium-size company take place in the shop, with the exception of minor and emergency repairs made on the jobs by the maintenance trucks. The large shops of the open-pit mining company can accommodate all the huge machines used in overburden and borax ore removal, with the exception of the 13-yd^3 capacity electric shovels.

The equipment of shops varies with the size and capabilities of the shops. Small or large, limited or unlimited in ability, the shops include some or all of the following equipment: bolt-and-nut machines; forming and bending machines; drilling and boring machines; boring and turning mills; pipe cutting and threading machines; planers; lathes; presses; gear cutters and hobbers; punching and shearing machines; shapers; rolls for bending and straightening; saws; slotters and key seaters; grinders; milling machines; hammer machines; forges; gas and electric cutting and welding machines; air compressors; painting outfits; steam cleaners; and wash racks.

In short, the well-equipped shop may have just about all the tools used by the manufacturer of the rock-earth machines.

Offices

Offices are for the people who manage the job and, if the job office is the same as the home office, it is for those who manage the job and for those who manage the company. In private works of a permanent nature, the office workers are all under the same roof, as is the management of the open-pit mine of the frontispiece and as is the direction of the centrally located company carrying on excavation in public and private works, as discussed in Figures 16-49 and 16-50.

Under other circumstances there is the field office, as shown in Figure 16-51. This office for highway building is 50 mi from the district office and 2000 mi from the home office of the major company. The people of the field office are concerned only with job operations, the details of payroll and purchasing being handled by the district office and the home office. The field office is made up of two portable units with a total of about 1000 ft^2 of office space. It houses managing, supervising, engineering, purchasing, and timekeeping personnel. The purchasing is confined to supplies for the job. Major purchases, such as machines, are handled through the district office and the home office.

The interlocking of policies and business methods of companies having field, district, and home offices varies generally according to longtime practices. However, it is axiomatic that the running of the job is the responsibility of the field office.

Figure 16-51 Field office for the building of a large mountain freeway. The two small, portable white buildings with about 1000 ft^2 of office space house all key company personnel concerned with grading, structures, and paving for the 2.1-mi eight-lane freeway, involving 3,120,000 yd^3 of rock excavation. These people include the project manager, superintendent, engineer, purchasing agent, timekeeper, bookkeeper, and secretary. All details of purchasing and payroll are handled by the district office, 50 mi away, and the home office, 2000 mi away. The records of foremen, grade checkers, service personnel, and mechanics are kept in files so that they are available at all times. The engineer's office is equipped with drafting tables, calculators, and surveying equipment.

The project is located geographically in the Merriam Mountains of San Diego County, California, an area with an equable year-round climate, so that such office housing is adequate. In an adverse climate such an economical minimum office setup would not be possible, as more personnel would need to be housed.

The 1978 cost for the two portable buildings with all their facilities is about $25,000. After this job of 2-yr duration they will be moved to another job or sold.

SUMMARY

The subjects discussed in this chapter have a direct bearing on the operation, the efficiency, and the resulting costs of an excavation job. Although seemingly of a minor nature, their selection, use, and care have an important bearing on the conduct and the economy of the excavation job. They actually are of major importance.

CHAPTER 17

Calculations of Quantities for Excavation

PLATE 17-1

Rock excavation in a 1,846,000-yd³ cut and the completed cut. *(a)* The work of tractor-rippers, push-loaded tractors, and scrapers is shown in the upper zone of rippable weathered rock and the work of track drills, bucket loaders, and rear-dump trucks is shown in the lower zone of semisolid and solid metavolcanic rocks. *(b)* The completed eight-lane freeway is shown after a year of work in excavation and paving. The location is Poway Junction, San Diego County, California.

The variable greenstones, tuffs, schists, and conglomerates of the cut of Plate 17-1 were fragmented by heavyweight tractor-rippers and by blasting. A total of 450,000 yd³ was blasted, 24 percent of the total 1,846,000 yd³. The maximum depth of the sidehill cut was about at the middle bench of the slope, as seen in *b*, where it was 90 ft. The average depth of rippable rock was 30 ft, as determined by seismic studies. The estimate for rock requiring blasting, based on Figure 17-11 (graph B), was 22 percent of the total yardage, or 406,000 yd³.

The estimate of the excavator for blasting rock was based on three determining factors: exploratory drilling by track drills, seismic studies, and familiarity with the complex rock formations of the area.

When the excavator prepares cost estimates for the work or prosecutes the work, the quantities involve basically volume or weight, horizontal and vertical haul distances, and the resultant inclined distance on a plus or minus grade.

1. *Volume:* The English unit is the cubic yard and the metric unit is the cubic meter, which equals 1.308 yd^3. Volume may be in terms of bank or in situ measurement, loose measurement as in a hauler or in a pile, or fill or compacted embankment measurement. The relationships among these measurements are given in Appendix 1.

2. *Weight:* The English unit of weight per unit volume is pounds per cubic yard and the metric unit is kilograms per cubic meter. The number of kilograms per cubic meter equals 0.593 times the number of pounds per cubic yard. The relationships among these units are given in Appendix 4. For haulage, weight rather than volume is often the unit of measurement, especially in borrow-pit excavation and in mining. The English units are the short or regular ton, 2000 lb, and the long ton of mining, 2240 lb. The metric ton equals 1000 kg or 2205 lb.

3. *Distance:* The English units are the foot, the station or 100 ft, and the mile. The meter equals 3.281 ft. The kilometer equals 3281 ft or 0.6214 mi.

4. *Area:* The English units are the acre, 43,560 ft^2, and the square mile, 640 acres. The square kilometer equals 0.3861 mi^2. The little used hectare, 10,000 m^2, equals 2.471 acres.

The main work of the excavator is to move a cubic yard of excavation from cut to fill or to another point of use. The work or energy for this task is in foot-pounds, a measure of force times distance. Force is a function of the weight of the rock-earth combined with that of the hauling machine per unit of rock-earth being hauled. The distance is three-dimensional from a point in one area to a point in another area lower or higher than the first point. Thus the estimator must determine the following quantities.

1. Dimensions, area, volume or weight, and center of gravity or mass of the cut.
2. Dimensions, area, volume or weight, and center of gravity or mass of the fill or point of deposit of the excavation, as in the case of a receiving hopper for a belt conveyor.
3. The swelling or shrinking factor for a cubic yard in the cut after it has been placed in a hauling unit or in a pile (where loose measurement applies) and after it has been placed in a fill. As both rock and earth swell on going from the cut to the loose condition, the factor is greater than unity. As rock swells from the cut to the fill, the factor is greater than unity. As earth shrinks from cut to fill, the factor is less than unity. Appendix 1 gives percentages of swell and shrink for the materials in the cut, in the loose condition, and in the fill. The swelling or shrinking factor for the material is given by:

Swelling factor = 1.00 + percentage of swell
Shrinking factor = 1.00 − percentage of shrink

4. The minimum, maximum, and average length of the haul and the grades of the haul road. These data are determined from those of items 1 and 2.

AREAS

There are three commonly used methods for determining areas.

1. By planimeter: This drafting instrument is used for determining the area of a figure by moving its tracing point around the perimeter of the plotted area, however irregular its shape. After the area in square inches has been determined, the value is multiplied by the square scale of the figure to calculate the scale value of the area. The precision of the planimeter depends on the size of the area and the skill of the draftsman and is about ±0.5 percent. This method is most commonly used.

2. By subdivision of the area into geometric figures with definite formulas for areas: The total area may be divided into such figures as rectangles, triangles, parallelograms, trapezoids, and trapezia. The precision depends on the degree of complete coverage of the total area by the small areas and the skill of the draftsman and is about ±0.5%.

3. By trapezoidal formula: The area is divided into a number of contiguous parallel strips of the same width W separated by ordinates or heights $h_0, h_1, h_2, \ldots, h_{n-1}, h_n$, and the following formula is applied:

$$\text{Area} = W\left(\frac{h_0}{2} + h_1 + h_2 + \ldots + h_{n-1} + \frac{h_n}{2}\right)$$

17-4 Calculations of Quantities for Excavation

The precision depends on the number of strips and the skill of the draftsman and it is about ± 0.5 percent.

The draftsman selects one of these methods in accordance with experience. For example, if the figure is a simple triangle, the area is calculated by method 2. On the other hand, if it is an irregular figure, the area should be calculated by the planimeter or by the trapezoidal formula.

VOLUMES

The term *solid* in this text is applied to both the whole of the cut and fill and to the parts into which the whole is subdivided so as to increase the precision of the volume calculations. The word *prism* is used often as an inclusive term for all solids. This usage is incorrect, as a prism is a solid whose two ends are parallel and equal and whose sides are parallelograms.

There are four basic methods for calculating a volume and the volume may be figured by any combination of these methods. Each method has its precision measure and its amount of work for a given calculation. As is almost always the case in estimating earthwork, the precision is inversely proportional to the amount of work in making the calculations.

Subdivision Method

This method involves subdivision of the main solid into smaller solids, the volumes of which are readily calculable. The borrow pit of Figure 17-2 is calculated by this method.

Figure 17-1 illustrates 15 solids with readily calculable volumes. Twelve of these are geometric solids with definite volume formulas. Each solid has one warped irregular face representing the ground surface, which is not a plane. It is this variable surface which makes accuracy an impossibility, and thus the volume must be calculated to within a certain precision. Formulas for the volumes are given, the precision being about ± 1.0 percent.

Three of the solids, the wedgelike prism, the inverted pyramid, and the irregular solid, are geometric with one warped irregular face. These solids represent those whose dimensions or whose volumes are not readily calculable. Approximate formulas are given, the precision being about $\pm 3.0\%$.

The subdivision method is sometimes called the borrow-pit method because of its frequent application to volumes of borrow pits. Figure 17-2 illustrates the subdivision of the excavation into 30 smaller solids. The volumes of these solids are calculated by the applicable formulas for the solids of Figure 17-1, as tabulated in the legend of Figure 17-2, and total 33,472 yd^3. When the method involves the use of a sufficient number of small solids, precisely or approximately calculated, the precision of measurement for the total volume is about ± 0.5 percent.

Prismoidal Formula

This method is accurate if the volume is a true prismoid. A prismoid is a solid having two polygonal faces connected by triangular faces. This is equivalent, if the connecting faces are taken as small enough, to the following description, which better defines the solids encountered in earthwork.

A prismoid is a solid having parallel plane ends with sides formed by moving a ruling line around the perimeters of the ends as directrices. Such a solid may be considered to be made up of prisms, pyramids, and wedges. The formula for the volume of a prismoid is given by the equation:

$$V = \frac{D_{1-2}}{6}(A_1 + A_m + A_2)$$

where V = volume, ft^3
  D_{1-2} = distance between end areas normal to their parallel planes, ft
  A_1 = area of end cross section at point 1, ft^2
  A_m = area of cross section at midpoint between end areas, ft^2
  A_2 = area of end cross section at point 2, ft^2

Note that A_m must be calculated because it is not the mean of A_1 and A_2.

Because of the tedious calculations for areas A_1, A_2, and A_m, the use of the prismoidal formula is generally confined to accurate calculations for volume. If the total excavation is made up of a large number of prismoids and solids like those of Figure 17-1, the calculations are involved and time-consuming. Accuracy, or a precision measure of ± 0.0 percent, is neither attainable or expected in earthwork computations. Accordingly, the prismoidal formula is rarely used, the subdivision method and the average-end-area method being sufficiently precise.

Average-End-Area Method

This simple and acceptably precise method is illustrated in Figure 17-3 and the accompanying tabulated data. The principle is that the volume of the solid between two parallel, or nearly parallel, cross-sectional areas is equal to the average of the two areas times the distance between the cross sections along their centerline. When the distance between cross sections is constant, the volume is given by the equation:

$$V = D \left(\frac{A_0}{2} + A_1 + A_2 + \ldots + A_{n-1} + \frac{A_n}{2} \right)$$

where V = volume, ft^3
D = distance between cross sections, ft
A_0–A_n = areas of cross sections, ft^2

The principle is not altogether true because the average of the two end areas is not the arithmetic mean of many intermediate areas. The method gives volumes generally slightly in excess of the actual volumes. The precision is about ± 1.0 percent.

Cross sections may be taken at any conservative intervals along the centerline, the distances being a matter of judgment depending on the relative areas of the cross sections and the irregularity of the ground surface. Generally the intervals are 25, 50, or 100 ft, but not more than 50 percent of the maximum height of the cross-sectional areas, so as to give good precision. The importance of spacings between cross sections is demonstrated in Figure 17-3 and the its accompanying tabulation.

Tables are available in civil and highway engineers' handbooks and in surveyors' handbooks giving volumes between cross sections at 50-ft intervals according to the sum of the end areas. These tables simplify calculations.

For illustrative purposes, in Figure 17-3 intervals are taken at 100 ft with the exception of those for the solids between stations $50+00$ and $51+00$ and between stations $55+00$ and $56+00$, where the intervals are 50 ft. When the intervals between $50+00$ and $51+00$ and between $55+00$ and $56+00$ are taken at 100 ft, the volume is 31,651 yd^3. However, when the intervals are taken at 50 ft, a more precise distance, the volume is 26,770 yd^3. The difference is 4881 yd^3. The error, based on the 258,474 yd^3 of the cut, is $+1.9$ percent. The exercise emphasizes the importance of close spacings for cross sections. Actually, good practice dictates 50-ft intervals for all the cross sections of Figure 17-3.

Contour Methods A variation of the conventional average-end-area method is computation from ground contours and finish-grade contours. The method is used for calculating volumes of cuts and fills of landscape grading and sometimes it is used for fine grading quantities following rough grading of excavation. The method and the calculations are illustrated in Figure 17-4 and the accompanying tabulated data.

The contour method may also be used to calculate the volume of a cut in a hill or ridge and the fill in a valley or ravine. Generally contour maps are available with contour intervals as small as 2 ft. These maps are available from the U.S. Geological Survey, the U.S. Army Corps of Engineers, or from a state, county, or some other public or private agency. In this method the end areas of the solids are horizontal plane figures at the elevations of the contours and the distance between areas is the contour interval. Thus the method, as contrasted to the conventional average-end-area method, may be considered as vertical rather than horizontal.

Another variation of the average-end-area method is the use of contours for calculating yardages in hills which are to be excavated either for materials or as part of a grading project. Figure 17-5 illustrates a 308-ft-high, 740-ft-long, 520-ft-wide hill which is to be leveled to a 275-ft elevation. The tabulation included in the legend gives calculations for

17-6 Calculations of Quantities for Excavation

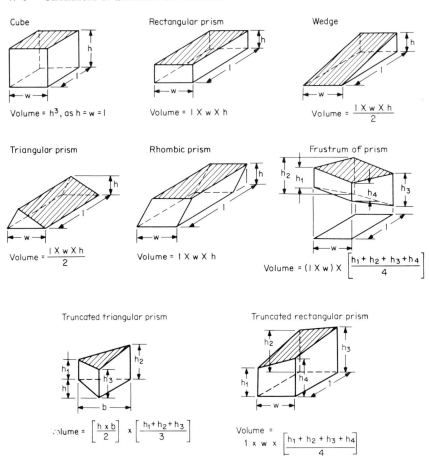

Figure 17-1 Solids with their formulas for calculating volume of excavation in cut and fill. The first 12 diagrams show solids which may be calculated precisely with a precision of about −1.0 percent. These solids have exact formulas for volume, provided the ground surface is a plane instead of a warped, irregular surface. The top surface generally is the ground surface in cuts and the bottom surface is the ground surface in fills. (Warped irregular ground surface is indicated by hachures—//////.)

The last three diagrams show solids which may be calculated approximately with a precision of about ±3.0 percent. These solids may or may not have definite formulas for volumes, but the measurements are difficult to ascertain. One surface is a warped, irregular ground surface. Calculations are based on the area of the ground surface projected onto a horizontal plane, and heights are vertical distances. The formulas for these irregular solids suggest the creation of other similar formulas for the approximate volumes of like solids.

Pyramid and cone

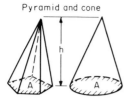

Volume = $\dfrac{A \times h}{3}$

A = area of base

Frustrum of pyramid and cone

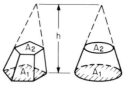

Volume = $\dfrac{h}{3} \times \left[A_1 + A_2 + (A_1 \times A_2)^{\frac{1}{2}} \right]$

A_1, A_2 = areas of bases in parallel planes

Wedgelike prism

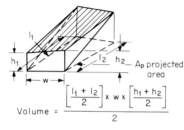

Volume = $\dfrac{\left[\dfrac{l_1 + l_2}{2}\right] \times w \times \left[\dfrac{h_1 + h_2}{2}\right]}{2}$

Pyramid, inverted

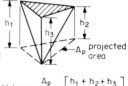

Volume = $\dfrac{A_p}{3} \times \left[\dfrac{h_1 + h_2 + h_3}{3}\right]$

Irregular solid

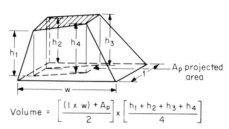

Volume = $\left[\dfrac{(l \times w) + A_p}{2}\right] \times \left[\dfrac{h_1 + h_2 + h_3 + h_4}{4}\right]$

Figure 17-1 (Continued)

17-8 Calculations of Quantities for Excavation

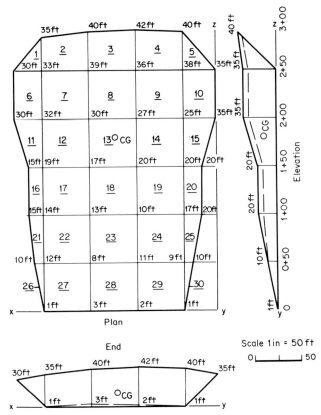

Figure 17-2 Views of borrow pit for calculating yardage by the method of subdivision of total excavation into solids. Calculated yardage of the borrow pit is 33,472 yd³. The yardages of solids 1 through 30 are calculated individually. The designation 35' and other elevations indicate elevations of natural ground above 0' base level of the floor of the borrow pit.

Calculations for the volume of the borrow pit by the method of subdividing the total excavation into 30 solids are tabulated below.

Figure 17-2 (Continued)

Prism	Kind of solid	Area of base, ft²	Calculations for volume, ft³	Volume, ft³
1	Pyramid, inverted	525	525 (35+33+30)/3/3	5,717
2	Wedgelike	1800	1800×37.5/2	33,750
3	Wedgelike	1875	1875×41.0/2	38,438
4	Wedgelike	1850	1850×41.0/2	37,925
5	Pyramid, inverted	700	700(40+35+38)/3/3	8,789
6	Wedgelike	1625	1625 30.0/2	24,375
7	Truncated rectangular	2500	2500(33+39+30+32)/4	83,750
8	Truncated rectangular	2500	2500(39+36+30+27)/4	82,500
9	Truncated rectangular	2500	2500(36+38+25+27)/4	78,750
10	Wedgelike	1575	1575×35.0/2	27,562
11	Wedgelike	1275	1275×22.5/2	14,344
12	Truncated rectangular	2500	2500(32+30+17+19)/4	61,250
13	Truncated rectangular	2500	2500(30+27+20+17)/4	58,750
14	Truncated rectangular	2500	2500(27+25+20+20)/4	57,500
15	Wedgelike	1125	1125×27.5/2	15,469
16	Wedgelike	825	825×15.0/2	6,188
17	Truncated rectangular	2500	2500(19+17+13+14)/4	39,375
18	Truncated rectangular	2500	2500(17+20+10+13)/4	37,500
19	Truncated rectangular	2500	2500(20+20+17+10)/4	41,875
20	Wedgelike	925	925×20.0/2	9,250
21	Wedgelike	650	650×12.5/2	4,062
22	Truncated rectangular	2500	2,500(14+13+8+12)/4	29,375
23	Truncated rectangular	2500	2,500(13+10+11+8)/4	26,250
24	Truncated rectangular	2500	2,500(10+17+9+11)/4	29,375
25	Wedgelike	650	650×15.0/2	4,875
26	Pyramid, approximate	60	60×50/3	1,000
27	Truncated rectangular	2500	2,500(12+8+3+1)/4	15,000
28	Truncated rectangular	2500	2,500(8+11+2+3)/4	15,000
29	Truncated rectangular	2500	2,500(11+9+1+3)/4	15,000
30	Pyramid, approximate	45	45×50/3	750
Total volume				903,744

Total volume = 903,744/27 = 33,472 yd³

The same method may be used to calculate the volumes in fills or embankment.

In this illustration of the principles, the grids are 50 ft square and the distances above the floor of the borrow pit are given in full feet. In practice, for depths of excavation up to about 40 ft the grids should be 20 ft square and the levels or elevations should be taken to within 0.1 ft.

It is always well to compare the results of one method of estimating with those of another when possible. An estimate by the average-end-area method, which is also discussed in this chapter, of the volume of this borrow pit gives 33,512 yd³. The difference between the two methods is 40 yd³, a remarkably close corroboration which demonstrates the high precisions of the two accepted methods for earthwork calculations.

17-10 Calculations of Quantities for Excavation

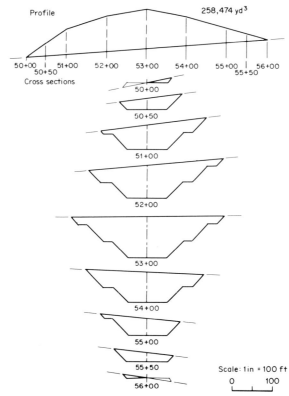

Figure 17-3 Profile and cross sections for calculating yardage in highway cut by average-end-area method.

Calculations for volume of highway cut by the average end area method are tabulated below.

Stations	Area	Areas, ft² Total areas	Average areas	Distance between sections, ft	Volume, yd³, between sections, distance × average area/27
50+00	300				
		3,810	1,905	50	3,528
50+50	3,510				
		13,900	6,950	50	12,871
51+00	10,390				
		28,170	14,085	100	52,171
52+00	17,780				
		40,670	20,335	100	75,321
53+00	22,890				
		36,530	18,265	100	67,654
54+00	13,640				
		19,740	9,870	100	36,558
55+00	6,100				
		8,500	4,250	50	7,871
55+50	2,400				
		2,700	1,350	50	2,500
56+00	300				
Total volume, yd³					258,474

The same average-end-area method is used to calculate the volume in a fill. In this example of principles, the distances between cross sections are taken at 50- and 100-ft intervals. A standard distance for high-precision measure in irregular topography is 50 ft.

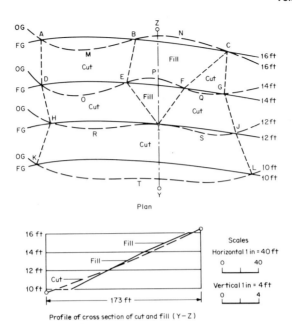

Figure 17-4 Contour method for calculating quantities in cut and fill for landscaping work. Contour intervals are 2 ft. Contours for original ground, OG, are shown by dashed lines. Contours for finished grade, FG, are shown by solid lines. Calculations for volumes of cuts and fills by the contour method in landscaping grading are tabulated below.

Contours, ft	Area	Areas, ft² Total areas	Average areas	Distance between contours, ft	Volume, yd³, between contours, distance × average area/27
In cut:					
10	2880				
		3820	1910	2	141
12	940				
12 Lt	600				
		1470	735	2	54
14 Lt	870				
		1920	960	2	71
16 Lt	1050				
12 Rt	340				
		430	215	2	16
14 Rt	90				
		90	45	2	3
16 Rt	0				
Total					285
In fill:					
12	0				
		310	155	2	11
14	310				
		770	385	2	29
16	460				
Total					40

There are two sets of cuts between contours 12 ft and 16 ft, separated by the fill. These are calculated separately, being labeled Lt, left, and Rt, right. The volumes of cuts and fills are in the area bounded by the perimeter designated by *ABNCGJLTKHDA*.

17-12　Calculations of Quantities for Excavation

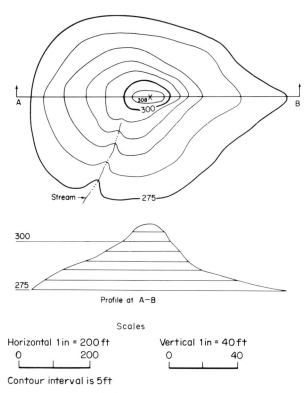

Figure 17-5　Contour method for calculating quantity in cut for leveling a hill. The contour interval is 5 ft. The calculated volume of the cut is 84,684 yd.³ The calculations involved in determining this volume are tabulated below.

Contours, elevations, ft	Area	Areas, ft² Total areas	Average areas	Distance between contours, ft	Volume, yd³, between contours, (Distance × average area)/27
275	265,500				
		428,000	214,000	5	39,630
280	162,500				
		241,500	120,750	5	22,361
285	79,000				
		128,000	64,000	5	11,852
290	49,000				
		71,000	35,500	5	6,574
295	22,000				
		31,500	15,750	5	2,917
300	9500				
		12,600	6300	5	1,167
305	3100				
		3100	1650	3	183
308	0				
Total volume, yd³					84,684

　The contour kind of the average-end-area method may be used for fills as well as cuts. It is especially appropriate for figuring volumes when contour maps are available, as no preliminary surveying is necessary.

this method. Again, this method may be described as vertical as opposed to the horizontal aspect of the conventional average-end-area method.

Average-End-Area Method With Prismoidal Adjustment

The error inherent in the average-end-area method may be corrected by means of the prismoidal adjustment. The equation for volume is:

$$V = \left(\frac{A_1+A_2}{2} \times D_{1-2}\right) - \left[\frac{D_{1-2}}{12}(c_1-c_2)(d_1-d_2)\right]$$

where V = volume of prism, ft^3
A_1 = area of cross section at point 1, ft^2
A_2 = area of cross section at point 2, ft^2
D_{1-2} = distance between cross sections along centerline, ft
c_1 = center height of cross section at point 1, ft
c_2 = center height of cross section at point 2, ft
d_1 = greatest width of cross section at point 1, ft
d_2 = greatest width of cross section at point 2, ft

In the discussion of Figure 17-3 the importance of taking cross sections at a maximum 50-ft centerline distance was emphasized. The quantity between stations $50+00$ and $51+00$ for 100-ft distance is 19,798 yd^3, but when the more precise 50-ft distances are used, the quantity is 16,399 yd^3, which gives a correction of -3399 yd^3, or -17.2 percent.

When the still more precise prismoidal correction is applied to the solid between stations $50+00$ and $51+00$, the volume is given by the equation:

$$V = \left(\frac{300+10{,}390}{2} \times 100\right) - \left[\frac{1}{12} \times 100 \times (0-59) \times (63-255)\right]$$
$$= 440{,}100 \text{ ft}^3 = 16{,}300 \text{ yd}^3$$

The volume of 16,300 yd^3 is 99 yd^3 less than that calculated by the use of 50-ft intervals between cross-sectional areas. This demonstrates two facts: (1) the tendency of the average-end-area method to give larger than actual volumes, and (2) the efficacy of the average-end-area method when corrected by the prismoidal adjustment.

Calculating Excavation Volumes According to Manner of Fragmentation

Chapter 11 deals with ripping and blasting. Figure 11-1 illustrates the three zones of rock-earth, fragmentation of which involves different methods and therefore different costs. Residuals require no work before loading and casting. Weathered rock or rippable rock calls for ripping. Semisolid or solid rock, so-called blasting rock, requires blasting. Chapter 3 discusses the weathering of different kinds of rocks and rock formations and this weathering is an important consideration in the appraisal of rippable and blasting rock. In a general sense the three groups of rocks weather in the following manners.

Igneous and metamorphic rocks weather rather uniformly downward from the ground surface so that the interfaces between residuals and weathered or rippable rock and between weathered rock and semisolid and solid, or blasting, rock tend to be parallel to the ground surface. The three zones merge almost imperceptibly into each other. Figure 3-7 illustrates the three zones in columnar basalt and Figure 11-14 shows the variable nature of ripping in the interface between rippable and blasting rock. These rocks, being amorphous, have no stratification or beddings and thus they have no defined interfaces.

Sedimentary rocks, on the other hand, are the result of deposition and have fairly definite interfaces between strata, as exhibited in Figure 3-3. In this limestone there is a top zone of residuals, a middle zone of rippable rock, and a bottom zone with considerable depth of blasting rock. Figure 17-6 illustrates the ripping of horizontally bedded limestone in the interface between rippable and nonrippable rock.

In order to prorate the relative costs of fragmentation (zero in the case of residuals) the estimator must determine the volumes of the rocks. Figure 17-7 illustrates ideally the contrast in spatial relationships between the igneous or metamorphic and the sedimentary rocks in a highway cut. These positions determine the respective volumes of residuals, rippable rock, and rock requiring blasting. One perceives that the idealistic Figure 17-7 may very well not represent natural weathering. For example, between stations

17-14   Calculations of Quantities for Excavation

Figure 17-6 Ripping sedimentary rock in an ideal interplane between rippable and blasting zones. The cross sections of this cut resemble those of station 52+00 of Figure 17-7. The tractor-ripper-bulldozer is working at a level between rippable rock and blasting rock. It is obvious that a change from ripping to blasting is imminent. The operator is having difficulty ripping out a slab of limestone about 18 in thick and weighing about 3 tons. Shortly blasting will be commenced at a notable saving in the cost for fragmentation. The location of the work is a loop highway around Bisbee, Arizona.

50+00 and 51+80 and between 54+20 and 56+00 there probably would be a thin layer of residuals in the case of the sedimentary rocks. However, such likelihoods do not obscure the sense of the comparisons.

The legend of Figure 17-7 includes the volumes of the three zones of rocks. One notes that in this cut the volume of igneous-metamorphic residuals is 2.79 times that for residuals of sedimentary rocks. In contrast, the volume of blasting rock in the igneous-metamorphic rocks is but 0.54 times that for the sedimentary formation. The volumes of rippable rock are about the same. One notes also that the thicknesses of residuals, weathered rocks, and semisolid and solid rocks are equal at station 53+00. Thus, the differences in volumes are due to beddings of the sedimentary formation.

The calculations for volumes in Figure 17-7 are based on determinations of average depths for the residuals and rippable rock throughout the length of the cut. If the estimator discovers highly variable depths, an alternative method is to divide the cut into sections, taking into account these differences.

It is apparent that the configuration of the cut also determines relative volumes. If the cut were a prism with the same trapezoidal cross section and if all interfaces were parallel to the base or grade, the volumes would be identical for both igneous-metamorphic and sedimentary rock formations.

Just as there is a marked difference in cost between ripping and blasting, so there is a sizable difference in cost for soft, medium, and hard ripping classifications, as set forth in Chapter 11. In large excavation jobs the rippable rock is sometimes divided into these three classifications in order to arrive at a more precise weighted-average direct job unit cost for the rippable rock.

Excavation Volumes and Weights According to Position or Condition

The volume and the weight of a cubic yard as measured in the cut or in situ change according to two later positions or locations. In the loose condition, as in a hauling unit or in a stockpile, the excavated material swells because of the creation of voids during excavation. As a result the loose cubic yard weighs less than the cubic yard in the cut. In the fill, as a result of compaction by gravity and by the compactors, material which occupied 1 yd^3 in the cut occupies more or less volume according to whether it swells or shrinks from cut to fill.

Appendix 1 gives weights per cubic yard for materials in the cut, percentage of swell and weight per cubic yard for the same material in the loose condition, and percentage of swell or shrink and weight per cubic yard for the same material in the fill. Data on the loose condition are important in calculating the production of loading, casting, and hauling machinery, as set forth in Chapters 12 and 13. Data for the compacted condition in the fill are important for calculations for the mass diagram, as described in this chapter. The swell and shrink percentages for materials in the fill are based on average compaction in the fill by both gravity and compactors. The degree of compaction by the combined agents is greater than that by gravity alone, as is the case when fills are built by the end- or side-dumping method.

In compiling Appendix 1 it was noticeable from the tables consulted that percentages of shrink from cut to fill have increased and percentages of swell have decreased over a period of 90 years. This continuous change is due to improvements in compacting methods and the gradual ending of the gravity or subsidence method of fill construction. One must be aware of this change if one uses older handbooks for data on swell and shrink percentages or factors. Awareness of changes in these values is especially important in the use of the mass diagram.

Residuals, consisting of clays, silts, sands, and gravels, shrink from cut to fill by an average of 12 percent. As 1 yd^3 in a cut becomes 0.88 yd^3 in the fill, 1.14 is the shrinking factor to be used in the mass diagram, the embankment measurement being thus converted to equivalent excavation measurement.

Rippable rock, generally weathered rock consisting of a mixture of rock particles and earth, may swell or shrink in the fill, depending on the relative amounts of rock particles and earth. These values are given in Appendix 1 for earth-rock and rock-earth mixtures. The term rippable rock implies a predominance of rock so that the average figure is a swell of +8 percent. The swelling factor for the mass diagram equals 1.00/1.08 or 0.93.

Blasting rock in the first stage of blasting contains 25 to 33 percent voids, depending on the amount of fines created during blasting. Thus the swelling after blasting varies from +33 to +50 percent. If compactors are not used on the fill, these swellings are still reduced by the compacting effects of haulers, tractor-bulldozers, water wagons, and other heavy moving machinery. Usually residuals and rippable rock, if available, are blended with the blasted rock so as to get a denser fill. By this process the overall percentage of swelling is reduced to a range +8 to +18 percent. The swelling factor for the mass diagram then ranges from 0.93 to 0.86.

One perceives that the changes in volumes and weights for rock-earth excavation call for careful calculations and, equally important, for common sense and experience. On this knowledge depend the productions of machines and the direct job unit cost of excavation.

CENTER OF GRAVITY OR CENTER OF MASS OF VOLUME IN EXCAVATION OR EMBANKMENT

The resultant of the weights, or parallel forces of gravity, on all the particles of a body always passes through a certain particle or point fixed with reference to the body no matter how the body is turned. This particle or point is the *center of gravity* of the body. In calculating average length of haul and difference in average elevations of cut and fill or point of deposit, the estimator must determine the centers of gravity of the volumes. As the volumes represent weights, the unit weight being assumed to be constant, the estimator may determine centers of gravity by distance and volume calculations.

The position of the center of gravity is determined by the static-moments method. The *static moment* of a volume or weight is the product of the volume or weight and its distance from a plane. The moment may be positive or negative according to whether the volume or weight is on the positive or negative side of the plane.

In order to calculate the center of gravity of an irregular volume, the volume is subdivided into parts whose centers of gravity are known or calculable. The static moments of these parts about a given plane are calculated and summed up. The position of the center of gravity of the whole volume with respect to the plane of reference is the summation of the moments divided by the whole volume. The calculated distance is from the center of gravity to the plane, the distance being normal to the plane. These calculations are presented below for three planes at right angles to each other, as the center of gravity is three-dimensional. The three calculations are made below with reference to the borrow pit of Figure 17-2, which gives a plot of the three positions of the center of gravity.

17-16 Calculations of Quantities for Excavation

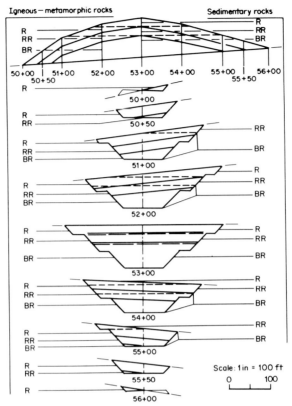

Figure 17-7 Calculations for volumes of rock-earth in cut according to manner of fragmentation.
Key:
  R—Soft residuals, earth not requiring ripping for excavation; 20-ft thickness at station $53+00$.
  RR—Rippable rock-earth, weathered soft to hard formation, which required ripping before excavation; 20-ft thickness at station $53+00$.
  BR—Rock requiring blasting for excavation, semisolid and solid formation; 60-ft thickness at station $53+00$.
  Igneous-metamorphic rocks—Rock formations which are not stratified or bedded and which weather downward from the surface in a fairly uniform manner in inverse proportion to depth. The interfaces between R and RR and between RR and BR are marked by solid lines.
  Sedimentary rocks—Rock formations which are stratified and bedded and which weather to generally plane interfaces. The interfaces between R and RR and between RR and BR are marked by dashed lines.

The cut is that of Figure 17-3, for which a total volume of 258,474 yd³ is calculated by the average-end-area method. When the total volume is subdivided into volumes for the R, RR, and BR classifications by the average-end-area method, the following distributions of volumes are obtained:

Classification of rock-earth	Igneous-metamorphic rocks		Sedimentary rocks	
	Volume, yd³	Percent of total volume	Volume, yd³	Percent of total volume
R (residuals)	102,407	40	36,666	14
RR (rippable rock)	69,999	27	63,019	24
BR (blasting rock)	86,068	33	158,789	62
Total rock-earth	258,474	100	258,474	100

Although the thicknesses of the R, RR, and BR classifications are the same at station $53+00$, the effects of the different attitudes and different weathering characteristics of the two kinds of rocks on the relative volumes of the three classifications are apparent in the tabulation.

Center of Gravity or Center of Mass of Volume in Excavation or Embankment 17-17

1. The center of gravity with respect to a vertical plane passing through points (x, y) at the lower edge of the borrow-pit cut is determined by taking the static moments of the volumes of 30 solids about plane xy:

Solids	Volume, ft^3	Horizontal Distance, ft^3	Moment, $ft^3 \cdot ft$
1, 2, 3, 4, and 5	124,619	270	33,600,000
6, 7, 8, 9, and 10	296,937	225	66,800,000
11, 12, 13, 14, and 15	207,313	175	36,300,000
16, 17, 18, 19, and 20	134,188	125	16,800,000
21, 22, 23, 24, and 25	93,937	75	7,000,000
26, 27, 28, 29, and 30	46,750	25	1,200,000
Totals	903,744		161,700,000

The distance from a vertical plane through points (x, y) to the center of gravity of the borrow-pit cut is given by the equation:

$$D = 161,700,000/903,744 = 179 \text{ ft}$$

One may quickly figure the distance by assuming the center of gravity to be two-thirds of 270, or 180 ft, from the vertical plane through points (xy). This is based on the rule that the center of gravity of a triangle is located at one-third of its altitude above the base, since the elevation view is approximately a triangle with its base at station $2+70$.

2. The center of gravity with respect to a vertical plane passing through points (y,z) at the right edge of the borrow-pit cut is determined by taking the static moments of the volumes of 30 solids about plane yz:

Solids	Volume, ft^3	Horizontal Distance, ft	Moment, $ft^3 \cdot ft$
5, 10, 15, 20, 25, and 30	66,695	25	1,700,000
4, 9, 14, 19, 24, and 29	260,425	60	15,600,000
3, 8, 13, 18, 23, and 28	258,438	110	28,400,000
2, 7, 12, 17, 22, and 27	262,500	160	42,000,000
1, 6, 11, 16, 21, and 26	55,686	195	10,900,000
Totals	903,744		98,600,000

The distance from a vertical plane through points (y,z) to the center of gravity of the borrow pit is given by the equation:

$$D = 98,600,000/903,744 = 109 \text{ ft}$$

Likewise, one may quickly figure the distance by assuming the center of the somewhat symmetrical volume to be in the middle of the width, or 110 ft from the vertical plane through points (y, z).

3. The center of gravity with respect to a plane corresponding to the floor or finished grade of the borrow-pit cut passing through points (x, y, z) is determined by taking static moments of volumes of 30 solids about the plane x,y,z:

Solids	Volume, ft^3	Vertical Distance, ft	Moment, $ft^3 \cdot ft$
1, 2, 3, 4, and 5	124,619	22	2,700,000
6, 7, 8, 9, and 10	296,937	16	4,800,000
11, 12, 13, 14, and 15	207,313	13	2,700,000
16, 17, 18, 19, and 20	134,188	8	1,100,000
21, 22, 23, 24, and 25	93,937	7	700,000
26, 27, 28, 29, and 30	46,750	5	200,000
Totals	903,744		12,200,000

The distance from the plane passing through points (x,y,z) to the center of gravity of the borrow pit cut is given by the equation:

$$D = 12,200,000/903,744 = 13 \text{ ft}$$

Again, one may figure quickly the distance by assuming the center of gravity to be located at $\frac{1}{3}(35+40+42+40)/4$, or 13 ft, above the plane. This is the rule that the center of

17-18   Calculations of Quantities for Excavation

gravity of a triangle is located at one-third of the altitude above the base of the triangle, as the elevation view is approximately a triangle with its base along line $y=z$.

The static-moment method is generally used, as well as the one involving formulas, to determine the locations of centers of gravity within the geometric solids. Estimation of the position of the center of gravity of the hilltop of Figure 17-5 by these two methods demonstrates the close results. The portion of the hill to be removed resembles a cone or pyramid, warranting the use of the formula for locating the center of gravity vertically in these two solids. According to the formula, the vertical height of the center of gravity of the solid is 0.25 of the altitude of the solid. In this case, as the height of the cone or pyramid is 33 ft above the base elevation of 275 ft, the elevation of the center of gravity is $(0.25 \times 33) + 275.0$, or 283.2 ft. When the static-moment method is applied, the moments of the volumes between contours are taken about the horizontal plane at an elevation of 275.0 ft. The sum of the moments is divided by the volume of the hill to be leveled, and the quotient is the distance of the center of gravity above the plane at a 275.0-ft elevation. The distance is calculated to be 7.5 ft and the resultant elevation of the center of gravity is 282.5 ft. The difference in the two elevations, 0.7 ft, is negligible.

Importance of Using Centers of Gravity When Estimating Hauling Costs

The relative influences of level, upgrade, and downgrade hauls on costs of hauling the same horizontal distance is emphasized in the following comparisons for the 50-ton-capacity rear-dump hauler of Figures 13-33 and 17-8. The specifications for this machine accompany the figure. Criteria for the three hauls are tabulated below.

Item	Level Haul	Upgrade Haul	Downgrade Haul
Lengths and grades of haul:			
In cut, level, ft	1000	1000	1000
On main haul road:			
Level, ft	5000		
Upgrade, +8%, ft	...	5000	
Downgrade, −8%, ft	...	...	5000
At primary crusher, level, ft	500	500	500
Difference in elevations, ft	0	+400	−400
Rolling resistance, %	3	3	3

The flywheel-brake horsepower of the engine is 537. The estimated horsepower absorption of the retarder or brake is 537. The wheel horsepower of the engine equals 0.76×537, or 408. The tare weight is 75,000 lb, the load is 100,000 lb, and the GVW is 175,000 lb.

The 1978 cost of the hauler is $187,000. The hourly cost of ownership and operation is estimated as indicated in the following tabulation.

50-ton-capacity off-the-road rear-dump truck (Table 8-3), $34\% \times \$187$	$63.60
Operator	11.80
Total hourly cost	$75.40

The hauling and returning times are calculated by the method including Table 13-12.

Hauling: Weight-to-power ratio $=175,000/537$ $\qquad = 326$ lb/hp
In cut, 1000 ft, level:

$$\text{Maximum speed} = \frac{408 \times 33{,}000}{0.03 \times 175{,}000} = 2565 \text{ ft/min} = 29.1 \text{ mi/h}$$

Average speed $=0.61 \times 2565 \qquad = 1565$ ft/min $= 17.8$ mi/h
Time $\qquad =1000/1565 \qquad\qquad\qquad = 0.64$ min
At crusher, 500 ft level:

$$\text{Maximum speed} = \frac{408 \times 33{,}000}{0.03 \times 175{,}000} = 2565 \text{ ft/min} = 29.1 \text{ mi/h}$$

Average speed $=0.51 \times 2565 \qquad = 1308$ ft/min $= 14.9$ mi/h
Time $\qquad =500/1308 \qquad\qquad\qquad = 0.38$ min
On main haul road:
5000 ft level:

Maximum speed = $\dfrac{408 \times 33{,}000}{0.03 \times 175{,}000}$ =2565 ft/min=29.1 mi/h

Average speed=0.92×2565 =2360 ft/min=26.8 mi/h
Time =5000/2360 =2.19 min
5000 ft, 8% upgrade:

Maximum speed = $\dfrac{408 \times 33{,}000}{(0.08+0.03) \times 175{,}000}$ =699 ft/min=7.9 mi/h

Average speed=maximum speed =699 ft/min=7.9 mi/h
Time=5000/699 =7.15 min
5000 ft, 8% downgrade

Maximum speed = $\dfrac{537 \times 33{,}000}{(0.08-0.03) \times 175{,}000}$ =2025 ft/min=23.0 mi/h

Average speed=maximum speed =2025 ft/min=23.0 mi/h
Time=500/2025 =2.47 min
Returning: Weight-to-power ratio=75,000/537 =140 lb/hp
At crusher, 500 ft level:

Maximum speed = $\dfrac{408 \times 33{,}000}{0.03 \times 75{,}000}$ =5984 ft/min=68.0 mi/h

Average speed=0.54×5984 =3231 ft/min=36.7 mi/h
This speed is not realistic; reduce to: 1616 ft/min=18.4 mi/h
Time =500/1616 =0.31 min
In cut, 1000 ft level:

Maximum speed = $\dfrac{408 \times 33{,}000}{0.03 \times 75{,}000}$ =5984 ft/min=68.0 mi/h

Average speed=0.67×5984 =4009 ft/min=45.6 mi/h
This speed is not realistic; reduce to 2004 ft/min=22.8 mi/h
Time =1000/2004 =0.50 min
On main haul road:
5000 ft level:

Maximum speed = $\dfrac{408 \times 33{,}000}{0.03 \times 75{,}000}$ =5984 ft/min=68.0 mi/h

Average speed=0.95×5984 =5685 ft/min=64.6 mi/h
This speed is not realistic; reduce to 2842 ft/min=32.2 mi/h
Time =5000/2842 =1.76 min
5000 ft, 8% downgrade:

Maximum speed = $\dfrac{537 \times 33{,}000}{(0.08-0.03) \times 75{,}000}$ =4726 ft/min=53.7 mi/h

Average speed=0.97×4726 =4584 ft/min=52.1 mi/h
This speed is not realistic; reduce to 2292 ft/min=26.0 mi/h
Time =5000/2292 =2.18 min
5000 ft, 8% upgrade:

Maximum speed = $\dfrac{408 \times 33{,}000}{(0.08+0.03) \times 75{,}000}$ =1632 ft/min=18.5 mi/h

Average speed=1.00×1632 =1632 ft/min=18.5 mi/h
Time=5000/1632 =3.06 min

Item	Level Haul	Upgrade Haul	Downgrade Haul
Hauling cycle, min:			
Loading 5 dippers from 10-yd³ capacity shovel	2.00	2.00	2.00
Two turnings and dumping	1.00	1.00	1.00
Hauling, 6500 ft:			
In cut, 1000 ft	0.64	0.64	0.64
On main haul road, 5000 ft	2.19	7.15	2.47
At crusher, 500 ft	0.38	0.38	0.38
Total time	3.21	8.17	3.49
Returning, 6500 ft:			

At crusher, 500 ft	0.31	0.31	0.31
On main haul road, 5000 ft	1.76	2.18	3.06
In cut, 1000 ft	0.50	0.50	0.50
Total time	2.57	2.99	3.87
Total cycle time, net, min	8.78	14.16	10.36
Average travel speeds, mi/h:			
Hauling, 6500 ft	23.0	9.0	21.2
Returning, 6500 ft	28.7	24.7	19.1
Hourly production:			
Loads hauled per 50-min working hour:			
50/8.78	5.69		
50/14.16	...	3.53	
50/10.36	...	...	4.83
Tons hauled per hour:			
50×5.69	284		
50×3.53	...	176	
50×4.83	...	...	242
Hourly cost of ownership and operation	$75.40	$75.40	$75.40
Direct job unit cost per ton:			
$75.40/284	$0.265		
$75.40/176	...	$0.428	
$75.40/242	...	...	$0.312

The unit cost is least for the level haul, intermediate for downgrade haul, and most for upgrade haul. Contrary to a sometimes held opinion, a downgrade haul costs more than a level haul. The difference between the unit costs for the upgrade haul with a 400-ft difference in elevation and the level haul is $0.163/ton. The additional unit cost for the upgrade haul compared with the equal-length level haul is $0.041/ton per 100 ft of vertical lift. The corresponding figure for the downgrade haul is $0.012/ton per 100 ft of vertical drop.

Often in the mining of ores from deep open pits, the estimator reckons cost in terms of a foot of vertical lift, as the length of haul is really a part of the vertical lift. In the example, involving an upgrade haul similar to those in mines, the direct job unit cost is $0.428/400 = $0.0011/ton per foot of vertical lift.

The analysis emphasizes forcefully the importance of calculating elevations of the center of gravity of the cut and the center of gravity of the fill or point of deposition when estimating costs of excavation.

MEASUREMENT OF HAUL AND THE MASS DIAGRAM

The mass diagram is a graphical means for measuring haul in terms of station yards or, more specifically, in terms of station cubic yards. A *station yard* is a measure of work, being the movement of 1 yd³ from the cut through a distance of one station, or 100 ft. The station is a unit of length in surveying. In addition to measuring the total haul in station yards, the mass diagram is used for the following purposes:

1. Calculating the amount of freehaul and overhaul in station yards: sometimes the units of measurement of haul are in terms of freehaul and overhaul rather than in terms of the one unit, the cubic yard, regardless of the distance moved. *Freehaul* is the movement of one cubic yard through a maximum distance. The maximum distance may be any length, but it is usually either 500 ft or 1000 ft. *Overhaul* is the movement of one cubic yard through any distance in excess of the freehaul distance. When the freehaul-overhaul system is used for bidding, a cost and a price must be established for freehaul and a cost and a price must be established for overhaul.

2. Making studies of the comparative costs of different schemes for hauling: these schemes generally involve the waste of fill from the cut and the borrowing of cut for the fill.

3. Determining quantities of excavation or embankment within a given length of cut or fill.

4. Determining the location of the centers of gravity of the cut and fill: these are generally determined horizontally along the centerline of the work, although they may be determined vertically by plotting a mass diagram in a vertical direction. The determination of a vertical center of gravity is rarely made.

Figure 17-8 Rear-dump truck of 50-ton capacity returning from long, steep downgrade haul. The truck is one of a fleet of oil-shale haulers, illustrated in Figure 13-33 and described in the accompanying text. The length of haul, one way, is 6 mi or 31,700 ft, and the average downgrade is −6.9 percent for the narrow, winding mountain haul road. The average net hauling cycle is 65 min.

The well-graded road with turnouts for passing is adequate for the production of the fleet of trucks used to serve the pilot plant, but it is too narrow and has too many short-radius curves for the speed and safety requirements of a large mining operation. Additionally, it offers an example of the gross error in applying the principles of Table 13-12 for estimating a hauling cycle on this unique job. The downgrade speed with use of the retarder is estimated to be 29.5 mi/h and the upgrade speed is estimated to be 20.6 mi/h, both of which are unrealistic. Based on an estimated time of 50 min for hauling and returning, it is probable that average speeds are about 18 mi/h downgrade and 12 mi/h upgrade.

The location of the pilot mining operation is in the Rocky Mountains near Anvil Point, Colorado. *(International Harvester Co.)*

The properties of a mass diagram are:

1. Ascending lines indicate that the excavation quantity exceeds the embankment quantity and descending lines indicate that the embankment quantity exceeds the excavation quantity between the stations represented by the line.

2. High and low points in the mass diagram curve occur at points of no cut and no fill on the accompanying centerline profile.

3. The maximum ordinates generally occur in the vicinity of the point of no cut and fill. The maximum ordinates occurring in the convex loops indicate a change from cut to fill and those occurring in the concave loops indicate a change from fill to cut.

4. The difference in length, algebraically, between any two ordinates is a measure of the total quantity of material between the stations at which the ordinates are drawn.

5. Excavation equals embankment between any two ordinates where a horizontal line intersects the curve of the mass diagram.

6. The direction of haul may be determined by this principle: Excavation represented

17-22   Calculations of Quantities for Excavation

by rising lines in the curve will be hauled toward a falling line, indicating embankment; or, convex loops indicate material to be hauled ahead, or upstation, and concave loops indicate material to be hauled back, or downstation.

7. The area between a horizontal line and the curve of the mass diagram is a measure of the haul in station yards between the two stations where the horizontal line intersects the curve.

The procedures for determining the station yards of total haul, freehaul, and overhaul from the mass diagram are explained in the following list.

Making Up Table 17-1, Calculations for Mass Diagram

1. Compute total excavation, cut, embankment, and fill, station by station for the work, including both main line and borrow pits.

2. Determine the ratio of cut yardage to fill yardage for the work, either for the entire yardage if the ratio is constant for the entire job, or for individual cuts and fills if the ratios vary according to the nature of the materials in the cuts. The weathered rock of Table 17-1 contains 75 percent rock fragments and 25 percent earth. From cut to fill it swells 12 percent. Thus the ratio is $1.00/1.12 = 0.89$. This is to say that 0.89 yd^3 in the cut becomes 1.00 yd^3 in the fill. Appendix 1 gives percentages of swell and shrink from cut to fill for different materials.

3. Convert embankment into equivalent excavation, using the determined ratio.

4. Tabulate excess cubic yards per station, excavation being a positive quantity and embankment being a negative quantity.

5. Calculate the algebraic total of excess cubic yards per station according to station.

Plotting Mass Diagram, Figure 17-9, from Data of Table 17-1

6. Plot the mass diagram, Figure 17-9, by using algebraic totals of Table 17-1 according to stations.

7. Divide the mass diagram into one or more sections where cut balances fill, as illustrated in area *DEFGHLK*.

8. Eliminate the freehaul of 500 ft from each of these areas, as illustrated by areas *ABCJ* and *KEFGL*.

9. Calculate the total haul in station yards and average haul in feet:
   a. Total haul, sta yd   = area *ABCDEFGHI*, in^2 × area scale
                                    = 12.40 × 50,000 = 620,000 sta yd
   b. Average haul, ft = total haul, sta yd/total excess excavation
                                    = 620,000/39,615 = 15.65 stations, or 1565 ft

10. Calculate the freehaul in station yards and average freehaul in feet:
   a. Freehaul, sta yd   = area *ABCJ* plus area *KEFGL*, in^2 × area scale
                                    = 3.82 × 50,000 = 191,000 sta yd
   b. Average freehaul, ft   = total freehaul, sta yd/total excess excavation
                                    = 191,000/39,615 = 4.82 stations or 482 ft

11. Calculate overhaul in station yards and average overhaul in feet:
   a. Overhaul, sta yd   = area *JCDKLHI*, in^2 × area scale
                                    = 8.58 × 50,000 = 429,000 sta yd
   b. Average overhaul, ft   = total overhaul, sta yd/total excess excavation
                                    = 429,000/39,615 = 10.83 stations or 1083 ft

Note that the sum of freehaul and overhaul must equal total haul in station yards, 620,000, and that the sum of average freehaul and average overhaul must equal average haul in feet, 1565.

Use of Mass Diagram for Calculating More Economical Haulage by Substituting Another Borrow Pit

The mass diagram of Figure 17-9 results from the plans of the contracting officer for the highway. It may be available to the excavator. If it is not a part of the plans, it may be

TABLE 17-1 Calculations for Mass Diagram
(Ratio of cut yardage to fill yardage of weathered rock, 0.89. Material swells 12% from cut to fill.)

Station to station	Excavation cut, yd³ (+)	Embankment fill, yd³ (−)	Embankment converted to equivalent excavation, yd³ (−)	Excess yd³ per station Excavation, yd³ (+)	Excess yd³ per station Embankment, yd³ (−)	Algebraic total, yd³
Borrow pit:						
0b–1b	8,000			8,000		+ 8,000
1b–2b	9,000			9,000		+17,000
2b–3b	1,500			1,500		+18,500
3b–4b	850			850		+19,350
4b–5b	13,530			13,530		+32,880

Distance from sta 5b, borrow pit, to sta 0, roadway, is 500 ft.

Station to station	Excavation cut, yd³ (+)	Embankment fill, yd³ (−)	Embankment converted to equivalent excavation, yd³ (−)	Excess yd³ per station Excavation, yd³ (+)	Excess yd³ per station Embankment, yd³ (−)	Algebraic total, yd³
Roadway:						
0–1		1,050	934		934	+31,946
1–2		1,200	1,068		1,068	+30,878
2–3	300	320	285	15		+30,893
3–4	1,500			1,500		+32,393
4–5	2,000			2,000		+34,393
5–6	1,720			1,720		+36,113
6–7	1,500			1,500		+37,613
7–8	100	420	374		274	+37,399
8–9		2,100	1,869		1,869	+35,470
9–10		5,250	4,673		4,673	+30,797
10–11		8,720	7,761		7,761	+23,036
11–12		13,120	11,678		11,678	+11,358
12–13		7,800	6,943		6,943	+ 4,415
13–14		4,010	3,569		3,569	+ 846
14–15		950	846		846	0
Totals	40,000	44,940	40,000	+39,615	−39,615	0

prepared by the excavator prior to or after the submission of his bid and acceptance of the bid. Sometimes the excavator may use acceptable material from a borrow pit located more favorably for the cost of haulage. The excavator then figures the relative costs for the specified pit and for the available pit and establishes his bid price accordingly. A typical analysis is given below.

Referring to Figure 17-9, the excavator locates an acceptable privately owned borrow pit located 500 ft from station 15+00 of the roadway in the same relative position as that of the mass diagram. The borrow pit is also 500 ft long and the material may be taken out to a uniform depth. The center of gravity of the pit is then 750 ft from station 15+00. A second mass diagram is plotted for this new plan. The resulting hauls from the mass diagram are:

Freehaul	182,500 sta yd
Overhaul	196,000
Total haul	378,500 sta yd

The significant saving in total haul afforded by the new borrow pit is 620,000−378,500, or 241,500 sta yd. The average haul is reduced from 1565 ft to 955 ft, a 610-ft saving. The excavator estimates that the 610-ft reduction in average haul by the scrapers will save $0.090/yd³, or $0.015/sta yd in direct job unit cost.

As 241,500 sta yd of haul is saved, the total cost saving is $3622.00. The material in the borrow pit may be purchased for $0.05/yd³. As 32,880 yd³ is required, the cost is $1644, giving a net saving of $1978. If the contracting officer approves the change in borrow pits, the excavator will use the privately owned pit rather than the one of the contracting agency.

17-24 Calculations of Quantities for Excavation

Use of Mass Diagram for Calculating Centers of Gravity in Cuts and Fills

The locations of the center of gravity in plan, elevation, and end views were calculated for the borrow pit of Figure 17-2 by the method of static moments.

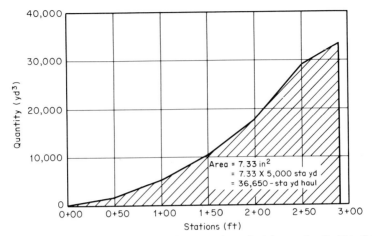

Mass diagram for borrow pit of Figure 17-2. The accumulative haul from station $0+00$ to $2+90$ is 36,659 sta yd. The yardage of excavation is 33,472 yd³. The center of gravity along a line parallel to a line normal to line $0+00-3+00$ at station $2+90$ may be determined by the equation for D, the distance downstation from station $2+90$.

$$D = 36,659/33,472 = 1.10 \text{ sta}$$

$$= 110 \text{ ft}$$

The distance from station $0+00 = 290 - 110 = 180$ ft. This 180-ft distance checks closely with the distance of 181 ft, determined previously by using the static-moment method for determining center of gravity.

The locations may also be calculated by means of three mass diagrams with a three-dimensional orientation. The center of gravity horizontal with respect to line xy is based on a mass diagram commencing at station $0+00$ and ending at station $2+90$. The center of gravity horizontal with respect to line yz is based on a mass diagram commencing at line yz and ending at the left edge of the borrow pit. The center of gravity vertical with respect to the plane xyz is based on a mass diagram commencing at the plane xyz and ending at point $42'$. These three determinants fix the location of the center of gravity within the borrow-pit excavation.

As an example of the method, the horizontal position of the center of gravity with respect to line $x-y$ in the plan view is calculated as shown below for the mass diagram of the borrow pit of Figure 17-2.

Station to Station	Excavation, yd³	Total Excavation, yd³
0+00–0+50	1,731	1,731
0+50–1+00	3,479	5,210
1+00–1+50	4,970	10,180
1+50–2+00	7,678	17,858
2+00–2+50	10,998	28,856
2+50–2+90	4,616	33,472

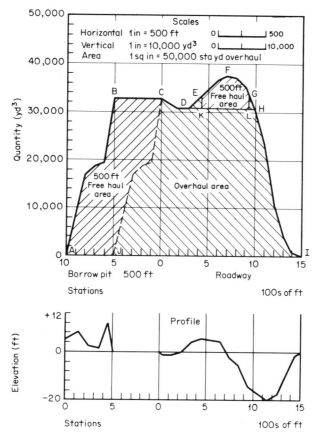

Figure 17-9 Mass diagram for calculating station yards of haul, freehaul, and overhaul in highway grading.

EXPEDIENT METHOD FOR APPROXIMATE VOLUMES OF RESIDUALS, RIPPABLE ROCK, AND BLASTING ROCK IN LINEAR CUT OF IGNEOUS OR METAMORPHIC ROCK

Sometimes the estimator must calculate yardages in a cut when precise results are not necessary and when only limited information is available. The limited information includes the total yardage of the cut, a centerline profile of the cut, and the estimated depths of residuals, rippable rock, and blasting rock. The approximate calculations may be desired for two purposes.

1. To determine the feasibility of doing the work. Some excavators specialize in either earth work or rock work and they want only an approximate estimate of the amount or percentage of excavation requiring blasting.

2. To obtain necessary quick, approximate estimate for bidding purposes. Occasionally and rarely, the estimator must come up with quantities when cross sections of a cut are not available or when time does not permit takeoffs of quantities from available cross sections.

Graph A, Figure 17-10, graph B, Figure 17-11, and graph C, Figure 17-12 offer a quick means for calculating volumes in a cut when only the following data are available:
    1. Total yardage in cut
    2. Maximum depth of cut at centerline
    3. Estimated average depths of residuals, rippable rock, and blasting rock
    4. Width of roadway

17-26   Calculations of Quantities for Excavation

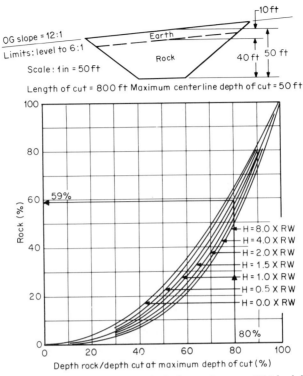

Figure 17-10 Through cuts with average 12:1 O. G. slope and average 1¼:1 backslopes (graph A). The graph shows percentage of rock vs. ratio of depth of rock to centerline depth of cut at maximum centerline depth of cut.

    Total volume by calculation   77,700 yd³
    Volume of rock by calculation   44,600 yd³
    Volume of rock by graph, 59% ×77,700   45,800 yd³
    Error of graph   +0.4%
H = maximum centerline depth of cut
RW = width of roadway

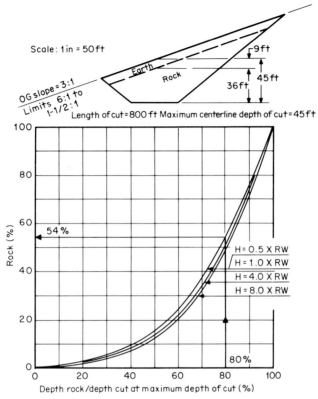

Figure 17-11 Through cuts with average 3:1 O. G. slope and average 1¼:1 backslopes (Graph B). The graph shows the percentage of rock versus the ratio of depth of rock to centerline depth of cut at maximum centerline depth of cut.

    Total volume by calculation   80,800 yd³
    Volume of rock by calculation   44,000 yd³
    Volume of rock by graph, 54% × 80,800   43,600 yd³
    Error of graph   −0.9%
H = maximum centerline depth of cut
RW = width of roadway

17-28 Calculations of Quantities for Excavation

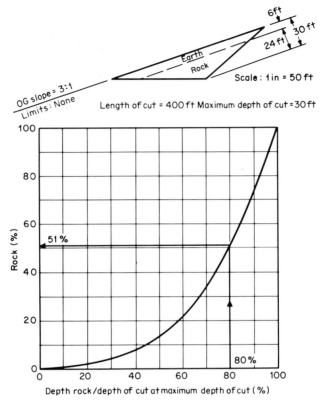

Figure 17-12 Sidehill cuts (Graph C). The graph shows the percentage of rock versus the ratio of depth of rock to the total depth of cut at maximum depth of cut.

Total volume by calculation	20,000 yd³
Volume of rock by calculation	10,200 yd³
Volume of rock by graph, 51% × 20,000	10,200 yd³
Error of graph	0.0%

The graphs are applicable only to rock formations without stratifications and beddings, which weather uniformly downward from the surface. These formations include igneous-metamorphic rocks and sometimes sedimentary rocks. The configurations of the cross sections of the three graphs correspond to those generally encountered in excavations of a linear nature such as those for canals, highways, and railroads.

The example of calculations is based on the desired volumes of residuals, rippable rock, and blasting rock of the cut of Figure 17-7. The criteria are:

Volume of cut	258,474 yd³
Maximum depth of cut along centerline, H	100 ft
Depth of residuals, R	20 ft
Depth of rippable rock, RR	20 ft
Depth of blasting rock, BR	60 ft
Width of roadway, RW	100 ft

For the calculations, the configuration of the cut suggests the use of graph A, Figure 17-10.

1. Total volume of rippable rock, RR, and blasting rock BR

$$H = 1.0 \times RW$$

$$\frac{\text{Depth of } (RR+BR)}{\text{Maximum depth of cut}} = \frac{80}{100} = 80\%$$

Entering the abscissa of the graph at 80%, intersecting the curve $H=1.0\times RW$, and leaving the graph at the ordinate 59% gives;

$$\text{Volume} = 59\% \times 258{,}474 = 152{,}500 \text{ yd}^3$$

2. Volume of blasting rock, BR

$$H = 1.0 \times RW$$

$$\frac{\text{Depth of } BR}{\text{Maximum depth of cut}} = \frac{60}{100} = 60\%$$

Entering the abscissa of the graph at 60%, intersecting the curve $H=1.0\times RW$, and leaving the graph at the ordinate 30% gives:

$$\text{Volume} = 30\% \times 258{,}474 = 77{,}542 \text{ yd}^3$$

3. Summary of classifications of rock-earth in the cut:

Classification	Yardage, yd³	Percentage yd³
Residuals, 258,474 − 152,500	105,974	41
Rippable rock, 152,500 − 77,542	74,958	29
Blasting rock	77,542	30
Total rock-earth	258,474	100

These same volumes have been calculated in Figure 17-7 by the average-end-area method. The comparison between this precise method and the approximate graphical method is tabulated below.

Classification	Average-End-Area Method, yd³	Graphical Method, yd³	Error of Graph Method, %
Residuals	102,407	105,974	+3.5
Rippable rock	69,999	74,958	+7.1
Blasting rock	86,068	77,542	−9.9
Totals	258,474	258,474	0.0

The average error for the three classifications of rock-earth is 6.8 percent. Of course, the error for the total excavation is 0 percent, as both methods are based on the same total 258,474 yd³. The average precision of the graphical method is ±5 percent, and it is acceptable because the several yardages always total up to the total yardage of the cut.

SUMMARY

Modern machines for rock-earth excavation accomplish work in terms of foot-pounds. A cubic yard with a given weight requires a combination of forces applied by loading and hauling machines to move it to a fill or to a point of deposit. Volumes of excavation and the corresponding weights, together with the weights of the moving machines, are two of the several factors determining the energy expended for the work. Costs are determined largely by the total foot-pounds of energy expended in the movement of rock-earth. Accordingly, a thorough knowledge of the principles set forth in this chapter is requisite for estimates of excavation costs.

CHAPTER 18

Preparation of Bid and Schedule of Work

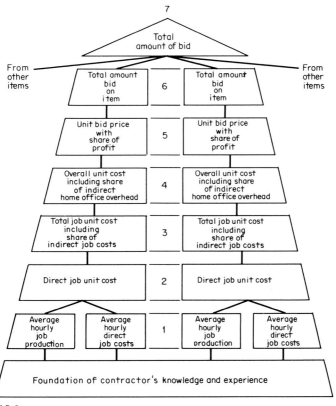

PLATE 18-1

Pyramid of the businesslike bid.

PREPARATION OF BID

There is no universally accepted plan for the preparation of a bid for excavation. Each excavator generally has a personal system, peculiar to that excavator's business methods and personal liking. The methods range from the simplest establishment of unit prices based on the experience of one person to the complex preparations of several persons for several months, as may be the case when bidding a large dam. The system suggested here for building a businesslike bid is simple and complete. It may be used in its entirety or it may be changed to suit the bookkeeping methods of the excavator, provided the fundamental procedures are carried out completely.

A bid is built like a pyramid. Like the Great Pyramid it must be based on a firm foundation. This conception is illustrated in Plate 18-1. The foundation is the excavator's knowledge and experience. The apex is the total amount of the bid. Each course of the pyramid rests squarely and solidly on the course below and supports in like manner the course above.

Hitherto in the text the discussions of cost of work have been confined largely to the direct job unit cost. This cost is secured by dividing the hourly cost of ownership and operation of the machine and of the operator(s) by the hourly production of the machine. These values make up courses 1 and 2 of the pyramid. The direct job unit cost must bear its share of the indirect job costs and the resultant total job unit cost is represented by course 3. The usual indirect job costs, chargeable to all the bid items, are set forth in Table 18-1. Likewise, the total job unit cost must bear its proportionate share of the cost of the indirect home office overhead, and the resultant overall unit cost is represented in course 4. The usual indirect home office expenses, chargeable to all the bid items, are set forth in Table 18-2. To the total job unit cost must be added a profit, either as percentages for all bid items or for certain individual bid items, and the unit bid price with its share of profit is course 5. The quantity of the bid item times the unit bid price gives the total amount bid on the item, and these amounts are course 6. Finally, the apex of the pyramid is the sum of all the amounts bid on the individual items. This is the capstone, 7.

A system of bidding must have a built-in facility for internal examination and for changes in unit bid prices at any time, and the pyramid system has this characteristic. Each bid item is calculated independently of all other bid items until it becomes part of the total amount bid at the apex or capstone. Thus it may be changed independently at any time without any necessary changes in the other bid items.

A bid is prepared around an invitation to bid a job. To illustrate the building of a

TABLE 18-1 Indirect Job Costs

1. General expenses
   a. Contract surety bond
   b. Special surety bond
   c. Nuisance insurance
   d. Protection of environment
   e. Local taxes
   f. Local legal charges
   g. Miscellany not covered by the items listed above
   h. Safety and traffic control
   i. Move-in and move-out charges
   j. Storage of machinery after completion of job
2. Field office
   a. Ownership or rental charges
   b. Light and heat
   c. Telephone and telegraph
   d. Equipment
   e. Supplies
   f. Cars
3. Office help
   a. Manager
   b. Bookkeepers
   c. Timekeepers
   d. Clerks and secretaries
   e. Traveling expenses
4. Shop and yard
   a. Ownership or rental charges
   b. Light, power, and heat
   c. Equipment
   d. Supplies
   e. Lubricating trucks
   f. Fueling trucks
   g. Maintenance trucks
   h. Pickups and cars
   i. Miscellany
5. Shop and yard help
   a. Master mechanic
   b. Mechanics
   c. Greasemen
   d. Guards
   e. Traveling expenses
6. Supervision
   a. Superintendent
   b. Engineers
   c. Foremen
   d. Cars and pickups
   e. Traveling expenses

18–4 Preparation of Bid and Schedule of Work

TABLE 18-2 Indirect Home Office Overhead

1. Property taxes	10. Office help
2. Business taxes	a. Manager
3. Property and liability insurance	b. Bookkeeper
4. Life and special insurance	c. Clerks and stenographers
5. Legal expenses	d. Cars
6. Accounting expenses	e. Traveling expenses
7. Miscellany	11. Shop and yard
8. Office	a. Rental or fixed charges
a. Rental or fixed charges	b. Equipment
b. Equipment	c. Supplies
c. Supplies	d. Light, heat, and power
d. Light and heat	e. Trucks
e. Telephone and telegraph	f. Cars and pickups
f. Miscellany	12. Shop help
9. Managerial help	a. Master mechanic
a. Officers	b. Mechanics
b. General superintendent	c. Helpers
c. Engineer	d. Guard
d. Cars	e. Traveling expenses
e. Traveling expenses	

businesslike bid, a step-by-step example of the procedure will be given for a simple rock excavation job. All costs for the bid preparation are for 1978.

Example of Bid Preparation

The owner has property suitable as a site for a quarry, aggregate plant, asphalt plant, and concrete plant, as illustrated by the maps and pictures of Figures 18-1 to 18-4. The owner is asking for bids for the removal of overburden from the quarry site and excavation of a channel change. The overburden is to be placed in embankments for the plant site and for the access roadway and in a stockpile for materials. The excavation from the channel change is to be placed in the stockpile for materials. The owner will build several culverts for the access roadway prior to the specified time for commencing work.

The invitation for bids includes the following pertinent conditions:

    1. Beginning of work is to be within 15 calendar days after receiving notice that the contract has been approved.

    2. Completion of work is to be within 270 calendar days after the beginning of work.

    3. Liquidated damages are $1000.00 per day for each and every calendar day of delay in finishing the work in excess of the number of calendar days allowed.

    4. A contract surety bond for the total amount of the bid is required.

    5. The quarry overburden is considered to be rippable. If for any reason blasting is necessary, the contingency is covered by a rock clause as part of the contract.

    6. The work will be done during the dry season when there is no water in Calavera Creek, the channel of which is to be changed.

Bid Items The individual items are as follows:

    1. Clearing and grubbing, 76.9 acres
    2. Development of water supply, lump sum
    3. Excavation of overburden for plant site embankment, 400,000 cybm
    4. Excavation of overburden for roadway embankment, 42,000 cybm
    5. Excavation of overburden for stockpile of material, 114,000 cybm
    6. Excavation of channel change for stockpile of material, 74,000 cybm
    7. Embankment of plant site, 400,000 cybm
    8. Embankment of roadway, 42,000 cybm
    9. Fill for stockpile from overburden excavation, 114,000 cybm
    10. Fill for stockpile from channel change excavation, 74,000 cybm

The unit for all excavation, embankments, and fills is the cubic yard bank measurement (cybm). It is believed that a cubic yard in the cut will make a cubic yard in the embankment and fill, there being no swell or shrink factor.

Direct Job Costs The estimator prepares a schedule of costs for each bid item. These schedules are based on estimates for the following controlling factors.

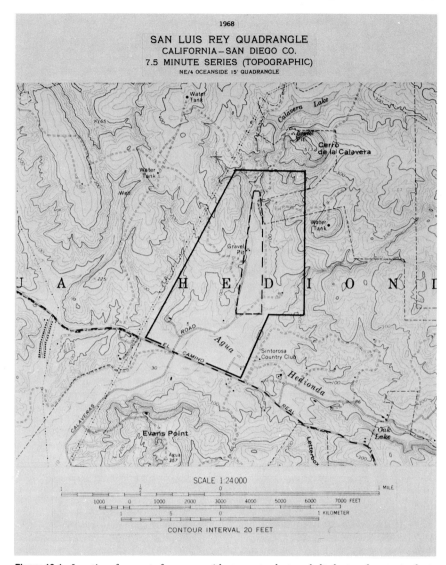

Figure 18-1 Location of property for quarry with aggregate plant, asphalt plant, and concrete plant. The property is outlined in solid lines and the location of plants is outlined in dashed lines. The rock is made up of thin residuals overlying weathered, semisolid, and solid granites. *(U.S. Geological Survey topographic map of San Luis Rey, 7½-minute quadrangle, San Diego County, California.)*

18–6 Preparation of Bid and Schedule of Work

Figure 18-2 General view of west side of property. Referring to Figure 18-1, the view is to the west from the water tank, across the site for the plants, and toward the gravel pit of the quarry site. The gravel pit is in the center of the picture. The area of the quarry to be worked first is on the far right side of the picture. Here overburden to a depth of 15 ft is to be excavated from a 23-acre area and to be used for embankment for the plant site, for embankment for the access roadway, and for fill of stockpile of materials. The overburden removal is 556,000 cybm. The excavation of the channel change of Calavera Creek is to the east of the plant site. The job is set by the owner for completion in 270 calendar days after the beginning of the work.

Figure 18-3 Small abandoned quarry located at site for large quarry. The small quarry shows up as the gravel pit of Figures 18-1 and 18-4. It is located at the midpoint of the western boundary of the site for plants. At this location the rippable granite formation is about 10 ft deep. At other locations the depth of rippable rock reaches 20 ft. The average depth of the overburden over the 23 acres is estimated to be 15 ft, giving about 556,000 yd^3 of overburden to be removed.

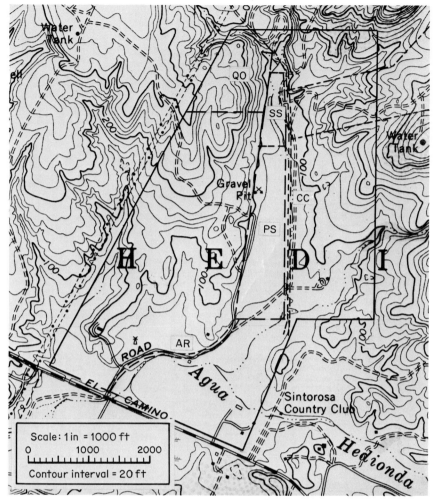

Figure 18-4 **Work areas for the preparation of the property.** Key: QO, quarry overburden; SS, site for stockpiles of materials; PS, site for aggregate, asphalt, and concrete plants; CC, channel change; AR, access road. Haul roads for quarry overburden are through areas QO, SS, PS, and AR; haul road for channel change excavation is through areas PS and SS. *(U.S. Geological Survey, topographic map of San Luis Rey, 7½-minute quadrangle, San Diego County, California.)*

1. Hourly production.
2. Time required in working hours for the work.
3. Machinery required for the work.
4. Hourly cost of ownership and operation of the machinery. This is sometimes called *hourly rental to the job*, but if this term is used it is not to be confused with rental of machinery from another owner.
5. Hourly cost of rental of machinery from another owner.
6. Hourly cost of labor, operator(s) and general, including all fringe benefits.

The estimates for the 10 bid items are discussed and tabulated below.

1. Clearing and Grubbing. The first work, along with development of water supply, includes removal of some 50 eucalyptus trees with up to 24-in diameters and light brush up to 4 ft in height from the 76.9-acre site. A clearing contractor will remove the trees and their stumps in exchange for the trees. The brush will be stacked and burned.

18-8 Preparation of Bid and Schedule of Work

Brushing will be done by three 300-hp crawler-tractor–brush rakes. The hourly cost of ownership and operation of each machine is $60.40. Production is 1.32 acres/h and the time for the work is 59 working hours.

The summary of the direct job costs is:

Summary of Direct Job Costs		Bid Item No. _1_ Unit _Acre_ Quantity _76.9_ Page No _1_					
		Working hours for completion _59_ Quantity per working hour _1.32_					
		1	2	3	4	5	6
Kind Of Cost	Operation And Machinery Data	Machinery O — Owned R — Rented	Labor Operators	Labor General	Materials	Sub-Contract	Total
H C	Hourly Costs						
	Brushing 3-300 hp crawler tractor brush rakes	O 181.20	46.80				228.00
	Burning burn men			11.50			11.50
	Totals	181.20	46.80	11.50			239.50
T C	Total Costs 59 h	10690.80	2761.20	678.50			14130.50
U C	Unit Costs 76.9 acre	139.01	35.91	8.82			183.74

2. *Development of Water Supply.* This item involves a lump-sum bid. Nearby is the utility company main line from the tank to the east of the property. The line passes within 300 ft of the southeast corner of the site for plants, Figure 18-4. The hourly cost of ownership and operation of the connecting line and the portable tank of 10,000-gal capacity is $7.20. The line will be in service for the 651 working hours required for the job from move-in to move-out of machinery. The total cost is $4687.20. The estimated cost of assembly and disassembly of line and tank is $1043.40. The total cost is $5730.60.

The summary of direct job costs is:

Summary of Direct Job Costs		Bid Item No. _2_ Unit _Lump Sum_ Quantity _1_						Page No _2_
		Working hours for completion _651_				Quantity per working hour		
			1	2	3	4	5	6
Kind Of Cost		Operation And Machinery Data	Machinery	Labor		Materials	Sub-Contract	Total
			O Owned R Rented	Operators	General			
H	C	Hourly Costs						
		Line from utility main to job	O 100					100
		Portable tank 10,000 gal capacity	O 620					620
		Totals	720					720
T	C	Total costs 651 h	468720					468720
		Estimated cost of assembly + disassembly						
		32 h of pickup truck @ $7.20	O 23040					23040
		64 h of pipe fitters @ $11.80			75520	5780		81300
		Totals Lump sum	491760		75520	5780		573060

3. Excavation of Overburden for Plant Site Embankment. For the required 400,000 cybm excavation, a total of 400,000 cybm of overburden will be removed from an area of 23 acres located in the northwest corner of the property, as illustrated in Figure 18-4. The average depth of the overburden is estimated to be 15 ft. The work will commence after completion of clearing and grubbing and development of water supply. The pioneering of haul roads is minimal and it will be done by the tractor-bulldozer-ripper and the motor grader during the initial stage of work when little or no ripping is necessary. The average haul is 2500 ft one way. The maximum grade for a short distance in the overburden area is −33 percent, calling for scraper operation on a 3:1 slope. The work will proceed at an average rate of 940 cybm/h, requiring a total of 426 working hours.

The machinery for the work is:

1. A 514-hp crawler-tractor-bulldozer-ripper at a cost of $128.90/h for ownership and operation.

18-10 Preparation of Bid and Schedule of Work

2. A 514-hp crawler-tractor–cushion bulldozer for push-loading scrapers at a cost of $103.30/h for ownership and operation.

3. Six 415-hp, 27-cysm push-loaded scrapers, five working and one standby. The hourly cost of ownership and operation for each machine is $84.70 when working and $42.30 when standby.

4. A 180-hp, 14-ft motor grader. The hourly cost of ownership and operation, $34.60, is divided equally between haul-road maintenance and embankment construction, as these are separate bid items. Thus, $17.30 is charged hourly to both item 3 and item 7.

5. A 250-hp, 2000-gal-capacity water wagon for haul-road maintenance. The hourly cost of ownership and operation is $15.90. Water is regarded as a material. Water consumption is estimated at 6 gal/h per cybm of excavation or 5640 gal/h. Water costs $0.225 per 100 ft^3, giving a cost of $1.70/h.

The summary of direct job costs is:

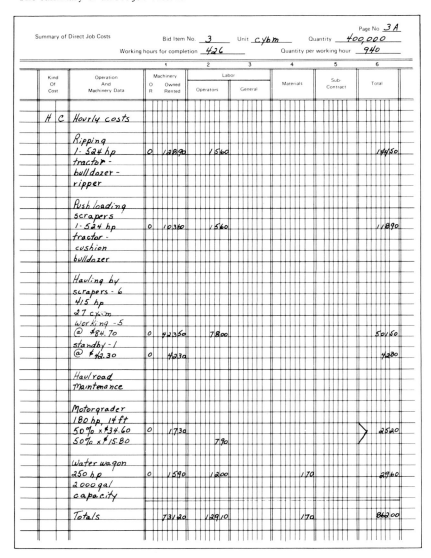

Preparation of Bid   **18-11**

Page No __3B__

Summary of Direct Job Costs   Bid Item No. __3__   Unit __Cybm__   Quantity __400,000__

Working hours for completion __426__   Quantity per working hour __940__

Kind Of Cost	Operation And Machinery Data	Machinery O/R Owned Rented (1)	Labor Operators (2)	Labor General (3)	Materials (4)	Sub-Contract (5)	Total (6)
TC	TOTAL COSTS 426 h	311491.20	54996.60		724.20		367212.01
UC	UNIT COSTS 400,000 cybm	0.779	0.137		0.002		0.918

18–12 Preparation of Bid and Schedule of Work

4. *Excavation of Overburden for Roadway Embankment.* The 42,000-cybm excavation will be handled similarly to that for the 400,000 cybm placed in the embankment for the plant site. It will follow the excavation for plant site. The average haul is 5600 ft one way. The work will proceed at the rate of 940 cybm per working hour, requiring 45 working hours for completion. The machinery and quantity of water for the work will be the same as those for item 3, except that a total of nine scrapers will be used because of the longer haul.

The summary of direct job costs is:

Summary of Direct Job Costs Bid Item No. 4 Unit cybm Quantity 42,000 Page No. 4
Working hours for completion 45 Quantity per working hour 940

Kind Of Cost	Operation And Machinery Data	Machinery		Labor		Materials	Sub-Contract	Total
		O	Owned	Operators	General			
		R	Rented					
H C	Hourly costs							
	Ripping, as item 3	O	12890	1560				14450
	Push loading scrapers, as item 3	O	10330	1560				11890
	Hauling by scrapers - 9 working - 8 @ $84.70	O	67760	12480				80240
	Standby - 1 @ $42.30	O	4230					4230
	Haul road maintenance, as item 3							
	Motor grader	O	1730	790				2520
	Water wagon	O	1590	1200		170		2960
	Totals		98530	17590		170		116290
T C	Total costs 45 h		4433850	791550		7650		5233050
U C	Unit Costs 42,000 cybm		1.056	0.188		0.002		1.246

5. *Excavation of Overburden for Stockpile of Materials.* The 114,000-cybm excavation will be handled similarly to the 400,000 cybm placed in the embankment for the plant site. It will follow the excavation for the roadway. The average haul is 900 ft one way. The work will proceed at a rate of 940 cybm per working hour, requiring 121

working hours for completion. The machinery and quantity of water for the work will be the same as those for item 3, except that a total of five scrapers will be used for the shorter haul.

The summary of direct job costs is:

					Labor				
	Kind Of Cost	Operation And Machinery Data	Machinery O Owned R Rented	Operators	General	Materials	Sub-Contract	Total	
H C		Hourly costs							
		Ripping, as item 3	O	128.90	15.60			144.50	
		Push loading scrapers, as item 3	O	103.40	15.60			118.90	
		Hauling by scrapers,-5 working -4 @ $84.76	O	338.80	62.40			401.20	
		Standby -1 @ $42.30	O	42.30				42.30	
		Haul road maintenance, as item 3							
		Motor grader	O	17.30	7.90			25.20	
		Water wagon	O	15.90	12.00		1.70	29.60	
		Totals		646.60	113.50		1.70	761.70	
T C		Total costs 121 h		78226.60	13733.50		205.70	92165.70	
U C		Unit costs 114,000 cybm		0.686	0.120		0.002	0.808	

Summary of Direct Job Costs — Bid Item No. 5 — Unit cybm — Quantity 114,000 — Page No. 5
Working hours for completion 121 — Quantity per working hour 940

6. *Excavation of Channel Change for Stockpile of Materials.* The 74,000-cybm excavation will be handled at the same time as that for overburden excavation for the plant site, roadway, and stockpile for materials. The haul averages 1300 ft one way on an average 2.0 percent upgrade. The work will proceed at an average rate of 472 cybm per working hour and it will require 157 working hours for completion.

The machinery for this work is:

1. Four 415-hp, 29-cysm self-loading scrapers. The hourly cost of ownership and operation of each machine is $92.90.

18-14 Preparation of Bid and Schedule of Work

2. A 300-hp crawler-tractor-bulldozer. The hourly cost of ownership and operation is $60.40. Of the total hourly cost, 50 percent is allocated to excavation of channel change and 50 percent is allocated to item 10, fill for stockpile from channel change excavation.

3. A 180-hp, 14-ft motor grader for haul road maintenance.

4. A 250-hp, 2000-gal water wagon for haul-road maintenance.

The hourly costs for the motor grader and the water wagon are not chargeable to this item, as they are a part of the costs for the simultaneous excavations of overburden for embankments for the plant site and roadway and for the fill of stockpile materials.

The summary of direct job costs is:

Kind of Cost	Operation And Machinery Data	Machinery O/R Owned Rented	Labor Operators	Labor General	Materials	Sub-Contract	Total
	Summary of Direct Job Costs Bid Item No. 6 Unit cybm Quantity 74,000 Working hours for completion 157 Quantity per working hour 472 Page No 6						
H C	Hourly costs						
	Hauling by self loading scrapers -4 415 hp 29 c.y.s m	O 37/60	62 40				43400
	Utility work by 300 hp crawler tractor- bulldozer 50% x $60.40 50% x $15.60		30 20	7 80			38 00
	Totals	40 180	70 20				472 00
T C	Total costs 157 h	6308 260	1102 140				741 0400
U C	Unit costs 74,000 cybm	0.852	0.149				1.001

7. *Embankment of Plant Site.* The 400,000 cybm of embankment will be placed at the rate of 940 cybm per working hour in a total of 426 working hours. The average depth of embankment is 6 ft. The machinery for this work is:

    1. One 300-hp, 67,000-lb, self-propelled tamping-foot compactor. The hourly cost of ownership and operation is $58.40.

    2. One 300-hp crawler-tractor-bulldozer. The hourly cost of ownership and operation is $60.40.

    3. One 180-hp, 14-ft motor grader. The hourly cost of ownership and operation is $34.60, 50 percent of which, $17.30, is allocated to this item and the remaining $17.30 to the simultaneous item 3.

    4. One 480-hp, 8000-gal water wagon. The hourly cost of ownership and operation is $81.90. Water is applied at the rate of 21 gal/cybm, and the hourly cost is $5.90.

The summary of direct job costs is:

Summary of Direct Job Costs — Bid Item No. 7 — Unit cybm — Quantity 400,000 — Page No. 7
Working hours for completion 426 — Quantity per working hour 940

Kind Of Cost	Operation And Machinery Data	Machinery O/R Owned/Rented	Labor Operators	Labor General	Materials	Sub-Contract	Total
H C	Hourly Costs						
	Spreading and shaping fill material						
	Motorgrader 180 hp, 14 ft. 50% × $34.60 50% × $15.80	O	17 30	7 90			25 20
	300 hp crawler tractor-bulldozer	O	60 40	15 60			76 00
	Wetting and compacting fill						
	Water wagon 480 hp. 8,000 gal capacity	O	81 90	12 00	5 90		99 80
	Compactor 300 hp 67,000 lb	O	58 40	15 60			74 00
	Totals		218 00	51 10	5 90		275 00
T C	Total costs 426 h		92,868 00	21,768 60	2513 40		117,150 00
U C	Unit costs 400,000 cybm		0.232	0.054	0.006		0.292

18–16 Preparation of Bid and Schedule of Work

8. Embankment of Roadway. The 42,000-cybm embankment will be placed at the rate of 940 cybm per working hour in a total of 45 working hours. The average depth of embankment is 6 ft. The machinery and quantity of water for this work will be the same as those for item 7, embankment for plant site. The hourly costs will be the same.

The summary of direct job costs is:

Summary of Direct Job Costs		Bid Item No. **8**	Unit **Cybm**	Quantity **42,000**	Page No **8**
		Working hours for completion **45**		Quantity per working hour **940**	
Kind Of Cost	Operation And Machinery Data	Machinery (O Owned / R Rented)	Labor — Operators	Labor — General	Materials	Sub-Contract	Total
HC	HOURLY COSTS, AS ITEM 7						
	TOTALS	21800	5110		590		27500
TC	TOTAL COSTS 45 h	981000	229900		26550		1237500
UC	UNIT COSTS 42,000 Cybm	0.234	0.055		0.006		0.295

9. Fill for Stockpile from Overburden Excavation. The 114,000-cybm stockpile will be placed in 121 working hours at the rate of 940 cybm per working hour. The average depth of fill is 15 ft. The machinery for this work is a 300-hp crawler-tractor-bulldozer. The hourly cost of ownership and operation is $60.40.

The summary of direct job costs is:

Kind Of Cost	Operation And Machinery Data	Machinery O Owned R Rented	Labor Operators	Labor General	Materials	Sub-Contract	Total
						Bid Item No. 9 Unit Cybm Quantity 114,000	Page No 9
						Working hours for completion 121 Quantity per working hour 940	
H C	Hourly Costs						
	Shaping fill						
	300 hp crawler tractor bulldozer	O	6040	1560			7600
	Totals		6040	1560			7600
T C	Total Costs 121 h		730840	188760			919600
U C	Unit Costs 114,000 cybm		0.064	0.017			0.081

18-18 Preparation of Bid and Schedule of Work

10. Fill for Stockpile from Channel Change Excavation. The stockpile, amounting to 74,000 cybm, will be placed at the rate of 472 cybm per working hour in a total of 157 working hours. The average depth of fill is 15 ft. The machinery for this work is a 300-hp crawler-tractor-bulldozer. The hourly cost of ownership and operation is $60.40. Of this, 50 percent, or $30.20, is allocated to this item and the remaining $30.20 is allocated to item 6, excavation of channel change for stockpile of materials.

The summary of direct job costs is:

Kind Of Cost	Operation And Machinery Data	Machinery O/R Owned/Rented	Labor Operators	Labor General	Materials	Sub-Contract	Total
H C	Hourly Costs						
	Shaping fill						
	300 hp crawler tractor bulldozer						
	50% × $60.40	O	30.20				38.00
	50% × $15.60			7.80			
	Totals		30.20	7.80			38.00
T C	Total Costs 157 h		4741.40	1224.60			5966.00
U C	Unit Costs 74,000 cybm		0.064	0.017			0.081

Page No 10
Bid Item No. 10 Unit cybm Quantity 74,000
Working hours for completion 157 Quantity per working hour 472

Indirect Job Costs The estimator prepares a summary of all indirect job costs, using a check list similar to Table 18-1. The summary is tabulated below. The total is $213,150, equivalent to 28.4 percent of the $749,359.80 direct job costs.

Summary Of Indirect Job Costs

Page No __1__
Total Direct Job Costs **$750,360.30**
Calendar Days Allowed For Job __180__

Item No	Kind of Cost	1 General	2 Field Office	3 Field Shop-Yard	4 Machinery	5 Salaries Wages	6 Travel Lodging
1	General expenses						
a	Contract surety bond	11000 —					
b	Special surety bond						
c	Nuisance insurance						
d	Protection of environment						
e	Local taxes						
f	Local legal charges						
g	Miscellany						
h	Safety and traffic control						
i	Move-in & move out	14400 —					
j	Storage of machinery						
2	Field office						
a	Ownership or rental charges		2970 —				
b	Light and heat		240 —				
c	Telephone and telegraph		360 —				
d	Equipment						
e	Supplies		1200 —				
f	Cars				1500		
3	Office help						
a	Manager						
b	Book-keepers					9000 —	
c	Time-keepers					7200 —	
d	Clerks and secretaries					5000 —	
e	Travelling expenses						1800 —

18-20　Preparation of Bid and Schedule of Work

Summary Of Indirect Job Costs　　　　　　　　　　　　　　　　Page No 2

Item No	Kind of Cost	1 General	2 Field Office	3 Field Shop-Yard	4 Machinery	5 Salaries Wages	6 Travel Lodging
4	Shop-Yard						
a	Ownership or rental charges			1080 -			
b	Light, power, and heat			1500 -			
c	Equipment			400 -			
d	Supplies			6000 -			
e	Lubricating trucks				6400 -		
f	Fueling trucks				8380 -		
g	Maintenance trucks				6440 -		
h	Pickups and cars				5150 -		
i	Miscellany						
5	Shop-Yard help						
a	Master Mechanic					12000 -	
b	Mechanics					8050 -	
c	Greasemen					14750 -	
d	Watchmen					10280 -	
e	Travelling expenses						1200 -
6	Supervision						
a	Superintendent					20000 -	
b	Engineers					10000 -	
c	Foremen					20000 -	
d	Cars and pickups				15100 -		
e	Travelling expenses						2000 -
	Totals	24400 -	4770 -	8680 -	53020 -	117280 -	5000 -
	Grand Totals						213150

Indirect Home Office Overhead The estimator is given a cost for a job's share of the home office overhead or expense. The cost is set by the home office and the methods for determining this cost vary with the policies of the company.

A logical method for determining the amount is to calculate the ratio of the estimated total cost of the job to the total costs of all work during an average period corresponding to the time allotted for the job. This ratio or percentage is then applied to the total job costs to determine the addition necessary to arrive at overall job costs. The cost for the home office overhead allocated to this particular job is $70,000.

The Bid Summary Sheet The bid summary sheet is generally a specially prepared sheet about 30 in wide and it contains columns similar to those of a bookkeeping sheet. It is worked horizontally bid item by bid item. In this text the sheet is divided into six pages for individual costs in order to simplify explanations. There are two repetitions of items which are eliminated in the large single sheet. They are the "bid item description" column for all but sheet A and the "percent allocation" column for all but sheet B.

Having data for direct job costs, indirect job costs, and indirect home office overhead costs, the estimator is ready to commence the bid summary sheet. The steps are explained with reference to the individual Sheets A, B, C, D, E, and F.

Sheet A. Sheet A is a recapitulation of the bid items of the invitation for bids.

Sheet B. Sheet B is a summary of the direct job cost pages for the individual bid items, together with calculated percentage distributions for the individual bid items. This is the sheet of basic data for the bid preparation.

Sheet C. Sheet C, by means of the percentage distributions of Sheet B, allocates total indirect job costs to each bid item. These indirect job costs are added to the direct job costs to give total job costs and resultant unit job costs.

Sheet D. Sheet D, also by means of the percentage distributions of Sheet B, allocates that percentage portion of home office overhead expense to each bid item. These indirect costs are added to the previous total job costs to give overall job costs and resultant unit overall costs.

Sheet E. Sheet E contains calculations for trial and revised bid prices, based on preliminary percentages of profit margins. Sheet E, for the sake of simplicity, contains only one column for revised bid prices. The large single-bid summary sheet generally contains two or more of these columns.

There are two ways of figuring profit margins. They may be calculated as a percentage addition to costs, in which case bid price equals overall job cost $\times (1.00 + \text{percentage profit margin})$. Or they may be expressed as a percentage of bid price, in which case bid price equals

$$\frac{\text{overall job cost}}{1.00 - \text{percentage profit margin}}$$

In this text the profit margin is expressed as a percentage of bid price or, equivalently, as a percentage of selling price.

Sheet F. Sheet F contains the submitted bid prices as they generally appear in the tendered bid. It represents the well-considered deliberations of the decision-making people. In this instance it is felt that a lowering of profit margin from 20 to 15 percent is necessary for the lowest total bid, and so the revised unit prices and total prices are used.

18-22   Preparation of Bid and Schedule of Work

BID SUMMARY SHEET A

Bidding Schedule   Page No __1__

Bid Item No.	Bid Item Description	Unit	Quantity
1	Cleaning & grubbing	Acre	169
2	Development of water supply	Lump Sum	1
3	Excavation of overburden for plant site embankment	cybm	400000
4	Excavation of overburden for roadway embankment	cybm	42000
5	Excavation of overburden for stockpile of material	cybm	114000
6	Excavation of channel change for stockpile of material	cybm	74000
7	Embankment of plant site	cybm	400000
8	Embankment of roadway	cybm	42000
9	Fill for stockpile from overburden excavation	cybm	114000
10	Fill for stockpile from channel change excavation	cybm	74000

BID SUMMARY SHEET B

Total And Unit Direct Job Costs With Percentage Allocations Of Total Costs

Page No __1__

Bid Item No	Bid Item Description	1. Total Direct Job Costs	2. Allocation (%)	3. Unit Direct Job Costs
1	C & G	1413050	1.9	1.374
2	DOWS	573060	0.8	573060
3	EOO FOR PSE	3672100	48.9	0.918
4	EOO FOR RE	5233050	7.0	1.246
5	EOO FOR SOM	9216570	12.3	0.808
6	EOCC FOR SOM	7410400	9.9	1.001
7	IOPS	11715000	15.6	0.292
8	EOR	1237500	1.6	0.295
9	FFSP FROM OE	919600	1.2	0.081
10	FFSP FROM CCE	596600	0.8	0.081
	Totals	75036030	100.0	

Preparation of Bid 18-23

18-24 Preparation of Bid and Schedule of Work

BID SUMMARY SHEET C

Total And Unit Job Costs Total Indirect Job Cost $213,150.00 Page No 1

Bid Item No	Bid Item Description	Allocation (%)	Indirect Job Cost Allocations	Direct Job Costs	Total Job Costs	Unit Job Costs
1	C & G	1.9	4049 85	141130 50	18180 35	236.42
2	DOWS	0.8	1705 20	5730 60	7435 80	7435.80
3	EOO FOR PSE	48.9	104230 35	367212 00	471442 35	1.179
4	EOO FOR RE	7.0	14920 50	52330 50	67251 00	1.601
5	EOO FOR SOM	12.3	26217 45	92165 70	118383 15	1.038
6	EOCC FOR SOM	9.9	21101 85	74104 00	95205 85	1.287
7	EOPS	15.6	33251 40	117150 00	150401 40	0.376
8	EOR	1.6	3410 40	12375 00	15785 40	0.376
9	FFSP FROM SE	1.2	2557 80	9196 00	11753 80	0.103
10	FFSP FROM CCE	0.8	1705 20	5966 00	7671 20	0.104
	Totals	100.0	213150 00	750360 30	963510 30	

Preparation of Bid 18–25

BID SUMMARY SHEET D

Total And Unit Overall Job Costs
Share Of Total Home Office Overhead Allocated To Job $70,000.00   Page No 1

Bid Item No	Bid Item Description	Allocation (%)	Home Office Overhead Allocations	Total Job Costs	Overall Job Costs	Unit Overall Job Costs
1	C & G	1.9	1425 00	1"035	1960.35	254.95
2	DOWS	0.8	600 00	7435 80	8035 80	8035 80
3	EOO FOR PSE	48.9	36675 00	47144235	50811735	1.270
4	EOO FOR RE	7.0	5250 00	67251 00	72501 00	1.726
5	EOO FOR SOM	12.3	9225 00	11838315	12760815	1.119
6	EOCC FOR SOM	9.9	7425 00	9520585	10263085	1.387
7	EOPS	15.6	11700 00	15040140	16210140	0.405
8	EOR	1.6	1200 00	1578540	1698540	0.404
9	FFSP FROM OE	1.2	900 00	1175380	1265380	0.111
10	FFSP FROM CCE	0.8	600 00	767120	827120	0.112
	Totals	100.0	75000 00	96351030	103851030	

18-26   Preparation of Bid and Schedule of Work

BID SUMMARY SHEET E

Trial And Revised Bid Prices

Page No 1

Bid Item No	Bid Item Description	Unit Overall Job Costs	Total Price Margin 20 %		Revised Prices Margin 15 %	
			Unit Price	Total Price	Unit Price	Total Price
1	C & G	254.95	319.00	24531.10	300.00	23070.00
2	DOWS	8035.80	10045.00	10045.00	9454.00	9454.00
3	EOO FOR PSE	1.270	1.59	636000.00	1.49	596000.00
4	EOO FOR RE	1.726	2.16	90720.00	2.03	85260.00
5	EOO FOR SOM	1.119	1.40	159600.00	1.32	150480.00
6	EOCC FOR SOM	1.387	1.73	128020.00	1.63	120620.00
7	EOPS	0.405	0.51	204000.00	0.48	192000.00
8	EOR	0.404	0.50	21000.00	0.48	20160.00
9	FFSP FROM OE	0.111	0.14	15960.00	0.13	14820.00
10	FFSP FROM CCE	0.112	0.14	10360.00	0.13	9620.00
	Totals			1300236.10		1221484.00
	Total costs			1038510.30		1038510.30
	Total profits			261725.80		182973.70
	% of profit			20.1		15.0

BID SUMMARY SHEET F

Submitted Bid Prices   Page No __1__

Bid Item No	Bid Item Description	Unit	Quantity	Unit Bid Price	Total Bid Price
1	C & G	Acre	769	300 00	230760 00
2	DOWS	Lump Sum	1	945400	945400
3	EOO FOR PSE	Cy bm	400000	1 49	596000 00
4	EOO FOR RE	Cy bm	42000	2 03	85260 00
5	EOO FOR SOM	Cy bm	114000	1 32	150480 00
6	EOCC FOR SOM	Cy bm	74000	1 63	120620 00
7	EOMS	Cy bm	400000	0 41	172000 00
8	EOR	Cy bm	42000	0 48	20160 00
9	FFSP FROM OE	Cy bm	114000	0 13	14820 00
10	FFSP FROM CCE	Cy bm	74000	0 13	9620 00

Total bid		1321604 00
Total costs		1085505030
Total profit		142371370
% of profit		15.0

The Unbalanced Bid

It is not uncommon to unbalance a bid. The contracting officer and the owner frown on this method of bidding. Nevertheless, it is done now and then for one or both of the following reasons.

    1. A desire to have large profits available during the early stages of the work. An example is to bid the excavation item high and the paving item low on a highway job because the excavation must be done and paid for some time before the paving work is done and paid for.

    2. A desire to have a large profit on a bid item which is apt to overrun beyond the quantity set up in the bid schedule. An example is excavation in a slide-prone area, where the total quantity might be 50 percent more than that of the bid schedule.

In the previous example of bid preparation there is little likelihood of an appreciable overrun in any bid item. However, there are four opportunities to unbalance the bid so as to have available initial profits which are larger than estimated. These opportunities

18-28 Preparation of Bid and Schedule of Work

arise with: item 1, clearing and grubbing; item 2, development of water supply; item 3, excavation of overburden for plant site embankment; and item 7, embankment of plant site.

The method is to increase the unit bid prices on these four items and to decrease the unit bid prices on the remaining six items so that the total bid price for the job remains the same. The calculations are simple. An example is the change of bid prices in the summary Sheet F of the example of bid preparation. The following are changes in total bid prices for items 1, 2, 3, and 7:

Bid item no.	Unit price changes		Total bid prices		
			Original bid	Unbalanced bid	Difference in bids
1	Unit price:	$300	$ 23,070		
		$500	. . .	$ 38,450	+$ 15,380
2	Lump sum	$9,454	9,454	18,908	+9,454
3	Unit price:	$1.49	596,000		
		$1.75	. . .	700,000	+104,000
7	Unit price:	$0.48	192,000		
		$0.56	. . .	224,000	+32,000
Totals			$820,524	$981,358	+$160,834

To calculate the resultant changes in total bid prices for items 4, 5, 6, 8, 9, and 10, the total, $160,834, subtraction from the original bid price is distributed over the remaining six items in proportion to the ratio of their original total bid prices to the sum of their original total bid prices. The calculations are:

Bid item no.	Adjustment to original total bid price	Subtraction from original total bid price	Original total bid price	Unbalanced total bid price
4	21.3%(×$160,834)	$ 34,258	$ 85,260	$ 51,002
5	37.5%	60,313	150,480	90,167
6	30.1%	48,410	120,620	72,210
8	5.0%	8,042	20,160	12,118
9	3.7%	5,951	14,820	8,869
10	2.4%	3,860	9,620	5,760
Totals	100.0%	$160,834	$400,960	$240,126

When the foregoing changes are made and when the unit bid prices are rounded out to the nearest cent, the unbalanced bid prices are as given in the following tabulation.

Bid item no.	Bid item description	Unit	Quantity	Unit bid price,$	Unit bid price,$
1	Clearing and grubbing	acre	76.9	500.00	38,450.00
2	Development of water supply	lump sum	1	18,908.00	18,908.00
3	Excavation of overburden for plant site embankment	cybm	400,000	1.75	700,000.00
4	Excavation of overburden for roadway embankment	cybm	42,000	1.21	50,820.00
5	Excavation of overburden for stockpile of material	cybm	114,000	0.79	90,060.00
6	Excavation of channel change for stockpile of material	cybm	74,000	0.98	72,520.00
7	Embankment of plant site	cybm	400,000	0.56	224,000.00
8	Embankment of roadway	cybm	42,000	0.29	12,180.00
9	Fill for stockpile from overburden excavation	cybm	114,000	0.08	9,120.00
10	Fill for stockpile from channel change excavation	cybm	74,000	0.08	5,920.00
Total bid					$1,221,978.00

The total cost for the job is $1,038,510.30. The profit for the unbalanced bid is $183,467.70, or 15.0 percent. The profit for the balanced bid is $182,973.70, or also 15.0 percent.

The financial effects of the unbalanced bid are both positive and negative.

1. During the first 2 weeks of work, during which clearing and grubbing and development of water supply will be completed, the engineer's estimate for work completed will be increased by $24,834.

2. During the next 11 weeks of work, during which the overburden will be placed in the embankment of the plant site, the combined total bids for items 3 and 7 show that the unbalanced bid exceeds the balanced bid by $136,000, and the engineer's estimate will reflect the additional amount.

3. Accordingly, the contractor will have total engineer's estimates exceeding by $161,000 those payable by a balanced bid. Explained in another way, an average of $80,500 has accumulated over a period of 3 months. At an 8 percent annual interest rate, the unbalanced bid creates an approximate subsidy of $1610 for the contractor. At the same time, the unbalanced bid minimizes the borrowing of money during the first stages of the work.

4. Of course, later there will be a corresponding adverse effect on items 4, 5, 6, 8, 9, and 10, which are being bid at total prices less than those of the balanced bid. These total bid prices amount to $240,126. However, the overall job costs on the same items are $340,650. Thus, these six items must take a loss of $100,524 in order to make possible the unbalanced bid.

Owners and contracting officers sometimes provide safeguards in the contract against use of unbalanced bids. One perceives that a skilled estimator can detect the unbalanced bid because of the absence of a normal similarity between the unit bid price and the prevalent unit bid price for the same work. Ethically, the unbalanced bid is not a businesslike bid.

Summary of Bid Preparation

Bid preparation is a systematic procedure based fundamentally on the hourly productions of workers and machines and on the hourly costs of labor and the ownership and operation of machines. The ensuing additional costs of indirect job expense and of the share of home office overhead are important in the sense that they must be kept to a minimum consistent with efficient management.

After the overall job cost has been determined, the percentage of profit for the job depends on several factors to be considered by the home office management. Among these considerations are:

1. Desirability of additional work
2. Desirability of the work as typical or atypical of the company's average work
3. Availability of labor and machines for the work
4. Imponderable risks of the job, such as weather conditions, accessibility, and labor relations
5. Competition for the work

Ultimately the owners, the company management, the job management, and the estimator arrive at a well-considered profit margin, which determines the final total bid price.

SCHEDULE OF WORK

The estimator generally prepares a time schedule for the several bid items or operations of the work. This program is especially important if the actions are mutually dependent or if they are apt to interfere with each other with respect to working space. An example of dependence is the laying of conduits for electricity, street lighting, and telephones in the same trench during site preparation for homes. An instance of interference is the excavations for storm drains and sewers in the same street, where space is not sufficient for efficient use of machinery.

In the previous example of bid preparation the operations or bid items could be worked independently. It was necessary only to schedule them in continuous order to arrive at a total time in work hours or calendar days for completion of the job. The following example of a work schedule is much more complex, as it involves 35 independent and

18-30 Preparation of Bid and Schedule of Work

interdependent bid items, for which separate and simultaneous times must be arranged.

The time schedules for the bid items are almost always set up as horizontal bar diagrams or charts, each bar representing a bid item or operation. The abscissa at the bottom of the chart is time in calendar days. There are several forms for the chart, all of which have two features in common: the horizontal bars and the bottom time schedule. Representative of these charts are the diagrams of the Critical-Path Method (CPM), which is used in the following pages to illustrate planning, scheduling, and completing work.

The Critical-Path Method (CPM)

Principles of CPM Job Scheduling The *critical path* is that path which the important controlling kinds of work must take in order that they may follow each other in logical sequence and may not interfere with each other in the orderly prosecution of the job. The path of the critical work elements is shown in the wide lines of Figure 18-5, which gives the CPM schedules of a site development project for which a detailed case analysis will be presented. The *noncritical paths* are those for kinds of work which may be carried on concurrently with the critical work because there are no mutual interferences. These paths are shown by the narrow lines of the CPM chart. *Float* is extra time for an operation beyond the actual estimated time, made possible because no other operation may take place logically or physically during the same period of time.

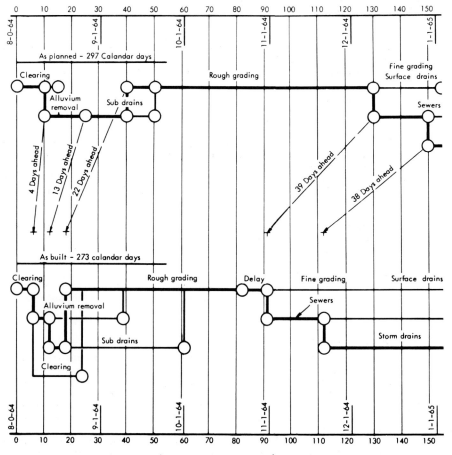

Figure 18-5 The critical-path method for scheduling construction as it was planned and as it was built. Construction was for a 22-acre mountainside development for 55 home sites. *(Horace K. Church, "How CPM Worked Out on a Grade-Sewer-Street Job," Roads and Streets, August 1966, p. 84.)*

By these plottings of horizontal bars, representing times for work, against the abscissa of calendar days at the bottom of the chart, one may visualize conveniently the following important factors in the conduct of a job.
1. The individual time schedules, usually based on bid items, for the operations
2. The relationships between the critical and the noncritical work elements
3. The overall time schedule for the job

Usually corrections are made to portray the "as built" schedule in contrast to the "as planned" schedule. The correction may be made on the "as planned" chart, but a better method is to plot a new "as built" schedule below the "as planned" one, as shown in Figure 18-5. The plotting of both charts is described in the ensuing discussion of "as planned" and "as built" schedules.

Example of CPM Job Scheduling The time- and money-saving value of CPM in preplanning a project is manifest. However, a case analysis showing just how well the preset work schedule was followed and how the CPM schedule was adjusted to meet unforeseen circumstances of the job is not often available.

The example is a typical development of a mountainside site for homes. Located in the Beaudry Mountains of Glendale, California, the tract included 22 acres with 55 home sites. The roughly square area represented the first stage of Tract 29116, which eventually totaled 41 acres.

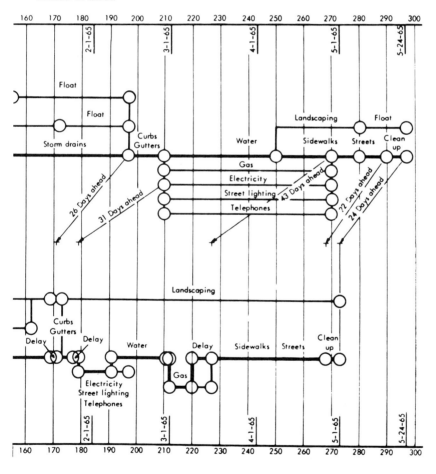

18–32 Preparation of Bid and Schedule of Work

The excavation was largely weathered and semisolid granite at high elevations and terrace deposits of conglomerates at low elevations. Of the 340,000 yd³ of rough excavation, only 4500 yd³ was blasted. Some 1800 yd³ in the street cut was blasted for utility excavation, and 27,000 yd³ of alluvia in the ravines was removed and compacted in the fill, bringing the total excavation to 367,000 yd³, or 16,700 yd³ acre and 6700 yd³ per home site or lot. The range of elevations between the top of the cut and the bottom of the fill was 320 ft in a lateral distance of 1200 ft, giving a maximum gradient for haulage of -27 percent at the beginning of the excavation and an average gradient of about -13 percent.

The writer prepared preliminary estimates for the job, made up the bid schedule and contract, selected low bidders, and supervised the construction. Individual contracts for site preparation were let to the lowest responsible bidders, this method of dividing the work lending itself to lowest overall cost. These contracts are a part of the tabulated costs for the complete home site development.

The round-figure costs of the site development, ready for the building of homes, were:

Land acquisition	18.7%	$110,000
Engineering	6.8	40,000
Supervision	2.5	15,000
Construction	72.0	424,000
Totals	100.0%	$589,000
Cost per acre		$26,800
Cost per lot or home site		$10,700

All this work was done in 1964–1965, and the costs set forth in the text are far too low for 1978. For equivalent 1978 costs, all costs should be multiplied by a factor of 2.3.

During the estimating for the job prior to receiving bids, the CPM analysis for scheduling the job was prepared, as illustrated in the upper part of Figure 18-5, "as planned." The worth of this front sight is to be appraised in terms of the back sight, which is diagramed in the lower, "as built" part of Figure 18-5. Naturally, the "as built" must differ from the "as planned" because initial planning is never perfect. The differences were in terms of management and machinery and, to a minor degree, in keeping with the weather. Figure 18-6 illustrates the rock excavation work at the beginning of the job and the completed job one year later when homes were being built.

An explanation of the "as planned" CPM schedule, the foresight, follows. The bid items for the job with their overall totals by classifications were:

Grading:	$195,000
Clearing and grubbing, acres	22
Excavation, including embankment, yd³	340,000
Alluvium removal, including embankment, yd³	27,000
Terraces and surface drains, lin ft	3000
Subsurface drains, lin ft	3500
Storm drains:	$48,000
R.C.P., 18-in to 54-in, lin ft	1600
C.M.P., 18-in, lin ft	100
Inlet structures, each	2
Catch basins, each	9
Manholes, each	10
Outlet structures, each	2
Sewers:	$28,000
V.C.P., 8-in, lin ft	3000
V.C.P., 6-in, lin ft	1500
Manholes, each	15
Streets, curbs and gutters, walks	$35,000
Curbs and gutters, concrete, 18-in, lin ft	5600
Sidewalks, 3-in thickness concrete, ft²	12,700
Cross gutters and aprons, ft²	2100
Pavement, asphalt, 3-in thickness, ft²	100,000
Utilities:	$112,000
Water:	
Mains, lin ft	3200
Laterals, lin ft	1500
Gas:	
Mains, lin ft	3200

Laterals, lin ft	1500
Electricity, underground:	
Mains, lin ft	3200
Laterals, lin ft	1500
Vaults and boxes, each	8
Street lighting, underground:	
Mains and laterals, lin ft	3600
Telephones, underground:	
Mains, lin ft	3200
Laterals, lin ft	1500
Vaults and boxes, each	7
Landscaping	$6,000
Total of contracts	$424,000

(a)

(b)

Figure 18-6 Beginning and ending of development of mountainside site for homes. *(a)* Two weeks after start of granite excavation by a scraper spread. The rock is being hauled downgrade to a narrow fill in the ravine which is visible just above the push-loading tractor. The completed fill will be 80 ft high. *(b)* Twelve months later the work has been completed and home building has been started. The two homes in the upper left corner of the picture are atop the high fill which was being placed in the ravine in the upper picture. Excavation was characterized by several deep sidehill cuts to be placed in one major high fill. The site is in the Beaudry Mountains of Los Angeles County, California.

18-34 Preparation of Bid and Schedule of Work

1. Clearing and grubbing were medium, involving light brush with perhaps 100 oak and eucalyptus trees with trunks of up to 24-in diameter. Ten critical days were allowed, with five extra float days for cleanup of the 22 acres. In all operations the assumed work schedule was a five-day, 45-h week, in keeping with local practice.

2. Alluvium removal and compaction were allowed 15 critical days for the 17,000 yd^3 of objectionable unconsolidated wash materials in the ravines.

3. Subdrains, 3500 lin ft, were allowed 15 critical days, at the end of which rough grading was scheduled to commence on the fortieth calendar day. The noncritical subdrains at high elevations were allowed an additional 10 days.

4. Rough grading with compaction in embankment, 340,000 yd^3, was scheduled for 90 critical days, 63 working days. The 27 idle days included Saturdays, Sundays, and holidays. It was assumed that the contractor would work a single spread of two to four scrapers with one or two push-loading tractors and one tractor-bulldozer-ripper. An average hourly production of 600 cybm was estimated.

5. Fine grading of streets and lots, part of the rough grading contract along with surface drains, 3000 lin ft, was allowed 25 noncritical days, along with 42 days of float time. The total time allowed for rough grading, including the times for other items included in the contract, was 197 calendar days.

6. Sewers, 4500 lin ft, were allowed 20 critical days, with 47 days of float time, and they were scheduled to commence on the 130th calendar day on the completion of rough grading.

7. Storm drains, 1700 lin ft, were allowed 47 critical days, commencing after 20 days of sewer work so as to allow time for the storm drain work to follow the sewer work. As a preventive measure against bank erosion, 100 lin ft of 18-in-diameter corrugated metal pipe was scheduled during the rough grading. This work was not regarded as critical to construction of other items and it was not included in the CPM schedule.

8. Curbs and gutters, 5600 lin ft, were allowed 13 critical days.

9. Underground utilities, 22,500 lin ft, including water, gas, electricity, street lighting, and telephone lines, were scheduled concurrently, with an allowance of 60 critical days before beginning of sidewalk construction.

10. Sidewalks, 12,700 ft^2, were allowed 10 critical days before street construction.

11. Streets, including cross gutters, 102,100 ft^2, were allowed 10 critical days before start of cleanup.

12. Cleanup was allowed seven critical days.

13. Landscaping, a noncritical item, was scheduled to start on the 125th day and to finish at the scheduled end of the project, the 297th calendar day.

Following is an explanation of the "as built" CPM schedule, the back sight. The "as built" part of the chart illustrates the difference between theory and practice. Expressed in another way, it is the difference between the ideas of the planner and the well-considered on-the-job changes of the contractor and the builder.

1. *Clearing and grubbing.* Instead of completing the work prior to alluvium removal, the grading contractor finished the heavy clearing in the ravines, began alluvium removal immediately thereafter, and completed light and medium clearing on the ridges later. The planned critical 10 days were reduced to 6 days, as built. The job was then four days ahead of schedule.

2. *Alluvium removal.* Instead of completing alluvium removal prior to subdrains, the contractor allowed only 6 critical days instead of 15 days as planned. The job was then 13 days ahead of schedule.

3. *Subdrains.* Instead of the planned 15 critical days the contractor scheduled only 6 days before starting rough grading. The job was then 22 days ahead of schedule.

4. *Rough grading.* Instead of using one set of two push-loading tractors, the contractor used two sets and increased the number of scrapers accordingly. Thus the 90 critical days, 63 working days, were reduced to 64 critical days, 47 working days. Instead of averaging 600 yd^3/h, production was 800 yd^3/h. Because of the unanticipated speed of rough grading, there was a delay of nine critical days before starting sewers. Nevertheless, construction was 39 days ahead of the scheduled 130 days. On November 1, 1964, rough grading was finished, ahead of the heavy winter rains.

5. *Fine grading* of streets and lots and surface drains, noncritical items, were completed 28 days ahead of schedule.

6. *Sewers.* Because of the feeling that separate contractors could schedule work

without interference on one street, sewers were allowed only 20 days of lead or critical time ahead of storm drains. This was a big mistake because of the crowded conditions. A minimum time of 21 days preceded the storm drains, and the total time for sewers was 71 days instead of the planned 42 days.

7. *Storm drains.* Forty-seven critical days after completion of sewers were allowed. Because of the aforementioned problem of interference, 57 calendar days were required. Thus, whereas construction was 39 days ahead of schedule at the completion of rough grading, it was only 26 days ahead at the completion of the storm drains.

8. *Curbs and gutters.* Starting two days after completion of the storm drains, high efficiency reduced critical time from 13 to 6 days. The work was then 31 days ahead of schedule.

9. *Utilities.* Water, gas, electricity, street lighting, and telephone lines were scheduled concurrently for 60 critical days. They were built in 41 days, so that at the end of the work the job was 50 days ahead of the planned 270 calendar days.

10. *Sidewalks and streets.* The contractor was delayed a week in moving in for the work so that sidewalks began 43 days ahead of schedule. The seasonal rains delayed sidewalks and streets so that 41 days were required instead of the planned 20 days. The job was then 22 days ahead of schedule.

11. *Cleanup.* This rather indefinite operation, along with landscaping, which had been started much ahead of schedule, required five days instead of the planned seven days. The project ended after completion of cleanup 24 days ahead of the CPM "as planned."

The overall time saving of 24 days under the planned 297 calendar days is 8 percent.

General Observations on the Value of Preplanning and Scheduling

The following observations apply specifically to the development of the mountainside site for the building of homes.

Estimating the Total Cost and the Total Time for the Work The estimator prepared the schedule for the work. In making up the chart he put in visual form the times and the sequences for the operations or the bid items. These times determined the necessary hourly productions on which cost estimates were based. Thus an easily changeable diagram was obtained which the owner and estimator could modify in order to secure the lowest total cost for the project in keeping with the desired total time for the work.

Bidding for the Work Prior to bidding, the selected bidders were able to visualize their required scheduled work and to see clearly their mutual time-motion relationships. Being practical contractors, they were able to give valuable time- and cost-saving suggestions for modifications in the schedule, which benefited both owner and contractors.

Prosecution of the Work The superintendent for the owner was able to see quickly the progress schedule and to discuss the facts with the contractors. At the same time he secured valuable advice from the contractors. The superintendent was able to keep all contractors informed concerning the best possible time sequences, and this facility was especially valuable to contractors about to come in on the job for their particular work. The contractors were able to visualize their own progress during the entire job, as well as the progress of their fellow contractors. This knowledge and understanding assisted in the solution of the problem of interference and congestion between the sewer and storm drain contractors.

Both owner and contractors were able to see mistakes in timing and to correct them on the job. And, most certainly, mistakes were recognized as avoidable on the next job. In this and in other senses, the charted schedule became a device for teaching correct procedures.

Financing the Work The owner was able to present a well-thought-out plan for the job to the lending institution. This, coupled with the well-known business acumen of the owner, contributed to a good understanding between borrower and lender. During construction the lending institution was able to see actual progress related to planned progress. This picture, as it was favorable to the owner, assisted the owner in borrowing money to finance the payroll of one of the contractors.

The project finished 24 days ahead of the planned 297 calendar days. This saving of time was attributable to the cooperation of contractors and owner, but it was also appar-

18-36 Preparation of Bid and Schedule of Work

ent that the CPM scheduling contributed to the cooperation. The resultant savings to the owner were estimated to be in accordance with the following tabulation.

Interest on money for land acquisition: $110,000, for 24 days at 5.0%, 1964–1965 interest rate	$ 361.64
Interest on money for engineering, supervision, and construction: average $239,500 borrowed, for 24 days at 5.0%.	787.40
Total saving during 24 days	$1149.04

SUMMARY

Preparation of bid and schedule for work are of equal importance. Getting the job with the lowest qualified bid must be backed up by the execution of the work so as to realize the intent of the bid insofar as costs and profit are concerned. Not only are they of equal significance but they are also mutual efforts in that bidding a job necessarily is based on a work schedule, as costs are reflected in the time allotted to the various bid items. Thus, a schedule for work is contemplated in the preparation of the bid. In a general sense the superintendent follows the schedule for work on which the bid was based but deviates from that schedule to adapt it to the problems encountered in the prosecution of the work. Obviously, mutuality of preparation of bid and schedule for work exists from the initial desire for the job to the final acceptance of the work.

Appendixes

APPENDIX 1 Approximate Material Characteristics
APPENDIX 2 Rock Clauses
APPENDIX 3 Depreciation Schedule for Machinery and Facilities
APPENDIX 4 Conversion Factors for Systems of Measurement
APPENDIX 5 Formulas Frequently Used in Calculations for Excavation Projects
APPENDIX 6 Swell Versus Voids of Materials and Hauling Machine Load Factors
APPENDIX 7 Approximate Angles of Repose of Materials
APPENDIX 8 Bearing Powers of Materials
APPENDIX 9 Abbreviations

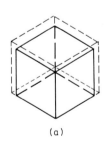 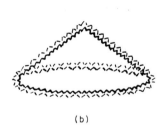 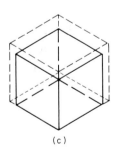

(a)   (b)   (c)

PLATE A-1

Relationship between cubic yard and cubic meter of semisolid and solid rock in bank, when transferred to hauler or stockpile, and when in compacted embankment. *(a)* Bank measurement: 1 yd³ = 0.76 m³; 1 m³ = 1.31 yd³. *(b)* Loose measurement in hauler or in stockpile; swell factor = 1.67. One unit bank measurement becomes 1.67 unit loose measurement; one unit loose measurement equals 0.60 unit bank measurement. *(c)* Compacted measurement in embankment; swell factor = 1.33. One unit bank measurement becomes 1.33 units compacted measurement; one unit compacted measurement equals 0.75 unit bank measurement. Cubic yards in solid and dashed lines; cubic meters in dashed and dotted lines.

 Note: For corresponding data for specific material see Appendix 1.

APPENDIX 1

Approximate Material Characteristics*

---

*I am indebted to some 25 authorities for the data in this Appendix. Approximately 1500 values for specific gravities, weights, and swell and shrinkage factors were analyzed, interpreted, and weighted for good averages.

For about 100 years, commencing with Trautwine's pioneering handbook of 1882, *Civil Engineer's Pocketbook,* authoritative sources in the United States have been publishing tables of material characteristics. Generally speaking, these tables include specific gravities, weights in natural bed, swell factors from the natural bed or cut to the loose condition, weights in the loose condition, swell or shrink factors from the natural bed or cut to uncompacted fills or compacted embankments, and weights in uncompacted fills or compacted embankments. Engineers, both public and private, contractors, mining companies, machinery manufacturers, and writers of handbooks have contributed to this array of data.

The following table in this appendix is a summary of existing data, commencing with Trautwine's tables based on his own meticulous laboratory and field work, and ending with personal data gathered during the past 50 years. The table is necessarily based on properly interpreted and weighted averages. It is therefore not absolute for a specific case, and engineering experience and judgment will guide the user in its proper application. Prior to examination of the table, the reader is referred to these explanatory notes.

Materials

Rock materials are noted to be I, igneous; S, sedimentary; or M, metamorphic. Materials marked by asterisks are ores in the mineral or near mineral state, and the weights do not allow for the containing gangues of the ore body. The weight of the mineral is constant, with a set specific gravity, but the weight of the gangue, such as the associated earthy materials contained in quartz, rhyolite, schist, and feldspar, varies considerably with respect to the weight of the mineral. In mining the engineer must estimate the unit weight of the ore body and the weight of the contained mineral.

For example, hematite, the iron mineral, weighs 8560 lb/yd^3. Associated gangue, however, varies with respect to the hematite. Suppose that the mineral hematite samples 40 percent by weight of the ore and that the gangue, weighing 4000 lb/yd^3, samples 60 percent by weight of ore. Then 1000 lb of ore in the natural bed occupies a volume of

$$\frac{40\% \times 1000}{8560} + \frac{60\% \times 1000}{4000} = 0.197 \text{ yd}^3$$

The ore, then, weighs 1000/0.197 = 5080 lb/yd^3, as contrasted to the weight of 8560 lb/yd^3 for the contained mineral hematite. At this juncture it is well to explain that miners sometimes use the word *hematite* for both the mineral and the ore.

Specific Gravity

When the value for specific gravity is in parentheses, it is an *apparent specific gravity* because the material is not in the solid state. Examples are gravel and rock-earth mixtures, which contain voids when in their natural bed.

Cubic Yard in Cut

The weight in the natural bed, or bank measurement, includes natural moisture. The average weight is subject to a maximum ±10 percent variation. Again, it is emphasized that ore weights are for the mineral only and not for an impure ore body containing gangue.

Cubic Yard in the Loose Condition

Percent swell from the natural bed to the loose condition is an average which is subject to a maximum 33 percent variation in both rock and earthy materials. Variations are multipliers and not percentages to be added to or subtracted from the given percent of swell. The swell factor of 67 percent, given for several rocks, is an average figure obtained from existing data for solid rock, and it has been applied to solidly bedded unweathered rocks for which no swell factors are available specifically. Percent swell factors for ores are in terms of the entire ore body rather than in terms of the contained mineral. Weights in the loose condition are averages, except when calculated on the basis of the aforementioned average 67 percent swell factor. All weights are subject to any adjusted value of the swell factor.

Cubic Yard in the Fill

In the table a cubic yard in a fill is a cubic yard in a compacted embankment. No values are given for ores in a fill as they are not construction materials. When they are in a fill, they are in a stockpile, and the values for a cubic yard in the loose condition are applicable. Percent swell or shrink from cut or natural bed to fill is an average, subject to a maximum 33 percent variation in both rock and earthy materials. Percentage variation is a multiplier.

It is absolutely necessary, especially in the case of rock materials, to distinguish between two methods of fill construction:

1. Natural or gravity compaction, which was common years ago before the development of compacting machinery, is little used now except in the building of waste fills and stockpiles of materials and ores. The swell and shrink factors from the cut or natural bed vary from 10 percent shrinkage for earthy materials to 67 percent swelling for rock materials. Because of different degrees of fragmentation in the cut and because of the wide variations of fill construction methods in natural or gravity compaction, no figures are tabulated.

2. Mechanical compaction by rollers, along with wetting of the fill, is today's accepted method for fill consolidation. The tabulated swell and shrink factors and weights are for this modern method of fill compaction.

Two other influences affect swell and shrink factors and resultant weights. First, crawler-tractor-rippers produce better fragmentation and better grading of both rock and earthy formations in the cut. Second, the average so-called rock job really consists of a rock-earth mixture which in itself is pretty well graded.

These three factors, nature of materials, use of tractor-rippers, and modern compacting methods, have made possible the prevalent high densities of fills, densities not in accordance with some previously tabulated data for swell and shrink factors from cut to fill. In the case of construction materials the writer has used swell and shrink factors and weights, including moisture, resulting from average compaction methods.

It is a fact that certain friable rocks in weathered and parent rock zones have low swell factors from cut to fill. These rocks are really equivalent to rock-earth mixtures in their behavior during excavation and compaction. Rock swell factors are in terms of solid rock in the cut and do not include allowances for overlain residual and weathered rocks or for earthy and friable materials, all of which would reduce greatly the swell factor from cut to fill.

Material	sp gr	Cubic yards, in cut— weight, lb	Cubic yards loose Percent swell	Weight, lb	Cubic yards in fill Swell or shrink, %	Weight, lb
Adobe, S	(1.91)	3230	35	2380	−10	3570
Andesite, I	2.94	4950	67	2970	33	3730
Asbestos	2.40	4040	67	2420		
Ashes, coal	(0.61)	1030	33	800	−50	2060
Asphaltum, S	1.28	2150	67	1390		
Asphalt rock, S	2.41	4050	62	2500		
Aragonite, calcium ore*	3.00	5050	67	3020		
Argentite, silver ore*	7.31	12300	67	7360		
Barite, barium ore*	4.48	7560	67	4520		
Basalt, I	2.94	4950	64	3020	36	3640
Bauxite, aluminum ore*	2.73	4420	50	2940		
Bentonite	1.60	2700	35	2000		
Biotite, mica ore*	2.88	4850	67	2900		
Borax, S	1.73	2920	75	1670		
Breccia, S	2.41	4050	33	3040	27	3190
Calcite, calcium ore*	2.67	4500	67	2700		
Caliche, S	(1.44)	2430	16	2100	−25	3200
Carnotite, uranium ore*	2.47	4150	50	2770		
Cassiterite, tin ore*	7.17	11380	67	6800		
Cement				2700		
Cerrusite, lead ore*	6.50	10970	67	6560		
Chalcocite, copper ore*	5.70	9600	67	5750		
Chalcopyrite, copper ore*	4.20	7060	67	4220		
Chalk, S	2.42	4060	50	2710	33	3050
Charcoal				1030		
Chat, mine tailings				2700		
Cinders	(0.76)	1280	33	960	−10	1420

A-6 Approximate Material Characteristics

Material	sp gr	Cubic yards, in cut—weight, lb	Cubic yards loose		Cubic yards in fill	
			Percent swell	Weight, lb	Swell or shrink, %	Weight, lb
Cinnabar, mercury ore*	8.10	13630	67	8170		
Clay, S:						
Dry	(1.91)	3220	35	2380	−10	3570
Damp	(1.99)	3350	40	2400	−10	3720
Clinker				2570		
Coal, S:						
Anthracite	1.55	2610	70	1530		
Bituminous	1.35	2280	67	1370		
Coke	(0.51)	860	0	860		
Colemanite, borax ore*	1.73	2920	75	1670		
Concrete:						
Stone	2.35	3960	72	2310	33	2910
Cyclopean	2.48	4180	72	2430	33	3150
Cinder	1.76	2970	72	1730	33	2240
Conglomerate, S	2.21	3720	33	2800	−8	4030
Decomposed rock:						
75% R, 25% E	(2.45)	4120	25	3300	12	3700
50% R, 50% E	(2.23)	3750	29	2900	−5	3940
25% R, 75% E	(2.01)	3380	26	2660	−8	3680
Diabase, I	3.00	5050	67	3010	33	3810
Diorite, I	3.10	5220	67	3130	33	3930
Diatomite, S:						
Ditomaceous earth	(0.87)	1470	62	910		
Dolomite, S	2.88	4870	67	2910	43	3400
Earth, loam, S:						
Dry	(1.84)	3030	35	2240	−12	3520
Damp	(2.00)	3370	40	2400	−4	3520
Wet, mud	(1.75)	2940	0	2940	−20	3520
Earth-rock mixtures:						
75% E, 25% R	(2.01)	3380	26	2660	−8	3680
50% E, 50% R	(2.23)	3750	29	2900	−5	3940
25% E, 75% R	(2.45)	4120	25	3300	12	3700
Feldspar, I	2.62	4410	67	2640	33	3320
Felsite, I	2.50	4210	67	2520	33	3170
Fluorite, S	3.10	5220	67	3130		
Gabbro, I	3.10	5220	67	3130	33	3940
Galena, lead ore*	7.51	12630	67	7570		
Gneiss, M	2.71	4550	67	2720	33	3420
Gob, mining refuse	(1.75)	2940	0	2940	−20	3520
Gravel, average graduation, S:						
Dry	(1.79)	3020	15	2610	−7	3240
Wet	(2.09)	3530	5	3350	−3	3640
Granite, I	2.69	4540	72	2640	33	3410
Gumbo, S:						
Dry	(1.91)	3230	50	2150	−10	3570
Wet	(1.99)	3350	67	2020	−10	3720
Gypsum, S	2.43	4080	72	2380		
Hematite, iron ore*	5.08	8560	75	4880		
Hessite, silver ore*	8.50	14300	67	8560		
Ice	0.93	1560	67	930		
Ilmenite, titanium ore*	4.75	8000	69	4730		
Kaolinite, S:						
Dry	(1.91)	3230	50	2150		
Wet	(1.99)	3350	67	2010		
Lignite	(1.25)	2100	65	1270		
Lime				2220		
Limestone, S	2.61	4380	63	2690	36	3220
Linnaeite, cobalt ore*	4.89	8230	67	4930		
Limonite, iron ore*	3.80	6400	55	4140		
Loam, earth, S:						
Dry	(1.84)	3030	35	2240	−12	3520

Approximate Material Characteristics   A–7

Material	sp gr	Cubic yards, in cut— weight, lb	Cubic yards loose		Cubic yards in fill	
			Percent swell	Weight, lb	Swell or shrink, %	Weight, lb
Damp	(2.00)	3370	40	2400	−4	3520
Wet, Mud	(1.75)	2940	0	2940	−20	3520
Loess, S:						
Dry	(1.91)	3220	35	2380	−10	3570
Wet	(1.99)	3350	40	2400	−10	3720
Magnesite, magnesium ore*	3.00	5050	50	3360		
Magnetite, iron ore*	5.04	8470	54	5520		
Marble, M	2.68	4520	67	2700	33	3400
Marl, S	2.23	3740	67	2240	33	2820
Masonry, rubble	2.33	3920	67	2350	33	2950
Millerite, nickel ore*	5.65	9530	67	5710		
Molybdenite, molybdenum ore*	4.70	7910	67	4750		
Mud, S	(1.75)	2940	0	2940	−20	3520
Muscovite, mica ore*	2.89	4860	67	2910		
Niccolite, nickel ore*	7.49	12600	67	7550		
Orpiment, arsenic ore*	3.51	5900	50	3940		
Pavement:						
Asphalt	1.93	3240	50	1940	0	3240
Brick	2.41	4050	67	2430	33	3050
Concrete	2.35	3960	67	2370	33	2980
Macadam	1.69	2840	67	1700	0	2840
Wood block	0.97	1630	72	950	33	1220
Peat	(0.70)	1180	33	890		
Phosphorite, phosphate rock, S	3.21	5400	50	3600		
Porphyry, I	2.74	4630	67	2770	33	3480
Potash, S	2.20	3700	50	2470		
Pumice, I	0.64	1080	67	650		
Pyrites, iron ore*	5.07	8540	67	5110		
Pyrolusite, manganese ore*	4.50	7560	50	5050		
Quartz, I	2.59	4360	67	2610	33	3280
Quartzite, M	2.68	4520	67	2710	33	3400
Realgar, arsenic ore*	3.51	5900	50	3930		
Rhyolite, I	2.40	4050	67	2420	33	3040
Riprap rock, average	2.67	4500	72	2610	43	3150
Rock-earth mixtures:						
75% R, 25% E	(2.45)	4120	25	3300	12	3700
50% R, 50% E	(2.23)	3750	29	2900	−5	3940
25% R, 75% E	(2.01)	3380	26	2660	−8	3680
Salt, rock, S	2.18	3670	67	2200		
Sand, average graduation, S:						
Dry	(1.71)	2880	11	2590	−11	3240
Wet	(1.84)	3090	5	3230	−11	3460
Sandstone, S	2.42	4070	61	2520	34	3030
Scheelite, tungsten ore*	5.98	10100	67	6050		
Schist, M	2.59	4530	67	2710	33	3410
Serpentine, asbestos ore*	2.62	4440	67	2650		
Shale, S	2.64	4450	50	2970	33	3350
Silt, S	(1.93)	3240	36	2380	−17	3890
Siltstone, S	2.42	4070	61	2520	−11	4560
Slag:						
Furnace	2.87	4840	98	2690	65	2930
Sand	(0.83)	1400	11	1260	−11	1570
Slate, M	2.68	4500	77	2600	33	3380
Smaltite, cobalt ore*	6.48	10970	67	6560		
Snow:						
Dry	(0.13)	220	0	220		
Wet	(0.51)	860	0	860		
Soapstone, talc ore*	2.70	4550	67	2720		
Sodium niter, chile saltpeter	2.20	2710	50	2470		
Stibnite, antimony ore*	4.58	7710	67	4610		
Sulfur	2.00	3450	50	2310		

A-8 Approximate Material Characteristics

Material	sp gr	Cubic yards, in cut— weight, lb	Cubic yards loose		Cubic yards in fill	
			Percent swell	Weight, lb	Swell or shrink, %	Weight, lb
Syenite, I	2.64	4460	67	2670	33	3350
Taconite, iron ore*	3.18	5370	60	3360		
Talc, M	2.70	4640	67	2780	33	3490
Topsoil, S	(1.44)	2430	56	1620	−26	3280
Trachyte, I	2.40	4050	67	2420	33	3050
Trap rock, igneous rocks, I	2.79	4710	67	2820	33	3540
Trash				400	−50	800
Tuff, S	2.41	4050	50	2700	33	3050
Witherite, barium ore*	4.29	7230	67	4320		
Wolframite, tungsten ore*	7.28	12280	67	7350		
Zinc blende, zinc ore*	4.02	6780	67	4060		
Zincite, zinc ore*	5.68	9550	67	5710		

Key to table:
I—igneous rock. S—sedimentary rock. M—metamorphic rock.
*—ores in the mineral state, with no gangues. Adjust for percentage of mineral bearing gangue or rock to estimate weight of entire ore body, as explained previously in text.
( )—apparent specific gravity, as material is not solid.

Weights per cubic yard in cut are subject to average ±10 percent variation. Swell and shrinkage factors for loose condition and embankment are subject to average ±33 percent variation. Weights in loose condition and in embankment are subject to adjustments in accordance with modified swell and shrinkage factors.

APPENDIX 2

Rock Clauses

When an owner and a contractor enter into a contract for rock excavation, the excavation may be described as either *classified* or *unclassified* in the bid schedule. If it is unclassified, the contractor must excavate all material in the cut at the agreed-upon unit bid price. On the other hand, if the material is classified as *common* and *rock*, the contractor will be paid accordingly once the kind of excavation is established. Naturally some means must be used to distinguish between common and rock, or else there will be disagreements. It is to the owner's interest to delay the change from common to rock because rock costs more than common. On the other hand, it is to the contractor's advantage to hasten the change if rock ripping becomes difficult because it may result in higher costs than drilling and blasting.

Before the advent of efficient tractor-rippers for fragmentation there was really no middle ground between common and rock. Common was described in specifications as earth, hardpan, and loose rock, which could be excavated by hand or by horse-drawn scrapers. Rock was described as solid rock or rock in place and boulders measuring ½ yd³ and up, the removal of which required drilling and blasting. The use of the tractor-ripper created a middle zone of weathered and semisolid rock between common and rock, which might be fragmented either by tractor-ripper or by blasting. It was then necessary to use some new impartial means to distinguish between common and rock. For some years the seismic timer described in Chapter 7 has been used to establish this line of demarcation.

The current 1978 rock clauses for surface and subsurface excavation are for the heavyweight crawler-tractor-rippers and for the heavyweight trenchers which are used regularly in rock excavation.

Rock Clauses

For Surface Excavation Normally Handled by Heavy Excavation Machinery Any surface excavation material with seismic shock-wave velocity greater than 6000 ft/s will be considered to be economically nonrippable by crawler-tractor-ripper and to require blasting prior to excavation. The contractor, when of the opinion that the velocity exceeds 6000 ft/s, will so advise the owner, and both owner and contractor will test the material to be excavated. If the velocity exceeds 6000 ft/s, the rock-earth will be classified as rock instead of common and the contractor will be paid accordingly. Both owner and contractor agree to expedite testing by seismic timer.

For Subsurface Excavation Normally Handled by Heavy Trenchers Any subsurface excavation with seismic shock-wave velocity greater than 4000 ft/s will be con-

A-10 Rock Clauses

sidered to be economically "nondiggable" by trenchers and to require blasting prior to excavation. The contractor, when of the opinion that the 4000 ft/s velocity is exceeded, will so advise the owner, and both owner and contractor will test the excavation. If the velocity exceeds 4000 ft/s, the rock will be classified as rock and the contractor will be paid accordingly. Both owner and contractor agree to expedite testing by seismic timer.

APPENDIX 3

Depreciation Schedule for Machinery and Facilities

Chapter 8 contains suggested rates of depreciation in years for machinery and facilities used in rock excavation. Table 8-3 is based on averages for the construction and mining industries.

For a long time an authoritative and useful table has been Schedule F of the U.S. Bureau of Internal Revenue. It has been used for all construction machinery and facilities since 1931, and has since been revised on several occasions. Below is the schedule, abridged to cover essentially machinery and facilities used in excavation.

Machinery and Facilities	Life, yr
Automobiles:	
Light	2
Medium	3
Heavy	5
Backfillers, power:	
Light	3
Medium	5
Heavy	6
Tractor-mounted	5
Barges:	
Steel	30
Wood	25
Bending machines:	
Angle	15
Pipe and rail	10
Bins:	
Steel and concrete	6
Steel	12
Wood	8
Bin frames, steel	6
Blacksmith shop, portable	4
Boilers:	
Upright	7
Locomotive	15
Stationary	20
Buckets:	
Cableway	6
Clamshell	6
Elevator	5
Orange-peel	6
Bail, pivot turnover	5
Scraper or dragline	6

A-12 Depreciation Schedule for Machinery and Facilities

Machinery and Facilities	Life, yr
Bulldozers:	
Grade builders	8
Tractor-mounted	4
Burner equipment, gas and oil	12
Cables, wire	4
Cableways, cable only	3
Cableway carriage	5
Capstans, electric	10
Cars:	
Ballast spreader	10
Boarding and tool	20
Dump, steel	8
Dump, wood	6
Flat, steel	12
Flat, wood	10
Skip, hoist	10
Tank	20
Carts, tool, steel	4
Chains:	
Hawsers and lines	6
Power, transmissions	5
Channelers, rock	6
Cleaning machines, steam or sand	15
Compressors:	
Belt-driven	10
Electric, portable	8
Gasoline, portable	6
Motor truck unit	5
Steam, portable	6
Controllers, motor	12
Conveyors:	
Belt elevating, portable	3
Belt elevating, stationary	6
Buckets:	
Cable-drag	6
Monorail	15
Chain, portable	6
Portable	5
Scraper	6
Cranes:	
Crawler, electric:	
2½–5 tons	5
10–15 tons	7
20 tons and over	9
Crawler, gasoline:	
2½–5 tons	5
10–15 tons	9
20 tons and over	12
Locomotive, gasoline	7
Steam:	
2½–5 tons	6
10–15 tons	10
20 tons and over	12
Locomotive	10
Dragline	10
Universal, gasoline 2½–5 tons, mounted on 10-ton truck	6
Dock or wharf, traveling	20
Crushers, rock:	
Portable	8
Stationary	10
Cutting and welding outfits	4
Diggers, clay, pneumatic	3
Draglines:	
Electric:	
½–¾ yd^3	6

Depreciation Schedule for Machinery and Facilities A–13

Machinery and Facilites	Life, yr
1–1½ yd³	8
2 yd³ and over	10
Gasoline:	
½–¾ yd³	5
1–1½ yd³	9
2 yd³ and over	12
Steam:	
½–¾ yd³	6
1–1½ yd³	10
2 yd³ and over	12
Dredges:	
Clamshell	16
Dipper	8
Hydraulic	20
Pipe	10
Drill boats	12
Drill points, well	5
Drills:	
Air drifter	3
Rock, electric	3
Jackhammer	3
Steam	5
Traction, well	7
Tripod	7
Tunnel, carriage	5
Well	10
Engines:	
Gasoline	10
Marine	20
Oil	20
Steam	11
Excavators:	
Cableway, complete	4
Trench, gasoline:	
Depth, 7–12 ft	6
Depth, 18 ft	8
Trench, steam	
Depth, 7–12 ft	8
Depth, 18 ft	10
Trench, vertical boom	5
Trench, wheel or ladder type	5
Generator sets:	
Steam engine	12
Turbine, headlight or floodlight	4
Graders:	
Blade, road:	
7- to 8-ft blade	4
9- to 10-ft blade	5
10-ft blade and over	8
Elevating	8
Rooters, wheel	5
Hoists:	
Air and electric	8
Gasoline	6
Steam	12
Hose:	
Fire, linen- or rubber-lined, cotton	5
Rubber, air, steam, or water	10
Levee construction:	
Draglines	8
Shovels	8
Tower excavators	12
Light plant	12
Lighters	22
Loaders, bucket:	

A-14 Depreciation Schedule for Machinery and Facilities

Machinery and Facilites	Life, yr
Crawler and portable	5
Stationary	6
Locomotives, industrial:	
Diesel	10
Electric	16
Gasoline:	
Up to 10 tons	8
10–20 tons	15
20 tons and over	20
Steam:	
Up to 10 tons	8
10–20 tons	18
20 tons and over	20
Locomotives, standard gauge	30
Motors:	
Electric, small	8
Electric, medium	10
Electric, large	12
Hydraulic	5
Pneumatic	5
Pipe, black or galvanized	4
Pipelines and fittings for floating dredges	10
Pit and quarry plants	6
Ploughs:	
Furrow	3
Rooter	6
Pumping units:	
Electric	6
Gasoline	6
Highway contractor's pump	4
Piston	5
Steam centrifugal	10
Pumps:	
Air lift	10
Centrifugal	6
Hydraulic	15
Oil	10
Steam piston	6
Rails, steel	10
Razing equipment for buildings	8
Rollers, road, gasoline- or steam-powered	10
Sandblast outfits	10
Saws, hand, electric and pneumatic	3
Scales, large-track and wagon	20
Scarifiers:	
Attachments	4
Drag, all steel	4
Grader-type	4
Scrapers:	
Blade, carryall	6
Fresno	4
Slip	2
Wheel	5
Shovel attachments	6
Shovels:	
Electric or gasoline, crawler- or wheel-mounted:	
½–¾ yd^3	5
1–1½ yd^3	6
2 yd^3 and over	8
Steam, crawler- or wheel-mounted:	
½–¾ yd^3	7
1–1½ yd^3	8
2 yd^3 and over	10
Railroad, steam	10
Tunnel	4

Machinery and Facilites	Life, yr
Spreaders, stone:	
Hopper wagon	5
Steel box	5
Switches:	
Portable	4
Stationary	5
Tampers, backfill, pneumatic	3
Tamping machines	10
Tanks, gasoline (storage, steel)	6
Tanks, water or air (storage, steel)	10
Towers, cableway:	
Steel	6
Wood	3
Steel boom with counterweights	5
Tractors, gasoline or steam:	
3 tons	4
5 tons	6
10 tons	8
20 tons	10
Trailers:	
Dump, steel or wood	10
Platform, wood	4
Drop platform, heavy duty	5
Trucks, automobile, dump:	
$1/3$–$2/3$ yd^3	3
1–$1 2/3$ yd^3	5
2 yd^3 and over	8
Tugs, screw-propelled, steam or gasoline	25
Wagons:	
Dump, steel or wood	6
Road oilers, steel tank	10
Tank or sprinkler:	
Steel	10
Wood	8
Welders, acetylene or electric	10
Winches, electric or pneumatic	10
Wire and cables, electric	6

NOTE 1: The depreciation schedule is not modern as far as kinds and sizes of excavation machinery are concerned. However, it is useful for the values for particular machines which are listed, and also for the selection of values for machines with characteristics similar to the machine under consideration.

NOTE 2: A supplementary pamphlet of the U.S. Internal Revenue Service is *Tax Information on Depreciation,* Publication 534, 1978. This pamphlet contains a guideline schedule of depreciation periods, not as complete as this table, and other helpful information concerning the general subject of depreciation of machinery and facilities.

APPENDIX 4

Conversion Factors for Systems of Measurement

ENGLISH UNITS TO ENGLISH UNITS

This unit	Multiplied by	Equals this unit
Feet (ft)	12	Inches (in)
Yards (yd)	3	Feet (ft)
Fathoms	6	Feet (ft)
Rods, poles, or perches	16.5	Feet (ft)
Furlongs (fur)	660	Feet (ft)
Miles, statute	5280	Feet (ft)
Miles, statute	1760	Yards (yd)
Miles, nautical	6080	Feet (ft)
Miles, nautical	2027	Yards (yd)
Square feet (ft^2)	144	Square inches (in^2)
Square yards (yd^2)	9	Square feet (ft^2)
Square miles (mi^2)	27,878,400	Square feet (ft^2)
Square miles (mi^2)	3,097,600	Square yards (yd^2)
Acres	43,560	Square feet (ft^2)
Acres	4840	Square yards (yd^2)
Gallons (gal)	231	Cubic inches (in^3)
Gallons, U.S. (gal)	0.833	Imperial gallon
Cubic feet (ft^3)	1728	Cubic inches (in^3)
Cubic feet (ft^3)	7.48	Gallons (gal)
Cubic yards (yd^3)	27	Cubic feet (ft^3)
Cubic yards (yd^3)	202	Gallons (gal)
Acre-feet (acre·ft)	43,560	Cubic feet (ft^3)
Acre-feet (acre·ft)	1613	Cubic yards (yd^3)
Acre-feet (acre·ft)	325,829	Gallons (gal)
Pounds, lb	16	Ounces (oz)
Tons, short	2000	Pounds (lb)
Tons, long	2240	Pounds (lb)
Miles per hour (mi/h)	88	Feet per minute (ft/min)
Miles per hour (mi/h)	1.47	Feet per second (ft/s)
Horsepower (hp)	33,000	Foot-pounds per minute (ft·lb/min)
Horsepower (hp)	550	Foot-pounds per second, standard (ft·lb/s)
Horsepower (hp)	0.746	Kilowatts (kW)
British thermal units (Btu)	778	Foot-pounds (ft·lb)
British thermal units (Btu)	0.0236	Horsepower (hp)
Equals this unit	Divided by	This unit

A–17

Conversion Factors for Systems of Measurement

METRIC UNITS TO METRIC UNITS

This unit	Multiplied by	Equals this unit
Meters (m)	100	Centimeters (cm)
Kilometers (km)	1000	Meters (m)
Square meters (m²)	10,000	Square centimeters (cm²)
Hectares	10,000	Square meters (m²)
Square kilometers (km²)	1,000,000	Square meters (m²)
Cubic meters (m³)	1,000,000	Cubic centimeters (cm³)
Cubic meters (m³)	1000	Liters (L)
Kilograms (kg)	1000	Grams (g)
Quintals	100	Kilograms (kg)
Tons, metric	1000	Kilograms (kg)
Kilometers per hour (km/h)	16.7	Meters per minute (m/min)
Kilometers per hour (km/h)	0.278	Meters per second (m/s)
Equals this unit	Divided by	This unit

ENGLISH UNITS TO METRIC UNITS AND METRIC UNITS TO ENGLISH UNITS

This unit	Multiplied by	Equals this unit
Inches (in)	2.54	Centimeters (cm)
Feet (ft)	0.305	Meters (m)
Yards, (yd)	0.914	Meters (m)
Miles, statute	1.609	Kilometers (km)
Square inches (in²)	6.45	Square centimeters (cm²)
Square feet (ft²)	0.0929	Square meters (m²)
Square yards (yd²)	0.836	Square meters (m²)
Acres	0.405	Hectares
Square miles (mi²)	2.590	Square kilometers (km²)
Cubic inches (in³)	16.4	Cubic centimeters (cm³)
Cubic feet (ft³)	0.0283	Cubic meters (m³)
Cubic yards (yd³)	0.765	Cubic meters (m³)
Gallons, U.S. (gal)	3.79	Liters (L)
Ounces (oz)	28.4	Grams (g)
Pounds (lb)	0.454	Kilograms (kg)
Tons, short	0.907	Tons, metric
Tons, long	1.016	Tons, metric
Feet per minute (ft/min)	0.305	Meters per minute (m/min)
Feet per second (ft/s)	0.305	Meters per second (m/s)
Miles per hour (mi/h)	1.609	Kilometers per hour (km/h)
Pounds per square inch (lb/in²)	0.0703	Kilograms per square centimeter (kg/cm²)
Pounds per square foot (lb/ft²)	4.887	Kilograms per square meter (kg/m²)
Pounds per cubic yard (lb/yd³)	0.593	Kilograms per cubic meter (kg/m³)
Foot-pounds or pound-feet (ft·lb)	0.138	Kilogram-meter (kg·m)
Horsepower (hp)	0.746	Kilowatts (kW)
British thermal units (Btu)	0.252	Calories (cal)
Equals this unit	Divided by	This unit

Degrees Fahrenheit, °F, to degrees, Celsius, °C:

$$°C = \tfrac{5}{9}(°F - 32) = 0.55(°F - 32)$$

Degrees Celsius, °C, to degrees Fahrenheit, °F:

$$°F = \tfrac{9}{5}\,°C + 32 = 1.80\,°C + 32$$

APPENDIX 5

Formulas Frequently Used in Calculations for Excavation Projects

ABBREVIATIONS USED IN FORMULAS

A	area	hp	horsepower
A_b	area of base	kWh	kilowatthour
a	acceleration	l	length
a_o	angle in degrees	m	mass
a_r	angle in radians, 1 radian=57.3°	min	minute
b	base	π	pi, 3.1416
b_u	upper base	r	radius of circle
b_l	lower base	s	distance
c	circumference of circle	s_p	perpendicular distance
d	diameter of circle	s	second
E	energy	t	time
f	force	V	volume
f_c	centrifugal force	v	velocity
g	acceleration of gravity, 32.2 ft/s²	W	work
h	height	w	width
h_s	slant height	wt	weight

FORMULAS

Circumference of circle: $\qquad c = \pi d$

Areas:
Square: $\qquad A = h^2$
Rectangle: $\qquad A = hb$
Parallelogram: $\qquad A = hb$

Trapezoid: $\qquad A = h \dfrac{b_u + b_l}{2}$

Triangle: $\qquad A = \dfrac{hb}{2}$

Trapezium: $\qquad A =$ sum of the areas of the two integral triangles
Circle: $\qquad A = \pi r^2 = 0.785 d^2$

Sector of circle: $\qquad A = \pi r^2 \dfrac{a_o}{360}$

Segment of circle: $\qquad A = 0.5 r^2 (a_r - \sin a_o)$

Ellipse: $\qquad A = \pi \dfrac{hw}{4}$

A–19

A-20 Formula Frequently Used in Calculations for Excavation Projects

Cone: $\quad A = A_b + (0.5ch_s)$
Sphere: $\quad A = 4\pi r^2$

Irregular area, approximately: Divide area into a number of strips with parallel sides and lengths $l_0, l_1, l_2, \ldots l_{n-1}, l_n$, separated by equal distances, s_p:

$$A = s_p\left[\frac{l_0}{2} + l_1 + l_2 + \ldots + l_{n-1} + \frac{l_n}{2}\right]$$

Volumes:
Cube: $\quad V = h^3$
Prism, rectangular: $\quad V = hwl$
Parallelopiped: $\quad V = As_p$
Wedge: $\quad V = 0.5(hwl)$
Cylinder: $\quad V = \pi r^2 h$
Cone or pyramid: $\quad V = 0.333 A_b h$
Sphere: $\quad V = 1.333\pi r^3 = 0.524 d^3$
Spherical sector: $\quad V = 0.667\pi r^2 h = 2.09\pi r^2 h$
Spherical segment: $\quad V = \dfrac{\pi h}{6}(3r^2 + h^2)$

Irregular volume, approximate: Divide volume into a number of zones with parallel sides of areas $A_0, A_1, A_2, \ldots, A_{n-1}, A_n$, separated by equal distances, s_p.

$$V = s_p\left(\frac{A_0}{2} + A_1 + A_2 + \ldots + A_{n-1} + \frac{A_n}{2}\right)$$

Motion:

$$s = vt$$
$$v = at$$
$$= gt$$
$$s = 0.5at^2$$
$$= 0.5gt^2$$
$$v^2 = 2as$$
$$= 2gs$$

Mass and weight:

$$wt = mg$$

Force:

$$f = ma$$
$$= mg$$
$$f_c = wt \times \frac{v^2}{rg}$$

Work and energy:

$$W = fs = fvt$$
$$E = fs = fvt$$

Power:

$$hp = \frac{W/s}{550}$$

$$hp = \frac{W/\min}{33{,}000}$$

$$kW = 1.341 \text{ hp}$$

Efficiency:

$$\text{Eff, \%} = \frac{\text{Output hp}}{\text{Input hp}} \times 100$$

Efficiency of overall work of a machine:

$$\text{Eff} = \frac{\text{min actually worked during hour}}{60}$$

The efficiency may be expressed as a working-minute hour, such as a 40- or 50-minute hour, or as a percentage, 67 and 83%, respectively. The efficiency factor is based on the machine's ability to work provided it is not held up by any extraneous delays.

APPENDIX 6

Swell versus Voids of Materials and Hauling Machine Load Factors

Swell, %	Voids, %	cybm/cylm
5	4.8	0.952
10	9.1	0.909
15	13.0	0.870
20	16.7	0.833
25	20.0	0.800
30	23.1	0.769
35	25.9	0.741
40	28.6	0.714
45	31.0	0.690
50	33.3	0.667
55	35.5	0.645
60	37.5	0.625
65	39.4	0.606
70	41.2	0.588
75	42.9	0.571
80	44.4	0.556
85	45.9	0.541
90	47.4	0.526
95	48.7	0.513
100	50.0	0.500

SOURCE: Caterpillar Tractor Co.

APPENDIX 7

Approximate Angles of Repose of Materials*

Material	Slope ratio, horizontal:vertical	Angle of repose, °
Ashes, coal	1.0:1	45
Cinders, coal	1.0:1	45
Clay:		
Dry	1.3:1	38
Damp	2.0:1	27
Coal, broken	1.4:1	36
Earth:		
Dry	1.3:1	38
Damp	2.0:1	27
Gravel:		
Round	1.7:1	30
Angular	1.3:1	38
Rock, broken:		
Soft	1.5:1	34
Hard	1.3:1	38
Rock, weathered:		
Residuals and weathered rock	1.5:1	34

SOURCE: Caterpillar Tractor Co.
*Angle of repose is the angle between the horizontal and the slope of a heaped pile of material.

APPENDIX 8

Bearing Powers of Materials

Material	English system		Metric system	
	lb/in²	tons/ft²	kg/cm²	tons/m²
Rock, solid	350	24	24.6	240
Rock, semishattered	70	5	4.9	50
Clay:				
Dry	55	4	3.9	40
Damp	27	2	1.9	20
Wet	14	1	1.0	10
Gravel, cemented	110	8	7.7	80
Sand:				
Dry, compacted	55	4	3.9	40
Dry, clean	27	2	1.9	20
Quicksand and alluvial soil	7	0.5	0.5	5

SOURCE: Caterpillar Tractor Co.

APPENDIX 9

Abbreviations

A	ampere
A	area
a	acceleration
ac	alternating current (adj.)
acre	acre
acre·ft	acre foot
A·h	ampere-hour
atm	atmosphere
avg	average
bhp	brake horsepower
bhp·h	brake horsepower-hour
Btu	British thermal unit
°C	degree Celsius
c	candle
cal	calorie
cg	centigram
cgs	centimeter-gram-second metric system
cif	cost, insurance, and freight
cm	centimeter
cm²	square centimeter
cm³	cubic centimeter
coef	coefficient
const	constant
cos	cosine
cp	candlepower
c to c	center to center
cwt	hundredweight
cybm	cubic yard bank measurement
cycm	cubic yard compacted measurement
cylm	cubic yard loose measurement
cysm	cubic yard struck measurement
...°	degree
d	diameter of circle
dbh	diameter breast high of tree
dc	direct current (adj.)
eff	efficiency
el	elevation
eq	equation
°F	degree Fahrenheit
f	force

Abbreviations

fbm	feet board measure or board feet
fc	footcandle
fob	free on board
ft	foot
ft^2	square foot
ft^3	cubic foot
ft·lb	foot pound
ft/min	feet per minute
ft^3/min	cubic feet per minute
ft/s	feet per second
ft^3/s	cubic feet per second
ft/s^2	foot per second per second
fur	furlong
g	acceleration of gravity, 32.2 ft/s^2
g	gram
gal	gallon
gal/min	gallons per minute
GR	grade resistance
GVW	gross vehicle weight (tare wt + payload wt)
h	hour
hp	horsepower
hp·h	horsepower hour
Hz	hertz (cycles per second)
ihp	indicated horsepower
in	inch
in^2	square inch
in^3	cubic inch
in·lb	inch-pound or pound-inch
kc	kilocycle
kg	kilogram
kg·m	kilogram-meter
kg/m^3	kilograms per cubic meter
kL	kiloliter
km	kilometer
km^2	square kilometer
kV	kilovolt
kVA	kilovolt-ampere
kW	kilowatt
kWh	kilowatt-hour
L	liter
lb	pound
lb/ft^2	pounds per square foot
lb/in^2	pounds per square inch
lin	linear
lin ft	linear feet
log	logarithm
lp	low pressure
lm	lumen
lm·h	lumen-hour
m	mass
m	meter
m^2	square meter
m^3	cubic meter
max	maximum
mg	milligram
mi	mile
mi^2	square mile
mi/h	miles per hour
mi/h/s	miles per hour per second
min	minimum
min	minute
mm	millimeter
oz	ounce

pt	pint
qt	quart
r/min	revolutions per minute
RR	rolling resistance
r/sec	revolutions per second
s	distance
s	second
scp	spherical candlepower
shp	shaft horsepower
sin	sine
sp gr	specific gravity
sta	station, 100 feet
sta yd	station cubic yard
tan	tangent
temp	temperature
ton/h	tons per hour
ton·mi	ton-mile
ton/min	tons per minute
V	volume
V	volt
v	velocity
VA	voltampere
W	work
W	watt
w/c	watts per candle
Wh	watthour
whp	wheel horsepower
wk	week
wt	weight
yd	yard
yd^2	square yard
yd^3	cubic yard
yr	year

NOTE: Abbreviations used in the fields of hydraulics and pneumatics are given at the beginning of Chapter 10.

Bibliography

Associated General Contractors of America: *Contractors' Equipment Manual,* Washington, D.C., 1974.
Bajpai, A. C., et al.: *Mathematics for Engineers and Scientists,* Wiley, New York, 1973.
Baumeister, Avallone, and Baumeister: *Marks' Standard Handbook for Mechanical Engineers,* 8th ed., McGraw-Hill, New York, 1978.
Beiser, Arthur, et al.: *The Earth,* Time-Life Books, New York, 1970.
California Division of Highways, *Standard Specifications,* Sacramento, 1960
Caterpillar Tractor Co.: *Handbook of Ripping,* Peoria, Ill., 1972.
Caterpillar Tractor Co.: *Performance Handbook,* Peoria, Ill., 1979.
Daugherty, Robert L., and Franzini, Joseph B.: *Fluid Mechanics with Engineering Applications,* 7th ed., McGraw-Hill, New York, 1977.
Dickenson, E. H., and Slager., T., Jr.: *Rock Drill Data,* Ingersoll-Rand Co., New York, 1960.
Dobrin, Milton D.: *Introduction to Geophysical Prospecting,* 3d ed., McGraw-Hill, New York, 1976.
E. I. du Pont de Nemours & Co.: *Blasters' Handbook,* Wilmington, Del., 1966.
Eshbach, O. W., and Souders., M.: *Handbook of Engineering Fundamentals,* 3d ed., Wiley, New York, 1975.
Fenton, Carroll L., and Fenton, Mildred. A.: *Rock Book,* Doubleday, Garden City, N.Y., 1970
Fink, Donald G., and Beaty., H. Wayne: *Standard Handbook for Electrical Engineers,* 11th ed., McGraw-Hill, New York, 1978.
Hyster Co.: *Compaction Handbook,* Kewanee, Ill., 1972.
Kent., R. T., et al.: *Mechanical Engineers' Handbook:* Wiley, New York, 1950.
Kissam, Philip: *Surveying for Civil Engineers,* 2d ed., McGraw-Hill, New York, 1976.
Kummel, Bernhard: *History of the Earth,* W. H. Freeman, San Francisco, 1970.
Longwell, C. R., Flint, R. F., and Skinner, B. J.: *Physical Geology,* 2d ed., Wiley, New York, 1977.
McGraw-Hill Encyclopedia of Science and Technology, Daniel N. Lapedes (ed.), McGraw-Hill, New York, 1979.
Nichols, Herbert L., Jr.: *Moving the Earth,* 3d ed., North Castle Books, Greenwich, Conn., 1976.
Peele, Robert: *Mining Engineers' Handbook,* Wiley, New York, 1951.
Peurifoy, R. L.: *Construction Planning, Equipment, and Methods,* 3d ed., McGraw-Hill, New York, 1979.
Pough, Frederick H.: *A Field Guide to Rocks and Minerals,* 4th ed., Houghton Mifflin, Boston, 1976.
Schultz, John R., and Cleaves, A. B.: *Geology in Engineering,* Wiley, New York, 1955.

Shelton, John S.: *Geology Illustrated*, W. H. Freeman, San Francisco, 1966.
Steel, Ernest W., and McGhee, Terrence: *Water Supply and Sewerage*, 5th ed., McGraw-Hill, New York, 1979.
Thornbury, William D.: *Principles of Geomorphology*, Wiley, New York, 1969.
Trauffer, Walter E. (ed.): *Pit and Quarry Handbook*, Pit and Quarry Publications, Chicago, Ill., 1975.
United States Mineral Resources, U.S. Bureau of Mines, Washington, 1973.
Urquhart, Leonard C.: *Civil Engineering Handbook*, 4th ed., McGraw-Hill, New York, 1959.
Wyckoff, Jerome: *Rock, Time, and Landforms*, Harper and Row, New York, 1966.

Glossary

Ablation   The formation of residual deposits by the washing away of loose or soluble materials.
Abrasion   The mechanical wear of rock on rock.
Acre   Unit for measuring land, equal to 43,560 ft² (4840 yd², or 160 square rod).
Acre-Foot   The amount of water required to cover 1 acre to a depth of 1 ft, equal to 1613 yd³ (1233 m³). Also used to measure materials in place, such as coal and gravel.
Adamantine Drill   Core drill using chilled shot as a cutting agent. Cores from 4 to 30 in (10 to 76 cm) in diameter are obtained.
Adit   A nearly horizontal or horizontal passage from the surface into a mine. Frequently called a drift.
Adobe   An impure calcareous clay which, when drying, breaks into roughly cubical blocks up to 20 in (51 cm) in dimension.
Agglomerate   A coarse-grained pyroclastic rock consisting largely of bombs and blocks.
Aggradation   The building up of any portion of the earth's surface toward a uniformity of grade or slope by the addition of materials.
Aggregate   Mineral material such as sand, gravel, shells, slag, or broken stone to be used in concrete.
Air Receiver   The air storage tank on an air compressor.
Air-Slaked   Wetted by exposure to moisture in the air.
Alkali Flat   A sterile plain, containing an excess of alkali, at the bottom of an undrained basin in an arid region. Sometimes called a playa.
Alluvial Fan   The outspread sloping deposit of boulders, gravel, and sand, left by a stream where it spreads from a gorge upon a plain or open valley floor.
Alluvium   The general name for all the sediment deposited in land environments by streams.
Altimeter   An aneroid barometer graduated to show elevation instead of pressure.
Amorphous   Without definite form. Applying to rocks and minerals having no definite crystalline structure.
Amortization   The repayment of debt, principal and interest, usually in equal installments. The process of gradually recovering the cost or value of an asset such as machinery.
Ampere   The intensity of electric current produced by 1 volt acting through a resistance of 1 ohm.
Amphibolite   A coarse-grained, mafic metamorphic rock containing more than 50 percent ferromagnesian minerals. Dark color. Hard and heavy.
Amygdule   A mineral that fills the vesicles or holes in igneous rock.
Andesite   An igneous extrusive rock of fine-grained texture, consisting chiefly of feldspar and ferromagnesian minerals. Medium color, hard and heavy.

Angle of Repose Measured from the horizontal, it is the minimum angle of plane, along which coarse particles in the material begin to fall under the influence of gravity. Also known as *critical slope*.

Angledozer A bulldozer with a blade which can be pivoted on a vertical center pin so as to cast its load to either side.

Anthracite Coal A hard, black lustrous coal containing 85 to 90 percent carbon. It produces an intense, hot fire. A metamorphic rock.

Anticline An upfold of layered rocks in the form of an arch, with the oldest strata in the center. Reverse of a syncline.

Apron The front gate of a scraper body, which is raised and lowered.

Archeozoic The second of the six groups of rocks. Corresponds to the same geologic era. From 2000 to 1200 million yr ago.

Area The amount of surface included between certain closed boundary lines. Any particular extent of surface, region, or tract.

Areal Geology Branch of geology that pertains to the distribution, position, and form of the areas of the Earth's surface occupied by different kinds of rocks or different rock formations.

Argillaceous Containing or consisting of clay, as the cementing agency in argillaceous limestones and sandstones.

Argillite Synonym of slate.

Arkose A sandstone containing at least 25 percent feldspar as well as quartz. Light to medium color. Medium hard and medium heavy.

Ash, Volcanic Extremely fine-grained particles ejected from a volcano.

Asset An owned value such as machinery.

Atmospheric Pressure Pressure of air enveloping the earth, averaged as 14.7 lb/in^2 at sea level or 29.92 in of mercury as measured by a standard barometer.

Auger A rotating drill having a screw thread that carries cuttings away from the surface being drilled.

Availability Factor A measure of the reliability of machines as regards freedom from mechanical failures. It is the actual working time divided by the available working time, in hours. An availability factor of 95 percent is excellent, 90 percent is good, and 85 percent is acceptable.

Average-End-Area Method A means of calculating the volume of earthwork between two cross sections. The volume equals the average of the cross-sectional areas times the distance between the cross sections.

Azoic The first of the six groups of rocks. Corresponds to the geologic era. From 4.5 billion to 2.0 billion yr ago.

Backfill Rock-earth used to replace excavation previously removed.

Backhoe A hoe-type or pull-type shovel.

Badlands A system of closely spaced ridges with little or no vegetation.

Baking The hardening of rocks through heating by magma or lava. It produces *contact metamorphism*.

Ballast Material such as water, sand, or metal, which increases the weight of a machine such as a compactor or dragline.

Bank Measure Measurement of material in its original place in the ground or in situ.

Bank Yards Cubic yards of material in its original place in the ground.

Barranca Spanish term meaning a gully with somewhat steep banks.

Basalt Extrusive igneous rock of dark color and fine structure. Sometimes it contains vesicles and the formation is columnar. Hard and heavy.

Base Line The main traverse or surveying line through the site of proposed construction, from which lines and grades for the work are plotted. Not to be confused with the centerline of construction.

Basement Complex Assemblage of igneous and metamorphic rocks lying beneath the oldest sedimentary rocks of the region.

Batholith Large body of intrusive igneous rock having an exposed area of more than 40 mi^2 (104 km^2).

Bauxite The clayey ore of aluminum. Light to medium color. Soft and medium-weight.

Bed A base for machinery. The bottom of the body of a hauling machine.

Bed Stratum of rock more than ½ in (1.2 cm) in thickness.

Bedding Plane The top or bottom surface of a bed or stratum of rock. Usually a separation and weakness in the rock structure.
Bedrock Continuous solid rock that everywhere underlies the regolith and which sometimes forms the Earth's surface, the regolith having been eroded away and the bedrock having become an outcrop or shield.
Belt Loader A movable or traveling belt conveyor with cutting edge or disk.
Bench A working level or step in a cut which is made in several layers.
Bench Mark A point of known or assumed elevation used as reference in determining and recording other elevations.
Bench Terrace A more or less level step in a slope of cut.
Berm An artificial ridge of earth.
Bid To submit a contract price. Also the tender or bidding price.
Bid, Unbalanced A bid in which some of the unit prices are abnormal, either too high or too low, but in which the total bid represents the total of a balanced bid.
Bit The part of a drill or excavating machine which cuts the rock-earth.
Bit, Carbide A bit having inserts of tungsten carbide.
Bit, Chopping A bit that is worked by raising and dropping.
Bit, Coring A bit that grinds the outside ring of the hole, leaving an inner core intact for sampling.
Bit, Diamond A rotary bit having diamonds set in its cutting surfaces.
Bit, Drag A diamond or fishtail bit. A bit that cuts by rotation of fixed cutting edges or points.
Bit, Fishtail A rotary bit having cutting edges or knives.
Bit, Multiuse A bit that is sharpened for new service when worn.
Bit, Plug A diamond bit that grinds out the full width of the hole.
Bit, Roller A bit that contains cutting elements that are rotated inside as it turns.
Bit, Throwaway A bit that is discarded when worn.
Bituminous Coal Black and firm soft coal that breaks into blocks and contains alternating layers having dull and bright luster. Its carbon content is between 75 and 85 percent. Lightweight sedimentary rock.
Black Powder Gunpowder used in blasting. Mixture of carbon, sodium or potassium nitrate, and sulfur.
Blade Usually, a part of an excavator which digs and pushes rock-earth but does not carry it.
Blast To loosen or move rock-earth by means of explosives or an explosion.
Blasthole A drill hole used for a charge of explosives.
Blasthole Drill A drill of any type used in the drilling of blastholes, usually air percussion, cable, rotary, or fusion-type.
Blasting Agent An inert explosive detonated by a high-strength primer.
Blasting Cap A copper shell closed at one end and containing a charge of detonating agent which is ignited from the spark of the fuse. Also, a fully enclosed copper shell containing a detonating agent which is ignited by an electrical current.
Blasting Fuse A slow-burning fuse used for igniting blasting charges.
Blasting Gelatin A jellylike high explosive made by dissolving nitrocotton in nitroglycerin.
Blasting Machine A battery- or hand-operated generator used to supply current to blasting circuits.
Blasting Mat A steel blanket composed of woven cable or interlocked steel rings to prevent the throw of blasted material.
Blockholing Blasting boulders by means of drilled holes and explosives.
Blowout A deflation basin excavated in shifting sand or other easily eroded regolith by the winds.
Blue Tops Grade stakes, the tops of which indicate finish grade level.
BM Bench mark.
Body The load-carrying part of a hauler or scraper.
Bog Soft spongy ground, usually wet and composed of more or less vegetable matter.
Bog Iron Ore A spongy variety of limonite. Found in layers or lumps on level sandy soils which have been covered by swamp or bog. Found also in existing swamps and bogs.
Bomb Rounded mass of lava expelled from a volcano by steam explosions.
Bonanza Mining term for a rich body of ore.

Booster Small amount of high explosive attached to a detonator for the purpose of increasing the rate of detonating the charge.

Borehole Hole drilled into rock for exploration or for blasting.

Boring Rotary drilling.

Borrow Pit An excavation from which material is taken for a nearby job.

Boulder Detrital material greater than 8 in (21 cm) in diameter.

Boulder Clay Stiff, hard, and usually unstratified clay of the glacial periods which contains boulders. Also called *till* or *hardpan.*

Bowl The body of a carrying scraper.

Brake Horsepower The horsepower output of an engine, motor, or mechanical device. Measured at the flywheel or belt.

Breccia General term for rock of any origin containing angular particles.

British Imperial Gallon The metric gallon, containing 277 in^3, equivalent to 1.20 U.S. gallons.

British Thermal Unit The quantity of heat required to raise the temperature of 1 lb of water 1°F at or near the temperature of maximum density, 39.1°F. Equivalent to 0.252 cal.

Brush Trees and shrubs of less than 4-in (10-cm) stump diameter. The Spanish term *chaparral* is a synonym.

Bucket The part of an excavator which digs, lifts, loads, and carries material.

Bucket Loader Usually a chain-bucket loader, but also a tractor-loader.

Buckshot A tough, tenacious earthen material which, when dry, shatters into irregularly shaped particles about the size of buckshot or larger, about ⅛ in (3 mm) or more.

Buffer Blasting Secondary fragmentation by crushing action, produced by setting off a row of holes nearest the face first and the other parallel rows behind it in succession. Successive crushing actions result, as material from each successive blast is thrown against that of the preceding one.

Bulldozer A term for the complete machine, tractor with bulldozer.

Bulking Increase in volume of fine materials due to moisture content.

Bumboat A small boat equipped with hoist for handling the pipes, lines, and anchors of a dredge.

Burden The distance of a blasthole from the face, or the volume of rock to be moved by the explosive in each blasthole.

Burn Cut A narrow section of rock pulverized by exploding heavy charges in parallel holes, as in tunneling.

Buttress Fill A compacted fill placed so as to prevent movement or sliding of an existing earthwork or foundation.

Cable Excavator A long-range cable-operated machine which works between a mast and an anchor or between a head tower and a foot tower.

Calcareous Composed essentially of calcium carbonate or cemented by calcium carbonate, as a calcareous sandstone.

Calcareous Tufa An open cellular deposit of calcium carbonate. Light color, medium hard, and medium weight.

Caliche Material composed essentially of soft limestone with varying percentages of clay. Generally soft to medium hard, although sometimes requiring blasting. Light in color. Found chiefly in deserts, where it occurs as a sedimentary rock in a shallow stratum.

Calorie The amount of heat required to raise the temperature of 1 gram of water 1°C at or near the temperature of maximum density, 4°C. Equivalent to 3.97 Btu.

Cambrian The first or oldest of the five rock systems of the Paleozoic group. Also the corresponding geologic period, from 570 to 500 million years ago.

Candle Unit of light intensity. At a distance of 1 ft, a candle produces 1 footcandle (fc) equivalent to 1 lumen per square foot on a surface normal to the beam.

Candlepower A light unit or that amount of light given by a sperm candle burning 120 grams per hour.

Cap A detonator set off by electric current or burning fuse.

Cap, Delay An electric blasting cap that explodes at a set interval after passage of the current.

Cap Rock The more impervious and harder rock found above the more weathered and softer rock.

Carbonaceous Coaly, containing carbon or coal. Used especially for shale or rock containing small particles of carbon distributed throughout the whole mass.
Carboniferous The last of the five systems of rocks of the Paleozoic group of rocks. Also the corresponding geologic period. Subdivided into the Mississippian, Pennsylvanian, and Permian series of rocks. From 340 to 230 million years ago.
Carbonite An explosive mixture of nitrobenzene, potassium nitrate, sulfur, and kieselguhr.
Carriage Cable The cable for traveling the carriage on a cable. Also the traction cable.
Cartridge A wrapped stick of dynamite or other explosive.
Casing A pipe lining for a drilled hole to prevent raveling of the surface.
Cassiterite An important ore of tin.
Cat A trademarked designation for any machine made by the Caterpillar Tractor Co. Widely used to indicate any crawler-tractor.
Catskinner The operator of a crawler-tractor.
Cave Collapse of an unstable bank.
Cellular Porous texture and fairly large cavities of certain volcanic rocks such as basalt and pumice.
Celsius Metric scale of temperature, sometimes called centigrade.
Cement Materials from solution which bind together the particles of a sedimentary rock.
Cementation The binding together of the particles of a sedimentary rock by such mineral cement as calcite, silica, and iron oxide.
Cenozoic The last of the six groups of rocks. Also the corresponding geologic era, from 63 million yr ago to the present.
Center of Gravity The point in a body about which all the weights of all the various parts balance each other. The center of mass of a cut or fill.
Center of Mass In a cut or fill, a cross-sectional line that divides the volume into its halves.
Center of Moments The point at which a body tends to rotate. A point arbitrarily selected for determining the resultant moment of a series of forces.
Centerpin In revolving machinery the fixed vertical shaft around which the machinery deck turns.
Center Stakes Stakes indicating the centerline of construction.
Centigrade See *Celsius*.
Centrifugal Force Outward force exerted by a body moving in a curved line.
Centripetal Force Inward force exerted by a body to keep it moving in a curved line. Force exerted as a result of a superelevated curve on haulage road.
Chain Bucket Loader A mobile loader using a series of buckets mounted on a chain to excavate and load material.
Chaining A term which originally meant measuring with a chain, but as now used denotes measuring with either a chain or tape.
Chalcocite An important copper ore, the mineral containing 80 percent copper.
Chalcopyrite An important copper ore, the mineral containing 35 percent copper.
Chalk A limestone that is weakly cohesive of white color and medium weight.
Chaparral In a general sense, brush that is stiff and thorny. Spanish term.
Chats The gangue which is found intimately mixed with some lead-zinc ores. Light to medium color. Medium weight.
Cherry Picker A small derrick or hoist, usually mounted on a truck or tractor.
Chert A sedimentary rock composed of silica which is either a precipitate or a replacement of calcium carbonate minerals. Light to medium color and medium weight.
Cheval-Vapeur The French term for horsepower.
Chip Blasting Blasting of shallow-depth ledge rock.
Chipping Loosening of shallow-depth rock by blasting or by air hammers.
Chips Small angular fragments of stone containing no dust.
Choker A short length of cable or chain with a noose, used to pull out stumps and for other similar purposes.
Chord The line joining two points on a curve.
Churn Drill A machine that drills a hole by raising and dropping a string of drilling tools suspended by a reciprocating cable. A cable drill.
Cinder Scoriaceous lava from a volcano. The residue from the burning of coal or other carbonaceous material.

Cinder Cone Small- to moderate-size cone built up from the cinders of a volcano.
Cinnabar The vermilion-colored, medium-weight common ore of mercury.
Circuit In electricity, the complete path of a current. In blasting, the path of the detonating electric circuit or detonating fuse such as Primacord.
Cirque A steep-walled amphitheatric recess in a mountainside, generally ascribed to glacial erosion.
Clamshell A bucket with two jaws, which clamp together to load by their own weight when lifted by the closing line.
Clastic Composed of broken fragments or grains cemented together. Sedimentary sandstone is an example.
Clay Fine argillaceous material which is more or less plastic when wet. It has grain size less than 0.0002 in (0.005 mm), light to medium color, and medium weight.
Claypan Stratum of stiff, compact, and relatively impervious clay. Not cemented.
Claystone Clastic rock consisting predominantly of clay-size particles.
Clearing Removing natural and artificial obstructions to excavation. Cutting down and removing trees and brush.
Cleavage Tendency of rock to split along definite, parallel, and closely spaced planes, which may be highly inclined to the bedding planes. Slate is an example.
Climate The total of all atmospheric and meteorological influences which combine to characterize a region and give it individuality by influencing the nature of its vegetation, soils, rocks, and landforms.
Clinometer An instrument for measuring the inclination of a line with respect to the horizontal. Used also for measuring the dip of a rock stratum or formation.
Clod Buster A drag that follows an excavating machine to break up lumps.
Close-Jointed Term applied to rock containing joints that are near together. Example is columnar basalt.
Coal Black sedimentary and metamorphic rock consisting chiefly of decomposed plant matter and containing less than 40 percent inorganic matter.
Coalescence The capacity for growing together, fusing, and binding.
Coarse and Fine Aggregates The term *coarse* is applied to graded mineral aggregate in which the largest particles have a diameter greater than ¼ in (6 mm). The term *fine* is applied to one in which the largest particles have a diameter less than ¼ in (6 mm).
Cobble Sediment particle having a diameter between 3 in (8 cm) and 8 in (21 cm).
Coefficient of Friction The ratio of the force required to move an object resting on a horizontal plane to the weight of the object. Expressed as a percentage as related to the natures of the object and the horizontal plane. It is also the tangent of the angle of inclination to the horizontal at which the object slides.
Cofferdam A barrier built so as to form an enclosure to keep water or soil from entering an area of excavation.
Cohesion The quality of a soil which makes it stick together.
Collar In drilling, the open end of the hole.
Colluvium Body of sediment consisting of alluvium in part and also containing angular fragments of the original rock.
Columnar Structure The contraction phenomenon in which prismatic columns commonly with six uniform faces form on the cooling of the magma. Examples are tabular bodies of igneous rocks such as andesites and basalts.
Common Excavation General term applied to soft excavation such as earth and residuals, as contrasted to hard excavation such as weathered rock and semisolid and solid rock.
Compacted Cubic Yard Measurement of material after it has been placed and compacted in the fill or embankment. Equivalent to 0.765 m².
Compaction Reduction in volume of fill or embankment by rolling, tamping, and wetting the material.
Compressor A machine for densifying air or gas from an initial intake pressure to a higher discharge pressure.
Conchoidal Shell-shaped. Compact rocks such as flint, argillite, and felsite break with concave or convex surfaces or conchoidal fracture.
Concretion A spheroidal aggregate formed by the segregation and precipitation of some soluble material such as quartz or calcite around a nucleus.
Concussion Shock waves or sharp air waves caused by an explosion or heavy blow.

Conglomerate Clastic sedimentary rock containing rounded pebbles or larger particles.
Contact Metamorphism Metamorphism in the vicinity of igneous rock resulting from the intense heating effect of the magma. Slate near granite is an example.
Continental Deposit Sedimentary rock deposit laid down within a general land area and in lakes and streams, as contrasted to marine deposits laid down in the sea.
Continental Shield An extensive area in which the Precambrian foundation rocks of a continental mass are exposed.
Contour Line An imaginary line on the surface of the ground, every point of which is at the same elevation.
Conveyor A machine that transports or hauls material by belts, cables, or chains.
Coquina An aggregation of shells and large shell fragments cemented by calcium carbonate. A form of limestone, with light color and light weight.
Core Cylindrical sample of rock recovered from an exploratory drill hole.
Core Drill A rotary drill, usually a diamond or shot drill, equipped with a hollow bit and core lifter.
Corrasion The mechanical detachment or wearing away of rock material by running water, glaciers, winds, waves, or mass movements.
Correlation In rocks and rock formations this term is used to mean determination of equivalence in geologic age and position in the sequence of strata in different areas.
Cost of Excavation The sum of direct job costs, indirect job expenses, and indirect home office expenses.
Cotton Rock Local name for soft, fine-grained siliceous magnesian limestone of the Silurian system.
Counterweight A nonworking load attached to one side or to the end of a machine to balance a working weight on the other side or end of the machine. An example is the counterweight on the back of a dragline in order to balance the weights of the long boom and the loaded bucket.
Country Rock A relative term used for the older rock into which the younger magma has intruded. The term is extended to mean the oldest rocks encountered.
Coyote Holes Horizontal tunnels into which explosives are placed for blasting a high rock face.
Crane A machine used for lifting and moving loads and for the handling of buckets for excavation.
Crawler One of a set of roller-chain tracks used to support and propel a machine, or a machine mounted on such tracks, known also as a *track layer*.
Creep The almost imperceptible slow downslope movement of the regolith of the Earth's surface.
Cretaceous Literally, of the nature of chalk. Also, the third and last system of Mesozoic rocks and the corresponding geologic period, 135 to 63 million years ago.
Crocus A term used in quarries to denote gneiss or other rock in contact with granite.
Crop Coal Coal of inferior quality near the surface, known also as *grass roots coal*.
Cross Bedding Laminations in sedimentary rocks confined to single beds and inclined to the general stratification, caused by actions of wind and water.
Cross Section A vertical section of the earthwork at right angles to the centerline. Used along with centerline distances to calculate earthwork quantities.
Crowd Forcing the bucket or dipper into the digging or the mechanism which does the forcing.
Crusher A machine which reduces rock to smaller and more uniform sizes.
Crust The topmost zone of the Earth, also called the *regolith*.
Crystalline Rock texture resulting from simultaneous growth of crystals.
Crystallization The process of development of crystals by condensation of materials in a gaseous state, by precipitation of materials in solution, or by solidification of materials in a melt.
Cubic Yard Unit of rock-earth excavation, equivalent to 0.765 m^3.
Cubic Yard Bank Measurement (cybm) Unit of excavation in cut or natural bed.
Cubic Yard Compacted Measurement (cycm) Unit of excavation in fill or embankment after compaction.
Cubic Yard Loose Measurement (cylm) Unit of excavation in machine, in stockpile, or in uncompacted fill or embankment.
Cubic Yard Struck Measurement (cysm) Unit of capacity of bucket, body, bowl, or

dipper of a machine. Measured by striking off the ends and sides of the container by a straight edge, excluding the teeth. It is sometimes called *water level measurement*.

Cuprite An important copper ore, the mineral containing 89 percent copper.

Cut Depth to which material is to be excavated. Also the volume of excavation for a given cut.

Cut and Cover A means of building a sanitary fill or dump. A cut is made, alternate compacted layers of trash and excavated material from a stockpile are placed in the area of the cut, and the entire area of the resulting fill is covered.

Cut and Fill A process of building earthworks by excavating the cuts and using the excavated material for the adjacent fills. In a balanced cut and fill, the volume of the excavated material equals the volume of the fill, with allowance for swell or shrink from cut to fill.

Cutterhead On a hydraulic or suction dredge, the set of revolving blades at the beginning of the suction line for fragmenting the hard materials.

Cutting An English term for excavating.

Cuttings The rock fragments from a drilling operation.

Cycle A complete set of individual operations which a machine performs before repetition, as the cycle for a shovel, which consists of load, swing, dump, and return. Frequent moving along the face of the cut is not part of the cycle.

Cycle of Erosion The sequence of landforms, essentially highlands and lowlands, through which a land mass evolves from the time it begins to be eroded until it is reduced to base level.

Dacite Igneous extrusive rock of the andesite family, containing quartz. Occurs in dikes as well as in lava flows. Medium to dark color. Heavy.

Data Facts, particularly those that can be expressed numerically.

Datum Any level surface taken as a plane of reference from which to measure elevations.

Deadheading Traveling without load, except from the dumping area to the loading point.

Dead Load The tare weight or unladen weight of a hauling machine.

Deadman The anchorage for a guy line or cable, usually embedded in the ground.

Decking Separating charges of explosives in the blasthole by inert material which prevents passing of concussion, each charge having a primer.

Decomposition The chemical alteration of rock minerals.

Deflagration Burning with sudden and startling combustion, as the explosion of black powder, contrasted to the more rapid detonation of dynamite.

Deflation The picking up and removal of loose rock particles by the wind.

Deformation Change of volume or shape or both in a rock body or a change in the original position of the rock body within the Earth's crust.

Degradation The lowering of any portion of the Earth's surface by erosion.

Delta A body of sediment deposited by a stream flowing into the standing or slow-current water of a lake, bay, or ocean.

Density *Weight per unit volume.* Generally used for rock-earth in terms of pounds per cubic yard or kilograms per cubic meter.

Denudation The wearing down and disintegration of rock masses by rain, frost, wind, running water, and other surficial agencies.

Deposit Anything laid down, such as rocks, minerals, and ores. Also the process of laying down.

Depreciation Loss of useful life for any cause. Specifically, in the case of machinery, loss resulting from wear, obsolescence, inadequacy, and other causes deemed to lessen usefulness. **Rate of depreciation** is an appraisal term defined as the percentage rate at which the value is being exhausted.

Desert Arid land where rainfall is less than 10 in (25 cm) annually. Deserts may have all kinds of topography and landforms at all elevations.

Detonation Practically instantaneous decomposition or combustion of an unstable compound with tremendous increase in volume.

Detonator Device to explode a charge of dynamite or other high explosive.

Detritus Collective term referring to broken pieces of older rocks, of minerals, or of skeletal remains of organisms.

Devonian The fourth system of rocks in the Paleozoic group. Also, the corresponding geologic period from 400 to 340 yr ago.
Dewatering Removing water by pumping, drainage, or evaporation.
Diameter Breast High Measurement of trees by diameter at the chest height of a man, about 5 ft (1.5 m).
Diamond Drill A rotary drill using a coring bit studded with black diamonds that is used chiefly in exploratory drilling.
Diastrophism The processes by which the crust of the Earth is deformed, producing continents and oceans, mountains and valleys, and flexures, folds, and faults in the rocks.
Differential Weathering The results of variations in the rate of weathering on different parts of a rock body.
Dike A mass of igneous rock which has intruded through a narrow fissure of existing country rock.
Diorite A coarse-grained igneous intrusive rock lacking quartz. Generally of medium color and medium weight.
Dip The angle of inclination of the plane of stratification of rock with a horizontal plane.
Dipper A digging bucket rigidly attached to the dipper stick, arm, or handle.
Disconformity See *Unconformity*.
Disintegration The mechanical breakup of rock.
Ditch Generally a long narrow trench or excavation.
Ditcher A machine used to excavate a ditch.
Dolerite A medium-grained, heavyweight, dark igneous rock of the basalt family.
Dolomite A heavyweight magnesian limestone, hard and generally of light bluish color.
Dolomitic Limestone A limestone containing dolomite, in which calcium carbonate predominates over magnesium carbonate.
Dome An uplift of rocks in which the beds dip outwards in all directions from the center.
Donkey A winch the drums of which are controlled separately by clutches and brakes.
Dozer Abbreviation for bulldozer or angledozer.
Dragscraper A digging and hauling device consisting of a bottomless bucket working on the cable of a mast and anchor, corresponding to headworks and tailworks. Also, a towed bottomless scraper used for leveling land and maintaining haul roads.
Drainage Basin The total area that contributes water to a stream.
Draw A topographical term referring to a natural depression or swale. Also a small watercourse.
Drawbar In a tractor the fixed or hinged bar extending to the rear and used as a fastening for lines and towed loads or machines.
Drawbar Horsepower The horsepower of the tractor engine, minus friction and slippage losses in the drive mechanisms and in the tracks or tires.
Drawbar Pull The force a tractor can exert on a load attached to the drawbar. The force depends on drawbar horsepower and speed if there is no slippage between tracks and ground. If there is a slippage the drawbar force is limited by the weight of the tractor and the coefficient of traction.
Dredge A machine for excavating material at the bottom or at the banks of a body of water. The material may be discharged on the bank or it may be emptied into a scow for transport to a distant point. Dredges are classified as: mechanical, of the grapple, dipper, and bucket types; and hydraulic of the self-propelled hopper, plain suction, and suction cutterhead types.
Drift In mining, a horizontal or nearly horizontal tunnel, generally following the vein of material.
Drifter An air percussion drill mounted on a column or crossbar and used for drilling underground.
Drillings The cuttings arising from the process of drilling.
Drilling Log In exploratory holes and blastholes, the detailed record of the rocks passed through in the drilling operation.
Drumlin Rounded oval hill of boulder clay formed by glacier deposition.
Dune Mound or ridge of sand deposited by wind action.
Dunite A hard and heavy intrusive igneous rock of medium color with granitic texture.
Dynamite A high explosive, made in three basic types. *Straight dynamite* contains

nitroglycerin as the principal or only explosive. *Extra dynamite* is, grade for grade, less sensitive to shock and friction than straight dynamite. *Gelatin dynamite* has an explosive base of nitrocotton-nitroglycerin gel, which is insoluble in water, for blasting under wet conditions.

Earth In this text the term is used both for the planet Earth, or the complete globe, and for the softer material making up part of the Earth's regolith, as contrasted to the harder material, rock.

Earth Drill An auger-type drill with bucket used for exploratory drilling in connection with rock-earth excavation.

Earthquake Powder An explosive mixture containing 79 percent nitrate and 21 percent charcoal.

Efficiency, Machine The ratio of time rate of energy output, horsepower, to time rate of energy input, horsepower.

Efficiency, Machine Operating The ratio of actual work time to available work time, expressed as minutes worked per hour, for example, the 50-min hour, or as a percentage, 83 percent.

Elevating Grader A mobile belt loader.

Elevation Vertical distance above or below a plane of reference.

Embankment A fill the top of which is higher than the surrounding surface. An embankment is usually compacted but it may be an uncompacted waste area such as overburden removed from a quarry.

Embayment A deep depression in a shoreline forming a large bay.

Emergence Rise of the land relative to the level of the sea. Generally a land of emergence is gentle and characterized by sandy soils, barrier beaches, and swamps. An example is southern New Jersey.

Energy The capacity to do work. Kinetic energy is that due to motion. Potential energy is that due to position or condition. Total energy is the sum of kinetic and potential energies.

Eocene The second series of rocks in the Tertiary system and the corresponding geologic epoch from 58 to 36 million yr ago.

Epoch That part of geologic time, a subdivision of a period, in which a series of rocks was deposited.

Era The largest subdivision of geologic time corresponding to the group of rocks deposited during the era. Six eras are recognized.

Erosion A general term describing mechanical disintegration, chemical decomposition, and movement of rock-earth materials from place to place on the Earth's surface.

Erratic A transported rock fragment, different from the bedrock beneath it. Agents of transport are water, ice, and wind.

Escarpment A steep slope or cliff. Sometimes called a *scarp,* especially in relation to a fault.

Esker A narrow long ridge, commonly sinuous, of sands and gravels formed by glacial meltwater.

Essential Minerals Those minerals which are necessary to the formation and identification of a rock.

Estimate A statement showing probable quantities and probable costs for proposed work.

Evaporate Nonclastic sedimentary rock precipitated from water solution as a result of evaporation. Rock salt is an example.

Excavation The cutting down of the natural ground surface, the material taken from the excavation, or the space formed by the removal of the material. Also known as *cutting.*

Excavation, Classified Excavation paid for at a unit price for common excavation and a unit price for rock excavation.

Excavation, Common Excavation in earthy materials not requiring blasting.

Excavation, Rock Excavation in rocky materials requiring blasting.

Excavation, Unclassified Excavation paid for at one unit price, whether common or rock excavation.

Exfoliation The peeling off of a rock formation in sheets, generally concentric, due to changes in temperature or to other causes.

Explosives Solid, liquid, or gaseous mixtures or chemical compounds which by chemical

action suddenly generate large volumes of heated gas. The energetic action of the explosives used in blasting rock depends largely on their chemical reactions.
Exposure A body of bedrock not covered by the regolith and forming part of the Earth's surface.
Extrusive Igneous Rock Rock that originated from solidification of molten lava. Also called *volcanic rock.*
Face The more or less vertical surface of rock exposed by excavation or blasting.
Facies Distinctive group of characteristics within a rock unit that differs as a group from those found elsewhere in the same unit.
Fan A fan-shaped body of alluvia built at the base of a steep slope by the action of water. Also known as an *alluvial fan.*
Fanglomerate The materials of a fan which are only slightly worn by the action of water. They are usually made up of coarse particles and characterized by a steep gradient down the slope.
Fault An abrupt break in the continuity of strata or formation of rock by the elevation or depression of the strata or formation on one side of the plane of the fault.
Fault Breccia A breccia consisting of irregular pieces of rock broken as the result of faulting.
Feeder A pushing device or short belt that supplies material to a crusher or belt conveyor.
Feldspar One of the most important groups of rock-forming minerals. Crystalline and light to medium color. Ultimately weathers to clay.
Felsite Extremely fine-grained medium- to dark-colored igneous intrusive rock with granitic texture. Heavy and hard.
Fill The height to which material is to be placed. Also the resulting earthwork. Also called *embankment,* especially if compacted.
Fill, Net In sidehill excavation, the yardage of fill required at any station, less the yardage of cut obtained at the same station.
Fill, Net Corrected Net fill after making allowance for swell or shrink from cut to fill.
Fine Grading The finish grading after rough grading in order to bring the grade to the necessary close elevation.
Fine-Grained Refers to those rocks in which the minerals are less than 0.040 in (1 mm) in diameter.
Fines The finer-grained particles of the soil, clays, silts, and sands.
Finish Grade The final grade called for by the specifications.
Fishing The operation of recovering an object left, lost, or dropped in a drill hole.
Fissile Capable of being split, as schist, slate, and shale.
Fissure A fracture of rocks along which the opposite walls have been pulled apart.
Flint A popular name for chert.
Float The loose fragments or particles of rock, as distinguished from the outcrop or bedrock.
Float Time In the time schedule for a job, the available time in addition to the necessary time for an operation due to possible interference with a simultaneous critical operation.
Floodplain That part of any stream valley which is inundated during floods.
Flow Rate of movement of water in a conduit, in terms of gallons or cubic feet per second or minute, or liters or cubic meters per second or minute.
Flow Breccia An extrusive or volcanic breccia caused by the breaking up of a hardened crust of extrusive rock as a result of a further flow of liquid lava.
Fluvial Of or pertaining to rivers, or produced by river action, as a *fluvial plain.*
Fold A pronounced bend in layers of rock.
Foliation Parallel or nearly parallel structure in metamorphic rocks along which the rock tends to split into flakes or thin slabs.
Fool's Gold Pyrites, a sulfide of iron. Often mistaken for gold because of its resemblance in color.
Foot In compactors, one of a number of projections from the drum which resemble the feet of sheep.
Footcandle (fc) The unit of illumination defined as the illumination on a surface which receives from all light sources, directly or by reflection, 1 lumen per square foot.

Foot Pound The unit of work done by a force of 1 pound when its point of application moves 1 foot in the direction of the force.

Force That which changes the state of rest or motion in matter, measured by the rate of change of momentum.

Formation The ordinary unit of geologic mapping consisting of a large and persistent stratum or strata of some kind of rock. Generally named after the locality where it was first identified.

Formula Any general equation. A rule or principle expressed in algebraic symbols. A method of reasoning stated in the form of an equation. An empirical formula is one derived from experience or experiments and often used in rock-earth excavation.

Fossil The naturally preserved remains or traces of an animal or a plant, generally associated with sedimentary rocks because igneous and metamorphic rocks destroy living things and their remains.

Fossiliferous Containing organic remains or fossils, said of rock-earth.

Foundation The material which supports a structure, cut or fill, whether strengthened or not by piles, mats, or other means to secure adequate bearing.

Fractures Cracks in rocks large enough to be distinctly visible to the naked eye, without regard to definite direction.

Fracture Zone A great linear system of breaks in the Earth's crust, sometimes called a *rift zone*, formed in connection with earthquake activity.

Fragment A rock or mineral particle larger than a grain (⅛ in, or 3 mm, in diameter).

Free Air Air at normal atmospheric conditions; the pressure exerted by a column of 76 cm (29.9 in) of mercury at sea level and 32.0°F (0°C), at standard acceleration of gravity, 32.2 ft/s² (9.80 m/s²).

Freehaul The distance within which material is moved without extra compensation, usually 1000 ft or less.

Freeway A limited-access highway devoted exclusively to the use of motor vehicles, in which all intersections are separated so that no traffic crosses at grade and opposing streams of traffic are separated. Distinguished by no stopping and free-flowing traffic.

Friction Resistance to motion when one body is sliding or tending to slide over another.

Fringe Benefits Those amounts which are added to a basic wage for work done, to take care of such costs as health and welfare, pensions, vacations, and education. As of 1978 they amount to an average 40 percent of basic wage.

Front The working attachment of an excavator, such as the boom of a dragline or the boom, handle, and dipper of a shovel.

Front-End Loader A bucket-type loader, in which the bucket operates at the front of the machine.

Frontal Apron The deposits of alluvia spread out in front of a glacier.

Frost Heaving The lifting of rock waste or human-built structures by the expansion of material during the freezing of the contained water.

Frost Wedging The pushing up or apart of rock particles by the action of ice.

Fumarole A volcano discharging gas nonexplosively. Indicative of latent geothermal energy.

Fusion Drill A drill which burns out a blasthole by means of fuel oil, oxygen, and cooling water delivered to a blowpipe within the hole. Also called a jet piercing drill.

Function A mathematical quantity which has value depending on the values of other quantities, called *independent variables* of the function.

Fuse A thin core of black powder surrounded by wrappings which, when ignited at one end, burns to the other firing end at a fixed speed.

Fuse, Detonating A stringlike core of PETN, a high explosive, contained within a waterproof reinforcing sheath. Primacord is an example.

Gabbro A medium- to dark-colored, generally coarse-grained intrusive igneous rock. Medium to hard and heavy.

Galena An important lead ore. The mineral contains 87 percent lead.

Gallon Unit of liquid volume. The U.S. standard gallon contains 231 in³ and holds 8.34 lb of water. The Imperial standard gallon contains 277 in³ (4.55 L) and 10.02 lb (4.55 kg) of water.

Gangue The nonvaluable minerals of an ore, or the waste materials.

Gantry An overhead structure of an excavator, which supports operating parts.

Gauge Thickness of wire or sheet metal; spacing of tracks, wheels, or tires; or spacing of rails.

Gauge Size The width of a drill bit along the cutting edge.

Geode A hollow nodule or concretion, the cavity of which is lined with crystals.

Geologic Column A composite diagram combining in a single column the succession of all known strata, fitted together on the basis of their fossils or other evidences or determinations of relative ages.

Geologic Cross Section A diagram showing the arrangement of rocks in a vertical plane.

Geologic Cycle The total of all internal and external processes working on the materials of the Earth's crust.

Geologic Map A map showing the distribution at the Earth's surface of rocks and rock formations of various kinds and various ages.

Geologic Record The archive of the Earth's history represented by regolith, bedrock, and the Earth's morphology.

Geologic Time Scale The time relationships established for the geologic column.

Geology The science of the Earth. *Historical geology:* The chronological order and events of the Earth's history, with special attention to organisms and fossils. *Physical geology:* The study of the composition and configuration of the Earth's physical features and rock masses, their relationships to each other, and the surficial and subsurface processes that operate on the Earth. Physical geology is sometimes divided into *structural* and *dynamic geology.*

Geomorphology Physical geology which deals with the form of the Earth, the general configuration of its surface, the distribution of land and water, and the changes that take place in the evolution of the landforms.

Geophysical Exploration Exploration to determine subsurface conditions based on the distribution within rocks of some physical property such as specific gravity, magnetic susceptibility, electrical conductivity, elasticity, or seismic characteristics.

Geyser An orifice in the ground that erupts steam, boiling water, and mud intermittently.

Giant A nozzle, manually or mechanically controlled, for directing a jet of water for hydraulicking. Also known as a *monitor.*

Gilsonite A hard, brittle native asphalt which is mined like coal.

Glacial Drift Any rock material, such as alluvia, transported by a glacier and deposited by the ice or by the meltwater from the glacier.

Glaciation The alteration of the land surface by the massive movement of glacial ice. The plucking, polishing, and striation of rocks by glacial action.

Glacier A body of ice, consisting mainly of crystallized snow, flowing on a land surface.

Glance A term used to describe various minerals with a splendent luster, such as silver glance and lead glance.

Glassy Texture The texture of some extrusive igneous rocks such as lavas which have cooled so quickly that they have not undergone crystallization. An example is obsidian.

Glory Hole A kind of open-pit mine in a large, vertical-sided pit. The material is fed by gravity through a funnel-shaped hole in the floor of the pit to hauling units located in an adit beneath the pit.

Gneiss A metamorphic coarse-grained rock breaking along irregular surfaces, generally with alternating layers of light- and dark-colored minerals. Hard and heavy. An example is granite gneiss, an altered granite.

Gorge A small canyon or narrow passage between hills.

Gossan A term referring to the weathered, iron-stained outcrop of a mineral deposit. An important indication of the presence of the ore.

Gouge Finely pulverized rock flour, sometimes claylike, caused by the grinding action between the walls of a fault.

Graben The depressed land between two faults, generally a long narrow debasement and typical of a basin range province.

Grade (1) The profile of the center of a roadway, canal, or like structure. (2) To prepare the ground by cutting and filling according to a definite plan. (3) The gradient, that is, the percentage of rise or fall to the horizontal distance.

Grade Assistance The assistance to a machine or body when descending a grade caused by the force of the inclined component of the weight of the machine or body.

Grade Resistance The resistance to a machine or body when ascending a grade caused by the force of the inclined component of the weight of the machine or body.

Grade Stake A stake indicating the amount of cut or fill in feet at a stated point.

Grain A mineral or rock particle having a diameter less than 1/8 in, or 3 mm.

Granite A coarse-grained light to dark intrusive igneous rock. Soft to hard and medium- to heavyweight.

Granite, Decomposed (Sometimes called DG.) Well-weathered granite in which some of the minerals have altered and softened, weakening the rock.

Granodiorite Coarse-grained light- to medium-colored rock intermediate between intrusive igneous granite and diorite. Soft to hard and medium- to heavyweight.

Granular Textural term referring to uniform size of grains or crystals of a rock, particles ranging in size from 0.002 in (0.05 mm) to 0.25 in (0.6 cm).

Grapple A clamshell-type bucket, sometimes with heavy tongs, having three or more jaws and generally used to handle rock such as riprap.

Gravel Inclusive term for materials resulting from the natural erosion of rock and ranging in size from coarse sand to cobbles, that is, from 1/8 in (3 mm) to 8 in (20 cm) in diameter.

Gravel, Pit Run The mixture of alluvia and foreign materials as they occur in any natural deposit. The source of prepared gravels.

Graywacke An old term applied to metamorphosed shaly sandstones that yield a tough, irregularly breaking rock different from both slate and quartzite.

Grid A set of surveyor's closely spaced reference lines laid out at right angles with elevations taken at intersections of the lines. The framework is used to calculate quantities of excavation.

Grit A coarse sand or sandstone formed mostly of hard angular quartz grains.

Grizzly A coarse screen or set of rails used to remove oversize pieces from excavation.

Ground Moraine The irregular sheet of till deposited partly beneath an advancing glacier and partly from the ice when it melts away.

Ground Pressure The weight of a machine divided by the area of the ground directly supporting the machine.

Group One of the six major divisions of rocks, corresponding to a geologic era.

Grouser A heavy lug attached to crawler tread plates and tractor wheels to obtain better traction in soft ground.

Grubbing Removing the stumps and roots from an area to be excavated.

Gypsite A dirty variety of gypsum with some impurities. A sedimentary rock, grayish in color and of medium hardness and medium weight.

Gypsum Hydrous calcium sulfate evaporate. Massive light-colored sedimentary rock. Medium to hard and heavy.

Gyratory Crusher A primary crusher having a central conical breaker with an eccentric motion contained in a chamber of circular cross section tapering from a wide top opening.

Half Track A heavy-duty truck with a high-speed crawler-track drive in the rear and steering wheels in the front.

Hammer, Air A machine hammer driven by compressed air, as a jackhammer.

Hand Level A small instrument consisting of a telescope and a leveling bubble tube for sighting a horizontal line.

Handle In a shovel or backhoe, the arm connecting the boom to the dipper. Same as *stick*.

Hanging Valley A valley the floor of which is notably higher than that of the valley or shore to which it leads. Hanging valleys are usually tributary to a glaciated main valley.

Hardness, Mineral A measure of the cohesion of the surface particles as determined by the capacity of the mineral to scratch another mineral. Minerals are referred to a scale of hardness of 10 minerals in ascending order of hardness. Each mineral will scratch the one below it on the scale, diamond being the hardest. The hardness ratings of this Mohs scale are: talc 1, gypsum 2, calcite 3, fluorite 4, apatite 5, orthoclase 6, quartz 7, topaz 8, corundum 9, diamond 10.

Hardness, Rock An arbitrary, relative term to define the resistance to fragmentation, either by ripping by heavyweight tractor-ripper or by blasting. The 1978 standards in terms of ripping by tractor-ripper in the 100,000-lb (45,000-kg) to the 160,000-lb (73,000-kg) weight class and in the 380- to 530-flywheel hp class are based on the range of seismic shock-wave velocities in the rock.

	Range of velocities				
	feet/second			meters/second	
No ripping	To		1500	To		460
Soft ripping	1500	to	4000	460	to	1260
Medium ripping	4000	to	5000	1220	to	1520
Hard ripping	5000	to	6000	1520	to	1830
Extremely hard ripping or blasting	6000	to	7000	1830	to	2130
Blasting	7000		Up	2130		Up

Hardpan A stratum of material accumulation that has been thoroughly cemented into a hard rocklike layer which will not soften when wet.

Hardwood Wood from a tree which is sturdy, with a texture of growth not given to breakage. Typical are oak, hard maple, and hickory.

Haul The distance material is moved from the loading point to the point of disposition.

Haul, Average The average distance material is hauled from the cut to the fill. This distance is sometimes calculated by means of a mass diagram.

Haul, Free See *Freehaul*.

Haul, Over See *Overhaul*.

Head The height of water above a specified level, consisting of the sum of static head, velocity head, and friction head.

Heading A collection of joints in a rock formation.

Heap The load of rock-earth carried above the sides of the hauler or the bucket of an excavator.

Hectare A metric measure of area, equivalent to 10,000 m^2, or 2.47 acres.

Hematite A common iron ore. The mineral contains 70 percent iron.

High Explosive An explosive which decomposes with extreme rapidity.

Highway A main road or thoroughfare with a mixture of traffic and without grade separations and free-flowing traffic as in the case of a freeway.

Hogback Ridge formed by the outcropping edge of tilted strata of rock.

Hoist A machine for lifting weights and loads, or the part of a machine which performs the hoisting operation, such as the hoist for the bucket of a dragline.

Holocene The second and latest series of rocks of the Quaternary system and also the corresponding geologic period, which commenced with the Neolithic age about 12,000 yr ago and extends to the present.

Hopper Body The body of a hauler which permits dumping the load through the bottom hopper doors or gates. Sometimes known as a bottom dump.

Hornfels A fine-grained, compact, highly metamorphosed rock, the original texture of which has been completely obliterated by contact metamorphism with nearby intruded magma. Derived from shale or other sedimentary rock; light to dark in color, heavy and hard.

Horse A large fragment of rock broken from one block and caught between the walls of the fault.

Horsepower A unit of power equal to a rate of 33,000 ft·lb/min (4560 m·kg/min) or 550 ft·lb/s (76 m·kg/s). The unit is actually about 1.5 times the output of a continuously working 1000-lb horse.

Horsepower, Engine Rated The average horsepower developed by an engine when operating at full throttle at rated speed and at specified altitude and temperature. Working horsepower is that delivered at the flywheel with full engine accessories for the working conditions.

Horsepower, Wheel The average horsepower delivered at the points of contact between the driving wheels and the ground surface. This horsepower equals the flywheel horsepower less the power lost through the power transmission assemblies. In a wheels-tires machine, such as a rear-dump truck, wheel horsepower equals approximately 0.76 times flywheel horsepower.

Horsepower, Drawbar The average horsepower delivered to the drawbar of a crawler-tractor. This horsepower equals the flywheel horsepower less power lost through the power transmission assemblies and to the rolling resistance of the tractor when tested under standard conditions of rolling resistance of 110 lb per ton of tractor weight. Drawbar horsepower equals approximately 0.69 times flywheel horsepower.

Horst The elevated land between two faults, generally a long narrow block and typical of a basin range province.

Humus The dark decomposed residue of plant and animal tissues.

Hydration The reaction of a compound with water to form a hydrate.

Hydraulic Fill A fill built by transporting the material by water. The material may be moved in an open natural channel or it may be pumped by a dredge.

Hydraulicking Use of water under high pressure directed against the face of the material to be moved. The dislodged material is carried away by the water, as in the case of overburden from deposits of rock and ore.

Ice Age The Pleistocene epoch, during which the four glacial periods occurred.

Igneous Rock Rock formed by the solidification of molten silicate materials.

Ilmenite An important iron and titanium ore. The mineral contains 36 percent iron and 32 percent titanium.

Impeller The member of a centrifugal pump using centrifugal force to discharge a fluid into the outlet passages.

Impervious Resistant to the percolation or absorption of water.

Incipient Erosion The early stages of erosion, especially with reference to gullying.

Inclined Plane A slope used to change the direction and the speed/power ratio of a force.

Indicated Horsepower The mechanical power developed in a steam cylinder by the steam working against the piston.

Indurated Hardened, as a rock hardened by heat, pressure, or the addition of some ingredient not commonly contained in the rock.

Inertia The property of matter by which it will remain at rest or in uniform motion in a straight line unless acted on by an external force.

Indigenous Rocks Rocks formed by magmas, the source of which has been the Earth's interior. Granite is an example. Same as intrusive igneous rocks.

Inhaul The line or mechanism by which a cable excavator bucket is pulled toward the dumping point.

Inlier An outcrop of rock surrounded on all sides by geologically younger rocks.

In situ As applied to rock or earth, in the natural position in which it was originally formed or deposited.

Intake That portion of a pipe, pump, or structure through which water or air enters from the source of supply.

Interface The boundary between two rocks or formations with different physical characteristics.

Intermontane Lying between two mountains.

Internal Friction The resistance of rock or earth particles to sliding over each other.

Intrusion A mass of igneous rock which, while molten, was forced between other rocks.

Intrusive Rocks Term applied to a rock which has been forced into other rocks. Sometimes called *plutonic*. Granodiorite is an example.

Invert The inside bottom of a pipe, trench, or tunnel.

Itacolumite A flexible sandstone which is composed of quartz grains, muscovite, talc, and a few other minerals. Peculiar to North Carolina and a few other places.

Jackhammer An air drill which can be operated by one person.

Jars A tool in the string of tools of a cable drill which contains slack to allow upward hammering to free a stuck drill bit.

Jaw Crusher A primary crusher with a fixed and an oscillating jaw widely spaced at the top and closely spaced at the bottom and a mechanism to move the oscillating jaw to and from the fixed jaw.

Jetty A long fill or structure extending into the water from the shore which serves to change the direction or velocity of flow of water.

Joint A fracture on which no appreciable movement parallel with the fracture has occurred. Common rock joints are caused by flexure in sedimentary rocks and by cooling in igneous rocks.

Joint System A combination of two or more intersecting sets of parallel joints.

Jurassic The middle of the three rock systems of the Mesozoic group and the corresponding geologic period, from 180 to 135 million years ago.
Kame A body of stratified drift material due to ice contact and shaped as a short, steep-sided knoll. Glacier-formed.
Kaolin A white nonplastic sedimentary rock formed from the residual minerals of granite. Soft and medium-weight.
Karst Topography Topography developed in a region of easily soluble limestone bedrock, characterized by sinkholes, caverns, and underground streams.
Kelly A square or fluted pipe which is turned by a rotary drill table while it is free to move up and down through the table. It carries the rotary bit.
Kettle A closed depression in drift material created by the melting of a mass of underlying ice.
Kieselguhr German name for *diatomaceous earth*, a filler or carrying agent in high explosives.
Kilogram A metric unit of weight, 1000 g, equivalent to 2.205 lb.
Kilometer A metric unit of length, 1000 m, equivalent to 0.621 mi.
Kilovolt-Ampere An electrical unit of power, equivalent to about 0.89 kW.
Kilowatt Metric and English unit of power, 1000 W, equivalent to 1.34 hp.
Kinetic Energy Energy due to motion.
Knife The cutting edge of some excavating machines.
Kyrock Bituminous silica sandstone from Kentucky, containing up to 12 percent natural bitumen. Dark, medium-hard, and heavy.
Laccolith Lenticular intrusion of igneous rock into a sedimentary rock formation, generally with a plane bottom and a domed top.
Ladder The digging-boom assembly of a hydraulic dredge or a chain-bucket-type excavator.
Lamina A stratum of less than ⅜ in (1 cm) thickness.
Laminations The banding of rocks caused by variations in different minerals, usually distinguishable by different colors. Sometimes laminations result in weakened planes between them.
Landforms Physical features of the surface of the land, such as hills and valleys, developed by the processes of aggradation and degradation.
Lapilli Particles of dust and small rock fragments ejected from volcanoes, with diameters ranging from ⅛ in (3 mm) to 1¼ in (32 mm).
Lapilli Tuff Medium-grained pyroclastic rock consisting of fused lapilli. Light to medium color, soft to hard, and medium-weight.
Laterite Reddish residual, product of weathering of soft rocks rich in iron and aluminum.
Latite A rock that is intermediate in properties among the igneous extrusive rocks such as rhyolite, andesite, and basalt. Medium to dark color, medium to hard, and medium-weight.
Lava General name for the magma outpourings of volcanoes; an igneous extrusive rock.
Leaching The continuous removal by water of soluble matter from regolith or bedrock.
Lead Wire In blasting circuits the heavy wire that connects the blasting machine or source of current with the connecting or cap wires.
Ledge A bedding or several beddings of rock, as in a quarry. An outcropping of horizontal or nearly horizontal rock.
Levee A rock or earth embankment to prevent inundation or erosion.
Leverman One who operates the controlling levers of a machine; an operator.
Life An appraisal term defined as the mean or expected duration of the life of a property.
Lift The depth of a cut taken out or of a fill placed during a cycle of excavation.
Lignite A brownish black soft coal, intermediate in composition between peat and bituminous coal. A sedimentary rock. Lightweight.
Lime Rock A natural sedimentary rock composed essentially of calcium carbonate with varying percentages of silica. Light to medium color, medium soft, and medium-weight.
Limestone A sedimentary rock consisting predominantly of calcium carbonate. Light to medium color, hard, and heavy.
Limonite An iron ore of hydrous oxide of iron. Brownish rust color when powdered. Soft and heavy. Mineral contains 86% iron.

Lip The cutting edge of a bucket or dipper. Applied chiefly to edges, including tooth sockets.
Liter Metric unit of capacity, equivalent to 1000 cm^3, 61.0 in^3, or 0.908 U.S. quart.
Lithification Rock formation. The conversion of sediments to sedimentary rock.
Load The placement of explosives in a blasthole.
Load Factor Average load carried by an engine, machine, or plant, expressed as a percentage of its maximum capacity.
Loam Earth having a relatively even mixture of clay, silt, and sand and generally containing a considerable portion of organic matter. Also called *topsoil*.
Local Metamorphism Contact metamorphism as distinguished from regional metamorphism.
Location The centerline and grade line of an engineering structure such as a highway, preparatory to its construction.
Loess Wind-deposited silt, usually accompanied by some clay and fine sand. Somewhat consolidated so that it maintains a steep bank of cut.
Log In geology and construction, the detailed record of the rocks passed through during a drilling operation.
Loose Yards Cubic yards loose measurement, a unit of volume of excavation.
Low Explosive An explosive which decomposes more slowly than high explosives and by deflagration rather than detonation.
Lumen (lm) The unit of light flux in terms of which the output of light sources is expressed, 1 lumen per square foot equaling 1 footcandle.
Luster The quality and the intensity of light reflected from a mineral or rock.
Machine An apparatus for applying mechanical power which has several parts, each with a different function.
Macrostructure A structural feature of rocks that can be discerned by the unassisted eye or with the help of a simple magnifier.
Mafic Rock A rock in which ferromagnesian minerals exceed 50 percent. Gabbro is an example.
Magazine A structure or container in which explosives are stored.
Magma Liquid molten rock, consisting of hot silicic solutions containing water and gases, from which igneous rocks are formed.
Magnetite An important magnetic iron ore. Black, hard, and heavy. The mineral contains 72 percent iron.
Malachite An important ore of copper. Greenish in color, medium hard, and medium-weight. The mineral contains 72 percent cupric oxide.
Mantle Rock The loose and more or less consolidated weathered rock resting on the bedrock.
Map Representation to a definite scale on a horizontal plane of the physical features of a portion of the Earth's surface.
Marble A hard, light-colored metamorphic rock derived from limestone or dolomite. Light color, hard, and heavy.
Marine Sediment Sediment deposited in the sea. Also the sedimentary rocks of the sea.
Marl An earthy mixture of calcium carbonate and clay. Light color, soft, and medium weight. A common lake deposit.
Mass Quantity of matter. Equal to weight divided by the acceleration of gravity.
Mass Diagram In earthwork calculations, a graphical representation of the algebraic cumulative quantities of cut and fill along the centerline, where cut is positive and fill is negative. Used to calculate haul in terms of station yards.
Massif A single mountainous mass of rock, which may be considered a unit.
Massive As applied to rock, homogeneous in structure, without stratification, flow-banding, foliation, schistosity, or the like.
Mass Shooting Simultaneous exploding of charges in all of a large number of blastholes, as contrasted with sequential firing with delay caps.
Mass Wasting The gravitative movement of rock debris downslope without the help of another agency such as water, ice, or wind.
Mean A quantity having an intermediate value among several others of which it expresses the average, obtained by dividing the sum of the quantities by their number.
Meander A looplike bend in the channel of a stream.
Mechanical Efficiency The ratio of the energy or work of the output of a machine to the energy of the input.

Mechanics The science of force and its effect on matter.
Medium-Grained Rock Rocks whose crystals range in diameter from 1/32 in (1 mm) to 3/16 in (5 mm).
Mesa A high, broad, flat tableland, bounded on at least one side by a steep cliff rising from the lower land.
Mesozoic The fifth of the groups of rocks preceding the present Cenozoic group. Also the corresponding geologic era, from 230 to 63 million yr ago.
Metamorphic Rock A rock which has been changed by temperature, pressure, and/or the chemical action of fluids into a new form which is more stable under the new conditions. Examples are gneiss from granite, quartzite from sandstone, and slate from shale.
Meteorite A particle of solid matter from outer space that has fallen to the ground from the atmosphere.
Meter A metric measure of length, equivalent to 3.28 ft.
Metric Ton A unit equivalent to 1000 kg, 2205 lb, or 1.102 short tons.
Mile Yard The movement of 1 yd^3 of excavation through a horizontal distance of 1 mi. A measure of payment for such excavation.
Millimeter A metric measure of length, equivalent to 0.0394 in.
Millisecond Delay A type of delay cap with a definite but extremely short interval between passage of current and detonation, usually in increments between 12 and 700 ms (0.012 and 0.700 s).
Mineral A naturally occurring inorganic substance of definite chemical composition and physical properties which is a constituent of rocks.
Mineralogy The science or study of minerals.
Miner's Inch An old measure of the flow of water as related to mining. The accepted value of one hundred years ago is the quantity of water flowing through an orifice 1 in square under a head of 6.5 in in 1 min. This flow is equivalent to 1.53 ft^3/min or 11.4 gal/min.
Mining The removal of rock or earth having value because of its chemical composition.
Miocene The fourth series of rocks in the Tertiary system. Also the corresponding geologic epoch, from 25 to 13 million yr ago.
Misfire Failure of all or part of an explosive charge to detonate.
Mississippian The first series of rocks of the Carboniferous system. Also the corresponding geologic epoch from 340 to 310 million yr ago.
Moil Point A short length of drill steel sharpened to a conical point used with a jackhammer for breaking rock or concrete or for hole-punching work.
Moisture Content The percentage weight of water in a given weight of rock, earth, or rock-earth.
Moldboard A curved surface of a plow, bulldozer blade, motorgrader blade, or other excavator which gives the material moving over it a rotary, spiral, or twisting movement.
Monadnock A conspicuous residual hill on a peneplane or leveled plain.
Monitor In hydraulicking, a high-pressure nozzle mounted in a swivel frame. Also called *giant*.
Monzonite A light- to medium-colored extrusive igneous rock intermediate between syenite and diorite. Medium-hard and medium-weight.
Moraine An accumulation of drift deposited by a glacier. If deposited along the side or sides of the glacier, it is a *lateral moraine*. If deposited at the end, it is a *terminal moraine*.
Motor In this text, a rotating machine which transforms electrical energy into mechanical energy. Not to be confused with an internal combustion engine, which transforms fossil fuel into mechanical energy.
Motor-Generator A generator propelled by an electric motor, known as an MG set. An example is the MG set of the Ward Leonard system for transforming ac power to dc power to mechanical power by means of a dc motor.
Mountain In a general sense, any land mass or landform which stands conspicuously above its surroundings. Geologically, it refers to parts of the Earth's crust having thick, crumpled strata, metamorphic rocks, and granitic batholiths.
Muck Soft mud containing vegetable matter. Also a general term referring to all kinds of excavation, including rock, with particular application to the excavation for a tunnel.
Mud Generally, any earthy material containing enough water to make it soft.

Mudflow Flow of a torrent so heavily charged with earth and debris that the mass is thick and viscous. Blocks of rock many feet in dimension can be transported vast distances.

Mudcapping Blasting boulders by means of explosives placed on the surface and covered with mud, the inertia of which intensifies the action of the explosives.

Net Cut In sidehill excavation, the cut required less the fill required at a particular station or part of the centerline.

Net Fill In sidehill excavation, the fill required less the cut required at a particular station or part of the centerline.

New Construction Term used in the economics of construction which applies to projects on entirely new locations or to reconstruction projects in which there is no salvage value for the existing construction.

Nitroglycerin A powerful liquid explosive, which is dangerously unstable unless combined with other materials such as kieselguhr (diatomaceous earth).

Normal Fault A generally steeply inclined fault in which the hanging wall block appears to have moved relatively downward.

Normal Haul A haul the cost of which has been included in the cost of excavation, so that no additional charge is made for it.

Obsidian An extrusive igneous rock that has cooled rapidly so as to produce a glasslike surface. Generally black. Hard and heavy.

Obsolescence The factor in depreciation of machinery resulting from changes in methods or design which makes the machinery less desirable or valuable for continuation of work.

Oil Shale A compact sedimentary rock containing organic matter, which yields oil when destructively distilled. Dark, medium-hard, and heavy.

Oligocene The third series of rocks in the Tertiary system. Also the corresponding geologic epoch, from 36 to 25 million yr ago.

Oolitic Limestone Limestone consisting largely of minute spherical grains of calcium carbonate resembling fish roe. Light in color and of medium hardness and medium weight.

Open-Pit Quarry or Mine An excavation in which the working area is kept open to the sky. Used to distinguish the method of quarrying or mining from underground work.

Operating Costs Machinery costs such as repairs and replacements, including labor, and fuel, oil, and lubricants. Sometimes operator's wages and fringe benefits are included in the cost but usually they are kept separate, as in this text.

Operator One whose work is to operate a machine.

Optimum Moisture Content The moisture content in percent by weight of dry rock-earth which results in the least voids or greatest density when the rock-earth is compacted.

Ordovician The second of the five systems of rocks in the Paleozoic group. Also the corresponding geologic period, from 500 to 430 million yr ago.

Ore A mineral or association of minerals that may under favorable circumstances be worked commercially for extraction of one or more of the minerals.

Original Horizontality The principle which states that most strata are nearly horizontal when originally deposited.

Orogeny Another term for mountain making. The deformation of the Earth's crust in the development of mountains.

Outcrop Underlying bedrock which comes to the surface of the ground and is exposed to view.

Outlet That portion of a pipe, pump, or structure through which water or air leaves.

Outlier An outcrop of rock surrounded on all sides by geologically older rocks.

Outwash Stratified drift deposited by streams of meltwater as they flow away from a glacier.

Overbreak Moving or loosening of rock beyond the intended line or plane of cut as a result of blasting.

Overburden Rock-earth mantle, waste material, or other matter found directly above the deposit of material to be excavated.

Overhang Projecting parts of a bank, face, or high wall.

Overhaul Transportation of excavation beyond certain specified limits known as *free-haul distances*.

Overhead Those indirect job and home-office expenses which cannot be charged to individual costs or bid items except by proration, as they are incurred in the maintenance of the business in general.

Overhead Shovel A bucket loader which loads at one end, swings the bucket overhead, and dumps at the other end. Used in closely confined areas, as in tunnels.

Overthrust The lateral thrusting of a mass of rock over or upon other rocks along a thrust fault.

Ownership Costs Machinery costs which are more or less fixed, such as depreciation, interest, taxes, insurance, storage, and replacement cost escalation.

Packsand A very fine-grained sandstone so loosely cemented by a slight calcareous cement as to be readily cut by a spade. A light- to medium-colored, soft, medium-weight sedimentary rock.

Pad Ground contact part of a crawler-type track. Also called a *shoe* or *plate*.

Paleocene The first of the five series of rocks of the Tertiary system. Also the corresponding geologic epoch, from 63 to 58 million yr ago.

Paleozoic The fourth of the six major groups of rocks. Also the corresponding geologic era, from 570 to 230 million yr ago.

Pan Name for a carrying scraper.

Parallel An arrangement of electric blasting caps in which the firing current passes through all of them at the same time.

Parallel Series Two or more series of electric blasting caps arranged in parallel.

Particle A general term referring to any size from microscopic mineral grains to huge rocks. Equivalent to an entity.

Parting A surface of separation within a rock body.

Pass A working trip or passage of an excavating machine.

Pass A defile or passageway through rough terrain.

Pay Formation A body of rock, earth, or ore the value of which is enough to justify excavation.

Pay Items Units in a contract which are covered by the specifications as bid items and listed as separate units for payment.

Payload The load of excavating machinery for which payment is made. In terms of cubic yards it may be bank measurement (cybm), loose measurement (cylm) or compacted measurement in the embankment (cycm). The load may be also in terms of short tons as in the case of borrow-pit excavation or in short or long tons as in the case of quarrying or mining.

Pea Gravel Clean gravel the particles of which equal peas in size, about $3/16$ in (5 mm) in diameter.

Peat A brownish lightweight mixture of decomposed plant tissues in which parts of the plant are easily recognized. The beginning of formation of coal.

Pebble Sediment particles having diameters between $1/16$ in (2 mm) and $2\frac{1}{2}$ in (64 mm).

Pediment A sloping surface, cut across bedrock, adjacent to the base of a highland in an arid country.

Pegmatite A very coarse-grained granite occurring in irregular dikes or lenses in granites and some other rocks. Light to dark in color, hard, and heavy.

Peneplane Almost a plain. A land surface worn down to very low relief by streams and mass wasting.

Penetration The rate of drilling at 100 percent efficiency with the drill bit working continuously. Expressed in feet per minute for fast drilling and inches per minute for slow drilling.

Pennsylvanian The second of the three series of rocks in the Carboniferous system. Also the corresponding geologic epoch, from 310 to 280 million yr ago.

Perched Water Table The upper surface of a body of free groundwater in a zone of saturation separated by unsaturated material from an underlying body of groundwater in a differing zone of saturation.

Peridotite An ultramafic dark intrusive igneous rock. Hard and heavy.

Perlite Glassy extrusive or volcanic igneous rock of rhyolitic composition. Light to medium color, hard, and lightweight.

Permian The third and last series of rocks in the Carboniferous system. Also the corresponding geologic epoch, from 280 to 230 million yr ago.

Petrology The study of rocks.

Phosphate Rock A sedimentary rock composed chiefly of calcium phosphate, along with impurities such as clay and lime. Medium color, medium hardness, and medium weight.

Phyllite A group of rocks associated with slate and sometimes called *leaf stone*. Of metamorphic origin. Medium to dark color, hard, and heavy.

Pi, π A number, approximately 3.1416, used in the determination of the properties of the circle and the sphere.

Pile A steel or wooden member usually driven or jetted into the ground and deriving its support from the underlying rock-earth and by the friction of the rock-earth on its surface. In earthwork its usual function is to form a wall, as in a cofferdam, to exclude water and soft material from the area of excavation.

Pillow Lava Ellipsoidal mass of extrusive igneous rock formed by the extrusion of lava under water. Dark, hard, and heavy.

Pioneering The first working over of rough or overgrown areas prior to excavation.

Pioneer Road A semipermanent road built along the job to provide means for moving workers and machinery.

Pit An open excavation, deep with respect to area.

Placer A deposit of heavy minerals concentrated mechanically, as by stream flow, wave action, or wind.

Placer Diggings Areas where placer mining has overturned or removed the earth and left a rough, eroded, and scarred surface.

Planation The bringing of a surface of the Earth to level or plane condition by natural or artificial means, as the complex process by which a stream develops its floodplain.

Planimeter An instrument for measuring the area of a plane figure by passing a tracer around the boundary of the area.

Plastic Limit The minimum amount of water, measured in percent based on the dry weight of material, that will make the material plastic.

Plastic Material A material that can be rolled into strings ⅛ in (3 mm) in diameter without crumbling.

Plateau An extensive upland underlain by essentially horizontal strata and having large, nearly flat areas.

Playa The dry bed of a lake on the nearly level floor of an intermontane basin.

Pneumatic Powered or inflated by compressed air.

Pleistocene The first of two series of rocks in the Quaternary system. Also the corresponding geologic epoch, from 2 million to 12,000 yr ago.

Pliocene The fifth and last series of rocks in the Tertiary system. Also the corresponding geologic epoch, from 13 to 2 million yr ago.

Plucking The process by which a glacier pulls away or quarries blocks of rock of considerable size from its channel walls.

Pluton Any body of intrusive igneous rock.

Point A general term to describe the teeth and the bits of excavating machinery.

Pontoon A float supporting part of a structure, as that which supports the pipe of a dredge from dredge to shore.

Poorly Sorted Sediments A sediment consisting of particles of many sizes.

Pore Spaces The spaces or voids in a body of rock or sediment which are unoccupied by solid materials.

Porphyritic Texture A texture of igneous rock in which some particles are conspicuously larger than the rest of the particles.

Porphyry An igneous rock with phenocrysts, the largest particles, making up more than 25 percent of the volume.

Potato Dirt A general term for any rock-earth which is easy to dig.

Pound-Foot The same as *foot-pound*.

Powder A general term for any low explosive such as black powder and earthquake powder.

Power The time rate at which work is done, expressed by horsepower or kilowatts.

Power Control Unit One or more winches mounted on the tractor and used to manipulate the working parts of bulldozers, loaders, scrapers, and other machines.

Power Train All moving parts connecting a prime mover with the point or points where work is performed.

Precambrian Rocks Rocks which are older than those of the Cambrian system. These rocks include the Azoic, Archeozoic, and Proterozoic groups. The demarcation marks the beginning of real fossil life, important in the classification and the correlation of rocks. From 4500 to 570 million yr ago.

Primacord Trademarked name for a a detonating fuse.

Prime Mover A machine to pull other machines. An engine or motor to drive a machine or machines.

Primer A high explosive and blasting cap used to initiate the explosion of a blasting agent. A primer is used in conjunction with ammonium nitrate explosives, which are in themselves inert.

Profile The intersection of a vertical plane through the centerline with surface of the ground and the plane or planes of the finished earthwork, or a drawing indicating the same, so as to indicate grades and distances, depth of cut, and height of fill for the earthwork.

Proterozoic The third of the six major groups of rocks. Also the corresponding geologic era, from 1200 to 570 million yr ago.

Province An area throughout which geologic history has been essentially the same over a long period of time or which is characterized by particular structural or physiographical features.

Puddingstone A sedimentary conglomerate rock in which the pebbles are rounded. Of light to medium color, hard, and medium-weight.

Puff Blowing Blowing chips out of a drill hole by means of exhaust air from the drill.

Pumice Extremely vesicular frothy natural glass. An extrusive igneous rock. Of light to medium color, sharp, and lightweight.

Push-Tractor A crawler-tractor equipped with a push block which loads push-loaded scrapers.

Pyroclastic Rock Fragmental extrusive igneous rocks, such as tuffs and breccias, produced by volcanic explosions. Sometimes classified as sedimentary rock. Light to dark in color, hard, and medium- to heavyweight.

Quantity The amount of material to be excavated and handled, expressed in the prescribed unit, usually cubic yard or ton.

Quarry A deposit of rock from which material is excavated for broken stone, crushed stone, or dimension stone.

Quartzite A metamorphic rock, derived from quartz sandstone. Light in color, hard, and heavyweight.

Quartz Monzonite A kind of granite in which the light-colored minerals such as the feldspars predominate. Light to medium in color, hard, and heavy.

Quaternary The last of the two systems of rocks in the Cenozoic group. Also the corresponding geologic period, from 2 million years ago to the present.

Quicksand A mass of silt and fine sand, thoroughly saturated with water, forming a semifluid.

Rake, Brush A rake blade for a crawler-tractor having a high top and light construction.

Rake, Rock A heavy-duty rock blade for a crawler-tractor equipped with teeth along the cutting edge.

Ramp An incline connecting two levels, such as a ramp connecting two working benches in a quarry.

Raveling The loosening and falling of materials from a bank or face.

Ravine A deep, more or less linear, depression or hollow worn by running water.

Receiver The air tank or reservoir on an air compressor.

Reclaiming Removing material from a stockpile.

Reef A rocky ledge or bar of sand under water in a stream channel.

Regional Metamorphism Extended metamorphism of rocks that, as contrasted to contact metamorphism, extends over a large area.

Regolith The noncemented rock fragments and materials which overlie the bedrock in most places. Regolith is of two kinds, residual and deposited.

Relief The difference in elevation between the high and the low parts of a land surface.

Relief Holes Holes drilled closely along a line which are not loaded but which serve to weaken the rock so that it breaks along the line when blasted.

Relocation A new alignment varying from the original location, generally of highways and railroads.

Residuals Earth and rock fragments formed in situ by the weathering of rocks and left as an overburden upon the country rock.

Reverse Fault A generally steeply inclined fault along which the hanging wall has moved relatively upward.
Revolving Shovel A shovel in which the upper works or deck revolves independently of the lower works about a centerpin.
Rhyolite A fine-grained extrusive igneous rock of light to medium color. Medium to hard and medium-weight.
Rift Zone The fractured and brecciated zone between the walls of a fault.
Right Bank That bank of a stream, reservoir, or dam which is on the right when one looks in the direction in which the current flows.
Right-of-Way The land or water rights necessary for a construction project.
Ripping The fragmentation of rock by a crawler-tractor equipped with ripper shanks and points.
Riprap Heavy rocks placed to form a revetment to prevent stream erosion, a seawall to prevent wave erosion, or a jetty to protect the entrance to a harbor or for some similar purpose. The rocks vary generally in size from 2 ft^3 to several cubic yards and in weight from about 300 lb to 10 tons.
Riverwash Alluvial deposits in streambeds and flood channels, subject to erosion and deposition during recurring flood periods.
Roadbed For a railroad, the finished surface of the roadway upon which the ballast rests. For a highway, the finished surface of the roadway between shoulder lines.
Roadway For a railroad, that part of the right-of-way prepared to receive construction of ditches, shoulders, and roadbed. For a highway, the entire construction area for the highway and its appurtenances.
Rocks A mass of material, loose or solid, which makes up an integral part of the Earth.
Rock Avalanche The extremely rapid downslope flow of a mass of dry rock particles.
Rock Cleavage Closely spaced partings in rocks controlled by platy particles that have been aligned in response to pressures within the Earth's crust.
Rock Cycle That part of the geologic cycle concerned with the creation, destruction, and alteration of rocks during erosion, transport, deposition, metamorphism, plutonism, and volcanism.
Rock Excavation The fragmentation, loading or casting, hauling, and depositing of rock or rock-earth.
Rock Fall The rapid descent of a rock mass, vertically from a cliff or by leaps down a slope.
Rock Slide The rapid descent of a rock mass down a slope.
Rolling Resistance A measure of the force which must be overcome to pull or roll tracks or wheels-tires over the ground. It is affected by ground conditions and the total weight of the machine, and is expressed for different ground conditions in terms of pounds per ton of weight of machine and in percentage of weight of machine.
Rotary Drill A drill using a rotary bit such as a simple fishtail type or a complex rotary-cone type.
Rotational Firing Blasting a block of rock nearest the face with a first explosion and afterwards timing other blasts successively so as to throw their burdens toward the space created by the preceding blast. Equivalent to *row shooting* or *buffer blasting*.
Rough Grading First stage of excavation, when the grade in the cut and fill is held to about ± 0.1 ft prior to the finish grading by a motor grader or other finishing machine.
Round A blast, including a succession of delay shots.
Row Shooting In a large blast, setting off the row of holes nearest the face first and the other rows behind it in succession. Same as *rotational firing*.
Rubble Rough rocks of irregular shapes and different sizes, broken from larger masses either naturally or artificially, as by geologic action or by blasting.
Rule-of-Thumb A statement or formula that is not accurate, but which is sufficiently precise for figuring an approximate value.
Run of Pit, Quarry, or Mine Material as it comes directly from the pit, quarry, or mine before any processing.
Runner The operator of some machines, such as a shovel or dragline.
Running An operating or producing machine, such as a drill.
Saddleback A hill or ridge having a concave outline along the top.
Sand Particles of sediment having diameters of more than $1/300$ in (0.08 mm) and less than $1/12$ in (2.0 mm).

Sandstone A clastic sedimentary rock consisting predominantly of sand-size particles. Light to medium in color, medium-weight, and soft to hard.
Sandy Loam Material containing much sand and silt, but having enough clay to make it somewhat coherent.
Scaling Prying loose pieces of rock off the face of a cut or roof of a tunnel to avoid danger of their falling unexpectedly.
Scalping The process of removing residuals and weathered rock of a cut prior to excavation of the hard rock.
Scarifier An accessory or tool of a motor grader, tractor, or other machine used chiefly for the loosening of materials of shallow depth.
Schist A well-foliated metamorphic rock having a tendency to split into thin layers. Medium to dark in color, hard, and heavy. An example is mica schist.
Scoria Lava in which the gas cavities are numerous and irregular in shape, producing a sharp rock. Light to dark in color, medium-weight, and hard.
Seams Parting planes in rocks, such as the cooling joints in igneous rocks and the separations between the strata of sedimentary rocks.
Secretions Rock materials which have been deposited from solution by infiltration into the rock cavities. The crystals of a geode are examples.
Sediment Regolith that has been transported and deposited by water, air, or ice; predecessor of sedimentary rocks.
Sedimentary Breccia A clastic sedimentary rock containing numerous angular particles of pebble size and larger. Of various colors and hardnesses and of medium weight.
Sedimentary Rock A rock formed by the cementation of sediment or by other processes acting at ordinary temperatures at or close beneath the surface.
Seismic Studies The analysis and appraisal of rock-earth excavation by seismic means by which shock-wave velocities at different depths are determined. The degree of consolidation of the material and, indirectly, the cost of the excavation are proportional to the shock-wave velocities.
Semiarid A term applied to a country neither entirely arid nor strictly humid but intermediate. Annual rainfall varies from about 10 in (25 cm) to about 30 in (76 cm).
Sequential Firing In a large blast, setting off the row of holes nearest the face first and the other rows behind it in succession. Same as *rotational firing* or *row shooting.*
Series An arrangement of electric blasting caps in which the firing current passes through them successively.
Series The third division of rocks, a part of the second division, or *system,* of rocks. Corresponds to a geologic *epoch*.
Series Parallel Two or more parallel circuits of blasting caps arranged in series.
Serpentine A metamorphic rock composed chiefly or wholly of the mineral serpentine. Greenish and massive, sometimes fibrous and lamellar. Medium to hard and medium-weight.
Shale A fine-grained sedimentary rock composed of clay-size and silt-size particles, generally with thinly laminated structure. Medium to dark in color, medium to hard, and medium- to heavyweight.
Shear Zone A zone of rock and in which shearing has occurred on a large scale so that the rock is crushed and brecciated.
Shoot Same as to *blast.*
Shooting Rock Material or rock which requires blasting.
Shot Rock Rock which has been blasted.
Shrinkage The diminution in dimensions and volume of a material. In excavation, the percentage loss in volume when 1 yd^3 in the cut is placed in the fill or embankment.
Sidehill Cut An excavation in a hill involving only one cut slope and, usually, one fill slope. It is sometimes considered to be any cut in a hillside.
Sill A relatively thin tabular sheet of magma which has penetrated a rock formation along approximately horizontal bedding planes.
Silt A sediment of particle size between 0.0002 in (0.005 mm) and 0.002 in (0.05 mm) in diameter.
Siltstone A clastic sedimentary rock consisting predominantly of silt-size particles.
Silurian The third system of rocks in the Paleozoic group and the corresponding geologic period, from 430 to 400 million yr ago.

Sink A large solution cavity open to the sky. A large shallow bowllike valley from which drain water escapes by evaporation or percolation.

Slab A large and flat but relatively thin mass of rock.

Slag Fused or partly fused compounds resulting in secondary products from the reduction of metallic ores.

Slate A fine-grained metamorphic rock derived from shale, in which pressure has produced very perfect cleavage planes. Sometimes called *argillite*. Medium to dark in color, hard, and heavy.

Slickensides Striated and polished surfaces of rocks abraded by movement along a fault.

Slide The movement of a part of the Earth's surface under the force of gravity.

Slide Rock Any loose fragmental rock lying on a slope. *Talus* is a term applied to slide rock.

Slip A fault in rock or a smooth joint or crack where the rocks have moved relative to each other.

Slope An incline or gradient as measured from the horizontal. Measured in degrees with the horizontal or as the ratio of horizontal distance to vertical distance, as 34° and 1.5:1 for the same inclination.

Slope Stake A surveying stake set at the point where the finished side slope of a cut or fill intersects the surface of the ground. Usually set along a line at right angles to the centerline and sometimes set a few feet from the actual point as a reference stake not to be lost in the prosecution of the work.

Sloughing Sliding of overlying material such as overburden upon rock.

Slump The downward slipping of a coherent body of earthy material along a surface of rupture.

Soapstone A metamorphic rock of interlocking fibrous texture and soft soapy feel, such as talc. Light to medium in color, soft, and medium- to heavyweight.

Softwood Wood which is light in texture, nonresistant, and easily worked. Typical are pines and cottonwoods.

Soil A mixture of earthy materials intermingled with organic matter so as to support rooted plants.

Sorting Selection by natural processes during transport of rock particles according to size, specific gravity, shape, durability, or other characteristics.

Specific Gravity A number stating a ratio of the weight of solid rock to the weight of an equal volume of pure water.

Specific Gravity, Apparent A number stating the ratio of the weight of the total volume of permeable rock with its included voids to the weight of an equal volume of pure water. Thus, the apparent specific gravity depends on the specific gravity of the rock and on the amount or volume of the contained voids.

Speed Indicators A measure of the travel speeds of a hauling machine. When laden it equals the engine(s) bhp divided by the GVW in 1000s of lb. When unladen it equals the engine(s) bhp divided by the tare weight in 1000s of lb.

Spoil Rock-earth wasted because it has no use. Overburden in a copper mine is an example.

Spoil Bank Bank or pile of spoil or wasted materials.

Spread The excavating machinery necessary to do the complete work for the whole or a part of the operation or job.

Square A unit of area equal to 100 ft^2 (equivalent to 9.29 m^2).

Stability The resistance of the material in a cut or fill to movement downslope due to inherent characteristics or to the weight of a superposed load.

Stack A small prominent island of bedrock, the remnant of a former narrow promontory eroded by wave action.

Station A distance of 100 ft measured along the centerline and designated by a stake bearing its number.

Station Yard A unit of quantity times distance, corresponding to 1 yd^3 moved horizontally through a distance of 100 ft.

Stemming Inert material such as cuttings placed in a blasthole instead of explosives, as in deck loading.

Stock A body of igneous rock intruded upward into an older rock formation.

Stockpile Material excavated and piled for future use.

Stone   Any natural rock deposit or formation of igneous, sedimentary, or metamorphic origin, either in its original or altered form.
Stratification   The deposition of sediment beds, layers, or strata. The arrangement of the rocks in such beds, layers, and strata. The parallel structure resulting from such deposition and arrangement.
Stratigraphy   The systematic study of stratified rocks.
Stratum   A bed or layer of rock.
Stream Terrace   A bench along the side of a valley, the upper surface of which was formerly the alluvial floor of the valley.
Strength   The ability of a rock body to resist stresses tending to change both shape and volume.
Striations   Scratches and grooves on bedrock surfaces, caused by grinding of rock against rock.
Strike   The direction of the line of intersection of the plane of rock stratification or bedding with the horizontal plane.
Stripping   Removing the undesirable material from a deposit intended to be be used.
Strip Mining   The open-pit mining of materials by removal of overburden.
Structural Geology   The study of rock deformation and the delineation of geologic structural features.
Submergence   Fall of the land relative to the level of the sea. A land of submergence has generally rough topography. The coast of Maine is an example.
Superelevation   Rise of the outside curve above the inside curve of a haul road to accommodate the centrifugal force of the hauling units and thus prevent skidding and overturning.
Surficial   A common word describing the unmoved surface of the earth.
Surge Bin   A compartment for temporary storage of materials which allows converting a variable rate of supply to an even rate of discharge of the same average amount. An example is the receiving hopper of a conveyor belt.
Swell   The increase in dimensions and volume of a material. In excavation, the percentage gain in volume when a cubic yard in the cut is placed in the fill or embankment.
Swing   In revolving shovels and like machines, to rotate the upper works or deck with respect to the lower works. In cable drills, to operate a string of drilling tools.
Swing-Return Angle   The angle for swing-return through which a revolving excavator must pass in a cycle of load-swing-dump-return.
Switchback   A hairpin turn of about 180° in a haul road.
Syenite   A granular intrusive igneous rock of generally dark color, containing little or no quartz. Medium to hard, and heavy.
Syncline   A downfold of rocks with troughlike form and having the youngest rocks in the center.
System   The second division of rocks within a major *group*. Corresponds to a geologic *period*.
Taconite   An important iron ore of ferruginous chert. Medium to dark, hard, and heavy.
Talc   A metamorphic rock with a soapy feel. Light to medium in color, soft, and medium- to heavyweight. Same as *soapstone*.
Talus   An apron of rock waste sloping outward from the cliff which supplies it.
Tamping   Compacting loose material by means of weights or weighted machines such as sheepsfoot or tamping-foot compactors.
Tamping Rollers   One or more steel drums fitted with projecting feet in a machine to be towed or self-propelled.
Tectonic   Pertaining to the rock structures and external forms resulting from the deformation of the Earth's crust.
Tephra   A collective term designating all particles ejected from volcanoes, irrespective of size, shape, and composition.
Terminal Moraine   The end moraine deposited by a glacier along its line of greatest advance.
Terrace   A relatively flat elongated surface, bounded by a steep ascending slope on one side and a less steep descending slope on the other side.
Tertiary   The first of the two systems of rocks in the Cenozoic group. Also the corresponding geologic period, from 63 to 2 million years ago.

Texture The sizes and shapes of the particles in a rock or sediment and the mutual relationships among them.
Throw Scattering of rock fragments from blasting.
Through Cut A cut with slopes on both sides. If there is little cut on one side, it is sometimes called a sidehill cut.
Thrust Fault A low-angle reverse fault with dip generally less than 45°.
Tight A term applied to rock-earth formations without weaknesses, which may require hard ripping or light blasting before excavation.
Till Nonsorted or ungraded glacial drift made up of all sizes of particles.
Tillite Sedimentary rock consisting of cemented till. Light to dark in color, medium to hard, and medium- to heavyweight.
Toe The lowest edge of a cut where the slope intersects the grade and the lowest edge of a fill where the slope intersects the ground.
Ton, Short A unit of weight equal to 2000 lb, 907 kg, or 0.907 metric ton.
Ton, Long A unit of weight used in mining, equal to 2240 lb.
Topographic Map A map that delineates surface forms, usually by contours.
Topography The relief and the form of a land surface.
Top Soil The good or productive soil as contrasted to the underlying soil. Ideally, a mixture of clay, silt, sand, and humus matter.
Track Drill A large percussion drill powered by compressed air or hydraulic oil. Mounted on crawler tracks.
Traction The friction of a machine on the surface on which it moves. Traction for excavating machinery is a measure of the weight of the machine on its driving tracks or wheels-tires and the coefficient of friction between the driving tracks or wheels-tires and the surface of the ground.
Trap Rock The dark-colored, fine-grained and dense igneous rocks with little or no quartz, such as basalt, diabase, and gabbro.
Travertine A collective variety of limestone, including dripstone, flowstone, and calcareous tufa.
Triassic The first system of rocks in the Mesozoic group. Also the corresponding geologic period, from 230 to 180 million yr ago.
Tuff A fine-grained pyroclastic rock consisting of ash and dust. Light to dark in color, soft to hard, and light- to medium-weight.
Ultramafic Rock Granular igneous rock consisting almost entirely of ferromagnesian minerals. Dark, hard, and heavy.
Unclassified Excavation Excavation not subdivided for bidding purposes into common, or earth, and rock. Regardless of nature, the excavation is bid at one unit bid price.
Unconformity A lack of continuity between units of rock in contact, corresponding to a gap in the geologic record, the gap being a period of erosion.
Uniformity Principle The concept that relationships established between processes and materials in the modern world can be applied as a basis for interpreting the geologic record and for reconstructing the history of the Earth.
Unit Cost In earthwork, the cost of producing a unit of work, such as a cubic yard of excavation or a cubic yard of compacted embankment.
Unwatering Same as *dewatering*.
Vein A tabular deposit of minerals, occupying a fracture, in which the particles have grown away from the walls toward the middle.
Vesicle A small cavity in an igneous rock formed by the expansion of a bubble of gas or steam during the solidification of the rock.
Voids The spaces between particles of a rock or rock-earth, expressed as a percentage of the entire volume.
Volcanic Ash The smallest of the particles ejected by the explosion of a volcano.
Volcanic Bomb The largest of the particles ejected by the explosion of a volcano.
Volcanic Mudflow A mudflow of water-saturated, predominantly fine-grained tephra.
Volcanism A term designating the aggregate of processes associated with the transfer of materials from the Earth's interior to the surface.
Volcano A vent or fissure through which molten and solid materials and hot gases pass upward to the Earth's surface.
Volume In excavation, the volume of cut or fill in cubic yards or cubic meters, as well

as the volume, loose measurement, in waste areas or stockpiles and in haulers measured in the same units.

Wacke   Residual sand, silt, and clay formed by the weathering of igneous rocks.

Walking Dragline   A huge dragline which moves itself by means of side-mounted shoes or pontoons actuated by overhead cams, the motion being akin to walking.

Wash Boring   An exploratory drill hole from which samples are brought up mixed with water.

Waste   Digging, hauling, and dumping valueless material to get it out of the way, as in the removal of overburden from an ore deposit. Also the material itself.

Water Gap   A pass in a ridge or mountain range through which a stream flows.

Watershed   Area which drains into a stream or other water passage.

Water Table   A surface of underground gravity-controlled water.

Watt   A standard unit of electrical energy, equivalent to 0.00134 hp.

Weathering   The mechanical disintegration and the chemical decomposition of rock materials during exposure to nature's destructive forces.

Weight, Gross Vehicle (GVW)   The combined tare weight of the hauling machine and payload weight.

Weight, Payload   The weight of the load for which payment is made.

Weight, Tare   The working weight of the hauling machine.

Welded Tuff   A fine-grained volcanic rock the particles of which were so hot when deposited that they fused together. Light to dark in color, hard, and medium- to heavyweight.

Well Drill   Name for the percussion cable drill.

Wellpoint   A pipe fitted with a driving point and a fine mesh screen, used to remove underground water.

Wellpoint System   Machinery used for the removal of underground water, consisting chiefly of centrifugal pump, header pipe, lateral pipes, riser pipes, valves, and wellpoints.

Well-Sorted Sediment   A sediment consisting of particles of about the same size.

Wind Gap   A former water gap through which the stream no longer flows.

Windrow   A ridge of loose material thrown up by a machine.

Work-Cost Indicator   A measure of the efficiency of a hauling machine. It equals the sum of gross vehicle weight plus tare weight divided by payload weight. It is an approximate expression of the ratio of total work done to pay work done during the cycle of a hauling machine.

Working Cycle   The complete set of productive operations of a machine. For a hauler of rock, the cycle includes loading, hauling, turning, dumping, returning, and turning.

Xenolith   A fragment of another rock, or earlier solidified portion of the same rock, in an igneous rock.

Yard   In excavation, the area in which the owner stores and repairs the machinery and usually maintains his offices. Also, an expression for a cubic yard.

Yard-Mile   Same as *Mile-Yard.*

Yard-Station   Same as *Station-Yard.*

Zone of Aeration   The zone of the regolith in which the open spaces or voids are filled mainly by air. Generally the weathered zone.

Zone of Fracture   The upper portion of the Earth's crust in which rocks are deformed mainly by fracture.

Zone of Rift   That volume of rock-earth between the walls of a fault which has been fractured, broken, brecciated, and pulverized by the long-term action of the fault. The zone may be 2 mi in width and several miles in depth.

Zone of Saturation   The subsurface zone below the water table in which all voids are filled with water.

Index

Abbreviations, **A**-27
Acadian series of rocks in geologic time scale, **1**-7
Aftercooler or intercooler:
   for air compressor, description of, **11**-67
   for engine; description of, **16**-52
      examples of, **16**-55 to **16**-57
      specifications for, **16**-58
Agglomerate (rock), description of, **2**-10
Aggradation of land:
   causes of, (*see* Crustal movements; Ice; Volcanics; Water; Wind)
   definition of, **4**-3
   example of, **1**-2
Airports, excavation for (*see* Open-cut excavations, for airports)
Alaska forest region:
   description of, **9**-4
   example of, **9**-10
   location of, **9**-1
Allied operations, machinery, and components (*see* Cofferdams; Engine-generator sets; Engines; Lighting plants; Motors, electric; Pumps, for unwatering for excavation; Servicing and maintaining machinery on the job; Tires; Transporting machinery; Weighing excavation; Wellpoint systems; Yards, shops, and offices)
Aluminum ore, description of, **2**-11
Andesite (rock):
   description of, **2**-9
   examples of, **4**-7, **5**-8, **7**-16, **11**-16
Angledozer:
   description of, **12**-9
   for exploring excavation, **7**-5
   for pioneering excavation, **12**-12
   specifications for, **12**-11

Angles of repose, **A**-23
Anticline:
   examples of, **4**-18, **4**-19
   occurrence of, **4**-18
Apatite (mineral), description of, **2**-12
Approximate material characteristics, **A**-3 to **A**-8
Archeozoic group of rocks:
   in geologic time scale, **1**-7
   in Grand Canyon of Colorado, **2**-2, **2**-4
Areas, calculations for, **17**-3
   by planimeter, **17**-3
   by subdivision of areas, **17**-3
   by trapezoidal formula, **17**-3
Argentite (mineral), description of, **2**-12
Argillite (rock):
   description of, **2**-11
   examples of, **2**-7, **2**-20, **2**-21, **16**-2
Asbestos ore, description of, **2**-11
Asphalt (rock), description of, **2**-10
Attitude and stratification of rock, **2**-20 to **2**-22
   dip and strike, **2**-20 to **2**-22
Auger drill (*see* Exploration of excavation, machinery, drills, mechanical, auger drill)
Automobiles and pickup trucks, costs hourly, **8**-7
Average-end-area method:
   description of, **17**-5
   with contour method (*see* Contour method for calculating volume)
   with prismoidal adjustment, description of, **17**-13
   use of, **17**-10
Azoic group of rocks in geologic time scale, **1**-7

Backhoes, **7**-4 to **7**-6, **12**-32 to **12**-41
  auxiliary machinery, **12**-41
  costs hourly, **8**-7
  cycle times, **12**-35, **12**-38
  dipper factors, **12**-37
  efficiency of, **12**-34
  examples of, **5**-17, **7**-6, **12**-36, **12**-38 to **12**-41
  for exploring excavation, **7**-4, **7**-5
    example of, **7**-6
    production of, **7**-6
  principles of operation, **12**-23
  production and costs of, **12**-37, **12**-40
  specifications for: crawler-mounted machines, **12**-34, **12**-35
    wheels-tires-mounted machines, **12**-39
  for spreading-mixing excavation: costs hourly, **8**-7
    example and production of, **14**-16
  summary, **12**-41
Balls, drop-swing: description of, **12**-49
  example of, **12**-43
Basalt (rock):
  description of, **2**-9
  examples of, **2**-6, **2**-15, **2**-16, **3**-8, **4**-2, **7**-7, **7**-25, **11**-13, **12**-25, **12**-54, **12**-81, **13**-66
Basin, example of, **4**-16
Bauxite (mineral), description of, **2**-11
Bearing powers of materials, **A**-25
Beddings and thicknesses of rocks, **2**-13
Belt conveyor systems, **13**-84 to **13**-113
  dynamics of, **13**-96 to **13**-111
    capacity of belt, **13**-96, **13**-98
    inclination of belt, **13**-100
    lengths of belt versus horizontal and vertical distances and degrees of inclination, **13**-100
    loading of belt, **13**-96, **13**-99
    number of drive pulleys, **13**-105
      tension equations, **13**-108
      tension factors, **13**-109
      tensions, allowable working, **13**-110
    speeds of belts, **13**-97
    troughing rollers, spacings for, **13**-111
  kinds of, **13**-84
    movable systems: costs hourly, **8**-9
      description of, **13**-84
      examples of, **13**-89 to **13**-91
      specifications for, **13**-92
    stationary systems: costs hourly, **8**-8
      description of, **13**-84
      examples of, **13**-84 to **13**-88, **13**-92
  parts for: belts, **13**-91 to **13**-94, **13**-101, **13**-103
    belt takeups, **13**-94
    brakes, **13**-94
    feeders, **13**-94, **13**-95
    grizzlies, **13**-92, **13**-94

Belt conveyor systems; parts for (*Cont.*):
    holdbacks, **13**-95
    prime movers, **13**-95
    pulleys, **13**-92
    return idlers, **13**-93
    supporting structures, **13**-92, **13**-93
    trippers, **13**-94, **13**-96
    troughing rollers, **13**-92, **13**-93
  power, **13**-101
    driving head pulleys by prime mover, **13**-105
      example of, **13**-105
      horsepower required, **13**-105
    moving empty belt, **13**-101
      equation and example of, **13**-103, **13**-104
      friction factors, **13**-102
      horsepower required, **13**-103, **13**-104
      weight of belts, **13**-103
      weight of moving parts, **13**-102
    moving load horizontally, **13**-104
      equation and example of, **13**-104, **13**-105
      horsepower required, **13**-104, **13**-106
    moving load vertically, **13**-105
      equation and example of, **13**-105
      horsepower required, vertical lifting, **13**-105, **13**-107
    total shaft horsepower of prime mover, **13**-105
    turning all pulleys, **13**-105
      example of, **13**-105
      horsepower required, **13**-108
  production and costs of, **13**-109, **13**-111, **13**-112
    compared to those of truck haulage, **13**-109, **13**-111 to **13**-113
  summary, **13**-113
Belt loaders, **12**-67 to **12**-78
  kinds and principles of, **12**-67
    belt loads, **12**-70
    belt speeds, **12**-70
    conversion factors for materials, **12**-70
    mobile belt loaders: integral-type, **12**-67 to **12**-72
      costs hourly, **8**-9
      examples of, **12**-71, **12**-72
      production of, **12**-68 to **12**-70, **12**-73
        and costs of, **12**-72
      specifications for, **12**-71
    pull-type, **12**-67 to **12**-70
      costs hourly, **8**-9
      examples of, **12**-67, **12**-68
      production of, **12**-68 to **12**-70
        and costs of, **12**-68, **12**-69
      specifications for, **12**-67, **12**-68

Belt loaders; kinds and principles of *(Cont.)*:
   movable belt loaders, **12**-72 to **12**-78
      costs hourly, **8**-9
      examples of, **12**-74, **12**-75, **12**-77, **12**-78
      principles of, **12**-72, **12**-73
      production of, **12**-73 to **12**-75
      and costs of, **12**-76 to **12**-78
      specifications for, **12**-74, **12**-75
   summary, **12**-78
Bernoulli's theorem, **10**-7
   general equation, **10**-7, **10**-8
Bid preparation, **18**-1 to **18**-29
   example of, **18**-4 to **18**-29
   principles of, **18**-3
   pyramid of businesslike bid, **18**-1, **18**-3
   unbalanced bid, example of, **18**-27
   summary, **18**-29
Big Muskie (world's largest full-revolving excavator):
   description of, **12**-2
   illustration of, **12**-1
   production of, **12**-2
   specifications for, **12**-2
Bit grinder, example of, **11**-104
Bits:
   for auger drill, **7**-8 to **7**-10
   for cable drill, **7**-27, **11**-90, **11**-91
   for core drill, **7**-19
   for down-the-hole drill, **11**-125, **11**-126
   for drifter drill (*see* Drifter drills, bits)
   for jackhammer drill, **7**-14, **11**-98
   for rotary drill (*see* Rotary drills, bits)
   for track drill (*see* Track drills, bits)
Black powder, description of, **11**-23
Blastholes, **11**-37 to **11**-59
   density of explosives per foot of blasthole, **11**-46
   drills for (*see* Drills, for blastholes)
   loading, **11**-55 to **11**-59
      by bulk loading, **11**-56, **11**-58, **11**-59
      by chamber loading, **11**-43
      by column loading, **11**-41
      by deck loading, **11**-41
      by manual loading, **11**-55 to **11**-57
      by pocket loading, **11**-43
      by use of stemming, **11**-41, **11**-42
   pattern of spacing and burden, **11**-44, **11**-45, **11**-47
   subdrilling, **11**-41
   trial-and-error spacings for, **11**-44, **11**-45, **11**-47
   yield per foot of blasthole, **11**-43 to **11**-45
Blasting, **11**-23 to **11**-139
   by buffer blasting: examples of, **11**-136, **11**-137
      purpose of, **11**-45
   by controlled blasting: by line drilling, **11**-48

Blasting; by controlled blasting *(Cont.)*:
   by preshearing or presplitting, **11**-48, **11**-49, **11**-51
   purpose of, **11**-45
   by smooth or contour blasting, **11**-48, **11**-51
   definition of, **11**-23
   by delay blasting: example of, **11**-37
   methods of, **11**-49, **11**-50
   dynamites (*see* Dynamites)
   examples of successful blasting: by buffer blasting, **11**-133, **11**-136
      by coyotes or tunnels, **11**-135, **11**-138, **11**-139
      by delay detonation, **11**-133, **11**-135
      by high face, **11**-133, **11**-134
      by sequential firing and buffer, **11**-133, **11**-137
   good fragmentation, examples of, **11**-54, **11**-55, **11**-135 to **11**-137, **11**-139
   methods of: description of, **11**-37
      by blastholes, **11**-37, **11**-39, **11**-40, **11**-119
      by combining blastholes and coyotes or tunnels, **11**-52, **11**-53, **11**-138, **11**-139
      by coyotes or tunnels, using blastholes for driving adits and laterals, **11**-40, **11**-51 to **11**-54
   noise from, **11**-59
   principles of, **11**-23
   safety precautions, **11**-60
      dos and don'ts, **11**-65 to **11**-67
      for transportation, storage, and use of explosives, **11**-61 to **11**-64
   secondary blasting: description of, **11**-55
      examples of, **11**-56, **11**-95, **11**-96
      explosives for, **11**-59
      significance of, **11**-55
   seismic shock-wave velocity and depth, relationship between, **7**-29, **7**-32, **7**-34, **7**-36, **11**-2, **11**-3, **11**-11, **11**-16, **11**-18
   stump removal, **9**-28
   summary, **11**-139
   vibrations from: calculations for intensity of, **11**-60
      destruction from, **11**-60
      permissible charges of explosives, **11**-60, **11**-61
   zone in excavation, **11**-2, **11**-3
Blasting agents, **11**-25 to **11**-28
   cartridged blasting agents: characteristics of, **11**-25
      description of, **11**-25
      example of, **11**-27
      performance of, **11**-25
   description and kinds of, **11**-25

4   Index

Blasting agents *(Cont.)*:
  free-running blasting agents: AN (ammonium nitrate prills), **11**-26, **11**-27
    AN/FO (ammonium nitrate/fuel oil prills), **11**-26 to **11**-28
    description and characteristics of, **11**-26 to **11**-28
    examples of, **11**-27
  plastic or viscous blasting agents:
    characteristics of, **11**-28
    description of, **11**-28
    example of, **11**-29
    performances of, **11**-28
Blasting circuits, **11**-32 to **11**-37
  description of, **11**-32
  detonating-fuse circuit, **11**-34, **11**-52
  electrical circuits, **11**-35
  safety-fuse circuit, **11**-32, **11**-33
Blasting initiating devices, **11**-28 to **11**-32
  for black powder, **11**-29, **11**-30
  for blasting agents, **11**-32, **11**-33
  description of, **11**-28, **11**-29
  for dynamites, **11**-29, **11**-31
Blasting machines:
  description of, **11**-36, **11**-37
  kinds of, **11**-36
Borax (mineral), description of, **2**-11
Boron ore:
  description of, **2**-11
  examples of, **2**-8, **11**-31, **13**-88
Bottom-dump haulers, **13**-51 to **13**-63
  body factors for loads, **13**-58
  comparison with scrapers as pure haulers, **13**-53, **13**-55
  costs hourly, **8**-25 to **8**-27
  cycle times, **13**-56, **13**-57
  examples of, **5**-11, **13**-52, **13**-54, **13**-57, **13**-58
  operation of, **13**-54
  production and costs of, **13**-56, **13**-59 to **13**-63
  specifications for: off-the-road machines, **13**-53, **13**-57, **13**-58
    off-the-road and on-the-road machines, **13**-55
  summary, **13**-63
  types and functions of, **13**-51
  (see also Scrapers, and haulers, wheels-tires-mounted)
Boulders:
  description of, **2**-10
  examples of, **2**-14, **3**-2 to **3**-4, **3**-7, **3**-10, **4**-5, **4**-6, **4**-10, **5**-17, **6**-2, **7**-8, **7**-12, **7**-16, **7**-28, **11**-3, **11**-16, **11**-33, **11**-95, **12**-20, **13**-48, **13**-125, **14**-1, **18**-6
Boyle's law, definition and equation, **11**-82
Breakage of rock:
  description of, **3**-9
  by earthquake and faults, **3**-10 to **3**-12
  by gravity, **3**-7, **3**-8
  by joints, **3**-6 to **3**-8

Breccia (rock):
  description of, **2**-9
  examples of, **2**-18, **4**-14, **6**-6
Brush-rake method for clearing and grubbing, **8**-24, **9**-18, **9**-22, **9**-29
Bucket drill for exploring excavation *(see* Exploration of excavation, machinery, drills, bucket drill)
Bucket factors, dipper and, **12**-25
Buckets:
  brush-type, for crawler-tractor: costs hourly, **8**-24
    examples of, **9**-20, **9**-21
  clamshell-type: costs hourly, **8**-9
    example of, **12**-44
    specifications for, **12**-50
  dragline-type: costs hourly, **8**-10
    examples of, **12**-2, **12**-54, **12**-56, **12**-58 to **12**-60, **13**-124, **13**-125
    specifications for, **13**-127
  orange-peel-type: costs hourly, as clamshell-type, **8**-10
    specifications for, as clamshell-type, **12**-50
Building sites and large foundation excavations:
  description of, **5**-4, **5**-5
  examples of, **5**-4, **13**-45
Buildings, portable: costs hourly, **8**-10
  examples of, **16**-87, **16**-89
Bulldozers, tractor: crawler-mounted *(see* Tractors, crawler-mounted, with bulldozer and angledozer)
  wheels-tires-mounted *(see* Spreading-mixing excavation, by tractor-bulldozers)
Burners, brush: costs hourly, **8**-9
  example of, **9**-24
Burrowing animals *(see* Plant roots and burrowing animals)
Butte, description of, **4**-11

Cable drills, **7**-24 to **7**-27, **11**-90 to **11**-94
  auxiliary machinery, **11**-92
  bits, **7**-27, **11**-92, **11**-93
  for blastholes: description of, **11**-90
    efficiency of, **11**-92
    example of, **11**-91
    explosives, **11**-93
    production and costs of, **11**-93
    speeds, drilling, **11**-92
  costs hourly, **8**-10
  for exploring excavation: costs hourly, **8**-10
    description of, **7**-24 to **7**-27
    drill tools, **7**-27
    efficiency of, **7**-25
    example of, **7**-25
    speeds, drilling, **7**-26
    use of, as wash-boring drill, **7**-26

Cable drills *(Cont.)*:
  principles of, **7**-24, **7**-25
  specifications for, **11**-91
  summary, **11**-94
  tools, **7**-27
Cableway scraper-bucket systems, **13**-113 to **13**-128
  dragscraper machines: characteristics and operation of, **13**-113 to **13**-115
    costs hourly, **8**-23
    crescent scrapers, specifications for, **13**-120
    examples of, **13**-113 to **13**-117
    hoists, **13**-114, **13**-116, **13**-117
    production of, **13**-120, **13**-121
      and costs of, **13**-122
    production conversion factors, **13**-121, **13**-122
    specifications for, **13**-118 to **13**-120
  summary, **13**-127, **13**-128
  track cable machine: characteristics and operation of, **13**-122
    costs hourly, **8**-24
    crescent scrapers, specifications for, **13**-120
    dragline buckets, specifications for, **13**-127
    examples of, **13**-123 to **13**-125, **13**-127
    hoists, **13**-117
    production of, **13**-123, **13**-124
      and costs of, **13**-124, **13**-125
    production conversion factors, **13**-123
    specifications for, **13**-126
    work-cost indicator, **13**-125
Calculations of quantities of excavation (*see* Areas; Center of gravity of mass or volume; Mass diagram for haul measurement; Volumes of earthwork)
Calyx or shot drill (*see* Exploration of excavation, machinery, drills, mechanical, core drills, calyx drill)
Cambrian system of rocks:
  in geologic time scale, **1**-7
  in Grand Canyon of Colorado, **1**-2 to **1**-4
Canadian series of rocks in geologic time scale, **1**-7
Canals, excavation (*see* Open-cut excavation for canals)
Carbonization of rock, description of, **3**-4
Cars, quarry: costs hourly, **8**-10
  examples of, **13**-128, **13**-130 to **13**-132
  specifications for, **13**-132
  (*see also* Railroads)
Cassiterite (mineral), description of, **2**-12
Casting excavation (*see* Loading and casting excavation)
Cayugan series of rocks in geologic time scale, **1**-7
Cenozoic group of rocks in geologic time scale, **1**-7

Center of gravity of mass or volume:
  calculations for, **17**-15, **17**-18, **17**-19
  definition of, **17**-15
  importance of, in estimating costs of haulage, **17**-19
Central hardwood forest region:
  description of, **9**-4
  example of, **9**-6
  location of, **9**-2
Chalcopyrite (mineral):
  description of, **2**-11
  examples of, **5**-2, **11**-132, **13**-78
Charles' law, definition and equation, **11**-82
Chippers, log: costs hourly, **8**-11
  example of, **9**-23
Chokers, chain, example of, **9**-26
Choppers, rolling: costs hourly, **8**-24
  example of, **9**-22
Chromite (mineral), description of, **2**-11
Chromium ore, description of, **2**-11
Chrysotile (mineral), description of, **2**-11
Cincinnatian series of rocks in geologic time scale, **1**-7
Cinnibar (mineral), description of, **2**-12
Cirque:
  description of, **4**-5
  example of, **1**-2
Clamshell bucket (*see* Buckets, clamshell-type)
Clay (rock):
  description of, **2**-10
  examples of, **4**-15, **7**-10, **14**-41, **14**-45
Clearing and grubbing, **9**-1 to **9**-34
  estimating costs of: per acre, **9**-12
    per cubic feet and board feet of trees, **9**-3, **9**-15
    data for, **9**-13
    sizes of trees, **9**-13
    timber cruise, **9**-13, **9**-14
  forest regions of the United States, **9**-4
    Alaska: description and trees, **9**-2, **9**-4
      example of, **9**-10
    central hardwood: description and trees, **9**-2, **9**-4
      example of, **9**-6
    Hawaii: description and trees, **9**-2, **9**-4
      example of, **9**-11
    northern: description and trees, **9**-2, **9**-4
      example of, **9**-5
    Pacific Coast: description and trees, **9**-2, **9**-4
      example of, **9**-9
    Rocky Mountain: description and trees, **9**-2, **9**-4
      example of, **9**-8
    southern: description and trees, **9**-2, **9**-4
      example of, **9**-7
    tropical: description and trees, **9**-2, **9**-4
      example of, **9**-12

Clearing and grubbing *(Cont.)*:
  machinery used for, 9-14 to 9-25
    brush buckets for crawler-tractor:
      costs hourly, 8-24
      examples of, 9-20, 9-21
    brush burners: costs hourly, 8-9
      example of, 9-24
    chippers, log: costs hourly, 8-11
      example of, 9-23
    chokers, chain, example of, 9-26
    choppers, rolling, for crawler-tractor:
      costs hourly, 8-24
      example of, 9-22
    cranes, crawler-mounted, with grapple: costs hourly, 8-11
      example of, 9-19
    cutters, stump: costs hourly, 8-12
      example of, 9-27
    explosives for stumping, 9-28
    haulers: brush: costs hourly, 8-25, 8-27
      example of, 9-21
      log: costs hourly, 8-25
      example of, 9-24
    plows, root, for crawler-tractor: costs hourly, 8-24
      example of, 9-29
    rakes, brush, for crawler-tractor: costs hourly, 8-24
      example of, 9-18
    saws: chain: costs hourly, 8-21
      examples of, 9-16
      circular: costs hourly, 8-21
      example of, 9-21
    tree pushers, crawler-tractor: costs hourly, 8-24
      example of, 9-17
    tree splitters, crawler-tractor: costs hourly, 8-24
      example of, 9-17
  methods of: brush-rake method, description of, 9-22, 9-29
    crane method, description of, 9-14, 9-15, 9-22
  payments for, 9-3
  principles of, 9-3, 9-4
  production and costs of, 9-20, 9-29 to 9-34
  root systems of trees, 9-13, 9-26, 9-28
  specifications for, 9-3
  summary, 9-34
  (*see also* Cranes, crawler-mounted machines; Loaders, bucket, crawler-mounted machines; Tractors, crawler-mounted, with bulldozer or angledozer: pulling scrapers and wagons)
Climate, effects of: on rock weathering, 3-13, 3-14
  on work, 13-13, 13-24, 13-44

Coal (rock):
  anthracite: description of, 2-10
    example of, 2-19
  bituminous: description of, 1-10
    examples of, 5-11, 12-2, 12-32, 12-40, 12-41, 12-80, 13-57, 13-58
Coefficients of traction:
  of crawler-tractors, 13-15
  of locomotive wheels, 13-134
  of wheels-tires-mounted scrapers and haulers, 13-28
Cofferdams, 16-7 to 16-13
  definition of, 16-7, 16-8
  design of, fundamentals, 16-9 to 16-12
  kinds of, 16-7, 16-8
  land cofferdams: description of, 16-8
    example of, 16-8
  water cofferdams: description of, 16-9
    examples of, 16-9, 16-10, 16-30
  stages of construction, 16-12, 16-13
  summary, 16-12
Common excavation, definition of, 11-5
Compaction of excavation (*see* Wetting-compacting excavation)
Compactors (*see* Wetting-compacting excavation)
Compressed air, uses of, in drilling, 11-62
Compressors, air, 11-62, 11-68 to 11-85
  auxiliary machinery, 11-62
  costs hourly, 8-5, 8-6
  description of, 11-62, 11-68
  friction losses in air transmission, 11-68, 11-69
    in pipe and hose fittings, 11-81
    in pipe and hoses, 11-74 to 11-80
  fundamental laws and definitions for air compression, 11-69, 11-82, 11-83
    Boyle's law, 11-82
    Charles' law, 11-82
    equation, characteristic, 11-83 to 11-85
  pipe, hose, and fittings, costs hourly, 8-7
  pipe sizes for air transmission, 11-82
  portable compressors: costs hourly, 8-5
    examples of, 7-16, 11-72
    specifications for, 11-68, 11-70
  stationary compressors: costs hourly, 8-5, 8-6
    examples of, 11-68, 11-72
    specifications for, 11-73
Conglomerate, rock: description of, 2-10
  examples of, 2-17, 5-17, 6-2, 7-7, 11-13, 12-10
Contour method for calculating volume, 17-5, 17-11, 17-12
Conversion factors for systems of measurements, A-17, A-18

Cooling of rocks, weathering effect of: description of, 3-6
   examples of, 3-7, 3-8
Copper ore, description of, 2-11
Core drill (*see* Exploration of excavation, machinery, drills, mechanical, core drills)
Correlation of rock formations for excavation (*see* Formations, rock)
Costs of machinery and facilities, 8-1 to 8-36
   cost of ownership and operation of machinery and facilities, 8-3, 8-4
   elements of tabulation, 8-3, 8-32 to 8-36
   tabulation, 8-5 to 8-31
   direct job unit cost, calculation of, 8-35, 8-36
   production of machinery: definition of, 8-34
      efficiency: and continuous production, 8-34, 8-35
      and intermittent production, 8-35
   summary, 8-36
Crane method for clearing and grubbing, 9-14, 9-15, 9-22
Cranes, 12-42 to 12-52
   buckets, 12-44, 12-49, 12-50
   capacity ratings: different standards, 12-43
      examples of, 12-46, 12-47
   crawler-mounted machines: costs hourly, 8-11
      examples of, 12-42
      specifications for, 12-45 to 12-47
   cycle times, 12-46 to 12-49
   drop-swing balls, 12-43, 12-49
   grapples, 9-19, 12-42, 12-49, 12-50
      costs hourly, 8-11
   operation of, 12-43
   production and costs of, 12-51
   selection of, 12-49 to 12-51
   slings, rock, 12-44, 12-49
   summary, 12-51, 12-52
   uses of, in excavation, 12-42
   wheels-tires-mounted machines: costs hourly, 8-11
      example of, 12-48
      specifications for, 12-45
Creep:
   description of, 4-14
   example of, 4-15
Cretaceous system and series of rocks:
   example of, 2-24
   in geologic time scale, 1-7
Crushers, primary, examples of rock dumping into, 13-74, 13-77, 13-78, 13-131

Crust of Earth, rocks and, 1-3, 2-5, 2-9 to 2-13
Crustal movements:
   description of, 4-17
   faults, 4-17, 4-18
   weathering by faults, examples of, 3-11, 4-17 to 4-19
Cubic yard of excavation:
   volume changes from bank to hauler or stockpile to embankment, A-2
   weight and volume changes from cut to loose condition to fill, A-5 to A-9
Cutters, stump: costs hourly, 8-12
   example of, 9-27

Dams and levees, excavation (*see* Open-cut excavations for dams and levees)
Decomposition of rock:
   by carbonization, description of, 3-4
   effects of climate on, 3-13
   by hydration: description of, 3-4
      example of, 2-8
   by hydrolysis: description of, 3-4
      example of, 3-1
   by oxidation, description of, 3-4
   by solution: description of, 3-4
      example of, 3-5
Deflation of rock:
   description of, 4-12
   example of, 3-7
Deformation of rock, examples of, 2-21, 4-19
Degradation of land:
   causes of (*see* Crustal movements; Gravity; Ice; Water; Wind)
   definition of, 4-3
Deposition of land (*see* Aggradation of land)
Depreciation of machinery and facilities (*see* Costs of machinery and facilities; Depreciation schedule for machinery and facilities)
Depreciation schedule for machinery and facilities, A-11
Development of water supply (*see* Water Supply)
Devonian system of rocks:
   in geologic time scale, 1-7
   in Grand Canyon of Colorado, 2-1 to 2-4
Diabase (rock), description of, 2-9
Diameter breast high (dbh) of tree, 9-13 to 9-15
Diamond drill (*see* Exploration of excavation, machinery, drills, mechanical, core drills, diamond drill)
Diatomaceous earth ore, description of, 2-11
Diatomite (mineral), description of, 2-11

Diorite (rock):
  description of, **2**-9
  example of, **3**-7
Dip and strike of rock formation, **2**-20 to **2**-22
  definition of, **2**-22
  example of, **2**-22
Dipper and bucket factors, **12**-25
Direct job unit cost:
  definition of, **8-35**, **8**-36
  equation for, **8**-36
Disintegration of rock, **3**-5
  by breakage: description of, **3**-9
    earthquakes and faults, **3**-10 to **3**-12
    gravity, **3**-7, **3**-9
    joints, **3**-6 to **3**-8
  by cooling: description of, **3**-6
    examples of, **3**-7, **3**-8
  effects of climate on: description of, **3**-11, **3**-12
    examples of, **3**-13, **3**-14
  by frost wedging: description of, **3**-5
    example of, **3**-13
  by plant roots and burrowing animals: description of, **3**-11
    example of, **3**-13
  by temperature changes: description of, **3**-5, **3**-6
    example of, **3**-7
Ditchers (*see* Trenchers)
Dolomite (rock):
  description of, **2**-10
  examples of, **2**-6, **2**-14
Down-the-hole drill, **11**-123 to **11**-129
  auxiliary machinery, **11**-124, **11**-125
  bits: examples of, **11**-125, **11**-126
    life of, **11**-125
  examples of, **11**-124, **11**-125
  explosives, **11**-125
  performances of, estimating, **11**-124
  principles of, **11**-123
  production and costs of, **11**-125
  specifications for, **11**-123
  speeds of, drilling, **11**-127
  summary, **11**-125
Draglines, **12**-52 to **12**-61
  in area mining, **12**-2, **12**-59, **12**-60
  crawler-mounted machines: costs hourly, **8**-12
    examples of, **12**-54, **12**-58
    specifications for, **12**-53, **12**-55
  cycle times, **12**-53, **12**-54, **12**-56
  pontoon-mounted walker-type machines: costs hourly, **8**-13
    examples of, **12**-2, **12**-56, **12**-59
    specifications for, **12**-55
  principles of, **12**-52
  production and costs of, **12**-54, **12**-57 to **12**-59
  summary, **12**-60, **12**-61

Dredges, **13**-145 to **13**-169
  classes and kinds of, **13**-145 to **13**-147
    bucket dredge: costs of, **13**-149
      description of, **13**-146, **13**-149
      production of, **13**-149
    dipper dredge: costs of, **13**-147, **13**-148
      description of, **13**-145, **13**-146, **13**-149
      production of, **13**-146, **13**-147
    grapple dredge: costs of, **13**-145, **13**-148
      description of, **13**-145, **13**-146, **13**-148
      production of, **13**-145, **13**-148
    self-propelled hopper dredge: description of, **13**-147, **13**-150
      production and costs of, **13**-150
    suction cutterhead dredge:
      capabilities of, **13**-150, **13**-151
      costs hourly, **8**-13
      description of, **13**-150, **13**-151
      dynamics of, **13**-157 to **13**-169
        friction head, **13**-158, **13**-159
        static head, **13**-158
        suction head, **13**-158
        total dynamic head, **13**-158, **13**-159
        velocity head, **13**-158
      example of, **13**-152
      friction losses: in pipe, **13**-159 to **13**-163
        in valves and fittings, **13**-163
      fuel costs, comparison of, **13**-163
      material or mixture pumped hourly, **13**-164
      operation of, **13**-151 to **13**-153
        centrifugal pump, **13**-153 to **13**-157
          performance curves, **13**-157
        controls, **13**-153
        cutterhead, **13**-152 to **13**-155
        pipeline, **13**-155, **13**-156
        pontoons, **13**-156
        suction head **13**-154
        working parts, **13**-152
      outputs of, different, **13**-151
      principles of, **13**-150, **13**-151
    suction dredge, plain, description of, **13**-147
  definition of, **13**-145
    power requirements, centrifugal pump, **13**-163, **13**-165 to **13**-168
  production and costs of, **13**-168, **13**-169
  specifications for, **13**-150
  weights and specific gravities of mixtures, **13**-158
  summary, **13**-169
Drifter drills, **11**-100 to **11**-106
  auxiliary machinery: description of, **11**-100, **11**-101
    selection of, **11**-101 to **11**-103

Drifter drills *(Cont.)*:
  bits: description of, **11**-103, **11**-104
    cost comparison of, **11**-104
    grinding, **11**-104
    life of, **11**-105
  costs hourly, **8**-6
  definition of, **11**-100
  drill rods, life, **11**-105
  examples of, **11**-101, **11**-103
  explosives, **11**-104
  production and costs of, **11**-104 to **11**-106
  specifications for, **11**-100
  speeds, drilling, **11**-101
  summary, **11**-106
Drills:
  for blastholes: for primary blasting *(see* Cable drills; Down-the hole drill; Drifter drills; Fusion drill; Jackhammer drills; Rotary drills; and Track drills)
    for secondary blasting *(see* Jackhammer drills)
  for exploration of excavation, **7**-3 to **7**-26
    manual drills: auger spiral drill, **7**-3, **7**-4
      description of, **7**-3
      sampling kit, **7**-3 to **7**-5
        split-tube sampler of, **7**-5
      for wash boring, **7**-3, **7**-4
    mechanical drills *(see* Exploration of excavation, machinery, drills, mechanical)
Drop-swing balls for cranes, **12**-43, **12**-49
Drumlins:
  description of, **4**-4
  example of, **4**-6
Dumping:
  and compacting excavation, **14**-1 to **14**-46
    example of, **14**-2
    sequential operations *(see* and placing excavation *below;* Spreading-mixing excavation; Wetting-compacting excavation)
    summary, **14**-46
  and placing excavations: for compacting embankments, example of, **14**-2
    description of, **14**-3
    for random fill, example of, **13**-142
    for rip-rap, example of, **14**-3
    for stockpile, example of, **13**-87
Dunes:
  description of, **4**-12
  example of, **4**-13
  excavation problems, **4**-12, **4**-13
Dynamites:
  characteristics of, **11**-25
  definition of, **11**-24
  kinds of, **11**-24, **11**-25
    extra dynamite: cartridges, **11**-26

Dynamites; kinds of; extra dynamite *(Cont.)*:
      description of, **11**-24
    gelatin dynamite: cartridges, **11**-26
      description of, **11**-24
    straight dynamite: cartridges, **11**-26
      description of, **11**-24
  properties of: density, description of, **11**-24
    strength, description of, **11**-24
    velocity, description of, **11**-24
    water resistance, description of, **11**-24

Earth (planet):
  and earth (material excavated), definition of, **1**-3
  and geology of excavation, **1**-1 to **1**-7
    crust or regolith of Earth, **1**-3
    depths of excavation, **1**-3
    excavation of Earth, **1**-3
    main divisions of geologic time for North America, **1**-7
    origin of Earth, **1**-3
    Principle of Uniformitarianism, **1**-6
    summary, **1**-7
    surficial changes of Earth: by aggradation, **1**-2 to **1**-5
      (see also Aggradation of land)
    by degradation, **1**-3 to **1**-6
      (see also Degradation of land)
    weathering of Earth, **1**-3
      (see also Rock weathering)
Earthquakes:
  degradation by, **1**-4
  rock breakage by, **2**-25
Efficiency of machines, **8**-35
Electro-osmosis method of unwatering excavation, description of, **16**-6, **16**-7
Embankment construction *(see* Dumping and placing excavation; Spreading-mixing excavation; Wetting-compacting excavation)
Engine-generator sets, **16**-53 to **16**-68
  applications of, in excavation, **16**-64
  costs hourly, **8**-15
  definitions, units of measurement, and principles of, **16**-53, **16**-61, **16**-64
  diesel engine-ac generator sets: cost estimate, feasibility, **16**-66, **16**-67
    example of, **16**-67
    kinds of, **16**-65, **16**-66
    specifications for, **16**-65 to **16**-68
  diesel engine-dc general sets, kinds of, **16**-64
  efficiencies of generators, **16**-64
Engines, internal-combustion, **13**-3 to **13**-7, **16**-50 to **16**-60
  characteristics and economics of, **13**-3
  costs hourly, **8**-13, **8**-14
  description of, **13**-3, **16**-50

Engines, internal-combustion (*Cont.*):
  diesel engines, **16**-52
    automotive-type, water-cooled: performance curves, **16**-57
      sectional view, **16**-56
      specifications for, **16**-58
    industrial-type: large, water-cooled: aftercooler or intercooler, **16**-52, **16**-54
      example of, **16**-61
      specifications for, **16**-63
    small, air-cooled: performance curves and view, **16**-59
      specifications for, **16**-60
  flywheel or brake horsepower of, **13**-5, **13**-6
  fuel consumptions and costs of, **13**-6, **13**-7
  gas engines, **16**-51
  gasoline engines, **16**-50, **16**-51
    sectional view and performance curves, **16**-51
    specifications for, **16**-52
  naturally aspirated four-cycle engine, **13**-3, **13**-4
  oil engine, **16**-50
  pressure-aspirated engine, **13**-4, **13**-5
  production and costs of, **16**-53
  wheel horsepower, **13**-28
Eocene series of rocks in geologic time scale, **1**-7
Erosion:
  by ice, **4**-3, **4**-7
  by water, **4**-7 to **4**-9
  by wind, **3**-7, **4**-12
Eskers, description and occurrence of, **4**-4, **4**-6
Examination of excavation, **6**-1 to **6**-20
  field work: in association with office work, **6**-10
    conclusions from field work and history of excavation in area, **6**-16, **6**-17
    correlations of vegetations and rocks, **6**-14
    equipment for, **6**-10, **6**-11
    example of: observations: general, **6**-12 to **6**-14
      when walking centerline, **6**-15, **6**-16
    progress and completion of work, subsequent, **6**-17, **6**-18
    seismic studies of excavation, **6**-11, **6**-12
    surveillance, long-range, **6**-14
    terrain and weather, **6**-13
    purpose of, **6**-11
  office work: action plan, **6**-3
    correlation of rock formations, **6**-4
    geologic maps, **6**-3, **6**-5, **6**-9, **6**-10
    history of excavation in area, **6**-3

Examination of excavation; office work (*Cont.*):
  master plan sheets, **6**-3
  plans and specifications for work, **6**-3
  soil survey sheets, **6**-8
  topographic maps, **6**-3, **6**-4, **6**-6
  summary, **6**-18
Excavations:
  classification of, **1**-3
  definition of, **1**-3
  examples of (*see* Open-cut excavations)
  excavation characteristics of rock, **2**-9 to **2**-11
Exploration of excavation, **7**-1 to **7**-37
  comparison of methods, **7**-36, **7**-37
  example of, **7**-2
  machinery, **7**-3
    backhoes (*see* Backhoes, for exploring excavation)
    bulldozers and angledozers, **7**-5, **7**-6
      examples of, **7**-7, **7**-35
      (*see also* Tractors, crawler-mounted, with bulldozer or angledozer)
    drills: manual auger spiral drill, **7**-3, **7**-4
      description of, **7**-3
      sampling kit, **7**-3, **7**-5
        split-tube sampler of, **7**-5
      wash-boring drill, **7**-3, **7**-4
    mechanical, **7**-6, **7**-8
      auger drill: costs hourly, **8**-7
        description of, **7**-8 to **7**-10
        drill assembly, **7**-10
        examples of, **7**-8, **7**-9
        speeds, drilling, **7**-9
      bucket drill: costs hourly, **8**-9
        description of, **7**-10 to **7**-12
        examples of, **7**-11, **7**-12
        speeds, drilling, **7**-10
      cable drill (*see* Cable drills, for exploring excavation)
      core drills, **7**-17, **7**-18
        calyx drill: cores, **7**-24
          costs hourly, **8**-11
          description of, **7**-21 to **7**-24
          drill head, **7**-22
          example of, **7**-23
          speeds, drilling, **7**-23
        diamond drill: bits, **7**-19
          cores, **7**-20
          costs hourly, **8**-11
          description of, **7**-18 to **7**-21
          example of, **7**-20
          performance of, **7**-19, **7**-21
          speeds, drilling, **7**-21
      jackhammer drill (*see* Jackhammer drills, for exploration of excavation)

Exploration of excavation; machinery; drills; mechanical *(Cont.):*
   rotary drill *(see* Rotary drills, for exploring excavation)
   track drill *(see* Track drills, for exploring excavation)
   wash-boring drill: description of, **7-26**
   principles of, **7-26**
  manual means, **7-3** to **7-5**
   soil sampling kit, example of, **7-5**
  seismic timer, **7-26, 7-35, 7-36**
   fragmentation methods according to seismic shock-wave velocities, **7-28**
   refraction studies: equipment for, **7-27**
    examples of, **7-31**
    graph and calculations, velocity versus depth, **7-29**
    principles of, **7-26, 7-27**
    Snell's law of refraction, **7-27**
   ripping and blasting zones according to seismic velocities, **7-32**
   uphole studies: equipment for, **7-33**
    graph and calculations, velocity versus depth, **7-34**
    principles of, **7-32**
   velocity versus depth relationships for the three classes of rock, **7-36**
  summary, **7-37**
Explosives:
  definition of, **11-23**
  high explosives, description of, **11-23**
  low explosives, description of, **11-23**
  theory of fragmentation of rock, **11-37** to **11-39**
  (see also Blastholes; Blasting; Blasting agents; Blasting circuits; Blasting initiating devices; Blasting machines; Dynamites)
Explosives loading trucks:
  costs hourly, **8-14**
  description of, **11-56, 11-58, 11-59**
  examples of, **11-58, 11-59**
Extrusive igneous rocks *(see* Igneous rocks, kinds of, extrusive or volcanic)

Fanglomerates:
  description of, **4-9, 4-10**
  examples of, **4-9, 4-10**
Faults:
  description of, **4-17**
  effects of, on excavation, **4-17, 4-19, 6-4**
  lateral fault, example of, **4-17**
  San Andreas fault, **3-10, 3-12, 4-19**
  thrust fault, example of, **4-18**
  vertical faults, examples of, **4-17**
Fittings, pipe, water and air *(see under* Water supply)
Flood lights *(see* Lighting plants)

Floodplains, description of, **4-9**
Flow of air *(see* Compressors, air)
Flow of water *(see under* Water supply)
Flywheel or brake horsepower, **13-5, 13-6**
Folds:
  description of, **4-18**
  examples of, **3-11, 4-19**
Forest regions *(see* Clearing and grubbing, forest regions of the United States)
Formations, rocks, **2-13** to **2-26**
  considerations for, **2-13**
  correlations for appraising excavation, **2-22** to **2-25**
   correlation chart, Southern California, **2-23**
   examples of, **2-24, 2-25**
  definition of, **2-13**
  examples of, **2-14** to **2-21**
  lineage of a formation and relationship to geologic divisions of rocks, **2-14**
  (see also Rocks)
Formulas frequently used in earthwork calculations, **A-19, A-20**
Fragmentation of rock, **11-1** to **11-139**
  considerations for, **11-4**
  definition of, **11-3**
  methods of, **11-3**
  specifications for work, **11-4**
  summary, **11-39**
  transition from ripping to blasting, **11-2**
  zones of, **11-3**
  (see also Blasting; Ripping)
Freezing water-bearing excavation, description of, **16-6**
Frost wedging:
  description of, **3-3**
  example of, **3-13**
Fusion drill, **11-61, 11-85** to **11-90**
  auxiliary machinery, **11-89**
  blowpipe, **11-86, 11-88**
  costs hourly, **8-15**
  description of, **11-61**
  examples of, **11-86, 11-87**
  explosives for, **11-89**
  penetration rates, **11-87** to **11-89**
  performance of, **11-87, 11-88**
  principles of, **11-85** to **11-87**
  production and costs of, **11-89, 11-90**
  specifications for, **11-85, 11-86**
  summary, **11-90**

Gabbro (rock), description of, **2-9**
Galena (mineral), description of, **2-12**
Gangue:
  definition of, **2-13**
  occurrence and removal of, **2-13**
Generators, electric *(see* Engine-generator sets)
Geologic time for North America, **1-7**

Geology:
  importance of earthmover's understanding of, 1-3
  maps, 6-3, 6-5, 6-9, 6-10
  Principle of Uniformitarianism, 1-6
  relationship to excavation, 1-3
  (*see also* Earth, and geology of excavation)
Geomorphology (*see* Landforms and geomorphology)
Georgian series of rocks in geologic time scale, 1-7
Glaciers:
  coverage in North America, 4-4
  example of, 4-6
  landforms created, 4-4 to 4-6
Gneiss (rock):
  description of, 2-11
  examples of, 2-7, 6-6, 13-44, 13-48
Gold (element), description of, 2-11
Gold ore, description of, 2-11
Gorge, example of, 4-8
Gouging rock by glacier:
  definition of, 4-5
  examples of, 1-2, 4-7
Graders:
  motor (*see* Motor graders)
  tow-type, 15-11, 15-14
    costs hourly, 8-5
    description of, 14-12
    examples of, 14-3, 15-13
    performance of, 15-13
    production of, 14-14
      and costs of, 14-14, 14-15, 15-14
    specifications for, 14-14
    working parts of, 14-13
Grand Canyon of Colorado:
  description of, 2-3
  geologic column, 2-3, 2-4
  history of, 2-2
  tour of rocks, 2-3 to 2-5
  view of formations, 2-2
Granite (rock):
  description of, 2-9
  examples of, 1-2, 2-2, 2-6, 2-14, 3-2, 3-7, 4-7, 4-8, 5-12, 7-6, 7-20, 7-35, 11-2, 11-8, 11-95, 11-107, 11-119, 11-137, 12-12, 12-22, 12-36, 12-44, 12-75, 13-38, 13-40, 13-41, 13-43, 13-48, 13-52, 13-70, 13-71, 13-73, 13-74, 14-3, 14-9, 14-12, 14-24, 14-33, 14-34, 15-8, 15-19, 15-20, 18-6, 18-33
Grapples (*see* Cranes, grapples; Dredges, grapple dredge)
Gravel (rock):
  description of, 2-10
  examples of, 4-3, 12-72, 13-89 to 13-90, 13-116, 13-148, 13-152
Gravity, agent for degradation: description of, 4-14
  examples of, 1-5, 1-6, 3-8, 3-9, 4-15

Graywacke (rock), example of, 2-20
Greenstone (rock):
  description of, 2-11
  example of, 2-8
Grooving rock by glacier:
  description of, 4-4
  example of, 4-7
Grubbing (*see* Clearing and grubbing)
Gypsum (mineral), description of, 2-11

Halite (mineral), description of, 2-12
Hard excavation, definition of, 2-13
Haul measurement (*see* Mass diagram for haul measurement)
Haul-road construction and maintenance, 15-1 to 15-21
  construction, 15-3 to 15-10
    alignment, 15-4, 15-5
    design standards, 15-5
    examples of, 15-6 to 15-10
    grades, 15-3, 15-4
    methods and machinery, 15-5, 15-6
    production and costs of, 15-10
    superelevation of curves, 15-4, 15-5
    (*see also* Clearing and grubbing; Dumping and compacting excavation; Fragmentation of rock; Hauling excavation; Loading and casting excavation; Water supply, development of)
  examples of haul road, 15-2, 15-19, 15-20
  maintenance, 15-10 to 15-21
    examples of, 15-12, 15-13, 15-15, 15-16, 15-18
    methods and machinery: motor graders, 15-11
      examples of, 15-2, 15-12
      performance of, 15-12
      production and costs of, 15-11
      (*see also* Motor graders)
    tow-type graders, 15-11, 15-14
      example of, 15-13
      performance of, 15-13
      production and costs of, 15-13
      (*See also* Graders, tow-type)
    water wagons, 15-14
      examples of, 15-15, 15-16, 15-18
      production of, 15-14 to 15-16
      and costs of, 15-16, 15-17
      (*see also* Water wagons)
  savings by good haul road, 15-21
  summary, 15-18, 15-21
Haulers:
  bottom-dump haulers (*see* Bottom-dump haulers)
  brush haulers: costs hourly, 8-15, 8-27
    example of, 9-21
  log haulers: costs hourly, 8-15
    example of, 9-24

Haulers *(Cont.)*:
  machinery haulers *(see* Transporting machinery)
  rear-dump haulers *(see* Rear-dump haulers)
  side-dump haulers *(see* Side-dump haulers)
Hauling excavation, **13**-1 to **13**-170
  engines for machines, characteristics and economics of, **13**-3 to **13**-7 *(see also* Engines)
  generalities for proper use of off-the-road machines, **13**-82 to **13**-84
  machines for *(see specific machine or system)*
  speed indicators of haulers, definition of, **13**-7, **13**-8 *(see also specifications of individual machines)*
  summary, **13**-169, **13**-170
  work-cost indicators of haulers, definition of, **13**-7 to **13**-9 *(see also specifications of individual machines)*
Hawaii forest region:
  description of, **9**-4
  example of, **9**-11
  location of, **9**-2
Hazen-Williams formula, **10**-8 to **10**-17
  coefficient C values for equation, **10**-14
  equation, **10**-8
  graph for flow of water by, **10**-14
  powers of numbers used as constants in flow formulas, **10**-15 to **10**-17
  table for flow of water by, **10**-10 to **10**-14
Hematite (mineral), description of, **2**-11
Highways, excavation for *(see* Open-cut excavations for highways)
Hoses:
  for air *(see* Compressors, air)
  for water *(see under* Water supply)
Huronian system of rocks in geologic time scale, **1**-7
Hydration of rock:
  description of, **3**-4
  example of, **2**-8
Hydraulickers, **13**-138 to **13**-144
  comparison of haulages by hydraulicker and scrapers, **13**-143, **13**-144
  costs hourly of components: monitors, **8**-16
    pipe and pipe fittings, **8**-17
    pumps, centrifugal, with power, **8**-17, **8**-18
  description of, **13**-138
  examples of, **13**-139 to **13**-142
  history of, **13**-138
  machinery and methods, **13**-139, **13**-140, **13**-142
    discharge of monitors or nozzles, **10**-24, **10**-25, **13**-143

Hydraulickers; machinery and methods *(Cont.)*:
  grades of flumes and water consumption, **13**-143
  water requirements for effective hydraulicking, **13**-143
  principles of, **13**-138, **13**-139
  production and costs of, **13**-142, **13**-143
  summary, **13**-143, **13**-144
  work-cost indicator, relative, **13**-144
Hydrolysis of rock:
  description of, **3**-4
  example of, **3**-1

Ice:
  action of, **4**-3 to **4**-7
  aggradation by, **4**-3
    examples of, **4**-5, **4**-6
  degradation by, **4**-3
    example of, **4**-7
  glaciers of North America, **4**-4
  rock weathering by, **3**-5
    example of, **3**-13
Igneous rocks:
  correlations of formations of, **2**-24
  definitions of, **2**-5, **2**-6, **2**-9
  descriptions of, **2**-9
  distribution of excavation, **5**-10
  excavation characteristic of, **2**-9
  kinds of: extrusive or volcanic: definition of, **2**-6, **2**-9
    descriptions of, **2**-9
    examples of, **1**-4, **2**-6, **2**-15 to **2**-17, **3**-8, **4**-2, **4**-14, **4**-15, **6**-2, **7**-16, **7**-23, **11**-10, **11**-13, **11**-16, **11**-118, **12**-8, **12**-9, **12**-12, **12**-25, **12**-36, **12**-54, **12**-81, **13**-66, **17**-2
    intrusive or plutonic: definition of, **2**-5, **2**-6, **2**-9
    descriptions of, **2**-9
    examples of, **1**-2, **2**-6, **3**-1, **3**-7, **4**-7, **5**-6, **5**-13, **7**-15, **7**-35, **10**-45, **11**-8, **11**-57, **11**-95, **11**-110, **11**-118, **11**-137, **12**-22, **12**-75, **13**-40, **13**-43, **13**-52, **13**-70, **13**-73, **14**-7, **14**-17, **14**-24, **14**-33 to **14**-35, **15**-8, **15**-13, **16**-74, **18**-6, **18**-33
  origin of, **2**-5, **2**-6, **2**-9
  seismic shock-wave velocities of, **7**-36
Illumination *(see* Lighting plants)
Initiating devices for explosives *(see* Blasting initiating devices)
Intercooler *(see* Aftercooler or intercooler)
Interest, taxes, insurance, and storage costs *(see* Costs of machinery and facilities)
Intrusive igneous rocks *(see* Igneous rocks, kinds of, intrusive or plutonic)

Jackhammer drills, 7-12 to 7-14, 11-94 to 11-100
  bits: examples of, 7-14
    life of, 11-98
  for blastholes (primary and secondary):
    auxiliary machinery, 11-96, 11-98
    description of, 11-94, 11-96
    examples of, 11-95
    explosives, 11-97
    production and costs of, 11-97, 11-98
    speeds, drilling, 11-96
    summary, 11-99, 11-100
  costs hourly, 8-7
  drill steel: description of, 7-13, 11-94
    life of, 11-98
  for exploration of excavation: auxiliary machinery, 7-13
    bits, 7-14
    description of, 7-12 to 7-14
    example of, 7-15
    speeds, drilling, 7-13
  multipliers for air consumption, 11-98
  specifications for, 11-96
Jet-piercing drill (*see* Fusion drill)
Job scheduling (*see* Schedule of work)
Jurassic system of rocks:
  example of, 2-20
  in geologic time scale, 1-7

Kaolin ore, description of, 2-12
Kaolinite (mineral), description of, 2-12
Keewatinian system of rocks in geologic time scale, 1-7
Keeweenawan system of rocks in geologic time scale, 1-7

Lambert's cosine law, 16-26
Landforms and geomorphology, 4-1 to 4-19
  definition of, 4-3
  forces and agents of aggradation and degradation: crustal movements (*see* Crustal movements)
    gravity (*see* Gravity)
    ice (*see* Ice)
    volcanics (*see* Volcanics)
    water (*see* Water)
    wind (*see* Wind)
  forms and kinds of: anticline: description of, 4-18
    example of, 4-19
    basin, examples of, 4-9, 4-16
    butte, description of, 4-11
    canyon, example of, 4-13
    creep: description of, 4-14
      example of, 4-15
    drumlin: description of, 4-4
      example of, 4-6

Landforms and geomorphology; forms and kinds of *(Cont.):*
    dune: description of, 4-12
      example of, 4-13
    esker, description of, 4-4
    fanglomerate: description of, 4-9, 4-10
      examples of, 4-9, 4-10
    fault (*see* Faults)
    floodplain, description of, 4-18
    lava flow: description of, 1-3, 4-13
      examples of, 1-4, 2-15, 2-17, 4-2, 11-16, 12-9
    mesa: description of, 4-11
      example of, 4-12
    moraine: description of, 4-4
      examples of, 4-5, 4-6
    mountain, examples of, 4-7, 4-9, 5-14
    mudflow: description of, 4-10, 4-11
      example of, 4-11
    peneplain: description of, 4-5
      example of, 4-6
    shore, examples of, 1-5, 4-2, 14-3
    slide: description of, 4-14, 4-16
      examples of, 1-5, 1-6
    slump: description of, 4-14
      example of, 4-15
    syncline, examples of, 4-18, 4-19
    talus slope: description of, 4-16
      examples of, 3-8, 3-9
    terrace: description of, 4-10
      example of, 4-3
    thrust, example of, 4-18
    valley: description of, 4-8, 4-9
      example of, 4-9
    volcano: description of, 4-13, 4-14
      examples of, 1-4, 4-14, 4-15
  summary, 4-19
Lava flow:
  description of, 1-3, 4-13
  examples of, 1-4, 2-15, 2-17, 4-2, 11-16, 12-9
Levees, excavation (*see* Open-cut excavations for dams and levees)
Lighting plants, 16-19 to 16-28
  calculations for, 16-22 to 16-26
  costs for floodlighting, 16-28
  costs hourly, 8-14
  description and kinds of, 16-19
  lamps, 16-27
  portable tower lighting plants: description of, 16-27
    examples of, 16-20 to 16-22
    specifications for, 16-27
  principles of illumination: definitions and complex calculations for, 16-22
    footcandle values for work, 16-23
    inverse cube rule, approximation, 16-16
    inverse square law, 16-26

Lighting plants; principles of
  illumination *(Cont.):*
    isofootcandle diagrams and utilization
      data, **16**-25, **16**-26
    Lambert's cosine law, **16**-26
    lightgrid, **16**-23
Lignite (rock), description of, **2**-10
Limestone (rock), **2**-10
  bituminous limestone, description of,
    2-10
  examples of, **2**-19, **3**-5, **3**-6, **3**-11, **11**-124,
    11-138, **11**-139, **12**-30, **12**-38, **13**-13,
    13-64, **13**-128, **13**-130 to **13**-132
Limonite (mineral), description of, **2**-11
Lineage of a formation, **2**-14
Load factors for hauling machines, **A**-21
Loaders, bucket, **12**-12 to **12**-21
  brush, crawler-tractor: costs hourly,
    8-24
  examples of, **9**-20, **9**-21
  bucket factors, **12**-15
  crawler-mounted machines: casting
    and hauling: cycle time,
      12-16, **12**-17
    example of, **12**-14
    production of, **12**-17
      and costs of, **12**-17
    costs hourly, **8**-16
    loading: cycle time, **12**-14
      example of, **12**-16
      production of, **12**-14, **12**-16
        and costs of, **12**-16
    specifications for, **12**-15
  principles of, **12**-12 to **12**-14
  summary, **12**-21
  wheels-tires-mounted machines: casting and hauling: cycle time,
      12-18, **12**-20
    production of, **12**-18, **12**-20
      and costs of, **12**-20
    costs hourly, **8**-16
    loading: cycle time, **12**-17, **12**-18
      examples of, **12**-15, **12**-20
      production of, **12**-18
        and costs of, **12**-18
    specifications for, **12**-13, **12**-19
Loading and casting excavation, **12**-1 to
  12-85
  Big Muskie: description of, **12**-2
    illustration of, **12**-1
  machines for *(see* Backhoes; Belt loaders; Cranes; Draglines; Loaders,
    bucket; Shovels; Tractors; Trenchers; Wheel-bucket excavators)
  summary, **12**-85
Locomotives:
  costs hourly, **8**-16
  examples of, **13**-128, **13**-130 to **13**-132
  selection procedure for, **13**-134

Locomotives *(Cont.):*
  specifications for, **13**-131
  (see also Railroads)
Loess, definition of, **4**-12

Magazines for storage of explosives:
  costs hourly, **8**-16
  distance for storage, **11**-63, **11**-64
  example of, **11**-64
Magnesite (mineral), description of, **2**-12
Magnesium ore, description of, **2**-12
Magnetite (mineral), description of, **2**-12
Maintaining machinery *(see* Servicing and
  maintaining machinery on the job)
Manganese ore, description of, **2**-12
Marble (rock):
  description of, **2**-11
  example of, **2**-8
Mass diagram for haul measurement:
  calculations for, **17**-25
  description of, **17**-20
  plotting, **17**-22
    table for, **17**-22, **17**-23
  properties of, **17**-21, **17**-22
  purposes of, **17**-20
  uses of, alternative, **17**-22 to **17**-24
Materials:
  approximate angles of repose of, **A**-23
  approximate characteristics of, **A**-3 to **A**-8
  bearing powers of, **A**-25
Medium excavation, definition of, **2**-13
Mercury ore, description of, **2**-12
Mesa:
  description of, **4**-11
  example of, **4**-12
Mesozoic group of rocks in geologic time
  scale, **1**-7
Metamorphic rocks:
  definition of, **2**-7
  descriptions of, **2**-10, **2**-11
  distribution in excavation, **5**-10
  examples of: amorphous, **2**-2, **2**-8, **2**-20,
    3-13, **7**-2, **12**-64
    laminated, **2**-2, **2**-7, **2**-8, **2**-20, **2**-21,
      6-19, **7**-2, **7**-7, **7**-16, **11**-12, **11**-16,
      11-96, **11**-118, **13**-9, **13**-44, **13**-48,
      15-18
  excavation characteristics of, **2**-10, **2**-11
  origin of, **2**-7, **2**-10
  seismic shock-wave velocities of, **7**-36
Meters, water, **10**-19, **10**-30, **10**-44
Minerals *(see* Ores and minerals)
Mines, excavation for *(see* Open-cut excavations, for pits and open mines)
Miocene series of rocks in geologic time
  scale, **1**-7
Mississippian series of rocks:
  in geologic time scale, **1**-7
  in Grand Canyon of Colorado, **2**-2 to **2**-4

Mixing excavation (*see* Spreading-mixing excavation)
Mohawkian series of rocks in geologic time scale, **1**-7
Molybdenite (mineral), description of, **2**-12
Monitors and nozzles, **10**-30
　examples of, **10**-2, **11**-13, **14**-26
　flow of water through, **10**-20, **10**-24, **10**-25, **13**-143
　for wetting excavation (*see* Wetting-compacting excavation, wetting excavation, by monitors or nozzles)
Moraine:
　description of, **4**-4
　lateral moraine, example of, **4**-5
　terminal moraine, examples of, **4**-5, **4**-6
Motor graders, **14**-8
　costs hourly, **8**-15
　specifications for, **14**-10
　(*see also* Haul-road construction and maintenance, maintenance, methods and machinery, motor graders; Spreading-mixing excavation)
Motors, electric, **16**-68 to **16**-76
　ac motors: dimensions and weights of, **16**-73
　　efficiencies of, **16**-74
　　example of, **16**-74
　　kinds of, **16**-69, **16**-70
　costs hourly, **8**-17
　dc motors: dimensions and weights of, **16**-70
　　efficiencies of, **16**-71
　　kinds of, **16**-68, **16**-69
　descriptions of, **16**-68
　NEMA design and classifications of, **16**-71, **16**-75, **16**-76
　production and costs of, **16**-71, **16**-72
Mountain examples of, **4**-7, **4**-9, **5**-14
Mudflows:
　description of, **4**-10, **4**-11
　example of, **4**-11

Niagaran series of rocks in geologic time scale, **1**-7
Niccolite (mineral), description of, **2**-12
Nickel ore, description of, **2**-12
Nitrate ore, description of, **2**-12
Northern forest region:
　description of, **9**-4
　example of, **9**-5
　location of, **9**-2
Nozzles (*see* Monitors and nozzles)

Obsidian (rock), description of, **2**-9
Offices (*see* Yards, shops, and offices)

Oligocene series of rocks in geologic time scale, **1**-7
Open-cut excavations, **5**-1 to **5**-18
　for airports: description of, **5**-3, **5**-4
　　examples of, **5**-3, **5**-4
　for building sites and large foundation excavations: description of, **5**-4, **5**-5
　　examples of, **5**-4, **13**-45
　for canals: description of, **5**-5, **5**-6
　　examples of, **5**-5, **5**-6
　for dams and levees: description of, **5**-6 to **5**-8
　　examples of, **5**-7, **5**-8, **12**-25, **12**-37, **12**-72, **12**-81
　distributions of rock in, **5**-10
　for highways: description of, **5**-8 to **5**-10
　　examples of, **5**-9, **6**-18, **6**-19, **13**-2
　for pits and open mines: description of, **5**-10, **5**-11
　　examples of, **5**-11, **11**-87, **12**-1, **12**-23, **13**-29, **13**-126
　for quarries: description of, **5**-11, **5**-12
　　examples of, **5**-12, **12**-15, **13**-128, **13**-132
　for railroads: description of, **5**-12, **5**-13
　　example of, **5**-14
　for sanitary fills: description of, **5**-13 to **5**-16
　　examples of, **5**-15, **14**-45
　summary, **5**-17, **5**-18
　for trenches and small foundation excavations: description of, **5**-16, **5**-17
　　examples of, **5**-17, **12**-38, **12**-62 to **12**-65
Operating expenses (*see* Costs of machinery and facilities)
Orange-peel bucket (*see* Buckets, orange-peel-type)
Ordovician system of rocks:
　example of, **2**-21
　in geologic time scale, **1**-7
Ores and minerals:
　classification of, **2**-13
　deposition of, **2**-13
　description of, **2**-11 to **2**-13
　(*see also specific ores and minerals*)
Oswegan series of rocks in geologic time scale, **1**-7
Overburden, examples of, **2**-19, **12**-33, **13**-58, **13**-123
Ownership and operation costs of machinery and facilities (*see* Costs of machinery and facilities)
Oxidation of rock, description of, **3**-4

Pacific coast forest region:
　description of, **9**-4
　example of, **9**-9
　location of, **9**-2

Paleocene series of rocks in geologic time scale, 1-7
Paleozoic group of rocks:
   in geologic time scale, 1-7
   in Grand Canyon of Colorado, 2-3, 2-4
Peneplain:
   description of, 4-5
   example of, 4-6
Penetration, rate of drilling, 7-6, 7-8,
Pennsylvanian series of rocks:
   examples of, 2-18, 2-25
   in geologic time scale, 1-7
   in Grand Canyon of Colorado, 2-2 to 2-4
Percussion drill:
   description of, 11-62
   (*see also* Cable drills; Drifter drills; Jackhammer drills; Track drills)
Permian series of rocks:
   in geologic time scale, 1-7
   in Grand Canyon of Colorado, 2-2 to 2-4
Phosphate ore:
   description of, 2-12
   examples of, 12-56, 13-91, 13-141, 13-142
Pickup trucks, costs hourly, 8-7
Pipe:
   for air (*see* Compressors, air)
   for water (*see* Water supply, development of)
Pits and open mines (*see* Open-cut excavations, for pits and open mines)
Placer gold, description of, 2-11
Planimeter, description of, 17-3
Plans and specifications for work (*see* Examination of excavation)
Plant roots and burrowing animals:
   description of, 3-11
   weathering by, 3-13
Pleistocene series of rocks:
   in geologic time scale, 1-7
   glaciation and land forms, 4-3, 4-4
Pliocene series of rocks in geologic time scale, 1-7
Plow, root, crawler-tractor: costs hourly, 8-24
     example of, 9-29
Plutonic igneous rocks (*see* Igneous rocks, kinds of, intrusive or plutonic)
Porphyry (rock):
   description of, 2-9
   example of, 5-2
Potassium ore, description of, 2-12
Powder, black, description of, 11-23
Precambrian groups and systems of rocks:
   examples of, 2-7, 2-25
   in geologic time scale, 1-7
   in Grand Canyon of Colorado, 2-1 to 2-4
Preparation of bid and schedule of work (*see* Bid preparation; Schedule of work)

Prismoidal formula, description of, 17-4, 17-5
Production of machinery:
   continuous production: efficiency of, 8-34, 8-35
     equations for, 8-35
     example of, 8-34
   definition of, 8-34
   intermittent production: efficiency of, 8-35
     equations for, 8-35
     example of, 8-35
     selection of efficiency, 8-35
   principles of, 8-34, 8-35
Proterozoic group of rocks:
   in geologic time scale, 1-7
   in Grand Canyon of Colorado, 2-2 to 2-4
Pumice, rock, description of, 2-9
Pumps:
   for unwatering for excavation, 16-12 to 16-19
   water, 10-30 to 10-39, 16-12 to 16-19
     centrifugal suction pumps, 16-13
     centrifugal submersible pumps, 16-13
     description of, 16-12
     diaphragm pumps, 16-13, 16-17
     for large unwatering jobs, 16-18, 16-19
     centrifugal submersible-type: costs hourly, 8-18
       description of, 16-13
       examples of, 16-16, 16-17
       performances of, 16-14
       production of, 16-16, 16-17
       specifications for, 16-14
     centrifugal turbine, deep-well-type:
       costs hourly, 8-19
       description of, 10-31, 10-38, 10-39
       example of, 10-38
       performance of, 10-39
       production of, 10-31, 10-38
       ratings for, 10-36, 10-37
       working parts for, 10-35
     centrifugal volute-type, 10-30, 16-13
       suction intake: costs hourly, 8-17, 8-18
       description of, 10-30, 10-31
       examples of, 10-32, 13-158, 16-4, 16-7, 16-14
       flow of water in, 10-32
       performance curves, 10-33
       production of, 16-14
       ratings for, 10-34, 10-36, 10-37
     diaphragm-type: costs hourly, 8-18
       description of, 16-13, 16-17
       examples of, 16-18, 16-19
       operation and working parts for, 16-18
       performance of, 16-15
       production of, 16-18
       and costs of, 16-17

Pumps; water; diaphragm-type *(Cont.)*:
    specifications for, **16**-15
    (*See also* Water supply, development of, Wellpoint systems)
Pyrolusite (mineral), description of, **2**-12

Quarries, excavation for (*see* Open-cut excavations for quarries)
Quartzite (rock):
    description of, **2**-11
    examples of, **2**-3, **3**-13
Quaternary system of rocks in geologic time scale, **1**-7

Railroads:
    excavation for (*see* Open-cut excavations for railroads)
    hauling, **13**-128 to **13**-138
        costs hourly: for cars, **8**-10
            for locomotives, **8**-16
        examples of, **13**-128, **13**-130 to **13**-132
        physical and economic criteria for, **13**-128
        principles and dynamics of: factors, **13**-129
            hauling cycle calculations for train, **13**-136, **13**-137
            locomotive selection procedure, **13**-134
                coefficients of adhesion or traction, **13**-134
                horsepower, **13**-135
                weight, **13**-135, **13**-136
            rail data: capacity, **13**-129
                gauge, **13**-129
                size, **13**-129
            resistance to train motion: by acceleration, **13**-133
                by air, **13**-133
                by curves, **13**-130
                by grades, **13**-130
                by total resistances, **13**-133
                by train, **13**-130
            roadbeds, **13**-129
        production and costs of, **13**-137
        specifications: for cars, **13**-132
            for locomotives, **13**-131
        summary, **13**-138
        work-cost indicator, **13**-138
Rake, brush, crawler-tractor, **9**-22, **9**-29
    costs hourly, **8**-24
    example of, **9**-18
Random fill, description of, **14**-3
Rear-dump haulers, **13**-63 to **13**-74
    body factors for loads, **13**-70
    costs hourly, **8**-26, **8**-28, **8**-29
    cycle time, **13**-70 to **13**-73

Rear-dump haulers *(Cont.)*:
    examples of, **5**-3, **5**-6, **5**-8, **13**-64 to **13**-67, **13**-70 to **13**-74, **14**-9, **15**-17, **15**-18, **15**-22
    production and costs of, **13**-72
    specifications for: off-the-road machines, **13**-68
        on-the-road and off-the-road machines, **13**-68, **13**-69
    summary, **13**-74
    types and applications of, **13**-63 to **13**-65
    (*See also* Scrapers, and haulers, wheels-tires-mounted)
Receiver, air compressor, **11**-62
Repairs and replacements, costs of (*see* Costs of machinery and facilities)
Replacement cost escalation (*see* Costs of machinery and facilities)
Reservoirs, water: description of, **10**-41
    example of, **10**-42
Residuals:
    description of, **3**-3
    examples of, **3**-3, **3**-5, **7**-7, **7**-8, **7**-11, **13**-123, **14**-41
Retarders, brake: description of, **13**-30
    use of, **13**-37
Rhyolite (rock), description of, **2**-9
Rift zone, examples of, **2**-25, **3**-12, **4**-19, **6**-4, **6**-5
Rip-rap:
    description of, **14**-3
    examples of, **12**-44, **12**-50, **14**-3
Ripper:
    attached-type (*see* Ripping)
    pull-type: costs hourly, **8**-19
        history of, **11**-5
Ripping, **11**-5 to **11**-22
    comfortable ripping, example of, **11**-8
    composite heavyweight tractor-bulldozer-ripper: idealized data for performance, **11**-10
        specifications for, **11**-7
    considerations for, **11**-4
    depths and spacings of furrows, **11**-9
    fragmentation by ripping and blasting, **11**-2, **11**-3
    kinds of rocks affecting ripping, **11**-9 to **11**-11
    methods of machinery for: controlling factors, **11**-11, **11**-12, **11**-14 to **11**-16
        examples of, **11**-2, **11**-8, **11**-9, **11**-11 to **11**-17
    principles of, **11**-5
    production and costs of: estimated costs of ownership and operation of machines, **11**-19, **11**-20
        estimated production and costs for ripping, **11**-18
        examples of, **11**-20, to **11**-22
        factors affecting, **11**-16, **11**-17

Ripping *(Cont.)*:
  ripper assembly of tractor: bulldozer-
      type, **11**-15
    hinged-type, **11**-8
    parallelogram-type, **11**-7
    parts for, **11**-5, **11**-6
    seismic shock-wave velocity relation-
      ship to ripping, examples of, **7**-29,
      7-30, **7**-32, **7**-34, **7**-36, **11**-2, **11**-3,
      11-8, **11**-10, **11**-11, **11**-14, **11**-16 to
      11-18
    speeds for, **11**-8 to **11**-10
    summary, **11**-139
  tractor-bulldozer-rippers for: classifica-
      tions of, **11**-5
    costs hourly, **8**-25
    history of, **11**-5
    specifications: for basic tractors, **12**-8,
      13-5
      for complete machines, **11**-5, **11**-7,
      11-11, **11**-17,
  transition from ripping to blasting, **11**-2,
      11-3, **11**-18
  uncomfortable ripping, example of, **11**-9
Rock clauses in contract, **A**-9
  for subsurface excavation, **A**-9, **A**-10
  for surface excavation, **A**-9
Rock excavation:
  definitions of, **1**-3, **11**-3, **11**-5
  percentage distribution of kinds of,
    5-10
Rock stream, example of, **4**-16
Rock weathering, **3**-1 to **3**-15
  definition of, **3**-4
  factors affecting, **3**-14
  forces for: chemical decomposition, **3**-4,
    3-5
    mechanical disintegration, **3**-5 to **3**-11
  rates of, **3**-12, **3**-14
  significance of, **3**-3, **3**-4
  summary, **3**-15
  (*See also* Decomposition of rock; Defla-
    tion of rock; Deformation of rock;
    Degradation of land; Disintegration
    of rock)
Rocks, **2**-1 to **2**-13
  classification of, **2**-5 to **2**-9
  definition of, **2**-5
  description of, **2**-9 to **2**-11
  in Grand Canyon of Colorado: forma-
    tions of, **2**-1 to **2**-4
    tour through rocks of, **2**-3 to **2**-5
  summary, **2**-26
  (*See also* Formations, rock; Igneous
    rocks; Metamorphic rocks; Ores and
    minerals; Sedimentary rocks)
Rocky Mountain forest region:
  description of, **9**-4
  example of, **9**-8
  location of, **9**-2

Rollers (*see* Wetting-compacting excava-
    tion)
Rolling resistances:
  of crawlers, **13**-16
  of trains, **13**-133
  of wheels-tires-mounted rock-earth
    haulers, **13**-16
Root plow, crawler-tractor (*see* Plow, root,
  crawler-tractor)
Roots, tree (*see* Clearing and grubbing)
Rotary drill, **7**-15 to **7**-17, **11**-125 to **11**-133
  auxiliary machinery, **11**-130
  bits, **7**-15
    cone bits: examples of, **7**-17, **11**-130
      life of, **11**-133
    drag bits: examples of, **7**-10
      life of, **11**-32
  for blast holes, **11**-125, **11**-126
    examples of, **11**-124, **11**-125, **11**-131,
      11-132
    explosives, **11**-132
    production and costs of, **11**-132,
      11-133
    speeds, drilling: with drag bits, **11**-131
      with tri-cone bits, **11**-131
    summary, **11**-133
  costs hourly, **8**-21
  description of, **11**-125, **11**-126
  for exploring excavation, **7**-15 to **7**-17
    bits, **7**-10; **7**-17
    costs hourly, **8**-21
    description of, **7**-15, to **7**-17
    example of, **7**-18
    speeds, drilling, **7**-16
  specifications for, **11**-130

Salt ore, description of, **2**-12
San Andreas fault:
  description of, **3**-10
  effects on excavation, **4**-19, **6**-4
  in fault system of California, **3**-12
  rift zone, **3**-10
Sand (rock):
  description of, **2**-10
  examples of, **4**-13, **5**-15, **12**-67, **14**-38,
    16-4, **16**-5, **16**-7
Sandstone (rock):
  description of, **2**-10
  examples of, **2**-6, **2**-17, **2**-18, **3**-9, **6**-2, **7**-7,
    7-12, **11**-11, **12**-65, **14**-23, **15**-12
Sanitary fills, excavation for (*see* Open-cut
  excavations for sanitary fills)
Saratoga series of rocks in geologic time
  scale, **1**-7
Saws:
  chain-type: costs hourly, **8**-21
    example of, **9**-16
  circular-type: costs hourly, **8**-21
    example of, **9**-21

Scales, platform (*see* Weighing excavation, scales)
Schedule of work, **18**-29 to **18**-36
    critical-path method: diagram of, **18**-30
        example of, **18**-31 to **18**-35
        principles of, **18**-30
    observations and values of, **18**-35
Schist (rock):
    description of, **2**-11
    examples of, **6**-19, **7**-7, **11**-12, **11**-12, **11**-16, **13**-9, **15**-8
Scoria (rock):
    description of, **2**-9
    examples of, **2**-17, **4**-15
Scrapers:
    and haulers, wheels-tires-mounted, **13**-31 to **13**-37
        cycle times, **13**-31
            hauling and returning times: fundamental calculations for, **13**-31 to **13**-33
            simple precise method for calculations for, **13**-33 to **13**-37
            loading times, **13**-31
            turnings and dumping times, **13**-31
        development of rock haulers, **13**-27
        kinds of machines, **13**-26
        principles of haulage, **13**-26 to **13**-22
            coefficients of traction, **13**-28
            dynamics of haulage, **13**-28 to **13**-31
            rolling resistances, **13**-28
            tractive force, **13**-28, **13**-29
            wheel horsepower versus engine horsepower, **13**-28
        selection considerations for scraper or hauler, **13**-80, **13**-81
        speed indicators, description and use of, **13**-7, **13**-8
        work-cost indicators, description and use of, **13**-7 to **13**-9
        (*See also* Bottom-dump haulers; Rear-dump haulers; Side-dump haulers)
    self-propelled, **13**-37 to **13**-51
        description and kinds of, **13**-37
        push-loaded scrapers: bowl factors for loads, **13**-20, **13**-42
            costs hourly, **8**-22
            cycle times, **13**-38, **13**-39, **13**-41 to **13**-43
            description of, **13**-37
            examples of, **5**-10, **13**-38, **13**-40, **13**-41, **13**-43 to **13**-45
            operation of, **13**-37, **13**-38, **13**-40, **13**-41
            production and costs of, **13**-42, **13**-43
            resolution of loading forces, **13**-38, **13**-39

Scrapers; push-loaded scrapers; self-propelled (*Cont.*):
            specifications for, **13**-39, **13**-40
        self-loaded scrapers: advantages and disadvantages of, **13**-44, **13**-45
            bowl factors for loads, **13**-20, **13**-49
            costs hourly, **8**-23
            cycle times, **11**-49
            description of, **13**-44
            examples of, **13**-46 to **13**-48
            fine grading by, **13**-45, **13**-48
            on-the-road haulage by, **13**-45, **13**-49
            operation of, **13**-44, **13**-45
            production and costs of, **13**-50, **13**-51
            specifications of, **13**-46
        summary, **13**-51
        tow-type (*see* Tractors, crawler-mounted, pulling scrapers and wagons and)
Sedimentary rocks:
    attitude and stratification, **2**-20, **2**-22
    correlation of formations, **2**-22 to **2**-25
    definitions of, **2**-6, **2**-10
    descriptions of, **2**-10
    dip and strike, **2**-22
    distribution in excavation, **5**-10
    excavation characteristics of, **2**-10
    forms of: consolidated, examples of, **1**-6, **2**-17 to **2**-19, **3**-5, **3**-9, **4**-12, **5**-3, **5**-14, **6**-14, **7**-9, **7**-18, **10**-4, **11**-14 to **11**-16, **11**-57, **11**-124, **11**-138, **12**-10, **12**-14, **12**-24, **12**-32, **12**-38, **12**-56, **12**-59, **12**-60, **12**-73, **12**-78, **13**-11 to **13**-13, **13**-24, **13**-47, **13**-58, **13**-128, **13**-131, **15**-20, **17**-16
        unconsolidated, examples of, **4**-3, **4**-6, **4**-10, **4**-13, **5**-15, **7**-9, **12**-67, **12**-80, **12**-83, **13**-54, **13**-89 to **13**-90, **13**-115, **13**-125, **13**-140 to **13**-142, **13**-152, **14**-13, **14**-31, **14**-38, **14**-41, **14**-45, **16**-29
    origin of, **2**-6, **2**-10
    seismic shock-wave velocities of, **7**-36
Seismic timer (*see* Exploration of excavation, seismic timer)
Semisolid rock, examples of, **3**-2, **3**-4, **3**-6
Serpentine (rock):
    description of, **2**-11
    example of, **2**-8
Servicing and maintaining machinery on the job, **16**-82 to **16**-87
    maintaining machinery: description of, **16**-85 to **16**-87
        examples of, **16**-85, **16**-86
        machinery and methods of, **16**-85
    servicing machinery: description of, **16**-82 to **16**-84
        lubrication van: costs hourly, **8**-28
            description of, **16**-82
            examples of, **16**-2, **16**-83
            production and costs of, **16**-83

Servicing and maintaining machinery on
    the job; servicing machinery
    (Cont.):
  specifications for, **16**-84
  refueling tanker: costs hourly, **8**-28
    description of, **16**-83
    production and costs of, **16**-84
  summary, **16**-90
  yard and buildings, example of, **16**-8
Shale (rock):
  description of, **2**-10
  examples of, **2**-17, **5**-3, **6**-6, **6**-14, **12**-73, **13**-67
Shops *(see* Yards, shops, and offices)
Shore, examples of, **1**-5, **4**-2, **14**-3
Shot or calyx drill *(see* Exploration of excavation, machinery, drills, mechanical, core drills, calyx drill)
Shovels, **12**-21 to **12**-33
  casting: examples of, **12**-30, **12**-32, **12**-33
    principles of, **12**-30
    production and costs of, **12**-31
  costs hourly, **8**-23
  cycle times, **12**-24 to **12**-29
  description of, **12**-21
  dipper factors, **12**-25
  loading: efficiency of, **12**-29
    examples of, **12**-22 to **12**-25
    production and costs of, **12**-29, **12**-30
  principals of, **12**-21, **12**-23
  specifications for, **12**-22, **12**-26 to **12**-28
  summary, **12**-31, **12**-32
Shrink and swell of earth-rock *(see* Swell, and shrink of earth-rock)
Side-dump haulers, **13**-74 to **13**-80
  costs hourly, **8**-26 to **8**-29
  cycle times, **13**-79
  description of, **13**-74, **13**-79
  examples and specifications for, **13**-75 to **13**-78
  loads, **13**-79
    body factors, **13**-80
  production and costs of, **13**-79, **13**-80
  summary, **13**-80
  (See also Scrapers, and haulers, wheels-tires-mounted)
Silt (rock):
  description of, **2**-10
  examples of, **5**-15, **14**-45
Siltstone (rock):
description of, **2**-10
examples of, **5**-55, **14**-23, **14**-39, **14**-45, **15**-12
Silurian system of rocks in geologic time scale, **1**-7
Silver ore, description of, **2**-12
Slate (rock):
  description of, **2**-11
  examples of, **2**-7, **2**-21

Slide:
  description and causes of, **1**-4, **1**-5
  examples, **1**-5, **1**-6
  prevention of, **1**-5
Slings, rock: description of, **12**-49
  example of, **12**-44
Slump:
  description of, **4**-14
  example of, **4**-15
Snell's law:
  definition of, **7**-28
  relationship of, to use of seismic timer, **7**-28, **7**-29
Soapstone (rock):
  description of, **2**-11
  example of, **12**-64
Soda niter (mineral), description of, **2**-12
Soft excavation, definition of, **2**-13
Soil survey sheet, example of, **6**-8
Solid rock, examples of, **3**-3, **3**-8, **11**-10, **11**-49, **11**-119
Solution of rock:
  description of, **3**-4
  example of, **3**-5
Southern forest region:
  description of, **9**-4
  example of, **9**-7
  location of, **9**-2
Specific gravities of materials, **A**-5 to **A**-8
Speed indicator, description and use of, **13**-7, **13**-8
  (See also specifications of individual machines)
Sphalerite (mineral), description of, **2**-13
Spheroidal weathering:
  description of, **3**-10
  examples of, **3**-3, **3**-7
Spreading-mixing excavation, **14**-5 to **14**-16
  by backhoes: costs hourly, **8**-7
    example and production of, **14**-16
    (See also Backhoes)
  by motor graders: costs hourly, **8**-15
    description of, **14**-8, **14**-12
    examples of, **14**-11, **14**-12
    production and costs of, **14**-5
    specifications for, **14**-10
    working parts for, **14**-10
  principals of, **14**-5
  by tow-type graders: costs hourly, **8**-15
    description of, **14**-12
    example of, **14**-13
    production of, **14**-14
      and costs of, **14**-14, **14**-15
    specifications for, **14**-14
    working parts for, **14**-13
  by tractor-bulldozers: crawler-mounted:
    costs hourly, **8**-24
    description of, **14**-5
    examples of, **14**-2, **14**-6

Spreading-mixing excavation; by
tractor-bulldozers *(Cont.):*
production of, **14**-5, **14**-7, **14**-8
specifications for, **12**-8, **13**-18
wheels-tires-mounted: costs hourly,
8-25
description of, **14**-5, **14**-7
examples of, **14**-7, **14**-9
production of, **14**-5, **14**-7, **14**-8
and costs of, **14**-7, **14**-8
specifications for, **14**-6
Stands, pipe, water, **10**-45, **10**-46
Stockpiles, examples of, **13**-88, **13**-152
Stratification of rocks:
attitude, dip and strike, **2**-20 to **2**-22
effect on economics of mining, **2**-21
effect on ripping, loading, and drilling,
2-21
thickness of beddings, **2**-13
Strike (*see* Dip and strike of rock formation)
Stumping or stump removal (*see* Clearing
and grubbing)
Submergency of land, example of, **4**-2
Sulfur (element), description of, **2**-12
Sulfur ore, description of, **2**-12
Swell:
and shrink of earth-rock, **17**-14, **17**-15,
A-2 to **A**-8, **A**-21
versus voids in materials and hauling
machine load factors, **A**-21
Syenite (rock), description of, **2**-12
Sylvite (mineral), description of, **2**-12
Syncline:
examples of, **4**-18, **4**-19
occurrence of, **4**-18

Taconite (mineral):
description of, **2**-12
examples of, **11**-55, **11**-86, **11**-87
Talus slope:
description of, **4**-16
examples of, **3**-8, **3**-9
Tanks, water (*see* Water supply, development of, machinery and methods,
tanks)
Temiskamian series of rocks in geologic
time scale, **1**-7
Temperature changes in rocks:
effects of, on rock disintegration, **3**-5, **3**-6
example of, **3**-3
Terrace, river: description of, **4**-10
example of, **4**-3
Tertiary system of rocks in geologic time
scale, **1**-7
Thrusts:
description of, **4**-18
example of, **4**-18

Till (rock):
description of, **4**-4
examples of, **4**-5, **4**-6
Tillite (rock), description of, **4**-5
Timber, merchantable: calculations for,
9-15
importance in clearing and grubbing,
9-13, **9**-14
Timber cruise:
description of, **9**-13, **9**-14
example of, **9**-14
Tin ore, description of, **2**-12
Tires, **16**-30 to **16**-50
construction of: nomenclature for, **16**-31
nylon-ply, **16**-30, **16**-32
radial-ply, with steel-ply carcass,
16-30, **16**-33
treads, basic, **16**-34
costs of, **16**-30, **16**-50
examples of, **16**-30, **16**-34, **16**-86
factors affecting life of: description of,
16-33, **16**-35, **16**-36
example of failures, **16**-36
three important factors, **16**-35
kinds of, **16**-30
life of, **16**-46, **16**-47
estimating: according to tire life factors, **16**-48, **16**-49
according to working conditions,
16-47
recapping, **16**-49
life and cost of, **16**-49
repairs to, estimating costs of, **16**-50
selection of: for specific or average work,
16-36, **16**-45
ton-mile-per-hour rating system for,
16-45, **16**-46
specifications for, **16**-36
standard-type, **16**-37 to **16**-41
wide-base-type, **16**-42 to **16**-44
treads, **16**-32, **16**-34
types of, **16**-30, **16**-31
Tonalite (rock):
description of, **2**-14
examples of, **2**-14, **3**-2
Topography of excavation:
examples of, **6**-2, **12**-67, **13**-2, **13**-130
maps, **6**-3, **6**-10, **18**-5, **18**-7
Track drills, **7**-14 to **7**-17, **11**-106 to **11**-123
auxiliary machinery, **11**-112
calculations for, **11**-112, **11**-120,
11-121
bits: cost comparisons, **11**-108
description of, **11**-103, **11**-121
examples of, **7**-14, **11**-103, **11**-126
grinding, **11**-104
life of, **11**-105
for blastholes: description of, **11**-106
efficiency of, **11**-106, **11**-110

Index   23

Track drills; for blastholes *(Cont.):*
    examples of, **11**-107, **11**-109, **11**-110, **11**-118, **11**-119
    explosives, **11**-121
    production of: and costs of, **11**-121, **11**-122
      estimating, **11**-110 to **11**-112
    speeds, drilling, **11**-109
      examples of, **11**-110
        methods for estimating, **11**-110 to **11**-112
  costs hourly, **8**-6
  drill rods: description of, **11**-121
    life of, **11**-105
  for exploring excavation, **7**-14, **7**-15
    bits, **7**-44, **11**-103
    costs hourly, **8**-6
    description, **7**-14 to **7**-16
    example of, **7**-16
    speeds, drilling, **7**-14
  specifications for, **11**-109
  summary, **11**-122, **11**-123
  working parts for, **11**-108
Tractors:
  crawler-mounted: with bulldozer or angledozer, **12**-3 to **12**-12, **13**-9 to **13**-11
    capacity of bulldozer or angledozer blades, **12**-4, **12**-6 to **12**-8
    casting: cycle times, **12**-10
      description of, **12**-9, **12**-10
      example of, **12**-12
      for exploration of excavation, **7**-5, **7**-7
      production and costs of, **12**-10, **12**-12
    costs hourly, **8**-24
    definition and description of, **12**-3, **12**-4
    hauling: cycle times, **12**-6, **13**-10
      description of, **13**-9
      examples of, **13**-9 to **13**-11
      production and costs of, **13**-9 to **13**-11
    loading: cycle times, **12**-6
      description of, **12**-7
      examples of, **12**-8 to **12**-10
      production and costs of, **12**-7, **12**-9
    resolution of forces, **12**-5
    specifications: for basic tractor, **13**-18
      for tractor-bulldozer, **12**-8
    speed indicator, **13**-11
    summary, **12**-12
    work-cost indicator, **13**-11
  pulling scrapers, wagons and, **13**-11 to **13**-35
    description of, **13**-12

Tractors; crawler-mounted *(Cont.):*
  scrapers: bowl factors for loads, **13**-20
    costs hourly, **8**-22
    cycle times, **13**-20
    examples of, **13**-12, **13**-21
    operation of, **13**-19
    principals of, **13**-11, **13**-12
    production and costs of, **13**-20 to **13**-22
    specifications for, **13**-19
  summary, **13**-35
  tractors: coefficient of traction, **13**-13 to **13**-15
    costs hourly, **8**-24
    drawbar horsepower, **13**-12 to **13**-14
    dynamics of haulage, **13**-15 to **13**-59
    elements of calculating forces and resistances, **13**-17
    rolling resistances of crawler- and wheels-tires-mounted haulers, **13**-16
    specifications for, **13**-18
  wagons: body factor loads, **13**-23
    costs hourly, **8**-29
    cycle times, **13**-24
    examples of, **13**-13
    operation of, **13**-22
    principles, **13**-11, **13**-12
    production and costs of, **13**-24, **13**-25
    specifications for, **13**-23
  with ripper (*see* Ripping)
  wheels-tires-mounted, with bulldozer (*see* Spreading-mixing excavation, by tractor-bulldozers)
Transporting machinery, **16**-72, **16**-76 to **16**-82
  costs hourly of machinery haulers, **8**-25, **8**-28
  legal loads on highways, example of, **16**-77 to **16**-78
  means for, **16**-72, **16**-76
  methods and machinery:
    carry-tow method, **16**-79
    considerations for, **16**-76
    gooseneck-type hauler, **16**-81
    tilt-type trailer, **16**-79
  production and costs of, **16**-81, **16**-82
  specifications for gooseneck-lowboy-type haulers, **16**-80
Trap rock, description of, **2**-9
Trapezoidal formula, **17**-3
Tree pusher, crawler-tractor: costs hourly, **8**-24
  example of, **9**-17

Tree splitter, crawler-tractor: costs hourly, 8-24
  example of, 9-17
Trees, (see Clearing and grubbing)
Trenchers, 12-61 to 12-66
  description of work, 12-61
  efficiency of, 12-61, 12-62
  ladder trencher: costs hourly, 8-27
    description of, 12-61
    examples of, 12-62 to 12-64
    production and costs of, 12-61
    specifications for, 12-63
  production of, 12-66
  summary, 12-66
  wheel trencher: costs hourly, 8-27
    description of, 12-62, 12-63
    example of, 12-65
    production and costs of, 12-66
    specifications for, 12-23
    working parts for, 12-65
Trenches:
  for exploring excavation, examples, 7-2, 7-6, 7-35
  (see also Open-cut excavations, for trenches and small foundation excavations)
Triassic system of rocks in geologic time scale, 1-7
Tropical forest region:
  description of, 9-4
  example of, 9-12
  location of, 9-2
Trucks (see Bottom-dump haulers; Rear-dump haulers; Side-dump haulers; Water wagons)
Tuff (rock):
  description of, 2-15
  examples of, 2-16, 11-13
  (See also Breccia)
Turbine pump (see Pumps, water, centrifugal turbine)

Uniformitarianism, Principal of, 1-6
  examples of use of, 2-22 to 2-25
Uraninite (mineral), description of, 2-13
Uranium ore, description of, 2-13

Valley:
  description of, 4-8, 4-9
  example of, 4-9
Valves, water: description of, 10-28
  example of, 10-18
  friction losses in, 13-163
Volcanic igneous rock (see Igneous rock, kinds of, extrusive or volcanic)
Volcanics:
  agent of aggradation, 4-13

Volcanics (Cont.):
  examples of, 1-3, 1-4, 2-15 to 2-17, 4-14, 4-15
Volumes of earthwork, calculations for, 17-4
  according to manner of fragmentation, 17-13, 17-14, 17-16, 17-17
  by approximate graphical method, 17-26 to 17-30
  by average-end-area method, 17-5, 17-10
    with prismoidal adjustment, 17-13
  by contour method, 17-5, 17-11 to 17-13
  by prismoidal formula, 17-4, 17-5
  by subdivision of volume, 17-4, 17-6 to 17-9
Volumes and weights of excavation:
  according to position or condition, 17-16, 17-17, A-2
  in cut, in loose condition, and in fill, A-3 to A-8
Volute pumps (see Pumps, water, centrifugal volute-type)

Wagons:
  rock (see Tractors, crawler-mounted, pulling scrapers and wagons and wagons)
  water (see Water wagons)
Wash-boring drills:
  cable-drill-type: costs hourly, 8-10
    description of, 7-26
  manual-type: description of, 7-3
    example of, 7-4
Waste fills, example of, iii
Water:
  action of, 4-7 to 4-12
  aggradation by: description of, 4-7, 4-8
    examples of, 4-9 to 4-11
  degradation by: description of, 4-7, 4-8
    example of, 4-8
  rock weathering by: description of, 3-4, 3-5
    examples of, 3-5, 3-9
Water supply, development of, 10-1 to 10-52
  Bernoulli's theorem, 10-7
    general equation, 10-7, 10-8
  C values in equations for flow of water, 10-14
  considerations for, 10-4, 10-6
  conversion formulas for, 10-3
  designs of, 10-6, 10-41 to 10-51
  equations for, fundamental, 10-6 to 10-8
  examples of, 10-2, 10-4, 10-5, 10-32, 10-38, 10-40, 10-42, 10-45, 10-46, 10-49, 10-50

Index 25

Water supply, development of *(Cont.):*
 flow and friction of water: in changes
   of pipe configuration, **10**-21 to **10**-23
  in hoses, **10**-9, **10**-19
  in meters, **10**-19
  in nozzles, **10**-20, **10**-24, **10**-25
  in pipe appurtenances, **10**-9, **10**-18,
   10-19, **10**-21
  in pipes, **10**-8, **10**-10 to **10**-14
  general equation for flow of water,
   10-7, **10**-8
   example of use of, **10**-22, **10**-23,
    10-25 to **10**-27
  Hazen-Williams formula, **10**-8 to **10**-17
   coefficient *C* values for equation,
    10-14
   equation, **10**-8
   graph for flow of water by, **10**-14
   powers of numbers used as constants in flow formulas, **10**-15 to
    10-17
   table for flow of water by, **10**-10 to
    10-14
 job factors affecting, **10**-6
 machinery and methods: considerations for, **10**-27
  hoses: costs hourly, **8**-7
   examples of, **10**-2, **11**-13, **14**-26
  meters: description of, **10**-30
   example of, **10**-44
  nozzles and monitors: description
   of, **10**-30
   examples of, **10**-2, **11**-13, **14**-26
  pipe: costs hourly, **8**-17
   description of, **10**-28, **10**-29
   examples of, **10**-29, **10**-45, **10**-46
   specifications for, **10**-30
  pipe fittings: costs hourly, **8**-17
   description of, **10**-28
   examples of, **10**-18, **10**-29, **10**-46
  pumps: description of, **10**-30
   kinds of: centrifugal volute, **10**-30,
    10-31
    centrifugal turbine, **10**-31,
     10-38
    (See also Pumps, water)
   reservoirs: description of, **10**-41
    example of, **10**-42
   stands, pipe, examples of, **10**-45
    10-46
   tanks: costs hourly, **8**-29
    description of, **10**-39
    economies of use, **10**-40, **10**-41
    portable tanks: examples of, **10**-5,
     10-40
     specifications for, **10**-39,
      10-40
    semiportable tanks: example of,
     10-32
    specifications for, **10**-40

Water supply; machinery and
  methods; tanks *(Cont.):*
 valves: description of, **10**-28
  example of, **10**-18
 powers of numbers used as constants
  in formulas, **10**-15 to **10**-17
 production and costs of, **10**-49 to **10**-52
 sources of water, **10**-6
 summary, **10**-52
 symbols for, **10**-3
 water requirements of, **10**-6
 weights and measures of, **10**-3
Water wagons, **14**-21 to **14**-25, **15**-14
 costs hourly, **8**-27, **8**-29
 description of, **14**-21
 examples of, **13**-21, **14**-2, **14**-23, **14**-24,
   14-45, **15**-15, **15**-16, **15**-18
 production of, **14**-22, **14**-23, **15**-14 to
  15-16
  and costs of, **14**-23, **14**-25, **15**-16,
   15-17
 specifications for, **14**-21, **14**-22
Weathering *(see* Rock weathering)
Weighing excavation, **16**-28 to **16**-30
 example of, **16**-29
 principles of, **16**-28
 production of, **16**-28, **16**-29
  and costs of, **16**-28, **16**-30
 scales, platform: costs hourly, **8**-22
  description of, **16**-28, **16**-29
  example of, **16**-29
  specifications for, **16**-29
 (See also Volumes and weights of excavation)
Welding equipment:
 costs hourly, **8**-30
 examples of, **16**-86, **16**-88
Wellpoint systems, **16**-3 to **16**-7
 costs hourly, **8**-30
 elements of, **16**-4
 parts of a wellpoint, **16**-3
 principles, methods, and machinery for,
  16-3 to **16**-5
 production and costs of, **16**-5, **16**-6
 schematic views of work, **16**-5
 stabilization methods, alternative, **16**-6
 three-stage system, example of, **16**-7
Wetting-compacting excavation, **14**-15 to
  14-46
 compacting excavation, **14**-26
  dynamics of machines for, **14**-26 to
   14-29
  flexibility of compactors, **14**-27, **14**-44
  forces of compactors, **14**-26
  operating characteristics of compactors, **14**-27
  by pneumatic-tires compactors:
   characteristics of, **14**-38 to **14**-39
   costs hourly, **8**-20
   examples of, **14**-40, **14**-41, **14**-44

Wetting-compacting excavation; by pneumatic-tires compactors *(Cont.)*:
  production of, **14**-35
    and costs of, **14**-43
  specifications for, **14**-42
  relative compaction work of machines, **14**-20
 by sheepsfoot compactors: characteristics of, **14**-29 to **14**-31
  costs hourly, **8**-19, **8**-20
  examples of, **14**-30 to **14**-31
  production of, **14**-26
    and costs of, **14**-43
  specifications for, **14**-30, **14**-32, **14**-33
 by tamping foot compactors: characteristics of, **14**-32, **14**-34
  costs hourly, **8**-19, **8**-20
  examples of, **14**-30, **14**-33 to **14**-35
  production of, **14**-26
    and costs of, **14**-34
  specifications for, **14**-30, **14**-33, **14**-37
 by vibratory compactors: characteristics of, **14**-34 to **14**-36
  costs hourly, **8**-20
  examples of, **14**-36, **14**-38, **14**-39
  production of, **14**-35
    and costs of, **14**-38
  specifications for, **14**-37, **14**-42
 field tests for, **14**-16 to **14**-17
 laboratory tests for, **14**-15 to **14**-18
 principles of compaction, **14**-15
 in sanitary land fills or dumps: description of, **14**-43, **14**-44
  examples of, **5**-15, **14**-45
  principles and methods of, **14**-43, **14**-44
 specifications for work, **14**-4, **14**-5, **14**-19, **14**-20
 summary, **14**-46
 wetting excavation: considerations for, **14**-20
  by monitors or nozzles: costs hourly, **8**-16
    description of, **14**-25, **14**-26
    examples of, **10**-2, **13**-140, **14**-26
    machines of water supply, **14**-25
    production and costs of, **14**-30
    (*See also* Monitors and nozzles)

Wetting-compacting excavation; wetting excavation *(Cont.)*:
  quantity of water, **14**-20
  by water wagons: costs hourly, **8**-27, **8**-29
    examples of, **14**-23, **14**-24
    production of, **14**-22, **14**-23
    and costs of, **14**-23, **14**-25
    (*See also* Water wagons)
Wheel-bucket excavators, **12**-78 to **12**-84
 circular-digging – linear-traveling:
  costs hourly, **8**-31
  examples of, **12**-81, **12**-83
  production and costs of, **12**-83, **12**-84
  specifications for, **12**-82
 linear-digging – linear-traveling: costs hourly, **8**-31
  examples of, **12**-80
  production and costs of, **12**-79, **12**-80, **12**-82
  specifications for, **12**-79
 principles of, **12**-78, **12**-79
 production, estimating, equation for, **12**-79
 summary, **12**-84, **12**-85
Wheel horsepower, **13**-28
Wind:
 action of, **4**-12
 aggradation by: description of, **4**-12
  example of, **4**-13
 degradation by: description of, **4**-12
  example of, **3**-7
Work-cost indicators:
 description and use of, **13**-7 to **13**-9
 (*See also* specifications of individual machines)

Yards, shops, and offices, **16**-87 to **16**-90
 location of, **16**-87
 offices: costs hourly, **8**-10
  description of, **16**-89
  example of, **16**-89
 shops: costs hourly, **8**-10
  description of, **16**-87, **16**-89
  example of, **16**-88
 yards: description of, **16**-87
  examples of, **16**-87

Zinc ore, description of, **2**-13

624.152
C47e

624.152 C47c1　　AAJ-0445
Church, Horace K.　　040101 001
Excavation handbook / Horace K

0 00003 0135222 5
Vermont Technical College

DATE DUE	
MAY 0 2 2014	

DEMCO, INC. 38-2931

DISCARD
Library
Vermont Technical College
Randolph Center, Vermont